Using sylvius with Neuroscience Second Edition

Throughout the text, *Sylvius* icons are near figures or passages of text that describe particular brain structures or pathways. Each icon includes a number that corresponds to the left column of the table below, which serves as the key to locating specific information within *Sylvius*. The second column in the table briefly describes the particular neuroanatomical structure (or structures) to be visualized in *Sylvius*. Since the program is organized into six sections or modules, the MODULE column indicates the location(s) within *Sylvius* where images and information for each icon can be found. Lastly, the SELECTION column contains further details for navigating to relevant content within a particular module.

The graphics above the four columns of the table show how to navigate from Icon 12, Structure of the Ventricular System.

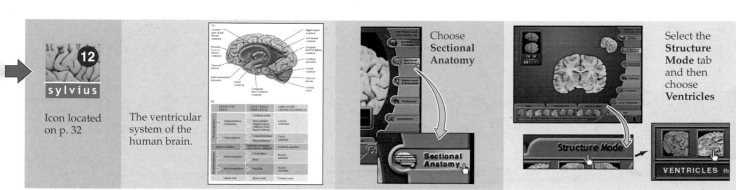

Icon located on p. 32 — The ventricular system of the human brain. — Choose **Sectional Anatomy** — Select the **Structure Mode** tab and then choose **Ventricles**

ICON	STRUCTURE(S) OF INTEREST	MODULE	SELECTION
1	Standard anatomical planes of section	Sectional anatomy Animations	Orientation tool Sagittal, coronal and axial MRI sections
2	Major subdivisions of CNS	Surface anatomy	Structure Mode: Embryonic divisions
3	Location of the cranial nerves	Spinal cord and brainstem Animations	Brainstem Model tab Brainstem rotation
4	Location of the cranial nerve nuclei	Spinal cord and brainstem Animations	Brainstem Model tab Brainstem rotation
5	Location of the cranial nerve nuclei	Spinal cord and brainstem Animations	Cross Sectional Atlas tab Brainstem rotation
6	Internal organization of the spinal cord	Spinal cord and brainstem	Cross Sectional Atlas tab
7	Surface features of the human brain	Surface anatomy Animations	Structure Mode: All Brain rotation
8	Surface features of the human brain	Surface anatomy Animations	Structure Mode: All Brain rotation
9	Surface features of the human brain	Surface anatomy Animations	Structure Mode: All Brain rotation
10	Internal organization of the brain	Sectional anatomy Animations	Structure Mode: All Sagittal, coronal, and axial MRI sections
11	Internal organization of the brain	Sectional anatomy Animations	Structure Mode: All Sagittal, coronal, and axial MRI sections
12	Structure of the ventricular system	Sectional anatomy Animations	Structure Mode: Ventricles Sagittal, coronal, and axial MRI sections

(continued on inside back cover)

NEUROSCIENCE
Second Edition

NEUROSCIENCE Second Edition

Edited by

DALE PURVES

GEORGE J. AUGUSTINE

DAVID FITZPATRICK

LAWRENCE C. KATZ

ANTHONY-SAMUEL LaMANTIA

JAMES O. McNAMARA

S. MARK WILLIAMS

Sinauer Associates, Inc.
Publishers
Sunderland, Massachusetts

THE COVER

Dorsal view of the human brain.
(Courtesy of S. Mark Williams.)

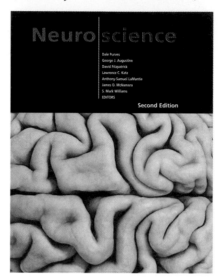

NEUROSCIENCE, SECOND EDITION

Copyright © 2001 by Sinauer Associates, Inc. All rights reserved.
This book may not be reproduced in whole or in part without permission.

Address inquiries and orders to
Sinauer Associates, Inc., P.O. Box 407
23 Plumtree Road, Sunderland, MA 01375 U.S.A.

Phone: 413-549-4300
FAX: 413-549-1118

email: orders@sinauer.com

www.sinauer.com

Library of Congress Cataloging-in-Publication Data

Neuroscience / edited by Dale Purves ... [et al.].-- 2nd ed.
    p. ; cm.
    Includes bibliographical references and index.
    ISBN 0-87893-742-0 (alk. paper)
    1. Neurosciences. 2. Neurophysiology. 3. Neurology. I. Purves, Dale.
    [DNLM: 1. Nervous System Physiology. 2. Neurochemistry. WL 102 N50588 2000]
    QP355.2 .N487 2000
    612.8--dc21

                                                              00-059496

Printed in U.S.A.

5   4   3

Contributors

George J. Augustine, Ph.D.
Dona M. Chikaraishi, Ph.D.
Michael D. Ehlers, M.D., Ph.D.
Gillian Einstein, Ph.D.
David Fitzpatrick, Ph.D.
William C. Hall, Ph.D.
Erich Jarvis, Ph.D.
Lawrence C. Katz, Ph.D.
Julie Kauer, Ph.D.
Anthony-Samuel LaMantia, Ph.D.
James O. McNamara, M.D.
Richard D. Mooney, Ph.D.
Miguel A. L. Nicolelis, M.D., Ph.D.
Dale Purves, M.D.
Peter H. Reinhart, Ph.D.
Sidney A. Simon, Ph.D.
J. H. Pate Skene, Ph.D.
Leonard E. White, Ph.D.
S. Mark Williams, Ph.D.

UNIT EDITORS

Unit I: George J. Augustine
Unit II: David Fitzpatrick
Unit III: William C. Hall and Miguel Nicolelis
Unit IV: Anthony-Samuel LaMantia and Larry Katz
Unit V: Dale Purves

NEW ART AND ACCOMPANYING CD-ROM

S. Mark Williams

SPECIAL CONSULTANTS

George R. Mangun, Ph.D.
Gregory McCarthy, Ph.D.

Contents in Brief

1 The Organization of the Nervous System 1

Unit I NEURAL SIGNALING

2 Electrical Signals of Nerve Cells 43
3 Voltage-Dependent Membrane Permeability 57
4 Channels and Transporters 77
5 Synaptic Transmission 99
6 Neurotransmitters 117
7 Neurotransmitter Receptors and Their Effects 141
8 Intracellular Signal Transduction 165

Unit II SENSATION AND SENSORY PROCESSING

9 The Somatic Sensory System 189
10 Pain 209
11 Vision: The Eye 223
12 Central Visual Pathways 251
13 The Auditory System 275
14 The Vestibular System 297
15 The Chemical Senses 317

Unit III MOVEMENT AND ITS CENTRAL CONTROL

16 Lower Motor Neuron Circuits and Motor Control 347
17 Upper Motor Neuron Control of the Brainstem and Spinal Cord 369
18 Modulation of Movement by the Basal Ganglia 391
19 Modulation of Movement by the Cerebellum 409
20 Eye Movements and Sensory Motor Integration 427
21 The Visceral Motor System 443

Unit IV THE CHANGING BRAIN

22 Early Brain Development 471
23 Construction of Neural Circuits 493
24 Modification of Brain Circuits as a Result of Experience 519
25 Plasticity of Mature Synapses and Circuits 535

Unit V COMPLEX BRAIN FUNCTIONS

26 The Association Cortices 565
27 Language and Lateralization 587
28 Sleep and Wakefulness 603
29 Emotions 625
30 Sex, Sexuality, and the Brain 645
31 Human Memory 665

Contents

Preface xvi

Acknowdgments xvii

Supplements to Accompany *NEUROSCIENCE* xviii

1. The Organization of the Nervous System 1

Overview 1

The Cellular Components of the Nervous System 1

Nerve Cells 3

Neuroglial Cells 6

Neural Circuits 8

Neural Systems 10

Some Anatomical Terminology 12

The Subdivisions of the Central Nervous System 13

    BOX A The Brainstem and Its Importance in Clinical Neuroanatomy 15

The External Anatomy of the Spinal Cord 18

The Internal Anatomy of the Spinal Cord 19

The External Anatomy of the Brain: Some General Points 20

The Lateral Surface of the Brain 21

The Dorsal and Ventral Surfaces of the Brain 22

The Midline Sagittal Surface of the Brain 23

The Internal Anatomy of the Brain 25

The Internal Anatomy of the Cerebral Hemispheres and Diencephalon 25

    BOX B Anatomical Brain Imaging Techniques 26

    BOX C Functional Brain Imaging: PET, SPECT, and *f*MRI 28

The Ventricular System 31

The Meninges 33

    BOX D Stroke 33

The Blood Supply of the Brain and Spinal Cord 35

    BOX E The Blood-Brain Barrier 38

Summary 39

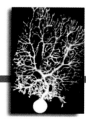

Unit I   NEURAL SIGNALING

2. Electrical Signals of Nerve Cells 43

Overview 43

Electrical Potentials Across Nerve Cell Membranes 43

How Ionic Movements Produce Electrical Signals 45

The Forces that Create Membrane Potentials 47

Electrochemical Equilibrium in an Environment with More Than One Permeant Ion 49

The Ionic Basis of the Resting Membrane Potential 51

The Ionic Basis of Action Potentials 52

    BOX A The Remarkable Giant Nerve Cells of Squid 53

    BOX B Action Potential Form and Nomenclature 55

Summary 56

3. Voltage-Dependent Membrane Permeability 57

Overview 57

Ionic Currents Across Nerve Cell Membranes 57

    BOX A The Voltage Clamp Method 58

Two Types of Voltage-Dependent Ionic Current 59

Two Voltage-Dependent Membrane Conductances 61

Reconstruction of the Action Potential 64

Long-Distance Signaling by Means of Action Potentials 65

    BOX B Threshold 66

    BOX C Passive Membrane Properties 68

The Refractory Period 72

Increased Conduction Velocity as a Result of Myelination 72

    BOX D Multiple Sclerosis 75

Summary 76

4. Channels and Transporters 77

Overview 77
Ion Channels Underlying Action Potentials 77

    BOX A The Patch Clamp Method 78

The Diversity of Ion Channels 81
Voltage-Gated Ion Channels 82

    BOX B Expression of Ion Channels in *Xenopus* Oocytes 85

Ligand-Gated Ion Channels 86
Stretch- and Heat-Activated Channels 86
The Molecular Structure of Ion Channels 86
Active Transporters Create and Maintain Ion Gradients 89

    BOX C Toxins That Poison Ion Channels 90

    BOX D Diseases Caused by Altered Ion Channels 92

Functional Properties of the Na^+/K^+ Pump 93
The Molecular Structure of the Na^+/K^+ Pump 95
Summary 96

5. Synaptic Transmission 99

Overview 99
Electrical Synapses 99
Chemical Synapses 101
Quantal Transmission at Neuromuscular Synapses 103
Release of Transmitters from Synaptic Vesicles 105
Local Recycling of Synaptic Vesicles 106
The Role of Calcium in Transmitter Secretion 107

    BOX A Diseases That Affect the Presynaptic Terminal 108

Molecular Mechanisms of Transmitter Secretion 111

    BOX B Toxins That Affect Transmitter Release 114

Summary 115

6. Neurotransmitters 117

Overview 117
What Defines a Neurotransmitter? 117

    BOX A Criteria That Define a Neurotransmitter 119

Two Major Categories of Neurotransmitters 120
Neurons Often Release More Than One Transmitter 122
Neurotransmitter Synthesis 123

Packaging Neurotransmitters 126
Neurotransmitter Release and Removal 126
Acetylcholine 127
Glutamate 129
GABA and Glycine 130

    BOX B Excitotoxicity in Acute Neuronal Injury 130

The Biogenic Amines 131
ATP and Other Purines 135
Peptide Neurotransmitters 135

    BOX C Biogenic Amine Neurotransmitters and Psychiatric Disorders 137

    BOX D Addiction 138

Summary 139

7. Neurotransmitter Receptors and Their Effects 141

Overview 141
Neurotransmitter Receptors Alter Postsynaptic Membrane Permeability 141
Principles Derived from Studies of the Neuromuscular Junction 141
Excitatory and Inhibitory Postsynaptic Potentials 147
Summation of Synaptic Potentials 149
Two Families of Postsynaptic Receptors 150
Cholinergic Receptors 151
Glutamate Receptors 153

    BOX A Neurotoxins That Act on Postsynaptic Receptors 156

    BOX B Myasthenia Gravis: An Autoimmune Disease of Neuromuscular Synapses 158

GABA and Glycine Receptors 159
Serotonin Receptors 161
Purinergic Receptors 161
Catecholamine Receptors 162
Peptide Receptors 162
Summary 162

8. Intracellular Signal Transduction 165

Overview 165
Strategies of Molecular Signaling 165
The Activation of Signaling Pathways 167
Receptor Types 168
G-Proteins and Their Molecular Targets 170
Second Messengers 172
Second Messenger Targets: Protein Kinases and Phosphatases 175
Nuclear Signaling 178
Examples of Neuronal Signal Transduction 181
Summary 185

Unit II 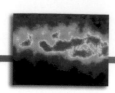 SENSATION AND SENSORY PROCESSING

9. The Somatic Sensory System 189
Overview 189
Cutaneous and Subcutaneous Somatic Sensory
  Receptors 189
Mechanoreceptors Specialized to Receive
  Tactile Information 192
Differences in Mechanosensory Discrimination
  Across the Body Surface 194

  **BOX A   Receptive Fields and Sensory Maps
  in the Cricket 195**

  **BOX B   Dynamic Aspects of Somatic Sensory
  Receptive Fields 196**

Mechanoreceptors Specialized for Proprioception 197
Active Tactile Exploration 199
The Major Afferent Pathway for Mechanosensory
  Information: The Dorsal Column–Medial
  Lemniscus System 199

  BOX C   Dermatomes 201

The Trigeminal Portion of the Mechanosensory
  System 202
The Somatic Sensory Components of the Thalamus
  203
The Somatic Sensory Cortex 204

  **BOX D   Patterns of Organization within the Sensory
  Cortices: Brain Modules 206**

Higher-Order Cortical Representations 207
Summary 207

10. Pain 209
Overview 209
Nociceptors 209
The Perception of Pain 210
Hyperalgesia and Sensitization 211
Central Pain Pathways: The Spinothalamic Tract 212
The Nociceptive Components of the Thalamus
  and Cortex 214

  BOX A   Referred Pain 215

  BOX B   Phantom Limbs and Phantom Pain 216

Central Regulation of Pain Perception 217
The Placebo Effect 219
The Physiological Basis of Pain Modulation 220
Summary 221

11. Vision: The Eye 223
Overview 223
Anatomy of the Eye 223
The Formation of Images on the Retina 224

  BOX A   Myopia and Other Refractive Errors 226

The Retina 227

Phototransduction 230

  BOX B   Retinitis Pigmentosa 231

Functional Specialization of the Rod and Cone
  Systems 234

  BOX C   Macular Degeneration 236

Anatomical Distribution of Rods and Cones 237
Cones and Color Vision 239

  **BOX D   The Importance of Context in Color
  Perception 240**

Retinal Circuits for Detecting Differences in
  Luminance 242

  BOX E   The Perception of Luminance 244

Contribution of Retinal Circuits to Light
  Adaptation 247
Summary 249

12. Central Visual Pathways 251
Overview 251
Central Projections of Retinal Ganglion Cells 251

  BOX A   The Blind Spot 254

The Retinotopic Representation of the
  Visual Field 255
Visual Field Deficits 258
The Functional Organization of the Striate
  Cortex 260
The Columnar Organization of the Striate
  Cortex 263

  **BOX B   Random Dot Stereograms and Related
  Amusements 264**

  **BOX C   Optical Imaging of Functional Domains in
  the Visual Cortex 266**

Parallel Streams of Information from Retina
  to Cortex 268
The Functional Organization of Extrastriate
  Visual Areas 270
Summary 273

13. The Auditory System 275
Overview 275
Sound 275
The Audible Spectrum 276
A Synopsis of Auditory Function 276

  BOX A   Four Causes of Acquired Hearing Loss 277

The External Ear 278
The Middle Ear 279
The Inner Ear 279
Hair Cells and the Mechanoelectrical Transduction
  of Sound Waves 282

BOX B **The Sweet Sound of Distortion 285**
Two Kinds of Hair Cells in the Cochlea 286
Tuning and Timing in the Auditory Nerve 287
How Information from the Cochlea Reaches
    Targets in the Brainstem 288
Integrating Information from the Two Ears 288
Monaural Pathways from the Cochlear Nucleus
    to the Lateral Lemniscus 292
Integration in the Inferior Colliculus 292
The Auditory Thalamus 293
The Auditory Cortex 294
Summary 295

14. The Vestibular System 297

Overview 297
The Vestibular Labyrinth 297
Vestibular Hair Cells 298
The Otolith Organs: The Utricle and Sacculus 300

BOX A **A Primer on (Vestibular) Navigation 300**

How Otolith Neurons Sense Linear Forces 302

BOX B **Adaptation and Tuning of Vestibular Hair
    Cells 304**

The Semicircular Canals 306
How Semicircular Canal Neurons Sense
    Angular Accelerations 307

BOX C **Throwing Cold Water on the Vestibular
    System 308**

Central Vestibular Pathways: Eye, Head, and Body
    Reflexes 310

BOX D **Mauthner Cells in Fish 312**

Vestibular Pathways to the Thalamus and Cortex
    314
Summary 315

15. The Chemical Senses 317

Overview 317
The Organization of the Olfactory System 317
Olfactory Perception in Humans 319
Physiological and Behavioral Responses to
    Odorants 320
The Olfactory Epithelium and Olfactory Receptor
    Neurons 321

BOX A **Olfaction, Pheromones, and Behavior
    in the Hawk Moth 322**

The Transduction of Olfactory Signals 323
Odorant Receptors and Olfactory Coding 324
The Olfactory Bulb 327

BOX B **Temporal "Coding" of Olfactory Information
    in Insects 328**

Central Projections of the Olfactory Bulb 329
The Organization of the Taste System 330
Taste Perception in Humans 332
The Organization of the Peripheral Taste System 334
Idiosyncratic Responses to Various Tastants 335
Taste Receptors and the Transduction of Taste
    Signals 336
Neural Coding in the Taste System 340
Central Processing of Taste Signals 340
Trigeminal Chemoreception 341

BOX C **Capsaicin 342**

Summary 343

Unit III   MOVEMENT AND ITS CENTRAL CONTROL

16. Lower Motor Neuron Circuits and
Motor Control 347

Overview 347
Neural Centers Responsible for Movement 347
Motor Neuron–Muscle Relationships 349
The Motor Unit 351
The Regulation of Muscle Force 353
The Spinal Cord Circuitry Underlying
    Muscle Stretch Reflexes 355
The Influence of Afferent Activity on Motor
    Behavior 357
Other Afferent Feedback that Affects Motor
    Performance 358
Flexion Reflex Pathways 361
Spinal Cord Circuitry and Locomotion 361

BOX A **Locomotion in the Leech and the
    Lamprey 362**

BOX B **The Autonomy of Central Pattern
    Generators: Evidence from the Lobster
    Stomatogastric Ganglion 364**

The Lower Motor Neuron Syndrome 366
Summary 367

BOX C **Amyotrophic Lateral Sclerosis 367**

17. Upper Motor Neuron Control of the
Brainstem and Spinal Cord 369

Overview 369
Descending Control of Spinal Cord Circuitry:
    General Information 369

Motor Control Centers in the Brainstem:
  Upper Motor Neurons That Maintain Balance
  and Posture 370
The Primary Motor Cortex: Upper Motor Neurons
  That Initiate Complex Voluntary Movements 375
Functional Organization of the Primary Motor
  Cortex 376

  BOX A Descending Projections to Cranial Nerve
  Motor Nuclei and Their Importance in Diagnosing the Cause of Motor Deficits 378

  BOX B What Do Motor Maps Represent? 380

  BOX C Sensory Motor Talents and Cortical Space
  383

The Premotor Cortex 384
Damage to Descending Motor Pathways:
  The Upper Motor Neuron Syndrome 386

  BOX A Muscle Tone 387

Summary 388

**18. Modulation of Movement by the
Basal Ganglia 391**
Overview 391
Projections to the Basal Ganglia 391
Projections from the Basal Ganglia to Other Brain
  Regions 396
Evidence from Studies of Eye Movements 397
Circuits within the Basal Ganglia System 400

  BOX A Huntington's Disease 400

  BOX B Parkinson's Disease: An Opportunity for
  Novel Therapeutic Approaches 403

  BOX C Basal Ganglia Loops and Non-Motor Brain
  Functions 406

Summary 407

**19. Modulation of Movement by the
Cerebellum 409**
Overview 409
Organization of the Cerebellum 409
Projections to the Cerebellum 412
Projections from the Cerebellum 414
Circuits within the Cerebellum 415

  BOX A Prion Diseases 418

Cerebellar Circuitry and the Coordination
  of Ongoing Movement 419
Consequences of Cerebellar Lesions 422

  BOX B Genetic Analysis of Cerebellar Function 423

Summary 425

**20. Eye Movements and Sensory Motor
Integration 427**
Overview 427
What Eye Movements Accomplish 427
The Actions and Innervation of Extraocular
  Muscles 428

  BOX A The Perception of Stabilized Retinal Images
  430

Types of Eye Movements and Their Functions 431
Neural Control of Saccadic Eye Movements 433

  BOX B Sensory Motor Integration in the Superior
  Colliculus 438

Neural Control of Smooth Pursuit Movements 440
Neural Control of Vergence Movements 440
Summary 440

21. The Visceral Motor System 443
Overview 443
Early Studies of the Visceral Motor System 443
The Sympathetic Division of the Visceral Motor
  System 444
The Parasympathetic Division of the Visceral
  Motor System 447
The Enteric Nervous System 450
Sensory Components of the Visceral Motor System
  453
Central Control of the Visceral Motor Functions 454
Neurotransmission in the Visceral Motor System 454

  BOX A The Hypothalamus 456

  BOX B Horner's Syndrome 458

Visceral Motor Reflex Functions 460
Autonomic Regulation of Cardiovascular Function
  460
Autonomic Regulation of the Bladder 462
Autonomic Regulation of Sexual Function 464
Summary 466

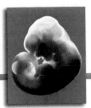

Unit IV   THE CHANGING BRAIN

22. Early Brain Development 471

Overview 471
The Initial Formation of the Nervous System:
  Gastrulation and Neurulation 471
The Molecular Basis of Neural Induction 473

   BOX A Retinoic Acid: Teratogen and Inductive
     Signal 474

Formation of the Major Brain Subdivisions 476
Genetic Abnormalities and Altered Human
  Brain Development 479
The Initial Differentiation of Neurons and Glia 480

   BOX B Homeotic Genes and Human Brain
     Development 480

   BOX C Rhombomeres 482

The Generation of Neuronal Diversity 483

   BOX D Neurogenesis and Neuronal Birthdating 486

Neuronal Migration 488
Summary 490

23. Construction of Neural Circuits 493

Overview 493
The Axonal Growth Cone 493
Non-Diffusible Signals for Axon Guidance 494
Diffusible Signals for Axon Guidance:
  Chemoattraction and Repulsion 498
The Formation of Topographic Maps 500
Selective Synapse Formation 502
Trophic Interactions and the Ultimate Size
  of Neuronal Populations 503

   BOX A Molecular Signals That Promote Synapse
     Formation 504

Further Competitive Interactions in the
  Formation of Neuronal Connections 506

   BOX B Why Do Neurons Have Dendrites? 509

Molecular Basis of Trophic Interactions 510
Neurotrophin Receptors 512

   BOX C The Discovery of BDNF and the
     Neurotrophin Family 514

The Effect of Neurotrophins on the Differentiation
  of Neuronal Form 515
Summary 516

24. Modification of Brain Circuits as a Result of Experience 519

Overview 519
Critical Periods 519

   BOX A Built-in Behaviors 520

The Development of Language: A Critical
  Period in Humans 521

   BOX B Birdsong 522

Critical Periods in Visual System Development 524
Effects of Visual Deprivation on Ocular Dominance
  525

   BOX C Transneuronal Labeling with Radioactive
     Amino Acids 527

Critical Periods, Cortical Plasticity, and
  Amblyopia in Humans 530
Mechanisms by which Neuronal Activity Affects
  the Development of Neural Circuits 531
Evidence for Critical Periods in Other Sensory
  Systems 533
Summary 533

25. Plasticity of Mature Synapses and Circuits 535

Overview 535
Mechanisms of Synaptic Plasticity in
  Relatively Simple Invertebrates 535
Mechanisms of Short-Term Synaptic Plasticity in
  the Mammalian Nervous System 539

   BOX A Genetics of Learning and Memory in the
     Fruit Fly 540

Mechanism of Long-Term Synaptic Plasticity in
  the Mammalian Nervous System 541
Long-Term Synaptic Potentiation 542
Molecular Mechanisms Underlying LTP 545

   BOX B Silent Synapses 548

Long-Term Synaptic Depression 550
Plasticity in the Adult Cerebral Cortex 553

   BOX C Epilepsy: The Effect of Pathological Activity
     on Neural Circuitry 554

Recovery from Neural Injury 556
Generation of Neurons in the Adult Brain 559

   BOX D Why Aren't We More Like Fish and Frogs?
     560

Summary 561

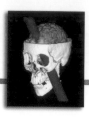

Unit V   COMPLEX BRAIN FUNCTIONS

26. The Association Cortices 565
Overview 565
The Association Cortices 565
An Overview of Cortical Structure 566
Specific Features of the Association Cortices 568

   **BOX A A More Detailed Look at Cortical
   Lamination 569**

Lesions of the Parietal Association Cortex:
   Deficits of Attention 571
Lesions of the Temporal Association Cortex:
   Deficits of Recognition 574
Lesions of the Frontal Association Cortex:
   Deficits of Planning 576
"Attention Neurons" in the Monkey Parietal
   Cortex 577

   BOX B Psychosurgery 579

"Recognition Neurons" in the Monkey Temporal
   Cortex 580
"Planning Neurons" in the Monkey Frontal
   Cortex 581

   BOX C Neuropsychological Testing 583

   BOX D Brain Size and Intelligence 584

Summary 585

27. Language and Lateralization 587
Overview 587
Language Is Both Localized and Lateralized 587

   BOX A Do Apes Have Language? 588

Aphasias 590

   BOX B Words, Syntax, and Meaning 591

A Dramatic Confirmation of Language
   Lateralization 593
Anatomical Differences between the Right
   and Left Hemispheres 595

   BOX C Handedness 596

Mapping Language Function 597
More on the Role of the Right Hemisphere
   in Language 599
Sign Language 600
Summary 601

28. Sleep and Wakefulness 603
Overview 603
Why Do Humans and Many Other Animals Sleep?
   603

   BOX A Styles of Sleep in Different Species 605

The Circadian Cycle of Sleep and Wakefulness 606

   **BOX B Molecular Mechanisms of Biological Clocks
   608**

Stages of Sleep 609
Physiological Changes in Sleep States 611

   BOX C Electroencephalography 612

The Possible Functions of REM Sleep and
   Dreaming 614
Neural Circuits Governing Sleep 615

   BOX D Consciousness 616

Thalamocortical Interactions 618
Sleep Disorders 619
Summary 622

29. Emotions 625
Overview 625
Physiological Changes Associated with
   Emotion 625
The Integration of Emotional Behavior 626

   **BOX A Facial Expressions: Pyramidal and
   Extrapyramidal Contributions 628**

The Limbic System 631
The Importance of the Amygdala 634

   BOX B The Anatomy of the Amygdala 634

   **BOX C The Reasoning Behind an Important
   Discovery 636**

   **BOX D Fear and the Human Amygdala: A Case
   Study 638**

The Relationship between Neocortex and
   Amygdala 639

   BOX E Affective Disorders 640

Cortical Lateralization of Emotional Functions 642
The Interplay of Emotion and Reason 643
Summary 644

30. Sex, Sexuality, and the Brain 645
Overview 645
Sexually Dimorphic Behavior 645

   **BOX A The Development of Male and Female
   Phenotypes 646**

What Is Sex? 647
Hormonal Influences on Sexual Dimorphism 649

   BOX B The Case of John/Joan 649

The Effect of Sex Hormones on Neural Circuitry 651

Central Nervous System Dimorphisms Related to Reproductive Behaviors 652

BOX C The Actions of Sex Hormones 654

Brain Dimorphisms Related to Cognitive Function 660

Hormone-Sensitive Brain Circuits in Adult Animals 663

Summary 663

31. Human Memory 665

Overview 665

Qualitative Categories of Human Memory 665

Temporal Categories of Memory 666

BOX A Phylogenetic Memory 667

The Importance of Association in Information Storage 668

BOX B Savant Syndrome 670

Forgetting 671

Brain Systems Underlying Declarative and Procedural Memories 673

BOX C Clinical Cases That Reveal the Anatomical Substrate for Declarative Memories 674

The Long-Term Storage of Information 676

BOX D Alzheimer's Disease 678

Memory and Aging 680

Summary 680

Glossary G-1

Illustration Credits IC-1

Index I-1

Preface

Whether judged in molecular, cellular, systemic, behavioral, or cognitive terms, the human nervous system is a stupendous piece of biological machinery. Given its accomplishments—all the artifacts of human culture, for instance—there is good reason for wanting to understand how the brain and the rest of the nervous system works. The debilitating and costly effects of neurological and psychiatric disease add a further sense of urgency to this quest. The aim of this book is to highlight the intellectual challenges and excitement—as well as the uncertainties—of what many see as the last great frontier of biological science. The information presented should serve as a starting point for undergraduates, medical students, graduate students in the neurosciences, and others who want to understand how the human nervous system operates. Like any other great challenge, neuroscience should be, and is, full of debate, dissension, and considerable fun. All these ingredients have gone into the construction of this book; we hope they will be conveyed in equal measure to readers at all levels.

Acknowledgments

We are grateful to numerous colleagues who provided helpful contributions, criticisms and suggestions. We particularly wish to thank David Amaral, Gary Banker, Marlene Behrmann, Ursula Bellugi, Dan Blazer, Bob Burke, Nell Cant, John Chapin, Milt Charlton, Michael Davis, Bob Desimone, Allison Doupe, Sascha du Lac, Jen Eilers, Peter Eimas, Everett Ellinwood, Robert Erickson, Howard Fields, Elizabeth Finch, Bob Fremeau, Michela Gallagher, Steve George, Pat Goldman-Rakic, Mike Haglund, Zach Hall, Matt Helms, Bill Henson, John Heuser, Jonathan Horton, Ron Hoy, Alan Humphrey, David Johnson, Jon Kaas, Herb Killackey, Len Kitzes, Arthur Lander, Story Landis, Darrell Lewis, Jeff Lichtman, Alan Light, Steve Lisberger, Donald Lo, Arthur Loewy, Eve Marder, Robert McCarley, Jim McIlwain, Chris Muly, Vic Nadler, Ron Oppenheim, Michael Platt, Scott Pomeroy, Rodney Radtke, Louis Reichardt, Marnie Riddle, Steve Roper, John Rubenstein, Ben Rubin, David Rubin, Josh Sanes, Cliff Saper, Lynn Selemon, Carla Shatz, Larry Squire, John Staddon, Peter Strick, Warren Strittmatter, Joe Takahashi, Jim Voyvodic, Jonathan Weiner, Christina Williams, and Fulton Wong. It is understood, of course, that any remaining errors are in no way attributable to our critics and advisors. We also thank the students at Duke University Medical School as well as many other students and colleagues who provided suggestions for improvement of the first edition. Finally, we owe special thanks to Polly Garner, Gayle Wood, Robert Reynolds, Doug Buchacek, and Nate O'Keefe, who labored long and hard to put the second edition together, and to Andy Sinauer, Joyce Zymeck, Joan Gemme, and the rest of the staff at Sinauer Associates for outstanding work and high standards.

FOR THE STUDENT

Sylvius 2.0 CD-ROM (ISBN 0-87893-917-2)

Every copy of the book includes *Sylvius 2.0: Fundamentals of Human Neural Structure*, an interactive CD-ROM atlas of the human nervous system. Many of the CD-ROM's brain images are incorporated into the art program of the book. Throughout the text, references to *Sylvius* are marked with an icon that corresponds to a *Sylvius* directory found inside the front and back covers. The directory includes book figure numbers, and descriptions of related *Sylvius* coverage.

Sylvius consists of six modules that allow users to:

- interact with a whole-brain specimen and examine its surface features from eight different vantage points.

- examine the brain's 3-D internal organization by "virtually" sectioning the whole-brain specimen in the axial, coronal, and sagittal planes.

- use a cross-sectional atlas or brainstem model to examine the location of spinal cord and brainstem structures.

- dynamically visualize the organization of all the major long-tract connections of the nervous system.

- rotate models of the brainstem and whole brain; view MRIs of a living brain.

- search the visual glossary for a brief description and the proper pronunciation of approximately 200 key anatomical structures.

FOR THE INSTRUCTOR

Instructor's Art CD-ROM (ISBN 0-87893-740-4)

This resource includes all the four-color illustrations from the book, reformatted and relabeled for maximum readability. Using either PowerPoint® (for PC) or Extensis® Portfolio™ Browser (Macintosh and PC-compatible), instructors can search for, view, and organize images for class presentations. For other users, external image files are available in folders arranged by chapter.

Overhead Transparencies (ISBN 0-87893-741-2)

A set of 100 key full-color figures from the book is available as overhead transparencies. These have been selected by the editors for teaching purposes, and resized and relabeled for optimal projection.

Chapter 1

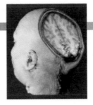

The Organization of the Nervous System

Overview

Perhaps the major reason that neuroscience remains such an exciting field is the wealth of unanswered questions about the fundamental structure and functions of the human brain. To understand this remarkable organ (and the rest of the nervous system), the myriad cell types that constitute the nervous system must be identified, their interconnections traced, and the physiological role of the resulting circuits defined. Adding to these several challenges is the fact that a specialized anatomical vocabulary has arisen to describe the structure of the nervous system, as well as a specialized set of physiological terms to describe its functions. In light of these conceptual and semantic difficulties, comprehending the brain and the rest of the nervous system is greatly facilitated by a general picture of the organization of the nervous system, and by a review of the basic terms and anatomical conventions used in discussing its structure and function.

The Cellular Components of the Nervous System

The fact that cells are the basic elements of living organisms was recognized early in the nineteenth century. It was not until well into the twentieth century, however, that neuroscientists agreed that nervous tissue, like all other organs, is made up of these fundamental units. In the early twentieth century, the first generation of "modern" neurobiologists had difficulty resolving the unitary nature of nerve cells with the available microscopes and cell staining techniques. The extraordinary shapes of individual nerve cells and the great extent of some of their branches was apparent, but this complexity actually tended to obscure their resemblance to the geometrically simpler cells of other tissues (Figure 1.1). Thus, some biologists concluded that each nerve cell was connected to its neighbors by protoplasmic links, forming a continuous nerve cell network or *reticulum.* This "reticular theory" of nerve cell communication, which was championed by the Italian neuropathologist Camillo Golgi (for whom the Golgi apparatus in cells is named), eventually fell from favor, primarily as a result of the work of the Spanish neuroanatomist Santiago Ramón y Cajal. The contrasting views represented by Golgi and Cajal occasioned one of the first spirited debates in modern neuroscience. Based on light microscopic examination of nervous tissue stained with silver salts according to a method pioneered by Golgi, Cajal argued persuasively that nerve cells are discrete entities, and that they communicate with one another by means of specialized contacts that eventually came to be called "synapses." The work that framed this debate was recognized by the award of the Nobel Prize for Physiology or Medicine in 1906 to both

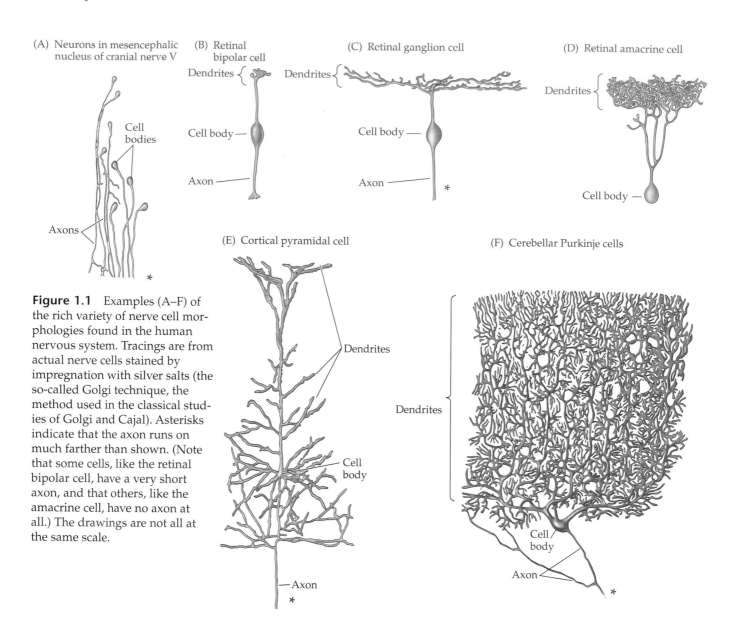

(A) Neurons in mesencephalic nucleus of cranial nerve V

Cell bodies

Axons

(B) Retinal bipolar cell

Dendrites

Cell body

Axon

(C) Retinal ganglion cell

Dendrites

Cell body

Axon

(D) Retinal amacrine cell

Dendrites

Cell body

(E) Cortical pyramidal cell

Dendrites

Cell body

Axon

(F) Cerebellar Purkinje cells

Dendrites

Cell body

Axon

Figure 1.1 Examples (A–F) of the rich variety of nerve cell morphologies found in the human nervous system. Tracings are from actual nerve cells stained by impregnation with silver salts (the so-called Golgi technique, the method used in the classical studies of Golgi and Cajal). Asterisks indicate that the axon runs on much farther than shown. (Note that some cells, like the retinal bipolar cell, have a very short axon, and that others, like the amacrine cell, have no axon at all.) The drawings are not all at the same scale.

Golgi and Cajal. The joint award indicated some ongoing concern about who was correct, despite Cajal's overwhelming evidence. Any lingering doubt was finally resolved with the advent of electron microscopy in the 1950s. The high-magnification, high-resolution pictures obtained with the electron microscope clearly established that nerve cells are indeed functionally independent units.

The nineteenth-century histological studies of Cajal, Golgi, and a host of successors led to the consensus that the cells of the nervous system can be divided into two broad categories: **nerve cells** (or **neurons**), and a variety of **supporting cells**. Nerve cells are specialized for electrical signaling over long distances, and understanding this process represents one of the more dramatic success stories in modern biology (the subject of Unit I). Supporting cells, in contrast, are not capable of electrical signaling, despite having some other important electrical properties. In the central nervous system (the brain and spinal cord), these supporting cells consist mostly of **neuroglial** cells. Although the cells of the human nervous system are in many ways similar to those of other organs, they are unusual in their extraordinary numbers (the human brain is estimated to contain 100 billion neurons and several times as many supporting cells). Most importantly, the nervous system has a greater range of distinct cell types—whether categorized by morphology, molecular identity, or physiological activity—than any other organ system. This rich structural and functional diversity, and the interconnection of nerve cells via **synapses** to form intricate ensembles, or **circuits**, is the foundation on which sensory processes, perception, and behavior are ultimately built.

For much of the twentieth century, neuroscientists relied on the same set of techniques developed by Cajal and Golgi to describe and categorize the diversity of cell types in the nervous system. From the late 1970s onward, however, remarkable new technologies made possible by the advances in cell biology and the revolution in molecular biology provided investigators with many additional tools to discern the properties of neurons (Figure 1.2). General cell staining methods showed mainly differences in cell sizes and distribution. Using antibody or mRNA probes, it became possible to appreciate distinctive features of neurons and glia in various regions of the nervous system, as well as the diversity of cell types within these regions. Moreover, new tracing methods allowed the interconnections of neurons to be explored much more fully. Tracer substances can be introduced into either living or fixed tissue, and are transported along nerve cell processes to reveal their origin and termination. Finally, ways of determining the molecular identity and morphology of nerve cells can be combined with measurements of their physiological activity, thus illuminating structure–function relationships.

Nerve Cells

Despite the specific molecular, morphological, and functional features of any particular nerve cell type, the basic structure of neurons resembles that of other cells. Thus, each nerve cell has a cell body containing a nucleus, endoplasmic reticulum, ribosomes, Golgi apparatus, mitochondria, and other organelles that are essential to the function of all cells (Figure 1.3). These features are best recognized using the high magnification and resolution afforded by the electron microscope. The distinguishing characteristic of nerve cells is their specialization for intercellular communication. This attribute is apparent in their overall morphology, in the specialization of their membranes for electrical signaling, and in the structural and functional intricacies of the synaptic contacts between them.

A particularly salient morphological feature of most nerve cells is the elaborate arborization of the dendrites (also called dendritic branches or dendritic processes) that arise from the neuronal cell body. The spectrum of neuronal geometries ranges from a small minority of cells that lack dendrites altogether to neurons with dendritic arborizations that rival the complexity of a mature tree (see Figure 1.1). The number of inputs that a particular neuron receives depends on the complexity of its dendritic arbor: Nerve cells that lack dendrites are innervated by just one or a few other nerve cells, whereas those with increasingly elaborate dendrites are innervated by a commensurately larger number of other neurons.

The dendrites (together with the cell body) provide the major site for **synaptic terminals** made by the axonal endings of other nerve cells. The synaptic contact itself is a special elaboration of the secretory apparatus found in most polarized epithelial cells. Typically, the **presynaptic terminal** is immediately adjacent to a **postsynaptic specialization** of the contacted cell. For the vast majority of synapses, there is no physical continuity between these pre- and postsynaptic elements. Instead, the pre- and postsynaptic compo-

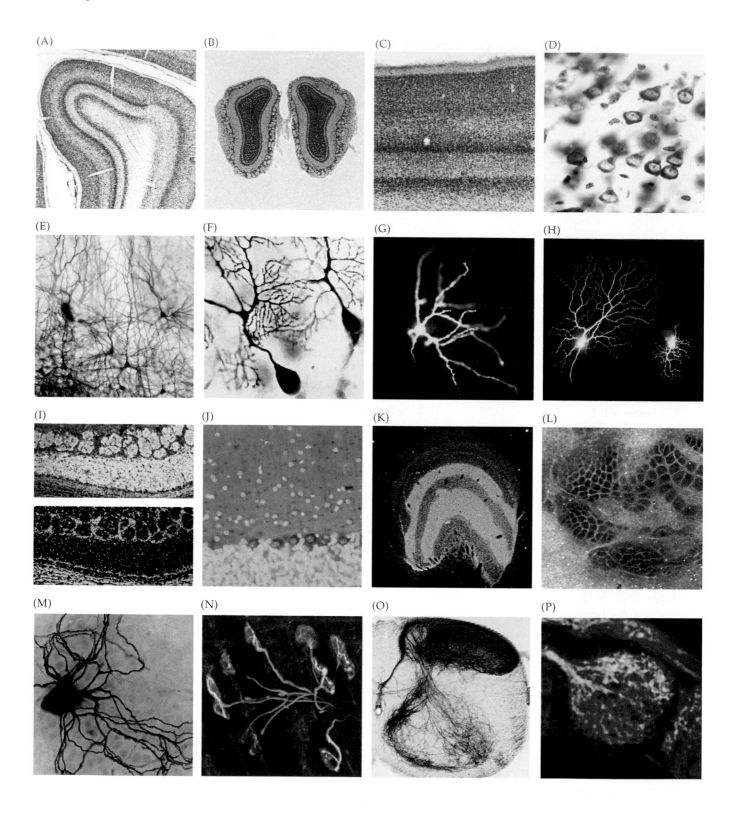

nents communicate via secretion of molecules from the presynaptic terminal that bind to receptors in the postsynaptic specialization. These molecules must traverse the extracellular space between pre- and postsynaptic elements; this interruption is called the synaptic cleft. The number of synaptic inputs received by each nerve cell in the human nervous system varies from 1 to about 100,000. This range of inputs reflects a fundamental purpose of nerve cells, namely to integrate information from other neurons. The num-

◀ **Figure 1.2** Structural diversity in the nervous system demonstrated with cellular and molecular markers. *First row:* Cellular organization of different brain regions demonstrated with Nissl stains, which label nerve and glial cell bodies. (A) The cerebral cortex at the boundary between the primary and secondary visual areas. (B) The olfactory bulbs. (C) Differences in cell density in cerebral cortical layers. (D) Individual Nissl-stained neurons and glia at higher magnification. *Second row:* Classical and modern approaches to seeing individual neurons and their processes. (E) Golgi-labeled cortical pyramidal cells. (F) Golgi-labeled cerebellar Purkinje cells. (G) Cortical interneuron labeled by intracellular injection of a fluorescent dye. (H) Retinal neurons labeled by intracellular injection of fluorescent dye. *Third row:* Cellular and molecular approaches to seeing neural connections and systems. (I) At top, an antibody that detects synaptic proteins in the olfactory bulb; at bottom, a fluorescent label shows the location of cell bodies. (J) Synaptic zones and the location of Purkinje cell bodies in the cerebellar cortex labeled with synapse-specific antibodies (green) and a cell body marker (blue). (K) The projection from one eye to the lateral geniculate nucleus in the thalamus, traced with radioactive amino acids (the bright label shows the axon terminals from the eye in distinct layers of the nucleus). (L) The map of the body surface of a rat in the somatic sensory cortex, shown with a marker that distinguishes zones of higher synapse density and metabolic activity. *Fourth row:* Peripheral neurons and their projections. (M) An autonomic neuron labeled by intracellular injection of an enzyme marker. (N) Motor axons (green) and neuromuscular synapses (orange) in transgenic mice genetically engineered to express fluorescent proteins. (O) The projection of dorsal root ganglia to the spinal cord, demonstrated by an enzymatic tracer. (P) Axons of olfactory receptor neurons from the nose labeled in the olfactory bulb with a vital fluorescent dye. (G courtesy of L. C. Katz; H courtesy of C. J. Shatz; K courtesy of P. Rakic; N,O courtesy of W. Snider and J. Lichtman; all others courtesy of A.-S. LaMantia and D. Purves.)

ber of inputs onto any particular cell is therefore an especially important determinant of neuronal function.

The information from the inputs that impinge on the neuronal dendrites is integrated and "read out" at the origin of the **axon**, the portion of the nerve cell specialized for signal conduction to the next site of synaptic interaction (see Figures 1.1 and 1.3). The axon is a unique extension from the neuronal cell body that may travel a few hundred micrometers or much farther, depending on the type of neuron and the size of the species. Many nerve cells in the human brain have axons no more than a few millimeters long, and a few have no axons at all (see, for example, the retinal amacrine cell in Figure 1.1; in fact, *amacrine* means "lacking a long process"). These short axons are a defining feature of local circuit neurons or interneurons throughout the brain. Many axons, however, extend to more distant targets. For example, the axons that run from the human spinal cord to the foot are about a meter long. The axonal mechanism that carries signals over such distances is called the **action potential**, a self-regenerating wave of electrical activity that propagates from its point of initiation at the cell body (called the axon hillock) to the terminus of the axon. At the axon ending, another set of synaptic contacts is made on yet other cells. The target cells of neurons include other nerve cells in the brain, spinal cord, and autonomic ganglia, and the cells of muscles and glands throughout the body.

The process by which information encoded by action potentials is passed on at synaptic contacts to the next cell in the pathway is called **synaptic transmission**. Presynaptic terminals (also called synaptic endings, axon terminals, or terminal boutons) and their postsynaptic specializations are typically **chemical synapses**, the most abundant type of synapse in the nervous system (another type, called electrical synapse, is described in Chapter 5). The secretory organelles in the presynaptic terminal of chemical synapses

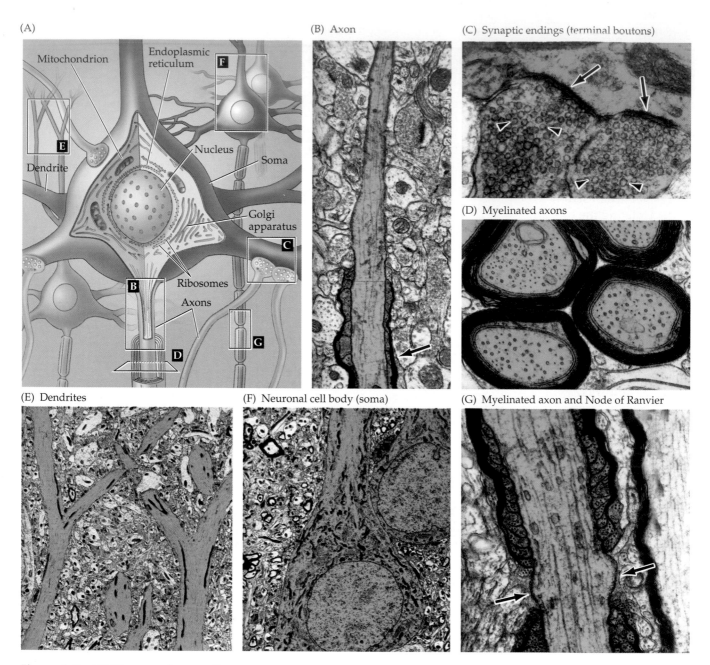

(A)

Mitochondrion

Endoplasmic reticulum

F

Nucleus

Soma

E

Dendrite

Golgi apparatus

C

Ribosomes

B

Axons

G

D

(B) Axon

(C) Synaptic endings (terminal boutons)

(D) Myelinated axons

(E) Dendrites

(F) Neuronal cell body (soma)

(G) Myelinated axon and Node of Ranvier

Figure 1.3 (A) Diagram of nerve cells and their component parts. (B) Axon initial segment (blue) entering a myelin sheath (gold). (C) Terminal boutons (blue) loaded with synaptic vesicles (arrowheads) forming synapses (arrows) with a dendrite (purple). (D) Transverse section of axons (blue) ensheathed by the processes of oligodendrocytes (gold). (E) Apical dendrites (purple) of cortical pyramidal cells. (F) Nerve cell bodies (purple) occupied by large round nuclei. (G) Portion of a myelinated axon (blue) illustrating the intervals between adjacent segments of myelin (gold) referred to as nodes of Ranvier (arrows). (Micrographs from Peters et al., 1991.)

are called **synaptic vesicles**, which are filled with **neurotransmitter** molecules. The neurotransmitters released from synaptic vesicles modify the electrical properties of the target cell by binding to **neurotransmitter receptors**, which are localized primarily at the postsynaptic specialization. Neurotransmitters, receptors, and the related transduction molecules are the machinery that allows nerve cells to communicate with one another, and with effector cells in muscles and glands.

Neuroglial Cells

Neuroglial cells—usually referred to simply as **glial cells** or **glia**—are quite different from nerve cells. The major distinction is that glia do not participate directly in synaptic interactions and electrical signaling, although their supportive functions help define synaptic contacts and maintain the signaling abilities of neurons. Glia are more numerous than nerve cells in the brain, outnumbering them by a ratio of perhaps 3 to 1. Although glial cells also have complex processes extending from their cell bodies, they are generally smaller than neurons, and they lack axons and dendrites (Figure 1.4). The term *glia* (from the Greek word meaning "glue") reflects the nineteenth-century presumption that these cells held the nervous system together in some way. The word has survived, despite the lack of any evidence that binding nerve cells together is among the many functions of glial cells. Glial roles that *are* well-established include maintaining the ionic milieu of nerve cells, modulating the rate of nerve signal propagation, modulating synaptic action by controlling the uptake of neurotransmitters, providing a scaffold for some aspects of neural development, and aiding in (or preventing, in some instances) recovery from neural injury.

There are three types of glial cells in the mature central nervous system: astrocytes, oligodendrocytes, and microglial cells (Figure 1.4A–C). **Astrocytes**, which are restricted to the brain and spinal cord, have elaborate local processes that give these cells a starlike appearance (hence the prefix "astro"). The major function of astrocytes is to maintain, in a variety of ways, an appropriate chemical environment for neuronal signaling. **Oligodendrocytes**, which are also restricted to the central nervous system, lay down a laminated, lipid-rich wrapping called **myelin** around some, but not all, axons. Myelin has important effects on the speed of action potential conduction (see Chapter 3). In the peripheral nervous system, the cells that elaborate myelin are called **Schwann cells**. As the name implies, **microglial cells** are smaller cells derived from hematopoietic stem cells (although some may be derived directly from neural stem cells). They share many properties with

Figure 1.4 Neuroglial cells. Tracings of an astrocyte (A), an oligodendrocyte (B), and a microglial cell (C) visualized by impregnation with silver salts. The images are at approximately the same scale. (D) Astrocytes in the brain labeled with an antibody against the astrocyte-specific protein (glial fibrillary acidic protein). (E) A scanning electron micrograph of a single oligodendroglial cell imaged in tissue culture. (F) A peripheral axon ensheated by the processes of Schwann cells (in green), except for at a distinct region called the node of Ranvier (labeled in red). (G) A microglial cell from the spinal cord, labeled with a cell-type-specific antibody. (A–C after Jones and Cowan, 1983; D courtesy of A.-S. LaMantia; E,F courtesy of B. Popko; G courtesy of A. Light.)

(A) Astrocyte (B) Oligodendrocyte (C) Microglial cell

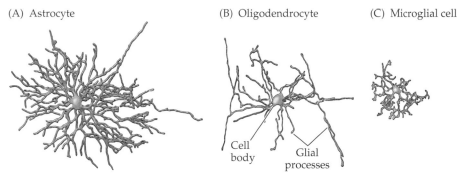

Cell body Glial processes

(D) (E) (F) (G)

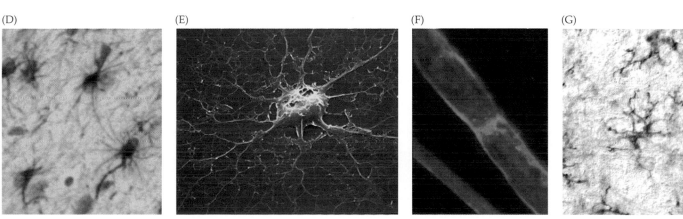

Figure 1.5 A simple reflex circuit, the knee-jerk response (more formally, the myotatic reflex), illustrates several points about the functional organization of neural circuits. Stimulation of peripheral sensors (a muscle stretch receptor in this case) initiates receptor potentials that trigger action potentials that travel centrally along the *afferent* axons of the sensory neurons. This information stimulates spinal motor neurons by means of synaptic contacts. The action potentials triggered by the synaptic potential in motor neurons travel peripherally in *efferent* axons, giving rise to muscle contraction and a behavioral response. One of the purposes of this particular reflex is to help maintain an upright posture in the face of unexpected changes.

tissue macrophages, and are primarily scavenger cells that remove cellular debris from sites of injury or normal cell turnover. Indeed, some neurobiologists prefer to categorize microglia as a type of macrophage. Following brain damage, the number of microglia at the site of injury increases dramatically. Some of these cells proliferate from microglia resident in the brain, while others come from macrophages that migrate to the injured area from the circulation.

Neural Circuits

Neurons never function in isolation; they are organized into ensembles or **circuits** that process specific kinds of information. Although the arrangement of neural circuits varies greatly according to the intended function, some features are characteristic of all such ensembles. The synaptic connections that define a circuit are typically made in a dense tangle of dendrites, axons terminals, and glial cell processes that together constitute **neuropil** (the suffix *-pil* comes from the Greek word *pilos*, meaning "felt"; see Figure 1.3). Thus, the neuropil between nerve cell bodies is the region where most synaptic connectivity occurs. The direction of information flow in any particular circuit is essential to understanding its function. Nerve cells that carry information toward the central nervous system (or farther centrally within the spinal cord and brain) are called **afferent neurons**; nerve cells that carry information away from the brain or spinal cord (or away from the cir-

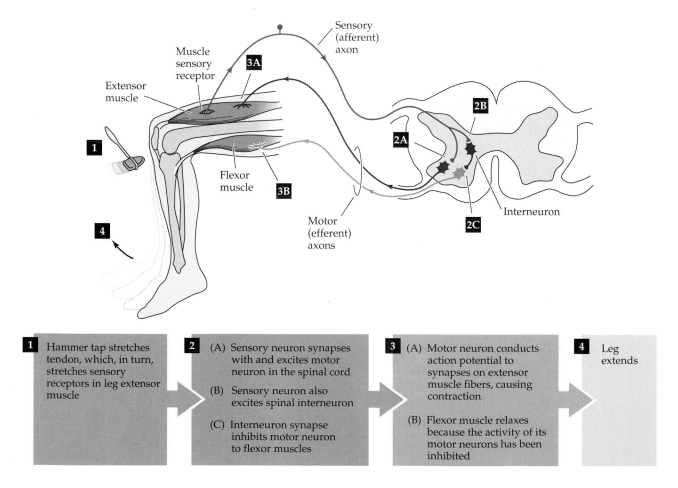

cuit in question) are called **efferent neurons**. Nerve cells that only participate in the local aspects of a circuit are called **interneurons** or **local circuit neurons**. These three classes—afferent neurons, efferent neurons, and interneurons—are the basic constituents of all neural circuits.

Neural circuits are both anatomical and functional entities. A simple example is the circuit that subserves the **myotatic** (or "knee-jerk") **spinal reflex** (Figure 1.5). The afferent limb of the reflex is **sensory neurons** of the dorsal root ganglion in the periphery. These afferents target neurons in the spinal cord. The efferent limb comprises **motor neurons** in the ventral horn of the spinal cord with different peripheral targets: One efferent group projects to flexor muscles in the limb, and the other to extensor muscles. The third element of this circuit is interneurons in the ventral horn of the spinal cord. The interneurons receive synaptic contacts from the sensory afferent neurons and make synapses on the efferent motor neurons that project to the flexor muscles. The synaptic connections between the sensory afferents and the extensor efferents are excitatory, causing the extensor muscles to contract; conversely, the interneurons activated by the afferents are inhibitory, and their activation by the afferents diminishes electrical activity in motor neurons and causes the flexor muscles to become less active (Figure 1.6). The result is a complementary activation and inactivation of the synergist and antagonist muscles that control the position of the leg.

A more detailed picture of the events underlying the myotatic or any other circuit can be obtained by electrophysiological recording (Figures 1.6 and 1.7). There are two basic approaches to measuring electrical activity: extracellular recording where an electrode is placed *near* the nerve cell of interest to detect activity, and intracellular recording where the electrode is placed *inside* the cell. Such recordings detect two basic types of signals. Extracellular recordings primarily detect action potentials, the all-or-nothing changes in the potential across nerve cell membranes that convey information from one point to another in the nervous system. Intracellular recordings can detect the smaller graded potential changes that serve to trigger action potentials. These graded triggering potentials can arise at either sensory receptors or synapses and are called receptor potentials or synaptic potentials, respectively. For the myotatic circuit, action potential activity can be measured from each element (afferents, efferents, and interneurons) before, during, and after a stimulus (see Figure 1.6). By comparing the onset,

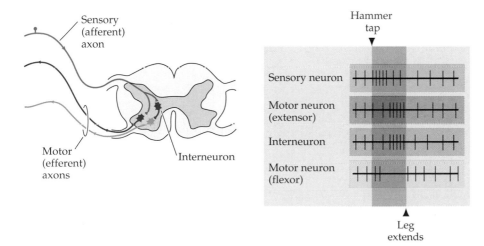

Figure 1.6 Relative frequency of action potentials in different components of the myotatic reflex as the reflex pathway is activated.

Figure 1.7 Intracellularly recorded responses underlying the myotatic reflex. (A) Action potential measured in a sensory neuron. (B) Postsynaptic triggering potential recorded in an extensor motor neuron. (C) Postsynaptic triggering potential in an interneuron. (D) Postsynaptic inhibitory potential in a flexor motor neuron. Such intracellular recordings are the basis for understanding the cellular mechanisms of action potential generation, and the sensory receptor and synaptic potentials that trigger these conducted signals.

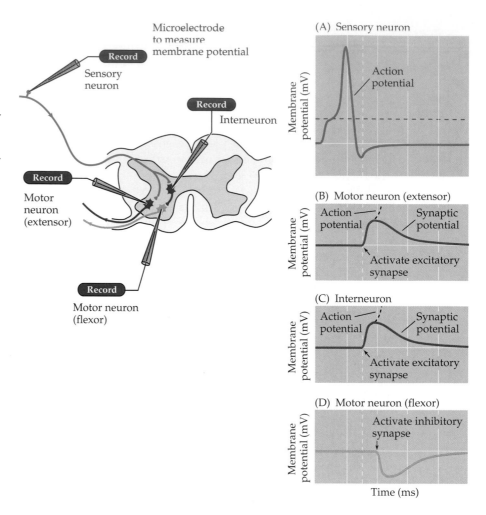

duration, and frequency of action potential activity in each cell, a functional picture of the circuit emerges. As a result of the stimulus, the sensory neuron is triggered to fire at higher frequency (i.e., more action potentials per unit time). This increase triggers in turn a higher frequency of action potentials in both the extensor motor neurons and the interneurons. Concurrently, the inhibitory synapses made by the interneurons onto the flexor motor neurons cause the frequency of action potentials in these cells to decline. Using intracellular recording (see Chapter 2), it is possible to observe directly the potential changes underlying the synaptic connections of the myotatic reflex circuit, as illustrated in Figure 1.7.

Neural Systems

Circuits that subserve similar functions are grouped in **neural systems** that serve broader behavioral purposes. The most general functional definition divides neural systems into **sensory systems** like vision or hearing that acquire and process information from the environment, and **motor systems** that allow the organism to respond to such information by generating movements. There are, however, large numbers of cells and circuits that lie between these relatively well defined input and output systems. These are collectively referred to as **associational systems**, and they carry out the most complex and least well characterized brain functions.

(A)

(B)

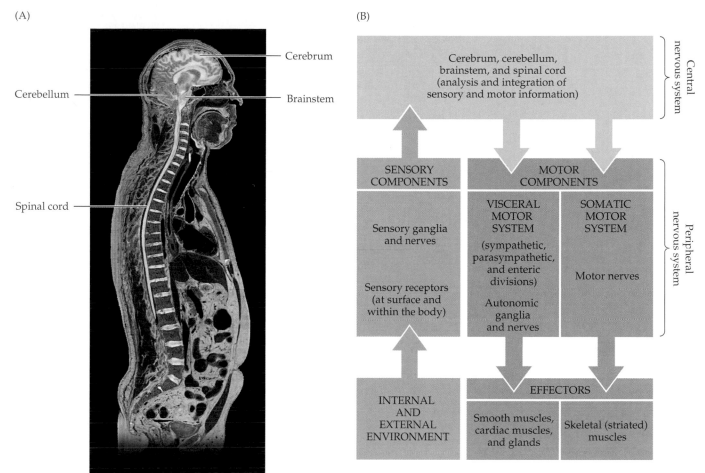

Figure 1.8 The major components of the nervous system and their functional relationships. (A) Digital 3-D reconstruction of the human body, illustrating the position of the brain, brainstem, and spinal cord in an intact human. (B) Diagram of the major components of the central and peripheral nervous systems and their functional relationships. Stimuli from the environment convey information to processing circuits within the brain and spinal cord, which in turn interpret their significance and send signals to peripheral effectors that move the body and adjust the workings of its internal organs. (A, modified from NPAC Visible Human Viewer, Northeast Parallel Architectures Center, Syracuse University.)

In addition to these broad functional distinctions, neuroscientists and neurologists have conventionally divided the vertebrate nervous system anatomically into central and peripheral components (Figure 1.8). The **central nervous system** comprises the **brain** (cerebrum, cerebellum, and brainstem) and the **spinal cord**. The **peripheral nervous system** includes sensory neurons, which link sensory receptors on the body surface, as well as in specialized receptor structures like the ear, with processing circuits in the central nervous system. The motor portion of the peripheral nervous system consists of two components. Motor axons that connect the brain and spinal cord to skeletal muscles make up the **somatic motor division** of the peripheral nervous system. The **visceral** or **autonomic motor division** consists of cells and axons that innervate smooth muscles, cardiac muscle, and glands.

In the peripheral nervous system, nerve cells are located in **ganglia**, which are simply local accumulations of nerve cell bodies (and supporting cells). Peripheral axons are gathered into **nerves**, which are bundles of axons, many of which are enveloped by the glial cells of the peripheral nervous system, the Schwann cells. In the central nervous system, nerve cells are arranged in two different configurations. **Nuclei** are compact accumulations of neurons having roughly similar connections and functions; these concentrated collections of nerve cells are found throughout the brain and spinal cord. In contrast, **cortex** (plural, cortices) describes sheet-like arrays of nerve cells. The cortices of the cerebral hemispheres and of the cerebellum provide the clearest example of this organizational principle. Axons in the central nervous system are gathered into **tracts**. Within a tract, glial cells of the central nervous system—astrocytes and oligodendrocytes—envelop the central axons. Finally, two gross histological terms applied to the central nervous system distinguish regions rich in neuronal cell bodies versus regions rich in axons. **Gray matter** refers to any accumulation of cell bodies and neuropil in the brain and spinal cord (e.g., nuclei or cortices), whereas **white matter** refers to axon tracts.

In the sensory portion of the peripheral nervous system, **sensory ganglia** lie adjacent to either the spinal cord (where they are referred to as **dorsal root ganglia;** see Figure 1.10) or the brainstem (where they are called **cranial nerve ganglia;** see Box A). The nerve cells in sensory ganglia send axons to the periphery that end in (or on) specialized receptors that transduce information about a wide variety of stimuli. The central processes of these sensory ganglion cells enter the spinal cord or brainstem. In the somatic motor portion of the peripheral nervous system, axons from motor neurons in the spinal cord give rise to peripheral motor axons that innervate the striated muscles to control skeletal movements and, consequently, most voluntary behaviors. The organization of the autonomic division of the peripheral nervous system is a bit more complicated. **Preganglionic visceral motor neurons** in the brainstem and spinal cord form synapses with peripheral motor neurons that lie in the **autonomic ganglia**. The motor neurons in autonomic ganglia innervate smooth muscle, glands, and cardiac muscle, thus controlling most involuntary (visceral) behavior. In the **sympathetic division** of the autonomic motor system, the ganglia are along or in front of the vertebral column and send their axons to a variety of peripheral targets. In the **parasympathetic division**, the ganglia are found within the organs they innervate. Another component of the visceral motor system, called the **enteric system**, is made up of small ganglia scattered throughout the wall of the gut.

Some Anatomical Terminology

To understand the spatial organization of these systems, some additional vocabulary employed to describe them needs to be defined. The terms used to specify location in the central nervous system are the same as those used for the gross anatomical description of the rest of the body (Figure 1.9). Thus, anterior and posterior indicate front and back; rostral and caudal, toward the head and tail; dorsal and ventral, top and bottom; and medial and lateral, the midline or to the side. Nevertheless, the comparison between these coordinates in the body versus the brain can be confusing. For the entire body these anatomical terms refer to the long axis, which is straight. The long axis of the central nervous system, however, has a bend in it. In human and other bipeds, a compensatory tilting of the rostral/caudal axis for the brain is necessary to properly compare body axes to brain axes. Once this adjustment has been made, the other axes for the brain can be easily assigned.

The proper assignment of these anatomical axes then dictates the standard planes for histological sections or tomographic images used to study the internal anatomy of the brain (see Figure 1.9C). **Horizontal sections** are taken parallel to the rostral/caudal axis of the brain. Sections taken in the plane dividing the two hemispheres are **sagittal**, and can be further categorized as **median** and **paramedian** according to whether the section is near the midline (median or midsagittal) or more lateral (paramedian). Sections in the plane of the face are called **frontal** or **coronal**. Different terms are usu-

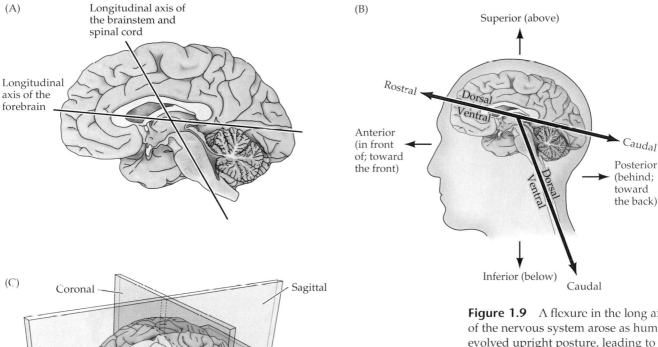

(A)

Longitudinal axis of the brainstem and spinal cord

Longitudinal axis of the forebrain

(B)

Superior (above)

Rostral

Dorsal
Ventral

Anterior (in front of; toward the front)

Caudal

Posterior (behind; toward the back)

Dorsal
Ventral

Inferior (below)

Caudal

(C)

Coronal

Sagittal

Horizontal

Figure 1.9 A flexure in the long axis of the nervous system arose as humans evolved upright posture, leading to an approximately 120° angle between the long axis of the brainstem and that of the forebrain (A). The consequences of this flexure for anatomical terminology are indicated in (B). The terms *anterior, posterior, superior,* and *inferior* refer to the long axis of the body, which is straight. Therefore, these terms indicate the same direction for both the forebrain and the brainstem. In contrast, the terms *dorsal, ventral, rostral,* and *caudal* refer to the long axis of the central nervous system. The dorsal direction is toward the back for the brainstem and spinal cord, but toward the top of the head for the forebrain. The opposite direction is ventral. The rostral direction is toward the top of the head for the brainstem and spinal cord, but toward the face for the forebrain. The opposite direction is caudal. (C) The major planes of section used in cutting or imaging the brain.

ally used to refer to sections of the spinal cord. The plane of section orthogonal to the long axis of the cord is called **transverse**, whereas sections parallel to the long axis of the cord are called **longitudinal**. In a transverse section through the human spinal cord, the dorsal and ventral axes and the anterior and posterior axes indicate the same directions. Tedious though this terminology may be, it is essential for understanding the basic subdivisions of the nervous system.

The Subdivisions of the Central Nervous System

The central nervous system (defined as the brain and spinal cord) is usually considered to have seven basic parts: the **spinal cord**, the **medulla**, the **pons**, the **cerebellum**, the **midbrain**, the **diencephalon**, and the **cerebral hemispheres** (Figure 1.10; see also Figure 1.8). The medulla, pons, and midbrain are collectively called the **brainstem** (Box A); the diencephalon and cerebral hemispheres are collectively called the **forebrain**. Within the brainstem are found **cranial nerve nuclei** that either receive input from **cranial sensory ganglia** via their respective **cranial sensory nerves** or give rise to axons that

(A)

(B)

(C)

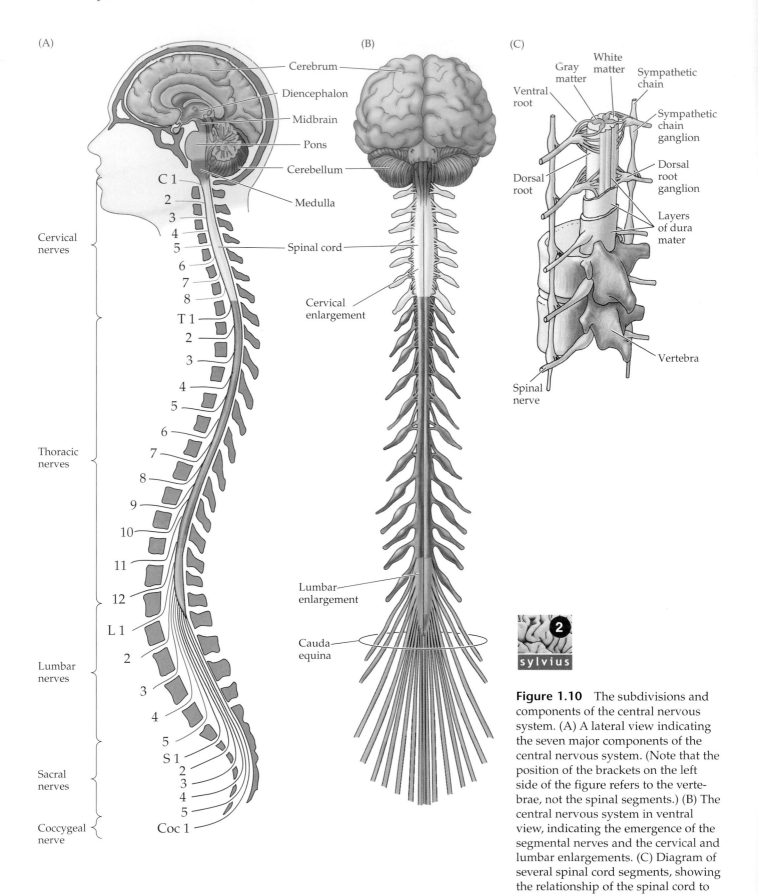

Cerebrum

Diencephalon

Midbrain

Pons

Cerebellum

Medulla

Spinal cord

Cervical enlargement

Lumbar enlargement

Cauda equina

Cervical nerves

Thoracic nerves

Lumbar nerves

Sacral nerves

Coccygeal nerve

C 1
2
3
4
5
6
7
8
T 1
2
3
4
5
6
7
8
9
10
11
12
L 1
2
3
4
5
S 1
2
3
4
5
Coc 1

Ventral root

Gray matter

White matter

Sympathetic chain

Sympathetic chain ganglion

Dorsal root

Dorsal root ganglion

Layers of dura mater

Vertebra

Spinal nerve

Figure 1.10 The subdivisions and components of the central nervous system. (A) A lateral view indicating the seven major components of the central nervous system. (Note that the position of the brackets on the left side of the figure refers to the vertebrae, not the spinal segments.) (B) The central nervous system in ventral view, indicating the emergence of the segmental nerves and the cervical and lumbar enlargements. (C) Diagram of several spinal cord segments, showing the relationship of the spinal cord to the bony canal in which it lies.

BOX A
The Brainstem and Its Importance in Clinical Neuroanatomy

Understanding the internal anatomy of the brainstem is generally regarded as essential for the practice of clinical medicine. The brainstem is the target or source for all cranial nerves that deal with sensory and motor function in the head and neck (figure A). Cranial nerve nuclei within the brainstem are the targets of cranial sensory nerves or the source of cranial motor nerves (figure B). Furthermore, the brainstem provides a "thruway" for all of the ascending sensory tracts from the spinal cord, the sensory tracts for the head neck (the trigeminal system), the descending motor tracts from the forebrain, and local pathways that link eye movement centers. All of these structures are compressed into a relatively small volume that has a regionally restricted vascular supply (see Figure

1.21). Thus, vascular accidents in the brainstem—which are common—result in distinctive, and often devastating, combinations of functional deficits (see Box D). These deficits can be used both for diagnosis and for better understanding of the intricate anatomy of the medulla, pons, and midbrain.

Cranial nerve nuclei that receive sensory input (analogous to the dorsal horns of the spinal cord) are located separately from those that give rise to motor output (which are analogous to the ventral horns; see figure B and Figure 1.11). The primary sensory neurons that innervate these nuclei are found in ganglia associated with the cranial nerves—similar to the relationship between dorsal root ganglia and the spinal cord. In general, sensory nuclei are found laterally in the

brainstem, whereas motor nuclei are located more medially (figure C). There are three types of brainstem motor nuclei: Somatic motor nuclei project to striated muscles; branchial motor nuclei project to muscles derived from embryonic structures called branchial arches (these arches give rise to the muscles—and bones—of the jaws and other craniofacial structures); and visceral motor nuclei project to peripheral ganglia that innervate smooth muscle or glandular targets, similar to preganglionic motor neurons in the spinal cord that innervate autonomic ganglia. Finally, the major ascending or descending tracts—carrying sensory or motor information to or from the brain—are found in the lateral and basal regions of the brainstem (see figure C).

Cranial nerves

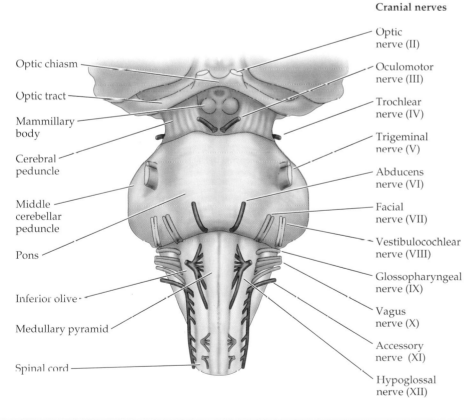

Optic chiasm
Optic tract
Mammillary body
Cerebral peduncle
Middle cerebellar peduncle
Pons
Inferior olive
Medullary pyramid
Spinal cord

Optic nerve (II)
Oculomotor nerve (III)
Trochlear nerve (IV)
Trigeminal nerve (V)
Abducens nerve (VI)
Facial nerve (VII)
Vestibulocochlear nerve (VIII)
Glossopharyngeal nerve (IX)
Vagus nerve (X)
Accessory nerve (XI)
Hypoglossal nerve (XII)

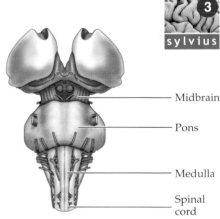

Midbrain
Pons
Medulla
Spinal cord

(A) At left, a ventral view of the brainstem showing the locations of the cranial nerves as they enter or exit the midbrain, pons and medulla. Exclusively sensory nerves are indicated in yellow, motor nerves in blue, and mixed sensory/motor nerves in green. At right, the territories included in each of the brainstem subdivisions (midbrain, violet; pons, green; medulla, pink) are indicated.

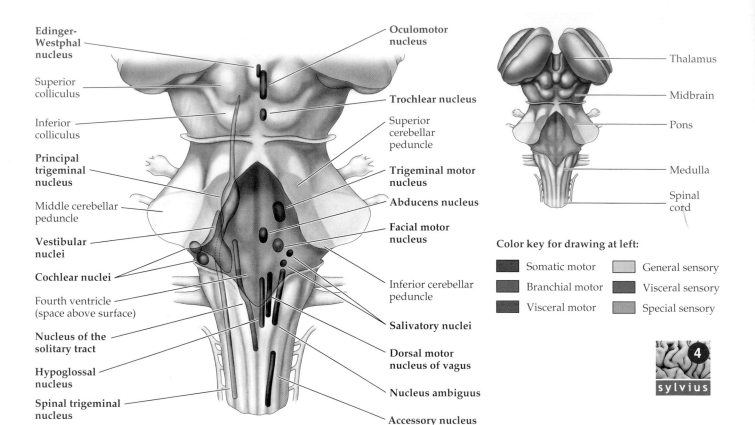

Edinger-Westphal nucleus
Superior colliculus
Inferior colliculus
Principal trigeminal nucleus
Middle cerebellar peduncle
Vestibular nuclei
Cochlear nuclei
Fourth ventricle (space above surface)
Nucleus of the solitary tract
Hypoglossal nucleus
Spinal trigeminal nucleus

Oculomotor nucleus
Trochlear nucleus
Superior cerebellar peduncle
Trigeminal motor nucleus
Abducens nucleus
Facial motor nucleus
Inferior cerebellar peduncle
Salivatory nuclei
Dorsal motor nucleus of vagus
Nucleus ambiguus
Accessory nucleus

Thalamus
Midbrain
Pons
Medulla
Spinal cord

Color key for drawing at left:

Somatic motor	General sensory
Branchial motor	Visceral sensory
Visceral motor	Special sensory

(B) At left, a "phantom" view of the dorsal surface of the brainstem showing the location of the brainstem cranial nerve nuclei that are either the target or the source of the cranial nerves (see Table 1.1 for the relationship between each cranial nerve and the cranial nerve nuclei). With the exception of the cranial nerve nuclei associated with the trigeminal nerve, there is fairly close correspondence between the location of the cranial nerve nuclei in the midbrain, pons, and medulla and the location of the associated cranial nerves. At right, the territories of the major brainstem subdivisions, viewed from the dorsal surface, are indicated.

The rostral/caudal organization of the cranial nerve nuclei (all of which are bilaterally symmetric) reflects the rostro-caudal distribution of head and neck structures (see figures A and B and Table 1.1). The more caudal the nucleus, the more caudally located the target structures in the periphery. This arrangement is most dramatically demonstrated by the dorsal column nuclei in the caudal medulla (see figure C), which are the targets of ascending spinal cord somatic sensory afferents. Similarly, the spinal accessory nuclei in the mid-medulla provide motor innervation for neck and shoulder muscles, and the motor nucleus of the vagus nerve provides pregan-

glionic innervation for many enteric and visceral targets. In the pons, the sensory and motor nuclei deal primarily with somatic sensation from the face (the principle trigeminal nuclei), movement of the jaws and the muscles of facial expression (the facial and trigeminal motor nuclei), and abduction of the eye (the abducens nuclei). Further rostral, in the mesencephalic portion of the brainstem, are nuclei concerned primarily with eye movements (the oculo-motor nuclei) and preganglionic parasympathetic innervation of the iris (the Edinger-Westphal nuclei). While this list is not complete, it indicates the basic order of the rostral/caudal organization of the brainstem.

Neurologists assess combinations of cranial nerve deficits to infer the location of brainstem lesions, or to place the source of brain dysfunction either in the spinal cord or brain. The most common brainstem lesions reflect the vascular territories that supply subsets of cranial nerve nuclei as well as ascending and descending tracts (see figure C). For example, an occlusion of the posterior inferior cerebellar artery (PICA), a branch of the vertebral artery that supplies the lateral region of the mid- and rostral medulla, results in damage to three cranial nerve nuclei and several tracts (see figure C). Accordingly, there are functional deficits that reflect the loss of the spinal trigeminal nucleus, the vestibular nucleus, and the nucleus ambiguus (which contains motor neurons that project to the larynx and pharynx) on the same side as the lesion. In addition, ascending pathways from the spinal cord that relay pain and temperature from the contralateral body surface are disrupted, leading to a contralateral loss of function. Finally, the inferior cerebellar peduncle, which contains projections that relay information about body position to the cerebellum for postural control, is damaged. This results in ataxia (clumsiness) on the side of the lesion. Anatomical relationships and shared vascularization, rather than any obvious functional similarity, unite these deficits and allow for anatomical localization of brainstem damage.

For both clinicians and neurobiologists, thinking about the brainstem requires integrating this sort of anatomical information with knowledge about functional organization and pathology.

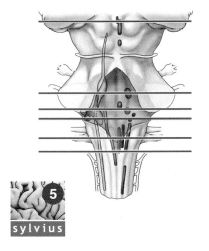

(C) Transverse sections through the brainstem and spinal cord showing internal organization along the rostral/caudal axis. The location of the cranial nerve nuclei, ascending, and descending tracts is indicated in each representative section. The identity of the nuclei (somatic sensory or motor, visceral sensory or motor, branchial sensory or motor) is indicated using the same color key as in figure B. In the section through the medulla, the shaded area indicates the vascular territory of the posterior inferior cerebral artery. Vascular occlusion of this artery results in the functional deficits corresponding to the combination of nuclei and tracts found in this region, described in the text of this Box.

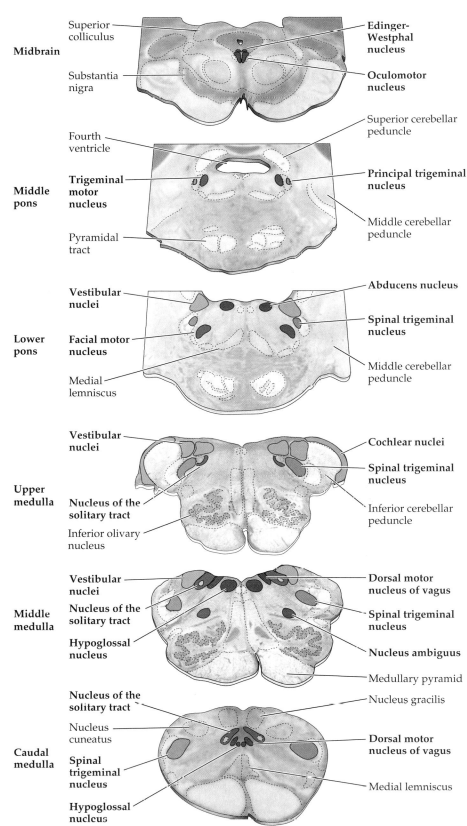

TABLE 1.1
The Cranial Nerves and Their Primary Functions

Cranial nerve	Name	Sensory and/or motor	Major function
I	Olfactory nerve	Sensory	Sense of smell
II	Optic nerve	Sensory	Vision
III	Oculomotor nerve	Motor	Eye movements; papillary constriction and accommodation; muscles of eyelid.
IV	Trochlear nerve	Motor	Eye movements
V	Trigeminal nerve	Sensory and motor	Somatic sensation from face, mouth, cornea; muscles of mastication
VI	Abducens nerve	Motor	Eye movements
VII	Facial nerve	Sensory and motor	Controls the muscles of facial expression; taste from anterior tongue; lacrimal and salivary glands
VIII	Auditory/vestibular nerve	Sensory	Hearing; sense of balance
IX	Glossopharyngeal nerve	Sensory and motor	Sensation from pharynx; taste from posterior tongue; carotid baroreceptors
X	Vagus nerve	Sensory and motor	Autonomic functions of gut; sensation from pharynx; muscles of vocal cords; swallowing
XI	Accessory nerve	Motor	Shoulder and neck muscles
XII	Hypoglossal nerve	Motor	Movements of tongue

constitute **cranial motor nerves** (Table 1.1). In addition, the brainstem is the conduit for several major tracts in the central nervous system. These tracts either relay sensory information from the spinal cord and brainstem to the midbrain and forebrain, or relay motor commands from the midbrain and forebrain back to motor neurons in the brainstem and spinal cord.

The External Anatomy of the Spinal Cord

The spinal cord extends caudally from the brainstem, running from the medullary-spinal junction at about the level of the first cervical vertebra to about the level of the twelfth thoracic vertebra (see Figure 1.10). The vertebral column (and the spinal cord within it) is divided into **cervical, thoracic, lumbar, sacral,** and **coccygeal** regions. The peripheral nerves (called the **spinal or segmental nerves**) that innervate much of the body arise from the spinal cord's 31 segmental pairs. The cervical region of the cord gives rise to eight cervical nerves (C1–C8), the thoracic to twelve thoracic nerves (T1–T12), the lumbar to five lumbar nerves (L1–L5), the sacral to five sacral nerves (S1–S5), and the coccygeal to one coccygeal nerve. The segmental spinal nerves leave the vertebral column through the intervertebral foramina that lie adjacent to the respectively numbered vertebral body. Sensory information carried by the afferent axons of the spinal nerves enters the cord via the **dorsal roots**, and motor commands carried by the efferent axons leave the cord via the **ventral roots** (see Figure 1.10C). Once the dorsal and ventral roots join, sensory and motor axons (with some exceptions) travel together in the segmental spinal nerves.

Location of cells whose axons form the nerve	Clinical test of function
Nasal epithelium	Test sense of smell with standard odor
Retina	Measure acuity and integrity of visual field
Oculomotor nucleus in midbrain; Edinger-Westphal nucleus in midbrain	Test eye movements (patient can't look up, down, or medially if nerve involved); look for ptosis, pupillary dilation
Trochlear nucleus in midbrain	Can't look downward when eye abducted
Trigeminal motor nucleus in pons; trigeminal sensory ganglion (the gasserian ganglion)	Test sensation on face; palpate masseter muscles and temporal muscle
Abducens nucleus in midbrain	Can't look laterally
Facial motor nucleus; superior salivatory nuclei in pons; trigeminal (gasserian) ganglion	Test facial expression plus taste on anterior tongue
Spiral ganglion; vestibular (Scarpa's) ganglion	Test audition with tuning fork; vestibular function with caloric test
Nucleus ambiguus; inferior salivatory	Test swallowing; pharyngeal gag reflex
Dorsal motor nucleus of vagus; vagal nerve ganglion	Test above plus hoarseness
Spinal accessory nucleus; nucleus ambiguus; intermediolateral column of spinal cord	Test sternocleidomastoid and trapezius muscles
Hypoglossal nucleus of medulla	Test deviation of tongue during protrusion (points to side of lesion)

Two regions of the spinal cord are enlarged to accommodate the greater number of nerve cells and connections needed to process information related to the upper and lower limbs (see Figure 1.10B). The spinal cord expansion that corresponds to the arms is called the **cervical enlargement** and includes spinal segments C5–T1; the expansion that corresponds to the legs is called the **lumbar enlargement** and includes spinal segments L2–S3. Because the spinal cord is considerably shorter than the vertebral column (see Figure 1.10A), lumbar and sacral nerves run for some distance in the vertebral canal before emerging, thus forming a collection of nerve roots known as the **cauda equina**. This region is the target for an important clinical procedure called a "lumbar puncture" that allows for the collection of cerebrospinal fluid by placing a needle into the space surrounding these nerves to withdraw fluid for analysis. In addition, local anesthetics can be safely introduced to produce spinal anesthesia; at this level, the risk of damage to the spinal cord from a poorly placed needle is minimized.

The Internal Anatomy of the Spinal Cord

The arrangement of gray and white matter in the spinal cord is relatively simple: The interior of the cord is formed by gray matter, which is surrounded by white matter (Figure 1.11A). In transverse sections, the gray matter is conventionally divided into dorsal (posterior) lateral and ventral (anterior) "horns." The neurons of the **dorsal horns** receive sensory information that enters the spinal cord via the dorsal roots of the spinal nerves. The **lateral** horns are present primarily in the thoracic region, and contain the preganglionic visceral

(A)

(B)

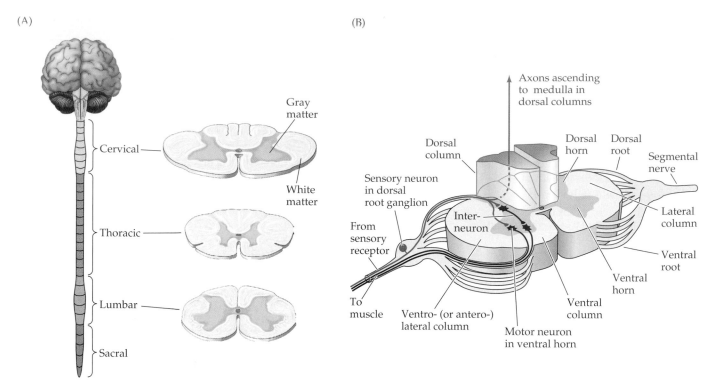

Figure 1.11 Internal structure of the spinal cord. (A) Transverse sections of the cord at three different levels, showing the characteristic arrangement of gray and white matter in the cervical, thoracic, and lumbar cord. (B) Diagram of the internal structure of the spinal cord.

motor neurons that project to the sympathetic ganglia (see Figure 1.10C). The **ventral horns** contains the cell bodies of motor neurons that send axons via the ventral roots of the spinal nerves to terminate on striated muscles. The white matter of the spinal cord is subdivided into dorsal (or posterior), lateral, and ventral (or anterior) columns, each of which contains axon tracts related to specific functions. The **dorsal columns** carry ascending sensory information from somatic mechanoreceptors (Figure 1.11B). The **lateral columns** include axons that travel from the cerebral cortex to contact spinal motor neurons. These pathways are also referred to as the **cortico-spinal tracts**. The **ventral** (and **ventrolateral** or **anterolateral**) **columns** carry both ascending information about pain and temperature, and descending motor information. Some general rules of spinal cord organization are (1) that neurons and axons that process and relay sensory information are found dorsally; (2) that preganglionic visceral motor neurons are found in an intermediate/lateral region; and (3) that somatic motor neurons and axons are found in the ventral portion of the cord.

The External Anatomy of the Brain: Some General Points

Three major structures are visible in most views of the human brain: the cerebral hemispheres, the cerebellum, and the caudal or medullary portion of the brainstem (see Figure 1.12). In addition to the large size of the cerebral hemispheres (about 85 percent of the brain by weight), their surface is highly convoluted. The ridges of these convolutions are known as **gyri** (singular, gyrus), and the valleys between them are called **sulci** (singular, sulcus) or, if they are especially deep, **fissures**. The convoluted surface of the cerebral hemispheres comprises a continuous layered or **laminated** sheet of neurons and supporting cells about 2 mm thick called the **cerebral cortex**. Neuroanatomists and evolutionary biologists have argued about the significance of cerebral convolutions, without clear resolution, for at least a century. All agree, however, that the infolding of the cerebral hemispheres allows a great deal more cortical surface area (about 1.6 m² on average) to exist within the confines of the cranium. Specific features of the brain are best appreciated in one of several different views.

The Lateral Surface of the Brain

A lateral view of the human brain is the best perspective from which to appreciate the lobes of the cerebral hemisphere (Figure 1.12A). Each hemisphere is conventionally divided into four **lobes**, named for the bones of the skull that overlie them: the **frontal**, **parietal**, **temporal**, and **occipital lobes**. The frontal lobe is the most anterior, and is separated from the parietal lobe by the **central sulcus** (Figure 1.12B). A particularly important feature of the frontal lobe is the **precentral gyrus**. (The prefix *pre-*, when used anatomically, refers to a structure that is in front of or anterior to another.) The cortex of the precentral gyrus is referred to as the **motor cortex** and contains

(A)

(B)

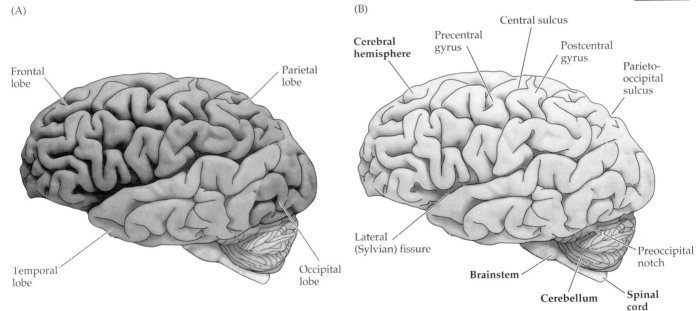

(C)

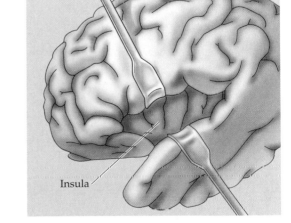

Figure 1.12 Lateral view of the human brain. (A) The four lobes of the brain. (B) Some of the major sulci and gyri evident from this perspective. (C) The banks of the lateral, or Sylvian, fissure have been pulled apart to expose the insula.

neurons whose axons project to the motor neurons in the brainstem and spinal cord that innervate the skeletal (striated) muscles. The temporal lobe extends almost as far anterior as the frontal lobe but is inferior to it, the two lobes being separated by the **lateral** (or **Sylvian**) **fissure**. The superior aspect of the temporal lobe contains cortex concerned with audition, and inferior portions deal with highly processed visual information. Hidden beneath the frontal and temporal lobes, the **insula** can be seen only if these two lobes are pulled apart or removed (Figure 1.12C). The insular cortex is largely concerned with visceral and autonomic function, including taste. The parietal lobe lies posterior to the central sulcus and superior to the lateral fissure. The **postcentral gyrus**, the most anterior gyrus in the parietal lobe, harbors cortex that is concerned with somatic (bodily) sensation; this area is therefore referred to as the **somatic sensory cortex**. The boundary between the parietal lobe and the occipital lobe, the most posterior of the hemispheric lobes, is somewhat arbitrary (a line from the parieto-occipital sulcus to the preoccipital notch). The occipital lobe, only a small part of which is apparent from the lateral surface of the brain, is primarily concerned with vision. In addition to their role in primary and sensory processing, each cortical lobe has characteristic cognitive functions. Thus, the frontal lobe is critical in planning responses to stimuli, the parietal lobe in attending to stimuli, the temporal lobe in recognizing stimuli, and the occipital lobe in vision.

The Dorsal and Ventral Surfaces of the Brain

Although the primary subdivisions of the cerebral hemispheres can be appreciated from a lateral view, other key landmarks are better seen from the dorsal and ventral surfaces. When viewed from the dorsal surface (Figure 1.13A), the approximate bilateral symmetry of the cerebral hemisphere is apparent. Although there is some variation, major landmarks like the central sulci and parieto-occipital sulci are usually very similar in arrangement on the two sides. If the cortical hemispheres are spread slightly apart, another major structure, the **corpus callosum**, can be seen bridging the two hemispheres. This tract contains axons that originate from neurons in both cerebral hemispheres that contact target nerve cells in the opposite hemisphere.

The external features of the brain best seen on its ventral surface are shown in Figure 1.13B. Extending along the inferior surface of the frontal lobe near the midline are the **olfactory tracts**, which arise from enlargements at their anterior ends called the **olfactory bulbs**. The olfactory bulbs receive input from neurons in the epithelial lining of the nasal cavity whose axons make up the first cranial nerve (cranial nerve I is therefore called the **olfactory nerve**; see Table 1.1). On the ventromedial surface of the temporal lobe, the **parahippocampal gyrus** conceals the **hippocampus**, a highly convoluted cortical structure that figures importantly in memory. Slightly more medial to the parahippocampal gyrus is the **uncus**, a slightly conical protrusion that includes the **pyriform cortex**. The pyriform cortex is the target of the lateral olfactory tract and processes olfactory information. At the most central aspect of the ventral surface of the forebrain is the **optic chiasm**, and immediately posterior, the ventral surface of the **hypothalamus**, including the **infundibular stalk** (the base of pituitary gland) and the **mammillary bodies**. Posterior to the hypothalamus are two large tracts, oriented roughly rostral/caudally, called the **cerebral peduncles**. These tracts contain axons from the cerebral hemispheres that project to the motor neurons in the brainstem

(A) **Dorsal view**

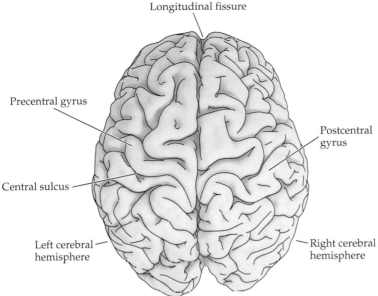

Longitudinal fissure

Precentral gyrus

Postcentral gyrus

Central sulcus

Left cerebral hemisphere

Right cerebral hemisphere

(B) **Ventral view**

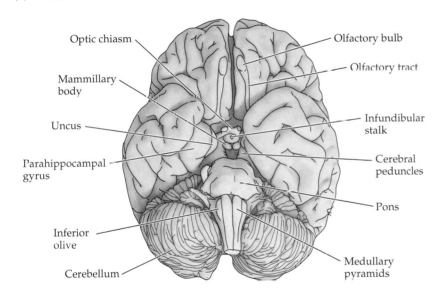

Optic chiasm

Olfactory bulb

Mammillary body

Olfactory tract

Uncus

Infundibular stalk

Parahippocampal gyrus

Cerebral peduncles

Pons

Inferior olive

Cerebellum

Medullary pyramids

(C)

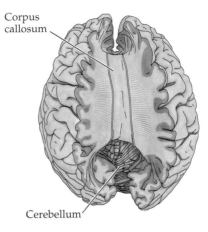

Corpus callosum

Cerebellum

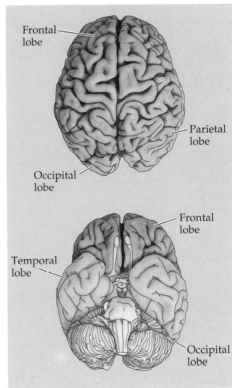

Frontal lobe

Parietal lobe

Occipital lobe

Frontal lobe

Temporal lobe

Occipital lobe

Figure 1.13 Dorsal view (A) and ventral view (B) of the human brain, indicating some of the major features visible from these perspectives. (C) The cerebral cortex has been removed in this dorsal view to reveal the underlying corpus callosum. (After Rohen et al., 1993.)

and into the lateral and ventral columns of the spinal cord. Finally, the ventral surfaces of the pons, medulla, and cerebellar hemispheres can be seen on the ventral surface of the brain (see also Box A).

The Midline Sagittal Surface of the Brain

When the brain is hemisected in the midsagittal plane, all of its major subdivisions plus a number of additional structures are visible on the cut surface (Figure 1.14). In this view, the cerebral hemispheres, because of their great size, are still the most obvious structures. The frontal lobe of each hemi-

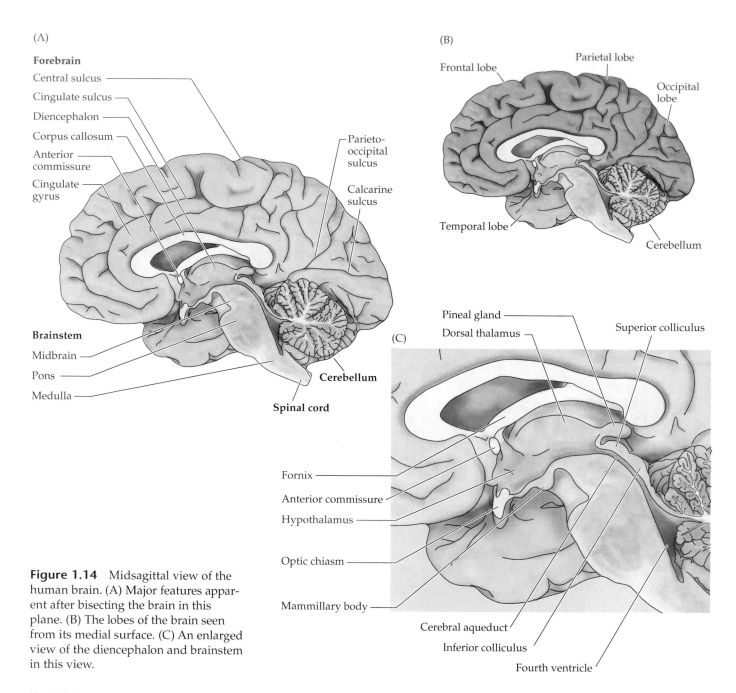

(A)

Forebrain

Central sulcus

Cingulate sulcus

Diencephalon

Corpus callosum

Anterior commissure

Cingulate gyrus

Parieto-occipital sulcus

Calcarine sulcus

Brainstem

Midbrain

Pons

Medulla

Cerebellum

Spinal cord

(B)

Frontal lobe

Parietal lobe

Occipital lobe

Temporal lobe

Cerebellum

(C)

Pineal gland

Dorsal thalamus

Superior colliculus

Fornix

Anterior commissure

Hypothalamus

Optic chiasm

Mammillary body

Cerebral aqueduct

Inferior colliculus

Fourth ventricle

Figure 1.14 Midsagittal view of the human brain. (A) Major features apparent after bisecting the brain in this plane. (B) The lobes of the brain seen from its medial surface. (C) An enlarged view of the diencephalon and brainstem in this view.

sphere extends forward from the central sulcus, the medial end of which can just be seen (Figure 1.14A,B). The **parieto-occipital sulcus**, running from the superior to the inferior aspect of the hemisphere, separates the parietal and occipital lobes. The **calcarine sulcus** divides the medial surface of the occipital lobe, running at very nearly a right angle from the parieto-occipital sulcus and marking the location of the primary visual cortex. A long, roughly horizontal sulcus, the **cingulate sulcus**, extends across the medial surface of the frontal and parietal lobes. The prominent gyrus below it, the **cingulate gyrus**, along with the cortex adjacent to it, is sometimes called the "limbic lobe" (the use of the term lobe here is used only loosely, as this region is not considered a fifth lobe of the brain). The "limbic lobe" (*limbic* means border or edge), which wraps around the corpus callosum, and the subcortical areas connected to it are referred to as the **limbic system**. These limbic structures are important in the regulation of visceral motor activity and emotional expression, among other functions. Finally, ventral to the cingulate gyrus is the midsagittal surface of the corpus callosum.

Although parts of the diencephalon, brainstem, and cerebellum are visible at the ventral surface of the brain, their overall structure is especially clear from the midsagittal surface (see Figure 1.14A). From this perspective, the diencephalon can be seen to consist of two parts. The **dorsal thalamus**, the largest component of the diencephalon, comprises a number of subdivisions, all of which relay information to the cerebral cortex from other parts of the brain. The **hypothalamus**, a small but especially crucial part of the diencephalon, is devoted to the control of homeostatic and reproductive functions. The hypothalamus is intimately related, both structurally and functionally, to the pituitary gland, a critical endocrine organ whose posterior part is attached to the hypothalamus by the pituitary stalk (or infundibulum; Figure 1.14C).

The midbrain, which can be seen only in this view, lies caudal to the thalamus, with the **superior and inferior colliculi** defining its dorsal surface or **tectum** (meaning "roof"); several midbrain nuclei, including the substantia nigra, lie in the ventral portion or **tegmentum** (meaning "covering") of the midbrain. The other prominent anatomical feature of the midbrain—the cerebral peduncles (also visible from the ventral surface)—are not apparent in a midsagittal view. The **pons** is caudal to the midbrain along the midsagittal surface, and the **cerebellum** lies over the pons just beneath the occipital lobe of the cerebral hemispheres. The major function of the cerebellum is coordination of motor activity, posture, and equilibrium. From the midsagittal surface, the most visible feature of the cerebellum is the **cerebellar cortex**, a continuous layered sheet of cells folded into ridges and valleys called **folia**. The most caudal structure seen from the midsagittal surface of the brain is the **medulla**, which merges into the spinal cord.

The Internal Anatomy of the Brain

A much more detailed neuroanatomical picture is apparent in gross or histological slices, or sections, through the brain. In these sections, deep structures that are not visible from any brain surface can be identified. In addition, relationships between brain structures seen from the surface can be appreciated more fully. The major challenge to understanding the internal anatomy of the brain is to integrate the rostral/caudal, dorsal/ventral, and medio-lateral landmarks seen on the brain surface with the position of structures seen in brain sections taken in the horizontal, frontal, or sagittal plane. This challenge is not only important for understanding brain function, it is also essential for interpreting noninvasive images of the brain (Boxes B and C), most of which are displayed as sections.

The Internal Anatomy of the Cerebral Hemispheres and Diencephalon

In any plane of section through the forebrain, the **cerebral cortex** is evident as a thin layer of neural tissue that covers the entire cerebrum. Most cerebral cortex is made up of six layers, and is referred to as **neocortex**. Phylo-

BOX B
Anatomical Brain Imaging Techniques

Until the early 1970s, the only technique available for imaging the structure of a living brain was X-ray technology. Conventional X-rays, however, have poor soft tissue contrast, involve relatively high radiation exposure, and provide only a two-dimensional view of brain structure or vasculature, making localization of lesions uncertain. As a result of these several problems, there was strong motivation in the 1960s and 1970s to discover better ways to image the brain.

A major advance was the development of computerized tomography (CT). CT uses a movable X-ray tube that is rotated around the patient's head. Opposite the tube (i.e., on the other side of the patient's head) are X-ray detectors far more sensitive than conventional film, thereby allowing much shorter exposure times (and less risk of radiation damage). Rather than acquiring a single image, as in conventional X-ray pictures, a CT scan gathers intensity information from many positions around the patient's head. These data are entered into a matrix, and the radiodensity at each point in the three-dimensional space of the head is calculated. With a sufficiently narrow X-ray beam, sensitive detectors, and digital signal processing techniques, small differences in radiodensity can be converted into an image. Since the information is gathered for the full volume of the head, the computed matrix contains information about the entire brain. It is therefore possible to generate "slices," or tomograms (*tomo* means "cut" or "slice") of various planes through the brain, visualizing internal structures at any desired level. CT scans readily distinguish gray matter and white matter, differentiate the ventricles quite well, and show many other brain structures with a spatial resolution of several millimeters.

Although computerized tomography opened a new era in brain imaging, it has been largely superseded by a technique called magnetic resonance imaging (MRI). How nuclear magnetic resonance produces an image is more difficult to explain than X-ray imaging. Magnetic resonance derives from the interaction of a magnet and a magnetic field. Consider,

for example, Earth's m... compass magnet. At r... needle points north. If tapped, however, it wil... forth (oscillate) at a fre... proportional to the ma... strength. The needle w... oscillate until friction... energy imparted by th... dle again points north... tion frequency is prop... netic field strength, kn... spatial variation of the oscillation frequency o... in principle be used to an image of) the locati... Earth's surface (albeit... atomic nuclei—princip... as the compass needle,... net plays the role of Ear... Graded spatial variatio... field are actually gener... gradient magnets orien... nal axes. If all the atom... by the magnetic field a... with a brief radiofrequ... emit energy in an oscil...

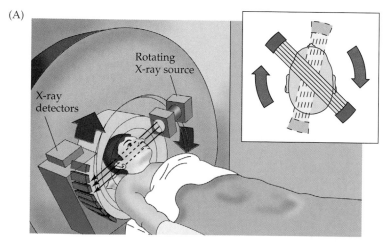

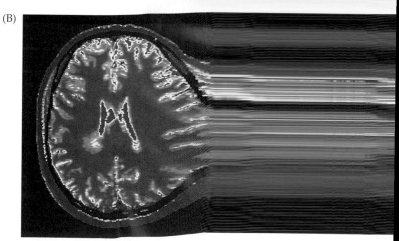

(A) In computerized tomography, the X-ray source and detectors are moved around the patient's head. This approach generates a matrix of intersecting points that have been obtained from several directions. The signal at each point can then be computed, allowing reconstruction of a "slice" through the brain that preserves three-dimensional relationships. (B) This CT scan shows a horizontal section of a normal adult brain.

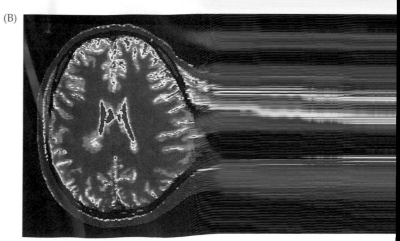

considered a fifth lobe of the brain). The"limbic lobe" (*limbic* means border or edge), which wraps around the corpus callosum, and the subcortical areas connected to it are referred to as the **limbic system**. These limbic structures are important in the regulation of visceral motor activity and emotional expression, among other functions. Finally, ventral to the cingulate gyrus is the midsagittal surface of the corpus callosum.

Although parts of the diencephalon, brainstem, and cerebellum are visible at the ventral surface of the brain, their overall structure is especially clear from the midsagittal surface (see Figure 1.14A). From this perspective, the diencephalon can be seen to consist of two parts. The **dorsal thalamus**, the largest component of the diencephalon, comprises a number of subdivisions, all of which relay information to the cerebral cortex from other parts of the brain. The **hypothalamus**, a small but especially crucial part of the diencephalon, is devoted to the control of homeostatic and reproductive functions. The hypothalamus is intimately related, both structurally and functionally, to the pituitary gland, a critical endocrine organ whose posterior part is attached to the hypothalamus by the pituitary stalk (or infundibulum; Figure 1.14C).

The midbrain, which can be seen only in this view, lies caudal to the thalamus, with the **superior and inferior colliculi** defining its dorsal surface or **tectum** (meaning "roof"); several midbrain nuclei, including the substantia nigra, lie in the ventral portion or **tegmentum** (meaning "covering") of the midbrain. The other prominent anatomical feature of the midbrain—the cerebral peduncles (also visible from the ventral surface)—are not apparent in a midsagittal view. The **pons** is caudal to the midbrain along the midsagittal surface, and the **cerebellum** lies over the pons just beneath the occipital lobe of the cerebral hemispheres. The major function of the cerebellum is coordination of motor activity, posture, and equilibrium. From the midsagittal surface, the most visible feature of the cerebellum is the **cerebellar cortex**, a continuous layered sheet of cells folded into ridges and valleys called **folia**. The most caudal structure seen from the midsagittal surface of the brain is the **medulla**, which merges into the spinal cord.

The Internal Anatomy of the Brain

A much more detailed neuroanatomical picture is apparent in gross or histological slices, or sections, through the brain. In these sections, deep structures that are not visible from any brain surface can be identified. In addition, relationships between brain structures seen from the surface can be appreciated more fully. The major challenge to understanding the internal anatomy of the brain is to integrate the rostral/caudal, dorsal/ventral, and medio-lateral landmarks seen on the brain surface with the position of structures seen in brain sections taken in the horizontal, frontal, or sagittal plane. This challenge is not only important for understanding brain function, it is also essential for interpreting noninvasive images of the brain (Boxes B and C), most of which are displayed as sections.

The Internal Anatomy of the Cerebral Hemispheres and Diencephalon

In any plane of section through the forebrain, the **cerebral cortex** is evident as a thin layer of neural tissue that covers the entire cerebrum. Most cerebral cortex is made up of six layers, and is referred to as **neocortex**. Phylo-

BOX B
Anatomical Brain Imaging Techniques

Until the early 1970s, the only technique available for imaging the structure of a living brain was X-ray technology. Conventional X-rays, however, have poor soft tissue contrast, involve relatively high radiation exposure, and provide only a two-dimensional view of brain structure or vasculature, making localization of lesions uncertain. As a result of these several problems, there was strong motivation in the 1960s and 1970s to discover better ways to image the brain.

A major advance was the development of computerized tomography (CT). CT uses a movable X-ray tube that is rotated around the patient's head. Opposite the tube (i.e., on the other side of the patient's head) are X-ray detectors far more sensitive than conventional film, thereby allowing much shorter exposure times (and less risk of radiation damage). Rather than acquiring a single image, as in conventional X-ray pictures, a CT scan gathers intensity information from many positions around the patient's head. These data are entered into a matrix, and the radiodensity at each point in the

three-dimensional space of the head is calculated. With a sufficiently narrow X-ray beam, sensitive detectors, and digital signal processing techniques, small differences in radiodensity can be converted into an image. Since the information is gathered for the full volume of the head, the computed matrix contains information about the entire brain. It is therefore possible to generate "slices," or tomograms (*tomo* means "cut" or "slice") of various planes through the brain, visualizing internal structures at any desired level. CT scans readily distinguish gray matter and white matter, differentiate the ventricles quite well, and show many other brain structures with a spatial resolution of several millimeters.

Although computerized tomography opened a new era in brain imaging, it has been largely superseded by a technique called magnetic resonance imaging (MRI). How nuclear magnetic resonance produces an image is more difficult to explain than X-ray imaging. Magnetic resonance derives from the interaction of a magnet and a magnetic field. Consider,

for example, Earth's magnetic field and a compass magnet. At rest, the compass needle points north. If the needle is tapped, however, it will swing back and forth (oscillate) at a frequency directly proportional to the magnetic field strength. The needle will continue to oscillate until friction dissipates the energy imparted by the tap and the needle again points north. Since the oscillation frequency is proportional to magnetic field strength, knowledge about the spatial variation of the field and the oscillation frequency of the needle could in principle be used to detect (and make an image of) the location of the needle on Earth's surface (albeit crudely). In MRI, atomic nuclei—principally hydrogen—act as the compass needle, and a strong magnet plays the role of Earth's magnetic field. Graded spatial variations in the magnetic field are actually generated by three sets of gradient magnets oriented along orthogonal axes. If all the atomic nuclei are aligned by the magnetic field and then "tapped" with a brief radiofrequency pulse, they emit energy in an oscillatory fashion as

(A) 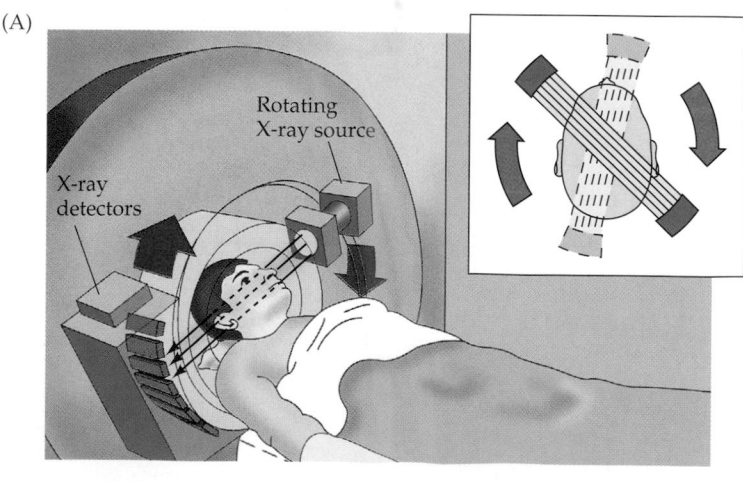 (B)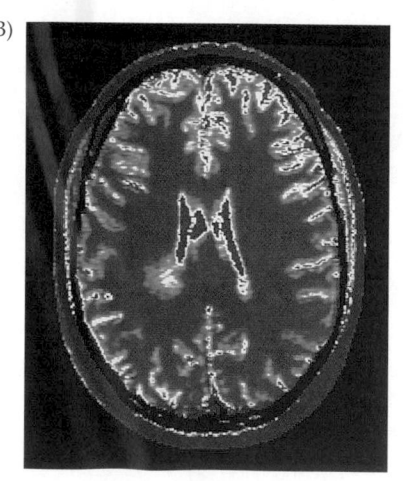

(A) In computerized tomography, the X-ray source and detectors are moved around the patient's head. This approach generates a matrix of intersecting points that have been obtained from several directions. The signal at each point can then be computed, allowing reconstruction of a "slice" through the brain that preserves three-dimensional relationships. (B) This CT scan shows a horizontal section of a normal adult brain.

they return to the alignment imposed by the field. By using detectors specifically sensitive to the radio frequencies emitted by the oscillating nuclei, together with computer techniques that convert signal intensity and magnetic gradient parameters into spatial locations, it is possible to construct extraordinarily detailed images of the brain. The resolution of MRI depends primarily on the strength of the magnetic field; currently, most clinical machines have field strengths of 1.5 Tesla, which provide resolution of under 1 mm. Higher field strength magnets (3–4 Tesla) are now also being used for human subjects to increase sensitivity and improve imaging resolution to fractions of a millimeter.

MRI has a number of features that have made it an especially valuable imaging tool for both diagnostic and research studies. First, it is entirely noninvasive; subjects are simply exposed to a strong magnetic field that is harmless (although accidents can happen if unsecured ferromagnetic objects are left in the vicinity). Second, unlike CT scans, views can be obtained from any angle. Since many brain structures are best seen in a particular plane, the ability to make "slices" from any point of view is a big advantage. Third, by varying the gradient and radio frequency pulse parameters, MRI scanners can be used to generate images based on a wide variety of different contrast mechanisms. For example, conventional MR images are designed to use certain properties of hydrogen nuclei that vary with tissue type to distinguish gray matter, white matter, and cerebrospinal fluid. By adjusting the pulse parameters, however, the same scanner can generate images in which gray and white matter are invisible but the brain vasculature stands out in sharp detail. Variants of this approach can also give information about the metabolic or biochemical state of selected brain regions. Moreover, recognition of the paramagnetic properties of hemoglobin has made MRI an important technique in the rapidly growing field of functional brain imaging (see Box C).

Safety and versatility have made MRI the technique of choice for imaging brain structure in most applications. It has not completely replaced CT imaging, however, as the latter is better for seeing bony or calcified structures in the head, and also remains the brain imaging technique of choice for patients who might have a problem with the high magnetic field (those with ferromagnetic metal plates or clips, for instance) or claustrophobia as a result of the confined space used in MRI. Together, CT and MRI have made it possible to see detailed structure of the living brain and have become invaluable tools for both diagnostic purposes and research.

References

CORMACK, A. M. (1980) Early two-dimensional reconstruction and recent topics stemming from it. Science 209: 1482–1486.

HOUNSFIELD, G. N. (1980) Computed medical imaging. Science 210: 22–28.

OLDENDORF, W. AND W. OLDENDORF JR. (1988) *Basics of Magnetic Resonance Imaging.* Boston: Kluwer Academic Publishers.

SCHILD, H. (1990) *MRI Made Easy (…Well, Almost).* Berlin: H. Heineman.

STARK, D. D. AND W. G. BRADLEY (1988) *Magnetic Resonance Imaging.* St. Louis, MO: Mosby Yearbook.

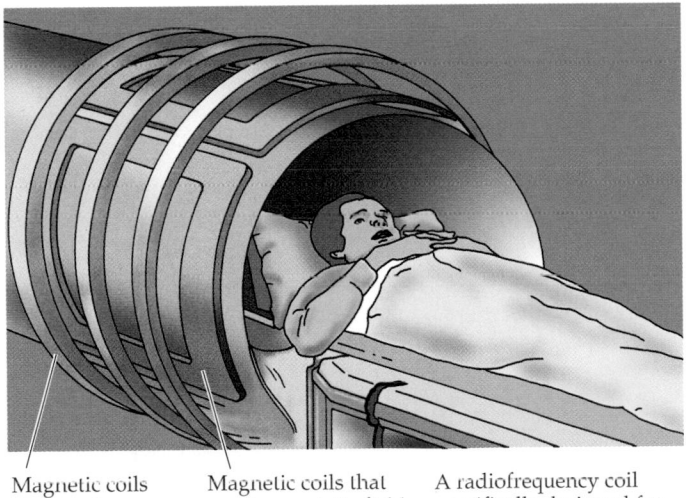

(C)

Magnetic coils that produce a static magnetic field in the long axis of the patient

Magnetic coils that produce a static field perpendicular to the long axis

A radiofrequency coil specifically designed for the head or other body part (not shown) perturbs the static fields to generate an MRI

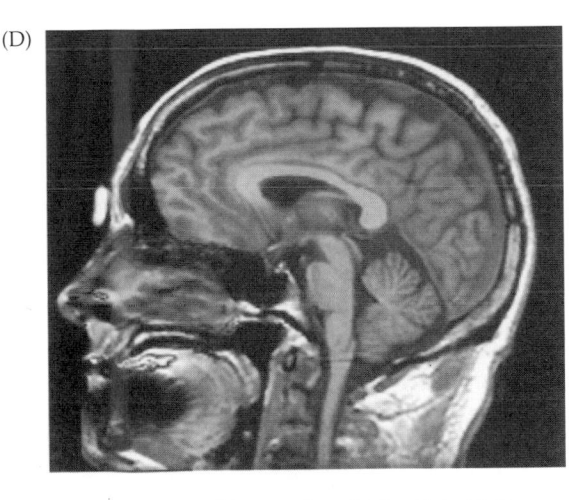

(D)

(C) Diagram of the machine used to obtain clinical MR images. Additional magnetic coils (not shown) produce linearly varying magnetic field gradients oriented along three orthogonal axes. (D) An MR image taken in the midsagittal plane. Note the extraordinary clarity with which all major brain components can be seen (compare with Figure 1.10).

BOX C
Functional Brain Imaging: PET, SPECT, and *f*MRI

The most informative brain imaging techniques for monitoring brain function now rely on detecting small changes in blood flow to visualize active areas of the brain. The brain utilizes a remarkably large fraction of the body's energy resources (about 20% of circulating glucose is consumed by the brain). Not surprisingly, at any given moment the most active nerve cells use more glucose and oxygen than relatively quiescent neurons. To meet the increased metabolic demands of particularly active neurons, the local flow of blood to the relevant brain area increases. Detecting and mapping these local changes in cerebral blood flow form the basis for three widely used functional brain imaging techniques: positron emission tomography (PET), single-photon emission computerized tomography (SPECT), and functional magnetic resonance imaging (*f*MRI). Because these techniques reveal patterns of activity in the intact brain, they have greatly enhanced the ability to understand both normal brain function and abnormal brain states associated with a variety of pathologies.

In PET scanning, unstable positron-emitting isotopes are synthesized in a cyclotron by bombarding nitrogen, carbon, oxygen, or fluorine with protons. Examples of the isotopes used include ^{15}O (half-life, 2 min), ^{18}F (110 min), and ^{11}C (20 min). These probes can be incorporated into many different reagents (including water, precursor molecules of specific neurotransmitters, or glucose) and used to analyze specific aspects of brain function. When the radiolabeled compounds are injected into the bloodstream, they distribute according to the physiological state of the brain. Thus, labeled oxygen and glucose accumulate in more metabolically active areas, and labeled transmitter probes are taken up

selectively by appropriate regions. As the unstable isotope decays, the extra proton breaks down into a neutron and a positron. The emitted positrons travel several millimeters, on average, until they collide with an electron. The collision of a positron with an electron destroys both particles, emitting two gamma rays from the site of the collision in directions that are exactly 180° apart. Gamma ray detectors placed around the head are therefore arranged to register a

"hit" only when two detectors 180° apart react simultaneously. By reconstructing the sites of the positron-electron collisions, the location of active regions can be imaged. The mean free path of the positrons in brain tissue limits the resolution of PET scanning to about 4 mm. Nonetheless, PET images can be superimposed on MRI images from the same subject (see Box B), providing detailed information about specific brain areas involved in a wide variety of functions. The elegance and power of

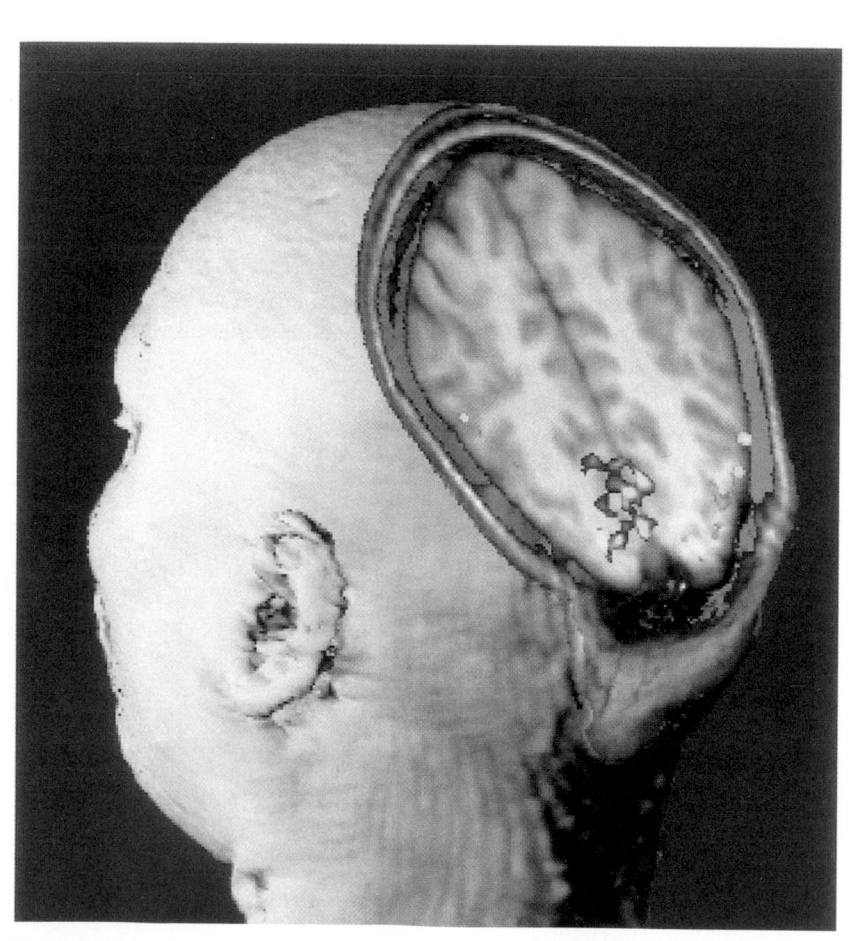

Example of functional magnetic resonance imaging. Regional changes in cerebral blood flow were measured during visual stimulation; the area of activated visual cortex (color) was then mapped onto the brain, a section of which is shown at the appropriate level in the head. (From Belliveau et al., 1991.)

this technique are evident in Figures 24.6 and 25.6 in Unit V of this book.

SPECT imaging is an outgrowth of older techniques for measuring regional cerebral blood flow. A radiolabeled compound with a relatively short half-life (for example, ^{133}Xe) is inhaled or injected into the circulation (in the latter case, ^{123}I-labeled iodoamphetamine is used); the probes bind to red blood cells and are carried throughout the body. As the label undergoes radioactive decay, it emits high-energy photons. The rate of clearance of the probes was initially detected using an array of sodium iodide photon detectors placed around the head. More recent approaches have used a gamma camera that can be rapidly moved around the head to collect photons from many different angles, thus permitting a more accurate three-dimensional image. The information gathered using SPECT can also be combined with structural information from other imaging techniques, such as CT scans and MRI scans, to provide better localization of the active areas. A limitation of SPECT scanning is its relatively low resolution (about 8 mm). Although this level is not sufficient to resolve the finer features of the brain, it reveals the major areas involved in normal processing or disease. SPECT imaging is neither as flexible nor as accurate as PET imaging, but it is much simpler, primarily because the radiolabeled probes are commercially available and do not require an on-site cyclotron (as does the synthesis of PET probes).

A variant of MRI, called functional MRI (fMRI), now offers the best approach to analyzing the brain at work (see figure). fMRI is based on the fact that oxyhemoglobin (the oxygen carrying form of hemoglobin) has a different magnetic resonance signal than deoxyhemoglobin (the oxygen-depleted form of hemoglobin) or the surrounding brain tissue. Brain areas activated by a specific task (e.g., the occipital cortex during visual behavior; see figure) utilize more oxygen. Initially, this activity decreases the levels of oxyhemoglobin and increases levels of deoxyhemoglobin. Within seconds, the brain microvasculature responds to this local oxygen depletion by increasing the flow of oxygen-rich blood to the active area. These changes in the concentration of oxyhemoglobin lead to localized blood oxygenation level dependent (BOLD) changes in the magnetic resonance signal, which form the basis for the fMRI signal. Thus, unlike PET or SPECT, fMRI uses signals intrinsic to the brain rather than signals originating from exogenous, radioactive probes; consequently, repeated observations can be made on the same individual, providing a major advantage over other imaging methods. fMRI also offers superior spatial localization (currently a few millimeters), as well as good temporal resolution (on the order of seconds or less under optimal circumstances, compared to minutes for other functional imaging techniques). As a result of these advantages, fMRI has emerged as the technology of choice for probing both the normal and abnormal functional architecture of the human brain.

References

BELLIVEAU, J. W. AND 7 OTHERS (1991) Functional mapping of the human visual cortex by magnetic resonance imaging. Science 254: 716–719.

COHEN, M. S. AND S. Y. BOOKHEIMER (1994) Localization of brain function using magnetic resonance imaging. Trends Neurosci. 17: 268–277.

KWONG, K. K. AND 9 OTHERS (1992) Dynamic magnetic resonance imaging of human brain activity during primary sensory stimulation. Proc. Natl. Acad. Sci. USA 89: 5675–5679.

OGAWA, S. AND 6 OTHERS (1992) Intrinsic signal changes accompanying sensory stimulation: Functional brain mapping with magnetic resonance imaging. Proc. Natl. Acad. Sci. USA 89: 5951–5955.

PETERSEN, S. E., P. T. FOX, A. Z. SNYDER AND M. E. RAICHLE (1990) Activation of extrastriate and frontal cortical areas by visual words and word-like stimuli. Science 249: 1041–1044.

RAICHLE, M. E. (1994) Images of the mind: Studies with modern imaging techniques. Ann. Rev. Psychol. 45: 333–356.

RAICHLE, M. E. AND M. I. POSNER (1994) *Images of Mind*. New York: Scientific American Library.

genetically older cortex (called paleocortex) with fewer cell layers occurs on the inferior and medial aspect of the temporal lobe within the parahippocampal gyrus. Cortex with even fewer layers (three), referred to as archicortex, occurs in the hippocampus and in the pyriform cortex within the uncus. The **hippocampal cortex** is folded into the medial aspect of the temporal lobe, and is therefore visible only in dissected brains or in sections (Figures 1.15 and 1.16). The largest structures embedded within the cerebral hemispheres are the **caudate** and **putamen nuclei** (together referred to as **striatum**) as well as the **globus pallidus** (Figure 1.16). Collectively these several structures are referred to as the **basal ganglia** (the term *ganglia* does

Figure 1.15 Major internal structures of the brain, shown after the upper half of the left hemisphere has been cut away.

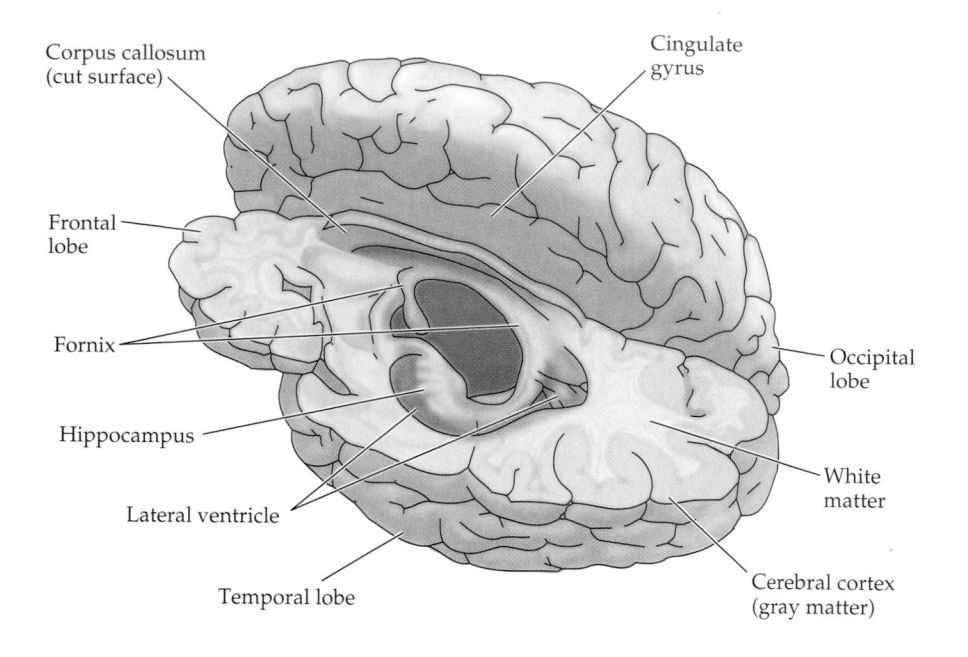

Corpus callosum (cut surface)

Cingulate gyrus

Frontal lobe

Fornix

Hippocampus

Lateral ventricle

Temporal lobe

Occipital lobe

White matter

Cerebral cortex (gray matter)

not usually refer to nuclei in the brain; the usage here is an exception). The basal ganglia are visible in horizontal sections through the mid-dorsal to mid-ventral portion of the forebrain, in frontal sections from just rostral to the uncus to the level of the diencephalon, and in paramedian sagittal sections. The neurons of these large nuclei receive input from the cerebral cortex and participate in the organization and guidance of complex motor functions. In the base of the forebrain, ventral to the basal ganglia are several smaller clusters of nerve cells known as the **septal** or **basal forebrain nuclei**. These nuclei are of particular interest because they are implicated in Alzheimer's disease. The other clearly discernible structure visible in sections through the cerebral hemispheres at the level of the uncus is the **amygdala**, a collection of nuclei important for emotional processing that lies in front of the hippocampus in the anterior pole of the temporal lobe.

In addition to these cortical and nuclear structures, the internal anatomy of the brain is characterized by a number of important axon tracts. As already mentioned, the two cerebral hemispheres and many of their component parts are interconnected by the **corpus callosum**; in some anterior sections, the smaller **anterior commissure** can also be seen (see Figure 1.14). Axons descending from (and ascending to) the cerebral cortex assemble into another large fiber bundle tract called the **internal capsule** (see Figure 1.16). The internal capsule lies just lateral to the diencephalon (forming a "capsule" around it), and many of its axons arise from or terminate in the dorsal thalamus. It is seen most clearly in frontal sections through the middle one-third of the rostral/caudal extent of forebrain, or in horizontal sections through the level of the thalamus. Other axons descending from the cortex in the internal capsule continue past the diencephalon to enter the cerebral peduncles of the midbrain. Axons in these tracts project to a number of targets in the brainstem and spinal cord. Thus, the internal capsule is the major pathway linking the cerebral cortex to the rest of the brain and spinal cord. Strokes or other injury to this structure interrupt the flow of ascending and descending nerve traffic, often with devastating consequences (Box D).

(A)

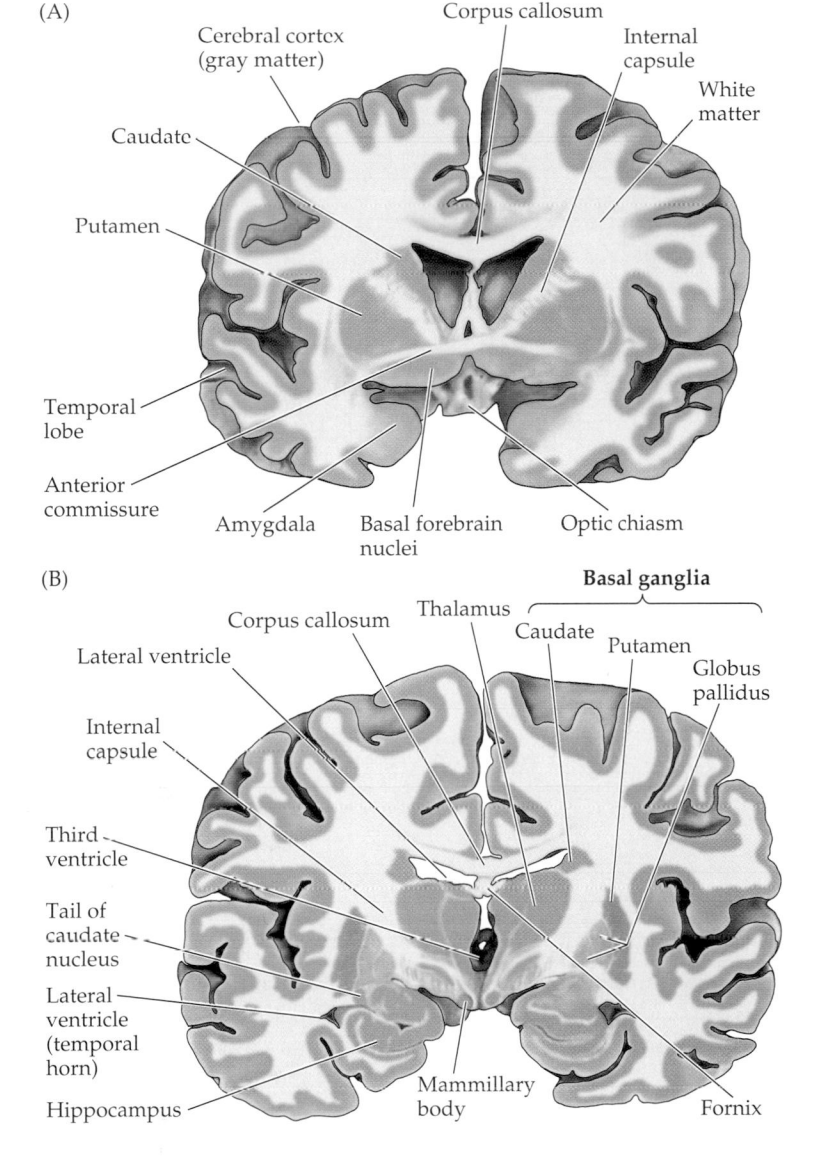

Corpus callosum

Cerebral cortex
(gray matter)

Internal
capsule

White
matter

Caudate

Putamen

Temporal
lobe

Anterior
commissure

Amygdala Basal forebrain
nuclei

Optic chiasm

(B)

Basal ganglia

Corpus callosum Thalamus

Caudate

Putamen

Globus
pallidus

Lateral ventricle

Internal
capsule

Third
ventricle

Tail of
caudate
nucleus

Lateral
ventricle
(temporal
horn)

Hippocampus

Mammillary
body

Fornix

(C)

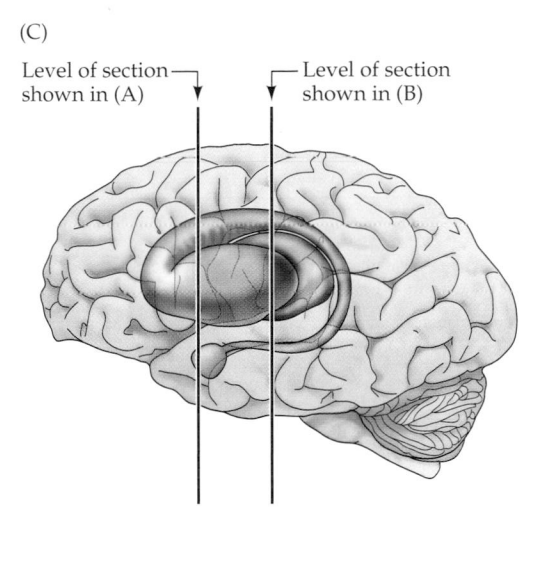

Level of section
shown in (A)

Level of section
shown in (B)

Figure 1.16 Internal structures of the brain seen in coronal section. (A) This plane of section runs through the basal ganglia. (B) A somewhat more posterior plane of section that includes the thalamus. (C) A transparent view of the basal ganglia showing the approximate location of the sections in (A) and (B). Notice that because the caudate nucleus has a tail that arcs into the temporal lobe, it may appear twice in the same section. The same is true of several other brain structures, including the lateral ventricles.

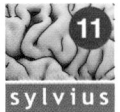

Finally, a smaller fiber bundle within each of the hemispheres, the **fornix,** interconnects the hippocampus and the hypothalamus.

The Ventricular System

The cerebral ventricles are a series of interconnected, fluid-filled spaces that lie in the core of the forebrain and brainstem (Figure 1.17). The presence of ventricular spaces in the various subdivisions of the brain reflects the fact that the ventricles are the adult derivatives of the open space or lumen of the embryonic neural tube (see Chapter 22). Although they have no unique function, the ventricular spaces present in sections through the brain provide another useful guide to location. The largest of these spaces are the **lateral ventricles** (one within each of the cerebral hemispheres). These particular ventricles are best seen in frontal sections, where their ventral surface is usually defined by the basal ganglia, their dorsal surface by the corpus callo-

Figure 1.17 The ventricular system of the human brain. (A) Location of the ventricles as seen in a transparent left lateral view. (B) Table showing the ventricular spaces associated with each of the major subdivisions of the brain. (See Chapter 22 for an account of brain development that more fully explains the origin of the ventricular spaces.)

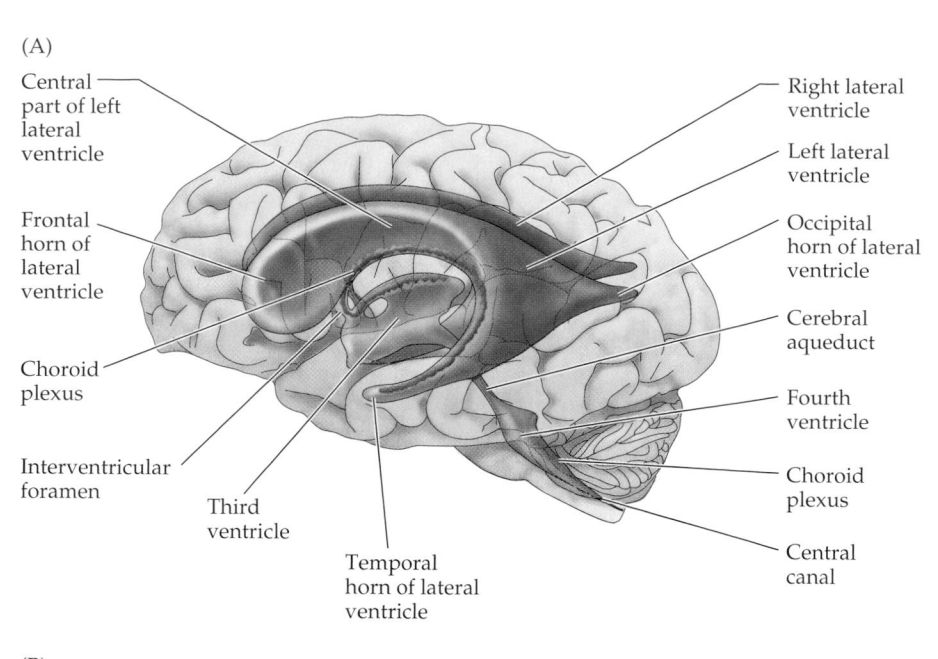

(A)

- Central part of left lateral ventricle
- Frontal horn of lateral ventricle
- Choroid plexus
- Interventricular foramen
- Third ventricle
- Temporal horn of lateral ventricle
- Right lateral ventricle
- Left lateral ventricle
- Occipital horn of lateral ventricle
- Cerebral aqueduct
- Fourth ventricle
- Choroid plexus
- Central canal

(B)

	EMBRYONIC BRAIN	ADULT BRAIN DERIVATIVES	ASSOCIATED VENTRICULAR SPACE
Prosencephalon	Telencephalon (forebrain)	Cerebral cortex	Lateral ventricles
		Basal ganglia Hippocampus Olfactory bulb Basal forebrain	
	Diencephalon	Dorsal thalamus	Third ventricle
		Hypothalamus	
	Mesencephalon	Midbrain (superior and inferior colliculi)	Cerebral aqueduct
Rhombencephalon	Metencephalon	Cerebellum	Fourth ventricle
		Pons	
	Myelencephalon	Medulla	Fourth ventricle
	Spinal cord	Spinal cord	Central canal

sum, and their medial surface by the **septum pellucidum**, a membranous tissue sheet that forms part of the midline sagittal surface of the cerebral hemispheres. The **third ventricle** forms a narrow midline space between the right and left thalamus, and communicates with the lateral ventricles through a small opening at the anterior end of the third ventricle (called the interventricular foramen). The third ventricle is continuous caudally with the **cerebral aqueduct**, which runs though the midbrain. At its caudal end, the aqueduct opens into the **fourth ventricle**, a larger space in the dorsal pons and medulla. The fourth ventricle narrows caudally to form the central canal of the spinal cord. The ventricles are filled with **cerebrospinal fluid**, and the lateral, third, and fourth ventricles are the site of the **choroid plexus**, which produces this fluid. The cerebrospinal fluid percolates through the

BOX D
Stroke

Stroke is the most common neurological cause for admission to a hospital, and is the third leading cause of death in the United States (after heart disease and cancer). The term "stroke" refers to the sudden appearance of a limited neurological deficit, such as weakness or paralysis of a limb, or the sudden inability to speak. The onset of the deficit within seconds, minutes, or hours marks the problem a vascular one. Brain function is exquisitely dependent on a continuous supply of oxygen, as evidenced by the onset of unconsciousness within about 10 seconds of blocking its blood supply (by cardiac arrest, for instance). The damage to neurons is at first reversible, but eventually becomes permanent if the blood supply is not promptly restored.

Strokes can be subdivided into three main types: thrombotic, embolic, and hemorrhagic. The thrombotic variety is caused by a local reduction of blood flow arising from an atherosclerotic buildup in one of the cerebral blood vessels that eventually occludes it. Alternatively, a reduction of blood flow can arise when an embolus (meaning an object loose in the bloodstream) dislodges from the

heart (or from an atherosclerotic plaque in the carotid or vertebral arteries) and travels to a cerebral artery (or arteriole) where it forms a plug. A hemorrhagic stroke occurs when a cerebral blood vessel ruptures, as can occur as a result of hypertension, a congenital aneurysm (bulging of a vessel), or a congenital arterio-venous malformation. The relative frequency of thrombotic, embolic, and hemorrhagic strokes is approximately 50%, 30%, and 20%, respectively.

The diagnosis of stroke relies primarily on an accurate history and a competent neurological examination. Indeed, the neurologist C. Miller Fisher, a master of bedside diagnosis, remarked that medical students and residents should learn neurology "stroke by stroke." Understanding the portion of the brain supplied by each of the major arteries (see text) enables an astute clinician to identify the occluded blood vessel.

More recently, imaging techniques such as CT scans and MRI (see Boxes B and C) have greatly facilitated the physician's ability to identify and localize small hemorrhages and regions of permanently damaged tissue. Moreover,

Doppler ultrasound, magnetic resonance angiography, and imaging of blood vessels by direct infusion of radio-opaque dye can now pinpoint atherosclerotic plaques, aneurysms, and other vascular abnormalities.

Several therapeutic approaches to strokes are feasible. Dissolving a thrombotic plug by tissue plasminogen activator and other compounds is now standard clinical practice for selected stroke victims. Furthermore, recent understanding of some of the mechanisms by which ischemia injures brain tissue has made pharmacological strategies to minimize neuronal injury after stroke a potentially effective possibility (see Box B in Chapter 6). Hemorrhagic strokes are of course treated neurosurgically, by finding and stopping the bleeding from the defective vessel when that is technically possible.

Reference

ADAMS, R. D., M. VICTOR AND A. H. ROPPER (1997) *Principles of Neurology*, 6th Ed. New York: McGraw-Hill, Ch. 34, pp. 777–873.

ventricular system and flows into the subarachnoid space through perforations in the thin covering of the fourth ventricle; it is eventually absorbed by specialized structures called arachnoid villi or granulations (see Figure 1.18), and returned to the venous circulation.

The Meninges

The cranial cavity is conventionally divided into three regions called the anterior, middle, and posterior cranial fossae. Surrounding and supporting the brain within this cavity are three protective tissue layers, which also extend down the brainstem and the spinal cord. Together these layers are called the **meninges** (Figure 1.18). The outermost layer of the meninges is called the **dura mater** because it is thick and tough. The middle layer is called the **arachnoid mater** because of spiderlike processes called arachnoid trabeculae that extend from it toward the third layer, the **pia mater**, a thin, delicate layer of cells that closely invests the surface of the brain. Since the pia closely adheres

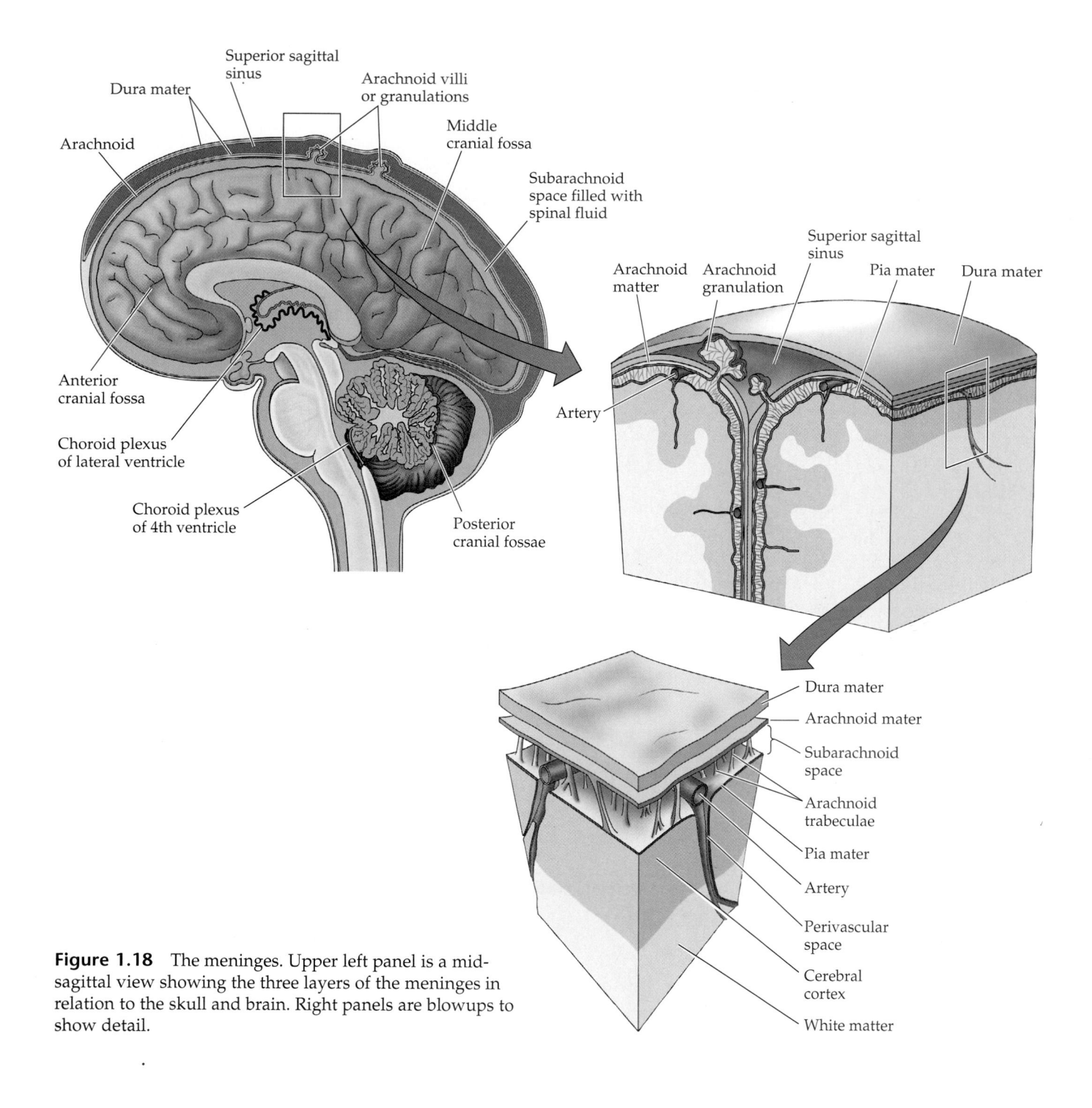

Figure 1.18 The meninges. Upper left panel is a mid-sagittal view showing the three layers of the meninges in relation to the skull and brain. Right panels are blowups to show detail.

to the brain as its surface curves and folds, whereas the arachnoid does not, there are places—called **cisterns**—where the subarachnoid space is especially large. The major arteries supplying the brain course through the subarachnoid space where they give rise to branches that penetrate the substance of the hemispheres. The subarachnoid space is therefore a frequent site of bleeding following trauma. A collection of blood between the meningeal layers is referred to as a subdural or subarachnoid hemorrhage, as distinct from bleeding within the brain itself.

The Blood Supply of the Brain and Spinal Cord

The entire blood supply of the brain and spinal cord depends on two sets of branches from the dorsal aorta. The **vertebral arteries** arise from the subclavian arteries, and the **internal carotid arteries** are branches of the common carotid arteries. The vertebral arteries and the ten **medullary arteries** that arise from segmental branches of the aorta provide the primary vascularization of the spinal cord. These medullary arteries join to form the **anterior** and **posterior spinal arteries** (Figure 1.19). If any of the medullary arteries are obstructed or damaged (during abdominal surgery, for example), the blood supply to specific parts of the spinal cord may be compromised. The pattern of resulting neurological damage differs according to whether the supply to the posterior or anterior artery is interrupted. As might be expected from the arrangement of ascending and descending neural pathways in the spinal cord, loss of the posterior supply generally leads to loss of sensory functions, whereas loss of the anterior supply more often causes motor deficits.

The brain receives blood from two sources: the **internal carotid arteries**, which arise at the point in the neck where the common carotid arteries bifurcate, and the **vertebral arteries** (Figure 1.20). The internal carotid arteries branch to form two major cerebral arteries, the **anterior** and **middle cerebral arteries**. The right and left vertebral arteries come together at the level of the pons on the ventral surface of the brainstem to form the midline **basilar artery**. The basilar artery joins the blood supply from the internal carotids in an arterial ring at the base of the brain (in the vicinity of the hypothalamus and cerebral peduncles) called the **circle of Willis**. The **posterior cerebral arteries** arise at this confluence, as do two small bridging arteries, the **anterior and posterior communicating arteries**. Conjoining the two major sources of cerebral vascular supply via the circle of Willis presumably improves the chances of any region of the brain continuing to receive blood if one of the major arteries becomes occluded (see Box D).

The major branches that arise from the internal carotid artery—the anterior and middle cerebral arteries—form the **anterior circulation** that sup-

Figure 1.19 Blood supply of the spinal cord. (A) View of the ventral (anterior) surface of the spinal cord. At the level of the medulla, the vertebral arteries give off branches that merge to form the anterior spinal artery. Approximately 10 to 12 segmental arteries (which arise from various branches of the aorta) join the anterior spinal artery along its course. These segmental arteries are known as medullary arteries. (B) The vertebral arteries (or the posterior inferior cerebellar artery) give rise to paired posterior spinal arteries that run along the dorsal (posterior) surface of the spinal cord. (C) Cross section through the spinal cord, illustrating the distribution of the anterior and posterior spinal arteries. The anterior spinal arteries give rise to numerous sulcal branches that supply the anterior two-thirds of the spinal cord. The posterior spinal arteries supply much of the dorsal horn and the dorsal columns. A network of vessels known as the vasocorona connects these two sources of supply and sends branches into the white matter around the margin of the spinal cord.

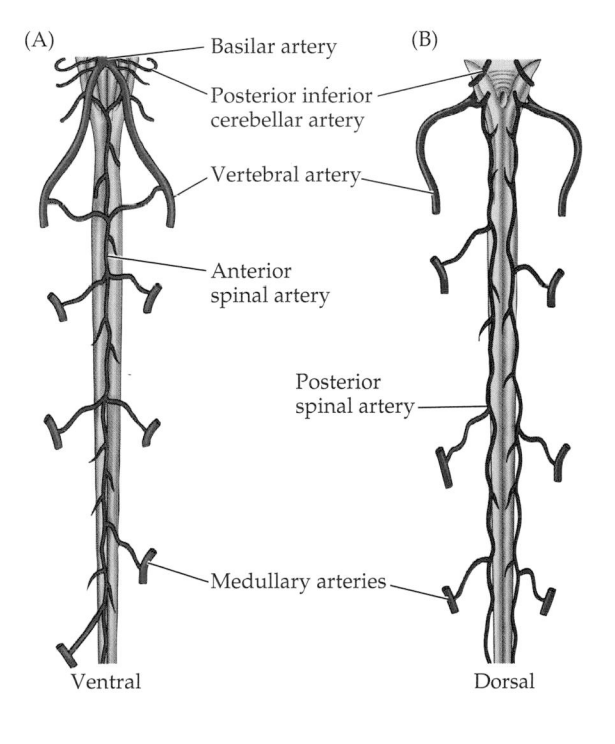

(A) Ventral

- Basilar artery
- Posterior inferior cerebellar artery
- Vertebral artery
- Anterior spinal artery

(B) Dorsal

- Posterior spinal artery
- Medullary arteries

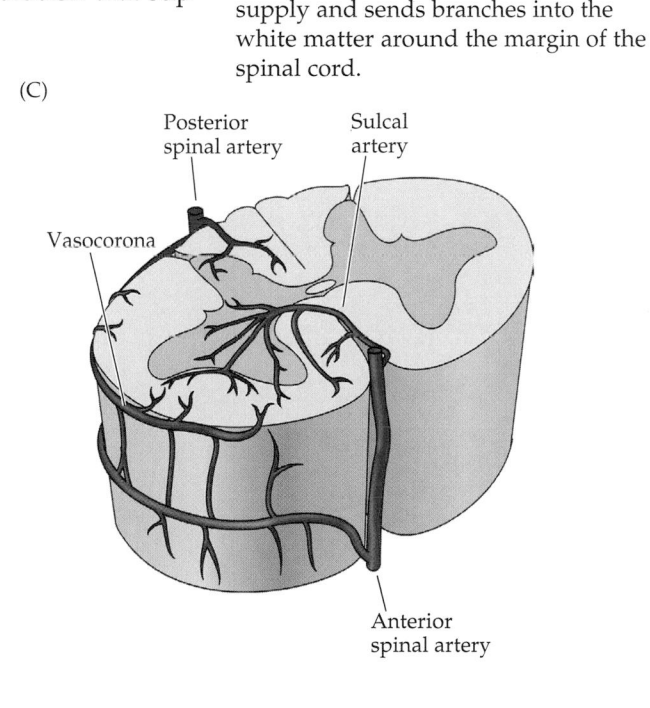

(C)

- Posterior spinal artery
- Sulcal artery
- Vasocorona
- Anterior spinal artery

(A)

Anterior cerebral artery

Internal carotid artery

Basilar artery

Anterior inferior cerebellar artery

Middle cerebral artery

Portion of temporal lobe removed

Posterior inferior cerebellar artery

Vertebral artery

Anterior communicating artery

Posterior communicating artery

Posterior cerebral artery (to midbrain)

Basilar artery (to pons)

(B)

Anterior cerebral artery

Posterior cerebral artery

Middle cerebral artery

(C)

Posterior cerebral artery

Anterior cerebral artery

(D)

Lenticulostriate arteries

Anterior cerebral artery

Middle cerebral artery

Internal carotid artery

Anterior communicating artery

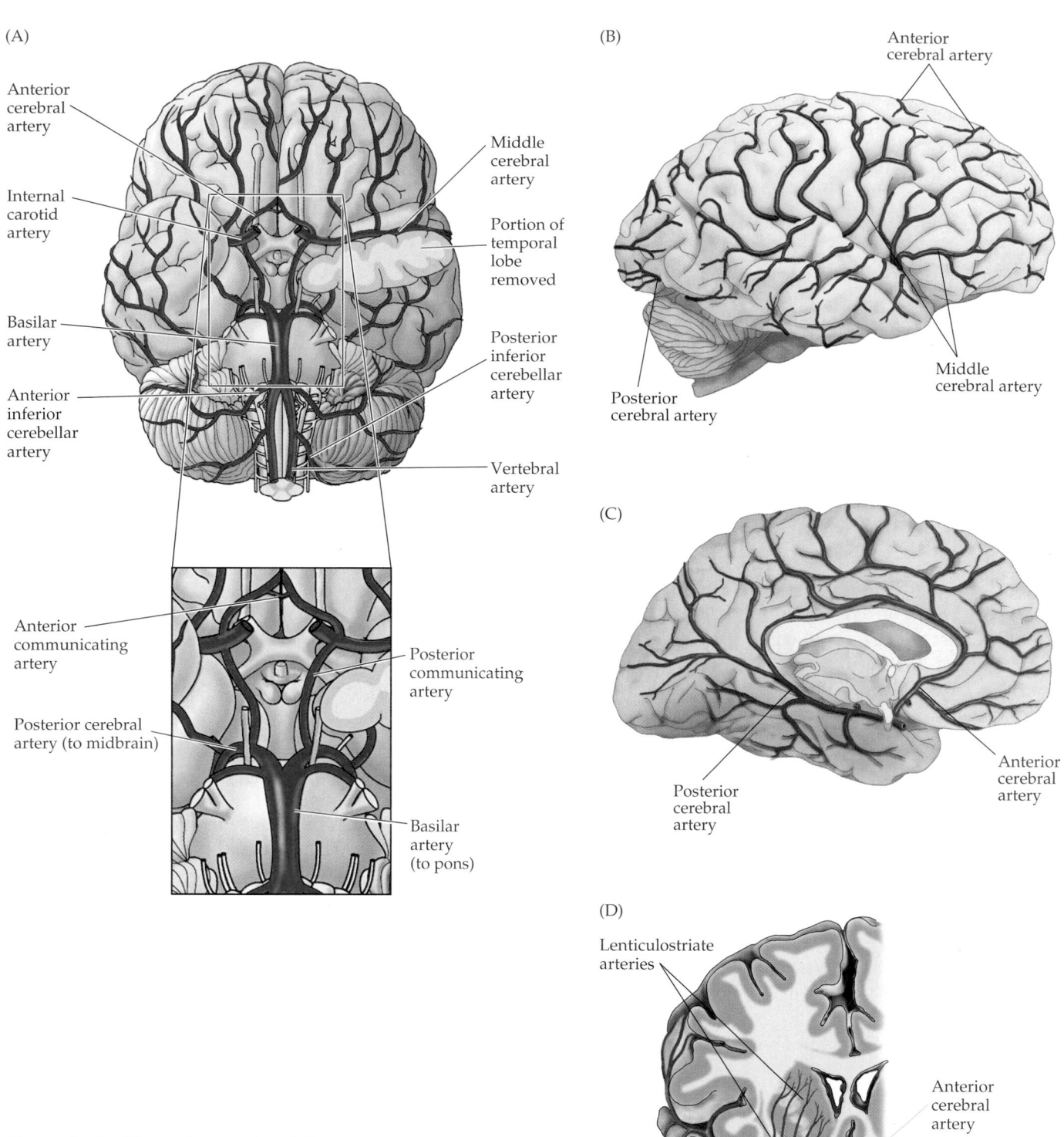

Figure 1.20 The major arteries of the brain. (A) Ventral view (compare with Figure 1.13B). The enlargement of the boxed area shows the circle of Willis. Lateral (B) and (C) midsagittal views showing anterior, middle, and posterior cerebral arteries. (D) Idealized frontal section showing course of middle cerebral artery.

plies the forebrain (Figure 1.20B). These arteries also originate from the circle of Willis. Each gives rise to branches that supply the cortex and branches that penetrate the basal surface of the brain, supplying deep structures such as the basal ganglia, thalamus, and internal capsule. Particularly prominent are the lenticulostriate arteries that branch from the middle cerebral artery. These arteries supply the basal ganglia and thalamus. The **posterior circulation** of the brain supplies the posterior cortex, the midbrain, and the brainstem; it comprises arterial branches arising from the **posterior cerebral**, **basilar**, and **vertebral arteries**. The pattern of arterial distribution is similar for all the subdivisions of the brainstem: Midline arteries supply medial structures, lateral arteries supply the lateral brainstem, and dorsal-lateral arteries supply dorsal-lateral brainstem structures and the cerebellum (Figures 1.20 and 1.21). Among the most important dorsal-lateral arteries (also called **long**

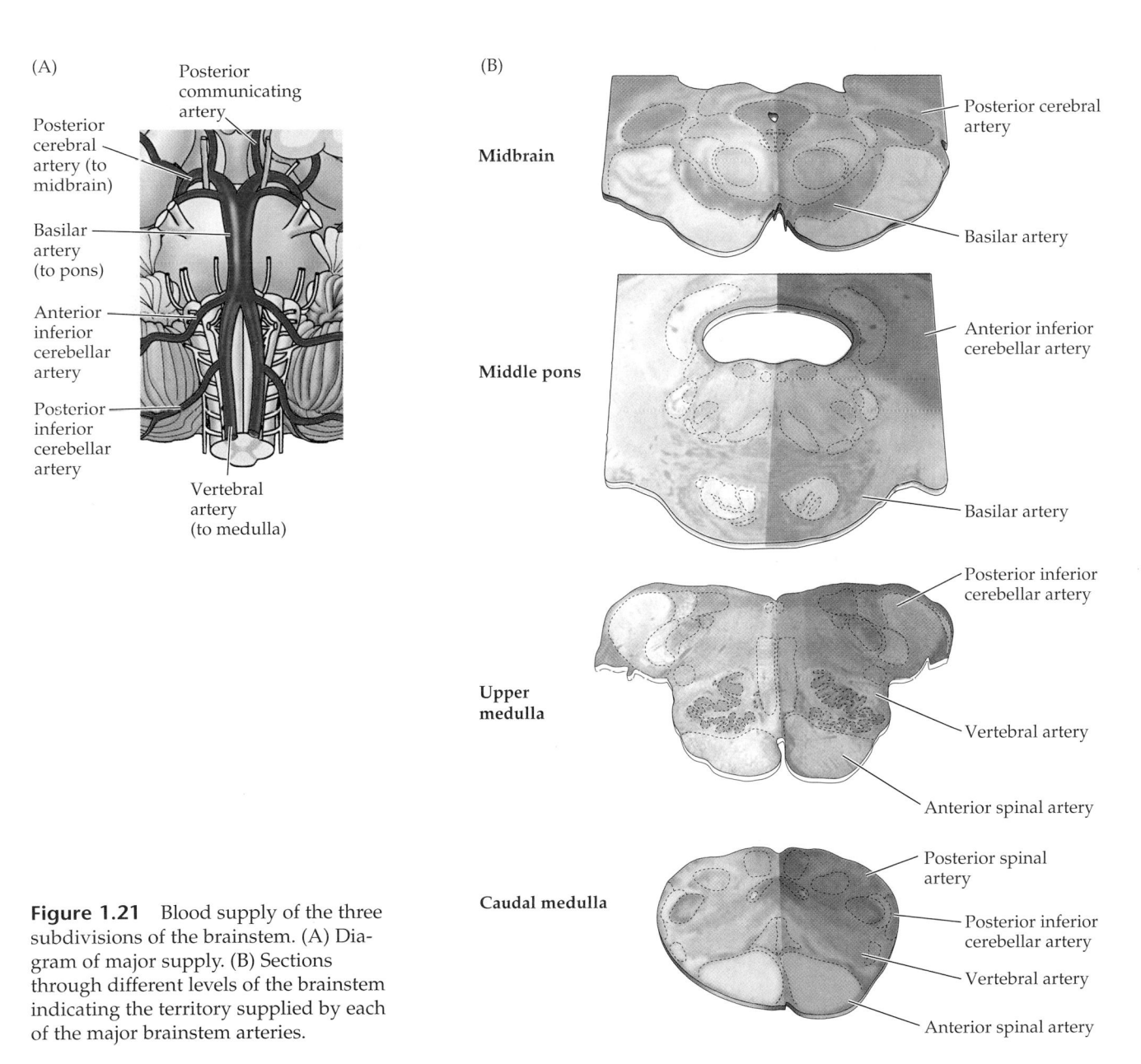

Figure 1.21 Blood supply of the three subdivisions of the brainstem. (A) Diagram of major supply. (B) Sections through different levels of the brainstem indicating the territory supplied by each of the major brainstem arteries.

BOX E
The Blood-Brain Barrier

The interface between the walls of capillaries and the surrounding tissue is important throughout the body, as it keeps vascular and extravascular concentrations of ions and molecules at appropriate levels in these two compartments. In the brain, this interface is especially significant and has been accorded an alliterative name, "the blood-brain barrier." The special properties of the blood-brain barrier were first observed by the nineteenth-century bacteriologist Paul Ehrlich, who noted that intravenously injected dyes leaked out of capillaries in most regions of the body to stain the surrounding tissues; the brain, however, remained unstained. Ehrlich wrongly concluded that the brain had a low affinity for the dyes; his student, Edwin Goldmann, showed that such dyes do not traverse the specialized walls of brain capillaries.

The restriction of large molecules like Ehrlich's dyes (and many smaller molecules) to the vascular space is the result of tight junctions between neighboring capillary endothelial cells in the brain. Such junctions are not found in capillaries elsewhere in the body, where the spaces between adjacent endothelial cells allow much more ionic and molecular traffic.

The structure of tight junctions was first demonstrated in the 1960s by Tom Reese, Morris Karnovsky, and Milton Brightman. Using electron microscopy after the injection of electron-dense intravascular agents such as lanthanum salts, they showed that the close apposition of the endothelial cell membranes prevented such ions from passing. Substances that traverse the walls of brain capillaries must move *through* the endothelial cell membranes. Accordingly, molecular entry into the brain should be determined by an agent's solubility in lipids, the major constituent of cell membranes. Nevertheless, many ions and molecules not readily soluble in lipids *do* move quite readily from the vascular space into brain tissue. A molecule like glucose, the primary source of metabolic energy for neurons and glial cells, is an obvious example. This paradox is explained by the presence of specific transporters for glucose and other critical molecules and ions.

In addition to tight junctions, astrocytic "end feet" (the terminal regions of astrocytic processes) surround the outside of capillary endothelial cells. The reason for this endothelial–glial allegiance is unclear, but may reflect an influence of astrocytes on the formation and maintenance of the blood-brain barrier.

The brain, more than any other organ, must be carefully shielded from abnormal variations in its ionic milieu, as well as from the potentially toxic molecules that find their way into the vascular space by ingestion, infection, or other means. The blood-brain barrier is thus important for protection and homeostasis. It also presents a significant problem for the delivery of drugs to the brain. Large (or lipid-insoluble) molecules can be introduced to the brain, but only by transiently disrupting the blood-brain barrier with hyperosmotic agents like mannitol.

References

BRIGHTMAN, M. W. AND T. S. REESE (1969) Junctions between intimately opposed cell membranes in the vertebrate brain. J. Cell Biol. 40: 648–677.

SCHMIDLEY, J. W. AND E. F. MAAS (1990) Cerebrospinal fluid, blood-brain barrier and brain edema. In *Neurobiology of Disease*, A. L. Pearlman and R.C. Collins (eds.). New York: Oxford University Press, Chapter 19, pp. 380–398.

REESE, T. S. AND M. J. KARNOVSKY (1967) Fine structural localization of a blood–brain barrier to exogenous peroxidase. J. Cell Biol. 34: 207–217.

(A)

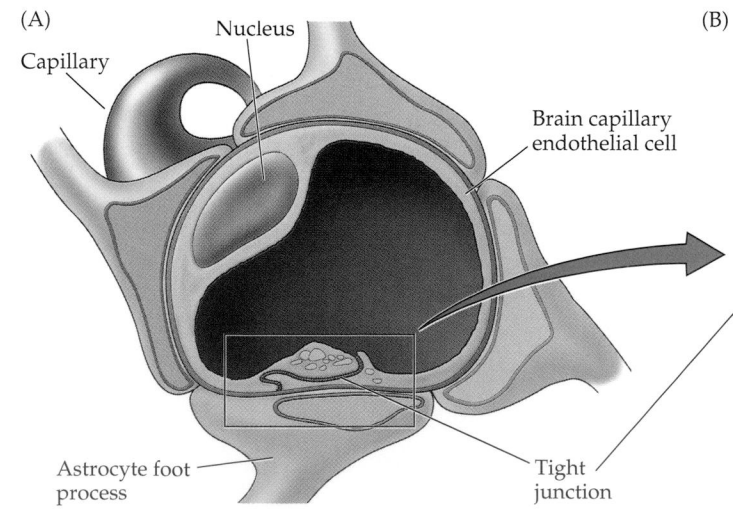

(B)

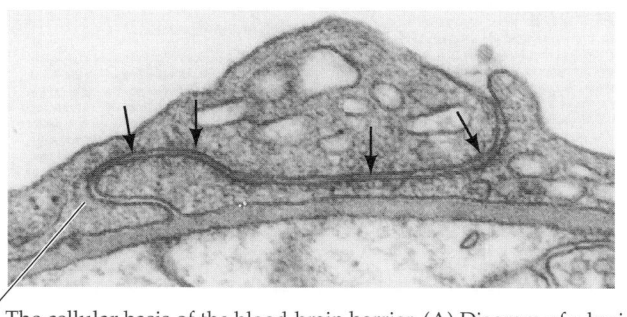

The cellular basis of the blood-brain barrier. (A) Diagram of a brain capillary in cross section and reconstructed views, showing endothelial tight junctions and the investment of the capillary by astrocytic end feet. (B) Electron micrograph of boxed area in (A), showing the appearance of tight junctions between neighboring endothelial cells (arrows). (A after Goldstein, Goldstein and Betz, 1986; B from Peters et al., 1991.)

circumferential arteries) are the **posterior inferior cerebellar artery** (**PICA**) and the **anterior inferior cerebellar artery** (**AICA**), which supply distinct regions of the medulla and pons. These arteries, as well as branches of the basilar artery that penetrate the brainstem from its ventral and lateral surfaces (called **paramedian** and **short circumferential** arteries), are especially common sites of occlusion and result in specific functional deficits of cranial nerve, somatic sensory, and motor function (see Boxes A and D).

The physiological demands served by the blood supply of the brain are particularly significant because neurons are more sensitive to oxygen deprivation than other kinds of cells with lower rates of metabolism. In addition, the brain is at risk from circulating toxins, and is specifically protected in this respect by the **blood-brain barrier** (Box E). As a result of the high metabolic rate of neurons, brain tissue deprived of oxygen and glucose as a result of compromised blood supply is likely to sustain transient or permanent damage. Brief loss of blood supply (referred to as ischemia) can cause cellular changes, which, if not quickly reversed, can lead to cell death. Sustained loss of blood supply leads much more directly to death and degeneration of the deprived cells. Strokes—an anachronistic term that refers to the death or dysfunction of brain tissue due to vascular disease—often follow the occlusion of (or hemorrhage from) the brain's arteries (see Box D). Historically, studies of the functional consequences of strokes, and their relation to vascular territories in the brain and spinal cord, provided information about the location of various brain functions. The location of the major language functions in the left hemisphere, for instance, was discovered in this way in the latter part of the nineteenth century (see Chapter 27). Now, noninvasive functional imaging techniques based on blood flow (see Box C) have largely supplanted the correlation of clinical signs and symptoms with the location of tissue damage observed at autopsy.

Summary

Although the human brain is often discussed as if it were a single organ, it includes a large number of cell types and circuits combined in a variety of systems and subsystems. Various types of neurons are assembled into richly interconnected circuits that relay and process electrical signals, which are the currency of all neural functions. Groups of circuits that process related information form distinct systems and subsystems in the brain and the rest of the nervous system. Knowledge about the organization and location of these systems is an essential first step toward understanding brain function. The human nervous system, like that of all vertebrates, comprises a central nervous system, which consists of the brain and spinal cord, and a peripheral nervous system, which includes the peripheral nerves (and their ganglia) extending to a wide array of targets (primarily muscles, glands, and specialized sensory receptors). Sensory components of the nervous system supply information to the central nervous system about the internal and external environment. The integrated effects of central processing are eventually translated into action by the motor components of the central and peripheral nervous systems. Different brain systems mediate an enormous range of functions, including perception, cognition, language, sleep, emotion, sexuality, and memory, to name but a few. The outline of brain structure and function described in this introductory chapter provides the basic framework for understanding these phenomena, and, together with the CD-ROM on neural structure that accompanies this book, should be referred to often as various aspects of brain function are explored in subsequent chapters.

ADDITIONAL READING

BRODAL, P. (1992) *The Central Nervous System: Structure and Function*. New York: Oxford University Press.

CARPENTER, M. B. AND J. SUTIN (1983) *Human Neuroanatomy*, 8th Ed. Baltimore, MD: Williams and Wilkins.

ENGLAND, M. A. AND J. WAKELY (1991) *Color Atlas of the Brain and Spinal Cord: An Introduction to Normal Neuroanatomy*. St. Louis: Mosby Yearbook.

HAINES, D. E. (1995) *Neuroanatomy: An Atlas of Structures, Sections, and Systems*, 2nd Ed. Baltimore: Urban and Schwarzenberg.

MARTIN, J. H. (1996) *Neuroanatomy: Text and Atlas*, 2nd Ed. Stamford, CT: Appleton and Lange.

NETTER, F. H. (1983) *The CIBA Collection of Medical Illustrations*, Vols. I and II. A. Brass and R. V. Dingle (eds.). Summit, NJ: CIBA Pharmaceutical Co.

PETERS, A., S. L. PALAY AND H. DE F. WEBSTER (1991) *The Fine Structure of the Nervous System: Neurons and Their Supporting Cells*, 3rd Ed. New York: Oxford University Press.

RAMÓN Y CAJAL, S. (1984) *The Neuron and the Glial Cell*. (Transl. by J. de la Torre and W. C. Gibson.) Springfield, IL: Charles C. Thomas.

RAMÓN Y CAJAL, S. (1990) *New Ideas on the Structure of the Nervous System in Man and Vertebrates*. (Transl. by N. Swanson and L. W. Swanson.) Cambridge, MA: MIT Press.

WAXMAN, S. G. AND J. DEGROOT (1995) *Correlative Neuroanatomy*, 22nd Ed. Norwalk, CT: Appleton and Lange.

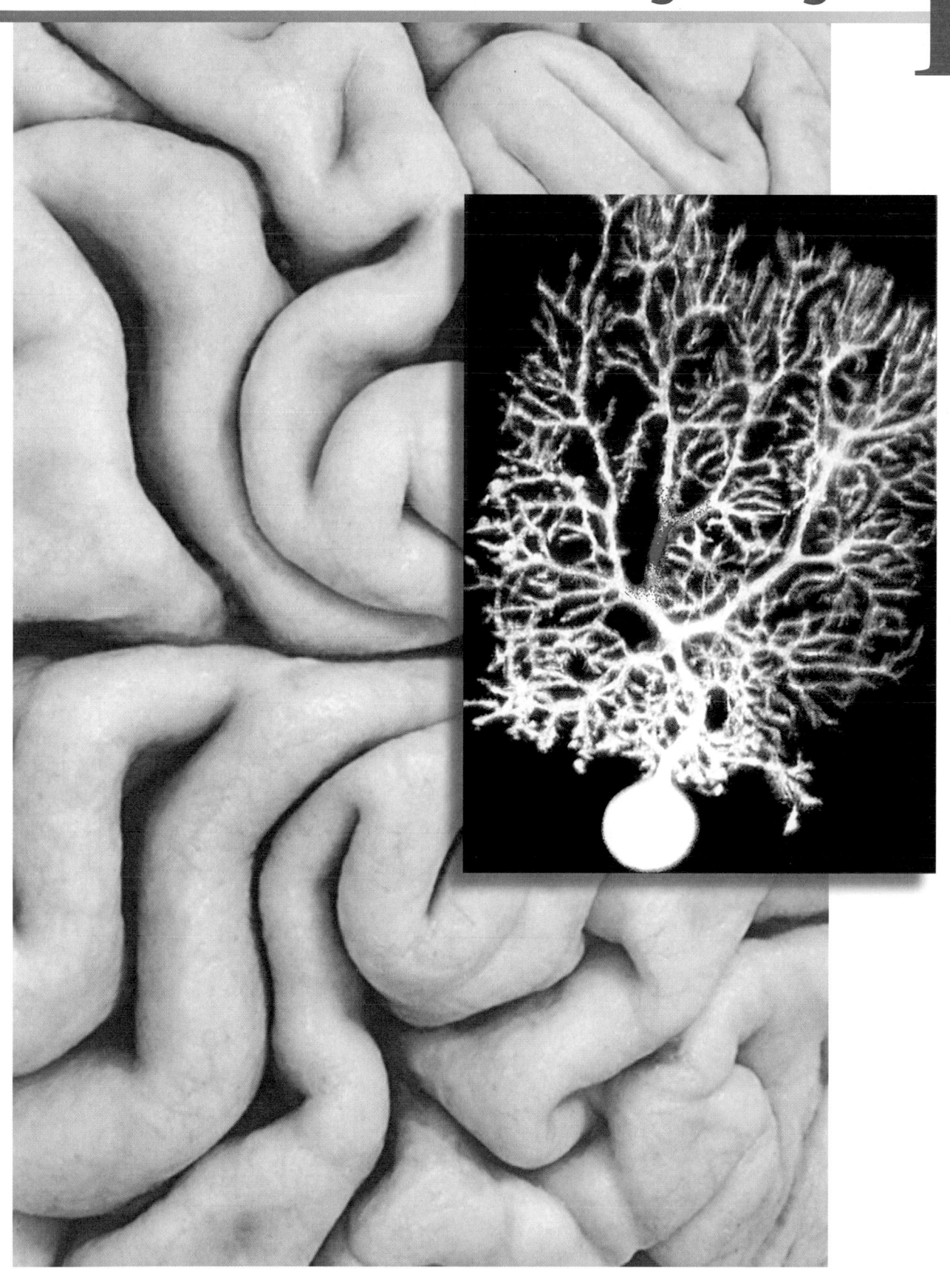

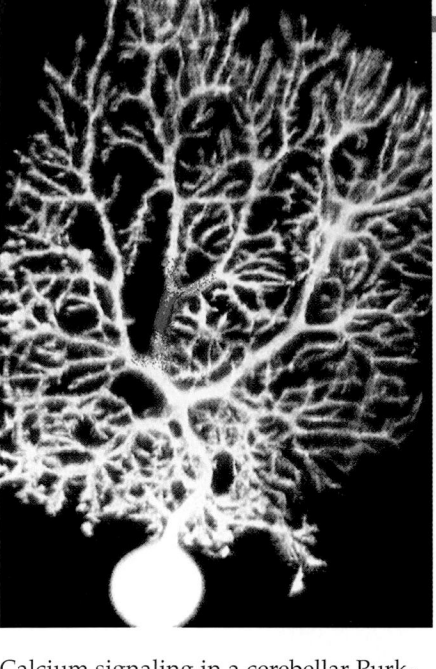

Calcium signaling in a cerebellar Purkinje neuron. The electrode shown in the lower left corner was used to fill the neuron with a fluorescent calcium indicator dye. This dye revealed the release of intracellular calcium ions (color) produced by the actions of the second messenger, IP_3. (Courtesy of Elizabeth A. Finch and George J. Augustine.)

Unit I
Neural Signaling

2   *Electrical Signals of Nerve Cells*
3   *Voltage-Dependent Membrane Permeability*
4   *Channels and Transporters*
5   *Synaptic Transmission*
6   *Neurotransmitters*
7   *Neurotransmitter Receptors and Their Effects*
8   *Intracellular Signal Transduction*

The primary purpose of the brain is to acquire, coordinate, and disseminate information about the body and its environment. To perform this task, neurons have evolved sophisticated means of generating electrical and chemical signals. This unit describes these signals and how they are produced. It also explains how one type of electrical signal, the action potential, allows information to travel along the length of a nerve cell, and how other types of signals—both electrical and chemical—are generated at synaptic connections between nerve cells. Synapses permit information transfer by interconnecting neurons to form the circuitry on which neural processing depends. These two types of signaling mechanisms—action potentials and synaptic signals—are the basis for the remarkable ability of the brain to sense, interpret, and ultimately act upon the environment.

The cellular and molecular mechanisms that give neurons their unique signaling abilities are also targets for disease processes that can compromise the function of the nervous system. A working knowledge of the cellular and molecular biology of neurons is therefore fundamental to understanding a variety of brain pathologies. An increasing number of diseases of the nervous system are beginning to be understood as discrete disorders of neuronal signaling molecules, and this information is now stimulating novel pharmacological and molecular biological approaches to diagnosing and treating these problems.

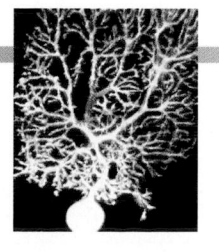

Chapter 2

Electrical Signals of Nerve Cells

Overview

Nerve cells generate electrical signals that transmit information. Although neurons are not intrinsically good conductors of electricity, they have evolved elaborate mechanisms for generating electrical signals based on the flow of ions across their plasma membranes. Ordinarily, neurons generate a negative potential, called the resting membrane potential, that can be measured by recording the voltage between the inside and outside of nerve cells. The action potential abolishes the negative resting potential and makes the transmembrane potential transiently positive. Action potentials are propagated along the length of axons and are the fundamental signal that carries information from one place to another in the nervous system. Generation of both the resting potential and the action potential can be understood in terms of the nerve cell's selective permeability to different ions, and of the normal distribution of these ions across the cell membrane.

Electrical Potentials Across Nerve Cell Membranes

Because electrical signals are the basis of information transfer in the nervous system, it is essential to understand how these signals arise. The use of electrical signals—as in sending electricity over wires to provide power or information—presents a series of problems in electrical engineering. A fundamental problem for neurons is that their axons, which can be quite long (remember that a spinal motor neuron can extend for a meter or more), are not good electrical conductors. Although neurons and wires are both capable of passively conducting electricity, the electrical properties of neurons compare poorly to even the most ordinary wire. To compensate for this deficiency, neurons have evolved a "booster system" that allows them to conduct electrical signals over great distances despite their intrinsically poor electrical characteristics. The electrical signals produced by this booster system are called **action potentials** (which are also referred to as "spikes" or "impulses").

The best way to observe an action potential is to use an intracellular microelectrode to record directly the electrical potential across the neuronal plasma membrane (Figure 2.1). A typical microelectrode is a piece of glass tubing pulled to a very fine point (with an opening of less than $1\,\mu$m diameter) and filled with a good electrical conductor, such as a concentrated salt solution. This conductive core can then be connected to a voltmeter, such as an oscilloscope, to record the transmembrane voltage of the nerve cell. When a microelectrode is inserted through the membrane of the neuron, it records a negative potential, indicating that the cell has a means of generating a con-

(A)

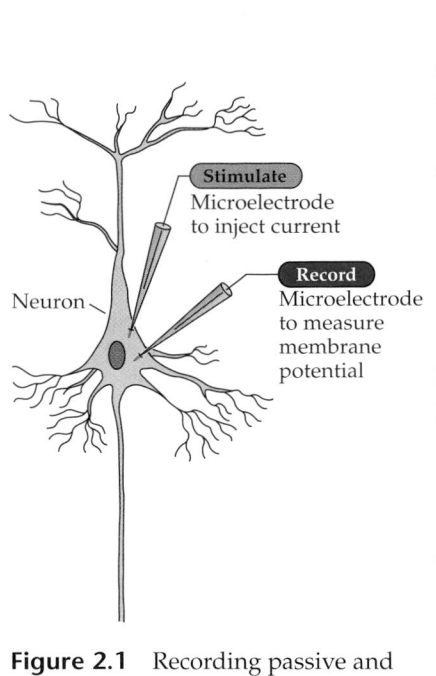

(B)

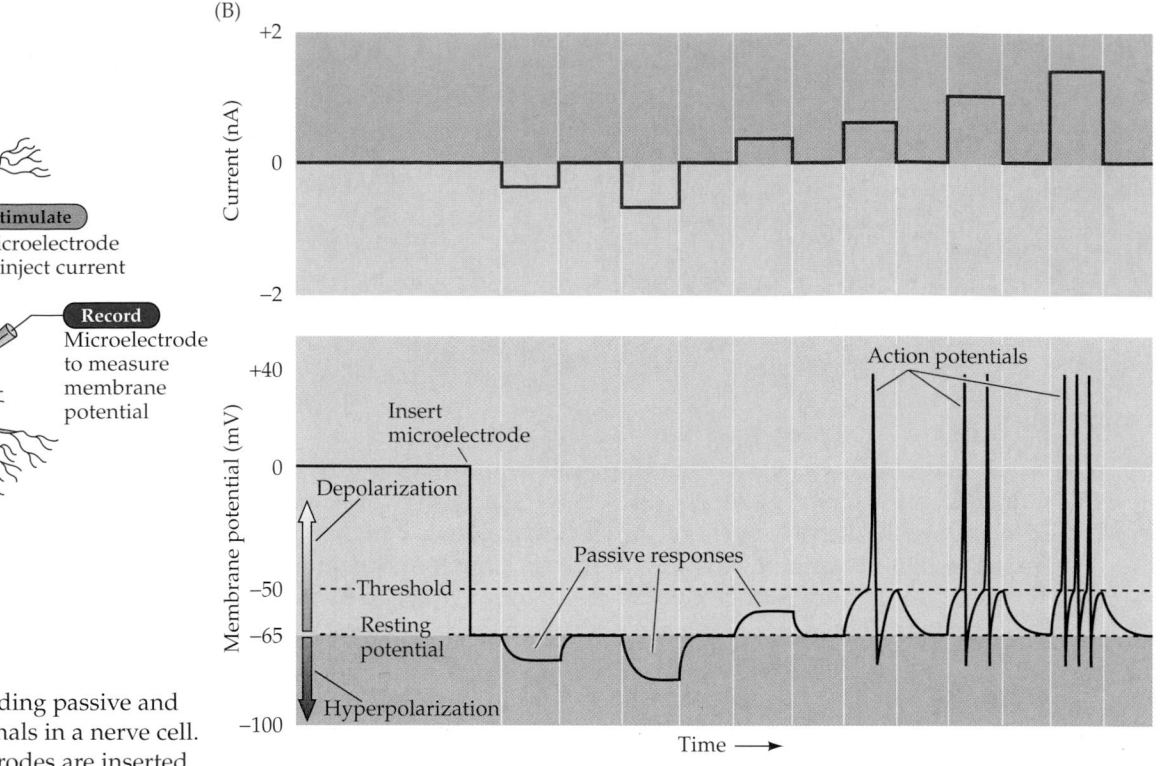

Figure 2.1   Recording passive and active electrical signals in a nerve cell. (A) Two microelectrodes are inserted into a neuron; one of these measures membrane potential while the other injects current into the neuron. (B) Inserting the voltage-measuring microelectrode into the neuron reveals a negative potential, the resting membrane potential. Injecting current through the current-passing microelectrode alters the neuronal membrane potential. Hyperpolarizing current pulses produce only passive changes in the membrane potential. While small depolarizing currents also elicit only passive responses, depolarizations that cause the membrane potential to meet or exceed threshold additionally evoke action potentials. Action potentials are active responses in the sense that they are generated by changes in the permeability of the neuronal membrane.

stant voltage across its membrane when it is at rest. This voltage, called the **resting membrane potential**, depends on the type of neuron being examined, but it is always a fraction of a volt (typically −40 to −90 mV).

Action potentials represent transient changes in the resting membrane potential of neurons. One way to elicit an action potential is to pass electrical current across the membrane of the neuron. In normal circumstances, this current would be generated by the action of neurotransmitters released by other neurons, or by the transduction of an external stimulus at specialized regions of sensory neurons (sensory receptors in the skin, for example; see Unit II). In the laboratory, however, electrical current suitable for initiating an action potential can be readily produced by inserting a second microelectrode into the same neuron and then connecting the electrode to a battery. If the current delivered in this way is such as to make the membrane potential more negative (**hyperpolarization**), nothing very dramatic happens. The membrane potential simply changes in proportion to the magnitude of the injected current. Such hyperpolarizing responses do not require any unique property of neurons and are therefore called passive electrical responses. A much more interesting phenomenon is seen if current of the opposite polarity is delivered, so that the membrane potential of the nerve cell becomes more positive than the resting potential (**depolarization**). In this case, at a certain level of membrane potential called the **threshold potential**, an action potential occurs (see Figure 2.1B).

The action potential, which is an active response generated by the neuron, appears on an oscilloscope as a brief (about 1 ms) change from negative to positive in the transmembrane potential. Importantly, the amplitude of the action potential is independent of the magnitude of the current used to evoke it; that is, larger currents do not elicit larger action potentials. The action potentials of a given neuron are therefore said to be all-or-none, because they occur fully or not at all. If the amplitude or duration of the stimulus current is increased sufficiently, multiple action potentials occur, as can be seen in the responses to the three different current intensities shown at the right of Figure 2.1B. It follows, therefore, that the intensity of a stimulus is encoded in the frequency of action potentials rather than in their amplitude.

This chapter addresses the underlying question of how nerve cells can generate electrical potentials by distributing ions across the neuronal membrane. Chapter 3 explores more specifically the means by which action potentials are produced and how these signals solve the problem of long-distance electrical conduction within nerve cells. Chapter 4 examines the properties of membrane molecules responsible for producing action potentials. Finally, Chapters 5–8 consider how electrical signals are transmitted from one nerve cell to another at synaptic contacts.

How Ionic Movements Produce Electrical Signals

Electrical potentials are generated across the membranes of neurons—and, indeed, all cells—because (1) there are *differences in the concentrations* of specific ions across nerve cell membranes, and (2) the membranes are *selectively permeable* to some of these ions. These two facts depend in turn on two different kinds of proteins in the cell membrane (Figure 2.2). The ion concentration gradients are established by proteins known as **active transporters**, which, as their name suggests, actively move ions into or out of cells against their concentration gradients. The selective permeability of membranes is due largely to **ion channels**, proteins that allow only certain kinds of ions to cross the membrane in the direction of their concentration gradients. Thus, channels and transporters basically work against each other, and in so doing they generate the resting membrane potential, action potentials, and the synaptic potentials and receptor potentials that trigger action potentials. The structure and function of these channels and transporters are described in Chapter 4.

To appreciate the role of ion gradients and selective permeability in generating a membrane potential, consider a simple system in which a membrane separates two compartments containing solutions of ions. In such a system, it is possible to determine the composition of the two solutions and,

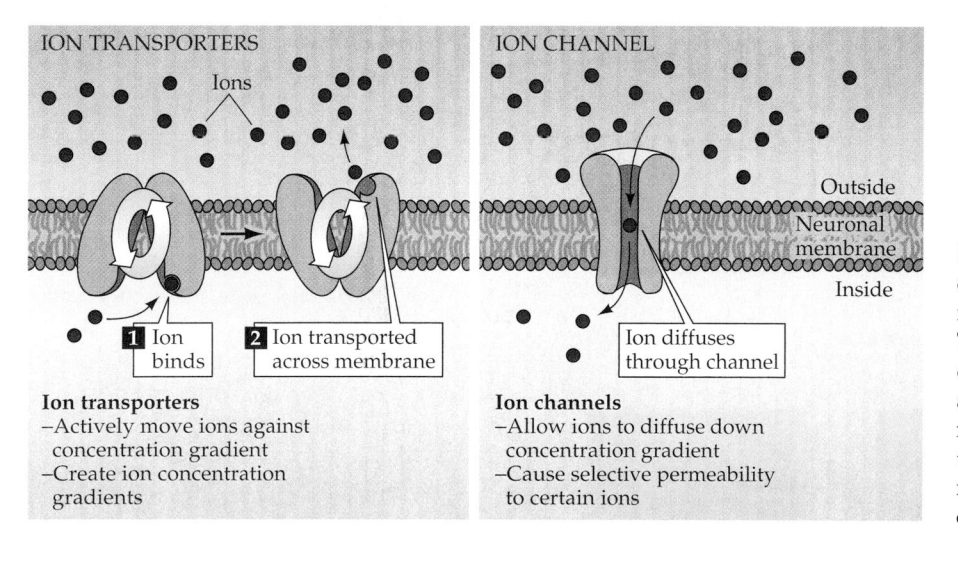

Ion transporters
–Actively move ions against concentration gradient
–Create ion concentration gradients

Ion channels
–Allow ions to diffuse down concentration gradient
–Cause selective permeability to certain ions

Figure 2.2 Ion transporters and ion channels are responsible for ionic movements across neuronal membranes. Transporters create ion concentration differences by actively transporting ions against their chemical gradients. Channels take advantage of these concentration gradients, allowing selected ions to move, via diffusion, down their chemical gradients.

thereby, control the ion gradients across the membrane. For example, take the case of a membrane that is permeable only to potassium ions (K^+). If the concentration of K^+ on each side of this membrane is equal, then no electrical potential will be measured across it (Figure 2.3A). However, if the concentration of K^+ is not the same on the two sides, then an electrical potential will be generated. For instance, if the concentration of K^+ on one side of the membrane (compartment 1) is 10 times higher than the K^+ concentration on the other side (compartment 2), then the electrical potential of compartment 1 will be negative relative to compartment 2 (Figure 2.3B). This difference in electrical potential is generated because the potassium ions flow down their concentration gradient and take their electrical charge (one positive charge per ion) with them as they go. Because neuronal membranes contain pumps that accumulate K^+ in the cell cytoplasm, and because potassium-permeable channels in the plasma membrane allow a transmembrane flow of K^+, an analogous situation exists in living nerve cells. A continual resting efflux of K^+ is therefore responsible for the resting membrane potential.

In the hypothetical case just described, an equilibrium will quickly be reached. As K^+ moves from compartment 1 to compartment 2 (the initial conditions on the left of Figure 2.3B), a potential is generated that tends to impede further flow of K^+. This impediment results from the fact that the potential gradient across the membrane tends to repel the positive potassium ions that would otherwise move across the membrane. Thus, as compartment 2 becomes positive relative to compartment 1, the increasing positivity makes compartment 2 less attractive to the positively charged K^+. The net movement (or flux) of K^+ will stop at the point (at equilibrium on the right of Figure 2.3B) where the potential change across the membrane (the relative positivity of compartment 2) exactly offsets the concentration gradient (the 10× excess of K^+ in compartment 1). At this **electrochemical equilibrium**, there is an exact balance between two opposing forces: (1) the concentration gradient that causes K^+ to move from compartment 1 to compartment 2, taking along positive charge, and (2) an opposing electrical gradient that increasingly

Figure 2.3 Electrochemical equilibrium. (A) A membrane permeable only to K^+ (yellow spheres) separates compartments 1 and 2, which contain the indicated concentrations of KCl. (B) Increasing the KCl concentration in compartment 1 to 10 mM initially causes a small movement of K^+ into compartment 2 (initial conditions) until the electromotive force acting on K^+ balances the concentration gradient, and the net movement of K^+ becomes zero (at equilibrium). (C) The relationship between the transmembrane concentration gradient ($[K^+]_2/[K^+]_1$) and the membrane potential. As predicted by the Nernst equation, this relationship is linear when plotted on semilogarithmic coordinates, with a slope of 58 mV per tenfold difference in the concentration gradient.

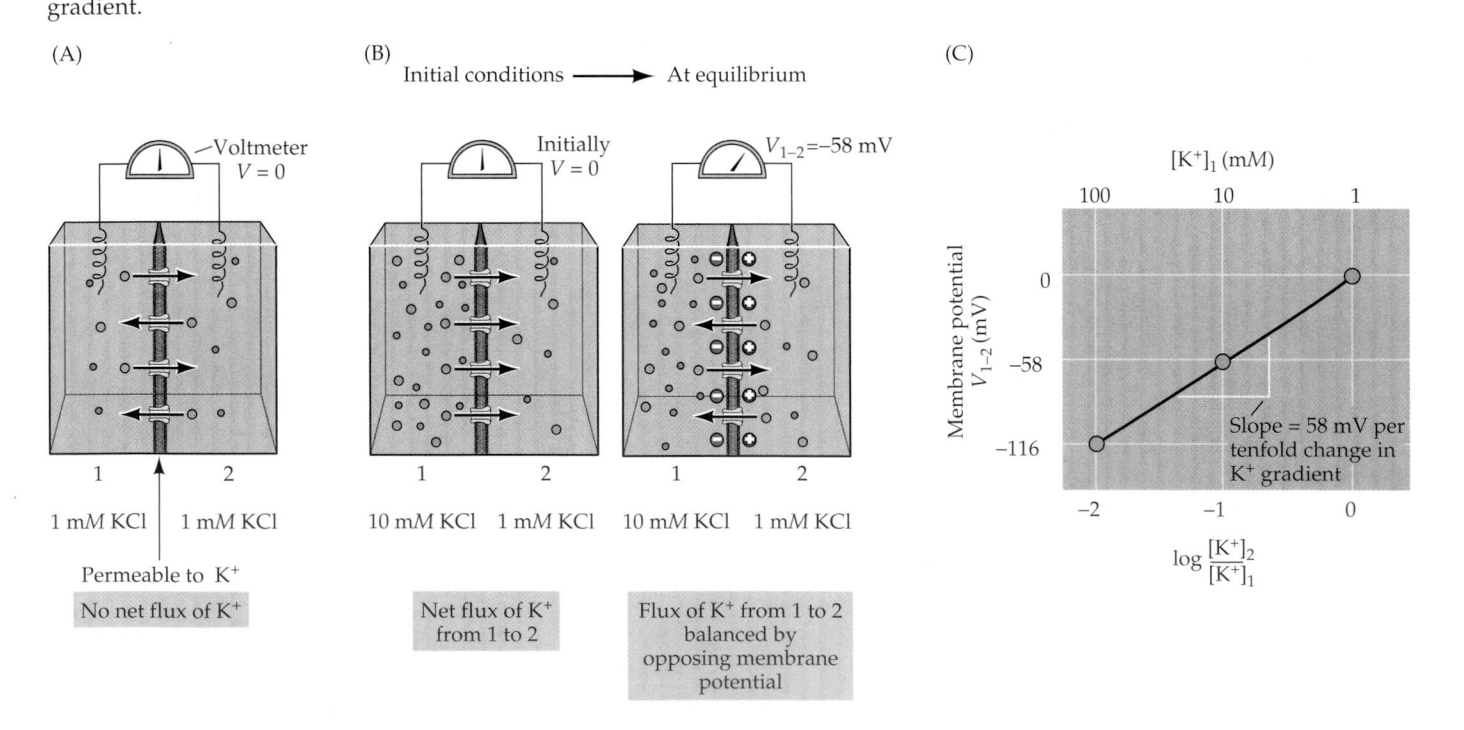

tends to stop K^+ from moving across the membrane (Figure 2.3B). The number of ions that needs to flow to generate this electrical potential is very small (per cm^2 of membrane, approximately 10^{-12} moles of K^+, or 10^{12} K^+ ions). This last fact is significant in two ways. First, it means that the concentrations of permeant ions on each side of the membrane remain essentially constant, even after the flow of ions has generated the potential. Second, the tiny fluxes of ions required to establish the membrane potential do not disrupt chemical electroneutrality because each ion has an oppositely charged counter ion (chloride ions in the example shown in Figure 2.3) to maintain the neutrality of the solutions on each side of the membrane. The concentration of K^+ remains equal to the concentration of Cl^- in the solutions in compartments 1 and 2, meaning that the separation of charge that creates the potential difference is restricted to the immediate vicinity of the membrane.

The Forces that Create Membrane Potentials

The electrical potential generated across the membrane at electrochemical equilibrium, the **equilibrium potential**, can be predicted by a simple formula called the **Nernst equation**. This relationship is generally expressed as

$$E_X = \frac{RT}{zF} \ln \frac{[X]_2}{[X]_1}$$

where E_X is the equilibrium potential for any ion X, R is the gas constant, T is the absolute temperature (in degrees on the Kelvin scale), z is the valence (electrical charge) of the permeant ion, and F is the Faraday constant (the amount of electrical charge contained in one mole of a univalent ion). The brackets indicate the concentrations of ion X on each side of the membrane and the symbol ln indicates the natural logarithm of the concentration gradient. Because it is easier to perform calculations using base 10 logarithms and to perform experiments at room temperature, this relationship is usually simplified to

$$E_X = \frac{58}{z} \log \frac{[X]_2}{[X]_1}$$

where log indicates the base 10 logarithm of the concentration ratio. Thus, for the example in Figure 2.3B, the potential across the membrane at electrochemical equilibrium is

$$E_K = \frac{58}{z} \log \frac{[K]_2}{[K]_1} = 58 \log \frac{1}{10} = -58 \text{ mV}$$

The equilibrium potential is conventionally defined in terms of the potential difference between the reference compartment, side 2 in Figure 2.3, and the other side. This approach is also applied to biological systems. In this case, the outside of the cell is the conventional reference point (defined as zero potential). Thus, when the concentration of K^+ is higher inside than out, an inside-negative potential is measured across the K^+-permeable neuronal membrane.

For a simple hypothetical system with only one permeant ion species, the Nernst equation allows the electrical potential across the membrane at equilibrium to be predicted exactly. For example, if the concentration of K^+ on side 1 is increased to 100 mM, the membrane potential will be −116 mV. More generally, if the membrane potential is plotted against the logarithm of the K^+ concentration gradient ($[K]_2/[K]_1$), the Nernst equation predicts a linear relationship with a slope of 58 mV (actually 58/z) per tenfold change in the K^+ gradient (Figure 2.3C).

To reinforce and extend the concept of electrochemical equilibrium, consider some additional experiments on the influence of ionic species and ionic permeability that could be performed on the simple model system in Figure 2.3. What would happen to the electrical potential across the membrane (the potential of side 1 relative to side 2) if the potassium on side 2 were replaced with 10 mM sodium (Na^+) and the K^+ in compartment 1 were replaced by 1 mM Na^+? No potential would be generated, because no Na^+ could flow across the membrane (which was defined as being permeable only to K^+). However, if under these ionic conditions (10 times more Na^+ in compartment 2) the K^+-permeable membrane were to be magically replaced by a membrane permeable only to Na^+, a potential of +58 mV would be measured at equilibrium. If 10 mM calcium (Ca^{2+}) were present in compartment 2 and 1 mM Ca^{2+} in compartment 1, and a Ca^{2+}-selective membrane separated the two sides, what would happen to the membrane potential? A potential of +29 mV would develop, because the valence of calcium is 2. Finally, what would happen to the membrane potential if 10 mM Cl^- were present in compartment 1 and 1 mM Cl^- were present in compartment 2, with the two sides separated by a Cl^--permeable membrane? Because the valence of this anion is −1, the potential would again be +58 mV.

The balance of chemical and electrical forces at equilibrium means that the electrical potential can determine ionic fluxes across the membrane, just as the ionic gradient can determine the membrane potential. To examine the influence of membrane potential on ionic flux, imagine connecting a battery across the two sides of the membrane to control the electrical potential across the membrane without changing the distribution of ions on the two sides (Figure 2.4). As long as the battery is off, things will be just as in Figure 2.3, with the

Figure 2.4 Membrane potential influences ion fluxes. (A) Connecting a battery across the K^+-permeable membrane allows direct control of membrane potential. When the battery is turned off (left), K^+ ions flow simply according to their concentration gradient. Setting the initial membrane potential (V_{1-2}) at the equilibrium potential for K^+ (center) yields no net flux of K^+, while making the membrane potential more negative than the K^+ equilibrium potential (right) causes K^+ to flow against its concentration gradient. (B) Relationship between membrane potential and direction of K^+ flux.

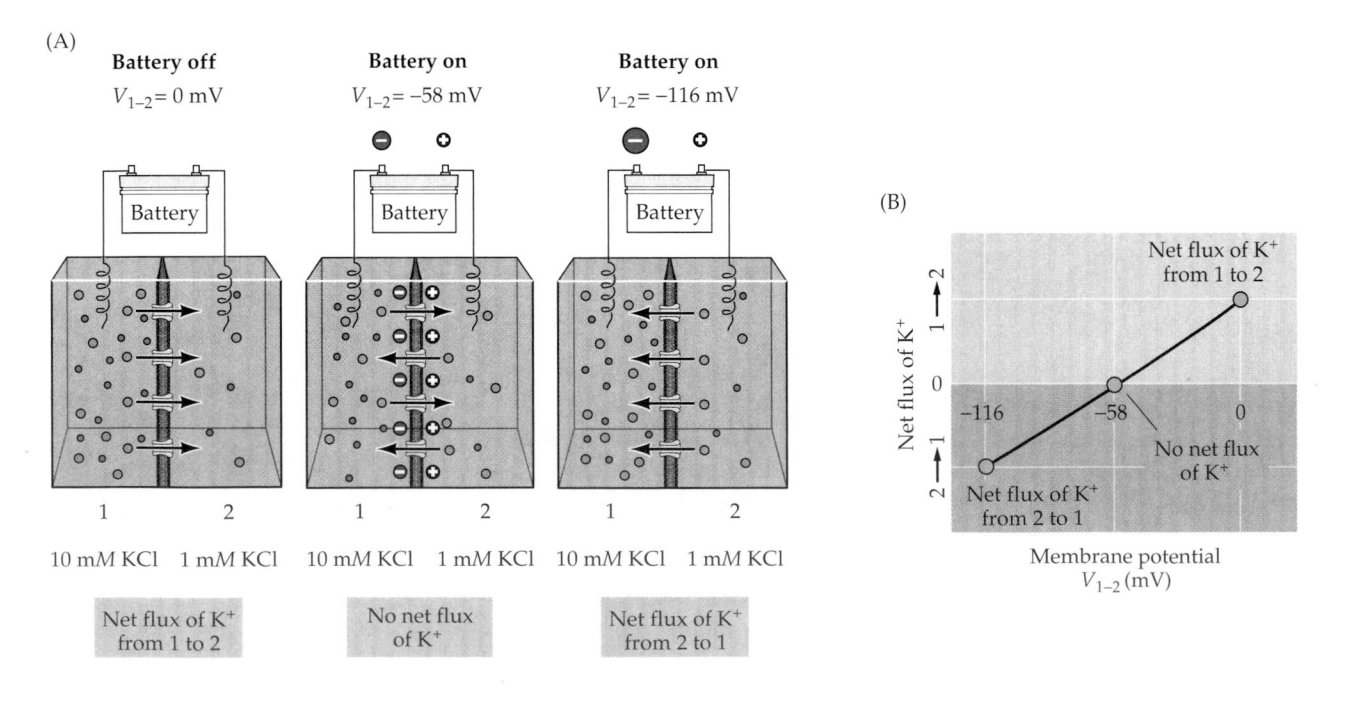

flow of K^+ from compartment 1 to compartment 2 causing a negative membrane potential (Figure 2.4A, left). However, if the battery is used to make compartment 1 initially more negative relative to compartment 2, there will be less K^+ flux, because the negative potential will tend to keep K^+ in compartment 1. How negative will side 1 need to be before there is no net flux of K^+? The answer is –58 mV, the voltage needed to counter the tenfold difference in K^+ concentrations on the two sides of the membrane (Figure 2.4A, center). If compartment 1 is initially made more negative than –58 mV, then K^+ will actually flow from compartment 2 into compartment 1, because the positive ions will be attracted to the more negative potential of compartment 1 (Figure 2.4A, right). This example demonstrates that both the direction and magnitude of ion flux depend on the membrane potential. Thus, in some circumstances the electrical potential can overcome an ionic concentration gradient.

The ability to alter ion flux experimentally by changing either the potential imposed on the membrane (Figure 2.4B) or the transmembrane concentration gradient for an ion (see Figure 2.3C) provides convenient tools for studying ion fluxes across the plasma membranes of neurons, as will be evident in many of the experiments described in the following chapters.

Electrochemical Equilibrium in an Environment with More Than One Permeant Ion

Now consider a somewhat more complex situation in which Na^+ and K^+ are unequally distributed across the membrane, as in Figure 2.5A. What would happen if 10 mM K^+ and 1 mM Na^+ were present in compartment 1, and 1 mM K^+ and 10 mM Na in compartment 2? If the membrane were permeable only to K^+, the membrane potential would be –58 mV; if the membrane were permeable only to Na^+, the potential would be +58 mV. But what would the potential be if the membrane were permeable to both K^+ and Na^+? In this case, the potential would depend on the relative permeability of the membrane to K^+ and Na^+. If it were more permeable to K^+, the potential would approach –58 mV, and if it were more permeable to Na^+, the potential would be closer to +58 mV. Because there is no permeability term in the Nernst equation, which only considers the simple case of a single permeant ion species, a more elaborate equation is needed that takes into account both

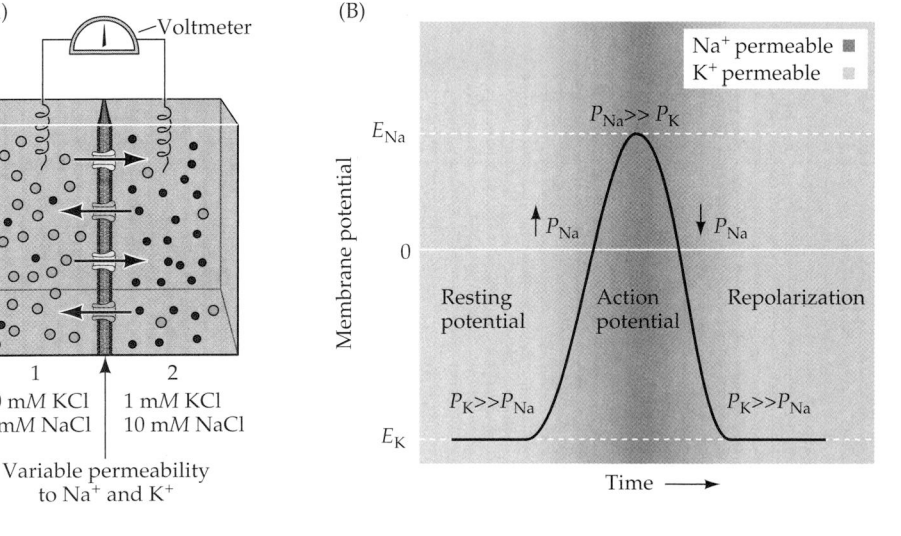

(A) Voltmeter

1
10 mM KCl
1 mM NaCl

2
1 mM KCl
10 mM NaCl

Variable permeability to Na^+ and K^+

(B)

Membrane potential

E_{Na}

0

E_K

Na$^+$ permeable
K$^+$ permeable

$P_{Na} \gg P_K$

$\uparrow P_{Na}$ $\downarrow P_{Na}$

Resting potential   Action potential   Repolarization

$P_K \gg P_{Na}$ $P_K \gg P_{Na}$

Time →

Figure 2.5 Resting and action potentials entail permeabilities to different ions. (A) Hypothetical situation in which a membrane variably permeable to Na^+ (red) and K^+ (yellow) separates two compartments that contain both ions. For simplicity, Cl^- ions are not shown in the diagram. (B) Schematic representation of the membrane ionic permeabilities associated with resting and action potentials. At rest, neuronal membranes are more permeable to K^+ (yellow) than to Na^+ (red); accordingly, the resting membrane potential is negative and approaches E_K. During an action potential, the membrane becomes very permeable to Na^+ (red); thus the membrane potential becomes positive and approaches E_{Na}. The rise in Na^+ permeability is transient, however, so that the membrane again becomes primarily permeable to K^+ (yellow), causing the potential to return to its negative resting value.

the concentration gradients of the permeant ions and the relative permeability of the membrane to each permeant species.

Such an equation was developed by David Goldman in 1943. For the case most relevant to neurons, in which K^+, Na^+, and Cl^- are the primary permeant ions, the **Goldman equation** is written

$$V = 58 \log \frac{P_K[K]_2 + P_{Na}[Na]_2 + P_{Cl}[Cl]_1}{P_K[K]_1 + P_{Na}[Na]_1 + P_{Cl}[Cl]_2}$$

where V is the voltage across the membrane (again, compartment 1 relative to the reference compartment 2) and P indicates the permeability of the membrane to each ion of interest. The Goldman equation is thus an extended version of the Nernst equation that takes into account the relative permeabilities of each of the ions involved. The relationship between the two equations becomes obvious in the situation where the membrane is permeable only to one ion, say, K^+; in this case, the Goldman expression collapses back to the simpler Nernst equation. In this context, it is important to note that the valence factor (z) in the Nernst equation has been eliminated; this is why the concentrations of negatively charged chloride ions, Cl^-, have been inverted relative to the concentrations of the positively charged ions [remember that $-\log (A/B) = \log (B/A)$].

If the membrane in Figure 2.5A is permeable to only K^+ and Na^+, the terms involving Cl^- drop out because P_{Cl} is 0. In this case, solution of the Goldman equation yields a potential of –58 mV when only K^+ is permeant, +58 mV when only Na^+ is permeant, and some intermediate value if both ions are permeant. For example, if K^+ and Na^+ were equally permeant, then the potential would be 0 mV.

With respect to neural signaling, it is particularly pertinent to ask what would happen if the membrane started out being permeable to K^+, and then temporarily switched to become most permeable to Na^+. In this circumstance, the membrane potential would start out at a negative level, become positive while the Na^+ permeability remained high, and then fall back to a negative level as the Na^+ permeability decreased again. As it turns out, this last case essentially describes what goes on in a neuron during the generation of an action potential. In the resting state, P_K of the neuronal plasma membrane is much higher than P_{Na}; since, as a result of the action of ion

TABLE 2.1
Extracellular and Intracellular Ion Concentrations

Ion	Concentration (mM)	
	Intracellular	*Extracellular*
Squid neuron		
Potassium (K^+)	400	20
Sodium (Na^+)	50	440
Chloride (Cl^-)	40–150	560
Calcium (Ca^{2+})	0.0001	10
Mammalian neuron		
Potassium (K^+)	140	5
Sodium (Na^+)	5–15	145
Chloride (Cl^-)	4–30	110
Calcium (Ca^{2+})	0.0001	1–2

transporters, there is always more K^+ inside the cell than outside (Table 2.1), the resting potential is negative (Figure 2.5B). As the membrane potential is depolarized (by synaptic action, for example), P_{Na} increases. The transient increase in Na^+ permeability causes the membrane potential to become even more positive (red region in Figure 2.5B), because Na^+ rushes in (remember that there is much more Na^+ outside a neuron than inside, again as a result of ion pumps). Because of this positive feedback loop, an action potential occurs. The rise in Na^+ permeability during the action potential is transient, however, as the membrane permeability to K^+ is restored, the membrane potential quickly returns to its resting level.

Armed with an appreciation of these simple electrochemical principles, it will be much easier to understand the more detailed account that follows of how neurons generate resting and action potentials.

The Ionic Basis of the Resting Membrane Potential

The action of ion transporters creates substantial transmembrane gradients for most ions. Table 2.1 summarizes the ion concentrations measured directly in an exceptionally large nerve cell found in the nervous system of the squid (Box A). Such measurements are the basis for stating that there is much more K^+ inside the neuron than out, and much more Na^+ outside than in. Similar concentration gradients occur in the neurons of most animals, including humans. However, because the ionic strength of mammalian blood is lower than that of sea-dwelling animals such as squid, in mammals the concentrations of each ion are several times lower. These transporter-dependent concentration gradients are the indirect source of the resting neuronal membrane potential and the action potential.

Once the ion concentration gradients across various neuronal membranes are known, the Nernst equation can be used to calculate that the equilibrium potential for K^+ and other major ions. Since the resting membrane potential of the squid neuron is approximately –65 mV, K^+ is the ion that is closest to being in electrochemical equilibrium when the cell is at rest. This fact implies that the resting membrane is more permeable to K^+ than to the other ions listed in Table 2.1, and that this permeability is the source of resting potentials.

It is possible to test this guess, as Alan Hodgkin and Bernard Katz did in 1949, by asking what happens to the resting membrane potential if the concentration of K^+ outside the neuron is altered. If the resting membrane were permeable only to K^+, then the Goldman equation (or even the simpler Nernst equation) predicts that the membrane potential will vary in proportion to the logarithm of the K^+ concentration gradient across the membrane. Assuming that the internal K^+ concentration is unchanged during the experiment, a plot of membrane potential against the logarithm of the external K^+ concentration should yield a straight line with a slope of 58 mV per tenfold change in external K^+ concentration at room temperature (see Figure 2.3C). (The slope becomes about 61 mV at mammalian body temperatures.)

When Hodgkin and Katz carried out this experiment on a living squid neuron, they found that the resting membrane potential did indeed change when the external K^+ concentration was modified, becoming less negative as external K^+ concentration was raised (Figure 2.6A). When the external K^+ concentration was raised high enough to equal the concentration of K^+ inside the neuron, thus making the K^+ equilibrium potential 0 mV, the resting membrane potential was also approximately 0 mV. In short, the resting membrane potential varied as predicted with the logarithm of the K^+ concentration, with a slope that approached 58 mV per tenfold change in K^+ concentration (Figure 2.6B). The value obtained was not exactly 58 mV

Figure 2.6 Experimental evidence that the resting membrane potential of a squid giant axon is determined by the K⁺ concentration gradient across the membrane. (A) Increasing the external K⁺ concentration makes the resting membrane potential more positive. (B) Relationship between resting membrane potential and external K⁺ concentration, plotted on a semilogarithmic scale. The straight line represents a slope of 58 mV per tenfold change in concentration, as given by the Nernst equation. (After Hodgkin and Katz, 1949.)

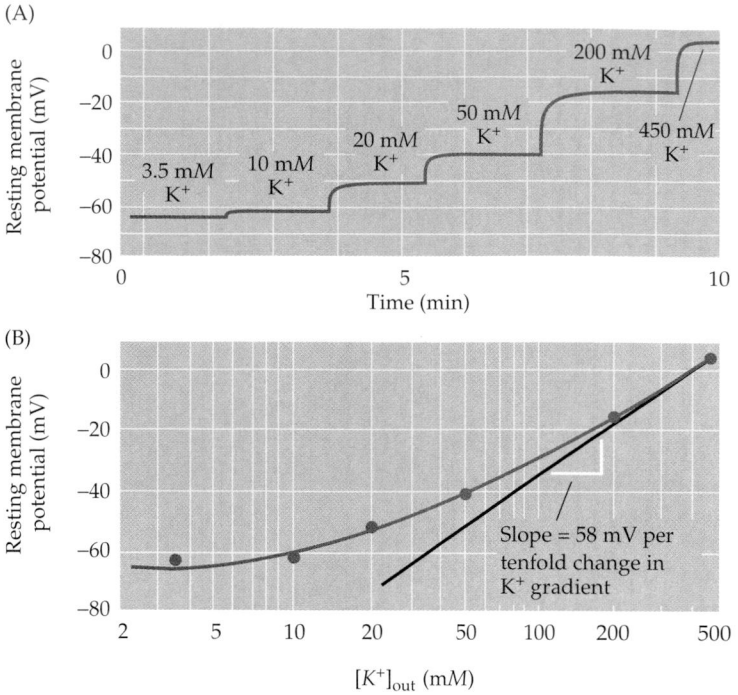

because other ions, such as Cl⁻ and Na⁺, are also slightly permeable, and thus influence the resting potential to a small degree. The contribution of these other ions is particularly evident at low external K⁺ levels, again as predicted by the Goldman equation. In general, however, manipulation of the external concentrations of these other ions has only a small effect, emphasizing that K⁺ permeability is indeed the primary source of the resting membrane potential.

In summary, Hodgkin and Katz showed that the inside-negative resting potential arises because (1) the membrane of the resting neuron is more permeable to K⁺ than to any of the other ions present, and (2) there is more K⁺ inside the neuron than outside. The selective permeability to K⁺ is caused by K⁺-permeable membrane channels that are open in resting neurons, and the large K⁺ concentration gradient is, as noted, produced by membrane transporters that selectively accumulate K⁺ within neurons. Many subsequent studies have confirmed the general validity of these principles.

The Ionic Basis of Action Potentials

What causes the membrane potential of a neuron to depolarize during an action potential? Although a general answer to this question has been given (increased permeability to Na⁺), it is well worth reasoning this out in more detail. Given the data presented in Table 2.1, one can use the Nernst equation to calculate that the equilibrium potential for Na⁺ (E_{Na}) in neurons, and indeed in most cells, is positive. Thus, if the membrane were to become highly permeable to Na⁺, the membrane potential would approach E_{Na}. Based on these considerations, Hodgkin and Katz hypothesized that the action potential arises because the neuronal membrane becomes temporarily permeable to Na⁺.

Taking advantage of the same style of ion substitution experiment they used to assess the resting potential, Hodgkin and Katz tested the role of Na⁺ in generating the action potential by asking what happens to the action

BOX A
The Remarkable Giant Nerve Cells of Squid

Many of the initial insights into how ion concentration gradients and changes in membrane permeability produce electrical signals came from experiments performed on the extraordinarily large nerve cells of the squid. The axons of these nerve cells can be up to 1 mm in diameter—100 to 1000 times larger than mammalian axons. Squid axons are large enough to allow experiments that would be impossible on most other nerve cells. For example, it is not difficult to insert simple wire electrodes inside these giant axons and make reliable electrical measurements. The relative ease of this approach yielded the first intracellular recordings of action potentials from nerve cells and, as will be discussed in the next chapter, the first experimental

measurements of the ionic currents that produce action potentials. It also is practical to extrude the cytoplasm from giant axons and measure its ionic composition (see Table 2.1). In addition, some giant nerve cells form synaptic contacts with other giant nerve cells, producing very large synapses that have been extraordinarily valuable in understanding the fundamental mechanisms of synaptic transmission (see Chapter 5).

Giant neurons evidently evolved in squid because they enhanced survival. These neurons participate in a simple neural circuit that activates the contraction of the mantle muscle, producing a jet propulsion effect that allows the squid to move away from predators at a remarkably fast speed. As discussed in

Chapter 3, larger axonal diameter allows faster conduction of action potentials. Thus, squid presumably have these huge nerve cells to escape more successfully from their numerous enemies. Today—more than 60 years after their discovery by John Z. Young at University College, London—the giant nerve cells of squid remain useful experimental systems for probing basic neuronal functions.

References

LLINÁS, R. (1982) Calcium in synaptic transmission. Sci. Am. 247 (April): 56–65.

YOUNG, J. Z. (1939) Fused neurons and synaptic contacts in the giant nerve fibres of cephalopods. Phil. Trans. R. Soc. Lond. 229(B): 465–503.

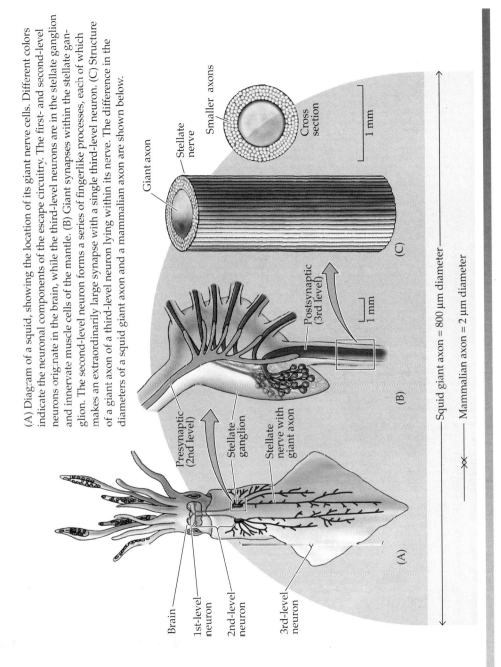

(A) Diagram of a squid, showing the location of its giant nerve cells. Different colors indicate the neuronal components of the escape circuitry. The first- and second-level neurons originate in the brain, while the third-level neurons are in the stellate ganglion and innervate muscle cells of the mantle. (B) Giant synapses within the stellate ganglion. The second-level neuron forms a series of fingerlike processes, each of which makes an extraordinarily large synapse with a single third-level neuron. (C) Structure of a giant axon of a third-level neuron lying within its nerve. The difference in the diameters of a squid giant axon and a mammalian axon are shown below.

Figure 2.7 The role of sodium in the generation of an action potential in a squid giant axon. (A) An action potential evoked with the normal ion concentrations inside and outside the cell. (B) The amplitude and rate of rise of the action potential diminish when external sodium concentration is reduced to one-third of normal, but (C) recover when the Na⁺ is replaced. (D) While the amplitude of the action potential is quite sensitive to the external concentration of Na⁺, the resting membrane potential (E) is little affected by changing the concentration of this ion. (After Hodgkin and Katz, 1949.)

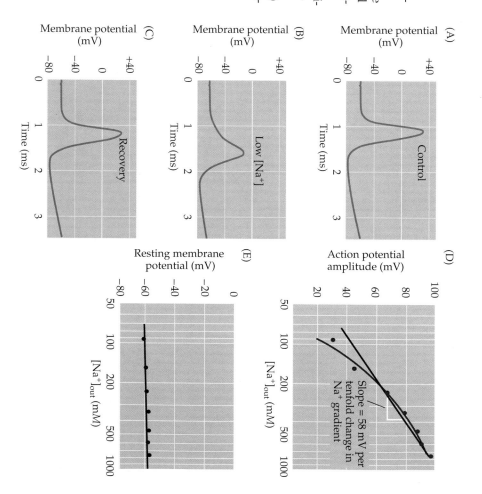

(A)

Membrane potential (mV)

Control

Time (ms)

(B)

Membrane potential (mV)

Low [Na⁺]

Time (ms)

(C)

Membrane potential (mV)

Recovery

Time (ms)

(D)

Action potential amplitude (mV)

[Na⁺]$_{out}$ (mM)

Slope = 58 mV per tenfold change in Na⁺ gradient

(E)

Resting membrane potential (mV)

[Na⁺]$_{out}$ (mM)

potential when Na⁺ is removed from the external medium. They found that lowering the external Na⁺ concentration reduces both the rate of rise of the action potential and its peak amplitude (Figure 2.7A–C). Indeed, when they examined this Na⁺ dependence quantitatively, they found a more-or-less linear relationship between the amplitude of the action potential and the logarithm of the external Na⁺ concentration (Figure 2.7D). The slope of this relationship approached a value of 58 mV per tenfold change in Na⁺ concentration, as expected for a membrane selectively permeable to Na⁺. In contrast, lowering Na⁺ concentration had very little effect on the resting membrane potential (Figure 2.7E). Thus, while the resting neuronal membrane is only slightly permeable to Na⁺, the membrane becomes extraordinarily permeable to Na⁺ during the **rising phase** and **overshoot phase** of the action potential (see Box B for an explanation of action potential nomenclature). This temporary increase in Na⁺ permeability results from the opening of Na⁺-selective channels that are essentially closed in the resting state. Membrane pumps maintain a large electrochemical gradient for Na⁺, which is in much higher concentration outside the neuron than in. When the Na⁺ channels open, Na⁺ flows into the neuron, causing the membrane potential to depolarize and approach E$_{Na}$.

The time that the membrane potential lingers near E$_{Na}$ (about +58 mV) during the overshoot phase of an action potential is brief because the increased membrane permeability to Na⁺ itself is short-lived. The membrane potential rapidly repolarizes to resting levels and is actually followed by a transient **undershoot**. As will be described in Chapter 3, these latter events in the action potential are due to an inactivation of the Na⁺ permeability and an increase in the K⁺ permeability of the membrane. During the undershoot,

BOX B
Action Potential Form and Nomenclature

The action potential of the squid giant axon has a characteristic shape, or waveform, with a number of different phases (figure A). During the rising phase, the membrane potential rapidly depolarizes. In fact, action potentials cause the membrane potential to depolarize so much that the membrane potential transiently becomes positive with respect to the external medium, producing an overshoot. The overshoot of the action potential gives way to a falling phase in which the membrane potential rapidly repolarizes. Repolarization takes the membrane potential to levels even more negative than the resting membrane potential for a short time; this brief period of hyperpolarization is called the undershoot.

Although the waveform of the squid action potential is typical, the details of the action potential form vary widely from neuron to neuron in different animals. In myelinated axons of vertebrate motor neurons (figure B), the action potential is virtually indistinguishable from that of the squid axon. However, the action potential recorded in the cell body of this same motor neuron (figure C) looks rather different. Thus, the action potential waveform can vary even within the same neuron. More complex action potentials are seen in other central neurons. For example, action potentials recorded from the cell bodies of neurons in the mammalian inferior olive (a region of the brainstem involved in motor control) last tens of milliseconds (figure D).

These action potentials exhibit a pronounced plateau during their falling phase, and their undershoot lasts even longer than that of the motor neuron. One of the most dramatic types of action potentials occurs in the cell bodies of cerebellar Purkinje neurons (figure E). These potentials have several complex phases that result from the summation of multiple, discrete action potentials.

The variety of action potential waveforms could mean that each type of neuron has a different mechanism of action potential production. Fortunately, however, these diverse waveforms all result from relatively minor variations in the scheme used by the squid giant axon. For example, plateaus in the repolarization phase result from the presence of ion channels that are permeable to Ca²⁺, and long-lasting undershoots result from the presence of extra types of K⁺ channels. The complex action potential of the Purkinje cell results from these extra features plus the fact that different types of action potentials are generated in various parts of the Purkinje neuron—cell body, dendrites, and axons—and are summed together in recordings from the cell body.

Thus, the lessons learned from the squid axon are applicable to, and indeed essential for, understanding action potential generation in all neurons.

References

BARRETT, E. F. AND J. N. BARRETT (1976) Separation of two voltage-sensitive potassium currents, and demonstration of a tetrodotoxin-resistant calcium current in frog motoneurones. J. Physiol. (Lond.) 255: 737–774.

DODGE, F. A. AND B. FRANKENHAEUSER (1958) Membrane currents in isolated frog nerve fibre under voltage clamp conditions. J. Physiol. (Lond.) 143: 76–90.

HODGKIN, A. L. AND A. F. HUXLEY (1939) Action potentials recorded from inside a nerve fibre. Nature 144: 710–711.

LLINÁS, R. AND M. SUGIMORI (1980) Electrophysiological properties of in vitro Purkinje cell dendrites in mammalian cerebellar slices. J. Physiol. (Lond.) 305: 197–213.

LLINÁS, R. AND Y. YAROM (1981) Electrophysiology of mammalian inferior olivary neurones *in vitro*. Different types of voltage-dependent ionic conductances. J. Physiol. (Lond.) 315: 549–567.

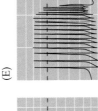
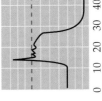
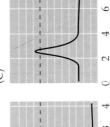
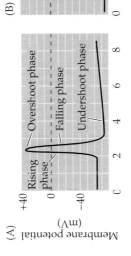

(A) The phases of an action potential of the squid giant axon. (B) Action potential recorded from a myelinated axon of a frog motor neuron. (C) Action potential recorded from the cell body of a frog motor neuron. The action potential is smaller and the undershoot prolonged in comparison to the action potential recorded from the axon of this same neuron (B). (D) Action potential recorded from the cell body of a neuron from the inferior olive of a guinea pig. This action potential has a pronounced plateau during its falling phase. (E) Action potential recorded from the cell body of a Purkinje neuron in the cerebellum of a guinea pig. (A after Hodgkin and Huxley, 1939; B after Dodge and Frankenhaeuser, 1958; C after Barrett and Barrett, 1976; D after Llinás and Yarom, 1981; E after Llinás and Sugimori, 1980.)

the membrane potential is transiently hyperpolarized because K+ permeability becomes even greater than it is at rest. The action potential ends when this phase of enhanced K+ permeability subsides, and the membrane potential thus returns to its normal resting level.

The ion substitution experiments carried out by Hodgkin and Katz provided convincing evidence that the resting membrane potential results from a high resting membrane permeability to K+, and that depolarization during an action potential results from a transient rise in membrane Na+ permeability. Although these experiments identified the ions that flow during an action potential, they did not establish *how* the neuronal membrane is able to change its ionic permeability to generate the action potential, or what mechanisms trigger this critical change. The next chapter addresses these issues, documenting the surprising conclusion that the neuronal membrane potential itself affects the membrane permeability.

Summary

Nerve cells generate electrical signals to convey information over substantial distances and to transmit it to other cells by means of synaptic connections. The action potential—the signal that conveys information along nerve cell axons—ultimately depends on the resting electrical potential across the neuronal membrane. A resting potential occurs because nerve cell membranes are permeable to one or more ion species subject to an electrochemical gradient. More specifically, a negative membrane potential at rest results from a net efflux of K+ across neuronal membranes that are predominantly permeable to K+. In contrast, an action potential occurs when a transient rise in Na+ permeability allows a net flow of Na+ in the opposite direction across the membrane that is now predominantly permeable to Na+. The brief rise in membrane Na+ permeability is followed by a secondary, transient rise in membrane K+ permeability that repolarizes the neuronal membrane and produces a brief undershoot of the action potential. As a result of these processes, the membrane is depolarized in an all-or-none fashion during an action potential. When these active permeability changes subside, the membrane potential returns to its resting level because of the high resting membrane permeability to K+.

ADDITIONAL READING

Reviews

HODGKIN, A. L. (1951) The ionic basis of electrical activity in nerve and muscle. Biol. Rev. 26: 339–409.

HODGKIN, A. L. (1958) The Croonian Lecture: Ionic movements and electrical activity in giant nerve fibres. Proc. R. Soc. Lond. (B) 148: 1–37.

Important Original Papers

BAKER, P. F., A. L. HODGKIN AND T. I. SHAW (1962) Replacement of the axoplasm of giant nerve fibres with artificial solutions. J. Physiol. (London) 164: 330–354.

COLE, K. S. AND H. J. CURTIS (1939) Electric impedence of the squid giant axon during activity. J. Gen. Physiol. 22: 649–670.

GOLDMAN, D. E. (1943) Potential, impedance, and rectification in membranes. J. Gen. Physiol. 27: 37–60.

HODGKIN, A. L. AND P. HOROWICZ (1959) The influence of potassium and chloride ions on the membrane potential of single muscle fibres. J. Physiol. (London) 148: 127–160.

HODGKIN, A. L. AND B. KATZ (1949) The effect of sodium ions on the electrical activity of the giant axon of the squid. J. Physiol. (London) 108: 37–77.

KEYNES, R. D. (1951) The ionic movements during nevous activity. J. Physiol. (London) 114: 119–150.

Books

HODGKIN, A. L. (1967) *The Conduction of the Nervous Impulse.* Springfield, IL: Charles C. Thomas.

HODGKIN, A. L. (1992) *Chance and Design.* Cambridge: Cambridge University Press.

JUNGE, D. (1992) *Nerve and Muscle Excitation,* 3rd Ed. Sunderland, MA: Sinauer Associates.

KATZ, B. (1966) *Nerve, Muscle, and Synapse.* New York: McGraw-Hill.

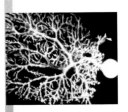

Chapter 3

Voltage-Dependent Membrane Permeability

Overview

The action potential, the primary electrical signal generated by nerve cells, reflects changes in membrane permeability to specific ions. Present understanding of membrane permeability is based on evidence obtained by the voltage clamp technique, which permits detailed characterization of permeability changes as a function of membrane potential and time. For most types of axons, these changes consist of a rapid and transient rise in sodium (Na^+) permeability, followed by a slower but more prolonged rise in potassium (K^+) permeability. Both permeabilities are voltage-dependent, increasing as the membrane potential depolarizes. The kinetics and voltage dependence of Na^+ and K^+ permeabilities provide a complete explanation of action potential generation. Depolarizing the membrane potential to the threshold level causes a rapid, self-sustaining increase in Na^+ permeability that produces the rising phase of the action potential; however, the Na^+ permeability increase is short-lived and is followed by a slower increase in K^+ permeability that restores the membrane potential to its usual negative resting level. A mathematical model that describes the behavior of these ionic permeabilities predicts virtually all of the observed properties of action potentials. Importantly, this same ionic mechanism also permits action potentials to be propagated along the length of neuronal axons, explaining how electrical signals are conveyed throughout the nervous system.

Ionic Currents Across Nerve Cell Membranes

The previous chapter introduced the idea that nerve cells generate electrical signals by virtue of a membrane that is differentially permeable to various ion species. In particular, a transient increase in the permeability of the neuronal membrane to Na^+ initiates the action potential. This chapter considers exactly how this increase in Na^+ permeability occurs. A key to understanding this phenomenon is the observation that action potentials are initiated *only* when the neuronal membrane potential becomes more positive than a certain threshold level. This relationship suggests that the mechanism responsible for the increase in Na^+ permeability is sensitive to the membrane potential. Therefore, if one could understand how a change in membrane potential activates Na^+ permeability, it should be possible to explain how action potentials are generated.

The fact that the Na^+ permeability that generates the membrane potential change is itself sensitive to the membrane potential presents both conceptual and practical obstacles to studying the mechanism of the action potential. A practical problem is the difficulty of systematically varying the mem-

BOX A
The Voltage Clamp Method

Breakthroughs in scientific research often rely on the development of new technologies. In the case of the action potential, detailed understanding came only after of the invention of the voltage clamp technique by Kenneth Cole in the 1940s. This device is called a voltage clamp because it controls, or clamps, membrane potential (or voltage) at any level desired by the experimenter. The method measures the membrane potential with a microelectrode (or other type of electrode) placed inside the cell (1), and electronically compares this voltage to the voltage to be maintained (called the command voltage) (2). The clamp circuitry then passes a current back into the cell though another intracellular electrode (3). This electronic feedback circuit holds the membrane potential at the desired level, even in the face of permeability changes that would normally alter the membrane potential (such as those generated during the action potential). Most importantly, the device permits the simultaneous measurement of the current needed to keep the cell at a given voltage (4). This current is exactly equal to the amount of current flowing across the neuronal membrane. Therefore, the voltage clamp technique can indicate how membrane potential influences ionic current flow across the membrane. This information gave Hodgkin and

Huxley the key insights that led to their model for action potential generation.

Today, the voltage clamp method remains widely used to study ionic currents in neurons and other cells. The most popular contemporary version of this approach is the patch clamp technique, a method that can be applied to virtually any cell and has a resolution high enough to measure the minute electrical currents flowing through single ion channels (see Box A in Chapter 4).

References

COLE, K. S. (1968) *Membranes, Ions and Impulses: A Chapter of Classical Biophysics.* Berkeley, CA: University of California Press.

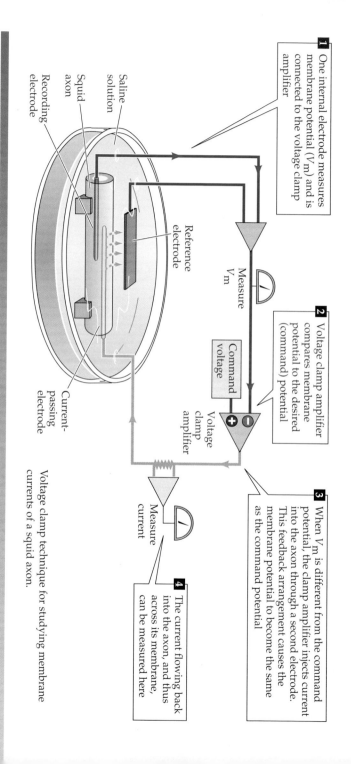

1 One internal electrode measures membrane potential (Vm) and is connected to the voltage clamp amplifier

Reference electrode

Measure Vm

Command voltage

2 Voltage clamp amplifier compares membrane potential to the desired (command) potential

Voltage clamp amplifier

3 When Vm is different from the command potential, the clamp amplifier injects current into the axon through a second electrode. This feedback arrangement causes the membrane potential to become the same as the command potential

Measure current

4 The current flowing back into the axon, and thus across its membrane, can be measured here

Saline solution

Squid axon

Recording electrode

Current-passing electrode

Voltage clamp technique for studying membrane currents of a squid axon.

brane potential to study the permeability change, because such changes in membrane potential will produce an action potential, which will cause further, uncontrolled changes in the membrane potential. Historically, then, it was not really possible to understand action potentials until a technique was developed that allowed experimenters to control membrane potential *and* simultaneously measure the underlying permeability changes. This technique, the **voltage clamp method** (Box A), provides the information needed to define the ionic permeability of the membrane at any level of membrane potential.

In the late 1940s, Alan Hodgkin and Andrew Huxley used the voltage clamp technique to work out the permeability changes underlying the action potential. They again chose to use the giant neuron of the squid because its large size (up to 1 mm in diameter; see Box A in Chapter 2) allowed insertion of the electrodes necessary for voltage clamping. Hodgkin and Huxley were the first investigators to test directly the hypothesis that potential-sensitive Na$^+$ and K$^+$ permeability changes are both necessary and sufficient for the production of action potentials.

Hodgkin and Huxley's first goal was to determine whether neuronal membranes do, in fact, have voltage-dependent permeabilities. To address this issue, they asked whether ionic currents flow across the membrane when its potential is changed. The result of one such experiment is shown in Figure 3.1. Figure 3.1A illustrates the currents produced by a squid axon when its membrane potential, V_m, is hyperpolarized from the resting level of –65 mV to –30 mV. The initial response of the axon results from the redistribution of charge across the axonal membrane. This capacitive current is nearly instantaneous, ending within a fraction of a millisecond. Aside from this brief event, very little current flows when the membrane is hyperpolarized. However, when the membrane potential is depolarized from –65 mV to 0 mV, the response is quite different (Figure 3.1B). Following the capacitive current. the axon produces a rapidly rising inward ionic current (inward refers to a positive charge entering the cell—that is, cations in or anions out), which gives way to a more slowly rising, delayed outward current. The fact that membrane depolarization elicits these ionic currents establishes that the membrane permeability of axons is indeed voltage-dependent.

Two Types of Voltage-Dependent Ionic Current

The results shown in Figure 3.1 demonstrate that the ionic permeability of neuronal membranes is voltage-sensitive, but the experiments do not identify how many types of permeability exist, or which ions are involved. As discussed in Chapter 2 (see Figure 2.4), varying the potential across a mem-

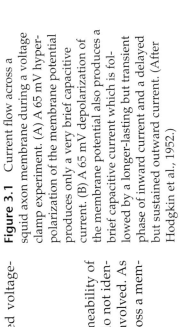

Figure 3.1 Current flow across a squid axon membrane during a voltage clamp experiment. (A) A 65 mV hyperpolarization of the membrane potential produces only a very brief capacitive current. (B) A 65 mV depolarization of the membrane potential also produces a brief capacitive current which is followed by a longer-lasting but transient phase of inward current and a delayed but sustained outward current. (After Hodgkin et al., 1952.)

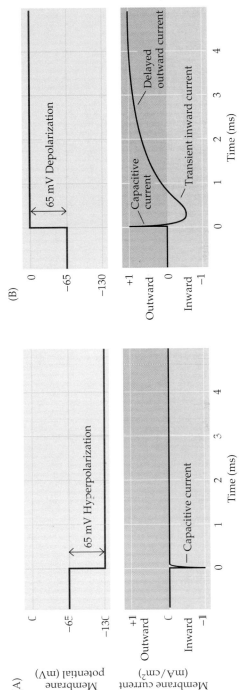

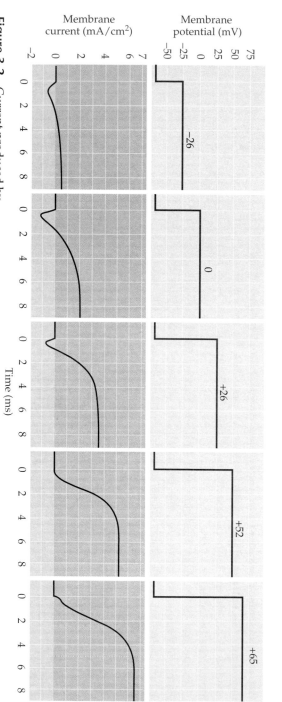

Figure 3.2   Current produced by membrane depolarizations to several different potentials. The early current first increases, then decreases in magnitude as the depolarization increases; note that this current is reversed in polarity at potentials more positive than about +55 mV. In contrast, the late current increases monotonically with increasing depolarization. (After Hodgkin et al., 1952.)

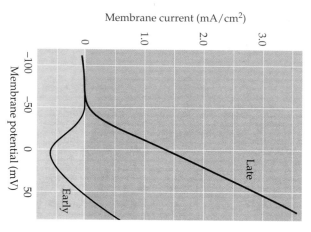

Figure 3.3   Relationship between current amplitude and membrane potential, taken from experiments such as the one shown in Figure 3.2. Whereas the late outward current increases steeply with increasing depolarization, the early inward current first increases in magnitude, but then decreases and reverses to outward current at about +55 mV (the sodium equilibrium potential). (After Hodgkin et al., 1952.)

brane makes it possible to deduce the equilibrium potential for the ionic fluxes through the membrane, and thus to identify the ions that are flowing. Since the voltage clamp method allows the membrane potential to be changed while ionic currents are being measured, it was a straightforward matter for Hodgkin and Huxley to determine ionic permeability by examining how the properties of the early inward and late outward currents changed as the membrane potential was varied (Figure 3.2). As already noted, no appreciable ionic currents flow at membrane potentials more negative than the resting potential. At more positive potentials, however, the currents not only flow but change in magnitude. The early current has a U-shaped dependence on membrane potential, increasing over a range of depolarizations up to approximately 0 mV but decreasing as the potential is depolarized further. In contrast, the late current increases monotonically with increasingly positive membrane potentials. These different responses to membrane potential can be seen more clearly when the magnitudes of the two current components are plotted as a function of membrane potential, as in Figure 3.3.

The voltage sensitivity of the early inward current gives an important clue about the nature of the ions carrying the current, namely, that no current flows when the membrane potential is clamped at +52 mV. For the squid neurons studied by Hodgkin and Huxley, the external Na^+ concentration is 440 mM, and the internal Na^+ concentration is 50 mM. For this concentration gradient, the Nernst equation predicts that the equilibrium potential for Na^+ should be +55 mV. Recall further from Chapter 2 that at the Na^+ equilibrium potential there is no net flux of Na^+ across the membrane, even if the membrane is highly permeable to Na^+. Thus, the experimental observation that no current flows at the membrane potential where Na^+ equilibrium potential flows is a strong indication that the early inward current is carried by entry of Na^+ into the axon.

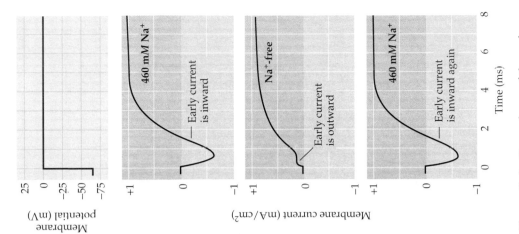

Figure 3.4 Dependence of the early inward current on sodium. In the presence of normal external concentrations of Na+, depolarization of a squid axon to 0 mV produces an inward initial current. However, removal of external Na+ causes the initial inward current to become outward, an effect that is reversed by restoration of external Na+. (After Hodgkin and Huxley, 1952a.)

An even more demanding way to test whether Na+ carries the early inward current is to examine the behavior of this current after removing external Na+. Removing the Na+ outside the axon makes E_{Na} negative; if the permeability to Na+ is increased under these conditions, current should flow outward as Na+ leaves the neuron, due to the reversed concentration gradient. When Hodgkin and Huxley performed this experiment, they obtained the result shown in Figure 3.4. Removing external Na+ caused the early inward current to reverse its polarity and become an outward current at a membrane potential that gave rise to an inward current when external Na+ was present. This result demonstrates convincingly that the early inward current measured when Na+ is present in the external medium must be due to Na+ entering the neuron.

Notice that removal of external Na+ in the experiment shown in Figure 3.4 has little effect on the outward current that flows after the neuron has been kept at a depolarized membrane voltage for several milliseconds. This result shows that the late outward current must be due to the flow of an ion other than Na+. Several lines of evidence presented by Hodgkin, Huxley, and others showed that this late outward current is caused by K+ exiting the neuron. Perhaps the most compelling demonstration of K+ involvement is that the amount of K+ efflux from the neuron, measured by loading the neuron with radioactive K+, is closely correlated with the magnitude of the late outward current.

Taken together, these experiments using the voltage clamp show that changing the membrane potential to a level more positive than the resting potential produces two effects: an early influx of Na+ into the neuron, followed by a delayed efflux of K+. The early influx of Na+ produces a transient inward current, whereas the delayed efflux of K+ produces a sustained outward current. The differences in the time course and ionic selectivity of the two fluxes suggest that two different ionic permeability mechanisms are activated by changes in membrane potential. Confirmation that there are indeed two distinct mechanisms has come from pharmacological studies of drugs that specifically affect these two currents (Figure 3.5). **Tetrodotoxin**, an alkaloid neurotoxin found in certain puffer fish, tropical frogs, and salamanders, blocks the Na+ current without affecting the K+ current. Conversely, **tetraethylammonium ions** block K+ currents without affecting Na+ currents. The differential sensitivity of Na+ and K+ currents to these drugs provides strong additional evidence that Na+ and K+ flow through independent permeability pathways. As discussed in Chapter 4, it is now known that these pathways are ion channels that are selectively permeable to either Na+ or K+. In fact, tetrodotoxin, tetraethylammonium, and other drugs that interact with specific types of ion channels have been extraordinarily useful tools in characterizing these channel molecules (see Chapter 4).

Two Voltage-Dependent Membrane Conductances

The next goal Hodgkin and Huxley set for themselves was to describe Na+ and K+ permeability changes mathematically. To do this, they assumed that the ionic currents are due to a change in **membrane conductance**, defined as the reciprocal of the membrane resistance. Membrane conductance is thus closely related, although not identical, to membrane permeability. When evaluating ionic movements from an electrical standpoint, it is convenient to describe them in terms of ionic conductances rather than ionic permeabilities. For present purposes, permeability and conductance can be considered

Figure 3.5 Pharmacological separation of Na⁺ and K⁺ currents into sodium and potassium components. Panel (1) shows the current that flows when the membrane potential of a squid axon is depolarized to 0 mV in control conditions. (2) Treatment with tetrodotoxin causes the early Na⁺ currents to disappear but spares the late K⁺ currents. (3) Addition of tetraethylammonium blocks the K⁺ currents without affecting the Na⁺ currents. (After Moore et al., 1967 and Armstrong and Binstock, 1965.)

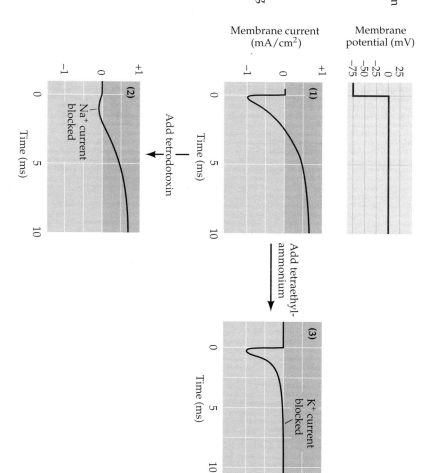

synonymous. If membrane conductance (g) obeys Ohm's Law (which states that voltage is equal to the product of current and resistance), then the ionic current that flows during an increase in membrane conductance is given by

$$I_{ion} = g_{ion}(V_m - E_{ion})$$

where I_{ion} is the ionic current, V_m is the membrane potential, and E_{ion} is the equilibrium potential for the ion flowing through the conductance, g_{ion}. The difference between V_m and E_{ion} is the electrochemical driving force acting on the ion.

Hodgkin and Huxley used this simple relationship to calculate the dependence of Na⁺ and K⁺ conductances on time and membrane potential. They knew V_m, which was set by their voltage clamp device (Figure 3.6A), and could determine E_{Na} and E_K from the ionic concentrations on the two sides of the axonal membrane (see Table 2.1). The currents carried by Na⁺ and K⁺—I_{Na} and I_K—could be determined separately from recordings of the membrane currents resulting from depolarization (Figure 3.6B), by measuring the difference between currents recorded in the presence and absence of external Na⁺ (as shown in Figure 3.4). From these parameters, Hodgkin and Huxley were able to calculate g_{Na} and g_K (Figure 3.6C,D) and drew two fundamental conclusions about these conductances. The first conclusion is that the Na⁺ and K⁺ conductances change over time. For example, both Na⁺ and K⁺ conductances require some time to **activate**, or turn on. In particular, the K⁺ conductance has a pronounced delay, requiring several milliseconds to reach its maximum (Figure 3.6D), whereas the Na⁺ conductance reaches its maximum more rapidly (Figure 3.6C). The more rapid activation of the Na⁺ con-

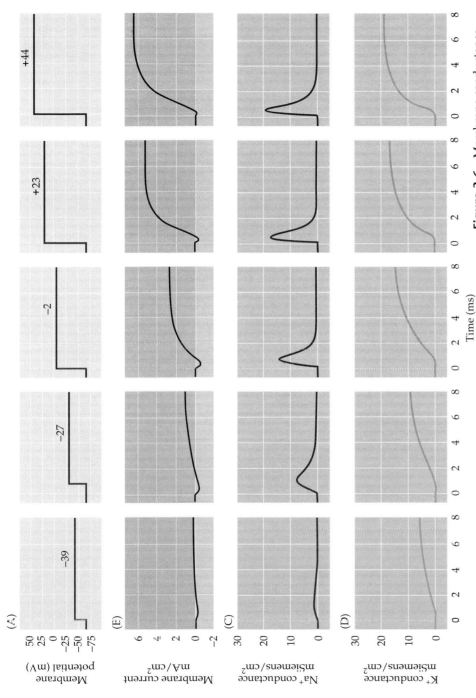

Figure 3.6 Membrane conductance changes underlying the action potential are time- and voltage-dependent. Depolarizations to various membrane potentials (A) elicit different membrane currents (B). Below are shown the Na$^+$ (C) and K$^+$ (D) conductances calculated from these currents. Both peak Na$^+$ conductance and steady-state K$^+$ conductance increase as the membrane potential becomes more positive. In addition, the activation of both conductances, as well as the rate of inactivation of the Na$^+$ conductance, occur more rapidly with larger depolarizations. (After Hodgkin and Huxley, 1952b.)

ductance allows the resulting inward Na$^+$ current to precede the delayed outward K$^+$ current (see Figure 3.6B). Although the Na$^+$ conductance rises rapidly, it quickly declines, even though the membrane potential is kept at a depolarized level. This fact shows that depolarization not only causes the Na$^+$ conductance to activate, but also causes it to decrease over time, or **inactivate**. The K$^+$ conductance of the squid axon does not inactivate in this way; thus, while the Na$^+$ and K$^+$ conductances share the property of time-dependent activation, only the Na$^+$ conductance inactivates. (Inactivating K$^+$ conductances have since been discovered in other types of nerve cells; see Chapter 4.) The time courses of the Na$^+$ and K$^+$ conductances are voltage-dependent, with the speed of both activation and inactivation increasing at more depolarized potentials. This finding accounts for more rapid time courses of membrane currents measured at more depolarized potentials.

The second conclusion derived from Hodgkin and Huxley's calculations is that both the Na$^+$ and K$^+$ conductances are voltage-dependent—that is, both conductances increase progressively as the neuron is depolarized. Figure 3.7 illustrates this fact by plotting the relationship between peak value of the conductances (from Figure 3.6C,D) against the membrane potential. Note the similar voltage dependence for each conductance; both conductances are quite small at negative potentials, maximal at very positive potentials, and exquisitely dependent on membrane voltage at intermediate potentials. The observation that these conductances are sensitive to changes

Figure 3.7 Depolarization increases Na⁺ and K⁺ conductances of the squid giant axon. The peak magnitude of Na⁺ conductance and steady-state value of K⁺ conductance both increase steeply as the membrane potential is depolarized. (After Hodgkin and Huxley, 1952b.)

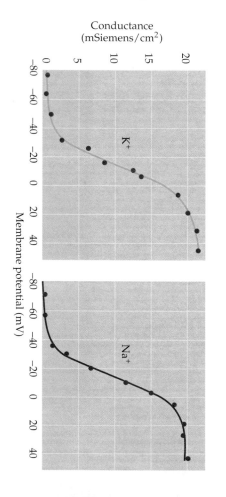

Figure 3.8 Mathematical reconstruction of the action potential. (A) Reconstruction of an action potential (black curve) together with the underlying changes in Na⁺ (red curve) and K⁺ (yellow curve) conductance. The size and time course of the action potential were calculated using only the properties of g_{Na} and g_K measured in voltage clamp experiments. Real action potentials evoked by brief current pulses of different intensities (B) are remarkably similar to those generated by the mathematical model (C). (After Hodgkin and Huxley, 1952d.)

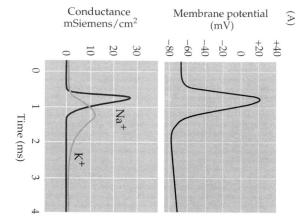

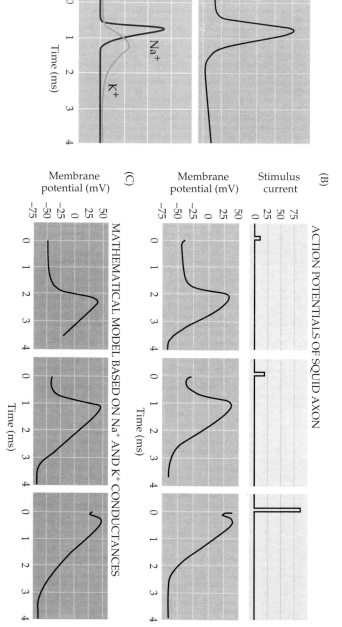

Reconstruction of the Action Potential

From their experimental measurements, Hodgkin and Huxley were able to construct a detailed mathematical model of the Na⁺ and K⁺ conductance changes. The goal of these modeling efforts was to determine whether the Na⁺ and K⁺ conductances alone are sufficient to produce an action potential. Using this information, they could in fact generate the form and time course of the action potential with remarkable accuracy (Figure 3.8A). Further, the Hodgkin-Huxley model predicted other features of action potential behavior

in membrane potential shows that the mechanism underlying the conductances somehow "senses" the voltage across the membrane.

All told, the voltage clamp experiments carried out by Hodgkin and Huxley showed that the ionic currents that flow when the neuronal membrane is depolarized are due to three different voltage-sensitive processes: (1) activation of Na⁺ conductance, (2) activation of K⁺ conductance, and (3) inactivation of Na⁺ conductance.

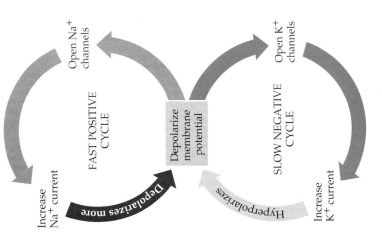

Figure 3.9 Feedback cycles responsible for membrane potential changes during an action potential. Membrane depolarization rapidly activates a positive feedback cycle fueled by the voltage-dependent activation of Na$^+$ conductance. This phenomenon is followed by the slower activation of a negative feedback loop as depolarization activates a K$^+$ conductance, which helps to repolarize the membrane potential and terminate the action potential.

in the squid axon, such as how the delay before action potential generation changes in response to stimulating currents of different intensities (Figure 3.8B,C). The model also predicted that the axon membrane would become refractory to further excitation for a brief period following an action potential, as was experimentally observed.

The Hodgkin-Huxley model also provided many insights into how action potentials are generated. Figure 3.8A shows a reconstructed action potential, together with the time courses of the underlying Na$^+$ and K$^+$ conductances. The coincidence of the initial increase in Na$^+$ conductance with the rapid rising phase of the action potential demonstrates that a selective increase in Na$^+$ conductance is responsible for action potential initiation. The increase in Na$^+$ conductance causes Na$^+$ to enter the neuron, thus depolarizing the membrane potential, which approaches E_{Na}. The rate of depolarization subsequently falls both because the electrochemical driving force on Na$^+$ decreases and because the Na$^+$ conductance inactivates. At the same time, depolarization slowly activates the voltage-dependent K$^+$ conductance, causing K$^+$ to leave the cell and repolarizing the membrane potential toward E_K. Because the K$^+$ conductance becomes temporarily higher than it is in the resting condition, the membrane potential actually becomes briefly more negative than the normal resting potential (the **undershoot**). The hyperpolarization of the membrane potential causes the voltage-dependent K$^+$ conductance (and any Na$^+$ conductance not inactivated) to turn off, allowing the membrane potential to return to its resting level.

This mechanism of action potential generation represents a positive feedback loop: Activating the voltage-dependent Na$^+$ conductance increases Na$^+$ entry into the neuron, which makes the membrane potential depolarize, which leads to the activation of still more Na$^+$ conductance, more Na$^+$ entry, and still further depolarization (Figure 3.9). Positive feedback continues unabated until Na$^+$ conductance inactivation and K$^+$ conductance activation restore the membrane potential to the resting level. Because this positive feedback loop, once initiated, is sustained by the intrinsic properties of the neuron—namely, the voltage dependence of the ionic conductances—the action potential is self-supporting, or **regenerative**. This regenerative quality explains why action potentials exhibit all-or-none behavior (see Figure 2.1), and why they have a threshold (Box B). The delayed activation of the K$^+$ conductance represents a negative feedback loop that eventually restores the membrane to its resting state.

Hodgkin and Huxley's reconstruction of the action potential and all its features shows that the properties of the voltage-sensitive Na$^+$ and K$^+$ conductances, together with the electrochemical driving forces created by ion transporters, are sufficient to explain action potentials. Their use of both empirical and theoretical methods brought an unprecedented level of rigor to a long-standing problem, setting a standard of proof that is achieved only rarely in biological research.

Long-Distance Signaling by Means of Action Potentials

The voltage-dependent mechanisms of action potential generation also explain the long-distance transmission of these electrical signals. Recall from Chapter 2 that neurons are relatively poor conductors of electricity, at least compared to a wire. Current conduction by wires, and by neurons in the absence of action potentials, is called **passive current flow** (Box C). The passive electrical properties of a nerve cell axon can be determined by measuring the voltage change resulting from a current pulse passed across the axonal membrane (Figure 3.10A). If this current pulse is not large enough to

BOX B
Threshold

An important—and potentially puzzling—property of the action potential is its initiation at a particular membrane potential, called threshold. Indeed, action potentials never occur without a depolarizing stimulus that brings the membrane to this level. The depolarizing "trigger" can be one of several events: a synaptic input, a receptor potential generated by specialized receptor organs, the endogenous pacemaker activity of cells that generate action potentials spontaneously, or the local current that mediates the spread of the action potential down the axon.

Why the action potential "takes off" at a particular level of depolarization can be understood by comparing the underlying events to a chemical explosion (figure A). As indicated, exogenous heat (analogous to the initial depolarization of the membrane potential) stimulates an exothermic chemical reaction, which produces more heat, which further enhances the reaction (figure B). As a result of this positive feedback loop, the rate of the reaction builds up exponentially—the definition of an explosion. In any such

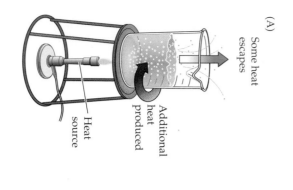

(A)

Some heat
escapes

Additional
heat
produced

Heat
source

process, however, there is a threshold, that is, a point up to which heat can be supplied without resulting in an explosion. The threshold for the chemical explosion diagrammed here is the point at which the amount of heat supplied exogenously is just equal to the amount of heat that can be dissipated by the circumstances of the reaction (such as escape of heat from the beaker).

The threshold of action potential initiation is, in principle, similar (figure C). There is a range of "subthreshold" depolarization, within which the rate of increased sodium entry is less than the rate of potassium exit (remember that the membrane at rest is highly permeable to K+, which therefore flows out as the membrane is depolarized). The point at which Na+ inflow just equals K+ outflow represents an unstable equilibrium analogous to the ignition point of an explosive mixture. The behavior of the membrane at threshold reflects this instability: The membrane potential may linger at the threshold level for a variable period before either returning to the resting level or flaring up into a full-blown

action potential. In theory at least, if there is a net internal gain of a single Na+ ion, an action potential occurs; conversely, the net loss of a single K+ ion leads to repolarization. A more precise definition of threshold, therefore, is that value of membrane potential, in depolarizing from the resting potential, at which the current carried by Na+ entering the neuron is exactly equal to the K+ current that is flowing out. Once the triggering event depolarizes the membrane beyond this point, the positive feedback loop of Na+ entry on membrane potential closes and the action potential "fires."

Because the Na+ and K+ conductances change dynamically over time, the threshold potential for producing an action potential also varies as a consequence of the previous activity of the neuron. For example, following an action potential, the membrane becomes temporarily refractory to further excitation because the threshold for firing an action potential transiently rises. There is, therefore, no specific value of membrane potential that defines the threshold for a given nerve cell in all circumstances.

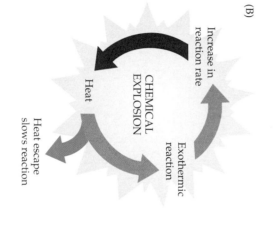

(B)

Increase in
reaction rate

Heat

CHEMICAL
EXPLOSION

Exothermic
reaction

Heat escape
slows reaction

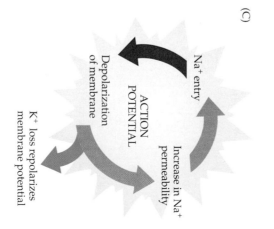

(C)

Na+ entry

Depolarization
of membrane

ACTION
POTENTIAL

Increase in Na+
permeability

K+ loss repolarizes
membrane potential

A positive feedback loop underlying the action potential explains the phenomenon of threshold.

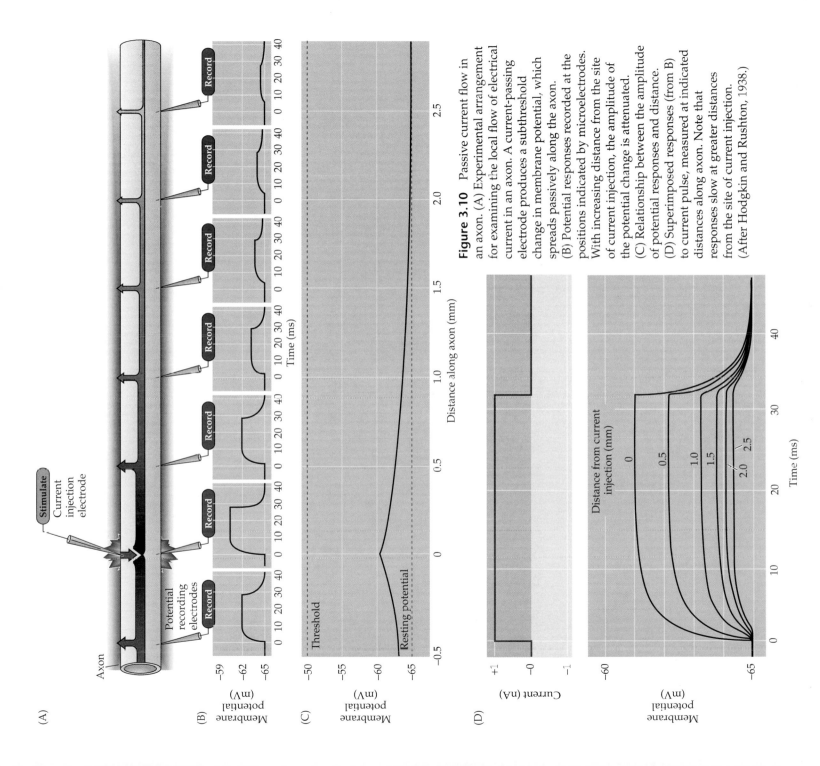

Figure 3.10 Passive current flow in an axon. (A) Experimental arrangement for examining the local flow of electrical current in an axon. A current-passing electrode produces a subthreshold change in membrane potential, which spreads passively along the axon. (B) Potential responses recorded at the positions indicated by microelectrodes. With increasing distance from the site of current injection, the amplitude of the potential change is attenuated. (C) Relationship between the amplitude of potential responses and distance. (D) Superimposed responses (from B) to current pulse, measured at indicated distances along axon. Note that responses slow at greater distances from the site of current injection. (After Hodgkin and Rushton, 1938.)

generate action potentials, the magnitude of the resulting potential change decays exponentially with increasing distance from the site of current injection (Figure 3.10B). Typically, the potential falls to a small fraction of its initial value at a distance no more than a couple of millimeters away from the site of injection (Figure 3.10C). The progressive decrease in the amplitude of

BOX C
Passive Membrane Properties

The passive flow of electrical current plays a central role in action potential propagation, synaptic transmission, and all other forms of electrical signaling in nerve cells. Therefore, it is worthwhile understanding in quantitative terms how passive current flow varies with distance along a neuron. For the case of a cylindrical axon, such as the one depicted in Figure 3.10, subthreshold current injected into one part of the axon spreads passively along the axon until the current is dissipated by leakage out across the axon membrane. The decrement in the current flow with distance (figure A) is described by a simple exponential function:

$$V_x = V_0 e^{-x/\lambda}$$

where V_x is the voltage response at any distance x along the axon, V_0 is the voltage change at the point where current is injected into the axon, e is the base of natural logarithms (approximately 2.7) and λ is the length constant of the axon. As evident in this relationship, the length constant is the distance where the initial voltage response (V_0) decays to $1/e$ (or 37%) of its value. The length constant is thus a way to characterize how far passive current flow spreads before it leaks out of the axon, with leakier axons having shorter length constants.

The length constant depends upon the physical properties of the axon, in particular the relative resistances of the plasma membrane (r_m), the intracellular axoplasm (r_i), and the extracellular medium (r_0). The relationship between these parameters is:

$$\lambda = \sqrt{\frac{r_m}{r_0 + r_i}}$$

Hence, to improve the passive flow of current along an axon, the resistance of the plasma membrane should be as high as possible and the resistances of the axoplasm and extracellular medium should be low.

Another important consequence of the passive properties of neurons is that currents flowing across a membrane do not immediately change the membrane potential. For example, when a rectangular current pulse is injected into the axon shown in the experiment illustrated in Figure 3.10A, the membrane potential depolarizes slowly over a few milliseconds and then repolarizes slowly over a similar time course when the current pulse ends (see Figure 3.10D). These delays in changing the membrane potential are due to the fact that the plasma membrane behaves as a capacitor, storing the initial charge that flows at the beginning and end of the current pulse. For the case of a cell whose membrane potential is spatially uniform, the change in the membrane potential at any time, V_t, after beginning the current

pulse (figure B) can also be described by an exponential relationship:

$$V_t = V_\infty (1 - e^{-t/\tau})$$

where V_∞ is the steady-state value of the membrane potential change, t is the time after the current pulse begins, and τ is the membrane time constant. The time constant is thus defined as the time when the voltage response (V) rises to $1 - (1/e)$ (or 63%) of V_∞. After the current pulse ends, the membrane potential change also declines exponentially according to the relationship:

$$V_t = V_\infty e^{-t/\tau}$$

During this decay, the membrane potential returns to $1/e$ of V_∞ by a time equal to τ. For cells with more complex geometries than the axon in Figure 3.10, the time courses of the changes in membrane potential are not simple exponentials, but nonetheless depend on the membrane time constant. Thus, the time constant characterizes how rapidly current

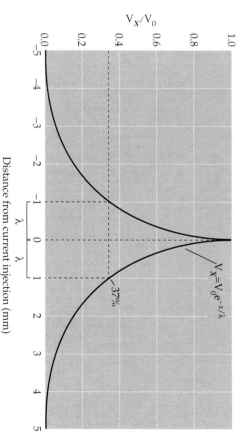

(A) Spatial decay of membrane potential along a cylindrical axon. A current pulse injected at one point in the axon (0 mm) produces voltage responses (V_x) that decay exponentially with distance. The distance where the voltage response is $1/e$ of its initial value (V_0) is the length constant, λ.

Graph axes: V_X/V_0 (vertical, 0.0 to 1.0); Distance from current injection (mm) (horizontal, −5 to 5). Curve labeled $V_x = V_0 e^{-x/\lambda}$, with 37% and λ markers.

rent flow changes the membrane potential. The membrane time constant also depends on the physical properties of the nerve cell, specifically on the resistance (r_m) and capacitance (c_m) of the plasma membrane such that:

$$\tau = r_m c_m$$

The values of r_m and c_m depend, in part, on the size of the neuron, with larger cells having lower resistances and larger capacitances. In general, small nerve cells tend to have long time constants and large cells brief time constants.

References

HODGKIN, A. L. AND W. A. H. RUSHTON (1938) The electrical constants of a crustacean nerve fibre. Proc. R. Soc. Lond. 133: 444–479.

JOHNSTON, D. AND S. M.-S. WU (1995) Foundations of Cellular Neurophysiology. Cambridge, MA: MIT Press.

RALL, W. (1977) Core conductor theory and cable properties of neurons. In Handbook of Physiology, Section 1: The Nervous System, Vol. 1: Cellular Biology of Neurons. E. R. Kandel (ed.). Bethesda, MD: American Physiological Society, pp. 39–98.

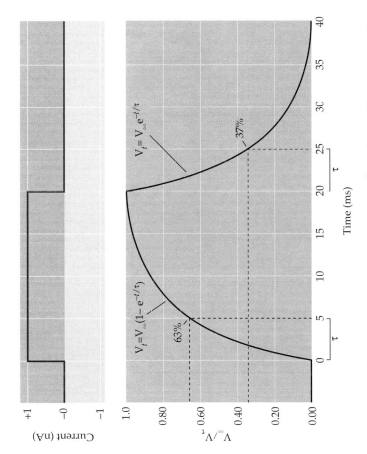

(B) Time course of potential changes produced in a spatially uniform cell by a current pulse. The rise and fall of the membrane potential (V_t) can be described as exponential functions, with the time constant, τ, defining the time required for the response to rise to $1-(1/e)$ of the steady-state value, V_∞, or to decline to $1/e$ of V_∞.

the induced potential change occurs because the injected current leaks out across the axonal membrane; accordingly, less current is available to change the membrane potential farther along the axon. Thus, the leakiness of the axonal membrane prevents effective passive transmission of electrical signals in all but the shortest axons (those 1 mm or less in length). Likewise, the leakiness of the membrane slows the time course of the responses measured at increasing distances from the site where current was injected (Figure 3.10D).

If the experiment shown in Figure 3.10 is repeated with a depolarizing current pulse sufficiently large to produce an action potential, the result is dramatically different (Figure 3.11A). In this case, an action potential occurs without decrement along the entire length of the axon, which may be a distance of a meter or more (Figure 3.11B). Thus, action potentials somehow circumvent the inherent leakiness of neurons.

How, then, do action potentials traverse great distances along such a poor passive conductor? The answer is in part provided by the observation that the amplitude of the action potentials recorded at different distances is constant. This all-or-none behavior indicates that more than simple passive flow of current must be involved in action potential propagation. A second clue comes from examination of the time of occurrence of the action potentials

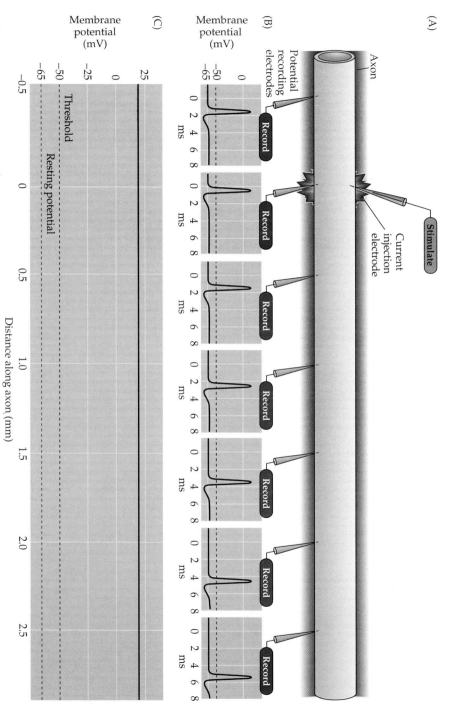

Figure 3.11 Propagation of an action potential. (A) Experimental arrangement: An electrode evokes an action potential by injecting a suprathreshold current. (B) Potential responses recorded at the positions indicated by microelectrodes. The amplitude of the action potential is constant along the length of the axon, although the time of appearance of the action potential is delayed with increasing distance. (C) Relationship between the amplitude of action potential responses and distance.

recorded at different distances from the site of stimulation: Action potentials occur later and later at greater distances along the axon (Figure 3.11B,C). Thus, the action potential has a measurable rate of transmission, called the **conduction velocity**. The delay in the arrival of the action potential at successively more distant points along the axon differs from the case shown in Figure 3.10, in which the electrical changes produced by passive current flow occur at more or less the same time at the successive points.

The mechanism of action potential propagation is easy to grasp once one understands how action potentials are generated and how current passively flows along an axon (Figure 3.12). A depolarizing stimulus—a synaptic potential or a receptor potential in an intact neuron, or an injected current pulse in an experiment—locally depolarizes the axon, thus opening the voltage-sensitive Na⁺ channels in that region. The opening of Na⁺ channels causes inward movement of Na⁺, and the resultant depolarization of the membrane potential generates an action potential at that site. Some of the local current generated by the action potential will then flow passively down the axon, in the same way that subthreshold currents spread along the axon (see Figure 3.10). Note that this passive current flow does not require the movement of Na⁺ along the axon but, instead, occurs by a shuttling of charge, similar to what happens when wires passively conduct electricity. This passive current flow depolarizes the membrane potential in the adjacent region of the axon, thus opening the Na⁺ channels in the neighboring membrane. The local depolarization triggers an action potential in this region, which then spreads again in a continuing cycle until the end of the axon is

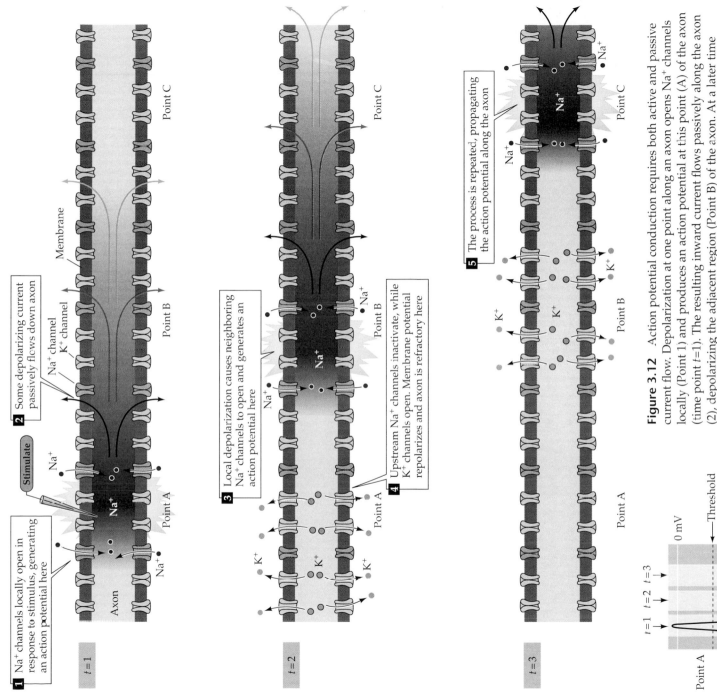

Figure 3.12 Action potential conduction requires both active and passive current flow. Depolarization at one point along an axon opens Na⁺ channels locally (Point 1) and produces an action potential at this point (A) of the axon (time point *t*=1). The resulting inward current flows passively along the axon (2), depolarizing the adjacent region (Point B) of the axon. At a later time (*t*=2), the depolarization of the adjacent membrane has opened Na⁺ channels at point B, resulting in the initiation of the action potential at this site and additional inward current that again spreads passively to an adjacent point (Point C) farther along the axon (3). At a still later time (*t*=3), the action potential has propagated even farther. This cycle continues along the full length of the axon (5). Note that as the action potential spreads, the membrane potential repolarizes due to K⁺ channel opening and Na⁺ channel inactivation, leaving a "wake" of refractoriness behind the action potential that prevents its backward propagation (4). Panel to the left of the figure legend shows the changing membrane potential as a function of time at the points indicated.

reached. Thus, action potential propagation requires the coordinated action of two forms of current flow—the passive flow of current as well as active currents flowing through voltage-dependent ion channels. The regenerative properties of Na^+ channel opening allow action potentials to propagate in an all-or-none fashion by acting as a booster at each point along the axon, thus ensuring the long-distance transmission of electrical signals.

The Refractory Period

The depolarization that produces Na^+ channel opening also causes delayed activation of K^+ channels and Na^+ channel inactivation, leading to repolarization of the membrane potential as the action potential sweeps along the length of an axon (see Figure 3.12). In its wake, the action potential leaves the Na^+ channels inactivated and K^+ channels activated for a brief time. These transitory changes make it harder for the axon to produce subsequent action potentials during this interval, which is called the **refractory period**. Thus, the refractory period limits the number of action potentials that a given nerve cell can produce per unit time. As might be expected, different types of neurons have different maximum rates of action potential firing due to different types and densities of ion channels. The refractoriness of the membrane in the wake of the action potential explains why action potentials do not propagate back toward the point of their initiation as they travel along an axon.

Increased Conduction Velocity as a Result of Myelination

The rate of action potential conduction limits the flow of information within the nervous system. It is not surprising, then, that various mechanisms have developed to optimize the propagation of action potentials along axons. Because action potential conduction requires passive and active flow of current (see Figure 3.12), the rate of action potential propagation is determined by both of these phenomena. One way of improving passive current flow is to increase the diameter of an axon, which effectively decreases the internal resistance to passive current flow (see Box C). The consequent increase in action potential conduction velocity presumably explains why giant axons evolved in invertebrates such as squid, and why rapidly conducting axons in all animals tend to be larger than slowly conducting ones.

Another strategy to improve the passive flow of electrical current is to insulate the axonal membrane, reducing the ability of current to leak out of the axon and thus increasing the distance along the axon that a given local current can flow passively (see Box C). This strategy is evident in the **myelination** of axons, a process by which oligodendrocytes in the central nervous system (and Schwann cells in the peripheral nervous system) wrap the axon in **myelin**, which consists of multiple layers of closely opposed glial membranes (Figure 3.13; see also Chapter 1). By acting as an electrical insulator, myelin greatly speeds up action potential conduction (Figure 3.14). For example, whereas unmyelinated axon conduction velocities range from about 0.5 to 10 m/s, myelinated axons can conduct at velocities up to 150 m/s. The major reason underlying this marked increase in speed is that the time-consuming process of action potential generation occurs only at specific points along the axon, called **nodes of Ranvier**, where there is a gap in the myelin wrapping (see Figure 1.4F). If the entire surface of an axon were insulated, there would be no place for current to flow out of the axon and action potentials could not be generated. As it happens, an action potential generated at one node of Ranvier elicits current that flows passively within the myelinated segment until the next node is reached. This local current flow

(A) Myelinated axon

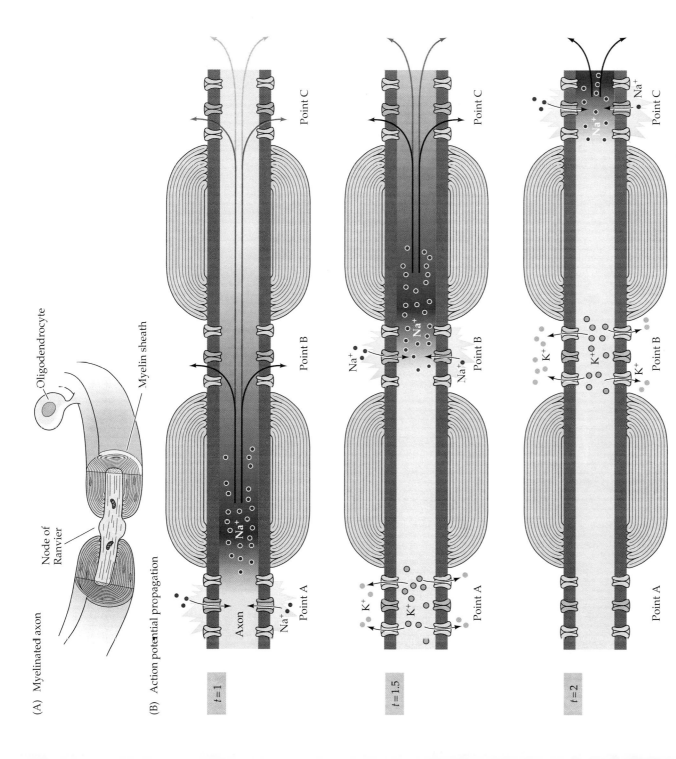

Oligodendrocyte

Myelin sheath

Node of Ranvier

(B) Action potential propagation

$t = 1$

Axon

Na⁺

Point A

Na⁺

Point B

Point C

$t = 1.5$

K⁺

K⁺

Point A

Na⁺

Na⁺

Na⁺

Point B

Point C

$t = 2$

Point A

K⁺

K⁺

K⁺

Point B

Na⁺

Point C

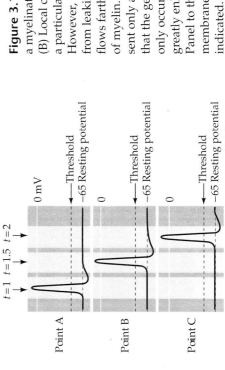

0 mV

Threshold
−65 Resting potential

0

Threshold
−65 Resting potential

0

Threshold
−65 Resting potential

$t=1$ $t=1.5$ $t=2$

Point A

Point B

Point C

Figure 3.13 Saltatory action potential conduction along a myelinated axon. (A) Diagram of a myelinated axon. (B) Local current in response to action potential initiation at a particular site flows locally, as described in Figure 3.12. However, the presence of myelin prevents the local current from leaking across the internodal membrane; it therefore flows farther along the axon than it would in the absence of myelin. Moreover, voltage-gated Na⁺ channels are present only at the nodes of Ranvier. This arrangement means that the generation of active, voltage-gated currents need only occur at these unmyelinated regions. The result is a greatly enhanced velocity of action potential conduction. Panel to the left of the figure legend shows the changing membrane potential as a function of time at the points indicated.

then generates an action potential in the neighboring segment, and the cycle is repeated along the length of the axon. Because current flows across the neuronal membrane only at the nodes (see Figure 3.13), this type of propagation is called **saltatory**, meaning that the action potential jumps from node to node. Not surprisingly, loss of myelin, as occurs in diseases such as multiple sclerosis, causes a variety of serious neurological problems (Box D).

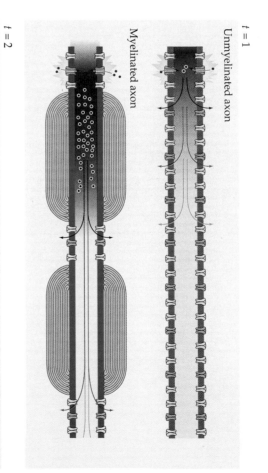

t = 1

Unmyelinated axon

Myelinated axon

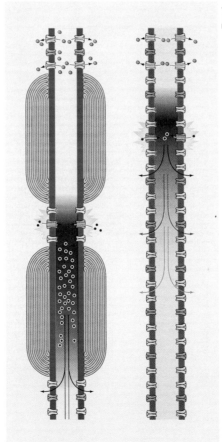

t = 2

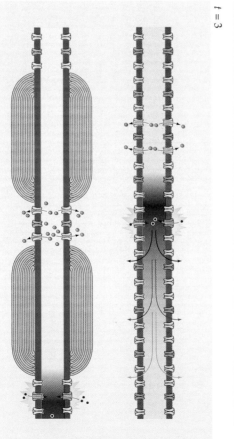

t = 3

Figure 3.14   Comparison of speed of action potential conduction in unmyelinated (upper) and myelinated (lower) axons.

BOX D
Multiple Sclerosis — Demyelination

Multiple sclerosis (MS) is a disease of the central nervous system characterized by a variety of clinical problems arising from multiple regions of demyelination and inflammation along axonal pathways. The disorder commonly begins between ages 20 and 40, and is characterized by the abrupt onset of neurological deficits that typically persist for days or weeks and then remit. The clinical course ranges from patients with no persistent neurological loss, some of whom experience only occasional later exacerbations, to others who progressively deteriorate as a result of extensive and relentless central nervous system involvement.

The signs and symptoms of MS are determined by the location of the affected regions. Particularly common are monocular blindness (due to lesions of the optic nerve), motor weakness or paralysis (due to lesions of the corticospinal tracts), abnormal somatic sensations (due to lesions of somatic sensory pathways, often in the posterior columns), double vision (due to lesions of medial longitudinal fasciculus), and dizziness (due to lesions of vestibular pathways). Abnormalities are often apparent in the cerebrospinal fluid, which usually contains an abnormal number of cells associated with inflammation and an increased content of antibodies (again, a sign of inflammation). The diagnosis of MS generally relies on the presence of a neurological problem that remits and then returns at an unrelated site. Confirmation can sometimes be obtained from magnetic resonance imaging (MRI), or functional evidence of lesions in a particular pathway by abnormal evoked potentials. The histological hallmark of MS at post-mortem exam is multiple lesions at different sites showing loss of myelin and infiltration by a characteristic complement of inflammatory cells.

The concept of MS as a demyelinating disease is deeply embedded in the clinical literature, although precisely how the demyelination translates into functional deficits is poorly understood. The loss of the myelin sheath surrounding many axons clearly compromises action potential conduction, and the abnormal patterns of nerve conduction that result presumably produce most of the clinical deficits in the disease. However, MS may have effects that extend beyond loss of the myelin sheath. It is clear that some axons are actually destroyed, probably as a result of the inflammation targeting the overlying myelin and/or loss of trophic support of the axon by oligodendrocytes. Thus, axon loss also contributes to the functional deficits in MS, especially in the chronic, progressive forms of the disease.

The ultimate cause of MS remains unclear. The immune system undoubtedly contributes to the damage and immunoregulatory therapies provide substantial benefits to some patients. Precisely how the immune system is activated to cause the injury is not known. The most popular hypothesis is that MS is an autoimmune disease. The fact that immunization of experimental animals with any one of several molecular constituents of the myelin sheath can induce a demyelinating disease (called experimental allergic encephalomyelitis) shows that an autoimmune attack on the myelin membrane is sufficient to produce a picture similar to MS. A possible explanation of the human disease is that a genetically susceptible individual becomes transiently infected (by a minor viral illness, for example) with a microorganism that expresses a molecule structurally similar to a component of myelin. An immune response to this antigen is mounted to attack the invader, but the failure of the immune system to discriminate between the foreign protein and self results in destruction of otherwise normal myelin. An alternative hypothesis is that MS is caused by a persistent infection with a virus or other microorganism; in this scenario, the immune system's ongoing efforts to get rid of the virus cause the damage to myelin. Tropical spastic paraparesis (TSP) provides a precedent for this idea. TSP is a disease characterized by the gradual progression of weakness of the legs and impaired control of bladder function associated with increased deep tendon reflexes and a positive Babinski sign (see Chapter 17). This clinical picture is similar to that of rapidly advancing MS. TSP is known to be caused by persistent infection with a retrovirus (human T lymphotropic virus–1). This precedent notwithstanding, proving the persistent viral infection hypothesis for MS requires unambiguous demonstration of the presence of a virus. Despite periodic reports of a virus associated with MS, convincing evidence has not been forthcoming. Thus, MS remains a daunting clinical challenge in which the neurologist has few effective treatment strategies.

References

ADAMS, R. D. AND M. VICTOR (1997) *Principles of Neurology*, 6th Ed. New York: McGraw-Hill, pp. 903–921.

JACOBS, L.D. ET AL. (1996) Intramuscular interferon β-1α for disease progression in relapsin multiple sclerosis. Ann. Neurol. 39: 285–294.

STEINMAN, L. (1996) Multiple sclerosis: A coordinated immunological attack against myelin in the central nervous system. Cell 85: 299–302.

TRAPP, B. D. AND 5 OTHERS (1998) Axonal transection in the lesions of multiple sclerosis. N. Engl. J. Med. 338: 278–285.

Summary

The action potential and all its complex properties can be explained by time- and voltage-dependent changes in the Na⁺ and K⁺ permeabilities of neuronal membranes. This conclusion derives primarily from evidence obtained by a device called the voltage clamp. The voltage clamp technique is an electronic feedback method that allows control of neuronal membrane potential and, simultaneously, direct measurement of the voltage-dependent fluxes of Na⁺ and K⁺ that produce the action potential. Voltage clamp experiments show that a transient rise in Na⁺ conductance activates rapidly and then inactivates during a sustained depolarization of the membrane potential. Such experiments also demonstrate a rise in K⁺ conductance that activates in a delayed fashion and, in contrast to the Na⁺ conductance, does not inactivate. Mathematical modeling of the properties of these conductances indicates that they, and they alone, are responsible for the production of all-or-none action potentials in the squid axon. Action potentials propagate along the nerve cell axons initiated by the voltage gradient between the active and inactive regions of the axon by virtue of the local current flow. In this way, action potentials compensate for the relatively poor passive electrical properties of nerve cells and enable neural signaling over long distances. These classical studies provide a solid basis for considering the functional and ultimately molecular variations on neural signaling taken up in the next chapter.

ADDITIONAL READING

Reviews

ARMSTRONG, C. M. AND B. HILLE (1998) Voltage-gated ion channels and electrical excitability. Neuron 20: 371–80.

NEHER, E. (1992) Ion channels for communication between and within cells. Science 256: 498–502.

Important Original Papers

ARMSTRONG, C. M. AND L. BINSTOCK (1965) Anomalous rectification in the squid giant axon injected with tetraethylammonium chloride. J. Gen. Physiol. 48: 859–872.

HODGKIN, A. L. AND A. F. HUXLEY (1952a) Currents carried by sodium and potassium ions through the membrane of the giant axon of *Loligo*. J. Physiol. 116: 449–472.

HODGKIN, A. L. AND A. F. HUXLEY (1952b) The components of membrane conductance in the giant axon of *Loligo*. J. Physiol. 116: 473–496.

HODGKIN, A. L. AND A. F. HUXLEY (1952c) The dual effect of membrane potential on sodium conductance in the giant axon of *Loligo*. J. Physiol. 116: 497–506.

HODGKIN, A. L. AND A. F. HUXLEY (1952d) A quantitative description of membrane current and its application to conduction and excitation in nerve. J. Physiol. 116: 507–544.

HODGKIN, A. L. AND W. A. H. RUSHTON (1938) The electrical constants of a crustacean nerve fibre. Proc. R. Soc. Lond. 133: 444–479.

HODGKIN, A. L., A. F. HUXLEY AND B. KATZ (1952) Measurements of current-voltage relations in the membrane of the giant axon of *Loligo*. J. Physiol. 116: 424–448.

MOORE, J. W., M. P. BLAUSTEIN, N. C. ANDERSON AND T. NARAHASHI (1967) Basis of tetrodotoxin's selectivity in blockage of squid axons. J. Gen. Physiol. 50: 1401–1411.

Books

AIDLEY, D. J. AND P. R. STANFIELD (1996) *Ion Channels: Molecules in Action*. Cambridge: Cambridge University Press.

JOHNSTON, D. AND S. M.-S. WU (1995) *Foundations of Cellular Neurophysiology*. Cambridge, MA: MIT Press.

JUNGE, D. (1992) *Nerve and Muscle Excitation*, 3rd Ed. Sunderland, MA: Sinauer Associates.

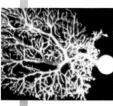

Chapter 4

Channels and Transporters

Overview

The generation of electrical signals in neurons requires both selective membrane permeability and specific ion concentration gradients across the plasma membrane. The membrane proteins that give rise to selective permeability are called ion channels, whereas other proteins called active transporters create and maintain ion gradients. As their name implies, ion channels have pores that permit particular ions to cross the neuronal membrane. Some of these channels are also able to sense the electrical potential across the membrane. Such voltage-gated channels open or close in response to the magnitude of the membrane potential, allowing the membrane permeability to be regulated by the potential. Some ion channels are gated by extracellular chemical signals such as neurotransmitters, others by intracellular signals such as second messengers, and still others respond to mechanical stimuli or temperature changes. Many types of ion channels have now been identified at the molecular level, and this diversity generates a wide spectrum of electrical characteristics among neuron types. In contrast to the functions of ion channels, active transporters are membrane proteins that produce and maintain ion concentration gradients. The most important of these is the Na^+ pump, which hydrolyzes ATP to regulate the intracellular concentrations of both Na^+ and K^+. Other active transporters produce concentration gradients for the full range of physiologically important ions, including Cl^-, Ca^{2+}, and H^+. From the perspective of electrical signaling, active transporters and ion channels are complementary: Transporters create the concentration gradients that drive ions through open ion channels, thus generating electrical signals.

Ion Channels Underlying Action Potentials

Although Hodgkin and Huxley had no knowledge of the physical nature of the conductance mechanisms underlying action potentials, they nonetheless proposed that nerve cell membranes have channels that allow ions to pass selectively from one side of the membrane to the other (see Chapter 3). Based on the ionic conductances and currents measured in voltage clamp experiments, the postulated channels had to have several properties. First, because the ionic currents are quite large, the channels had to be capable of allowing ions to move across the membrane at high rates. Second, because the ionic currents depend on the electrochemical gradient across the membrane, the channels had to make use of these gradients. Third, because Na^+ and K^+ flow across the membrane independently of each other, different channel types had to be capable of discriminating between Na^+ and K^+, allowing only one of these ions to flow across the membrane under the rele-

BOX A
The Patch Clamp Method

A wealth of new information about ion channels resulted from the invention of the patch clamp method in the 1970s. This technique is based on a very simple idea. A glass pipette with a very small opening is used to make tight contact with a tiny area, or patch, of neuronal membrane. After the application of a small amount of suction to the back of the pipette, the seal between pipette and membrane becomes so tight that no ions can flow between the pipette and the membrane. Thus, all the ions that flow when a single ion channel opens must flow into the pipette. The resulting electrical current, though small, can be measured with an ultrasensitive electronic amplifier connected to the pipette. Based on the geometry involved, this arrangement usually is called the *cell-attached patch clamp recording method*. As with the conventional voltage clamp method, the patch clamp method allows experimental control of the membrane potential to characterize the voltage dependence of membrane currents.

Although the ability to record currents flowing through single ion channels is an important advantage of the cell-attached patch clamp method, minor technical modifications yield still other advantages. For example, if the membrane patch within the pipette is disrupted by briefly applying strong suction, the interior of the pipette becomes continuous with the cytoplasm of the cell. This arrangement allows measurements of electrical potentials and currents from the entire cell and is therefore called the *whole-cell recording method*. The whole-cell configuration also allows diffusional exchange between the pipette and the cytoplasm, producing a convenient way to inject substances into the interior of a "patched" cell.

Two other variants of the patch clamp method originate from the finding that

once a tight seal has formed between the membrane and the glass pipette, small pieces of membrane can be pulled away

Four configurations in patch clamp measurements of ionic currents.

from the cell without disrupting the seal; this yields a preparation that is free of the complications imposed by the rest of the

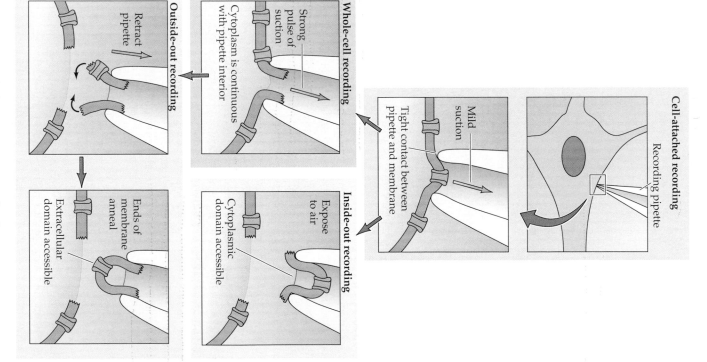

Cell-attached recording

Recording pipette

Tight contact between pipette and membrane

Mild suction

Whole-cell recording

Cytoplasm is continuous with pipette interior

Strong pulse of suction

Inside-out recording

Cytoplasmic domain accessible

Expose to air

Outside-out recording

Retract pipette

Ends of membrane anneal

Extracellular domain accessible

cell. Simply retracting a pipette that is in the cell-attached configuration causes a small vesicle of membrane to remain attached to the pipette. By exposing the tip of the pipette to air, the vesicle opens to yield a small patch of membrane with its (former) intracellular surface exposed. This arrangement, called the *inside-out patch recording configuration*, allows the measurement of single-channel currents with the added benefit of making it possible to change the medium to which the intracellular surface of the membrane is exposed. Thus, the inside-out configuration is particularly valuable when study-

ing the influence of intracellular molecules on ion channel function. Alternatively, if the pipette is retracted while it is in the whole-cell configuration, a membrane patch is produced that has its extracellular surface exposed. This arrangement, called the *outside-out recording configuration*, is optimal for studying how channel activity is influenced by extracellular chemical signals, such as neurotransmitters (see Chapter 7). This range of possible configurations makes the patch clamp method an unusually versatile technique for studies of ion channel function.

References

HAMILL, O. P., A. MARTY, E. NEHER, B. SAKMANN AND F. J. SIGWORTH (1981) Improved patch-clamp techniques for high-resolution current recording from cells and cell-free membrane patches. Pflügers Arch. 391: 85–100.

LEOIS, R. A. AND J. L. RAE (1998) Low-noise patch-clamp techniques. Meth. Enzym. 293: 218–266.

SIGWORTH, F. J. (1986) The patch clamp is more useful than anyone had expected. Fed. Proc. 45: 2673–2677.

vant conditions. Finally, given that the conductances are voltage-dependent, the channels had to be able to sense the voltage drop across the membrane, opening only when the voltage reached appropriate levels. While this concept of channels was highly speculative in the 1950s, later experimental work established beyond any doubt that transmembrane proteins called voltage-sensitive ion channels indeed exist and are responsible for all of the ionic conductance phenomena described in Chapter 3.

The first direct evidence for the presence of voltage-sensitive, ion-selective channels in nerve cell membranes came from measurements of the ionic currents flowing through individual ion channels. The voltage-clamp apparatus used by Hodgkin and Huxley could only resolve the *aggregate* current resulting from the flow of ions through many thousands of channels. A technique capable of measuring the currents flowing through single channels was devised in 1976 by Erwin Neher and Bert Sakmann at the Max Planck Institute in Goettingen. This remarkable approach, called patch clamping (Box A), revolutionized the study of membrane currents. In particular, the patch clamp method provided the means to test directly Hodgkin and Huxley's proposals about the characteristics of ion channels.

Currents flowing through Na^+ channels are best examined in experimental circumstances that prevent the flow of current through other types of channels that are present in the membrane (e.g., K^+ channels). Under such conditions, depolarizing a patch of membrane from a squid giant axon causes tiny inward currents to flow, but only occasionally (Figure 4.1). The size of these currents is minuscule—approximately 1–2 pA (i.e., 10^{-12} ampere), which is orders of magnitude smaller than the Na^+ currents measured by voltage clamping the entire axon. The currents flowing through single channels are called **microscopic currents** to distinguish them from the **macroscopic currents** flowing through a large number of channels distributed over a much more extensive region of surface membrane. Although microscopic currents are certainly small, a current of 1 pA nonetheless reflects the flow of thousands of ions per millisecond. Thus, as predicted, a single channel can let many ions pass through the membrane in a very short time.

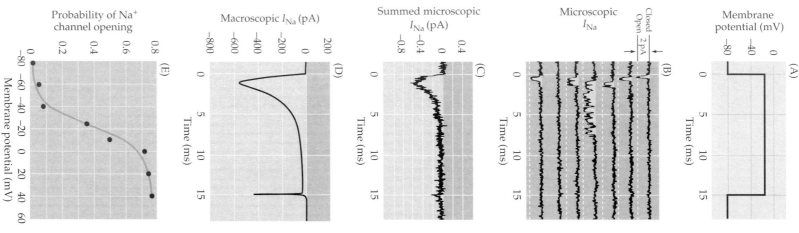

Figure 4.1 Patch clamp measurements of ionic currents flowing through single Na⁺ channels in a squid giant axon. In these experiments, Cs⁺ was applied to the axon to block voltage-gated K⁺ channels. Depolarizing voltage pulses (A) applied to a patch of membrane containing a single Na⁺ channel result in brief currents (B, downward deflections) in the seven successive recordings of membrane current (I_{Na}). (C) The sum of many such current records shows that most channels open in the initial 1–2 ms following depolarization of the membrane, after which the probability of channel openings diminishes because of channel inactivation. (D) A macroscopic current measured from another axon shows the close correlation between the time courses of microscopic and macroscopic Na⁺ currents. (E) The probability of an Na⁺ channel opening depends on the membrane potential, increasing as the membrane is depolarized. (B,C after Bezanilla and Correa, 1995; D after Vanden-burg and Bezanilla, 1991; E after Correa and Bezanilla, 1994.)

Several observations further proved that the microscopic currents in Figure 4.1B are due to the opening of single, voltage-activated Na⁺ channels. First, the currents are carried by Na⁺; thus, they are directed inward at potentials more negative than E_{Na}, reverse their polarity at potentials more positive than E_{Na}, are outward at more positive potentials, and are reduced in size when the Na⁺ concentration of the external medium is decreased. This behavior exactly parallels that of the macroscopic Na⁺ currents described in Chapter 3. Second, the channels have a time course of opening, closing, and inactivating that matches the kinetics of macroscopic Na⁺ currents. This correspondence is difficult to appreciate in the measurement of microscopic currents flowing through a single open channel, because individual channels open and close in a stochastic (random) manner, as can be seen by examining the individual traces in Figure 4.1B. However, repeated depolarization of the membrane potential causes each Na⁺ channel to open and close many times. When the current responses to a large number of such stimuli are averaged together, the collective response has a time course that looks much like the macroscopic Na⁺ current (Figure 4.1C). In particular, the channels open mostly at the beginning of a prolonged depolarization, showing that they subsequently inactivate, as predicted from the macroscopic Na⁺ current (compare Figures 4.1C and 4.1D). Third, both the opening and closing of the channels are voltage-dependent; thus, the channels are closed at –80 mV but open when the membrane potential is depolarized. In fact, the probability that any given channel will be open varies with membrane potential (Figure 4.1E), again as predicted from the macroscopic Na⁺ conductance (see Figure 3.7). Finally, tetrodotoxin, which blocks the macroscopic Na⁺ current (see Box C), also blocks microscopic Na⁺ currents. Taken together, these results show that the macroscopic Na⁺ current measured by Hodgkin and Huxley does indeed arise from the aggregate effect of many thousands of microscopic Na⁺ currents, each representing the opening of a single voltage-sensitive Na⁺ channel.

Patch clamp experiments have also revealed the properties of the channels responsible for the macroscopic K⁺ currents associated with action potentials. When the membrane potential is depolarized (Figure 4.2A), microscopic outward currents (Figure 4.2B) can be observed under conditions that block Na⁺ channels. The microscopic outward currents exhibit all the features expected for currents flowing through action-potential-related K⁺ channels. Thus, the microscopic currents (Figure 4.2C), like their macroscopic counterparts (Figure 4.2D), fail to inactivate during brief depolarizations. Moreover, these single-channel currents are sensitive to ionic manipulations and drugs that affect the macroscopic K⁺ currents and, like the macroscopic K⁺ currents, are voltage-dependent (Figure 4.2E). This and other evidence

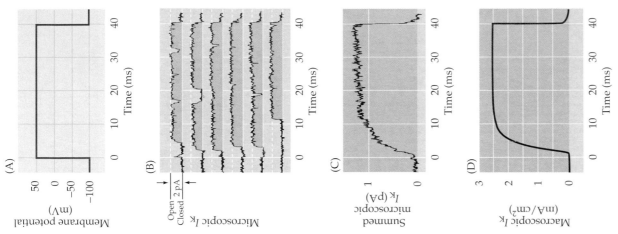
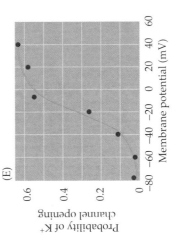

Figure 4.2 Patch clamp measurements of ionic currents flowing through single K+ channels in a squid giant axon. In these experiments, tetrodotoxin was applied to the axon to block voltage-gated Na+ channels. Depolarizing voltage pulses (A) applied to a patch of membrane containing a single K+ channel results in brief currents (B, upward deflections) whenever the channel opens. (C) The sum of such current records shows that most channels open with a delay, but remain open for the duration of the depolarization. (D) A macroscopic current measured from another axon shows the correlation between the time courses of microscopic and macroscopic K+ currents. (E) The probability of a K+ channel opening depends on the membrane potential, increasing as the membrane is depolarized. (B and C after Augustine and Bezanilla, in Hille 1992; D after Augustine and Bezanilla, 1990; E after Perozo et al., 1991.)

shows that macroscopic K+ currents associated with action potentials arise from the opening of many voltage-sensitive K+ channels.

In summary, patch clamping has allowed direct observation of microscopic ionic currents flowing through single ion channels, confirming that voltage-sensitive Na+ and K+ channels are responsible for the macroscopic conductances and currents that underlie the action potential. Measurements of the behavior of single ion channels also indicate many of the molecular attributes of these channels. For example, they show that the membrane of the squid axon contains at least two types of channels—one selectively permeable to Na+ and a second selectively permeable to K+. Both channel types are **voltage-gated**, meaning that their opening is influenced by membrane potential (see Figures 4.1 and 4.2). For each channel, depolarization increases the probability of channel opening, whereas hyperpolarization closes them. Thus, both channel types have a **voltage sensor** that detects the membrane potential (Figure 4.3). However, these channels differ in important respects. In addition to their different ion selectivities, depolarization also inactivates the Na+ channel but not the K+ channel, causing Na+ channels to pass into a nonconducting state. The Na+ channel must therefore have an additional molecular mechanism responsible for **inactivation**. And, as expected from the macroscopic behavior of the Na+ and K+ currents described in Chapter 3, the kinetic properties of the gating of the two channels differs. This detailed information about the physiology of single channels set the stage for subsequent studies of the molecular characteristics of ion channels.

The Diversity of Ion Channels

Molecular genetic studies, in conjunction with the patch clamp method and other techniques, have led to many additional advances in understanding ion channels. Genes encoding Na+ and K+ channels, as well as many other channel types, have now been identified and cloned. A surprising fact that has emerged from these molecular studies is the diversity of genes that code for ion channels. More than 100 ion channel genes have now been discovered, a number that could not have been anticipated from early studies of ion channel function. To understand the functional significance of this multitude of ion channel genes, the channels can be selectively expressed in well-defined experimental systems, such as in cultured cells or frog oocytes (Box B), and then studied with patch clamping and other physiological techniques. Such studies have found many genes encoding voltage-gated channels that respond to membrane potential in much the same way as the Na+ and K+ channels that underlie the action potential. Other channels, however, are

Figure 4.3 Functional states of voltage-gated Na⁺ and K⁺ channels. The gates of both channels are closed when the membrane potential is hyperpolarized. When the potential is depolarized, voltage sensors (indicated by +) allow the channel gates to open—first the Na⁺ channels and then the K⁺ channels. Na⁺ channels also inactivate during prolonged depolarization, whereas many types of K⁺ channels do not.

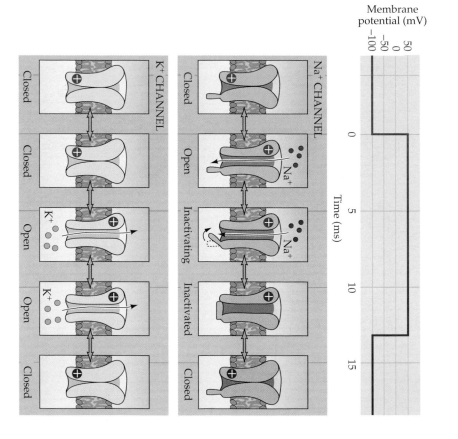

gated by chemical signals that bind to extracellular or intracellular domains on these proteins and are insensitive to membrane voltage. Still others are sensitive to mechanical displacement, or to changes in temperature.

Because channel genes often contain one or more sites for splicing, multiple forms of channel subunits can be generated by a single gene. Differences arising from subunit composition can have a dramatic effect on the functional properties of the channels. Subunit proteins can also undergo posttranslational modifications, such as phosphorylation by protein kinases (see Chapter 8), which may further change their functional characteristics. Thus, although the basic electrical signals of the nervous system are relatively stereotyped, the proteins responsible for their generation are remarkably diverse, conferring distinct signaling properties to the many neuronal cell types that populate the nervous system, as well as providing the basis for a broad range of neurological diseases.

Voltage-Gated Ion Channels

Voltage-gated ion channels that are selectively permeable to each of the major physiological ions—Na⁺, K⁺, Ca²⁺, and Cl⁻—have now been discovered (Figure 4.4 A–D). Indeed, many different genes have been discovered for each type of voltage-gated ion channel. For example, 10 human Na⁺ channel genes have been identified. This finding was unexpected because Na⁺ channels from many different cell types have similar functional properties, consistent with their origin from a single gene. It is now clear, however, that all of these Na⁺ channel genes produce proteins that differ in their structure, function, and distribution in specific tissues. For example, in addi-

VOLTAGE-GATED CHANNELS

LIGAND-GATED CHANNELS

(A) Na$^+$ channel (B) Ca^{2+} channel (C) K$^+$ channel (D) Cl$^-$ channel (E) Neurotransmitter receptor (F) Ca^{2+}-activated K$^+$ channel (G) Cyclic nucleotide gated channel

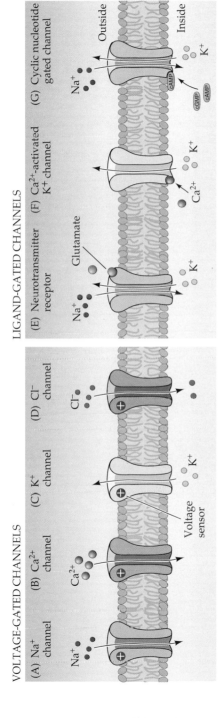

Outside

Inside

Na$^+$

Ca^{2+}

K$^+$

Cl$^-$

Voltage sensor

Glutamate

Na$^+$

K$^+$

Ca^{2+}

K$^+$

Na$^+$

K$^+$

cAMP

cGMP

cGMP

cAMP

Figure 4.4 Types of voltage-gated ion channels. Examples of voltage-gated channels include those selectively permeable to Na$^+$ (A), Ca^{2+} (B), K$^+$ (C), and Cl$^-$ (D). Ligand-gated ion channels include those activated by the extracellular presence of neurotransmitters, such as glutamate (E). Other ligand-gated channels are activated by intracellular second messengers, such as Ca^{2+} (F) or the cyclic nucleotides, cAMP and cGMP (G).

tion to the rapidly inactivating Na$^+$ channels discovered by Hodgkin and Huxley in squid axon, a voltage-sensitive Na$^+$ channel that does *not* inactivate has been identified in mammalian axons. As might be expected, this channel gives rise to action potentials of long duration and is one of the targets of local anesthetics such as benzocaine and lidocaine.

Other electrical responses in neurons are due to the activation of voltage-gated Ca^{2+} channels (Figure 4.4B). In some neurons, voltage-gated Ca^{2+} channels give rise to action potentials in much the same way as voltage-sensitive Na$^+$ channels. In many other neurons, Ca^{2+} channels can control the shape of action potentials generated primarily by Na$^+$ conductance changes. By affecting intracellular Ca^{2+} concentrations, the activity of Ca^{2+} channels regulates an enormous range of biochemical processes within cells (see Chapter 5). Perhaps the most important of the processes regulated by voltage-sensitive Ca^{2+} channels is the release of neurotransmitters at synapses (see Chapter 5). Given these crucial functions, it is perhaps not surprising that 16 different Ca^{2+} channel genes have been identified. Like Na$^+$ channels, different Ca^{2+} channels differ in their activation and inactivation properties, allowing subtle variations in both electrical and chemical signaling processes mediated by Ca^{2+}. As a result, drugs that block voltage-gated Ca^{2+} channels are especially valuable in treating a variety of conditions ranging from heart disease to anxiety disorders.

The largest and most diverse class of voltage-gated ion channels are the K$^+$ channels (Figure 4.4C). Nearly 100 K$^+$ channel genes are now known, and these fall into several distinct groups that differ substantially in their gating properties (Figure 4.5). Some take minutes to inactivate, as in the case of squid axon K$^+$ channels studied by Hodgkin and Huxley. Others inactivate within milliseconds, as is typical of most voltage-gated Na$^+$ channels. These properties influence the duration and rate of action potential firing, with important consequences for axonal conduction and synaptic transmission. Perhaps the most important function of K$^+$ channels is the part they play in generating the resting membrane potential (see Chapter 2). At least two families of K$^+$ channels that are open at hyperpolarized membrane potentials contribute to setting the resting membrane potential.

Finally, several types of voltage-gated Cl$^-$ channel also have been identified (see Figure 4.4D). These channels are present in every type of neuron, where they control excitability, contribute to the resting membrane potential, and help regulate cell volume.

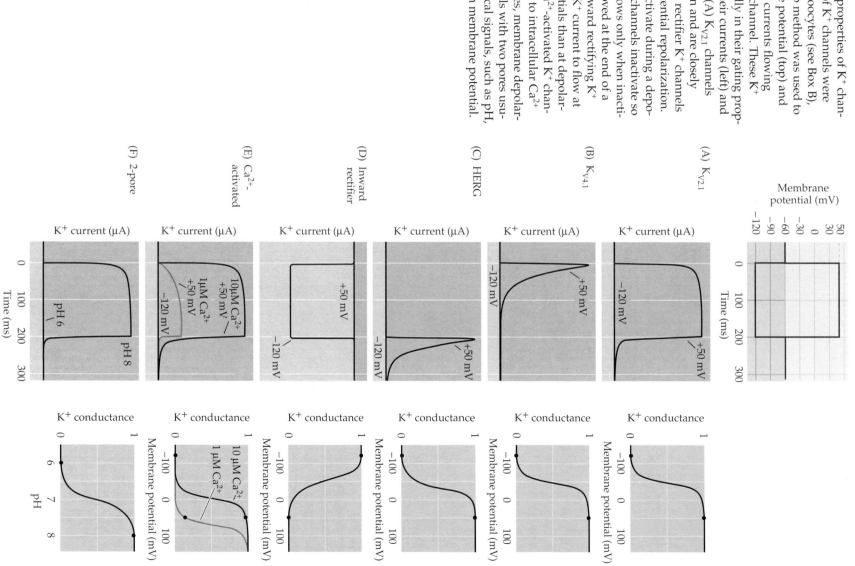

Figure 4.5 Diverse properties of K+ channels. Different types of K+ channels were expressed in *Xenopus* oocytes (see Box B), and the voltage clamp method was used to change the membrane potential (top) and measure the resulting currents flowing through each type of channel. These K+ channels vary markedly in their gating properties, as evident in their currents (left) and conductances (right). (A) K$_{V2.1}$ channels show little inactivation and are closely related to the delayed rectifier K+ channels involved in action potential repolarization. (B) K$_{V4.1}$ channels inactivate during a depolarization. (C) HERG channels inactivate so rapidly that current flows only when inactivation is rapidly removed at the end of a depolarization. (D) Inward rectifying K+ channels allow more K+ current to flow at hyperpolarized potentials than at depolarized potentials. (E) Ca^{2+}-activated K+ channels open in response to intracellular Ca^{2+} ions and, in some cases, membrane depolarization. (F) K+ channels with two pores usually respond to chemical signals, such as pH, rather than changes in membrane potential.

BOX B
Expression of Ion Channels in *Xenopus* Oocytes

Bridging the gap between the sequence of an ion channel gene and understanding channel function is a challenge. To meet this challenge, it is essential to have an experimental system in which the gene product can be expressed efficiently, and in which the function of the resulting channel can be studied with methods such as the patch clamp technique. Ideally, the vehicle for expression should be readily available, have few endogenous channels, and be large enough to permit mRNA and DNA to be microinjected with ease. Oocytes (immature eggs) from the clawed African frog, *Xenopus laevis* (figure A), fulfill all these demands. These huge cells (approximately 1 mm in diameter; figure B) are easily harvested from the female *Xenopus*. Work performed in the 1970s by John Gurdon, a developmental biologist, showed that injection of exogenous mRNA into frog oocytes causes them to synthesize foreign protein in prodigious quantities. In the early 1980s, Ricardo Miledi, Eric Barnard, and other neurobiologists demonstrated that *Xenopus* oocytes could express exogenous ion channels, and that physiological methods could be used to study the ionic currents generated by the newly-synthesized channels (figure C).

As a result of these pioneering studies, heterologous expression experiments have now become a standard way of studying ion channels. The approach has been especially valuable in deciphering the relationship between channel structure and function. In such experiments, defined mutations (often affecting a single nucleotide) are made in the part of the channel gene that encodes a structure of interest; the resulting channel proteins are then expressed in oocytes to assess the functional consequences of the mutation.

The ability to combine molecular and physiological methods in a single cell system has made *Xenopus* oocytes a powerful experimental tool. Indeed, this system has been as valuable to contemporary studies of voltage-gated ion channels as the squid axon was to such studies in the 1950s and 1960s.

References

GUNDERSEN, C. B., R. MILEDI AND I. PARKER (1984) Slowly inactivating potassium channels induced in *Xenopus* oocytes by messenger ribonucleic acid from *Torpedo* brain. J. Physiol. (Lond.) 353: 231–248.

GURDON, J. B., C. D. LANE, H. R. WOODLAND AND G. MARBAIX (1971) Use of frog eggs and oocytes for the study of messenger RNA and its translation in living cells. Nature 233: 177–182.

STÜHMER, W. (1998) Electrophysiological recordings from *Xenopus* oocytes. Meth. Enzym. 293: 280–300.

SUMIKAWA, K., M. HOUGHTON, J. S. EMTAGE, B. M. RICHARDS AND E. A. BARNARD (1981) Active multi-subunit ACh receptor assembled by translation of heterologous mRNA in *Xenopus* oocytes. Nature 292: 862–864.

(A)

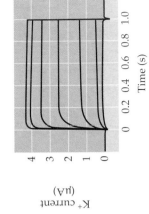

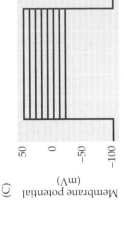

(B)

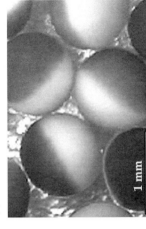

(C)

(A) The clawed African frog, *Xenopus laevis*. (B) Several oocytes from *Xenopus* highlighting the dark coloration of the original pole and the lighter coloration of the vegetal pole. (C) Results of a voltage clamp experiment showing K⁺ currents produced following injection of K⁺ channel mRNA into an oocyte. (A,B courtesy of P. Reinhart; C after Gundersen et al., 1984.)

Ligand-Gated Ion Channels

Many types of ion channels respond to chemical signals (ligands) rather than to changes in the membrane potential (Figure 4.4E–G). The most important of these **ligand-gated ion channels** in the nervous system is the class activated by binding neurotransmitters (Figure 4.4E). These channels are essential for synaptic transmission and other forms of cell-cell signaling phenomena discussed in Chapters 5–8. Whereas the voltage-gated ion channels underlying the action potential typically allow only one type of ion to permeate, channels activated by extracellular ligands are usually less selective, allowing two or more types of ions to pass through the channel pore.

Other ligand-gated channels are sensitive to chemical signals from within the cytoplasm of neurons (see Chapter 8). These channels have ligand-binding domains on their *intracellular* surfaces that interact with second messengers such as Ca^{2+} (Figure 4.4F) and the cyclic nucleotides cAMP and cGMP (Figure 4.4G). Such channels can be selective for specific ions such as K^+ or Cl^-, or can be permeable to all physiological cations. The main function of these channels is to convert intracellular chemical signals into electrical information. This process is particularly important in sensory transduction, where channels gated by cyclic nucleotides convert odors and light into electrical signals. Some intracellularly activated ion channels are in the cell surface membrane, but others are in intracellular membranes such as the endoplasmic reticulum. These latter channels are selectively permeable to Ca^{2+} and regulate the release of Ca^{2+} from the lumen of the endoplasmic reticulum into the cytoplasm. The Ca^{2+} released can then trigger a spectrum of cellular responses.

Stretch- and Heat-Activated Channels

Some ion channels respond to heat or membrane deformation. Heat-activated ion channels contribute to the sensations of pain and temperature and help mediate inflammation (see Chapter 10). Other ion channels respond to mechanical distortion of the plasma membrane and are the basis of stretch receptors and neuromuscular stretch reflexes (see Chapters 9, 16 and 17). A specialized form of these channels enables hearing by allowing auditory hair cells to respond to sound waves (see Chapter 13).

In summary, this tremendous variety of ion channels allows neurons to generate electrical signals in response to changes in membrane potential, synaptic input, intracellular second messengers, light, odors, heat, sound, touch, and many other stimuli.

The Molecular Structure of Ion Channels

Understanding the physical structure of ion channels is obviously the key to sorting out how they actually work. For instance, much insight into the detailed operation of ion channels has come from recent X-ray crystallographic studies of a bacterial K^+ channel (Figure 4.6). This particular molecule was chosen for analysis because the large quantity of channel protein needed for crystallography could be obtained by growing vast numbers of bacteria. The channel is formed by subunits that each cross the plasma membrane twice; between these two membrane-spanning structures is a loop that inserts into the plasma membrane (Figure 4.6A). Four of these subunits are assembled together to form a channel (Figure 4.6B). In the center of the assembled channel is a narrow opening through the protein that allows K^+ to flow across the membrane. The **pore**, as this tunnel is usually called, is formed

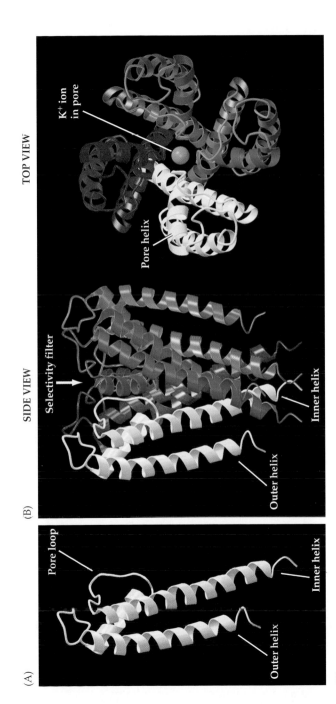

TOP VIEW

K⁺ ion in pore

SIDE VIEW

Selectivity filter

Inner helix

Outer helix

(B)

(A)

Pore loop

Inner helix

Outer helix

Pore helix

(C)

Selectivity filter

K⁺ ions

Negatively charged pore helix

Pore

Water-filled cavity

Figure 4.6 Structure of a simple bacterial K⁺ channel determined by crystallography. (A) Structure of one subunit of the channel, which consists of two membrane-spanning domains and a pore loop that inserts into the membrane. (B) Three-dimensional arrangement of four subunits (each in a different color) to form a K⁺ channel. The top view illustrates a K⁺ ion (green) within the channel pore. (C) The permeation pathway of the K⁺ channel consists of a large aqueous cavity connected to a narrow selectivity filter. Helical domains of the channel point negative charges (red) toward this cavity, allowing K⁺ ions (green) to become dehydrated and then move through the selectivity filter. (A,B from Doyle et al., 1998; C after Doyle et al., 1998.)

by the protein loop, as well as by the membrane-spanning domains. The structure of the pore is well suited for conducting K⁺ ions (Figure 4.6C). The narrowest part is near the outside mouth of the channel and is so narrow that only a nonhydrated K⁺ can fit through this bottleneck. Larger cations, such as Cs⁺, cannot traverse this region of the pore and smaller cations such as Na⁺ cannot enter the pore because the "walls" of the pore are too far apart to stabilize a dehydrated Na⁺ ion. This part of the channel complex is thus responsible for the selective permeability to K⁺ and is therefore called the **selectivity filter**. Deeper within the channel is a water-filled cavity that connects to the interior of the cell via the pore. This cavity evidently collects K⁺ from the interior of the cell and, utilizing negative charges from the protein, allows K⁺ ions to become dehydrated. These "naked" ions are then able to move through the selectivity filter (recall that the normal concentration gradient drives K⁺ out of cells). In sum, the physical structure of this particular K⁺ channel provides a detailed picture of how ions are conducted from one side of the plasma membrane to the other, and how a channel can be selectively permeable to K⁺. It is likely that other K⁺ channels, as well as other types of ion channels, use similar structures to achieve ion selectivity.

Although this bacterial K⁺ channel nicely illustrates the structural principles responsible for conduction and selectivity in ion channels, the channel does not open or close in response to changes in the membrane potential. Its structure cannot, therefore, indicate much about the mechanisms involved in the voltage gating of ion channels. Understanding this aspect of ion channel function has come from less direct studies of other channel proteins. Although such channels have not yet proven amenable to crystallographic study, much about their ionic selectivity and gating properties has been learned by exploring the functions of particular amino acids within the proteins using mutagenesis and the expression of such channels in *Xenopus* oocytes. The general transmembrane architecture of all the major ion channel families is consistent with that of the bacterial K⁺ channel (Figure 4.7).

Thus, these molecules are all integral membrane proteins that span the plasma membrane repeatedly. K⁺ channel subunits typically span the membrane six times (Figure 4.7C,E), though there are some K⁺ channels that, like the bacterial channel, span the membrane four times (Figure 4.7D), and others that span the membrane only twice (Figure 4.7F) or seven times (Figure 4.7E). Four of these K⁺ channel subunits aggregate together to form a single functional ion channel. Na⁺ (and Ca²⁺) channel subunits cross the membrane 24 times (Figure 4.7A,B), and the channel is evidently formed by just one of these Na⁺ or Ca²⁺ channel subunits. Like the bacterial K⁺ channel, the membrane-spanning domains of all ion channels appear to form a central pore through

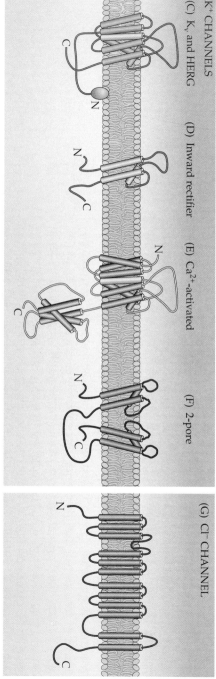

Figure 4.7 Topology of the principal subunits of voltage-gated Na⁺, Ca²⁺, K⁺, and Cl⁻ channels. Repeating motifs of Na⁺ (A) and Ca²⁺ (B) channels are labeled I, II, III, and IV; (C–F) K⁺ channels are more diverse. In all cases, four subunits combine to form a functional channel. (G) Chloride channels are structurally distinct from all other voltage-gated channels.

(A) Na⁺ CHANNEL

I II III IV

β subunit

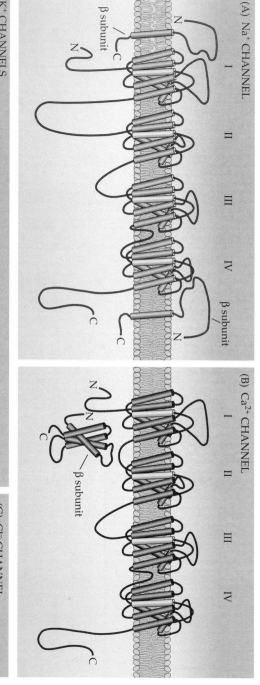

(B) Ca²⁺ CHANNEL

I II III IV

β subunit

K⁺ CHANNELS
(C) Kᵥ and HERG
(D) Inward rectifier
(E) Ca²⁺-activated
(F) 2-pore

(G) Cl⁻ CHANNEL

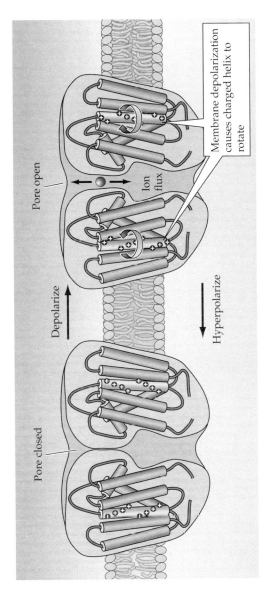

Figure 4.8 A charged voltage sensor permits voltage-dependent gating of ion channels. The process of voltage activation involves the rotation of a positively charged transmembrane domain. This movement causes a change in the conformation of the pore loop, enabling the channel to conduct specific ions.

which ions can diffuse (Figure 4.8). This pore typically arises from the same sort of protein loop that lines the pore of the bacterial channel and has a selectivity filter that determines the permeant ions. As expected, the amino acid composition of the pore loop differs among channels, presumably allowing each channel type to conduct a specific ion. These distinct structural features of channel subunits also provide unique binding sites for drugs and also for various neurotoxins known to block specific subclasses of ion channels (Box C).

A special feature of voltage-activated channels is a sensor that detects the electrical potential across the membrane. This sensor is a transmembrane domain that contains positive charges (see Figure 4.8). The domain is a helical structure with charged amino acids along one face of the helix. Evidently, membrane depolarization influences the charged amino acids such that the helix rotates, thereby allowing the channel pore to open. Similarly, channels that inactivate as a function of membrane voltage also have sequences of amino acids that plug the channel pore during prolonged depolarization.

In short, ion channels are integral membrane proteins with characteristic features that allow them to assemble into multimolecular aggregates. Collectively, these structures allow channels to conduct ions, sense the transmembrane potential, inactivate, and bind to various neurotoxins. Mutations in ion channel genes can have profound effects on channel structure and function, thereby leading to a variety of neurological disorders (Box D).

Active Transporters Create and Maintain Ion Gradients

Up to this point, the discussion of the molecular basis of electrical signaling has taken for granted the fact that nerve cells maintain ion concentration gradients across their surface membranes. However, none of the ions of physiological importance (Na^+, K^+, Cl^-, and Ca^{2+}) are in electrochemical equilibrium. Because channels produce electrical effects by allowing one or more of these ions to diffuse down their electrochemical gradients, there would be a gradual dissipation of these concentration gradients unless nerve cells could restore ions displaced during the current flow that occurs as a result of both neural signaling and the continual ionic leakage that occurs at rest. The work of generating and maintaining ionic concentration gradients for particular ions is carried out by a group of plasma membrane proteins known as **active transporters**.

BOX C
Toxins That Poison Ion Channels

Given the importance of Na⁺ and K⁺ channels for neuronal excitation, it is not surprising that a number of organisms have evolved channel-specific toxins as mechanisms for self-defense or for capturing prey. A rich collection of toxins selectively target the ion channels of neurons and other cells. These toxins are valuable not only for survival, but for studying the function of cellular ion channels. The best-known channel toxin is *tetrodotoxin*, which is produced by certain puffer fish and other animals. Tetrodotoxin produces a potent and specific obstruction of the Na⁺ channels responsible for action potential generation, thereby paralyzing the animals unfortunate enough to ingest it. *Saxitoxin*, a chemical homologue of tetrodotoxin produced by dinoflagellates, has a similar action on Na⁺ channels. The potentially lethal effects of eating shellfish that have ingested these "red tide" dinoflagellates are due to the potent neuronal actions of saxitoxin.

Scorpions paralyze their prey by injecting a potent mix of peptide toxins that also affect ion channels. Among these are the *α-toxins*, which slow the inactivation of Na⁺ channels (figure A1); exposure of neurons to these toxins prolongs the action potential (figure A2), thereby scrambling information flow within the nervous system of the soon-to-be-devoured victim. Other peptides in scorpion venom, called *β-toxins*, shift the voltage dependence of Na⁺ channel activation (figure B). These toxins cause Na⁺ channels to open at potentials much more negative than normal, disrupting action potential generation. Some alkaloid toxins combine these actions, both removing inactivation *and* shifting activation of Na⁺ channels. One such toxin is *batrachotoxin*, produced by a species of frog; some tribes of South American

Indians use this poison on their arrow tips. A number of plants produce similar toxins, including *aconitine*, from buttercups; *veratridine*, from lilies; and a number of insecticidal toxins produced by plants such as chrysanthemums and rhododendrons.

Potassium channels have also been targeted by toxin-producing organisms. Peptide toxins affecting K⁺ channels include *dendrotoxin*, from wasps; *apamin*, from bees; and *charybdotoxin*, yet another toxin produced by scorpions. All of these toxins block K⁺ channels as their primary action; no toxin is known to affect the activation or inactivation of these chan-

nels, although such agents may simply be awaiting discovery.

References

CAHALAN, M. (1975) Modification of sodium channel gating in frog myelinated nerve fibers by *Centruroides sculpturatus* scorpion venom. J. Physiol. (Lond.) 244: 511–534.

SCHMIDT, O. AND H. SCHMIDT (1972) Influence of calcium ions on the ionic currents of nodes of Ranvier treated with scorpion venom. Pflügers Arch. 333: 51–61.

NARAHASHI, T. (2000) Neuroreceptors and ion channels as the basis for drug action: Present, and future. J. Pharmacol. Exptl. Therapeutics 294: 1–26.

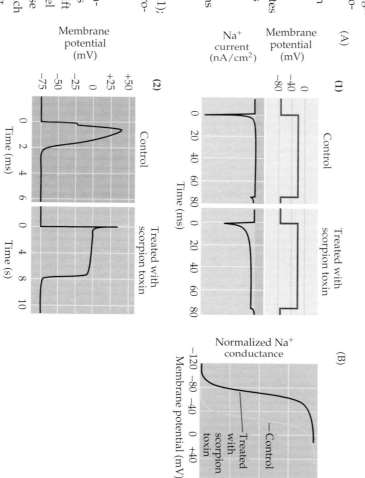

(A) Effects of toxin treatment on frog axons. (1) α-Toxin from the scorpion *Leiurus quinquestriatus* prolongs Na⁺ currents recorded with the voltage clamp method. (2) As a result of the increased Na⁺ current, α-toxin greatly prolongs the duration of the axonal action potential. Note the change in timescale after treating with toxin. (B) Treatment of a frog axon with β-toxin from another scorpion, *Centruroides sculpturatus*, shifts the activation of Na⁺ channels, so that Na⁺ conductance begins to increase at potentials much more negative than usual. (A after Schmidt and Schmidt, 1972; B after Cahalan, 1975.)

Active transporters carry out this task by forming complexes with the ions that they are translocating. The process of ion binding and unbinding for transport typically requires several milliseconds. As a result, ion translocation by active transporters is much slower than ion movement through channels: Recall that ion channels can conduct thousands of ions across a membrane each millisecond. In short, active transporters effectively store energy in the form of ion concentration gradients, whereas the opening of ion channels rapidly dissipates this stored energy during relatively brief electrical signaling events.

Several types of active transporter have now been identified (Figure 4.9). Although the specific jobs of these transporters differ, all must translocate ions against their electrochemical gradients. Moving ions uphill requires the consumption of energy, and neuronal transporters fall into two classes based on their energy sources. Some transporters acquire energy directly from the hydrolysis of ATP and are called **ATPase pumps** (Figure 4.9, left). The most prominent example of an ATPase pump is the **Na$^+$ pump** (or, more properly, the Na$^+$/K$^+$ ATPase pump), which is responsible for maintaining transmembrane concentration gradients for both Na$^+$ and K$^+$ (Figure 4.9A). Another is the Ca^{2+} pump, which provides one of the main mechanisms for removing Ca^{2+} from cells (Figure 4.9B). The second class of active transporter does not use ATP directly, but depends instead on the electrochemical gradients of other ions as an energy source. This type of transporter carries one or more ions *up* its electrochemical gradient while simultaneously taking another ion (most often Na$^+$) *down* its gradient. Because at least two species of ions are involved in such transactions, these transporters are usually called **ion exchangers** (Figure 4.9, right). An example of such a transporter is the Na$^+$/Ca^{2+} exchanger, which shares with the Ca^{2+} pump the important job of keeping intracellular Ca^{2+} concentrations low (Figure 4.9C). Another exchanger in this category regulates both intracellular Cl$^-$ concentration and pH by swapping intracellular Cl$^-$ for another extracellular anion, bicarbonate (Figure 4.9D). Other ion exchangers, such as the Na$^+$/H$^+$ exchanger (Figure 4.9E), also regulate intracellular pH, in this case by acting directly on the concentration of H$^+$. Yet other ion exchangers are involved in transporting neurotransmitters into synaptic terminals (Figure 4.9F), as described in the following chapter. Although the electrochemical gradient of Na$^+$ (or other counter ions) is the proximate source of energy for ion exchangers, these gradients ultimately depend on the hydrolysis of ATP by ATPase pumps, such as the Na$^+$/K$^+$ ATPase pump.

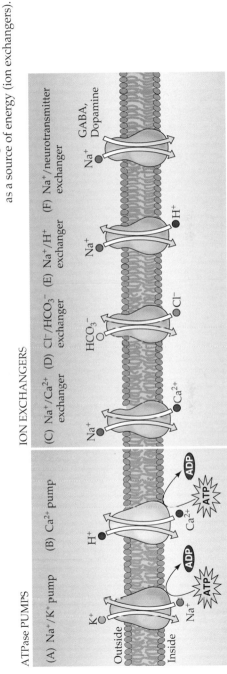

Figure 4.9 Examples of ion transporters found in cell membranes. (A,B) Some transporters are powered by the hydrolysis of ATP (ATPase pumps), whereas others (C–F) use the electrochemical gradients of co-transported ions as a source of energy (ion exchangers).

BOX D
Diseases Caused by Altered Ion Channels

Several genetic diseases, collectively called "channelopathies," result from small but critical alterations in ion channel genes. The best-characterized of these diseases are those that affect skeletal muscle cells. In these disorders, alterations in ion channel proteins produce either myotonia (muscle stiffness due to excessive electrical excitability) or paralysis (due to insufficient muscle excitability). Other disorders arise from ion channel defects in heart, kidney, and the inner ear.

Channelopathies associated with ion channels localized in brain are much more difficult to study. Nonetheless, voltage-gated Ca^{2+} channels have recently been implicated in a range of neurological diseases. These include episodic ataxia, spinocerebellar degeneration, night blindness, and migraine headaches. Familial hemiplegic migraine (FHM) is characterized by migraine attacks that typically last one to three days. During such episodes, patients experience severe headaches and vomiting. Several mutations in a human Ca^{2+} channel have been identified in families with FHM, each having different clinical symptoms. For example, a mutation in the pore-forming region of the channel produces hemiplegic migraine with progressive cerebellar ataxia, whereas other mutations cause only the usual FHM symptoms. How these altered Ca^{2+} channel properties lead to migraine attacks is not known.

Genetic mutations in (A) Ca^{2+} channels, (B) Na^+ channels, (C) K^+ channels, and (D) Cl^- channels that result in diseases. Red regions indicate the sites of these mutations; the red circles indicate the mutations in single amino acids. (After Lehmann-Horn and Jurkat-Kott 1999.)

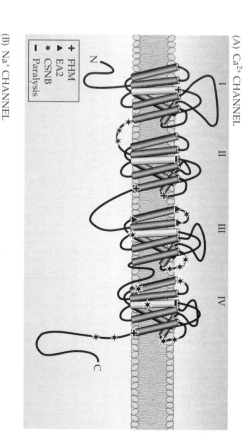

(A) Ca^{2+} CHANNEL

+ FHM
▲ EA2
★ CSNB
▬ Paralysis

(B) Na^+ CHANNEL

▼ GEFS
● Myotonia
■ Paralysis

β subunit

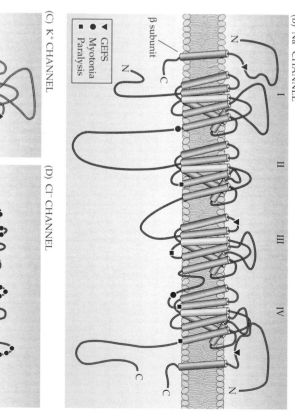

(C) K^+ CHANNEL

× EA1
◆ BFNC

(D) Cl^- CHANNEL

● Myotonia

Episodic ataxia type 2 (EA2) is a neurological disorder in which affected individuals suffer recurrent attacks of abnormal limb movements and severe ataxia. These problems are sometimes accompanied by vertigo, nausea, and headache. Usually, attacks are precipitated by emotional stress, exercise, or alcohol and last for a few hours. The mutations in EA2 cause Ca^{2+} channels to be truncated at various sites, which may cause the clinical manifestations of the disease by preventing the normal assembly of Ca^{2+} channels in the membrane.

X-linked congenital stationary night blindness (CSNB) is a recessive retinal disorder that causes night blindness, decreased visual acuity, myopia, nystagmus, and strabismus. Complete CSNB causes retinal rod photoreceptors to be nonfunctional. Incomplete CSNB causes subnormal (but measurable) functioning of both rod and cone photoreceptors. Like EA2, the incomplete type of CSNB is caused by mutations producing truncated Ca^{2+} channels. Abnormal retinal function may arise from decreased Ca^{2+} currents and neurotransmitter release from photoreceptors (see Chapter 11).

A defect in brain Na^+ channels causes generalized epilepsy with febrile seizures (GEFS) that begins in infancy and usually continues through early puberty. This defect has been mapped to two mutations: one on chromosome 2 that

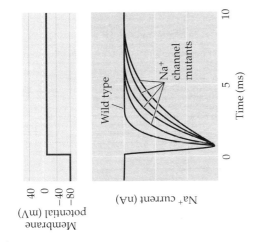

Mutations in Na^+ channels slow the rate of inactivation of Na^+ currents. (After Barchi, 1995.)

encodes an α-subunit for a voltage-gated Na^+ channel, and the other on chromosome 19 that encodes a Na^+ channel β subunit. These mutations cause a slowing of Na^+ channel inactivation (see figure above), which may explain the neuronal hyperexcitability underlying GEFS. Another type of seizure, benign familial neonatal convulsion, is due to K^+ channel mutations. This disease is characterized by frequent brief seizures commencing within the first week of life and disappearing spontaneously within a few months. The mutation has been mapped to at least two voltage-gated K^+ channel genes. A reduction in K^+ current

flow through the mutated channels probably accounts for the hyperexcitability associated with this defect. A related disease, episodic ataxia type 1 (EA1), has been linked to a defect in another type of voltage-gated K^+ channel. EA1 is characterized by brief episodes of ataxia. Mutant channels inhibit the function of other, nonmutant K^+ channels and may produce clinical symptoms by impairing action potential repolarization.

References

BARCHI, R. L. (1995) Molecular pathology of the skeletal muscle sodium channel. Ann. Rev. Physiol. 57: 355–385.

BERKOVIC, S. F. AND I. E. SCHEFFER (1997) Epilepsies with single gene inheritance. Brain Develop. 19 :13–28.

LEHMANN-HORN, F. AND K. JURKAT-ROTT (1999) Voltage-gated ion channels and hereditary disease. Physiol. Rev. 79: 1317–1372.

OPHOFF, R. A., G. M. TERWINDT, R. R. FRANTS AND M. D. FERRARI (1998) P/Q-type Ca^{2+} channel defects in migraine, ataxia and epilepsy. Trends Pharm. Sci. 19: 121–127.

COOPER, E. C. AND L. Y. JAN (1999) Ion channel genes and human neurological disease: Recent progress, prospects, and challenges. Proc. Natl. Acad. Sci. USA 96: 4759–4766.

DAVIES, N. P. AND M. G. HANNA (1999) Neurological channelopathies: Diagnosis and therapy in the new millennium. Ann. Med. 31: 406–420.

JEN, J. (1999) Calcium channelopathies in the central nervous system. Curr. Op. Neurobiol. 9: 274–280.

Functional Properties of the Na^+/K^+ Pump

Of these various transporters, the best understood is the Na^+/K^+ pump. The activity of this pump is estimated to account for 20–40% of the brain's energy consumption, indicating its importance for brain function. The Na^+ pump was first discovered in neurons in the 1950s, when Richard Keynes at Cambridge University used radioactive Na^+ to demonstrate the energy-dependent efflux of Na^+ from squid giant axons. Keynes and his collaborators found that this efflux ceased when the supply of ATP in the axon was interrupted by treatment with metabolic poisons (Figure 4.10A, point 4). Other conditions that lower intracellular ATP also prevent Na^+ efflux. These experiments showed that removing intracellular Na^+ requires cellular metabolism. Further studies

Figure 4.10 Ionic movements due to the Na+/K+ pump. (A) Measurement of radioactive Na+ efflux from a squid giant axon. This efflux depends on external K+ and intracellular ATP. (B) A model for the movement of ions by the Na+/K+ pump. Uphill movements of Na+ and K+ are driven by ATP, which phosphorylates the pump. These fluxes are asymmetrical, with three Na+ carried out for every two K+ brought in. (A after Hodgkin and Keynes, 1955; B after Lingrel et al., 1994.)

(A)

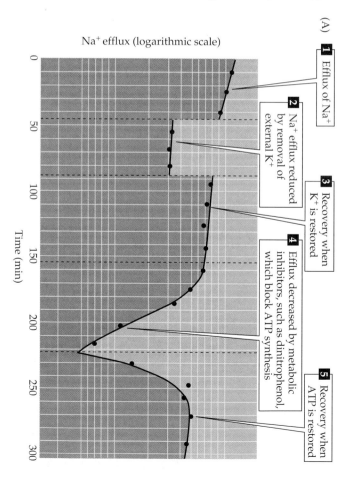

Na+ efflux (logarithmic scale)

1 Efflux of Na+

2 Na+ efflux reduced by removal of external K+

3 Recovery when K+ is restored

4 Efflux decreased by metabolic inhibitors, such as dinitrophenol, which block ATP synthesis

5 Recovery when ATP is restored

Time (min)

(B)

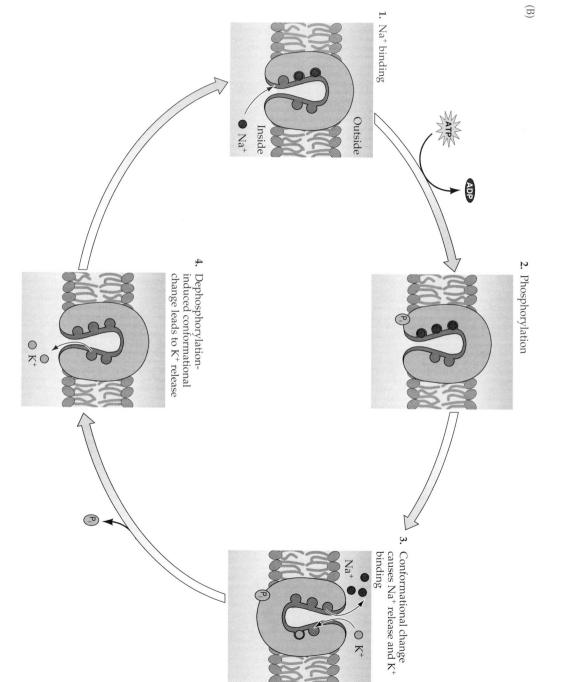

1. Na+ binding

Outside

Inside

Na+

ATP

ADP

2. Phosphorylation

Pi

3. Conformational change causes Na+ release and K+ binding

Na+

K+

Pi

4. Dephosphorylation-induced conformational change leads to K+ release

K+

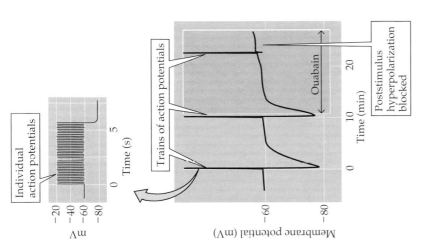

Figure 4.11 The electrogenic transport of ions by the Na^+/K^+ pump can influence membrane potential. Measurements of the membrane potential of a small unmyelinated axon show that a train of action potentials is followed by a long-lasting hyperpolarization. This hyperpolarization is blocked by ouabain, indicating that it results from the activity of the Na^+/K^+ pump. (After Rang and Ritchie, 1968.)

with radioactive K^+ demonstrated that Na^+ efflux is associated with simultaneous, ATP-dependent influx of K^+. These opposing fluxes of Na^+ and K^+ are operationally inseparable: Removal of external K^+ greatly reduces Na^+ efflux (Figure 4.10, point 2) and vice versa. These energy-dependent movements of Na^+ and K^+ implicated an ATP-hydrolyzing Na^+/K^+ pump in the generation of the transmembrane gradients of both Na^+ and K^+. The exact mechanism responsible for these fluxes of Na^+ and K^+ is still not entirely clear, but the pump is thought to alternately shuttle these ions across the membranes in a cycle fueled by the transfer of a phosphate group from ATP to the pump protein (Figure 4.10B).

Additional quantitative studies of the movements of Na^+ and K^+ indicate that the two ions are not pumped at identical rates: The K^+ influx is only about two-thirds the Na^+ efflux. Thus, the pump apparently transports two K^+ into the cell for every three Na^+ that are removed (see Figure 4.10B). This stoichiometry causes a net loss of one positively charged ion from inside of the cell during each round of pumping, meaning that the pump generates an electrical current that can hyperpolarize the membrane potential. For this reason, the Na^+/K^+ pump is said to be **electrogenic**. Because pumps act much more slowly than ion channels, the current produced by the Na^+/K^+ pump is quite small. For example, in the squid axon, the net current generated by the pump is less than 1% of the current flowing through voltage-gated Na^+ channels and affects the resting membrane potential by only a millivolt or less.

Although the electrical current generated by the activity of the Na^+/K^+ pump is small, under special circumstances the pump can significantly influence the membrane potential. For instance, prolonged stimulation of small unmyelinated axons produces a substantial hyperpolarization (Figure 4.11). During the period of stimulation, Na^+ enters through voltage-gated channels and accumulates within the axons. As the pump removes this extra Na^+, the resulting current generates a long-lasting hyperpolarization. Support for this interpretation comes from the observation that conditions that block the Na^+/K^+ pump—for example, treatment with ouabain, a plant glycoside that specifically inhibits the pump—prevent the hyperpolarization. The electrical contribution of the Na^+/K^+ pump is particularly significant in these small-diameter axons because their large surface-to-volume ratio causes intracellular Na^+ concentration to rise to higher levels than it would in other cells. Nonetheless, it is important to emphasize that, in most circumstances, the Na^+/K^+ pump plays no part in generating the action potential and has very little *direct* effect on the resting potential.

The Molecular Structure of the Na^+/K^+ Pump

The results described above indicate that the Na^+ and K^+ pump must exhibit several molecular properties: (1) It must bind both Na^+ and K^+; (2) it must possess sites that bind ATP and receive a phosphate group from this ATP; and (3) it must bind ouabain, the toxin that blocks this pump (Figure 4.12A). Molecular studies have identified protein molecules that can account for these properties of the Na^+/K^+ pump. This pump is a large, integral membrane protein made up of at least two subunits, called α and β. The primary sequence shows that the α subunit spans the membrane 10 times, with most of the molecule found on the cytoplasmic side, whereas the β subunit spans the membrane once and is predominantly extracellular. Although a detailed account of the functional domains of the Na^+/K^+ pump is not yet available, some parts of the amino acid sequence have identified functions (Figure 4.12B). One intracellular domain of the protein is required for ATP binding and hydrolysis, and the amino acid phosphorylated by

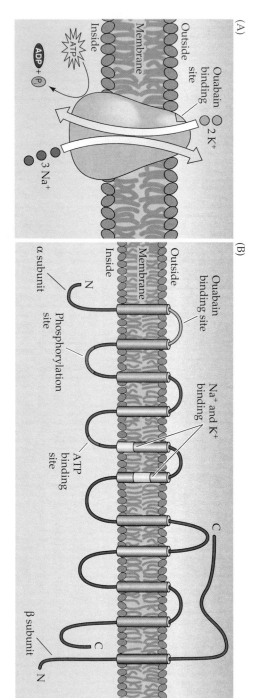

(A)

Ouabain binding site

2 K⁺

Outside

Membrane

Inside

ATP

ADP + Ⓟ

3 Na⁺

(B)

Ouabain binding site

Na⁺ and K⁺ binding

Outside

Membrane

Inside

α subunit

N

Phosphorylation site

ATP binding site

β subunit

C

C

N

Figure 4.12 Molecular structure of the Na⁺/K⁺ pump. (A) General features of the pump. (B) The molecule spans the membrane 10 times. Amino acid residues thought to be important for binding of ATP, K⁺, and ouabain are highlighted. (After Lingrel et al., 1994.)

ATP has been identified. Another extracellular domain may represent the binding site for ouabain. However, the sites involved in the most critical function of the pump—the movement of Na⁺ and K⁺—are not yet defined. Nonetheless, altering certain membrane-spanning domains (red in Figure 4.12B) impairs ion translocation; moreover, kinetic studies indicate that both ions bind to the pump at the same site. Because these ions move across the membrane, it is likely that this site traverses the plasma membrane; it is also likely that the site has a negative charge, since both Na⁺ and K⁺ are positively charged. The observation that removing negatively charged residues in a membrane-spanning domain of the protein (pale orange in Figure 4.12B) greatly reduces Na⁺ and K⁺ binding provides at least a hint about the ion-translocating domain of the transporter molecule.

Summary

Ion transporters and channels have complementary functions. The primary purpose of transporters is to generate transmembrane concentration gradients, which are then exploited by ion channels to generate electrical signals. Ion channels are responsible for the voltage-dependent conductances of nerve cell membranes. The channels underlying the action potential are integral membrane proteins that open or close ion-selective pores in response to the membrane potential, allowing specific ions to diffuse across the membrane. The flow of ions through single open channels can be detected as tiny electrical currents; the synchronous opening of many such channels generates the macroscopic currents that produce action potentials. Molecular studies show that such voltage-gated channels have conserved structures that are responsible for features such as ion permeation and voltage sensing, as well as the features that specify ion selectivity and toxin sensitivity. Other types of channels are sensitive to chemical signals, such as neurotransmitters or second messengers, or to heat or membrane deformation. A large number of ion channel genes create channels with a correspondingly wide range of functional characteristics, thus allowing different types of neurons to have a remarkable spectrum of electrical properties. Ion transporter proteins are quite different in both structure and function. The energy needed for ion movement against a concentration gradient (e.g., in the maintenance of the resting potential) is provided either by the hydrolysis of ATP or by the electrochemical gradient of co-transported ions. The Na⁺/K⁺ pump produces and maintains the transmembrane gradients of Na⁺ and K⁺, while other transporters are responsible for the electrochemical gradients for other physiologically important ions, such as Cl⁻, Ca²⁺, and H⁺. Together, transporters and channels provide a reasonably comprehensive molecular explanation for the ability of neurons to generate electrical signals.

ADDITIONAL READING

Reviews

ARMSTRONG, C. M. AND B. HILLE (1998) Voltage-gated ion channels and electrical excitability. Neuron 20: 371–380.

BEZANILLA, F. AND A. M. CORREA (1995) Single-channel properties and gating of Na⁺ and K⁺ channels in the squid giant axon. In *Cephalopod Neurobiology*, N. J. Abbott, R. Williamson and L. Maddock (eds.). New York: Oxford University Press, pp. 131–151.

CATTERALL, W. A. (1988) Structure and function of voltage-sensitive ion channels. Science 242: 50–61.

ISOM, L. L., K. S. DE JONGH AND W. A. CATTERALL (1994) Auxiliary subunits of voltage-gated ion channels. Neuron 12: 1183–1194.

JAN, L. Y. AND Y. N. JAN (1997) Voltage-gated and inwardly rectifying potassium channels. J. Physiol. 505. 267–282.

JENTSCH, T. J., T. FRIEDRICH, A. SCHRIEVER AND H. YAMADA (1999) The ClC chloride channel family. Pflugers Archiv. 437: 783–795.

LINGREL, J. B., J. VAN HUYSSE, W. O'BRIEN, E. JEWELL-MOTZ, R. ASKEW AND P. SCHULTHEIS (1994) Structure-function studies of the Na, K-ATPase. Kidney Internat. 45: S32–S39.

NEHER, E. (1992) Nobel lecture: Ion channels for communication between and within cells. Neuron 8: 605–612.

SKOU, J. C. (1988) Overview: The Na,K pump. Meth. Enzymol. 156: 1–25.

Important Original Papers

ANTZ, C. AND 7 OTHERS (1997) NMR structure of inactivation gates from mammalian voltage-dependent potassium channels. Nature 385: 272–275.

BEZANILLA, F., E. PEROZO, D. M. PAPAZIAN AND E. STEFANI (1991) Molecular basis of gating charge immobilization in Shaker potassium channels. Science 254: 679–683.

BOULTER, J. AND 6 OTHERS (1990) Molecular cloning and functional expression of glutamate receptor subunit genes. Science 249: 1033–1037.

CATERINA, M. J., M. A. SCHUMACHER, M. TOMINAGA, T. A. ROSEN, J. D. LEVINE AND D. JULIUS (1997) The capsaicin receptor: A heat-activated ion channel in the pain pathway. Nature 389: 816–824.

CHA, A., G. E. SNYDER, P. R. SELVIN AND F. BEZANILLA (1999) Atomic scale movement of the voltage-sensing region in a potassium channel measured via spectroscopy. Nature 402: 809–813.

DOYLE, D. A. AND 7 OTHERS (1998) The structure of the potassium channel: Molecular basis of K⁺ conduction and selectivity. Science 280: 69–77.

FAHLKE, C., H. T. YU, C. L. BECK, T. H. RHODES AND A. L. GEORGE JR. (1997) Pore-forming segments in voltage-gated chloride channels. Nature 390: 529–532.

HO, K. AND 6 OTHERS (1993) Cloning and expression of an inwardly rectifying ATP-regulated potassium channel. Nature 362: 31–38.

HODGKIN, A. L. AND R. D. KEYNES (1955) Active transport of cations in giant axons from *Sepia* and *Loligo*. J. Physiol. 128: 28–60.

HOSHI, T., W. N. ZAGOTTA AND R. W. ALDRICH (1990) Biophysical and molecular mechanisms of Shaker potassium channel inactivation. Science 250: 533–538.

LLANO, L., C. K. WEBB AND F. BEZANILLA (1988) Potassium conductance of squid giant axon. Single-channel studies. J. Gen. Physiol. 92: 179–196.

MIKAMI, A. AND 7 OTHERS (1989) Primary structure and functional expression of the cardiac dihydropyridine-sensitive calcium channel. Nature 340: 230–233.

NODA, M. AND 6 OTHERS (1986) Expression of functional sodium channels from cloned cDNA. Nature 322: 826–828.

NOWYCKY, M. C., A. P. FOX AND R. W. TSIEN (1985) Three types of neuronal calcium channel with different calcium agonist sensitivity. Nature 316: 440–443.

PAPAZIAN, D. M., T. L. SCHWARZ, B. L. TEMPEL, Y. N. JAN AND L. Y. JAN (1987) Cloning of genomic and complementary DNA from Shaker, a putative potassium channel gene from *Drosophila*. Science 237: 749–753.

RANG, H. P. AND J. M. RITCHIE (1968) On the electrogenic sodium pump in mammalian non-myelinated nerve fibres and its activation by various external cations. J. Physiol. 196: 183–221.

SIGWORTH, F. J. AND E. NEHER (1980) Single Na⁺ channel currents observed in cultured rat muscle cells. Nature 287: 447–449.

THOMAS, R. C. (1969) Membrane current and intracellular sodium changes in a snail neurone during extrusion of injected sodium. J. Physiol. 201: 495–514.

TOYOSHIMA, C., M. NAKASAKO, H. NOMURA AND H. OGAWA (2000) Crystal structure of the calcium pump of sarcoplasmic reticulum at 2.6 Å resolution. Nature 405: 647–655.

VANDERBERG, C. A. AND F. BEZANILLA (1991) A sodium channel model based on single channel, macroscopic ionic, and gating currents in the squid giant axon. Biophys. J. 60: 1511–1533.

WALDMANN, R., G. CHAMPIGNY, F. BASSILANA, C. HEURTEAUX AND M. LAZDUNSKI (1997) A proton-gated cation channel involved in acid-sensing. Nature 386: 173–177.

WEI, A. M., A. COVARRUBIAS, A. BUTLER, K. BAKER, M. PAK AND L. SALKOFF (1990) K⁺ current diversity is produced by an extended gene family conserved in *Drosophila* and mouse. Science 248: 599–603.

YANG, N., A. L. GEORGE JR. AND R. HORN (1996) Molecular basis of charge movement in voltage-gated sodium channels. Neuron 16: 113–22.

Books

AIDLEY, D. J. AND P. R. STANFIELD (1996) *Ion Channels: Molecules in Action*. Cambridge: Cambridge University Press.

HILLE, B. (1992) *Ionic Channels of Excitable Membranes*, 2nd Ed. Sunderland, MA: Sinauer Associates.

JUNGE, D. (1992) *Nerve and Muscle Excitation*, 3rd Ed. Sunderland, MA: Sinauer Associates.

NICHOLLS, D. G. (1994) *Proteins, Transmitters and Synapses*. Oxford: Blackwell Scientific

SIEGEL, G. J., B. W. AGRANOFF, R. W. ALBERS, S. K. FISHER AND M. D. UHLER (1999) *Basic Neurochemistry*. Philadelphia: Lippincott-Raven.

Chapter 5

Synaptic Transmission

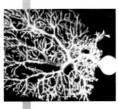

Overview

The human brain contains at least 100 billion neurons, each with the ability to influence many other cells. Clearly, highly sophisticated and efficient mechanisms are needed to enable communication among this astronomical number of elements. Such communication is made possible by synapses, the functional contacts between neurons. Although there are many kinds of synapses within the brain, they can be divided into two general classes: electrical synapses and chemical synapses. Electrical synapses permit direct, passive flow of electrical current from one neuron to another. The current flows through gap junctions, which are specialized membrane channels that connect the two cells. In contrast, chemical synapses enable cell-to-cell communication via the secretion of neurotransmitters; the chemical agents released by the presynaptic neurons produce secondary current flow in postsynaptic neurons by activating specific receptor molecules. The secretion of neurotransmitters is triggered by the influx of Ca^{2+} through voltage-gated channels, which gives rise to a transient increase in Ca^{2+} concentration within the presynaptic terminal. The rise in Ca^{2+} concentration causes synaptic vesicles—the presynaptic organelles that store neurotransmitters—to fuse with the presynaptic plasma membrane and release their contents into the space between the pre- and postsynaptic cells. Although it is not yet understood exactly how Ca^{2+} triggers exocytosis, specific proteins on the surface of the synaptic vesicle and elsewhere in the presynaptic terminal evidently mediate this process.

Electrical Synapses

Although they are a distinct minority, electrical synapses are found in all nervous systems, including the human brain. The structure of an electrical synapse is shown schematically in Figure 5.1A. The membranes of the two communicating neurons come extremely close at the synapse and are actually linked together by an intercellular specialization called a **gap junction**. Gap junctions contain precisely aligned, paired channels in the membrane of the pre- and postsynaptic neurons, such that each channel pair forms a pore (Figure 5.2A). The pore of a gap junction channel is much larger than the pores of the voltage-gated ion channels described in the previous chapter. As a result, a variety of substances can simply diffuse between the cytoplasm of the pre- and postsynaptic neurons. In addition to ions, substances that diffuse through gap junction pores include molecules with molecular weights as great as several hundred daltons. This permits ATP and other important intracellular metabolites, such as second messengers (see Chapter 8), to be transferred between neurons.

(A)

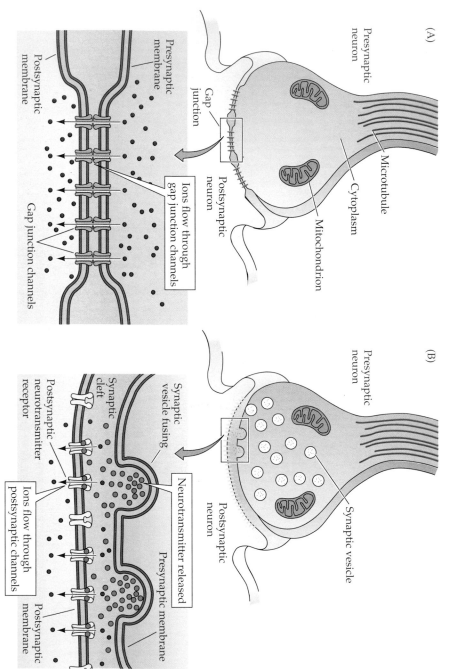

Presynaptic
neuron

Postsynaptic
neuron

Gap
junction

Microtubule

Cytoplasm

Mitochondrion

Postsynaptic
membrane

Presynaptic
membrane

Ions flow through
gap junction channels

Gap junction channels

(B)

Presynaptic
neuron

Postsynaptic
neuron

Synaptic vesicle

Synaptic
vesicle fusing

Synaptic
cleft

Postsynaptic
neurotransmitter
receptor

Neurotransmitter released

Presynaptic membrane

Ions flow through
postsynaptic channels

Postsynaptic
membrane

Figure 5.1　Electrical and chemical synapses differ fundamentally in their transmission mechanisms. (A) At electrical synapses, gap junctions between pre- and postsynaptic membranes permit current to flow passively through intercellular channels (see blowup). This current flow changes the postsynaptic membrane potential, initiating (or in some instances inhibiting) the generation of postsynaptic action potentials. (B) At chemical synapses, there is no direct flow of current from pre- to postsynaptic cell. Synaptic current flows across the postsynaptic membrane only in response to the secretion of neurotransmitters which open or close postsynaptic ion channels after binding to receptor molecules (see blowup).

Electrical synapses thus work by allowing ionic current to flow passively through the gap junction pores from one neuron to another. The usual source of this current is the potential difference generated locally by the action potential (see Chapter 3). The "upstream" neuron, which is the source of current, is called the **presynaptic** element, and the "downstream" neuron into which this current flows is termed **postsynaptic**. This arrangement has a number of interesting consequences. One is that transmission can be bidirectional; that is, current can flow in either direction across the gap junction, depending on which member of the coupled pair is invaded by an action potential (although some types of gap junctions have special features that render their transmission unidirectional). Another important feature of the electrical synapse is that transmission is extraordinarily fast: Because passive current flow across the gap junction is virtually instantaneous, communication can occur without the delay that is characteristic of chemical synapses.

These features are apparent in the operation of the first electrical synapse to be discovered in the crayfish nervous system. A postsynaptic electrical signal is observed at this synapse within a fraction of a millisecond after the generation of a presynaptic action potential (Figure 5.2B). In fact, at least part of this brief synaptic delay is caused by propagation of the action potential into the presynaptic terminal, so that there may be essentially no delay at all in the transmission of electrical signals across the synapse. Such synapses interconnect many of the neurons that allow the crayfish to escape from its predators, thus minimizing the time between the presence of a threatening stimulus and a potentially life-saving motor response.

A more general purpose of electrical synapses is to synchronize electrical activity among populations of neurons. For example, certain hormone-

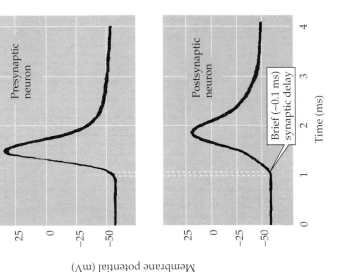

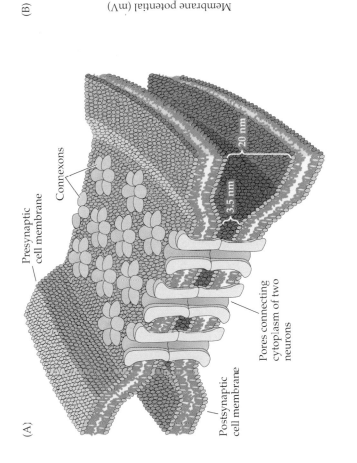

(A)

Presynaptic cell membrane

Connexons

Pores connecting cytoplasm of two neurons

Postsynaptic cell membrane

3.5 nm

20 nm

(B)

Membrane potential (mV)

Presynaptic neuron

Postsynaptic neuron

Brief (~0.1 ms) synaptic delay

Time (ms)

Figure 5.2 Structure and function of gap junctions at electrical synapses. (A) Gap junctions consist of hexameric complexes formed by the coming together of subunits called connexons, which are present in both the pre- and postsynaptic membranes. The pores of the channels connect to one another, creating electrical continuity between the two cells. (B) Rapid transmission of signals at an electrical synapse in the crayfish. An action potential in the presynaptic neuron causes the postsynaptic neuron to be depolarized within a fraction of a millisecond. (B after Furshpan and Potter, 1959.)

secreting neurons within the mammalian hypothalamus are connected by electrical synapses. This arrangement ensures that all cells fire action potentials at about the same time, thus facilitating a burst of hormone secretion into the circulation. The fact that gap junction pores are large enough to allow molecules such as ATP and second messengers to diffuse intercellularly also permits electrical synapses to coordinate the intracellular signaling and metabolism of coupled neurons.

Chemical Synapses

The general structure of a chemical synapse is shown schematically in Figure 5.1B. The space between the pre- and postsynaptic neurons is substantially greater at chemical synapses than at electrical synapses and is called the **synaptic cleft**. However, the key feature of all chemical synapses is the presence of small, membrane-bounded organelles called **synaptic vesicles** within the presynaptic terminal. These spherical organelles are filled with one or more **neurotransmitters**, the chemical signals secreted from the presynaptic neuron, and it is these chemical agents acting as messengers between the communicating neurons that gives this type of synapse its name. There are many kinds of neurotransmitters (see Chapter 6), the best studied example being acetylcholine, the transmitter employed at peripheral neuromuscular synapses, in autonomic ganglia, and at some central synapses.

Transmission at chemical synapses is based on the elaborate sequence of events depicted in Figure 5.3. The process is initiated when an action potential invades the terminal of the presynaptic neuron. The change in membrane potential caused by the arrival of the action potential leads to the opening of voltage-gated calcium channels in the presynaptic membrane. Because of the steep concentration gradient of Ca^{2+} across the presynaptic membrane (the external Ca^{2+} concentration is approximately 10^{-3} M, whereas the internal Ca^{2+} concentration is approximately 10^{-7} M), the opening of these channels causes a rapid influx of Ca^{2+} into the presynaptic terminal,

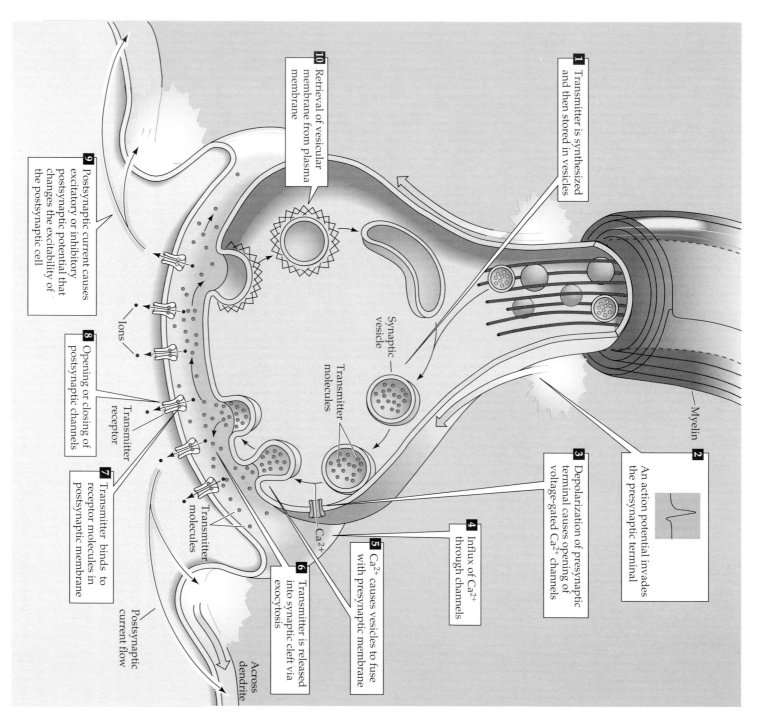

Figure 5.3   Sequence of events involved in transmission at a typical chemical synapse.

1 Transmitter is synthesized and then stored in vesicles

2 An action potential invades the presynaptic terminal

—Myelin

3 Depolarization of presynaptic terminal causes opening of voltage-gated Ca^{2+} channels

4 Influx of Ca^{2+} through channels

5 Ca^{2+} causes vesicles to fuse with presynaptic membrane

6 Transmitter is released into synaptic cleft via exocytosis

7 Transmitter binds to receptor molecules in postsynaptic membrane

8 Opening or closing of postsynaptic channels

9 Postsynaptic current causes excitatory or inhibitory postsynaptic potential that changes the excitability of the postsynaptic cell

10 Retrieval of vesicular membrane from plasma membrane

Synaptic vesicle

Transmitter molecules

Ca^{2+}

Ions

Transmitter receptor

Transmitter molecules

Postsynaptic current flow

Across dendrite

with the result that the Ca^{2+} concentration of the cytoplasm in the terminal transiently rises to a much higher value. Elevation of the presynaptic Ca^{2+} concentration, in turn, allows synaptic vesicles to fuse with the plasma membrane of the presynaptic neuron. The Ca^{2+}-dependent fusion of synaptic vesicles with the terminal membrane causes their contents, most importantly neurotransmitters, to be released into the synaptic cleft.

Following exocytosis, transmitters diffuse across the synaptic cleft and bind to specific receptors on the membrane of the postsynaptic neuron (see Chapter 7). The binding of neurotransmitter to the receptors causes channels in the postsynaptic membrane to open (or sometimes to close), thus changing the ability of ions to flow into (or out of) the postsynaptic cells. The resulting neurotransmitter-induced current flow alters the conductance and usually the membrane potential of the postsynaptic neuron, increasing or decreasing the probability that the neuron will fire an action potential. In this way, information is transmitted from one neuron to another.

Quantal Transmission at Neuromuscular Synapses

Much of the evidence leading to the present understanding of chemical synaptic transmission was obtained from experiments examining the release of acetylcholine at neuromuscular junctions. These synapses between spinal motor neurons and skeletal muscle cells are simple, large, and peripherally located, making them particularly amenable to experimental analysis. Such synapses occur at specializations called **end plates** because of the saucerlike appearance of the site on the muscle fiber where the presynaptic axon elaborates its terminals (Figure 5.4A). Most of the pioneering work on neuromuscular transmission was performed by Bernard Katz and his collaborators at University College London during the 1950s and 1960s, and Katz has been widely recognized for his remarkable contributions to understanding synaptic transmission. Though he worked primarily on the frog neuromuscular junction, numerous subsequent experiments have confirmed the applicability of data about events at this particular synapse to transmission at chemical synapses throughout the nervous system.

When an intracellular microelectrode is used to record the membrane potential of a muscle cell, an action potential in the presynaptic motor neuron can be seen to elicit a transient depolarization of the postsynaptic muscle fiber. This change in membrane potential, called an **end plate potential** (**EPP**), is normally large enough to bring the membrane potential of the muscle cell well above the threshold for producing a postsynaptic action potential (Figure 5.4B). The postsynaptic action potential triggered by the EPP causes the muscle fiber to contract.

One of Katz's seminal findings, in studies carried out with Paul Fatt in 1951, was that spontaneous changes in muscle cell membrane potential occur even in the absence of stimulation of the presynaptic motor neuron (Figure 5.4C). These changes have the same shape as EPPs but are much smaller (typically less than 1 mV in amplitude, compared to an EPP of perhaps 40 or 50 mV). Both EPPs and these small, spontaneous events are sensitive to pharmacological agents that block postsynaptic acetylcholine

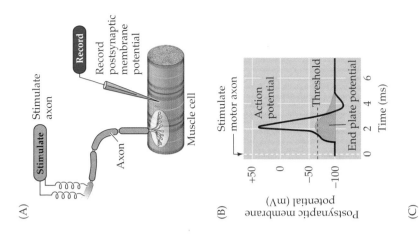

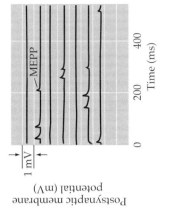

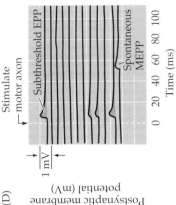

Figure 5.4 Synaptic transmission at the neuromuscular junction. (A) Experimental arrangement, typically using the muscle of a frog or rat. The axon of the motor neuron innervating the muscle fiber is stimulated with an extracellular electrode, while an intracellular microelectrode is inserted into the postsynaptic muscle cell to record its electrical responses. (B) End plate potentials (EPPs) evoked by stimulation of a motor neuron are normally above threshold and therefore produce an action potential in the postsynaptic muscle cell. (C) Spontaneous miniature EPPs (MEPPs) occur in the absence of presynaptic stimulation. (D) When the neuromuscular junction is bathed in a solution that has a low concentration of Ca^{2+}, stimulating the motor neuron evokes EPPs whose amplitudes are reduced to about the size of MEPPs. (After Fatt and Katz, 1952.)

receptors, such as curare (see Box A in Chapter 7). These and other parallels between EPPs and the spontaneously occurring depolarizations led Katz and his colleagues to call these spontaneous events **miniature end plate potentials**, or **MEPPs**.

The relationship between the full-blown end plate potential and MEPPs was clarified by careful analysis of the EPPs. The magnitude of the EPP provides a convenient electrical assay of neurotransmitter secretion from a motor neuron terminal; however, measuring it is complicated by the need to prevent muscle contraction from dislodging the microelectrode. The usual means of eliminating muscle contractions is either to lower Ca^{2+} concentration in the extracellular medium or to partially block the postsynaptic transmitter receptors with the drug curare. As expected from the scheme illustrated in Figure 5.3, lowering the Ca^{2+} concentration reduces neurotransmitter secretion, thus reducing the magnitude of the EPP below the threshold for postsynaptic action potential production and allowing it to be measured more precisely. Under such conditions, stimulation of the motor neuron produces very small EPPs that fluctuate in amplitude from trial to

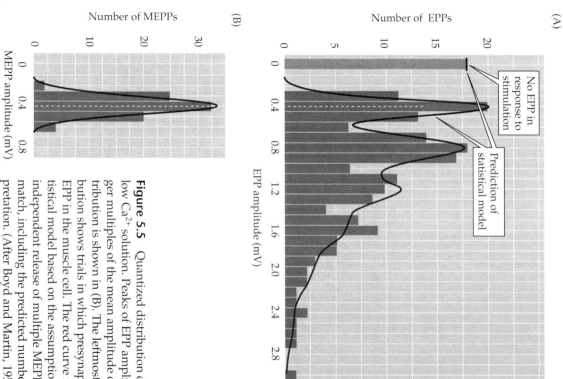

Figure 5.5 Quantized distribution of EPP amplitudes evoked in a low Ca^{2+} solution. Peaks of EPP amplitudes (A) tend to occur in integer multiples of the mean amplitude of MEPPs, whose amplitude distribution is shown in (B). The leftmost bar in the EPP amplitude distribution shows trials in which presynaptic stimulation failed to elicit an EPP in the muscle cell. The red curve indicates the prediction of a statistical model based on the assumption that the EPPs result from the independent release of multiple MEPP-like quanta. The observed match, including the predicted number of failures, supports this interpretation. (After Boyd and Martin, 1955.)

(A)

Number of EPPs

No EPP in response to stimulation

Prediction of statistical model

EPP amplitude (mV)

(B)

Number of MEPPs

MEPP amplitude (mV)

trial (Figure 5.4D). These fluctuations give considerable insight into the mechanisms responsible for neurotransmitter release. In particular, the evoked response in low Ca^{2+} is now known to result from the release of unit amounts of neurotransmitter by the presynaptic nerve terminal. Indeed, the amplitude of the smallest evoked response is strikingly similar to the size of single MEPPs (compare Figure 5.4C and D). Further supporting this similarity, increments in the EPP response (Figure 5.5A) occur in units about the size of single MEPPs (Figure 5.5B). These "quantal" fluctuations in the amplitude of EPPs indicated to Katz and colleagues that EPPs are made up of individual units, each equivalent to a MEPP.

The idea that EPPs represent the simultaneous release of many MEPP-like units can be tested statistically. A method of statistical analysis based on the independent occurrence of unitary events (called Poisson statistics) predicts what the distribution of EPP amplitudes would look like during a large number of trials of motor neuron stimulation, under the assumption that EPPs are built up from unitary events like MEPPs (see Figure 5.5A). The distribution of EPP amplitudes determined experimentally was found to be just that expected if transmitter release from the motor neuron is indeed quantal (the red curve in Figure 5.5A). Such analyses confirmed the idea that release of acetylcholine does indeed occur in discrete packets, each equivalent to a MEPP. In short, a presynaptic action potential causes a postsynaptic EPP because it synchronizes the release of many transmitter quanta.

Release of Transmitters from Synaptic Vesicles

The discovery of the quantal release of packets of neurotransmitter immediately raised the question of how such quanta are formed and discharged into the synaptic cleft. At about the time Katz and his colleagues were using physiological methods to discover quantal release of neurotransmitter, electron microscopy revealed, for the first time, the presence of synaptic vesicles in presynaptic terminals. Putting these two discoveries together, Katz and others proposed that synaptic vesicles loaded with transmitter are the source of the quanta. Subsequent biochemical studies confirmed that synaptic vesicles are the repositories of transmitters. These studies have shown that acetylcholine is highly concentrated in the synaptic vesicles of motor neurons, where it is present at a concentration of about 100 mM. Given the diameter of a synaptic vesicle (~50 nm), approximately 10,000 molecules of neurotransmitter are contained in a single vesicle. This number corresponds quite nicely to the amount of acetylcholine that must be applied to a neuromuscular junction to mimic a MEPP, providing further support for the idea that quanta arise from discharge of the contents of single synaptic vesicles.

As noted, synaptic vesicles secrete their contents into the synaptic cleft by fusing with the plasma membrane of the presynaptic terminal. To prove that fusion actually occurs, it is necessary to show that each fused vesicle causes a single quantal event to be recorded postsynaptically. This challenge was met in the late 1970s, when John Heuser, Tom Reese, and colleagues correlated measurements of vesicle fusion with the quantal content of EPPs at the neuromuscular junction. In their experiments, the number of vesicles that fused with the presynaptic plasma membrane was measured by electron microscopy in terminals that had been treated with a drug (4-aminopyridine, or 4-AP) that enhances the number of vesicle fusion events produced by single action potentials (Figure 5.6A). Parallel electrical measurements were made of the quantal content of the EPPs elicited in this way. A comparison of the number of synaptic vesicle fusions observed with the electron

(A)

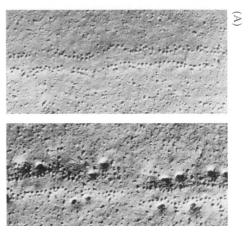

(B)

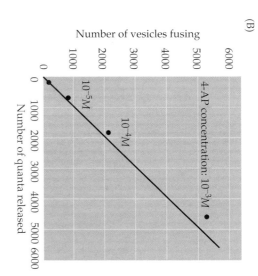

Figure 5.6 Relationship of synaptic vesicle exocytosis and quantal transmitter release. (A) A special electron microscopical technique called freeze-fracture microscopy was used to visualize the fusion of synaptic vesicles in presynaptic terminals of frog motor neurons. Left: Image of the plasma membrane of an unstimulated presynaptic terminal. Right: Image of the plasma membrane of a terminal stimulated by an action potential. Stimulation causes the appearance of dimple-like structures that represent the fusion of synaptic vesicles with the presynaptic membrane. The view is as if looking down on the release sites from outside the presynaptic terminal. (B) Comparison of the number of observed vesicle fusions to the number of quanta released by a presynaptic action potential. Transmitter release was varied by using a drug (4-AP) that affects the duration of the presynaptic action potential, thus changing the amount of calcium that enters during the action potential. The diagonal line is the 1:1 relationship expected if each vesicle that opened released a single quantum of transmitter. (From Heuser et al., 1979.)

microscope and the number of quanta released at the synapse showed a good correlation between these two measures (Figure 5.6B). These results remain one of the strongest lines of support for the idea that a quantum of transmitter release is due to a synaptic vesicle fusing with the presynaptic membrane. More recent evidence, based on other means of measuring vesicle fusion, has left no doubt about the validity of this general interpretation of chemical synaptic transmission.

Local Recycling of Synaptic Vesicles

The fusion of synaptic vesicles causes new membrane to be added to the plasma membrane of the presynaptic terminal, but the addition is not permanent. Although a bout of exocytosis can dramatically increase the surface area of presynaptic terminals, this extra membrane is removed within a few minutes. Heuser and Reese performed another important set of experiments showing that the fused vesicle membrane is actually retrieved and taken back into the cytoplasm of the nerve terminal (a process called endocytosis). The experiments, again carried out at the frog neuromuscular junction, were based on filling the synaptic cleft with horseradish peroxidase (HRP), an enzyme that can be made to produce a dense reaction product that is visible in an electron microscope. Under appropriate experimental conditions, endocytosis could then be visualized by the uptake of HRP into the nerve terminal (Figure 5.7). To activate endocytosis, the presynaptic terminal was stimulated with a train of action potentials, and the subsequent fate of the HRP was followed by electron microscopy. Immediately following stimulation, the HRP was found within special endocytotic organelles called coated vesicles (Figure 5.7A,B). A few minutes later, however, the coated vesicles had disappeared and the HRP was found in a different organelle, the endosome (Figure 5.7C). Finally, approximately an hour after stimulating the terminal, the HRP reaction product appeared inside synaptic vesicles (Figure 5.7D).

These observations indicate that synaptic vesicle membrane is recycled within the presynaptic terminal via the sequence summarized in Figure 5.7E. In this process, called the **synaptic vesicle cycle**, the retrieved vesicular membrane passes through diverse intracellular compartments (coated vesicles and endosomes) and is eventually used to make new synaptic vesicles. The newly made vesicles are refilled with neurotransmitter, docked at the presynaptic plasma membrane, and primed to participate in exocytosis once again.

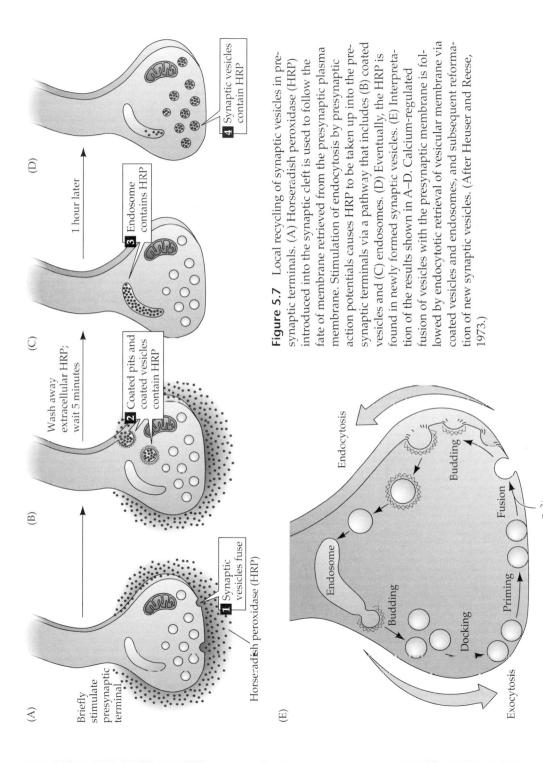

Figure 5.7 Local recycling of synaptic vesicles in presynaptic terminals. (A) Horseradish peroxidase (HRP) introduced into the synaptic cleft is used to follow the fate of membrane retrieved from the presynaptic plasma membrane. Stimulation of endocytosis by presynaptic action potentials causes HRP to be taken up into the presynaptic terminals via a pathway that includes (B) coated vesicles and (C) endosomes. (D) Eventually, the HRP is found in newly formed synaptic vesicles. (E) Interpretation of the results shown in A–D. Calcium-regulated fusion of vesicles with the presynaptic membrane is followed by endocytotic retrieval of vesicular membrane via coated vesicles and endosomes, and subsequent reformation of new synaptic vesicles. (After Heuser and Reese, 1973.)

The precursors to synaptic vesicles *originally* are produced in the endoplasmic reticulum and Golgi apparatus in the neuronal cell body. Because of the long distance between the cell body and the presynaptic terminal in most neurons, transport of vesicles from the soma would not permit rapid replenishment of synaptic vesicles during continuous neural activity. Thus, local recycling is well suited to the peculiar anatomy of neurons, giving nerve terminals the means to provide a continual supply of synaptic vesicles. As might be expected, defects in synaptic vesicle recycling can cause severe neurological disorders, some of which are described in Box A.

The Role of Calcium in Transmitter Secretion

As was apparent in the experiments of Katz and others described in the preceding sections, lowering the concentration of Ca^{2+} outside a presynaptic motor nerve terminal reduces the size of the EPP (compare Figure 5.4B and D). Moreover, measurement of the number of transmitter quanta released under such conditions shows that the reason the EPP gets smaller is that lowering Ca^{2+} concentration decreases the number of vesicles that fuse with the plasma membrane of the terminal. An important insight into *how* Ca^{2+} regulates the fusion of synaptic vesicles was the discovery that presynaptic terminals have voltage-sensitive Ca^{2+} channels in their plasma membranes (see Chapter 4).

BOX A
Diseases That Affect the Presynaptic Terminal

Various steps in the exocytosis and endocytosis of synaptic vesicles are targets of a number of rare but debilitating neurological diseases. Many of these are myasthenic syndromes, in which abnormal transmission at neuromuscular synapses leads to weakness and fatigability of skeletal muscles (see Box B in Chapter 7). One of the best-understood examples of such disorders is the Lambert-Eaton myasthenic syndrome (LEMS), a frequent complication in patients with certain kinds of cancers. Biopsies of muscle tissue removed from LEMS patients allow intracellular recordings identical to those shown in Figure 5.4. Such recordings have shown that when a motor neuron is stimulated, the number of quanta contained in individual EPP is greatly reduced, although the amplitude of spontaneous MEPPs is normal. Thus, LEMS impairs evoked neurotransmitter release, but does not affect the size of individual quanta.

Several lines of evidence indicate that this reduction in neurotransmitter release is due to a loss of voltage-gated Ca^{2+} channels in the presynaptic terminal of motor neurons (see figure). Thus, the defect in neuromuscular transmission can be overcome by increasing the extracellular concentration of Ca^{2+}, and anatomical studies indicate a lower density of Ca^{2+} channel proteins in the presynaptic plasma membrane. The loss of presynaptic Ca^{2+} channels in LEMS apparently arises from a defect in the immune system. The blood of LEMS patients has a very high concentration of antibodies that bind to Ca^{2+} channels, and it seems likely that these antibodies are the primary cause of LEMS. For example, removal of Ca^{2+} channel antibodies from the blood of LEMS patients by plasma exchange reduces muscle weakness. Similarly, immunosuppressant drugs also can alleviate LEMS symptoms. Perhaps most telling, injecting these anti-

bodies into experimental animals elicits muscle weakness and abnormal neuromuscular transmission. Why the immune system generates antibodies against Ca^{2+} channels is not clear. Most LEMS patients have small-cell carcinoma, a form of lung cancer that may somehow initiate the immune response to Ca^{2+} channels. Whatever the origin, the binding of antibodies to Ca^{2+} channels causes a reduction in Ca^{2+} channel currents. It is this antibody-induced defect in presynaptic Ca^{2+} entry that accounts for the muscle weakness associated with LEMS.

Congenital myasthenic syndromes are genetic disorders that also cause muscle weakness by affecting neuromuscular transmission. Some of these syndromes affect the acetylcholinesterase that degrades acetylcholine in the synaptic cleft, whereas others arise from autoimmune attack of acetylcholine receptors (see Box B in Chapter 7). However, a number of congenital myasthenic syndromes arise from defects in acetylcholine release due to altered synaptic vesicle traffic within the motor neuron terminal. Neuromuscular synapses in some of these patients have EPPs with reduced quantal content, a deficit that is especially prominent when the synapse is activated repeatedly. Electron microscopy shows that presynaptic motor nerve terminals have a greatly reduced number of synaptic vesicles. The defect in neurotransmitter release evidently results from an inadequate number of synaptic vesicles available for release during sustained presynaptic activity. The origins of this shortage of synaptic vesicles is not clear, but could result either from an impairment in endocytosis in the nerve terminal (see figure) or from a reduced supply of vesicles from the motor neuron cell body.

Still other patients suffering from familial infantile myasthenia appear to have neuromuscular weakness that arises from reductions in the size of individual

quanta, rather than the number of quanta released. Motor nerve terminals from these patients have synaptic vesicles that are normal in number, but smaller in diameter. This finding suggests a different type of genetic lesion that somehow alters formation of new synaptic vesicles following endocytosis, thereby leading to less acetylcholine in each vesicle.

Another disorder of synaptic transmitter release results from poisoning by anaerobic *Clostridium* bacteria. This genus of microorganisms produces some of the most potent toxins known, including several botulinum toxins and tetanus toxin. Both botulinum and tetanus are potentially deadly disorders.

Botulism can occur by consuming food containing *Clostridium* bacteria or by infection of wounds with the spores of

Clostridium. In either case, the presence of the bacteria can cause paralysis of skeletal muscles and the syndrome of botulism. Botulinum and tetanus toxins both act by blocking the exocytosis of synaptic vesicles. Both block neurotransmitter release by cleaving the SNARE proteins involved in fusion of synaptic vesicles with the presynaptic plasma membrane (see figure).

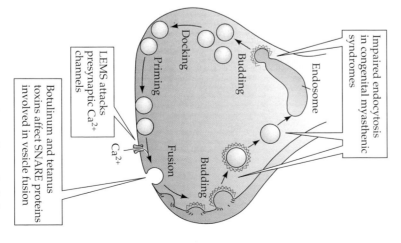

Impaired endocytosis in congenital myasthenic syndromes

Endosome

LEMS attacks presynaptic Ca^{2+} channels

Budding

Docking

Priming

Fusion

Budding

Ca^{2+}

Botulinum and tetanus toxins affect SNARE proteins involved in vesicle fusion

Presynaptic targets of several neurological disorders.

these ubiquitous organisms. In either case, the presence of the toxin can cause paralysis of peripheral neuromuscular synapses due to abolition of neurotransmitter release. This interference with neuromuscular transmission causes skeletal muscle weakness, in extreme cases producing respiratory failure due to paralysis of the diaphragm and other muscles required for breathing. Botulinum toxins also block synapses innervating the smooth muscles of several organs, giving rise to visceral motor dysfunction.

Tetanus typically results from the contamination of puncture wounds by *Clostridium* bacteria that produce tetanus toxin. In contrast to botulism, tetanus poisoning blocks the release of inhibitory transmitters from interneurons in the spinal cord. This effect causes a loss of

synaptic inhibition on spinal motor neurons, producing hyperexcitation of skeletal muscle and tetanic contractions in affected muscles (hence the name of the disease).

Although their clinical consequences are dramatically different, clostridial toxins have a common mechanism of action (see figure). Tetanus toxin and botulinum toxins work by cleaving the SNARE proteins involved in fusion of synaptic vesicles with the presynaptic plasma membrane (see Box B). This proteolytic action presumably accounts for the block of transmitter release at the afflicted synapses. The different actions of these toxins on synaptic transmission at excitatory motor versus inhibitory synapses apparently results from the fact that these toxins are taken up by different types of

neurons: Whereas the botulinum toxins are taken up by motor neurons, tetanus toxin is preferentially targeted to interneurons. The basis for this differential uptake of toxins is not known, but presumably arises from the presence of different types of toxin receptors on the two types of neurons.

References

ENGEL, A. G. (1991) Review of evidence for loss of motor nerve terminal calcium channels in Lambert-Eaton myasthenic syndrome. Ann. N.Y. Acad. Sci. 635: 246–258.

ENGEL, A. G. (1994) Congenital myasthenic syndromes. Neurol. Clin. 12: 401–437.

LANG, B. AND 13 OTHERS (1998) The role of autoantibodies in Lambert-Eaton myasthenic syndrome. Ann. N.Y. Acad. Sci. 841: 596–605.

MASELL, R. A. (1998) Pathogenesis of human botulism. Ann. N.Y. Acad. Sci. 841: 122–139.

The first indication of presynaptic Ca^{2+} channels was provided by Katz and Ricardo Miledi. They observed that presynaptic terminals treated with tetrodotoxin (which blocks Na^+ channels; see Chapter 7) could still produce a peculiarly prolonged type of action potential. The explanation for this surprising finding was that current was still flowing through Ca^{2+} channels, substituting for the current ordinarily carried by Na^+ channels. Subsequent voltage clamp experiments, performed by Rodolfo Llinás and others at a giant presynaptic terminal of the squid (Figure 5.8A), confirmed the presence of voltage-gated Ca^{2+} channels in the presynaptic terminal (Figure 5.8B). Such experiments showed that the amount of neurotransmitter released is very sensitive to the exact amount of Ca^{2+} that enters. Further, blockade of these Ca^{2+} channels with drugs also inhibits transmitter release (Figure 5.8B, right). That Ca^{2+} entry into presynaptic terminals causes a rise in the concentration of Ca^{2+} within the terminal has also been documented by microscopic imaging of terminals filled with Ca^{2+}-sensitive fluorescent dyes (Figure 5.9A). These observations all confirm that the voltage-gated Ca^{2+} channels are directly involved in neurotransmission. Thus, presynaptic action potentials open voltage-gated Ca^{2+} channels, with a resulting influx of Ca^{2+}.

The consequences of the rise in presynaptic Ca^{2+} concentration for neurotransmitter release has been directly shown in two ways. First, microinjection of Ca^{2+} into presynaptic terminals triggers transmitter release in the absence of presynaptic action potentials (Figure 5.9B). Second, presynaptic microinjection of Ca^{2+} chelators (chemicals that bind Ca^{2+} and keep its concentration buffered at low levels) prevents presynaptic action potentials from causing transmitter secretion (Figure 5.9C). These results prove beyond any doubt that a rise in presynaptic Ca^{2+} concentration is both necessary and sufficient for neurotransmitter release.

(A)

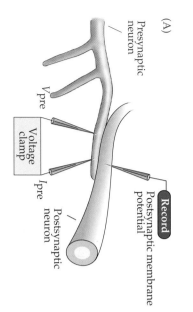

Presynaptic neuron

V pre

Voltage clamp

I pre

Record

Postsynaptic neuron

Postsynaptic membrane potential

(B)

CONTROL

CADMIUM ADDED

Presynaptic membrane potential (mV)
0
−25
−50
−75

Presynaptic calcium current (μA/cm²)
200
0

Postsynaptic membrane potential (mV)
0
−25
−50

Time (ms)
−3 0 3 6 9 12

−3 0 3 6 9 12

Figure 5.8 The entry of Ca²⁺ through the specific voltage-dependent calcium channels in the presynaptic terminals causes transmitter release. (A) Experimental setup using an extraordinarily large synapse in the squid. The voltage clamp method detects currents flowing across the presynaptic membrane when the membrane potential is depolarized. (B) Pharmacological agents that block currents flowing through Na⁺ and K⁺ channels reveal a remaining inward current flowing through Ca²⁺ channels. This influx of calcium triggers transmitter secretion, as indicated by a change in the postsynaptic membrane potential. Treatment of the same presynaptic terminal with cadmium, a calcium channel blocker, eliminates both the presynaptic calcium current and the postsynaptic response. (After Augustine and Eckert, 1984.)

Figure 5.9 Further evidence that calcium entry triggers transmitter release from presynaptic terminals. (A) Fluorescence microscopy measurements of presynaptic Ca²⁺ concentration at the squid giant synapse (see Figure 5.8A). A train of presynaptic action potentials causes a rise in Ca²⁺ concentration, as revealed by a rise in Ca²⁺ concentration, as revealed by a dye (called fura-2) that fluoresces more strongly when the Ca²⁺ concentration increases. (B) Microinjection of Ca²⁺ into a squid giant presynaptic terminal triggers transmitter release, measured as a depolarization of the postsynaptic membrane potential. (C) Microinjection of BAPTA, a Ca²⁺ chelator, into a squid giant presynaptic terminal prevents transmitter release. (A from Smith et al., 1993; B after Miledi, 1971; C after Adler et al., 1991.)

(A)

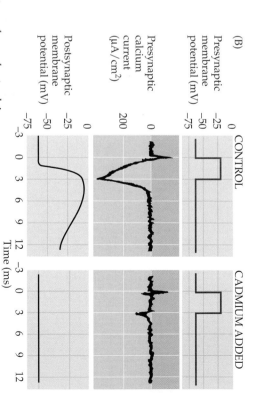

250 μm

Ca²⁺

(B)

Postsynaptic membrane potential (mV)
−64
−65

Ca²⁺ injection

Time (s)
0 1 2 3 4

(C)

Presynaptic membrane potential (mV)
25
0
−25
−50
−75

Postsynaptic membrane potential (mV)
25
0
−25
−50
−75

CONTROL

INJECT Ca²⁺ BUFFER

Time (ms)
0 1 2 3 4 5

0 1 2 3 4 5

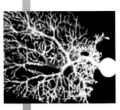

Chapter 6

Neurotransmitters

Overview

For the most part, neurons in the human brain communicate with one another by releasing chemical messengers called neurotransmitters. All neurotransmitter molecules undergo a similar cycle of use involving (1) synthesis and packaging into vesicles in the presynaptic cell; (2) release from the presynaptic cell and binding to receptors on one or more postsynaptic cells; and (3) rapid removal and/or degradation. The total number of neurotransmitters is not known, but is well over 100. Despite this diversity, these agents can be classified into two broad categories: small-molecule neurotransmitters and neuropeptides. In general, small-molecule neurotransmitters mediate rapid synaptic actions, whereas neuropeptides tend to modulate slower, ongoing synaptic functions. Abnormalities of neurotransmitter function contribute to a wide range of neurological and psychiatric disorders. As a result, altering aspects of neurotransmitter release, binding, and reuptake or removal by pharmacological or other means is central to many therapeutic strategies.

What Defines a Neurotransmitter?

As briefly described in the preceding chapter, neurotransmitters are chemical signals released from presynaptic nerve terminals into the synaptic cleft. The subsequent binding of neurotransmitters to specific receptors on postsynaptic neurons (or other classes of target cells) transiently changes the electrical properties of the target cells, leading to an enormous variety of postsynaptic effects (see Chapters 7 and 8).

The notion that electrical information can be transferred from one neuron to the next by means of chemical signaling was the subject of intense debate through the first half of the twentieth century. A key experiment that supported this idea was performed in 1926 by German physiologist Otto Loewi. Acting on an idea that allegedly came to him in the middle of the night, Loewi proved that electrical stimulation of the vagus nerve slows the heartbeat by releasing a chemical signal. He isolated and perfused the hearts of two frogs, monitoring the rates at which they were beating (Figure 6.1). The gist of his experiment was to collect the perfusate flowing through the stimulated heart and transfer it to the second heart. Even though the second heart had not been stimulated, its beat also slowed, showing that the vagus nerve regulates the heart rate by releasing a chemical that accumulates in the perfusate. Originally referred to as "vagus substance," the agent was later shown to be **acetylcholine (ACh)**, which over the years has become the most thoroughly studied neurotransmitter. ACh acts not only in the heart but at a variety of postsynaptic targets in the central and peripheral nervous systems,

(A)

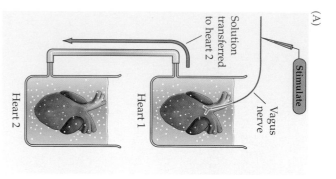

(B)

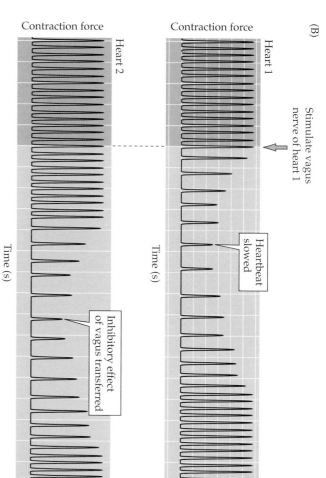

Figure 6.1 Loewi's experiment demonstrating chemical neurotransmission. (A) Diagram of experimental setup. (B) Where the vagus nerve of an isolated frog's heart was stimulated, the heart rate decreased (upper panel). If the perfusion fluid from the stimulated heart was transferred to a second heart, its rate decreased as well (lower panel).

preeminently at the neuromuscular junction of striated muscles and in the visceral motor system (see Chapters 5 and 21).

Over the years, a number of formal criteria have emerged that definitively identify a substance as a neurotransmitter (Box A). Nonetheless, identifying the neurotransmitters active at any particular synapse remains a difficult undertaking, and for many synapses (particularly in the brain), the nature of the neurotransmitter is not well established. Substances that have not met all the criteria outlined in Box A are referred to as "putative" neurotransmitters.

The distinctive characteristics of neurotransmitters, compared to other signaling molecules, are made clearer by comparison with the actions of the hormones secreted by the endocrine system. Hormones typically influence target cells far removed from the hormone-secreting cell (see Chapter 8). This "action at a distance" is achieved by the release of hormones into the bloodstream. In contrast, the distance over which neurotransmitters act is miniscule. At many synapses, transmitters bind only to receptors on the postsynaptic cell that directly underlies the presynaptic terminal (Figure 6.2A); in such cases, the transmitter acts over distances less than a micrometer. Even when neurotransmitters diffuse locally to alter the electrical properties of multiple postsynaptic (and sometimes presynaptic) cells in the vicinity (Figure 6.2B), they act only over distances of tens to hundreds of micrometers. While the elongated axonal processes of neurons allow neurotransmitters to be released as much as a meter away from the neuronal cell body, these transmitters still act only near the presynaptic site of release (Figure 6.2C).

While the distinction between neurotransmitters and hormones is generally clear-cut, a substance can act as a neurotransmitter in one region of the brain while serving as a hormone elsewhere. For example, vasopressin and

BOX A
Criteria That Define a Neurotransmitter

Three primary criteria have been used over the years to confirm that a molecule acts as a neurotransmitter at a given chemical synapse.

1. *The substance must be present within the presynaptic neuron.* Clearly, a chemical cannot be secreted from a presynaptic neuron unless it is present there. Because elaborate biochemical pathways are required to produce neurotransmitters, showing that the enzymes and precursors required to synthesize the substance are present in presynaptic neurons provides additional evidence that the substance is used as a transmitter. Note, however, that since the transmitters glutamate, glycine, and aspartate are also needed for protein synthesis and other metabolic reactions in all neurons, their presence is *not* sufficient evidence to establish them as neurotransmitters.

2. *The substance must be released in response to presynaptic depolarization, and*

the release must be Ca^{2+}-dependent. Another essential criterion for identifying a neurotransmitter is to demonstrate that it is released from the presynaptic neuron in response to presynaptic electrical activity, and that this release requires Ca^{2+} influx into the presynaptic terminal. Meeting this criterion is technically challenging, not only because it may be difficult to selectively stimulate the presynaptic neurons, but also because enzymes and transporters efficiently remove the secreted neurotransmitters.

3. *Specific receptors for the substance must be present on the postsynaptic cell.* A neurotransmitter cannot act on its target unless specific receptors for the transmitter are present in the postsynaptic membrane. One way to demonstrate that application of exogenous transmitter mimics the post-

synaptic effect of presynaptic stimulation. A more rigorous demonstration is to show that agonists and antagonists that alter the normal postsynaptic response have the same effect when the substance in question is applied exogenously. High-resolution histological methods can also be used to show that specific receptors are present in the postsynaptic membrane (by detection of radioactively labeled receptor antibodies, for example).

Fulfilling these criteria establishes unambiguously that a substance is used as a transmitter at a given synapse. Practical difficulties, however, have prevented these standards from being applied at many types of synapses. It is for this reason that so many substances must be referred to as "putative" neurotransmitters.

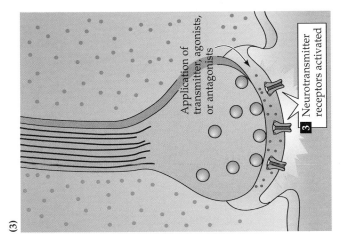

(2)

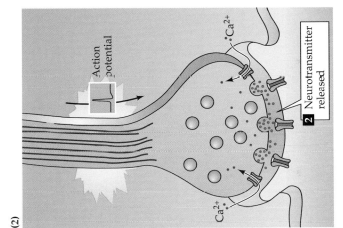

(1)

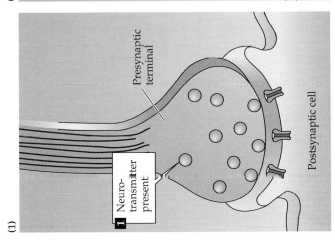

(3)

Demonstrating the identity of a neurotransmitter at a synapse requires showing (1) its presence, (2) its release, and (3) the postsynaptic presence of specific receptors.

(A)

(B)

(C)

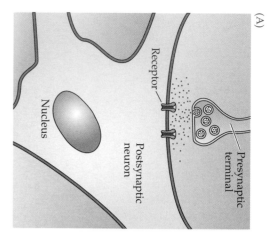

Receptor

Nucleus

Postsynaptic
neuron

Presynaptic
terminal

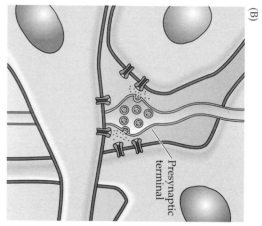

Presynaptic
terminal

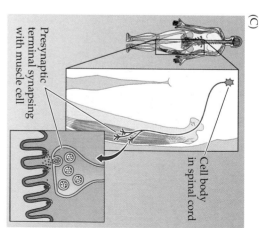

Presynaptic
terminal synapsing
with muscle cell

Cell body
in spinal cord

Figure 6.2   Localization of neurotransmitter action. Neurotransmitters in
general act either locally (A), by altering
the electrical excitability of a small
region of a single postsynaptic cell, or
more diffusely (B), by altering the electrical excitability of a few postsynaptic
cells. (C) Neurons can exert their
actions over greater distances by having
long axons that locally release neurotransmitters onto distant targets.

oxytocin, two peptide hormones that are released into the circulation from
the posterior pituitary, also function as neurotransmitters at a number of
central synapses. A number of other peptides also serve as both hormones
and neurotransmitters.

Two Major Categories of Neurotransmitters

By the 1950s, the list of neurotransmitters (defined by the criteria described
in Box A) had expanded to include four amines—epinephrine, norepinephrine, dopamine, and serotonin—in addition to ACh. Over the following
decade, three amino acids—glutamate, γ-aminobutyric acid (GABA), and
glycine—were also shown to be neurotransmitters. Subsequently, other small
molecules were added to the list, and considerable evidence now suggests that
histamine, aspartate, and ATP should be included (Figure 6.3). The most
recent class of molecules discovered to be transmitters are a large number of
polypeptides; since the 1970s, more than 100 such molecules have been
shown to meet at least some of the criteria outlined in Box A.

For purposes of discussion, it is useful to separate this variety of agents
into two broad categories based simply on their size (Figure 6.3). **Neuropeptides** are relatively large transmitter molecules composed of 3 to 36 amino
acids (Figure 6.4). Individual amino acids, such as glutamate and GABA, as
well as the transmitters acetylcholine, serotonin, and histamine, are much
smaller than neuropeptides and have therefore come to be called **small-mol-**

Figure 6.3   Examples of small-molecule and peptide neurotransmitters. Small-
molecule transmitters can be subdivided into acetylcholine, the amino acids,
purines, and biogenic amines. The catecholamines, so named because they all share
the catechol moiety (i.e., a hydroxylated benzene ring), make up a distinctive sub-
group within the biogenic amines. Serotonin and histamine contain an indole ring and
an imidazole ring, respectively. Size differences between the small-molecule neurotransmitters and the peptide neurotransmitters are indicated by the space-filling
models for glycine, norepinephrine, and methionine enkephalin. (Carbon atoms are
black, nitrogen atoms blue, and oxygen atoms red.)

▼

SMALL-MOLECULE NEUROTRANSMITTERS

Acetylcholine $(CH_3)_3\overset{+}{N}-CH_2-CH_2-O-\overset{O}{\underset{\|}{C}}-CH_3$

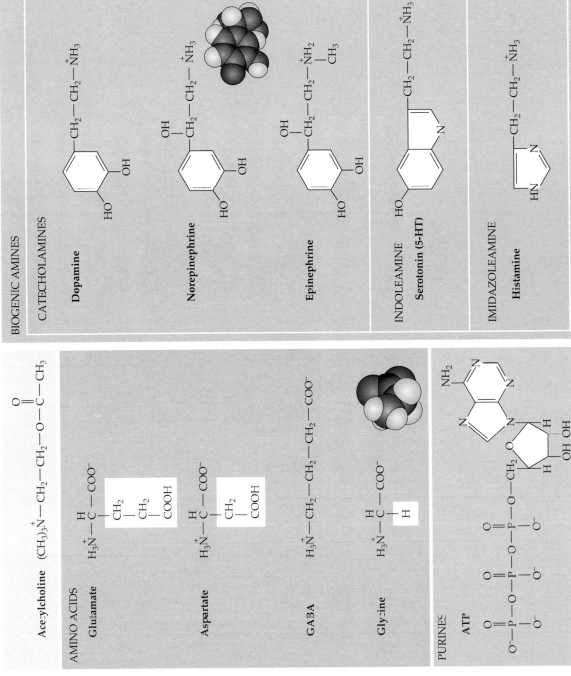

AMINO ACIDS

Glutamate

Aspartate

GABA

Glycine

PURINES

ATP

BIOGENIC AMINES

CATECHOLAMINES

Dopamine

Norepinephrine

Epinephrine

INDOLEAMINE

Serotonin (5-HT)

IMIDAZOLEAMINE

Histamine

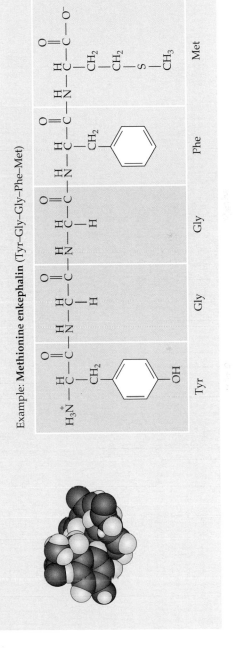

PEPTIDE NEUROTRANSMITTERS (more than 100 peptides, usually 3–30 amino acids long)

Example: **Methionine enkephalin (Tyr–Gly–Gly–Phe–Met)**

Tyr	Gly	Gly	Phe	Met

Methionine enkephalin	Tyr Gly Gly Phe Met
Leucine enkephalin	Tyr Gly Gly Phe Leu

β-Endorphin	Tyr Gly Gly Phe Met Thr Ser Glu Lys Ser Gln Thr Pro Leu Val Thr Leu Phe Lys Asn Ala Ile Val Lys Asn Ala His Lys Gly Gln
α-Endorphin	Tyr Gly Gly Phe Met Thr Ser Glu Lys Ser Gln Thr Pro Leu Val Thr

Amino acid properties
- Hydrophobic
- Polar, uncharged
- Acidic
- Basic

Substance P	Arg Pro Lys Pro Gln Gln Phe Phe Gly Leu Met
Somatostatin-14	Ala Gly Cys Lys Asn Phe Phe Trp Lys Thr Phe Thr Ser Cys
Thyrotropin releasing hormone (TRH)	Glu His Pro
Leutinizing hormone-releasing hormone (LHRH)	Glu His Trp Ser Tyr Gly Leu Arg Pro Gly
Angiotensin-II	Asp Arg Val Tyr Ile His Pro Phe
Vasopressin	Cys Tyr Phe Gln Asn Cys Pro Arg Gly
Oxytocin	Cys Tyr Ile Gln Asn Cys Pro Leu Gly
Cholecystokinin octapeptide (CCK-8)	Asp Tyr Met Gly Trp Met Asp Phe
Vasoactive intestinal peptide (VIP)	His Asp Ala Val Phe Thr Asp Asn Tyr Thr Arg Leu Arg Lys Gln Met Ala Val Lys Lys Tyr Leu Asn Ser Ile Leu Asn
Neuropeptide-Y	Tyr Pro Ser Lys Pro Asp Asn Pro Gly Glu Asp Ala Pro Ala Glu Asp Leu Ala Arg Tyr Tyr Ser Ala Leu Arg His Tyr Ile Asn Leu Ile Thr Arg Gln Arg Tyr
Neurotensin	Glu Leu Tyr Glu Asn Lys Pro Arg Arg Pro Tyr Ile Leu
Bombesin (BBS-14)	Glu Gln Arg Leu Gly Asn Gln Trp Ala Val Gly His Leu Met

Figure 6.4   Neuropeptides vary in length, but usually contain between 3 and 36 amino acids. Note that one peptide can include the sequence of other neuroactive peptides. For example, β-endorphin contains both α-endorphin and methionine enkephalin (olive box).

ecule neurotransmitters. Within the category of small-molecule neurotransmitters, the **biogenic amines** (dopamine, norepinephrine, epinephrine, serotonin, and histamine) are often discussed separately because of their similar chemical properties and postsynaptic actions.

Neurons Often Release More Than One Transmitter

Until relatively recently, it was believed that a given neuron produced only a single type of neurotransmitter. There is now convincing evidence, however, that many types of neurons contain and release two or more different neurotransmitters. There are now numerous examples of different peptides being present in the same terminal, as well as cases in which two small-molecule neurotransmitters are found within the same presynaptic ending, or in which a peptide neurotransmitter is found along with the same presynaptic ending. When more than one transmitter is present within a nerve terminal, the molecules are called **co-transmitters.** Because each class of transmitter tends to be packaged in a separate population of synaptic vesicles, co-transmitters typically are segregated within a presynaptic terminal (there are, however, instances in which two or more co-transmitters are present in the same synaptic vesicle).

Having more than one transmitter lends considerable versatility to synaptic transmission. In particular, since co-transmitters are often packaged in different types of vesicles, the transmitters need not be released simultaneously. For

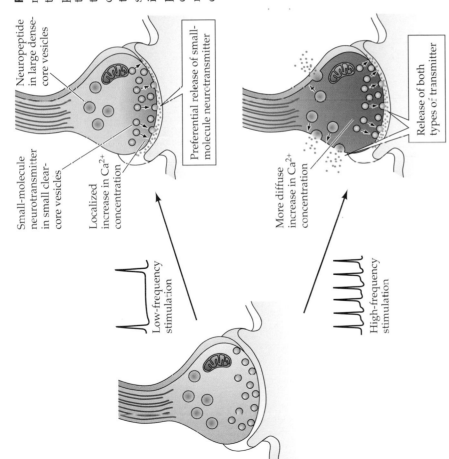

Figure 6.5 Differential release of neuropeptide and small-molecule co-transmitters. Low-frequency stimulation preferentially raises the Ca²⁺ concentration close to the membrane, favoring the release of transmitter from small clear-core vesicles docked at presynaptic specializations. High-frequency stimulation leads to a more general increase in Ca²⁺, causing the release of peptide neuro transmitters from large dense-core vesicles as well as small-molecule neurotransmitters from small clear-core vesicles.

Neuropeptide in large dense-core vesicles

Small-molecule neurotransmitter in small clear-core vesicles

Localized increase in Ca²⁺ concentration

Low-frequency stimulation

Preferential release of small-molecule neurotransmitter

More diffuse increase in Ca²⁺ concentration

High-frequency stimulation

Release of both types of transmitter

example, low-frequency stimulation often releases only small neurotransmitters, whereas high-frequency stimulation is required to release neuropeptides from the same presynaptic terminals (Figure 6.5). As a result, the chemical signaling properties of a synapse often change according to the level of presynaptic activity.

This differential release of co-transmitters is probably based on the distribution of Ca²⁺ and vesicles in presynaptic terminals. Whereas the vesicles containing small-molecule transmitters are typically docked at the plasma membrane in advance of Ca²⁺ entry, vesicles containing peptide transmitters are farther away from the plasma membrane (see Figures 6.5 and 6.7). At low firing frequencies, the concentration of Ca²⁺ may increase only in the vicinity of presynaptic Ca²⁺ channels, limiting release to small-molecule transmitters because only the vesicles containing these agents are immediately adjacent to the channels. High-frequency stimulation increases the Ca²⁺ concentration more evenly throughout the presynaptic terminal, thereby inducing release of neuropeptides as well.

Neurotransmitter Synthesis

Effective synaptic transmission requires close control of the concentration of neurotransmitters within the synaptic cleft. Neurons have therefore developed

(A) LIFE CYCLE OF NEUROTRANSMITTER

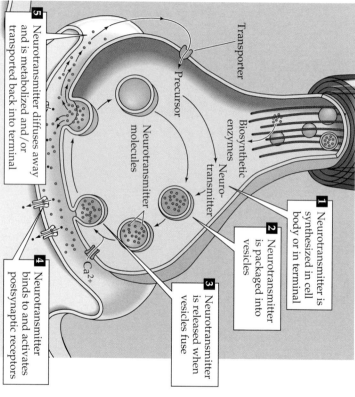

1 Neurotransmitter is synthesized in cell body or in terminal

2 Neurotransmitter is packaged into vesicles

3 Neurotransmitter is released when vesicles fuse

4 Neurotransmitter binds to and activates postsynaptic receptors

5 Neurotransmitter diffuses away and is metabolized and/or transported back into terminal

Transporter

Precursor

Biosynthetic enzymes

Neuro-transmitter

Neurotransmitter molecules

Ca²⁺

(B) SMALL-MOLECULE TRANSMITTERS

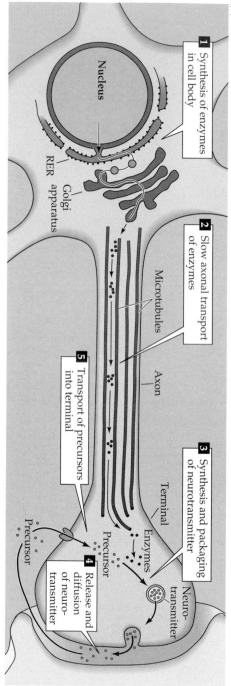

1 Synthesis of enzymes in cell body

2 Slow axonal transport of enzymes

3 Synthesis and packaging of neurotransmitter

4 Release and diffusion of neuro-transmitter

5 Transport of precursors into terminal

Nucleus

Golgi apparatus

RER

Microtubules

Axon

Terminal

Enzymes

Precursor

Neuro-transmitter

Precursor

(C) PEPTIDE TRANSMITTERS

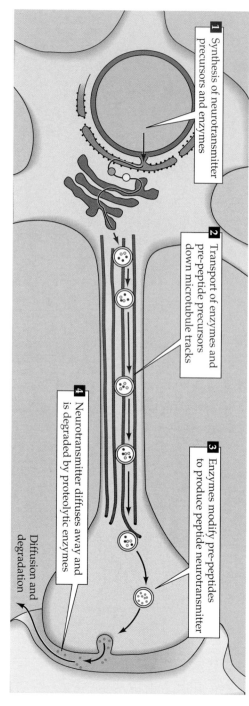

1 Synthesis of neurotransmitter precursors and enzymes

2 Transport of enzymes and pre-peptide precursors down microtubule tracks

3 Enzymes modify pre-peptides to produce peptide neurotransmitter

4 Neurotransmitter diffuses away and is degraded by proteolytic enzymes

Diffusion and degradation

▼ **Figure 6.6** The synthesis, packaging, secretion, and removal of neurotransmitters. (A) The life cycle of transmitter agents entails (1) neurotransmitter synthesis, (2) packaging into vesicles, (3) fusion of vesicles resulting in neurotransmitter release, and (4) activation of postsynaptic receptors. Neurotransmitters are then removed from the synaptic cleft (5). In many cases, the neurotransmitter and/or a breakdown product is reused for neurotransmitter synthesis. (B) Small-molecule neurotransmitters are synthesized at nerve terminals. The enzymes necessary for neurotransmitter synthesis are made in the cell body of the presynaptic cell (1) and are transported down the axon by slow axonal transport (2). Precursors are taken up into the terminals by specific transporters, and neurotransmitter synthesis and packaging take place within the nerve endings (3). After vesicle fusion and release (4), the neurotransmitter may be enzymatically degraded. The reuptake of the neurotransmitter (or its metabolites) starts another cycle of synthesis, packaging, release, and removal (5). (C) Peptide neurotransmitters, as well as the enzymes that modify their precursors, are synthesized in the cell body (1). Enzymes and propeptides are packaged into vesicles in the Golgi apparatus. During fast axonal transport of these vesicles to the nerve terminals (2), the enzymes modify the propeptides to produce one or more neurotransmitter peptides (3). After vesicle fusion and exocytosis, the peptides diffuse away and are degraded by proteolytic enzymes (4).

a sophisticated ability to regulate the synthesis, packaging, release, and degradation (or removal) of neurotransmitters to achieve the desired levels of transmitter molecules (Figure 6.6A). In general, each of these component processes is specific to the transmitter involved, requiring enzymes found only in (or in association with) neurons that use the transmitter at their synapses.

As a rule, the synthesis of small-molecule neurotransmitters occurs within presynaptic terminals (Figure 6.6B). The enzymes needed for transmitter synthesis are synthesized in the neuronal cell body and transported to the nerve terminal cytoplasm at 0.5–5 millimeters a day by a mechanism called **slow axonal transport**. The precursor molecules used by these synthetic enzymes are usually taken into the nerve terminal by transporter proteins found in the plasma membrane of the terminal. The enzymes generate a cytoplasmic pool of neurotransmitter that must then be loaded into synaptic vesicles by transport proteins in the vesicular membrane (see Chapter 4). For some small-molecule neurotransmitters, the final synthetic steps actually occur inside the synaptic vesicles.

The mechanisms responsible for the synthesis and packaging of peptide transmitters are fundamentally different from those used for the small-molecule neurotransmitters (Figure 6.6C). Peptide-secreting neurons generally synthesize polypeptides in their cell bodies that are much larger than the final, "mature" peptide. Processing these polypeptides, which are called **pre-propeptides** (or pre-proproteins), takes place by a sequence of reactions in several intracellular organelles. Pre-propeptides are synthesized in the rough endoplasmic reticulum, where the signal sequence of amino acids—that is, the sequence indicating that the peptide is to be secreted—is removed. The remaining polypeptide, called a **propeptide** (or proprotein), then traverses the Golgi apparatus and is packaged into vesicles in the *trans*-Golgi network. The final stages of peptide neurotransmitter processing occur after packaging into vesicles and involve proteolytic cleavage, modification of the ends of the peptide, glycosylation, phosphorylation, and disulfide bond formation.

Neuropeptide synthesis is, therefore, much like the synthesis of proteins secreted from non-neuronal cells (pancreatic enzymes, for instance). A major difference, however, is that the neuronal axon often presents a very long distance between the site of a peptide's synthesis and its ultimate secretion. The

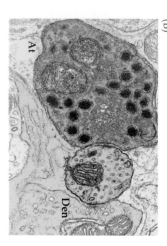
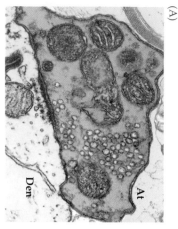

Figure 6.7 Different types of synaptic vesicles. (A) Small clear-core vesicles at a synapse between an axon terminal (At) and a dendritic spine (Den) in the central nervous system. Such vesicles typically contain small-molecule neurotransmitters. (B) Large dense-core vesicles in another type of central axon terminal (At) synapsing onto a dendrite (Den). Such vesicles typically contain neuropeptides (or in some cases biogenic amines). (From Peters, Palay, and Webster, 1991.)

peptide-filled vesicles must therefore be transported along the axon to the synaptic terminal. The mechanism responsible for such movement, known as **fast axonal transport**, carries vesicles at rates up to 400 mm/day along cytoskeletal elements called microtubules (in contrast to the slow axonal transport of the enzymes that synthesize small-molecule transmitters). Microtubules are long, cylindrical filaments, 25 nm in diameter, present throughout neurons and other cells. Peptide-containing vesicles are moved along these microtubule "tracks" by ATP-requiring "motor" proteins such as kinesin.

Packaging Neurotransmitters

Following their synthesis, neurotransmitters are stored for future use in synaptic vesicles. The nature of these vesicles varies for different transmitters. Some of the small-molecule neurotransmitters—acetylcholine and the amino acid transmitters—are packaged in vesicles 40 to 60 nm in diameter, the centers of which appear clear in electron micrographs; accordingly, these vesicles are referred to as **small clear-core vesicles** (Figure 6.7A). The transmitters are concentrated in synaptic vesicles by transporter proteins in the vesicle membrane using an energy-requiring mechanism. Neuropeptides, in contrast, are packaged into larger synaptic vesicles that range from 90 to 250 nm in diameter. These vesicles are electron-dense in electron micrographs—hence, they are referred to as **large dense-core vesicles** (Figure 6.7B). Finally, the biogenic amines can be packaged in either small (40–60 nm diameter) dense-core vesicles or larger (60–120 nm diameter), irregularly shaped, dense-core vesicles, depending on the particular class of neuron.

Neurotransmitter Release and Removal

Once loaded with transmitter molecules, vesicles associate with the presynaptic membrane and fuse with it in response to Ca²⁺ influx, as described in Chapter 5. The mechanisms of vesicle release are similar for all transmitters, although there are differences in the speed of this process. In general, small-molecule transmitters are secreted more rapidly than peptides. For example, while secretion of ACh from motor neurons requires only a fraction of a millisecond, many neuroendocrine cells, such as those in the hypothalamus, require high-frequency bursts of action potentials for many seconds to release peptide hormones from their nerve terminals. These differences in the rate of transmitter release make neurotransmission rapid at synapses employing small-molecule transmitters and relatively slow at synapses that use peptides. As already mentioned, these differences in the rate of release probably arise from spatial differences in vesicle localization and presynaptic Ca²⁺ signaling (see Figure 6.5). Thus, the small clear-core vesicles used to store small-molecule transmitters are often docked at active zones (specialized regions of the presynaptic membrane; see Chapter 5), whereas the large dense-core vesicles used to store peptides are not (compare Figure 6.7A and B). Since biogenic amines are sometimes packaged into small vesicles that dock at active zones and are sometimes packaged and released much like peptides, the speed of their release can vary greatly.

When the neurotransmitter has been secreted into the synaptic cleft, it binds to specific receptors on the postsynaptic cell, thereby generating a postsynaptic electrical signal, as described in much more detail in Chapter 7. The transmitter must then be removed rapidly to enable the postsynaptic cell to engage in another cycle of neurotransmitter release, binding, and signal generation.

The mechanisms by which neurotransmitters are removed vary but always involve diffusion in combination with reuptake into nerve terminals or surrounding glial cells, degradation by transmitter-specific enzymes, or in some cases a combination of these mechanisms. For most of the small-molecule neurotransmitters, specific transporter proteins remove the transmitters (or their metabolites) from the synaptic cleft, ultimately delivering them back to the presynaptic terminal for reuse (see Figure 6.6A).

The particulars of synthesis, packaging, release and removal differ for each neurotransmitter. These variations are elaborated for the major neurotransmitters in the following sections, and are summarized in Table 6.1.

Acetylcholine

Acetylcholine is the neurotransmitter at neuromuscular junctions, at synapses in the ganglia of the visceral motor system, and at a variety of sites within the central nervous system. Whereas a great deal is known about the function of cholinergic transmission at the neuromuscular junction and at ganglionic synapses, the actions of ACh in the central nervous system are not as well understood.

Acetylcholine is synthesized in nerve terminals from acetyl coenzyme A (acetyl CoA, which is synthesized from glucose) and choline, in a reaction catalyzed by choline acetyltransferase (CAT) (Figure 6.8). The presence of CAT in a neuron is thus a strong indication that ACh is used as one of its transmitters. Choline is present in plasma at a concentration of about 10

TABLE 6.1
Functional Features of the Major Neurotransmitters

Neurotransmitter	Postsynaptic effect[a]	Precursor(s)	Rate-limiting step in synthesis	Removal mechanism	Type of vesicle
ACh	Excitatory	Choline + acetyl CoA	CAT	AChEase	Small, clear
Glutamate	Excitatory	Glutamine	Glutaminase	Transporters	Small, clear
GABA	Inhibitory	Glutamate	GAD	Transporters	Small, clear
Glycine	Inhibitory	Serine	Phosphoserine	Transporters	Small, clear
Catecholamines (epinephrine, norepinephrine, dopamine)	Excitatory	Tyrosine	Tyrosine hydroxylase	Transporters, MAO, COMT	Small dense-core, or large irregular dense-core
Serotonin (5-HT)	Excitatory	Tryptophan	Tryptophan hydroxylase	Transporters, MAO	Large, dense-core
Histamine	Excitatory	Histidine	Histidine decarboxylase	Transporters	Large, dense-core
ATP	Excitatory	ADP	Mitochondrial oxidative phosphorylation; glycolysis	Hydrolysis to AMP and adenosine	Small, clear
Neuropeptides	Excitatory and inhibitory	Amino acids (protein synthesis)	Synthesis and transport	Proteases	Large, dense-core

[a]The most common postsynaptic effect is indicated; the same transmitter can elicit postsynaptic excitation *or* inhibition depending on the nature of the ion channels affected by transmitter binding (see Chapter 7).

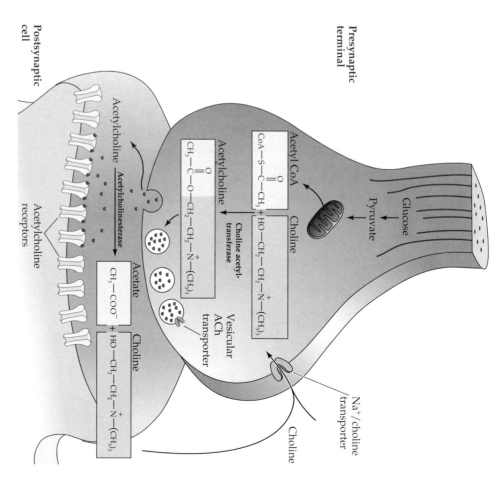

Presynaptic terminal

Glucose

Pyruvate

Acetyl CoA

$$CoA - S - C - CH_3$$
$$\overset{O}{\underset{\|}{}}$$

Choline

Na⁺/choline transporter

Choline

Acetylcholine

Choline acetyl-transferase

$$CH_3 - C - O - CH_2 - CH_2 - \overset{+}{N} - (CH_3)_3$$
$$\overset{O}{\underset{\|}{}}$$

Vesicular ACh transporter

Acetylcholinesterase

Acetylcholine

Acetate

$$CH_3 - COO^-$$

Choline

$$HO - CH_2 - CH_2 - \overset{+}{N} - (CH_3)_3$$

Acetylcholine receptors

Postsynaptic cell

Figure 6.8 Acetylcholine metabolism in cholinergic nerve terminals. The synthesis of acetylcholine from choline and acetyl CoA requires choline acetyltransferase. Acetyl CoA is derived from pyruvate generated by glycolysis, while choline is transported into the terminals via a Na⁺-dependent transporter. After release, acetylcholine is rapidly metabolized by acetylcholinesterase and choline is transported back into the terminal.

mM, and is taken up into cholinergic neurons by a high-affinity Na⁺/choline transporter. About 10,000 molecules of ACh are packaged into each vesicle by a vesicular ACh transporter.

In contrast to most other small-molecule neurotransmitters, the postsynaptic action of ACh at many cholinergic synapses (the neuromuscular junction in particular) are not terminated by reuptake but by a powerful hydrolytic enzyme, acetylcholinesterase (AChE). This enzyme is concentrated in the synaptic cleft, ensuring a rapid decrease in ACh concentration after its release from the presynaptic terminal. AChE has a very high catalytic activity (about 5000 molecules of ACh per AChE molecule per second) and hydrolyzes ACh into acetate and choline. As already mentioned, cholinergic nerve terminals typically contain a high-affinity, Na⁺-choline transporter that takes up the choline produced by ACh hydrolysis.

Among the many interesting drugs that interact with cholinergic enzymes are the organophosphates. This group also includes some potent chemical warfare agents. One such compound is the nerve gas "Sarin," which was made notorious a few years ago after a group of terrorists released this gas in Tokyo's underground rail system. Organophosphates can be lethal to humans (and insects) because they inhibit AChE, causing ACh to accumulate at cholinergic synapses This build-up of ACh depolarizes the postsynaptic cell and renders it refractory to subsequent ACh release, causing, among other effects, neuro-muscular paralysis.

Glutamate

Glutamate is generally acknowledged to be the most important transmitter for normal brain function. Nearly all excitatory neurons in the central nervous system are glutamatergic, and it is estimated that over half of all brain synapses release this agent. Glutamate plays an especially important role in clinical neurology because elevated concentrations of extracellular glutamate, released as a result of neural injury, are toxic to neurons (Box B).

Glutamate is a nonessential amino acid that does not cross the blood-brain barrier and must be synthesized in neurons from local precursors. The most prevalent glutamate precursor in synaptic terminals is glutamine. Glutamine is released by glial cells and, once within presynaptic terminals, is metabolized to glutamate by the mitochondrial enzyme glutaminase (Figure 6.9). Glutamate can also be synthesized by transamination of 2-oxoglutarate, an intermediate of the tricarboxylic acid (TCA) cycle. Hence, some of the glucose metabolized by neurons can also be used for glutamate synthesis.

Following its packaging into synaptic vesicles by a Mg^{2+}/ATP-dependent transport process, glutamate-filled vesicles are ready to dock and be released from presynaptic sites. Glutamate is removed from the synaptic cleft by several high-affinity glutamate transporters present in both glial cells and presynaptic terminals. Glial cells contain the enzyme glutamine synthetase, which converts glutamate into glutamine; glutamine is then transported out of the glial cells and into nerve terminals. In this way, synaptic terminals cooperate with glial cells to maintain an adequate supply of the neurotransmitter. This overall sequence of events is referred to as the **glutamate-glutamine cycle.**

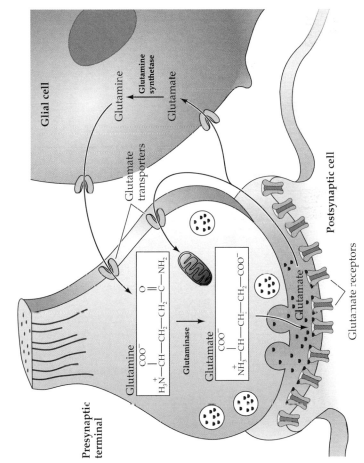

Figure 6.9 Glutamate synthesis and cycling between neurons and glia. The action of glutamate released into the synaptic cleft is terminated by uptake into neurons and surrounding glial cells via specific transporters. Within the nerve terminal, the glutamine released by glial cells and taken up by neurons is converted back to glutamate.

BOX B
Excitotoxicity in Acute Neuronal Injury

Excitotoxicity refers to the ability of glutamate and related compounds to destroy neurons by prolonged excitatory synaptic transmission. Normally, the concentration of glutamate released into the synaptic cleft rises to high levels (approximately 1 mM), but it remains at this level for only a few milliseconds. If abnormally high concentrations of glutamate accumulate in the cleft, the excessive activation of neuronal glutamate receptors can literally excite neurons to death.

The phenomenon of excitotoxicity was discovered in 1957 when D. R. Lucas and J. P. Newhouse serendipitously found that feeding sodium glutamate to infant mice destroys neurons in the retina. Roughly a decade later, John Olney extended this discovery by showing that regions of glutamate-induced neuronal loss can occur throughout the brain. The damage was evidently restricted to the postsynaptic cells—the dendrites of the target neurons were grossly swollen—while the presynaptic terminals were spared. Olney also examined the relative potency of glutamate analogs and found that their neurotoxic actions paralleled their ability to activate postsynaptic glutamate receptors. Furthermore, glutamate receptor antagonists were effective in blocking the neurotoxic effects of glutamate. In light of this evidence, Olney postulated that glutamate destroys neurons by a mechanism similar to transmission at excitatory glutamatergic synapses, and coined the term *excitotoxic* to refer to this pathological effect.

Evidence that excitotoxicity is an important cause of neuronal damage after brain injury has come primarily from studying the consequences of oxygen deprivation. The most common cause of reduced blood flow to the brain (ischemia) is the occlusion of a cerebral blood vessel (i.e., a stroke; see Box D in Chapter 1). The idea that excessive synaptic activity contributes to ischemic injury emerged from the observation that concentrations of glutamate and aspartate in extracellular space increase during ischemia. Moreover, microinjection of glutamate receptor antagonists in experimental animals protects neurons from ischemia-induced damage.

Together, these findings imply that extracellular accumulation of glutamate during ischemia activates glutamate receptors excessively, and that this somehow triggers a chain of events that leads to neuronal death. The reduced supply of oxygen presumably elevates extracellular glutamate levels by slowing the energy-dependent uptake of glutamate at synapses.

Excitotoxic mechanisms have now been shown to be involved in other acute forms of neuronal insult, including hypoglycemia, trauma, and repeated intense seizures (called status epilepticus). Understanding excitotoxicity therefore has important implications for treating a variety of neurological disorders. For instance, a blockade of glutamate receptors could, in principle, protect neurons from injury due to stroke, trauma, or other causes. Unfortunately, clinical trials of glutamate receptor antagonists have

not led to much improvement in the outcome of stroke. The ineffectiveness of this quite logical treatment is probably due to several factors, one of which is that substantial excitotoxic injury occurs quite soon after ischemia, prior to the typical initiation of treatment. It is also likely that excitotoxicity is only one of several mechanisms by which ischemia damages neurons, other candidates being the triggering of genetically controlled cell death programs (apoptosis) and damage secondary to inflammation. Pharmacological interventions that target all these mechanisms nonetheless hold considerable promise for minimizing brain injury after stroke and trauma.

References

LEE, J., G. J. ZIPFEL AND D. W. CHOI (1999) The changing landscape of ischaemic brain injury mechanisms. Nature 399(Supp.): 7–14.

LUCAS, D. R. AND J. P. NEWHOUSE (1957) The toxic effects of sodium l-glutamate on the inner layers of the retina. Arch. Opthalmol. 58: 193–201.

MOTT, D. D. AND 7 OTHERS (1998) Phenylethanolamines inhibit NMDA receptors by enhancing proton inhibition. Nature Neurosci. 1: 659–667.

OLNEY, J. W. (1969) Brain lesions, obesity and other disturbances in mice treated with monosodium glutamate. Science 164: 719–721.

OLNEY, J. W. (1971) Glutamate-induced neuronal necrosis in the infant mouse hypothalamus: An electron microscopic study. J. Neuropathol. Exp. Neurol. 30: 75–90.

ROTHMAN, S. M. (1983) Synaptic activity mediates death of hypoxic neurons. Science 220: 536–537.

GABA and Glycine

Most inhibitory neurons in the brain and spinal cord use either γ-aminobutyric acid (GABA) or glycine as a neurotransmitter. Like glutamate, GABA was identified in brain tissue during the 1950s, and the details of its synthesis and degradation were worked out shortly thereafter. David Curtis and Jeffrey

Watkins were the first to show several decades ago that GABA inhibits the ability of mammalian neurons to fire action potentials. It is now known that as many as one-third of the synapses in the brain use GABA as their neurotransmitter. Unlike glutamate, GABA is not an essential metabolite, nor is it incorporated into protein. Thus, the presence of GABA in neurons and terminals is a good initial indication that the cells in question use GABA as a neurotransmitter. GABA is most commonly found in local circuit interneurons, although the Purkinje cells of the cerebellum provide an example of a GABAergic projection neuron (see Chapter 19).

The predominant precursor for GABA synthesis is glucose, which is metabolized to glutamate by the tricarboxylic acid cycle enzymes, although pyruvate and glutamine can also act as precursors. The enzyme glutamic acid decarboxylase (GAD), which is found almost exclusively in GABAergic neurons, catalyzes the conversion of glutamate to GABA (Figure 6.10A). GAD requires a cofactor, pyridoxal phosphate, for activity. Because pyridoxal phosphate is derived from vitamin B_6, a B_6 deficiency can lead to diminished GABA synthesis. The significance of this became clear after a disastrous series of infant deaths was linked to the omission of vitamin B_6 from infant formula. The lack of B_6 resulted in a large reduction in the GABA content of the brain, and the subsequent loss of synaptic inhibition caused seizures that in some cases were fatal.

The mechanism of GABA removal is similar to that for glutamate: Both neurons and glia contain high-affinity transporters for GABA. Most GABA is eventually converted to succinate, which is metabolized further in the tricarboxylic acid cycle that mediates cellular ATP synthesis. The enzymes required for this degradation, GABA aminotransferase and succinic semialdehyde dehydrogenase, are both mitochondrial enzymes. Inhibition of GABA breakdown causes a rise in tissue GABA content and an increase in the activity of inhibitory neurons. Drugs that act as agonists or modulators on postsynaptic GABA receptors, such as benzodiazepines and barbiturates, are used clinically to treat epilepsy and are effective sedatives and anesthetics.

The distribution of the neutral amino acid glycine in the central nervous system is more localized than that of GABA. About half of the inhibitory synapses in the spinal cord use glycine; most other inhibitory synapses use GABA. Glycine is synthesized from serine by the mitochondrial isoform of serine hydroxymethyltransferase (Figure 6.10B). Once released from the presynaptic cell, glycine is rapidly removed from the synaptic cleft by specific membrane transporters. Mutations in the genes coding for some of these enzymes result in hyperglycinemia, a devastating neonatal disease characterized by lethargy, seizures, and mental retardation.

The Biogenic Amines

There are five established biogenic amine neurotransmitters: the three **catecholamines**—**dopamine, norepinephrine (noradrenaline)**, and **epinephrine (adrenaline)**—and **histamine** and **serotonin** (see Figure 6.3). In terms of synthesis, packaging, release, and degradation, the amine neurotransmitters fall somewhere between the properties of the other small-molecule neurotransmitters and those of the neuropeptides.

All the catecholamines (so named because they share the catechol moiety) are derived from a common precursor, the amino acid tyrosine (Figure 6.11). The first step in catecholamine synthesis is catalyzed by tyrosine hydroxylase in a reaction requiring oxygen as a co-substrate and tetrahydrobiopterin as a cofactor to synthesize dihydroxyphenylalanine (DOPA). Because tyrosine hydroxylase is rate-limiting for the synthesis of all three transmitters, its presence is a valuable criterion for identifying catecholaminergic neurons.

Figure 6.10 Synthesis, release, and reuptake of the inhibitory neurotransmitters GABA and glycine. (A) GABA is synthesized from glutamate by the enzyme glutamic acid decarboxylase, which requires pyridoxal phosphate. (B) Glycine can be synthesized by a number of metabolic pathways; in the brain, the major precursor is serine. High-affinity transporters terminate the actions of these transmitters and return GABA or glycine to the synaptic terminals for reuse.

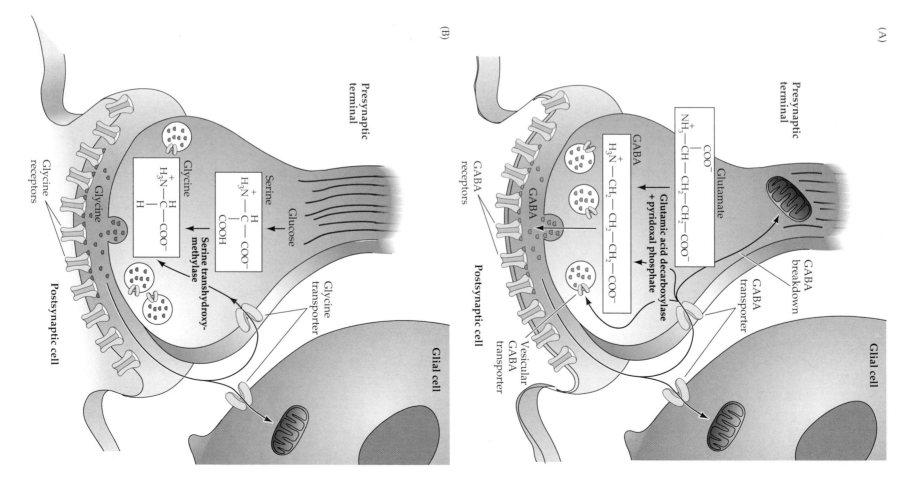

(A)

(B)

Tyrosine

O_2 → Tyrosine hydroxylase

Dihydroxyphenylalanine (DOPA)

CO_2 → DOPA decarboxylase

Dopamine

O_2 → Dopamine-β hydroxylase

Norepinephrine

RCH_3 → R Phenylethanolamine *N*-methyl-transferase

Epinephrine

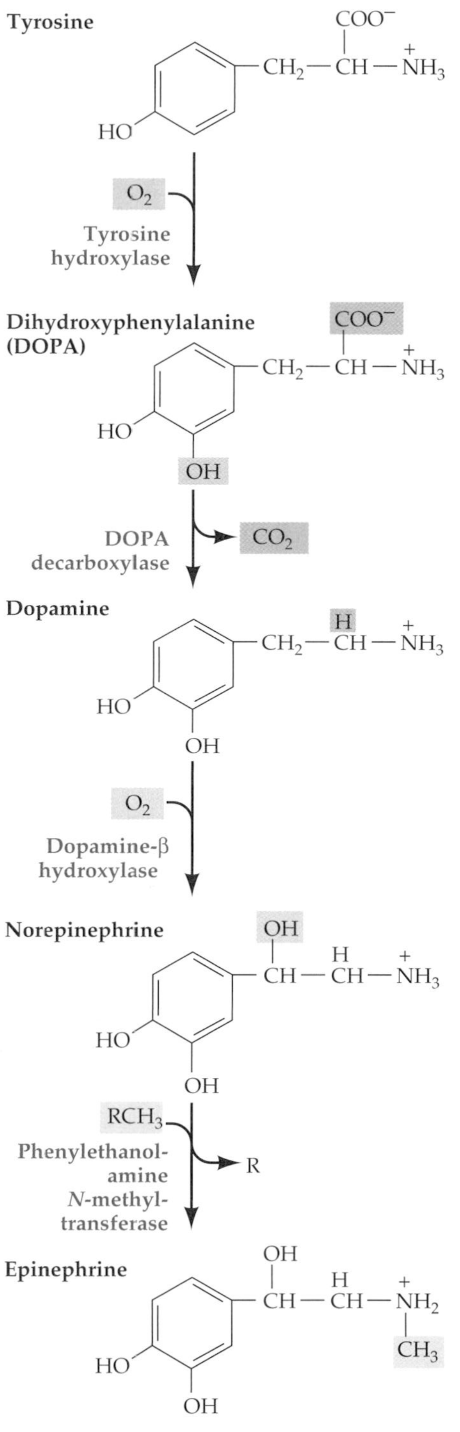

Figure 6.11 The biosynthetic pathway for the catecholamine neurotransmitters. The amino acid tyrosine is the precursor for all three catecholamines. The first step in this reaction pathway, catalyzed by tyrosine hydroxylase, is rate-limiting.

• *Dopamine* is produced by the action of DOPA decarboxylase on DOPA (see Figure 6.11). Although present in several brain regions (Figure 6.12A), the major dopamine-containing area of the brain is the corpus striatum, which receives major input from the substantia nigra and plays an essential role in the coordination of body movements. In Parkinson's disease, for instance, the dopaminergic neurons of the substantia nigra degenerate, leading to a characteristic motor dysfunction (see Box B in Chapter 18). Although dopamine does not readily cross the blood-brain barrier, its precursor, levodopa, does. Levodopa is absorbed in the small bowel but is rapidly catabolized in the GI tract and in peripheral tissues. Hence, the disease can be treated by administering levodopa together with carbidopa, a dopamine decarboxylase inhibitor, and selegiline, a monoamine oxidase inhibitor. Dopamine is also believed to be involved in motivation, reward, and reinforcement. For example, cocaine and other addictive drugs act by stimulating the release of dopamine from specific brain areas (see Box D). Once released, dopamine binds to specific dopamine receptors, as well as to some β-adrenergic receptors. It not only acts as a neurotransmitter in the central nervous system but also plays a poorly understood role in some sympathetic ganglia. Dopamine is also used clinically to treat shock because it dilates renal arteries by activating dopamine receptors and increases cardiac output by activating β-adrenergic receptors in the heart.

• *Norepinephrine* (also called noradrenaline) synthesis requires dopamine β-hydroxylase, which catalyzes the production of norepinephrine from dopamine (see Figure 6.11). Dopamine is transported by vesicles into adrenergic terminals, where it is converted to norepinephrine. The most prominent class of neurons that synthesize norepinephrine is sympathetic ganglion cells, since norepinephrine is the major peripheral transmitter in this division of the visceral motor system (see Chapter 21). Norepinephrine is also the transmitter used by the locus coeruleus, a brainstem nucleus that projects diffusely to a variety of forebrain targets (Figure 6.12B), where it influences sleep and wakefulness, attention, and feeding behavior.

• *Epinephrine* (also called adrenaline) is present in the brain at lower levels than the other catecholamines. The enzyme that synthesizes epinephrine, phenylethanolamine-*N*-methyltransferase (see Figure 6.11), is present only in epinephrine-secreting neurons. Epinephrine-containing neurons in the central nervous system are found in two groups in the rostral medulla, the function of which is not known.

All three catecholamines are removed by reuptake into nerve terminals or surrounding glial cells by a Na^+-dependent transporter. The two major enzymes involved in the catabolism of catecholamines are monoamine oxidase (MAO) and catechol *O*-methyltransferase (COMT). Both neurons and glia contain mitochondrial MAO and cytoplasmic COMT. Inhibitors of these enzymes, such as phenelzine and tranylcypromine, are used clinically as antidepressants (see Box C).

• *Histamine* is produced from the amino acid histidine by a histidine decarboxylase and is metabolized by the combined actions of histamine methyltransferase and MAO. (Figure 6.13A). High concentrations of histamine and histamine decarboxylase are found in neurons in the hypothalamus that send sparse but widespread projections to almost all regions of the brain and spinal cord (see Figure 6.12C). The central histamine projections mediate arousal and attention, similar to central ACh and norepinephrine projections. This partly explains why antihistamines that cross the blood-brain barrier, such as diphenhydramine (Benadryl®), act as sedatives. Histamine also is released from mast cells in response to allergic reactions or tissue damage.

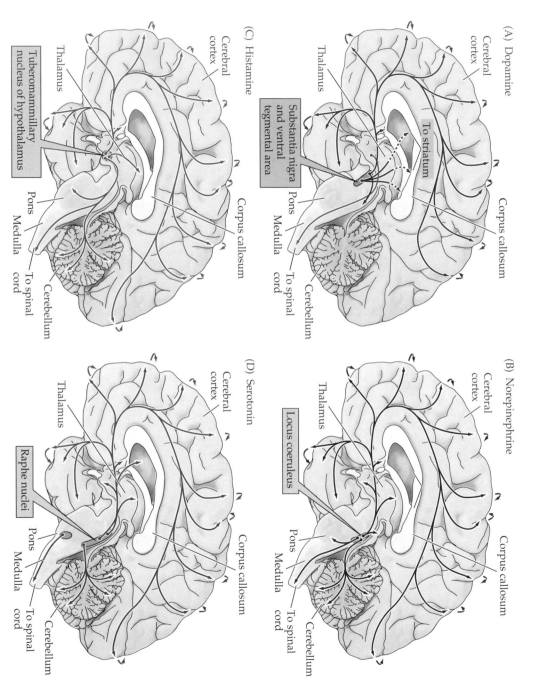

(A) Dopamine

Cerebral cortex

Thalamus

Substantia nigra and ventral tegmental area

To striatum

Corpus callosum

Pons
Medulla
Cerebellum
To spinal cord

(B) Norepinephrine

Cerebral cortex

Thalamus

Locus coeruleus

Corpus callosum

Pons
Medulla
Cerebellum
To spinal cord

(C) Histamine

Cerebral cortex

Tuberomammillary nucleus of hypothalamus

Thalamus

Corpus callosum

Pons
Medulla
Cerebellum
To spinal cord

(D) Serotonin

Cerebral cortex

Thalamus

Raphe nuclei

Corpus callosum

Pons
Medulla
Cerebellum
To spinal cord

Figure 6.12 The distribution in the human brain of neurons and their projections (arrows) containing biogenic amine neurotransmitters. Curved arrows along the perimeter of the cortex indicate the innervation of lateral cortical regions not shown in this midsagittal plane of section.

The close proximity of mast cells to blood vessels, together with the potent actions of histamine on blood vessels, raises the possibility that histamine may influence brain blood flow.

• *Serotonin*, or 5-hydroxytryptamine (5-HT), was initially thought to increase vascular tone by virtue of its presence in serum (hence the name serotonin). 5-HT is synthesized from the amino acid tryptophan, which is an essential dietary requirement. Tryptophan is taken up into neurons by a plasma membrane transporter and hydroxylated in a reaction catalyzed by the enzyme tryptophan-5-hydroxylase (Figure 6.13B), the rate-limiting step for 5-HT synthesis. As in the case of other biogenic amines, the synaptic effects of serotonin are terminated by transport back into serotonergic nerve terminals. The primary catabolic pathway is mediated by MAO. Serotonin is located in groups of neurons in the raphe region of the pons and upper brainstem, which have widespread projections to the forebrain (see Figure 6.12D) and have been implicated in the regulation of sleep and wakefulness (see Chapter 28). A number of antipsychotic drugs used in the treatment of depression and anxiety are thought to act specifically on serotonergic neurons.

Because biogenic amines are implicated in such a wide range of behaviors (ranging from central homeostatic functions to cognitive phenomena such as attention), it is not surprising that drugs affecting the synthesis, receptor binding, or catabolism of these neurotransmitters are among the most important in the armamentarium of modern pharmacology (Box C).

ATP and Other Purines

Interestingly, all synaptic vesicles contain ATP, which is co-released with one or more "classical" neurotransmitters. This observation raises the possibility that ATP acts as a co-transmitter. It has been known since the 1920s that the extracellular application of ATP (or its breakdown products AMP and adenosine) to neurons can elicit electrical responses. The idea that some purines (so named because all these compounds contain a purine ring; see Figure 6.3) are also neurotransmitters has now received considerable experimental support. ATP acts as an excitatory neurotransmitter in motor neurons of the spinal cord, as well as sensory and autonomic ganglia. Postsynaptic actions of ATP have also been demonstrated in the central nervous system, specifically for dorsal horn neurons and in a subset of hippocampal neurons. Adenosine, however, cannot be considered a classical neurotransmitter because it is not stored in synaptic vesicles or released in a Ca^{2+}-dependent manner (see Box A). Rather, it is generated from ATP by the action of extracellular enzymes. A number of enzymes, such as apyrase and ecto-5' nucleotidase, as well as nucleoside transporters are involved in the rapid catabolism and removal of purines from extracellular locations. Despite the relative novelty of this evidence, it suggests that excitatory transmission via purinergic synapses is widespread in the mammalian brain.

Peptide Neurotransmitters

Many peptides known to be hormones also act as neurotransmitters, and often these are co-released with small-molecule neurotransmitters. The biological activity of the peptide neurotransmitters depends on their amino acid sequence (see, for example, Figure 6.4). As already described, propeptide precursors are typically larger than their active peptide products and can give rise to more than one species of neuropeptide (Figure 6.14). Hence, the release of multiple neuroactive peptides from a single vesicle often elicits complex postsynaptic responses. Peptides are catabolized into inactive amino acid fragments by enzymes called peptidases, usually located on the extracellular surface of the plasma membrane. Some peptide transmitters have been implicated in modulating emotions (see Chapter 29). Others, such as substance P and the opioid peptides, are involved in the perception of pain (see Chapter 10). Still other peptides, such as melanocyte-stimulating hormone, adrenocorticotropin, and β-endorphin, regulate complex responses to stress. The large number of neuropeptide transmitters have been loosely grouped into five categories: the brain/gut peptides, opioid peptides, pituitary peptides, hypothalamic releasing hormones, and a catchall category containing all other peptides not easily classified.

Substance P is an example of the first of these categories. The study of neuropeptides actually began more than 60 years ago with the accidental discovery of substance P, a powerful hypotensive agent. (The peculiar name derives from the fact that this molecule was an unidentified component of *powder* extracts from brain and intestine.) This 11-amino-acid peptide (see Figure 6.4) is present in high concentrations in the human hippocampus, neocortex, and also in the gastrointestinal tract; hence its classification as a brain/gut peptide. It is also released from C fibers, the small-diameter afferents in peripheral nerves that convey information about pain and tempera-

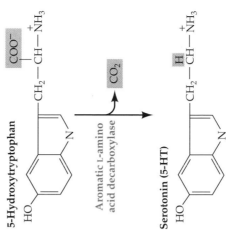

Figure 6.13 Synthesis of histamine and serotonin. (A) Histamine is synthesized from the amino acid histidine. (B) Serotonin is derived from the amino acid tryptophan by a two-step process that requires the enzymes tryptophan-5-hydroxylase and a decarboxylase.

Figure 6.14 Proteolytic processing of the pre-propeptides, pre-proopiomelanocortin (A) and pre-proenkephalin A (B). For each pre-propeptide, the signal sequence is indicated in orange at the left; the locations of active peptide products are indicated by different colors. The maturation of the pre-propeptides involves cleaving the signal sequence and other proteolytic processing. Such processing can result in a number of different neuroactive peptides such as ACTH, γ-lipotropin, and β-endorphin (A), or multiple copies of the same peptide, such as met-enkephalin (B).

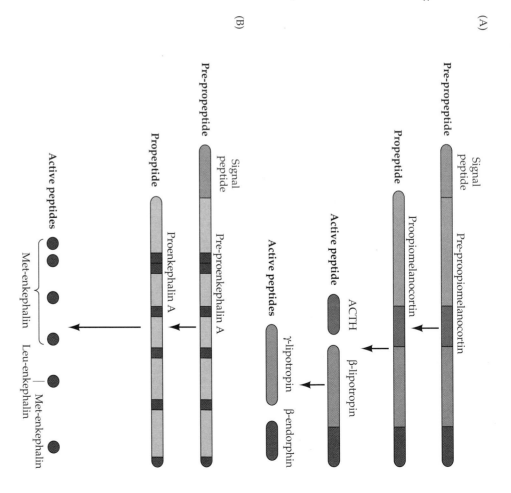

(A)

Pre-propeptide

Signal peptide

Pre-proopiomelanocortin

Propeptide

Proopiomelanocortin

Active peptide

ACTH

β-lipotropin

Active peptides

γ-lipotropin

β-endorphin

(B)

Pre-propeptide

Signal peptide

Pre-proenkephalin A

Propeptide

Proenkephalin A

Active peptides

Met-enkephalin

Met-enkephalin

Leu-enkephalin

Met-enkephalin

ture (as well as postganglionic autonomic signals). Substance P is also a sensory neurotransmitter in the spinal cord, where its release can be inhibited by opioid peptides released from spinal cord interneurons, resulting in the suppression of pain (see Chapter 10). The diversity of neuropeptides is highlighted by the finding that the gene coding for substance P also encodes a number of other neuroactive peptides including neurokinin A, neuropeptide K, and neuropeptide γ.

An especially important category of peptide neurotransmitters is the family of opioids. These peptides are so named because they bind to the same postsynaptic receptors activated by opium (see Chapter 10). The opium poppy has been cultivated for at least 5000 years, and has been used as an analgesic since at least the Renaissance. The active ingredients in opium are a variety of plant alkaloids, predominantly morphine. Morphine, named for the Greek god of dreams, Morpheus, is still one of the most effective analgesics in use today, despite its addictive potential (see Box D). Synthetic opiates such as meperidine and methadone are also used as analgesics, and fentanyl, a drug with 80 times the analgesic potency of morphine, is widely used in clinical anesthesiology.

The opioid peptides were discovered in the 1970s during a search for endogenous compounds that mimicked the actions of morphine. It was hoped that such compounds would be analgesic, and that understanding them would shed light on drug addiction. The endogenous ligands of the opioid receptors have now been identified as a family of more than 20 opioid peptides that fall into three classes: the endorphins, the enkephalins, and the dynor-

BOX C
Biogenic Amine Neurotransmitters and Psychiatric Disorders

The regulation of the biogenic amine neurotransmitters is altered in a variety of psychiatric disorders. Indeed, most psychotropic drugs (defined as drugs that alter behavior, mood, or perception) selectively affect one or more steps in the synthesis, packaging, or degradation of biogenic amines. Sorting out how these drugs work has been extremely useful in beginning to understand the molecular mechanisms underlying some of these diseases.

Based on their effects on humans, psychotherapeutic drugs can be divided into several broad categories: antipsychotics, antianxiety drugs, antidepressants, and stimulants. The first antipsychotic drug used to ameliorate disorders such as schizophrenia was reserpine. Reserpine was developed in the 1950s and initially used as an antihypertensive agent; it blocks the uptake of norepinephrine into synaptic vesicles and therefore depletes the transmitter at aminergic terminals, diminishing the ability of the sympathetic division of the visceral motor system to cause vasoconstriction (see Chapter 21). A major side effect in hypertensive patients treated with reserpine—behavioral depression—suggested the possibility of using it as an antipsychotic agent in patients suffering from agitation and pathological anxiety. (Its ability to cause depression in mentally healthy individuals also suggested that aminergic transmitters are involved in mood disorders; see Box E in Chapter 29.)

Although reserpine is no longer used as an antipsychotic agent, its initial success stimulated the development of antipsychotic drugs such as chlorpromazine, haloperidol, and benperidol, which over the last several decades have radically changed the approach to treating psychotic disorders. Prior to the discovery of these drugs, psychotic patients were typically hospitalized for long peri-

ods, sometimes indefinitely, and in the 1940s were subjected to desperate measures such as frontal lobotomy (see Box C in Chapter 26). Modern antipsychotic drugs now allow most patients to be treated on an outpatient basis after a brief hospital stay. Importantly, the clinical effectiveness of these drugs is correlated with their ability to block brain dopamine receptors, implying that excessive dopamine release is responsible for some types of psychotic illness. A great deal of effort continues to be expended on developing more effective antipsychotic drugs with fewer side effects, and on discovering the mechanism and site of action of these medications.

The second category of psychotherapeutic drugs is the antianxiety agents. Anxiety disorders are estimated to afflict between 10% and 35% of the population, making them the most common psychiatric problem. The two major forms of pathological anxiety—panic attacks and generalized anxiety disorder—both respond to drugs that affect aminergic transmission. The agents used to treat panic disorders include inhibitors of the enzyme monoamine oxidase (MAO inhibitors) required for the catabolism of the amine neurotransmitters, and blockers of serotonin receptors. The most effective drugs in treating generalized anxiety disorder have been benzodiazepines, such as chlordiazepoxide (Librium®), and diazepam (Valium®). In contrast to most other psychotherapeutic drugs, these agents increase the efficacy of transmission at GABA$_A$ synapses rather than acting at aminergic synapses.

Antidepressants and stimulants also affect aminergic transmission. A large number of drugs are used clinically to treat depressive disorders, and in the laboratory to explore the neurotransmitter abnormalities underlying these common conditions. The three major classes of

antidepressants—MAO inhibitors, tricyclic antidepressants, and serotonin uptake blockers such as fluoxetine (Prozac®) and trazodone—all influence various aspects of aminergic transmission. MAO inhibitors such as phenelzine block the breakdown of amines, whereas the tricyclic antidepressants such as desipramine block the reuptake of norepinephrine and other amines. The extraordinarily popular antidepressant fluoxetine (Prozac®) selectively blocks the reuptake of serotonin without affecting the reuptake of catecholamines. Stimulants such as amphetamine are also used to treat some depressive disorders. Amphetamine stimulates the release of norepinephrine from nerve terminals; the transient "high" resulting from taking amphetamine is presumably the emotional opposite of the depression that sometimes follows reserpine-induced norepinephrine depletion.

Despite the relatively small number of aminergic neurons in the brain, this litany of pharmacological actions emphasizes that these neurons are critically important in the maintenance of mental health.

References

BRUNELLO, N. C., C. MASOTTO, L. STEARDO, R. MARKSTEIN AND G. RACAGNI (1995) New insights into the biology of schizophrenia through the mechanism of action of clozapine. Neuropsychopharmacol. 13: 177–213.

CARLSSON, A. (1993) Thirty years of dopamine research. Adv. Neurol. 60: 1–10.

FIBIGER, H. C. (1995) Neurobiology of depression: Focus on dopamine. Adv. Biochem. Psychopharmacol. 49: 1–17.

PERRY, P. J. (1995) Clinical use of the newer antipsychotic drugs. Am. J. Health Syst. Pharm. 52: S9–S14.

SEIDEN, L. S., K. E. SABOL AND G. A. RICAURTE (1993) Amphetamine: Effects on catecholamine systems and behavior. Ann. Rev. Pharmacol. Toxicol. 33: 639–677.

BOX D
Addiction

Drug addiction is a chronic, relapsing disease with obvious medical, social, and political consequences. Addiction (also called substance dependence) is ultimately a pharmacological problem that reflects a drug-induced imbalance of the release, response, and/or reuptake of neurotransmitters. The diagnostic manual of the American Psychiatric Association defines addiction in terms of both *physical* dependence (essentially a disorder of homeostasis; see Chapter 21) and *psychological* dependence (in which an individual continues the drug-taking behavior despite obviously maladaptive consequences). The range of substances that can generate this sort of dependence is wide; the primary agents of abuse at present are opioids, cocaine, amphetamines, marijuana, alcohol, and nicotine. Addiction to more "socially acceptable" agents such as alcohol and nicotine are sometimes regarded as less problematic, but in fact involve medical and behavioral consequences that are at least as great as for drugs of abuse that are considered more dangerous. Importantly, the phenomenon of addiction is not limited to human behavior, but is demonstrable in laboratory animals: Most of these same agents are self-administered if primates, rodents, or other species are provided with the opportunity to do so.

In addition to a compulsion to take the agent of abuse, a major feature of addiction for many drugs is a constellation of negative physiological and emotional features, loosely referred to as "withdrawal syndrome," that occur when the drug is not taken. The symptoms of withdrawal are different for each agent of abuse, but in general are characterized by effects opposite those of the positive experience induced by the drug itself. Consider, as an example, cocaine, a

drug that was estimated to be in regular use by 5 to 6 million Americans during the decade of the 1990s, with about 600,000 regular users either addicted or at high risk for addiction. The positive effects of the drug smoked or inhaled as a powder in the form of the alkaloidal free base is a "high" that is nearly immediate but generally lasts only a few minutes, typically leading to a desire for additional drug in as little as ten minutes to half an hour. The "high" is described as a feeling of well-being, self-confidence, and satisfaction. Conversely, when the drug is not available, frequent users experience depression, sleepiness, fatigue, drug-craving, and a general sense of malaise. Another aspect of addiction to cocaine or other agents is tolerance, defined as a reduction in the response to the drug upon repeated administration. Tolerance is a widespread phenomenon in pharmacology but is particularly significant in drug addiction, since it progressively increases the dose needed to experience the desired effects.

Although it is fair to say that the neurobiology of addiction is poorly understood, for cocaine and many other agents of abuse the addictive effects are based on increased levels of dopamine in critical brain regions involved in motivation and emotional reinforcement (see Chapter 29). The most important of these areas is the midbrain dopamine system, especially its projections from the ventral-tegmental area to the nucleus accumbens. Agents such as cocaine appear to act by raising dopamine levels in these areas, making this transmitter more available to receptors by interfering with re-uptake by the dopamine transporter. The reinforcement and motivation of drug-taking behaviors is thought to be related to the projections to the nucleus accumbens.

The most common opioid drug of abuse is heroin. Heroin is a derivative of the opium poppy and is not legally available for clinical purposes in the United States. The number of heroin addicts in the United States is estimated to be between 750,000 and 1 million individuals. The positive feelings produced by heroin, generally described as the "rush," are often compared to the feeling of sexual orgasm and begin in less than a minute after intravenous injection. There is then a feeling of general well-being (referred to as "on the nod") that lasts about an hour. The symptoms of withdrawal can be intense; these are restlessness, irritability, nausea, muscle pain, depression, sleeplessness, and a sense of anxiety and malaise. The reinforcing aspects of the drug entail the same dopaminergic circuitry in the ventral-tegmental area and nucleus accumbens as does cocaine, although additional areas are certainly involved, particularly the sites of opioid receptors described in Chapter 10. Interestingly, addiction to heroin or any other agent is not an inevitable consequence of drug use, but depends critically on the environment. For instance, returning veterans who were heroin addicts in Vietnam typically lost their addiction upon returning to the United States. Likewise, patients given other opioids (e.g., morphine) for painful conditions rarely become addicts.

The treatment of any form of addiction is difficult, and must be tailored to the circumstances of the individual. In addition to treating acute problems of withdrawal and "detoxification," patterns of behavior must be changed that may take months or years. Addiction is thus a chronic disease state that requires continual monitoring during the lifetime of susceptible individuals.

References

AMERICAN PSYCHIATRIC ASSOCIATION (1994) Diagnostic and Statistical Manual of Mental Disorders, 4th Ed. (DSM IV). Washington, D.C.

KOOB, G. F., P. P. SANNA AND F. E. BLOOM (1998) Neuroscience of addiction. Neuron 21: 467–476.

LESHNER, A. I. (1997) Addiction is a brain disease, and it batters. Science 278: 45–47.

NESTLER, E. J. AND G. K. AGHAJANIAN (1997) Molecular and cellular basis of addiction. Science 278: 58–63.

O'BRIEN, C. P. (1996) *Goodman and Gilman's The Pharmaceutical Basis of Therapeutics*, 9th Ed. New York: McGraw-Hill, Chapter 24, pp. 557–577.

RITZ, M. C., R. J. LAMB, S. R. GOLDBERG AND M. J. KUHAR (1987) Cocaine receptors on dopamine transporters are related to self-administration of cocaine. Science 237: 1219–1223.

phins. Each of these classes are liberated from an inactive pre-propeptide (pre-proopiomelanocortin, pre-proenkephalin A, and pre-prodynorphin), derived from distinct genes. Opioid precursor processing is carried out by tissue-specific processing enzymes packaged into vesicles along with the precursor peptide in the Golgi apparatus (see Figure 6.14).

Opioid peptides are widely distributed throughout the brain and are often co-ocalized with other small-molecule neurotransmitters such as GABA and 5-HT. In general, these peptides tend to be depressants. When injected intracerebrally, they act as analgesics and have been shown to be involved in the mechanisms underlying acupuncture-induced analgesia. Opioids are also involved in complex behaviors such as sexual attraction and aggressive/submissive behaviors. They have also been implicated in psychiatric disorders such as schizophrenia and autism, although the evidence for this is debated.

Unfortunately, the repeated administration of opioids leads to tolerance and addiction. A better understanding of these neurotransmitters and their actions will be essential for developing strategies to deal with this extraordinary social and medical problem (Box D).

Summary

The large number of neurotransmitters in the nervous system can be divided into two broad classes: small-molecule transmitters and neuropeptides. Neurotransmitters are synthesized from defined precursors by regulated enzymatic pathways, packaged into one of several vesicle types, and released into the synaptic cleft in a Ca^{2+}-dependent manner. Many synapses release more than one type of neurotransmitter, and multiple transmitters can be packaged in the same synaptic vesicle. The postsynaptic effects of neurotransmitters are terminated by the degradation of the transmitter in the synaptic cleft, by transport of the transmitter back into cells, or by diffusion out of the synaptic cleft. Glutamate is the major excitatory neurotransmitter in the brain, whereas GABA and glycine are the major inhibitory neurotransmitters. The actions of these small-molecule neurotransmitters are typically faster than those of the neuropeptides. Thus, most small-molecule transmitters mediate synaptic transmission when a rapid response is essential, whereas the neuropeptide transmitters, as well as the biogenic amines and some small-molecule neurotransmitters, tend to modulate ongoing activity in the brain or in peripheral target tissues in a more gradual and ongoing way. Drugs that influence transmitter actions have enormous importance in the treatment of neurological and psychiatric disorders, as well as in a broad spectrum of other medical problems and conditions ranging from hypertension to shock.

ADDITIONAL READING

Reviews

BECKER, C.-D. (1995) Glycine receptors: Molecular heterogeneity and implications for disease. Neuroscientist 1: 130–141.

BOURIN, M., G. B. BAKER AND J. BRADWEIN (1998) Neurobiology of panic disorder. J. Psychosomatic Res. 44: 163–180.

CARLSSON, A. (1987) Perspectives on the discovery of central monoaminergic neurotransmission. Ann. Rev. Neurosci. 10: 19–40.

CIVELLI, O. (1998) Functional genomics: The search for novel neurotransmitters and neuropeptides. FEBS Letters 430: 55–58.

EMSON, P. C. (1979) Peptides as neurotransmitter candidates in the CNS. Prog. Neurobiol. 13: 61–116

HÖKFELT, T. D. AND 10 OTHERS (1987) Coexistence of peptides with classical neurotransmitters. Experientia Suppl. 56: 154–179.

HYLAND, K. (1999) Neurochemistry and defects of biogenic amine neurotransmitter metabolism. J. Inher. Metab. Dis. 22: 353–363.

INESTROSSA, N. C. AND A. PERELMAN (1989) Distribution and anchoring of molecular forms of acetylcholinesterase. Trends Pharmacol. Sci. 10: 325–329.

JUNG, L. H. AND R. H. SCHELLER (1991) Peptide processing and targeting in the neuronal secretory pathway. Science 251: 1330–1335.

KOOB, G. F., P. P. SANNA AND F. E. BLOOM (1998) Neuroscience of addiction. Neuron 21: 467–476.

MASSON, J., C. SAGN, M. HAMON AND S. E. MESTIKAWY (1999) Neurotransmitter transporters in the central nervous system. Pharmacol. Rev. 51: 439–464.

MELDRUM, B. AND J. GARTHWAITE (1990) Glutamate neurotoxicity may underlie slowly pro-gressive degenerative diseases such as Huntington's Disease and ongoing way. Drugs Pharmacol. Sci. 11: 379–387.

PERRY, E., M. WALKER, J. GRACE AND R. PERRY (1999) Acetylcholine in mind: A neurotransmitter correlate of consciousness? Trends Neurosci. 22: 273–280.

SCHWARTZ, J. C., J. M. ARRANG, M. GARBARG, H. POLLARD AND M. RUAT (1991) Histaminergic transmission in the mammalian brain. Physiol. Rev. 71: 1–51.

SHAFQAT, S., M. VELAZ-FAIRCLOTH, A. GUNANO-FERRAZ AND R. T. FREMEAU (1993) Molecular characterization of neurotransmitter transporters. Molec. Endocrinol. 7: 1517–1529.

TUCEK, S., J. RICNY AND V. DOLEZAL (1990) Advances in the biology of cholinergic neurons. Adv. Neurol. 51: 109–115.

Important Original Papers

CHEN, Z. P., A. LEVY AND S. L. LIGHTMAN (1995) Nucleotides as extracellular signalling molecules. J. Neuroendocrinol. 7: 83–96.

CURTIS, D. R., J. W. PHILLIS AND J. C. WATKINS (1959) Chemical excitation of spinal neurons. Nature 183: 611–612.

DALE, H. H., W. FELDBERG AND M. VOGT (1936) Release of acetylcholine at voluntary motor nerve endings. J. Physiol. 86: 353–380.

HÖKFELT, T., O. JOHANSSON, A. LJUNGDAHL, J. M. LUNDBERG AND M. SCHULTZBERG (1980) Peptidergic neurons. Nature 284: 515–521.

HUGHES, J., T. W. SMITH, H. W. KOSTERLITZ, L. A. FOTHERGILL, B. A. MORGAN AND H. R. MORRIS (1975) Identification of two related pentapeptides from the brain with potent opiate agonist activity. Nature 258: 577–580.

JONAS, P., J. BISCHOFBERGER AND J. SANDKUHLER (1998) Corelease of two fast neurotransmitters at a central synapse. Science 281: 419–424.

KUPERMANN, I. (1991) Functional studies of cotransmission. Physiol. Rev. 71: 683–732.

LOEWI, O. (1921) Über humorale übertragbarheit der herznervenwirkung. Pflügers Arch. 189: 239–242.

SOSSIN, W. S., A. SWEET-CORDERO AND R. H. SCHELLER (1990) Dale's hypothesis revisited: Different neuropeptides derived from a common prohormone are targeted to different processes. Proc. Natl. Acad. Sci. U.S.A. 87: 4845–4548.

THOMAS, S. A. AND R. D. PALMITER (1995) Targeted disruption of the tyrosine hydroxylase gene reveals that catecholamines are required for mouse fetal development. Nature 374: 640–643.

WANG, Y.M. AND 8 OTHERS (1997) Knockout of the vesicular monoamine transporter 2 gene results in neonatal death and supersensitivity to cocaine and amphetamine. Neuron 19: 1285–1296.

Books

BRADFORD, H. F. (1986) Chemical Neurobiology. New York: W. H. Freeman.

COOPER, J. R., F. E. BLOOM AND R. H. ROTH (1991) The Biochemical Basis of Neuropharmacology. New York: Oxford University Press.

HALL, Z. (1992) An Introduction to Molecular Neurobiology. Sunderland, MA: Sinauer Associates, Chapters 3–7.

NICHOLLS, D. G. (1994) Proteins, Transmitters, and Synapses. Boston: Blackwell Scientific.

SIEGEL, G.J., B. W. AGRANOFF, R. W. ALBERS, S. K. FISHER AND M. D. UHLER (1999) Basic Neurochemistry. Philadelphia: Lippincott-Raven.

FELDMAN, R. S., J. S. MEYER AND L. F. QUENZER (1997) Principles of Neuropharmacology. Sunderland, MA: Sinauer Associates.

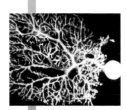

Chapter 7

Neurotransmitter Receptors and Their Effects

Overview

Neurotransmitters evoke postsynaptic electrical responses by binding to members of a diverse group of proteins called neurotransmitter receptors. The receptors then give rise to electrical signals by opening or closing ion channels in the postsynaptic membrane. Whether the postsynaptic actions of a particular neurotransmitter are excitatory or inhibitory is determined by the class of ion channel affected by the transmitter, and by the concentration of permeant ions inside and outside the cell. There are two major classes of receptors: those in which the receptor and ion channel are also an ion channel, and those in which the receptor and ion channel are separate molecules. The former are called ionotropic receptors or ligand-gated ion channels, and give rise to fast postsynaptic responses that typically last only a few milliseconds. The latter are called metabotropic receptors, and they produce slower postsynaptic effects that may endure much longer. Numerous drugs affect the central nervous system by activating or blocking neurotransmission.

Neurotransmitter Receptors Alter Postsynaptic Membrane Permeability

In 1907, the British physiologist John N. Langley introduced the concept of **receptor molecules** to explain the specific and potent actions of certain chemicals on muscle and nerve cells. Much subsequent work has shown that receptor molecules do indeed account for the ability of neurotransmitters, hormones, and drugs to alter the functional properties of neurons. While it has been clear since Langley's day that receptors are important for synaptic transmission, their identity and detailed mechanism of action remained a mystery until quite recently. It is now known that neurotransmitter receptors are proteins embedded in the plasma membrane of postsynaptic cells. Domains of receptor molecules that extend into the synaptic cleft bind neurotransmitters that are released into this space by the presynaptic neuron. The binding of neurotransmitters, either directly or indirectly, causes ion channels in the postsynaptic membrane to open or close (Figure 7.1). Typically, the resulting ion fluxes change the membrane potential of the postsynaptic cell, thus mediating the transfer of information across the synapse.

Principles Derived from Studies of the Neuromuscular Junction

A particularly accessible system for understanding neurotransmitter receptors has been the neuromuscular junction. The binding of the neurotransmit-

Figure 7.1   Receptors that mediate the postsynaptic actions of neurotransmitters have two functions. First, specific binding sites on the extracellular side of receptors allow these proteins to detect the presence of neurotransmitters in the synaptic cleft. Second, transmitter-bound receptors alter the ionic permeability of the postsynaptic membrane by virtue of being coupled, directly or indirectly, to ion channels in the postsynaptic membrane. Opening or closing these channels as a result of transmitter binding allows ionic currents to flow, thus changing the postsynaptic membrane potential.

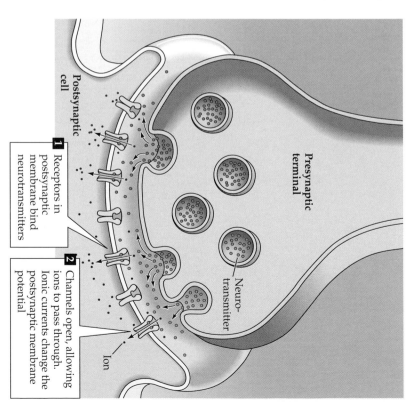

Postsynaptic cell

Presynaptic terminal

Neuro-transmitter

Ion

1 Receptors in postsynaptic membrane bind neurotransmitters

2 Channels open, allowing ions to pass through. Ionic currents change the postsynaptic membrane potential

ter acetylcholine (ACh) to postsynaptic receptors opens ion channels in the muscle fiber membrane. This effect can be demonstrated directly by using the patch clamp method (see Box A in Chapter 4) to measure the minute postsynaptic currents that flow when two molecules of individual ACh bind to receptors. Exposure of the extracellular surface of a patch of postsynaptic membrane to ACh causes single-channel currents to flow for a few milliseconds (Figure 7.2A). Even if ACh is continually applied such that the transmitter is nearly always bound to receptors, the channels open and close repeatedly. Thus, ACh opens ion channels stochastically when it binds to receptors.

The electrical consequences of ACh binding to a single receptor are greatly multiplied when an action potential in a presynaptic motor neuron causes the approximately simultaneous release of several million molecules of ACh into the synaptic cleft. In this more physiological case, the transmitter molecules bind to many thousands of ACh receptors packed in a dense array on the postsynaptic membrane, transiently opening a very large number of postsynaptic ion channels. Although individual ACh receptors only open briefly, (Figure 7.2B1), the opening of a large number of channels is synchronized by the brief duration during which ACh is secreted from presynaptic terminals (Figure 7.2B2,3). The macroscopic current resulting from the summed opening of many ion channels is called the **end plate current**, or **EPC**. Because the current flowing during the EPC is normally inward, it causes the postsynaptic membrane potential to depolarize. This depolarizing change in potential—the end plate potential, or EPP (Figure 7.2C)—typically triggers a postsynaptic action potential due to the opening of voltage-activated Na^+ and K^+ channels (see Chapter 5).

(A) Patch clamp measurement of single ACh receptor current

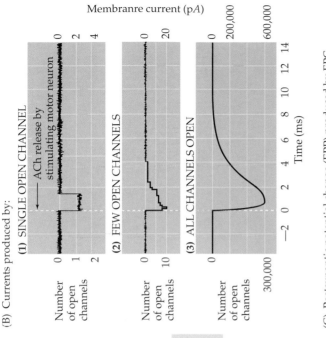

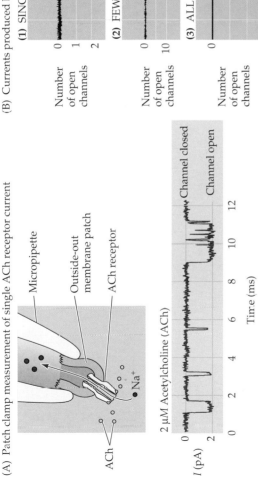

(B) Currents produced by:

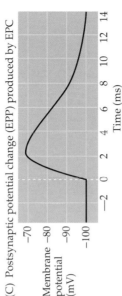

(1) SINGLE OPEN CHANNEL

ACh release by stimulating motor neuron

(2) FEW OPEN CHANNELS

(3) ALL CHANNELS OPEN

(C) Postsynaptic potential change (EPP) produced by EPC

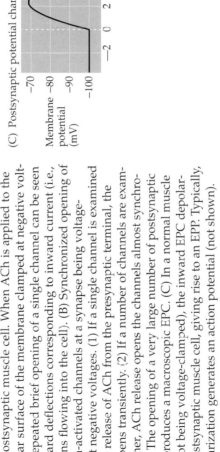

Figure 7.2 Activation of ACh receptors at neuromuscular synapses. (A) Outside-out patch clamp measurement of single ACh receptor currents from a patch of membrane removed from the postsynaptic muscle cell. When ACh is applied to the extracellular surface of the membrane clamped at negative voltages, the repeated brief opening of a single channel can be seen as downward deflections corresponding to inward current (i.e., positive ions flowing into the cell). (B) Synchronized opening of many ACh-activated channels at a synapse being voltage-clamped at negative voltages. (1) If a single channel is examined during the release of ACh from the presynaptic terminal, the channel opens transiently. (2) If a number of channels are examined together, ACh release opens the channels almost synchronously. (3) The opening of a very large number of postsynaptic channels produces a macroscopic EPC. (C) In a normal muscle cell (i.e., not being voltage-clamped), the inward EPC depolarizes the postsynaptic muscle cell, giving rise to an EPP. Typically, this depolarization generates an action potential (not shown).

The ions that flow when ACh is bound to postsynaptic receptors can be identified using the principles of ion permeation that were introduced in Chapters 2–4. In particular, the identity of the ions that are carrying the current can be determined by knowing the membrane potential at which no current flows in response to transmitter binding. When the potential of the postsynaptic muscle cell is controlled by the voltage clamp method (Figure 7.3A), the magnitude of the membrane potential clearly affects the amplitude and polarity of EPCs (Figure 7.3B). Thus, when the postsynaptic membrane potential is made more negative than the resting potential, the amplitude of the EPC becomes larger, whereas this current is reduced when the membrane potential is made more positive. At approximately 0 mV, no EPC is detected, and at even more positive potentials, the current reverses its polarity, becoming outward rather than inward (Figure 7.3C). The potential where the EPC reverses, about 0 mV in the case of the neuromuscular junction, is called the **reversal potential**.

From Ohm's Law (see Chapter 3), the magnitude of the EPC at any membrane potential is given by the product of the ionic conductance activated by ACh (g_{ACh}) and the electrochemical driving force on the ions flowing through ligand-gated channels (remember that conductance is the inverse of

(A) Scheme for voltage clamping postsynaptic muscle fiber

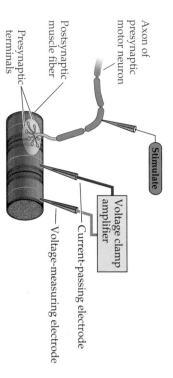

(B) Effect of membrane voltage on postsynaptic end plate currents

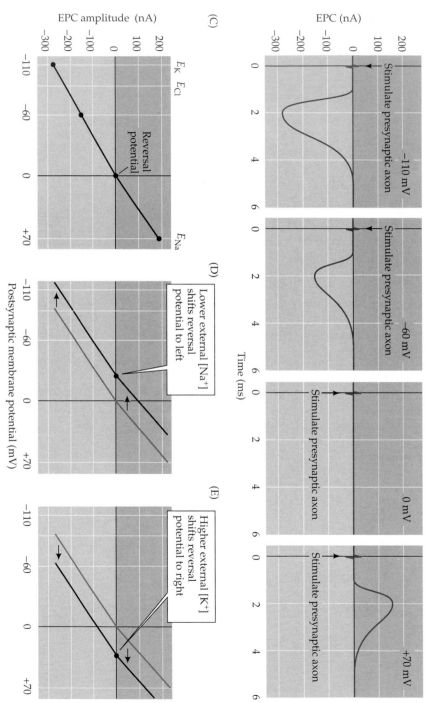

Figure 7.3   The influence of the postsynaptic membrane potential on end plate currents. (A) A postsynaptic muscle fiber is voltage clamped using two electrodes, while the presynaptic neuron is electrically stimulated to cause the release of ACh from presynaptic terminals. This experimental arrangement allows the recording of macroscopic EPCs produced by ACh. (B) Amplitude and time course of EPCs generated by stimulating the presynaptic motor neuron while the postsynaptic cell is voltage clamped at four different membrane potentials. (C) The relationship between the peak amplitude of EPCs and postsynaptic membrane potential is nearly linear, with a reversal potential (the voltage at which the direction of the current changes from inward to outward) close to 0 mV. Also indicated on this graph are the equilibrium potentials of Na⁺, K⁺, and Cl⁻ ions. (D) Lowering the external Na⁺ concentration causes EPCs to reverse at more negative potentials. (E) Raising the external K⁺ concentration makes the reversal potential more positive. (After Takeuchi and Takeuchi, 1960.)

resistance). The driving force is simply the difference between the postsynaptic membrane potential, V_m, and the reversal potential for the EPC, E_{rev}. Thus, the value of the EPC is given by the relationship

$$EPC = g_{ACh}(V_m - E_{rev})$$

This relationship predicts that the EPC will be an inward current at potentials more negative than E_{rev} because the electrochemical driving force, $V_m - E_{rev}$, is a negative number; thus, the EPC becomes smaller at potentials approaching E_{rev} because the driving force is reduced. At potentials more positive than E_{rev}, the EPC is outward because the driving force is reversed in direction (that is, positive).

Because the channels opened by ACh are not sensitive to membrane voltage, g_{ACh} should depend only on the number of channels opened by ACh (which depends in turn on the concentration of ACh in the synaptic cleft). Thus, the magnitude and polarity of the postsynaptic membrane potential determines the direction and amplitude of the EPC solely by altering the driving force on ions flowing through the receptor channels opened by ACh.

When V_m is at the reversal potential, $V_m - E_{rev}$ is equal to 0 and there is no net driving force on the ions that can permeate the receptor-activated channel. As a result, the identity of the ions that flow during the EPC can be deduced by observing how the reversal potential of the EPC compares to the equilibrium potential for various ion species (Figure 7.4). For example, if ACh were to open an ion channel permeable only to K^+, then the reversal potential of the EPC would be at the equilibrium potential for K^+, which for a muscle cell is close to −100 mV (Figure 7.4A). If the ACh-activated channels were permeable only to Na^+, then the reversal potential of the current would be approximately +70 mV, the Na^+ equilibrium potential of muscle cells (Figure 7.4B); if these channels were permeable only to Cl^-, then the reversal potential would be approximately −50 mV (Figure 7.4C). By this reasoning, ACh-activated channels cannot be permeable to only one of these ions, because the reversal potential of the EPC is not near the equilibrium potential for any of them (see Figure 7.3C). However, if these channels were permeable to both Na^+ and K^+, then the reversal potential of the EPC would be between −70 mV and −100 mV (Figure 7.4D).

The fact that EPCs reverse at approximately 0 mV is therefore consistent with the idea that ACh-activated ion channels are almost equally permeable to both Na^+ and K^+. This implication has been tested by experiments in which the extracellular concentration of these two ions is altered. As expected, the magnitude and reversal potential of the EPC are changed by altering the concentration gradient of each ion. Lowering the external Na^+ concentration, which makes E_{Na} more negative, produces a negative shift in E_{rev} (Figure 7.3D), whereas elevating external K^+ concentration, which makes E_K more positive, causes E_{rev} to shift to a more positive potential (Figure 7.3E). Such experiments confirm that the ACh-activated ion channels are in fact permeable to both Na^+ and K^+.

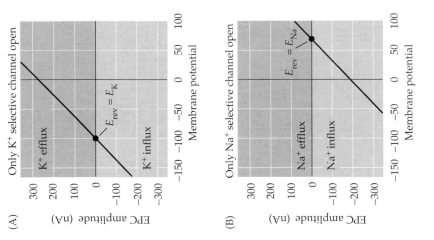
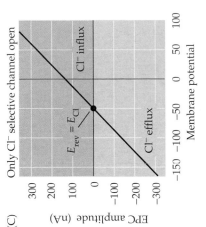
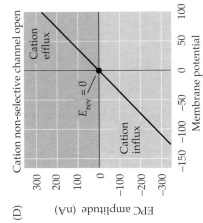

Figure 7.4 The effect of ion channel selectivity on the reversal potential. Voltage clamping a postsynaptic cell while activating presynaptic neurotransmitter release reveals the identity of the ions permeating the postsynaptic receptors being activated. (A) The activation of postsynaptic channels permeable only to K^+ results in currents reversing at E_K, near −100 mV. (B) The activation of postsynaptic Na^+ channels results in currents reversing at E_{Na}, near +70 mV. (C) Cl^- selective currents reverse at E_{Cl}, near −50 mV. (D) Ligand-gated channels that are about equally permeable to both K^+ and Na^+ show a reversal potential near 0 mV.

Even though the channels opened by the binding of ACh to its receptors are permeable to both Na⁺ and K⁺, at the resting membrane potential the EPC is generated primarily by Na⁺ influx (Figure 7.5). If the membrane potential is kept at E_K, the EPC arises entirely from an influx of Na⁺ because at this potential there is no driving force on K⁺ (Figure 7.5A). At the usual muscle fiber

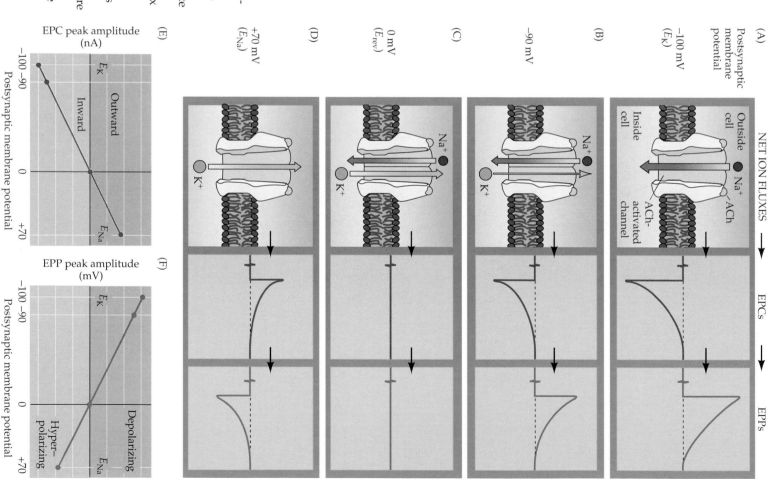

(A) Postsynaptic membrane potential

−100 mV (E_K)

(B) −90 mV

(C) 0 mV (E_{rev})

(D) +70 mV (E_{Na})

NET ION FLUXES → EPCs → EPPs

(E) EPC peak amplitude (nA)

Outward / Inward

E_K −100 −90 +70 E_{Na}

Postsynaptic membrane potential

(F) EPP peak amplitude (mV)

Depolarizing / Hyper-polarizing

E_K −100 −90 +70 E_{Na}

Postsynaptic membrane potential

Figure 7.5 Na⁺ and K⁺ movements during EPCs and EPPs. (A–D) Each of the postsynaptic potentials (V_{post}) indicated at the left results in different relative fluxes of net Na⁺ and K⁺ (ion fluxes). These ion fluxes determine the amplitude and polarity of the EPCs, which in turn determine the EPPs. Note that at about 0 mV the Na⁺ flux is exactly balanced by an opposite K⁺ flux resulting in no net current flow, and hence, no change in the membrane potential. (E) EPCs are inward currents at potentials more negative than E_{rev} and outward currents at potentials more positive than E_{rev}. (F) EPPs depolarize the postsynaptic cell at potentials more negative than E_{rev}. At potentials more positive than E_{rev}, EPPs hyperpolarize the cell.

resting membrane potential of –90 mV, there is a small driving force on K+, but a much greater one on Na+. Thus, during the EPC, much more Na+ flows into the muscle cell than K+ flows out (Figure 7.5B); it is the net influx of positively charged Na+ that constitutes the inward current measured as the EPC. At the reversal potential of about 0 mV, Na+ influx and K+ efflux are exactly balanced, so no current flows during the opening of channels by ACh binding (Figure 7.5C). At potentials more positive than E_{rev}, the balance reverses; for example, at E_{Na} there is no influx of Na+ and a large efflux of K+ because of the large driving force on Na+ (Figure 7.5D). Even more positive potentials cause efflux of both Na+ and K+ and produce an even larger outward EPC.

Were it possible to measure the end plate potential charge at the same time as the end plate current (the voltage clamp technique prevents this by keeping membrane potential constant), the EPP would be seen to vary in parallel with the amplitude and polarity of the EPC (Figures 7.5E,F). At the usual postsynaptic resting membrane potential of –90 mV, the large inward EPC causes the postsynaptic membrane potential to become more depolarized (see Figure 7.5F). However, at 0 mV, the EPP reverses its polarity, and at more positive potentials, the EPP is hyperpolarizing. Thus, the polarity and magnitude of the EPP depend on the electrochemical driving force, which determines the polarity and magnitude of the EPC. EPPs will depolarize when the membrane potential is more negative than E_{rev}, and hyperpolarize when the membrane potential is more positive than E_{rev}. The general rule, then, is that *the action of a transmitter drives the postsynaptic membrane potential toward E_{rev} for the particular ion channels being activated.*

Although this discussion has focused on the neuromuscular junction, similar mechanisms generate postsynaptic responses at all chemical synapses. A principle that emerges is that transmitter binding to postsynaptic receptors produces a postsynaptic conductance change as ion channels are opened (or sometimes closed). The postsynaptic conductance is increased if—as at the neuromuscular junction—channels are opened, and decreased if channels are closed. This conductance change typically generates an electrical current, the **postsynaptic current (PSC)**, which in turn changes the postsynaptic membrane potential to produce a **postsynaptic potential (PSP)**. As in the case of the EPP at the neuromuscular junction, the PSP will be depolarizing if its reversal potential is more positive than the postsynaptic membrane potential and hyperpolarizing if its reversal potential is more negative.

The conductance changes and the PSPs that typically accompany them are the ultimate outcome of most chemical synaptic transmission, concluding a sequence of electrical and chemical events that begins with the invasion of an action potential into the terminals of a presynaptic neuron. In many ways, the events that produce PSPs at synapses are similar to those that generate action potentials in axons; in both cases, conductance changes produced by ion channels lead to ionic current flow that changes the membrane potential (see Figure 7.5).

Excitatory and Inhibitory Postsynaptic Potentials

Postsynaptic conductance changes and the potential changes that accompany them alter the probability that an action potential will be produced in the postsynaptic cell. At the neuromuscular junction, synaptic action increases the probability that an action potential will occur in the postsynaptic muscle cell; indeed, the large amplitude of the EPP ensures that an action potential always is triggered. At many other synapses, PSPs actually *decrease* the probability that the postsynaptic cell will generate an action potential. PSPs are called **excitatory** (or **EPSPs**) if they increase the likeli-

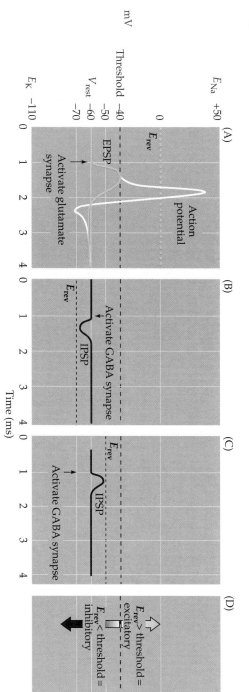

Figure 7.6 Reversal potentials and threshold potentials determine postsynaptic excitation and inhibition. (A) If the reversal potential for a PSP (0 mV) is more positive than the action potential threshold (−40 mV), the effect of a transmitter is excitatory, and it generates EPSPs. (B) If the reversal potential for a PSP is more negative than the action potential threshold, the transmitter is inhibitory and generate IPSPs. (C) IPSPs can nonetheless depolarize the postsynaptic cell if their reversal potential is between the resting potential and the action potential threshold. (D) The general rule of postsynaptic action is: If the reversal potential is more positive than threshold, excitation results; inhibition occurs if the reversal potential is more negative than threshold.

hood of a postsynaptic action potential occurring, and **inhibitory** (or **IPSPs**) if they decrease this likelihood. Given that most neurons receive inputs from both excitatory and inhibitory synapses, it is important to understand more precisely the mechanisms that determine whether a particular synapse excites or inhibits its postsynaptic partner.

The principles of excitation just described for the neuromuscular junction are pertinent to all excitatory synapses. The principles of postsynaptic inhibition are much the same as for excitation, and are also general. In both cases, neurotransmitters binding to receptors open or close ion channels in the postsynaptic cell. Whether a postsynaptic response is an EPSP or an IPSP depends on the type of channel that is coupled to the receptor, and on the concentration of permeant ions inside and outside the cell. In fact, the only factor that distinguishes postsynaptic excitation from inhibition is the reversal potential of the PSP in relation to the threshold voltage for generating action potentials in the postsynaptic cell.

Consider, for example, a neuronal synapse that uses glutamate as the transmitter. Many such synapses have receptors that, like the ACh receptors at neuromuscular synapses, open ion channels that are nonselectively permeable to cations. When these glutamate receptors are activated, both Na^+ and K^+ flow across the postsynaptic membrane. The reversal potential (E_{rev}) for the postsynaptic current is approximately 0 mV, whereas the resting potential of neurons is approximately −60 mV. The resulting EPSP will depolarize the postsynaptic membrane potential, bringing it toward 0 mV. For the particular neuron shown in Figure 7.6A, the action potential threshold voltage is −40 mV. Thus, the EPSP increases the probability that the postsynaptic neuron will produce an action potential, defining this synapse as excitatory.

As an example of inhibitory postsynaptic action, consider a neuronal synapse that uses GABA as its transmitter. At such synapses, the GABA receptors typically open channels that are selectively permeable to Cl^-. When these channels open, negatively charged chloride ions can flow across the membrane. Assume that the postsynaptic neuron has a resting potential of −60 mV and an action potential threshold of −40 mV, as in the previous example. If E_{Cl} is −70 mV, as is typical for many neurons, transmitter release at this synapse will inhibit the postsynaptic cell (because E_{Cl} is more negative than the action potential threshold). In this case, the electrochemical driving force ($V_m - E_{rev}$) causes Cl^- to flow into the cell, generating an outward PSC (because Cl^- is negatively charged) and consequently a hyperpolarizing IPSP

(Figure 7.6B). Because E_{Cl} is more negative than the action potential threshold, the conductance change arising from the binding of GABA keeps the postsynaptic membrane potential more negative than threshold, thereby reducing the probability that the postsynaptic cell will fire an action potential.

However, not all inhibitory synapses produce hyperpolarizing IPSPs. For instance, in the neuron just described, if E_{Cl} were –50 mV instead of –70 mV, then the synapse would still be inhibitory because the reversal potential of the IPSP remains more negative than the action potential threshold (–40 mV). Because the electrochemical driving force now causes Cl⁻ to flow out of the cell, however, the IPSP is actually depolarizing (Figure 7.6C). Nonetheless, this depolarizing IPSP inhibits the postsynaptic cell because the cell's membrane potential is kept more negative than the threshold potential for action potential initiation. Another way to think about this peculiarity is that if another depolarizing input were to bring the cell's resting potential to –41 mV, just below threshold for firing an action potential, the opening of these GABA-activated channels would result in a hyperpolarizing current, bringing the membrane potential closer to –50 mV, the reversal potential for these channels. Thus, while EPSPs depolarize the postsynaptic cell, IPSPs can hyperpolarize or depolarize; indeed, an inhibitory conductance change may produce no potential change at all and still exert an inhibitory effect.

Although the particulars of postsynaptic action can be complex, a simple rule distinguishes postsynaptic excitation from inhibition: An EPSP has a reversal potential more positive than the action potential threshold, whereas an IPSP has a reversal potential more negative than threshold (Figure 7.6D). Intuitively, this rule can be understood by realizing that an EPSP will tend to depolarize the membrane potential so that it exceeds threshold, whereas an IPSP will always act to keep the membrane potential more negative than the threshold potential.

Summation of Synaptic Potentials

The postsynaptic effects of most synapses in the brain are not nearly as large as those at the neuromuscular junction; indeed, PSPs due to the activity of individual synapses are usually well below the threshold for generating postsynaptic action potentials, and may be only a fraction of a millivolt. How, then, can neurons in the brain transmit information from presynaptic to postsynaptic cells if most central synaptic effects are subthreshold? The answer is that neurons in the central nervous system are typically innervated by thousands of synapses, and the PSPs produced by each active synapse can *sum together*—in space and in time—to determine the behavior of the postsynaptic neuron.

Consider the highly simplified case of a neuron that is innervated by two excitatory synapses, each generating a subthreshold EPSP, and an inhibitory synapse that produces an IPSP (Figure 7.7A). While activation of either one of the excitatory synapses alone (E1 or E2 in Figure 7.7B) produces a subthreshold EPSP, activation of both excitatory synapses at about the same time causes the two EPSPs to sum together. If the sum of the two EPSPs (E1 + E2) depolarizes the postsynaptic neuron sufficiently to reach the threshold potential, a postsynaptic action potential results. **Summation** thus allows subthreshold EPSPs to influence action potential production. Likewise, an IPSP generated by an inhibitory synapse (I) can sum (algebraically speaking) with a subthreshold EPSP to reduce its amplitude (E1 + I) or can sum with supra-threshold EPSPs to prevent the postsynaptic neuron from reaching threshold (E1 + I + E2).

Catecholamine Receptors

Catecholamines act exclusively by activating G-protein-coupled receptors. Many of these metabotropic catecholamine receptors contribute to complex behaviors. For example, administration of dopamine receptor agonists elicits hyperactivity and repetitive, stereotyped behavior in laboratory animals. Activation of another type of dopamine receptor in the medulla inhibits vomiting. Thus, antagonists to these receptors are used to induce vomiting after poisoning or a drug overdose. Dopamine receptor antagonists also can elicit catalepsy, a state in which it is difficult to initiate voluntary motor movement. Most dopamine receptor subtypes (see Figure 7.13B) act by either activating or inhibiting adenylyl cyclase.

The other two catecholamines, NE and epinephrine, each act on α- and β-adrenergic receptors (Figure 7.13B). Two subclasses of α-adrenergic receptors are known. Activation of α_1-receptors usually results in a slow depolarization linked to the inhibition of K^+ channels, while activation of α_2-receptors produces a slow hyperpolarization due to the activation of a different type of K^+ channel. There are three subtypes of β-adrenergic receptor, two of which are expressed in many types of neurons. Agonists and antagonists of adrenergic receptors, such as the β-blocker propanolol (Inderol®), are widely used clinically. However, most of their actions are on smooth muscle receptors, particularly the cardiovascular and respiratory systems (see Chapter 21).

Peptide Receptors

Virtually all neuropeptides mediate their effects by activating G-protein-coupled receptors. The study of these metabotropic peptide receptors in the brain has been difficult because few specific agonists and antagonists are known. Peptides activate their receptors at low (nM to μM) concentrations compared to the concentrations required to activate receptors for small-molecule neurotransmitters (see Chapter 6). These properties allow the postsynaptic targets of peptides to be quite far removed from presynaptic terminals and to modulate the electrical properties of neurons that are simply in the vicinity of the site of peptide release. Neuropeptide receptor activation is particularly important in regulating the postganglionic output from sympathetic ganglia and the activity of the gut (see Chapter 21). Peptide receptors, particularly the neuropeptide Y receptor, are also implicated in the initiation and maintenance of feeding behavior leading to satiety or obesity.

Other behaviors ascribed to peptide receptor activation include anxiety and panic attacks, and antagonists of cholecystokinin receptors are clinically useful in the treatment of these afflictions. Other useful drugs have been developed by targeting the opiate receptors. Three well-defined opioid receptor subtypes (μ, δ, and κ) play a role in reward mechanisms as well as addiction (see Box D in Chapter 6). The μ-opiate receptor has been specifically identified as the primary site for drug reward mediated by opiate drugs.

Summary

The actions of neurotransmitters at synapses throughout the brain arise from the tremendous diversity of postsynaptic neurotransmitter receptors, which are proteins embedded in the plasma membranes of postsynaptic cells. These receptors translate chemical signals into electrical signals by binding neurotransmitter molecules secreted by presynaptic neurons, which leads in turn to opening or closing of postsynaptic ion channels. The postsynaptic currents produced by the synchronous opening or closing of ion channels changes the conductance of the postsynaptic cell, thus increasing or decreasing its excitability. Conductance changes that increase the probability of firing an action potential are excitatory, whereas those that decrease the probability of generating an action potential are inhibitory. Because postsynaptic neurons are usually innervated by many different inputs, the integrated effect of the conductance changes underlying all EPSPs and IPSPs produced in a postsynaptic cell at any moment determines whether or not the cell fires an action potential. Two broadly different families of neurotransmitter receptors have evolved to carry out the postsynaptic signaling actions of neurotransmitters. Ionotropic or ligand-gated ion channels combine the neurotransmitter receptor and ion channel in one molecular entity, and therefore give rise to rapid postsynaptic electrical responses. Metabotropic receptors regulate the activity of postsynaptic ion channels indirectly, usually via G-proteins, and induce slower and longer-lasting electrical responses. Metabotropic receptors are especially important in regulating behavior, and drugs targeting these receptors have been clinically valuable in treating a wide range of behavioral disorders. The postsynaptic response at a given synapse is determined by the combination of receptor subtypes, G-protein subtypes, and ion channels that are expressed in the postsynaptic cell. Because each of these features can vary both within and among neurons, a tremendous diversity of transmitter-mediated effects is possible.

ADDITIONAL READING

Reviews

BARNES, N. M. AND T. SHARP (1999) A review of central 5-HT receptors and their function. Neuropharm. 38(8): 1083–1152.

BROWN, A. M. AND L. BIRNBAUMER (1990) Ionic channels and their regulation by G-proteins. Ann. Rev. Physiol. 52: 197–213.

BURNSTOCK, G. (1999) Current status of purinergic signalling in the nervous system. Prog. Brain Res. 120: 3–10.

CHANGEUX, J.-P. (1993) Chemical signaling in the brain. Sci. Am. 269 (May): 58–62.

DINGLEDINE, R. (1991) Molecular properties of AMPA/kainate receptors. Trends Pharmacol. Sci. 12: 360–362.

FREDHOLM, B. B. (1995) Adenosine, adenosine receptors and the actions of caffeine. Pharmacol. Toxicol. 76: 93–101.

HILLE, B. (1994) Modulation of ion channel function by G-protein-coupled receptors. Trends Neurosci. 17: 531–535.

KAUPMANN, K. AND 10 OTHERS (1997) Expression cloning of GABAβ receptors uncovers similarity to metabotropic glutamate receptors. Nature 386: 239–246.

LOVINGER, D. M. (1999) 5-HT3 receptors and the neural actions of alcohols: An increasingly exciting topic. Neurochem. Internat. 35(2): 125–30.

MACKENZIE, A. B., A. SURPRENANT AND R. A. NORTH (1999) Functional and molecular diversity of purinergic ion channel receptors. Ann. NY Acad. Sci. 868: 716–729.

NAKANISHI, S. (1992) Molecular diversity of glutamate receptors and implication for brain function. Science 258: 597–603.

ROSENMUND, C., Y. STERN-BACH AND C. F. STEVENS (1998) The tetrameric structure of a glutamate receptor channel. Science 280: 1596–1599.

SCHWARTZ, M. W., S. C. WOODS, D. PORTE JR., R. J. SEELEY AND D. G. BASKIN (2000) Central nervous system control of food intake. Nature 404: 661–671.

WEBB, T. E. AND E. A. BARNARD (1999) Molecular biology of P2Y receptors expressed in the nervous system. Prog. Brain Res. 120: 23–31.

Important Original Papers

BRICKLEY, S. G., S. G. CULL-CANDY AND M. FARRANT (1999) Single-channel properties of synaptic and extrasynaptic GABA$_A$ receptors suggest differential targeting of receptor subtypes. J. Neurosci. 19: 2960–2973.

DAVIES, P. A. AND 6 OTHERS (1999) The 5-HT3B subunit is a major determinant of serotonin-receptor function. Nature 397: 359–363.

GU, J. G. AND A. B. MACDERMOTT (1997) Activation of ATP P2X receptors elicits glutamate release from sensory neuron synapses. Nature 389: 749–753.

HARRIS, B. A., J. D. ROBISHAW, S. M. MUMBY AND A. G. GILMAN (1985) Molecular cloning of complementary DNA for the alpha subunit of the G protein that stimulates adenylate cyclase. Science 229: 1274–1277.

HOLLMANN, M., C. MARON AND S. HEINEMANN (1994) N-glycosylation site tagging suggests a three transmembrane domain topology for the glutamate receptor GluR1. Neuron 13: 1331–1343.

LEDEBT, C. AND 9 OTHERS (1997) Aggressiveness, hypoalgesia and high blood pressure in mice lacking the adenosine A2a receptor. Nature 388: 674–678.

MATTERA, R. AND 8 OTHERS (1989) Splice variants of the alpha subunit of the G-protein G$_s$ activate both adenylyl cyclase and calcium channels. Science 243: 804–807.

NAVEILHAN, P. AND 10 OTHERS (1999) Normal feeding behavior, body weight and leptin response require the neuropeptide Y Y2 receptor. Nature Med. 5: 1188–1193.

UNWIN, N. (1995) Acetylcholine receptor channels imaged in the open state. Nature 373: 37–43.

WICKMAN, K. AND 7 OTHERS (1994) Recombinant G$_{\beta\gamma}$ activates the muscarinic-gated atrial potassium channel I$_{KACh}$. Nature 368: 255–257.

Books

FELDMAN, R. S., J. S. MEYER AND L. F. QUENZER (1997) *Principles of Neuropharmacology*. Sunderland, MA: Sinauer Associates.

HALL, Z. (1992) *An Introduction to Molecular Neurobiology*. Sunderland, MA: Sinauer Associates, Chapters 3–7.

HILLE, B. (1992) *Ionic Channels of Excitable Membranes*. Sunderland, MA: Sinauer Associates, Chapters 1–8, 16–20.

MYCEK, M. J., R. A. HARVEY AND P. C. CHAMPE (2000) *Pharmacology*, 2nd Ed. Philadelphia/NY: Lippincott–Williams and Wilkins Publishers.

NICHOLLS, D. G. (1994) *Proteins, Transmitters, and Synapses*. Boston: Blackwell Scientific.

SIEGEL, G. J. AND 5 OTHERS (1999) *Basic Neurochemistry*, 6th Ed. Philadelphia/NY: Lippincott–Raven Publishers.

Chapter 8

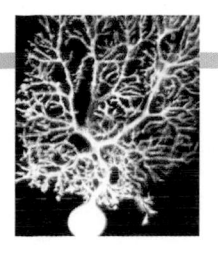

Intracellular Signal Transduction

Overview

As is apparent in the preceding chapters, electrical signaling mechanisms allow one nerve cell to receive and transmit information to another. This chapter considers the related processing of biochemical information that occurs specifically *within* neurons and other cells. This intracellular processing typically begins when extracellular chemical signals, such as neurotransmitters, hormones, and trophic factors, bind to specific receptors expressed by cellular targets. Such binding activates the receptors and stimulates cascades of intracellular reactions involving GTP-binding proteins, second messenger molecules, protein kinases, ion channels, and many other effector proteins that temporarily change the physiological state of the target cell. These various intracellular signal transduction pathways can also cause longer-lasting changes by altering the transcription of genes, thus affecting the protein composition of the target cells. The large number of components involved in intracellular signaling pathways allows precise temporal and spatial control over the function of individual neurons and other target cells, at the same time allowing coordination of the electrical and chemical activity of neuronal populations.

Strategies of Molecular Signaling

Chemical communication coordinates the behavior of individual nerve and glial cells in physiological processes that range from neural differentiation to learning and memory. Indeed, molecular signaling ultimately mediates all brain functions. To carry out such communication, a series of extraordinarily diverse and complex chemical signaling pathways has evolved. The preceding chapters have described in some detail the electrical signaling mechanisms that allow neurons to generate action potentials for conduction of information, and synaptic transmission, a special form of chemical signaling that transfers this information from one neuron to another. Chemical signaling is not, however, limited to synapses (Figure 8.1A). Other well-characterized forms of chemical communication include **paracrine** signaling, which acts over a longer range than synaptic transmission and involves the secretion of chemical signals onto a group of nearby target cells, and **endocrine** signaling, which refers to the secretion of hormones into the bloodstream where they can affect targets throughout the body.

Chemical signaling of any sort requires three components: a molecular *signal* that transmits information from one cell to another, a *receptor* molecule that transduces the information provided by the signal, and a *target* molecule that mediates the cellular response (Figure 8.1B). The part of this process

(A) **Synaptic** **Paracrine** **Endocrine** (B)

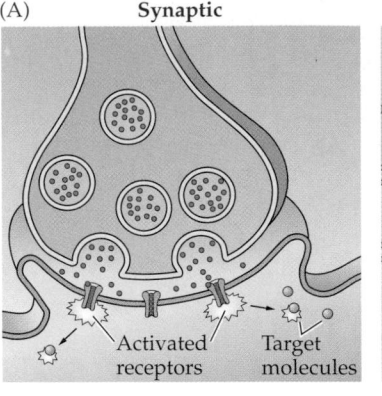 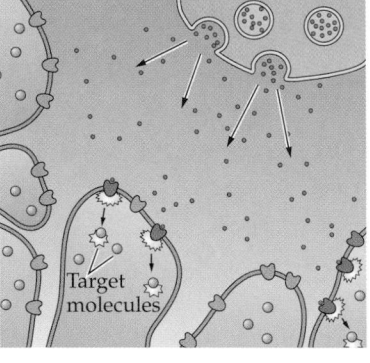 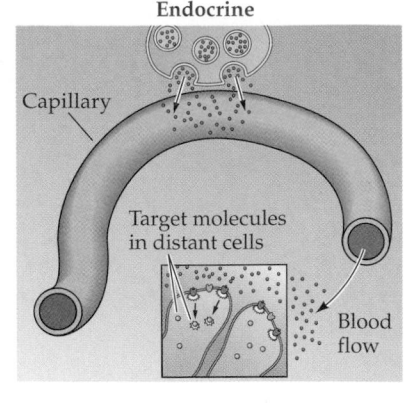 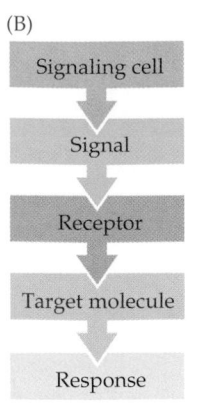

Figure 8.1 Chemical signaling mechanisms. (A) Forms of chemical communication include synaptic transmission, paracrine signaling, and endocrine signaling. (B) The essential components of chemical signaling are: cells that initiate the process by releasing signaling molecules; specific receptors on target cells; second messenger target molecules; and subsequent cellular responses.

that take place within the confines of the target cell is called **intracellular signal transduction**. A good example of transduction is the sequence of events triggered by chemical synaptic transmission (see Chapter 5): Neurotransmitters serve as the signal, neurotransmitter receptors serve as the transducing receptor, and ion channels are the target molecule that is altered to cause the electrical response of the postsynaptic cell. In many cases, however, synaptic transmission activates additional intracellular pathways that have a variety of functional consequences. For example, the binding of the neurotransmitter norepinephrine to its receptor activates GTP-binding proteins, which produces second messengers within the postsynaptic target, activates enzyme cascades, and eventually changes the chemical properties of numerous target molecules.

A major advantage of such chemical signaling schemes is **signal amplification**. Amplification occurs because individual signaling reactions can produce a large number of products. In the case of norepinephrine signaling, a single norepinephrine molecule binding to its receptor can generate many thousands of second messenger molecules (such as cyclic AMP), yielding an amplification of tens of thousands of phosphates transferred to target proteins (Figure 8.2). Similar amplification occurs in all signal transduction pathways. Furthermore, because the transduction processes often are mediated by a sequential set of enzymatic reactions, each with its own amplification factor, a small number of signal molecules ultimately can activate a very large number of target molecules. Such amplification guarantees that a physiological response is evoked in the face of other, potentially countervailing, influences.

Another rationale for complex signal transduction schemes is to permit precise control of cell behavior over a wide range of times. Some molecular interactions allow information to be transferred rapidly, while others are slower and longer lasting. For example, the signaling cascades associated with synaptic transmission at neuromuscular junctions allow an observer to respond to rapidly changing cues, such as the trajectory of a pitched ball, while the slower responses triggered by adrenal medullary hormones secreted during a challenging game (epinephrine and norepinephrine) produce slower (and longer-lasting) effects on muscle metabolism (see Chapter 21). To encode information that varies so widely over time, the concentration of the relevant signaling molecules must be precisely controlled. On one hand, the concentration of every signaling molecule within the signaling cascade must return to subthreshold values before the arrival of another stimulus. On the other hand, keeping the intermediates in a signaling pathway activated is critical for a sustained response. Having multiple levels of molecular interactions facilitates the timing of these events.

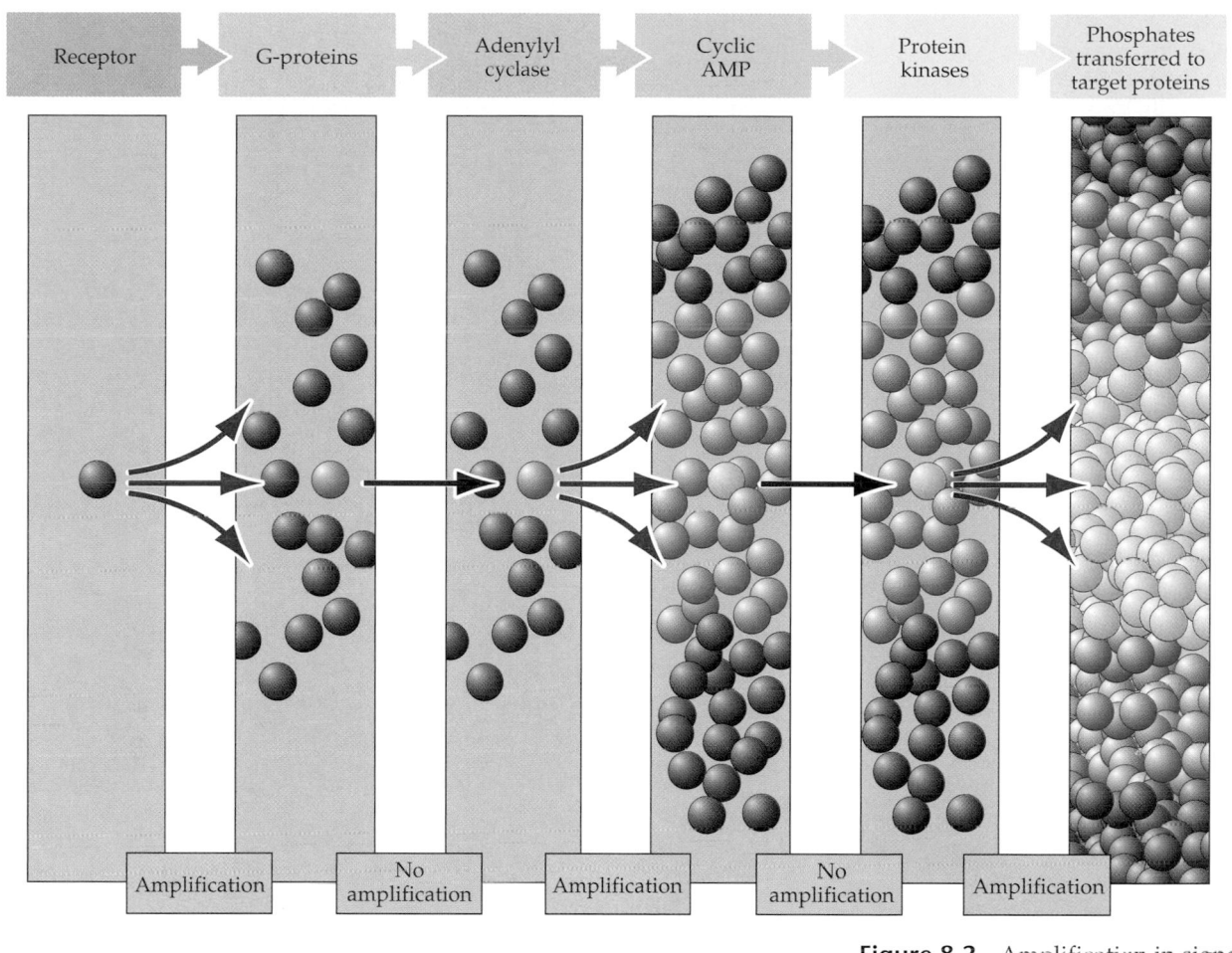

Figure 8.2 Amplification in signal transduction pathways. The activation of a single receptor by a signaling molecule, such as the neurotransmitter norepinephrine, can lead to the activation of numerous G-proteins inside cells. These activated proteins can bind to other signaling molecules, such as the enzyme adenylyl cyclase. Each activated enzyme molecule can then generate a large number of cAMP molecules. cAMP binds to and activates another family of enzymes, protein kinases. These enzymes can then phosphorylate many target proteins. While not every step in this signaling pathway involves amplification, overall the cascade results in a tremendous increase in the potency of the initial signal.

The Activation of Signaling Pathways

The molecular components of these signal transduction pathways are always activated by a chemical signaling molecule. Signaling molecules can be grouped into three classes: **cell-impermeant**, **cell-permeant**, and **cell-associated signaling molecules** (Figure 8.3). The first two classes are secreted molecules and thus can act on target cells removed from the site of signal synthesis or release. Cell-impermeant signaling molecules typically bind to receptors associated with cell membranes. Hundreds of secreted molecules have now been identified, including the neurotransmitters discussed in Chapter 6, as well as proteins such as neurotrophic factors (see Chapter 23), and peptide hormones such as glucagon, insulin, and various reproductive hormones. These signaling molecules are typically short-lived, either because they are rapidly metabolized or because they are internalized by endocytosis once bound to their receptors.

Cell-permeant signaling molecules can cross the plasma membrane to directly act on receptors that are inside the cell. Examples include numerous steroid (glucocorticoids, estradiol, and testosterone) and thyroid (thyroxin) hormones, and retinoids. These signaling molecules are relatively insoluble in aqueous solutions and are often transported in blood and other extracellular fluids by binding to specific carrier proteins. In this form, they may persist in the bloodstream for hours or even days.

The third group of chemical signaling molecules, cell-associated signaling molecules, are arrayed on the extracellular side of the plasma membrane.

Figure 8.3   Three classes of cell signaling molecules. (A) Cell-impermeant molecules, such as neurotransmitters, cannot readily traverse the plasma membrane of the target cell and must bind to the extracellular portion of transmembrane receptor proteins. (B) Cell-permeant molecules are able to cross the plasma membrane and bind to receptors in the cytoplasm or nucleus of target cells. (C) Cell-associated molecules are presented on the extracellular surface of the plasma membrane. These signals activate receptors on target cells only if they are directly adjacent to the signaling cell.

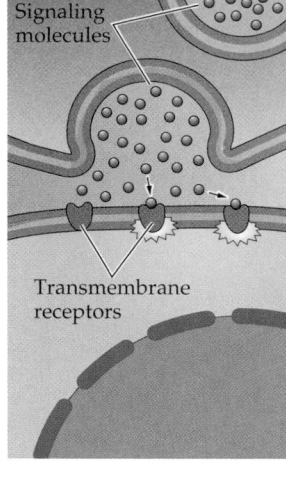

(A) Cell-impermeant molecules

Signaling molecules

Transmembrane receptors

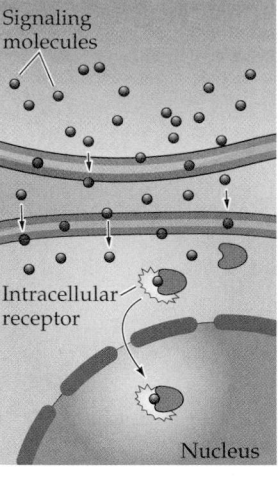

(B) Cell-permeant molecules

Signaling molecules

Intracellular receptor

Nucleus

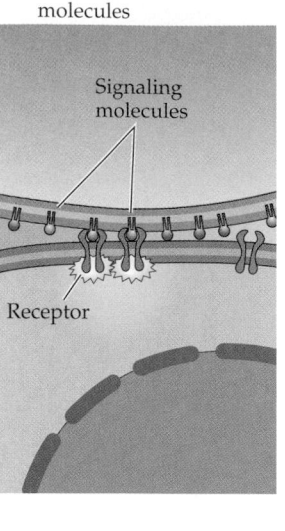

(C) Cell-associated molecules

Signaling molecules

Receptor

These molecules act only on cells physically in contact with the cell possessing such signals. Examples include proteins such as the integrins and neural cell adhesion molecules (NCAMs) that influence axonal growth (see Chapter 23). Membrane-bound signaling molecules are more difficult to study, but are clearly important in neuronal development and other circumstances where physical contact between cells provides information about cellular identities.

Receptor Types

Regardless of the nature of the initiating signal, the cellular responses are determined by the presence of receptors that specifically bind the signaling molecules. Binding of signal molecules causes a conformational change in the receptor, which then triggers the subsequent signaling cascade. Given that chemical signals can act either at the plasma membrane or within the cytoplasm (or nucleus) of the target cell, it is not surprising that receptors are found on both sides of the plasma membrane.

The receptors for impermeant signal molecules are membrane-spanning proteins that include components both outside and inside the cell surface. The extracellular domain of such receptors includes the binding site for the signal, while the intracellular domain activates intracellular signaling cascades after the signal binds. A large number of these receptors have been identified and are grouped into three families defined by the mechanism used to transduce signal binding into a cellular response (Figure 8.4A–C).

Channel-linked receptors (also called ligand-gated ion channels) have the receptor and transducing functions as part of the same protein molecule. Interaction of the chemical signal with the binding site of the receptor causes the opening or closing of an ion channel pore in another part of the same molecule. The resulting ion flux changes the membrane potential of the target cell and, in some cases, can also lead to entry of Ca^{2+} ions that serve as a second messenger signal within the cell. Good examples of such receptors are the neurotransmitter receptors described in Chapter 7.

Enzyme-linked receptors also have an extracellular binding site for chemical signals. The intracellular domain of such receptors is an enzyme whose catalytic activity is regulated by the binding of an extracellular signal. The great majority of these receptors are **protein kinases,** often tyrosine kinases, that phosphorylate intracellular target proteins, thereby changing the physiological function of the target cells. Noteworthy members of this

(A) Channel-linked receptors

(B) Enzyme-linked receptors

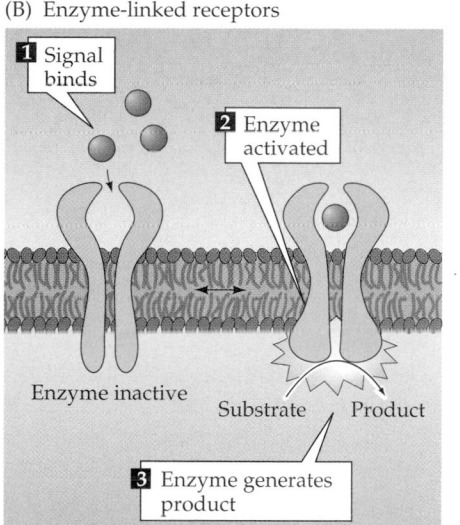

Figure 8.4 Categories of cellular receptors. Membrane-impermeant signaling molecules can bind to and activate either channel-linked receptors (A), enzyme-linked receptors (B), or G-protein-coupled receptors (C). Membrane permeant signaling molecules activate intracellular receptors (D).

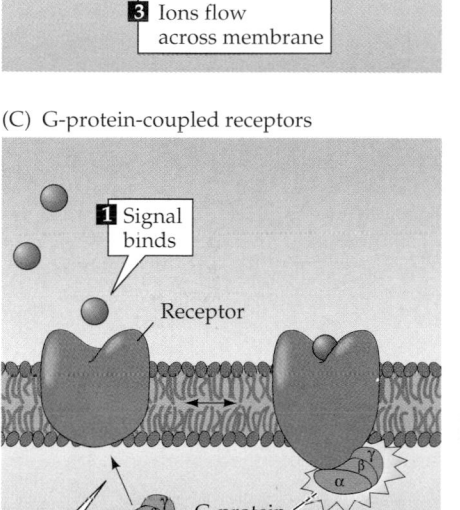

(C) G-protein-coupled receptors

(D) Intracellular receptors

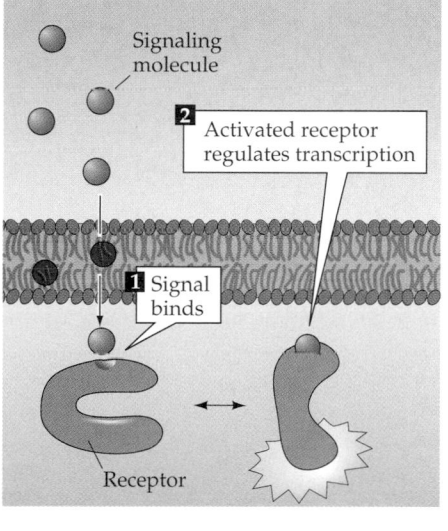

group of receptors include the Trk family of neurotrophin receptors (see Chapter 23) and other receptors for growth factors.

G-protein-coupled receptors regulate intracellular reactions by an indirect mechanism involving an intermediate transducing molecule, called the **GTP-binding proteins** (or **G-proteins**). Because these receptors all share the structural feature of crossing the plasma membrane seven times, they are also referred to as 7-transmembrane receptors (or metabotropic receptors; see Chapter 7). Hundreds of different G-protein-linked receptors have been identified. Well-known examples include the β-adrenergic receptor, the muscarininc type of acetylcholine receptor, metabotropic glutamate receptors, receptors for odorants in the olfactory system, and many types of receptors for peptide hormones. Rhodopsin, a light-sensitive 7-transmembrane protein in retinal photoreceptors, is another form of G-protein-linked receptor (see Chapter 11).

Intracellular receptors are activated by cell-permeant or lipophilic signaling molecules (Figure 8.4D). Many of these receptors lead to the activation of signaling cascades that produce new mRNA and protein within the target cell. Often such receptors comprise a receptor protein bound to an inhibitory protein complex. When the signaling molecule binds to the receptor, the inhibitory complex dissociates to expose a DNA-binding domain on the receptor. This activated form of the receptor can then move into the nucleus and directly interact with nuclear DNA, resulting in altered transcription. Some intracellular receptors are located primarily in the cytoplasm, while others are in the nucleus. In either case, once these receptors are activated they can affect gene expression by altering DNA transcription.

G-Proteins and Their Molecular Targets

Both G-protein-linked receptors and enzyme-linked receptors can activate biochemical reaction cascades that ultimately modify the function of target proteins. For both these receptor types, the coupling between receptor activation and their subsequent effects are the GTP-binding proteins. There are two general classes of GTP-binding protein (Figure 8.5). **Heterotrimeric G-proteins** are composed of three distinct subunits (α, β, and γ). There are many different α, β, and γ subunits, allowing a bewildering number of G-protein permutations. Regardless of the specific composition of the heterotrimeric G-protein, its α subunit binds to guanine nucleotides, either GTP or GDP. Binding of GDP allows the α subunit to bind to the β and γ subunits to form an inactive trimer. Binding of an extracellular signal to a G-protein-coupled receptor allows the G-protein to bind to the receptor and causes GDP to be replaced with GTP (Figure 8.5A). When GTP is bound to the G-protein, the α subunit dissociates from the $\beta\gamma$ complex and activates the G-protein. Following activation, both the GTP-bound α subunit and the free $\beta\gamma$ complex can bind to downstream effector molecules and mediate a variety of responses in the target cell.

The second class of GTP-binding proteins are **monomeric G-proteins** (also called **small G-proteins**). These monomeric GTPases also relay signals from activated cell surface receptors to intracellular targets such as the cytoskele-

Figure 8.5 Types of GTP-binding protein. (A) Heterotrimeric G-proteins are composed of three distinct subunits (α, β, and γ). Receptor activation causes the binding of the G-protein and the α subunit to exchange GDP for GTP, leading to a dissociation of the α and $\beta\gamma$ subunits. The biological actions of these G-proteins are terminated by hydrolysis of GTP, which is enhanced by GTPase-activating (GAP) proteins. (B) Monomeric G-proteins use similar mechanisms to relay signals from activated cell surface receptors to intracellular targets. Binding of GTP stimulates the biological actions of these G-proteins, and their activity is terminated by hydrolysis of GTP, which is also regulated by GAP proteins.

(A) Heterotrimeric G-proteins

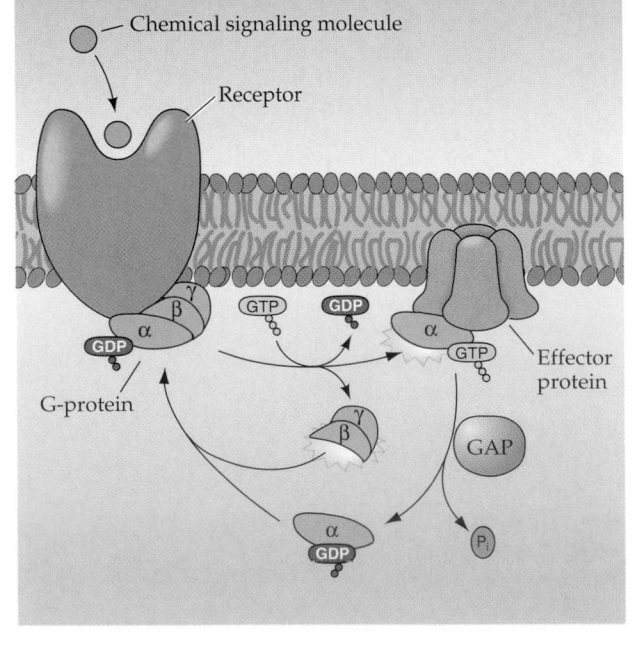

(B) Monotrimeric G-proteins

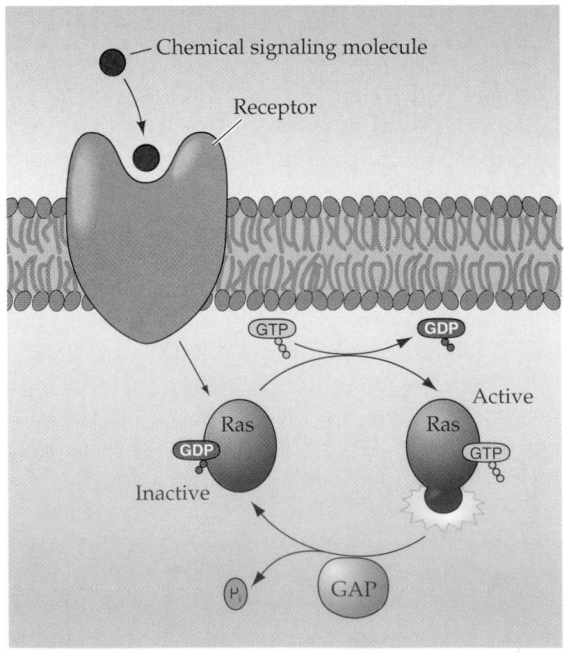

ton and the vesicle trafficking apparatus of the cell. The first small G-protein was discovered in a virus that causes *rat* sarcoma tumors and was therefore called **ras**. Ras is a molecule that helps regulate cell differentiation and proliferation by relaying signals from receptor kinases to the nucleus; the viral form of ras is defective, which accounts for the ability of the virus to cause the uncontrolled cell proliferation that leads to tumors. Since then, a large number of small GTPases have been identified and can be sorted into five different subfamilies with different functions. For instance, some are involved in vesicle trafficking in the presynaptic terminal or elsewhere in the neuron, while others play a central role in protein and RNA trafficking in and out of the nucleus.

Termination of signaling by both heterotrimeric and monomeric G-proteins is determined by hydrolysis of GTP to GDP. The rate of GTP hydrolysis is an important property of a particular G-protein that can be regulated by other proteins, termed GTPase-activating proteins (GAPs). By replacing GTP with GDP, GAPs return G-proteins to their inactive form. GAPs were first recognized as regulators of small G-proteins, but recently similar proteins have been found to regulate the α subunits of heterotrimeric G-proteins. Hence, monomeric and trimeric G-proteins function as molecular timers that are active in their GTP-bound state, and become inactive when they have hydrolized the bound GTP to GDP (Figure 8.5B).

Activated G-proteins alter the function of many downstream effectors. Most of these effectors are enzymes that produce intracellular second messengers. Effector enzymes include adenylyl cyclase, guanylyl cyclase, phospholipase C, and others (Figure 8.6). The second messengers produced by these enzymes trigger the complex biochemical signaling cascades discussed in the next section. Because each of these cascades is activated by specific G-protein subunits, the pathways activated by a particular receptor are determined by the specific identity of the G-protein subunits associated with it.

Figure 8.6 Effector pathways associated with G-protein-coupled receptors. In all three examples shown here, binding of a neurotransmitter to such a receptor leads to activation of a G-protein and subsequent recruitment of second messenger pathways. G_s, G_q and G_i refer to three different types of heterotrimeric G-protein.

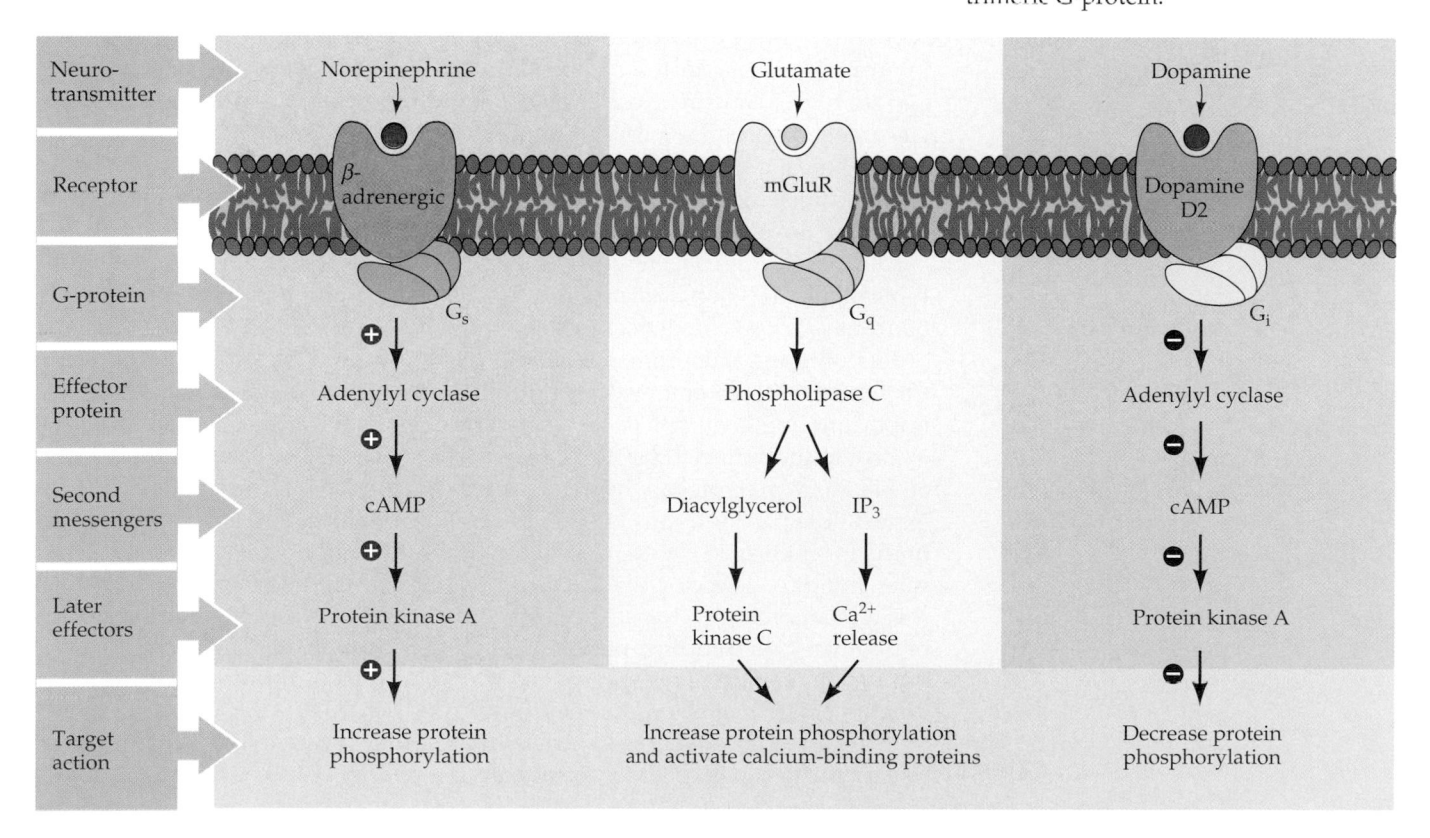

As well as activating effector molecules, G-proteins can also directly bind to and activate ion channels. For example, some neurons, as well as heart muscle cells, have G-protein-coupled receptors that bind acetylcholine. Because these receptors are also activated by the agonist muscarine, they are usually called muscarinic receptors (see Chapters 7 and 21). Activation of muscarinic receptors can open K^+ channels, thereby inhibiting the rate at which the neuron fires action potentials, or slowing the heartbeat of muscle cells. These inhibitory responses are believed to be the result of $\beta\gamma$ subunits of G proteins binding to the K^+ channels. The activation of α subunits can also lead to the rapid closing of voltage-gated Ca^{2+} and Na^+ channels. Because these channels carry inward currents involved in generating action potentials, closing them makes it more difficult for target cells to fire (see Chapters 3 and 4).

In summary, the binding of chemical signals to their receptors activates cascades of signal transduction events in the cytosol of target cells. Within such cascades, G-proteins serve a pivotal function as the molecular transducing elements that couple membrane receptors to their molecular effectors within the cell. The diversity of G-proteins and their downstream targets leads to many types of physiological responses. By directly regulating the gating of ion channels, G-proteins can influence the membrane potential of target cells.

Second Messengers

Neurons use many different second messengers as intracellular signals. These messengers differ in the mechanism by which they are produced and removed, as well as their downstream targets and effects (Figure 8.7A). This section summarizes the attributes of some of the principal second messengers.

• *Calcium.* The calcium ion (Ca^{2+}) is perhaps the most common intracellular messenger in neurons. Indeed, few neuronal functions are immune to the influence—direct or indirect—of Ca^{2+}. In all cases, information is transmitted by a transient rise in the cytoplasmic calcium concentration, which allows Ca^{2+} to bind to a large number of Ca^{2+}-binding proteins that serve as molecular targets. One of the most thoroughly studied targets of Ca^{2+} is **calmodulin**, a Ca^{2+}-binding protein abundant in the cytosol of all cells. Binding of Ca^{2+} to calmodulin activates this protein, which then initiates its effects by binding to still other downstream targets, such as protein kinases.

Ordinarily the concentration of Ca^{2+} ions in the cytosol is extremely low, typically 50–100 nanomolar (10^{-9} M). The concentration of Ca^{2+} ions outside neurons—in the bloodstream or cerebrospinal fluid, for instance—is several orders of magnitude higher, typically several millimolar (10^{-3} M). This steep Ca^{2+} gradient is maintained by a number of mechanisms (Figure 8.7B). Most important in this maintenance are two proteins that translocate Ca^{2+} from the cytosol to the extracellular medium: an ATPase called the **calcium pump**, and an **Na^+/Ca^{2+} exchanger**, which is a protein that replaces intracellular Ca^{2+} with extracellular sodium ions (see Chapter 4). In addition to these plasma membrane mechanisms, Ca^{2+} is also pumped into the endoplasmic reticulum and mitochondria. These organelles can thus serve as storage depots of Ca^{2+} ions that are later released to participate in signaling events. Finally, nerve cells contain other Ca^{2+}-binding proteins—such as **calbindin**—that serve as Ca^{2+} buffers. Such buffers reversibly bind Ca^{2+} and thus blunt the magnitude and kinetics of Ca^{2+} signals within neurons.

The Ca^{2+} ions that act as intracellular signals enter cytosol by means of one or more types of Ca^{2+}-permeable ion channels (see Chapter 4). These

(A)

Second messenger	Sources	Intracellular targets	Removal mechanisms
Ca^{2+}	Plasma membrane: Voltage-gated Ca^{2+} channels, Various ligand-gated channels; Endoplasmic reticulum: IP_3 receptors, Ryanodine receptors	Calmodulin, Protein kinases, Protein phosphatases, Ion channels, Synaptotagmin, Many other Ca^{2+}-binding proteins	Plasma membrane: Na^+/Ca^{2+} exchanger, Ca^{2+} pump; Endoplasmic reticulum: Ca^{2+} pump; Mitochondria
Cyclic AMP	Adenylyl cyclase acts on ATP	Protein kinase A, Cyclic nucleotide-gated channels	cAMP phosphodiesterase
Cyclic GMP	Guanylyl cyclase acts on GTP	Protein kinase G, Cyclic nucleotide-gated channels	cGMP phosphodiesterase
IP_3	Phospholipase C acts on PIP_2	IP_3 receptors on endoplasmic reticulum	Phosphatases
Diacylglycerol	Phospholipase C acts on PIP_2	Protein kinase C	Various enzymes
Nitric oxide	Nitric oxide synthase acts on arginine	Guanylyl cyclase	Spontaneous oxidation

Figure 8.7 Neuronal second messengers. (A) Mechanisms responsible for producing and removing second messengers, as well as the downstream targets of these messengers. (B) Proteins involved in delivering calcium to the cytoplasm and in removing calcium from the cytoplasm. (C) Mechanisms of production and degradation of cyclic nucleotides. (D) Pathways involved in production and removal of diacylglycerol (DAG) and IP_3. (E) Nitric oxide synthesis and degradation.

(B)

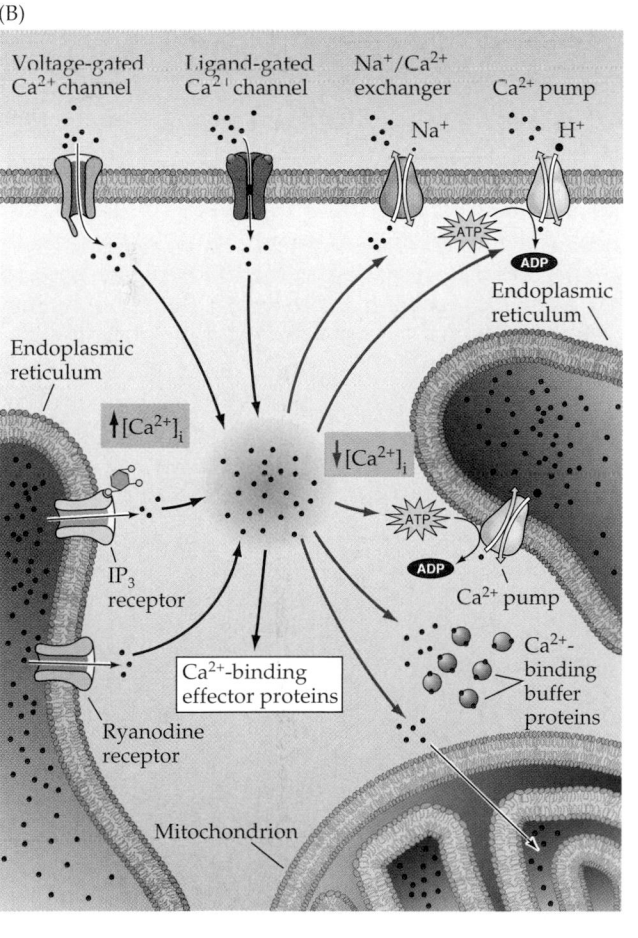

(C)

(D)

(E)

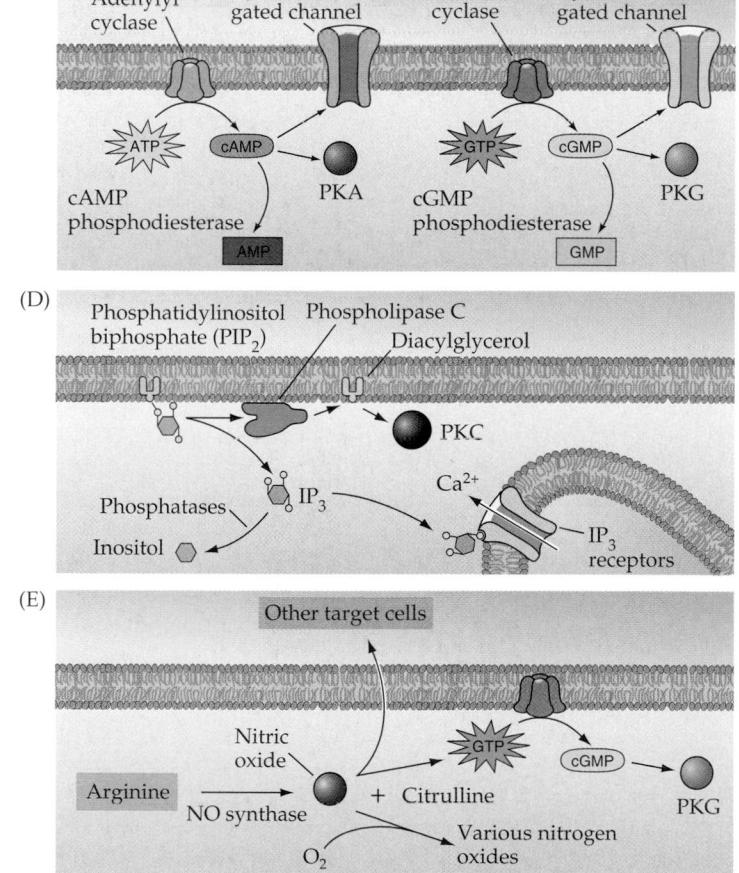

can be voltage-gated Ca^{2+} channels or ligand-gated channels in the plasma membrane, both of which allow Ca^{2+} to flow down the Ca^{2+} gradient and into the cell from the extracellular medium. In addition, other channels allow Ca^{2+} to be released from the interior of the endoplasmic reticulum into the cytosol. These intracellular Ca^{2+}-releasing channels are gated, so they can be opened or closed in response to various intracellular signals. One such channel is the **inositol trisphosphate (IP$_3$) receptor.** As the name implies, these channels are regulated by IP$_3$, a second messenger described in more detail below. A second type of intracellular Ca^{2+}-releasing channel is the **ryanodine receptor,** named after a drug that binds to and partially opens these receptors. Among the biological signals that activate ryanodine receptors are cytoplasmic Ca^{2+} and, at least in muscle cells, depolarization of the plasma membrane.

These various mechanisms for elevating and removing Ca^{2+} ions allows precise control of both the timing and location of Ca^{2+} signaling within neurons, which in turn allows Ca^{2+} to control many different signaling events. For example, voltage-gated Ca^{2+} channels allow Ca^{2+} concentrations to rise very rapidly and locally within presynaptic terminals to trigger neurotransmitter release, as already described in Chapters 5 and 6. Slower and more widespread rises in Ca^{2+} concentration regulate a wide variety of other responses, including gene expression in the cell nucleus.

• *Cyclic nucleotides.* Another important group of second messengers are the cyclic nucleotides, specifically cyclic adenosine monophosphate (cAMP) and cyclic guanosine monophosphate (cGMP) (Figure 8.7C). Cyclic AMP is a derivative of the common cellular energy storage molecule, ATP. Cyclic AMP is produced when G-proteins activate adenylyl cyclase in the plasma membrane. This enzyme converts ATP into cAMP by removing two phosphate groups from the ATP. Cyclic GMP is similarly produced from GTP by the action of guanylyl cyclase. Once the intracellular concentration of cAMP or cGMP is elevated, these nucleotides can bind to two different classes of targets. The most common targets of cyclic nucleotide action are protein kinases, either the cAMP-dependent protein kinase (PKA) or the cGMP-dependent protein kinase (PKG). These enzymes mediate many physiological responses by phosphorylating target proteins, as described in the following section. In addition, cAMP and cGMP can bind to certain ion channels, thereby influencing neuronal signaling. These cyclic-nucleotide gated channels are ligand-gated (see Chapter 4); they are particularly important in phototransduction and other sensory transduction processes, such as olfaction. Cyclic nucleotide signals are degraded by phosphodiesterases, enzymes that cleave phosphodiester bonds and convert cAMP into AMP or cGMP into GMP.

• *Diacylglycerol and IP$_3$.* Remarkably, membrane lipids can also be converted into intracellular second messengers (Figure 8.7D). The two most important messengers of this type are produced from phosphatidylinositol bisphosphate (PIP$_2$). This lipid component is cleaved by phospholipase C, an enzyme activated by certain G-proteins and by calcium ions. Phospholipase C splits the PIP$_2$ into two smaller molecules that each act as second messengers. One of these messengers is diacylglycerol (DAG), a molecule that remains within the membrane and activates protein kinase C, which phosphorylates substrate proteins in both the plasma membrane and elsewhere. The other messenger is inositol trisphosphate (IP$_3$), a molecule that leaves the cell membrane and diffuses within the cytosol. IP$_3$ binds to IP$_3$ receptors, channels that release calcium from the endoplasmic reticulum. Thus, the action of IP$_3$ is to produce yet another second messenger (perhaps a third messenger, in this case!) that triggers a whole spectrum of reactions in the cytosol. The actions of DAG and IP$_3$

are terminated by enzymes that convert these two molecules into inert forms that can be recycled to produce new molecules of PIP_2.

• *Nitric oxide.* An unusual, but especially interesting, second messenger is nitric oxide (NO; Figure 8.7E). NO is produced by the action of nitric oxide synthase, an enzyme that converts the amino acid arginine into a metabolite, citrulline, and simultaneously generates NO. The nitric oxide synthase found in neurons is regulated by calcium binding to calmodulin and is coupled to a variety of neurotransmitter systems. In comparison to other second messengers used by neurons, NO is unusual in that it is a gas. NO also permeates the plasma membrane, meaning that NO generated within one cell can travel through the extracellular medium and act within other nearby cells. Thus, NO has a range of influence that extends well beyond the cell of origin. This property of NO makes it a useful signal for coordinating the activities of multiple cells in a very localized region; indeed, NO often is considered a neurotransmitter rather than a second messenger and may provide certain forms of synaptic plasticity in small networks of neurons (NO can diffuse only a few tens of micrometers from its site of production before it decays). At least some of the biological actions of NO are due to the activation of guanylyl cyclase, which then produces cGMP in target cells. NO reacts nonspecifically with many other molecules and decays spontaneously by reacting with oxygen to produce inactive nitrogen oxides. As a result, NO signals last for only a short time, on the order of seconds or less. NO may also be involved in some neurological diseases; for example, an emerging hypothesis is that an imbalance between nitric oxide and superoxide generation underlies some neurodegenerative diseases.

Second Messenger Targets: Protein Kinases and Phosphatases

Second messengers typically regulate neuronal functions by modulating the phosphorylation state of intracellular proteins (Figure 8.8). Phosphorylation (the addition of phosphate groups) rapidly and reversibly changes protein function. Proteins are phosphorylated by a wide variety of **protein kinases**; phosphate groups are removed by other enzymes called **protein phosphatases**. The degree of phosphorylation of a target protein thus reflects a balance between the competing actions of protein kinases and phosphatases, integrating a host of cellular signaling pathways. The substrates of protein kinases and phosphatases include enzymes, neurotransmitter receptors, ion channels, and structural proteins.

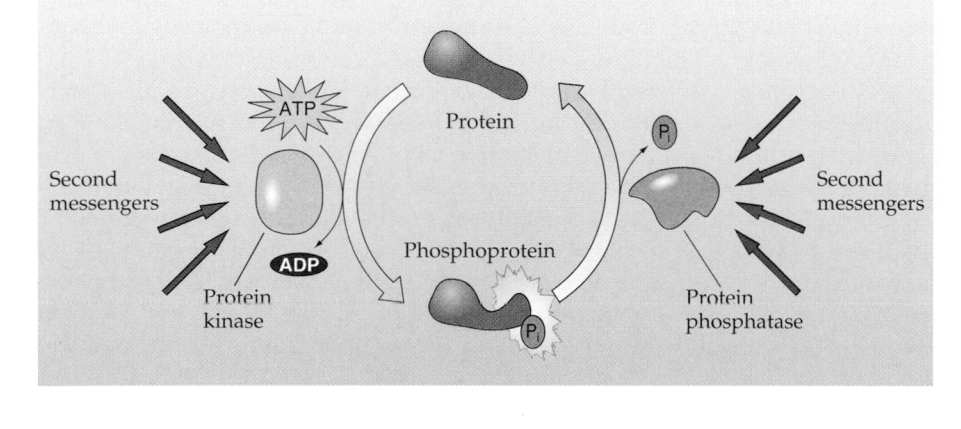

Figure 8.8 Regulation of cellular proteins by phosphorylation. Protein kinases transfer phosphate groups (P_i) from ATP to serine, threonine, or tyrosine residues on substrate proteins. This phosphorylation reversibly alters the structure and function of cellular proteins. Removal of the phosphate groups is catalyzed by protein phosphatases. Both kinases and phosphatases are regulated by a variety of intracellular second messengers.

Protein kinases and phosphatases typically act either on the serine and threonine residues (Ser/Thr kinases or phosphatases) or the tyrosine residues (Tyr kinases or phosphatases) of their substrates. Some of these enzymes act specifically on only one or a handful of protein targets, while others are multifunctional and have a broad range of substrate proteins. The activity of protein kinases and phosphatases can be regulated either by second messengers, such as cAMP or Ca^{2+}, or by extracellular chemical signals, such as growth factors (see Chapter 23). Typically, second messengers activate Ser/Thr kinases, whereas extracellular signals activate Tyr kinases. Although thousands of protein kinases are expressed in the brain, a relatively small number function as regulators of neuronal signaling.

• *cAMP-dependent protein kinase (PKA).* The primary effector of cAMP is the cAMP-dependent protein kinase (PKA). PKA is a tetrameric complex of two catalytic subunits and two inhibitory (regulatory) subunits. cAMP activates PKA by binding to the regulatory subunits and causing them to release active catalytic subunits. Such displacement of inhibitory domains is a general mechanism for activation of several protein kinases by second messengers (Figure 8.9A). The catalytic subunit of PKA phosphorylates serine and threonine residues of many different target proteins. Although this subunit is similar to the catalytic domains of other protein kinases, distinct amino acids allow the PKA to bind to specific target proteins, thus allowing only those targets to be phosphorylated in response to intracellular cAMP signals.

• *Ca^{2+}/calmodulin-dependent protein kinase type II (CaMKII).* Ca^{2+} ions binding to calmodulin can regulate protein phosphorylation/dephosphorylation. In neurons, the most abundant Ca^{2+}/calmodulin-dependent protein kinase is CaMKII, a multifunctional Ser/Thr protein kinase. CaMKII is composed of approximately 12 subunits, which in the brain are the α and β types. Each subunit contains a catalytic domain and a regulatory domain, as well as other domains that allow the enzyme to oligomerize and target to the proper region within the cell. Ca^{2+}/calmodulin activates CaMKII by displacing the inhibitory domain from the catalytic site (Figure 8.9B). CaMKII phosphorylates a large number of substrates, including ion channels and other proteins involved in intracellular signal transduction.

• *Protein kinase C (PKC).* Another important group of Ser/Thr protein kinases is protein kinase C (PKC). PKCs are diverse monomeric kinases activated by the second messengers DAG and Ca^{2+}. DAG causes PKC to move from the cytosol to the plasma membrane, where it also binds Ca^{2+} and phosphatidylserine, a membrane phospholipid (Figure 8.9C). These events relieve autoinhibition and cause PKC to phosphorylate various protein substrates. PKC also diffuses to sites other than the plasma membrane—such as the cytoskeleton, perinuclear sites, and the nucleus—where it phosphorylates still other substrate proteins. Prolonged activation of PKC can be accomplished with phorbol esters, tumor-promoting compounds that activate PKC by mimicking DAG.

• *Protein tyrosine kinases.* Two classes of protein kinases transfer phosphate groups to tyrosine residues on substrate proteins. Receptor tyrosine kinases are transmembrane proteins with an extracellular domain that binds to protein ligands (growth factors, neurotrophic factors, or cytokines) and an intracellular catalytic domain that phosphorylates the relevant substrate proteins. Non-receptor tyrosine kinases are cytoplasmic or membrane-associated enzymes that are indirectly activated by extracellular signals. Tyrosine phosphorylation is less common than Ser/Thr phosphorylation, and it often serves to recruit signaling molecules to the phosphorylated protein. Tyrosine

(A) PKA

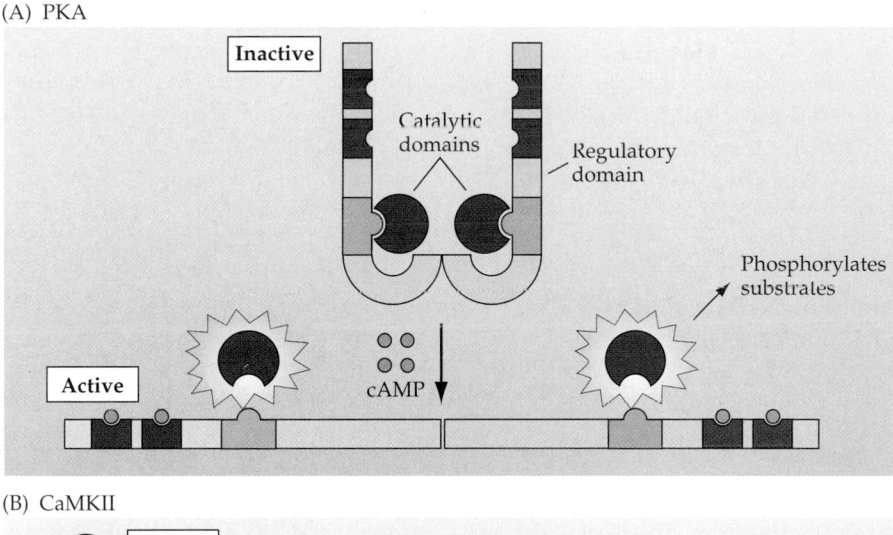

(B) CaMKII

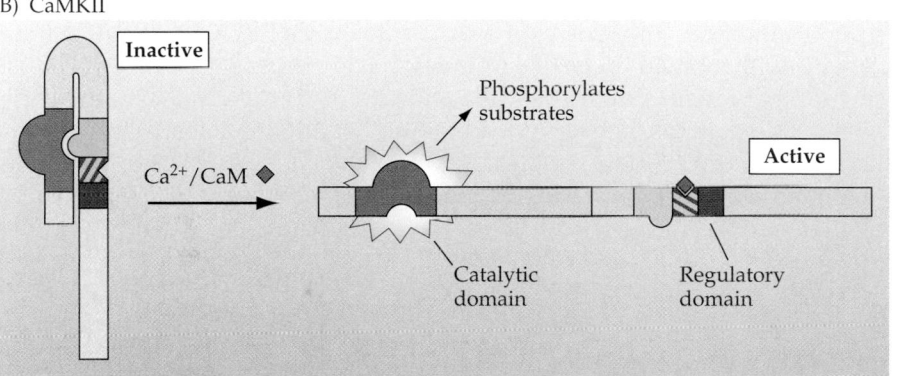

(C) PKC

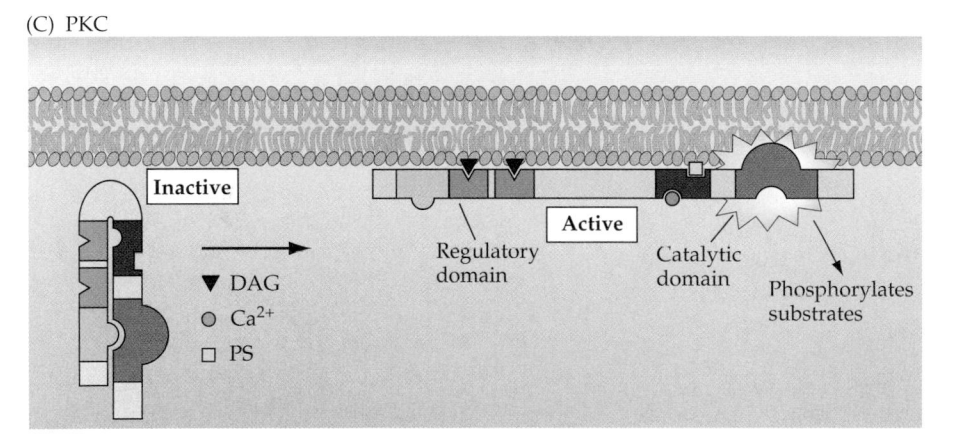

Figure 8.9 Mechanism of activation of protein kinases. Protein kinases contain several specialized domains with specific functions. Each of the kinases has homologous catalytic domains responsible for transferring phosphate groups to substrate proteins. These catalytic domains are kept inactive by the presence of an autoinhibitory domain that occupies the catalytic site. Binding of second messengers, such as cAMP, DAG, and Ca^{2+}, to the appropriate regulatory domain of the kinase removes the autoinhibitory domain and allows the catalytic domain to be activated. For some kinases, such as PKC and CaMKII, the autoinhibitory and catalytic domain are part of the same molecule. For other kinases, such as PKA, the autoinhibitory domain is a separate subunit.

kinases are particularly important for cell growth and differentiation (see Chapters 22 and 23).

• *Mitogen-activated protein kinase (MAPK)*. In addition to protein kinases that are directly activated by second messengers, some of these molecules can be activated by other signals, such as phosphorylation by another protein kinase. Important examples of such protein kinases are the mitogen-activated protein kinases (MAPKs), also called extracellular signal-regulated kinases (ERKs). MAPKs were first identified as participants in the control of cell growth and are now known to have many other signaling functions. MAPKs are normally inactive in neurons but become activated when they are phosphorylated by other kinases. In fact, MAPKs are part of a kinase cascade in which one protein kinase phosphorylates and activates the next protein kinase in the cascade. The extracellular signals that trigger these kinase cascades are often extracellular growth factors that bind to receptor tyrosine kinases that, in turn, activate monomeric G proteins such as ras. Once activated, MAPKs can phosphorylate transcription factors, proteins that regulate gene expression. Among the wide variety of other MAPK substrates are various enzymes, including other protein kinases, and cytoskeletal proteins.

The best characterized protein phosphatases are the Ser/Thr phosphatases PP1, PP2A, and PP2B (also called calcineurin). In general, protein phosphatases display less substrate specificity than protein kinases. Their limited specificity may arise from the fact that the catalytic subunits of the three major protein phosphatases are highly homologous, though each still associates with specific targeting or regulatory subunits. PP1 dephosphorylates a wide array of substrate proteins and is probably the most prevalent Ser/Thr protein phosphatase in mammalian cells. PP1 activity is regulated by several inhibitory proteins expressed in neurons. PP2A is a multisubunit enzyme with a broad range of substrates that overlap with PP1. PP2B, or calcineurin, is present at high levels in neurons. A distinctive feature of this phosphatase is its activation by Ca^{2+}/calmodulin. PP2B is composed of a catalytic and a regulatory subunit. Ca^{2+}/calmodulin activates PP2B primarily by binding to the catalytic subunit and displacing the inhibitory regulatory domain. PP2B generally does not have the same molecular targets as CaMKII, even though both enzymes are activated by Ca^{2+}/calmodulin.

In summary, activation of membrane receptors elicits complex cascades of enzyme activation, often resulting in second messenger production and protein phosphorylation or dephosphorylation. These cytoplasmic signals produce a variety of rapid physiological responses by transiently regulating enzyme activity, ion channels, cytoskeletal proteins, and many other cellular processes. In addition, these signals can propagate to the nucleus to cause long-lasting changes in gene expression.

Nuclear Signaling

Second messengers can elicit prolonged changes in neuronal function by promoting the synthesis of new RNA and protein. The resulting accumulation of new proteins requires at least 30–60 minutes, a time frame that is orders of magnitude slower than the responses mediated by ion fluxes or phosphorylation. Likewise, the reversal of such events requires hours to days. In some cases, genetic "switches" can be thrown to permanently alter a neuron, as in neuronal differentiation (see Chapter 22).

The amount of protein present in cells is determined primarily by the rate of transcription of DNA into RNA (Figure 8.10). The first step in RNA synthesis is the decondensation of the structure of chromatin to provide binding sites for the RNA polymerase complex and for **transcriptional activator proteins**, also called **transcription factors**. Transcriptional activator proteins attach to binding sites that are present on the DNA molecule near the start of the target gene sequence; they also bind to other proteins that promote unwrapping of DNA. The net result of these actions is to allow RNA polymerase, an enzyme complex, to assemble on the **promoter** region of the DNA and begin transcription. In addition to clearing the promoter for RNA polymerase, activator proteins can stimulate transcription by interacting with the RNA polymerase complex or by interacting with other activator proteins that influence the polymerase.

Intracellular signal transduction cascades regulate gene expression by converting transcriptional activator proteins from an inactive state to an

(A)   (B)

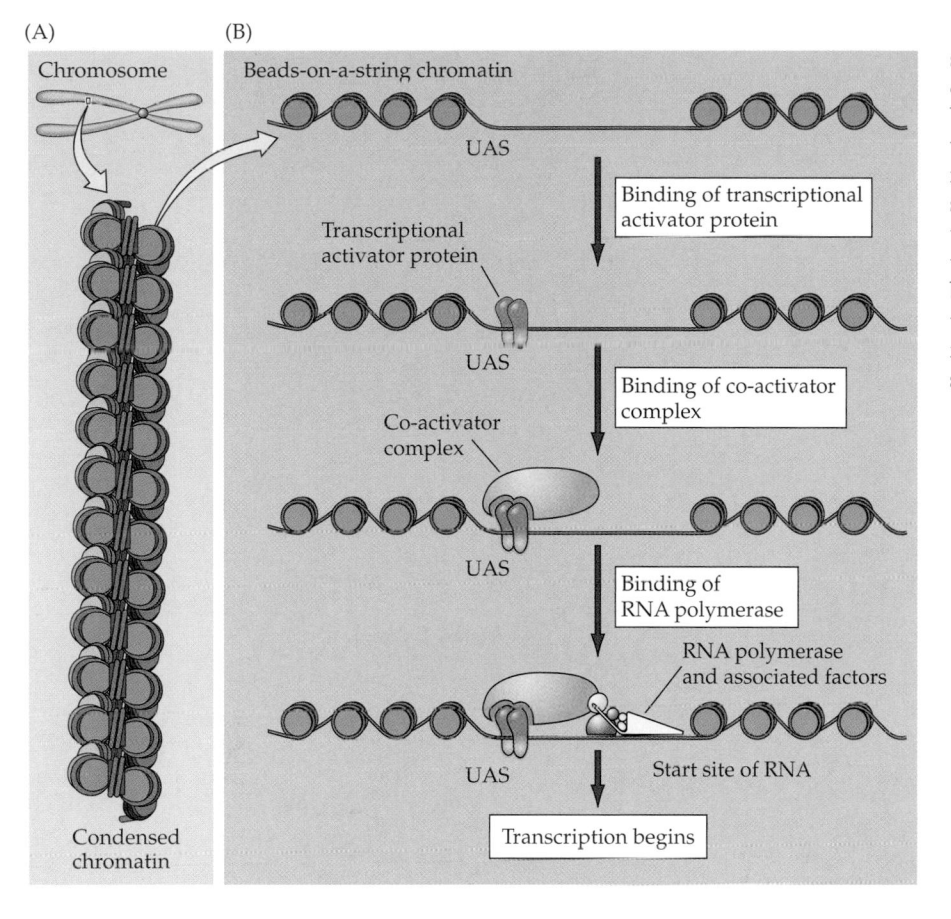

Figure 8.10 Steps involved in transcription of DNA into RNA. Condensed chromatin (A) is decondensed into a beads-on-a-DNA-string array (B) in which an upstream activator site (UAS) is free of proteins and is bound by a sequence-specific transcriptional activator protein (transcription factor). The transcriptional activator protein then binds co-activator complexes that enable the RNA polymerase with its associated factors to bind at the start site of transcription and initiate RNA synthesis.

active state in which they are able to bind to DNA. This conversion comes about in several ways. The key activator proteins and the mechanisms that allow them to regulate gene expression in response to signaling events are briefly summarized in the following sections.

• *CREB.* The *c*AMP *r*esponse *e*lement *b*inding protein, usually abbreviated **CREB**, is a ubiquitous transcriptional activator (Figure 8.11). CREB is normally bound to its binding site on DNA (called the cAMP response element, or CRE), either as a homodimer or bound to another, closely related transcription factor. In unstimulated cells, CREB is not phosphorylated and has little or no transcriptional activity. However, phosphorylation of CREB greatly potentiates transcription. Several signaling pathways are capable of causing CREB to be phosphorylated. Both PKA and the ras pathway, for example, can phosphorylate CREB. CREB can also be phosphorylated in response to increased intracellular calcium, in which case the CRE site is also called the CaRE (calcium response element) site. The calcium-dependent phosphorylation of CREB is primarily caused by Ca^{2+}/calmodulin kinase IV, a relative of CaMKII. CREB phosphorylation must be maintained long enough for transcription to ensue, even though neuronal electrical activity only transiently raises intracellular calcium concentration. Such signaling cascades can potentiate CREB-mediated transcription by inhibiting a protein phosphatase that dephosphorylates CREB. CREB is thus an example of the convergence of multiple signaling pathways onto a single transcriptional activator.

Many genes whose transcription is regulated by CREB have been identified. CREB-sensitive genes include the immediate early gene, *c-fos* (see below), the neurotrophin BDNF (see Chapter 23), the enzyme tyrosine hydroxylase (which

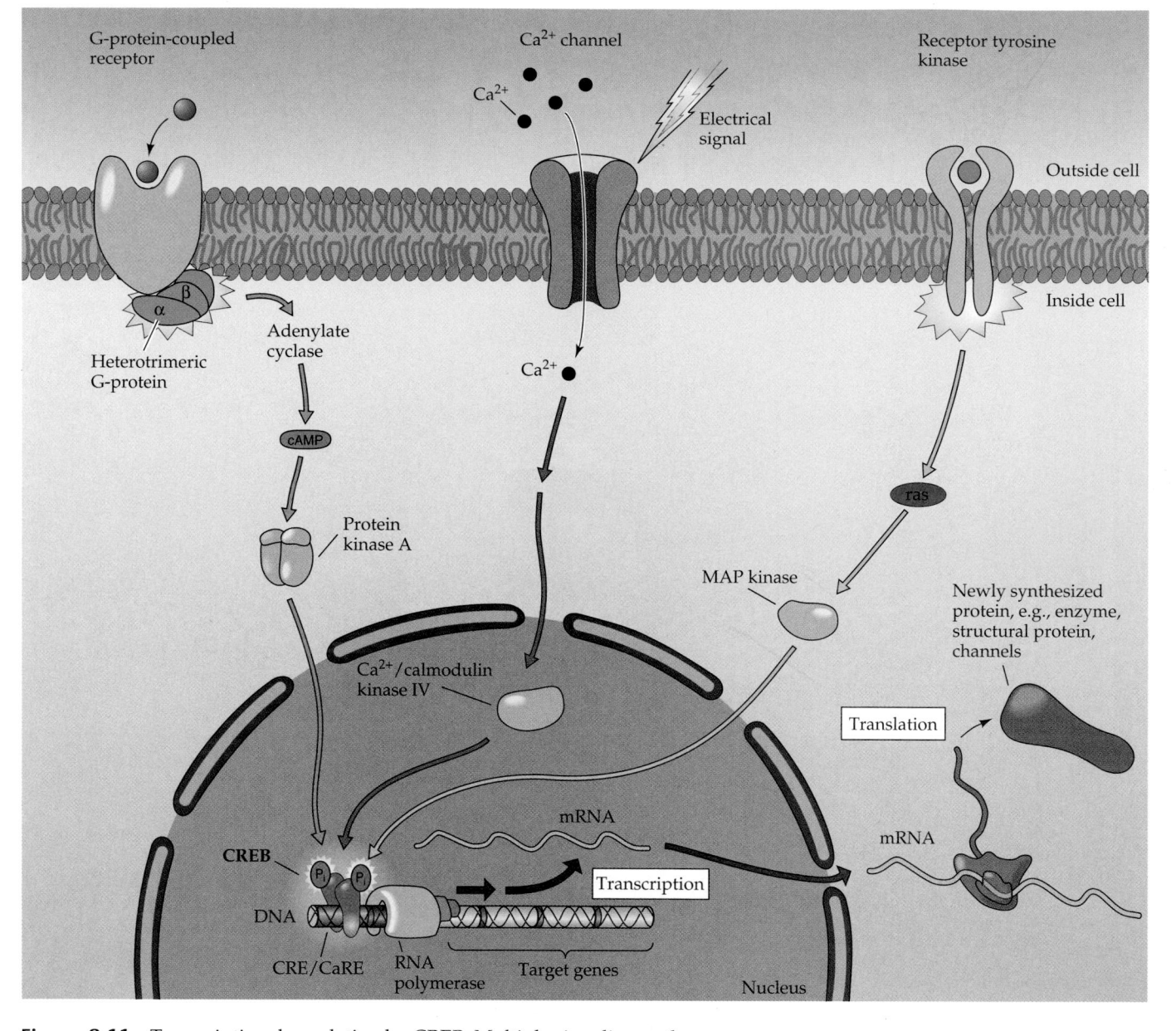

Figure 8.11 Transcriptional regulation by CREB. Multiple signaling pathways converge by activating kinases that phosphorylate CREB. These include PKA, Ca²⁺/calmodulin kinase IV, and MAP kinase. Phosphorylation of CREB allows it to bind co-activators (not shown in the figure), which then stimulate RNA polymerase to begin synthesis of RNA. RNA is then processed and exported to the cytoplasm, where it serves as mRNA for translation into protein.

is important for synthesis of catecholamine neurotransmitters; see Chapter 6), and many neuropeptides (including somatostatin, enkephalin, and corticotropin releasing hormone). CREB also is thought to mediate long-lasting changes in brain function. For example, CREB has been implicated in spatial learning, behavioral sensitization, long-term memory of odorant-conditioned behavior, and long-term synaptic plasticity (see Chapter 25).

• *Nuclear receptors.* Nuclear receptors for membrane-permeant ligands also are transcriptional activators. The receptor for glucocorticoid hormones illustrates one mode of action of such receptors. In the absence of glucocorticoid hormones, the receptors are located in the cytoplasm. Binding of glucocorticoids causes the receptor to unfold and move to the nucleus, where it binds a specific recognition site on the DNA. This DNA binding activates the relevant RNA polymerase complex to initiate transcription and subsequent gene expression. Thus, a critical regulatory event for steroid receptors is their translocation to the nucleus to allow DNA binding.

The receptors for thyroid hormone (TH) and other non-steroid nuclear receptors illustrate a second mode of regulation. In the absence of TH, the receptor is bound to DNA and serves as a potent repressor of transcription. Upon binding TH, the receptor undergoes a conformational change that ultimately opens the promoter for polymerase binding. Hence, TH binding switches the receptor from being a repressor to being an activator of transcription.

• *c-fos*. A different strategy of gene regulation is apparent in the function of the transcriptional activator protein, **c-fos**. In resting cells, c-fos is present at a very low concentration. However, stimulation of the target cell causes c-fos to be synthesized, and the amount of this protein rises dramatically over 30–60 minutes. Therefore, c-fos is considered to be an **immediate early gene** because its synthesis is directly triggered by the stimulus. Once synthesized, c-fos can act as a transcriptional activator to induce synthesis of second-order genes. These are termed **delayed response genes** because their activity is delayed by the fact that an immediate early gene—c-fos in this case—needs to be synthesized first.

Multiple signals converge on c-fos by activating different transcription factors that bind to at least three distinct sites in the promoter region of the c-fos gene. The regulatory region of the c-fos gene contains a binding site that mediates transcriptional induction by cytokines and ciliary neurotropic factor. Another site is targeted by growth factors such as neurotrophins through ras and protein kinase C, and a CRE/CaRE that can bind to CREB and thereby respond to cAMP or calcium entry resulting from electrical activity. In addition to synergistic interactions among these c-fos sites, transcriptional signals can be integrated by converging on the same activator, such as CREB.

Nuclear signaling events typically result in the generation of a huge and relatively stable complex composed of a functional transcriptional activator protein, additional proteins that bind to the activator protein, and the RNA polymerase and associated proteins bound at the start site of transcription. Most of the relevant signaling events act to "seed" this complex by generating an active transcriptional activator protein by phosphorylation, by inducing a conformational change in the activator upon ligand binding, by fostering nuclear localization, by removing an inhibitor, or simply by making more activator protein.

Examples of Neuronal Signal Transduction

Understanding the general properties of signal transduction processes at the plasma membrane, in the cytosol, and within the nucleus make it possible to consider how these processes work in concert to mediate specific functions in the brain. Three important signal transduction pathways can illustrate some of the roles of intracellular signal transduction processes in the nervous system.

• *NGF/TrkA*. The first of these is signaling by the **nerve growth factor** (**NGF**). This protein is a member of the neurotrophin growth factor family and is required for the differentiation, survival, and synaptic connectivity of sympathetic and sensory neurons (see Chapter 23). NGF works by binding to a high-affinity tyrosine kinase receptor, TrkA, found on the plasma membrane of these target cells (Figure 8.12). NGF binding causes TrkA receptors

Figure 8.12   Mechanism of action of NGF. NGF binds to a high-affinity tyrosine kinase receptor, TrkA, on the plasma membrane to induce phosphorylation of TrkA at two different tyrosine residues. These phosphorylated tyrosines serve to tether various adapter proteins or phospholipase C (PLC), which, in turn, activate three major signaling pathways: the PI 3 kinase pathway leading to activation of Akt kinase, the ras pathway leading to MAP kinases, and the PLC pathway leading to release of intracellular Ca^{2+} and activation of PKC. The ras and PLC pathways primarily stimulate processes responsible for neuronal differentiation, while the PI 3 kinase pathway is primarily involved in cell survival.

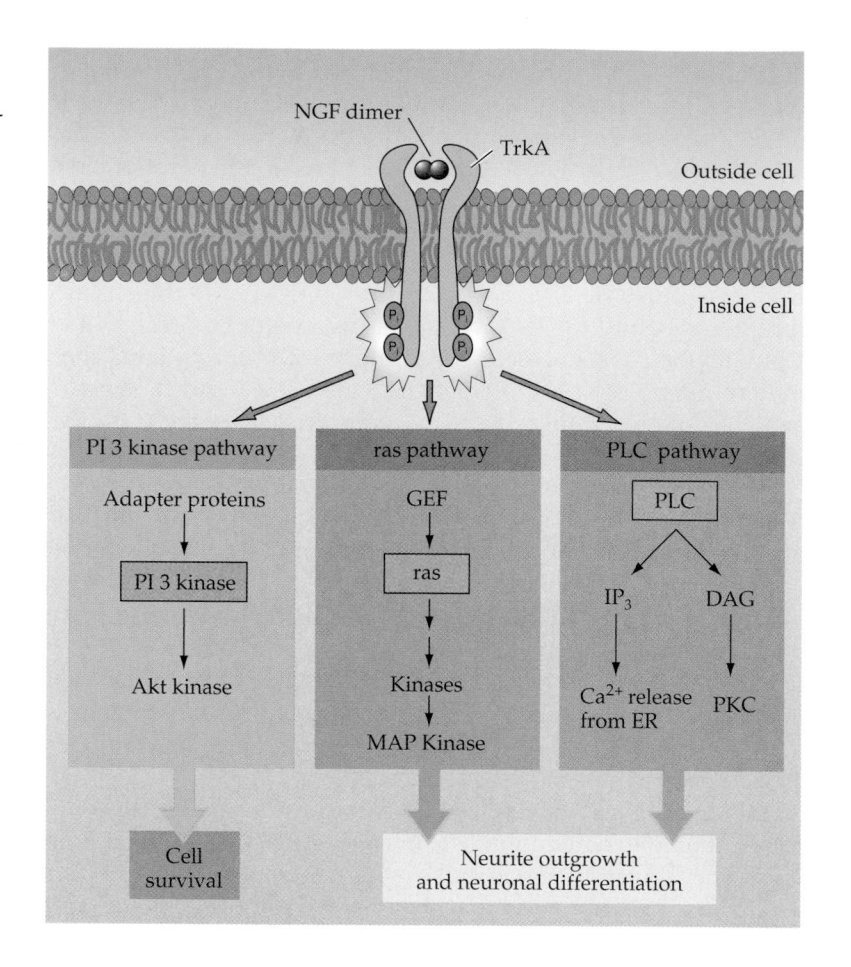

to dimerize, and the intrinsic tyrosine kinase activity of each receptor then phosphorylates its partner receptor. Phosphorylated TrkA receptors trigger the ras cascade, resulting in the activation of multiple protein kinases. Some of these kinases translocate to the nucleus to activate transcriptional activators, such as CREB. This ras-based component of the NGF pathway is primarily responsible for inducing and maintaining differentiation of NGF-sensitive neurons. Phosphorylation of TrkA also causes this receptor to stimulate the activity of phospholipase C, which increases production of IP_3 and DAG. IP_3 induces release of Ca^{2+} from the endoplasmic reticulum, and diacylglycerol activates PKC. These two second messengers appear to target many of the same downstream effectors as ras. Finally, activation of TrkA receptors also causes activation of other protein kinases (such as Akt kinase) that inhibit cell death. This pathway, therefore, primarily mediates the NGF-dependent survival of sympathetic and sensory neurons described in Chapter 23.

• *Long-term depression (LTD).* Another example of intracellular signaling can be observed during synaptic transmission between parallel fibers and Purkinje cells in the cerebellum. This synapse is central to information flow through the cerebellar cortex, which in turn helps coordinate motor movements (see Chapter 19). When parallel fibers are active, they release the neurotransmitter glutamate onto the dendrites of their Purkinje cell targets. This activates AMPA-type receptors, which are ligand-gated ion channels (see Chapter 7), causing a small inward current that depolarizes the Purkinje cell for a few milliseconds. In addition to this electrical signal, parallel fiber transmission also generates two postsynaptic second messengers (Figure

8.13). The glutamate released by parallel fibers activates metabotropic glutamate receptors, which are G-protein-linked receptors that cause activation of phospholipase C. This enzyme produces IP_3 and DAG within the Purkinje cell. The IP_3 causes Ca^{2+} to be released from the endoplasmic reticulum, transiently elevating cytoplasmic Ca^{2+} concentration in the dendrites until the IP_3 is metabolized and the Ca^{2+} is pumped out of the cytoplasm.

Given appropriate patterns of electrical activity, the calcium released by IP_3 can produce long-term depression (LTD), a form of synaptic plasticity that causes the parallel fiber synapses to become less effective (see Chapter 25). How the Ca^{2+} produced by parallel fiber activity produces LTD is unclear. One possibility is that Ca^{2+} activates protein kinase C, which is also activated by the DAG that is produced in Purkinje cells by parallel fibers. Thus, these two messengers may converge on PKC, which then phosphorylates an unknown substrate. Ultimately, these signaling processes change the postsynaptic, AMPA-type glutamate receptors, so that these receptors produce smaller electrical signals in response to the glutamate released from the parallel fibers. This weakening of the parallel fiber synapse is the final cause of LTD.

In summary, parallel fiber synaptic transmission produces brief electrical signals and chemical signals that last much longer. The actions of IP_3 and DAG also are restricted to small parts of the Purkinje cell dendrite, which is

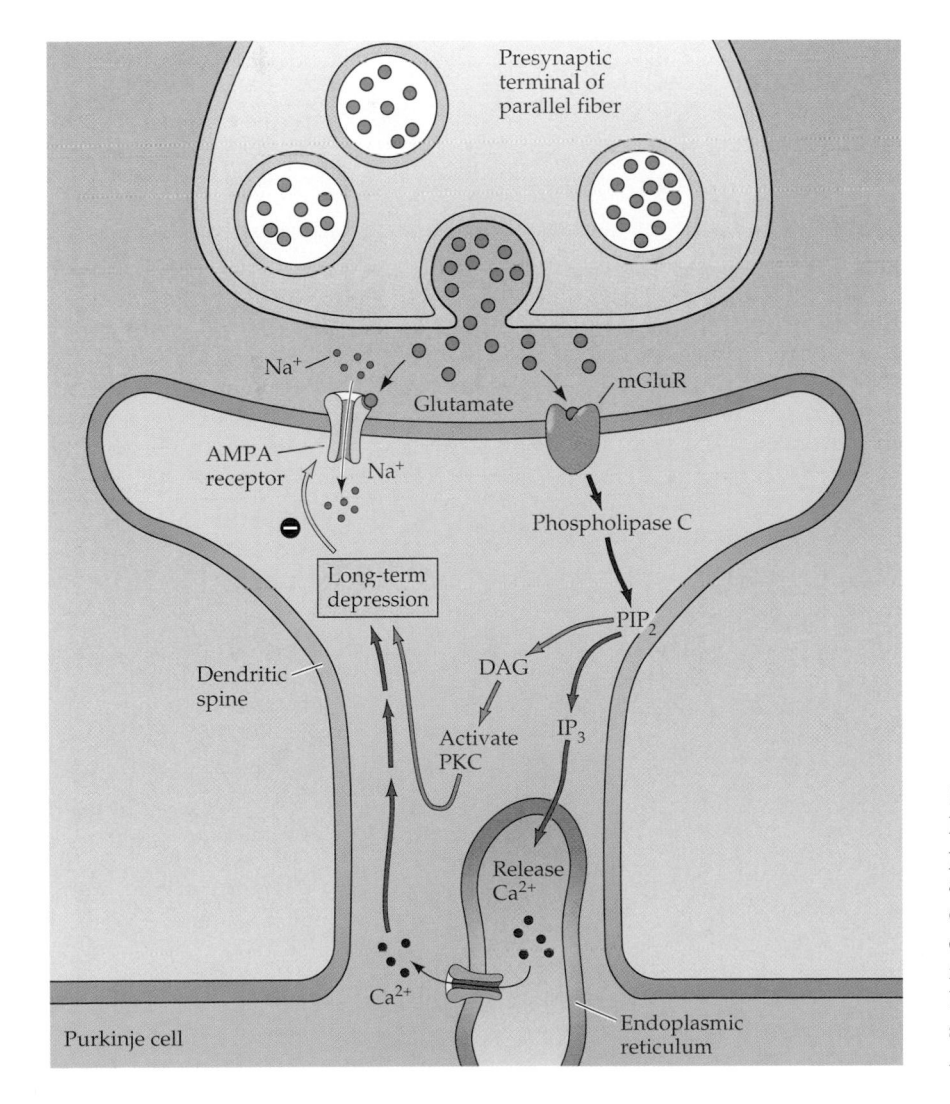

Figure 8.13 Signaling at cerebellar parallel fiber synapses. Glutamate released by parallel fibers activates both AMPA-type and metabotropic receptors. The latter produces IP_3 and DAG within the Purkinje cell. The IP_3 causes Ca^{2+} to be released from the endoplasmic reticulum, while DAG activates protein kinase C. These signals together change the properties of AMPA receptors to produce LTD.

a more limited spatial range than the parallel fiber EPSP, which spreads throughout the entire dendrite and cell body of the Purkinje cell. Thus, in contrast to the electrical signals, the second messenger signals can impart precise information about the location of active synapses, potentially influencing synapses in the vicinity of active parallel fibers.

• *Phosphorylation of tyrosine hydroxylase.* A third example of intracellular signaling in the nervous system is the regulation of the enzyme tyrosine hydroxylase. Tyrosine hydroxylase governs the synthesis of the catecholamine neurotransmitters: dopamine, norepinephrine, and epinephrine (see Chapter 6). A number of signals, including electrical activity, other neurotransmitters, and NGF, increase the rate of catecholamine synthesis by increasing the catalytic activity of tyrosine hydroxylase (Figure 8.14). The rapid increase of tyrosine hydroxylase activity is largely due to phosphorylation of this enzyme.

Tyrosine hydroxylase is a substrate for several protein kinases, including PKA, CaMKII, MAP kinase, and PKC. Phosphorylation causes conformational changes that increase the catalytic activity of tyrosine hydroxylase. Stimuli that elevate cAMP, Ca^{2+}, or DAG can all increase tyrosine hydroxylase activity and thus increase the rate of catecholamine biosynthesis. This regulation by several different signals allows for close control of tyrosine hydroxylase activity, and illustrates how several different pathways can converge to influence a key enzyme involved in synaptic transmission.

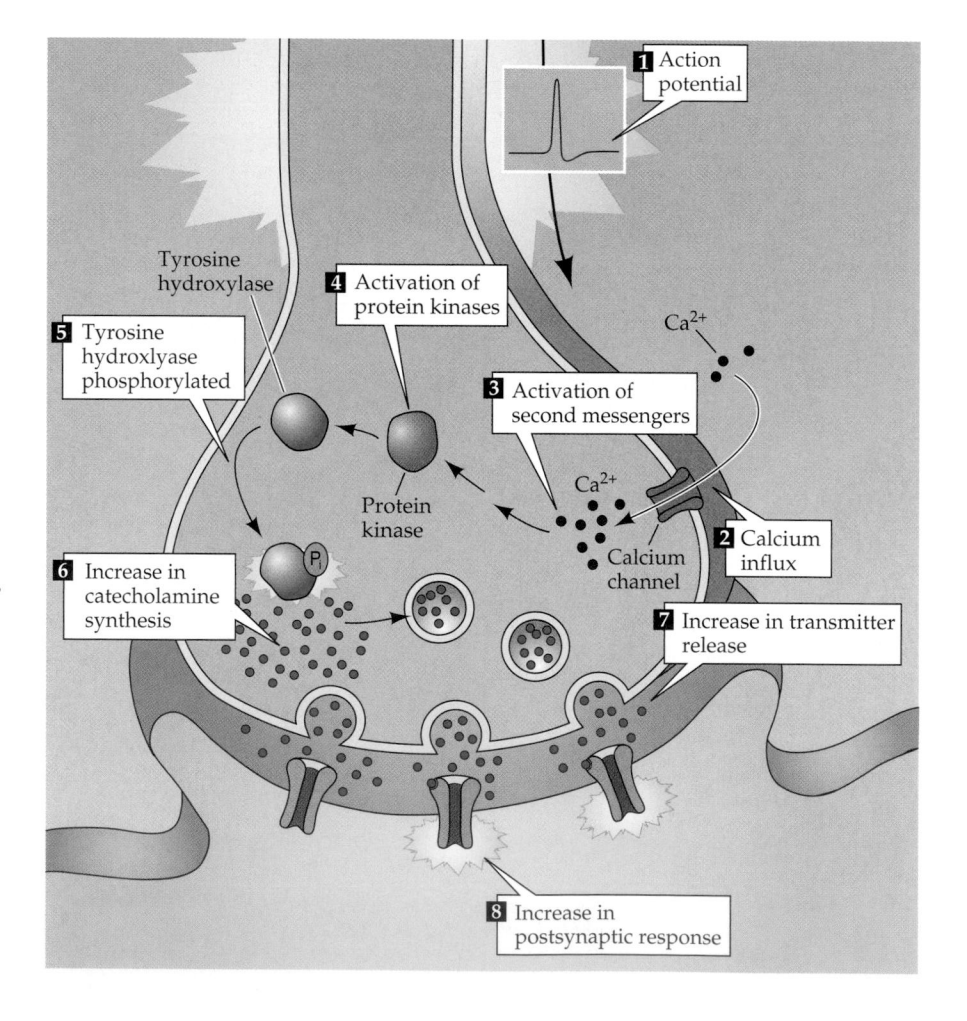

Figure 8.14 Regulation of tyrosine hydroxylase by protein phosphorylation. This enzyme governs the synthesis of the catecholamine neurotransmitters and is stimulated by a number of intracellular signals. In the example shown here, neuronal electrical activity (1) causes influx of Ca^{2+} (2). The resultant rise in intracellular Ca^{2+} concentration (3) activates protein kinases (4), which phosphorylates tyrosine hydroxylase (5) to stimulate catecholamine synthesis (6). This, in turn, increases release of catecholamines (7) and enhances the postsynaptic response produced by the synapse (8).

Summary

A rich diversity of signal transduction pathways exists within all neurons. Activation of these pathways typically is initiated by chemical signals such as neurotransmitters and hormones. These molecules bind to receptors that include ligand-gated ion channels, G-protein-coupled receptors and tyrosine kinase receptors. Many of these receptors activate either heterotrimeric or monomeric G-proteins that regulate intracellular enzyme cascades and/or ion channels. A common outcome of the activation of these receptors is the production of second messengers, such as cAMP, Ca^{2+}, and IP_3, that bind to effector enzymes. Particularly important effectors are protein kinases and phosphatases that regulate the phosphorylation state of their substrates, and thus their function. These substrates can be metabolic enzymes or other signal transduction molecules, such as ion channels, protein kinases, or transcription factors that regulate gene expression. Examples of transcription factors include CREB, steroid hormone receptors, and c-fos. This plethora of molecular components allows intracellular signal transduction pathways to generate responses over a wide range of times and distances, greatly augmenting and refining the information-processing ability of neuronal circuits.

ADDITIONAL READING

Reviews

EXTON, J. H. (1998) Small GTPases. J. Biol. Chem. 273: 19923.

FISCHER, E. H. (1999) Cell signaling by protein tyrosine phosphorylation. Adv. Enzyme Regul. Review 39: 359–369.

FRIEDMAN, W. J. AND L. A. GREENE (1999) Neurotrophin signaling via Trks and p75. Exp. Cell Res. 253: 131–142.

GILMAN, A. G. (1984) G proteins and dual control of adenylate cyclase. Cell 36: 577–579.

GRAVES J. D. AND E. G. KREBS (1999) Protein phosphorylation and signal transduction. Pharmacol. Ther. 82: 111–121.

KUMER, S. AND K. VRANA (1996) Intricate regulation of tyrosine hydroxylase activity and gene expression. J. Neurochem. 67: 443–462.

LEVITAN, I. B. (1999) Modulation of ion channels by protein phosphorylation. How the brain works. Adv. Second Mess. Phosphoprotein Res. 33: 3–22.

NEER, E. J. (1995) Heterotrimeric G proteins: Organizers of transmembrane signals. Cell 80: 249–257.

RODBELL, M. (1995) Nobel Lecture. Signal transduction: Evolution of an idea. Bioscience Reports 15: 117–133.

STAMLER, J. S., E. J. TOONE, S. A. LIPTON AND N. J. SUCHER (1997) (S)NO Signals: Translocation, Regulation, and a Consensus Motif. Neuron 18: 691–696.

Important Original Papers

BACSKAI, B. J. AND 6 OTHERS (1993) Spatially resolved dynamics of cAMP and protein kinase A subunits in *Aplysia* sensory neurons. Science 260: 222–226.

BURGESS, G. M., P. P. GODFREY, J. S. MCKINNEY, M. J. BERRIDGE, R. F. IRVINE AND J.W. PUTNEY JR. (1984) The second messenger linking receptor activation to internal Ca release in liver. Nature 309: 63–66.

CONNOR, J. A. (1986) Digital imaging of free calcium changes and of spatial gradients in growing processes in single, mammalian central nervous system cells. Proc. Natl. Acad. Sci. USA 83: 6179–6183.

DE KONINCK, P. AND H. SCHULMAN (1998) Sensitivity of CaM kinase II to the frequency of Ca^{2+} oscillations. Science 279: 227–230.

FINCH, E. A. AND G. J. AUGUSTINE (1998) Local calcium signaling by IP_3 in Purkinje cell dendrites. Nature 396: 753–756.

KAMMERMEIER, P. J. AND S. R. IKEDA (1999) Expression of RGS2 alters the coupling of metabotropic glutamate receptor 1a to M-type K^+ and N-type Ca^{2+} channels. Neuron 22: 819–829.

KRAFT, A. S. AND W. B. ANDERSON (1983) Phorbol esters increase the amount of Ca^{2+}, phospholipid-dependent protein kinase associated with plasma membrane. Nature 301: 621–623.

LINDGREN, N. AND 8 OTHERS (2000) Regulation of tyrosine hydroxylase activity and phosphorylation at ser(19) and ser(40) via activation of glutamate NMDA receptors in rat striatum. J. Neurochem. 74: 2470–2477.

MILLER, S. G. AND M. B. KENNEDY (1986) Regulation of brain type II Ca^{2+}/calmodulin-dependent protein kinase by autophosphorylation: A Ca^{2+}-triggered molecular switch. Cell. 44: 861–870.

NORTHUP, J. K., P. C. STERNWEIS, M. D. SMIGEL, L. S. SCHLEIFER, E. M. ROSS AND A. G. GILMAN (1980) Purification of the regulatory component of adenylate cyclase. Proc. Natl. Acad. Sci. USA 77: 6516–6520.

SAITOH, T. AND J. H. SCHWARTZ (1985) Phosphorylation-dependent subcellular translocation of a Ca^{2+}/calmodulin-dependent protein kinase produces an autonomous enzyme in *Aplysia* neurons. J. Cell Biol. 100: 835–842.

SHEN, K. AND T. MEYER (1999) Dynamic control of CaMKII translocation and localization in hippocampal neurons by NMDA receptor stimulation. Science 284: 162–166.

SU, Y. AND 7 OTHERS (1995) Regulatory subunit of protein kinase A: Structure of deletion mutant with cAMP binding domains. Science 269: 807–813.

TAO, X., S. FINKBEINER, D. B. ARNOLD, A. J. SHAYWITZ AND M. E. GREENBERG (1998). Ca^{2+} influx regulates BDNF transcription by a CREB family transcription factor-dependent mechanism. Neuron 20: 709–726.

TESMER, J. J., R. K. SUNAHARA, A. G. GILMAN AND S. R. SPRANG (1997) Crystal structure of the catalytic domains of adenylyl cyclase in a complex with Gsα.GTPγS. Science 278: 1907–1916.

ZHANG, G., M. G. KAZANIETZ, P. M. BLUMBERG AND J. H. HURLEY (1995) Crystal structure of the cys2 activator-binding domain of protein kinase C delta in complex with phorbol ester. Cell 81: 917–924.

Books

ALBERTS, B., D. BRAY, J. LEWIS, M. RAFF, K. ROBERTS AND J. D. WATSON (1996) *Molecular Biology of the Cell*, 3rd Ed. New York: Garland Publishing, Inc.

CARAFOLI, E. AND C. KLEE (1999) *Calcium as a Cellular Regulator*. New York: Oxford University Press.

Sensation and Sensory Processing

II

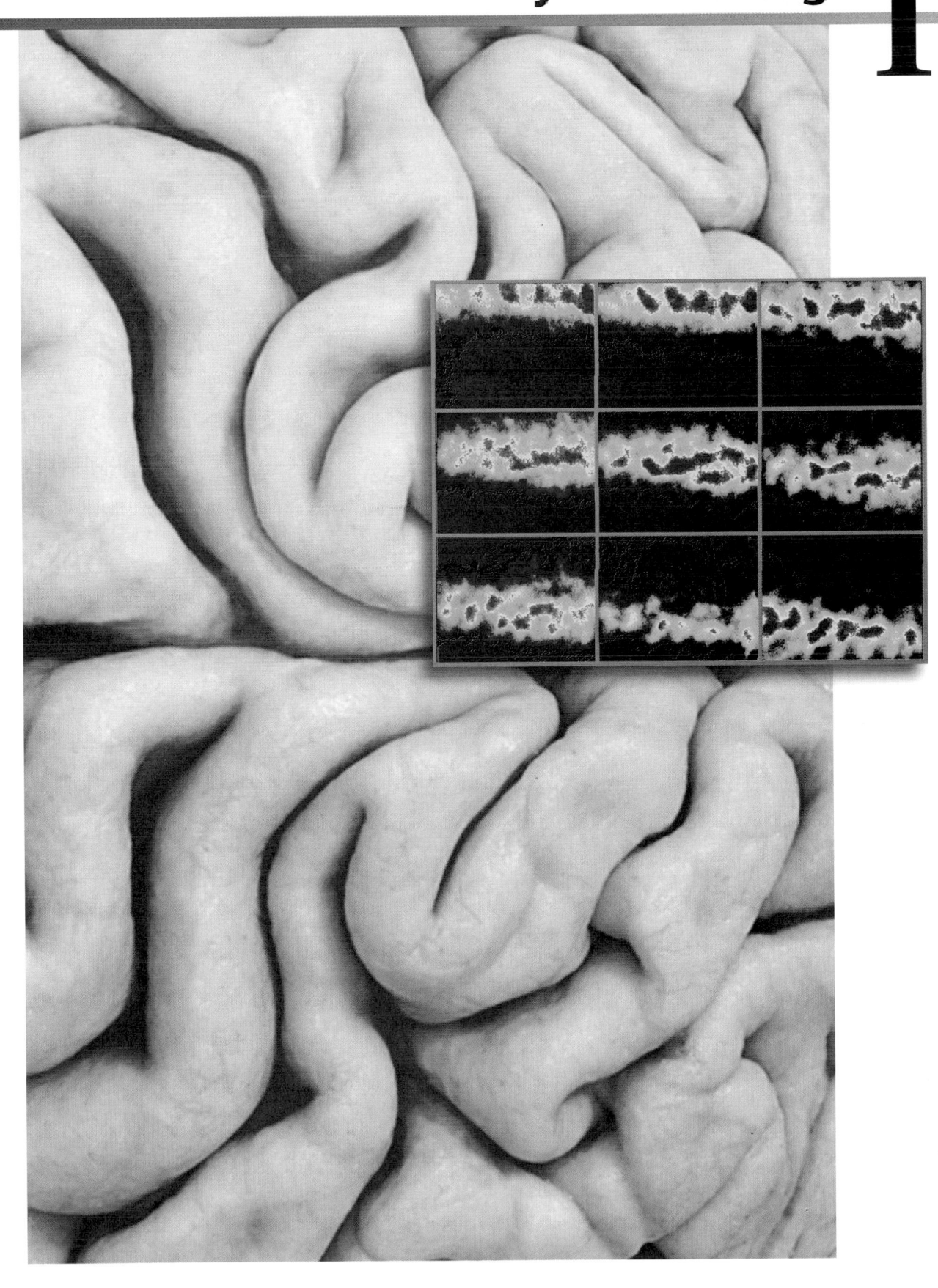

Surface view of the primary visual cortex illustrating patterns of neural activity visualized with intrinsic signal optical imaging techniques (see Box C in Chapter 12). Each panel illustrates the activity evoked by viewing a single thin vertical line. The smooth progression of the activated region from the upper left to the lower right panel illustrates the orderly mapping of visual space. The patchy appearance of the activated region in each panel reflects the columnar mapping of orientation preference. Red regions are the most active, black the least. (Courtesy of Bill Bosking, Justin Crowley, Tom Tucker, and David Fitzpatrick.)

Unit II
Sensation and Sensory Processing

9 *The Somatic Sensory System*
10 *Pain*
11 *Vision: The Eye*
12 *Central Visual Pathways*
13 *The Auditory System*
14 *The Vestibular System*
15 *The Chemical Senses*

Sensation entails the ability to transduce, encode, and ultimately perceive information generated by stimuli arising from both the external and internal environment, and much of the brain is devoted to these tasks. Although the basic senses—somatic sensation, vision, audition, vestibular sensation, and the chemical senses—are very different from one another, a few fundamental rules govern the way the nervous system deals with each of these diverse modalities. Highly specialized nerve cells called receptors convert the energy associated with mechanical forces, light, sound waves, odorant molecules, or ingested chemicals into neural signals that convey information about the stimulus to the brain. These afferent sensory signals activate central neurons capable of representing both the qualitative and quantitative aspects of the stimulus (what it is and how strong it is), and in some modalities—somatic sensation, vision, and audition—the location of the stimulus in space (where it is).

The clinical evaluation of patients routinely requires an assessment of the sensory systems to infer the nature and location of potential neurological problems. Knowledge of where and how the different sensory modalities are transduced, relayed, represented, and further processed to generate appropriate behavioral responses is therefore essential to understanding and treating a wide variety of diseases. Accordingly, these chapters on the neurobiology of sensation also serve to introduce some of the major structure/function relationships in the sensory components of the nervous system.

Chapter 9

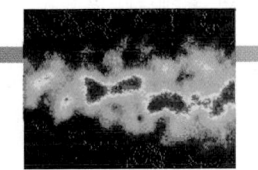

The Somatic Sensory System

Overview

The somatic sensory system has two major components: a subsystem for the detection of mechanical stimuli (e.g., light touch, vibration, pressure, and cutaneous tension), and a subsystem for the detection of painful stimuli and temperature. Together, these two subsystems give humans and other animals the ability to identify the shapes and textures of objects, to monitor the internal and external forces acting on the body at any moment, and to detect potentially harmful circumstances. This chapter focuses on the mechanosensory subsystem; the pain and temperature subsystem is taken up in the following chapter.

Mechanosensory processing of external stimuli is initiated by the activation of a diverse population of cutaneous and subcutaneous mechanoreceptors at the body surface that relays information to the central nervous system for interpretation and ultimately action. Additional receptors located in muscles, joints, and other deep structures monitor mechanical forces generated by the musculoskeletal system and are called proprioceptors. Mechanosensory information is carried to the brain by several ascending pathways that run in parallel through the spinal cord, brainstem, and thalamus to reach the primary somatic sensory cortex in the postcentral gyrus of the parietal lobe. The primary somatic sensory cortex projects in turn to higher-order association cortices in the parietal lobe, and back to the subcortical structures involved in mechanosensory information processing.

Cutaneous and Subcutaneous Somatic Sensory Receptors

The specialized sensory receptors in the cutaneous and subcutaneous tissues are dauntingly diverse (Table 9.1). They include free nerve endings in the skin, nerve endings associated with specializations that act as amplifiers or filters, and sensory terminals associated with specialized transducing cells that influence the ending by virtue of synapse-like contacts. Based on function, this variety of receptors can be divided into three groups: **mechanoreceptors**, **nociceptors**, and **thermoceptors**. On the basis of their morphology, the receptors near the body surface can also be divided into **free** and **encapsulated** types. Nociceptor and thermoceptor specializations are referred to as **free nerve endings** because the unmyelinated terminal branches of these neurons ramify widely in the upper regions of the dermis and epidermis; their role in pain and temperature sensation is discussed in Chapter 10. Most other cutaneous receptors show some degree of **encapsulation**, which helps determine the nature of the stimuli to which they respond.

Despite their variety, all somatic sensory receptors work in fundamentally the same way: Stimuli applied to the skin deform or otherwise change the

TABLE 9.1
The Major Classes of Somatic Sensory Receptors

Receptor type	Anatomical characteristics	Associated axons[a] (and diameters)	Axonal conduction velocities	Location	Function	Rate of adaptation	Threshold of activation
Free nerve endings	Minimally specialized nerve endings	C, Aδ	2–20 m/s	All skin	Pain, temperature, crude touch	Slow	High
Meissner's corpuscles	Encapsulated; between dermal papillae	Aβ 6–12 μm		Principally glabrous skin	Touch, pressure (dynamic)	Rapid	Low
Pacinian corpuscles	Encapsulated; onionlike covering	Aβ 6–12 μm		Subcutaneous tissue, interosseous membranes, viscera	Deep pressure, vibration (dynamic)	Rapid	Low
Merkel's disks	Encapsulated; associated with peptide-releasing cells	Aβ		All skin, hair follicles	Touch, pressure (static)	Slow	Low
Ruffini's corpuscles	Encapsulated; oriented along stretch lines	Aβ 6–12 μm		All skin	Stretching of skin	Slow	Low
Muscle spindles	Highly specialized (see Figure 9.5 and Chapter 15)	Ia and II		Muscles	Muscle length	Both slow and rapid	Low
Golgi tendon organs	Highly specialized (see Chapter 15)	Ib		Tendons	Muscle tension	Slow	Low
Joint receptors	Minimally specialized	—		Joints	Joint position	Rapid	Low

[a]In the 1920s and 1930s, there was a virtual cottage industry classifying axons according to their conduction velocity. Three main categories were discerned, called A, B, and C. A comprises the largest and fastest axons, C the smallest and slowest. Mechanoreceptor axons generally fall into category A. The A group is further broken down into subgroups designated α (the fastest), β, and δ (the slowest). To make matters even more confusing, muscle afferent axons are usually classified into four additional groups—I (the fastest), II, III, and IV (the slowest)—with subgroups designated by lowercase roman letters!

nerve endings, which in turn affects the ionic permeability of the receptor membrane. Changes in permeability generate a depolarizing current in the nerve ending, thus producing a **receptor** (or **generator**) **potential** that triggers action potentials, as described in Chapters 2 and 3. This overall process, in which the energy of a stimulus is converted into an electrical signal in the sensory neuron, is called **sensory transduction** and is the critical first step in all sensory processing.

The *quality* of a mechanosensory (or any other) stimulus (i.e., what it represents and where it is) is determined by the properties of the relevant receptors and the location of their central targets (Figure 9.1). The quantity or strength of the stimulus is conveyed by the rate of action potential discharge triggered by the receptor potential (although this relationship is nonlinear and often quite complex). Some receptors fire rapidly when a stimulus is first presented and then fall silent in the presence of continued stimulation (which is to say they "adapt" to the stimulus), whereas others generate a sustained discharge in the presence of an ongoing stimulus (Figure 9.2). The usefulness of having some receptors that adapt quickly and others that do not is to provide information about both the *dynamic* and *static* qualities of a stimulus. Receptors that initially fire in the presence of a stimulus and then become quiescent are particularly effective in conveying information about changes in

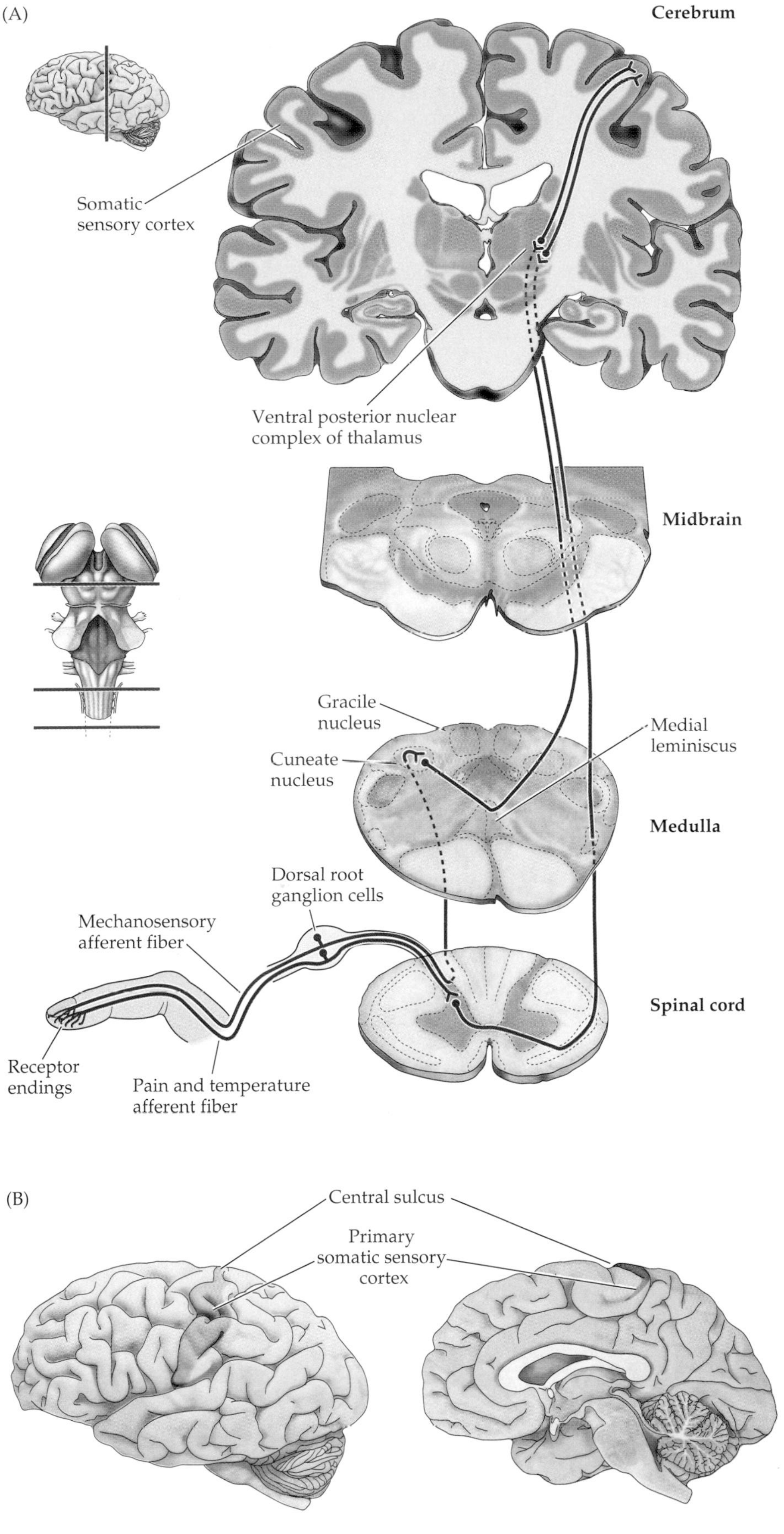

(A)

Somatic
sensory cortex

Cerebrum

Ventral posterior nuclear
complex of thalamus

Midbrain

Gracile
nucleus

Cuneate
nucleus

Medial
leminiscus

Medulla

Dorsal root
ganglion cells

Mechanosensory
afferent fiber

Spinal cord

Receptor
endings

Pain and temperature
afferent fiber

Figure 9.1 General organization of the somatic sensory system. (A) Mechanosensory information about the body reaches the brain by way of a three-neuron relay (shown in red). The first synapse is made by the terminals of the centrally projecting axons of dorsal root ganglion cells onto neurons in the brainstem nuclei (the local branches involved in segmental spinal reflexes are not shown here). The axons of these second-order neurons synapse on third-order neurons of the ventral posterior nuclear complex of the thalamus, which in turn send their axons to the primary somatic sensory cortex. Information about pain and temperature takes a different course (shown in blue; the anterolateral system), and is discussed in the following chapter. (B) Lateral and midsagittal views of the human brain, illustrating the approximate location of the primary somatic sensory cortex in the anterior parietal lobe, just posterior to the central sulcus.

(B)

Central sulcus

Primary
somatic sensory
cortex

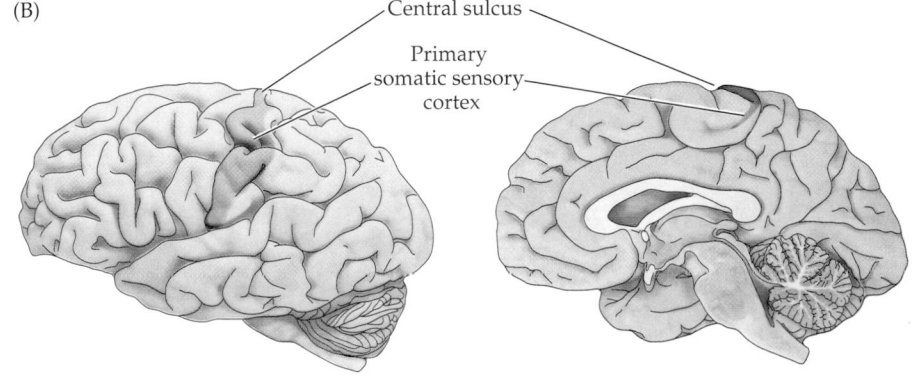

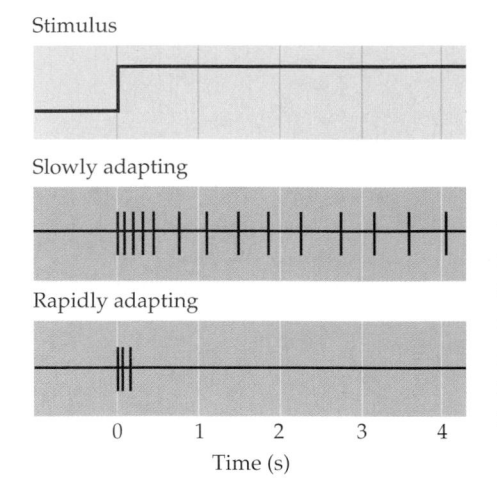

Stimulus

Slowly adapting

Rapidly adapting

Time (s)

Figure 9.2   Slowly adapting mechanoreceptors continue responding to a stimulus, whereas rapidly adapting receptors respond only at the onset (and often the offset) of stimulation. These functional differences allow the mechanoreceptors to provide information about both the static (via slowly adapting receptors) and dynamic (via rapidly adapting receptors) qualities of a stimulus.

the information the receptor reports; conversely, receptors that continue to fire convey information about the persistence of a stimulus. Accordingly, somatic sensory receptors and the neurons that give rise to them are usually classified into rapidly or slowly adapting types (see Table 9.1). **Rapidly adapting**, or **phasic**, receptors respond maximally but briefly to stimuli; their response decreases if the stimulus is maintained. Conversely, **slowly adapting**, or **tonic**, receptors keep firing as long as the stimulus is present.

Mechanoreceptors Specialized to Receive Tactile Information

Four major types of encapsulated mechanoreceptors are specialized to provide information to the central nervous system about touch, pressure, vibration, and cutaneous tension: Meissner's corpuscles, Pacinian corpuscles, Merkel's disks, and Ruffini's corpuscles (Figure 9.3 and Table 9.1). These receptors are referred to collectively as **low-threshold** (or high-sensitivity) mechanoreceptors because even weak mechanical stimulation of the skin induces them to produce action potentials. All low-threshold mechanoreceptors are innervated by relatively large myelinated axons (type Aβ; see Table 9.1), ensuring the rapid central transmission of tactile information.

Meissner's corpuscles, which lie between the dermal papillae just beneath the epidermis of the fingers, palms, and soles, are elongated recep-

Figure 9.3   The skin harbors a variety of morphologically distinct mechanoreceptors. This diagram represents the smooth, hairless (also called glabrous) skin of the fingertip. The major characteristics of the various receptor types are summarized in Table 9.1. (After Darian-Smith, 1984.)

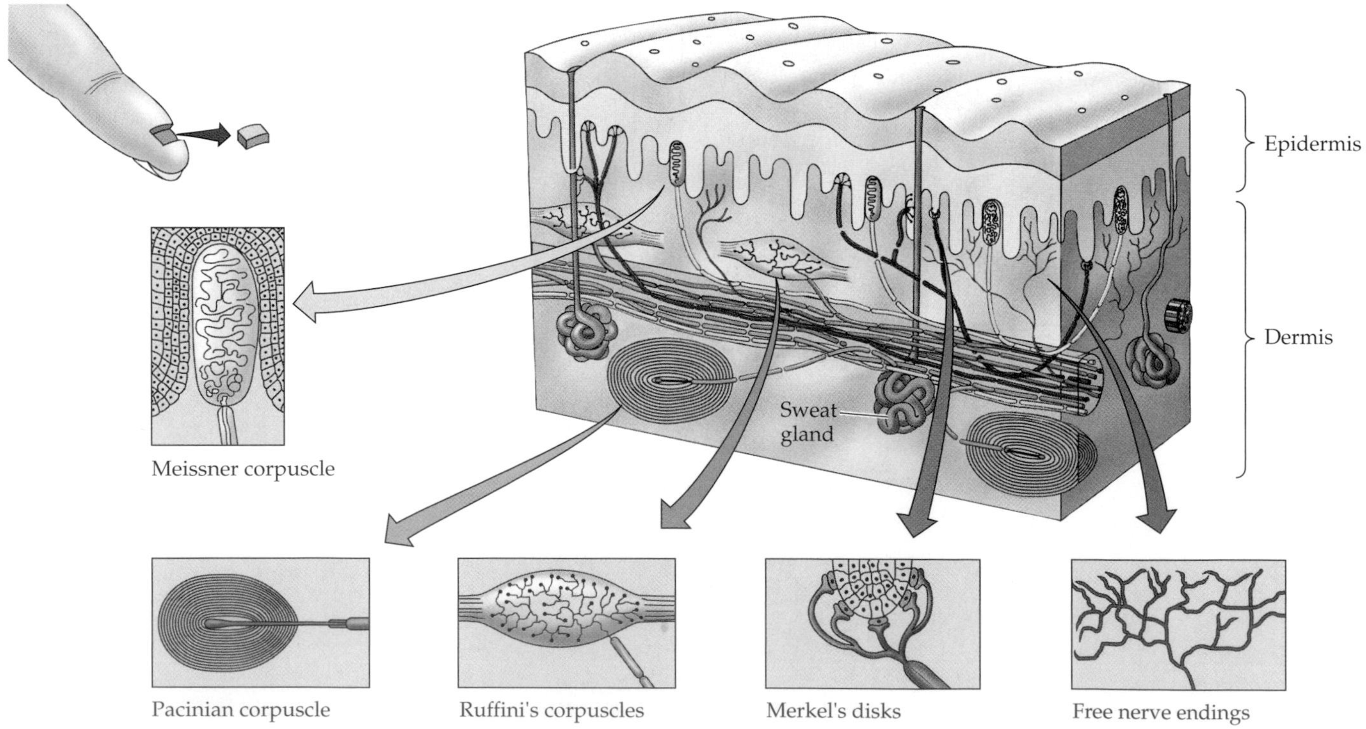

Epidermis

Dermis

Sweat gland

Meissner corpuscle

Pacinian corpuscle

Ruffini's corpuscles

Merkel's disks

Free nerve endings

tors formed by a connective tissue capsule that comprises several lamellae of Schwann cells. The center of the capsule contains one or more afferent nerve fibers that generate rapidly adapting action potentials following minimal skin depression. Meissner's corpuscles are the most common mechanoreceptors of "glabrous" (smooth, hairless) skin (the fingertips, for instance), and their afferent fibers account for about 40% of the sensory innervation of the human hand. These corpuscles are particularly efficient in transducing information about the relatively low-frequency vibrations (30–50 Hz) that occur when textured objects are moved across the skin.

Pacinian corpuscles are large encapsulated endings located in the subcutaneous tissue (and more deeply in interosseous membranes and mesenteries of the gut). These receptors differ from Meissner's corpuscles in their morphology, distribution, and response threshold. The Pacinian corpuscle has an onionlike capsule in which the inner core of membrane lamellae is separated from an outer lamella by a fluid-filled space. One or more rapidly adapting afferent axons lie at the center of this structure. The capsule again acts as a filter, in this case allowing only transient disturbances at high frequencies (250–350 Hz) to activate the nerve endings. Pacinian corpuscles adapt more rapidly than Meissner's corpuscles and have a lower response threshold. These attributes suggest that Pacinian corpuscles are involved in the discrimination of fine surface textures or other moving stimuli that produce high-frequency vibration of the skin. In corroboration of this supposition, stimulation of Pacinian corpuscle afferent fibers in humans induces a sensation of vibration or tickle. They make up 10–15% of the cutaneous receptors in the hand. Pacinian corpuscles located in interosseous membranes probably detect vibrations transmitted to the skeleton. Structurally similar endings found in the bills of ducks and geese and in the legs of cranes and herons detect vibrations in water; such endings in the wings of soaring birds detect vibrations produced by air currents. Because they are rapidly adapting, Pacinian corpuscles, like Meissner's corpuscles, provide information primarily about the dynamic qualities of mechanical stimuli.

Slowly adapting cutaneous mechanoreceptors include **Merkel's disks** and **Ruffini's corpuscles** (see Figure 9.3 and Table 9.1). Merkel's disks are located in the epidermis, where they are precisely aligned with the papillae that lie beneath the dermal ridges. They account for about 25% of the mechanoreceptors of the hand and are particularly dense in the fingertips, lips, and external genitalia. The slowly adapting nerve fiber associated with each Merkel's disk enlarges into a saucer-shaped ending that is closely applied to another specialized cell containing vesicles that apparently release peptides that modulate the nerve terminal. Selective stimulation of these receptors in humans produces a sensation of light pressure. These several properties have led to the supposition that Merkel's disks play a major role in the static discrimination of shapes, edges, and rough textures.

Ruffini's corpuscles, although structurally similar to other tactile receptors, are not well understood. These elongated, spindle-shaped capsular specializations are located deep in the skin, as well as in ligaments and tendons. The long axis of the corpuscle is usually oriented parallel to the stretch lines in skin; thus, Ruffini's corpuscles are particularly sensitive to the cutaneous stretching produced by digit or limb movements. They account for about 20% of the receptors in the human hand and do not elicit any particular tactile sensation when stimulated electrically. Although there is still some question as to their function, they probably respond primarily to internally generated stimuli (see the section on proprioception below).

Differences in Mechanosensory Discrimination Across the Body Surface

The accuracy with which tactile stimuli can be sensed varies from one region of the body to another, a phenomenon that illustrates some further principles of somatic sensation. Figure 9.4 shows the results of an experiment in which variation in tactile ability across the body surface was measured by **two-point discrimination**. This technique measures the minimal interstimulus distance required to perceive two simultaneously applied stimuli as distinct (the indentations of the points of a pair of calipers, for example). When applied to the skin, such stimuli of the fingertips are discretely perceived if they are only 2 mm apart. In contrast, the same stimuli applied to the forearm are not perceived as distinct until they are at least 40 mm apart! This marked regional difference in tactile ability is explained by the fact that the encapsulated mechanoreceptors that respond to the stimuli are three to four times more numerous in the fingertips than in other areas of the hand, and many times more dense than in the forearm. Equally important in this regional difference are the sizes of the neuronal receptive fields. The **receptive** field of a somatic sensory neuron is the region of the skin within which a tactile stimulus evokes a sensory response in the cell or its axon (Boxes A and B). Analysis of the human hand shows that the receptive fields of mechanosensory neurons are 1–2 mm in diameter on the fingertips but 5–10 mm on the palms. The receptive fields on the arm are larger still. The importance of receptive field size is easy to envision. If, for instance, the receptive fields of all cutaneous receptor

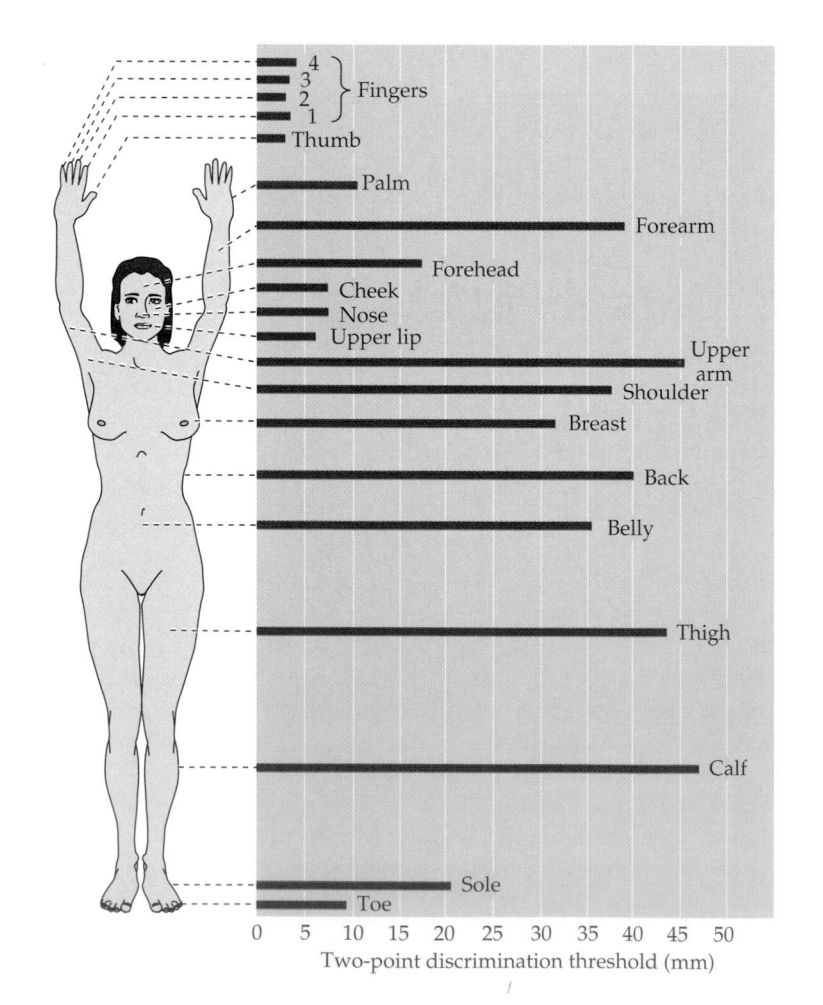

Figure 9.4 Variation in the sensitivity of tactile discrimination as a function of location on the body surface, measured here by two-point discrimination. (After Weinstein, 1969.)

BOX A
Receptive Fields and Sensory Maps in the Cricket

Two principles of somatic sensory organization have emerged from studies of the mammalian brain: (1) individual neurons are tuned to particular aspects of complex stimuli; and (2) these stimulus qualities are represented in an orderly fashion in relevant regions of the nervous system. These principles apply equally well to invertebrates, including the equivalent of the somatic sensory system in insects such as crickets, grasshoppers, and cockroaches.

In the cricket, the salient tactile stimulation for the animal comes from air currents that displace sensory hairs of bilaterally symmetric sensory structures called cerci (sing., *cercus*). The location and structure of specific cercal hairs allow them to be displaced by air currents having different directions and speeds. Accordingly, the peripheral sensory neurons associated with the hairs represent the full range of air current directions and velocities impinging on the animal. This information is carried centrally and is systematically represented in a region of the cricket central nervous system called the terminal ganglion.

Individual neurons in this ganglion correspond to the cercal hairs, and have receptive fields and response properties that represent a full range of directions and speeds for extrinsic mechanical forces, including air currents (see figure). For the cricket, the significance of this information is most likely detection of the direction and speed of oncoming objects to execute motor programs for escape (this also likely the significance of this representation for the cockroach, who can therefore escape the oncoming blow from a stamping foot).

Much like the somatic sensory system in mammals, the primary sensory afferents project to the terminal ganglion in an orderly fashion, such that there is a somatotopic map of air current directions. And, like mammals, individual neurons within this representation are tuned to specific aspects of the mechanical forces acting on the cricket. The major function of this system for the cricket is detection of the direction and speed of predators, which allows the cricket to initiate an appropriate escape response.

These facts about insects' mechanosensory system emphasize that somatic sensory functions are basically similar across a wide range of animals.

Accordingly, regardless of sensory modality, nervous system organization, or the identity of the organism, it is likely that stimulus specificity will be reflected in receptive fields of individual neurons and there will be orderly mapping of those receptive fields into either a topographic or computational map in the animal's brain.

References

JACOBS, G. A. AND F. E. THEUNISSEN (1996) Functional organization of a neural map in the cricket cercal sensory system. J. Neurosci. 16: 769–784.

MILLER, J. P., G. A. JACOBS AND F. E. THEUNISSEN (1991) Representation of sensory information in the cricket cercal sensory system. I. Response properties of the primary interneurons. J. Neurophys. 66: 1680–1689.

MURPHEY, R. K. (1981) The structure and development of somatotopic map in crickets: The cercal afferent projection. Dev. Biol. 88: 236–246.

MURPHEY, R. K. AND H. V. B. HIRSCH (1982) From cat to cricket: The genesis of response selectivity of interneurons. Curr. Topics Dev. Biol. 17: 241–256.

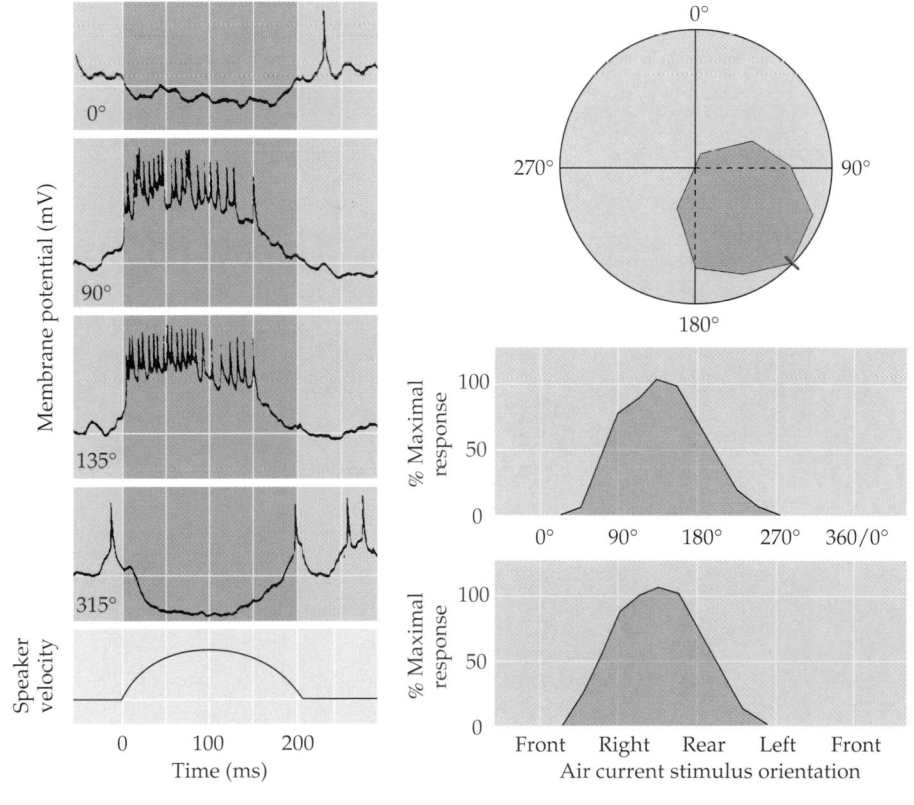

Intracellular recording of action potential activity of an individual sensory neuron's responses to different directions of wind current. The plots at right indicate this neuron's receptive field for wind direction and the tuning curve for the neuron's selective firing to its preferred direction. (After Miller et al., 1991.)

BOX B
Dynamic Aspects of Somatic Sensory Receptive Fields

When we manually explore objects, multiple contacts between the skin and the object surface generate extraordinarily complex patterns of tactile stimuli. As a consequence, the somatic sensory system must process subtle signals that change continuously in time. Nonetheless, we routinely discriminate the size, texture and shape of objects with great accuracy. Until recently, the temporal structure of such stimuli was not considered a major variable in characterizing the physiological properties of somatic sensory neurons. For instance, the classical definition of the receptive field of a somatic sensory neuron takes into account only the overall area of the body surface that elicits significant variation in the neuron's firing rate (see text). By the same token, the topographic maps in the somatic sensory system have been interpreted as evi-

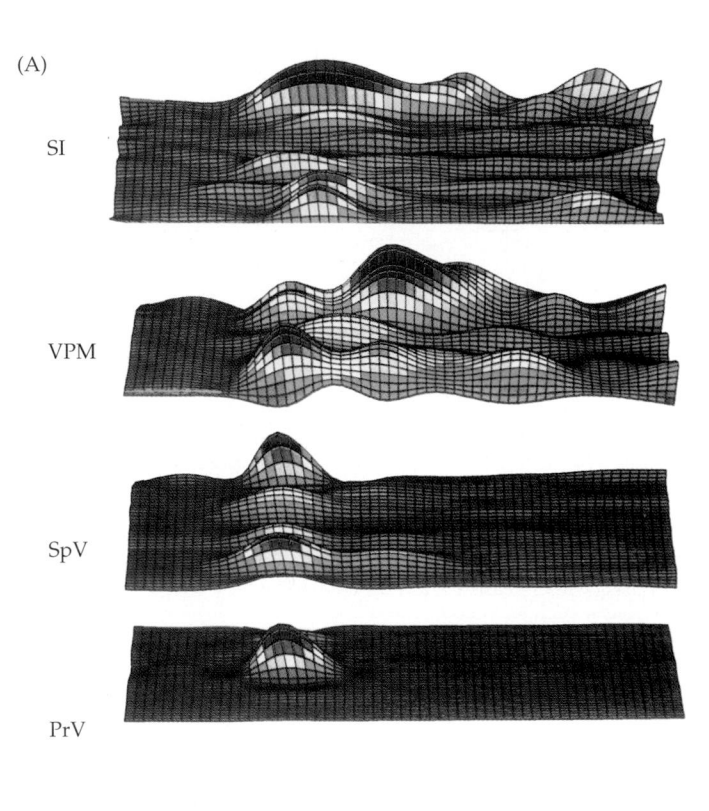

(A)

SI

VPM

SpV

PrV

Dynamic aspects of receptive fields. (A) Simultaneous electrode recordings in behaving rats allow monitoring of the spatiotemporal spread of neuronal activation across several levels of the somatic sensory system following stimulation (of a single facial whisker, in this example). These 3-D graphs represent patterns of neuronal ensemble activity at each level of the pathway (the x axis represents the poststimulus time in ms, the y axis the number of neurons recorded at each level; the color-coded gradient in the z axis shows the response of the neurons, with red the highest firing and green the lowest). SI, somatic sensory cortex; VPM, ventral posterior medial nucleus of the thalamus; SpV, spinal nucleus of the trigeminal brainstem complex; and PrV, the principle nucleus of the brainstem trigeminal complex. (B) Cortical receptive fields (from two different animals). Each panel represents the matrix of whiskers on the animals' snout (whisker columns are on the x axis and whisker rows on the y axis) for a 4 ms epoch of poststimulus time. Within a particular time period, the center of the receptive field is defined as the whisker eliciting the greatest response magnitude (yellow). Note that the receptive field centers shift as a function of time. (A from Nicholelis et al., 1997; B from Ghazanfar and Nicholelis, 1999.)

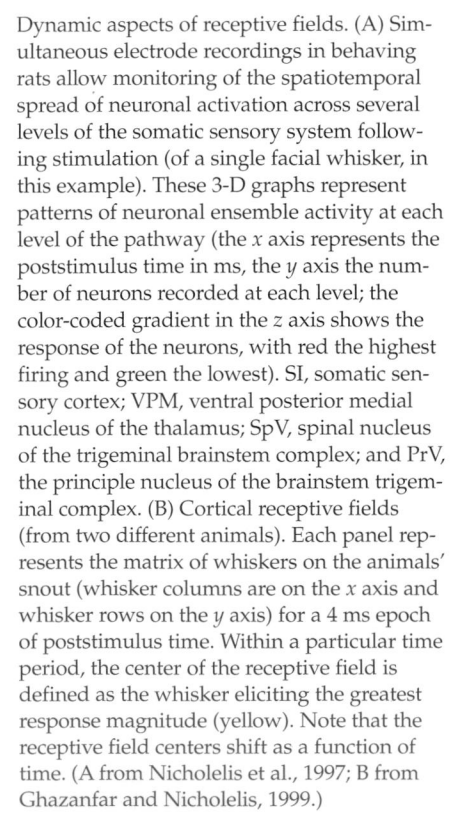

(B)

dence that tactile information processing involves primarily spatial criteria.

The advent of multiple electrode recording to simultaneously monitor the activity of large populations of single neurons has begun to change this "static" view of the somatic sensory system. In both primates and rodents, this approach has shown that the receptive fields of cortical and subcortical neurons vary as a function of time: The neuron responds differently to a spatially defined stimulus as the period of stimulation proceeds (see figure).

This coupling of space and time can also be demonstrated at level of somatotopic maps (see text). By recording the activity of single neurons located in different regions of the map simultaneously, it is apparent that the stimulation of a small area of the skin tends to excite more and more neurons as time goes by. Thus, many more neurons than those located in the area of the map directly representing the stimulated skin actually respond to the stimulus, albeit at longer latencies. The end result of these more complex neuronal responses is the emergence of spatiotemporal representations at all levels of the somatic sensory system. Thus, contrary to the classical notion of receptive fields, the somatic sensory system processes information in a dynamic way. Such processing is not only relevant for the normal operation of the system, but may also account for some aspects of adult plasticity (see Chapter 25).

References

GHAZANFAR, A. A. AND M. A. L. NICOLELIS (1999) Spatiotemporal properties of layer V neurons of the rat primary somatosensory cortex. Cereb. Cortex 4: 348–361.

NICOLELIS, M. A. L., A. A. GHAZANFAR, B. FAGGIN, S. VOTAW AND L. M. O. OLIVEIRA (1997) Reconstructing the engram: Simultaneous, multiple site, many single neuron recordings. Neuron 18: 529–537.

NICOLELIS, M. A. L. AND 7 OTHERS (1998) Simultaneous encoding of tactile information by three primate cortical areas. Nature Neurosci. 1: 621–630.

neurons covered the entire digital pad, it would be impossible to discriminate two spatially separate stimuli applied to the fingertip (since all the receptive fields would be returning the same spatial information).

Receptor density and receptive field sizes in different regions are not the only factors determining somatic sensation. Psychophysical analysis of tactile performance suggests that something more than the cutaneous periphery is needed to explain variations in tactile perception. For instance, sensory thresholds in two-point discrimination tests vary with practice, fatigue, and stress. The contextual significance of stimuli is also important in determining what we feel; even though we spend most of the day wearing clothes, we usually ignore the tactile stimulation that they produce. Some aspect of the mechanosensory system allows us to filter out this information and pay attention to it only when necessary. The fascinating phenomenon of "phantom limb" sensations after amputation (see Box B in Chapter 10) provides further evidence that tactile perception is not fully explained by the peripheral information that travels centrally. The central nervous system clearly plays an active role in determining the perception of the mechanical forces that act on us.

Mechanoreceptors Specialized for Proprioception

Whereas cutaneous mechanoreceptors provide information derived from external stimuli, another major class of receptors provides information about mechanical forces arising from the body itself, the musculoskeletal system in particular. These are called **proprioceptors**, roughly meaning "receptors for self." The purpose of proprioceptors is primarily to give detailed and continuous information about the position of the limbs and other body parts in space (specialized mechanoreceptors also exist in the heart and major vessels to provide information about blood pressure, but these neurons are considered to be part of the visceral motor system; see Chapter 21). Low-threshold mechanoreceptors, including muscle spindles, Golgi tendon organs, and joint receptors, provide this kind of sensory information, which is essential to the accurate performance of complex movements. Information about the position and motion of the head is particularly important; in this case, proprioceptors are integrated with the highly specialized vestibular system, which is considered separately in Chapter 14.

The most detailed knowledge about proprioception derives from studies of **muscle spindles**, which are found in all but a few striated (skeletal) muscles. Muscle spindles consist of four to eight specialized **intrafusal muscle fibers** surrounded by a capsule of connective tissue. The intrafusal fibers are distributed among the ordinary (extrafusal) fibers of skeletal muscle in a parallel arrangement (Figure 9.5). In the largest of the several intrafusal fibers, the nuclei are collected in an expanded region in the center of the fiber called a bag; hence the name *nuclear bag fibers*. The nuclei in the remaining two to six smaller intrafusal fibers are lined up single file, with the result that these fibers are called *nuclear chain fibers*. Myelinated sensory axons belonging to group Ia innervate muscle spindles by encircling the middle portion of both types of intrafusal fibers (see Figure 9.5 and Table 9.1). The Ia axon terminal is known as the **primary sensory ending** of the spindle. Secondary innervation is provided by group II axons that innervate the nuclear chain fibers and give off a minor branch to the nuclear bag fibers. The intrafusal muscle fibers contract when commanded to do so by motor axons derived from a pool of specialized motor neurons in the spinal cord (called γ **motor neurons**). The major function of muscle spindles is to provide information about muscle length (that is, the degree to which they are being stretched). A detailed account of how these important receptors function during movement is given in Chapters 16 and 17.

The density of spindles in human muscles varies. Large muscles that generate coarse movements have relatively few spindles; in contrast, extraocular muscles and the intrinsic muscles of the hand and neck are richly supplied with spindles, reflecting the importance of accurate eye movements, the need to manipulate objects with great finesse, and the continuous demand for precise positioning of the head. This relationship between receptor density and muscle size is consistent with the generalization that the sensory motor apparatus at all levels of the nervous system is much richer for the

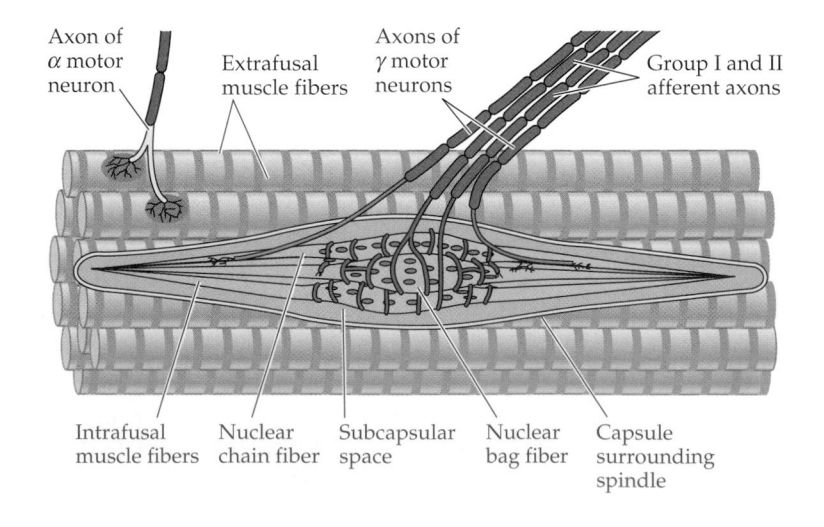

Figure 9.5 A muscle spindle and several extrafusal muscle fibers. See text for description. (After Matthews, 1964.)

hands, head, speech organs, and other parts of the body that are used to perform especially important and demanding tasks. Spindles are lacking altogether in a few muscles, such as those of the middle ear, which do not require the kind of feedback that these receptors provide.

Whereas muscle spindles are specialized to signal changes in muscle *length*, low-threshold mechanoreceptors in tendons inform the central nervous system about changes in muscle *tension*. These mechanoreceptors, called **Golgi tendon organs**, are innervated by branches of group Ib afferents and are distributed among the collagen fibers that form the tendons (see Chapter 16).

Finally, rapidly adapting mechanoreceptors in and around joints gather dynamic information about limb position and joint movement. The function of these **joint receptors** is not well understood.

Active Tactile Exploration

Tactile discrimination—that is, perceiving the detailed shape or texture of an object—normally entails active exploration. In humans, this is typically accomplished by using the hands to grasp and manipulate objects, or by moving the fingers across a surface so that a sequence of contacts between the skin and the object of interest is established. Psychophysical evidence indicates that relative movement between the skin and a surface is the single most important requirement for accurate discrimination of texture. Animal experiments confirm the dependence of tactile discrimination on active exploration. Rats, for instance, discriminate the details of texture by rhythmically brushing their facial whiskers across surfaces. Active touching, which is called **haptics**, involves the interpretation of complex spatiotemporal patterns of stimuli that are likely to activate many classes of mechanoreceptors. Haptics also requires dynamic interactions between motor and sensory signals, which presumably induce sensory responses in central neurons that differ from the responses of the same cells during passive stimulation of the skin (see Box B).

The Major Afferent Pathway for Mechanosensory Information: The Dorsal Column–Medial Lemniscus System

The action potentials generated by tactile and other mechanosensory stimuli are transmitted to the spinal cord by afferent sensory axons traveling in the peripheral nerves. The neuronal cell bodies that give rise to these first-order axons are located in the **dorsal root** (or **sensory**) **ganglia** associated with each segmental spinal nerve (see Figure 9.1 and Box C). Dorsal root ganglion cells are also known as **first-order neurons** because they initiate the sensory process. The ganglion cells thus give rise to long peripheral axons that end in the somatic receptor specializations already described, and shorter central axons that reach the dorsolateral region of the spinal cord via the **dorsal (sensory) roots** of each spinal cord segment. The large myelinated fibers that innervate low-threshold mechanoreceptors are derived from the largest neurons in these ganglia, whereas the smaller ganglion cells give rise to smaller afferent nerve fibers that end in the high-threshold nociceptors and thermoceptors (see Table 9.1).

Depending on whether they belong to the mechanosensory system or to the pain and temperature system, the first-order axons carrying information from somatic receptors have different patterns of termination in the spinal cord and define distinct somatic sensory pathways within the central nervous system (see Figure 9.1). **The dorsal column–medial lemniscus pathway**

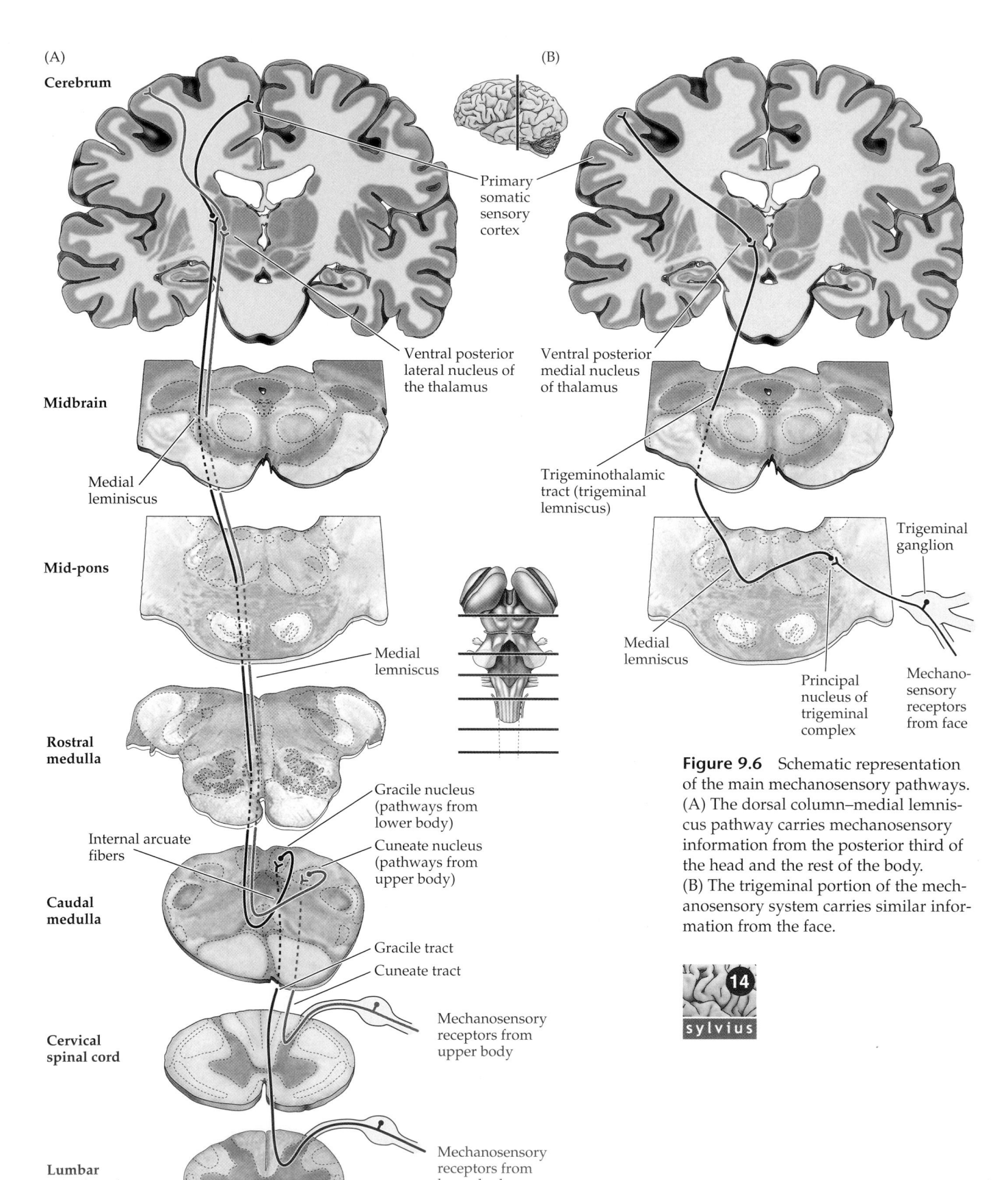

(A)

Cerebrum

(B)

Primary somatic sensory cortex

Ventral posterior lateral nucleus of the thalamus

Ventral posterior medial nucleus of thalamus

Midbrain

Medial leminiscus

Trigeminothalamic tract (trigeminal lemniscus)

Mid-pons

Trigeminal ganglion

Medial lemniscus

Medial lemniscus

Principal nucleus of trigeminal complex

Mechanosensory receptors from face

Rostral medulla

Gracile nucleus (pathways from lower body)

Cuneate nucleus (pathways from upper body)

Internal arcuate fibers

Caudal medulla

Gracile tract

Cuneate tract

Cervical spinal cord

Mechanosensory receptors from upper body

Lumbar spinal cord

Mechanosensory receptors from lower body

Figure 9.6 Schematic representation of the main mechanosensory pathways. (A) The dorsal column–medial lemniscus pathway carries mechanosensory information from the posterior third of the head and the rest of the body. (B) The trigeminal portion of the mechanosensory system carries similar information from the face.

sylvius 14

BOX C
Dermatomes

Each dorsal root (or sensory) ganglion and associated spinal nerve arises from an iterated series of embryonic tissue masses called somites. This fact of development explains the overall segmental arrangement of somatic nerves (and the targets they innervate) in the adult (see figure). The territory innervated by each spinal nerve is called a dermatome. In humans, the cutaneous area of each dermatome has been defined in patients in whom specific dorsal roots were affected (as in herpes zoster, or "shingles") or after surgical interruption (for relief of pain or other reasons). Such studies show that dermatomal maps vary among individuals. Moreover, dermatomes also overlap substantially, so that injury to an individual dorsal root does not lead to complete loss of sensation in the relevant skin region, the overlap being more extensive for touch, pressure, and vibration than for pain and temperature. Thus, testing for pain sensation provides a more precise assessment of a segmental nerve injury than does testing for responses to touch, pressure, or vibration. Finally, the segmental distribution of proprioceptors does not follow the dermatomal map but is more closely allied with the pattern of muscle innervation. Despite these limitations, knowledge of dermatomes is essential in the clinical evaluation of neurological patients, particularly in determining the level of a spinal lesion.

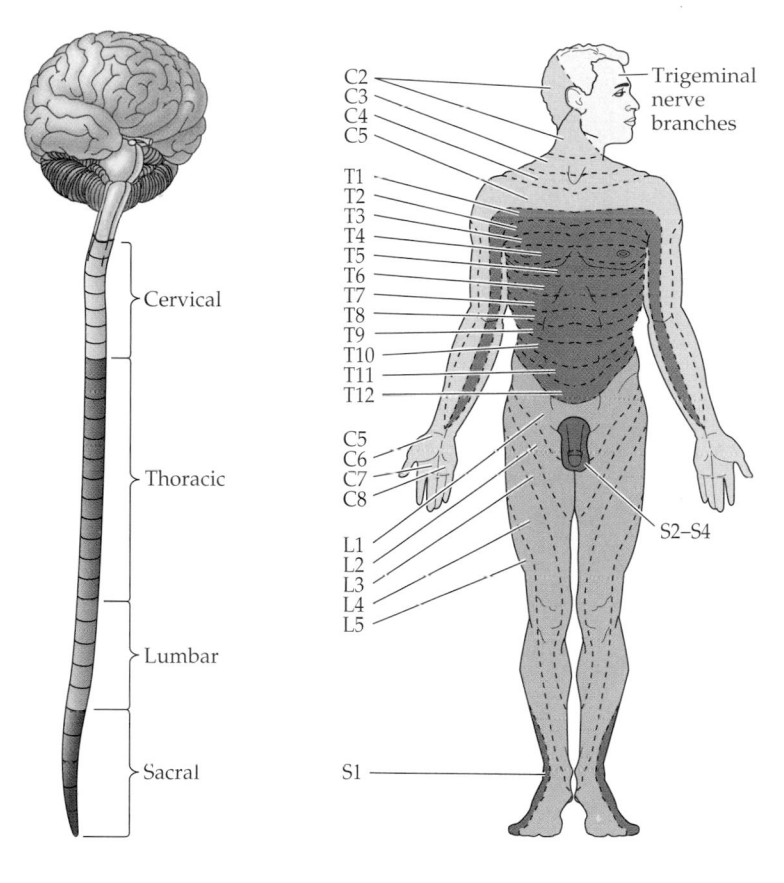

The innervation arising from a single dorsal root ganglion and its spinal nerve is called a dermatome. The full set of sensory dermatomes is shown here for a typical adult. Knowledge of this arrangement is particularly important in defining the location of suspected spinal (and other) lesions. The numbers refer to the spinal segments by which each nerve is named.

carries the majority of information from the mechanoreceptors that mediate tactile discrimination and proprioception (Figure 9.6); the **spinothalamic (anterolateral) pathway** mediates pain and temperature sensation and is described in Chapter 10. This difference in the afferent pathways of these modalities is one of the reasons that pain and temperature sensation is treated separately here.

Upon entering the spinal cord, the first-order axons carrying information from peripheral mechanoreceptors bifurcate into ascending and descending branches, which in turn send collateral branches to several spinal segments. Some collateral branches penetrate the dorsal horn of the cord and synapse on neurons located mainly in a region called Rexed's laminae III–V. These synapses mediate, among other things, segmental reflexes such as the "knee-jerk" or myotatic reflex described in Chapter 1, and are further considered in Chapters 16 and 17. The major branch of the incoming axons, however, ascends ipsilaterally through the **dorsal columns** (also called the posterior funiculi) of the cord, all the way to the lower medulla, where it terminates by contacting **second-order neurons** in the **gracile** and **cuneate nuclei** (together referred to as the **dorsal column nuclei**; see Figures 9.1 and 9.6A). Axons in the dorsal columns are topographically organized such that the fibers that convey information from lower limbs are in the medial subdivision of the dorsal columns, called the **gracile tract**, a fact of some significance in the clinical localization of neural injury. The lateral subdivision, called the **cuneate tract**, contains axons conveying information from the upper limbs, trunk, and neck. At the level of the upper thorax, the dorsal columns account for more than a third of the cross-sectional area of the human spinal cord.

Despite their size, lesions limited to the dorsal columns of the spinal cord in both humans and monkeys have only a modest effect on the performance of simple tactile tasks. They do, however, impede the ability to detect the direction and speed of tactile stimuli and degrade the ability to sense the position of the limbs in space. Dorsal column lesions may also reduce a patient's ability to initiate active movements related to tactile exploration. For instance, such individuals have difficulty recognizing numbers and letters drawn on the skin. The relatively mild deficit that follows dorsal column lesions is presumably explained by the fact that some axons responsible for cutaneous mechanoreception also run in the spinothalamic (pain and temperature) pathway, as described in Chapter 10.

The second-order relay neurons in the dorsal column nuclei send their axons to the somatic sensory portion of the thalamus (see Figure 9.6A). The axons from dorsal column nuclei project in the dorsal portion of each side of the lower brainstem, where they form the **internal arcuate tract**. The internal arcuate axons subsequently cross the midline to form another named tract that is elongated dorsoventrally, the **medial lemniscus**. (The crossing of these fibers is called the decussation, or crossing, of the medial lemniscus; the word *lemniscus* means "ribbon.") In a cross-section through the medulla, such as the one shown in Figure 9.6A, the medial lemniscal axons carrying information from the lower limbs are located ventrally, whereas the axons related to the upper limbs are located dorsally (again, a fact of some clinical importance). As the medial lemniscus ascends through the pons and midbrain, it rotates 90° laterally, so that the upper body is eventually represented in the medial portion of the tract, and the lower body in the lateral portion. The axons of the medial lemniscus thus reach the ventral posterior lateral (VPL) nucleus of the thalamus, whose cells are the **third-order neurons** of the dorsal column-medial lemniscus system (see Figure 9.7).

The Trigeminal Portion of the Mechanosensory System

The dorsal column-medial lemniscus pathway described in the preceding section carries somatic information from the upper and lower body and from the posterior third of the head. To make matters even more complicated, tactile and proprioceptive information from the face is conveyed from the periphery to the thalamus by a different route. Information derived from the face is transmitted to the central nervous system via the **trigeminal somatic sensory system** (Figure 9.6B). Low-threshold mechanoreception in the face is mediated by first-order neurons in the trigeminal (cranial nerve V) ganglion. The peripheral processes of these neurons form the three main subdivisions of **the trigeminal nerve** (the **ophthalmic**, **maxillary**, and **mandibular**

branches), each of which innervates a well-defined territory on the face and head, including the teeth and the mucosa of the oral and nasal cavities. The central processes of trigeminal ganglion cells form the sensory roots of the trigeminal nerve; they enter the brainstem at the level of the pons to terminate on neurons in the subdivisions of the **trigeminal brainstem complex**.

The trigeminal complex has two major components: the **principal nucleus** (responsible for processing mechanosensory stimuli), and the **spinal nucleus** (responsible for processing painful and thermal stimuli). Thus, most of the axons carrying information from low-threshold cutaneous mechanoreceptors in the face terminate in the principal nucleus. In effect, this nucleus corresponds to the dorsal column nuclei that relay mechanosensory information from the rest of the body. The spinal nucleus corresponds to a portion of the spinal cord that contains the second-order neurons in the pain and temperature system for the rest of the body (see Chapter 10). The second-order neurons of the trigeminal brainstem nuclei give off axons that cross the midline and ascend to the ventral posterior medial (VPM) nucleus of the thalamus by way of the **trigeminothalamic tract** (also called the trigeminal lemniscus).

The Somatic Sensory Components of the Thalamus

Each of the several ascending somatic sensory pathways originating in the spinal cord and brainstem converge on the thalamus (Figure 9.7). The **ventral posterior complex** of the thalamus, which comprises a lateral and a medial nucleus, is the main target of these ascending pathways. As already noted, the more laterally located **ventral posterior lateral (VPL) nucleus** receives projections from the medial lemniscus carrying all somatosensory information from the body and posterior head, whereas the more medially located **ventral posterior medial (VPM) nucleus** receives axons from the trigeminal lemniscus (that is, mechanosensory and nociceptive information from the face). Accordingly, the ventral posterior complex of the thalamus contains a complete representation of the somatic sensory periphery.

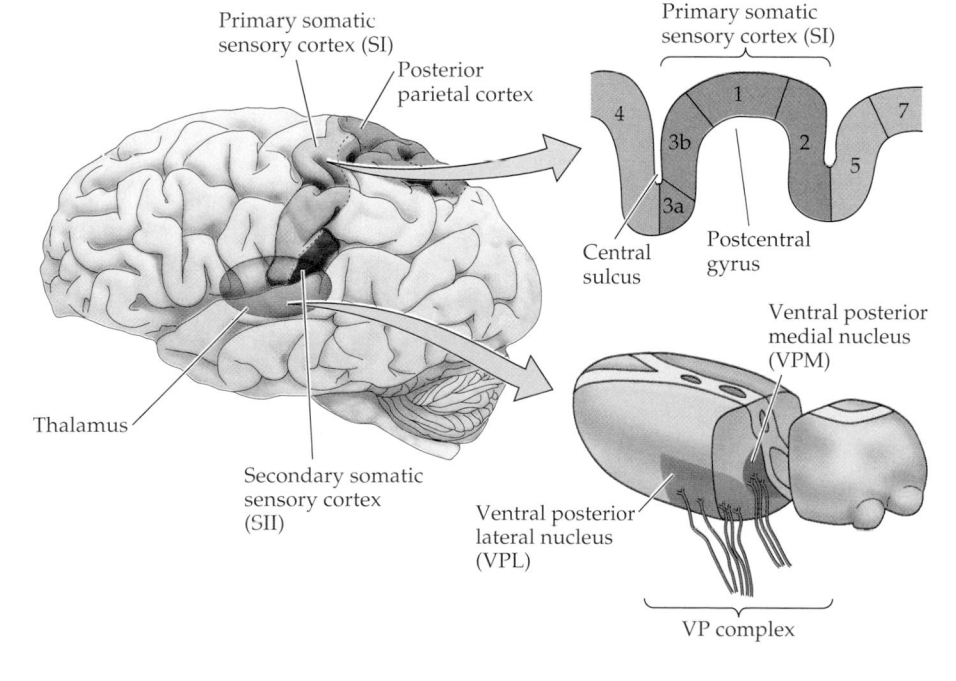

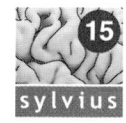

Figure 9.7 Diagram of the somatic sensory portions of the thalamus and their cortical targets in the postcentral gyrus. The ventral posterior nuclear complex comprises the VPM, which relays somatic sensory information carried by the trigeminal system from the face, and the VPL, which relays somatic sensory information from the rest of the body. Inset above shows organization of the primary somatosensory cortex in the postcentral gyrus, shown here in a section cutting across the gyrus. (After Brodal, 1992, and Jones et al., 1982.)

The Somatic Sensory Cortex

The axons arising from neurons in the ventral posterior complex of the thalamus project to cortical neurons located primarily in layer IV of the somatic sensory cortex (see Figure 9.7; also see Box A in Chapter 26 for a more detailed description of cortical lamination). The **somatic sensory cortex** in humans, which is located in the parietal lobe, comprises four distinct regions, or fields, known as **Brodmann's areas 3a, 3b, 1**, and **2**. Although area 3b is generally known as the **primary somatic sensory cortex** (also called SI), all four areas are involved in processing tactile information. Experiments carried out in nonhuman primates indicate that neurons in areas 3b and 1 respond primarily to cutaneous stimuli, whereas neurons in 3a respond mainly to stimulation of proprioceptors; area 2 neurons process both tactile and proprioceptive stimuli. Mapping studies in humans and other primates show further that each of these four cortical areas contains a separate and complete representation of the body. In these **somatotopic maps**, the foot, leg, trunk, forelimbs, and face are represented in a medial to lateral arrangement, as shown in Figures 9.8A,B and 9.9.

Although the topographic organization of the several somatic sensory areas is similar, the functional properties of the neurons in each region and their organization are distinct (Box D). For instance, the neuronal receptive fields are relatively simple in area 3b; the responses elicited in this region are

Figure 9.8 Somatotopic order in the human primary somatic sensory cortex. (A) Diagram showing the region of the human cortex from which electrical activity is recorded following mechanosensory stimulation of different parts of the body. The patients in the study were undergoing neurosurgical procedures for which such mapping was required. Although modern imaging methods are now refining these classical data, the human somatotopic map first defined in the 1930s has remained generally valid. (B) Diagram along the plane in (A) showing the somatotopic representation of body parts from medial to lateral. (C) Cartoon of the homunculus constructed on the basis of such mapping. Note that the amount of somatic sensory cortex devoted to the hands and face is much larger than the relative amount of body surface in these regions. A similar disproportion is apparent in the primary motor cortex, for much the same reasons (see Chapter 17). (After Penfield et al., 1953, and Corsi, 1991.)

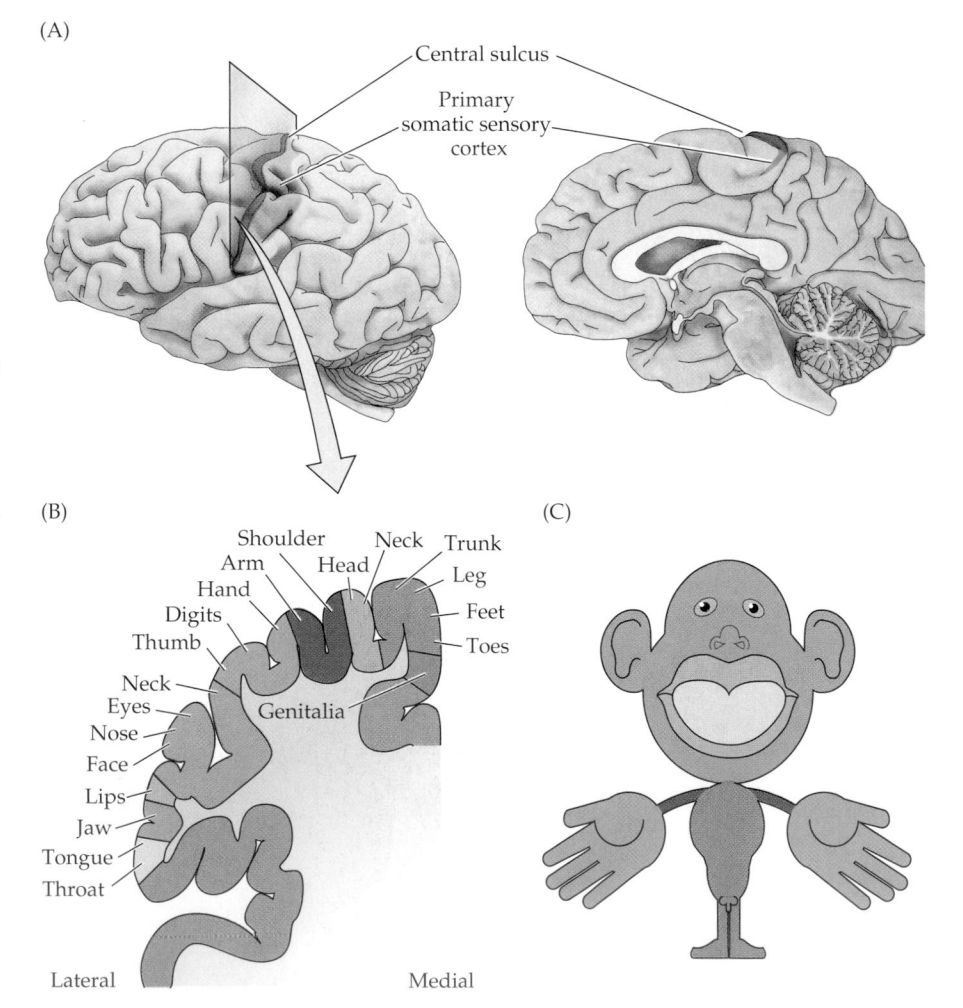

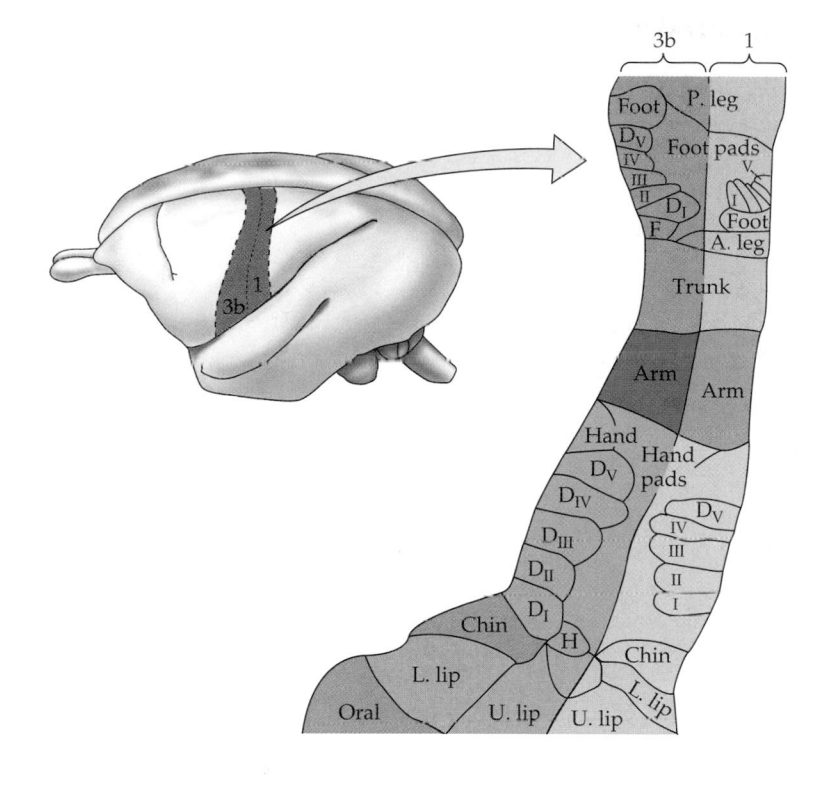

Figure 9.9 The primary somatic sensory map in the owl monkey based, as in Figure 9.8, on the electrical responsiveness of the cortex to peripheral stimulation. Much more detailed mapping is possible in experimental animals than in neurosurgical patients. The enlargement on the right shows areas 3b and 1, which process most cutaneous mechanosensory information. The arrangement is generally similar to that determined in humans. (After Kaas, 1983.)

generally to stimulation of a single finger. In areas 1 and 2, however, the majority of the receptive fields respond to stimulation of multiple fingers. Furthermore, neurons in area 1 respond preferentially to particular directions of skin stimulation, whereas many area 2 neurons require complex stimuli to activate them (such as a particular shape). Lesions restricted to area 3b produce a severe deficit in both texture and shape discrimination. In contrast, damage confined to area 1 affects the ability of monkeys to perform accurate texture discrimination. Area 2 lesions tend to produce deficits in finger coordination, and in shape and size discrimination.

A salient feature of cortical maps, recognized soon after their discovery, is their failure to represent the body in actual proportion. When neurosurgeons determined the representation of the human body in the primary sensory (and motor) cortex, the homunculus (literally, "little man") defined by such mapping procedures had a grossly enlarged face and hands compared to the torso and proximal limbs (Figure 9.8C). These anomalies arise because manipulation, facial expression, and speaking are extraordinarily important for humans, requiring more central (and peripheral) circuitry to govern them. Thus, in humans, the cervical spinal cord is enlarged to accommodate the extra circuitry related to the hand and upper limb, and as stated earlier, the density of receptors is greater in regions such as the hands and lips. Such distortions are also apparent when topographical maps are compared across species. In the rat brain, for example, an inordinate amount of the somatic sensory cortex is devoted to representing the large facial whiskers that provide a key component of the somatic sensory input for rats and mice (see Boxes B and D), while raccoons overrepresent their paws and the platypus its bill. In short, the sensory input (or motor output) that is particularly significant to a given species gets relatively more cortical representation.

BOX D
Patterns of Organization within the Sensory Cortices: Brain Modules

Observations over the last 40 years have made it clear that there is an iterated substructure within the somatic sensory (and many other) cortical maps. This substructure takes the form of units called *modules*, each involving hundreds or thousands of nerve cells in repeating patterns. The advantages of these iterated patterns for brain function remain largely mysterious; for the neurobiologist, however, such iterated arrangements have provided important clues about cortical connectivity and the mechanisms by which neural activity influences brain development (see Chapters 23 and 24).

The observation that the somatic sensory cortex comprises elementary units of vertically linked cells was first noted in the 1920s by the Spanish neuroanatomist Rafael Lorente de Nó, based on his studies in the rat. The potential importance of cortical modularity remained largely unexplored until the 1950s, however, when electrophysiological experiments indicated an arrangement of repeating units in the brains of cats and, later, monkeys. Vernon Mountcastle, a neurophysiologist at Johns Hopkins, found that vertical microelectrode penetrations in the primary somatosensory cortex of these animals encountered cells that responded to the same sort of mechanical stimulus presented at the same location on the body surface. Soon after Mountcastle's pioneering work, David Hubel and Torsten Wiesel discovered a similar arrangement in the cat primary visual cortex. These and other observations led Mountcastle to the general view that "the elementary pattern of organization of the cerebral cortex is a vertically oriented column or cylinder of cells capable of input-output functions of considerable complexity." Since these discoveries in the late 1950s and early 1960s, the view that modular circuits represent a fundamental feature of the mammalian cerebral cortex has gained wide acceptance, and many

such entities have now been described in various cortical regions (see figure).

This wealth of evidence for patterned circuits has led many neuroscientists to conclude, like Mountcastle, that modules are a fundamental feature of the cerebral cortex, essential for perception, cognition, and perhaps even consciousness. Despite the prevalence of iterated modules, there are some problems with the view that modular units are universally important in cortical function. First, although modular circuits of a given class are readily seen in the brains of some species, they have not been found in other, sometimes closely related, animals. Second, not all regions of the mammalian cortex are organized in a modular fashion. And third, no clear function of such modules has been discerned, much effort and speculation notwithstanding. This salient feature of the organization of the somatic sensory cortex and other cortical (and some subcortical) regions therefore remains a tantalizing puzzle.

References

HUBEL, D. H. (1988) *Eye, Brain, and Vision*. Scientific American Library. New York: W. H. Freeman.

LORENTE DE NÓ, R. (1949) The structure of the cerebral cortex. *Physiology of the Nervous System*, 3rd Ed. New York: Oxford University Press.

MOUNTCASTLE, V. B. (1957) Modality and topographic properties of single neurons of cat's somatic sensory cortex. J. Neurophysiol. 20: 408–434.

MOUNTCASTLE, V. B. (1998) *Perceptual Neuroscience: The Cerebral Cortex*. Cambridge: Harvard University Press.

PURVES, D., D. RIDDLE AND A. LaMANTIA (1992) Iterated patterns of brain circuitry (or how the cortex gets its spots). Trends Neurosci. 15: 362–368.

WOOLSEY, T. A. AND H. VAN DER LOOS (1970) The structural organization of layer IV in the somatosensory region (SI) of mouse cerebral cortex. The description of a cortical field composed of discrete cytoarchitectonic units. Brain Res. 17: 205–242.

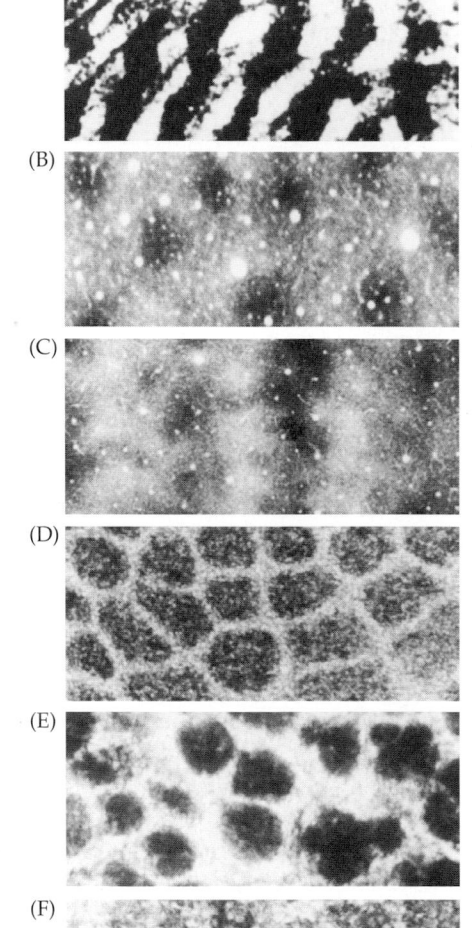

(A)

(B)

(C)

(D)

(E)

(F)

Examples of iterated, modular substructures in the mammalian brain. (A) Ocular dominance columns in layer IV in the primary visual cortex (V1) of a rhesus monkey. (B) Repeating units called "blobs" in layers II and III in V1 of a squirrel monkey. (C) Stripes in layers II and III in V2 of a squirrel monkey. (D) Barrels in layer IV in primary somatic sensory cortex of a rat. (E) Glomeruli in the olfactory bulb of a mouse. (F) Iterated units called "barreloids" in the thalamus of a rat. These and other examples indicate that modular organization is commonplace in the brain. (From Purves et al., 1992.)

Higher-Order Cortical Representations

Somatic sensory information is distributed from the primary somatic sensory cortex to "higher-order" cortical fields (as well as to subcortical structures). One of these higher-order cortical centers, the secondary somatosensory cortex (sometimes called SII and adjacent to the primary cortex; see Figure 9.7), receives convergent projections from the primary somatic sensory cortex and sends projections in turn to limbic structures such as the amygdala and hippocampus (see Chapters 29 and 31). This latter pathway is believed to play an important role in tactile learning and memory. Neurons in motor cortical areas in the frontal lobe also receive tactile information from the anterior parietal cortex and, in turn, provide feedback projections to several cortical somatic sensory regions. Such integration of sensory and motor information is considered in Chapters 20 and 26, where the role of these "association" regions of the cerebral cortex are discussed in more detail.

Finally, a fundamental but often neglected feature of the somatic sensory system is the presence of massive descending projections. These pathways originate in sensory cortical fields and run to the thalamus, brainstem, and spinal cord. Indeed, descending projections from the somatic sensory cortex outnumber ascending somatic sensory pathways! Although their physiological role is not well understood, it is generally assumed (with some experimental support) that descending projections modulate the ascending flow of sensory information at the level of the thalamus and brainstem.

Summary

The components of the somatic sensory system considered in this chapter process information conveyed by mechanical stimuli that impinge upon the body surface or that are generated within the body itself (proprioception) (the pain and temperature components of this diverse system are considered in Chapter 10). This processing is performed by neurons distributed across several brain structures that are connected by both ascending and descending pathways. Transmission of afferent mechanosensory information from the periphery to the brain begins with a variety of receptor types that initiate action potentials. This activity is conveyed centrally via a chain of neurons, referred to as the first-, second-, and third-order cells. First-order neurons are located in the dorsal root and cranial nerve ganglia. Second-order neurons are located in brainstem nuclei. Third-order neurons are found in the thalamus, from whence they project to the cerebral cortex. These pathways are topographically arranged throughout the system, the amount of cortical and subcortical space allocated to various body parts being proportional to the density of peripheral receptors. Studies of non-human primates show that specific cortical regions correspond to each functional submodality; area 3b, for example, processes information from low-threshold cutaneous receptors, and area 3a from proprioceptors. Thus, at least two broad criteria operate in the organization of the somatic sensory system: modality and somatotopy. The end result of this complex interaction is the unified perceptual representation of the body and its ongoing interaction with the environment.

ADDITIONAL READING

Reviews

CHAPIN, J. K. (1987) Modulation of cutaneous sensory transmission during movement: Possible mechanisms and biological significance. In *Higher Brain Function: Recent Explorations of the Brain's Emergent Properties*. S. P. Wise (ed.). New York: John Wiley and Sons, pp. 181–209.

DARIAN-SMITH, I. (1982) Touch in primates. Ann. Rev. Psychol. 33: 155–194.

JOHANSSON, R. S. AND A. B. VALLBO (1983) Tactile sensory coding in the glabrous skin of the human. Trends Neurosci. 6: 27–32.

KAAS, J. H. (1990) Somatosensory system. In *The Human Nervous System*. G Paxinos (ed.). San Diego: Academic Press, pp. 813–844.

KAAS, J. H. (1993) The functional organization of somatosensory cortex in primates. Ann. Anat. 175: 509–518.

MOUNTCASTLE, V. B. (1975) The view from within: Pathways to the study of perception. Johns Hopkins Med. J. 136: 109–131.

WOOLSEY, C. (1958) Organization of somatic sensory and motor areas of the cerebral cortex. In *Biological and Biochemical Bases of Behavior*. H. F. Harlow and C. N. Woolsey (eds.). Madison, WI: University of Wisconsin Press, pp. 63–82.

Important Original Papers

ADRIAN, E. D. AND Y. ZOTTERMAN (1926) The impulses produced by sensory nerve endings. Part II. The response of a single end organ. J. Physiol. 61: 151–171.

JOHANSSON, R. S. (1978) Tactile sensibility of the human hand: Receptive field characteristics of mechanoreceptive units in the glabrous skin. J. Physiol. (Lond.) 281: 101–123.

JOHNSON, K. O. AND G. D. LAMB (1981) Neural mechanisms of spatial tactile discrimination: Neural patterns evoked by Braille-like dot patterns in the monkey. J. Physiol. (London) 310: 117–144.

JONES, E. G. AND D. P. FRIEDMAN (1982) Projection pattern of functional components of thalamic ventrobasal complex on monkey somatosensory cortex. J. Neurophysiol. 48: 521–544.

JONES, E. G. AND T. P. S. POWELL (1969) Connexions of the somatic sensory cortex of the rhesus monkey. I. Ipsilateral connexions. Brain 92: 477–502.

LaMOTTE, R. H. AND M. A. SRINIVASAN (1987) Tactile discrimination of shape: Responses of rapidly adapting mechanoreceptive afferents to a step stroked across the monkey fingerpad. J. Neurosci. 7: 1672–1681.

LAUBACH, M., J. WESSBER AND M. A. L. NICOLELIS (2000) Cortical ensemble activity increasingly predicts behavior outcomes during learning of a motor task. Nature 405: 567–571.

MOORE, C. I. AND S. B. NELSON (1998) Spatiotemporal subthreshold receptive fields in the vibrissa representation of rat primary somatosensory cortex. J. Neurophysiol. 80: 2882–2892.

MOORE, C. I., S. B. NELSON AND M. SUR (1999) Dynamics of neuronal processing in rat somatosensory cortex. TINS 22: 513–520.

NICOLELIS, M. A. L., L. A. BACCALA, R. C. S. LIN AND J. K. CHAPIN (1995) Sensorimotor encoding by synchronous neural ensemble activity at multiple levels of the somatosensory system. Science 268: 1353–1358.

SUR, M. (1980) Receptive fields of neurons in areas 3b and 1 of somatosensory cortex in monkeys. Brain Res. 198: 465–471.

WALL, P. D. AND W. NOORDENHOS (1977) Sensory functions which remain in man after complete transection of dorsal columns. Brain 100: 641–653.

ZHU, J. J. AND B. CONNORS (1999) Intrinsic firing patterns and whisker-evoked synaptic responses of neurons in the rat barrel cortex. J. Neurophysiol. 81: 1171–1183.

Chapter 10

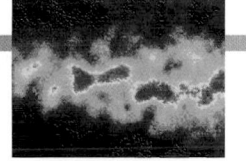

Pain

Overview

A natural assumption is that the sensation of pain arises from excessive stimulation of the same receptors that generate other somatic sensations (i.e., those discussed in Chapter 9). This is not the case. Although similar in some ways to the sensory processing of ordinary mechanical stimulation, the perception of pain (called nociception) depends on specifically dedicated receptors and pathways. Since alerting the brain to the dangers implied by noxious stimuli differs substantially from informing it about innocuous somatic sensory stimuli, it makes good sense that a special subsystem be devoted to the perception of potentially threatening circumstances. The overriding importance of pain in clinical practice, as well as the many aspects of pain physiology and pharmacology that remain imperfectly understood, continue to make nociception an extremely active area of research.

Nociceptors

The relatively unspecialized nerve cell endings that initiate the sensation of pain are called **nociceptors** (*noci-* is derived from the Latin for "hurt") (see Figure 9.2). Like other cutaneous and subcutaneous receptors, they transduce a variety of stimuli into receptor potentials, which in turn trigger afferent action potentials. Moreover, nociceptors, like other somatic sensory receptors, arise from cell bodies in dorsal root ganglia (or in the trigeminal ganglion) that send one axonal process to the periphery and the other into the spinal cord or brainstem (see Figure 9.1).

Because peripheral nociceptive axons terminate in unspecialized "free endings," it is conventional to categorize nociceptors according to the properties of the axons associated with them (see Table 9.1). As described in the previous chapter, the somatic sensory receptors responsible for the perception of innocuous mechanical stimuli are associated with myelinated axons that have relatively rapid conduction velocities. The axons associated with nociceptors, in contrast, conduct relatively slowly, being only lightly myelinated or, more commonly, unmyelinated. Accordingly, axons conveying information about pain fall into either the Aδ group of myelinated axons, which conduct at about 20 m/s, or into the C fiber group of unmyelinated axons, which conduct at velocities generally less than 2 m/s. Thus, even though the conduction of all nociceptive information is relatively slow, there are fast and slow pain pathways.

In general, the faster-conducting Aδ nociceptors respond either to dangerously intense mechanical or to mechanothermal stimuli, and have receptive fields that consist of clusters of sensitive spots. Other unmyelinated nocicep-

(A)

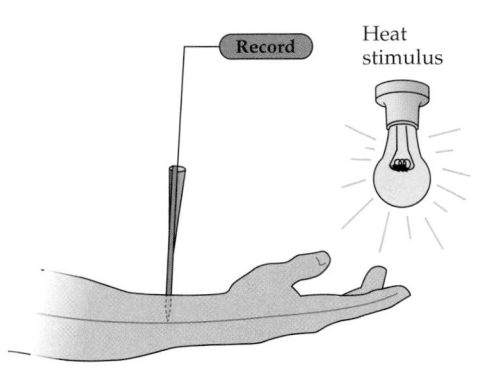

(B)

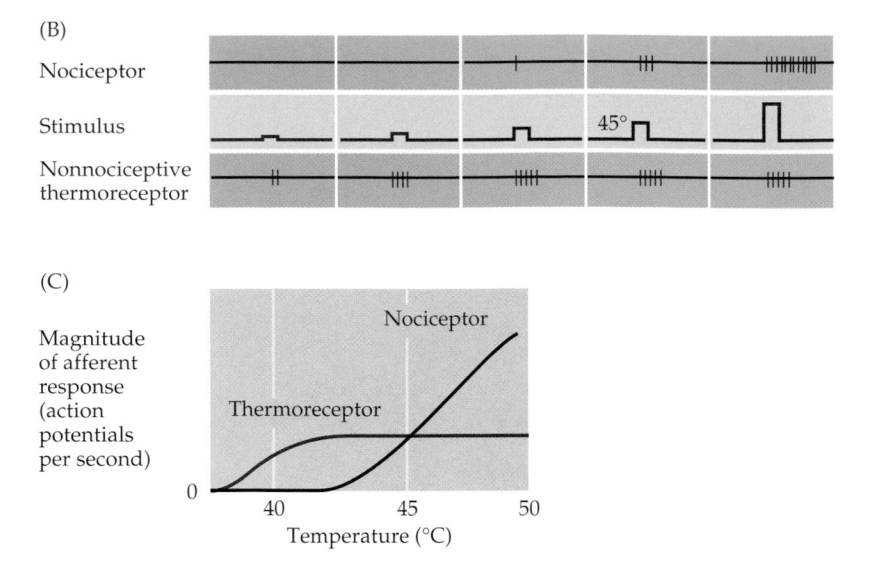

(C)

Figure 10.1 Experimental demonstration that nociception involves specialized neurons, not simply greater discharge of the neurons that respond to normal stimulus intensities. (A) Arrangement for transcutaneous nerve recording. (B) In the painful stimulus range, the axons of thermoreceptors fire action potentials at the same rate as at lower temperatures; the number and frequency of action potential discharge in the nociceptive axon, however, continues to increase. (Note that 45°C is the approximate threshold for pain.) (C) Summary of results. (After Fields, 1987.)

tors tend to respond to thermal, mechanical, and chemical stimuli, and are therefore said to be *polymodal*. In short, there are three major classes of nociceptors in the skin: **Aδ mechanosensitive nociceptors, Aδ mechanothermal nociceptors**, and **polymodal nociceptors,** the latter being specifically associated with C fibers. The receptive fields of all pain-sensitive neurons are relatively large, particularly at the level of the thalamus and cortex, presumably because the detection of pain is more important than its precise localization.

Studies carried out in both humans and experimental animals demonstrated some time ago that the rapidly conducting axons that subserve somatic sensory sensation are not involved in the transmission of pain. A typical experiment of this sort is illustrated in Figure 10.1. The peripheral axons responsive to nonpainful mechanical or thermal stimuli do not discharge at a greater rate when painful stimuli are delivered to the same region of the skin surface. The nociceptive axons, on the other hand, begin to discharge only when the strength of the stimulus (a thermal one in the example in Figure 10.1) reaches high levels; at this same stimulus intensity, other thermoreceptors discharge at a rate no different from the maximum rate already achieved within the nonpainful temperature range, indicating that there are both nociceptive and nonnociceptive thermoreceptors. Equally important, direct stimulation of the large-diameter somatic sensory afferents at any frequency in humans does not produce sensations that are described as painful. In contrast, the smaller-diameter, more slowly conducting Aδ and C fibers are active when painful stimuli are delivered; and when stimulated electrically in human subjects, they produce pain.

The Perception of Pain

How, then, do these different classes of nociceptors lead to the perception of pain? As mentioned, one way of determining the answer has been to stimulate different nociceptors in human volunteers while noting the sensations reported. In general, two categories of pain perception have been described: a sharp **first pain** and a more delayed (and longer-lasting) sensation that is generally called **second pain** (Figure 10.2A). Stimulation of the large, rapidly conducting Aα and Aβ axons in peripheral nerves does not elicit the sensation of pain. When the stimulus intensity is raised to a level that activates a

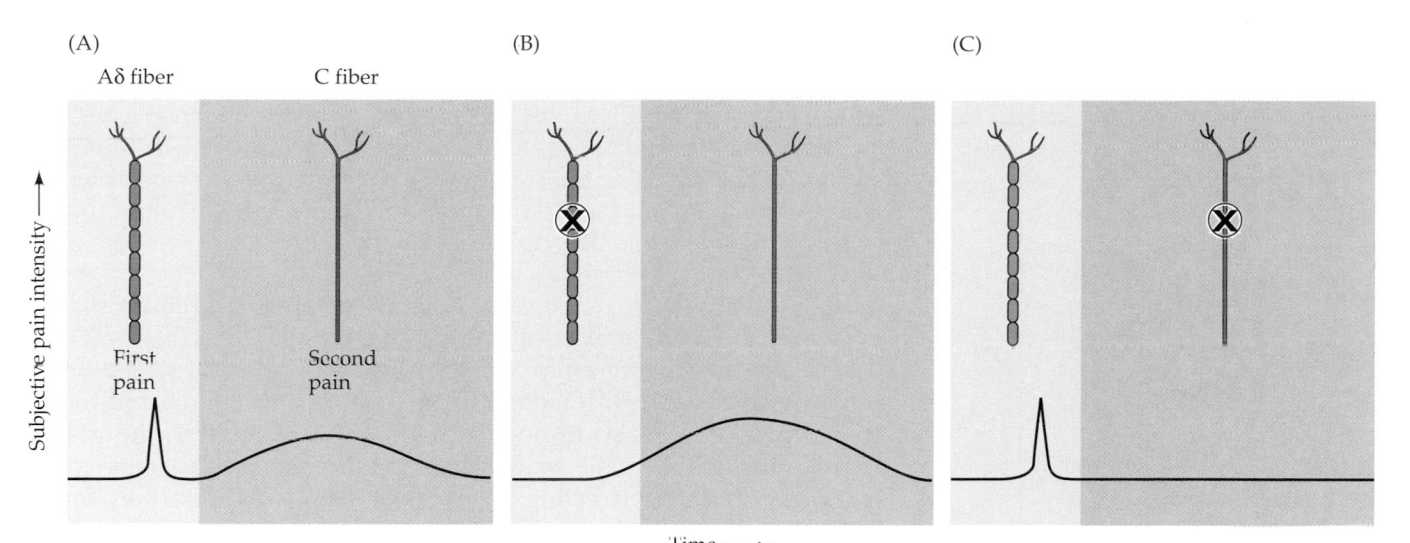

Figure 10.2 Pain can be separated into an early perception of sharp pain and a later sensation that is described as having a duller, burning quality. (A) First and second pain, as these sensations are called, are carried by different axons, as can be shown by (B) the selective blockade of the more rapidly conducting myelinated axons that carry the sensation of first pain, or (C) blockade of the more slowly conducting C fibers that carry the sensation of second pain. (After Fields, 1990.)

subset of Aδ fibers, however, a tingling sensation or, if the stimulation is intense enough, a feeling of sharp pain is reported. If the stimulus intensity is increased still further, so that the small-diameter, slowly conducting C fiber axons are brought into play, then a duller, longer-lasting sensation of pain is experienced. It is also possible to selectively anesthetize C fibers and Aδ fibers; in general, these selective blocking experiments confirm that the Aδ fibers are responsible for first pain, and that C fibers are responsible for the duller, longer-lasting second pain (Figure 10.2B,C).

Hyperalgesia and Sensitization

Painful stimuli are usually associated with tissue damage (e.g., cuts, scrapes, and bruises). The familiar phenomenon of **hyperalgesia** is defined as the enhanced sensitivity and responsivity to stimulation of the area around the damaged tissue. Thus, in the region surrounding an injury, stimuli that would not normally cause pain are perceived as painful, and stimuli that would ordinarily be painful are significantly more so (therefore, *hyper*algesia). The cause of this phenomenon is the **sensitization** of nociceptors by various substances released when tissue is damaged (Table 10.1). Evidently, the release of bradykinin, histamine, prostaglandins, and other agents from the site of injury enhances the responsiveness of nociceptive endings. Electrical activity in the nociceptors themselves also stimulates the local release of chemical substances (such as substance P) that cause vasodilation, swelling, and the release of histamine from mast cells. Injury and pain are thus intertwined in a complex cascade of local signals. The involvement of these substances in the production of pain has also provided clues about how some analgesics may work, suggesting strategies for pain relief. Aspirin (salicylic acid), for example, evidently acts by inhibiting cyclooxygenase, an enzyme important in the biosynthesis of prostaglandins.

The presumed purpose of the complex chemical signaling arising from local damage is not only to protect the injured area (as a result of the painful perceptions produced by ordinary stimuli close to the site of damage), but also to promote healing and guard against infection by means of local effects such as increased blood flow and inflammation.

TABLE 10.1
Substances Released Following Tissue Damage

Substance	Source
Potassium	Damaged cells
Serotonin	Platelets
Bradykinin	Plasma
Histamine	Mast cells
Prostaglandins	Damaged cells
Leukotrienes	Damaged cells
Substance P	Primary afferent fibers

Source: Modified from Fields, 1987.

Central Pain Pathways: The Spinothalamic Tract

The pathways that carry information about noxious stimuli to the brain, as might be expected for such an important and multifaceted system, are complex. The major pathways are summarized in Figure 10.3, which omits some of the less well understood subsidiary routes. Because projections from non-nociceptive temperature-sensitive neurons follow the same anatomical route, they are included in this description, even though they are not part of the pain system.

Like the other sensory neurons in dorsal root ganglia, the central axons of nociceptive nerve cells enter the spinal cord via the dorsal roots (Figure 10.3A). Axons carrying information from pain and temperature receptors are generally found in the most lateral division of the dorsal roots, but the cell bodies of these neurons are not discretely localized within the ganglia (although they are generally smaller than the mechanosensory nerve cells). When these centrally projecting axons reach the dorsal horn, they branch into ascending and descending collaterals, forming the **dorsolateral tract of Lissauer** (named after the German neurologist who first described this pathway in the late nineteenth century). Axons in Lissauer's tract run up and down for one or two spinal cord segments before they penetrate the gray matter of the dorsal horn. Once within the dorsal horn, the axons give off branches that contact neurons located in several of Rexed's laminae (these laminae are the descriptive divisions of the spinal gray matter in cross section, again named after the neuroanatomist who described these details in the 1950s). Both Aδ and C fibers send branches to innervate neurons in Rexed's lamina I (also called the marginal zone) and lamina II (called the substantia gelatinosa).

Information from Rexed's lamina II is transmitted to second-order projection neurons in laminae IV, V, and VI, the neurons of which also receive some direct innervation from the terminals of the first-order neurons. The axons of these second-order neurons in laminae IV–VI (which are collectively known as the **nucleus proprius**) cross the midline and ascend all the way to the brainstem and thalamus in the anterolateral (also called ventrolateral) quadrant of the contralateral half of the spinal cord. These fibers, together with axons from second-order lamina I neurons, form the **spinothalamic tract**, the major ascending pathway for information about pain and temperature. This overall pathway is also referred to as the **anterolateral system**, much as the mechanosensory pathway is referred to as the dorsal column–medial lemniscus system.

The location of the spinothalamic tract is particularly important clinically because of the characteristic sensory deficits that follow certain spinal cord injuries. Since the mechanosensory pathway ascends ipsilaterally in the cord, a unilateral spinal lesion will produce sensory loss of touch, pressure, vibration, and proprioception below the lesion on the same side. The pathways for pain and temperature, however, cross the midline to ascend on the opposite side of the cord. Therefore, diminished sensation of pain below the lesion will be observed on the side *opposite* the mechanosensory loss (and the lesion). This pattern is referred to as a **dissociated sensory loss** and (together with local dermatomal signs; see Box C in Chapter 9) helps define the level of the lesion (Figure 10.4).

As is the case of the mechanosensory pathway, information about noxious and thermal stimulation of the face follows a separate route to the thalamus (Figure 10.3B). First-order axons originating from the trigeminal ganglion cells and from ganglia associated with nerves VII, IX, and X carry information from facial nociceptors and thermoreceptors into the brainstem. After

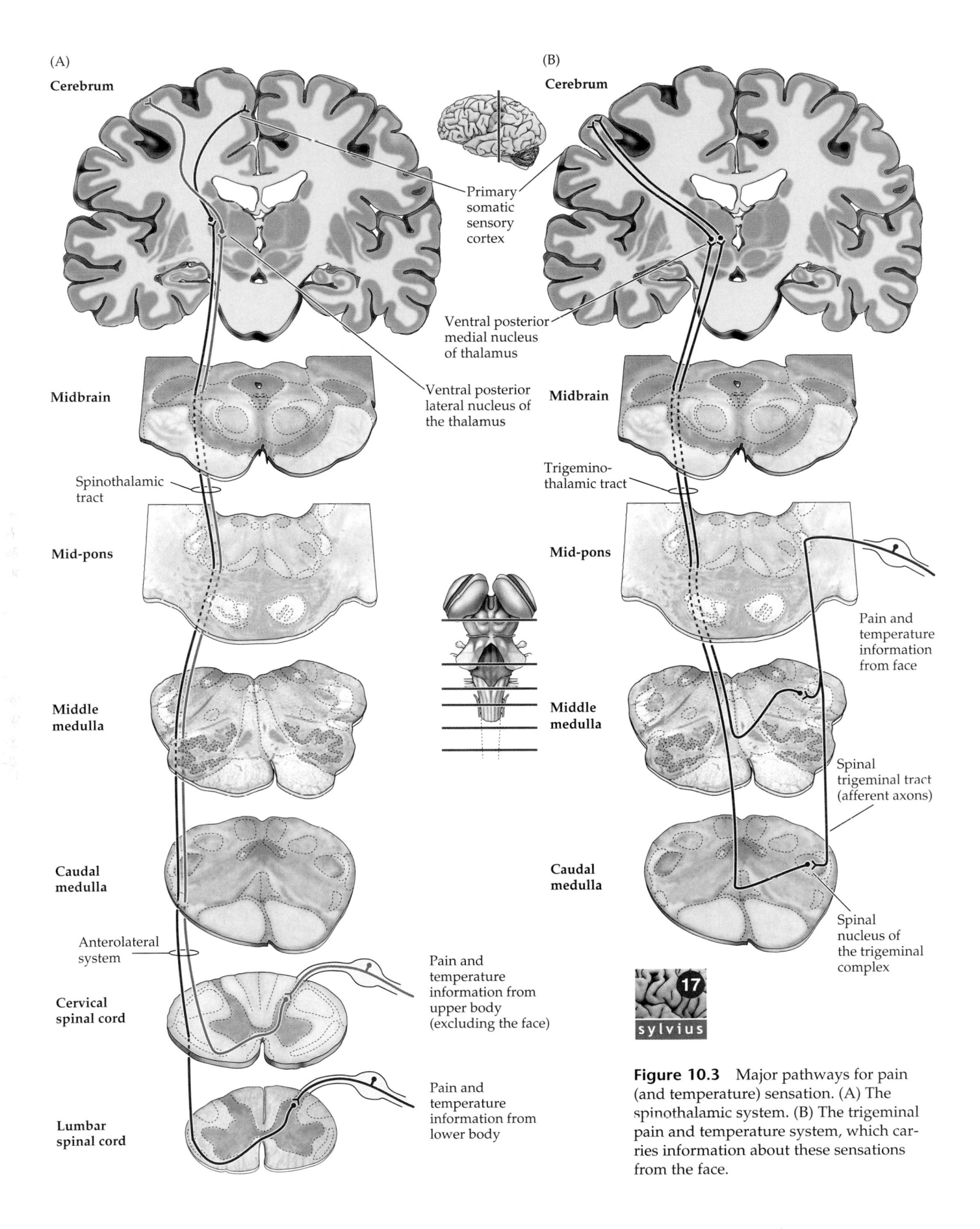

(A)

Cerebrum

Primary somatic sensory cortex

Ventral posterior medial nucleus of thalamus

Ventral posterior lateral nucleus of the thalamus

Midbrain

Spinothalamic tract

Mid-pons

Middle medulla

Caudal medulla

Anterolateral system

Cervical spinal cord

Pain and temperature information from upper body (excluding the face)

Lumbar spinal cord

Pain and temperature information from lower body

(B)

Cerebrum

Midbrain

Trigemino-thalamic tract

Mid-pons

Pain and temperature information from face

Middle medulla

Spinal trigeminal tract (afferent axons)

Caudal medulla

Spinal nucleus of the trigeminal complex

Figure 10.3 Major pathways for pain (and temperature) sensation. (A) The spinothalamic system. (B) The trigeminal pain and temperature system, which carries information about these sensations from the face.

Figure 10.4 Pattern of "dissociated" sensory loss following a spinal cord hemisection at the 10th thoracic level on the left side. This pattern, together with motor weakness on the same side as the lesion, is sometimes referred to as the Brown-Séquard syndrome.

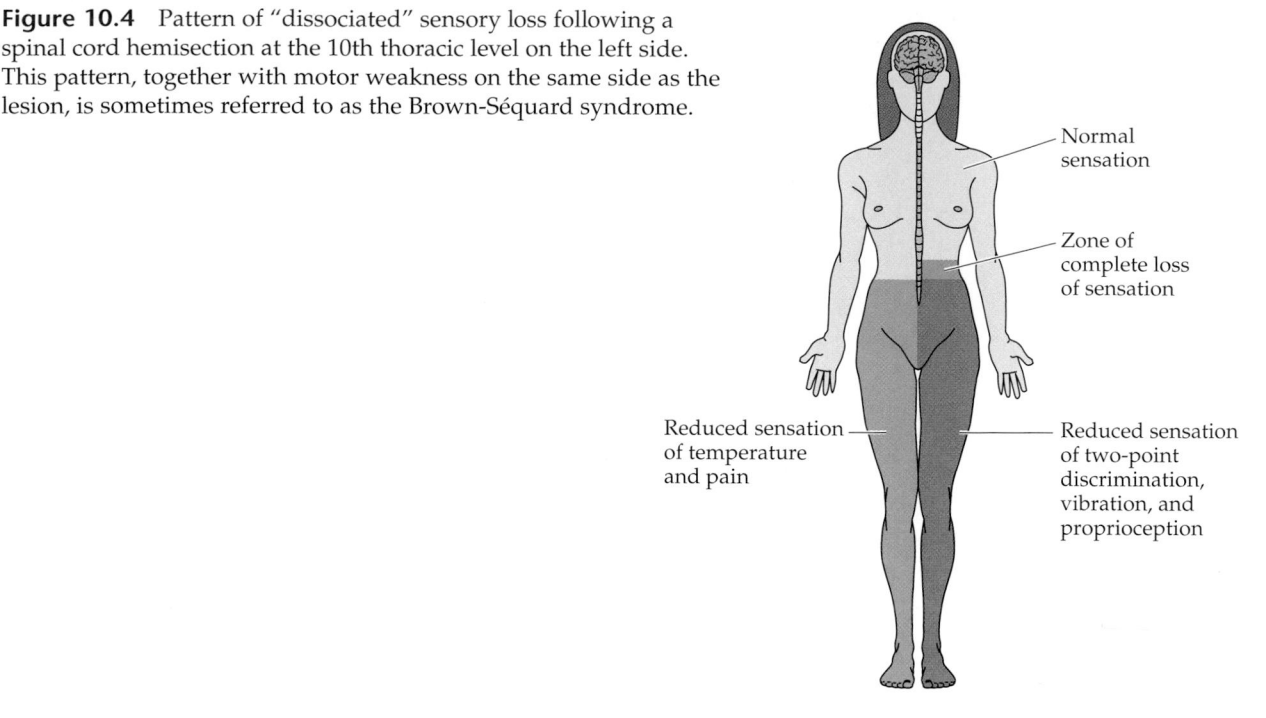

Normal sensation

Zone of complete loss of sensation

Reduced sensation of temperature and pain

Reduced sensation of two-point discrimination, vibration, and proprioception

entering the pons, these small myelinated and unmyelinated trigeminal fibers *descend* to the medulla, forming the **spinal trigeminal tract** (or spinal tract of cranial nerve V), and terminate in two subdivisions of the spinal trigeminal complex: the pars interpolaris and pars caudalis. Axons from the second-order neurons in these two trigeminal nuclei, like their counterparts in the spinal cord, cross the midline and ascend to the contralateral thalamus in the **trigeminal lemniscus** (also called the trigemino-thalamic tract).

The complexity of the pain pathways (recall that several minor routes are omitted in this account) often makes the origin of a patient's complaints about pain difficult to assess (Box A). For the same reason, chronic pain is often difficult to treat. Such pain can arise from inflammation (as in neuritis), injury to nerve endings and scar formation (as in the pain that can follow surgical amputation; Box B), or nerve invasion by cancer. Injuries to the central nervous system structures that process nociceptive information can also lead to intractable pain. The common denominator of conditions that cause chronic pain is irritation of nociceptive endings, axons, or processing circuits causing abnormal activity that is interpreted as pain. Surgical interruption of a particular tract to abolish chronic pain is not usually effective; the pain, although initially alleviated, tends to return. Indeed, there is often no completely successful treatment for these unfortunate patients.

The Nociceptive Components of the Thalamus and Cortex

In the thalamus, the major target nuclei of the ascending pain and temperature axons are, like the targets for mechanosensory axons, in the ventral posterior nuclear complex. The ventral posterior medial (VPM) and ventral posterior lateral (VPL) nuclei receive the bulk of these axons. Neurons in the VPM nucleus receive nociceptive information from the face, while neurons in the VPL nucleus receive nociceptive information from the rest of the body.

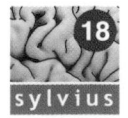
sylvius

BOX A
Referred Pain

Surprisingly, there are few, if any, neurons in the dorsal horn of the spinal cord that are specialized solely for the transmission of *visceral* pain. Obviously, we recognize such pain, but it is conveyed centrally via dorsal horn neurons that are also concerned with *cutaneous* pain. As a result of this economical arrangement, the disorder of an internal organ is sometimes perceived as cutaneous pain. A patient may therefore present to the physician with the complaint of pain at a site other than its actual source, a potentially confusing phenomenon called referred pain. The most common clinical example is anginal pain (pain arising from heart muscle that is not being adequately perfused with blood) referred to the upper chest wall, with radiation into the left arm and hand. Other important examples are gallbladder pain referred to the scapular region, esophageal pain referred to the chest wall, ureteral pain (e.g., from passing a kidney stone) referred to the lower abdominal wall, bladder pain referred to the perineum, and the pain from an inflamed appendix referred to the anterior abdominal wall around the umbilicus. Understanding referred pain can lead to an astute diagnosis that might otherwise be missed.

References

CAPPS, J. A. AND G. H. COLEMAN (1932) *An Experimental and Clinical Study of Pain in the Pleura, Pericardium, and Peritoneum.* New York: Macmillan.

HEAD, H. (1893) On disturbances of sensation with special reference to the pain of visceral disease. Brain 16: 1–32.

KELLGREW, J. H. (1939–1942) On the distribution of pain arising from deep somatic structures with charts of segmental pain areas. Clin. Sci. 4: 35–46.

Examples of pain arising from a visceral disorder referred to a cutaneous region (color).

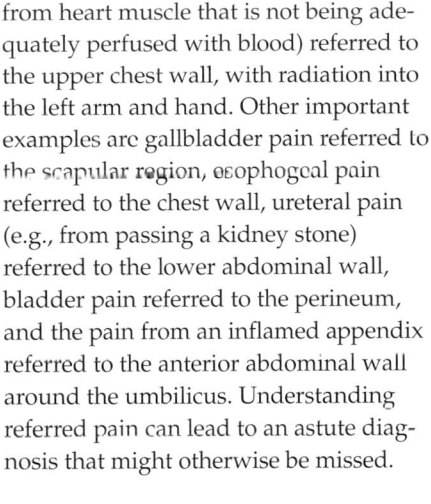
Esophagus

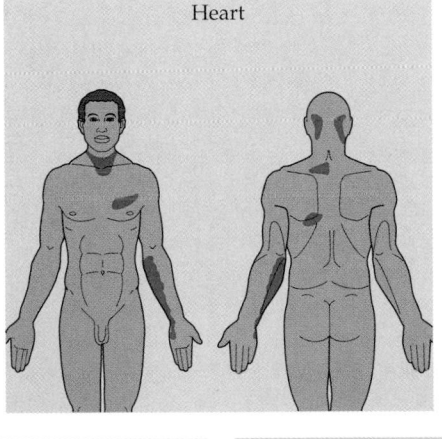
Heart

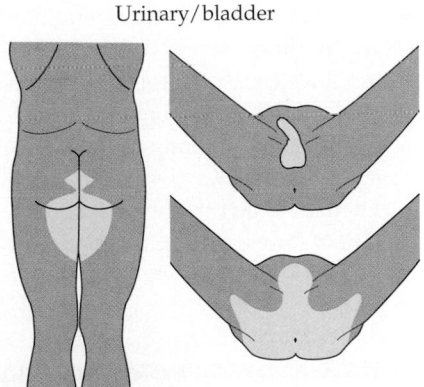
Urinary/bladder

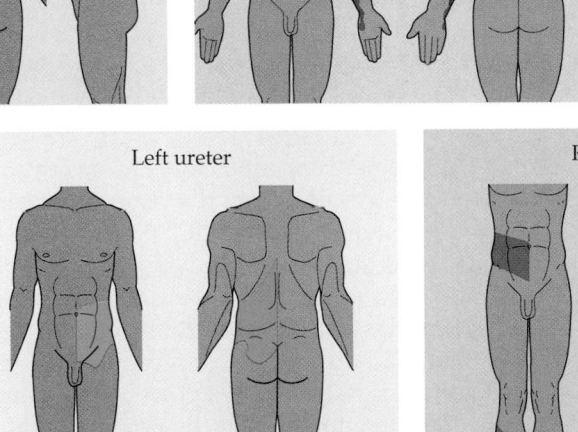
Left ureter

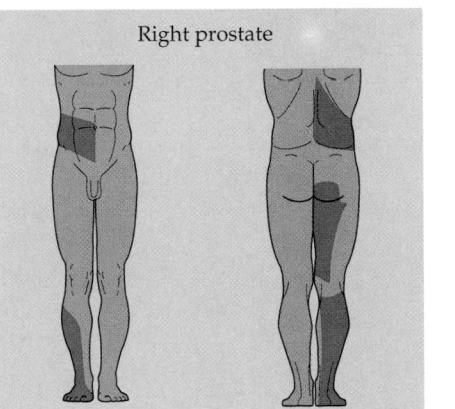
Right prostate

BOX B
Phantom Limbs and Phantom Pain

Following the amputation of an extremity, nearly all patients have an illusion that the missing limb is still present. Although this illusion usually diminishes over time, it persists in some degree throughout the amputee's life and can often be reactivated by injury to the stump or other perturbations. Such phantom sensations are not limited to amputated limbs; phantom breasts following mastectomy, phantom genitalia following castration, and phantoms of the entire lower body following spinal cord transection have all been reported. Phantoms are also common after local nerve block for surgery. During recovery from brachial plexus anesthesia, for example, it is not unusual for the patient to experience a phantom arm, perceived as whole and intact, but displaced from the real arm. When the real arm is viewed, the phantom appears to jump "into" the arm and may emerge and reenter intermittently as the anesthesia wears off. These sensory phantoms demonstrate that the central machinery for processing somatic sensory information is not idle in the absence of peripheral stimuli; apparently, the central sensory processing apparatus continues to operate independently of the periphery, giving rise to these bizarre sensations.

Phantoms might simply be a curiosity—or a provocative clue about higher-order somatic sensory processing—were it not for the fact that a substantial number of amputees also develop phantom pain. This common problem is usually described as a tingling or burning sensation in the missing part. Sometimes, however, the sensation becomes a more serious pain that patients find increasingly debilitating. Phantom pain is, in fact, one of the more common causes of chronic pain syndromes and is extraordinarily difficult to treat. Because of the widespread nature of central pain processing, ablation of the spinothalamic tract, portions of the thalamus, or even primary sensory cortex does not generally relieve the discomfort felt by these patients.

Indeed, considerable functional reorganization of somatotopic maps in the primary somatosensory cortex occurs in amputees (see Chapter 25). This reorganization starts immediately after the amputation and tends to evolve for several years. One of the effects of this process is that neurons that have lost their original inputs (i.e., from the removed limb) respond to tactile stimulation of other body parts. A surprising consequence is that stimulation of the face, for example, can be experienced as if the missing limb had been touched.

Further evidence that the phenomenon of phantom limb is the result of a central representation is the experience of children born without limbs. Such individuals have rich phantom sensations, despite the fact that a limb never developed. This observation suggests that a full represenation of the body exists independently of the peripheral elements that are mapped. Based on these results, Ronald Melzack proposed that the loss of a limb generates an internal mismatch between the brain's representation of the body and the pattern of peripheral tactile input that reaches the neocortex. The consequence would be an illusory sensation that the missing body

The similar arrangement for mechanosensory and noxious stimuli is presumably responsible for discriminative aspects of pain (the ability to locate a pain and judge its intensity). A parallel projection to the reticular formation of the medulla, pons, and midbrain is probably responsible for the general arousal that pain causes, and for the autonomic activation that follows a noxious stimulus (the classic fight-or-flight reaction; see Chapter 21 and Figure 10.5). Other thalamic nuclei, such as the central lateral nucleus and the intralaminar complex, receive projections from the anterolateral system and the reticular formation and also participate in the arousal response evoked by a noxious stimulus. Despite the location of nociceptive neurons in the same general regions of the thalamus as the mechanosensory neurons, at a more detailed level they comprise an essentially separate system.

The cortical representation of pain is the least well documented aspect of the central pathways for nociception. Although the thalamic neurons that relay noxious sensations via the ventral posterior nuclear complex project to the primary somatic sensory cortex, ablations of the relevant regions of the

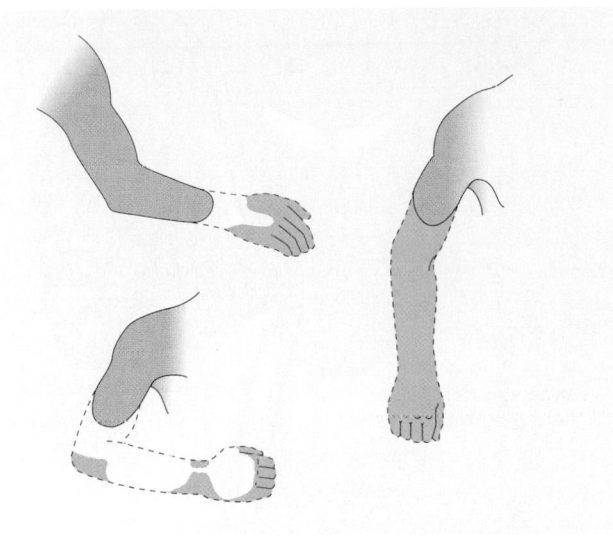

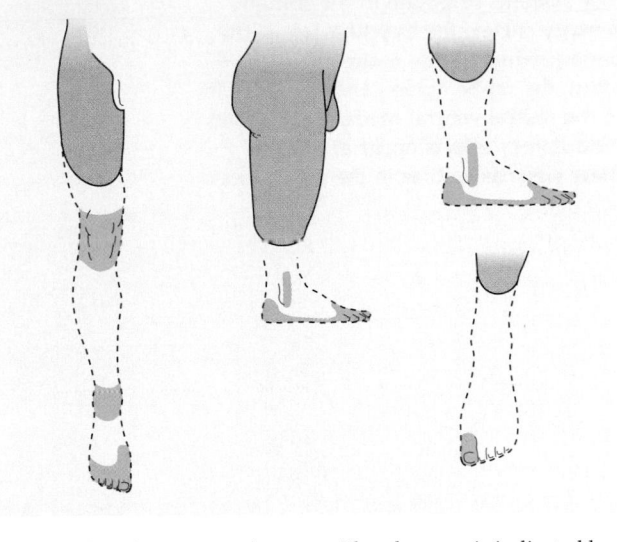

Drawings of phantom arms and legs, based on patients' reports. The phantom is indicated by a dashed line, with the colored regions showing the most vividly experienced parts. Note that some phantoms are telescoped into the stump. (After Solonen, 1962.)

part is still present and functional. With time, the brain may adapt to this loss and alter its intrinsic somatic representation to better accord with the new configuration of the body. This change could explain why the phantom sensation appears almost immediately after limb loss, but usually decreases in intensity over time.

References

MELZACK, R. (1989) Phantom limbs, the self and the brain. The D.O. Hebb Memorial Lecture. Canad. Psychol. 30: 1–14.

MELZACK, R. (1990) Phantom limbs and the concept of a neuromatrix. Trends Neurosci. 13: 88–92.

NASHOLD, B. S., JR. (1991) Paraplegia and pain. In *Deafferentation Pain Syndromes: Pathophysiology and Treatment*. B. S. Nashold, Jr. and J. Ovelmen-Levitt (eds.). New York: Raven Press, pp. 301–319.

RAMACHANDRAN, V. S. AND S. BLAKESLEE (1998) *Phantoms in the Brain*. New York: William Morrow & Co.

parietal cortex do not generally alleviate chronic pain (although they impair contralateral mechanosensory perception, as expected). Perhaps this is because widespread cortical activation, mediated by projections from the central lateral nucleus and the intralaminar complex, occurs in response to a noxious stimulus. Whatever the explanation, the cortical processing of pain remains something of a mystery.

Central Regulation of Pain Perception

With respect to the *interpretation* of pain, observers have long commented on the difference between the objective reality of a painful stimulus and the subjective response to it. Modern studies of this discrepancy have provided considerable insight into how circumstances affect pain perception and, ultimately, into the anatomy and pharmacology of the pain system.

During World War II, Henry Beecher and his colleagues at Harvard Medical School made a fundamental observation. In the first systematic study of

Figure 10.5 The descending systems that modulate the transmission of ascending pain signals. These modulatory systems originate in the somatic sensory cortex, the hypothalamus, the periaqueductal gray matter of the midbrain, the raphe nuclei, and other nuclei of the rostral ventral medulla. Complex modulatory effects occur at each of these sites, as well as in the dorsal horn.

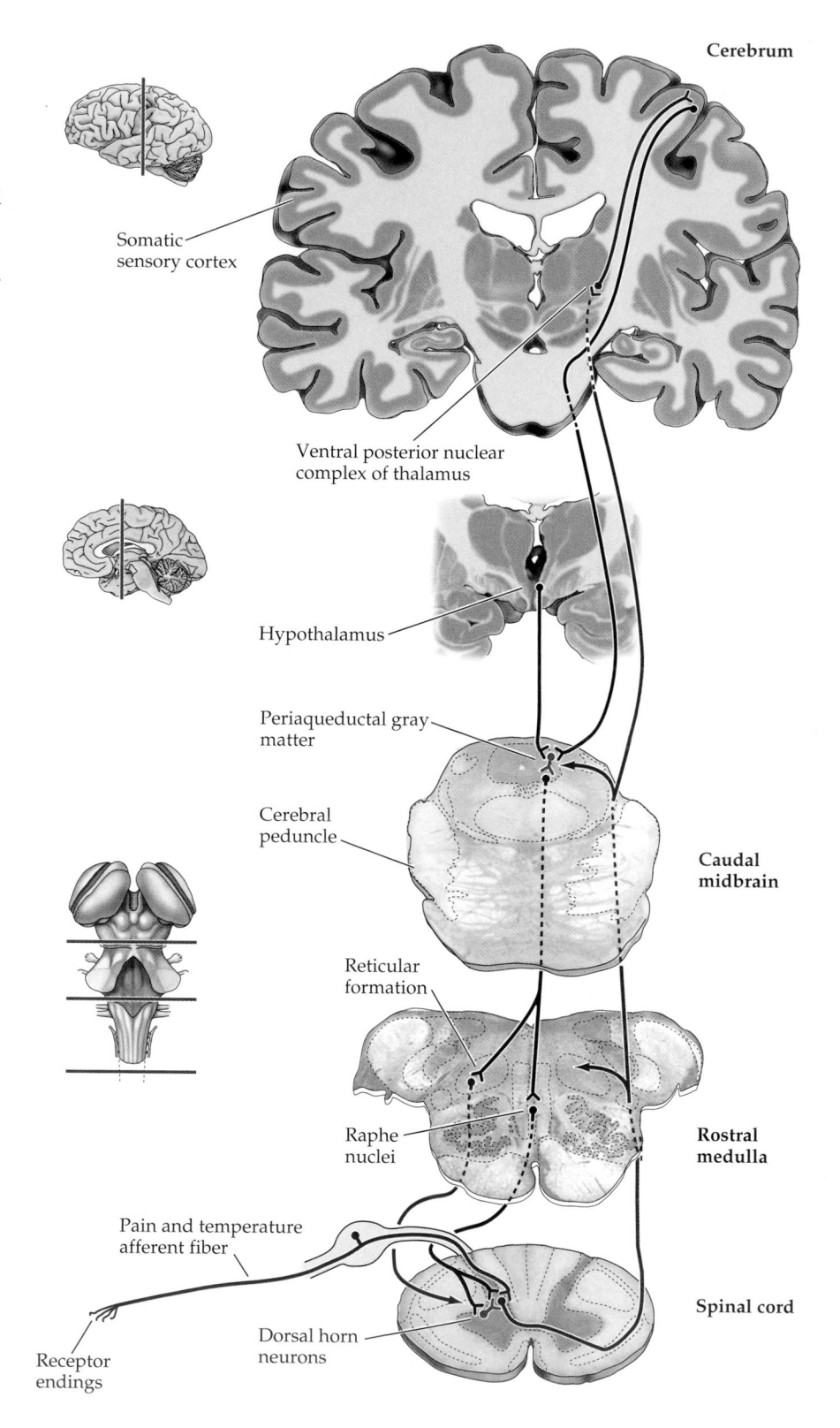

Cerebrum

Somatic sensory cortex

Ventral posterior nuclear complex of thalamus

Hypothalamus

Periaqueductal gray matter

Cerebral peduncle

Caudal midbrain

Reticular formation

Raphe nuclei

Rostral medulla

Pain and temperature afferent fiber

Spinal cord

Dorsal horn neurons

Receptor endings

its kind, they found that soldiers suffering from severe battle wounds often experienced little or no pain. Indeed, many of the wounded expressed surprise at this odd dissociation. Beecher, an anesthesiologist, concluded that the perception of pain depends on its context. For instance, the pain of an injured soldier on the battlefield would presumably be mitigated by the imagined benefits of being removed from danger, whereas a similar injury in a domestic setting would present quite a different set of circumstances that could exacerbate the pain (loss of work, financial liability, and so on). Such observations, together with the well-known placebo effect (discussed in the next section), make clear that the perception of pain is subject to central modulation (although all sensations are subject to at least some degree of this kind of modification). This statement, incidentally, should not be taken as a vague acknowledgment about the importance of psychological or "top-down" influences on sensory experience. On the contrary, there has been a gradual realization among neuroscientists and neurologists that such "psychological" effects are as real and important as any other neural phenomenon. This appreciation has provided a much more rational view of psychosomatic problems in general, and pain in particular.

The Placebo Effect

The placebo effect is defined as a physiological response following the administration of a pharmacologically inert "remedy." The word *placebo* means "I will please," and the placebo effect has a long history of use (and abuse) in medicine. The reality of the effect is undisputed. In one classic study, medical students were given one of two different pills, one said to be a sedative and the other a stimulant. In fact, both pills contained only inert ingredients. Of the students who received the "sedative," more than two-thirds reported that they felt drowsy, and students who took two such pills felt sleepier than those who had taken only one. Conversely, a large fraction of the students who took the "stimulant" reported that they felt less tired. Moreover, about a third of the entire group reported side effects ranging from headaches and dizziness to tingling extremities and a staggering gait! Only 3 of the 56 students studied reported that the pills they took had no appreciable effect.

In another study of this general sort, 75% of patients suffering from postoperative wound pain reported satisfactory relief after an injection of sterile saline. The researchers who carried out this work noted that the responders were indistinguishable from the nonresponders, both in the apparent severity of their pain and psychological makeup. Most tellingly, this placebo effect in postoperative patients could be blocked by naloxone, a competitive antagonist of opiate receptors, indicating a substantial pharmacological basis for the pain relief experienced (see the next section). A common misunderstanding about the placebo effect is the view that patients who respond to a therapeutically meaningless reagent are not suffering real pain, but only "imagining" it; this is certainly not the case.

Among other things, the placebo effect probably explains the efficacy of acupuncture anesthesia and the analgesia that can sometimes be achieved by hypnosis. In China, surgery has often been carried out under the effect of a needle (often carrying a small electrical current) inserted at locations dictated by ancient acupuncture charts. Before the advent of modern anesthetic techniques, operations such as thyroidectomies for goiter were commonly done without extraordinary discomfort, particularly among populations where stoicism was the cultural norm.

The mechanisms of pain amelioration on the battlefield, in acupuncture anesthesia, and with hypnosis are presumably related. Although the mechanisms by which the brain affects the perception of pain are only beginning to be understood, the effect is neither magical nor a sign of a suggestible intellect. In short, the placebo effect is quite real.

The Physiological Basis of Pain Modulation

Understanding the central modulation of pain perception (on which the placebo effect is presumably based) was greatly advanced by the finding that electrical or pharmacological stimulation of certain regions of the midbrain produces relief of pain (see Figure 10.5). This analgesic effect arises from activation of descending pain-modulating pathways that project, via the medulla, to neurons in the dorsal horn—particularly in Rexed's lamina II—that control the ascending information in the nociceptive system. The major brainstem regions that produce this effect are located in poorly defined nuclei in the periaqueductal gray matter and the rostral medulla. Electrical stimulation at each of these sites in experimental animals not only produces analgesia by behavioral criteria, but also demonstrably inhibits the activity of nociceptive projection neurons in the dorsal horn of the spinal cord.

A quite ordinary example of the modulation of painful stimuli is the ability to reduce the sensation of sharp pain by activating low-threshold mechanoreceptors: If you crack your shin or stub a toe, a natural (and effective) reaction is to vigorously rub the site of injury for a minute or two. Such observations, buttressed by experiments in animals, led Ronald Melzack and Patrick Wall to propose that the flow of nociceptive information through the spinal cord is modulated by concomitant activation of the large myelinated fibers associated with low-threshold mechanoreceptors. Even though further investigation led to modification of some of the original propositions in Melzack and Wall's **gate theory of pain**, the idea stimulated a great deal of work on pain modulation.

The most exciting advance in this long-standing effort has been the discovery of **endogenous opioids**. For centuries it had been apparent that opium derivatives such as morphine are powerful analgesics—indeed, they remain a mainstay of analgesic therapy today. Modern animal studies have shown that a variety of brain regions are susceptible to the action of opiate drugs, particularly—and significantly—the periaqueductal gray matter and the rostral ventral medulla. There are, in addition, opiate-sensitive regions at the level of the spinal cord. In other words, the areas that produce analgesia when stimulated are also responsive to exogenously administered opiates. It seems likely, then, that opiate drugs act at most or all of the sites shown in Figure 10.5 in producing their dramatic pain-relieving effects.

The analgesic action of opiates implied the existence of specific brain and spinal cord receptors for these drugs long before the receptors were actually found during the 1960s and 1970s. Since such receptors are unlikely to exist for the purpose of responding to the administration of opium and its derivatives, the conviction grew that there must be *endogenous* compounds for which these receptors had evolved (see Chapter 6). Several categories of endogenous opioids have now been isolated from the brain and intensively studied (Table 10.2). These agents are found in the same regions that are involved in the modulation of nociceptive afferents, although each of the families of endogenous opioid peptides has a somewhat different distribution. All three of the major groups (**enkephalins, endorphins,** and **dynor-**

TABLE 10.2
Endogenous Opioids[a]

Name	Amino acid sequence[b]
Leucine-enkephalin	*Tyr-Gly-Gly-Phe*-Leu-OH
Methionine-enkephalin	*Tyr-Gly-Gly-Phe*-Met-OH
β-Endorphin	*Tyr-Gly-Gly-Phe*-Met-Thr-Ser-Glu-Lys-Ser-Gln- Thr-Pro-Leu-Val-Thr-Leu-Phe-Lys-Asn-Ala-Ile- Val-Lys-Asn-Ala-His-Lys-Gly-Gln-OH
α-Neoendorphin	*Tyr-Gly-Gly-Phe*-Leu-Arg-Lys-Tyr-Pro-Lys
Dynorphin	*Tyr-Gly-Gly-Phe*-Leu-Arg-Arg-Ile-Arg-Pro-Lys- Leu-Lys-Trp-Asp-Asn-Gln-OH

[a]The role of these agents as neurotransmitters is discussed in Chapter 7.
[b]Note the initial homology, indicated by italics.

phins) are present in the periaqueductal gray matter. The enkephalins and dynorphins have also been found in the rostral ventral medulla and in the spinal cord regions involved in the modulation of pain.

An impressive aspect of this story is the wedding of physiology, pharmacology, and clinical research to yield a much richer understanding of the intrinsic modulation of pain. This information has finally begun to explain the subjective variability of painful stimuli and the striking dependence of pain perception on the context of the experience. Precisely how pain is modulated is being explored in many laboratories at present, motivated by the tremendous clinical (and economic) benefits that would accrue from still deeper knowledge of the pain system and its molecular underpinnings.

Summary

Whether from a structural or functional perspective, pain is an extraordinarily complex sensory modality. Because of the importance of warning an animal about dangerous circumstances, the mechanisms and pathways that subserve nociception are widespread and redundant. The major nociceptive pathway, like other somatic sensory modalities, comprises a three-neuron relay from periphery to cortex. This arrangement differs from the mechanosensory pathway primarily in that the central axons of dorsal root ganglion cells synapse on second-order neurons in the spinal cord, which then cross the midline and project to brainstem and thalamic nuclei in the contralateral spinal cord. The thalamic neurons, in turn, project to the same cortical areas as other somatic sensory modalities. The molecular basis of pain modulation is particularly intricate and is only beginning to be deciphered. The major features are the modulation of pain peripherally by the release of a variety of agents at the injury site, and the central modulation of afferent pain pathways by endogenous opioids that act at the level of both the spinal cord and the brainstem. Tremendous progress in understanding pain has been made in the last 25 years, and much more seems likely, given the importance of the problem. No patients are more distressed—or more difficult to treat—than those with chronic pain. Indeed, some aspects of pain seem much more destructive to the sufferer than required by any physiological purposes; consider, for example, the pain of a chronic illness such as invasive cancer. Perhaps such seemingly excessive effects are a necessary but unfortunate by-product of the protective benefits of this vital sensory modality.

ADDITIONAL READING

Reviews

DUBNER, R. AND M. S. GOLD (1999) The neurobiology of pain. Proc. Natl. Acad. Sci. 96: 7627–7630.

FIELDS, H. L. AND A. I. BASBAUM (1978) Brain stem control of spinal pain transmission neurons. Ann. Rev. Physiol. 40: 217–248.

LEVINE, J. D. (1998) New directions in pain reearch molecules to maladies. Neuron 20: 649–654.

MCLESKY, E. W. AND M. S. GOLD (1999) Ion channels of nociception. Ann. Rev. Physiol. 61: 835–856.

NASHOLD, B. S., JR. AND J. OVELMEN-LEVITT (1993) Chronic pain. In *Neuroscience Year*, B. Smith and G. Edelman (eds.). Boston: Birkhäuser, pp. 35–39.

Important Original Papers

BASBAUM, A. I. AND H. L. FIELDS (1979) The origin of descending pathways in the dorsolateral funiculus of the spinal cord of the cat and rat: Further studies on the anatomy of pain modulation. J. Comp. Neurol. 187: 513–522.

BEECHER, H. K. (1946) Pain in men wounded in battle. Ann. Surg. 123: 96.

BLACKWELL, B., S. S. BLOOMFIELD AND C. R. BUNCHER (1972) Demonstration to medical students of placebo response and non-drug factors. Lancet 1: 1279–1282.

CATERINA, M. J. AND 8 OTHERS (2000) Impaired nociception and pain sensation in mice lacking the capsaicin receptor. Science 288: 306–313.

CRAIG, A. D., M. C. BUSHNELL, E.-T. ZHANG AND A. BLOMQVIST (1994) A thalamic nucleus specific for pain and temperature sensation. Nature 372: 770–773.

CRAIG, A. D., E. M. REIMAN, A. EVANS AND M. C. BUSHNELL (1996) Functional imaging of an illusion of pain. Nature 384: 258–260.

LEVINE, J. D., H. L. FIELDS AND A. I. BASBAUM (1993) Peptides and the primary afferent nociceptor. J. Neurosci. 13: 2273–2286.

MOGIL, J. S. AND J. E. GRISEL (1998) Transgenic studies of pain. Pain 77: 107–128.

Books

FIELDS, H. L. (1987) *Pain*. New York: McGraw-Hill.

FIELDS, H. L. (ed.) (1990) *Pain Syndromes in Neurology*. London: Butterworths.

KOLB, L. C. (1954) *The Painful Phantom*. Springfield, IL: Charles C. Thomas.

SKRABANEK, P. AND J. MCCORMICK (1990) *Follies and Fallacies in Medicine*. New York: Prometheus Books.

WALL, P. D. AND R. MELZACK (1989) *Textbook of Pain*. New York: Churchill Livingstone.

Chapter 11

Vision: The Eye

Overview

The human visual system is extraordinary in the quantity and quality of information it supplies about the world. A glance is sufficient to describe the location, size, shape, color, and texture of objects and, if the objects are moving, their direction and speed. Equally remarkable is the fact that visual information can be discerned over a wide range of stimulus intensities, from the faint light of stars at night to bright sunlight. The next two chapters describe the molecular, cellular, and higher-order mechanisms that allow us to see. The first steps in the process of seeing are determined by the optics of the eye, the molecular mechanisms by which light energy is transduced into electrical signals in the retina, and the retinal circuitry that determines the information relayed from the eye to the lateral geniculate nucleus of the thalamus, and ultimately to the primary visual cortex in the occipital lobe.

Anatomy of the Eye

The eye is a fluid-filled sphere enclosed by three layers of tissue (Figure 11.1). Most of the outer layer is composed of a tough white fibrous tissue, the **sclera**. At the front of the eye, however, this opaque outer layer is transformed into the cornea, a specialized transparent tissue that permits light rays to enter the eye. The middle layer of tissue includes three distinct but continuous structures: the iris, the ciliary body, and the choroid. The **iris** is the colored portion of the eye that can be seen through the cornea. It contains two sets of muscles with opposing actions, which allow the size of the pupil (the opening in its center) to be adjusted under neural control. The **ciliary body** is a ring of tissue that encircles the lens and includes a muscular component that is important for adjusting the refractive power of the lens, and a vascular component (the so-called ciliary processes) that produces the fluid that fills the front of the eye. The **choroid** is composed of a rich capillary bed that serves as the main source of blood supply for the photoreceptors of the retina. Only the innermost layer of the eye, the **retina**, contains neurons that are sensitive to light and are capable of transmitting visual signals to central targets.

En route to the retina, light passes through the cornea, the lens, and two distinct fluid environments. The **anterior chamber**, the space between the lens and the cornea, is filled with **aqueous humor**, a clear, watery liquid that supplies nutrients to these structures as well as to the lens. Aqueous humor is produced by the ciliary processes in the **posterior chamber** (the region between the lens and the iris) and flows into the anterior chamber through the pupil. A specialized meshwork of cells that lies at the junction of the iris

Figure 11.1 Anatomy of the human eye.

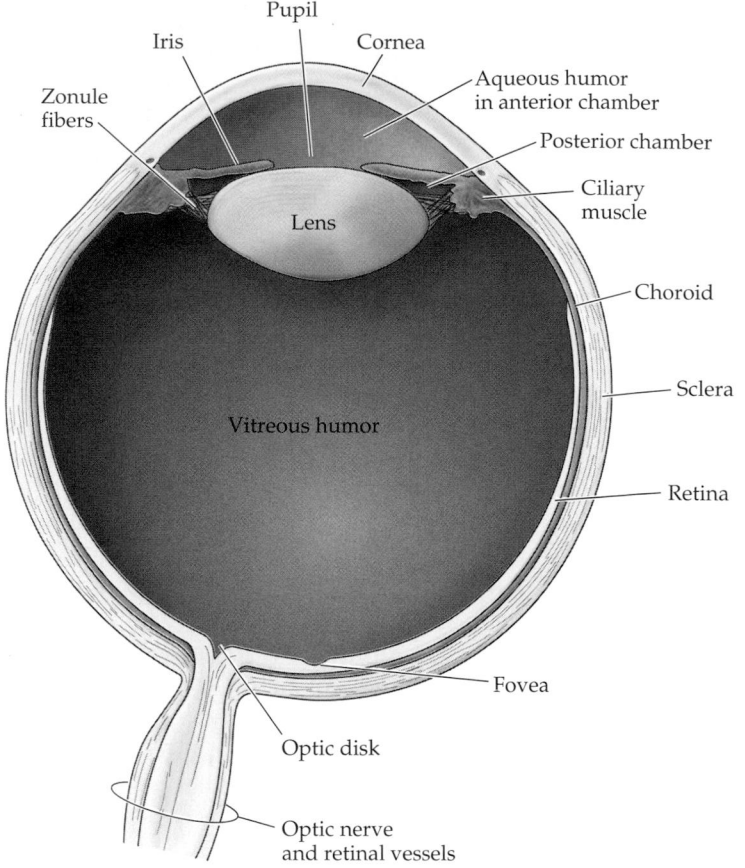

and the cornea is responsible for its uptake. Under normal conditions, the rates of aqueous humor production and uptake are in equilibrium, ensuring a constant intraocular pressure. Abnormally high levels of intraocular pressure, which occur in glaucoma, can reduce the blood supply to the eye and eventually damage retinal neurons.

The space between the back of the lens and the surface of the retina is filled with a thick, gelatinous substance called the **vitreous humor**, which accounts for about 80% of the volume of the eye. In addition to maintaining the shape of the eye, the vitreous humor contains phagocytic cells that remove blood and other debris that might otherwise interfere with light transmission. The housekeeping abilities of the vitreous humor are limited, however, as a large number of middle-aged and elderly individuals with vitreal "floaters" will attest. Floaters are collections of debris too large for phagocytic consumption that therefore remain to cast annoying shadows on the retina; they typically arise when the aging vitreous membrane pulls away from the overly long eyeball of myopic individuals (Box A).

The Formation of Images on the Retina

Because light rays diverge in all directions from their source, the set of rays from each point in space that reach the pupil must be focused. The formation of focused images on the photoreceptors of the retina depends on the refraction (bending) of light by the **cornea** and the **lens** (Figure 11.2). The cornea is responsible for most of the necessary refraction, a contribution easily appreciated by considering the hazy out-of-focus images experienced

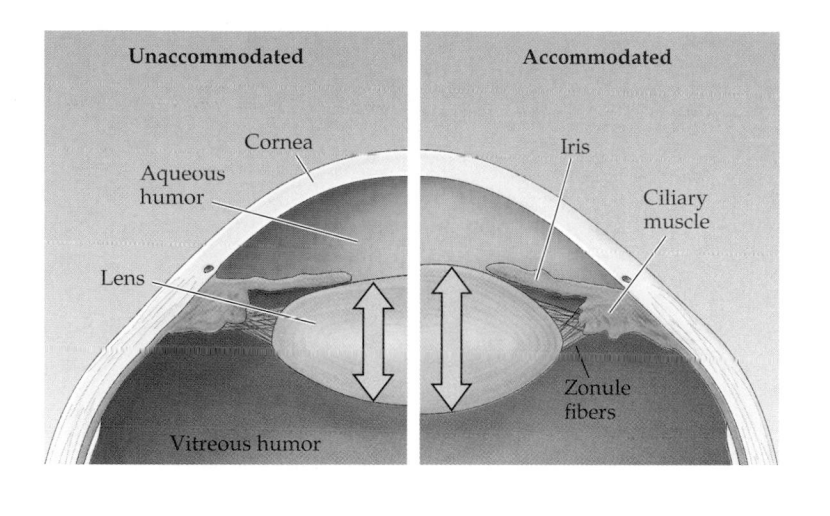

Figure 11.2 Diagram showing the anterior part of the human eye in the unaccommodated (left) and accommodated (right) state. Accommodation for focusing on near objects involves the contraction of the ciliary muscle, which reduces the tension in the zonule fibers and allows the elasticity of the lens to increase its curvature.

when swimming underwater. Water, unlike air, has a refractive index close to that of the cornea; as a result, immersion in water virtually eliminates the refraction that normally occurs at the air/cornea interface. The lens has considerably less refractive power than the cornea; however, the refraction supplied by the lens is adjustable, allowing objects at various distances from the observer to be brought into sharp focus on the retinal surface.

Dynamic changes in the refractive power of the lens are referred to as **accommodation**. When viewing distant objects, the lens is made relatively thin and flat and has the least refractive power. For near vision, the lens becomes thicker and rounder and has the most refractive power (see Figure 11.2). These changes result from the activity of the **ciliary muscle** that surrounds the lens. The lens is held in place by radially arranged connective tissue bands (called zonule fibers) that are attached to the ciliary muscle. The shape of the lens is thus determined by two opposing forces: the elasticity of the lens, which tends to keep it rounded up (removed from the eye, the lens becomes spheroidal), and the tension exerted by the zonule fibers, which tends to flatten it. When viewing distant objects, the force from the zonule fibers is greater than the elasticity of the lens, and the lens assumes the flatter shape appropriate for distance viewing. Focusing on closer objects requires relaxing the tension in the zonule fibers, allowing the inherent elasticity of the lens to increase its curvature. This relaxation is accomplished by contraction of the ciliary muscle. Because the ciliary muscle forms a ring around the lens, when the muscle contracts, the attachment points of the zonule fibers move toward the central axis of the eye, thus reducing the tension on the lens. Unfortunately, changes in the shape of the lens are not always able to produce a focused image on the retina, in which case a sharp image can be focused only with the help of additional corrective lenses (see Box A).

Adjustments in the size of the **pupil** (i.e., the circular opening in the iris) also contribute to the clarity of images formed on the retina. Like the images formed by other optical instruments, those generated by the eye are affected by spherical and chromatic aberrations, which tend to blur the retinal image. Since these aberrations are greatest for light rays that pass farthest from the center of the lens, narrowing the pupil reduces both spherical and chromatic aberration, just as closing the iris diaphragm on a camera lens improves the sharpness of a photographic image. Reducing the size of the pupil also increases the depth of field—that is, the distance within which objects are seen without blurring. However, a small pupil also limits the amount of light that reaches the retina, and, under conditions of dim illumination, visual acuity becomes limited by

BOX A
Myopia and Other Refractive Errors

Optical discrepancies between the various components of the eye cause a majority of the human population to have some form of refractive error, called ametropia. People who are unable to bring distant objects into clear focus are said to be nearsighted, or myopic (figures A,B). Myopia can be caused by the corneal surface being too curved, or by the eyeball being too long. In either case, with the lens as flat as it can be, the image of distant objects focuses in front of, rather than on, the retina. People who are unable to focus on near objects are said to be farsighted, or hyperopic. Hyperopia can be caused by the eyeball being too short or the refracting system too weak (figure C). Even with the lens in its most rounded-up state, the image is out of focus on the retinal surface (focusing at some point behind it). Both myopia and hyperopia are correctable by appropriate lenses, concave (minus) and convex (plus), respectively, or by the increasingly popular technique of corneal surgery.

Myopia, or nearsightedness, is by far the most common ametropia (an estimated 50% of the population in the United States is affected). Given the large number of people who need glasses, contact lenses, or surgery to correct this refractive error, one naturally wonders how nearsighted people coped before spectacles were invented only a few centuries ago. From what is now known about myopia, most people's vision may have been considerably better in ancient times. The basis for this assertion is the surprising finding that the growth of the eyeball is strongly influenced by focused light falling on the retina. This phenomenon was first described in 1977 by Torsten Wiesel and Elio Raviola at Harvard Medical School, who studied monkeys reared with their lids sutured (the same approach used to demonstrate the effects of visual deprivation on cortical

connections in the visual system; see Chapter 24), a procedure that deprives the eye of focused retinal images. They found that animals growing to maturity under these conditions show an elongation of the eyeball. The effect of focused light deprivation appears to be a local one, since the abnormal growth of the eye occurs in experimental animals even if the optic nerve is cut. Indeed, if only a portion of the retinal surface is deprived of focused light, then only that region of the eyeball grows abnormally.

Although the mechanism of light-mediated control of eye growth is not fully understood, many experts now believe that the prevalence of myopia is due to some aspect of modern civilization—perhaps learning to read and write at an early age—that interferes with the normal feedback control of vision on eye development, leading to abnormal elongation of the eyeball. A corollary of this

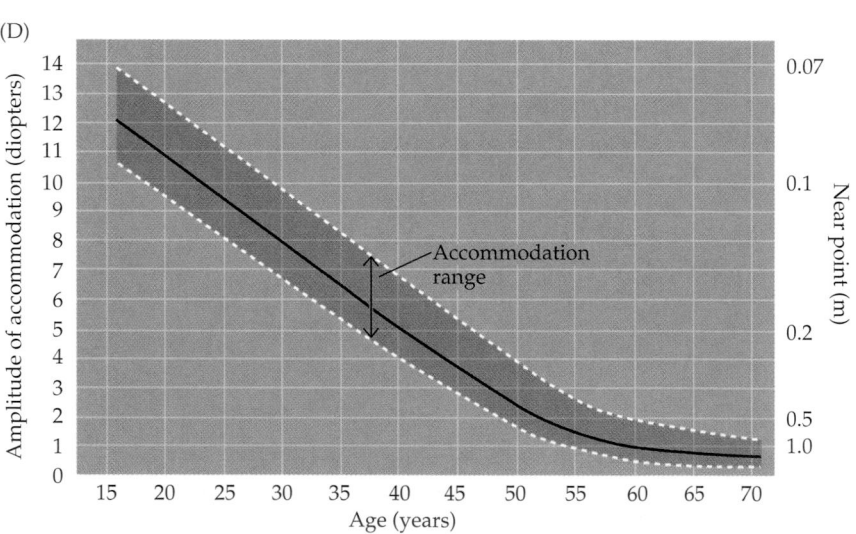

(A) Emmetropia (normal)

(B) Myopia (nearsighted)

(C) Hyperopia (farsighted)

(D)

Refractive errors. (A) In the normal eye, with ciliary muscles relaxed, an image of a distant object is focused on the retina. (B) In myopia, light rays are focused in front of the retina. (C) In hyperopia, images are focused at a point beyond the retina. (D) Changes in the ability of the lens to round up (accommodate) with age. The graph also shows how the near point (the closest point to the eye that can be brought into focus) changes. Accommodation, which is an optical measurement of the refractive power of the lens, is given in diopters. (D after Westheimer, 1974.)

hypothesis is that if children (or, more likely, their parents) wanted to improve their vision, they might be able to do so by practicing far vision to counterbalance the near work "overload." Practically, of course, most people would probably choose wearing glasses or contacts rather than indulging in the onerous daily practice that would presumably be required. Not everyone agrees, however, that such a remedy would be effective, and a number of investigators (and drug companies) are exploring the possibility of pharmacological intervention during the period of childhood when abnormal eye growth is presumed to occur. In any event, it is a remarkable fact that deprivation of focused light on the retina causes a compensatory growth of the eye and that this feedback loop is so easily perturbed.

Even people with normal (emmetropic) vision as young adults eventually experience difficulty focusing on near objects. One of the many consequences of aging is that the lens loses its elasticity; as a result,

the maximum curvature the lens can achieve when the ciliary muscle contracts is gradually reduced. The near point (the closest point that can be brought into clear focus) thus recedes, and objects (such as this book) must be farther and farther away from the eye in order to focus them on the retina. At some point, usually during early middle age, the accommodative ability of the eye is so reduced that near vision tasks like reading become difficult or impossible (figure D). This condition is referred to as presbyopia, and can be corrected by convex lenses for near-vision tasks, or by bifocal lenses if myopia is also present (which requires a negative correction). Bifocal correction presents a particular problem for those who prefer contact lenses. Because contact lenses float on the surface of the cornea, having the distance correction above and the near correction below (as in conventional bifocal glasses) doesn't work (although "omnifocal" contact lenses have recently been used with success). A solution to this problem for some contact lens wearers has been to put

a near correcting lens in one eye and a distance correcting lens in the other! The success of this approach is another testament to the remarkable ability of the visual system to adjust to a wide variety of unusual circumstances.

References

Bock, G. and K. Widdows (1990) *Myopia and the Control of Eye Growth.* Ciba Foundation Symposium 155. Chichester: Wiley.

Coster, D. J. (1994) *Physics for Ophthalmologists.* Edinburgh: Churchill Livingston.

Hart, W. M. Jr. (ed.) (1992) *Adler's Physiology of the Eye: Clinical Application*, 9th Ed. St. Louis, MO: Mosby Year Book.

Sherman, S. M., T. T. Norton and V. A. Casagrande (1977) Myopia in the lid sutured tree shrew. Brain Res. 124: 154–157.

Wallman, J., J. Turkel and J. Tractman (1978) Extreme myopia produced by modest changes in early visual experience. Science 201: 1249–1251.

Wiesel, T. N. and E. Raviola (1977) Myopia and eye enlargement after neonatal lid fusion in monkeys. Nature 266: 66–68.

the number of available photons rather than by optical aberrations. An adjustable pupil thus provides an effective means of limiting optical aberrations, while maximizing depth of field to the extent that different levels of illumination permit. The size of the pupil is controlled by innervation from both sympathetic and parasympathetic divisions of the visceral motor system, which are in turn modulated by several brainstem centers (see Chapters 20 and 21).

The Retina

Despite its peripheral location, the **retina** or neural portion of the eye, is actually part of the central nervous system. During development, the retina forms as an outpocketing of the diencephalon, called the optic vesicle, which undergoes invagination to form the optic cup (Figure 11.3; see also Chapter 22). The inner wall of the optic cup gives rise to the retina, while the outer wall gives rise to the **pigment epithelium**. This epithelium is a melanin-containing structure that reduces backscattering of light that enters the eye; it also plays a critical role in the maintenance of photoreceptors, renewing photopigments and phagocytosing the photoreceptor disks, whose turnover at a high rate is essential to vision.

Consistent with its status as a full-fledged part of the central nervous system, the retina comprises complex neural circuitry that converts the

(B) 4.5-mm embryo

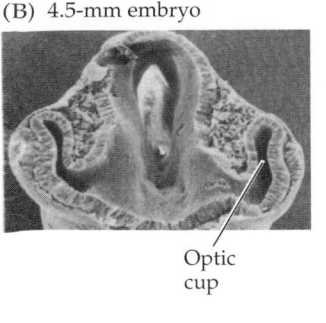

Optic
cup

(C) 5-mm embryo

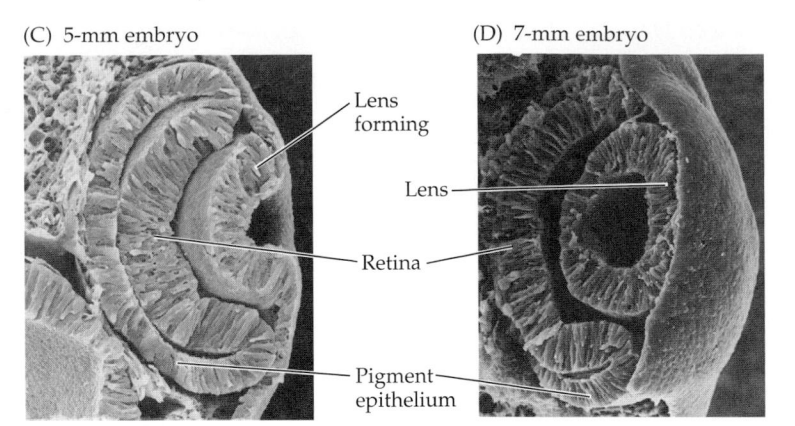

Lens
forming

Lens

Retina

Pigment
epithelium

(D) 7-mm embryo

opment of the
retina develops as
m the neural tube,
cle. (B) The optic
to form the optic
er wall of the optic
ural retina, while the
s the pigment epithe-
ilfer and Yang, 1980;
osney.)

graded electrical activity of photoreceptors into action potentials that travel to the brain via axons in the optic nerve. Although it has the same types of functional elements and neurotransmitters found in other parts of the central nervous system, the retina comprises only a few classes of neurons, and these are arranged in a manner that has been less difficult to unravel than the circuits in other areas of the brain. There are five types of neurons in the retina: **photoreceptors, bipolar cells, ganglion cells, horizontal cells**, and **amacrine cells**. The cell bodies and processes of these neurons are stacked in five alternating layers, with the cell bodies located in the inner nuclear, outer nuclear, and ganglion cell layers, and the processes and synaptic contacts located in the inner plexiform and outer plexiform layers (Figure 11.4). A direct three-neuron chain—photoreceptor cell to bipolar cell to ganglion cell—is the major route of information flow from photoreceptors to the optic nerve.

There are two types of light-sensitive elements in the retina: **rods** and **cones**. Both types of photoreceptors have an outer segment that is composed of membranous disks that contain photopigment and lies adjacent to the pigment epithelial layer, and an inner segment that contains the cell nucleus and gives rise to synaptic terminals that contact bipolar or horizontal cells. Absorption of light by the photopigment in the outer segment of the photoreceptors initiates a cascade of events that changes the membrane potential of the receptor, and therefore the amount of neurotransmitter released by the photoreceptor synapses onto the cells they contact. The synapses between photoreceptor terminals and bipolar cells (and horizontal cells) occur in the outer plexiform layer; more specifically, the cell bodies of photoreceptors make up the outer nuclear layer, whereas the cell bodies of bipolar cells lie in the inner nuclear layer. The short axonal processes of bipolar cells make synaptic contacts in turn on the dendritic processes of ganglion cells in the inner plexiform layer. The much larger axons of the ganglion cells form the **optic nerve** and carry information about retinal stimulation to the rest of the central nervous system.

The two other types of neurons in the retina, **horizontal cells** and **amacrine cells**, have their cell bodies in the inner nuclear layer and are primarily responsible for lateral interactions within the retina. These lateral interactions between receptors, horizontal cells, and bipolar cells in the outer plexiform layer are largely responsible for the visual system's sensitivity to luminance contrast over a wide range of light intensities. The processes of amacrine cells, which extend laterally in the inner plexiform layer, are post-synaptic to bipolar cell terminals and presynaptic to the dendrites of ganglion cells (see Figure 11.4). The processes of horizontal cells ramify in the

hypothesis is that if children (or, more likely, their parents) wanted to improve their vision, they might be able to do so by practicing far vision to counterbalance the near work "overload." Practically, of course, most people would probably choose wearing glasses or contacts rather than indulging in the onerous daily practice that would presumably be required. Not everyone agrees, however, that such a remedy would be effective, and a number of investigators (and drug companies) are exploring the possibility of pharmacological intervention during the period of childhood when abnormal eye growth is presumed to occur. In any event, it is a remarkable fact that deprivation of focused light on the retina causes a compensatory growth of the eye and that this feedback loop is so easily perturbed.

Even people with normal (emmetropic) vision as young adults eventually experience difficulty focusing on near objects. One of the many consequences of aging is that the lens loses its elasticity; as a result,

the maximum curvature the lens can achieve when the ciliary muscle contracts is gradually reduced. The near point (the closest point that can be brought into clear focus) thus recedes, and objects (such as this book) must be farther and farther away from the eye in order to focus them on the retina. At some point, usually during early middle age, the accommodative ability of the eye is so reduced that near vision tasks like reading become difficult or impossible (figure D). This condition is referred to as presbyopia, and can be corrected by convex lenses for near-vision tasks, or by bifocal lenses if myopia is also present (which requires a negative correction). Bifocal correction presents a particular problem for those who prefer contact lenses. Because contact lenses float on the surface of the cornea, having the distance correction above and the near correction below (as in conventional bifocal glasses) doesn't work (although "omnifocal" contact lenses have recently been used with success). A solution to this problem for some contact lens wearers has been to put

a near correcting lens in one eye and a distance correcting lens in the other! The success of this approach is another testament to the remarkable ability of the visual system to adjust to a wide variety of unusual circumstances.

References

BOCK, G. AND K. WIDDOWS (1990) *Myopia and the Control of Eye Growth.* Ciba Foundation Symposium 155. Chichester: Wiley.

COSTER, D. J. (1994) *Physics for Ophthalmologists.* Edinburgh: Churchill Livingston.

HART, W. M. JR. (ed.) (1992) *Adler's Physiology of the Eye: Clinical Application,* 9th Ed. St. Louis, MO: Mosby Year Book.

SHERMAN, S. M., T. T. NORTON AND V. A. CASAGRANDE (1977) Myopia in the lid sutured tree shrew. Brain Res. 124: 154–157.

WALLMAN, J., J. TURKEL AND J. TRACTMAN (1978) Extreme myopia produced by modest changes in early visual experience. Science 201: 1249–1251.

WIESEL, T. N. AND E. RAVIOLA (1977) Myopia and eye enlargement after neonatal lid fusion in monkeys. Nature 266: 66–68.

the number of available photons rather than by optical aberrations. An adjustable pupil thus provides an effective means of limiting optical aberrations, while maximizing depth of field to the extent that different levels of illumination permit. The size of the pupil is controlled by innervation from both sympathetic and parasympathetic divisions of the visceral motor system, which are in turn modulated by several brainstem centers (see Chapters 20 and 21).

The Retina

Despite its peripheral location, the **retina** or neural portion of the eye, is actually part of the central nervous system. During development, the retina forms as an outpocketing of the diencephalon, called the optic vesicle, which undergoes invagination to form the optic cup (Figure 11.3; see also Chapter 22). The inner wall of the optic cup gives rise to the retina, while the outer wall gives rise to the **pigment epithelium**. This epithelium is a melanin-containing structure that reduces backscattering of light that enters the eye; it also plays a critical role in the maintenance of photoreceptors, renewing photopigments and phagocytosing the photoreceptor disks, whose turnover at a high rate is essential to vision.

Consistent with its status as a full-fledged part of the central nervous system, the retina comprises complex neural circuitry that converts the

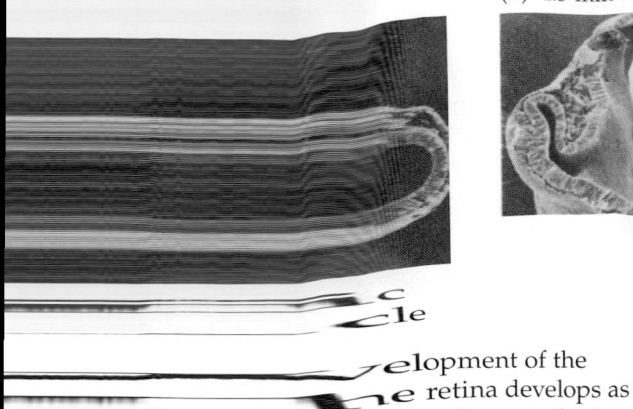

(B) 4.5-mm embryo

Optic
cup

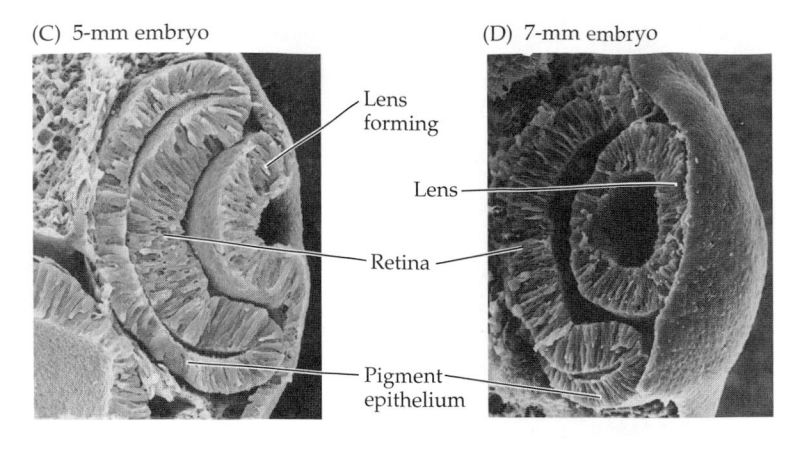

(C) 5-mm embryo

Lens
forming

Lens

Retina

Pigment
epithelium

(D) 7-mm embryo

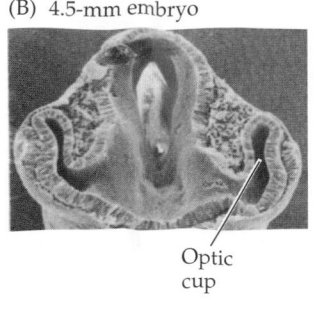

c
cle

elopment of the
e retina develops as
om the neural tube,
sicle. (B) The optic
s to form the optic
ner wall of the optic
neural retina, while the
es the pigment epithe-
Hilfer and Yang, 1980;
Tosney.)

graded electrical activity of photoreceptors into action potentials that travel to the brain via axons in the optic nerve. Although it has the same types of functional elements and neurotransmitters found in other parts of the central nervous system, the retina comprises only a few classes of neurons, and these are arranged in a manner that has been less difficult to unravel than the circuits in other areas of the brain. There are five types of neurons in the retina: **photoreceptors, bipolar cells, ganglion cells, horizontal cells**, and **amacrine cells**. The cell bodies and processes of these neurons are stacked in five alternating layers, with the cell bodies located in the inner nuclear, outer nuclear, and ganglion cell layers, and the processes and synaptic contacts located in the inner plexiform and outer plexiform layers (Figure 11.4). A direct three-neuron chain—photoreceptor cell to bipolar cell to ganglion cell—is the major route of information flow from photoreceptors to the optic nerve.

There are two types of light-sensitive elements in the retina: **rods** and **cones**. Both types of photoreceptors have an outer segment that is composed of membranous disks that contain photopigment and lies adjacent to the pigment epithelial layer, and an inner segment that contains the cell nucleus and gives rise to synaptic terminals that contact bipolar or horizontal cells. Absorption of light by the photopigment in the outer segment of the photoreceptors initiates a cascade of events that changes the membrane potential of the receptor, and therefore the amount of neurotransmitter released by the photoreceptor synapses onto the cells they contact. The synapses between photoreceptor terminals and bipolar cells (and horizontal cells) occur in the outer plexiform layer; more specifically, the cell bodies of photoreceptors make up the outer nuclear layer, whereas the cell bodies of bipolar cells lie in the inner nuclear layer. The short axonal processes of bipolar cells make synaptic contacts in turn on the dendritic processes of ganglion cells in the inner plexiform layer. The much larger axons of the ganglion cells form the **optic nerve** and carry information about retinal stimulation to the rest of the central nervous system.

The two other types of neurons in the retina, **horizontal cells** and **amacrine cells**, have their cell bodies in the inner nuclear layer and are primarily responsible for lateral interactions within the retina. These lateral interactions between receptors, horizontal cells, and bipolar cells in the outer plexiform layer are largely responsible for the visual system's sensitivity to luminance contrast over a wide range of light intensities. The processes of amacrine cells, which extend laterally in the inner plexiform layer, are post-synaptic to bipolar cell terminals and presynaptic to the dendrites of ganglion cells (see Figure 11.4). The processes of horizontal cells ramify in the

hypothesis is that if children (or, more likely, their parents) wanted to improve their vision, they might be able to do so by practicing far vision to counterbalance the near work "overload." Practically, of course, most people would probably choose wearing glasses or contacts rather than indulging in the onerous daily practice that would presumably be required. Not everyone agrees, however, that such a remedy would be effective, and a number of investigators (and drug companies) are exploring the possibility of pharmacological intervention during the period of childhood when abnormal eye growth is presumed to occur. In any event, it is a remarkable fact that deprivation of focused light on the retina causes a compensatory growth of the eye and that this feedback loop is so easily perturbed.

Even people with normal (emmetropic) vision as young adults eventually experience difficulty focusing on near objects. One of the many consequences of aging is that the lens loses its elasticity; as a result,

the maximum curvature the lens can achieve when the ciliary muscle contracts is gradually reduced. The near point (the closest point that can be brought into clear focus) thus recedes, and objects (such as this book) must be farther and farther away from the eye in order to focus them on the retina. At some point, usually during early middle age, the accommodative ability of the eye is so reduced that near vision tasks like reading become difficult or impossible (figure D). This condition is referred to as presbyopia, and can be corrected by convex lenses for near-vision tasks, or by bifocal lenses if myopia is also present (which requires a negative correction). Bifocal correction presents a particular problem for those who prefer contact lenses. Because contact lenses float on the surface of the cornea, having the distance correction above and the near correction below (as in conventional bifocal glasses) doesn't work (although "omnifocal" contact lenses have recently been used with success). A solution to this problem for some contact lens wearers has been to put

a near correcting lens in one eye and a distance correcting lens in the other! The success of this approach is another testament to the remarkable ability of the visual system to adjust to a wide variety of unusual circumstances.

References

BOCK, G. AND K. WIDDOWS (1990) *Myopia and the Control of Eye Growth.* Ciba Foundation Symposium 155. Chichester: Wiley.

COSTER, D. J. (1994) *Physics for Ophthalmologists.* Edinburgh: Churchill Livingston.

HART, W. M. JR. (ed.) (1992) *Adler's Physiology of the Eye: Clinical Application,* 9th Ed. St. Louis, MO: Mosby Year Book.

SHERMAN, S. M., T. T. NORTON AND V. A. CASAGRANDE (1977) Myopia in the lid sutured tree shrew. Brain Res. 124: 154–157.

WALLMAN, J., J. TURKEL AND J. TRACTMAN (1978) Extreme myopia produced by modest changes in early visual experience. Science 201: 1249–1251.

WIESEL, T. N. AND E. RAVIOLA (1977) Myopia and eye enlargement after neonatal lid fusion in monkeys. Nature 266: 66–68.

the number of available photons rather than by optical aberrations. An adjustable pupil thus provides an effective means of limiting optical aberrations, while maximizing depth of field to the extent that different levels of illumination permit. The size of the pupil is controlled by innervation from both sympathetic and parasympathetic divisions of the visceral motor system, which are in turn modulated by several brainstem centers (see Chapters 20 and 21).

The Retina

Despite its peripheral location, the **retina** or neural portion of the eye, is actually part of the central nervous system. During development, the retina forms as an outpocketing of the diencephalon, called the optic vesicle, which undergoes invagination to form the optic cup (Figure 11.3; see also Chapter 22). The inner wall of the optic cup gives rise to the retina, while the outer wall gives rise to the **pigment epithelium**. This epithelium is a melanin-containing structure that reduces backscattering of light that enters the eye; it also plays a critical role in the maintenance of photoreceptors, renewing photopigments and phagocytosing the photoreceptor disks, whose turnover at a high rate is essential to vision.

Consistent with its status as a full-fledged part of the central nervous system, the retina comprises complex neural circuitry that converts the

(A) 4-mm embryo

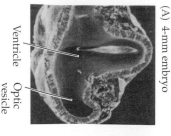

Ventricle

Optic
vesicle

(B) 4.5-mm embryo

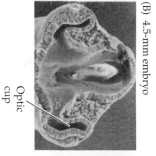

Optic
cup

(C) 5-mm embryo

(D) 7-mm embryo

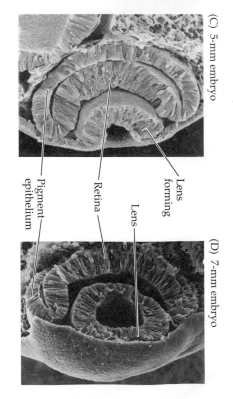

Pigment
epithelium

Retina

Lens
forming

Lens

Figure 11.3 Development of the human eye. (A) The retina develops as an outpocketing from the neural tube, called the optic vesicle. (B) The optic vesicle invaginates to form the optic cup. (C, D) The inner wall of the optic cup becomes the neural retina, while the outer wall becomes the pigment epithelium. (A–C from Hilfer and Yang, 1980; D courtesy of K. Tosney.)

graded electrical activity of photoreceptors into action potentials that travel to the brain via axons in the optic nerve. Although it has the same types of functional elements and neurotransmitters found in other parts of the central nervous system, the retina comprises only a few classes of neurons, and these are arranged in a manner that has been less difficult to unravel than the circuits in other areas of the brain. There are five types of neurons in the retina: **photoreceptors**, **bipolar cells**, **ganglion cells**, **horizontal cells**, and **amacrine cells**. The cell bodies and processes of these neurons are stacked in five alternating layers, with the cell bodies located in the inner nuclear, outer nuclear, and ganglion cell layers, and the processes and synaptic contacts located in the inner plexiform and outer plexiform layers (Figure 11.4). A direct three-neuron chain—photoreceptor cell to bipolar cell to ganglion cell—is the major route of information flow from photoreceptors to the optic nerve.

There are two types of light-sensitive elements in the retina: **rods** and **cones**. Both types of photoreceptors have an outer segment that is composed of membranous disks that contain photopigment and lies adjacent to the pigment epithelial layer, and an inner segment that contains the cell nucleus and gives rise to synaptic terminals that contact bipolar or horizontal cells. Absorption of light by the photopigment in the outer segment of the photoreceptors initiates a cascade of events that changes the membrane potential of the receptor, and therefore the amount of neurotransmitter released by the photoreceptor synapses onto the cells they contact. The synapses between photoreceptor terminals and bipolar cells (and horizontal cells) occur in the outer plexiform layer; more specifically, the cell bodies of photoreceptors make up the outer nuclear layer, whereas the cell bodies of bipolar cells lie in the inner nuclear layer. The short axonal processes of bipolar cells make synaptic contacts in turn on the dendritic processes of ganglion cells in the inner plexiform layer. The much larger axons of the ganglion cells form the **optic nerve** and carry information about retinal stimulation to the rest of the central nervous system.

The two other types of neurons in the retina, **horizontal cells** and **amacrine cells**, have their cell bodies in the inner nuclear layer and are primarily responsible for lateral interactions within the retina. These lateral interactions between receptors, horizontal cells, and bipolar cells in the outer plexiform layer are largely responsible for the visual system's sensitivity to luminance contrast over a wide range of light intensities. The processes of amacrine cells, which extend laterally in the inner plexiform layer, are post-synaptic to bipolar cell terminals and presynaptic to the dendrites of ganglion cells (see Figure 11.4). The processes of horizontal cells ramify in the

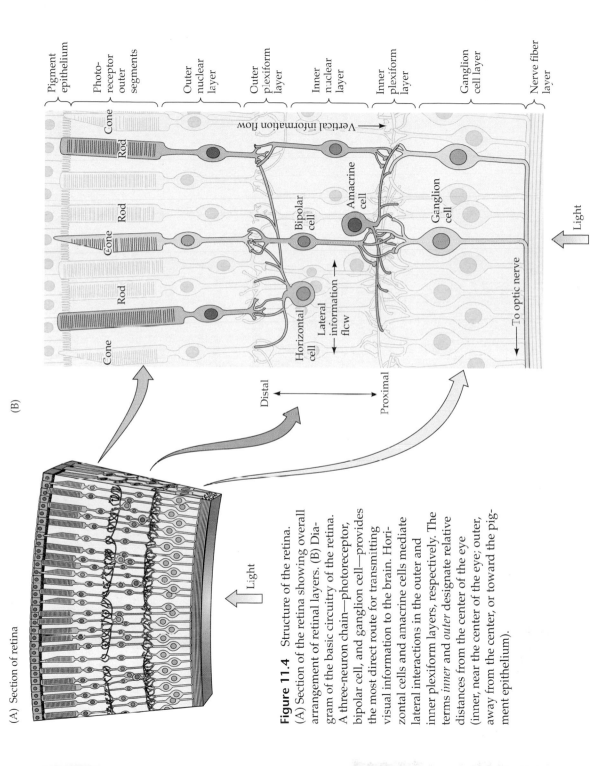

Figure 11.4 Structure of the retina. (A) Section of the retina showing overall arrangement of retinal layers. (B) Diagram of the basic circuitry of the retina. A three-neuron chain—photoreceptor, bipolar cell, and ganglion cell—provides the most direct route for transmitting visual information to the brain. Horizontal cells and amacrine cells mediate lateral interactions in the outer and inner plexiform layers, respectively. The terms *inner* and *outer* designate relative distances from the center of the eye (inner, near the center of the eye; outer, away from the center, or toward the pigment epithelium).

outer plexiform layer. Several subclasses of amacrine cells that make distinct contributions to visual function. One class of amacrine cells, for example, plays an important role in transforming the persistent responses of bipolar cells to light into the brief transient responses exhibited by some types of ganglion cells. Another type serves as an obligatory step in the pathway that transmits information from rod photoreceptors to retinal ganglion cells. The variety of amacrine cell subtypes illustrates the more general rule that although there are only five basic retinal cell types, there can be considerable diversity within a given cell type. This diversity is the basis for pathways that convey different sorts of information to central targets in a parallel manner.

At first glance, the spatial arrangement of retinal layers seems counterintuitive, since light rays must pass through the non-light-sensitive elements of the retina (and retinal vasculature!) before reaching the outer segments of the photoreceptors, where photons are absorbed (see Figure 11.4). The reason for this curious feature of retinal organization lies in the special relationship that exists between the outer segments of the photoreceptors and the pigment epithelium. The outer segments contain membranous disks that house the light-sensitive photopigment and other proteins involved in the transduction process. These disks are formed near the inner segment of the photoreceptor and move toward the tip of the outer segment, where they are shed. The pigment epithelium plays an essential role in removing the expended receptor disks; this is no small task, since all the disks in the outer segments are replaced every 12 days. In addition, the pigment epithelium contains the biochemical machinery that is required to regenerate photopigment molecules after they have been exposed to light. It is presumably the demands of the photoreceptor disk life cycle and photopigment recycling that explain why rods and cones are found in the outermost rather than the innermost layer of the retina. Disruptions in the normal relationships between pigment epithelium and retinal photoreceptors such as those that occur in retinitis pigmentosa have severe consequences for vision (Box B).

Phototransduction

In most sensory systems, activation of a receptor by the appropriate stimulus causes the cell membrane to depolarize, ultimately stimulating an action potential and transmitter release onto the neurons it contacts. In the retina, however, photoreceptors do not exhibit action potentials; rather, light activation causes a graded change in membrane potential and a corresponding change in the rate of transmitter release onto postsynaptic neurons. Indeed, much of the processing within the retina is mediated by graded potentials, largely because action potentials are not required to transmit information over the relatively short distances involved.

Perhaps even more surprising is that shining light on a photoreceptor, either a rod or a cone, leads to membrane *hyperpolarization* rather than depolarization (Figure 11.5). In the dark, the receptor is in a depolarized state, with a membrane potential of roughly −40 mV (including those portions of the cell that release transmitters). Progressive increases in the intensity of illumination cause the potential across the receptor membrane to become more negative, a response that saturates when the membrane potential reaches about −65 mV. Although the sign of the potential change may seem odd, the only logical requirement for subsequent visual processing is a consistent relationship between luminance changes and the rate of transmitter release from the photoreceptor terminals. As in other nerve cells, transmitter release from the synaptic terminals of the photoreceptor is dependent on voltage-sensitive Ca^{2+} channels in the terminal membrane. Thus, in the dark, when photoreceptors are relatively depolarized, the number of open Ca^{2+} channels in the synaptic terminal is high, and the rate of transmitter release

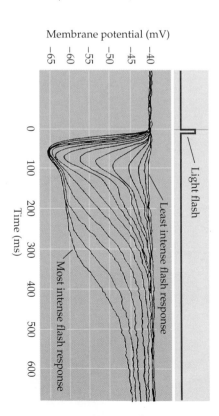

Figure 11.5 An intracellular recording from a single cone stimulated with different amounts of light (the cone has been taken from the turtle retina, which accounts for the relatively long time course of the response). Each trace represents the response to a brief flash that was varied in intensity. At the highest light levels, the response amplitude saturates (at about −65 mV). The hyperpolarizing response is characteristic of vertebrate photoreceptors; interestingly, some invertebrate photoreceptors depolarize in response to light. (After Schnapf and Baylor, 1987.)

BOX B
Retinitis Pigmentosa

Retinitis pigmentosa (RP) refers to a heterogeneous group of hereditary eye disorders characterized by progressive vision loss due to a gradual degeneration of photoreceptors. An estimated 100,000 people in the United States have RP. In spite of the name, inflammation is not a prominent part of the disease process. Classification of this group of disorders under one rubric is based on the clinical features most commonly observed in these patients. The hallmarks of RP are night blindness and reduction of peripheral vision, narrowing of the retinal vessels, and the migration of pigment from disrupted retinal pigment epithelium into the retina, forming clumps of various sizes, often next to the retinal blood vessels (see figure).

Typically, patients first notice difficulty seeing at night due to the loss of rod photoreceptors; the remaining cone photoreceptors then become the mainstay of visual function. Over years and decades, however, the cones also degenerate, leading to a progressive loss of vision. In most RP patients, visual field defects begin in the midperiphery, between 30° and 50° from fixation. The defective regions gradually enlarge, leaving islands of vision in the periphery and a constricted central field (called tunnel vision). When the visual field contracts to 20° or less and/or central vision is 20/200 or worse, the patient becomes legally blind.

Inheritance patterns indicate that RP can be transmitted in X-linked (XLRP), autosomal dominant (ADRP), or recessive (ARRP) modes. In the United States, the percentage of each genetic type is estimated to be 9%, 16%, and 41%, respectively. When only one member of a pedigree has RP, the case is classified as "simplex," which accounts for 34% of all cases. Among the three genetic types of RP, ADRP is the mildest. These patients often retain good central vision to 60 years of age and beyond. In contrast, patients with the XLRP form of the disease are usually legally blind by 30 to 40 years of age. However, the severity and the age of onset of the symptoms varies greatly among patients with the same genetic type of RP. This variation is apparent even within the same family when presumably all the affected members have the same genetic mutation.

Many RP-inducing mutations have now been described. To date, 6 genomic loci have been mapped for XLRP and two genes cloned. Similarly, 11 loci and 5 genes for ADRP and 13 loci and 7 genes for ARRP have been identified. Of the genes identified so far, many encode photoreceptor-specific proteins, several being associated with phototransduction in the rods, such as rhodopsin, subunits of the cGMP phosphodiesterase, and the cGMP-gated Ca^{2+} channel. Multiple mutations in each of the cloned genes have been found. For example, in the case of the rhodopsin gene, 90 different mutations have been identified among ADRP patients.

The heterogeneity of RP at all levels, from the genetic mutations to the clinical symptoms, has important implications for understanding its pathogenesis and designing therapies. Given the complex molecular etiology of RP, it is unlikely that a single cellular mechanism will explain the pathogenesis for all cases. On the other hand, regardless of the specific mutation, the vision loss that is most critical to RP patients is due to the gradual degeneration of cones. In many cases, the protein that the RP-causing mutation affects is not even expressed in the cones; the prime example is rhodopsin—the rod-specific visual pigment. Therefore, the loss of cones may be an indirect consequence of a rod-specific mutation. Identifying the cellular mechanisms that directly cause the degeneration of cones should lead to a better understanding of RP.

References

WELEBER, R. G. (1994) Retinitis pigmentosa and allied disorders. In *Retina*, 2nd Ed., Volume 1: *Basic Science and Inherited Retinal Diseases*. S. J. Ryan and T. E. Ogden (eds.), St. Louis, MO: Mosby Year Book, pp. 335–466.

The Foundation Fighting Blindness of Hunt Valley, MD, maintains a web site that provides updated information about many forms of retinal degeneration: **www.blindness.org**

RetNet provides updated information, including references to original articles, on genes and mutations associated with retinal diseases: **www.sph.uth.tmc.edu/RetNet**

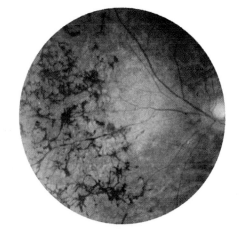

Characteristic appearance of the retina in patients with retinitis pigmentosa. Note dark clumps of pigment that are the hallmark of this disorder.

is correspondingly great; in the light, when receptors are hyperpolarized, the number of open Ca^{2+} channels is reduced, and the rate of transmitter release is also reduced. The reason for this unusual arrangement compared to other sensory receptor cells is not known.

The relatively depolarized state of photoreceptors in the dark depends on the presence of ion channels in the outer segment membrane that permit Na^+ and Ca^{2+} ions to flow into the cell, thus reducing the degree of inside negativity (Figure 11.6). The probability of these channels in the outer segment being open or closed is regulated in turn by the levels of the nucleotide cyclic guanosine monophosphate (cGMP) (as in many other second messenger systems; see Chapter 8). In darkness, high levels of cGMP in the outer segment keep the channels open. In the light, however, cGMP levels drop and some of the channels close, leading to hyperpolarization of the outer segment membrane, and ultimately the reduction of transmitter release at the photoreceptor synapse.

The series of biochemical changes that ultimately leads to a reduction in cGMP levels begins when a photon is absorbed by the photopigment in the receptor disks. The photopigment contains a light-absorbing chromophore (**retinal**, an aldehyde of vitamin A) coupled to one of several possible proteins called **opsins** that tune the molecule's absorption of light to a particular region of the spectrum. Indeed, it is the different protein component of the photopigment in rods and cones that contributes to the functional specialization of these two receptor types. Most of what is known about the molecular events of phototransduction has been gleaned from experiments in rods, in which the photopigment is **rhodopsin** (Figure 11.7A). When the retinal moiety in the rhodopsin molecule absorbs a photon, its configuration changes from the 11-*cis* isomer to all-*trans* retinal; this change then triggers a series of alterations in the protein component of the molecule (Figure 11.7B). The changes lead, in turn, to the activation of an intracellular messenger called **transducin**, which activates a phosphodiesterase that hydrolyzes cGMP. All of these events take place within the disk membrane. The hydrolysis by phosphodiesterase at the disk membrane lowers the concentration of cGMP throughout the outer segment, and thus reduces the number of cGMP

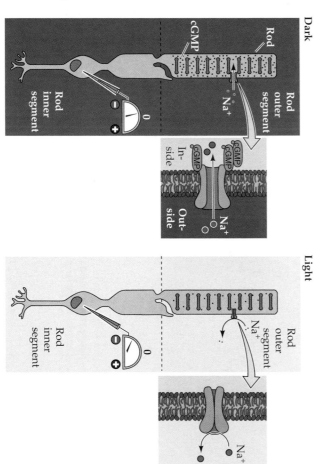

Figure 11.6 Cyclic GMP-gated channels in the outer segment membrane are responsible for the light-induced changes in the electrical activity of photoreceptors (a rod is shown here, but the same scheme applies to cones). In the dark, cGMP levels in the outer segment are high; this molecule binds to the Na^+-permeable channels in the membrane, keeping them open and allowing sodium (and other cations) to enter, thus depolarizing the cell. Exposure to light leads to a decrease in cGMP levels, and receptor closing of the channels, and receptor hyperpolarization.

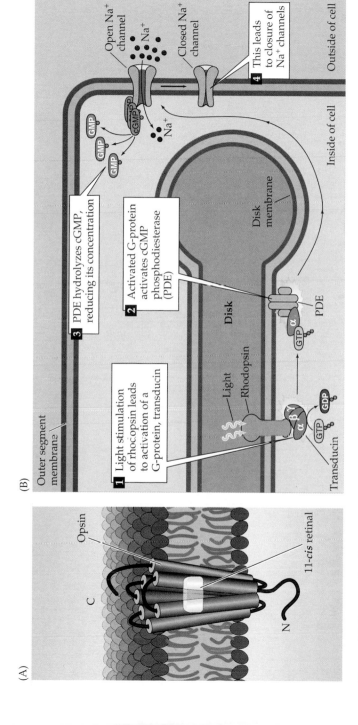

Figure 11.7 Details of phototransduction in rod photoreceptors. (A) The molecular structure of rhodopsin, the pigment in rods. (B) The second messenger cascade of phototransduction. Light stimulation of rhodopsin in the receptor disks leads to the activation of a G-protein (transducin), which in turn activates a phosphodiesterase (PDE). The phosphodiesterase hydrolyzes cGMP, reducing its concentration in the outer segment and leading to the closure of sodium channels in the outer segment membrane.

molecules that are available for binding to the channels in the surface of the outer segment membrane, leading to channel closure.

One of the important features of this complex biochemical cascade initiated by photon capture is that it provides enormous signal amplification. It has been estimated that a single light-activated rhodopsin molecule can activate 800 transducin molecules, roughly eight percent of the molecules on the disk surface. Although each transducin molecule activates only one phosphodiesterase molecule, each of these is in turn capable of catalyzing the breakdown of as many as six cGMP molecules. As a result, the absorption of a single photon by a rhodopsin molecule results in the closure of approximately 200 ion channels, or about 2% of the number of channels in each rod that are open in the dark. This number of channel closures causes a net change in the membrane potential of about 1 mV.

Equally important is the fact that the magnitude of this amplification varies with the prevailing levels of illumination, a phenomenon known as **light adaptation**. At low levels of illumination, photoreceptors are the most sensitive to light. As levels of illumination increase, sensitivity decreases, preventing the receptors from saturating and thereby greatly extending the range of light intensities over which they operate. The concentration of Ca^{2+} in the outer segment appears to play a key role in the light-induced modulation of photoreceptor sensitivity. The cGMP-gated channels in the outer segment are permeable to both Na^+ and Ca^{2+}; thus, light-induced closure of these channels leads to a net decrease in the internal Ca^{2+} concentration. This decrease triggers a number of changes in the phototransduction cascade, all of which tend to reduce the sensitivity of the receptor to light. For example, the decrease in Ca^{2+} increases the activity of guanylate cyclase, the cGMP synthesizing enzyme, leading to an increase in cGMP levels. Likewise, the decrease in Ca^{2+} increases the affinity of the cGMP-gated channels for cGMP, reducing the impact of the light-induced reduction of cGMP levels. The regulatory effects of Ca^{2+} on the phototransduction cascade are only one part of the mechanism that adapts retinal sensitivity to background levels of illumination; another important contribution comes from neural interactions between horizontal cells and photoreceptor terminals.

Once initiated, additional mechanisms limit the duration of this amplifying cascade and restore the various molecules to their inactivated states. The protein **arrestin**, for instance, blocks the ability of activated rhodopsin to activate transducin, and facilitates the breakdown of activated rhodopsin. The all-*trans* retinal then dissociates from the opsin, diffuses into the cytosol of the outer segment, and is transported out of the outer segment and into the pigment epithelium, where appropriate enzymes ultimately convert it to 11-*cis* retinal. After it is transported back into the outer segment, the 11-*cis* retinal recombines with opsin in the receptor disks. The recycling of rhodopsin is critically important for maintaining the light sensitivity of photoreceptors. Even under intense levels of illumination, the rate of regeneration is sufficient to maintain a significant number of active photopigment molecules.

Functional Specialization of the Rod and Cone Systems

The two types of photoreceptors, rods and cones, are distinguished by shape (from which they derive their names), the type of photopigment they contain, distribution across the retina, and pattern of synaptic connections (Figure 11.8). These properties reflect the fact that the rod and cone systems (the receptors and their connections within the retina) are specialized for different aspects of vision. The rod system has very low spatial resolution but is extremely sensitive to light; it is therefore specialized for sensitivity at the expense of resolution. Conversely, the cone system has very high spatial resolution but is relatively

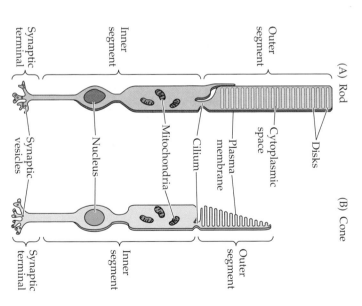

Figure 11.8 Structural differences between rods and cones. Although generally similar in structure, rods (A) and cones (B) differ in their size and shape, as well as in the arrangement of the membranous disks in their outer segments.

(A) Rod

Disks
Cytoplasmic space
Outer segment
Plasma membrane
Cilium
Mitochondria
Inner segment
Nucleus
Synaptic vesicles
Synaptic terminal

(B) Cone

Outer segment
Inner segment
Synaptic terminal

insensitive to light; it is therefore specialized for acuity at the expense of sensitivity. The properties of the cone system also allow us to see color.

The range of illumination over which the rods and cones operate is shown in Figure 11.9. At the lowest levels of light, only the rods are activated. Such rod-mediated perception is called **scotopic vision**. The difficulty of making visual discriminations under very low light conditions where only the rod system is active is obvious. The problem is primarily the poor resolution of the rod system (and, to a lesser degree, the fact that there is no perception of color in dim light because the cones are not involved to a significant degree). Although cones begin to contribute to visual perception at about the level of starlight, spatial discrimination is still very poor. As illumination increases, cones become more and more dominant in determining what is seen, and they are the major determinant of perception under relatively bright conditions such as normal indoor lighting or sunlight. The contributions of rods to vision drops out nearly entirely in so-called **photopic vision** because their response to light saturates—that is, the membrane potential of individual rods no longer varies as a function of illumination because all of the membrane channels are closed (see Figure 11.5). **Mesopic vision** occurs in levels of light at which both rods and cones contribute—at twilight, for example. From these considerations it should be clear that most of what we think of as "seeing" is mediated by the cone system, and that loss of cone function is devastating, as occurs in elderly individuals suffering from macular degeneration (Box C). Individuals who have lost cone function are legally blind, whereas those who have lost rod function only experience difficulty seeing at low levels of illumination (night blindness; see Box B).

Differences in the transduction mechanisms of the two receptor types also contribute to the ability of rods and cones to respond to different ranges of light intensity. For example, rods produce a reliable response to a single photon of light, whereas more than 100 photons are required to produce a comparable response in a cone. It is not, however, that cones fail to effectively capture photons. Rather, the change in current produced by single photon capture in cones is comparatively small and difficult to distinguish from noise. Another difference is that the response of an individual cone does not saturate at high levels of steady illumination, as does the rod response. Although both rods and cones adapt to operate over a range of luminance values, the adaptation mechanisms of the cones are more effective. This difference in adaptation is apparent in the time course of the response of rods and cones to light flashes. The response of a cone, even to a bright light flash that produces the maximum change in photoreceptor current, recovers in about 200 milliseconds, more than four times faster than rod recovery.

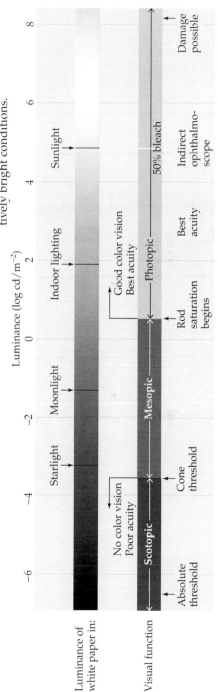

Figure 11.9 The range of luminance values over which the visual system operates. At the lowest levels of illumination, only rods are activated. Cones begin to contribute to perception at about the level of starlight and are the only receptors that function under relatively bright conditions.

BOX C
Macular Degeneration

An estimated six million Americans suffer from a condition known as age-related macular degeneration (AMD) which causes a progressive loss of central vision. Since central vision is critical for most visual tasks, diseases that affect the macula (see Figure 12.1) severely limit the ability to perform visual tasks. Indeed, AMD is the most common cause of vision loss in people over age 55, and the incidence continues to rise due to the increasing percentage of elderly individuals in the population.

The underlying problem, which remains poorly understood, is degeneration of the photoreceptors. Usually, patients first notice a blurring of central vision when performing tasks that require detailed vision such as reading. Images may also appear distorted. A graph paper–like chart, known as the Amsler grid, is actually used as a simple test for early signs of AMD. By focusing on a marked spot in the middle of the grid, the patient can assess whether the parallel and perpendicular lines on the grid appear blurred or distorted. Blurred central vision may progress to having blind spots within central vision. In most cases, both eyes are eventually involved. The risk of developing AMD increases with age, but the causes of the disease are not known. Various studies have implicated hereditary factors, cardiovascular disease, environmental factors such as smoking and light exposure, and nutritional causes; indeed, it may be that all these contribute to the risk of developing AMD.

Macular degeneration is broadly divided into two types. In the exudative-neovascular form, or "wet" AMD, which accounts for 10% of all cases, abnormal blood vessel growth occurs under the macula. These blood vessels leak fluid and blood into the retina and thus cause

damage to the photoreceptors. Wet AMD tends to progress rapidly and can cause severe damage; rapid loss of central vision may occur over just a few months. The treatment for this form of disease is laser therapy. By transferring thermal energy, the laser beam destroys the leaky blood vessels under the macula, which slows the rate of vision loss. A disadvantage of this approach, however, is that the high thermal energy delivered by the beam also destroys healthy tissue nearby. An improvement in the laser treatment of AMD involves a light-activated drug to target abnormal blood vessels. Once the drug is administered, low energy laser pulses aimed at the abnormal blood vessels are delivered to stimulate the drug, which in turn destroys the abnormal blood vessels with minimal damage to the surrounding tissue.

The remaining 90% of AMD cases are the nonexudative, or "dry" form. In these patients there is a gradual disappearance of the retinal pigment epithelium (RPE), resulting in circumscribed areas of atrophy. Since photoreceptor loss follows the disappearance of RPE, the affected retinal areas have little or no visual function. Vision loss from dry AMD occurs more gradually over the course of many years. These patients usually retain some central vision, although the loss can be severe enough to compromise performance of tasks that require seeing details. Unfortunately, there is no treatment for dry AMD.

Occasionally, macular degeneration occurs at a much earlier age. Many of these cases are caused by genetic mutations. There are many forms of hereditary macular degeneration, each with its own clinical manifestations and genetic cause. The most common form of juvenile macular degeneration is known as Stargardt disease, which is inherited as

an autosomal recessive. Patients are usually diagnosed under the age of 20. Although the progression of vision loss is variable, most of these patients are legally blind by age 50. Mutations that cause Stargardt disease have been identified in the ABCR gene, which codes for a protein that transports retinoids across the photoreceptor membrane. Thus, the visual cycle of photopigment regeneration may be disrupted in this form of macular degeneration, presumably by the dysfunctional proteins encoded by the abnormal gene. Interestingly, the ABCR gene is expressed only in rods, suggesting that the cones may have their own visual cycle enzymes.

References

SARKS, S. H. AND SARKS, J. P. (1994) Age-related macular degeneration—atrophic form. In *Retina*, 2nd Ed., Volume 2: *Medical Retina*. S. J. Ryan, A. P. Schachat and R. P. Murphy (eds.), St. Louis, MO: Mosby Year Book, pp. 1071–1102.

ELMAN, M. J. AND S. L. FINE (1994) Exudative age-related macular degeneration. In *Retina*, 2nd Ed., Volume 2: *Medical Retina*. S. J. Ryan, A. P. Schachat and R. P. Murphy (eds.), St. Louis, MO: Mosby Year Book, pp. 1186–1240.

DEUTMAN, A. F. (1994) Macular dystrophies. In *Retina*, 2nd Ed., Volume 2: *Medical Retina*. S. J. Ryan, A. P. Schachat and R. P. Murphy (eds.), St. Louis, MO: Mosby Year Book, pp. 1103–114.

The Foundation Fighting Blindness of Hunt Valley, MD, maintains a web site that provides updated information about many forms of retinal degeneration: **www.blind-ness.org**

RetNet provides updated information, including references to original articles, on genes and mutations associated with retinal diseases: **www.sph.uth.tmc.edu/RetNet**

The arrangement of the circuits that transmit rod and cone information to retinal ganglion cells also contributes to the different characteristics of scotopic and photopic vision. In most parts of the retina, rod and cone signals converge on the same ganglion cells; i.e., individual ganglion cells respond to both rod and cone inputs, depending on the level of illumination. The early stages of the pathways that link rods and cones to ganglion cells, however, are largely independent. For example, the pathway from rods to ganglion cells involves a distinct class of bipolar cell (called rod bipolar) that, unlike cone bipolar cells, does not contact retinal ganglion cells. Instead, rod bipolar cells synapse with the dendritic processes of a specific class of amacrine cell that makes gap junctions and chemical synapses with the terminals of cone bipolars; these processes, in turn, make synaptic contacts on the dendrites of ganglion cells in the inner plexiform layer.

Finally, the rod and cone systems differ dramatically in their degree of convergence, a factor that contributes greatly to their distinct properties. Each rod bipolar cell is contacted by a number of rods, and many rod bipolar cells contact a given amacrine cell. In contrast, the cone system is much less convergent. Thus, each retinal ganglion cell that dominates central vision (called midget ganglion cells) receives input from only one cone bipolar cell, which, in turn, is contacted by a single cone. Convergence makes the rod system a better detector of light, because small signals from many rods are pooled to generate a large response in the bipolar cell. At the same time, convergence reduces the spatial resolution of the rod system, since the source of a signal in a rod bipolar cell or retinal ganglion cell could have come from anywhere within a relatively large area of the retinal surface. The one-to-one relationship of cones to bipolar and ganglion cells is, of course, just what is required to maximize acuity.

Anatomical Distribution of Rods and Cones

The distribution of rods and cones across the surface of the retina also has important consequences for vision (Figure 11.10). Despite the fact that perception in typical daytime light levels is dominated by cone-mediated vision, the total number of rods in the human retina (91 million) far exceeds the number of cones (roughly 4.5 million). As a result, the density of rods is much greater

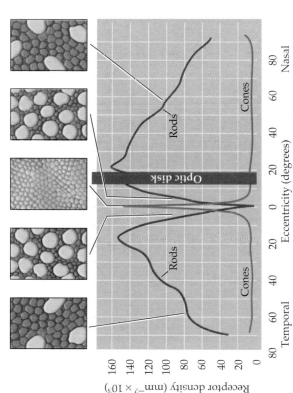

Figure 11.10 Distribution of rods and cones in the human retina. Graph illustrates that cones are present at a low density throughout the retina, with a sharp peak in the center of the fovea. Conversely, rods are present at high density throughout most of the retina, with a sharp decline in the fovea. Boxes at top illustrate the appearance of cross sections through the outer segments of the photoreceptors at different eccentricities. The increased density of cones in the fovea is accompanied by a striking reduction in the diameter of their outer segments.

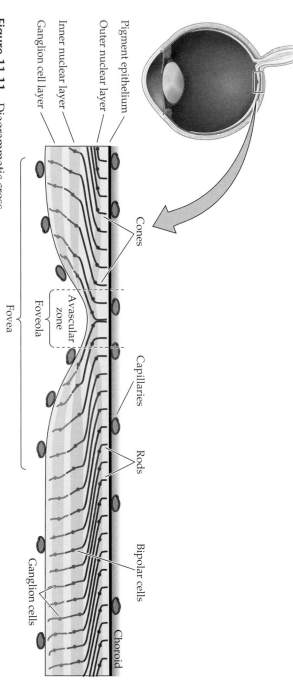

Figure 11.11 Diagrammatic cross section through the human fovea. The overlying cellular layers and blood vessels are displaced so that light rays are subject to a minimum of scattering before they strike the outer segments of the cones in the center of the fovea, called the foveola.

Labels: Pigment epithelium · Outer nuclear layer · Inner nuclear layer · Ganglion cell layer · Cones · Fovea · Foveola · Avascular zone · Capillaries · Rods · Bipolar cells · Ganglion cells · Choroid

than cones throughout most of the retina. However, this relationship changes dramatically in the **fovea**, a highly specialized region of the central retina that measures about 1.2 millimeters in diameter (Figure 11.11). In the fovea, cone density increases almost 200-fold, reaching, at its center, the highest receptor packing density anywhere in the retina. This high density is achieved by decreasing the diameter of the cone outer segments such that foveal cones resemble rods in their appearance. The increased density of cones in the fovea is accompanied by a sharp decline in the density of rods. In fact, the central 300 μm of the fovea, called the **foveola**, is totally rod-free.

The extremely high density of cone receptors in the fovea, and the one-to-one relationship with bipolar cells and retinal ganglion cells (see earlier), endows this region (and the cone system generally) with the capacity to mediate high visual acuity. As cone density declines with eccentricity and the degree of convergence onto retinal ganglion cells increases, acuity is markedly reduced. Just 6° eccentric to the line of sight, acuity is reduced by 75%, a fact that can be readily appreciated by trying to read the words on any line of this page beyond the word being fixated on. The restriction of highest acuity vision to such a small region of the retina is the main reason humans spend so much time moving their eyes (and heads) around—in effect directing the foveas of the two eyes to objects of interest (see Chapter 20). It is also the reason why disorders that affect the functioning of the fovea have such devastating effects on sight (see Box C). Conversely, the exclusion of rods from the fovea, and their presence in high density away from the fovea, explain why the threshold for detecting a light stimulus is lower outside the region of central vision. It is easier to see a dim object (such as a faint star) by looking away from it, so that the stimulus falls on the region of the retina that is richest in rods (see Figure 11.10).

Another anatomical feature of the fovea (which literally means "pit") that contributes to the superior acuity of the cone system is that the layers of cell bodies and processes that overlie the photoreceptors in other areas of the retina are displaced around the fovea, and especially the foveola (see Figure 11.11). As a result, light rays are subjected to a minimum of scattering before they strike the photoreceptors. Finally, another potential source of optical distortion that lies in the light path to the receptors—the retinal blood vessels—are diverted away from the foveola. This central region of the fovea is therefore dependent on the underlying choroid and pigment epithelium for oxygenation and metabolic sustenance.

Cones and Color Vision

A special property of the cone system is color vision. Perceiving color allows humans (and many other animals) to discriminate objects on the basis of the distribution of the wavelengths of light that they reflect to the eye. While differences in luminance are often sufficient to distinguish objects, color adds another perceptual dimension that is especially useful when differences in luminance are subtle or nonexistent. Color obviously gives us a quite different way of perceiving and describing the world we live in.

Unlike rods, which contain a single photopigment, there are three types of cones that differ in the photopigment they contain. Each of these photopigments has a different sensitivity to light of different wavelengths, and for this reason are referred to as "blue," "green," and "red," or, more appropriately, short (S), medium (M), and long (L) wavelength cones, terms that more or less describe their spectral sensitivities (Figure 11.12). This nomenclature implies that individual cones provide color information for the wavelength of light that excites them best. In fact, individual cones, like rods, are entirely color blind in that their response is simply a reflection of the number of photons they capture, regardless of the wavelength of the photon (or, more properly, its vibrational energy). It is impossible, therefore, to determine whether the change in the membrane potential of a particular cone has arisen from exposure to many photons at wavelengths to which the receptor is relatively insensitive, or fewer photons at wavelengths to which it is most sensitive. This ambiguity can only be resolved by *comparing* the activity in different classes of cones. Based on the responses of individual ganglion cells, and cells at higher levels in the visual pathway (see Chapter 12), comparisons of this type are clearly involved in how the visual system extracts color information from spectral stimuli. Despite these insights, understanding of the neural mechanisms that underlie color perception has been elusive (Box D).

Much additional information about color vision has come from studies of individuals with abnormal color detecting abilities. Color vision deficiencies result either from the inherited failure to make one or more of the cone pig-

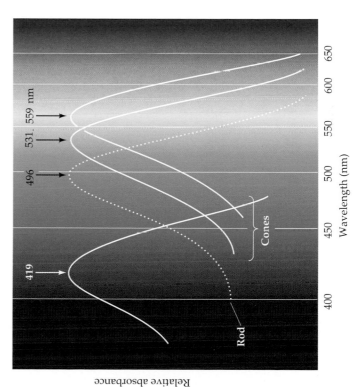

Figure 11.12 Color vision. The absorption spectra of the four photopigments in the normal human retina. The solid curves indicate the three kinds of cone opsins; the dashed curve shows rod rhodopsin for comparison. Absorbance is defined as the log value of the intensity of incident light divided by intensity of transmitted light.

BOX D
The Importance of Context in Color Perception

Seeing the luminance of objects (that is, their brightness) can presumably be signaled by simply increasing or decreasing the overall firing rate of the relevant retinal ganglion cells, properly adapted to the overall level of ambient light (see text and Box E). Seeing color, however, logically demands that retinal responses to different wavelengths in some way be *compared*. The discovery of three human cone types and their different absorption spectra is correctly regarded, therefore, as the basis for human color vision. Nevertheless, how the three human cone types and the higher-order neurons they contact (see Chapter 12) produce the sensations of color is still unclear. Indeed, this issue has been debated by some of the greatest minds in science (Hering, Helmholtz, Maxwell, Schroedinger, Mach, and Land, to name only a few) since Thomas Young first proposed that humans must have three different receptive "particles," i.e., cone types. A fundamental problem has been that, although the relative activities of three cone types

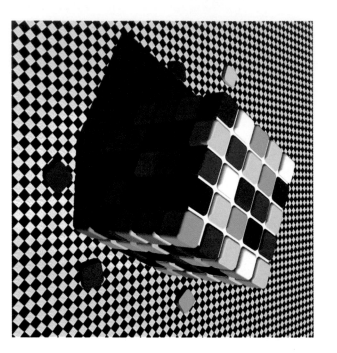

can more or less explain the colors perceived in color matching experiments performed in the laboratory, the perception of color is strongly influenced by context. For example, a patch returning the exact same spectrum of wavelengths to the eye can appear quite different depending on its surround, a phenomenon called *color contrast* (see figure). Moreover, test patches returning different spectra to the eye can appear to be the same color, an effect called *color constancy*. Although these phenomena were well known in the nineteenth century, they were not accorded a central place in color vision theory until Edwin Land's work in the 1950s. In his most famous demonstration, Land (who among other achievements founded the Polaroid company and became a billionaire) used a collage of colored papers that have been referred to as "the Land Mondrians" because of their similarity to the work of the Dutch artist Piet Mondrian. Using a telemetric photometer and three adjustable illuminators generating short,

middle, and long wavelength light, Land showed that two patches that in white light appeared quite different in color (e. g., green and brown) continued to look their respective colors even when the three illuminators were adjusted so that the light being returned from the "green" surfaces produced exactly the same readings on the three telephotometers as had previously come from the "brown" surface—a striking demonstration of color constancy!

The phenomena of color contrast and color constancy have led to a heated debate about how color percepts are generated that now spans several decades. For Land, the answer lay in a series of ratiometric equations that could integrate the spectral returns of different regions over the entire scene. It was recognized even before Land's death in 1991, however, that his so-called retinex theory did not work in all circumstances and was in any event a description rather than an explanation. An alternative explanation of these contextual aspects of color vision is that color, like brightness, is generated empirically according to what spectral stimuli have typically signified (see Box E).

The brown tile at the center of the illuminated upper face of the cube and the orange tile at the center of the shadowed face are actually returning the same spectral light to the eye (as is the tan tile lying on the ground-plane in the foreground). Readers who find this hard to believe can convince themselves by cutting holes in a sheet of paper such that the rest of the scene is masked out, in which case the two tiles on the faces of the cube look identical in both color and brightness. This illustration provides a dramatic example of the influence of context on the color perceived. (From Lotto and Purves, 1999.)

References

LAND, E. (1986) Recent advances in Retinex theory. *Vis. Res.* 26: 7–21.

LOTTO, R. B. AND D. PURVES (1999) The effects of color on brightness. *Nature Neurosci.* 2(11): 1010–1014.

ments or from an alteration in the absorption spectra of cone pigments (or, rarely, from lesions in the central stations that process color information; see Chapter 12). Under normal conditions, most people can match any color in a test stimulus by adjusting the intensity of three superimposed light sources generating long, medium, and short wavelengths. The fact that only three such sources are needed to match (nearly) all the perceived colors is strong confirmation of the fact that color sensation is based on the relative levels of activity in three sets of cones with different absorption spectra. That color vision is **trichromatic** was first recognized by Thomas Young at the beginning of the nineteenth century (thus, people with normal color vision are called *trichromats*). For about 5–6% of the male population in the United States and a much smaller percentage of the female population, however, color vision is more limited. Only two colors of light are needed to match all the colors that these individuals can perceive; the third color category is simply not seen. Such **dichromacy**, or "color blindness" as it is commonly called, is inherited as a recessive, sex-linked characteristic and exists in two forms: *protanopia*, in which all color matches can be achieved by using only green and blue light, and *deuteranopia*, in which all matches can be achieved by using only blue and red light. In another major class of color deficiencies, all three light sources (i.e., short, medium, and long wavelengths) are needed to make all possible color matches, but the matches are made using values that are significantly different from those used by most individuals. Some of these *anomalous trichromats* require more red than normal to match other colors (protanomalous trichromats); others require more green than normal (deuteranomalous trichromats).

Jeremy Nathans and his colleagues at Johns Hopkins University have provided a deeper understanding of these color vision deficiencies by identifying and sequencing the genes that encode the three human cone pigments (Figure 11.13). The genes that encode the red and green pigments show a high degree of sequence homology and lie adjacent to each other on the X chromosome, thus explaining the prevalence of color blindness in males. In contrast, the blue-sensitive pigment gene is found on chromosome 7 and is quite different in its amino acid sequence. These facts suggest that the red and green pigment genes evolved relatively recently, perhaps as a result of the duplication of a single ancestral gene; they also

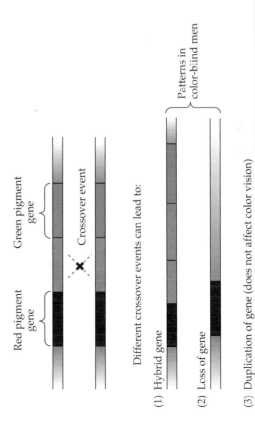

Figure 11.13 Many deficiencies of color vision are the result of genetic alterations in the red or green cone pigments due to the crossing over of chromosomes during meiosis. This recombination can lead to the loss of a gene, the duplication of a gene, or the formation of a hybrid with characteristics distinct from those of normal genes.

explain why most color vision abnormalities involve the red and green cone pigments.

Human dichromats lack one of the three cone pigments, either because the corresponding gene is missing or because it exists as a hybrid of the red and green pigment genes (see Figure 11.13). For example, some dichromats lack the green pigment gene altogether, while others have a hybrid gene that is thought to produce a red-like pigment in the "green" cones. Anomalous trichromats also possess hybrid genes, but these elaborate pigments whose spectral properties lie between those of the normal red and green pigments. Thus, although most anomalous trichromats have two distinct sets of long-wavelength cones (one normal, one hybrid), there is more overlap in their absorption spectra than in normal trichromats, and thus less difference in how the two sets of cones respond to a given wavelength (with resulting anomalies in color perception).

Retinal Circuits for Detecting Differences in Luminance

Despite the esthetically pleasing nature of color vision, most of the information in visual scenes consists of spatial variations in light intensity (a black and white movie, for example, has most of the information a color version has, although it is deficient in some respects and certainly less fun to watch). How the spatial patterns of light and dark that fall on the photoreceptors are deciphered by central targets has also been a vexing problem (Box E). To understand what is accomplished by the complex neural circuits within the retina during this process, it is useful to start by considering the responses of individual retinal ganglion cells to small spots of light. Stephen Kuffler, working at Johns Hopkins University in the 1950s, pioneered this approach by characterizing the responses of single ganglion cells in the cat retina. He found that each ganglion cell responds to stimulation of a small circular patch of the retina, which defines the cell's receptive field (see Chapter 9 for discussion and definition of receptive fields). Based on these responses, Kuffler distinguished two classes of ganglion cells, "on"-center and "off"-center. Turning on a spot of light in the center of an **on-center ganglion cell** receptive field produces a burst of electrical activity (an "on response") (Figure 11.14). Turning the light on in the center of an **off-center ganglion cell** receptive field has the opposite effect: The spontaneous rate of firing decreases, and when the spot of light is turned off, the cell responds with a burst of action potentials (an "off response"). On- and off-center ganglion cells are present in roughly equal numbers. The receptive fields have overlapping distributions, so that every point on the retinal surface (that is, every part of visual space) is analyzed by several on-center and off-center ganglion cells. A rationale for having these two distinct types of retinal ganglion cells was suggested by Peter Schiller and his colleagues, who examined the effects of pharmacologically inactivating on-center ganglion cells on a monkey's ability to detect a variety of visual stimuli. After silencing on-center ganglion cells, the animals showed a deficit in their ability to detect stimuli that were brighter than the background; however, they could still see objects that were darker than the background.

These observations imply that information about increases or decreases in luminance is carried separately to the brain by the axons of these two different types of retinal ganglion cells. Having separate luminance "channels" means that changes in light intensity, whether increases or decreases, are always conveyed to the brain by an increased number of action potentials. Because ganglion cells rapidly adapt to changes in luminance, their "resting" discharge

Figure 11.14 The responses of on-center and off-center retinal ganglion cells to stimulation of different regions of their receptive fields. See text for explanation.

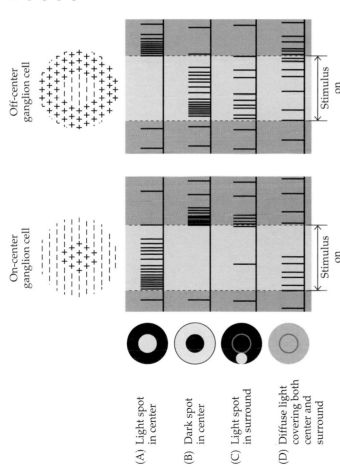

On-center ganglion cell

Off-center ganglion cell

(A) Light spot in center

(B) Dark spot in center

(C) Light spot in surround

(D) Diffuse light covering both center and surround

Stimulus on

Stimulus on

rate in constant illumination is relatively low. Although an increase in discharge rate above resting level serves as a reliable signal, a decrease in firing rate from an initially low rate of discharge might not. Thus, having luminance changes signaled by two classes of adaptable cells provides unambiguous information about both luminance increments and decrements.

The functional differences between these two ganglion cell types can be understood in terms of both their anatomy and their physiological properties and relationships. On- and off-center ganglion cells have dendrites that arborize in separate strata of the inner plexiform layer, forming synapses selectively with the terminals of bipolar cells that respond to luminance increases and decreases, respectively. As mentioned previously, the principal difference between ganglion cells and bipolar cells lies in the nature of their electrical response. Like most other cells in the retina, bipolar cells have graded potentials rather than action potentials. Graded depolarization of on-center bipolar cells leads to an increase in transmitter release at their synapses and consequent depolarization of the on-center ganglion cells that they contact. Likewise, depolarization of off-center bipolar cells leads to an increase in transmitter at their synapses and depolarization of the off-center ganglion cells that they contact.

The selective response of on- and off-center bipolar cells to light increments and decrements is explained by the fact that they express different types of glutamate receptors. Off-center bipolar cells have ionotropic (AMPA) receptors that cause the cells to depolarize in response to glutamate released from photoreceptor terminals. In contrast, on-center bipolar cells express a G-protein-coupled metabotropic glutamate receptor. When bound to glutamate, these receptors activate an intracellular cascade that closes cAMP-gated Na^+ channels, reducing inward current and hyperpolarizing the cell. Thus, glutamate has opposite effects on these two classes of cells, depolarizing off-center bipolar cells and hyperpolarizing on-center cells. In order to understand the response of on- and off-center bipolar cells to changes in light intensity, recall that photoreceptors hyperpolarize in response to light increments, decreasing

BOX E
The Perception of Luminance

Understanding the link between retinal stimulation and what we see (perception) is arguably the central problem in vision), and the perception of luminance—that is, the sensation we call brightness—is probably the simplest place to consider this problem.

As indicated in the text, how we see the brightness differences between adjacent territories with distinct luminances (i.e., contrast) depends in the first instance on the relative firing rate of retinal ganglion cells, modified by lateral interactions. However, there is a problem with the assumption that the central nervous system simply "reads out" these relative rates of ganglion cell activity to sense brightness. The difficulty, as in perceiving color, is that the brightness of a given target is markedly affected by its context in ways that are difficult or impossible to explain in terms of the retinal output as such. The accompanying figures, which illustrate two simultaneous brightness contrast illusions, help make this point. In figure A, two photometrically identical (equiluminant) gray squares look differently bright as a function of the background in which they are presented. The standard interpretation of this phenomenon is that the receptive field properties illustrated in Figures 11.14 through 11.16 cause ganglion cells to fire differently depending on whether the surround of the equiluminant target is dark or light. The demonstration in figure B, however, undermines this explanation since in this case the target surrounded by more dark area actually looks *darker* than the same target surrounded by a larger light area.

An alternative interpretation of luminance perception that accounts for these puzzling phenomena is that perceptions of brightness are generated empirically. In this conception, it is not local contrast that causes the perceived differences in

the brightness of the two identical test patches, but rather what the luminances of the test patches in relation to the luminances in the rest of the scene have typically turned out to be in the past. The gist of this explanation is illustrated in figure C. The "scene" in the standard

presentation of simultaneous brightness contrast shown in figure A is deeply ambiguous with respect to the underlying significance of the luminances in the stimulus: They could represent any of the different combinations illustrated in in figure C (i.e., two differently painted

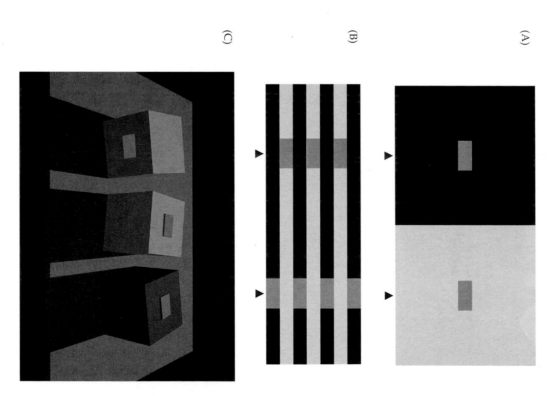

(A)

(B)

(C)

(A) Standard illusion of simultaneous brightness contrast. (B) Another illusion of simultaneous brightness contrast that is difficult to explain in conventional terms. (C) Cartoons of some possible sources of the standard simultaneous brightness contrast illusion in (A). (Courtesy of R. Beau Lotto and Dale Purves.)

surfaces in different illuminants, as in a comparison of the target patches on the left and middle cubes, or two similarly reflecting surfaces in similar amounts of light, as in a comparison of the target patches on the middle and right cubes). In consequence, an expedient—and perhaps the only—way the visual system can cope with this ambiguity (which is present in all visual stimuli) is to generate the perception of the stimulus in A (and B) empirically, i.e., based on what this particular luminance profile has turned out to signify in the past. Statistical information about the meanings of luminance profiles in representative visual scenes has never been determined.

However, the dark and light surrounds in figure A will not *always* have represented surface reflectances in an evenly illuminated scene, the only circumstance in which the equiluminant test diamonds would actually have been equiluminant objects. On the contrary, as indicated in figure C, the stimulus will often have arisen from a variety of other possible sources. As a result, the stimulus in figure A is seen not as the luminance profile it actually represents, but as a statistical construct determined by its past significance. The advantage of seeing luminance according to the relative probabilities of the possible sources of the stimulus is that percepts generated in this way give the observer the best chance of making appropriate behavioral responses to the profoundly ambiguous visual world in which we live.

References

ADELSON, E. H. (1999) Light perception and lightness illusions. In *The Cognitive Neurosciences*, 2nd Ed. M. Gazzaniga (ed.). Cambridge, MA: MIT Press, pp. 339–351.

CORNSWEET, T. (1970) *Visual Perception*. New York: Academic Press.

PURVES, D., A. SHIMPI AND R. B. LOTTO (1999) An empirical explanation of the Cornsweet effect. J. Neurosci. 19: 8542–8551.

WILLIAMS, S. M., A. N. MCCOY AND D. PURVES (1998) An empirical explanation of brightness. Proc. Natl. Acad. Sci. 95: 13301–13306.

their release of neurotransmitter. Under these conditions, on-center bipolar cells contacted by the photoreceptors are freed from the hyperpolarizing influence of the photoreceptor's transmitter, and they depolarize. In contrast, for off-center cells, the reduction in glutamate represents the withdrawal of a depolarizing influence, and these cells hyperpolarize. Decrements in light intensity naturally have the opposite effect on these two classes of bipolar cells, hyperpolarizing on-center cells and depolarizing off-center ones.

Kuffler's work also called attention to the fact that retinal ganglion cells do not act as simple "photodetectors." Indeed, most ganglion cells are relatively poor at signaling differences in the level of diffuse illumination. Instead, they are sensitive to *differences* between the level of illumination that falls on the receptive field center and the level of illumination that falls on the surround—that is, to **luminance contrast** (see Figure 11.14). The center of a ganglion cell receptive field is surrounded by a concentric region that, when stimulated, antagonizes the response to stimulation of the receptive field center. For example, as a spot of light is moved from the center of the receptive field of an on-center cell toward its periphery, the response of the cell to the spot of light decreases (Figure 11.15). When the spot falls completely outside the center (that is, in the surround), the response of the cell falls below its resting level; the cell is effectively inhibited until the distance from the center is so great that the spot no longer falls on the receptive field at all, in which case the cell returns to its resting level of firing. Off-center cells also show an antagonistic surround. Light stimulation of the surround of an off-center cell increases the firing rate of the cell, a response that opposes the decrease in firing rate that occurs when the center is stimulated (see Figure 11.14C). Because of their antagonistic surrounds, ganglion cells respond much more vigorously to small spots of light confined to their receptive field centers than to large spots, or to uniform illumination of the visual field (see Figure 11.14D).

Figure 11.15   Rate of discharge of an on-center ganglion cell to a spot of light as a function of the distance of the spot from the receptive field center. Zero on the *x* axis corresponds to the center; at a distance of 5°, the spot falls outside the receptive field.

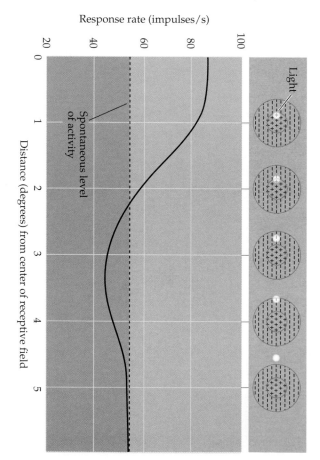

Figure 11.16   Responses of a hypothetical population of on-center ganglion cells whose receptive fields (A–E) are distributed across a light-dark edge. Those cells whose activity is most affected have receptive fields that lie along the light-dark edge.

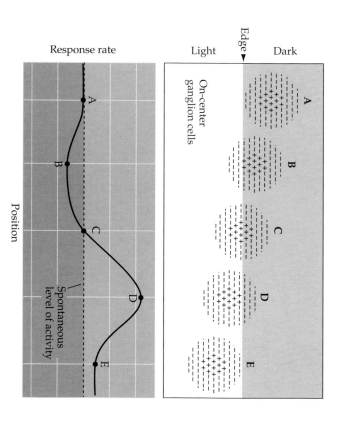

To appreciate how center-surround antagonism makes the ganglion cell sensitive to luminance contrast, consider the activity levels in a hypothetical population of on-center ganglion cells whose receptive fields are distributed across a retinal image of a light-dark edge (Figure 11.16). The neurons whose firing rates are most affected by this stimulus—either increased (neuron D) or decreased (neuron B)—are those with receptive fields that lie along the light-dark border; those with receptive fields completely illuminated (or completely darkened) are less affected (neurons A and E). Thus, the information supplied by the retina to central visual stations for further processing does not give equal weight to all regions of the visual scene; rather, it emphasizes the regions where there are differences in luminance.

Contribution of Retinal Circuits to Light Adaptation

In addition to making ganglion cells especially sensitive to light-dark borders in the visual scene, center-surround mechanisms contribute to the process of **light adaptation**. As illustrated for an on-center cell in Figure 11.17, the response rate of a ganglion cell to a small spot of light turned on in its receptive field center varies as a function of the spot's intensity. In fact, response rate is proportional to the spot's intensity over a range of about one log unit. However, the intensity of spot illumination required to evoke a given discharge rate is dependent on the background level of illumination. Increases in background level of illumination are accompanied by adaptive shifts in the cell's operating range such that greater stimulus intensities are required to achieve the same discharge rate. Thus, firing rate is not an absolute measure of light intensity, but rather signals the difference from background level of illumination.

Because the range of light intensities over which we can see is enormous compared to the narrow range of ganglion cell discharge rates (see Figure 11.9), adaptational mechanisms are essential. By scaling the ganglion cell's response to ambient levels of illumination, the entire dynamic range of a neuron's firing rate is used to encode information about intensity differences over the range of luminance values that are relevant for a given visual scene. Due to the antagonistic center-surround organization of retinal ganglion cells, the signal sent to the brain from the retina downplays the background level of illumination (see Figure 11.14). This arrangement presumably explains why the relative brightness of objects remains much the same over a wide range of lighting conditions. In bright sunlight, for example, the print on this page reflects considerably more light to the eye than it does in room light. In fact, the *print* reflects more light in sunlight than the *paper* reflects in room light; yet it continues to look black and the page white, indoors or out.

Like the mechanism responsible for generating the on- and off-center response, the antagonistic surround of ganglion cells is a product of interactions that occur at the early stages of retinal processing (Figures 11.18 and 11.19). Much of the suppression is thought to arise via lateral connections established by horizontal cells and receptor terminals. Horizontal cells receive synaptic inputs from photoreceptor terminals and are linked via gap junctions with a vast network of other horizontal cells distributed over a

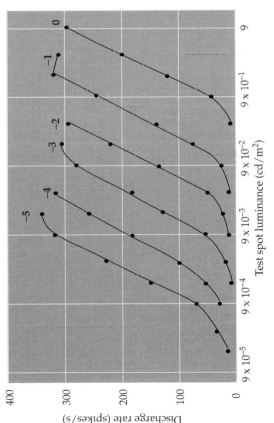

Figure 11.17 A series of curves illustrating the discharge rate of a single on-center ganglion cell to the onset of a small test spot of light in the center of its receptive field. Each curve represents the discharge rate evoked by spots of varying intensity at a constant background level of illumination, which is given by the red numbers at the top of each curve (the highest background level is 0, the lowest –5). The response rate is proportional to stimulus intensity over a range of 1 log unit, but the operating range shifts to the right as the background level of illumination increases.

Figure 11.18 Circuitry responsible for generating receptive field center responses of retinal ganglion cells. In this case, the centers of the on- and off-center ganglion cell receptive fields are illuminated without illumination of the surrounds. Light falling on the receptor leads to a reduction in transmitter release that has opposite effects on the two populations of bipolar cells: on-center bipolar cells depolarize, and off-center bipolar cells hyperpolarize. This, in turn, leads to an increase in the firing rate of on-center ganglion cells and a decrease in the firing rate of off-center ganglion cells.

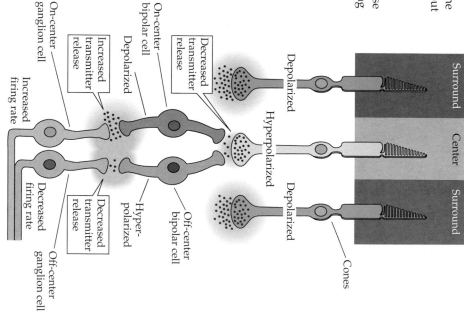

wide area of the retinal surface. As a result, the activity in horizontal cells reflects levels of illumination over a broad area of the retina. Although the details of their actions are not clear, horizontal cells exert their suppressive influence directly on photoreceptor terminals, regulating the amount of transmitter that the receptors release onto bipolar cell dendrites. The net effect of inputs from this horizontal cell network is to oppose changes in membrane potential that are induced by phototransduction events in the outer segment of the photoreceptor. Thus, the most effective stimulus for a cone is a small spot of light centered on its inner segment. Under these conditions, the cone's membrane potential is largely determined by its phototransduction cascade. As the stimulus size increases, the strength of the horizontal cell network's influence on the membrane potential of the cone increases, and there is a reduction in the light-evoked response. This reduction is reflected in changes in transmitter release onto bipolar cells and, ultimately, in the responses of retinal ganglion cells.

Thus, even at the earliest stages in visual processing, neural signals do not represent the absolute numbers of photons that are captured by receptors, but rather the relative intensity of stimulation—how much the current level of stimulation differs from ambient levels. While it may seem that the actions of horizontal cells decrease the sensitivity of the retina, they play a critical role in allowing the full range of the photoreceptor's electrical response (about 30 millivolts) to be applied to the limited range of stimulus

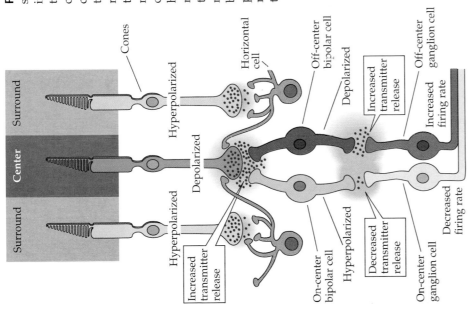

Figure 11.19 Circuitry responsible for generating the antagonistic surrounds of retinal ganglion cell receptive fields. In this case, an increase in illumination is confined to photoreceptors that contribute to the surround of the ganglion cells' receptive field; there is no change in the illumination of the photoreceptors that contribute to the center. Onset of light in the receptive field surround causes the photoreceptors in this location to hyperpolarize, reducing glutamate release onto horizontal cell processes. Horizontal cells contacting these photoreceptors hyperpolarize and decrease their rate of transmitter release onto the synaptic terminals of the photoreceptors that contribute to the receptive field center. The transmitter utilized by the horizontal cells hyperpolarizes photoreceptor terminals, and its reduction leads to a net depolarization of the photoreceptor terminals that contribute to the receptive field center. As a result, these terminals increase their rate of transmitter release onto the processes of bipolar cells causing the on-center bipolar cells they contact to hyperpolarize and the off-center bipolar cells to depolarize. Changes in the release of transmitter from the bipolar cells then lead to changes in the spike rate of retinal ganglion cells as shown.

intensities that are present at any given moment. The network mechanisms of adaptation described here function in conjunction with cellular mechanisms in the receptor outer segments that regulate the sensitivity of the phototransduction cascade at different light levels. Together, they allow retinal circuits to convey the most salient aspects of luminance changes to the central stages of the visual system described in the following chapter.

Summary

The light that falls on photoreceptors is transformed by retinal circuitry into a pattern of action potentials that ganglion cell axons convey to the visual centers in the brain. This process begins with phototransduction, a biochemical cascade that ultimately regulates the opening and closing of ion channels in the membrane of the photoreceptor's outer segment, and thereby the amount of neurotransmitter the photoreceptor releases. Two systems of photoreceptors—rods and cones—allow the visual system to meet the conflicting demands of sensitivity and acuity, respectively. Retinal ganglion cells operate quite differently from the photoreceptor cells. The center-surround arrangement of ganglion cell receptive fields makes these neurons particularly sensitive to luminance contrast and relatively insensitive to the overall level of illumination. It also allows the retina to adapt, such that it can respond effectively over the enormous range of illuminant intensities in the world. The underlying organization is generated by the synaptic interactions between photoreceptors, horizontal cells, and bipolar cells in the outer plexiform layer. As a result, the signal sent to the visual centers in the brain is already highly processed when it leaves the retina, emphasizing those aspects of the visual scene that convey the most information.

ADDITIONAL READING

Reviews

NATHANS, J. (1987) Molecular biology of visual pigments. Ann. Rev. Neurosci. 10: 163–194.

SCHNAPF, J. L. AND D. A. BAYLOR (1987) How photoreceptor cells respond to light. Sci. Am. 256 (April): 40–47.

STERLING, P. (1990) Retina. In *The Synaptic Organization of the Brain*, G. M. Shepherd (ed.). New York: Oxford University Press, pp. 170–213.

STRYER, L. (1986) Cyclic GMP cascade of vision. Ann. Rev. Neurosci. 9: 87–119.

Important Original Papers

BAYLOR, D. A., M. G. F. FUORTES AND P. M. O'BRYAN (1971) Receptive fields of cones in the retina of the turtle. J. Physiol. (Lond.) 214: 265–294.

DOWLING, J. E. AND F. S. WERBLIN (1969) Organization of the retina of the mud puppy, *Necturus maculosus*. I. Synaptic structure. J. Neurophysiol. 32: 315–338.

FASENKO, E. E., S. S. KOLESNIKOV AND A. L. LYUBARSKY (1985) Induction by cyclic GMP of cationic conductance in plasma membrane of retinal rod outer segment. Nature 313: 310–313.

KUFFLER, S. W. (1953) Discharge patterns and functional organization of mammalian retina. J. Neurophysiol. 16: 37–68.

NATHANS, J., D. THOMAS AND D. S. HOGNESS (1986) Molecular genetics of human color vision: The genes encoding blue, green and red pigments. Science 232: 193–202.

NATHANS, J., T. P. PIANTANIDA, R. EDDY, T. B. SHOWS AND D. S. HOGNESS (1986) Molecular genetics of inherited variation in human color vision. Science 232: 203–210.

SCHILLER, P. H., J. H. SANDELL AND J. H. R. MAUNSELL (1986) Functions of the "on" and "off" channels of the visual system. Nature 322: 824–825.

WERBLIN, F. S. AND J. E. DOWLING (1969) Organization of the retina of the mud puppy, *Necturus maculosus*. II. Intracellular recording. J. Neurophysiol. 32: 339–354.

Books

BARLOW, H. B. AND J. D. MOLLON (1982) *The Senses*. London: Cambridge University Press.

DOWLING, J. E. (1987) *The Retina: An Approachable Part of the Brain*. Cambridge, MA: Belknap Press.

HART, W. M. J. (ed.) (1992) *Adler's Physiology of the Eye: Clinical Application*, 9th Ed. St. Louis, MO: Mosby Year Book.

HELMHOLTZ, H. L. F. VON (1924) *Helmholtz's Treatise on Physiological Optics*, Vol. I–III. Transl. from the Third German Edition by J. P. C. Southall. Menasha, WI: George Banta Publishing Company.

HOGAN, M. J., J. A. ALVARADO AND J. E. WEDDELL (1971) *Histology of the Human Eye: An Atlas and Textbook*. Philadelphia: Saunders.

HUBEL, D. H. (1988) *Eye, Brain, and Vision*, Scientific American Library Series. New York: W.H. Freeman.

HURVICH, L. (1981) *Color Vision*. Sunderland, MA: Sinauer Associates, pp. 180–194.

OGLE, K. N. (1964) *Researches in Binocular Vision*. Hafner: New York.

OYSTER, C. (1999) *The Human Eye: Structure and Function*. Sunderland, MA: Sinauer Associates.

POLYAK, S. (1957) *The Vertebrate Visual System*. Chicago: The University of Chicago Press.

RODIECK, R. W. (1973) *The Vertebrate Retina*. San Francisco: W. H. Freeman.

RODIECK, R. W. (1998) *First Steps in Seeing*. Sunderland, MA: Sinauer Associates.

WANDELL, B. A. (1995) *Foundations of Vision*. Sunderland, MA: Sinauer Associates.

Chapter 12

Central Visual Pathways

Overview

Information supplied by the retina initiates interactions between multiple subdivisions of the brain that eventually lead to conscious perception of the visual scene, at the same time stimulating more conventional reflexes such as adjusting the size of the pupil, directing the eyes to targets of interest, and regulating homeostatic behaviors that are tied to the day/night cycle. The pathways and structures that mediate this broad range of functions are necessarily diverse. Of these, the primary visual pathway from the retina to the dorsal lateral geniculate nucleus in the thalamus and on to the primary visual cortex is the most important and certainly the most thoroughly studied component of the visual system. Different classes of neurons within this pathway encode the varieties of visual information—luminance, spectral differences, orientation, and motion—that we ultimately see. The parallel processing of different categories of visual information continues in cortical pathways that extend beyond primary visual cortex, supplying a variety of visual areas in the occipital, parietal, and temporal lobes. Visual areas in the temporal lobe are primarily involved in object recognition, whereas those in the parietal lobe are concerned with motion. Normal vision depends on the integration of information in all these cortical areas. The processes underlying visual perception are not understood and remain one of the central challenges of modern neuroscience.

Central Projections of Retinal Ganglion Cells

Ganglion cell axons exit the retina through a circular region in its nasal part called the **optic disk** (or optic papilla), where they bundle together to form the **optic nerve**. This region of the retina contains no photoreceptors and, because it is insensitive to light, produces the perceptual phenomenon known as the **blind spot** (Box A). The optic disk is easily identified as a whitish circular area when the retina is examined with an ophthalmoscope; it also is recognized as the site from which the ophthalmic artery and veins enter (or leave) the eye (Figure 12.1). In addition to being a conspicuous retinal landmark, the appearance of the optic disk is a useful gauge of intracranial pressure. The subarachnoid space surrounding the optic nerve is continuous with that of the brain; as a result, increases in intracranial pressure—a sign of serious neurological problems such as a space-occupying lesion—can be detected as a swelling of the optic disk (called papilledema).

Axons in the optic nerve run a straight course to the **optic chiasm** at the base of the diencephalon. In humans, about 60% of these fibers cross in the chiasm, while the other 40% continue toward the thalamus and midbrain

Figure 12.1 The retinal surface of the right eye, viewed with an ophthalmoscope. The optic disk is the region where the ganglion cell axons leave the retina to form the optic nerve; it is also characterized by the entrance and exit, respectively, of the ophthalmic arteries and veins that supply the retina. The macula lutea can be seen as a distinct area at the center of the optical axis (the optic disk lies nasally); the macula is the region of the retina that has the highest visual acuity. The fovea is a depression or pit about 1.5 mm in diameter that lies at the center of the macula (see Chapter 11).

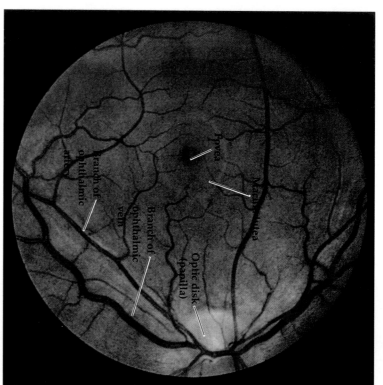

targets on the same side. Once past the chiasm, the ganglion cell axons on each side form the **optic tract**. Thus, the optic tract, unlike the optic nerve, contains fibers from *both* eyes. The partial crossing (or decussation) of ganglion cell axons at the optic chiasm allows information from corresponding points on the two retinas to be processed by approximately the same cortical site in each hemisphere, an important issue that is considered in the next section.

The ganglion cell axons in the optic tract reach a number of structures in the diencephalon and midbrain (Figure 12.2). The major target in the diencephalon is the **dorsal lateral geniculate nucleus** of the thalamus. Neurons in the lateral geniculate nucleus, like their counterparts in the thalamic relays of other sensory systems, send their axons to the cerebral cortex via the internal capsule. These axons pass through a portion of the internal capsule called the **optic radiation** and terminate in the **primary visual (or striate) cortex** (also referred to as **Brodmann's area 17 or V1**), which lies largely along and within the calcarine fissure in the occipital lobe. The **retinogeniculostriate pathway**, or **primary visual pathway**, conveys information that is essential for most of what is thought of as seeing. Thus, damage anywhere along this route results in serious visual impairment.

A second major target of the ganglion cell axons is a collection of neurons that lies between the thalamus and the midbrain in a region known as the **pretectum**. Although small in size compared to the lateral geniculate nucleus, the pretectum is particularly important as the coordinating center for the **pupillary light reflex** (i.e., the reduction in the diameter of the pupil that occurs when sufficient light falls on the retina) (Figure 12.3). The initial component of the pupillary light reflex pathway is a bilateral projection from the retina to the pretectum. Pretectal neurons, in turn, project to the

Figure 12.2 Central projections of retinal ganglion cells. Ganglion cell axons terminate in the lateral geniculate nucleus of the thalamus, the superior colliculus, the pretectum, and the hypothalamus. For clarity, only the crossing axons of the right eye are shown.

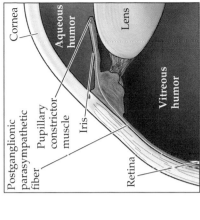

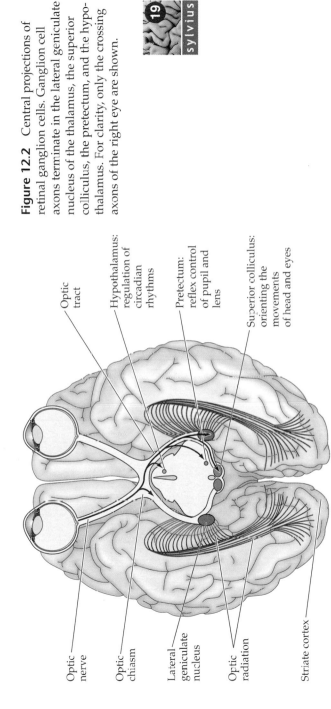

Optic
tract

Hypothalamus:
regulation of
circadian
rhythms

Pretectum:
reflex control
of pupil and
lens

Superior colliculus:
orienting the
movements
of head and eyes

Optic
nerve

Optic
chiasm

Lateral
geniculate
nucleus

Optic
radiation

Striate cortex

Edinger-Westphal nucleus, a small group of nerve cells that lies close to the nucleus of the oculomotor nerve (cranial nerve III) in the midbrain. The Edinger-Westphal nucleus contains the preganglionic parasympathetic neurons that send their axons via the oculomotor nerve to terminate on neurons in the ciliary ganglion (see Chapter 20). Neurons in the ciliary ganglion innervate the constrictor muscle in the iris, which decreases the diameter of the pupil when activated. Shining light in the eye thus leads to an increase in the activity of pretectal neurons, which stimulates the Edinger-Westphal

Figure 12.3 The circuitry responsible for the pupillary light reflex. This pathway includes bilateral projections from the retina to the pretectum and projections from the pretectum to the Edinger-Westphal nucleus. Neurons in the Edinger-Westphal nucleus terminate in the ciliary ganglion, and neurons in the ciliary ganglion innervate the pupillary constrictor muscles. Notice that the afferent axons activate both Edinger-Westphal nuclei via the neurons in the pretectum.

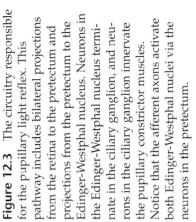

Cornea

Aqueous
humor

Lens

Iris

Vitreous
humor

Retina

Postganglionic
parasympathetic
fiber

Pupillary
constrictor
muscle

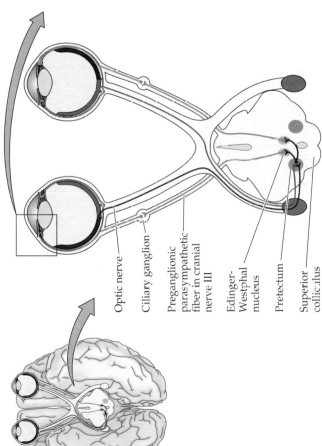

Optic nerve

Ciliary ganglion

Preganglionic
parasympathetic
fiber in cranial
nerve III

Edinger-
Westphal
nucleus

Pretectum

Superior
colliculus

BOX A
The Blind Spot

It is logical to suppose that a visual field defect (called a *scotoma*) arising from damage to the retina or central visual pathways would be obvious to the individual suffering from such pathology. When the deficit involves a peripheral region of the visual field, however, a scotoma often goes unnoticed until a car accident or some other mishap all too dramatically reveals the sensory loss. In fact, all of us have a physiological scotoma of which we are quite unaware, the so-called "blind spot." The blind spot is the substantial gap in each monocular visual field that corresponds to the location of the optic disk, the receptor-free region of the retina where the optic nerve leaves the eye (see Figure 12.1).

To find the "blind spot" of the right eye, close the left eye and fixate on the X shown in the figure here, holding the book about 30–40 centimeters away. Now take a pencil in your right hand and, without breaking fixation, move the tip slowly toward the X from the right side of the page. At some point, the tip of the pencil (indeed the whole end of the pencil) will disappear; mark this point and continue to move the pencil to the left until it reappears; then make another mark. The borders of the blind spot along the vertical axis can be determined in the same way by moving the pencil

up and down so that its path falls between the two horizontal marks. To prove that information from the region of visual space bounded by the marks is really not perceived, put a penny inside the demarcated area. When you fixate the X with both eyes and then close the left eye, the penny will disappear, seemingly magical event that amazed the French royal court when it was first reported by the natural philosopher Edmé Mariotte in 1668.

How can we be unaware of such a large defect in the visual field (typically about 5°–8°)? The optic disk is located in the nasal retina of each eye. With both eyes open, information about the corresponding region of visual space is, of course, available from the temporal retina of the other eye. But this fact does not explain why the blind spot remains undetected with one eye closed. When the world is viewed monocularly, the visual system appears to "fill-in" the missing part of the scene based on the information supplied by the regions surrounding the optic disk. To observe this phenomenon, notice what happens when a pencil or some other object lies *across* the optic disk representation. Remarkably, the pencil looks complete! Although electrophysiological recordings have shown that neurons in the visual

cortex whose receptive fields lie in the optic disk representation can be activated by stimulating the regions that surround the optic disk of the contralateral eye, suggesting that "filling-in" the blind spot is based on cortical mechanisms that integrate information from different points in the visual field, the mechanism of this striking phenomenon is not clear. Herman von Helmholtz pointed out in the nineteenth century that it may just be that this part of the visual world is ignored, the pencil being completed across the blind spot because the rest of the scene simply "collapses" around it.

References

FIORANI, M., M. G. P. ROSA, R. GATTASS AND C. E. ROCHA-MIRANDA (1992) Dynamic surrounds of receptive fields in striate cortex: A physiological basis for perceptual completion? Proc. Natl. Acad. Sci. USA 89: 8547–8551.

GILBERT, C. D. (1992) Horizontal integration and cortical dynamics. Neuron 9: 1–13.

RAMACHANDRAN, V. S. AND T. L. GREGORY (1991) Perceptual filling in of artificially induced scotomas in human vision. Nature 350: 699–702.

VON HELMHOLTZ, H. (1968) *Helmholtz's Treatise on Physiological Optics*, Vols. I–III (Translated from the 3rd German Ed. published in 1910). J. P. C. Southall (ed.), New York: Dover Publications. See pp. 204ff in Vol. III.

neurons and the ciliary ganglion neurons they innervate, thus constricting the pupil.

In addition to its normal role in regulating the amount of light that enters the eye, the pupillary reflex provides an important diagnostic tool that allows the physician to test the integrity of the visual sensory apparatus, the motor outflow to the pupillary muscles, and the central pathways that mediate the reflex. Under normal conditions, the pupils of both eyes respond identically, regardless of which eye is stimulated; that is, light in one eye produces constriction of both the stimulated eye (the direct response) and the unstimulated eye (the consensual response; see Figure 12.3). Comparing the response in the two eyes is often helpful in localizing a lesion. For example, a direct response in the left eye without a consensual response in the right eye suggests a problem with the visceral motor outflow to the right eye, possibly damage to the oculomotor nerve or Edinger-Westphal nucleus in the brainstem. Failure to elicit a response (either direct or indirect) to stimulation of the left eye if both eyes respond normally to stimulation of the right eye suggests damage to the sensory input from the left eye, possibly to the left retina or optic nerve.

There are two other important targets of retinal ganglion cell axons. One is the **suprachiasmatic nucleus** of the hypothalamus, a small group of neurons at the base of the diencephalon (see Figure 28.4). The **retinohypothalamic pathway** is the route by which variation in light levels influences the broad spectrum of visceral functions that are entrained to the day/night cycle (see Chapters 21 and 28). The other target is the **superior colliculus**, a prominent structure visible on the dorsal surface of the midbrain (see Figure 1.14). The superior colliculus coordinates head and eye movements; its functions are considered in Chapter 20.

The Retinotopic Representation of the Visual Field

The spatial relationships among the ganglion cells in the retina are maintained in their central targets as orderly representations or "maps" of visual space. Importantly, information from the left half of the visual world is represented in the right half of the brain, and vice versa.

Understanding the neural basis for this arrangement requires considering how images are projected onto the two retinas, and which parts of the two retinas cross at the optic chiasm. Each eye sees a part of visual space that defines its **visual field** (Figure 12.4A). For descriptive purposes, each retina and its corresponding visual field are divided into quadrants. In this scheme, the surface of the retina is subdivided by vertical and horizontal lines that intersect at the center of the fovea (Figure 12.4B). The vertical line divides the retina into **nasal** and **temporal divisions** and the horizontal line divides the retina into **superior** or **inferior divisions**. Corresponding vertical and horizontal lines in visual space (also called meridians) intersect at the **point of fixation** (the point in visual space that the fovea is aligned with) and define the quadrants of the visual field. The crossing of light rays diverging from different points on an object at the pupil causes the images of objects in the visual field to be inverted and left-right reversed on the retinal surface as the rays are focused. As a result, objects in the temporal part of the visual field are seen by the nasal part of the retina, and objects in the superior part of the visual field are seen by the inferior part of the retina. (It may help in understanding Figure 12.4B to imagine that you are looking at the back surfaces of the retinas, with the corresponding visual fields projected onto them.)

With both eyes open, the two foveas are normally aligned on a single target in visual space, causing the visual fields of both eyes to overlap exten-

Figure 12.4 Projection of the visual fields onto the left and right retinas. (A) Projection of an image onto the surface of the retina. The passage of light rays through the optical elements of the eye results in images that are inverted and left-right reversed on the retinal surface. (B) Retinal quadrants and their relation to the organization of monocular and binocular visual fields, as viewed from the back surface of the eyes. Vertical and horizontal lines drawn through the center of the fovea define retinal quadrants (bottom). Comparable lines drawn through the point of fixation define visual field quadrants (center). Color coding illustrates corresponding retinal and visual field quadrants. The overlap of the two monocular visual fields is shown at the top.

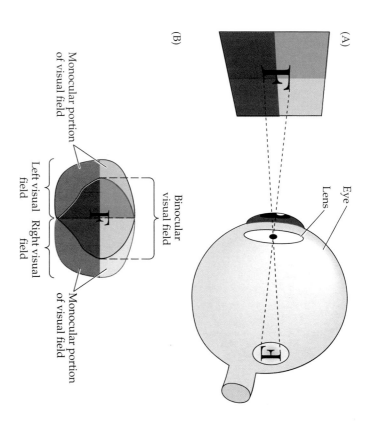

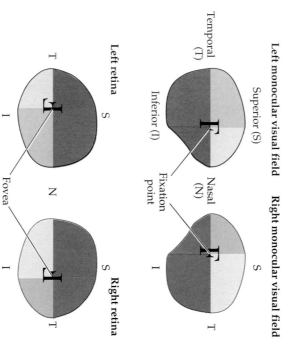

sively (see Figure 12.4B and Figure 12.5). This **binocular field** of view consists of two symmetrical visual hemifields (left and right). The left binocular hemifield includes the nasal visual field of the right eye and the temporal visual field of the left eye; the right hemifield includes the temporal visual field of the right eye and the nasal visual field of the left eye. The temporal visual fields are more extensive than the nasal visual fields, reflecting the size of the nasal and temporal retinas respectively. As a result, vision in the periphery of the field of view is strictly monocular, mediated by the most medial portion of the nasal retina. Most of the rest of the field of view can be seen by both eyes; i.e., individual points in visual space lie in the nasal visual field of the other. It is worth noting, however, that the shape of the face and nose impact the extent of this region of binocular vision. In particular, the inferior nasal visual fields are less extensive than the superior nasal fields,

Figure 12.5 Projection of the binocular field of view onto the two retinas and its relation to the crossing of fibers in the optic chiasm. Points in the binocular portion of the left visual field (B) fall on the nasal retina of the left eye and the temporal retina of the right eye. Points in the binocular portion of the right visual field (C) fall on the nasal retina of the right eye and the temporal retina of the left eye. Points that lie in the monocular portions of the left and right visual fields (A and D) fall on the left and right nasal retinas, respectively. The axons of ganglion cells in the nasal retina cross in the optic chiasm, whereas those from the temporal retina do not. As a result, information from the left visual field is carried in the right optic tract, and information from the right visual field is carried in the left optic tract.

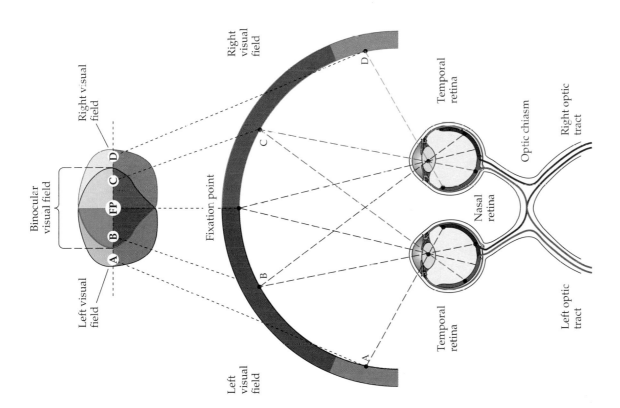

and consequently the binocular field of view is smaller in the lower visual field than in the upper (see Figure 12.4B).

Ganglion cells that lie in the nasal division of each retina give rise to axons that cross in the chiasm, while those that lie in the temporal retina give rise to axons that remain on the same side (see Figure 12.5). The boundary (or line of decussation) between contralaterally and ipsilaterally projecting ganglion cells runs through the center of the fovea and defines the border between the nasal and temporal hemiretinas. Images of objects in the left visual hemifield (such as point B in Figure 12.5) fall on the nasal retina of the left eye and the temporal retina of the right eye, and the axons from ganglion cells in these regions project through the right optic tract. Objects in the right visual hemifield (such as point C in Figure 12.5) fall on the nasal retina of the right eye and the temporal retina of the left eye; the axons from ganglion cells in these regions project through the left optic tract. As mentioned previously, objects in the monocular portions of the visual hemifields (points A and D in Figure 12.5) are seen only by the most peripheral nasal retina of each eye; the axons of ganglion cells in these regions (like the rest of the nasal retina) run in the contralateral optic tract. When the

(A)

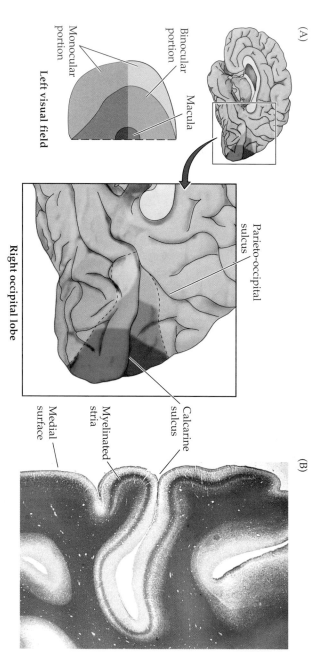

Left visual field

Binocular portion

Monocular portion

Macula

Parieto-occipital sulcus

Right occipital lobe

Calcarine sulcus

Myelinated stria

Medial surface

(B)

sylvius 20

Figure 12.6 Visuotopic organization of the striate cortex in the right occipital lobe, as seen in mid-sagittal view. (A) The primary visual cortex occupies a large part of the occipital lobe. The area of central vision (the fovea) is represented over a disproportionately large part of the caudal portion of the lobe, whereas peripheral vision is represented more anteriorly. The upper visual field is represented below the calcarine sulcus, the lower field above the calcarine sulcus. (B) Photomicrograph of a coronal section of the human striate cortex, showing the characteristic myelinated band, or stria, that gives this region of the cortex its name. The calcarine sulcus on the medial surface of the occipital lobe is indicated. (B courtesy of T. Andrews and D. Purves.)

axons in the optic tract reach the lateral geniculate nucleus, they terminate in an orderly map of the contralateral hemifield (albeit in separate right and left eye layers; see Figure 12.14).

Lateral geniculate neurons, in turn, maintain this topographic order in their projection to the striate cortex (Figure 12.6). The fovea is represented in the posterior part of the striate cortex, whereas the more peripheral regions of the retina are represented in progressively more anterior parts of the striate cortex. The upper visual field is mapped below the calcarine sulcus, and the lower visual field above it. As in the somatic sensory system, the amount of cortical area devoted to each unit area of the sensory surface is not uniform, but reflects the density of receptors and sensory axons that supply the peripheral region. Thus, like the representation of the hand region in the somatic sensory cortex, the representation of the macula is disproportionately large, occupying most of the caudal pole of the occipital lobe.

Visual Field Deficits

A variety of retinal or more central pathologies can cause visual field deficits that are limited to particular regions of visual space. Because the spatial relationships in the retinas are maintained in central visual structures, a careful analysis of the visual fields can often indicate the site of neurological damage. Relatively large visual field deficits are called **anopsias** and smaller ones are called **scotomas** (see Box A). The former term is combined with various prefixes to indicate the specific region of the visual field from which sight has been lost (Figures 12.7 and 12.8).

Damage to the retina or one of the optic nerves before it reaches the chiasm results in a loss of vision that is limited to the eye of origin. In contrast, damage in the region of the optic chiasm—or more centrally—results in specific types of deficits that involve the visual fields of both eyes (Figure 12.8). Damage to structures that are central to the optic chiasm, including the optic tract, lateral geniculate nucleus, optic radiation, and visual cortex,

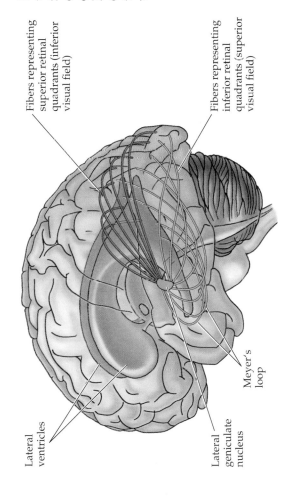

Lateral
ventricles

Fibers representing
superior retinal
quadrants (inferior
visual field)

Fibers representing
inferior retinal
quadrants (superior
visual field)

Lateral
geniculate
nucleus

Meyer's
loop

Figure 12.7 Course of the optic radiation to the striate cortex. Axons carrying information about the superior portion of the visual field sweep around the lateral horn of the ventricle in the temporal lobe (Meyer's loop) before reaching the occipital lobe. Those carrying information about the inferior portion of the visual field travel in the parietal lobe.

results in deficits that are limited to the contralateral visual hemifield. For example, interruption of the optic tract on the right results in a loss of sight in the left visual field (that is, blindness in the temporal visual field of the left eye and the nasal visual field of the right eye). Because such damage affects corresponding parts of the visual field in each eye, there is a complete loss of vision in the affected region of the binocular visual field, and the deficit is referred to as a homonymous hemianopsia (in this case, a left homonymous hemianopsia).

In contrast, damage to the optic chiasm results in visual field deficits that involve noncorresponding parts of the visual field of each eye. For example, damage to the middle portion of the optic chiasm (which is often the result of pituitary tumors) can affect the fibers that are crossing from the nasal retina of each eye, leaving the uncrossed fibers from the temporal retinas intact. The resulting loss of vision is confined to the temporal visual field of each eye and is known as bitemporal hemianopsia. It is also called heteronomous hemianopsia to emphasize that the parts of the visual field that are lost in each eye do not overlap. Individuals with this condition are able to see in both left and right visual fields, provided both eyes are open. However, all information from the most peripheral parts of visual fields (which are seen only by the nasal retinas) is lost.

Damage to central visual structures is rarely complete. As a result, the deficits associated with damage to the chiasm, optic tract, optic radiation, or visual cortex are typically more limited than those shown in Figure 12.8. This is especially true for damage along the optic radiation, which fans out under the temporal and parietal lobes in its course from the lateral geniculate nucleus to the striate cortex. Some of the optic radiation axons run out into the temporal lobe on their route to the striate cortex, an anomaly called **Meyer's loop** (see Figure 12.7). Meyer's loop carries information from the superior portion of the contralateral visual field. More medial parts of the optic radiation, which pass under the cortex of the parietal lobe, carry information from the inferior portion of the contralateral visual field. Damage to parts of the temporal lobe with involvement of Meyer's loop can thus result in a superior homonymous quadrantanopsia; damage to the optic radiation underlying the parietal cortex results in an inferior homonymous quadrantanopsia.

Injury to central visual structures can also lead to a phenomenon called *macular sparing*, i.e., the loss of vision throughout wide areas of the visual field, with the exception of foveal vision. Macular sparing is commonly

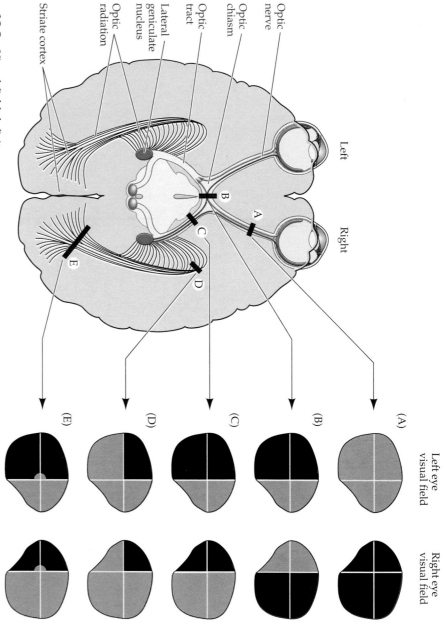

Figure 12.8 Visual field deficits resulting from damage at different points along the primary visual pathway. The diagram on the left illustrates the basic organization of the primary visual pathway and indicates the location of various lesions. The right panels illustrate the visual field deficits associated with each lesion. (A) Loss of vision in right eye. (B) Bitemporal (heteronomous) hemianopsia. (C) Left homonymous hemianopsia. (D) Left superior quadrantanopsia. (E) Left homonymous hemianopsia with macular sparing.

found with damage to the cortex, but can be a feature of damage anywhere along the length of the visual pathway. Although several explanations for macular sparing have been offered, including overlap in the pattern of crossed and uncrossed ganglion cells supplying central vision, the basis for this selective preservation is not clear.

The Functional Organization of the Striate Cortex

Much in the same way that Stephen Kuffler explored the response properties of individual retinal ganglion cells (see Chapter 11), David Hubel and Torsten Wiesel used microelectrode recordings to examine the properties of neurons in more central visual structures.

The responses of neurons in the lateral geniculate nucleus were found to be remarkably similar to those in the retina, with a center-surround receptive field organization and selectivity for luminance increases or decreases. However, the small spots of light that were so effective stimulating neurons in the retina and lateral geniculate nucleus were largely ineffective in visual cortex. Instead, most cortical neurons in cats and monkeys were found to respond vigorously to light-dark bars or edges, and only if the bars were presented at a particular range of orientations within the cell's receptive field (Figure 12.9). The responses of cortical neurons are thus tuned to the orientation of edges, much like cone receptors are tuned to the wavelength of light; the peak in the tuning curve (the orientation to which a cell is most responsive) is referred to as the neuron's preferred orientation. By sampling the responses of a large number of single cells, Hubel and Weisel demonstrated that all edge orientations were roughly equally represented in visual cortex.

(A) Experimental setup

Light bar stimulus
projected on screen

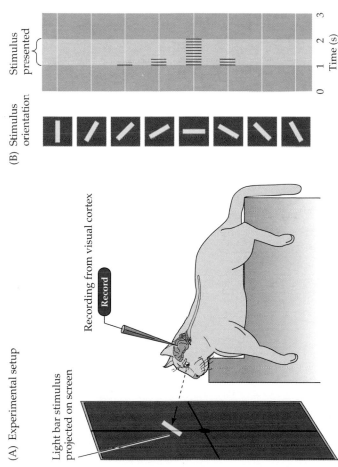

Recording from visual cortex

Record

(B) Stimulus
orientation

Stimulus
presented

0   1   2   3
Time (s)

Figure 12.9 Neurons in the primary visual cortex respond selectively to oriented edges. (A) An anesthetized animal is fitted with contact lenses to focus the eyes on a screen, where images can be projected; an extracellualr electrode records the neuronal responses. (B) Neurons in visual cortex typically respond vigorously to a bar of light oriented at a particular angle and weakly—or not at all—to other orientations.

As a result, a given orientation in a visual scene appears to be "encoded" in the activity of a distinct population of **orientation-selective neurons**.

Hubel and Wiesel also found that there are subtly different subtypes within a class of neurons that preferred the same orientation. For example, the receptive fields of some cortical cells, which they called **simple cells**, were composed of spatially separate "on" and "off" response zones, as if the "on" and "off" centers of lateral geniculate cells that supplied these neurons were arrayed in separate parallel bands. Other neurons, referred to as **complex cells**, exhibited mixed "on" and "off" responses throughout their receptive field, as if they received their inputs from a number of simple cells. Further analysis uncovered cortical neurons sensitive to the *length* of the bar of light that was moved across their receptive field, decreasing their rate of response when the bar exceeded a certain length. Still other cells responded selectively to the *direction* in which an edge moved across their receptive field. Although the mechanisms responsible for generating these selective responses are still not well understood, there is little doubt that the specificity of the receptive field properties of neurons in the striate cortex (and beyond) plays an important role in determining the basic attributes of visual scenes.

Another feature that distinguishes the responses of neurons in the striate cortex from those at earlier stages in the primary visual pathway is **binocularity**. Although the lateral geniculate nucleus receives inputs from both eyes, the axons terminate in separate layers, so that individual geniculate neurons are monocular, driven by either the left or right eye but not by both (Figure 12.10; see also Figure 12.14). In some species, including most (but not all) primates, inputs from the left and right eyes remain segregated to some degree even beyond the geniculate because the axons of geniculate neurons terminate in alternating eye-specific columns within cortical layer IV—the so-called **ocular dominance columns** (see the next section). Beyond this point, the signals from the two eyes are combined at the cellular level. Thus, most cortical neurons have binocular receptive fields, and these fields are almost identical, having the same size, shape, preferred orientation, and roughly the same position in the visual field of each eye.

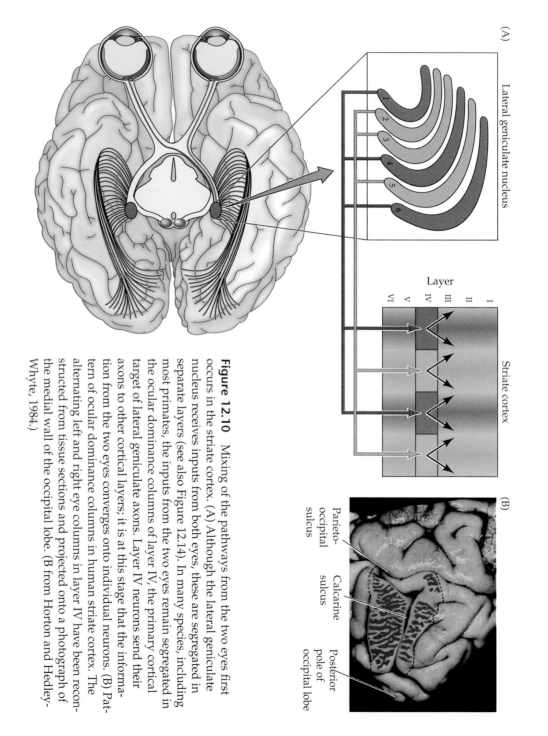

Figure 12.10 Mixing of the pathways from the two eyes first occurs in the striate cortex. (A) Although the lateral geniculate nucleus receives inputs from both eyes, these are segregated in separate layers (see also Figure 12.14). In many species, including most primates, the inputs from the two eyes remain segregated in the ocular dominance columns of layer IV, the primary cortical target of lateral geniculate axons. Layer IV neurons send their axons to other cortical layers; it is at this stage that the information from the two eyes converges onto individual neurons. (B) Pattern of ocular dominance columns in human striate cortex. The alternating left and right eye columns in layer IV have been reconstructed from tissue sections and projected onto a photograph of the medial wall of the occipital lobe. (B from Horton and Hedley-Whyte, 1984.)

Bringing together the inputs from the two eyes at the level of the striate cortex provides a basis for **stereopsis**, the special sensation of depth that arises from viewing nearby objects with two eyes instead of one. Because the two eyes look at the world from slightly different angles, objects that lie in front of or behind the plane of fixation project to noncorresponding points on the two retinas. To convince yourself of this fact, hold your hand at arm's length and fixate on the tip of one finger. Maintain fixation on the finger as you hold a pencil in your other hand about half as far away. At this distance, the image of the pencil falls on noncorresponding points on the two retinas and will therefore be perceived as two separate pencils (a phenomenon called double vision, or *diplopia*). If the pencil is now moved toward the finger (the point of fixation), the two images of the pencil fuse and a single pencil is seen in front of the finger. Thus, for a small distance on either side of the plane of fixation, where the disparity between the two views of the world remains modest, a single image is perceived; the disparity between the two eye views of objects nearer or farther than the point of fixation is interpreted as depth (Figure 12.11).

Although the neurophysiological basis of stereopsis is not understood, some neurons in the striate cortex and in other visual cortical areas have receptive field properties that make them good candidates for extracting information about binocular disparity. Unlike many binocular cells whose monocular receptive fields sample the same region of visual space, these neurons have monocular receptive fields that are slightly displaced (or perhaps differ

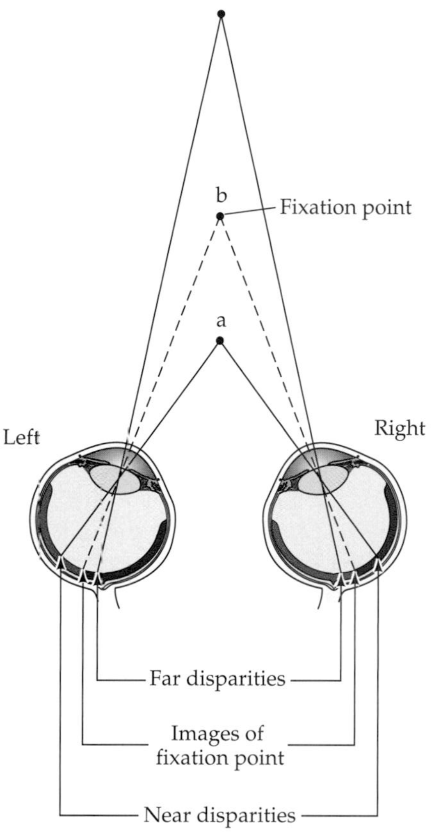

Figure 12.11 Binocular disparities are generally thought to be the basis of stereopsis. When the eyes are fixated on b, points that lie beyond the plane of fixation (point c) or in front of the point of fixation (point a) project to noncorresponding points on the two retinas. When these disparities are small, the images are fused and the disparity is interpreted by the brain as small differences in depth. When the disparities are greater, double vision occurs (although this normal phenomenon is generally unnoticed).

in their internal organization) so that the cell is maximally activated by stimuli that fall on noncorresponding parts of the retinas. Some of these neurons (so-called **far cells**) discharge to disparities beyond the plane of fixation, while others (**near cells**) respond to disparities in front of the plane of fixation. The pattern of activity in these different classes of neurons seems likely to contribute to sensations of stereoscopic depth (Box B).

Interestingly, the preservation of the binocular responses of cortical neurons is contingent on the normal activity from the two eyes during early postnatal life. Anything that creates an imbalance in the activity of the two eyes—for example, the clouding of one lens or the abnormal alignment of the eyes during infancy (strabismus)—can permanently reduce the effectiveness of one eye in driving cortical neurons, and thus impair the ability to use binocular information as a cue for depth. Early detection and correction of visual problems is therefore essential for normal visual function in maturity (see Chapter 24).

The Columnar Organization of the Striate Cortex

The variety of response properties exhibited by cortical neurons raises the question of how neurons with different receptive fields are arranged within striate cortex. For the most part, the responses of neurons are qualitatively similar at any one point in primary visual cortex, but tend to shift smoothly across its surface. With respect to orientation, for example, all the neurons encountered in an electrode penetration perpendicular to the surface at a particular point will very likely have the same orientation preference, forming a "column" of cells with similar response properties. Adjacent columns, however, usually have slightly different orientation preferences; the sequence of orientation preferences encountered along a tangential electrode penetration gradually shifts as the electrode advances (Figure 12.12). Thus, orientation preference is mapped in the cortex, much like receptive field

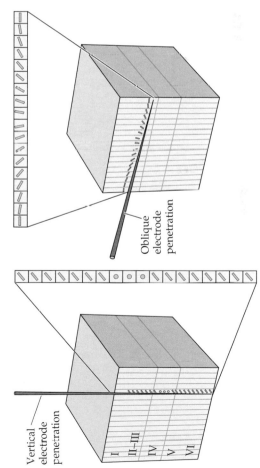

Figure 12.12 Columnar organization of orientation selectivity in the monkey striate cortex. Vertical electrode penetrations encounter neurons with the same preferred orientations, whereas oblique penetrations show a systematic change in orientation across the cortical surface. The circles denote the lack of orientation-selective cells in layer IV.

BOX B
Random Dot Stereograms and Related Amusements

An important advance in studies of stereopsis was made in 1959 when Bela Julesz, then working at the Bell Laboratories in Murray Hill, New Jersey, discovered an ingenious way of showing that stereoscopy depends on matching information seen by the two eyes without any prior recognition of what object(s) such matching might generate. Julesz, a Hungarian whose background was in engineering and physics, was working on the problem of how to "break" camouflage. He surmised that the brain's ability to fuse the slightly different views of the two eyes to bring out new information would be an aid in overcoming military camouflage. Julesz also realized that, if his hypothesis was correct, a hidden figure in a random pattern presented to the two eyes should emerge when a portion of the otherwise identical pattern was shifted horizontally in the view of one eye or the other. A horizontal shift in one direction would cause the hidden object to appear in front of the plane of the background, whereas a shift in the other direction would cause the hidden object to appear in back of the plane. Such a figure, called a random dot stereogram, and the method of its creation are shown in figures A and B. The two images can be easily fused in a stereoscope (like the familiar Viewmaster® toy) but can also be fused simply by allowing the eyes to diverge. Most people find it easiest to do this by imagining that they are looking "through" the figure; after some seconds, during which the brain tries to make sense of what it is presented with, the two images merge and the hidden figure appears (in this case, a square that occupies the middle portion of the figure). The random dot stereogram has been widely used in stereoscopic research for about 40 years, although how such stimuli elicit depth remains very much a matter of dispute.

An impressive—and extraordinarily popular—derivative of the random dot stereogram is the autostereogram (figure C). The possibility of autostereograms was first discerned by the nineteenth-century British physicist David Brewster. While staring at a Victorian wallpaper with an iterated but offset pattern, he noticed that when the patterns were fused, he perceived two different planes. The plethora of autostereograms that can be seen today in posters, books, and newspapers are close cousins of the random dot stereogram in that computers are used to shift patterns of iterated information with respect to each other. The result is that different planes emerge from what appears to be a meaningless array of visual information (or, depending on the taste of the creator, an apparently "normal" scene in which the iterated and displaced information is hidden). Some autostereograms are designed to reveal the hidden figure when the eyes diverge, and others when they converge. (Looking at a plane more distant than the plane of the surface causes divergence; looking at a plane in

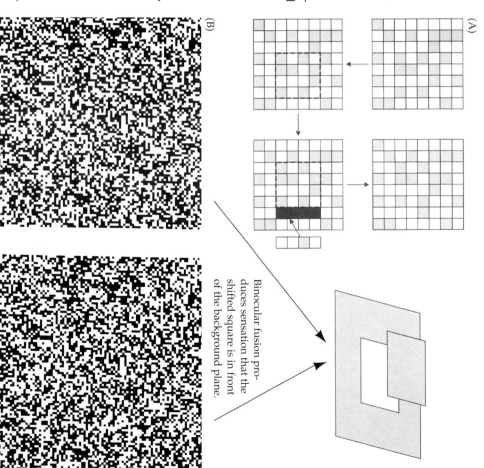

(A)

(B)

Binocular fusion produces sensation that the shifted square is in front of the background plane.

front of the picture causes the eyes to converge; see Figure 12.11.)

The elevation of the autostereogram to a popular art form should probably be attributed to Chris W. Tyler, a student of Julesz's and a visual psychophysicist, who was among the first to create commercial autostereograms. Numerous graphic artists, preeminently in Japan,

where the popularity of the autostereogram has been enormous, have generated many of such images. As with the random dot stereogram, the task in viewing the autostereogram is not clear to the observer. Nonetheless, the hidden figure emerges, often after minutes of effort in which the brain automatically tries to make sense of the occult information.

References

Julesz, B. (1971) *Foundations of Cyclopean Perception*. Chicago: The University of Chicago Press.

Julesz, B. (1995) *Dialogues on Perception*. Cambridge, MA: MIT Press.

N. E. Thing Enterprises (1993) *Magic Eye: A New Way of Looking at the World*. Kansas City: Andrews and McMeel.

(C)

Random dot stereograms and autostereograms. (A) To construct a random dot stereogram, a random dot pattern is created to be observed by one eye. The stimulus for the other eye is created by copying the first image, displacing a particular region horizontally, and then filling in the gap with a random sample of dots. (B) When the right and left images are viewed simultaneously but independently by the two eyes (by using a stereoscope or fusing the images by converging or diverging the eyes), the shifted region (a square) appears to be in a different plane from the other dots. (C) An autostereogram. The hidden figure (three geometrical forms) emerges by diverging the eyes in this case. (A after Wandell, 1995; C courtesy of Jun Oi.)

location (Box C). Unlike the map of visual space, however, the map of orientation preference is iterated many times, such that the same orientation preference is repeated at approximately 1-mm intervals across the striate cortex. This iteration presumably ensures that there are neurons for each region of visual space that represent the full range of orientation values. The orderly

BOX C
Optical Imaging of Functional Domains in the Visual Cortex

The recent availability of optical imaging techniques has made it possible to visualize how response properties, such as the selectivity for edge orientation or ocular dominance, are mapped across the cortical surface. These methods generally rely on intrinsic signals (changes in the amount of light reflected from the cortical surface) that correlate with levels of neural activity. Such signals are thought to arise at least in part from local changes in the ratio of oxyhemoglobin and deoxyhemoglobin that accompany such activity; more active areas having a higher deoxyhemoglobin/oxyhemoglo-

bin ratio (see also Box D in Chapter 1). This change can be detected when the cortical surface is illuminated with red light (605–700 nm). Under these conditions, active cortical regions absorb more light than less active ones. With the use of a sensitive video camera, and averaging over a number of trials (the changes are small, 1 or 2 parts per thousand), it is possible to visualize these differences and use them to map cortical patterns of activity (figure A).

This approach has now been successfully applied to both striate and extrastriate areas in both experimental animals

and human patients undergoing neurosurgery. The results emphasize that maps of stimulus features are a general principle of cortical organization. For example, orientation preference is mapped in a continuous fashion such that adjacent positions on the cortical surface tend to have only slightly shifted orientation preferences. However, there are points where continuity breaks down. Around these points, orientation preference is represented in a radial pattern resembling a pinwheel, covering the whole 180° of possible orientation values (figure B).

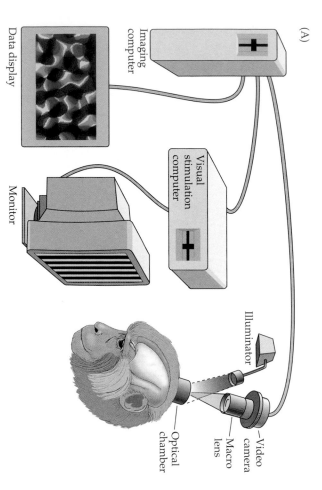

(A)

Imaging computer

Data display

Monitor

Visual stimulation computer

Illuminator

Video camera

Macro lens

Optical chamber

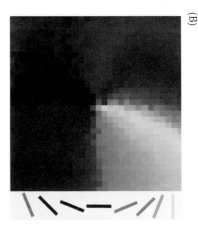

(B)

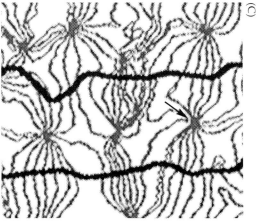

(C)

(A) The technique of optical imaging. A sensitive video camera is used to record changes in light absorption that occur as the animal views various stimuli presented on a video monitor. Images are digitized and stored in a computer in order to construct maps that compare patterns of activity associated with different stimuli. (B) Maps of orientation preference in the visual cortex visualized with optical imaging. Each color represents the angle of an edge that was most effective in activating the neurons at a given site. Orientation preference changes in a continuous fashion, rotating around pinwheel centers. (C) Comparison of optical image maps of orientation preference and ocular dominance in monkey visual cortex. The thin gray lines represent the borders between ocular dominance columns. The thick black lines represent the borders of orientation preference and ocular dominance bands at right angles. Iso-orientation contours, which converge at orientation pinwheel centers (arrow). Iso-orientation contour lines generally intersect the borders of ocular dominance bands at right angles. (B from Bonhoeffer and Grinvald, 1993; C from Obermeyer and Blasdel, 1993.)

This powerful technique can also be used to determine how maps for different stimulus properties are arranged relative to one another, and to detect additional maps such as that for direction of motion. A comparison of ocular dominance bands and orientation preference maps, for example, shows that pinwheel centers are generally located in the center of ocular dominance bands, and that the iso-orientation contours that emanate from the pinwheel centers run orthogonal to the borders of ocular dominance bands (figure C). An orderly relationship between maps of orientation selectivity and direction selectivity has also been demonstrated. These systematic relationships between the functional maps that coexist within primary visual cortex are thought to ensure that all combinations of stimulus features (orientation, direction, ocular dominance, and spatial frequency) are analyzed for all regions of visual space.

References

BLASDEL, G. G. AND G. SALAMA (1986) Voltage sensitive dyes reveal a modular organization in monkey striate cortex. Nature 321: 579–585.

BONHOEFFER, T. AND A. GRINVALD (1993) The layout of iso-orientation domains in area 18 of the cat visual cortex: Optical imaging reveals a pinwheel-like organization. J. Neurosci 13: 4157–4180.

BONHOEFFER, T. AND A. GRINVALD (1996) Optical imaging based on intrinsic signals: The methodology. In *Brain Mapping: The Methods*, A. Toge (ed.). New York: Academic Press.

OBERMAYER, K. AND G. G. BLASDEL (1993) Geometry of orientation and ocular dominance columns in monkey striate cortex. J. Neurosci. 13: 4114–4129.

WELIKY, M., W. H. BOSKING AND D. FITZPATRICK (1996) A systematic map of direction preference in primary visual cortex. Nature 379: 725–728.

progression of orientation preference (as well as other properties that are mapped in this systematic way) is accommodated within the orderly map of visual space by the fact that the mapping is relatively coarse. Each small region of visual space is represented by a set of neurons whose receptive fields cover the full range of orientation preferences, the set being distributed over several millimeters of the cortical surface.

The columnar organization of the striate cortex is equally apparent in the binocular responses of cortical neurons. Although most neurons in the striate cortex respond to stimulation of both eyes, the relative strength of the inputs from the two eyes varies from neuron to neuron. At the extremes of this continuum are neurons that respond almost exclusively to the left or right eye; in the middle are neurons that respond equally well to both eyes. As in the case of orientation preference, vertical electrode penetrations tend to encounter neurons with similar ocular preference (or **ocular dominance**, as it is usually called), whereas tangential penetrations show gradual shifts in ocular dominance. And, like the arrangement of orientation preference, a movement of about a millimeter across the surface is required to sample the full complement of ocular dominance values (Figure 12.13). These shifts in ocular dominance result from the ocular segregation of the inputs from lateral geniculate nucleus within cortical layer IV (see Figure 12.10).

Although the modular arrangement of the visual cortex was first recognized on the basis of these orientation and ocular dominance columns, further work has shown that other stimulus features such as color, direction of motion, and spatial frequency also tend to be distributed in iterated patterns that are systematically related to each other (for example, orientation columns tend to intersect ocular dominance columns at right angles). In short, the striate cortex is composed of repeating units, or modules, that contain all the neuronal machinery necessary to analyze a small region of visual space for a variety of different stimulus attributes. As described in Box D in Chapter 9, a number of other cortical regions show a similar columnar arrangement of their processing circuitry.

Figure 12.13 Columnar organization of ocular dominance. (A) Cortical neurons in all layers vary in the strength of their response to the inputs from the two eyes, from complete domination by one eye to equal influence of the two eyes. (B) Tangential electrode penetration across the superficial cortical layers reveals a gradual shift in the ocular dominance of the recorded neurons from one eye to the other. In contrast, all neurons encountered in a vertical electrode penetration (other than those neurons that lie in layer IV) tend to have the same ocular dominance.

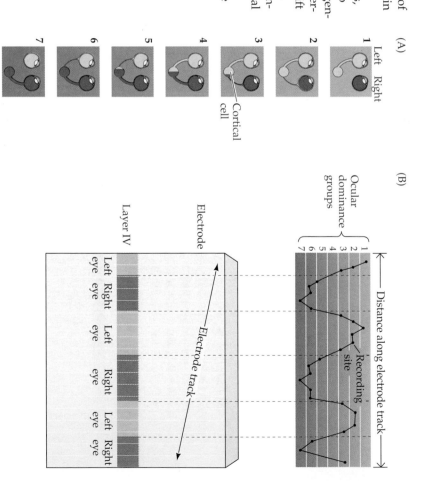

(A)

Left   Right

1

2

3   Cortical cell

4

5

6

7

(B)

Ocular dominance groups

1 2 3 4 5 6 7

Distance along electrode track

Recording site

Electrode

Electrode track

Layer IV

Left Right eye eye

Left Left eye eye

Right Right eye eye

Left Left eye eye

Right Right eye eye

Parallel Streams of Information from Retina to Cortex

The information passed on by the retina to central visual structures has been treated so far as if it were derived from a relatively uniform population of ganglion cells that differ only in sign ("on"-center or "off"-center). In fact, there are several functionally distinct populations of retinal ganglion cells, each of which has "on"- and "off"-center subtypes distributed across the surface of the retina. In primates, two that are of particular interest are called P and M ganglion cells (because of their relationship to the parvocellular and magnocellular layers of the geniculate, respectively). M ganglion cells have larger cell bodies and dendritic fields, and larger-diameter axons than P cells (Figure 12.14A). These differences are expressed in their response properties; M ganglion cells have larger receptive fields than P cells, and their axons have faster conduction velocities.

M and P cells also differ in ways that are not so obviously related to their morphology. M cells respond transiently to the presentation of visual stimuli, while P cells respond in a sustained fashion. Moreover, P ganglion cells can transmit information about color, whereas M cells cannot. P cells convey color information because their receptive field centers and surrounds are driven by different classes of cones (i.e., cones responding with greatest sensitivity to short-, medium-, or long-wavelength light). For example, some P ganglion cells have centers that receive inputs from long-wavelength ("red") cones and surrounds that receive inputs from medium-wavelength ("green") cones. Others have centers that receive inputs from "green cones" and surrounds from "red cones" (see Chapter 11). As a result, P cells are sensitive to differences in the wavelengths of light striking their receptive field center and surround. Although M ganglion cells also receive inputs from cones, there is

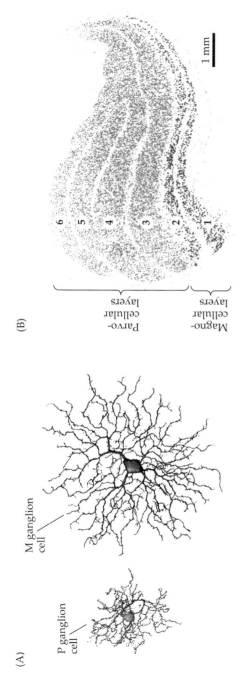

(A)

P ganglion
cell

M ganglion
cell

(B)

6
5
4
3
2
1

Parvo-
cellular
layers

Magno-
cellular
layers

1 mm

Figure 12.14 Magno- and parvocellular streams. (A) Tracings of M and P ganglion cells as seen in flat mounts of the retina after staining by the Golgi method. M cells have large-diameter cell bodies and large dendritic fields. They supply the magnocellular layers of the lateral geniculate nucleus. P cells have smaller cell bodies and dendritic fields. They supply the parvocellular layers of the lateral geniculate nucleus. (B) Photomicrograph of the human lateral geniculate nucleus showing the magnocellular and parvocellular layers. (A after Watanabe and Rodieck, 1989; B courtesy of Tim Andrews and Dale Purves.)

no difference in the type of cone input to the receptive field center and surround; the center and surround of each M cell receptive field is driven by all cone types. The absence of cone specificity to center-surround antagonism makes M cells largely insensitive to differences in the wavelengths of light that strike their receptive field centers and surrounds, and they are thus unable to transmit color information to their central targets.

M and P ganglion cells terminate in different layers of the lateral geniculate nucleus (Figure 12.14B). In addition to being specific for input from one eye or the other, the geniculate layers are also distinguished on the basis of cell size: M cells terminate selectively in the **magnocellular layers of the lateral geniculate nucleus**, while P cells terminate in the **parvocellular layers.** Because the information from the magnocellular and parvocellular layers of the LGN remains separate at least in the initial stages of cortical processing, the terms **magnocellular stream** and **parvocellular stream** are often used to signify the central pathways that convey information derived from M and P ganglion cells.

These differences in the response properties of M and P ganglion cells suggest that the magno- and parvocellular streams make different contributions to visual perception. This idea has been tested experimentally by examining the visual capabilities of monkeys after selectively damaging either the magno- or parvocellular layers of the lateral geniculate nucleus. Damage to the magnocellular layers has little effect on visual acuity or color vision, but sharply reduces the ability to perceive quickly moving stimuli. In contrast, damage to the parvocellular layers has no effect on motion perception but severely impairs visual acuity and color perception. These observations suggest that the visual information conveyed by the parvocellular stream is particularly important for high-resolution vision—the detailed analysis of the shape, size, and color of objects. The magnocellular stream, on the other hand, is particularly concerned with information about the movement of objects in space.

The Functional Organization of Extrastriate Visual Areas

Anatomical and electrophysiological studies in monkeys have led to the discovery of a multitude of areas in the occipital, parietal, and temporal lobes that are involved in processing visual information (Figure 12.15). Each of these areas contains a map of visual space, and each is largely dependent on the primary visual cortex for its activation. The response properties of the neurons in some of these areas suggest that they are specialized for different aspects of the visual scene. For example, the **middle temporal area (MT)** contains neurons that respond selectively to the direction of a moving edge without regard to its color. In contrast, neurons in another cortical area called **V4** respond selectively to the color of a visual stimulus without regard to its direction of movement. These physiological findings are supported by behavioral evidence; thus, damage to area MT leads to a specific impairment in a monkey's ability to perceive the direction of motion in a stimulus pattern, while other aspects of visual perception remain intact.

Recent functional imaging studies have indicated a similar arrangement of visual areas within human extrastriate cortex. Using retinotopically restricted stimuli, it has been possible to localize at least 10 separate representations of the visual field (Figure 12.16). One of these areas exhibits a large motion-selective signal, suggesting that it is the homologue of the motion-selective middle temporal area described in monkeys. Another area exhibits color-selective responses, suggesting that it may be similar to V4 in

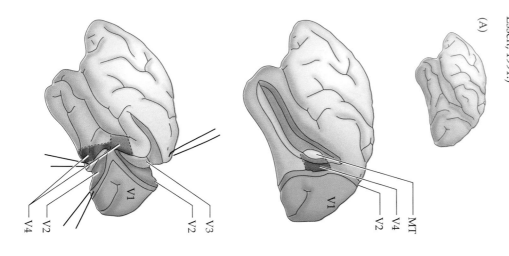

(A)

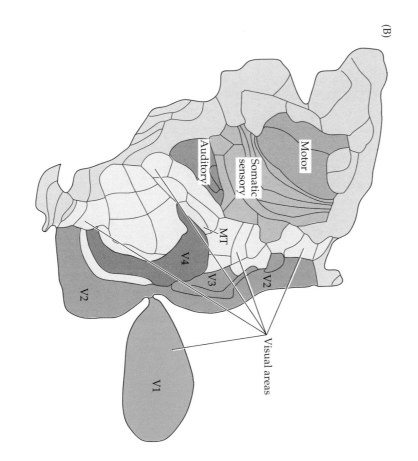

(B)

Figure 12.15 Subdivisions of the extrastriate cortex in the macaque monkey. (A) Each of the subdivisions indicated in color contains neurons that respond to visual stimulation. Many are buried in sulci, and the overlying cortex must be removed in order to expose them. Some of the more extensively studied extrastriate areas are specifically identified (V2, V3, V4, and MT). V1 is the primary visual cortex; MT is the middle temporal area. (B) The arrangement of extrastriate and other areas of neocortex in a flattened view of the monkey neocortex. There are at least 25 areas that are predominantly or exclusively visual in function, plus 7 other areas suspected to play a role in visual processing. (A after Maunsell and Newsome, 1987; B after Felleman and Van Essen, 1991.)

nonhuman primates. A role for these areas in the perception of motion and color, respectively, is further supported by evidence for increases in activity not only during the presentation of the relevant stimulus, but also during periods when subjects experience motion or color afterimages.

The clinical description of selective visual deficits after localized damage to various regions of extrastriate cortex also supports functional specialization of extrastriate visual areas in humans. For example, a well-studied patient who suffered a stroke that damaged the extrastriate region thought to be comparable to area MT in the monkey was unable to appreciate the motion of objects. The neurologist who treated her noted that she had difficulty in pouring tea into a cup because the fluid seemed to be "frozen." In addition, she could not stop pouring at the right time because she was unable to perceive when the fluid level had moved to the brim. The patient also had trouble following a dialogue because she could not follow the movements of the speaker's mouth. Crossing the street was potentially terrifying because she couldn't judge the movement of approaching cars. As the patient related, "When I'm looking at the car first, it seems far away. But then, when I want

Figure 12.16 Localization of multiple visual areas in the human brain using *f*MRI. (A,B) Lateral and medial views (respectively) of the human brain, illustrating the location of primary visual cortex (V1) and additional visual areas V2. V3, VP (ventral posterior area), V4, MT (middle temporal area), and MST (medial superior temporal area). (C) Unfolded and flattened view of retinotopically defined visual areas in the occipital lobe. Dark grey areas correspond to cortical regions that were embedded in sulci; light regions correspond to regions that were located on the surface of gyri. Visual areas in humans show a close resemblance to visual areas originally defined in monkeys (compare with Figure 12.15). (After Sereno et al., 1995.)

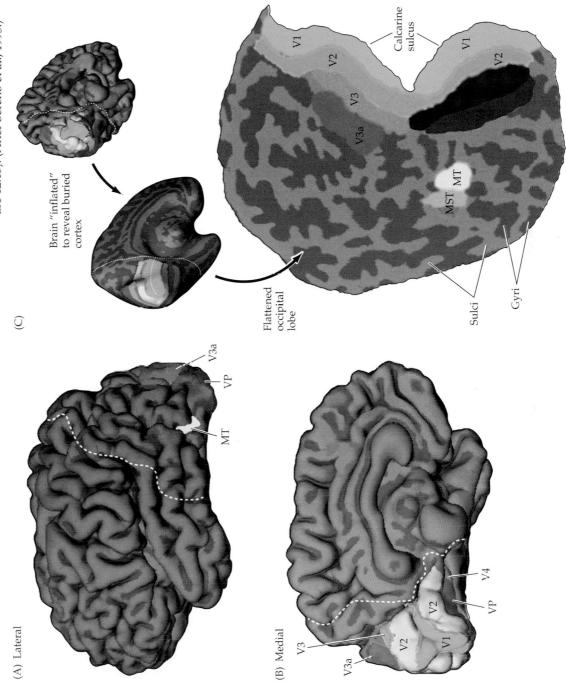

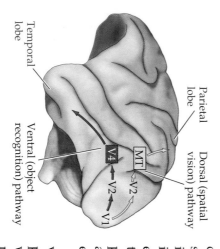

Parietal lobe

Dorsal (spatial vision) pathway

Temporal lobe

Ventral (object recognition) pathway

V4

MT

→V2

V2

V1

Figure 12.17 The visual areas beyond the striate cortex are broadly organized into two pathways: a ventral pathway that leads to the temporal lobe, and a dorsal pathway that leads to the parietal lobe. The ventral pathway plays an important role in object recognition, the dorsal pathway in spatial vision.

to cross the road, suddenly the car is very near." Her ability to perceive other features of the visual scene, such as color and form, was intact.

Another example of a specific visual deficit as a result of damage to extrastriate cortex is **cerebral achromatopsia**. These patients lose the ability to see the world in color, whereas other aspects of vision remain in good working order. The normal colors of a visual scene are described as being replaced by "dirty" shades of gray, much like looking at a poor quality black-and-white movie. Achromatopsic individuals know the normal colors of objects—that a school bus is yellow, an apple red—but can no longer see them. Thus, when asked to appropriately color the objects they have represented. It is important to distinguish this condition from the color blindness that arises from the congenital absence of one or more cone pigments in the retina (see Chapter 11). In achromatopsia, the three types of cones are functioning normally; it is damage to specific extrastriate cortical areas that renders the patient unable to use the information supplied by the retina.

Based on the anatomical connections between visual areas, differences in electrophysiological response properties, and the effects of cortical lesions, a consensus has emerged that extrastriate cortical areas are organized into two largely separate systems that eventually feed information into cortical association areas in the temporal and parietal lobes (see Chapter 26). One system, called the ventral stream, includes area V4 and leads from the striate cortex into the inferior part of the temporal lobe. This system is thought to be responsible for high-resolution form vision and object recognition. The dorsal stream, which includes the middle temporal area, leads from striate cortex into the parietal lobe. This system is thought to be responsible for spatial aspects of vision, such as the analysis of motion, and positional relationships between objects in the visual scene (Figure 12.17).

The functional dichotomy between these two streams is supported by observations on the response properties of neurons and the effects of selective cortical lesions. Neurons in the ventral stream exhibit properties that are important for object recognition, such as selectivity for shape, color, and texture. At the highest levels in this pathway, neurons exhibit even greater selectivity, responding preferentially to faces and objects (see Chapter 26). In contrast, those in the dorsal stream are not tuned to these properties, but show selectivity for direction and speed of movement. Consistent with this interpretation, lesions of the parietal cortex severely impair an animal's ability to distinguish objects on the basis of their position, while having little effect on its ability to perform object recognition tasks. In contrast, lesions of the inferotemporal cortex produce profound impairments in the ability to perform recognition tasks but no impairment in spatial tasks. These effects are remarkably similar to the syndromes associated with damage to the parietal and temporal lobe in humans (see Chapters 26 and 27).

What, then, is the relationship between these "higher order" extrastriate visual pathways and the magno- and parvocellular pathways that supply the primary visual cortex? Not long ago, it seemed that these intracortical pathways were simply a continuation of the geniculostriate pathways—that is, the magnocellular pathway provided input to the dorsal stream and the parvocellular pathway provided input to the ventral stream. However, more recent work has indicated that the situation is more complicated. The temporal pathway clearly has access to the information conveyed by both the magno- and parvocellular streams; and the parietal pathway, while dominated by inputs from the magnocellular stream, also receives inputs from the parvocellular stream. Thus, interaction and cooperation between the magno- and parvocellular streams appear to be the rule in complex visual perceptions.

Summary

Retinal ganglion cells send their axons to a number of central visual structures that serve different functions. The most important projections are to the pretectum for mediating the pupillary light reflex, to the hypothalamus for the regulation of circadian rhythms, to the superior colliculus for the regulation of eye and head movements, and—most important of all—to the lateral geniculate nucleus for mediating vision and visual perception. The retinogeniculostriate projection (the primary visual pathway) is arranged topographically such that central visual structures contain an organized map of the contralateral visual field. Damage anywhere along the primary visual pathway, which includes the optic nerve, optic tract, lateral geniculate nucleus, optic radiation, and striate cortex, results in a loss of vision confined to a predictable region of visual space. Compared to retinal ganglion cells, neurons at higher levels of the visual pathway become increasingly selective in their stimulus requirements. Thus, most neurons in the striate cortex respond to light-dark edges only if they are presented at a certain orientation; some are selective for the length of the edge, and others to movement of the edge in a specific direction. Indeed, a point in visual space is related to a set of cortical neurons, each of which is specialized for processing a limited set of the attributes in the visual stimulus. The neural circuitry in the striate cortex also brings together information from the two eyes; most cortical neurons (other than those in layer IV, which are segregated into eye-specific columns) have binocular responses. Binocular convergence is presumably essential for the detection of binocular disparity, an important component of depth perception. Finally, the visual system shows a remarkable degree of parallel function that begins at the retina. Separate classes of retinal ganglion cells supply the magno- and parvocellular layers of the lateral geniculate nucleus; these pathways are specialized for the detection of rapidly moving stimuli (magnocellular stream) and color and spacial detail (parvocellular stream). Parceling of function continues in the ventral and dorsal streams that lead from the striate cortex to the extrastriate and association areas in the temporal and parietal lobes, respectively. Areas in the inferotemporal cortex are especially important in object recognition, whereas areas in the parietal lobe are critical for understanding the spatial relations between objects in the visual field.

ADDITIONAL READING

Reviews

COURTNEY, S. M. AND L. G. UNGERLEIDER (1997) What fMRI has taught us about human vision. Curr. Op. Neurobiol. 7: 554–561.

FELLEMAN, D. J. AND D. C. VAN ESSEN (1991) Distributed hierarchical processing in primate cerebral cortex. Cerebral Cortex 1: 1–47.

HORTON, J. C. (1992) The central visual pathways. In *Adler's Physiology of the Eye.* W. M. Hart (ed.) St. Louis: Mosby Yearbook.

HUBEL, D. H. AND T. N. WIESEL (1977) Functional architecture of macaque monkey visual cortex. Proc. R. Soc. (Lond.) 198: 1–59.

LIVINGSTONE, E. M. AND D. H. HUBEL (1988) Segregation of form, color, movement, and depth: Anatomy, physiology, and perception. Science 240: 740–749.

MAUNSELL, J. H. R. (1992) Functional visual streams. Curr. Opin. Neurobiol. 2: 506–510.

SCHILLER, P. H. AND N. K. LOGOTHETIS (1990) The color-opponent and broad-band channels of the primate visual system. Trends Neurosci. 13: 392–398.

TOOTELL, R.B., A. M. DALE, M. I. SERENO AND R. MALACH (1996) New images from human visual cortex. Trends Neurosci. 19:481–489.

UNGERLEIDER, J. G. AND M. MISHKIN (1982) Two cortical visual systems. In *Analysis of Visual Behavior.* D. J. Ingle, M. A. Goodale and R. J. W. Mansfield (eds.). Cambridge, MA: MIT Press, pp. 549–586.

Important Original Papers

HUBEL, D. H. AND T. N. WIESEL (1962) Receptive fields, binocular interaction and functional architecture in the cat's visual cortex. J. Physiol. (Lond.; 160: 106–154.

HUBEL, D. H. AND T. N. WIESEL (1968) Receptive fields and functional architecture of monkey striate cortex. J. Physiol. (Lond.) 195: 215–243.

ZIHL, J., D. CRAMON AND D. VON N MAI (1983) Selective disturbance of movement vision after bilateral brain damage. Brain 106: 313–340.

SERENO, M. I. AND 7 OTHERS (1995) Borders of multiple visual areas in humans revealed by functional magnetic resonance imaging. Science 268: 889–893.

Books

HUBEL, D. H. (1988) *Eye, Brain, and Vision.* New York: Scientific American Library.

ZEKI, S. (1993) *A Vision of the Brain.* Oxford: Blackwell Scientific Publications.

RODIECK, R. W. (1998) *The First Steps in Seeing.* Sunderland, MA: Sinauer Associates.

Chapter 13

The Auditory System

Overview

The auditory system is one of the engineering masterpieces of the human body. At the heart of the system is an array of miniature acoustical detectors packed into a space no larger than a pea. These detectors can faithfully transduce vibrations as small as the diameter of an atom, and they can respond a thousand times faster than visual photoreceptors. Such rapid auditory responses to acoustical cues facilitate the initial orientation of the head and body to novel stimuli, especially those that are not initially within the field of view. Although humans are highly visual creatures, much human communication is mediated by the auditory system; indeed, loss of hearing can be more socially debilitating than blindness. From a cultural perspective, the auditory system is essential not only to language, but also to music, one of the most aesthetically sophisticated forms of human expression. For these and other reasons, audition represents a fascinating and especially important aspect of sensation, and more generally of brain function.

Sound

When people speak of sound, they are usually referring to pressure waves generated by vibrating air molecules. Sound waves are much like the ripples that radiate outward when a rock is thrown in a pool of water. However, instead of occurring across a two-dimensional surface, sound waves propagate in three dimensions, creating spherical shells of alternating compression and rarefaction. Like all wave phenomena, sound waves have four major features: **waveform, phase, amplitude** (usually expressed in log units known as decibels, abbreviated dB), and **frequency** (expressed in cycles per second or Hertz, abbreviated Hz). For the human listener, the amplitude and frequency of a sound roughly correspond to **loudness** and **pitch**, respectively.

The waveform of a sound is its amplitude plotted against time. It helps to begin by visualizing an acoustical waveform as a sine wave. At the same time, it must be kept in mind that sounds composed of single sine waves (i.e., pure tones) are extremely rare in nature; most sounds in speech, for example, consist of acoustically complex waveforms. Interestingly, such complex waveforms can often be modeled as the sum of sinusoidal waves of varying amplitudes, frequencies, and phases. In engineering applications, an algorithm called the Fourier transform decomposes a complex signal into its sinusoidal components. In the auditory system, as will be apparent later in the chapter, the inner ear acts as a sort of acoustical prism, decomposing complex sounds into a myriad of constituent tones.

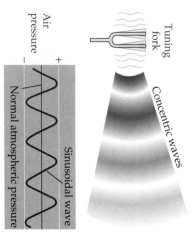

Figure 13.1 Diagram of the periodic condensation and rarefaction of air molecules produced by the vibrating tines of a tuning fork. The molecular disturbance of the air is pictured as if frozen at the instant the constituent molecules responded to the resultant pressure wave. Shown below is a plot of the air pressure versus distance from the fork. Note its sinusoidal quality.

Figure 13.1 diagrams the behavior of air molecules near a tuning fork that vibrates sinusoidally when struck. The vibrating tines of the tuning fork produce local displacements of the surrounding molecules, such that when the tine moves in one direction, there is molecular condensation; when it moves in the other direction, there is rarefaction. These changes in density of the air molecules are equivalent to local changes in air pressure.

Such regular, sinusoidal cycles of compression and rarefaction can be thought of as a form of circular motion, with one complete cycle equivalent to one full revolution (360°). This point can be illustrated with two sinusoids of the same frequency projected onto a circle, a strategy that also makes it easier to understand the concept of phase (Figure 13.2). Imagine that two tuning forks, both of which resonate at the same frequency, are struck at slightly different times. At a given time, t = 0, one wave is at position P and the other at position Q. By projecting P and Q onto the circle, their respective phase angles, θ_1 and θ_2, are apparent. The sine wave that starts at P reaches a particular point on the circle, say 180°, at time t_1, whereas the wave that starts at Q reaches 180° at time t_2. Thus, phase differences have corresponding time differences, a concept that is important in appreciating how the auditory system locates sounds in space.

The human ear is extraordinarily sensitive to sound. At the threshold of hearing, air molecules are displaced an average of only 10 picometers (10^{-11} m), a distance 10,000 times smaller than the wavelength of visible light. The intensity of such a sound is about one-trillionth of a watt per square meter! This means a listener on an otherwise noiseless planet could hear a 1-watt, 3-kHz sound source located over 300 miles away (consider that very dim lightbulbs consume more than 1 watt of power). Even dangerously high sound pressure levels exert power on the eardrum only in the milliwatt range (Box A).

The Audible Spectrum

Humans can detect sounds in a frequency range from about 20 Hz to 20 kHz. (Human infants can actually hear frequencies slightly higher than 20 kHz, but lose some high-frequency sensitivity as they mature; the upper limit in average adults is often closer to 15–17 kHz.) Not all mammalian species are sensitive to the same range of frequencies. Most small mammals are sensitive to very high frequencies, but not to low frequencies. For instance, some species of bats are sensitive to tones as high as 200 kHz, but their lower limit is around 20 kHz—the upper limit for young people with normal hearing. One reason for these differences is that small objects, including the auditory structures of these small mammals, are better resonators for high frequencies, whereas large objects are better for low frequencies (which also explains why the violin has a higher pitch than the cello).

A Synopsis of Auditory Function

The auditory system transforms sound waves into distinct patterns of neural activity, which are then integrated with information from other sensory systems to guide behavior, including orienting movements to acoustical stimuli and intraspecies communication. The first stage of this transformation occurs at the external and middle ears, which collect sound waves and amplify their pressure, so that the sound energy in the air can be successfully transmitted to the fluid-filled cochlea of the inner ear. In the inner ear, a series of biomechanical processes occur that break up the signal into simpler, sinusoidal components, with the result that the frequency, amplitude,

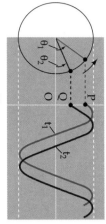

Figure 13.2 A sine wave and its projection as circular motion. The two sinusoids shown are at different phases, such that point P corresponds to phase angle θ_1 and point Q corresponds to phase angle θ_2.

BOX A
Four Causes of Acquired Hearing Loss

Acquired hearing loss is an increasingly common sensory deficit that can often lead to impaired oral communication and social isolation. Four major causes of acquired hearing loss are acoustical trauma, infection of the inner ear, ototoxic drugs, and presbyacusis (literally, the hearing of the old). The exquisite sensitivity of the auditory periphery, combined with the direct mechanical linkage between the acoustical stimulus and the receptor cells, make the ear especially susceptible to acute or chronic acoustical trauma. Extremely loud, percussive sounds, such as those generated by explosives or gunfire, can rupture the eardrum and so severely distort the inner ear that the organ of Corti is torn. The resultant loss of hearing is abrupt and often quite severe. Less well appreciated is the fact that repeated exposure to less dramatic but nonetheless loud sounds, including those produced by industrial or household machinery or by amplified musical instruments, can also damage the inner ear. Although these sounds leave the eardrum intact, specific dam-

age is done to the hair bundle itself; the stereocilia of cochlear hair cells of animals exposed to loud sounds shear off at their pivot points with the hair cell body, or fuse together in a platelike fashion that impedes movement. In humans, the mechanical resonance of the ear to stimulus frequencies centered about 3 kHz means that exposure to loud, broadband noises (such as those generated by jet engines) results in especially pronounced deficits near this resonant frequency.

Ototoxic drugs include aminoglycoside antibiotics (such as gentamycin and kanamycin), which directly affect hair cells, and ethacrynic acid, which poisons the potassium-extruding cells of the stria vascularis that generate the endocochlear potential. In the absence of these ion pumping cells, the endocochlear potential, which supplies the energy to drive the transduction process, is lost. Although still a matter of some debate, the relatively nonselective transduction channel apparently affords a means of entry for aminoglycoside antibiotics, which then poison hair cells

by disrupting phosphoinositide metabolism. In particular, outer hair cells and those inner hair cells that transduce high-frequency stimuli are more affected, simply because of their greater energy requirements. Finally, presbyacusis, the hearing loss associated with aging, may in part stem from atherosclerotic damage to the especially fine microvasculature of the inner ear, as well as from genetic predispositions to hair cell damage. Recent advances in understanding the genetic transmission of acquired hearing loss, in both humans and mice, point to mutations in myosin isoforms unique to hair cells as a likely culprit.

References

HOLT, J. R. AND D. P. COREY (1999) Ion channel defects in hereditary hearing loss. Neuron 22(2): 217–219.

KEATS, B. J. AND D. P. COREY (1999) The usher syndromes. Am. J. Med. Gen. 89(3): 158–166.

PRIUSKA, E. M. AND J. SCHACT (1997) Mechanism and prevention of aminoglycoside ototoxicity: Outer hair cells as targets and tools. Ear, Nose, and Throat J. 76(3): 164–171.

and phase of the original signal are all faithfully transduced by the sensory **hair cells** and encoded by the electrical activity of the **auditory nerve fibers.**

One product of this process of acoustical decomposition is the systematic representation of sound frequency along the length of the cochlea, referred to as **tonotopy,** which is an important feature preserved throughout the central auditory pathways. The earliest stage of central processing occurs at the cochlear nucleus, where the peripheral auditory information diverges into a number of parallel central pathways. Accordingly, the output of the cochlear nucleus has several targets. One of these is the superior olivary complex, the first place that information from the two ears interacts and the site of the initial processing of the cues that allow us to localize sound in space. The cochlear nucleus also projects to the inferior colliculus of the midbrain, a major integrative center and the first place where auditory information can interact with the motor system. The inferior colliculus is an obligatory relay for information traveling to the thalamus and cortex, where additional integrative aspects of sound that are especially germane to speech (such as sound combinations that vary over time) are processed.

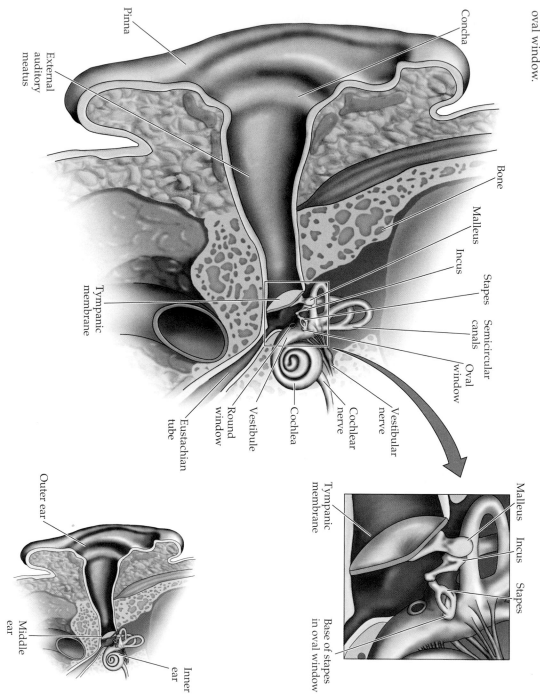

Figure 13.3 The human ear. Note the large surface area of the tympanic membrane (eardrum) relative to the oval window.

The External Ear

The external ear, which consists of the **pinna, concha,** and **auditory meatus,** gathers sound energy and focuses it on the eardrum, or **tympanic membrane** (Figure 13.3). One consequence of the configuration of the external ear is that it selectively boosts the sound pressure 30- to 100-fold for frequencies around 3 kHz. This amplification makes humans especially sensitive to frequencies in this range—and also explains why they are particularly prone to acoustical injury and hearing loss near this frequency (see Box A). Not surprisingly, most human speech sounds are distributed in the bandwidth around 3 kHz. Most vocal communication occurs in the low-kHz range because transmission of airborne sound is less efficient at higher frequencies, and the detection of lower frequencies is difficult for animals the size of humans.

A second important function of the pinna and concha is to selectively filter different sound frequencies in order to provide cues about the elevation of the sound source. The convolutions of the pinna are shaped so that the external ear transmits more high-frequency components from an elevated source than from the same source at ear level. This effect can be demonstrated by recording sounds from different elevations after they have passed through an artificial external ear; when the recorded sounds are played back via earphones, so that the whole series is at the same elevation relative to the listener, the recordings from higher elevations are perceived as coming from positions higher in space than the recordings from lower elevations.

The Middle Ear

Sounds impinging on the external ear are airborne; however, the environment within the inner ear, where the sound-induced vibrations are converted to neural impulses, is aqueous. The major function of the middle ear is to match relatively low-impedance airborne sounds to the higher-impedance fluid of the inner ear. The term "impedance" in this context describes a medium's resistance to movement. Normally, when sound travels from a low-impedance medium like air to a much higher-impedance medium like water, almost all (more than 99.9%) of the acoustical energy is reflected. The middle ear (see Figure 13.3) overcomes this problem and ensures transmission of the sound energy across the air-fluid boundary by boosting the pressure measured at the tympanic membrane almost 200-fold by the time it reaches the inner ear.

Two mechanical processes occur within the middle ear to achieve this large pressure gain. The first and major boost is achieved by focusing the force impinging on the relatively large-diameter tympanic membrane on to the much smaller-diameter **oval window**, the site where the bones of the middle ear contact the inner ear. A second and related process relies on the mechanical advantage gained by the lever action of the three small interconnected middle ear bones, or **ossicles** (i.e., the malleus, incus, and stapes; see Figure 13.3), which connect the tympanic membrane to the oval window.

Bony and soft tissue, including that surrounding the inner ear, have impedances close to that of water. Therefore, even without an intact tympanic membrane or middle ear ossicles, acoustical vibrations can still be transferred directly through the bones and tissues of the head to the inner ear. In the clinic, bone conduction can be exploited to determine the source of a patient's hearing loss. For example, when a tuning fork is applied directly to the vertex of a patient's head, it is perceived as being equally loud in the two ears when there is either uni- or bilateral mechanical damage to the middle ear (**conductive hearing loss**) but not when there is unilateral damage to the hair cells of the inner ear or to the auditory nerve itself (**sensorineural hearing loss**; see Box A).

The Inner Ear

The **cochlea** of the inner ear is the most critical structure in the auditory pathway, for it is there that the energy from sonically generated pressure waves is transformed into neural impulses. The cochlea not only amplifies sound waves and converts them into neural signals, but it also acts as a mechanical frequency analyzer, decomposing complex acoustical waveforms into simpler elements. Many features of auditory perception derive directly from the physical properties of the cochlea; hence, it is important to consider this structure in some detail.

The cochlea (from the Latin for "snail") is a small (about 10 mm wide) coiled structure, which, were it uncoiled, would form a tube about 35 mm long (Figures 13.4 and 13.5). Both the oval window and the **round window**

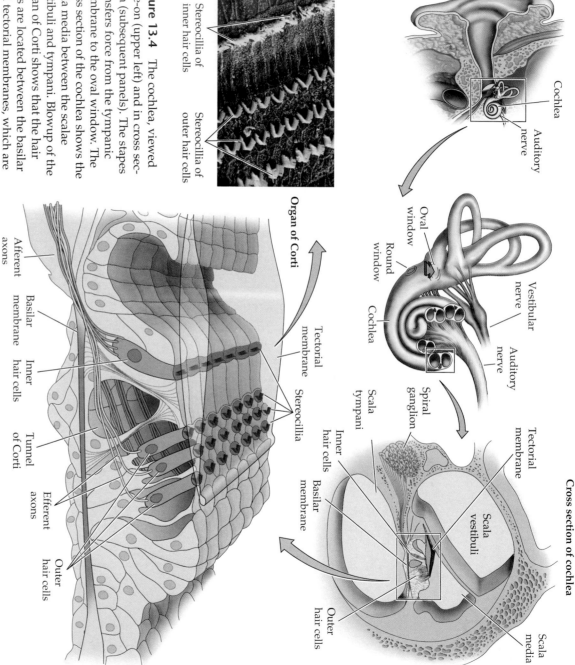

Figure 13.4 The cochlea, viewed face-on (upper left) and in cross section (subsequent panels). The stapes transfers force from the tympanic membrane to the oval window. The cross section of the cochlea shows the scala media between the scalae vestibuli and tympani. Blowup of the organ of Corti shows that the hair cells are located between the basilar and tectorial membranes, which are rendered transparent in the line drawing and removed in the scanning electron micrograph. The hair cells are named for their tufts of stereocilia; inner hair cells receive afferent inputs from cranial nerve VIII, whereas outer hair cells receive mostly efferent input. (Micrograph from Kessel and Kardon, 1979.)

are at the basal end of this tube. The cochlea is bisected by the cochlear partition, which is a flexible structure that supports the **basilar membrane** and the **tectorial membrane**. There are fluid-filled spaces on each side of the cochlear partition, named the **scala vestibuli** and the **scala tympani**; a distinct channel, the **scala media**, runs within the cochlear partition. The cochlear partition does not extend all the way to the apical end of the cochlea; instead there is an opening, known as the **helicotrema**, that joins the scala vestibuli to the scala tympani. As a result of this structural arrangement, inward movement of the oval window displaces the fluid of the inner ear, which causes the round window to bulge out slightly and deforms the basilar membrane.

The manner in which the basilar membrane vibrates in response to sound is the key to understanding cochlear function. Measurements of the vibration of different parts of the basilar membrane, as well as the discharge rates of individual auditory nerve fibers, show that both these features are highly tuned; that is, they respond most intensely to a sound of a specific frequency. Frequency tuning within the inner ear is attributable in part to the geometry of the basilar membrane, which is wider and more flexible at the apical end and narrower and stiffer at the basal end. One feature of such a system is that regardless of where energy is supplied to it, movement always begins at the stiff end (i.e., the base), and then propagates to the more flexible end (i.e., the apex). Georg von Békésy, working at Harvard University, showed that a membrane that varies systematically in its width and flexibility vibrates maximally at different positions as a function of the stimulus frequency (Figure 13.5). Using tubular models and human cochleas taken from cadavers, he found that an acoustical stimulus initiates a traveling wave of the same frequency in the cochlea, which propagates from the base toward the apex of the basilar membrane, growing in amplitude and slowing in velocity until a point of maximum displacement is reached. This point of maximal displacement is determined by the sound frequency. The points responding to high frequencies are at the base of the basilar membrane, and the points responding to low frequencies are at the apex, giving rise to a topographical mapping of frequency (that is, to **tonotopy**). An important and striking feature of the tonotopically organized basilar membrane is that complex sounds cause a pattern of vibration equivalent to the superposition of the vibrations generated by the individual tones making up that complex sound, thus accounting for the decompositional aspects of cochlear function mentioned earlier.

Von Békésy's model of cochlear mechanics was a passive one, resting on the premise that the basilar membrane acts like a series of linked resonators, much as a concatenated set of tuning forks. Each point on the basilar membrane was postulated to have a characteristic frequency at which it vibrated most efficiently; because it was physically linked to adjacent areas of the membrane, each point also vibrated (if somewhat less readily) at other fre-

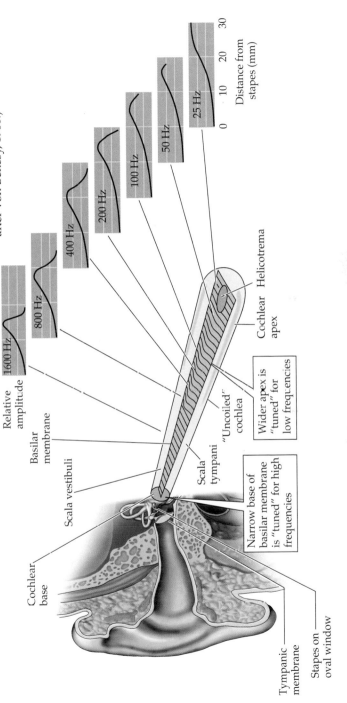

Figure 13.5 Traveling waves along the cochlea. A traveling wave is shown at a given instant along the cochlea, which has been uncoiled for clarity. The graphs profile the amplitude of the traveling wave along the basilar membrane for different frequencies and show that the position where the traveling wave reaches its maximum amplitude varies directly with the frequency of stimulation. (Drawing after Dallos, 1992; graphs after von Békésy, 1960.)

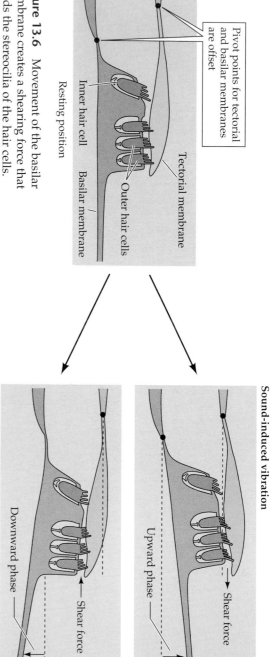

Figure 13.6 Movement of the basilar membrane creates a shearing force that bends the stereocilia of the hair cells. The pivot point of the basilar membrane is offset from the pivot point of the tectorial membrane, so that when the basilar membrane is displaced, the tectorial membrane moves across the tops of the hair cells, bending the stereocilia.

quencies, thus permitting propagation of the traveling wave. It is now clear, however, that the tuning of the auditory periphery, whether measured at the basilar membrane or recorded as the electrical activity of auditory nerve fibers, is too sharp to be explained by passive mechanics alone. At very low sound intensities, the basilar membrane vibrates much more than would be predicted by linear extrapolation from the motion measured at high intensities. Therefore, the ear's sensitivity arises from an active biomechanical process, as well as from its passive resonant properties (Box B). The outer hair cells, which together with the inner hair cells comprise the sensory cells of the inner ear, are the most likely candidates for driving this active process. The details of this process are poorly understood.

The motion of the traveling wave initiates sensory transduction by displacing the hair cells that sit atop the basilar membrane. Because these structures are anchored at different positions, the vertical component of the traveling wave is translated into a shearing motion between the basilar membrane and the overlying tectorial membrane (Figure 13.6). This motion bends the tiny processes, called **stereocilia**, that protrude from the apical ends of the hair cells, leading to voltage changes across the hair cell membrane. How the bending of stereocilia leads to receptor potentials in hair cells is considered in the following section.

Hair Cells and the Mechanoelectrical Transduction of Sound Waves

The hair cell is an evolutionary triumph that solves the problem of transforming vibrational energy into an electrical signal. The scale at which the hair cell operates is truly amazing: At the limits of human hearing, hair cells can faithfully detect movements of atomic dimensions and respond in the tens of microseconds! Furthermore, hair cells can adapt rapidly to constant stimuli, thus allowing the listener to extract signals from a noisy background.

The hair cell is a flask-shaped epithelial cell named for the bundle of hair-like processes that protrude from its apical end into the scala media. Each hair bundle contains anywhere from 30 to a few hundred hexagonally arranged stereocilia, with one taller **kinocilium** (Figure 13.7A). Despite their names, only the kinocilium is a true ciliary structure, with the characteristic

Figure 13.7 The structure and function of the hair bundle. The vestibular hair bundles shown here resemble those of cochlear hair cells, except for the presence of the kinocilium, which disappears in the mammalian cochlea shortly after birth. (A) The hair bundle of a guinea pig vestibular hair cell. This view shows the increasing height leading to the kinocilium (arrow). (B) Cross section through the vestibular hair bundle shows the 9 + 2 array of microtubules in the kinocilium (on right), which contrasts with the simpler actin filament structure of the stereocilia. (C) Scanning electron micrograph of a guinea pig cochlear outer hair cell bundle viewed along the plane of mirror symmetry. Note the graded lengths of the stereocilia, and the absence of a kinocilium. (D) Diagram of the stereocilia and tip links, which, when stretched by movement toward the kinocilium, open channels that generate a depolarizing current. When compressed, these same structures lead to hyperpolarization of the hair cell. (A from Lindeman, 1973; B from Hudspeth, 1983; C from Pickles, 1988; D after Pickles et al., 1984.)

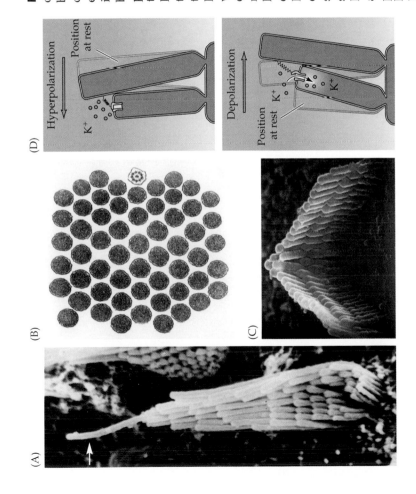

two central tubules surrounded by nine doublet tubules (Figure 13.7B). The function of the kinocilium is unclear, and in the cochlea of humans and other mammals it actually disappears shortly after birth (Figure 13.7C). The stereocilia are simpler, containing only an actin cytoskeleton. Each stereocilium tapers where it inserts into the apical membrane, forming a hinge about which each stereocilium pivots (Figure 13.7D). The stereocilia are graded in height and are arranged in a bilaterally symmetric fashion (in vestibular hair cells, this plane runs through the kinocilium). Displacement of the hair bundle parallel to this plane toward the tallest stereocilia depolarizes the hair cell, while movements parallel to this plane toward the shortest stereocilia cause hyperpolarization. In contrast, displacements perpendicular to the plane of symmetry do not alter the hair cell's membrane potential. The hair bundle movements at the threshold of hearing are approximately 0.3 nm, about the diameter of an atom of gold!

Hair cells can convert the displacement of the stereociliary bundle into an electrical potential in as little as 10 microseconds; indeed, such speed is required to faithfully transduce high-frequency signals and enable the accurate localization of the source of the sound. The need for microsecond resolution places certain constraints on the transduction mechanism, ruling out the relatively slow second messenger pathways used in visual and olfactory transduction (see Chapters 8, 11 and 15); a direct, mechanically gated transduction channel is needed to operate this quickly. Evidently the filamentous structures that connect the tips of adjacent stereocilia, known as **tip links**, directly open cation-selective transduction channels when stretched, allowing K^+ ions to flow into the cell (see Figure 13.7D). As the linked stereocilia pivot from side to side, the tension on the tip link varies, modulating the ionic flow and resulting in a graded receptor potential that follows the movements of the stereocilia (Figure 13.8). The tip link model also explains

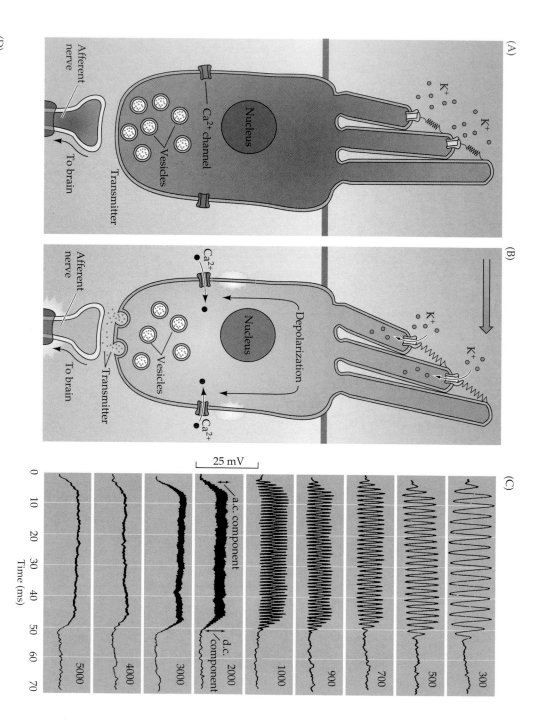

(A)

(B)

(C)

25 mV

a.c. component

d.c. component

5000
4000
3000
2000
1000
900
700
500
300

Time (ms)
0 10 20 30 40 50 60 70

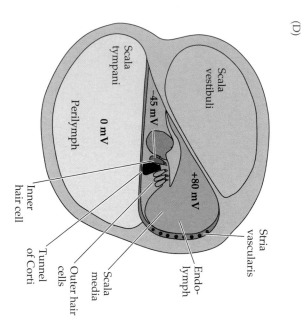

(D)

Scala
vestibuli

Scala
tympani

Perilymph

0 mV

−45 mV

+80 mV

Perilymph

Scala
media

Endo-
lymph

Stria
vascularis

Outer
hair
cells

Tunnel
of Corti

Inner
hair cell

Figure 13.8 Mechanoelectrical transduction mediated by hair cells. (A,B) When the hair bundle is deflected toward the tallest stereocilium, cation-selective channels open near the tips of the stereocilia, allowing K⁺ ions to flow into the hair cell down their electrochemical gradient (see text on next page for the explanation of this peculiar situation). The resulting depolarization of the hair cell opens voltage-gated Ca²⁺ channels in the cell soma, allowing calcium entry and release of neurotransmitter onto the nerve endings of the auditory nerve. (C) Receptor potentials generated by an individual hair cell in the cochlea in response to pure tones (indicated in Hz at the right of the tracings). Note that the hair cell potential faithfully follows the waveform of the stimulating sinusoids for low frequencies (<3kHz), but still responds with a DC offset to higher frequencies. (D) The stereocilia of the hair cells protrude into the endolymph, which is high in K⁺ and has an electrical potential of +80 mV relative to the perilymph. (A,B after Lewis and Hudspeth, 1983; C after Palmer and Russell, 1986.)

BOX B
The Sweet Sound of Distortion

As early as the first half of the eighteenth century, musical composers such as Giuseppe Tartini and W. A. Sorge discovered that upon playing pairs of tones, other tones not present in the original stimulus are also heard. These combination tones, fc, are mathematically related to the played tones, f_1 and f_2 ($f_2 > f_1$), by the formula

$$fc = mf_1 \pm nf_2$$

where m and n are positive integers. Combination tones have been used for a variety of compositional effects, as they can strengthen the harmonic texture of a chord. Furthermore, organ builders sometimes use the difference tone ($f_2 - f_1$) created by two smaller organ pipes to produce the extremely low tones that would otherwise require building one especially large pipe.

Modern experiments indicate that this distortion product is actually due to the nonlinear properties of the inner ear. M. Ruggero and his colleagues placed small glass beads (10–30 μm in diameter) on the basilar membrane of an anesthetized animal and then deter-

mined the velocity of the basilar membrane in response to different combinations of tones by measuring the Doppler shift of laser light reflected from the beads. When two tones were played into the ear, the basilar membrane vibrated not only at those two frequencies, but also at other frequencies predicted by the above formula.

Related experiments on hair cells studied in vitro suggest that these nonlinearities result from the properties of the mechanical linkage of the transduction apparatus. By moving the hair bundle sinusoidally with a metal-coated glass fiber, A. J. Hudspeth and his coworkers found that the hair bundle exerts a force at the same frequency. However, when two sinusoids were applied simultaneously, the forces exerted by the hair bundle occurred not only at the primary frequencies, but at several combination frequencies as well. These distortion products are due to the transduction apparatus, since blocking the transduction channels causes the forces exerted at the combination frequencies to disappear, even though the

forces at the primary frequencies remain unaffected. It seems that the tip links add a certain extra springiness to the hair bundle in the small range of motions over which the transduction channels are changing between closed and open states. If nonlinear distortions of basilar membrane vibrations arise from the properties of the hair bundle, then it is likely that hair cells can indeed influence basilar membrane motion, thereby accounting for the cochlea's extreme sensitivity. Apparently, when we hear difference tones, we are paying the price in distortion for an exquisitely fast and sensitive transduction mechanism.

References

PLANCHART, A. E. (1960) A study of the theories of Giuseppe Tartini. J. Music Theory 4(1): 32–61.

ROBLES, L., M. A. RUGGERO AND N. C. RICH (1991) Two-tone distortion in the basilar membrane of the cochlea. Nature 439: 413–414.

JARAMILLO, F., V. S. MARKIN AND A. J. HUDSPETH (1993) Auditory illusions and the single hair cell. Nature 364: 527–529.

why only deflections along the axis of the hair bundle activate transduction channels, since tip links join adjacent stereocilia along the axis directed toward the tallest stereocilia (see also Box B in Chapter 14).

Understanding the ionic basis of hair cell transduction has been greatly advanced by intracellular recordings made from these tiny structures. The hair cell has a resting potential between −45 and −60 mV relative to the fluid that bathes the basal end of the cell. At the resting potential, only a small fraction of the transduction channels are open. When the hair bundle is displaced in the direction of the tallest stereocilium, more transduction channels open, causing depolarization as K$^+$ enters the cell. Depolarization in turn opens voltage-gated calcium channels in the hair cell membrane, and the resultant Ca^{2+} influx causes transmitter release from the basal end of the cell onto the auditory nerve endings (Figure 13.8A,B). Such calcium-dependent exocytosis is similar to chemical neurotransmission elsewhere in the central and peripheral nervous system (see Chapters 5 and 6). Because some of the transduction channels are open at rest, the receptor potential is biphasic: Movement toward the tallest stereocilia depolarizes the cell, while movement in the opposite direction leads to hyperpolarization. This situation allows the hair cell to generate a sinusoidal receptor potential in response to a sinusoidal stimulus, thus preserving the temporal information present in the original signal up to frequencies of around 3 kHz (Figure 13.8C).

The high speed demands of mechanoelectrical transduction have resulted in some fascinating ionic specializations within the inner ear. An unusual adaptation of the hair cell in this regard is that K^+ serves both to depolarize and repolarize the cell, enabling the hair cell's K^+ gradient to be largely maintained by passive ion movement alone. As with other epithelial cells, the basal and apical surfaces of the hair cell are separated by tight junctions, allowing separate extracellular ionic environments at these two surfaces. The apical end is exposed to the K^+-rich, Na^+-poor **endolymph**, which is produced by dedicated ion pumping cells in the **stria vascularis** (Figure 13.8D). The basal end is bathed in the same fluid that fills the scala tympani, known as **perilymph**, which resembles other extracellular fluids in that it is K^+-poor and Na^+-rich. In addition, the compartment containing endolymph is about 80 mV more positive than the perilymph compartment (this difference is known as the endocochlear potential), while the inside of the hair cell is about 45 mV more negative than the perilymph (and 125 mV more negative than the endolymph). The resulting electrical gradient across the membrane of the stereocilia (about 125 mV) drives K^+ through open transduction channels into the hair cell, even though these cells already have a high internal K^+ concentration. K^+ entry via the transduction channels leads to depolarization of the hair cell, which in turn opens voltage-gated Ca^{2+} and K^+ channels located in the membrane of the hair cell soma (see Box B in Chapter 14). The opening of *somatic* K^+ channels favors K^+ efflux, and thus repolarization; the efflux occurs because the perilymph surrounding the basal end is low in K^+ relative to the cytosol, and because the equilibrium potential for K^+ is more negative than the hair cell's resting potential. Repolarization of the hair cell via K^+ efflux is also facilitated by Ca^{2+} entry. In addition to modulating the release of neurotransmitter, Ca^{2+} entry opens Ca^{2+}-dependent K^+ channels, which provide another avenue for K^+ to enter the perilymph. Indeed, the interaction of Ca^{2+} influx and Ca^{2+}-dependent K^+ efflux can lead to electrical resonances that enhance the tuning of response properties within the inner ear (also explained in Box B in Chapter 14). In essence, the hair cell operates as two distinct compartments, each dominated by its own Nernst equilibrium potential for K^+; this arrangement ensures that the hair cell's ionic gradient will not run down, even during prolonged stimulation. At the same time, compounds such as ethacrynic acid (see Box A), which selectively poison the ion-pumping cells of the stria vascularis, can cause the endocochlear potential to dissipate, resulting in a sensorineural hearing deficit. In short, the hair cell exploits the different ionic milieus of its apical and basal surfaces to provide extremely fast and energy-efficient repolarization.

Two Kinds of Hair Cells in the Cochlea

The cochlear hair cells in humans consist of one row of **inner hair cells** and three rows of **outer hair cells** (see Figure 13.4). The inner hair cells are the actual sensory receptors, and 95% of the fibers of the auditory nerve that project to the brain arise from this subpopulation. The terminations on the outer hair cells are almost all from efferent axons that arise from cells in the brain.

A clue to the significance of this efferent pathway was provided by the discovery that basilar membrane motion is influenced by an active process within the cochlea, as already noted. First, it was found that the cochlea actually emits sound under certain conditions. These otoacoustical emissions can be detected by placing a sensitive microphone at the eardrum and monitoring the response after briefly presenting a tone. Such emissions can also occur spontaneously. These observations clearly indicate that a process within the cochlea is capable of producing sound. Second, stimulation of the crossed olivocochlear bundle, which supplies efferent input to the outer hair cells, can broaden eighth nerve tuning curves. Finally, isolated outer hair cells move in response to small electrical currents, apparently due to the transduction process being driven in reverse. Thus, it seems likely that the outer hair cells sharpen the frequency-resolving power of the cochlea by actively contracting and relaxing, thus changing the stiffness of the tectorial membrane at particular locations. An active process of this sort is necessary in any event to explain the nonlinear vibration of the basilar membrane at low sound intensities (Box B).

Tuning and Timing in the Auditory Nerve

The rapid response time of the transduction apparatus allows the membrane potential of the hair cell to follow deflections of the hair bundle up to quite high frequencies of oscillation. In humans, the receptor potentials of certain hair cells and the action potentials of their associated auditory nerve fiber can follow stimuli of up to about 3 kHz in a one-to-one fashion. Such real-time encoding of stimulus frequency by the pattern of action potentials in the auditory nerve is known as the "volley theory" of auditory information transfer. Even these extraordinarily rapid processes, however, fail to follow frequencies above 3 kHz (see Figure 13.8C). Accordingly, some other mechanism must be used to transmit auditory information at higher frequencies. The tonotopically organized basilar membrane provides an alternative to temporal coding, namely a "labeled-line" coding mechanism. In this case, frequency information is specified by preserving the tonotopy of the cochlea at higher levels in the auditory pathway. Because the auditory nerve fibers associate with the inner hair cells in approximately a one-to-one ratio, each auditory nerve fiber transmits information about only a small part of the audible frequency spectrum. As a result, auditory nerve fibers related to the apical end of the cochlea respond to low frequencies, and fibers that are related to the basal end respond to high frequencies (see Figure 13.5). The limitations of specific fibers can be seen in electrophysiological recordings of responses to sound (Figure 13.9). These threshold functions are called **tuning curves**, and the lowest threshold of the tuning curve is called the **characteristic frequency**. Since the topographical order of the characteristic frequency of neurons is retained throughout the system, information about frequency is also preserved.

Cochlear implants, which consist of threadlike multisite microstimulation electrodes driven by digital signal processors, exploit the tonotopic organization of the cochlea, and particularly its eighth nerve afferents, to roughly recreate the patterns of eighth nerve activity elicited by sounds. In patients with damaged hair cells, such implants can effectively bypass the impaired transduction apparatus, and thus restore some degree of auditory function.

The other prominent feature of hair cells—their ability to follow the waveform of low-frequency sounds—is also important in other more subtle aspects of auditory coding. As mentioned earlier, hair cells have biphasic response properties. Because hair cells release transmitter only when depolarized, auditory nerve fibers fire only during the positive phases of low-frequency sounds (Figure 13.10). The resultant "phase locking" that results pro-

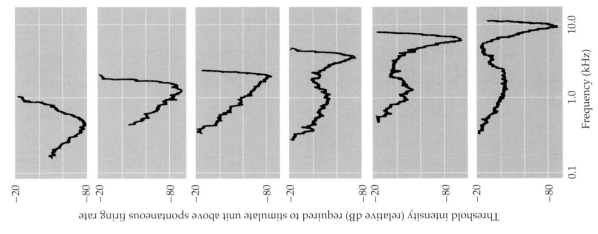

Figure 13.9 Frequency tuning curves of six different fibers in the auditory nerve. Each graph plots, across all frequencies to which the fiber responds, i.e., the minimum sound level required to increase the fiber's firing rate above its spontaneous firing level. The lowest point in the plot is the weakest sound intensity to which the neuron will respond. The frequency at this point is called the neuron's characteristic frequency. (After Kiang and Moxon, 1972.)

Threshold intensity (relative dB) required to stimulate unit above spontaneous firing rate

Frequency (kHz)

Figure 13.10 Temporal response patterns of a low-frequency axon in the auditory nerve. The stimulus waveform is indicated beneath the histograms, which show the phase-locked responses to a 50-ms tone pulse of 260 Hz. Note that the spikes are all timed to the same phase of the sinusoidal stimulus. (After Kiang, 1984.)

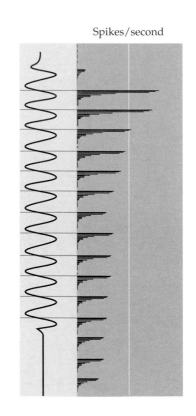

Spikes/second

vides temporal information from the two ears to neural centers that compare interaural time differences. The evaluation of interaural time differences provides a critical cue for sound localization, by means of which a person is able to perceive auditory "space." That auditory space can be perceived in this way is especially remarkable, given that the cochlea, unlike the retina, cannot represent space directly.

How Information from the Cochlea Reaches Targets in the Brainstem

A hallmark of the ascending auditory system is its parallel organization. This arrangement becomes evident as soon as the auditory nerve enters the brainstem, where it branches to innervate the three divisions of the cochlear nucleus. The auditory nerve (the major component of cranial nerve VIII) comprises the central processes of the bipolar spiral ganglion cells in the cochlea (see Figure 13.4); each of these cells sends a peripheral process to contact one or more hair cells and a central process to innervate the cochlear nucleus. Within the cochlear nucleus, each auditory nerve fiber branches, sending an ascending branch to the anteroventral cochlear nucleus and a descending branch to the posteroventral cochlear nucleus and the dorsal cochlear nucleus (Figure 13.11). The tonotopic organization of the cochlea is maintained in the three parts of the cochlear nucleus, each of which contains different populations of cells with quite different properties. In addition, the patterns of termination of the auditory nerve axons differ in density and type; thus, there are several opportunities at this level for transformation of the information from the hair cells.

Integrating Information from the Two Ears

Just as the auditory nerve branches to innervate several different targets in the cochlear nuclei, the neurons in these nuclei give rise to several different pathways (see Figure 13.11). One clinically relevant feature of the ascending projections of the auditory brainstem is a high degree of bilateral connectivity, which means that damage to central auditory structures is almost never manifested as a monaural hearing loss. Indeed, monaural hearing loss strongly implicates unilateral peripheral damage, either to the middle or inner ear, or to the eighth nerve itself. Given the relatively byzantine organization already present at the level of the auditory brainstem, it is useful to consider these pathways in the context of their functions.

The best-understood function mediated by the auditory brainstem nuclei, and certainly the one most intensively studied, is sound localization. Humans use at least two different strategies to localize the horizontal position of sound

Figure 13.11 Diagram of the major auditory pathways. Although many details are missing from this diagram, two important points are evident: (1) the auditory system entails several parallel pathways, and (2) information from each ear reaches both sides of the system, even at the level of the brainstem.

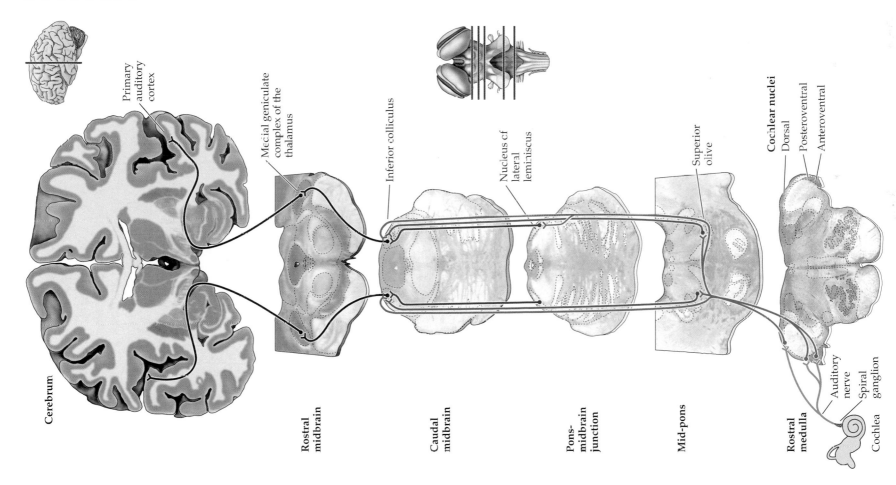

Cerebrum

Primary auditory cortex

Medial geniculate complex of the thalamus

Inferior colliculus

Nucleus of lateral lemniscus

Superior olive

Cochlear nuclei
Dorsal
Posteroventral
Anteroventral

Auditory nerve
Spiral ganglion
Cochlea

Rostral midbrain

Caudal midbrain

Pons-midbrain junction

Mid-pons

Rostral medulla

sources, depending on the frequencies in the stimulus. For frequencies below 3 kHz (which can be followed in a phase-locked manner), interaural *time* differences are used to localize the source; above these frequencies, interaural *intensity* differences are used as cues. Parallel pathways originating from the cochlear nucleus serve each of these strategies for sound localization.

The human ability to detect interaural time differences is remarkable. The longest interaural time differences, which are produced by sounds arising directly lateral to one ear, are on the order of only 700 microseconds (a value given by the width of the head divided by the speed of sound in air, about 340 m/s). Psychophysical experiments show that humans can actually detect interaural time differences as small as 10 microseconds; two sounds presented through earphones separated by such small interaural time differences are perceived as arising from the side of the leading ear. This sensitivity translates into an accuracy for sound localization of about 1°.

How is timing in the 10 microseconds range accomplished by neural components that operate in the millisecond range? The neural circuitry that computes such tiny interaural time differences consists of binaural inputs to the **medial superior olive (MSO)** that arise from the right and left anteroventral cochlear nuclei (Figure 13.12; see also Figure 13.11). The medial superior olive contains cells with bipolar dendrites that extend both medially and laterally. The lateral dendrites receive input from the ipsilateral anteroventral cochlear nucleus, and the medial dendrites receive input from the contralateral anteroventral cochlear nucleus (both inputs are excitatory). As might be

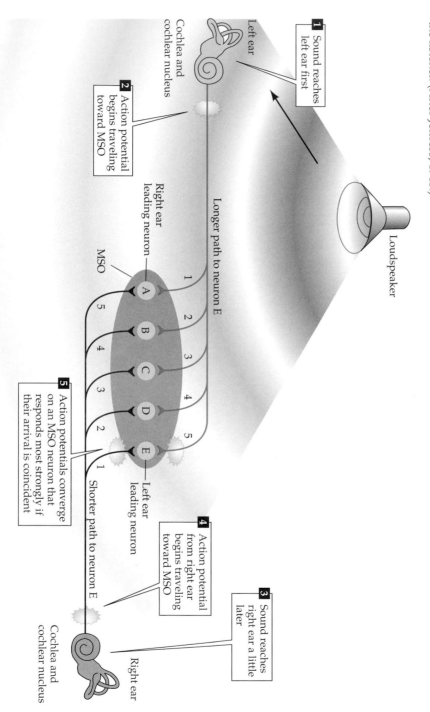

Figure 13.12 Diagram illustrating how the MSO computes the location of a sound by interaural time differences. A given MSO neuron responds most strongly when the two inputs arrive simultaneously, as occurs when the contralateral and ipsilateral inputs precisely compensate (via their different lengths) for differences in the time of arrival of a sound at the two ears. The systematic (and inverse) variation in the delay lengths of the two inputs creates a map of sound location: In this model, E would be most sensitive to sounds located to the left, and A to sounds from the right; C would respond best to sounds coming from directly in front of the listener. (After Jeffress, 1948.)

Loudspeaker

Left ear

Cochlea and cochlear nucleus

1 Sound reaches left ear first

2 Action potential begins traveling toward MSO

Longer path to neuron E

Right ear leading neuron

MSO

A B C D E

1 2 3 4 5

5 Action potentials converge on an MSO neuron that responds most strongly if their arrival is coincident

4 Action potential begins traveling toward MSO

3 Sound reaches right ear a little later

Left ear leading neuron

Shorter path to neuron E

Right ear

Cochlea and cochlear nucleus

expected, the MSO cells work as **coincidence detectors**, responding when both excitatory signals arrive at the same time. For a coincidence mechanism to be useful in localizing sound, different neurons must be maximally sensitive to different interaural time delays. The axons that project from the anteroventral cochlear nucleus evidently vary systematically in length to create delay lines. (Remember that the length of an axon multiplied by its conduction velocity equals the conduction time.) These anatomical differences compensate for sounds arriving at slightly different times at the two ears, so that the resultant neural impulses arrive at a particular MSO neuron simultaneously, making each cell especially sensitive to sound sources in a particular place.

Sound localization perceived on the basis of interaural time differences requires phase-locked information from the periphery, which, as already emphasized, is available to humans only for frequencies below 3 kHz. (In barn owls, the reigning champions of sound localization, phase locking occurs at up to 9 kHz.) Therefore, a second mechanism must come into play at higher frequencies. At frequencies higher than about 2 kHz, the human head begins to act as an acoustical obstacle because the wavelengths of the sounds are too short to bend around it. As a result, when high-frequency sounds are directed toward one side of the head, an acoustical "shadow" of lower intensity is created at the far ear. These intensity differences provide a second cue about the location of a sound. The circuits that compute the position of a sound source on this basis are found in the **lateral superior olive (LSO)** and the **medial nucleus of the trapezoid body (MNTB)**. Excitatory axons project directly from the ipsilateral anteroventral cochlear nucleus to the LSO (as well as to the MSO; see Figure 13.11). Note that the LSO also receives inhibitory input from the contralateral ear, via an inhibitory neuron in the MNTB (Figure 13.13). This excitatory/inhibitory interaction results in a net excitation of the LSO on the same side of the body as the sound source.

Figure 13.13 Lateral superior olive neurons encode sound location through interaural intensity differences. (A) LSO neurons receive direct excitation from the ipsilateral cochlear nucleus; input from the contralateral cochlear nucleus is relayed via inhibitory interneurons in the MNTB. (B) This arrangement of excitation–inhibition makes LSO neurons fire most strongly in response to sounds arising directly lateral to the listener on the same side as the LSO, because excitation from the ipsilateral input will be great and inhibition from the contralateral input will be small. In contrast, sounds arising from in front of the listener, or from the opposite side, will silence the LSO output, because excitation from the ipsilateral input will be minimal, but inhibition driven by the contralateral input will be great. Note that LSOs are paired and bilaterally symmetrical; each LSO only encodes the location of sounds arising on the same side of the body as its location.

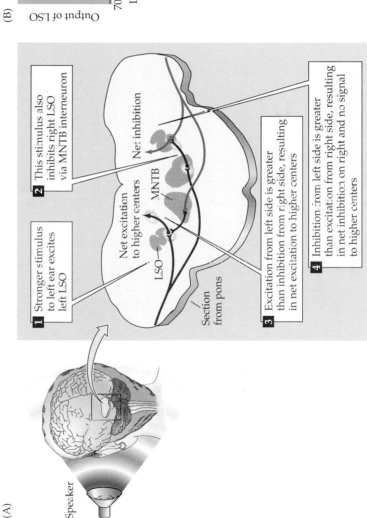

(A)

Speaker

1 — Stronger stimulus to left ear excites left LSO

2 — This stimulus also inhibits right LSO via MNTB interneuron

Net excitation to higher centers

Net inhibition

LSO

MNTB

Section from pons

3 — Excitation from left side is greater than inhibition from right side, resulting in net excitation to higher centers

4 — Inhibition from left side is greater than excitation from right side, resulting in net inhibition on right and no signal to higher centers

(B)

Output of LSO

Left LSO output

Right LSO output

Left > right   Right > left

70   40   20   0   −20   −40   −70

Relative loudness

For sounds arising directly lateral to the listener, firing rates will be highest in the LSO on that side; in this circumstance, the excitation via the ipsilateral anteroventral cochlear nucleus will be maximal, and inhibition from the contralateral MNTB minimal. In contrast, sounds arising closer to the listener's midline will elicit lower firing rates in the ipsilateral LSO because of increased inhibition arising from the contralateral MNTB. For sounds arising at the midline, or from the other side, the increased inhibition arising from the MNTB is powerful enough to completely silence LSO activity. Note that each LSO only encodes sounds arising in the ipsilateral hemifield; it therefore takes both LSOs to represent the full range of horizontal positions.

In summary, there are two separate pathways—and two separate mechanisms—for localizing sound. Interaural time differences are processed in the medial superior olive, and interaural intensity differences are processed in the lateral superior olive. These two pathways are eventually merged in the midbrain auditory centers.

Monaural Pathways from the Cochlear Nucleus to the Lateral Lemniscus

The binaural pathways for sound localization are only part of the output of the cochlear nucleus. This fact is hardly surprising, given that auditory perception involves much more than locating the position of the sound source. A second major set of pathways from the cochlear nucleus bypasses the superior olive and terminates in the **nuclei of the lateral lemniscus** on the contralateral side of the brainstem (see Figure 13.11). These particular pathways respond to sound arriving at one ear only and are thus referred to as monaural. Some cells in the lateral lemniscus nuclei signal the onset of sound, regardless of its intensity or frequency. Other cells in the lateral lemniscus nuclei process other temporal aspects of sound, such as duration. The precise role of these pathways in processing temporal features of sound is not yet known. As with the outputs of the superior olivary nuclei, the pathways from the nuclei of the lateral lemniscus converge at the midbrain.

Integration in the Inferior Colliculus

Auditory pathways ascending via the olivary and lemniscal complexes, as well as other projections that arise directly from the cochlear nucleus, project to the midbrain auditory center, the **inferior colliculus** (see Figure 13.11). In examining how integration occurs in the inferior colliculus, it is again instructive to turn to the most completely analyzed auditory mechanism, the binaural system for localizing sound. As already noted, space is not mapped on the auditory receptor surface; thus the perception of auditory space must somehow be synthesized by circuitry in the lower brainstem and midbrain. Experiments in the barn owl, an extraordinarily proficient animal at localizing sounds, show that the convergence of binaural inputs in the midbrain produces something entirely new relative to the periphery—namely, a computed topographical representation of auditory space. Neurons within this **auditory space map** in the colliculus respond best to sounds originating in a specific region of space and thus have both a preferred elevation and a preferred horizontal location, or azimuth. Although comparable maps of auditory space have not yet been found in mammals, humans have a clear perception of both the elevational and azimuthal components of a sound's location, suggesting that we have a similar auditory space map.

Another important property of the inferior colliculus is its ability to process sounds with complex temporal patterns. Many neurons in the inferior colliculus respond only to frequency-modulated sounds, while others respond only to sounds of specific durations. Such sounds are typical components of biologically relevant sounds, such as those made by predators, or intraspecific communication sounds, which in humans include speech. The inferior colliculus is evidently the first stage in a system, continued in the auditory thalamus and cortex, that analyzes sounds that have particular significance.

The Auditory Thalamus

Despite the parallel pathways in the auditory stations of the brainstem and midbrain, the **medial geniculate complex (MGC)** in the thalamus is an obligatory relay for all ascending auditory information destined for the cortex (see Figure 13.11). Most input to the MGC arises from the inferior colliculus, although a few auditory axons from the lower brainstem bypass the inferior colliculus to reach the auditory thalamus directly. The MGC has several divisions, including the ventral division, which functions as the major thalamocortical relay, and the dorsal and medial divisions, which are organized like a belt around the ventral division.

In some mammals, the strictly maintained tonotopy of the lower brainstem areas is exploited by convergence onto MGC neurons, generating specific responses to certain spectral combinations. The original evidence for this statement came from research on the response properties of cells in the MGC of echolocating bats. Some cells in the so-called belt areas of the bat MGC respond only to combinations of widely spaced frequencies that are specific components of the bat's echolocation signal and of the echoes that are reflected from objects in the bat's environment. In the mustached bat, where this phenomenon has been most thoroughly studied, the echolocation pulse has a changing frequency (frequency-modulated, or FM) component that includes a fundamental frequency and one or more harmonics. The fundamental frequency (FM_1) has low intensity and sweeps from 30 kHz to 20 kHz. The second harmonic (FM_2) is the most intense component and sweeps from 60 kHz to 40 kHz. Note that these frequencies do not overlap. Most of the echoes are from the intense FM_2 sound, and virtually none arise from the weak FM_1, even though the emitted FM_1 is loud enough for the bat to hear. Apparently, the bat assesses the distance to an object by measuring the delay between the FM_1 emission and the FM_2 echo. Certain MGC neurons respond when FM_2 follows FM_1 by a specific delay, providing a mechanism for sensing such frequency combinations. Because each neuron responds best to a particular delay, a range of distances is encoded by the population of MGC neurons.

Bat sonar illustrates two important points about the function of the auditory thalamus. First, the MGC is the first station in the auditory pathway where selectivity for combinations of frequencies is found. The mechanism responsible for this selectivity is presumably the ultimate convergence of inputs from cochlear areas with different spectral sensitivities. Second, cells in the MGC are selective not only for frequency combinations, but also for specific time intervals between the two frequencies. The principle is the same as that described for binaural neurons in the medial superior olive, but in this instance, two monaural signals with different frequency sensitivity coincide, and the time difference is in the millisecond rather than the microsecond range.

In summary, neurons in the medial geniculate complex receive convergent inputs from spectrally and temporally separate pathways. This complex, by virtue of its convergent inputs, mediates the detection of specific spectral and temporal combinations of sounds. In many species, including humans, varying spectral and temporal cues are especially important features of communication sounds. It is not known whether cells in the human medial geniculate are selective to combinations of sounds, but the processing of speech certainly requires both spectral and temporal combination sensitivity.

The Auditory Cortex

The ultimate target of afferent auditory information is the auditory cortex. Although the auditory cortex has a number of subdivisions, a broad distinction can be made between a primary area and peripheral, or belt, areas. The **primary auditory cortex (A1)** is located on the superior temporal gyrus in the temporal lobe and receives point-to-point input from the ventral division of the medial geniculate complex; thus, it contains a precise tonotopic map. The **belt areas** of the auditory cortex receive more diffuse input from the belt areas of the medial geniculate complex and therefore are less precise in their tonotopic organization.

The primary auditory cortex (A1) has a topographical map of the cochlea (Figure 13.14), just as the primary visual cortex (V1) and the primary somatic sensory cortex (S1) have topographical maps of their respective sensory epithelia. Unlike the visual and somatic sensory systems, however, the cochlea has already decomposed the acoustical stimulus so that it is arrayed tonotopically along the length of the basilar membrane. Thus, A1 is said to comprise a tonotopic map, as do most of the ascending auditory structures between the cochlea and the cortex. Orthogonal to the frequency axis of the tonotopic map is a striped arrangement of binaural properties. The neurons in one stripe are excited by both ears (and are therefore called EE cells), while the neurons in the next stripe are excited by one ear and inhibited by the other ear (EI cells). The EE and EI stripes alternate, an arrangement that is reminiscent of the ocular dominance columns in V1 (see Chapter 12). The sorts of sensory processing that occur in the other divisions of the auditory cortex are not well understood, but they are likely to be important to higher-order processing of natural sounds, including those used for communication. It appears that some areas are specialized for processing combinations of frequencies, while others are specialized for processing modulations of amplitude or frequency.

Sounds that are especially important for intraspecific communication often have a highly ordered temporal structure. In humans, the best example

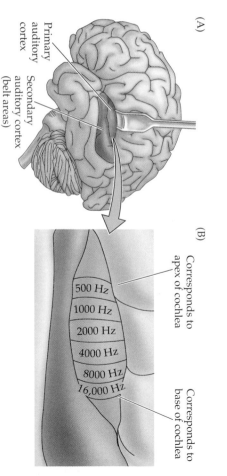

Figure 13.14 The human auditory cortex. (A) Diagram showing the brain in left lateral view, including the depths of the lateral sulcus, where part of the auditory cortex occupying the superior temporal gyrus normally lies hidden. The primary auditory cortex (A1) is shown in blue; the surrounding belt areas of the auditory cortex are in red. (B) The primary auditory cortex has a tonotopic organization, as shown in this blowup diagram of a segment of A1.

(A)

Primary
auditory
cortex

Secondary
auditory cortex
(belt areas)

(B)

Corresponds to
apex of cochlea

500 Hz
1000 Hz
2000 Hz
4000 Hz
8000 Hz
16,000 Hz

Corresponds to
base of cochlea

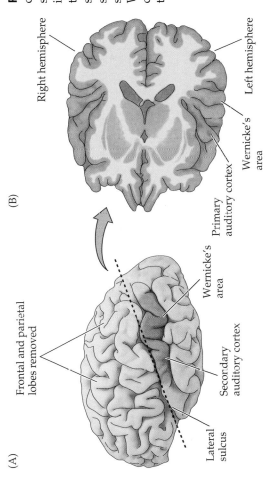

(A)

Frontal and parietal lobes removed

Wernicke's area

Secondary auditory cortex

Lateral sulcus

(B)

Right hemisphere

Primary auditory cortex

Wernicke's area

Left hemisphere

Figure 13.15 The human auditory cortical areas related to processing speech sounds. (A) Diagram of the brain in left lateral view, showing locations in the intact hemisphere. (B) An oblique section (plane of dashed line in A) shows the cortical areas on the superior surface of the temporal lobe. Note that Wernicke's area, a region important in comprehending speech, is just posterior to the primary auditory cortex.

of such time-varying signals is speech, where different phonetic sequences are perceived as distinct syllables and words. Behavioral studies in cats and monkeys show that the auditory cortex is especially important for processing temporal sequences of sound. If the auditory cortex is ablated in these animals, they lose the ability to discriminate between two complex sounds that have the same frequency components but which differ in temporal sequence. Thus, without the auditory cortex, monkeys cannot discriminate one conspecific communication sound from another. Studies of human patients with bilateral damage to the auditory cortex also reveal severe problems in processing the temporal order of sounds. It seems likely, therefore, that specific regions of the human auditory cortex are specialized for processing elementary speech sounds, as well as other temporally complex acoustical signals, such as music. Indeed, Wernicke's area, which is critical to the comprehension of human language, lies within the secondary auditory area (Figure 13.15; see also Chapter 27).

Summary

Sound waves are transmitted via the external and middle ear to the cochlea of the inner ear, which exhibits a traveling wave when stimulated. For high-frequency sounds, the amplitude of the traveling wave reaches a maximum at the base of the cochlea; for low-frequency sounds, the traveling wave reaches a maximum at the apical end. The associated motions of the basilar membrane are transduced primarily by the inner hair cells, while the basilar membrane motion is itself actively modulated by the outer hair cells. Damage to the outer or middle ear results in conductive hearing loss, while hair cell damage results in a sensorineural hearing deficit. The tonotopic organization of the cochlea is retained at all levels of the central auditory system. Projections from the cochlea travel via the eighth nerve to the three main divisions of the cochlear nucleus. The targets of the cochlear nucleus neurons include the superior olivary complex and nuclei of the lateral lemniscus, where the binaural cues for sound localization are processed. The inferior colliculus is the target of nearly all of the auditory pathways in the lower brainstem and carries out important integrative functions, such as processing sound frequencies and integrating the cues for localizing sound in space. The primary auditory cortex, which is also organized tonotopically, is essential for basic auditory functions, such as frequency discrimination and sound localization. The belt areas of the auditory cortex have a less strict tonotopic organization and probably process complex sounds, such as those that mediate communication. In the human brain, the major speech comprehension areas are located in the zone immediately adjacent to the auditory cortex.

ADDITIONAL READING

Reviews

COREY, D. P. AND A. J. HUDSPETH (1979) Ionic basis of the receptor potential in a vertebrate hair cell. Nature 281: 675–677.

DALLOS, P. (1992) The active cochlea. J. Neurosci. 12: 4575–4585.

GARCIA-ANOVEROS, J. AND D. P. COREY (1997) The molecules of mechanosensation. Ann. Rev. Neurosci. 20: 567–597.

HEFFNER, H. E. AND R. S. HEFFNER (1990) Role of primate auditory cortex in hearing. In *Comparative Perception, Volume II: Complex Signals*. W. C. Stebbins and M. A. Berkley (eds.). New York: John Wiley.

HUDSPETH, A. J. (1997) How hearing happens. Neuron 19(5): 947–950.

KIANG, N. Y. S. (1984) Peripheral neural processing of auditory information. In *Handbook of Physiology*, Section 1: *The Nervous System*, Volume III. *Sensory Processes*, Part 2. J. M. Brookhart, V. B. Mountcastle, I. Darian-Smith and S. R. Geiger (eds.), Bethesda, MD: American Physiological Society.

LEWIS, R. S. AND A. J. HUDSPETH (1983) Voltage- and ion-dependent conductances in solitary vertebrate hair cells. Nature 304: 538–541.

MIDDLEBROOKS, J. C., A. E. CLOCK, L. XU AND D. M. GREEN (1994) A panoramic code for sound location by cortical neurons. Science 264: 842–844.

NEFF, W. D., I. T. DIAMOND AND J. H. CASSEDAY (1975) Behavioral studies of auditory discrimination. In *Handbook of Sensory Physiology*, Volumes V–II. W. D. Keidel and W. D. Neff (eds.). Berlin: Springer-Verlag.

SUGA, N. (1990) Biosonar and neural computation in bats. Sci. Am. 262 (June): 60–68.

Important Original Papers

CRAWFORD, A. C. AND R. FETTIPLACE (1981) An electrical tuning mechanism in turtle cochlear hair cells. J. Physiol. 312: 377–412.

FITZPATRICK, D. C., J. S. KANWAL, J. A. BUTMAN AND N. SUGA (1993) Combination-sensitive neurons in the primary auditory cortex of the mustached bat. J. Neurosci. 13: 931–940.

KNUDSEN, E. I. AND M. KONISHI (1978) A neural map of auditory space in the owl. Science 200: 795–797.

JEFFRESS, L. A. (1948) A place theory of sound localization. J. Comp. Physiol. Psychol. 41: 35–39.

SUGA, N., W. E. O'NEILL AND T. MANABE (1978) Cortical neurons sensitive to combinations of information-bearing elements of biosonar signals in the mustache bat. Science 200: 778–781.

VON BÉKÉSY, G. (1960) *Experiments in Hearing*. New York: McGraw-Hill. (A collection of von Békésy's original papers.)

Books

PICKLES, J. O. (1988) *An Introduction to the Physiology of Hearing*. London: Academic Press.

YOST, W. A. AND G. GOUREVITCH (eds.) (1987) *Directional Hearing*. Berlin: Springer-Verlag.

YOST, W. A. AND D. W. NIELSEN (1985) *Fundamentals of Hearing*. Fort Worth: Holt, Rinehart and Winston.

Chapter 14

The Vestibular System

Overview

The vestibular system provides the sense of balance and the information about body position that allows rapid compensatory movements in response to both self-induced and externally generated forces. The peripheral portion of the vestibular system is a part of the inner ear that acts as a miniaturized accelerometer and inertial guidance device, continually reporting information about the motions and position of the head and body to integrative centers located in the brainstem, cerebellum, and somatic sensory cortices. Although we are normally unaware of its function, the vestibular system is a key component in both postural reflexes and eye movements when the head is moving, and sense of orientation in space are all adversely affected. These manifestations of vestibular damage are especially important in the evaluation of brainstem injury. The circuitry of the vestibular system extends through a large part of the brainstem, and simple clinical tests of vestibular function can be performed to determine brainstem involvement, even on comatose patients.

The Vestibular Labyrinth

The main peripheral component of the vestibular system is an elaborate set of interconnected canals—the **labyrinth**—that has much in common, and is in fact continuous with, the cochlea. Like the cochlea (see Chapter 13), the vestibular system is derived from the otic placode of the embryo, and it uses the same specialized set of sensory cells—hair cells—to transduce physical motion into neural impulses. In the cochlea, the motion is due to airborne sounds; in the vestibular system, the motions transduced arise from head movements, inertial effects due to gravity, and ground-borne vibrations (Box A).

The labyrinth is buried deep in the temporal bone and consists of the two **otolith organs** (the **utricle** and the **sacculus**) and the **semicircular canals** (Figure 14.1). The elaborate and tortuous architecture of these components explains why this part of the vestibular system is called the labyrinth. The utricle and sacculus are specialized primarily to respond to *linear accelerations* of the head and *static head position*, whereas the semicircular canals, as their shapes suggest, are specialized for responding to *rotational accelerations* of the head.

The intimate relationship between the cochlea and the labyrinth goes beyond their common embryonic origin. Indeed, the cochlear and vestibular spaces are actually joined (see Figure 14.1), and the specialized ionic environments of the vestibular end organ parallel those of the cochlea. The membranous sacs within the bone are filled with fluid (endolymph) and are collectively called the membranous labyrinth. The endolymph (like the cochlear

Figure 14.1   The labyrinth and its innervation. The vestibular and auditory portions of the eighth nerve are shown; the small connection from the vestibular nerve to the cochlea contains auditory efferent fibers. General orientation in head is shown in Figure 13.3; see also Figure 14.8.

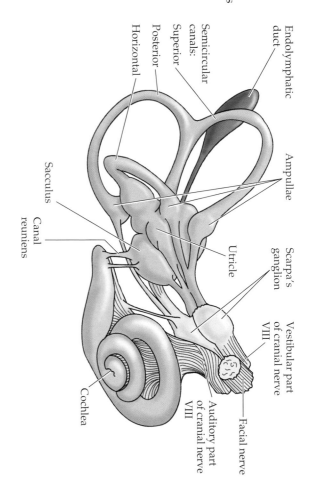

Endolymphatic duct

Semicircular canals:
Superior
Posterior
Horizontal

Ampullae

Sacculus

Canal reuniens

Scarpa's ganglion

Utricle

Vestibular part of cranial nerve VIII

Auditory part of cranial nerve VIII

Facial nerve

Cochlea

endolymph) is similar to intracellular solutions in that it is high in K⁺ and low in Na⁺. Between the bony walls (the osseous labyrinth) and the membranous labyrinth is another fluid, the perilymph, which is similar in composition to cerebrospinal fluid (i.e., low in K⁺ and high in Na⁺; see Chapter 13).

The vestibular hair cells are located in the utricle and sacculus and in three juglike swellings, called **ampullae**, located at the base of the semicircular canals next to the utricle. Within each of these structures, the vestibular hair cells extend their hair bundles into the endolymph of the membranous labyrinth. As in the cochlea, tight junctions seal the apical surfaces of the vestibular hair cells, ensuring that endolymph selectively bathes the hair cell bundle, while remaining separate from the perilymph surrounding the basal portion of the hair cell.

Vestibular Hair Cells

The vestibular hair cells, which, like cochlear hair cells, transduce minute displacements into behaviorally relevant receptor potentials, provide the basis for vestibular function. Vestibular and auditory hair cells are quite similar; a detailed description of hair cell structure and function has already been given in Chapter 13. As in the case of auditory hair cells, movement of the stereocilia toward the kinocilium in the vestibular end organs opens mechanically gated transduction channels located at the tips of the stereocilia, depolarizing the hair cell and causing neurotransmitter release onto (and excitation of) the vestibular nerve fibers. Movement of the stereocilia in the direction away from the kinocilium closes the channels, hyperpolarizing the hair cell and thus reducing vestibular nerve activity. The biphasic nature of the receptor potential means that some transduction channels are open in the absence of stimulation, with the result that hair cells tonically release transmitter, thereby generating considerable spontaneous activity in vestibular nerve fibers (Box B). One consequence of these spontaneous action potentials is that the firing rates of vestibular fibers can increase or

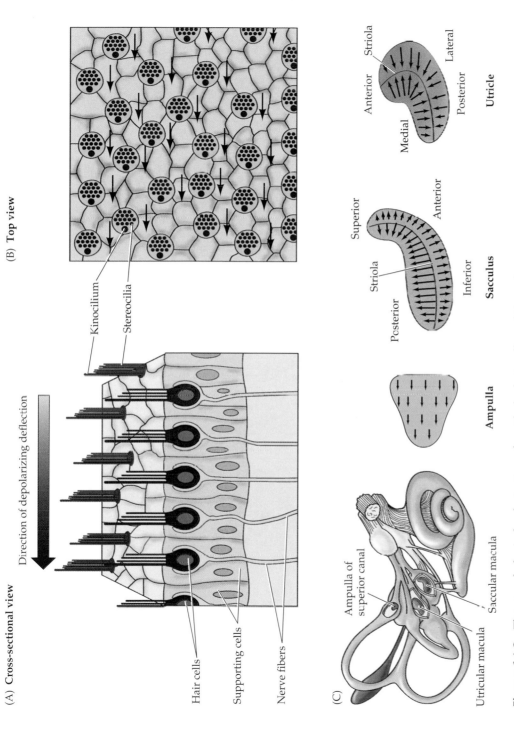

(A) Cross-sectional view

Direction of depolarizing deflection

Kinocilium

Stereocilia

Hair cells

Supporting cells

Nerve fibers

(B) Top view

(C)

Ampulla of
superior canal

Utricular macula

Saccular macula

Superior

Striola

Posterior

Anterior

Inferior

Sacculus

Ampulla

Anterior

Striola

Medial

Lateral

Posterior

Utricle

Figure 14.2 The morphological polarization of vestibular hair cells and the polarization maps of the vestibular organs. (A) A cross section of hair cells shows that the
kinocilia of a group of hair cells are all located on the same side of the hair cell. The
arrow indicates the direction of deflection that depolarizes the hair cell. (B) View
looking down on the hair bundles. (C) In the ampulla located at the base of each
semicircular canal, the hair bundles are oriented in the same direction. In the sacculus and utricle, the striola divides the hair cells into populations with opposing hair
bundle polarities.

decrease in a manner that faithfully mimics the receptor potentials produced
by the hair cells.

Importantly, the hair cell bundles in each vestibular organ have specific
orientations (Figure 14.2). As a result, the organ as a whole is responsive to
displacements in all directions. In a given semicircular canal, the hair cells in
the ampulla are all polarized in the same direction (Figure 14.2C). In the utricle and sacculus, a specialized area called the **striola** divides the hair cells
into two populations with opposing polarities (Figures 14.2C; see also Figure
14.4C). The directional polarization of the receptor surfaces is a basic principle of organization in the vestibular system, as will become apparent in the
following descriptions of the individual vestibular organs.

BOX A
A Primer on (Vestibular) Navigation

The function of the vestibular system can be simplified by remembering some basic terminology of classical mechanics. All bodies moving in a three-dimensional framework have six degrees of freedom: three of these are translational and three are rotational. The translational elements refer to linear movements in the x, y, and z axes (the horizontal and vertical planes). Translational motion in these planes (linear acceleration and static displacement of the head) is the primary concern of the otolith organs. The three degrees of rotational freedom refer to a body's rotation relative to the x, y, and z axes and are commonly referred to as roll, pitch, and yaw. The semicircular canals are primarily responsible for sensing rotational accelerations around these three axes.

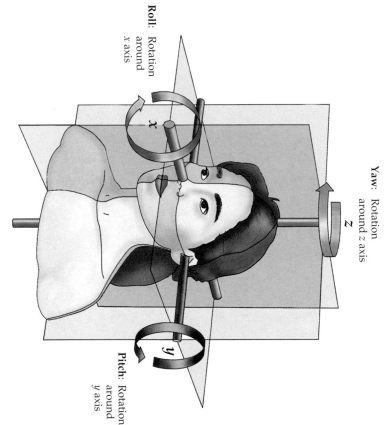

Yaw: Rotation around z axis

Roll: Rotation around x axis

Pitch: Rotation around y axis

Figure 14.3   Scanning electron micrograph of calcium carbonate crystals (otoconia) in the utricular macula of the cat. Each crystal is about 50 mm long. (From Lindeman, 1973.)

The Otolith Organs: The Utricle and Sacculus

Displacements and linear accelerations of the head, such as those induced by tilting or translational movements (see Box A), are detected by the two otolith organs: the sacculus and the utricle. Both of these organs contain a sensory epithelium, the **macula**, which consists of hair cells and associated supporting cells. Overlying the hair cells and their hair bundles is a gelatinous layer, and above this is a fibrous structure, the **otolithic membrane**, in which are embedded crystals of calcium carbonate called **otoconia** (Figures 14.3 and 14.4A). The crystals give the otolith organs their name (*otolith* is Greek for "ear stones"). The otoconia make the otolithic membrane considerably heavier than the structures and fluids surrounding it; thus, when the head tilts, gravity causes the membrane to shift relative to the sensory epithelium (Figure 14.4B). The resulting shearing motion between the otolithic membrane and the macula displaces the hair bundles, which are embedded in the lower, gelatinous surface of the membrane. This displacement of the hair bundles generates a receptor potential in the hair cells. A shearing motion between the macula and the otolithic membrane also occurs when the head undergoes linear accelerations (see Figure 14.5); the greater relative mass of the otolithic membrane causes it to lag behind the macula temporarily, leading to transient displacement of the hair bundle. The similar effects exerted on otolithic hair cells by certain head tilts and linear accelerations

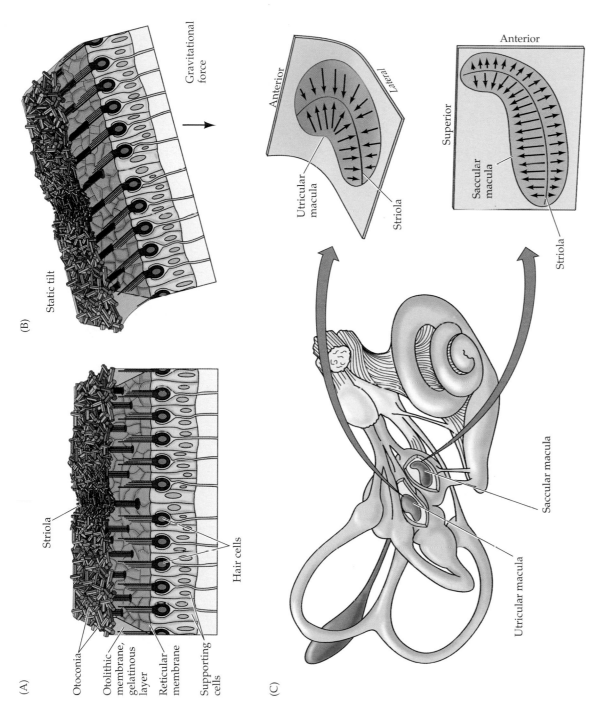

(A)

Striola

Otoconia

Otolithic membrane, gelatinous layer

Reticular membrane

Supporting cells

Hair cells

(B)

Static tilt

Gravitational force

(C)

Anterior

Lateral

Utricular macula

Striola

Anterior

Superior

Saccular macula

Striola

Utricular macula

Saccular macula

Figure 14.4 Morphological polarization of hair cells in the utricular and saccular maculae. (A) Cross section of the utricular macula showing hair bundles projecting into the gelatinous layer when the head is level. (B) Cross section of the utricular macula when the head is tilted. (C) Orientation of the utricular and saccular maculae in the head; arrows show orientation of the kinocilia, as in Figure 14.2. The *saccules* on either side are oriented more or less vertically, and the *utricles* more or less horizontally. The striola is a structural landmark consisting of small otoconia arranged in a narrow trench that divides each otolith organ. In the utricular macula, the kinocilia are directed toward the striola. In the saccular macula, the kinocilia point away from the striola. Note that, given the utricle and sacculus on both sides of the body, there is a continuous representation of all directions of body movement.

explains the perceptual equivalency of these different stimuli when visual feedback is absent, as occurs in the dark or when the eyes are closed.

As already mentioned, the orientation of the hair cell bundles is organized relative to the striola, which demarcates the overlying layer of otoconia (see Figure 14.4A). The striola forms an axis of mirror symmetry such that hair cells on opposite sides of the striola have opposing morphological polarizations. Thus, a tilt along the axis of the striola will excite the hair cells on one side while inhibiting the hair cells on the other side. The saccular macula is oriented vertically and the utricular macula horizontally, with a continuous variation in the morphological polarization of the hair cells located in each macula (as shown in Figure 14.4C, where the arrows indicate the direction of movement that produces excitation). Inspection of the excitatory orientations in the macula indicates that the utricle responds to movements of the head in the horizontal plane, such as sideways head tilts and rapid lateral displacements, whereas the sacculus responds to movements in the vertical plane (up–down and forward–backward movements in the sagittal plane). Note that the saccular and utricular maculae on one side of the head are mirror images of those on the other side. Thus, a tilt of the head to one side has opposite effects on corresponding hair cells of the two utricular maculae. This concept is important in understanding how the central connections of the vestibular periphery mediate the interaction of inputs from the two sides of the head (see next section).

How Otolith Neurons Sense Linear Forces

The structure of the otolith organs enables them to sense both static displacements, as would be caused by tilting the head, and transient displacements caused by translational movements of the head. The mass of the otolithic membrane relative to the surrounding endolymph, as well as the otolithic membrane's physical uncoupling from the underlying macula, means that hair bundle displacement will occur transiently in response to linear accelerations and tonically in response to tilting of the head. Therefore, both tonic and transient information can be conveyed by these sense organs. Figure 14.5 illustrates some of the forces produced by head tilt and linear accelerations on the utricular macula.

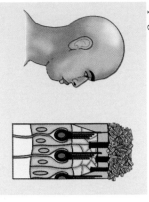

Upright

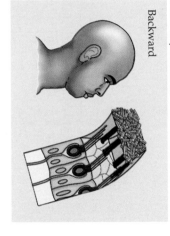

Head tilt; sustained

Backward

No head tilt; transient

Forward acceleration

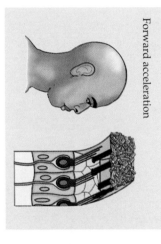

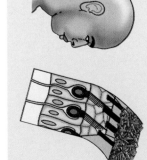

Forward

Deceleration

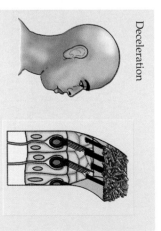

Figure 14.5 Forces acting on the head and the resulting displacement of the otolithic membrane of the utricular macula. For each of the positions and accelerations due to translational movements, some set of hair cells will be maximally excited, whereas another set will be maximally inhibited. Note that head tilts produce displacements similar to certain accelerations.

Figure 14.6 Response of a vestibular nerve axon from an otolith organ (the utricle in this example). (A) The stimulus (top) is a change that causes the head to tilt. The histogram shows the neuron's response to tilt. (B) A response of the same fiber to tilting in the opposite direction. (After Goldberg and Fernandez, 1976.)

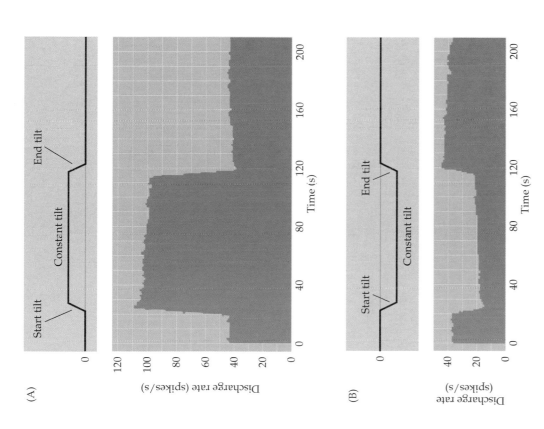

These properties of hair cells are reflected in the responses of the vestibular nerve fibers that innervate the otolith organs. These vestibular nerve fibers have a steady and relatively high firing rate when the head is upright. The change in firing rate in response to a given movement can be either sustained or transient, decaying back to a tonic level, thereby signaling either absolute head position or linear acceleration. An example of the sustained response of a vestibular nerve fiber innervating the utricle is shown in Figure 14.6. These responses are from axons in a monkey that was seated in a chair that could be tilted for several seconds to produce a steady force. It is important to note that prior to the tilt, the axon has a high firing rate, which increases or decreases depending on the direction of the tilt. Notice also that the response remains at a high level as long as the tilting force remains constant; thus, such neurons faithfully encode the static force being applied to the head (Figure 14.6A). When the head is returned to the original position, the firing level of the neurons returns to baseline value. Conversely, when the tilt is in the opposite direction, the neurons respond by decreasing their firing rate below the resting level (Figure 14.6B) and remain depressed as long as the static force continues. In a similar fashion, transient increases or decrease in firing rate from spontaneous levels signal the direction of linear accelerations of the head. In summary, the otolith organs detect linear forces acting on the head,

BOX B
Adaptation and Tuning of Vestibular Hair Cells

Hair Cell Adaptation

The minuscule movement of the hair bundle at sensory threshold has been compared to the displacement of the top of the Eiffel Tower by a thumb's breadth! Despite its great sensitivity, the hair cell can still adapt quickly and continuously to static displacements of the hair bundle caused by large movements. Such adjustments are especially useful in the otolith organs, where adaptation permits hair cells to maintain sensitivity to small linear and angular accelerations of the head despite the constant input from gravitational forces that are over a million times greater. In other receptor cells, such as photoreceptors, adaptation is accomplished by regulating the second messenger cascade induced by the initial transduction event. The hair cell has to depend on a different strategy, however, because there is no second messenger system between the initial transduction event and the subsequent receptor potential (as might be expected for receptors that respond so rapidly).

Adaptation occurs in both directions in which the hair bundle displacement generates a receptor potential, albeit at different rates for each direction. When the hair bundle is pushed toward the kinocilium, tension is initially increased in the gating spring. During adaptation, tension decreases back to the resting level, perhaps because one end of the gating spring repositions itself along the shank of the stereocilium. When the hair bundle is displaced in the opposite direction, away from the kinocilium, tension in the spring initially decreases; adaptation then involves an increase in spring tension. One theory is that a calcium-regulated motor such as a myosin ATPase climbs along actin filaments in the stereocilium and actively resets the tension in the transduction spring. During sustained depolarization, some Ca^{2+} enters through the transduction channel, along with K^+, Ca^{2+} then causes the motor to spend a greater fraction of its time unbound from the actin, resulting in slippage of the spring down the side of the stereocilium. During sustained hyperpolarization (figure A), Ca^{2+} levels drop below normal resting levels, and the motor spends more of its time bound to the actin, thus climbing up the actin filaments and increasing the spring tension. As tension increases, though, some of the previously closed transduction channels open, admitting Ca^{2+} and thus slowing the motor's progress until a balance is struck between the climbing and slipping of the motor. In support of this model, when internal Ca^{2+} is reduced artificially, spring tension increases. This model of hair cell adaptation presents an elegant molecular solution to the regulation of a mechanical process.

Electrical Tuning

Although mechanical tuning plays an important role in generating frequency selectivity in the cochlea, there are other mechanisms that contribute to this process in vestibular and auditory nerve cells. These other tuning mechanisms are especially important in the otolith organs,

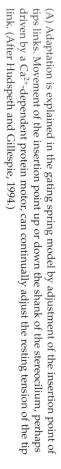

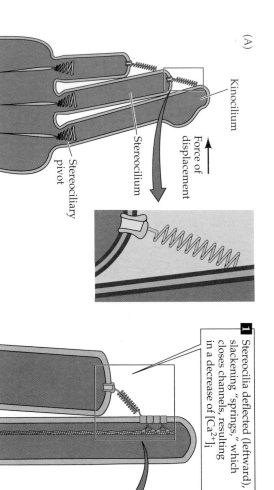

(A)

Kinocilium

Stereocilium

Stereociliary pivot

Force of displacement

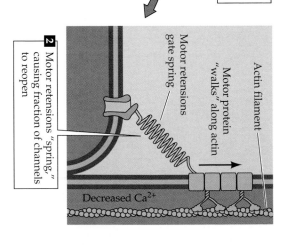

1 Stereocilia deflected (leftward), slackening "springs," which closes channels, resulting in a decrease of $[Ca^{2+}]_i$

2 Motor retensions "spring," causing fraction of channels to reopen

Motor retensions gate spring

Motor protein "walks" along actin

Actin filament

Decreased Ca^{2+}

(A) Adaptation is explained in the gating spring model by adjustment of the insertion point of tips links. Movement of the insertion point up or down the shank of the stereocilium, perhaps driven by a Ca^{2+}-dependent protein motor, can continually adjust the resting tension of the tip link. (After Hudspeth and Gillespie, 1994.)

where, unlike the cochlea, there are no obvious macromechanical resonances to selectively filter and/or enhance biologically relevant movements. One such mechanism is an electrical resonance displayed by hair cells in response to depolarization. The membrane potential of a hair cell undergoes damped sinusoidal oscillations at a specific frequency in response to the injection of depolarizing current pulses (figure B).

The ionic mechanism of this process involves two major types of ion channels located in the membrane of the hair cell soma. The first of these is a voltage-activated Ca^{2+} conductance, which lets Ca^{2+} into the cell soma in response to depolarization, such as that generated by the transduction current. The second is a Ca^{2+}-activated K^+ conductance, which is triggered by the rise in internal Ca^{2+} concentration. These two currents produce an interplay of depolarization and repolarization that results in electrical resonance (figure C). Activation of the hair cell's calcium-activated K^+ conductance

occurs 10 to 100 times faster than that of similar currents in other cells. Such rapid kinetics allow this conductance to generate an electrical response that usually requires the fast properties of a voltage-gated channel.

Although a hair cell responds to hair bundle movement over a wide range of frequencies, the resultant receptor potential is largest at the frequency of electrical resonance. The resonance frequency represents the characteristic frequency of the hair cell, and transduction at that frequency will be most efficient. This electrical resonance has important implications for structures like the utricle and sacculus, which may encode a range of characteristic frequencies based on the different resonance frequencies of their constituent hair cells. Thus, electrical tuning in the otolith organs can generate enhanced tuning to biologically relevant frequencies of stimulation, even in the absence of macromechanical resonances within these structures.

References

ASSAD, J. A. AND D. P. COREY (1992) An active motor model for adaptation by vertebrate hair cells. J. Neurosci. 12: 3291–3309.

CRAWFORD, A. C. AND R. FETTIPLACE (1981) An electrical tuning mechanism in turtle cochlear hair cells. J. Physiol. 312: 377–412.

HUDSPETH, A. J. (1985) The cellular basis of hearing: The biophysics of hair cells. Science 230: 745–752.

HUDSPETH, A. J. AND P. G. GILLESPIE (1994) Pulling strings to tune transduction: Adaptation by hair cells. Neuron 12: 1–9.

LEWIS, R. S. AND A. J. HUDSPETH (1988) A model for electrical resonance and frequency tuning in saccular hair cells of the bull-frog, *Rana catesbeiana*. J. Physiol. 400: 275–297.

LEWIS, R. S. AND A. J. HUDSPETH (1983) Voltage- and ion-dependent conductances in solitary vertebrate hair cells. Nature 304: 538–541.

SHEPHERD, G. M. G. AND D. P. COREY (1994) The extent of adaptation in bullfrog saccular hair cells. J. Neurosci. 14: 6217–6229.

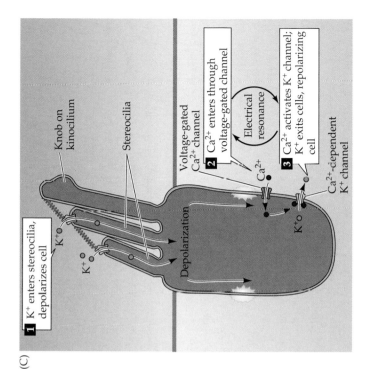

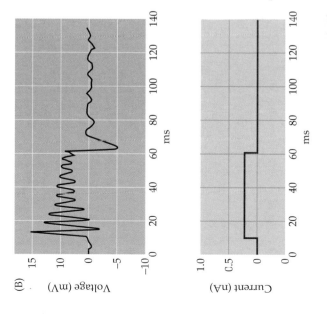

(B) Voltage oscillations (upper trace) in an isolated hair cell in response to a depolarizing current injection (lower trace).

(C) Proposed ionic basis for electrical resonance in hair cells. (B after Lewis and Hudspeth, 1983; C after Hudspeth, 1985.)

whether by static displacement of hair bundles due to gravity or by transient displacement of hair bundles due to linear accelerations, and this information is preserved in the firing rate of vestibular axons.

The range of orientations of hair bundles within the otolith organs enables them to transmit information about linear forces in every direction the body moves (see Figure 14.4C). The utricle, which is primarily concerned with motion in the horizontal plane, and the sacculus, which is concerned with vertical motion, combine to effectively gauge in three dimensions the linear forces acting on the head at any instant. Tilts of the head off the horizontal plane and translational movements of the head in any direction stimulate a distinct subset of hair cells in the saccular and utricular maculae, while simultaneously suppressing the responses of other hair cells in these organs. Ultimately, variations in hair cell polarity within the otolith organs produce patterns of vestibular nerve fibers activity that, at a population level, can unambiguously encode head position and the forces acting to influence it.

The Semicircular Canals

Whereas the otolith organs are primarily concerned with translational movements, the semicircular canals sense head *rotations*, arising either from self-induced movements or from angular accelerations of the head imparted by external forces. Each of the three semicircular canals has at its base a bulbous expansion called the **ampulla** (Figure 14.7), which houses the sensory epithelium, or **crista**, that contains the hair cells. The structure of the canals suggests how they detect the angular accelerations that arise through rotation of the head. The hair bundles extend out of the crista into a gelatinous mass, the **cupula**, that bridges the width of the ampulla, forming a fluid barrier through which endolymph cannot circulate. As a result, the compliant cupula is distorted by movements of the endolymphatic fluid. When the head turns in the plane of one of the semicircular canals, the inertia of the endolymph produces a force across the cupula, distending it away from the direction of head movement and causing a displacement of the hair bundles within the crista (Figure 14.8A,B). In contrast, linear accelerations of the head produce equal forces on the two sides of the cupula, so the hair bundles are not displaced.

Unlike the saccular and utricular maculae, all of the hair cells in the crista within each semicircular canal are organized with their kinocilia pointing in the same direction (see Figure 14.2C). Thus, when the cupula moves in the appropriate direction, the entire population of hair cells is depolarized and activity in all of the innervating axons increases. When the cupula moves in the opposite direction, the population is hyperpolarized and neuronal activity decreases. Deflections orthogonal to the excitatory-inhibitory direction produce little or no response.

Each semicircular canal works in concert with a partner located on the other side of the head, which has its hair cells aligned oppositely. There are three such pairs: the two pairs of horizontal canals, and the superior canal on each side working with the posterior canal on the other side (Figure 14.8C). Head rotation deforms the cupula in opposing directions for the two partners, resulting in changes in their firing rates (Box C). For example, the orientation of the horizontal canals makes them selectively sensitive to rotation in the horizontal plane. More specifically, the hair cells in the canal towards which the head is turning are depolarized, while those on the other side are hyperpolarized. For example, when the head turns to the left, the

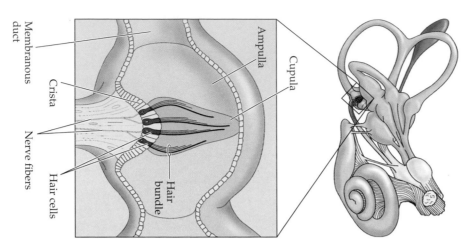

Figure 14.7 The ampulla of the posterior semicircular canal showing the crista, hair bundles, and cupula. The cupula is distorted by the fluid in the membranous canal when the head rotates.

Membranous duct

Ampulla

Cupula

Crista

Nerve fibers

Hair cells

Hair bundle

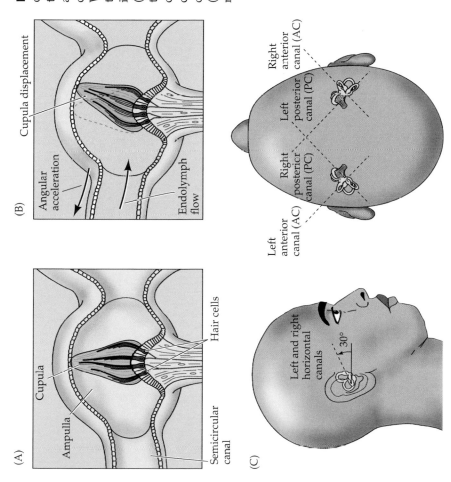

Figure 14.8 Functional organization of the semicircular canals. (A) The position of the cupula without angular acceleration. (B) Distortion of the cupula during angular acceleration. When the head is rotated in the plane of the canal (arrow outside canal), the inertia of the endolymph creates a force (arrow inside the canal) that displaces the cupula. (C) Arrangement of the canals in pairs. The two horizontal canals form a pair; the right anterior canal (AC) and the left posterior canal (PC) form a pair; the left AC and the right PC form a pair.

cupula is pushed toward the kinocilium in the left horizontal canal, and the firing rate of the relevant axons in the left vestibular nerve increases. In contrast, the cupula in the right horizontal canal is pushed away from the kinocilium, with a concomitant decrease in the firing rate of the related neurons. If the head movement is to the right, the result is just the opposite. This push-pull arrangement operates for all three pairs of canals; the pair whose activity is modulated is in the rotational plane, and the member of the pair whose activity is increased is on the side toward which the head is turning. The net result is a system that provides information about the rotation of the head in any direction.

How Semicircular Canal Neurons Sense Angular Accelerations

Like those that innervate the otolith organs, the vestibular fibers originating from bipolar neurons in Scarpa's ganglion that innervate the semicircular canals exhibit a high and steady spontaneous firing rate. As a result, they can transmit information by either increasing or decreasing their firing rate. As already suggested, such bidirectional responses enable the afferent nerves to faithfully follow the receptor potentials, and thus to more effectively encode head movements. The bidirectional responses of fibers innervating the hair cells of the semicircular canal have been studied by recording the axonal firing rates in a monkey's vestibular nerve. Seated in a chair that

BOX C
Throwing Cold Water on the Vestibular System

Testing the integrity of the vestibular system can indicate much about the condition of the brainstem, particularly in comatose patients.

Normally, when the head is not being rotated, the output of the nerves from the right and left sides are equal; thus, no eye movements occur. When the head is rotated in the horizontal plane, the vestibular afferent fibers on the side toward the turning motion increase their firing rate, while the afferents on the opposite side decrease their firing rate (figures A and B). The net difference in firing rates then leads to slow move-

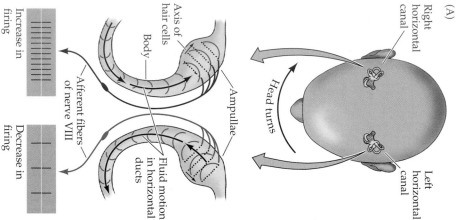

(A)
Right horizontal canal
Axis of hair cells
Body
Head turns
Ampullae
Left horizontal canal
Afferent fibers of nerve VIII
Fluid motion in horizontal ducts
Increase in firing
Decrease in firing

ments of the eyes counter to the turning motion. This reflex response generates the slow component of a normal eye movement pattern called physiological nystagmus (figure B1). (The fast compo-

nent is a saccade that resets the eye position; see Chapter 20.)

Unwanted and deleterious nystagmus (which means "nodding" or oscillatory movements of the eyes) can occur if there is unilateral damage to the vestibular system. In this case, the silencing of the spontaneous output from the damaged side will result in a pathological difference in firing rate because the spontaneous discharge from the intact side remains (figure B2). The difference in firing rates will cause spontaneous nystagmus, even though no head movements are being made.

Such responses can thus be used to assess the integrity of the brainstem in patients when damage is suspected. If a patient is placed on his or her back and the head is elevated to about 30° above horizontal, the horizontal semicircular canals lie in an almost vertical orientation. Irrigating one ear with cold water will then lead to spontaneous eye movements because convection currents in the

(B)
(1) **Physiological nystagmus**

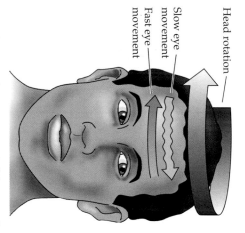

Head rotation
Slow eye movement
Fast eye movement

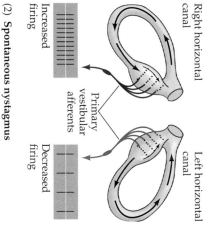

Right horizontal canal
Primary vestibular afferents
Left horizontal canal
Increased firing
Decreased firing

(2) **Spontaneous nystagmus**

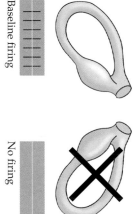

Baseline firing
No firing

(A) View looking down on the top of a person's head illustrates the fluid motion generated in the left and right horizontal canals, and the changes in vestibular nerve firing rates when the head turns to the right. (B) In normal individuals, rotating the head elicits physiological nystagmus (1), which consists of a slow eye movement counter to the direction of head turning. The slow component of the eye movements is due to the net differences in left and right vestibular nerve firing rates acting via the central circuit diagrammed in Figure 14.10. Spontaneous nystagmus (2), where the eyes move rhythmically from side to side in the absence of any head movements, occurs when one of the head movements, occurs when one of the canals is damaged. In this situation, net differences in vestibular nerve firing rates exist even when the head is stationary because the vestibular nerve innervating the intact canal fires steadily when at rest, in contrast to a lack of activity on the damaged side.

(C) Caloric testing of vestibular function is possible because irrigating an ear with water slightly warmer than body temperature generates convection currents in the canal that mimic the endolymph movement induced by turning the head to the irrigated side. Irrigation with cold water induces the opposite effect. These currents result in changes in the firing rate of the associated vestibular nerve, with an increased rate on the warmed side and a decreased rate on the chilled side. As in head rotation and spontaneous nystagmus, net differences in firing rates generate eye movements.

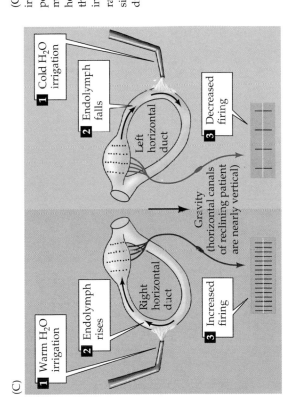

canal mimic rotatory head movements away from the irrigated ear (figure C). In normal individuals, these eye movements consist of a slow movement toward the irrigated ear and a fast movement away from it. The fast movement is most readily detected by the observer, and the significance of its direction can be kept in

mind by using the mnemonic COWS (Cold Opposite, Warm Same). This test can also be used in comatose patients, but because saccadic movements are no longer made, the response consists of only the slow movement component. When cold water is used, the slow movement will be toward the irrigated ear

unless damage to the brainstem has occurred (figure D). Lesions in the caudal pons or rostral medulla (the site of the vestibular nuclei) can abolish or alter these responses.

(D) Ocular reflexes in unconscious patients

(1) Condition: Brainstem intact

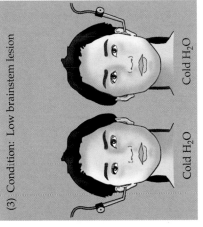

(2) Condition: MLF lesion (bilateral)

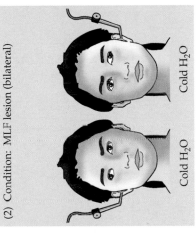

(3) Condition: Low brainstem lesion

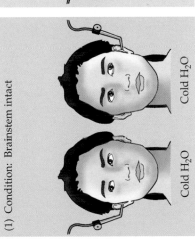

(D) Caloric testing can be used to test the function of the brainstem in an unconscious patient. The figures show the slow eye movements resulting from cold water irrigation in one ear for three different conditions: (1) with the brainstem intact; (2) with a lesion of the medial longitudinal fasciculus (MLF; note that irrigation in this case results in movement of the eye only on the irrigated side); and (3) with a low brainstem lesion (see Figure 14.10).

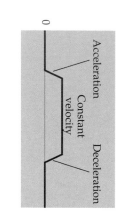

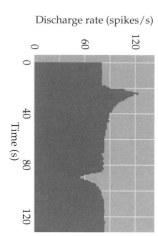

Discharge rate (spikes/s)

Figure 14.9   Response of a vestibular nerve axon from the semicircular canal to angular acceleration. The stimulus (top) is a rotation that first accelerates, then maintains constant velocity, and then decelerates the head. The axon increases its firing above resting level in response to the acceleration, returns to resting level during constant velocity, then decreases its firing rate below resting level during deceleration; these changes in firing rate reflect inertial effects on the displacement of the cupula. (After Goldberg and Fernandez, 1971.)

could be rotated, the monkey was first rotated at an accelerated rate, then at constant velocity for several seconds, and finally the chair was decelerated to a stop (Figure 14.9). The maximum firing rates observed correspond to the period of acceleration; the maximum inhibition corresponds to the period of deceleration. During the constant-velocity phase, the response adapts so that the firing rate subsides to resting level; after the movement stops, the neuronal activity decreases transiently before returning to the resting level. Neurons innervating paired canals have a complimentary response pattern. Note that the rate of adaptation (on the order of tens of seconds) corresponds to the time it takes the cupula to return to its undistorted state (and for the hair bundles to return to their undeflected position); adaption therefore can occur even while the head is still turning, as long as a constant angular velocity is maintained. Such constant forces are rare in nature, although they are encountered on ships, airplanes, and space vehicles, where prolonged accelleratory arcs are sometimes described.

Central Vestibular Pathways: Eye, Head, and Body Reflexes

The vestibular end organs communicate via the vestibular branch of the eighth cranial nerve with targets in the brainstem and the cerebellum that perform much of the processing necessary to compute head position and motion. As with the cochlear nerve, the vestibular nerves arise from a population of bipolar neurons, the cell bodies of which in this instance reside in the **vestibular nerve ganglion** (also called **Scarpa's ganglion**; see Figure 14.1). The distal processes of these cells innervate the semicircular canals and the otolith organs, while the central processes project via the vestibular portion of cranial nerve VIII to the vestibular nuclei (and also directly to the cerebellum) (Figure 14.10). Because vestibular and auditory fibers run together in the eighth nerve, damage to this structure often results in both auditory and vestibular disturbances. The **vestibular nuclei** are important centers of integration, receiving input from the vestibular nuclei of the opposite side, as well as from the cerebellum and the visual and somatic sensory systems.

One of the main functions of the vestibular system is to coordinate head and eye movements (other functions include protective or escape reactions; see Box D). The **vestibulo-ocular reflex (VOR)** in particular is a mechanism for producing eye movements that counter head movements, thus permitting the gaze to remain fixed on a particular point (Box C; see also Chapter 20). For example, activity in the left horizontal canal induced by leftward rotation of the head excites neurons in the left vestibular nucleus and results in reflexive eye movements to the right. This effect is due to excitatory projections from the vestibular nucleus to the contralateral nucleus abducens that, along with the oculomotor nucleus, help execute conjugate eye movements. For instance, horizontal movement of the two eyes toward the right requires contraction of the left medial and right lateral rectus muscles. Vestibular nerve fibers originating in the left horizontal semicircular canal project to the medial and lateral vestibular nuclei (see Figure 14.10). Excitatory fibers from the medial vestibular nucleus cross to the contralateral abducens nucleus, which has two outputs. One of these is a motor pathway that causes the lateral rectus of the right eye to contract; the other is an excitatory projection that crosses the midline and ascends via the **medial longitudinal fasciculus** to the left oculomotor nucleus, where it activates neurons that cause the medial rectus of the left eye to contract. Finally, inhibitory neurons project from the medial vestibular nucleus to the left abducens nucleus, directly causing the motor drive on the lateral rectus of the left eye to decrease and also indirectly causing the right medial rectus to relax. The consequence of these several connections is that excitatory input from the horizontal canal on one side produces eye movements toward the opposite side. Therefore, turning the head to the left causes eye movements to the right. In a similar fashion, head turns in other planes activate other semicircular canals, causing other appropriate compensatory eye movements. The rostro-caudal set of cranial nerve nuclei involved in the VOR (i.e., the vestibular, abducens and oculomotor nuclei), as well as the VOR's persistence in the unconscious state, make this reflex especially useful for detecting brainstem damage in the comatose patient (see Box C).

Loss of the VOR can have severe consequences. A patient with vestibular damage finds it difficult or impossible to fixate on visual targets while the

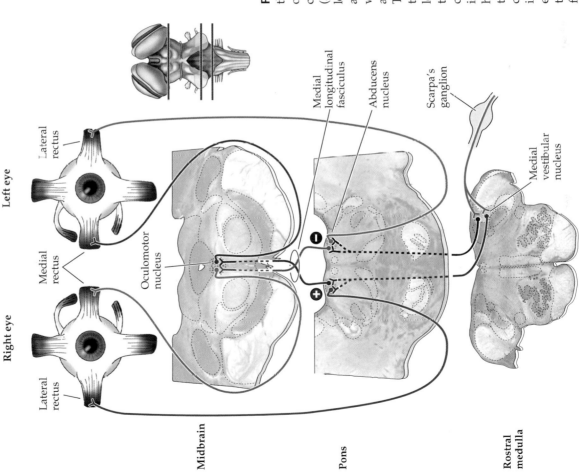

Figure 14.10 Connections underlying the vestibulo-ocular reflex. Projections of the vestibular nucleus to the nuclei of cranial nerves III (oculomotor) and VI (abducens). The connections to the oculomotor nucleus and to the contralateral abducens nucleus are excitatory (red), whereas the connections to ipsilateral abducens nucleus are inhibitory (black). There are connections from the oculomotor nucleus to the medial rectus of the left eye and from the adbucens nucleus to the lateral rectus of the right eye. This circuit moves the eyes to the right, that is, in the direction away from the left horizontal canal, when the head rotates to the left. Turning to the right, which causes increased activity in the right horizontal canal, has the opposite effect on eye movements. The projections from the right vestibular nucleus are omitted for clarity.

BOX D
Mauthner Cells in Fish

One of the primary functions of the vestibular system is to provide information about the direction and speed of ongoing movements, ultimately enabling rapid, coordinated reflexes to compensate for both self-induced and externally generated forces. One of the most impressive and speediest vestibular-mediated reflexes is the tail-flip escape behavior of fish (and larval amphibians), a stereotyped response that allows a potential prey to elude its predators (figure A). In response to a perceived risk (tap on the side of a fish tank if you want to observe the reflex), fish flick their tail and are thus propelled laterally away from the approaching threat.

The circuitry underlying the tail-flip escape reflex includes a pair of giant medullary neurons called Mauthner cells, their vestibular inputs, and the spinal cord motor neurons to which the

Mauthner cells project. (In most fish, there is one pair of Mauthner cells in a stereotypic location. Thus, these cells can be consistently visualized and studied from animal to animal.) Movements in the water, such as might be caused by an approaching predator, excite saccular hair cells in the vestibular labyrinth. These receptor potentials are transmitted via the central processes of vestibular ganglion cells in cranial nerve VIII to the two Mauthner cells in the brainstem. As in the vestibulospinal pathway in humans, the Mauthner cells project directly to spinal motor neurons. The small number of synapses intervening between the receptor cells and the motor neurons is one of the ways that this circuit has been optimized for speed by natural selection. The large size of the Mauthner axons is another; the axons from these cells in a goldfish are about 50 μm in diameter! (See Chapter 4.)

The optimization for speed and direction in the escape reflex also is reflected in the synapses vestibular nerve afferents make on each Mauthner cell (figure B). These connections are electrical synapses that allow rapid and faithful transmission of the vestibular signal.

An appropriate direction of escape is promoted by two features: (1) each Mauthner cell projects only to contralateral motor neurons; and (2) a local network of bilaterally projecting interneurons inhibits activity in the Mauthner cell away from the side on which the vestibular activity originates. In this way, the Mauthner cell on the appropriate side faithfully generates action potentials that command contractions of contralateral tail musculature, thus moving the fish out of the path of the oncoming predator (figure C). Conversely, the Mauthner cell on the opposite side is

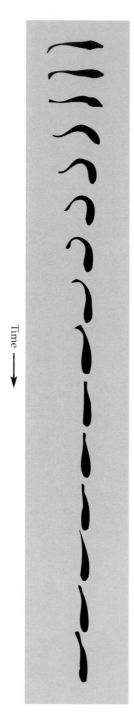

Time ⟶

(A) Bird's eye view of the sequential body orientations of a fish engaging in a tail-flip escape behavior, with time progressing from left to right. This behavior is largely mediated by vestibular inputs onto Mauthner cells.

head is moving, a condition called **oscillopsia**. If the damage is unilateral, the patient usually recovers the ability to fixate objects during head movements. However, a patient with bilateral loss of vestibular function has the persistent and disturbing sense that the world is moving when the head moves. The underlying problem in such cases is that information about head and body movements normally generated by the vestibular organs is not available to the oculomotor centers, so that corrective eye movements cannot be made.

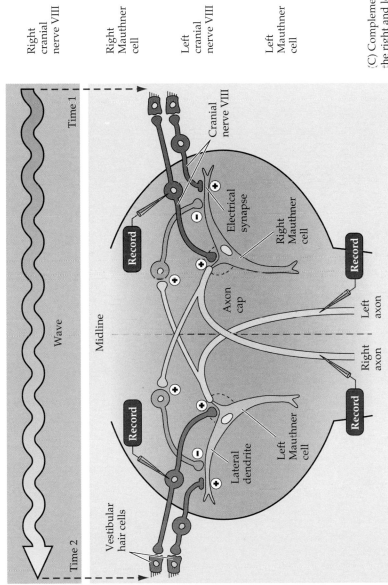

Right cranial nerve VIII

Right Mauthner cell

Left cranial nerve VIII

Left Mauthner cell

Time 1   Time 2

Time →

(C) Complementary responses of the right and left Mauthner cells mediating the escape response. Times 1 and 2 correspond to those indicated in figure B. (After Furshpan and Furukuwa, 1962.)

Wave

Time 1

Cranial nerve VIII

Record

Electrical synapse

Right Mauthner cell

Axon cap

Left axon

Record

Right axon

Midline

Record

Lateral dendrite

Left Mauthner cell

Vestibular hair cells

Time 2

(B) Diagram of synaptic events in the Mauthner cells of a fish in response to a disturbance in the water coming from the right.

silenced by the local inhibitory network during the response (figure C).

The Mauthner cells in fish are analogous to the reticulospinal and vestibulospinal pathways that control balance, posture, and orienting movements in mammals. The equivalent behavioral responses in humans are evident in a friendly game of tag.

References

EATON, R. C., R. A. BOMBARDIERI, AND D. L. MEYER (1977) The Mauthner-initiated startle response in teleost fish. J. Exp. Biol. 56: 65–81.

FURSHPAN, E. J. AND T. FURUKAWA (1952) Intracellular and extracellular responses of the several regions of the Mauthner cell of the goldfish. J. Neurophysiol. 25:732–771.

JONTES, J. D., J. BUCHANAN AND S. J. SMITH (2000) Growth cone and dendrite dynamics

in zebrafish embryos: Early events in synaptogenesis imaged in vivo. Nature Neurosci. 3: 231–237.

O'MALLEY, D. M., Y. H. KAO AND J. R. FETCHO (1996) Imaging the functional organization of zebrafish hindbrain segments during escape behaviors. Neuron 17: 1145–1155.

Descending projections from the vestibular nuclei are essential for postural adjustments of the head and body. As with the VOR, these postural reflexes are extremely fast, in part due to the small number of synapses interposed between the vestibular organ and the relevant motor neurons (Box D). Axons from the medial vestibular nucleus descend in the medial longitudinal fasciculus to reach the upper cervical levels of the spinal cord (Figure 14.11). This pathway regulates head position by reflex activity of neck muscles in response to stimulation of the semicircular canals from rota-

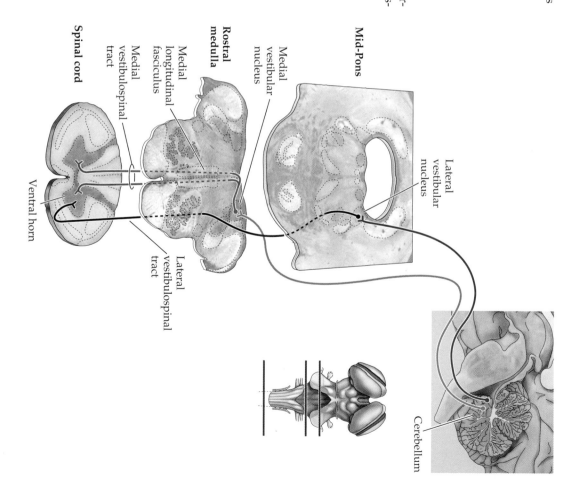

Mid-Pons

Lateral vestibular nucleus

Medial vestibular nucleus

Rostral medulla

Medial longitudinal fasciculus

Medial vestibulospinal tract

Lateral vestibulospinal tract

Spinal cord

Ventral horn

Cerebellum

Figure 14.11   Descending projections from the medial and lateral vestibular nuclei to the spinal cord. The medial vestibular nuclei project bilaterally in the medial longitudinal fasciculus to reach the medial part of the ventral horns and mediate head reflexes in response to activation of semicircular canals. The lateral vestibular nucleus sends axons via the lateral vestibular tract to contact anterior horn cells innervating the axial and proximal limb muscles. Neurons in the lateral vestibular nucleus receive input from the cerebellum, allowing the cerebellum to influence posture and equilibrium.

tional accelerations of the head. For example, during a downward pitch of the body (e.g., tripping), the superior canals are activated and the head muscles reflexively pull the head up. The dorsal flexion of the head initiates other reflexes, such as forelimb extension and hindlimb flexion, to stabilize the body and protect against a fall (see Chapter 17).

The inputs from the otolith organs project mainly to the lateral vestibular nucleus, which in turn sends axons in the lateral vestibulospinal tract to the spinal cord (see Figure 14.11). The input from this tract exerts a powerful excitatory influence on the extensor (antigravity) muscles. When hair cells in the otolith organs are activated, signals reach the medial part of the ventral horn. By activating the ipsilateral pool of motor neurons innervating extensor muscles in the trunk and limbs, this pathway mediates balance and the maintenance of upright posture. Decerebrate rigidity, which is characterized by rigid extension of the limbs, arises when the brainstem is transected above the level of the vestibular nucleus. The tonic activation of extensor muscles in this instance suggests that the vestibulospinal pathway is normally strongly suppressed by descending projections from higher levels of the brain, especially the cerebral cortex (see also Chapter 17).

Vestibular Pathways to the Thalamus and Cortex

In addition to these several descending projections, the superior and lateral vestibular nuclei send axons to the ventral posterior nuclear complex of the thalamus, which projects to two cortical areas relevant to vestibular sensations (Figure 14.12). One cortical target is just posterior to the primary somatosensory cortex, near the representation of the face; the other is at the transition between the somatic sensory cortex and the motor cortex (Brodmann's area 3a; see Chapter 9). Electrophysiological studies of individual neurons in these areas show that the relevant cells respond to proprioceptive and visual stimuli as well as to vestibular stimuli. Many of these neurons are activated by moving visual stimuli as well as by rotation of the body (even with the eyes closed), suggesting that these cortical regions are involved in the perception of body orientation in extrapersonal space.

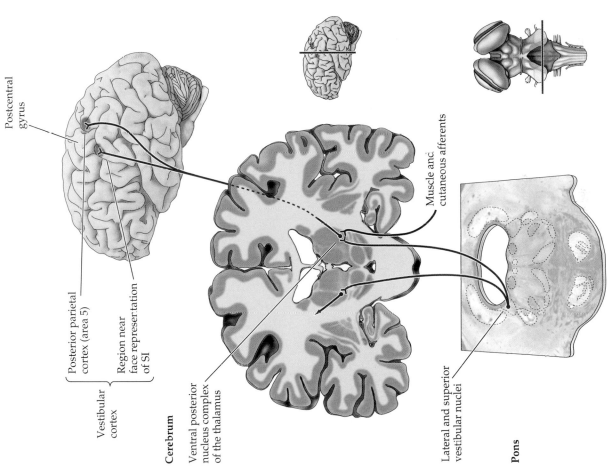

Postcentral
gyrus

Posterior parietal
cortex (area 5)

Vestibular
cortex

Region near
face representation
of SI

Cerebrum

Ventral posterior
nucleus complex
of the thalamus

Muscle and
cutaneous afferents

Lateral and superior
vestibular nuclei

Pons

Figure 14.12   Thalamocortical pathways carrying vestibular information. The lateral and superior vestibular nuclei project to the thalamus. From the thalamus, the vestibular neurons project to the vicinity of the central sulcus near the face representation. Sensory inputs from the muscles and skin also converge on thalamic neurons receiving vestibular input (see Chapter 9).

Summary

The receptor cells of the vestibular system are located in the otolith organs and the semicircular canals of the inner ear and provide information about the motion and position of the body in space. The otolith organs provide information necessary for postural adjustments of the somatic musculature, particularly the axial musculature, when the head tilts in various directions or undergoes linear accelerations. This information represents linear forces acting on the head that arise through static effects of gravity or from translational movements. The semicircular canals, in contrast, provide information about rotational accelerations of the head. This latter information generates reflex movements that adjust the eyes, head, and body during motor activities. Among the best studied of these reflexes are eye movements that compensate for head movements, thereby stabilizing the visual scene when the head turns. Input from all the vestibular organs is integrated with input from the visual and somatic sensory systems to provide perceptions of body position and orientation in extrapersonal space.

ADDITIONAL READING

Reviews

BENSON, A. (1982) The vestibular sensory system. In *The Senses*; H. B. Barlow and J. D. Mollon (eds.). New York: Cambridge University Press.

BRANDT, T. (1991) Man in motion: Historical and clinical aspects of vestibular function. A review. Brain 114: 2159–2174.

FURMAN, J. M. AND R. W. BALOH (1992) Otolith-ocular testing in human subjects. Ann. New York Acad. Sci. 656: 431–451.

GOLDBERG, J. M. (1991) The vestibular end organs: Morphological and physiological diversity of afferents. Curr. Opin. Neurobiol. 1: 229–235.

GOLDBERG, J. M. AND C. FERNANDEZ (1984) The vestibular system. In *Handbook of Physiology*, Section 1: *The Nervous System*, Volume III: *Sensory Processes*, Part II, J. M. Brookhart, V. B. Mountcastle, I. Darian-Smith and S. R. Geiger (eds.). Bethesda, MD: American Physiological Society.

Important Original Papers

GOLDBERG, J. M. AND C. FERNANDEZ (1971) Physiology of peripheral neurons innervating semicircular canals of the squirrel monkey. Parts 1, 2, 3. J. Neurophysiol. 34: 635–684.

GOLDBERG, J. M. AND C. FERNANDEZ (1976) Physiology of peripheral neurons innervating otolith organs of the squirrel monkey. Parts 1, 2, 3. J. Neurophysiol. 39: 970–1008.

LINDEMAN, H. H. (1973) Anatomy of the otolith organs. Adv. Oto.-Rhino.-Laryng. 20: 405–433.

Book

BALOH, R. W. AND V. HONRUBIA (1990) *Clinical Neurophysiology of the Vestibular System*, 2nd Ed. Philadelphia: F. A. Davis Co.

Chapter 15

The Chemical Senses

Overview

Three sensory systems associated with the nose and mouth—olfaction, taste, and the trigeminal chemosensory system—are dedicated to the detection of chemicals in the environment. The olfactory system detects airborne molecules called odors. In humans, odors provide information about food, self, other people, animals, plants, and many other aspects of the environment. Olfactory information can influence feeding behavior, social interactions and, in many animals, reproduction. The taste (or gustatory) system detects ingested, primarily water-soluble molecules called tastants. Tastants provide information about the quality, quantity, pleasantness, and safety of ingested food. The trigeminal chemosensory system provides information about irritating or noxious molecules that come into contact with skin or mucous membranes of the eyes, nose and mouth. All three of these chemosensory systems rely on receptors in the nasal cavity, mouth, or on the face that interact with the relevant molecules and generate receptor and action potentials, thus transmitting the effects of chemical stimuli to appropriate regions of the central nervous system.

The Organization of the Olfactory System

From an evolutionary perspective, the chemical senses—particularly olfaction—are deemed the "oldest" sensory systems; nevertheless, they remain in many ways the least understood of the sensory modalities. The olfactory system (Figure 15.1) is the most thoroughly studied component of the chemosensory triad and processes information about the identity, concentration, and quality of a wide range of chemical stimuli. These stimuli, called **odorants**, interact with olfactory receptor neurons in an epithelial sheet—the **olfactory epithelium**—that lines the interior of the nose (Figure 15.1A,B). The axons arising from the receptor cells project directly to neurons in the **olfactory bulb**, which projects in turn to the pyriform cortex in the temporal lobe (Figure 15.1C). The olfactory system is thus unique among the sensory systems in that it does not entail a thalamic relay en route to the primary cortical region that processes the sensory information. The olfactory tract also projects to a number of other targets in the forebrain, including the hypothalamus and amygdala. Projections from the pyriform cortex and other forebrain regions, via the thalamus, provide olfactory information to several additional regions of the cerebral cortex (see Figure 15.1C). The further processing that occurs in these various regions identifies the odorant and initiates appropriate motor, visceral, and emotional reactions to olfactory stimuli.

Despite its phylogenetic "age" and unusual trajectory to the cortex, the olfactory system abides by the basic principle that governs other sensory

(A)

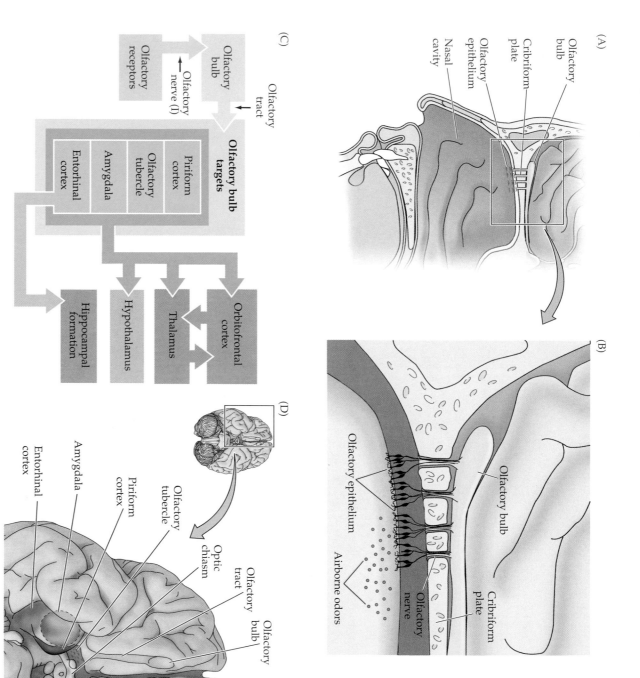

Olfactory
bulb

Cribriform
plate

Olfactory
epithelium

Nasal
cavity

(B)

Olfactory bulb

Cribriform
plate

Olfactory epithelium

Olfactory
nerve

Airborne odors

(C)

Olfactory
bulb

Olfactory
tract

Olfactory
nerve (I)

Olfactory
receptors

**Olfactory bulb
targets**

Entorhinal
cortex

Amygdala

Olfactory
tubercle

Piriform
cortex

Hippocampal
formation

Hypothalamus

Thalamus

Orbitofrontal
cortex

(D)

Entorhinal
cortex

Amygdala

Piriform
cortex

Olfactory
tubercle

Optic
chiasm

Olfactory
tract

Olfactory
bulb

Figure 15.1 Organization of the human olfactory system. (A) Peripheral and central components of the olfactory pathway. (B) Enlargement of region boxed in (A) showing the relationship between the olfactory epithelium, containing the olfactory receptor neurons, and the olfactory bulb (the central target of olfactory receptor neurons). (C) Diagram of the basic pathways for processing olfactory information. (D) Central components of the olfactory system.

modalities: Interactions of chemical stimuli with receptors at the periphery are transduced and encoded into electrical signals, which are then transmitted to higher-order centers. Nevertheless, less is known about the central organization of the olfactory system than other sensory pathways. For example, the somatic sensory and visual cortices described in the preceding chapters all feature spatial maps of the relevant receptor surface, and the auditory cortex features frequency maps. Whether any analogous maps exist in the pyriform cortex (or the olfactory bulb) is not yet known. Indeed, until recently it has been difficult to imagine on what sensory qualities an olfactory map would be based, or what features might be processed in parallel (as occurs in other sensory systems).

Olfactory Perception in Humans

In humans, olfaction is often considered the least acute of the senses, and a number of animals are obviously superior to humans in their olfactory abilities. This difference is probably explained by the larger number of olfactory receptor neurons (and odorant receptor molecules; see below) in the olfactory epithelium in many species and the relatively larger area of cortex devoted to olfaction. In a 70-kg human, the surface area of the olfactory epithelium is approximately 10 cm². In contrast, a 3-kg cat has about 20 cm² of olfactory epithelium. Humans are nonetheless quite good at detecting and identifying airborne molecules in the environment (Figure 15.2). For instance, the major aromatic constituent of bell pepper (2-isobutyl-3-methoxypyrazine) can be detected at a concentration of 0.01 nM. The threshold concentrations for odorant detection and identification nonetheless vary greatly. Ethanol, for example, cannot be identified until its concentration reaches approximately 2 mM. Small changes in molecular structure can also lead to large perceptual differences: The molecule *d-carvone* smells like caraway seeds, whereas *l-carvone* smells like spearmint!

Since the number of odorants is very large, there have been several attempts to classify them in groups. One useful classification was developed in the 1950s by John Amoore, who divided odors into categories based on their perceived quality, molecular structure, and the fact that some people, called anosmics, have difficulty smelling one or another group. The categories Amoore described were pungent, floral, musky, earthy, ethereal, camphor, peppermint, ether, and putrid, and these are still used to describe

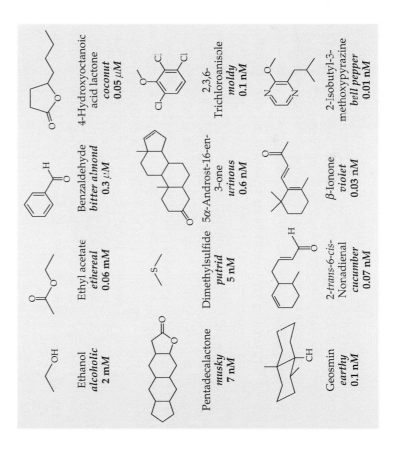

Figure 15.2 Chemical structure and human perceptual threshold for 12 common odorants. Molecules perceived at low concentrations are more lipid-soluble, whereas those with higher thresholds are more water-soluble. (After Pelosi, 1994.)

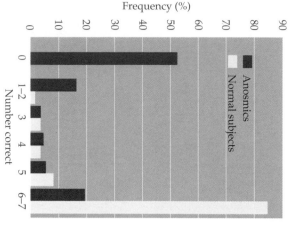

Figure 15.3 Anosmia is the inability to identify common odors. When subjects are presented with seven common odors (a test frequently used by neurologists), the vast majority of "normal" individuals can identify all seven odors correctly (in this case, baby powder, chocolate, cinnamon, coffee, mothballs, peanut butter, and soap). Some people, however, have difficulty identifying even these common odors. When individuals previously identified as anosmics were presented with the same battery of odors, only a few could identify all of the odors (less than 15%), and more than half could not identify any of the odors. (After Cain and Gent, in Meiselman and Rivlin, 1986.)

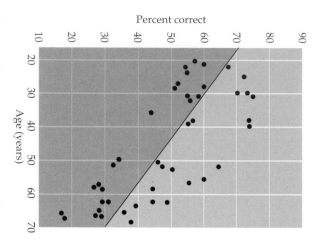

Figure 15.4 Normal decline in olfactory sensitivity with age. The ability to identify 80 common odorants declines markedly between 20 and 70 years of age. (After Murphy, 1986.)

odors, to study the cellular mechanisms of olfactory transduction, and to discuss the central representation of olfactory information. Nevertheless, this classification remains entirely empirical. A further complication in rationalizing the perception of odors is that their quality may change with concentration. For example, at low concentrations indole has a floral odor, whereas at higher concentrations it smells putrid. Despite these problems, the longevity of Amoore's scheme makes clear that the olfactory system can identify odorant classes that have distinct perceptual qualities. Humans can, of course, perceive individual odorant molecules. Thus, coconuts, violets, cucumbers, and bell peppers all have a unique odor generated by a particular molecule. Most naturally occurring odors, however, are blends of several odorant molecules, even though they are typically experienced as a single smell (such as the perceptions elicited by perfumes or the bouquet of a wine).

Psychologists and neurologists have developed a variety of tests that measure the ability to detect common odors. Although most people are able to consistently identify a broad range of test odorants, others fail to identify one or more common smells (Figure 15.3). Such chemosensory deficits, called anosmias, are often restricted to a single odorant, suggesting that a specific element in the olfactory system, mostly likely an olfactory receptor type, is missing. For example, about 1 person in 1000 is insensitive to butyl mercaptan, the foul-smelling odorant released by skunks. More serious is the inability to detect hydrogen cyanide (1 in 10 people), which can be lethal, or ethyl mercaptan, the chemical added to natural gas to aid in its detection from leaks.

The ability to identify odors normally decreases with age. If otherwise healthy subjects are challenged to identify a large battery of common odorants, people between 20 and 40 years of age can typically identify about 50–75% of the odors, whereas those between 50 and 70 correctly identify only about 30–45% (Figure 15.4). A more radically diminished or distorted sense of smell can accompany eating disorders, psychotic disorders, diabetes, taking certain medications, and Alzheimer's disease (all for reasons that remain obscure). Although the loss of human olfactory sensitivity is not usually a source of great concern, it can diminish the enjoyment of food and, if severe, can affect the ability to identify and respond appropriately to potentially dangerous odors such as spoiled food, smoke, or natural gas (see above).

Physiological and Behavioral Responses to Odorants

In addition to olfactory perceptions, odorants can elicit a variety of physiological responses. Examples are the visceral motor responses to the aroma of appetizing food (salivation and increased gastric motility) or to a noxious smell (gagging and, in extreme cases, vomiting). Olfaction can also influence reproductive and endocrine functions. Women housed in single-sex dormitories, for instance, have menstrual cycles that tend to be synchronized. This phenomenon appears to be mediated by olfaction. Thus, volunteers exposed to gauze pads from the underarms of women at different stages of their menstrual cycles also tend to experience synchronized menses. Olfaction also influences maternal/child interactions. Infants recognize their mothers within hours after birth by smell, preferentially orienting toward their mothers' breasts and showing increased rates of suckling when fed by their mother compared to being fed by other lactating females (see also Chapter 24). By the same token, mothers can discriminate their own infant's odor when challenged with a range of odor stimuli from infants of similar age.

In other animals, including many mammals, species-specific odorants called pheromones play important roles in behavior, by influencing social, reproductive, and parenting behaviors (Box A). In rats and mice, odorants thought to be pheromones are detected by G-protein-coupled receptors located at the base of the nasal cavity in distinct, encapsulated chemosensory structures called vomeronasal organs (VNO). Despite many attempts to identify pheromones in humans, there is little evidence for them. Indeed VNOs are found bilaterally in only 8% of adults and, there is no clear indication that these structures have any significant function in humans.

The Olfactory Epithelium and Olfactory Receptor Neurons

The transduction of olfactory information occurs in the olfactory epithelium, the sheet of neurons and supporting cells that lines approximately half of the nasal cavities. (The remaining surface is lined by respiratory epithelium, which lacks neurons and serves primarily as a protective surface.) The olfactory epithelium includes several distinct cell types (Figure 15.5A). The most important of these is the **olfactory receptor neuron**, a bipolar cell that gives rise to a small-diameter, unmyelinated axon at its basal surface that transmits olfactory information centrally. At its apical surface, the receptor neuron gives rise to a single process that expands into a knoblike protrusion from which several microvilli, called **olfactory cilia**, extend into a thick layer

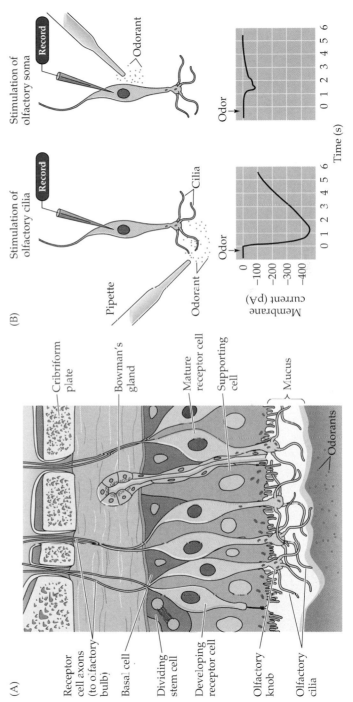

Figure 15.5 Structure and function of the olfactory epithelium. (A) Diagram of the olfactory epithelium showing the major cell types: olfactory receptor neurons and their cilia, sustentacular cells (that detoxify potentially dangerous chemicals), and basal cells. Bowman's glands produce mucus. Nerve bundles of unmyelinated neurons and blood vessels run in the basal part of the mucosa (called the lamina propria). Olfactory receptor neurons are generated continuously from basal cells. (B) Generation of receptor potentials in response to odors takes place in the cilia of receptor neurons. Thus, odorants evoke a large inward (depolarizing) current when applied to the cilia (left), but only a small current when applied to the cell body (right). (A after Anholt, 1987; B after Firestein et al., 1991.)

BOX A
Olfaction, Pheromones, and Behavior in the Hawk Moth

Olfactory information guides essential behaviors in virtually all species. The importance of olfactory cues in reproductive behaviors has been particularly well characterized in the hawk moth, *Manduca sexta*. In *Manduca*, males identify potential mates by following a plume of pheromones exuded by the female. Similarly, the female uses an olfactory cue—a molecule made by tobacco plants—to identify an appropriate site to lay eggs. These olfactory functions in the moths are sexually dimorphic: Only males respond to female pheromones, and only females detect the olfactory stimulus from the tobacco plant needed for egg laying.

These abilities are mediated by an olfactory system that shares some remarkable similarities with mammalian systems. Male and female moths have different olfactory receptor cells (and

associated structures) on their antennae which generate receptor potentials in response to the female-specific pheromones or the tobacco plant odorants. These peripheral receptors project to olfactory recipient structures that are reminiscent of the mammalian olfactory bulb (see figure). The target structure in the moth—called the antennal lobe—is comprised of an array of iterated circuits that are referred to as glomeruli and are surprisingly similar in both structure and function to glomeruli in the mammalian olfactory bulb. In males, the antennal receptor neurons sensitive to the female pheromone project to a distinct subset of glomeruli called the macroglomerular complex. These glomeruli are specifically active in the presence of female pheromone and, if absent, prevent any behavioral response to the female scent. Finally, the develop-

ment of these sexually dimorphic central circuits is controlled by the periphery. If a male antenna is transplanted to a genotypically female moth, a macroglomerular complex develops in the antennal lobe. The female-specific pheromone has been identified, as have several receptor molecules specifically associated with the male or female olfactory pathway. Not surprisingly, pheromone receptors in the male are members of a special class of seven transmembrane odorant receptors found in other invertebrates and vertebrates.

The matching of identified glomeruli with receptor cells expressing specific receptor molecules may be a general rule in olfactory systems. If so, the neurobiology of a sexually dimorphic olfactory behavior in the moth provides an ideal model system in which to study chemosensory processing of specific odorants.

Antennal nerve

Glomeruli

Antennal
lobe neurons

Macroglomerular
complex

Male

Female

Glomeruli

Male and female olfactory glomeruli in the antennal lobe are specialized for odorant-mediated behaviors. The male-specific macroglomerular complex (MCG) is essential for processing the female pheromone.

References

FARKAS, S. R. AND H. H. SHOREY (1972) Chemical trial following by flying insects: A mechanism for orientation to a distant odor source. Science 178: 67–68.

MATSUMOTO, S. G. AND J. G. HILDEBRAND (1981) Olfactory mechanisms in the moth *Manduca sexta*: Response characteristics and morphology of central neurons in the antennal lobe. Proc. Roy. Soc. London B. 213: 249–277.

SCHNEIDERMAN, A. M., S. G. MATSUMOTO AND J. G. HILDEBRAND (1982) Trans-sexually grafted antennae influence development of sexually dimorphic neurons in moth brain. Nature 298: 844–846.

SCHNEIDERMAN, A. M., J. G. HILDEBRAND, M. M. BRENNAN AND J. H. TUMLINSON (1986) Trans-sexually grafted antennae alter pheromone-directed behavior in a moth. Nature 323: 801–803.

STRAUSFELD, N. J. AND J. G. HILDEBRAND (1999) Olfactory systems: Common design, uncommon origin. Curr. Opin. Neurobiol. 9: 634–639.

of mucus. The mucus that lines the nasal cavity and controls the ionic milieu of the olfactory cilia is produced by secretory specializations (called Bowman's glands) distributed throughout the epithelium. Two other cell classes, basal cells and sustentacular (supporting) cells, are also present in the olfactory epithelium. This entire apparatus—mucus layer and epithelium with neural and supporting cells—is called the **nasal mucosa.**

The superficial location of the nasal mucosa allows the olfactory receptor neurons direct access to odorant molecules. Another consequence, however, is that these neurons are exceptionally exposed. Airborne pollutants, allergens, microorganisms, and other potentially harmful substances subject the olfactory receptor neurons to more or less continual damage. Several mechanisms help maintain the integrity of the olfactory epithelium in the face of this trauma. The respiratory epithelium, a non-neural epithelium found at the most external aspect of the nasal cavity, warms and moistens the inspired air. In addition, it secretes mucus, which traps and neutralizes potentially harmful particles. In both the respiratory and olfactory epithelium, immunoglobulins are secreted into the mucus, providing an initial defense against harmful antigens, and the sustentacular cells contain enzymes (cytochrome P450s and others) that catabolize organic chemicals and other potentially damaging molecules that enter the nasal cavity. The ultimate solution to this problem, however, is to replace olfactory receptor neurons in a normal cycle of degeneration and regeneration. In rodents, the entire population of olfactory neurons is renewed every 6 to 8 weeks. This feat is accomplished by maintaining among the basal cells a population of precursors (stem cells) that divide to give rise to new receptor neurons (see Figure 15.5A). This naturally occurring regeneration of olfactory receptor cells provides an opportunity to investigate how neural precursor cells can successfully produce new neurons and reconstitute function in the mature central nervous system, a topic of broad clinical interest. Recent evidence suggests that many of the signaling molecules that influence neuronal differentiation, axon outgrowth, and synapse formation in development elsewhere in the nervous system (see Chapters 22 and 23) perform similar functions for regenerating olfactory receptor neurons in the adult. Understanding how the new olfactory receptor neurons extend axons to the brain and reestablish appropriate functional connections is obviously relevant to stimulating the regeneration of functional connections elsewhere in the brain after injury or disease (see Chapter 25).

The Transduction of Olfactory Signals

The cellular and molecular machinery for olfactory transduction is located in the olfactory cilia (Figure 15.5B). Odorant transduction begins with odorant binding to specific receptors on the external surface of cilia. Binding may occur directly, or by way of proteins in the mucus (called odorant binding proteins) that sequester the odorant and shuttle it to the receptor. Several additional steps then generate a receptor potential by opening ion channels. In mammals, the principal pathway involves cyclic nucleotide-gated ion channels, similar to those found in rod photoreceptors (see Chapter 11). The olfactory receptor neurons contain an olfactory-specific G-protein (G_{olf}), which activates an olfactory-specific adenylate cyclase (Figure 15.6A). The resulting increase in cyclic AMP (cAMP) opens channels that permit Na^+ and Ca^{2+} entry (mostly Ca^{2+}), thus depolarizing the neuron. This depolarization, amplified by a Ca^{2+}-activated Cl^- current, is conducted passively from the cilia to the axon hillock region of the olfactory receptor neuron, where action potentials are generated and transmitted to the olfactory bulb.

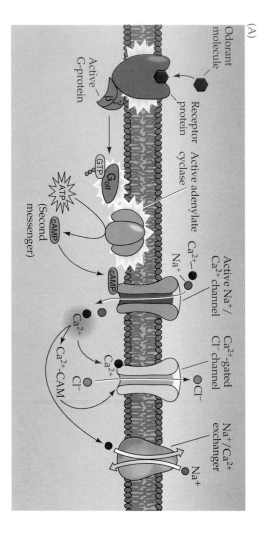

(A)

Odorant molecule

Active G-protein

Receptor protein

Active adenylate cyclase

Active Na+/ Ca2+ channel

GTP

Golf

ATP (Second messenger)

cAMP

cAMP

Ca2+

Ca2+

Ca2+-CAM

Cl−

Na+

Na+, Ca2+

Ca2+-gated Cl− channel

Na+/Ca2+ exchanger

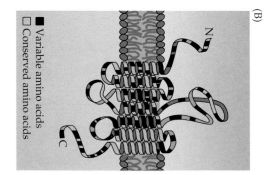

(B)

N

C

■ Variable amino acids
□ Conserved amino acids

Figure 15.6 Olfactory transduction and olfactory receptor molecules. (A) Odorants in the mucus bind directly (or are shuttled via odorant binding proteins) to one of many receptor molecules located in the membranes of the cilia. This association activates an odorant-specific G-protein (G$_{olf}$) that, in turn, activates an adenylate cyclase, resulting in the generation of cyclic AMP (cAMP). One target of cAMP is a cation-selective channel that, when open, permits the influx of Na+ and Ca2+ into the cilia, resulting in depolarization. The ensuing increase in intracellular Ca2+ opens Ca2+-gated Cl− channels that provide most of the depolarization of the olfactory receptor potential. The receptor potential is reduced in magnitude when cAMP is broken down by specific phosphodiesterases to reduce its concentration. At the same time, Ca2+ complexes with calmodulin (Ca2+–CAM) and binds to the channel, reducing its affinity for cAMP. Finally, Ca2+ is extruded through the Ca2+/Na+ exchange pathway. (B) The generic structure of putative olfactory odorant receptors. These proteins have seven transmembrane domains, plus a variable cell surface region and a cytoplasmic tail that interacts with G-proteins. As many as 1000 genes encode proteins of similar inferred structure in several mammalian species, including humans. Each gene presumably encodes an odorant receptor that detects distinct sets of odorant molecules. (Adapted from Menini, 1999.)

Olfactory receptor neurons are especially efficient at extracting a signal from chemosensory noise. Fluctuations in the cAMP concentration in an olfactory receptor neuron could, in theory, cause the receptor cell to be activated in the absence of odorants. Such nonspecific responses do not occur, however, because the cAMP-gated channels are blocked at the resting potential by the high Ca2+ and Mg2+ concentrations in mucus. To overcome this voltage-dependent block, several channels must be opened at once. This requirement ensures that olfactory receptor neurons fire only in response to stimulation by odorants. Moreover, changes in the odorant concentration change the latency of response, the duration of the response, and/or the firing frequency of individual neurons, each of which provides additional information about the environmental circumstances to the central stations in the system.

Finally, like other sensory receptors, olfactory neurons adapt in the continued presence of a stimulus. Adaptation is apparent subjectively as a decreased ability to identify or discriminate odors during prolonged exposure (e.g., decreased awareness of being in a "smoking" room at a hotel as the minutes pass). Physiologically, olfactory receptor neurons indicate adaptation by a reduced rate of action potentials in response to the continued presence of an odorant. Adaptation occurs because of: (1) increased Ca2+ binding by calmodulin, which decreases the sensitivity of the channel to cAMP; and (2) the extrusion of Ca2+ through the activation of Na+/Ca2+ exchange proteins, which reduces the amplitude of the receptor potential.

Odorant Receptors and Olfactory Coding

Olfactory receptor molecules (Figure 15.6B) are homologous to a large family of other G-protein-linked receptors that includes β-adrenergic receptors and the photopigment rhodopsin. Odorant receptor proteins have seven membrane-spanning hydrophobic domains, potential odorant binding sites in the extracellular domain of the protein, and the ability to interact with G-proteins at the carboxyl terminal region of their cytoplasmic domain. The amino acid sequences for these molecules also show substantial variability, particularly in regions that code for the membrane-spanning domains.

The specificity of olfactory signal transduction is presumably the result of this variety of odorant receptor molecules present in the nasal epithelium. In rodents (the mouse has been the animal of choice for such studies because of its well-established genetics), genes identified from an olfactory epithelium cDNA library have defined about 1000 different odorant receptors, making this the largest known gene family. In humans, the number of olfactory receptor genes is smaller (about 500–750). Since approximately 75% of these genes do not encode full-length proteins, the number of functional human receptors is about 100–200. This relatively small number of odorant receptor types may reflect our poor sense of smell compared to other species. Nevertheless, the combined activity of this number of receptors is easily large enough to account for the number of distinct odors that can be discriminated by the human olfactory system (estimated to be about 10,000).

Messenger RNAs for different olfactory receptor genes are expressed in subsets of olfactory neurons that occur in bilaterally symmetric patches of olfactory epithelium defined by the expression of receptors. Genetic analysis shows that each olfactory receptor neuron expresses only one or at most a few of the 1000 or so odorant receptor genes. Thus, different odors activate molecularly and spatially distinct subsets of olfactory receptor neurons. In short, individual odorants can activate multiple receptors, and individual receptors can be activated by multiple odorants.

Like other sensory receptor cells, olfactory receptor neurons are sensitive to a subset of chemical stimuli that define a "tuning curve." Depending on the particular olfactory receptor molecules they contain, some olfactory receptor neurons exhibit marked selectivity to particular chemical stimuli, whereas others are activated by a number of different odorant molecules (Figure 15.7A). In addition, olfactory receptor neurons can exhibit different thresholds for a particular odorant. That is, receptor neurons that are inactive at concentrations sufficient to stimulate some neurons are activated

Figure 15.7 Responses of olfactory receptor neurons to selected odorants. (A) Neuron 1 responds similarly to three different odorants. In contrast, neuron 2 responds to only one of these odorants. Neuron 3 responds to two of the three stimuli. The responses of these receptor neurons were recorded by whole-cell patch clamp recording; downward deflections represent inward currents measured at a holding potential of −55 mV. (B) Responses of a single olfactory receptor neuron to changes in the concentration of a single odorant, isoamyl acetate. The upper trace in each panel (red) indicates the duration of the odorant stimulus; the lower trace the neuronal response. The frequency and number in each panel of action potentials increases as the odorant concentration increases. (A after Firestein et al., 1992; B after Getchell, 1986.)

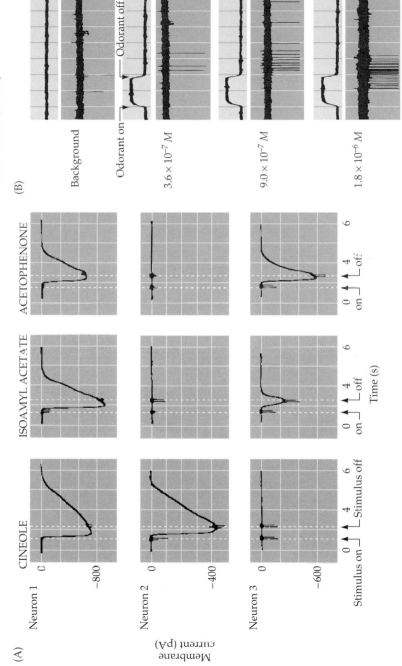

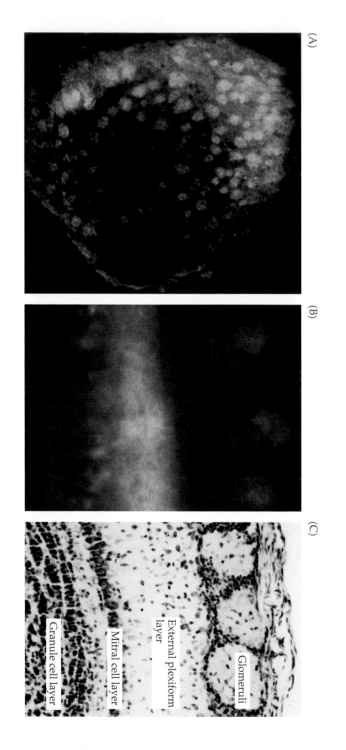

(A)

(B)

(C)

Granule cell layer

Mitral cell layer

External plexiform layer

Glomeruli

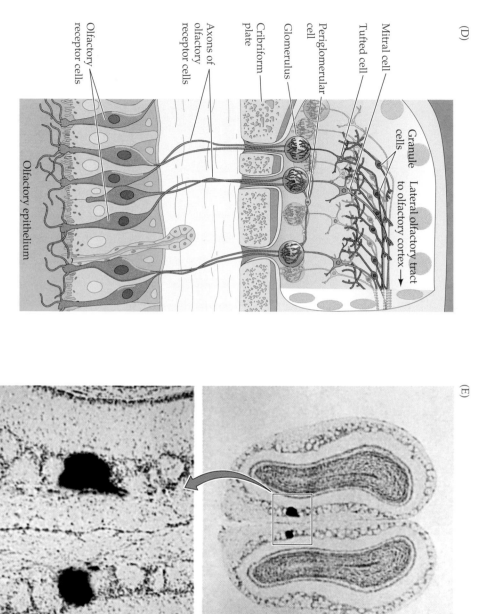

(D)

Olfactory receptor cells

Olfactory epithelium

Axons of olfactory receptor cells

Cribriform plate

Glomerulus

Periglomerular cell

Tufted cell

Mitral cell

Granule cells

Lateral olfactory tract to olfactory cortex

(E)

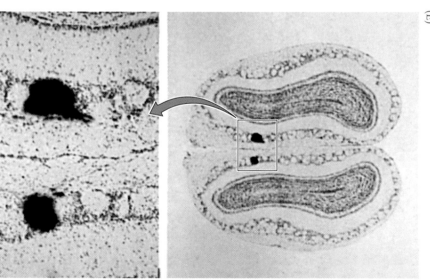

▼ **Figure 15.8**   The organization of the mammalian olfactory bulb. (A) When the bulb is viewed from its dorsal surface (visualized here in a living mouse in which the overlying bone has been removed), olfactory glomeruli can be seen. The dense accumulation of dendrites and synapses that constitute glomeruli are stained here with a vital fluorescent dye that recognizes neuronal processes. (B) Among the major neuronal components of each glomerulus are the apical tufts of mitral cells, which project to the pyriform cortex and other bulb targets (see Figure 15.1C). In this image of a coronal section through the bulb, they have been labeled retrogradely by placing the lipophilic tracer Di-I in the lateral olfactory tract. (C) The cellular structure of the olfactory bulb, shown in a Nissl-stained coronal section. The five layers of the bulb are indicated. The glomerular layer includes the tufts of mitral cells, the axon terminals of olfactory receptor neurons, and periglomerular cells that define the margins of each glomerulus. The external plexiform layer is made up of lateral dendrites of mitral cells, cell bodies and lateral dendrites of tufted cells, and dendrites of granule cells that make dendro-dendritic synapses with the other dendritic elements. The mitral cell layer is defined by the cell bodies of mitral cells, and mitral cell axons are found in the internal plexiform layer. Finally, granule cell bodies are densely packed into the granule cell layer. (D) A schematic of the laminar and circuit organization of the olfactory bulb, shown in a cutaway view from its medial surface. Olfactory receptor cell axons synapse with mitral cell apical dendritic tufts and periglomerular cell processes within glomeruli. Granule cells and mitral cell lateral dendrites constitute the major synaptic elements of the external plexiform layer. (E) Axons from olfactory receptor neurons that express a single odorant receptor gene converge on a small subset of bilaterally symmetric glomeruli. These glomeruli indicated in the boxed area in the upper panel are shown at higher magnification in the lower panel. The projections from the olfactory epithelium have been labeled by a reporter transgene inserted by homologous recombination ("knocked in") into the genetic locus that encodes the particular receptor. (A from LaMantia et al., 1992; B,C from Pomeroy et al., 1990; E from Mombaerts et al., 1996.)

when exposed to higher concentrations of an odorant. These characteristics suggest why the perception of an odor can change as a function of its concentration (Figure 15.7B).

How these olfactory responses convey the type and concentration of a given odorant is a complex issue that is unlikely to be explained at the level of the primary neurons. Nevertheless, neurons with specific receptors are located in particular parts of the olfactory epithelium. These neurons project to specific subsets of glomeruli in the olfactory bulb. Thus, the regions of the olfactory epithelium and bulb that are stimulated by particular odorants are clearly significant (Figure 15.8). As in other sensory systems, this topographical arrangement is referred to as **space coding**, although the meaning of this phrase in the olfactory system is much less clear than in vision, for example (where a topographical map correlates with visual space). The coding of olfactory information also has a temporal dimension. Sniffing, for example, is a periodic event that elicits trains of action potentials and synchronous activity of populations of neurons. Information conveyed by timing is called **temporal coding** and occurs in a variety of species (Box B). How, and whether, spatial or temporal coding contributes to olfactory perception is just beginning to be elucidated.

The Olfactory Bulb

Transducing and relaying odorant information centrally from olfactory receptor neurons are only the first steps in processing olfactory signals. As the olfactory receptor axons leave the olfactory epithelium, they coalesce to form a large number of bundles that together make up the **olfactory nerve**

BOX B
Temporal "Coding" of Olfactory Information in Insects

Most studies of olfaction in mammals have emphasized the spatial patterns of receptors in the nose and glomeruli in the bulb that are activated by specific odorants. However, beginning with Edgar Adrian's study of the hedgehog olfactory bulb in 1942, odor-induced temporal oscillations have been described in species as diverse as turtles and primates. A variety of functions have been proposed for these oscillatory phenomena, including identification of odor type and perception of odor intensity.

Gilles Laurent and colleagues at California Institute of Technology have recently found that olfaction in insects does show an important temporal component related to behavior. By recording intracellularly from neurons in the antennal lobe in crickets (a structure analogous to the olfactory bulb in mammals; see also Box A) and extracellularly in the mushroom body (analogous to the mammalian pyriform cortex), they found that the projection neurons in the antennal lobe (corresponding to mammalian mitral cells) respond to a given odor with a variety of temporal patterns that differ from odor to odor but are reproducible for the same odor. The figure here shows a schematic representation of these temporal aspects of the odor response of four such projection neurons. The upper panel shows a local field potential recording from the mushroom body (MB) that represents the synaptic activity of many neurons. During presentation of the odor, a pattern of activity is generated by the synchronized firing of many projection neurons. Interestingly, this

oscillation at 20–30 Hz is independent of the odor. Each small sphere in the lower four panels represents the state of one of the four neurons before, during, and after the application of an odorant. White balls represents a silent or inhibited state, blue balls an active but unsynchronized state, and orange balls an active *and* synchronized state. The figure shows that at different times during the odor presentation, various neurons are in synchrony and thus contribute at different times to the field potential recorded in the mushroom body. Desynchronizing the neurons has the effect of eliminating the 20–30 Hz oscillation. Desynchronization does not modify the insects' responses to odors, but eliminates their ability to distinguish among similar odors.

These observations suggest that coherent firing among neurons is an important component of olfactory processing in this species, and raise the possibility that temporal coding is a more

important aspect of mammalian olfaction than has so far been imagined.

References

ADRIAN, E. D. (1942) Olfactory reactions in the brain of the hedgehog. J. Physiol. (Lond.) 100: 459–473.

FREEMAN, W. J. AND K. A. GRADISKI (1987) Relation of olfactory EEG to behavior: Factor analysis. Behav. Neurosci. 101: 766–777.

KAY L. M. AND G. LAURENT (1999) Odor- and context-dependent modulation of mitral cell activity in behaving rats. Nature Neurosci. 2(11): 1003–1009.

LAM, Y.-W., L. B. COHEN, M. WACHOWIAK AND M. R. ZOCHOWSKI (2000) Odors elicit three different oscillations in the turtle olfactory bulb. J. Neurosci. 202: 749–762.

LAURENT, G. (1999) A systems perspective on early olfactory coding. Science 286(5440): 723–728.

LAURENT, G., M. WEHR AND H. DAVIDOWITZ (1996) Temporal representation of odors in an olfactory network. J. Neurosci. 15: 3837–3847.

STOPFER, M. AND G. LAURENT (1999) Short-term memory in olfactory network dynamics. Nature 402(6762): 664–668.

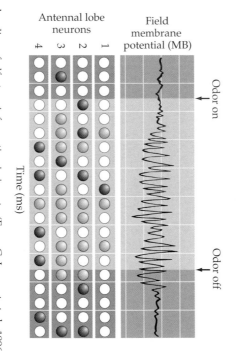

Temporal coding of olfactory information in insects. (From G. Laurent et al., 1996.)

(cranial nerve I). Each olfactory nerve projects ipsilaterally. The target of the olfactory nerve on each side is the **olfactory bulb**, which lies on the ventral anterior aspect of the ipsilateral forebrain.

The most distinctive feature of the olfactory bulb is an array of more or less spherical accumulations of neuropil 100–200 μm in diameter called

glomeruli, which lie just beneath the surface of the bulb and receive the primary olfactory axons (Figure 15.8A–C). In addition to these structures, the bulb comprises several cell and neuropil layers that receive, process, and relay olfactory information.

Within each glomerulus, the axons of the receptor neurons contact the apical dendrites of **mitral cells,** which are the principal projection neurons of the olfactory bulb (Figure 15.8B). The cell bodies of the mitral cells are located in a distinct layer deep to the olfactory glomeruli (Figure 15.8C) and, in adults, extend a primary dendrite into a single glomerulus, where the dendrite gives rise to an elaborate tuft of branches onto which the primary olfactory axons synapse. Each glomerulus in the mouse (where glomerular connectivity has been studied quantitatively) includes the apical dendrites of approximately 25 mitral cells, which receive innervation from approximately 25,000 olfactory receptor axons. This degree of convergence presumably serves to increase the sensitivity of mitral cells to ensure odor detection, and to increase the signal strength by averaging out uncorrelated noise. Each glomerulus also includes dendritic processes from two other classes of local circuit neurons: tufted cells and periglomerular cells (approximately 50 tufted cells and 25 periglomerular cells contribute to each glomerulus) (see Figure 15.8B). Although it is generally assumed that these neurons sharpen the sensitivity of individual glomeruli, their function is unclear. Finally, granule cells in the olfactory bulb synapse primarily on the basal dendrites of mitral cells within the external plexiform layer (Figure 15.8C,D). These cells make dendrodendritic synapses on mitral cells, and are important for establishing local lateral inhibitory circuits in the olfactory bulb.

Olfactory receptor neurons that express a distinct odorant receptor molecule project to bilaterally symmetrical subsets of glomeruli (Figure 15.8E). Thus there is a special zone-to-zone projection between individual glomeruli in the olfactory bulb and groups of olfactory receptor neurons. As already mentioned, however, there is no obvious systematic representation in this arrangement as there is, for example, in the somatic sensory or visual system. Rather, there is an affinity between widely distributed cells in the olfactory epithelium and ensembles of target glomeruli. This arrangement suggests that individual glomeruli respond specifically (or at least selectively) to distinct odorants.

Many investigations have confirmed the selective (but not uniquely specific) responsiveness of glomeruli to particular odorants using electrophysiological methods, voltage-sensitive dyes, and, most recently, intrinsic signals that depend on blood flow (Figure 15.9). Such studies have also shown that increasing the odorant concentration increases the activity of individual glomeruli, as well as the number of glomeruli activated.

Central Projections of the Olfactory Bulb

Glomeruli in the olfactory bulb are the sole target of olfactory receptor neurons, and thus the only relay—via the axons of mitral and tufted cells—for olfactory information from the periphery to the rest of the brain. The mitral cell axons form a bundle—the **lateral olfactory tract**—that projects to the accessory olfactory nuclei, the olfactory tubercle, the entorhinal cortex, and portions of the amygdala (see Figure 15.1A). The major target of the olfactory tract is the three-layered **pyriform cortex** in the ventromedial aspect of the temporal lobe near the optic chiasm. Although neurons in pyriform cortex respond to odors, there is no evidence of the predictable arrangement between receptor types and glomerular distribution found in the olfactory bulb. The further processing that occurs in this region is not well understood.

Figure 15.9 Glomerular activity recorded by optical imaging (see Box C in Chapter 12). Dorsal surface of the olfactory bulb in a living rat monitored as increasing concentrations of amyl acetate are presented to the animal. The higher the concentration, the more intense the activity in the particular glomeruli that respond to the odor. The column at left shows the entire dorsal surface of the olfactory bulb; the column at right shows a higher magnification of the individual glomeruli. (From Rubin and Katz, 1999.)

Amyl actetate concentration

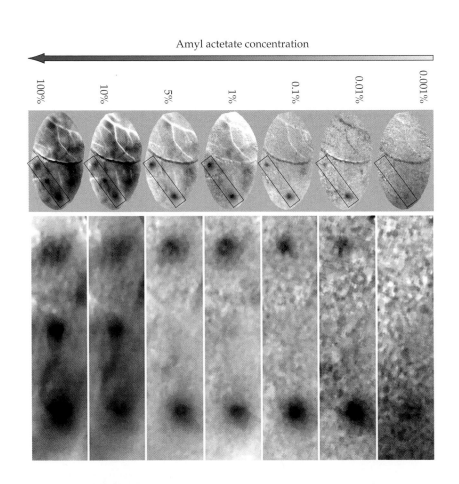

100%

10%

5%

1%

0.1%

0.01%

0.001%

The axons of pyramidal cells in the pyriform cortex project in turn to several thalamic and hypothalamic nuclei and to the hippocampus and amygdala. Some neurons from pyriform cortex also innervate a region in the orbitofrontal cortex comprising multimodal neurons that respond to olfactory and gustatory stimuli. Information about odors thus reaches a variety of forebrain regions, allowing olfactory cues to influence cognitive, visceral, emotional, and homeostatic behaviors

The Organization of the Taste System

The taste system, acting in concert with the olfactory and trigeminal systems, indicates whether food should be ingested. Once in the mouth, the chemical constituents of food interact with receptors on taste-cell-containing epithelial specializations called **taste buds** in the tongue. The taste cells transduce these stimuli and provide additional information about the identity, concentration, and pleasant or unpleasant quality of the substance. This information also prepares the gastrointestinal system to receive food by causing salivation and swallowing (or gagging and regurgitation if the substance is noxious). Information about the temperature and texture of food is transduced and relayed from the mouth via somatic sensory receptors from the trigeminal and other sensory cranial nerves to the thalamus and somatic sensory cortices (see Chapters 9 and 10). Of course, food is not simply eaten for nutritional value; "taste" also depends on cultural and psychological factors. How else can one explain why so many people enjoy consuming hot peppers or bitter-tasting liquids such as beer?

Like the olfactory system, the taste system includes both peripheral receptors and a number of central pathways (Figure 15.10). Taste cells (the peripheral receptors) are found in taste buds distributed on the dorsal surface of the tongue, soft palate, pharynx, and the upper part of the esophagus (Figure 15.10A; see also Figure 15.11). These cells make synapses with primary sensory axons that run in the chorda tympani and greater superior petrosal branches of the facial nerve (cranial nerve VII), the lingual branch of the glossopharyngeal nerve (cranial nerve IX), and the superior laryngeal branch of the vagus nerve (cranial nerve X) to innervate the taste buds in the tongue, palate, epiglottis, and esophagus, respectively. The central axons of these pri-

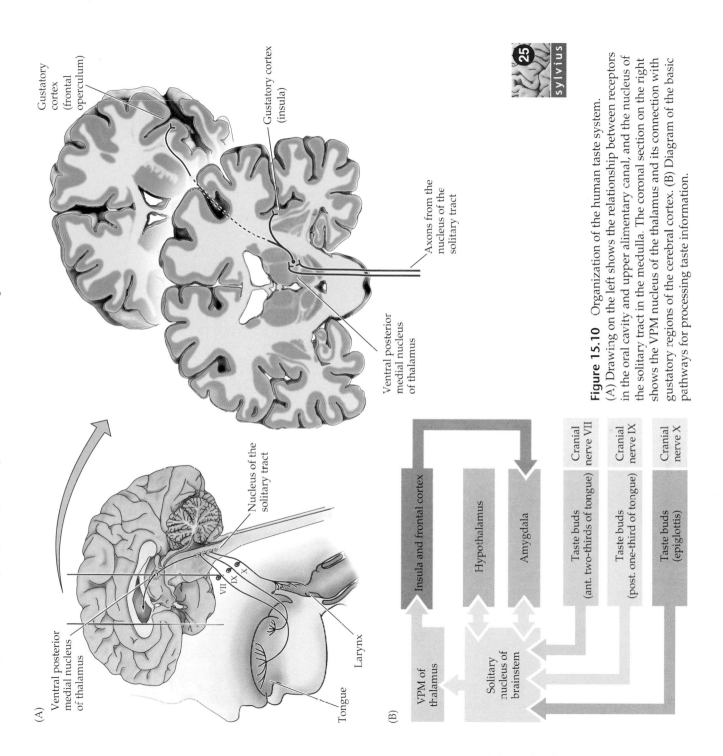

(A)

Ventral posterior medial nucleus of thalamus

Nucleus of the solitary tract

VII

IX

X

Tongue

Larynx

Gustatory cortex (frontal operculum)

Gustatory cortex (insula)

Axons from the nucleus of the solitary tract

Ventral posterior medial nucleus of thalamus

(B)

VPM of thalamus

Solitary nucleus of brainstem

Insula and frontal cortex

Hypothalamus

Amygdala

Taste buds (ant. two-thirds of tongue)

Cranial nerve VII

Taste buds (post. one-third of tongue)

Cranial nerve IX

Taste buds (epiglottis)

Cranial nerve X

Figure 15.10 Organization of the human taste system. (A) Drawing on the left shows the relationship between receptors in the oral cavity and upper alimentary canal, and the nucleus of the solitary tract in the medulla. The coronal section on the right shows the VPM nucleus of the thalamus and its connection with gustatory regions of the cerebral cortex. (B) Diagram of the basic pathways for processing taste information.

mary sensory neurons in the respective cranial nerve ganglia project to rostral and lateral regions of the **nucleus of the solitary tract** in the medulla (Figure 15.10B), which is also known as the **gustatory nucleus** of the solitary tract complex (recall that the posterior region of the solitary nucleus is the main target of afferent visceral sensory information related to the sympathetic and parasympathetic divisions of the visceral motor system; see Chapter 21).

The distribution of these cranial nerves and their branches in the oral cavity is topographically represented along the rostral-caudal axis of the rostral portion of the gustatory nucleus; the terminations from the facial nerve are most rostral, the glossopharyngeal are intermediate, and those from the vagus nerve are more caudal. Integration of taste and visceral sensory information is presumably facilitated by this arrangement. The caudal part of the nucleus of the solitary tract also receives innervation from subdiaphragmatic branches of the vagus nerve, which control gastric motility. Interneurons connecting the rostral and caudal regions of the nucleus represent the first interaction between visceral and gustatory stimuli. This close relationship of gustatory and visceral information makes good sense, since an animal must quickly recognize if it is eating something that is likely to make it sick, and respond accordingly.

Axons from the rostral (gustatory) part of the solitary nucleus project to the ventral posterior complex of the thalamus, where they terminate in the medial half of the **ventral posterior medial nucleus.** This nucleus projects in turn to several regions of the cortex, including the anterior insula in the temporal lobe and the operculum of the frontal lobe. There is also a secondary cortical taste area in the caudolateral orbitofrontal cortex, where neurons respond to combinations of visual, somatic sensory, olfactory, and gustatory stimuli. Interestingly, when a given food is consumed to the point of satiety, specific orbitofrontal neurons in the monkey diminish their activity to that tastant, suggesting that these neurons are involved in the motivation to eat (or not to eat) particular foods. Finally, reciprocal projections connect the nucleus of the solitary tract via the pons to the hypothalamus and amygdala (see Figure 15.10B). These projections presumably influence appetite, satiety, and other homeostatic responses associated with eating (recall that the hypothalamus is the major center governing homeostasis; see Chapter 21).

Taste Perception in Humans

Most taste stimuli are nonvolatile, hydrophilic molecules soluble in saliva. Examples include salts such as NaCl needed for electrolyte balance; essential amino acids such as glutamate needed for protein synthesis; sugars such as glucose needed for energy; and acids such as citric acid that indicate the palatability of various foods (oranges, in the case of citrate). Bitter-tasting molecules include plant alkaloids, such as atropine, quinine, and strychnine, that may be poisonous. Placing bitter compounds in the mouth usually deters ingestion unless one "acquires a taste" for the substance, as for quinine in tonic water.

The taste system encodes information about the quantity as well as the identity of stimuli. In general, the higher the stimulus concentration, the greater the perceived intensity of taste. Threshold concentrations for most ingested tastants are quite high, however. For example, the threshold concentration for citric acid is about 2 mM; for salt (NaCl), 10 mM; and for sucrose, 20 mM. Since the body requires substantial concentrations of salts and carbohydrates, taste cells may respond only to relatively high concentrations of these essential substances to promote an adequate intake. Clearly, it is advantageous for the taste system to detect potentially dangerous substances (e.g., bitter-tasting plant compounds) at much lower concentrations.

Thus, the threshold concentration for quinine is 0.008 m*M*, and for strychnine 0.0001 m*M*. As in olfaction, gustatory sensitivity declines with age. Adults tend to add more salt and spices to food than children. The decreased sensitivity to salt can be problematic for older people with electrolyte and/or fluid balance problems. Unfortunately, a safe and effective substitute for NaCl has not yet been developed.

There are at least two common misconceptions about taste perception. The first is that sweet is perceived at the tip of the tongue, salt along its posterolateral edges, sour along the mediolateral edges, and bitter on the back of the tongue. This arrangement was initially proposed in 1901 by Deiter Hanig, who measured taste thresholds for NaCl, sucrose, quinine, and hydrochloric acid (HCl). Hanig never said that other regions of the tongue were *insensitive* to these chemicals, but only indicated which regions were the *most* sensitive. People missing the anterior part of their tongue (or who have facial nerve lesions) can still taste sweet and salty stimuli. In fact, all of these tastes can be detected over the full surface of the tongue (Figure 15.11A). However, different regions of the tongue do have different thresholds. Because the tip of the tongue is most responsive to sweet-tasting compounds, and because these compounds produce pleasurable sensations, information from this region acti-

Figure 15.11 Taste buds and the peripheral innervation of the tongue. (A) Distribution of taste papillae on the dorsal surface of the tongue. Different responses to sweet, salty, sour, and bitter tastants recorded in the three cranial nerves that innervate the tongue and epiglottis are indicated at left. The size of the circles representing sucrose, NaCl, HCl, quinine, and water corresponds to the relative response of the papillae to these stimuli. (B) Diagram of a circumvallate papilla showing location of individual taste buds. (C) Light micrograph of a taste bud. (D) Diagram of a taste bud, showing various types of taste cells and the associated gustatory nerves. The apical surface of the receptor cells have microvilli that are oriented toward the taste pore. (C from Ross, Rommell and Kaye, 1995.)

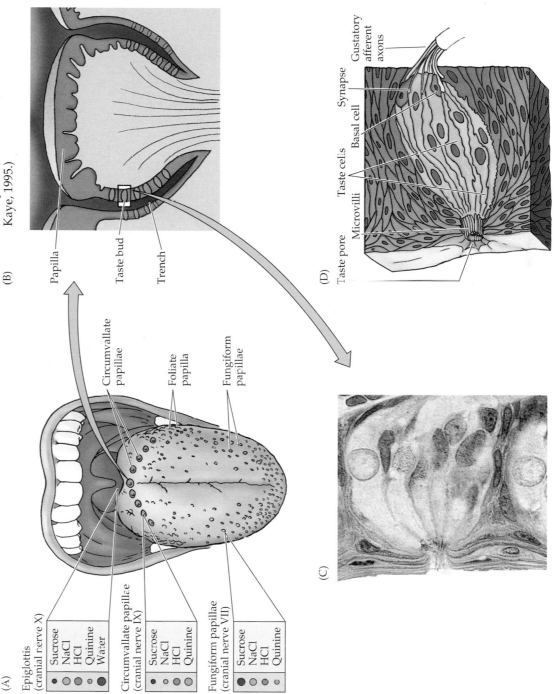

vates feeding behaviors such as mouth movements, salivary secretion, insulin release, and swallowing. In contrast, responses to bitter compounds are indeed greatest on the back of the tongue. Activation of this region by bitter-tasting substances elicits protrusion of the tongue and other protective reactions that prevent ingestion. Sour-tasting compounds elicit grimaces, puckering responses, and massive salivary secretion to dilute the tastant.

A second misconception about taste perception is that there are only four "primary" tastes: salt, sweet, sour, and bitter. If this were true, then all tastes could be represented as a combination of these "primaries." Although these four tastes do indeed represent distinct perceptions, this classification is obviously limited. People experience a variety of additional taste sensations, including astringency (cranberries and tea), pungency (hot pepper and ginger), fat, starchy, and various metallic tastes (to name but a few). None of these, however, fits into these four categories. Moreover, some cultures consider other tastes to be "primary." For example, the Japanese consider the taste of monosodium glutamate to be distinct from that of salt, and even give it a different name ("umami," which means delicious). Finally, mixtures of various chemicals may elicit entirely new taste sensations. While it is possible to estimate the number of perceived odors (approximately 10,000), these uncertainties have made it difficult to estimate the number of tastes. In neither taste nor olfaction is there a clear relationship between "primary" perceptual classes and the cellular and molecular machinery of sensory transduction.

The Organization of the Peripheral Taste System

Approximately 4000 taste buds in humans are distributed throughout the oral cavity and upper alimentary canal. Taste buds are about 50 mm wide at their base and approximately 80 mm long, each containing 30 to 100 taste cells (the sensory receptor cells), plus a few basal cells (Figure 15.11B–D). About 75% percent of all taste buds are found on the dorsal surface of the tongue in small elevations called **papillae** (see Figure 15.11A). There are three types of papillae: **fungiform** (which contain about 25% of the total number of taste buds), **circumvallate** (which contain 50% of the taste buds), and **foliate** (which contain 25%). Fungiform papillae are found only on the anterior two-thirds of the tongue; the highest density (about 30/cm²) is at the tip. They have a mushroom-like structure (hence their name) and typically have about 3 taste buds at their apical surface. There are 9 circumvallate papillae arranged in a chevron at the rear of the tongue. Each consists of a circular trench containing about 250 taste buds along the trench walls. Two foliate papillae are present on the posterolateral tongue, each having about 20 parallel ridges with about 600 taste buds in their walls. Thus, chemical stimuli on the tongue first stimulate receptors in the fungiform papillae and then in the foliate and circumvallate papillae. Tastants subsequently stimulate scattered taste buds in the pharynx, larynx, and upper esophagus.

Taste cells in individual taste buds (see Figure 15.11C,D) synapse with primary afferent axons from branches of three cranial nerves: the facial (VII), glossopharyngeal (IX), and vagus (X) nerves (see Figure 15.10). The taste cells in fungiform papillae on the anterior tongue are innervated exclusively by the chorda tympani branch of the facial nerve; in circumvallate papillae, the taste cells are innervated exclusively by the lingual branch of the glossopharyngeal nerve; and in the palate they are innervated by the greater superior petrosal branch of the facial nerve. Taste buds of the epiglottis and esophagus are innervated by the superior laryngeal branch of the vagus nerve.

The initiating events of chemosensory transduction occur in the taste cells, which have receptors on microvilli that emerge from the apical surface of the taste cell (see Figure 15.11D). The synapses that relay the receptor activity are made onto the afferent axons of the various cranial nerves at the basal surface. The apical surfaces of individual taste cells in taste buds are clustered in a small opening (about 1 mm) near the surface of the tongue called a **taste pore**. Like olfactory receptor neurons (and presumably for the same reasons), taste cells have a lifetime of only about 2 weeks and are normally regenerated from basal cells.

Idiosyncratic Responses to Various Tastants

Taste responses vary among individuals. For example, many people (about 30–40% of the U.S. population) cannot taste the bitter compound phenylthiocarbamide (PTC) but can taste molecules such as quinine and caffeine that also produce bitter sensations. Indeed, humans can be divided into two groups with quite different thresholds for bitter compounds containing the N—C=S group found in PTC. The difference between these individuals is the presence of a single autosomal gene (*Ptc*) with a dominant (tasters) and a recessive (nontasters) allele. Interestingly, people who are extremely sensitive to PTC or its analogues, called "supertasters," have more taste buds than normal and tend to avoid certain foods such as grapefruit, green tea, and broccoli, all of which contain bitter-tasting compounds. Thus, an individual's genetic makeup with respect to taste receptors has implications for diet, and even health.

In the same vein, a number of quite different compounds taste sweet to humans. These include saccharides (glucose, sucrose, and fructose), organic anions (saccharin), amino acids (aspartame, or Nutrasweet®), L-phenylalanine methyl ester, and proteins (monellin and thaumatin). People can distinguish among different sweeteners, and some find that saccharin has a bitter-tasting component. One reason for such discriminability is that some of these compounds activate separate receptors. For example, saccharides activate cAMP pathways, whereas nonsaccharide sweeteners such as amino acids activate IP$_3$ pathways (see next section). Thus, the perceptual experience of "sweet" encompasses much more than the taste of sucrose, can be elicited by various sensory transduction mechanisms, and may generate sensory qualities different from those generated by sucrose.

Taste sensitivity for salt also relies on a number of mechanisms. Not all salts, or even all monovalent chloride salts, activate the same pathway. Psychophysical studies have shown that amiloride, a diuretic that blocks Na$^+$ entry through amiloride-sensitive Na$^+$ channels, decreases the taste intensity of NaCl and LiCl, but not KCl. Although LiCl, like NaCl, tastes salty, it cannot be used as a salt Na$^+$ substitute because of its profound effects on the central nervous system (it is used clinically to treat manic-depressive disorders). Sodium succinate, NH$_4$Cl, and CsCl do not taste exclusively salty. Indeed, CsCl has a bitter or salty-bitter taste that probably arises from the inhibition of K$^+$ channels. Additional evidence for a distinct receptor for NaCl comes from developmental studies. Infants up to 4 months old can distinguish between water and sucrose (and lactose), water and acid, and water and bitter tastants, but they cannot distinguish between water and a 0.2 *M* NaCl solution. Thus, either the receptor for Na$^+$ has not yet been expressed, or, if expressed, it is not yet functional. Infants between the ages of 4 and 6 months, however, can discriminate between NaCl solutions and water, and children can detect the full salty taste of NaCl at about 4 years of age.

Sour taste is produced by relatively high concentrations of acid. At the same H$^+$ activity, weak organic acids such as tartaric acid and citric acid exhibit distinctly different tastes from HCl, whereas strong inorganic acids such as HCl, HNO$_3$, and H$_2$SO$_4$ have similar tastes. The accompanying anion influences the activity of gustatory responses to acids by virtue of its ability to diffuse across the tight junctions between taste cells, thus creating a variety of sour tastes.

In short, the taste system uses many mechanisms to distinguish among the various chemicals placed in the mouth. The four words "sweet," "sour," "salty," and "bitter" do not accurately or fully describe tastes, which encompass a far more subtle range of sensations and may mean different things to different people.

Taste Receptors and the Transduction of Taste Signals

Although receptor molecules that bind various tastants are found primarily on the apical microvilli of the taste cells, the transduction machinery involves ion channels on both the apical and basolateral membranes (Figure 15.12).

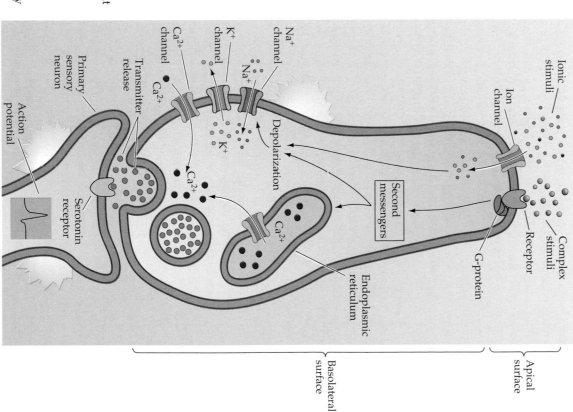

Figure 15.12 Transduction mechanisms in a generic taste cell. The apical and basolateral surfaces of the cell are separated by tight junctions. The apical surface contains both channels and G-protein-coupled receptors that are activated by chemical stimuli. The basolateral surface contains voltage-gated Na⁺, K⁺, and Ca²⁺ channels, as well as all the machinery for synaptic transmission mediated by serotonin. Also shown are the relevant second messenger systems and intracellular compartments that store Ca²⁺. The increase in intracellular Ca²⁺ either by the activation of voltage-gated Ca²⁺ channels or via the release from intracellular stores causes synaptic vesicles to fuse and release their transmitter onto receptors on primary sensory neurons.

Figure 15.13 Examples of various channels and G-protein-coupled receptors that activate taste transduction in response to various compounds. When stimulated, each of these channels or receptors changes neurotransmitter release via either direct changes in depolarization or second messenger-mediated changes in intracellular Ca^{2+} concentration.

Salt

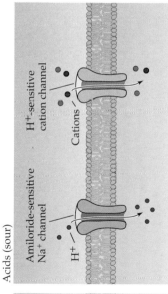

Amiloride-sensitive
Na^+ channel

Na^+

Acids (sour)

H$^+$-sensitive
cation channel

Amiloride-sensitive
Na^+ channel

Cations

H^+

Bitter

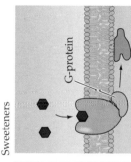

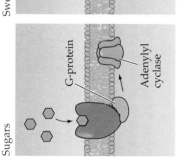

G-protein

PDE

cAMP   cGMP

AMP

GMP

G-protein

Phospholipase C

IP$_3$

Divalent salt/
quinine-sensitive
K$^+$ channel

K^+

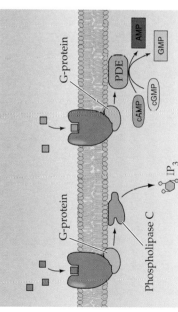

Monosodium
glutamate ("Unami")

MGluR-4
(Metabotropic
glutamate receptor)

G-protein

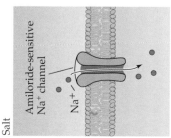

Sugars

G-protein

Adenylyl
cyclase

Sweeteners

G-protein

Phospholipase C

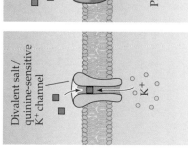

Channels typically found in axonal membranes are located on the basolateral aspect of taste cells. These include voltage-gated Na^+, K^+, and Ca^{2+} channels that produce depolarizing potentials when taste cells interact with chemical stimuli. The resulting receptor potentials raise Ca^{2+} to levels sufficient for synaptic vesicle fusion and synaptic transmission, thus eliciting action potentials in the afferent axons. In general, the greater the tastant concentration, the greater the depolarization of the taste cell.

The molecular identity of taste receptors has been examined in several experimental animals, including nonhuman primates. The "receptor" for salt (NaCl) is apparently an epithelial-type Na^+ channel on the apical membrane of some taste cells (Figure 15.13). In general, the larger the NaCl concentration applied to the tongue, the larger the depolarization in the relevent taste cells. These Na^+ channels are regulated by hormones involved in water and electrolyte balance (for example, antidiuretic hormone and aldosterone), which mediate Na^+-specific appetite and intake. Protons (H^+) can also diffuse through this channel, albeit more slowly than Na$^+$; this fact may explain why the addition of acids like lemon juice to salty foods reduces their salty taste. Protons, which are primarily responsible for sour taste, also interact with distinct channels on the apical membranes of a subset of taste cells (see Figure 15.13). These cations activate proton-gated cation and Cl$^-$ channels (see Figure 15.12). Thus, several mechanisms underlie the reception and transduction of acidic stimuli (Figure 15.14).

The transduction of sweet-tasting compounds involves the activation of G-protein-coupled receptors (GPCRs) on the apical surface of taste cells (see Figure 15.13). The particulars of the cascade depend on a number of factors, including the specifics of the stimulus. In the case of sweeteners such as the saccharides, activation of GPCRs depolarizes taste cells by activating adenylate cyclase, which in turn increases the cAMP concentration that will either directly or indirectly close basolateral K$^+$ channels. Synthetic sweeteners, such as saccharine, activate different GPCRs that in turn activate phospholipase C (PLC) to produce IP$_3$ and DAG. An increase in IP$_3$ raises intracellular Ca^{2+} concentration, leading to transmitter release. An increase in DAG activates PKA, and PKA in turn phosphorylates and closes basolateral K$^+$ channels, further contributing to this effect. Both of these pathways for the perception of sweetness can co-exist in the same taste cell.

There are many chemically distinct classes of bitter-tasting compounds (see Figure 15.13). Some of these are alkaloids, like quinine and caffeine; others are L-amino acids, urea, and even salts like MgSO$_4$. Again, not all of these bitter tastants use the same receptor or transduction pathways. Indeed, about forty new receptors for bitter tastants have been recently cloned, and many are found in the same cells. Bitter- tasting organic compounds typically bind to GPCRs that activate **gustducin** (a G-protein found in taste cells homologous to transducin in photoreceptors), which in turn activates phosphodiesterase, thus lowering the cyclic nucleotide concentration and closing cyclic nucleotide-gated channels on the basolateral membranes of taste cells. Gustducin-knockout mice exhibit impaired responses to bitter compounds, suggesting that this second messenger is involved in the transduction of bitter tastes. Many bitter tastants bind directly to GCPRs that activate PLC and the production of IP$_3$, leading to an increase in intracellular Ca^{2+}. Given the wide variety of chemical structures that evoke a bitter taste, it is not surprising that a variety of transduction mechanisms are employed.

The taste of amino acids fall into both the sweet (D-amino acids) and bitter categories (L-isomers), and also uses a variety of transduction mechanisms. An exception is the amino acid L-glutamate (and its sodium salt), which elicits a quite different taste (see above). The effects of L-glutamate on taste cells involves both ionotropic receptors that activate ion channels, and unusual taste-specific metabotropic glutamate receptors (mGluR4) that are less sensitive to glutamate and that close ion channels through a cAMP-dependent pathway.

The overall picture that emerges from these admittedly complicated details is that taste cells have a variety of transduction mechanisms. In general, individual taste cells respond to several types of chemical stimuli. Nevertheless, taste cells also exhibit gustatory selectivity. Like olfactory cells, the lower the threshold concentration for detecting a single tastant, the greater the selectivity of the relevant taste cell. Finally, taste receptor mechanisms also adapt to the ongoing presence of a stimulus, although the mechanisms are not understood. If a chemical is left on the tongue for a sufficient time, it ceases to be perceived (consider saliva, for example). Thus, to obtain the full

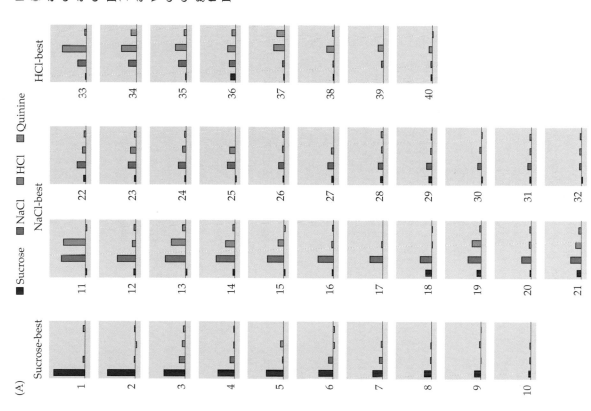

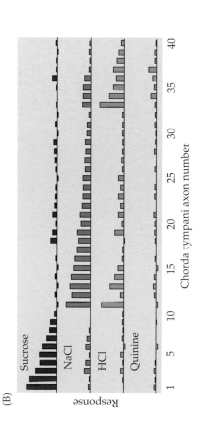

Figure 15.14 Different ways of encoding taste. (A) Response profiles of individual chorda tympani axons to four different stimuli (indicated by the four different colors). The numbers indicate individual axons. The responses reflect the net activity for 5 seconds after application of each tastant. The response patterns suggest a "labeled line" mechanism; axons 1–10 are sucrose-best, axons 11–32 are NaCl-best, and axons 33–40 are HCl-best. (B) When plotted another way, however, the responses of the same fibers are consistent with an "across neuron" scheme of taste coding. Each row depicts the pattern elicited by a single tastant in the full population of 40 axons. Each taste has its own distinct pattern. (After Smith and Frank, 1993.)

taste of foods, one must either frequently change the types of foods placed in the mouth or wait a sufficient time between helpings, facts that have long been appreciated by restauranteurs and gourmets.

Neural Coding in the Taste System

Taste "coding" refers to the way that the identity, concentration, and "hedonic" (pleasurable or unpleasurable) value of tastants is represented in the patterns of action potentials relayed to the brain. As already noted, psychophysical studies indicate that different regions of the tongue and oral cavity have different sensitivities to various tastants. This same selectivity is observed in recordings from the primary afferent axons in the cranial nerves of experimental animals. Specifically, the chorda tympani branch of the facial nerve responds best to NaCl and sucrose (that is, for a given concentration a greater number of action potentials and axons are activated by these substances); the glossopharyngeal nerve responds best to acid and quinine; and the superior laryngeal branch of the vagus nerve responds best to acid and water. These observations imply that a single peripheral neuron might respond best to sucrose, for example, but could also be activated by fructose, NaCl, and acetic acid. The information contained in the patterns of action potentials elicited by each of these tastants evidently arises from the activation of the several taste cells that synapse with the primary afferent neurons. This peripheral convergence has the effect of increasing the sensitivity of the taste system to a given stimulus.

Two competing ideas have been proposed about how this sensory information for taste is encoded: the **labeled line hypothesis** and the **ensemble** or **"across-neuron" hypothesis** (a similar argument has been made for olfaction). The labeled line paradigm categorizes individual taste cells and associated neurons into classes such as "NaCl-best," "sugar-best," and so on, meaning that one of these stimulus categories will evoke the greatest number of action potentials per unit time in any given gustatory neuron (Figure 15.14A). Thus the phrase "labeled line" implies that the activity in one neuron type is both necessary and sufficient to represent a given sensory attribute. A corollary is that distinct types of gustatory information are transmitted throughout the central components of the taste system along pathways that preserve the "best" fiber categories. In contrast, the ensemble or "across neuron" hypothesis proposes that the pattern of responses to a particular stimulus across all fibers is the central feature of coding (Figure 15.14B). In this model, the taste of sucrose, for example, is "computed" from the response evoked from all 40 taste cells illustrated in Figure 15.14, regardless of the specific subset that is active or silent.

Central Processing of Taste Signals

If the labeled line theory is correct, the physiological properties of distinct taste receptor cells in the periphery should be faithfully projected to the central relay stations of the taste system. There is, in fact, some evidence for "best" taste responses in the nucleus of the solitary tract, thalamus, and cortex. For instance, the drug amiloride has been shown to inhibit the responses to NaCl only from NaCl-best neurons in the central nervous system, as it does for the NaCl-best primary afferents. It is important to note, however, that "best" types are determined by evaluating a limited number of taste stimuli at a limited number of concentrations. In addition, gustatory information contains information about quality, intensity, and hedonic value, all of which are unlikely to be encoded in a single type of neuron. In short, given present evidence, it is difficult to prove the validity of the labeled line theory.

If the ensemble or across neuron model is correct, patterns of activation across subsets of neurons should be apparent, and these patterns should be associated with stimulation by particular tastants (see Figure 15.14B). In the presence of amiloride, the patterns for NaCl and KCl are very similar, and rats cannot distinguish between them. The pattern of activity elicited by a single stimulus in a set of taste-responsive neurons does indeed remain relatively stable at a particular stimulus concentration, although increasing the concentration recruits new neurons, thus changing the pattern and the taste. In keeping with this ensemble model, mixtures of stimuli tend to have unique patterns of activity, and unique tastes are experienced that cannot easily be described as a combination of the alleged four or five "primary" tastes. Markedly changing the pattern changes the taste and, the closer the patterns, the closer the tastes.

The tuning curves for neurons in the solitary nucleus, thalamus, and cortex become ever more broadly tuned to stimuli that are nonetheless perceptually distinct, suggesting that ensembles of neurons extract specific information about chemicals placed in the oral cavity. Moreover, many of the neurons in the gustatory cortices are multimodal: They respond to thermal, mechanical, olfactory, and visual stimuli pertaining to food, as well as to taste stimuli.

In summary, neural coding for taste, as for olfaction, involves information obtained from ensembles of neurons from many interacting neural subsystems and cannot, based on present evidence, be reduced to a single conceptual scheme.

Trigeminal Chemoreception

The third of the major chemosensory systems, the trigeminal chemosensory system, consists of polymodal nociceptive neurons and their axons in the trigeminal nerve (cranial nerve V) and, to a lesser degree, nociceptive neurons whose axons run in the glossopharyngeal and vagus nerves (IX and X). These neurons and their associated endings are typically activated by chemicals classified as irritants, including air pollutants (e.g., sulfur dioxide), ammonia (smelling salts), ethanol (liquor), acetic acid (vinegar), carbon dioxide (in soft drinks), menthol (in various inhalants), and capsaicin (the compound in chili peppers) that elicits the characteristic burning sensation) (Box C). Irritant-sensitive polymodal nociceptors alert the organism to potentially harmful chemical stimuli that have been ingested, respired, or come in contact with the face, and are closely tied to the trigeminal pain system discussed in Chapter 10.

Trigeminal chemosensory information from the face, scalp, cornea, and mucous membranes of the oral and nasal cavities is relayed via the three major sensory branches of the trigeminal nerve: the ophthalmic, maxillary, and mandibular (Figure 15.15). The central target of these afferent axons is the spinal component of the trigeminal nucleus, which relays this information to the ventral posterior medial nucleus of the thalamus and thence to the somatic sensory cortex and other cortical areas that process facial irritation and pain (see Chapter 10).

Many compounds classified as irritants can also be recognized as odors or tastes; however, the threshold concentrations for trigeminal chemoreception are much higher than those for olfaction or taste. When potentially irritating compounds are presented to people who have lost their sense of smell, perceptual thresholds are found to be approximately 100 times higher than those of normal subjects who perceive the compounds as odors

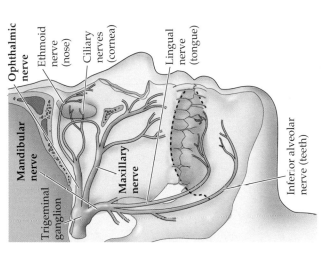

Figure 15.15 Diagram of the branches of the trigeminal nerve that innervate the oral, nasal, and ocular cavities. The chemosensitive structures innervated by each trigeminal branch are indicated in parentheses.

Figure labels:
Ophthalmic nerve
Ethmoid nerve (nose)
Ciliary nerves (cornea)
Mandibular nerve
Lingual nerve (tongue)
Maxillary nerve
Trigeminal ganglion
Inferior alveolar nerve (teeth)

BOX C
Capsaicin

Capsaicin, the principle ingredient responsible for the pungency of hot peppers, is eaten daily by over a third of the world's population. Capsaicin activates responses in a subset of nociceptive C fibers (polymodal nociceptors; see Chapter 10) by opening ligand-gated ion channels that permit the entry of Na^+ and Ca^{2+}. One of these channels (VR-1) has been cloned and has been found to be activated by capsaicin, acid, and anandamide (an endogeneous compound that also activates cannabinoid receptors), and by heating the tissue to about 43°C. It follows that anandamide and temperature are probably the endogenous activators of these channels. Mice whose VR-1 receptors have been knocked out drink capsaicin solutions as if they were water. Receptors for capsaicin have been found in polymodal nociceptors of all mammals, but are not present in birds (leading to the

production of squirrel-proof birdseed laced with capsaicin!).

When applied to the mucus membranes of the oral cavity, capsaicin acts as an irritant, producing protective reactions. When injected into skin, it produces a burning pain and elicits hyperalgesia to thermal and mechanical stimuli. Repeated applications of capsaicin also desensitize pain fibers and prevent neuromodulators such as substance P, VIP, and somatostatin from being released by peripheral and central nerve terminals. Consequently, capsaicin is used clinically as an analgesic and anti-inflammatory agent; it is usually applied topically in a cream (0.075%) to relieve the pain associated with arthritis, postherpetic neuralgia, mastectomy, and trigeminal neuralgia. Thus, this remarkable chemical irritant not only gives gustatory pleasure on an enormous scale, but is also a useful pain reliever!

References

CATERINA, M. J., M. A. SCHUMACHER, M. TOMINAGA, T. A ROSEN, J. D. LEVINE AND D. JULIUS (1997) The capsaicin receptor: A heat-activated ion channel in the pain pathway. Nature 389, 816–766.

ZYGMUNT, P. M. AND 7 OTHERS (1999) Vanilloid receptors on sensory nerves mediate the vasodilator action of anandamide. Nature 400(6743): 452–457.

TOMINAGA, M. AND 8 OTHERS (1998) The cloned capsaicin receptor integrates multiple pain-producing stimuli. Neuron 21(3): 531–543.

CATERINA, M. J. AND 8 OTHERS (2000) Impaired nociception and pain sensation in mice lacking the capsaicin receptor. Science 288: 306–313.

SZALLASI, A. AND P. M. BLUMBERG (1999) Vanilloid (capsaicin) receptors and mechanisms. Pharm. Reviews 51(2): 159–212.

(A)

Habañero

Jalapeño

Red chile

(B) Capsaicin

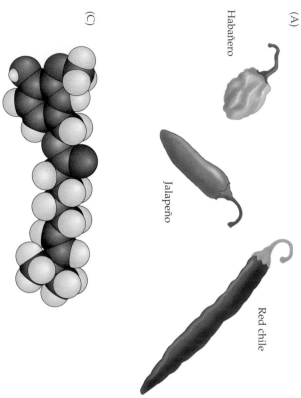

(C)

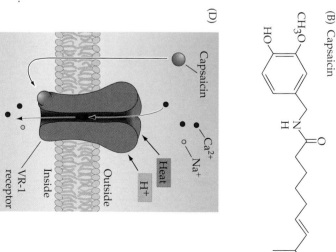

(D)

Capsaicin

Ca²⁺

Na⁺

Outside

Inside

Heat

H⁺

VR-1 receptor

(A) Some popular peppers that contain capsaicin. (B) The chemical structure of capsaicin. (C) The capsaicin molecule. (D) Schematic of the VR-1/capsaicin receptor channel. This channel can be activated by capsaicin intercellularly, or by heat or protons (H⁺) at the cell surface.

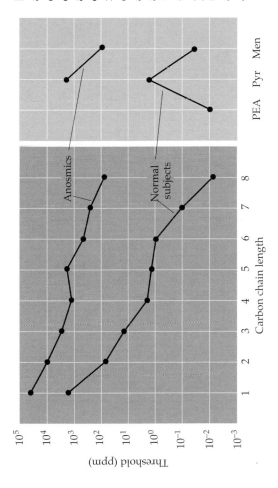

Figure 15.16 Perceptual thresholds in anosmic and normal subjects for related organic chemicals. In anosmics, these chemicals are only detected as irritants at relatively high concentrations (indicated here in parts per million—ppm); in normal subjects, they are first detected at much lower concentrations as odors. The numbers 1–8 stand for the aliphatic alcohols from methanol to 1-octanol. Perceptual thresholds for three additional common irritants—phenylethyl alcohol (PEA), pyridine (Pyr), and menthol (Men)—are shown at the far right. (After Commetto-Muniz and Cain, 1990.)

(Figure 15.16). Similar differences occur in identifying chemicals as tastes rather than irritants. Thus, 0.1 *M* NaCl has a salty taste, but 1.0 *M* NaCl is perceived as an irritant. Another common irritant is ethanol. When placed on the tongue at moderate temperatures and high concentrations—as in drinking vodka "neat"—ethanol produces a burning sensation.

A variety of physiological responses mediated by the trigeminal chemosensory system are triggered by exposure to irritants. These include increased salivation, vasodilation, tearing, nasal secretion, sweating, decreased respiratory rate, and bronchoconstriction. Consider, for instance, the experience that follows the ingestion of capsaicin, the irritant in hot chili peppers (see Box C). These reactions are generally protective in that they dilute the stimulus (tearing, salivation, sweating) and prevent inhaling or ingesting more of it.

The receptors for irritants are primarily on the terminal branches of polymodal nociceptive neurons, as described for the pain and temperature systems in Chapter 10. Although these receptors respond to many of the same stimuli as olfactory receptor neurons (e.g., aldehydes, alcohols), they are probably not activated by the same mechanism; for instance, the G-protein-coupled receptors for odorants are found only in olfactory receptor neurons. With the exception of capsaicin and acidic stimuli, both of which activate cation-selective ion channels, little is known about the transduction mechanisms for irritants, or their central processing.

Summary

The chemical senses—olfaction, taste, and the trigeminal chemosensory system—all contribute to an awareness of airborne or soluble molecules from a variety of sources. Humans and other mammals rely on this information for behaviors as diverse as attraction, avoidance, reproduction, feeding, and avoiding potentially dangerous circumstances. Receptor neurons in the olfactory epithelium transduce chemical stimuli into neuronal activity via the stimulation of G-protein-linked receptors; this interaction leads to elevated levels of second messengers such as cAMP, which in turn open cation-selective channels. These events generate receptor potentials in the membrane of the olfactory receptor neuron, and ultimately action potentials in the afferent axons of these cells. Taste receptor cells, in contrast, use a variety of mechanisms for transducing chemical stimuli. These include ion channels that are directly activated by salts and amino acids, and G-protein-linked receptors that activate second messengers. For both smell and taste, the spatial and temporal patterns of action potentials provide information about the identity and intensity of chemical stimuli. Each of the approximately 10,000 odors that humans recognize (and an undetermined number of tastes and irritant molecules) is evidently encoded by the activity of a distinct population of receptor cells in the nose, tongue, and oral cavity. Olfaction, taste, and trigeminal chemosensation all are relayed via specific pathways in the central nervous system. Receptor neurons in the olfactory system project directly to the olfactory bulb. In the taste system, information is relayed centrally via the solitary nucleus in the brainstem. In the trigeminal chemosensory system, information is relayed via the spinal trigeminal nucleus. Each of these structures project in turn to many sites in the brain that process chemosensory information in ways that remain poorly understood but nevertheless give rise to some of the most sublime pleasures that humans experience.

ADDITIONAL READING

Reviews

BUCK, L. B. (2000) The molecular architecture of odor and pheromone sensing in mammals. Cell 100: 611–618.

ERICKSON, R. P. (1985) Definitions: A matter of taste. In *Taste, Olfaction, and the Central Nervous System*. D. W. Pfaff (ed.). New York: Rockefeller University Press, p. 129.

HERNESS, M. S. AND T. A. GILBERTSON (1999) Cellular mechanisms of taste transduction. Ann. Rev. Physiol. 61: 873–900.

HILDEBRAND, J. G. AND G. M. SHEPHERD (1997) Mechanisms of olfactory discrimination: Converging evidence for common principles across phyla. Ann. Rev. Neurosci. 20: 595–631.

KRUGER, L. AND P. W. MANTYH (1989) Gustatory and related chemosensory systems. In *Handbook of Chemical Neuroanatomy*, Vol. 7, *Integrated Systems of the CNS*, Part II. A. Björkland, T. Hökfelt and L. W. Swanson (eds.). New York: Elsevier Science, pp. 323–410.

LAURENT, G. (1999) A systems perspective on early olfactory coding. Science 286: 723–728.

LINDEMANN, B. (1996) Taste reception. Physiol. Rev. 76: 719–766.

MENINI, A. (1999) Calcium signaling and regulation in olfactory neurons. Curr. Opin. Neurobiol. 9: 419–426.

YAMAMOTO, T., T. NAGAI, T. SHIMURA AND Y. YASOSHIMA (1998) Roles of chemical mediators in the taste system. Jpn. J. Pharmacol 76: 325–348.

ZUTALL, F AND T. LEINDERS-ZUTALL (2000) The cellular and molecular basis of odor adoption. Chem. Senses 25: 473–481.

Important Original Papers

ADLER, E., M. A HOON, K. L MUELLER, J. CHRANDRASHEKAR, N. J. P. RYBA AND C. S. ZUKER (2000) A novel family of mammalian taste receptors. Cell 100: 693–702.

ASTIC, L. AND D. SAUCIER (1986) Analysis of the topographical organization of olfactory epithelium projections in the rat. Brain Res. Bull. 16(4): 455–462.

AVENET, P. AND B. LINDEMANN (1988) Amiloride-blockable sodium currents in isolated taste receptor cells. J. Memb. Biol. 105: 245–255.

BUCK, L. AND R. AXEL (1991) A novel multigene family may encode odorant receptors: A molecular basis for odor recognition. Cell 65: 175–187.

CATERINA, M. J. AND 8 OTHERS (2000) Impaired nociception and pain sensation in mice lacking the capsaicin receptor. Science 288: 306–313.

CHAUDHARI, N., A. M. LANDIN AND S. D. ROPER (2000) A metabotropic glutamate receptor variant functions as a taste receptor. Nature Neurosci. 3: 113–119.

GRAZIADEI, P. P. C. AND G. A. MONTI-GRAZIADEI (1980) Neurogenesis and neuron regeneration in the olfactory system of mammals. III. Deafferentation and reinnervation of the olfactory bulb following section of the fila olfactoria in rat. J. Neurocytol. 9: 145–162.

KAY, L. M. AND G. LAURENT (2000) Odor- and context-dependent modulation of mitral cell activity in behaving rats. Nature Neurosci. 2: 1003–1009.

MALNIC, B., J. HIRONO, T. SATO AND L. B. BUCK (1999) Combinatorial receptor codes for odors. Cell 96: 713–723.

MOMBAERTS, P. AND 7 OTHERS (1996) Visualizing an olfactory sensory map. Cell 87(4): 675–686.

ROLLS, E. T. AND L. L. BAYLIS (1994) Gustatory, olfactory and visual convergence within primate orbitofrontal cortex. J. Neurosci. 14: 5437–5452.

SCHIFFMAN, S. S., E. LOCKHEAD AND F. W. MAES (1983) Amiloride reduces taste intensity of salts and sweeteners. Proc. Natl. Acad. Sci. USA 80: 6136–6140.

VASSAR, R., S. K. CHAO, R. SITCHERAN, J. M. NUNEZ, L. B. VOSSHALL AND R. AXEL (1994) Topographic organization of sensory projections to the olfactory bulb. Cell 79(6): 981–991.

WONG, G. T., K. S. GANNON AND R. F. MARGOLSKEE (1996) Transduction of bitter and sweet taste by gustducin. Nature 381: 796–800.

Books

BARLOW, H. B. AND J. D. MOLLON (1989) *The Senses.* Cambridge: Cambridge University Press, Chapters 17–19.

DOTY, R. L. (ED.) (1995) *Handbook of Olfaction and Gustation.* New York: Marcel Dekker.

FARBMAN, A. I. (1992) *Cell Biology of Olfaction.* New York: Cambridge University Press.

GETCHELL, T. V., L. M. BARTOSHUK, R. L. DOTY AND J. B. SNOW JR. (1991) *Smell and Taste in Health and Disease.* New York: Raven Press.

SIMON, S. A. AND S. D. ROPER (1993) *Mechanisms of Taste Transduction.* Boca Raton: CRC Press, Chapters 2, 6, 9, 10, 12, 13, and 14.

Movement and Its Central Control

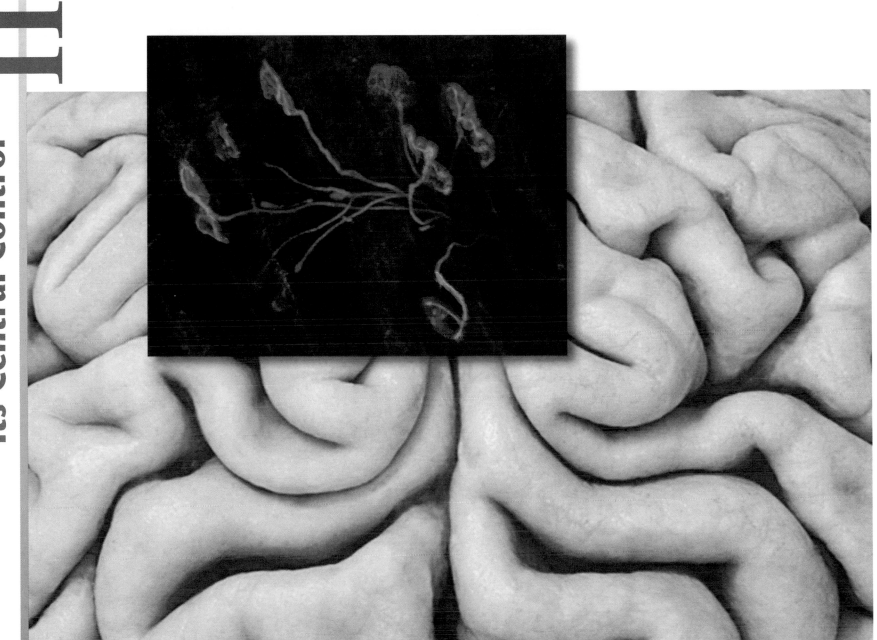

Unit III
Movement and Its Central Control

16 *Lower Motor Neuron Circuits and Motor Control*
17 *Upper Motor Neuron Control of the Brainstem and Spinal Cord*
18 *Modulation of Movement by the Basal Ganglia*
19 *Modulation of Movement by the Cerebellum*
20 *Eye Movements and Sensory-Motor Integration*
21 *The Visceral Motor System*

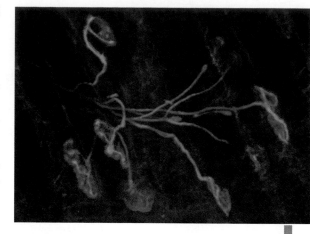

Fluorescence photomicrograph showing motor axons (green) and neuromuscular synapses (orange) in transgenic mice that have been genetically engineered to express fluorescent proteins. (Courtesy of Bill Snider and Jeff Lichtman.)

Movements, whether voluntary or involuntary, are produced by spatial and temporal patterns of muscular contractions orchestrated by the brain and spinal cord. Analysis of these circuits is fundamental to an understanding of both normal behavior and the etiology of a variety of neurological disorders. This unit considers the brainstem and spinal cord circuitry that make elementary reflex movements possible, as well as the circuits that organize the intricate patterns of neural activity responsible for more complex motor acts. Ultimately, all movements produced by the skeletal musculature are initiated by "lower" motor neurons in the spinal cord and brainstem that directly innervate skeletal muscles (the innervation of visceral smooth muscles is separately organized by the autonomic divisions of the visceral motor system). The lower motor neurons are controlled directly by local circuits within the spinal cord and brainstem that coordinate individual muscle groups, and also indirectly by "upper" motor neurons in higher centers that regulate those local circuits, thus enabling and coordinating complex sequences of movements. Especially important are circuits in the basal ganglia and cerebellum that regulate the upper motor neurons, ensuring that movements are performed with spatial and temporal precision.

Specific disorders of movement often signify damage to a particular brain region. For example, clinically important and intensively studied neurodegenerative disorders such as Parkinson's disease, Huntington's disease, and amyotrophic lateral sclerosis result from pathological changes in different parts of the motor system. Knowledge of the various levels of motor control is essential for understanding, diagnosing, and treating these diseases.

Chapter 16

Lower Motor Neuron Circuits and Motor Control

Overview

Skeletal (striated) muscle contraction is initiated by "lower" motor neurons in the spinal cord and brainstem. The cell bodies of the lower neurons are located in the ventral horn of the spinal cord gray matter and in the motor nuclei of the cranial nerves in the brainstem. These neurons (also called α motor neurons) send axons directly to skeletal muscles via the ventral roots and spinal peripheral nerves, or via cranial nerves in the case of the brainstem nuclei. The spatial and temporal patterns of activation of lower motor neurons are determined primarily by local circuits located within the spinal cord and brainstem. Descending pathways comprising the axons of "upper" motor neurons modulate the activity of lower motor neurons by influencing this local circuitry. The cell bodies of upper motor neurons are located either in the cortex or in brainstem centers, such as the vestibular nucleus, the superior colliculus, and the reticular formation. The axons of the upper motor neurons typically contact the local circuit neurons in the brainstem and spinal cord, which, via relatively short axons, contact in turn the appropriate combinations of lower motor neurons. The local circuit neurons also receive direct input from sensory neurons, thus mediating important sensory motor reflexes that operate at the level of the brainstem and spinal cord. Lower motor neurons, therefore, are the final common pathway for transmitting neural information from a variety of sources to the skeletal muscles.

Neural Centers Responsible for Movement

The neural circuits responsible for the control of movement can be divided into four distinct but highly interactive subsystems, each of which makes a unique contribution to motor control (Figure 16.1). The first of these subsystems is the local circuitry within the gray matter of the spinal cord and the analogous circuitry in the brainstem. The relevant cells include the **lower motor neurons** (which send their axons out of the brainstem and spinal cord to innervate the skeletal muscles of the head and body, respectively) and the **local circuit neurons** (which are the major source of synaptic input to the lower motor neurons). All commands for movement, whether reflexive or voluntary, are ultimately conveyed to the muscles by the activity of the lower motor neurons; thus they comprise, in the words of the great British neurophysiologist Charles Sherrington, the "final common path" for movement. The local circuit neurons receive sensory inputs as well as descending projections from higher centers. Thus, the circuits they form provide much of the coordination between different muscle groups that is essential for organized movement. Even after the spinal cord is disconnected from the

Figure 16.1 Overall organization of neural structures involved in the control of movement. Four systems—local spinal cord and brainstem circuits, descending modulatory pathways, the basal ganglia, and the cerebellum—make essential and distinct contributions to motor control.

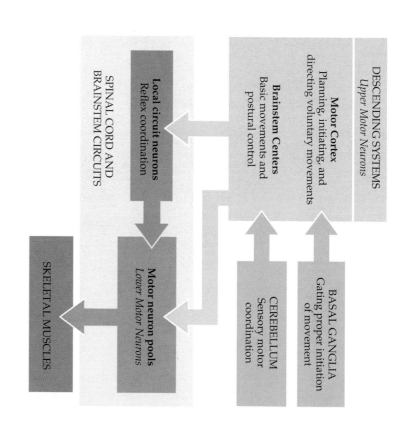

DESCENDING SYSTEMS
Upper Motor Neurons

Motor Cortex
Planning, initiating, and directing voluntary movements

Brainstem Centers
Basic movements and postural control

BASAL GANGLIA
Gating proper initiation of movement

CEREBELLUM
Sensory motor coordination

SPINAL CORD AND BRAINSTEM CIRCUITS

Local circuit neurons
Reflex coordination

Motor neuron pools
Lower Motor Neurons

SKELETAL MUSCLES

brain in an experimental animal such as a cat, appropriate stimulation of local spinal circuits elicits involuntary but highly coordinated movements of the four limbs that resemble walking.

The second motor subsystem consists of neurons whose cell bodies lie in the brainstem or cerebral cortex. The axons of these higher-order or **upper motor neurons** descend to synapse with the local circuit neurons or, more rarely, with the lower motor neurons directly. The upper motor neuron pathways that arise in the cortex are essential for the initiation of voluntary movements and for complex temporal sequences of movement. In particular, descending projections from cortical areas in the frontal lobe, including Brodmann's area 4 (the **primary motor cortex**), the lateral part of area 6 (the **lateral premotor cortex**), and the medial part of area 6 (the **medial premotor cortex**) are essential for planning, initiating, and directing temporal sequences of voluntary movements. Upper motor neurons originating in the brainstem are responsible for regulating muscle tone and for orienting the eyes, head, and body with respect to vestibular, somatic, auditory, and visual sensory information. Their contributions are thus critical for basic navigational movements of the body, and in the control of posture.

The third and fourth subsystems are structures (or groups of structures) that have no direct access to either the local circuit neurons or the lower motor neurons; instead, they control movement by regulating the activity of the upper motor neurons. The third and larger of these subsystems, the **cerebellum**, is located on the dorsal surface of the pons (see Chapter 1). The cerebellum acts via its efferent pathways to the upper motor neurons as a servomechanism, detecting the difference, or "motor error," between an intended movement and the movement actually performed (see Chapter 19). The cerebellum uses this information about discrepancies to mediate both real-time and long-term reductions in these motor errors (the latter being a form of motor learning). As might be expected from this account, patients with cerebellar damage exhibit persistent errors in movement. The fourth subsystem, embedded in the depths of the forebrain, consists of a group of structures collectively referred to as the **basal ganglia** (see Chapter 1). The basal ganglia suppress unwanted movements and prepare (or "prime") upper motor neuron circuits for the initiation of movements. The problems associated with disorders of basal ganglia, such as Parkinson's disease and Huntington's disease, attest to the importance of this complex in the initiation of voluntary movements (see Chapter 18).

Despite much effort, the sequence of events that leads from thought to movement is still poorly understood. The picture is clearest, however, at the level of control of the muscles themselves. It therefore makes sense to begin an account of motor behavior by considering the anatomical and physiological relationships between lower motor neurons and the muscle fibers they innervate.

Motor Neuron–Muscle Relationships

By injecting into muscle groups visible tracers that are transported by the axons of the lower motor neurons back to their cell bodies, the lower motor neurons that innervate the body's skeletal muscles can be seen in histological sections of the ventral horns of the spinal cord. Each lower motor neuron innervates muscle fibers within a single muscle, and all the motor neurons innervating a single muscle (called the **motor neuron pool** for that muscle) are grouped together into rod-shaped clusters that run parallel to the long axis of the cord for one or more spinal cord segments (Figure 16.2).

An orderly relationship between the location of the motor neuron pools and the muscles they innervate is evident both along the length of the spinal cord and across the mediolateral dimension of the cord, an arrangement that in effect provides a spatial map of the body's musculature. For example, the motor neuron pools that innervate the arm are located in the cervical enlargement of the cord and those that innervate the leg in the lumbar enlargement (see Chapter 1). The mapping, or topography, of the same motor neuron pools in the mediolateral dimension can be appreciated in a cross section through the cervical enlargement (the level illustrated in Figure 16.3). Thus, neurons that innervate the axial musculature (i.e., the postural muscles of the trunk) are located medially in the cord. Lateral to these cell groups are motor neuron pools innervating muscles located progressively more laterally in the body. Neurons that innervate the muscles of the shoulders (or pelvis, if one were to look at a similar section in the lumbar enlargement; see Figure 16.2) are the next most lateral group, whereas those that innervate the proximal muscles of the arm (or leg) are located laterally to these. The motor neuron pools that innervate the distal parts of the extremities, the fingers or toes, lie farthest from the midline. This spatial organization provides clues about the functions of the descending upper motor neuron pathways described in the following chapter; some of these pathways terminate primarily in the medial region of the spinal cord, which is concerned with postural muscles, whereas other pathways terminate more laterally, where they have access to the lower motor neurons that control fine movements of the fingers.

Two types of lower motor neuron are found in these neuronal pools. Small *γ* **motor neurons** innervate specialized muscle fibers that, in combina-

(A)

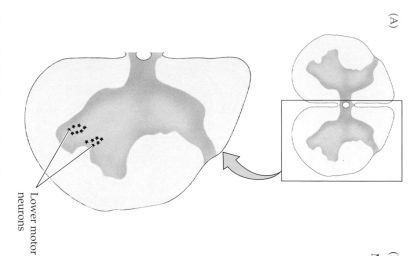

Lower motor neurons

(B) Medial gastrocnemius injection · Soleus injection

(C)

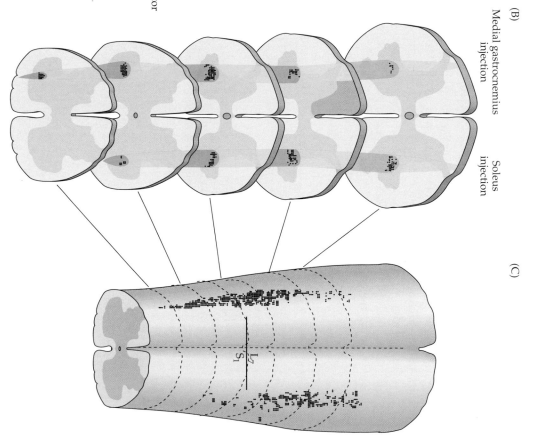

Figure 16.2 Organization of lower motor neurons in the ventral horn of the spinal cord demonstrated by retrograde labeling from individual muscles. Neurons were identified by placing a retrograde tracer into the medial gastrocnemius or soleus muscle of the cat. (A) Section through the lumbar level of the spinal cord showing the distribution of labeled cell bodies. Lower motor neurons form distinct clusters (motor pools) in the ventral horn. Spinal cord cross sections (B) and a reconstruction seen from the dorsal surface (C) illustrate the distribution of motor neurons innervating individual skeletal muscles in the long axis of the cord. The cylindrical shape and distinct distribution of different pools are especially evident in the dorsal view of the reconstructed cord. The dashed lines in (C) represent individual lumbar and sacral spinal cord segments. (After Burke et al., 1977.)

tion with the nerve fibers that innervate them, are actually sensory receptors called muscle spindles (see Chapter 9). The muscle spindles are embedded within connective tissue capsules in the muscle, and are thus referred to as intrafusal muscle fibers (*fusal* means capsular). The intrafusal muscle fibers are also innervated by sensory axons that send information to the brain and spinal cord about the length and tension of the muscle. The function of the γ motor neurons is to regulate this sensory input by setting the intrafusal muscle fibers to an appropriate length (see next section). The second type of lower motor neuron, called **α motor neurons**, innervates the extrafusal muscle fibers, which are the striated muscle fibers that actually generate the forces needed for movement.

Although the following discussion focuses on the lower motor neurons in the spinal cord, comparable sets of motor neurons responsible for the control of muscles in the head and neck are located in the brainstem. The latter neurons are distributed in the eight motor nuclei of the cranial nerves in the medulla, pons, and midbrain (see Chapter 1). Somewhat confusingly, but quite appropriately, these motor neurons in the brainstem are also called lower motor neurons.

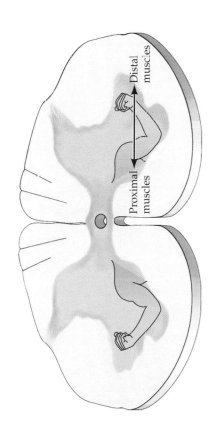

Figure 16.3 Somatotopic organization of lower motor neurons in a cross section of the ventral horn at the cervical level of the spinal cord. Motor neurons innervating axial musculature are located medially, whereas those innervating the distal musculature are located more laterally.

The Motor Unit

Most mature extrafusal skeletal muscle fibers in mammals are innervated by only a single α motor neuron. Since there are more muscle fibers by far than motor neurons, individual motor axons branch within muscles to synapse on many different fibers that are typically distributed over a relatively wide area within the muscle, presumably to ensure that the contractile force of the motor unit is spread evenly (Figure 16.4). In addition, this arrangement

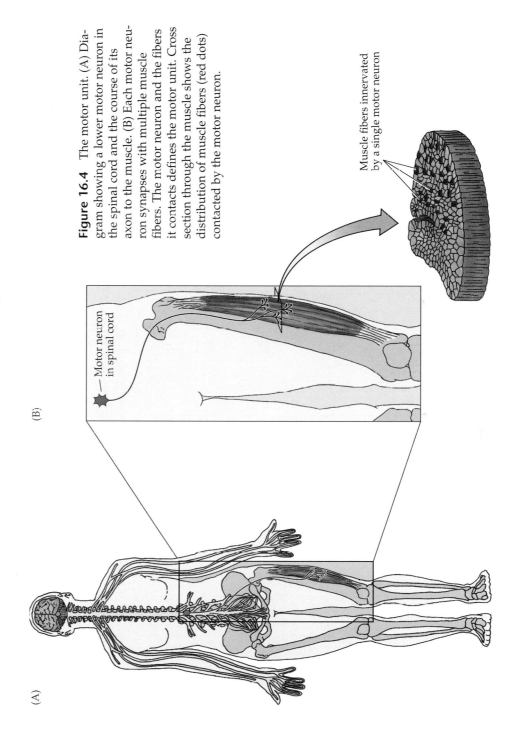

Figure 16.4 The motor unit. (A) Diagram showing a lower motor neuron in the spinal cord and the course of its axon to the muscle. (B) Each motor neuron synapses with multiple muscle fibers. The motor neuron and the fibers it contacts defines the motor unit. Cross section through the muscle shows the distribution of muscle fibers (red dots) contacted by the motor neuron.

reduces the chance that damage to one or a few α motor neurons will significantly alter a muscle's action. Because an action potential generated by a
motor neuron normally brings to threshold all of the muscle fibers it contacts, a single α motor neuron and its associated muscle fibers together constitute the smallest unit of force that can be activated to produce movement.
Sherrington was the first to recognize this fundamental relationship between
an α motor neuron and the muscle fibers it innervates, for which he coined
the term **motor unit**.

Both motor units and the α motor neurons themselves vary in size. Small
α motor neurons innervate relatively few muscle fibers and form motor
units that generate small forces, whereas large motor neurons innervate
larger, more powerful motor units. Motor units also differ in the types of
muscle fibers that they innervate. In most skeletal muscles, the small motor
units innervate small "red" muscle fibers that contract slowly and generate
relatively small forces; but, because of their rich myoglobin content, plentiful
mitochondria, and rich capillary beds, such small red fibers are resistant to
fatigue. These small units are called **slow (S) motor units** and are especially
important for activities that require sustained muscular contraction, such as
the maintenance of an upright posture. Larger α motor neurons innervate
larger, pale muscle fibers that generate more force; however, these fibers
have sparse mitochondria and are therefore easily fatigued. These units are
called **fast fatigable (FF) motor units** and are especially important for brief
exertions that require large forces, such as running or jumping. A third class
of motor units has properties that lie between those of the other two. These
fast fatigue-resistant (FR) motor units are of intermediate size and are not
quite as fast as FF units. As the name implies, they are substantially more
resistant to fatigue, and generate about twice the force of a slow motor unit
(Figure 16.5).

These distinctions among different types of motor units indicate how the
nervous system produces movements appropriate for different circumstances. In most muscles, small, slow motor units have lower thresholds for
activation than the larger units and are tonically active during motor acts
that require sustained effort (standing, for instance). The threshold for the

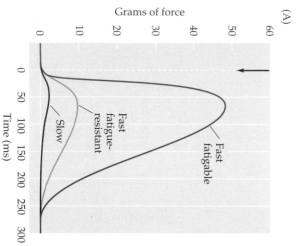
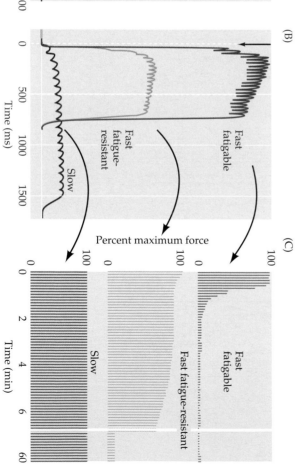
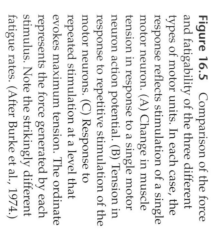

(A)

Grams of force

Time (ms)

Slow

Fast
fatigue-
resistant

Fast
fatigable

(B)

Time (ms)

Fast
fatigue-
resistant

Fast
fatigable

Slow

Percent maximum force

(C)

Time (min)

Slow

Fast fatigue-resistant

Fast
fatigable

Figure 16.5 Comparison of the force
and fatigability of the three different
types of motor units. In each case, the
response reflects stimulation of a single
motor neuron. (A) Change in muscle
tension in response to a single motor
neuron action potential. (B) Tension in
response to repetitive stimulation of the
motor neurons. (C) Response to
repeated stimulation at a level that
evokes maximum tension. The ordinate
represents the force generated by each
stimulus. Note the strikingly different
fatigue rates. (After Burke et al., 1974.)

large, fast motor units is reached only when rapid movements requiring great force are made, such as jumping. The functional distinctions between the various classes of motor units also explain some structural differences among muscle groups. For example, a motor unit in the soleus (a muscle important for posture that comprises mostly small, slow units) has an average innervation ratio of 180 muscle fibers for each motor neuron. In contrast, the gastrocnemius, a muscle that comprises both small and larger units, has an innervation ratio of 1000–2000 muscle fibers per motor neuron, and can generate forces needed for sudden changes in body position. More subtle variations are present in athletes on different training regimens. Thus, muscle biopsies show that sprinters have a larger proportion of powerful but rapidly fatiguing pale fibers in their legs than do marathoners. Other differences are related to the highly specialized functions of particular muscles. For instance, the eyes require rapid, precise movements but little strength; in consequence, extraocular muscle motor units are extremely small (with an innervation ratio of only 3!) and have a very high proportion of muscle fibers capable of contracting with maximal velocity.

The Regulation of Muscle Force

Increasing or decreasing the number of motor units active at any one time changes the amount of force produced by a muscle. In the 1960s, Elwood Henneman and his colleagues at Harvard Medical School found that steady increases in muscle tension could be produced by progressively increasing the activity of axons that provide input to the relevant pool of lower motor neurons. This gradual increase in tension results from the recruitment of motor units in a fixed order according to their size. By stimulating in an experimental animal either sensory nerves or upper motor pathways that project to a lower motor neuron pool while measuring the tension changes in the muscle, Henneman found that the smallest motor neurons in the pool are the only units activated by weak synaptic stimulation. When synaptic input increases, progressively larger motor neurons are recruited: As the synaptic activity driving a motor neuron pool increases, low threshold S units are recruited first, then FR units, and finally, at the highest levels of activity, the FF units. Since these original experiments, evidence for the orderly recruitment of motor units has been found in a variety of voluntary and reflexive movements. As a result, this systematic relationship has come to be known as the **size principle.**

An illustration of how the size principle operates for the motor units of the medial gastrocnemius muscle in the cat is shown in Figure 16.6. When the animal is standing quietly, the force measured directly from the muscle tendon is only a small fraction (about 5%) of the total force that the muscle can generate. The force is provided by the S motor units, which make up about 25% of the motor units in this muscle. When the cat begins to walk, larger forces are necessary: locomotor activities that range from slow walking to fast running require up to 25% of the muscle's total force capacity. This additional need is met by the recruitment of FR units. Only movements such as galloping and jumping, which are performed infrequently and for short periods, require the full power of the muscle; such demands are met by the recruitment of the FF units. Thus, the size principle provides a simple solution to the problem of grading muscle force: The combination of motor units activated by such orderly recruitment optimally matches the physiological properties of different motor unit types with the range of forces required to perform different motor tasks.

Figure 16.6 The recruitment of motor neurons in the cat medial gastrocnemius muscle under different behavioral conditions. Slow (S) motor units provide the tension required for standing. Fast fatigue-resistant (FR) units provide the additional force needed for walking and running. Fast fatigable (FF) units are recruited for the most strenuous activities. (After Walmsley et al., 1978.)

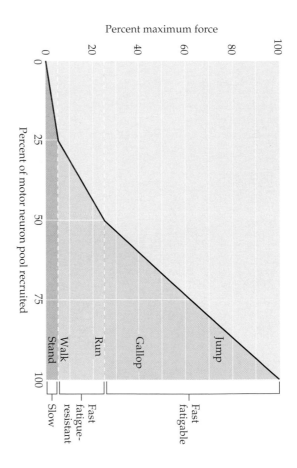

Figure 16.7 The effect of stimulation rate on muscle tension. (A) At low frequencies of stimulation, each action potential in the motor neuron results in a single twitch of the related muscle fibers. (B) At higher frequencies, the twitches sum to produce a force greater than that produced by single twitches. (C) At a still higher frequency of stimulation, the force produced is greater, but individual twitches are still apparent. (D) At the highest rates of motor neuron activation, individual twitches are no longer apparent (a condition called fused tetanus).

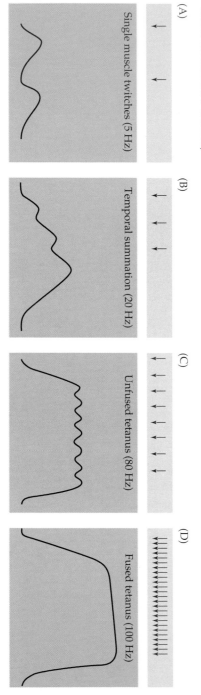

Single muscle twitches (5 Hz)

Temporal summation (20 Hz)

Unfused tetanus (80 Hz)

Fused tetanus (100 Hz)

(A)

(B)

(C)

(D)

The frequency of the action potentials generated by motor neurons also contributes to the regulation of muscle tension. The increase in force that occurs with increased firing rate reflects the summation of successive muscle contractions: The muscle fibers are activated by the next action potential before they have had time to completely relax, and the forces generated by the temporally overlapping contractions are summed (Figure 16.7). The lowest firing rates during a voluntary movement are on the order of 8 per second (Figure 16.8). As the firing rate of individual units rises to a maximum of about 20–25 per second in the muscle being studied here, the amount of force produced increases. At the highest firing rates, individual muscle fibers are in a state of "fused tetanus"—that is, the tension produced in individual motor units no longer has peaks and troughs that correspond to the individual twitches evoked by the motor neuron's action potentials. Under normal conditions, the maximum firing rate of motor neurons is less than that required for fused tetanus (see Figure 16.8). However, the asynchronous firing of different lower motor neurons provides a steady level of input to the muscle that causes the contraction of a relatively constant number of motor units and averages out the changes in tension due to contractions and relaxations of individual motor units. All this allows the resulting movements to be executed smoothly.

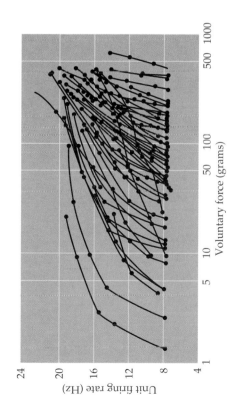

Figure 16.8 Motor units recorded in a muscle of the human hand as the amount of voluntary force produced is progressively increased. Motor units (represented by the lines between the dots) are initially recruited at a low frequency of firing (8 Hz); the rate of firing for each unit increases as the subject generates more and more force. (After Monster and Chan, 1977.)

The Spinal Cord Circuitry Underlying Muscle Stretch Reflexes

The local circuitry within the spinal cord mediates a number of sensory motor reflex actions. The simplest of these reflex arcs entails the response to muscle stretch, which provides direct excitatory feedback to the motor neurons innervating the muscle that has been stretched (Figure 16.9). As already mentioned, the sensory signal for the **stretch reflex** originates in **muscle spindles**, sensory receptors embedded within most muscles (see previous section and Chapter 9). The spindles comprise 8–10 intrafusal fibers arranged in parallel with the extrafusal fibers that make up the bulk of the muscle (Figure 16.9A). Large-diameter sensory fibers, called Ia afferents, are coiled around the central part of the spindle. These afferents are the largest axons in peripheral nerves and, since action potential conduction velocity is a direct function of axon diameter (see Chapters 2 and 3), they allow for very rapid adjustments in this reflex arc when the muscle is stretched. The stretch imposed on the muscle deforms the intrafusal muscle fibers, which in turn initiate action potentials by activating mechanically gated ion channels in the afferent axons coiled around the spindle. The centrally projecting branch of the sensory neuron forms monosynaptic excitatory connections with the α motor neurons in the ventral horn of the spinal cord that innervate the same (homonymous) muscle and, via local circuit neurons, inhibitory connections with the α motor neurons of antagonistic (heteronymous) muscles. This arrangement is an example of what is called reciprocal innervation and results in rapid contraction of the stretched muscle and simultaneous relaxation of the antagonist muscle. All of this leads to especially rapid and efficient responses to changes in the length or tension in the muscle (Figure 16.9B). The excitatory pathway from a spindle to the α motor neurons innervating the same muscle is unusual in that it is a monosynaptic reflex; in most cases, sensory neurons from the periphery do not contact the lower motor neuron directly but exert their effects through local circuit neurons.

This monosynaptic reflex arc is variously referred to as the "stretch," "deep tendon," or "myotatic reflex," and it is the basis of the knee, ankle, jaw, biceps, or triceps responses tested in a routine neurological examination. The tap of the reflex hammer on the tendon stretches the muscle and therefore excites a volley of activity from the muscle spindles in the afferent axons. The afferent volley is relayed to the α motor neurons in the brainstem or spinal cord, and an efferent volley returns to the muscle (see Figure 1.5). Since muscles are always under some degree of stretch, this reflex circuit is

(A) Muscle spindle

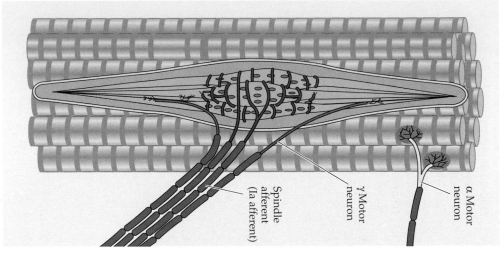

α Motor
neuron

γ Motor
neuron

Spindle
afferent
(Ia afferent)

(B)

α Motor
neuron
Ia afferent

Muscle spindle
Homonymous
muscle

Synergist
Antagonist

Passive
stretch

Resistance

Inhibited

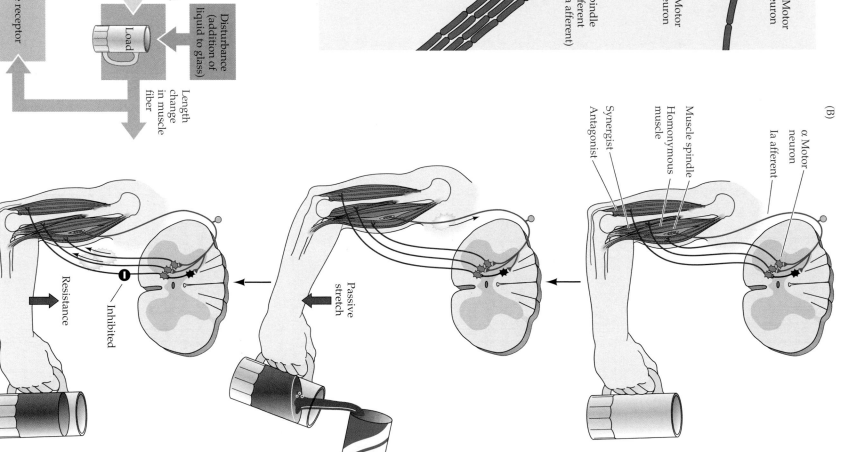

(C)

Descending
facilitation
and inhibition

α Motor
neuron

Muscle

Force
required
to hold
glass

Load

Disturbance
(addition of
liquid to glass)

Length
change
in muscle
fiber

Spindle receptor

Increase spindle
afferent discharge

▼ **Figure 16.9** Stretch reflex circuitry. (A) Diagram of muscle spindle, the sensory receptor that initiates the stretch reflex. (B) Stretching a muscle spindle leads to increased activity in Ia afferents and an increase in the activity of α motor neurons that innervate the same muscle. Ia afferents also excite the motor neurons that innervate the same muscle, and inhibit the motor neurons that innervate antagonists (see also Figure 1.5). (C) The stretch reflex operates as a negative feedback loop to regulate muscle length.

normally responsible for the steady level of tension in muscles called **muscle tone.** Changes in muscle tone occur in a variety of pathological conditions, and it is these changes that are assessed by examination of tendon reflexes.

In terms of engineering principles, the stretch reflex arc is a negative feedback loop used to maintain muscle length at a desired value (Figure 16.9C). The appropriate muscle length is specified by the activity of descending pathways that influence the motor neuron pool. Deviations from the desired length are detected by the muscle spindles, since increases or decreases in the stretch of the intrafusal fibers alter the level of activity in the sensory fibers that innervate the spindles. These changes lead in turn to adjustments in the activity of the α motor neurons, returning the muscle to the desired length by contracting the stretched muscle and relaxing the opposed muscle group, and by restoring the level of spindle activity to what it was before.

The smaller γ motor neurons control the functional characteristics of the muscle spindles by modulating their level of excitability. As already described, when the muscle is stretched, the spindle is also stretched and the rate of discharge in the afferent fibers increased. When the muscle shortens, however, the spindle is relieved of tension, or "unloaded," and the sensory axons that innervate the spindle might therefore be expected to fall silent during contraction. However, they remain active. The γ motor neurons terminate on the contractile poles of the intrafusal fibers, and the activation of these neurons causes intrafusal fiber contraction—in this way maintaining the tension on the middle (or equatorial region) of the intrafusal fibers where the sensory axons terminate. Thus, co-activation of the α and γ motor neurons allows spindles to function (i.e., send information centrally) at all muscle lengths during movements and postural adjustments.

The Influence of Afferent Activity on Motor Behavior

The level of γ activity often is referred to as γ bias, or gain, and can be adjusted by upper motor neuron pathways as well as by local reflex circuitry. The larger the gain of the stretch reflex, the greater the change in muscle force that results from a given amount of stretch applied to the intrafusal fibers. If the gain of the reflex is high, then a small amount of stretch applied to the intrafusal fibers will produce a large increase in the number of α motor neurons recruited and a large increase in their firing rates; this in turn leads to a large increase in the amount of tension produced by the extrafusal fibers. If the gain is low, a greater stretch is required to generate the same amount of tension in the extrafusal muscle fibers. In fact, the gain of the stretch reflex is continuously adjusted to meet different functional requirements. For example, when standing in a moving bus, the gain of the stretch reflex can be modulated by upper motor neuron pathways to compensate for the variable changes that occur as the bus stops and starts or progresses relatively smoothly. During voluntary movements, α and γ motor

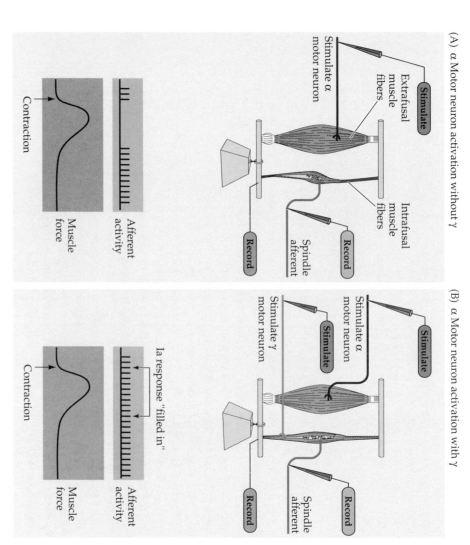

(A) α Motor neuron activation without γ

Stimulate α motor neuron

Stimulate

Extrafusal muscle fibers

Intrafusal muscle fibers

Spindle afferent

Record

Record

Afferent activity

Contraction

Muscle force

(B) α Motor neuron activation with γ

Stimulate α motor neuron

Stimulate

Stimulate γ motor neuron

Stimulate

Spindle afferent

Record

Record

Afferent activity

Ia response "filled in"

Contraction

Muscle force

Figure 16.10 The role of γ motor neuron activity in regulating the responses of muscle spindles. (A) When α motor neurons are stimulated without activation of γ motor neurons, the response of the Ia fiber decreases as the muscle contracts. (B) When both α and γ motor neurons are activated, there is no decrease in Ia firing during muscle shortening. Thus, the γ motor neurons can regulate the gain of muscle spindles so they can operate efficiently at any length of the parent muscle. (After Hunt and Kuffler, 1951.)

neurons are often co-activated by higher centers to prevent muscle spindles from being unloaded (Figure 16.10).

In addition, the level of γ motor neuron activity can be modulated independently of α activity if the context of a movement requires it. In general, the baseline activity level of γ motor neurons is high if a movement is relatively difficult and demands rapid and precise execution. For example, recordings from cat hindlimb muscles show that γ activity is high when the animal has to perform a difficult movement such as walking across a narrow beam. Unpredictable conditions, as when the animal is picked up or handled, also lead to marked increases in γ activity and greatly increased spindle responsiveness.

Gamma motor neuron activity, however, is not the only factor that sets the gain of the stretch reflex. The gain also depends on the level of excitability of the α motor neurons that serve as the effector side of this reflex loop. Thus, in addition to the influence of descending upper motor neuron projections, other local circuits in the spinal cord can change the gain of the stretch reflex by excitation or inhibition of either α or γ motor neurons.

Other Afferent Feedback that Affects Motor Performance

Another sensory receptor that is important in the reflex regulation of motor unit activity is the **Golgi tendon organ**. Golgi tendon organs are encapsulated afferent nerve endings located at the junction of the muscle and tendon (Figure 16.11A; see also Table 9.1). Each tendon organ is related to a single group Ib sensory axon (the Ib axons being slightly smaller than the Ia axons that innervate the muscle spindles). In contrast to the parallel arrangement

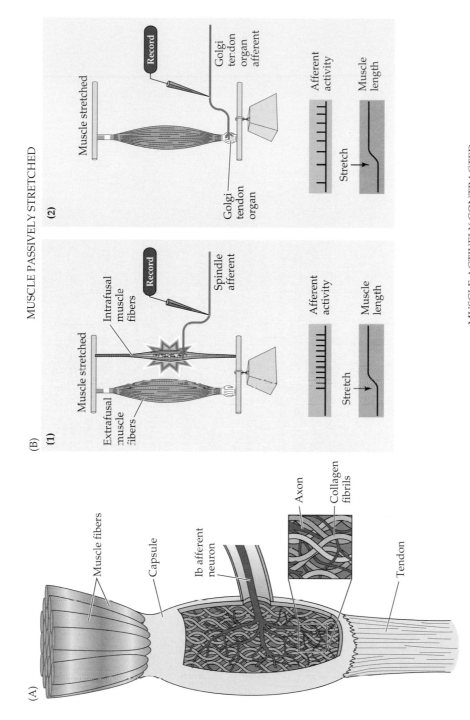

(A)

MUSCLE PASSIVELY STRETCHED

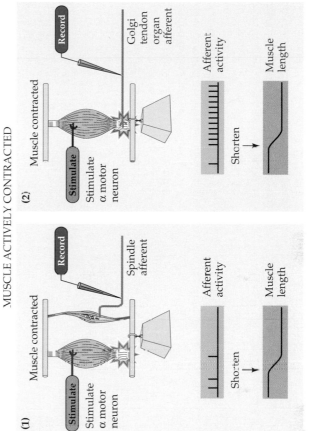

MUSCLE ACTIVELY CONTRACTED

Figure 16.11   Comparison of the function of muscle spindles and Golgi tendon organs. (A) Golgi tendon organs are arranged in series with extrafusal muscle fibers because of their location at the junction of muscle and tendon. (B) The two types of muscle receptors, the muscle spindles (1) and the Golgi tendon organs (2), have different responses to passive muscle stretch (*top*) and active muscle contraction (*bottom*). Both afferents discharge in response to passively stretching the muscle, although the Golgi tendon organ discharge is much less than that of the spindle. When the extrafusal muscle fibers are made to contract by stimulation of their motor neurons, however, the spindle is unloaded and therefore falls silent, whereas the rate of Golgi tendon organ firing increases. (B after Patton, 1965.)

of extrafusal muscle fibers and spindles, Golgi tendon organs are in series with the extrafusal muscle fibers. When a muscle is passively stretched, most of the change in length occurs in the muscle fibers, since they are more elastic than the fibrils of the tendon. When a muscle actively contracts, however, the force acts directly on the tendon, leading to an increase in the tension of the collagen fibrils in the tendon organ and compression of the intertwined sensory receptors. As a result, Golgi tendon organs are exquisitely sensitive to increases in muscle *tension* that arise from muscle contraction but, unlike spindles, are relatively insensitive to *passive stretch* (Figure 16.11B).

The Ib axons from Golgi tendon organs contact inhibitory local circuit neurons in the spinal cord (called Ib inhibitory interneurons) that synapse, in turn, with the α motor neurons that innervate the same muscle. The Golgi tendon circuit is thus a negative feedback system that regulates muscle tension; it decreases the activation of a muscle when exceptionally large forces are generated and this way protects the muscle's integrity. This reflex circuit also operates at reduced levels of muscle force, counteracting small changes in muscle tension by increasing or decreasing the inhibition of α motor neurons. Under these conditions, the Golgi tendon system tends to maintain a steady level of force, counteracting effects that diminish muscle force (such as fatigue). In short, if the muscle spindle system is considered a feedback system that monitors and maintains muscle *length*, then the Golgi tendon system is a feedback system that monitors and maintains muscle *force*.

Like the muscle spindle system, the Golgi tendon organ system is not a closed loop. The Ib inhibitory interneurons also receive synaptic inputs from a variety of other sources, including cutaneous receptors, joint receptors,

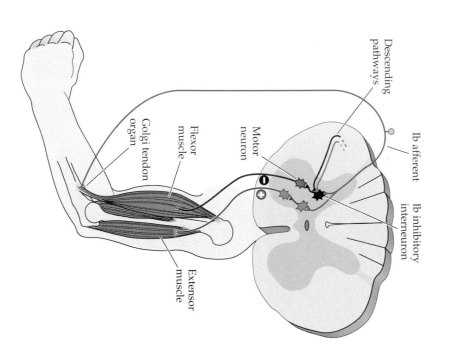

Figure 16.12 Negative feedback regulation of muscle tension by Golgi tendon organs. The Ib afferents from tendon organs contact inhibitory interneurons that decrease the activity of α motor neurons innervating the same muscle. The Ib inhibitory interneurons also receive input from other sensory fibers, as well as from descending pathways. This arrangement prevents muscles from generating excessive tension.

Descending pathways

Ib afferent

Ib inhibitory interneuron

Motor neuron

Flexor muscle

Golgi tendon organ

Extensor muscle

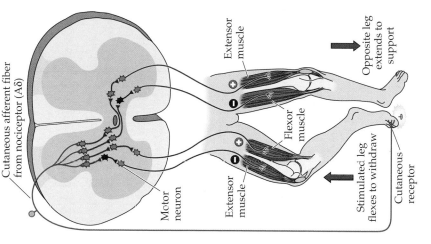

Figure 16.13 Spinal cord circuitry responsible for the flexion reflex. Stimulation of cutaneous receptors in the foot leads to activation of spinal cord local circuits that withdraw (flex) the stimulated extremity and extend the other extremity to provide compensatory support.

muscle spindles, and descending upper motor neuron pathways (Figure 16.12). Acting in concert, these inputs regulate the responsiveness of Ib interneurons to activity arising in Golgi tendon organs.

Flexion Reflex Pathways

So far, the discussion has focused on reflexes driven by sensory receptors located within muscles or tendons. Other reflex circuitry, however, mediates the withdrawal of a limb from a painful stimulus, such as a pinprick or the heat of a flame. Contrary to what might be imagined, given the speed with which we are able to withdraw from painful stimuli, this **flexion reflex** involves several synaptic links (Figure 16.13). As a result of activity in this circuitry, stimulation of nociceptive sensory fibers leads to excitation of ipsilateral flexor muscles and reciprocal inhibition of ipsilateral extensor muscles. Flexion of the stimulated limb is also accompanied by an opposite reaction in the contralateral limb (i.e., the contralateral extensor muscles are excited while flexor muscles are inhibited). This **crossed extension reflex** serves to enhance postural support during withdrawal of the affected limb from the painful stimulus.

Like the other reflex pathways, local circuit neurons in the flexion reflex pathway receive converging inputs from several different sources, including cutaneous receptors, other spinal cord interneurons, and upper motor neuron pathways. Although the functional significance of this complex pattern of connectivity is unclear, changes in the character of the reflex following damage to descending pathways provides some insight. Under normal conditions, a noxious stimulus is required to evoke the flexion reflex; following damage to descending pathways, however, other types of stimulation, such as squeezing a limb, can sometimes produce the same response. This observation suggests that the descending projections to the spinal cord modulate the responsiveness of the local circuitry to a variety of sensory inputs.

Spinal Cord Circuitry and Locomotion

The contribution of local circuitry to motor control is not, of course, limited to reflexive responses to sensory inputs. Studies of rhythmic movements such as locomotion and swimming in animal models (Box A) have demonstrated that local circuits in the spinal cord are fully capable of controlling the timing and coordination of such complex patterns of movement, and of adjusting them in response to altered circumstances (Box B).

A good example is locomotion (walking, running, etc.). The movement of a single limb during locomotion can be thought of as a cycle consisting of two phases: a stance phase, during which the limb is extended and placed in contact with the ground to propel humans or other bipeds forward; and a swing phase, during which the limb is flexed to leave the ground and then brought forward to begin the next stance phase (Figure 16.14A). Increases in the speed of locomotion reduce the amount of time it takes to complete a cycle, and most of the change in cycle time is due to a shortening of the stance phase; the swing phase remains relatively constant over a wide range of locomotor speeds.

In quadrupeds, changes in locomotor speed are also accompanied by changes in the sequence of limb movements. At low speeds, for example, there is a back-to-front progression of leg movements, first on one side and then on the other. As the speed increases to a trot, the movements of the right forelimb and left hindlimb are synchronized (as are the movements of

BOX A
Locomotion in the Leech and the Lamprey

All animals must coordinate body movements so they can navigate successfully in their environment. All vertebrates, including mammals, use local circuits in the spinal cord (pattern generators) to control the coordinated movements associated with locomotion. The cellular basis of organized locomotor activity, however, has been most thoroughly studied in an invertebrate, the leech, and a simple vertebrate, the lamprey.

Both the leech and the lamprey lack peripheral appendages for locomotion possessed by many vertebrates (limbs, flippers, fins, or their equivalent). Furthermore, their bodies comprise repeating muscle segments (as well as repeating skeletal elements in the lamprey). Thus, in order to move through the water, both animals must coordinate the movement of each segment. They do this by orchestrating a sinusoidal displacement of each body segment in sequence, so that the animal is propelled forward through the water (figure A).

The leech is particularly well-suited for studying the circuit basis of coordinated movement. The nervous system in the leech consists of a series of interconnected segmental ganglia, each with motor neurons that innervate the corresponding segmental muscles (see figure A). These segmental ganglia facilitate electrophysiological studies, because there is a limited number of neurons in each and each neuron has a distinct identity. The neurons can thus be recognized and studied from animal to animal, and their electrical activity correlated with the sinusoidal swimming movements.

A central pattern generator circuit similar to the generator for crustacean gut movement (see Box B) coordinates this undulating motion. In the leech, the relevant neural circuit is an ensemble of sensory neurons, interneurons, and

motor neurons repeated in each segmental ganglion that controls the local sequence of contraction and relaxation in each segment of the body wall musculature (figure B). The sensory neurons detect the stretching and contraction of the body wall associated with the sequential swimming movements. Dorsal and ventral motor neurons in the circuit provide innervation to dorsal and ventral muscles, whose phasic contractions propel the leech forward. The sensory information and motor neuron signals are coordinated by interneurons that fire rhythmically, and thus set up phasic patterns of activity in the dorsal and ventral cells that lead to sinusoidal movement (see figure B). The intrinsic swimming rhythm is established by a variety of membrane conductances that mediate periodic bursts of suprathreshold action potentials followed by well-defined periods of hyperpolarization.

The lamprey, one of the simplest vertebrates, is distinguished by its clearly segmented musculature and by its lack of bilateral fins or other appendages. In order to move through the water, the lamprey contracts and relaxes each muscle segment in sequence (figure C), which produces a sinusoidal motion, much like that of the leech. Again, a central pattern generator coordinates this sinusoidal movement. Unlike the leech with its segmental ganglia, the lamprey has a continuous spinal cord that innervates its muscle segments. The lamprey spinal cord is simpler than that of other vertebrates, and several classes of identified neurons occupy stereotyped positions. This orderly arrangement facilitates the identification and analysis of neurons that constitute the central pattern generator circuit.

In the lamprey spinal cord, the intrinsic firing pattern of a set of interconnected sensory neurons, interneurons

and motor neurons establishes the pattern of undulating muscle contractions that underlie swimming (figure D). The patterns of connectivity between neurons, the neurotransmitters used by each class of cell, and the physiological properties of the elements in the lamprey pattern generator are now known. Different neurons in the circuit fire with distinct rhythmicity, thus controlling specific aspects of the swim cycle (figure E). Particularly important are reciprocal inhibitory connections across the midline that coordinate the pattern generating circuitry on each side of the spinal cord. This circuitry in the lamprey thus provides a basis for understanding the circuits that control locomotion in more complex vertebrates.

These observations on pattern generating circuits for locomotion in these relatively simple animals have stimulated parallel studies of terrestrial mammals in which central pattern generators in the spinal cord also coordinate locomotion. Although different in detail, terrestrial locomotion ultimately relies on the sequential movements similar to those that propel the leech and the lamprey through aquatic environments.

References

GRILLNER, S., D. PARKER AND A. EL MANIRA (1998) Vertebrate locomotion: A lamprey perspective. Ann. N.Y. Acad. Sci. 860: 1–18.

MARDER, E. AND R. M. CALABRESE (1996) Principles of rhythmic motor pattern generation. Physiol. Rev. 76: 687–717.

STENT, G. S., W. B. KRISTAN, W. O. FRIESEN, C. A. ORT, M. POON AND R. M. CALABRESE (1978) Neural generation of the leech swimming movement. Science 200: 1348–1357.

(A) LEECH

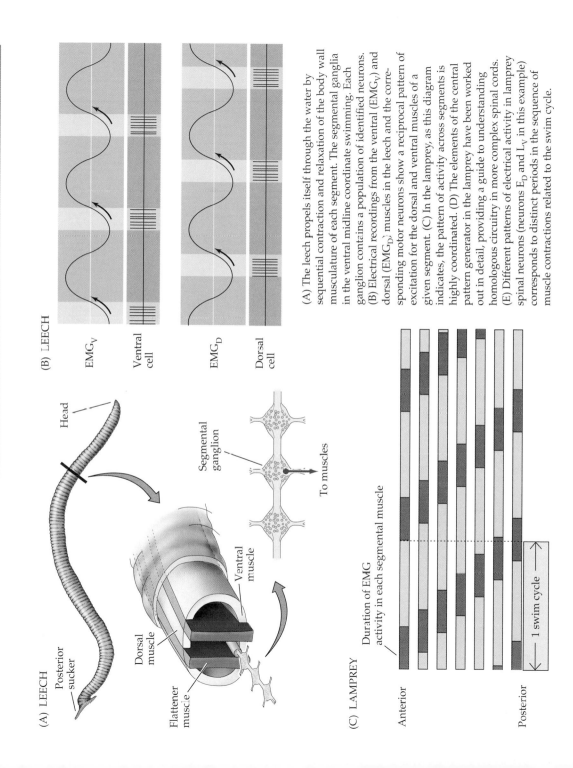

Head

Posterior
sucker

Dorsal
muscle

Flattener
muscle

Ventral
muscle

Segmental
ganglion

To muscles

(B) LEECH

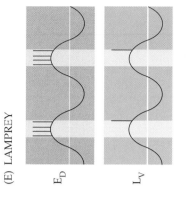

EMG_V

Ventral
cell

EMG_D

Dorsal
cell

(A) The leech propels itself through the water by sequential contraction and relaxation of the body wall musculature of each segment. The segmental ganglia in the ventral midline coordinate swimming. Each ganglion contains a population of identified neurons. (B) Electrical recordings from the ventral (EMG_V) and dorsal (EMG_D) muscles in the leech and the corresponding motor neurons show a reciprocal pattern of excitation for the dorsal and ventral muscles of a given segment. (C) In the lamprey, as this diagram indicates, the pattern of activity across segments is highly coordinated. (D) The elements of the central pattern generator in the lamprey have been worked out in detail, providing a guide to understanding homologous circuitry in more complex spinal cords. (E) Different patterns of electrical activity in lamprey spinal neurons (neurons E_D and L_V in this example) corresponds to distinct periods in the sequence of muscle contractions related to the swim cycle.

(C) LAMPREY

Duration of EMG
activity in each segmental muscle

Anterior

Posterior

1 swim cycle

(D) LAMPREY

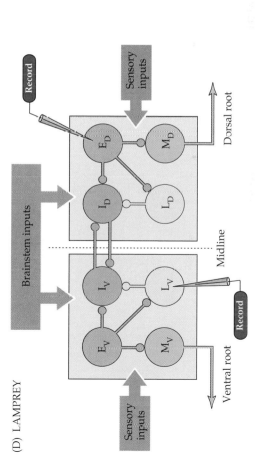

Record

E_D

M_D

Dorsal root

I_D

L_D

Sensory
inputs

Brainstem inputs

Midline

I_V

L_V

Record

E_V

M_V

Ventral root

Sensory
inputs

(E) LAMPREY

E_D

L_V

BOX B
The Autonomy of Central Pattern Generators: Evidence from the Lobster Stomatogastric Ganglion

A principle that has emerged from studies of central pattern generators is that rhythmic patterns of firing elicit complex motor responses without need of ongoing sensory stimulation. A good example is the behavior mediated by a small group of nerve cells in lobsters and other crustaceans called the stomatogastric ganglion (STG) that controls the muscles of the gut (figure A). This ensemble of 30 motor neurons and interneurons in the lobster is perhaps the most completely characterized neural circuit known. Of the 30 cells, defined subsets are essential for two distinct rhythmic movements: gastric mill movements that mediate grinding of food by "teeth" in the lobster's foregut, and pyloric movements that propel food into the hindgut. Phasic firing patterns of the motor neurons and interneurons of the STG are directly correlated with these two rhythmic movements. Each of the relevant cells has now been identified based on its position in the ganglion, and its electrophysiological and neuropharmacological properties characterized (figures B,C).

Patterned activity in the motor neurons and interneurons of the ganglion begins only if the appropriate neuromodulatory input is provided by sensory axons that originate in other ganglia. Depending upon the activity of the sensory axons, neuronal ensembles in the STG produce one of several characteristic rhythmic firing patterns. Once activated, however, the intrinsic membrane properties of identified cells within the ensemble sustain the rhythmicity of the circuit in the absence of further sensory input.

Another key fact that has emerged from this work is that the same neurons

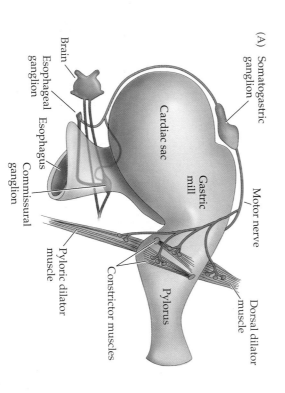

(A) Somatogastric ganglion
Brain
Esophageal ganglion
Esophagus
Commissural ganglion
Cardiac sac
Gastric mill
Motor nerve
Pyloric dilator muscle
Pylorus
Dorsal dilator muscle
Constrictor muscles

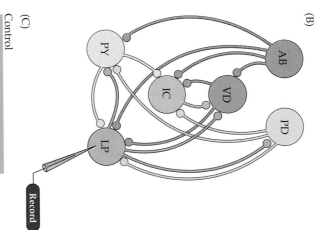

(B)

(C)

Control

Pilocarpine

Serotonin

Proctolin

Dopamine

(A) Location of the lobster stomatogastric ganglion in relation to the gut. (B) Subset of identified neurons in the stomatogastric ganglion that generates gastric mill and pyloric activity. The abbreviations indicate individual identified neurons, all of which project to different pyloric muscles (except for the AB neuron, which is an interneuron). (C) Recording from one of the neurons, the lateral pyloric or LP neuron, in this circuit showing the different patterns of activity elicited by several neuromodulators known to be involved in the normal behavior of this ganglion.

can participate in different programmed motor activity, as circumstances demand. For example, the subset of neurons producing gastric mill activity overlaps the subset that generates pyloric activity. This economic use of neuronal subsets has not yet been described in the central pattern generators of mammals, but seems likely to be a feature of all such circuits.

References

HARTLINE, D. K. AND D. M. MAYNARD (1975) Motor patterns in the stomatogastric ganglion of the lobster, *Panulirus argus*. J. Exp. Biol. 62: 405–420.

MARDER, E. AND R. M. CALABRESE (1996) Principles of rhythmic motor pattern generation. Physiol. Rev. 76: 687–717.

SELVERSTON, A. I., D. F. RUSSELL AND J. P. MILLER (1976) The stomatogastric nervous system: Structure and function of a small neural network. Progress in Neurobiology 7: 215–290.

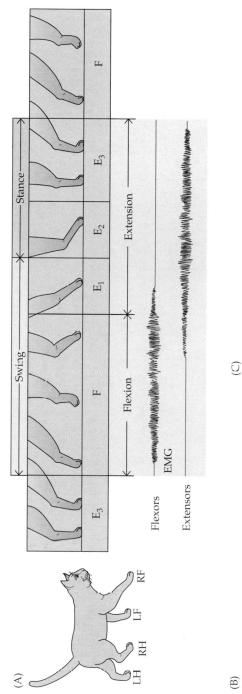

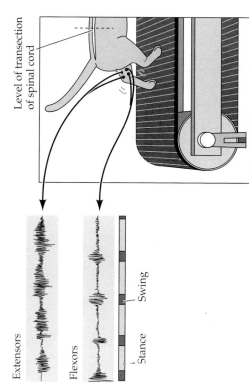

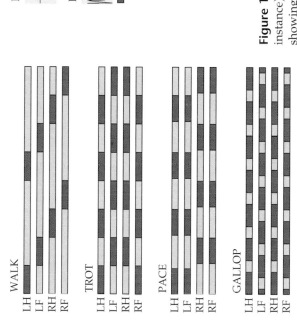

Figure 16.14 The cycle of locomotion for terrestrial mammals (a cat in this instance) is organized by central pattern generators. (A) The step cycle, showing leg flexion (F) and extension (E) and their relation to the swing and stance phases of locomotion. EMG indicates electromyographic recordings. (B) Comparison of the stepping movements for different gaits. Brown bars, foot lifted (swing phase); gray bars, foot planted (stance phase). (C) Transection of the spinal cord at the thoracic level isolates the hindlimb segments of the cord. The hindlimbs are still able to walk on a treadmill after recovery from surgery, and reciprocal bursts of electrical activity can be recorded from flexors during the swing phase and from extensors during the stance phase of walking. (After Pearson, 1976.)

the left forelimb and right hindlimb). At the highest speeds (gallop), the movements of the two front legs are synchronized, as are the movements of the two hindlimbs (Figure 16.14B).

Given the precise timing of the movement of individual limbs and the coordination between limbs that is required in this process, it is natural to assume that locomotion is accomplished by higher centers that organize the spatial and temporal activity patterns of the individual limbs. However, following transection of the spinal cord at the thoracic level, a cat's hindlimbs will still make coordinated locomotor movements if the animal is supported and placed on a moving treadmill (Figure 16.14C). Under these conditions, the speed of locomotor movements is determined by the speed of the treadmill, suggesting that the movement is nothing more than a reflexive response to stretching the limb muscles. This possibility is ruled out, however, by experiments in which the dorsal roots are also sectioned. Although the speed of walking is slowed and the movements are less coordinated than under normal conditions, appropriate locomotor movements are still observed. These and other observations in experimental animals show that the basic rhythmic patterns of limb movement during locomotion are not dependent on sensory input; nor are they dependent on input from descending projections from higher centers. Rather, each limb appears to have its own **central pattern generator**—an oscillatory spinal cord local circuit responsible for the alternating flexion and extension of the limb during locomotion. Under normal conditions, the central pattern generators for the limbs are variably coupled to each other by additional local circuits in order to achieve the different sequences of movements that occur at different speeds.

Although some locomotor movements can also be elicited in humans following damage to descending pathways, these are considerably less effective than the movements seen in the cat. The reduced ability of the transected spinal cord to mediate rhythmic stepping movements in humans presumably reflects an increased dependence of local circuitry on upper motor neuron pathways. Perhaps bipedal locomotion carries with it requirements for postural control greater than can be accommodated by spinal cord circuitry alone. Whatever the explanation, the basic oscillatory circuits that control such rhythmic behaviors as flying, walking, and swimming in many animals also play an important part in human locomotion.

The Lower Motor Neuron Syndrome

The symptoms that arise from damage to the lower motor neurons of the brainstem and spinal cord are referred to as the "lower motor neuron syndrome." In clinical neurology, this constellation of problems must be distinguished from the "upper motor neuron syndrome" that results from damage to the descending upper motor neuron pathways (see Chapter 17 for a discussion of the signs and symptoms associated with damage to upper motor neurons).

Damage to lower motor neuron cell bodies or their peripheral axons results in paralysis (loss of movement) or paresis (weakness) of the affected muscles. In addition to paralysis and/or paresis, the lower motor neuron syndrome includes a loss of reflexes (areflexia) due to interruption of the efferent (motor) limb of the sensory motor reflex arcs. Damage to lower motor neurons also entails a loss of muscle tone, since tone is in part dependent on the monosynaptic reflex arc that links the muscle spindles to the lower motor neurons (see also Box D in Chapter 17). A somewhat later effect is atrophy of the affected muscles due to denervation and disuse. The mus-

BOX C
Amyotrophic Lateral Sclerosis

Amyotrophic lateral sclerosis (ALS) is a neurodegenerative disease that affects an estimated 0.05% of the population. It is also called Lou Gehrig's disease, after the New York Yankees baseball player who died of the disorder in 1936. ALS is characterized by the slow but inexorable degeneration of α motor neurons in the ventral horn of the spinal cord and brainstem, and eventually of neurons in the motor cortex. Affected individuals show progressive weakness and wasting of skeletal muscles and usually die within 5 years of onset. Sadly, these patients are condemned to watch their own demise, since the intellect remains intact. There is no effective therapy.

Approximately 10% of ALS cases are familial. Familial ALS (FALS) is usually inherited as an autosomal dominant trait, and is largely indistinguishable from other forms of the disease. The familial form, however, has provided an opportunity for geneticists and neurologists to decipher the genetic basis of the disease in this subset of patients: The defect is a mutation in the long arm of

chromosome 21. Interestingly, this same chromosomal region contains the gene that encodes the cytosolic antioxidant enzyme copper/zinc superoxide dismutase (SOD). Evidence that superoxide radicals can destroy nerve cells suggested that a mutation of the SOD1 gene may cause FALS. This finding led in turn to the identification of mutations of SOD1 in roughly 20% of families with FALS. Further confirmation of this causal sequence in the familial form of ALS is that transgenic mice that overexpress mutant SOD1 protein develop a neurological disease that mimics ALS both behaviorally and pathologically. Overexpression of normal SOD, however, is not associated with motor neuron disease in mice. Together, these data suggest that the mutant SOD molecule is in some way cytotoxic.

How the mutant SOD1 damages motor neurons is not known. Possibilities include generation of damaging free radicals and death by oxidative stress, enhanced release of copper from the mutant enzyme and neuronal death by

copper toxicity, and abnormal folding of mutant SOD1 and neuronal death as a consequence of aberrant protein-protein interactions. It is also unclear what accounts for the curious predilection of the disease for motor neurons. Despite these uncertainties, identification of mutant SOD1 as the cause of familial ALS has given scientists a valuable clue to the molecular pathogenesis of at least some forms of this tragic disorder.

References

BROWN, R. G. (1998) SOD1 aggregates in ALS: Cause, correlate, or consequence? Nature Medicine 4: 1362–1363.

DENG, H. X. AND 19 OTHERS (1993) Amyotrophic lateral sclerosis and structural defects in Cu,Zn superoxide dismutase. Science 261: 1047–1051.

MULDER, D. W., L. T. KURLAND, K. P. OFFORD AND C. M. BEARD (1986) Familial adult motor neuron disease: Amyotrophic lateral sclerosis. Neurol. 36: 511–517.

ROSEN, D. R. AND 32 OTHERS (1993) Mutations in Cu/Zn superoxide dismutase gene are associated with familial amyotrophic lateral sclerosis. Nature 362: 59–62.

cles involved may also exhibit fibrillations and fasciculations, which are spontaneous twitches characteristic of single denervated muscle fibers or motor units, respectively. These phenomena arise from changes in the excitability of denervated muscle fibers in the case of fibrillation, and from abnormal activity of injured α motor neurons in the case of fasciculations. These spontaneous contractions can be readily recognized in an electromyogram, providing an especially helpful clinical tool in diagnosing lower motor neuron disorders (Box C).

Summary

Four distinct but highly interactive motor subsystems—local circuits in the spinal cord and brainstem, descending upper motor neuron pathways that control these circuits, the basal ganglia, and the cerebellum—all make essential contributions to motor control. Alpha motor neurons located in the spinal cord and in the cranial nerve nuclei in the brainstem directly link the nervous system and muscles, with each motor neuron and its associated muscle fibers constituting a functional entity called the motor unit. Motor units vary in size, amount of tension produced, speed of contraction, and degree of fatigability. Graded increases in muscle tension are mediated by both the orderly recruitment of different types of motor units and an increase in motor neuron firing frequency. Local circuitry involving sensory inputs, local circuit neurons, and α and γ motor neurons are especially important in the reflexive control of muscle activity. The stretch reflex is a monosynaptic circuit with connections between sensory fibers arising from muscle spindles and the α motor neurons that innervate the same or synergistic muscles. Gamma motor neurons regulate the gain of the stretch reflex by adjusting the level of tension in the intrafusal muscle fibers of the muscle spindle. This mechanism sets the baseline level of activity in α motor neurons and helps to regulate muscle length and tone. Other reflex circuits provide feedback control of muscle tension and mediate essential functions such as the rapid withdrawal of limbs from painful stimuli. Much of the spatial coordination and timing of muscle activation required for complex rhythmic movements such as locomotion are provided by specialized local circuits called central pattern generators. Because of their essential role in all of these circuits, damage to lower motor neurons leads to paralysis of the associated muscle and to other changes, including the loss of reflex activity, the loss of muscle tone, and eventually muscle atrophy.

ADDITIONAL READING

Reviews

BURKE, R. E. (1981) Motor units: Anatomy, physiology and functional organization. In *Handbook of Physiology*, V. B. Brooks (ed.). Section 1: *The Nervous System*. Volume 1, Part 1. Bethesda, MD: American Physiological Society, pp. 345–422.

BURKE, R. E. (1990) Spinal cord: Ventral horn. In *The Synaptic Organization of the Brain*, 3rd Ed. G. M. Shepherd (ed.). New York: Oxford University Press, pp. 88–132.

GRILLNER, S. AND P. WALLEN (1985) Central pattern generators for locomotion, with special reference to vertebrates. Ann. Rev. Neurosci. 8: 233–261.

HENNEMAN, E. (1990) Comments on the logical basis of muscle control. In *The Segmental Motor System*, M. C. Binder and L. M. Mendell (eds). New York: Oxford University Press, pp. 7–10.

HENNEMAN, E. AND L. M. MENDELL (1981) Functional organization of the motoneuron pool and its inputs. In *Handbook of Physiology*, V. B. Brooks (ed.). Section 1: *The Nervous System*. Volume 1, Part 1. Bethesda, MD: American Physiological Society, pp. 423–507.

LUNDBERG, A. (1975) Control of spinal mechanisms from the brain. In *The Nervous System*, Volume 1: *The Basic Neurosciences*. D. B. Tower (ed.). New York: Raven Press, pp. 253–265.

PATTON, H. D. (1965) Reflex regulation of movement and posture. In *Physiology and Biophysics*, 19th Ed., T. C. Ruch and H. D. Patton (eds.). Philadelphia: Saunders, pp. 181–206.

PEARSON, K. (1976) The control of walking. Sci. Am. 235: 72–86.

PROCHAZKA, A., M. HULLIGER, P. TREND AND N. DURMULLER (1988) Dynamic and static fusimotor set in various behavioral contexts. In *Mechanoreceptors: Development, Structure, and Function*. P. Hnik, T. Soukup, R. Vejsada and J. Zelena (eds.). New York: Plenum, pp. 417–430.

SCHMIDT, R. F. (1983) Motor systems. In *Human Physiology*. R. F. Schmidt and G. Thews (eds.). Berlin: Springer Verlag, pp. 81–110.

Important Original Papers

BURKE, R. E., D. N. LEVINE, M. SALCMAN AND P. TSAIRES (1974) Motor units in cat soleus muscle: Physiological, histochemical, and morphological characteristics. J. Physiol. (Lond.) 238: 503–514.

BURKE, R. E., P. L. STRICK, K. KANDA, C. C. KIM AND B. WALMSLEY (1977) Anatomy of medial gastrocnemius and soleus motor nuclei in cat spinal cord. J. Neurophysiol. 40: 667–680.

HENNEMAN, E., E. SOMJEN, AND D. O. CARPENTER (1965) Excitability and inhibitability of motoneurons of different sizes. J. Neurophysiol. 28: 599–620.

HUNT, C. C. AND S. W. KUFFLER (1951) Stretch receptor discharges during muscle contraction. J. Physiol. (Lond.) 113: 298–315.

LIDDELL, E. G. T. AND C. S. SHERRINGTON (1925) Recruitment and some other factors of reflex inhibition. Proc. R. Soc. London 97: 488–518.

LLOYD, D. P. C. (1946) Integrative pattern of excitation and inhibition in two-neuron reflex arcs. J. Neurophysiol. 9: 439–444.

MONSTER, A. W. AND H. CHAN (1977) Isometric force production by motor units of extensor digitorum communis muscle in man. J. Neurophysiol. 40: 1432–1443.

WALMSLEY, B., J. A. HODGSON AND R. E. BURKE (1978) Forces produced by medial gastrocnemius and soleus muscles during locomotion in freely moving cats. J. Neurophysiol. 41: 1203–1216.

Books

BRODAL, A. (1981) *Neurological Anatomy in Relation to Clinical Medicine*, 3rd Ed. New York: Oxford University Press.

SHERRINGTON, C. (1947) *The Integrative Action of the Nervous System*, 2nd Ed. New Haven: Yale University Press.

Chapter 17

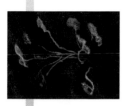

Upper Motor Neuron Control of the Brainstem and Spinal Cord

Overview

The axons of upper motor neurons descend from higher centers to influence the local circuits in the brainstem and spinal cord that organize movements. The sources of these upper motor neuron pathways include several brainstem centers and a number of cortical areas in the frontal lobe. The motor control centers in the brainstem are especially important in ongoing postural control. Each center has a distinct influence. Two of these centers, the vestibular nuclear complex and the reticular formation, have widespread effects on body position. Another brainstem center, the red nucleus, controls movements of the arms. The motor and "premotor" areas of the frontal lobe, in contrast, are responsible for the planning and precise control of complex sequences of voluntary movements. Most upper motor neurons, regardless of their source, influence the generation of movements by directly affecting the activity of the local organizing circuits in the brainstem and spinal cord (see Chapter 16). Upper motor neurons in the cortex also control movement indirectly, via pathways that project to the brainstem motor control centers, which, in turn, project to the local organizing circuits in the brainstem and cord. A major function of these indirect pathways is to maintain the body's posture during cortically initiated voluntary movements.

Descending Control of Spinal Cord Circuitry: General Information

Some insight into the functions of the different sources of the upper motor neurons is provided by the way the lower motor neurons and local circuit neurons—the ultimate targets of the upper motor neurons—are arranged within the spinal cord. As described in Chapter 16, lower motor neurons in the ventral horn of the spinal cord are organized in a somatotopic fashion: The most medial part of the ventral horn contains lower motor neuron pools that innervate axial muscles or proximal muscles of the limbs, whereas the more lateral parts contain lower motor neurons that innervate the distal muscles of the limbs. The local circuit neurons, which lie primarily in the intermediate zone of the spinal cord and supply much of the direct input to the lower motor neurons, are also topographically arranged. Thus, the medial region of the intermediate zone of the spinal cord gray matter contains the local circuit neurons that synapse with lower motor neurons in the medial part of the ventral horn, whereas the lateral regions of the intermediate zone contain local circuit neurons that synapse primarily with lower motor neurons in the lateral ventral horn.

369

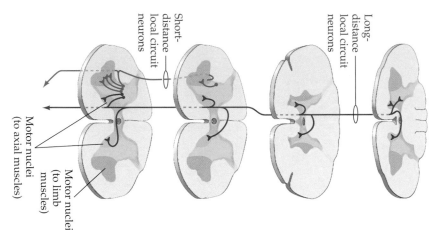

Long-distance local circuit neurons

Short-distance local circuit neurons

Motor nuclei (to axial muscles)

Motor nuclei (to limb muscles)

Figure 17.1 Local circuit neurons that supply the medial region of the ventral horn are situated medially in the intermediate zone of the spinal cord gray matter and have axons that extend over a number of spinal cord segments and terminate bilaterally. In contrast, local circuit neurons that supply the lateral parts of the ventral horn are located more laterally, have axons that extend over a few spinal cord segments, and terminate only on the same side of the cord. Descending pathways that contact the medial parts of the spinal cord gray matter are involved primarily in the control of posture; those that contact the lateral parts are involved in the fine control of the distal extremities.

The patterns of connections made by local circuit neurons in the medial region of the intermediate zone are different from the patterns made by those in the lateral region, and these differences are related to their respective functions (Figure 17.1). The medial local circuit neurons, which supply the lower motor neurons in the medial ventral horn, have axons that project to many spinal cord segments; indeed, some project to targets along the entire length of the cord. Moreover, many of these local circuit neurons also have axonal branches that cross the midline in the commissure of the spinal cord to innervate lower motor neurons in the medial part of the contralateral hemicord. This arrangement ensures that groups of axial muscles on both sides of the body act in concert to maintain and adjust posture. In contrast, local circuit neurons in the lateral region of the intermediate zone have shorter axons that typically extend fewer than five segments and are predominantly ipsilateral. This more restricted pattern of connectivity underlies the finer and more differentiated control that is exerted over the muscles of the distal extremities, such as that required for the independent movement of individual fingers during manipulative tasks.

Differences in the way upper motor neuron pathways from the cortex and brainstem terminate in the spinal cord conform to these functional distinctions between the local circuits that organize the activity of axial and distal muscle groups. Thus, most upper motor neurons that project to the medial part of the ventral horn also project to the medial region of the intermediate zone; the axons of the neurons have collateral branches that terminate over many spinal cord segments, reaching medial cell groups on both sides of the spinal cord. The sources of these projections are primarily the vestibular nuclei and the reticular formation (see next section); as their terminal zones in the medial spinal cord gray matter suggest, they are concerned primarily with postural mechanisms (Figure 17.2). In contrast, descending axons from the motor cortex generally terminate in lateral parts of the spinal cord gray matter and have terminal fields that are restricted to only a few spinal cord segments. These corticospinal pathways are primarily concerned with precise movements involving more distal parts of the limbs.

Two additional brainstem structures, the superior colliculus and the red nucleus, also contribute upper motor neuron pathways to the spinal cord. The axons arising from the superior colliculus project to medial cell groups in the cervical cord, where they influence the lower motor neuron circuits that control axial musculature of the neck (see Figure 17.2). These projections are particularly important in generating orienting movements of the head (the role of the superior colliculus in the generation of head and eye movements is covered in detail in Chapter 20). The red nucleus projections are also limited to the cervical level of the cord, but these terminate in lateral regions of the ventral horn and intermediate zone (see Figure 17.2). In humans, the axons arising from the red nucleus control primarily the proximal muscles of the arm. The limited distribution of rubrospinal projections may seem surprising, given the large size of the red nucleus in humans (*rubro-* means red; the adjective is derived from the rich capillary bed that gives the nucleus a reddish color in fresh tissue). In fact, the bulk of the red nucleus in humans is a subdivision that does not project to the spinal cord at all, but relays information from the cortex to the cerebellum (see Chapter 19).

Motor Control Centers in the Brainstem: Upper Motor Neurons That Maintain Balance and Posture

As described in Chapter 14, the **vestibular nuclei** are the major destination of the axons that form the vestibular division of the eighth cranial nerve; as

(A) MEDIAL BRAINSTEM PATHWAYS

(B) LATERAL BRAINSTEM PATHWAYS

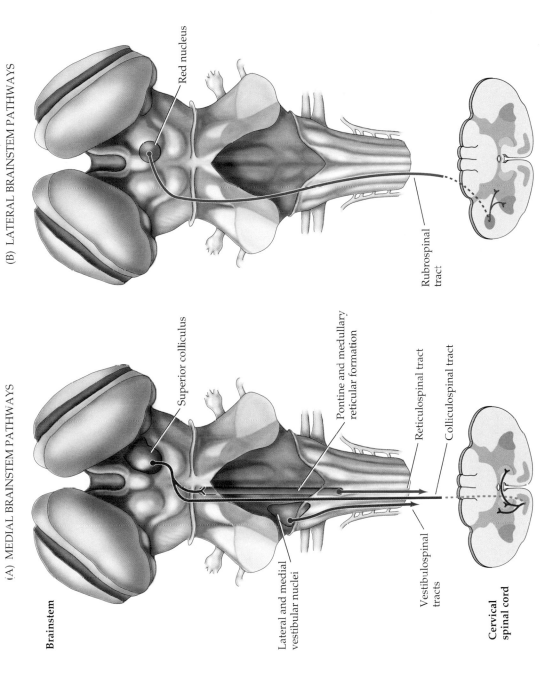

Brainstem

Superior colliculus

Red nucleus

Lateral and medial vestibular nuclei

Pontine and medullary reticular formation

Reticulospinal tract

Colliculospinal tract

Vestibulospinal tracts

Rubrospinal tract

Cervical spinal cord

Figure 17.2 Descending projections from the brainstem to the spinal cord. Pathways that influence motor neurons in the medial part of the ventral horn originate in the reticular formation, vestibular nucleus, and superior colliculus. Those that influence motor neurons that control the arm muscles originate in the red nucleus and terminate in more lateral parts of the ventral horn.

such, they receive sensory information from the semicircular canals and the otolith organs that specifies the position and angular acceleration of the head. Many of the cells in the vestibular nuclei that receive this information are upper motor neurons with descending axons that terminate in the medial region of the spinal cord gray matter, although some extend more laterally to contact the neurons that control the proximal muscles of the limbs. The projections from the vestibular nuclei that control axial muscles and those that influence proximal limb muscles originate from different cells and take different routes (called the medial and lateral vestibulospinal tracts).

The **reticular formation** is a complicated network of circuits located in the core of the brainstem and extending from the rostral midbrain to the caudal medulla (Figure 17.3). Unlike the well-defined sensory and motor nuclei of the cranial nerves, the reticular formation comprises clusters of neurons scattered among a welter of interdigitating axon bundles; it is therefore difficult to subdivide anatomically. The neurons within the reticular formation have a variety of functions, including cardiovascular and respiratory control (see Chapter 21), governance of myriad sensory motor reflexes (see Chapter 20), regulation of sleep and wakefulness (see Chapter 28), and, most important for present purposes, motor control.

Figure 17.3 The location of the reticular formation in relation to some other major landmarks at different levels of the brainstem. Neurons in the reticular formation are scattered among the axon bundles that course through the medial portion of the midbrain, pons, and medulla.

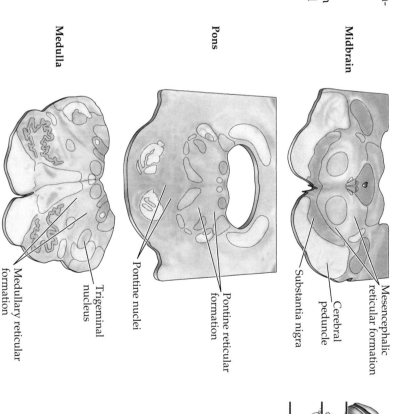

Midbrain

Mesencephalic reticular formation

Cerebral peduncle

Substantia nigra

Pons

Pontine reticular formation

Pontine nuclei

Medulla

Trigeminal nucleus

Medullary reticular formation

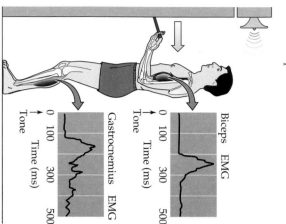

Figure 17.4 Anticipatory maintenance of body posture. At the onset of a tone, the subject pulls on a handle, contracting the biceps muscle. To ensure postural stability, contraction of the gastrocnemius muscle precedes that of the biceps.

Biceps   EMG

Tone

0   100   300   500
Time (ms)

Gastrocnemius   EMG

Tone

0   100   300   500
Time (ms)

The descending motor control pathways from the reticular formation to the spinal cord are similar to those of the vestibular nuclei: They terminate primarily in the medial parts of the gray matter where they influence the local circuit neurons that coordinate axial and proximal limb muscles (see Figure 17.2).

Both the vestibular nuclei and the reticular formation provide information to the spinal cord that maintains posture in response to environmental (or self-induced) disturbances of body position and stability. As expected, the vestibular nuclei make adjustments in posture and equilibrium in response to information from the inner ear. Direct projections from the vestibular nuclei to the spinal cord ensure a rapid compensatory response to any postural instability detected by the inner ear (see Chapter 14). In contrast, the motor centers in the reticular formation are controlled largely by other motor centers in the cortex or brainstem. The relevant neurons in the reticular formation initiate adjustments that stabilize posture during ongoing movements.

The way the upper motor neurons of the reticular formation maintain posture can be appreciated by analyzing their activity during voluntary movements. Even the simplest movements are accompanied by the activation of muscles that at first glance seem to have little to do with the primary purpose of the movement. For example, Figure 17.4 shows the pattern of muscle activity that occurs as a subject uses his arm to pull on a handle in response to an auditory tone. Activity in the biceps muscle begins about 200 ms after the tone. However, as the records show, the contraction of the biceps is accompanied by a significant increase in the activity of a proximal leg muscle, the gastrocnemius (as well as many other muscles not monitored in the experiment). In fact, contraction of the gastrocnemius muscle begins well before contraction of the biceps.

These observations show that postural control entails an anticipatory, or feedforward, mechanism (Figure 17.5). As part of the motor plan for moving

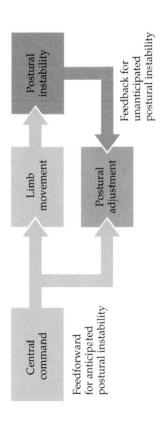

Figure 17.5 Feedforward and feedback mechanisms of postural control. Feedforward postural responses are "preprogrammed" and typically precede the onset of limb movement (see Figure 17.4). Feedback responses are initiated by sensory inputs that detect postural instability.

the arm, the effect of the impending movement on body stability is "evaluated" and used to generate a change in the activity of the gastrocnemius muscle. This change actually precedes and provides postural support for the movement of the arm. In the example given here, contraction of the biceps would tend to pull the entire body forward, an action that is opposed by the contraction of the gastrocnemius muscle. In short, this feedforward mechanism "predicts" the resulting disturbance in body stability and generates an appropriate stabilizing response.

The importance of the reticular formation for feedforward mechanisms of postural control has been explored in more detail in cats trained to use a forepaw to strike an object. As expected, the forepaw movement is accompanied by feedforward postural adjustments in the other legs to maintain the animal upright. These adjustments shift the animal's weight from an even distribution over all four feet to a diagonal pattern, in which the weight is carried mostly by the contralateral, nonreaching forelimb and the ipsilateral hindlimb. Lifting of the forepaw and postural adjustments in the other limbs can also be induced in an alert cat by electrical stimulation of the motor cortex. After pharmacological inactivation of the reticular formation, however, electrical stimulation of the motor cortex evokes only the forepaw movement, without the feedforward postural adjustments that normally accompany them.

The results of this experiment can be understood in terms of the fact that the upper motor neurons in the motor cortex influence the spinal cord circuits by two routes: **direct projections** to the spinal cord and **indirect projections** to brainstem centers under consideration here that in turn project to the spinal cord (Figure 17.6). The reticular formation is one of the major destinations of these latter projections from the motor cortex; thus, cortical upper motor neurons initiate both the reaching movement of the forepaw and also the postural adjustments in the other limbs necessary to maintain body stability. The forepaw movement is initiated by the direct pathway from the cortex to the spinal cord (and possibly by the red nucleus as well), whereas the postural adjustments are mediated via pathways from the motor cortex that reach the spinal cord indirectly, after an intervening relay in the reticular formation (the corticoreticulospinal pathway).

Further evidence for the contrasting functions of the direct and indirect pathways from the motor cortex and brainstem to the spinal cord has come from experiments carried out by the Dutch neurobiologist Hans Kuypers, who examined the behavior of rhesus monkeys that had the direct pathway transected at the level of the medulla, leaving the indirect descending upper motor neuron pathways to the spinal cord via the brainstem centers intact. Immediately after the surgery, the animals were able to use axial and proximal muscles to stand, walk, run, and climb, but they had great difficulty using the distal parts of their limbs (especially their hands) independently of other body movements. For example, the monkeys could cling to the cage

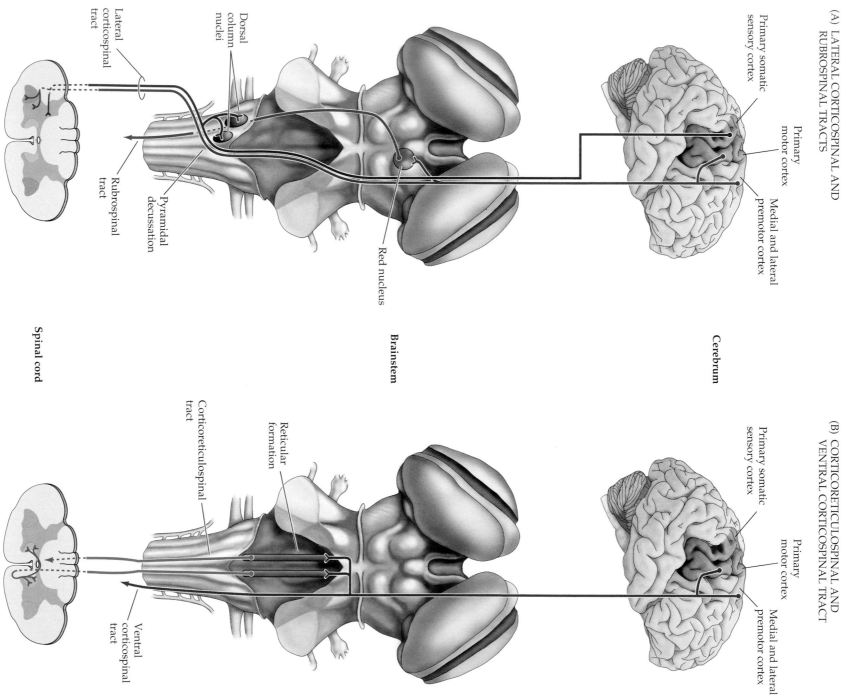

(A) LATERAL CORTICOSPINAL AND RUBROSPINAL TRACTS

Lateral corticospinal tract

Dorsal column nuclei

Rubrospinal tract

Pyramidal decussation

Red nucleus

Primary somatic sensory cortex

Primary motor cortex

Medial and lateral premotor cortex

Spinal cord

Brainstem

Cerebrum

(B) CORTICORETICULOSPINAL AND VENTRAL CORTICOSPINAL TRACT

Corticoreticulospinal tract

Reticular formation

Ventral corticospinal tract

Primary somatic sensory cortex

Primary motor cortex

Medial and lateral premotor cortex

▼ **Figure 17.6** Direct and indirect pathways from the motor cortex to the lateral (A) and medial (B) gray matter of the spinal cord. Neurons in the motor cortex that supply the lateral part of the ventral horn also terminate on neurons in the red nucleus. Neurons in the motor cortex that supply the medial part of the ventral horn also terminate on neurons in the reticular formation. Thus, the motor cortex has both direct and indirect routes by which it can influence the activity of spinal cord neurons.

but were unable to reach toward and pick up food with their fingers; rather, they used the entire arm to sweep the food toward them. After several weeks, the animals recovered some independent use of their hands and were again able to pick up objects of interest, but this action still involved the concerted closure of all of the fingers. The ability to make independent, fractionated movements of the fingers, as in opposing the movements of the fingers and thumb to pick up an object, never returned. These observations show that following damage to the direct corticospinal pathway at the level of the medulla, the indirect projections from the motor cortex via the brainstem centers (or from brainstem centers alone) are capable of sustaining motor behavior that involves primarily the use of proximal muscles. In contrast, the direct projections from the motor cortex to the spinal cord enable the speed and agility of movements, providing a higher degree of precision in fractionated finger movements than is possible using the indirect pathways alone.

Selective damage to the corticospinal tract (i.e., the direct pathway) in humans is rarely seen in the clinic. Nonetheless, this evidence in nonhuman primates showing that direct projections from the cortex to the spinal cord are essential for the performance of discrete finger movements helps explain the limited recovery in humans after damage to the motor cortex or to the internal capsule. Immediately after such an injury, such patients are typically paralyzed. With time, however, some ability to perform voluntary movements reappears. These movements, which are presumably mediated by the brainstem centers, are crude for the most part, and the ability to perform discrete finger movements such as those required for writing, typing, or buttoning typically remains impaired.

The Primary Motor Cortex: Upper Motor Neurons That Initiate Complex Voluntary Movements

The upper motor neurons in the cerebral cortex reside in several adjacent and highly interconnected areas in the frontal lobe, which together mediate the planning and initiation of complex temporal sequences of voluntary movements. These cortical areas all receive regulatory input from the basal ganglia and cerebellum via relays in the ventrolateral thalamus (see Chapters 18 and 19), as well as inputs from the somatic sensory regions of the parietal lobe (see Chapter 9). Although the phrase "motor cortex" is sometimes used to refer to these frontal areas collectively, more commonly it is restricted to the **primary motor cortex**, which is located in the precentral gyrus (Figure 17.7). The primary motor cortex can be distinguished from the adjacent "premotor" areas both cytoarchitectonically (it is area 4 in Brodmann's nomenclature) and by the low intensity of current necessary to elicit movements by electrical stimulation in this region. The low threshold for eliciting movements is an indicator of a relatively large and direct pathway from the primary area to the lower motor neurons of the brainstem and spinal cord. This section and the next focus on the organization and functions of the primary motor cortex and its descending pathways, whereas the subsequent section addresses the contributions of the adjacent premotor areas.

Figure 17.7 The primary motor cortex and the premotor area in the human cerebral cortex as seen in lateral (A) and medial (B) views. The primary motor cortex is located in the precentral gyrus; the premotor area is more rostral.

sylvius 27

(A) Lateral view

Lateral premotor cortex

Medial premotor cortex

Primary motor cortex

(B) Medial view

Medial premotor cortex

Primary motor cortex

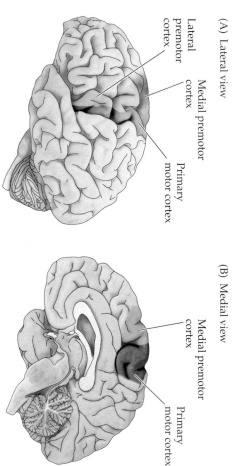

The pyramidal cells of cortical layer V (also called Betz cells) are the upper motor neurons of the primary motor cortex. Their axons descend to the brainstem and spinal motor centers in the **corticobulbar** and **corticospinal tracts**, passing through the internal capsule of the forebrain to enter the cerebral peduncle at the base of the midbrain (Figure 17.8). They then run through the base of the pons, where they are scattered among the transverse pontine fibers and nuclei of the pontine gray matter, coalescing again on the ventral surface of the medulla where they form the **medullary pyramids**. The components of this upper motor neuron pathway that innervate cranial nerve nuclei, the reticular formation, and the red nucleus (that is, the corticobulbar tract) leave the pathway at the appropriate levels of the brainstem (see Figure 17.8 and Box A). At the caudal end of the medulla, most, but not all, of the axons in the pyramidal tract cross (or "decussate") to enter the lateral columns of the spinal cord, where they form the **lateral corticospinal tract**. A smaller number of axons enters the spinal cord without crossing; these axons, which comprise the **ventral corticospinal tract**, terminate either ipsilaterally or contralaterally, after crossing in the midline (via spinal cord commissure). The ventral corticospinal pathway arises primarily from regions of the motor cortex that serve axial and proximal muscles (see Figure 17.6).

The lateral corticospinal tract forms the direct pathway from the cortex to the spinal cord and terminates primarily in the lateral portions of the ventral horn and intermediate gray matter (see Figures 17.6 and 17.8). The indirect pathway to lower motor neurons in the spinal cord runs, as already described, from the motor cortex to two of the sources of upper motor neurons in the brainstem: the red nucleus and the reticular formation. In general, the axons to the reticular formation originate from the parts of the motor cortex that project to the medial region of the spinal cord gray matter, whereas the axons to the red nucleus arise from the parts of the motor cortex that project to the lateral region of the spinal cord gray matter (see Figure 17.6).

Functional Organization of the Primary Motor Cortex

Clinical observations and experimental work dating back a hundred years or more have provided a reasonably coherent picture of the functional organization of the motor cortex. By the end of the nineteenth century, experimental work in animals by the German physiologists G. Theodor Fritsch and Eduard Hitzig had shown that electrical stimulation of the motor cortex elicits contractions of muscles on the contralateral side of the body. At about the

Figure 17.8 The corticospinal tract. Neurons in the motor cortex give rise to axons that travel through the internal capsule and coalesce on the ventral surface of the midbrain, within the cerebral peduncle. These axons continue through the pons and come to lie on the ventral surface of the medulla, giving rise to the pyramids. Most of these pyramidal fibers cross in the caudal part of the medulla to form the lateral corticospinal tract in the spinal cord. Those axons that do not cross (not illustrated) descend on the same side and form the ventral corticospinal tract (see Figure 17.6). The axons that terminate in the reticular formation of the pons and medulla comprise components of the corticobulbar tract.

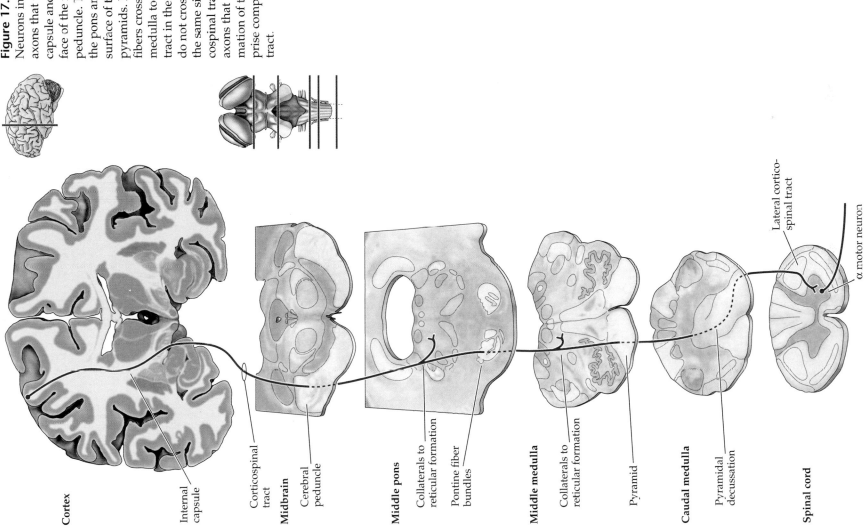

Cortex

Internal capsule

Corticospinal tract

Midbrain

Cerebral peduncle

Middle pons

Collaterals to reticular formation

Pontine fiber bundles

Middle medulla

Collaterals to reticular formation

Pyramid

Caudal medulla

Pyramidal decussation

Spinal cord

Lateral corticospinal tract

α motor neuron

BOX A
Descending Projections to Cranial Nerve Motor Nuclei and Their Importance in Diagnosing the Cause of Motor Deficits

The symptoms pertinent to the cranial nerves and their nuclei are of special importance to physicians seeking to identify the location of neurological lesions that produce motor deficits. An especially good example is the cranial nerves that innervate the face. Axons descending from upper motor neurons in the face representation of the primary motor cortex leave the corticospinal pathway at various levels of the brainstem to innervate the somatic sensory and motor nuclei of the relevant cranial nerves, thus forming the corticobulbar pathways. These axons terminate either contralaterally or bilaterally, depending on the nucleus; this pattern of termination has special significance for understanding the neurologic deficits following cortical damage. The projections to most of the motor nuclei (including the trigeminal nucleus, facial nucleus, nucleus ambiguus, and spinal accessory nucleus) are bilateral. In contrast, those to the hypoglossal nucleus and a part of the

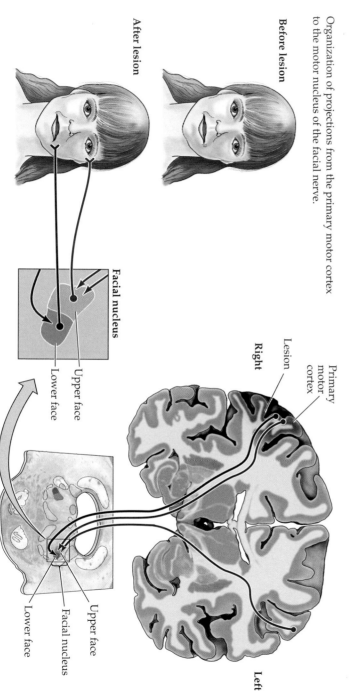

After lesion

Before lesion

Facial nucleus
Upper face
Lower face

Primary motor cortex
Lesion
Right
Left

Upper face
Facial nucleus
Lower face

Organization of projections from the primary motor cortex to the motor nucleus of the facial nerve.

facial nucleus are contralateral. The control of muscles innervated by motor nuclei that receive *bilateral* projections from the cortex is relatively unaffected by *unilateral* damage to the cortex or the descending fiber pathways; the remaining intact projection is sufficient for normal (or near normal) muscle control. This sparing of function is not apparent, however, for those nuclei that receive descending projections only from the contralateral cortex.

The descending projections to the facial nucleus deserve special comment in this regard. Motor neurons in the part of the nucleus that innervates the upper face receive bilateral projections from the motor cortex, while those that innervate the lower face receive projections only from the contralateral cortex. Thus, damage of the descending projections to the facial nucleus has different effects on muscles of the upper and lower parts of the face; upper facial muscles (frontalis, orbicularis oculi) continue to operate nor-

mally, while lower facial muscles (orbicularis oris, buccinator) are weakened. Accordingly, patients with damage to the face representation in the motor cortex on one side can wrinkle the forehead and close the eyes voluntarily, but lose voluntary control of the contralateral muscles of the mouth (see figure). The most obvious sign is a sagging of the corner of the mouth and a stilted smile. This arrangement can be especially informative in distinguishing damage to the brainstem (or to the facial nerve itself) from damage to higher centers. Damage to the brainstem or facial nerve affects *all* the muscles of facial expression (both upper and lower face) on the side of the lesion. Damage to the motor cortex, however, affects only the control of the lower part of the face on the side contralateral to the lesion.

Reference

BRODAL, A. (1981) *Neurological Anatomy in Relation to Clinical Medicine*, 3rd Ed. New York: Oxford University Press.

same time, the British neurologist John Hughlings Jackson surmised that the motor cortex contains a complete representation, or map, of the body's musculature. Jackson reached this conclusion from the fact that the abnormal movements during some types of epileptic seizures "march" systematically from one part of the body to another. For instance, partial motor seizures may start with abnormal movements of a finger, progress to involve the entire hand, then the forearm, the arm, the shoulder, and, finally, the face.

This early evidence for motor maps in the cortex was confirmed shortly after the turn of the nineteenth century when Charles Sherrington published his classical studies of the organization of the motor cortex in great apes, using focal electrical stimulation. During the 1930s, one of Sherrington's students, the American neurosurgeon Wilder Penfield, extended this work by demonstrating that the human motor cortex also contains a spatial map of the body's musculature. By correlating the location of muscle contractions with the site of electrical stimulation on the surface of the motor cortex (the same method used by Sherrington), Penfield mapped the representation of the muscles in the precentral gyrus in over 400 neurosurgical patients (Figure 17.9). He found that this motor map shows the same disproportions observed in the somatic sensory maps in the postcentral gyrus (see Chapter 9). Thus, the musculature used in tasks requiring fine motor control (such as movements of the face and hands) occupies a greater amount of space in the

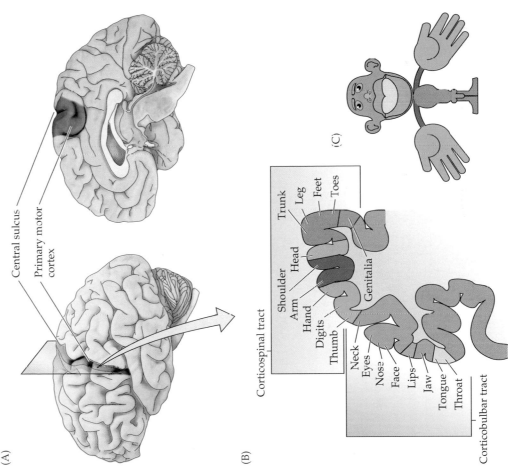

(A)

Central sulcus

Primary motor cortex

(B)

Corticospinal tract

Shoulder
Trunk
Arm   Head   Leg
Hand   Feet
Digits   Genitalia   Toes
Thumb

Neck
Eyes
Nose
Face
Lips
Jaw
Tongue
Throat

Corticobulbar tract

(C)

Figure 17.9 Topographic map of the body musculature in the primary motor cortex. (A) Location of primary motor cortex in the precentral gyrus. (B) Section along the precentral gyrus, illustrating the somatotopic organization of the motor cortex. The most medial parts of the motor cortex are responsible for controlling muscles in the legs; the most lateral portions are responsible for controlling muscles in the face. (C) Disproportional representation of various portions of the body musculature in the motor cortex. Representations of parts of the body that exhibit fine motor control capabilities (such as the hands and face) occupy a greater amount of space than those that exhibit less precise motor control (such as the trunk).

BOX B
What Do Motor Maps Represent?

Electrical stimulation studies carried out by the neurosurgeon Wilder Penfield and his colleagues in human patients (and by Clinton Woolsey and his colleagues in experimental animals) clearly demonstrated a systematic map of the body's musculature in the primary motor cortex (see text). The fine structure of this map, however, has been a continuing source of controversy. Is the map in the motor cortex a "piano keyboard" for the control of individual muscles, or is it a map of movements, in which specific sites control multiple muscle groups that contribute to the generation of particular actions? Initial experiments implied that the map in the motor cortex is a fine-scale representation of individual muscles. Thus, stimulation of small regions of the map activated single muscles, suggesting that vertical columns of cells in the motor cortex were responsible for controlling the actions of particular muscles, much as columns in the somatic sensory map are thought to analyze particular types of stimulus information (see Chapter 9).

More recent studies using anatomical and physiological techniques, however, have shown that the map in the motor cortex is far more complex than a columnar representation of particular muscles. Individual pyramidal tract axons are now known to terminate on sets of spinal motor neurons that innervate different muscles. This relationship is evident even for neurons in the hand representation of the motor cortex, the region that controls the most discrete, fractionated movements. Furthermore, cortical microstimulation experiments have shown that a single muscle is represented multiple times over a wide region of the motor cortex (about 2–3 mm in primates) in a complex, mosaic fashion. It seems likely that horizontal connections within the motor cortex create ensembles of neurons that coordinate the pattern of firing in the population of ventral horn cells that ultimately generate a given movement.

Thus, while the somatotopic maps in the motor cortex generated by early

studies are correct in their overall topography, the fine structure of the map is far more intricate. Unraveling these details of motor maps still holds the key to understanding how patterns of activity in the motor cortex generate a given movement.

References

BARINAGA, M. (1995) Remapping the motor cortex. Science 268: 1696–1698.

LEMON, R. (1988) The output map of the primate motor cortex. Trends Neurosci. 11: 501–506.

PENFIELD, W. AND E. BOLDREY (1937) Somatic motor and sensory representation in the cerebral cortex of man studied by electrical stimulation. Brain 60: 389–443.

SCHIEBER, M. H. AND L. S. HIBBARD (1993) How somatotopic is the motor cortex hand area? Science 261: 489–491.

WOOLSEY, C. N. (1958) Organization of somatic sensory and motor areas of the cerebral cortex. In Biological and Biochemical Bases of Behavior, H. F. Harlow and C. N. Woolsey (eds.), Madison, WI: University of Wisconsin Press, pp. 63–81.

Electrical stimulation studies carried out by the neurosurgeon Wilder Penfield and his colleagues in human patients (and by Clinton Woolsey and his colleagues in experimental animals) clearly demonstrated a systematic map of the body's musculature in the primary motor cortex (see text). The fine structure of this map, however, has been a continuing source of controversy. Is the map in the motor cortex a "piano keyboard" for the control of individual muscles, or is it a map of movements, in which specific sites control multiple muscle groups that contribute to the generation of particular actions? Initial experiments implied that the map in the motor cortex is a fine-scale representation of individual muscles. Thus, stimulation of small regions of the map activated single muscles, suggesting that vertical columns of cells in the motor cortex were responsible for controlling the actions of particular muscles, much as columns in the somatic sensory map are thought to analyze particular types of stimulus information (see Chapter 9).

The introduction in the 1960s of intracortical microstimulation (a more refined method of cortical activation) allowed a more detailed understanding of motor maps. Microstimulation entails the delivery of electrical currents an order of magnitude smaller than those used by Sherrington and Penfield. By passing the current through the sharpened tip of a metal microelectrode inserted into the cortex, the upper motor neurons in layer V that project to lower motor neuron circuitry can be stimulated focally. Although intracortical stimulation generally confirmed Penfield's spatial map in the motor cortex, it also showed that the finer organization of the map is rather different than most neuroscientists imagined. For example, when microstimulation was combined with recordings of muscle electrical activity, even the smallest currents capable of eliciting a response initiated the excitation of several muscles (and the simultaneous inhibition of others), suggesting that organized movements rather than individual muscles are represented in the map (see Box B). Furthermore, within major subdivisions of the map (e.g., arm, forearm, or finger regions), a particular movement could be elicited by

map than does the musculature requiring less precise motor control (such as that of the trunk). The behavioral implications of cortical motor maps are considered in Boxes B and C.

stimulation of widely separated sites, indicating that nearby regions are linked by local circuits to organize specific movements. This interpretation has been supported by the observation that the regions responsible for initiating particular movements overlap substantially.

About the same time that these studies were being undertaken, Ed Evarts and his colleagues at the National Institutes of Health were pioneering a technique in which implanted microelectrodes were used to record the electrical activity of individual motor neurons in awake, behaving monkeys (Figure 17.10). In these experiments, the monkeys were trained to perform a variety of motor tasks, thus providing a means of correlating neuronal activity with voluntary movements. Evarts and his group found that the force generated by contracting muscles changed as a function of the firing rate of upper motor neurons. Moreover, the firing rates of the active neurons often changed *prior* to movements involving very small forces. Evarts therefore proposed that the primary motor cortex contributes to the initial phase of recruitment of lower motor neurons involved in the generation of finely controlled movements. Additional experiments showed that the activity of primary motor neurons is correlated not only with the magnitude, but also with the direction of the force produced by muscles. Thus, some neurons show progressively less activity as the direction of movement deviates from the neuron's "preferred direction."

A further advance was made in the mid-1970s by the introduction of spike-triggered averaging (Figure 17.11). By correlating the timing of the cortical neuron's discharges with the onset times of the contractions generated by the various muscles used in a movement, this method provides a way of measuring the influence of a single cortical motor neuron on a population of lower motor neurons in the spinal cord. In recording such activity from different muscles as monkeys performed wrist flexion or extension, it became apparent that the activity of a number of different muscles is directly facilitated by the discharges of a given upper motor neuron. This peripheral muscle group was called the "muscle field" of the upper motor neuron. On average, the size of the muscle field in the wrist region was two to three muscles per upper motor neuron. These observations confirmed that single upper motor neurons contact several lower motor neuron pools;

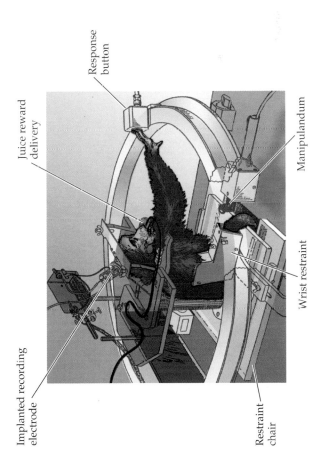

Implanted recording electrode

Juice reward delivery

Response button

Manipulandum

Wrist restraint

Restraint chair

Figure 17.10 Experimental apparatus developed to record the activity of single neurons in awake primates trained to perform specific movements.

Figure 17.11 The influence of single cortical upper motor neurons on muscle activity. (A) Diagram illustrates the spike triggering average method for correlating muscle activity with the discharges of single upper motor neurons. (B) The response of a thumb muscle (bottom trace) follows by a fixed latency the single spike discharge of a pyramidal tract neuron (top trace). This technique can be used to determine all the muscles that are influenced by a given motor neuron (see text). (After Porter and Lemon, 1993.)

(A) Detection of postspike facilitation

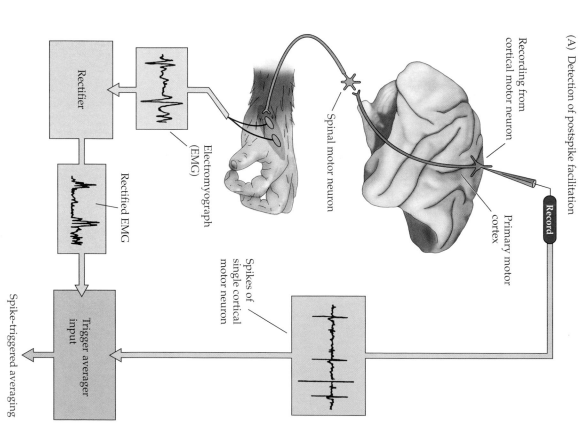

(B) Postspike facilitation by cortical motor neuron

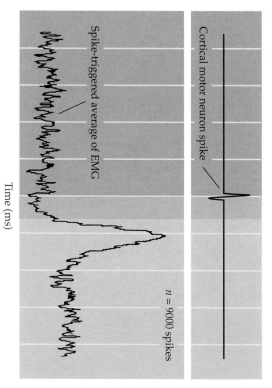

Spike-triggered averaging

Cortical motor neuron spike

Spike-triggered average of EMG

n = 9000 spikes

Time (ms)

BOX C
Sensory Motor Talents and Cortical Space

Are special sensory motor talents, such as the exceptional speed and coordination displayed by talented athletes, ballet dancers, or concert musicians, visible in the structure of the nervous system? The widespread use of noninvasive brain imaging techniques (see Boxes B and C in Chapter 1) has generated a spate of studies that have tried to answer this and related questions. Most of these studies have sought to link particular sensory motor skills to the amount of brain space devoted to such talents. For example, a study of professional violinists, cellists, and classical guitarists purported to show that representations of the "fingering" digits of the left hand in the right primary somatic sensory cortex are larger than the corresponding representations in nonmusicians.

Although such studies in humans remain controversial (the techniques are only semiquantitative), the idea that greater motor talents (or any other ability) will be reflected in a greater amount of brain space devoted to that task makes good sense. In particular, comparisons across species show that special talents are invariably based on commensurately sophisticated brain circuitry, which means more neurons, more synaptic contacts between neurons, and more supporting glial cells—all of which occupy

more space within the brain. The size and proportion of bodily representations in the primary somatic sensory and motor cortices of various animals reflects species-specific nuances of mechanosensory discrimination and motor control. Thus, the representations of the paws are disproportionately large in the sensorimotor cortex of raccoons; rats and mice devote a great deal of cortical space to representations of their prominent facial whiskers; and a large fraction of the sensorimotor cortex of the star-nosed mole is given over to representing the elaborate nasal appendages that provide critical mechanosensory information for this burrowing species. The link between behavioral competence and the allocation of space is equally apparent in animals in which a particular ability has diminished, or has never developed fully, during the course of evolution.

Nevertheless, it remains uncertain how—or if—this principle applies to variations in behavior among members of the same species, including humans. For example, there does not appear to be any average hemisphere asymmetry in the allocation of space in either the primary sensory or motor area, as measured cytoarchitectonically. Some asymmetry might be expected simply because 90% of humans prefer to use the right

hand when they perform challenging manual tasks. It seems likely that individual sensory motor talents among humans will be reflected in the allocation of an appreciably different amount of space to those behaviors, but this issue is just beginning to be explored with quantitative methods that are adequate to the challenge.

References

CATANIA, K. C. AND J. H. KAAS (1995) Organization of the somatosensory cortex of the star-nosed mole. J. Comp. Neurol. 351: 549–567.

ELBERT, T., C. PANTEV, C. WIENBRUCH, B. ROCKSTROH AND E. TAUB (1995) Increased cortical representation of the fingers of the left hand in string players. Science 270: 305–307.

WELKER, W. I. AND S. SEIDENSTEIN (1959) Somatic sensory representation in the cerebral cortex of the raccoon (*Procyon lotos*). J. Comp. Neurol. 111: 469–501.

WHITE, L. E., T. J. ANDREWS, C. HULETTE, A. RICHARDS, M. GROELLE, J. PAYDARFAR AND D. PURVES (1997) Structure of the Human Sensorimotor System II. Lateral symmetry. Cereb. Cortex 7: 31–47.

WOOLSEY, T. A. AND H. VAN DER LOOS (1970) The structural organization of layer IV in the somatosensory region (SI) of mouse cerebral cortex. The description of a cortical field composed of discrete cytoarchitectonic units. Brain Res. 17: 205–242.

the results are also consistent with the general conclusion that *movements*, rather than individual muscles, are encoded by the activity of the upper motor neurons in the cortex (see Box B).

Finally, the relative amount of activity across large populations of neurons appears to encode the direction of visually-guided movements. Thus, the direction of movements in monkeys could be predicted by calculating a "neuronal population vector" derived simultaneously from the discharges of many "broadly tuned" upper motor neurons (Figure 17.12). These observations showed that the discharges of individual upper motor neurons cannot specify the direction of an arm movement, simply because they are tuned too broadly;

Chapter 18

Modulation of Movement by the Basal Ganglia

Overview

As described in the preceding chapter, motor regions of the cortex and brainstem contain upper motor neurons that initiate movement by projecting to local circuit and lower motor neurons in the brainstem and spinal cord. This chapter and the next discuss two additional regions of the brain that are important in motor control: the basal ganglia and the cerebellum. In contrast to the components of the motor system that harbor upper motor neurons, the basal ganglia and cerebellum do not project directly to either the local circuit or lower motor neurons; instead, they influence movement by regulating the activity of upper motor neurons. The term "basal ganglia" refers to a large and functionally diverse set of nuclear structures that lie deep within the cerebral hemispheres. The subset of these nuclei relevant to motor function includes the caudate, putamen, and the globus pallidus. Two additional structures, the substantia nigra in the base of the midbrain and the subthalamic nucleus in the ventral thalamus, are closely associated with the motor functions of these basal ganglia nuclei and are included in the discussion. The motor components of the basal ganglia, together with the substantia nigra and the subthalamic nucleus, effectively make a subcortical loop that links most areas of the cortex with pools of upper motor neurons in the primary motor and premotor cortex and in the brainstem. The neurons in this loop respond in anticipation of and during movements, and their effects on upper motor neurons are required for the normal initiation of voluntary movements. When one of these components of the basal ganglia or associated structures is compromised, the patient cannot switch smoothly between commands that initiate a movement and those that terminate the movement. The disordered movements that result can be understood as a consequence of abnormal upper motor neuron activity in the absence of the supervisory control normally provided by the basal ganglia.

Projections to the Basal Ganglia

The basal ganglia are divided into several functionally distinct groups of nuclei (Figure 18.1). The first and larger of these groups is called the **corpus striatum**, which includes the **caudate** and **putamen**. These two subdivisions of the corpus striatum are the *input* zone of the basal ganglia, their neurons being the targets of most of the pathways that reach this complex from other parts of the brain (Figure 18.2). The name (which means "striped body") is given because the axon fascicles that pass through the caudate and putamen give them a striped appearance when cut in cross section. The destination of the incoming axons from the cortex are onto the dendrites of a class of cells called **medium spiny neurons** in the corpus striatum (Figure 18.3). The large

(A)

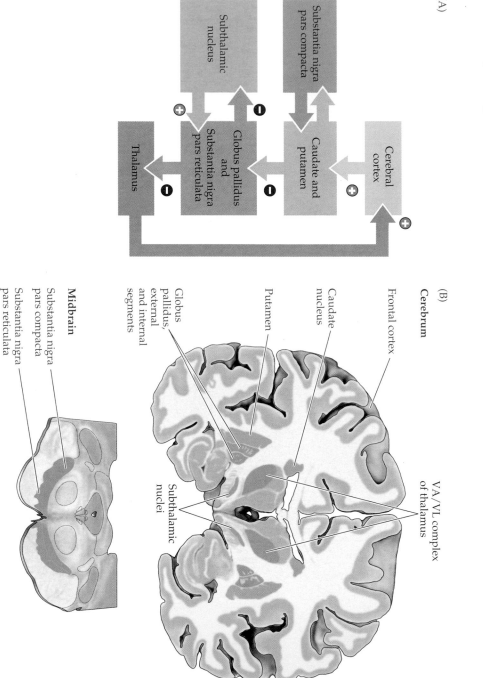

(B)
Cerebrum

Frontal cortex

VA/VL complex
of thalamus

Caudate
nucleus

Putamen

Globus
pallidus,
external
and internal
segments

Subthalamic
nuclei

Midbrain

Substantia nigra
pars compacta

Substantia nigra
pars reticulata

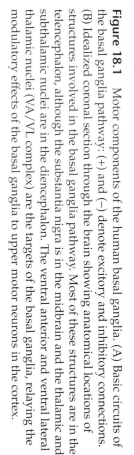

Figure 18.1 Motor components of the human basal ganglia. (A) Basic circuits of the basal ganglia pathway; (+) and (–) denote excitory and inhibitory connections. (B) Idealized coronal section through the brain showing anatomical locations of structures involved in the basal ganglia pathway. Most of these structures are in the telencephalon, although the substantia nigra is in the midbrain and the thalamic and subthalamic nuclei are in the diencephalon. The ventral anterior and ventral lateral thalamic nuclei (VA/VL complex) are the targets of the basal ganglia, relaying the modulatory effects of the basal ganglia to upper motor neurons in the cortex.

dendritic trees of these neurons allow them to integrate inputs from a variety of cortical, thalamic, and brainstem structures. The axons arising in turn from the medium spiny neurons converge on neurons in the globus pallidus and the substantia nigra pars reticulata. The globus pallidus and substantia nigra pars reticulata are the main sources of *output* from the basal ganglia complex.

Nearly all regions of the neocortex project directly to the corpus striatum, making the cerebral cortex the largest input to the basal ganglia by far. Indeed, the only cortical areas that do not project to the corpus striatum are the primary visual and primary auditory cortices (Figure 18.4). Of those cortical areas that do innervate the striatum, the heaviest projections are from association areas in the frontal and parietal lobes, with substantial contributions from the temporal, insular, and cingulate cortices as well. All of these projections, referred to collectively as the **corticostriatal pathway**, travel through the internal capsule to reach the caudate and putamen directly (see Figure 18.2).

The cortical inputs to the caudate and putamen are not equivalent, however, a fact that reflects functional differences between these two nuclei. The

Cerebrum

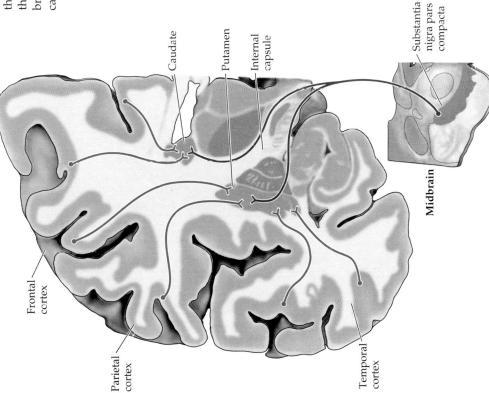

Figure 18.2 Anatomical organization of the inputs to the basal ganglia. An idealized coronal section through the human brain, showing the projections from the cerebral cortex and the substantia nigra pars comparta to the caudate and putamen.

caudate nucleus receives cortical projections primarily from multimodal association cortices, and from motor areas in the frontal lobe that control eye movements. As the name implies, the association cortices do not process any one type of sensory information; rather, they receive inputs from a number of primary and secondary sensory cortices and associated thalamic nuclei (see Chapter 26). The putamen, on the other hand, receives input from the primary and secondary somatic sensory cortices in the parietal lobe, the secondary (extrastriate) visual cortices in the occipital and temporal lobes, the premotor and motor cortices in the frontal lobe, and the auditory association areas in the temporal lobe. The fact that different cortical areas project to different regions of the striatum implies that the corticostriatal pathway consists of multiple parallel pathways serving different functions. This interpretation is supported by the observation that the segregation is maintained in the structures that receive projections from the striatum, and in the pathways that project from the basal ganglia to other brain regions.

There are other indications that the corpus striatum is functionally subdivided according to its inputs. For example, visual and somatic sensory cortical projections are topographically mapped within different regions of the putamen. Moreover, the cortical areas that are functionally interconnected at the level of the cortex give rise to projections that overlap extensively in the striatum. Anatomical studies by Ann Graybiel and her colleagues at Massa-

(A)

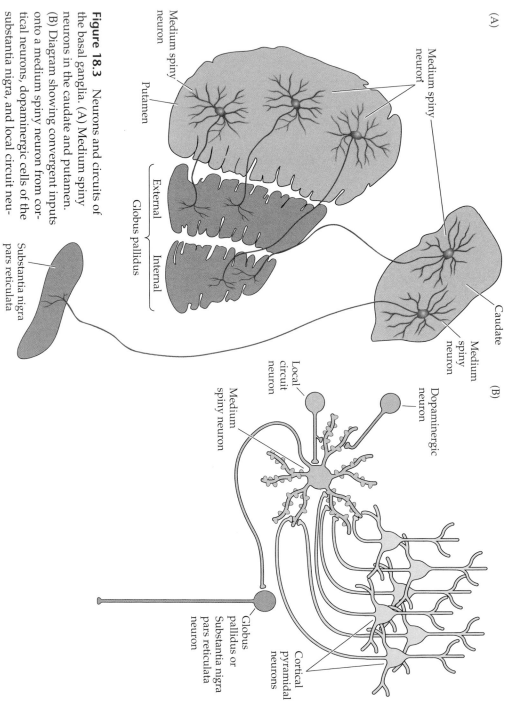

(B)

Figure 18.3 Neurons and circuits of the basal ganglia. (A) Medium spiny neurons in the caudate and putamen. (B) Diagram showing convergent inputs onto a medium spiny neuron from cortical neurons, dopaminergic cells of the substantia nigra, and local circuit neurons. The primary output of the medium spiny cells is to the globus pallidus and to the substantia nigra pars reticulata.

chusetts Institute of Technology have shown that cortical regions concerned with the hand (see Chapter 9) converge in specific rostrocaudal bands within the striatum; conversely, regions in the same cortical areas concerned with the leg converge in other striatal bands. These rostrocaudal bands, therefore, appear to be functional units concerned with the movement of

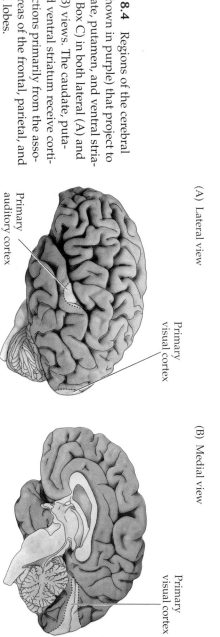

(A) Lateral view   (B) Medial view

Figure 18.4 Regions of the cerebral cortex (shown in purple) that project to the caudate, putamen, and ventral striatum (see Box C) in both lateral (A) and medial (B) views. The caudate, putamen, and ventral striatum receive cortical projections primarily from the association areas of the frontal, parietal, and temporal lobes.

particular body parts. Another study by the same group showed that the more extensively cortical areas are interconnected, the greater the overlap in their projections to the striatum.

A further indication of functional subdivision within the striatum is the spatial distribution of different types of medium spiny neurons. Although medium spiny neurons are distributed throughout the striatum, they occur in clusters of cells called "patches" or "striosomes" and in a surrounding "matrix" of neurochemically distinct cells. Whereas the distinction between the patches and matrix was originally based only on differences in the types of neuropeptides contained by the medium spiny cells in the two regions, the cell types are now known to differ in the sources of their inputs from the cortex and in the destinations of their projections to other parts of the basal ganglia. For example, even though most cortical areas project to medium spiny neurons in both these compartments, limbic areas of the cortex (such as the cingulate gyrus) project more heavily to the patches, whereas motor and somatic sensory areas project preferentially to the neurons in the matrix. These differences in the connectivity of medium spiny neurons in the patches and matrix further support the conclusion that functionally distinct pathways project in parallel from the cortex to the striatum.

The nature of the signals transmitted to the caudate and putamen from the cortex is not understood. It is known, however, that collateral axons of the corticocortical, corticothalamic, and corticospinal pathways all make excitatory glutamatergic synapses on the dendritic spines of medium spiny neurons (see Figure 18.3B). The arrangement of these cortical synapses is such that the number of contacts established between an individual cortical axon and a single medium spiny cell is very small, whereas the number of spiny neurons contacted by a single axon is extremely large. This divergence of cortical axon terminals allows a single medium spiny neuron to integrate the influences of thousands of cortical cells.

The medium spiny cells also receive noncortical inputs from interneurons, from the midline and intralaminar nuclei of the thalamus, and from most of the brainstem aminergic nuclei. In contrast to the cortical inputs, the motor circuit neuron and thalamic synapses are made on the dendritic shafts and close to the cell soma, where they can modulate the effectiveness of cortical synaptic activation arriving from the more distal dendrites. The aminergic ones are dopaminergic synapses from a subdivision of the substantia nigra called **pars compacta** because of its densely packed cells. These synapses are, in contrast, located on the base of the spines in close proximity to the cortical synapses, where they more directly modulate cortical input (see Figure 18.3B). As a result, inputs from both the cortex and the substantia nigra pars compacta are relatively far from the initial segment of the medium spiny neuron axon, where the nerve impulse is generated. Accordingly, the medium spiny neurons must simultaneously receive many excitatory inputs from cortical and nigral neurons in order to become active. The medium spiny neurons are, therefore, usually silent.

When the medium spiny neurons do become active, their activity is associated with the imminent occurrence of a movement. Extracellular recordings show that these neurons typically increase their rate of discharge just before an impending movement. Neurons in the putamen tend to discharge in anticipation of body movements, whereas caudate neurons fire prior to eye movements. These anticipatory discharges are evidently part of a movement selection process; in fact, they can precede the initiation of movement by as much as several seconds. Similar recordings have also shown that the discharges of some striatal neurons vary according to the location in space of the *target* of a movement, rather than with the starting position of the limb relative to the target. Thus, the activity of these cells may encode the decision to reach toward the target, rather than simply the direction and amplitude of a movement as such.

Projections from the Basal Ganglia to Other Brain Regions

The medium spiny neurons of the caudate and putamen give rise to inhibitory GABAergic projections that terminate in another pair of nuclei in the basal ganglia complex: the **internal division of the globus pallidus** and a specific region of the substantia nigra called **pars reticulata** (because, unlike the pars compacta, axons passing through give it a reticulated appearance). These nuclei are in turn major sources of the output from the basal ganglia (Figure 18.5). The globus pallidus and substantia nigra pars reticulata have similar output functions. In fact, developmental studies show that pars reticulata is actually part of the globus pallidus, although the two eventually become separated by fibers of the internal capsule. The striatal projections to these two nuclei resemble the

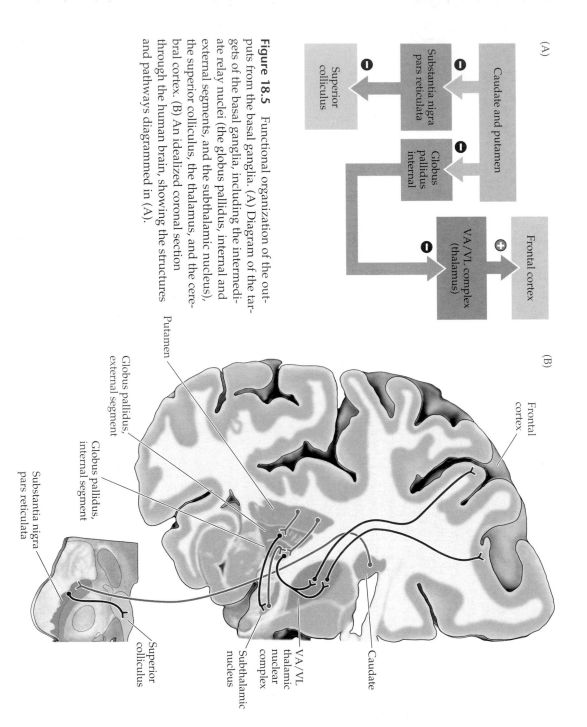

(A)

(B)

Figure 18.5 Functional organization of the outputs from the basal ganglia. (A) Diagram of the targets of the basal ganglia, including the intermediate relay nuclei (the globus pallidus, internal and external segments, and the subthalamic nucleus), the superior colliculus, the thalamus, and the cerebral cortex. (B) An idealized coronal section through the human brain, showing the structures and pathways diagrammed in (A).

corticostriatal pathways in that they terminate in rostrocaudal bands, the locations of which vary with the locations of their source in the striatum.

A striking feature of the projections from the medium spiny neurons to the globus pallidus and substantia nigra is the degree of their convergence onto pallidal and reticulata cells. In humans, for example, the corpus striatum contains approximately 100 million neurons, about 75% of which are medium spiny neurons. In contrast, the main target of these neurons, the globus pallidus, comprises only about 700,000 cells. Thus, on average, more than 100 medium spiny neurons innervate each pallidal cell.

The efferent neurons of the internal globus pallidus and substantia nigra pars reticulata together give rise to the major pathways that link the basal ganglia with upper motor neurons located in the cortex and in the brainstem (see Figure 18.5). The pathway to the motor cortex arises primarily in the internal globus pallidus and is relayed via the **ventral anterior and ventral lateral nuclei** of the dorsal thalamus. These thalamic nuclei project directly to motor areas of the cortex, thus completing a vast loop that originates in multiple cortical areas and terminates, after relays in the basal ganglia and thalamus, back in the motor areas of the frontal lobe. In contrast, the axons from substantia nigra pars reticulata synapse on upper motor neurons in the superior colliculus that command eye movements, without any intervening relay in the thalamus (see Figure 18.5 and Chapter 20). This difference between the globus pallidus and substantia nigra pars reticulata is not absolute, however, since many reticulata axons also project to the thalamus where they contact relay neurons that project to the frontal eye fields of the premotor cortex.

Because the efferent cells of both the globus pallidus and substantia nigra pars reticulata are GABAergic, the main output of the basal ganglia is *inhibitory*. In contrast to the quiescent medium spiny neurons, the neurons in both these output zones have high levels of spontaneous activity that tend to prevent unwanted movements by tonically inhibiting the superior colliculus and thalamus. Since the medium spiny neurons of the striatum also are GABAergic and inhibitory, the net effect of the excitatory inputs that reach the striatum from the cortex is to inhibit the tonically active inhibitory cells of the globus pallidus and substantia nigra pars reticulata (Figure 18.6). Thus, in the absence of body movements, the globus pallidus neurons, for example, provide tonic inhibition to the relay cells in the ventral lateral and anterior nuclei of the thalamus. When the pallidal cells are inhibited by activity of the medium spiny neurons, the thalamic neurons are *disinhibited* and can relay signals from other sources to the upper motor neurons in the cortex. This **disinhibition** is what normally allows the upper motor neurons to send commands to local circuit and lower motor neurons that initiate movements. Conversely, an abnormal reduction in the tonic inhibition as a consequence of basal ganglia dysfunction leads to excessive excitability of the upper motor neurons, and thus to the involuntary movement syndromes that are characteristic of basal ganglia disorders such as **Huntington's disease** (Box A; see also Figure 18.9B).

Evidence from Studies of Eye Movements

The permissive role of the basal ganglia in the initiation of movement is perhaps most clearly demonstrated by studies of eye movements carried out by Okihide Hikosaka and Robert Wurtz at the National Institutes of Health (Figure 18.7). As described in the previous section, the substantia nigra pars

Figure 18.6   A chain of nerve cells arranged in a disinhibitory circuit. Top: Diagram of the connections between two inhibitory neurons, A and B, and an excitatory neuron, C. Bottom: Pattern of the action potential activity of cells A, B, and C when A is at rest, and when neuron A fires transiently as a result of its excitatory inputs. Such circuits are central to the gating operations of the basal ganglia.

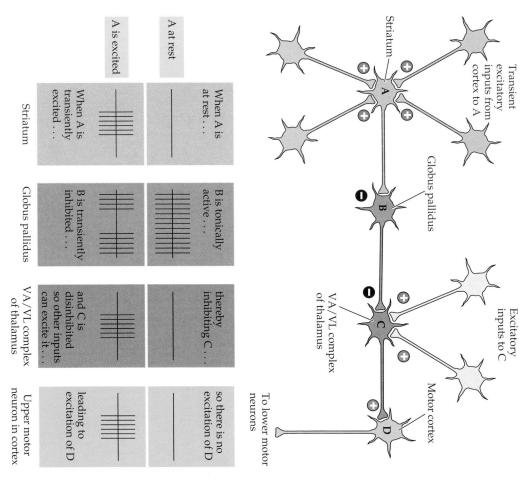

reticulata is part of the output circuitry of the basal ganglia. Instead of projecting to the cortex, however, it sends axons mainly to the deep layers of the superior colliculus. The upper motor neurons in these layers command the rapid orienting movements of the eyes called "saccades" (see Chapter 20). When the eyes are not scanning the environment, these upper motor neurons are tonically inhibited by the spontaneously active reticulata cells to prevent unwanted saccades. Shortly before the onset of a saccade, the tonic discharge rate of the reticulata neurons is sharply reduced by input from the GABAergic medium spiny neurons of the caudate, which have been activated by signals from the cortex. The subsequent reduction in the tonic discharge from reticulata neurons disinhibits the upper motor neurons of the superior colliculus, allowing them to generate the bursts of action potentials that command the saccade. Thus, the projections from substantia nigra pars reticulata to the upper motor neurons act as a physiological "gate" that must be "opened" to allow either sensory or other, more complicated signals from cognitive centers to activate the upper motor neurons and initiate a saccade. Upper motor neurons in the cortex are similarly gated by the basal ganglia but, as discussed earlier, the tonic inhibition is mediated mainly by the GABAergic projection from the internal division of the globus pallidus to the relay cells in the ventral lateral and anterior nuclei of the thalamus (see Figures 18.5 and 18.6).

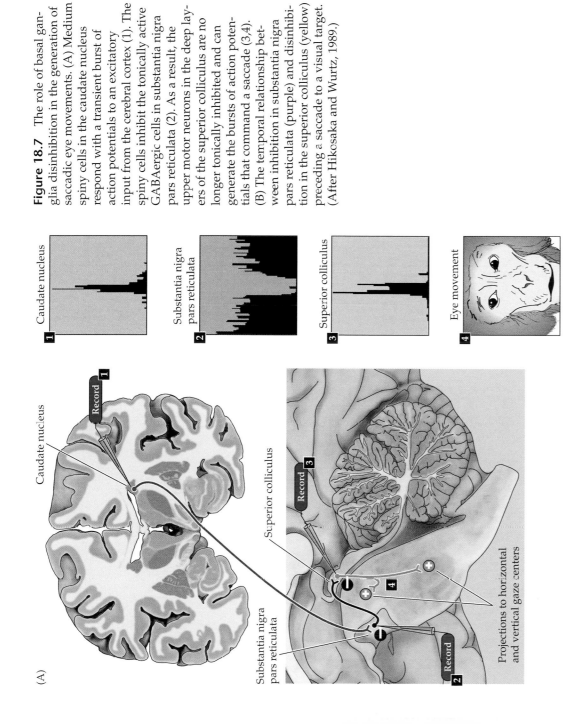

Figure 18.7   The role of basal ganglia disinhibition in the generation of saccadic eye movements. (A) Medium spiny cells in the caudate nucleus respond with a transient burst of action potentials to an excitatory input from the cerebral cortex (1). The spiny cells inhibit the tonically active GABAergic cells in substantia nigra pars reticulata (2). As a result, the upper motor neurons in the deep layers of the superior colliculus are no longer tonically inhibited and can generate the bursts of action potentials that command a saccade (3,4). (B) The temporal relationship between inhibition in substantia nigra pars reticulata (purple) and disinhibition in the superior colliculus (yellow) preceding a saccade to a visual target. (After Hikosaka and Wurtz, 1989.)

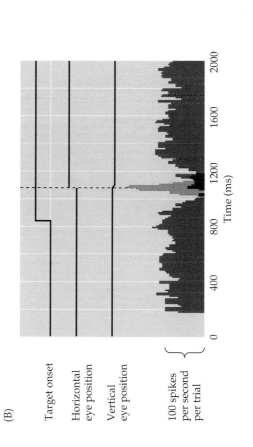

BOX A
Huntington's Disease

In 1872, a physician named George Huntington described a group of patients seen by his father and grandfather in their practice in East Hampton, Long Island. The disease he defined, which became known as Huntington's disease (HD), is characterized by the gradual onset of defects in behavior, cognition, and movement beginning in the fourth and fifth decades of life. The disorder is inexorably progressive, resulting in death within 10 to 20 years. HD is inherited in an autosomal dominant pattern, a feature that has led to a much better understanding of its cause in molecular terms.

One of the more common of the neurodegenerative diseases, HD usually presents as an alteration in mood (especially depression) or a change in character that often takes the form of increased irritability, suspiciousness, and impulsive or eccentric behavior. Defects of memory and attention may also occur. The hallmark of the disease, however, is a movement disorder consisting of rapid, jerky motions with no clear purpose; these *choreiform* movements may be confined to a finger or may involve a whole extremity, the facial musculature, or even the vocal apparatus. The movements themselves are involuntary, but the patient often incorporates them into apparently deliberate actions, presumably in an effort to obscure the problem. There is no weakness, ataxia, or deficit of sensory function. Occasionally, the disease begins in childhood or adolescence. The clinical manifestations in juveniles include rigidity, seizures, more marked dementia, and a rapidly progressive course.

A distinctive neuropathology is associated with these clinical manifestations: a profound but selective atrophy of the caudate and putamen, with some associated degeneration of the frontal and temporal cortices (see Figure 18.9A). This known as Huntington's disease (HD), is pattern of destruction is thought to explain the disorders of movement, cognition, and behavior, as well as the sparing of other neurological functions.

The availability of extensive HD pedigrees has allowed geneticists to decipher the molecular cause of this disease. HD was one of the first human diseases in which DNA polymorphisms were used to localize the mutant gene, which in 1983 was mapped to the short arm of chromosome 4. This discovery led to an intensive effort to identify the HD gene within this region by positional cloning. Ten years later, these efforts culminated in identification of the gene (named Huntingtin) responsible for the disease. In contrast to previously recognized forms of mutations such as point mutations, deletions, or insertions, the mutation of Huntingtin is an unstable triplet repeat. In normal individuals Huntingtin contains between 15 and 34 repeats, whereas the gene in HD patients contains from 42 to over 66 repeats.

HD is one of a growing number of diseases attributed to unstable DNA segments. Other examples are fragile X syndrome, myotonic dystrophy, spinal and bulbar muscular atrophy, and spinocerebellar ataxia type 1. In the latter two and HD, the repeats consist of a DNA segment (CAG) that codes for the amino acid glutamine and is present within the coding region of the gene.

The mechanism by which the increased number of polyglutamine repeats injures neurons is not clear. The leading hypothesis is that the increased numbers of glutamines alter protein folding, which somehow triggers a cascade of molecular events culminating in dysfunction and neuronal death. Interestingly, although Huntingtin is expressed predominantly in the expected neurons in the basal ganglia, it is also present in regions of the brain that are not affected in HD. Indeed, the gene is expressed in many organs outside the nervous system. How and why the mutant Huntingtin uniquely injures striatal neurons is unclear. Continuing to elucidate this molecular pathogenesis will no doubt provide further insight into this and other triplet repeat diseases.

References

GUSELLA, J. F. AND 13 OTHERS (1983) A polymorphic DNA marker genetically linked to Huntington's disease. Nature 306: 234–238.

HUNTINGTON, G. (1872) On chorea. Med. Surg. Reporter 26: 317.

HUNTINGTON'S DISEASE COLLABORATIVE RESEARCH GROUP (1993) A novel gene containing a trinucleotide repeat that is expanded and unstable on Huntington's disease chromosomes. Cell 72: 971–983.

PRICE D. L., S. S. SISODIA AND D. R. BORCHELT (1999) Genetic neurodegenerative diseases: The human illness and transgenic models. Science 282: 1079–1083.

WEXLER, A. (1995) *Mapping Fate: A Memoir of Family, Risk and Genetic Research.* New York: Times Books.

Circuits within the Basal Ganglia System

The projections from the medium spiny neurons of the caudate and putamen to the internal segment of the globus pallidus and substantia nigra pars reticulata are part of a "direct pathway" and, as just described, serve to release the upper motor neurons from tonic inhibition. This pathway is sum-

(A) Direct pathway

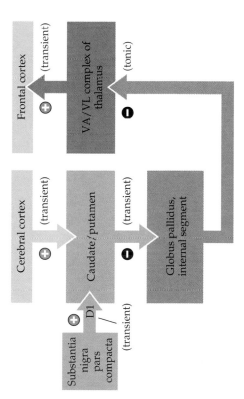

(B) Indirect and direct pathways

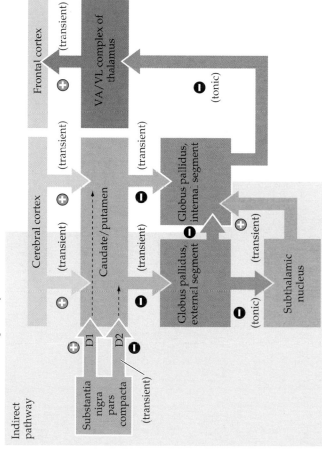

Figure 18.8 Disinhibition in the direct and indirect pathways through the basal ganglia. (A) In the direct pathway, transiently inhibitory projections from the caudate and putamen project to tonically active inhibitory neurons in the *internal* segment of the globus pallidus, which project in turn to the VA/VL complex of the thalamus. Transiently excitatory inputs to the caudate and putamen from the cortex and substantia nigra are also shown, as is the transiently excitatory input from the thalamus back to the cortex. (B) In the indirect pathway (shaded by yellow), transiently active inhibitory neurons from the caudate and putamen project to tonically active inhibitory neurons of the *external* segment of the globus pallidus. Note that the influence of nigral dopaminergic input to neurons in the indirect pathway is inhibitory. The globus pallidus (external segment) neurons project to the subthalamic nucleus, which also receives a strong excitatory input from the cortex. The subthalamic nucleus in turn projects to the globus pallidus (internal segment), where its transiently excitatory drive acts to oppose the disinhibitory action of the direct pathway. In this way, the indirect pathway modulates the effects of the direct pathway.

marized in Figure 18.8A. A second pathway serves to increase the level of tonic inhibition and is called the "indirect pathway" (Figure 18.8B). This pathway provides a second route linking the corpus striatum with the internal globus pallidus and substantia nigra pars reticulata. In the indirect pathway, another population of medium spiny neurons projects to the lateral or **external segment of the globus pallidus**. This external division sends projections to both the internal segment of the globus pallidus and the **subthalamic nucleus** of the ventral thalamus (see Figure 18.1). But, instead of projecting to structures *outside* of the basal ganglia, the subthalamic nucleus projects back to the internal segment of the globus pallidus and to the substantia nigra pars reticulata. As already described, these latter two nuclei project out of the basal ganglia, which thus allows the indirect pathway to influence the activity of the upper motor neurons.

The indirect pathway through the basal ganglia apparently serves to modulate the disinhibitory actions of the direct pathway. The subthalamic nucleus neurons that project to the internal globus pallidus and substantia nigra pars reticulata are excitatory. Normally, when the indirect pathway is activated by signals from the cortex, the medium spiny neurons discharge and inhibit the tonically active GABAergic neurons of the external globus pallidus. As a result, the subthalamic cells become more active and, by virtue of their excitatory synapses with cells of the internal globus pallidus and reticulata, they increase the inhibitory outflow of the basal ganglia. Thus, in contrast to the direct pathway, which when activated releases tonic inhibition, the net effect of activity in the indirect pathway is to increase inhibitory influences on the upper motor neurons. The indirect pathway can thus be regarded as a "brake" on the normal function of the direct pathway. Indeed, many neural systems achieve fine control of their output by a similar interplay between excitation and inhibition.

The consequences of imbalances in this fine control mechanism are apparent in diseases that affect the subthalamic nucleus. These disorders remove a source of excitatory input to the internal globus pallidus and reticulata, and thus abnormally reduce the inhibitory outflow of the basal ganglia. A basal ganglia syndrome called **hemiballismus**, which is characterized by violent, involuntary movements of the limbs, is the result of damage to the subthalamic nucleus. The involuntary movements are initiated by abnormal discharges of upper motor neurons that are receiving less tonic inhibition from the basal ganglia.

Another circuit within the basal ganglia system entails the dopaminergic cells in the pars compacta subdivision of substantia nigra and modulates the output of the corpus striatum. The medium spiny neurons of the corpus striatum project directly to substantia nigra pars compacta, which in turn sends widespread dopaminergic projections back to the spiny neurons. These dopaminergic influences on the spiny neurons are complex: The same nigral neurons can provide excitatory inputs mediated by D1 type dopaminergic receptors on the spiny cells that project to the internal globus pallidus (the direct pathway), and inhibitory inputs mediated by D2 type receptors on the spiny cells that project to the external globus pallidus (the indirect pathway). Since the actions of the direct and indirect pathways on the output of the basal ganglia are antagonistic, these different influences of the nigrostriatal axons produce the same effect, namely a decrease in the inhibitory outflow of the basal ganglia.

The modulatory influences of this second internal circuit help explain many of the manifestations of basal ganglia disorders. For example, **Parkinson's Disease** is caused by the loss of the nigrostriatal dopaminergic neurons (Figure 18.9A and Box B). As mentioned earlier, the normal effects of

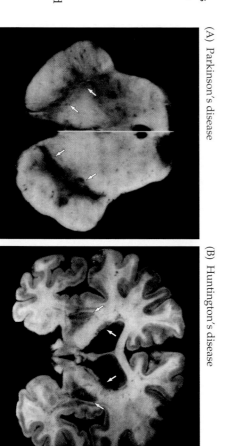

(A) Parkinson's disease

(B) Huntington's disease

Figure 18.9 The pathological changes in certain neurological diseases provide insights about the function of the basal ganglia. (A) Left: The midbrain from a patient with Parkinson's disease. The substantia nigra (pigmented area) is largely absent in the region above the cerebral peduncles (arrows). Right: The mesencephalon from a normal subject, showing intact substantia nigra (arrows). (B) The size of the caudate and putamen (the striatum) (arrows) is dramatically reduced in patients with Huntington's disease. (From Bradley et al., 1991.)

BOX B
Parkinson's Disease: An Opportunity for Novel Therapeutic Approaches

Parkinson's disease is one of the major degenerative diseases of the nervous system. Described by James Parkinson in 1817, this disorder is characterized by tremor at rest, slowness of movement (bradykinesia), rigidity of the extremities and neck, and minimal facial expressions. Walking entails short steps, stooped posture, and a paucity of associated movements (e. g., arm swinging). To make matters worse, in some patients these abnormalities of motor function are associated with dementia. Following a gradual onset between the ages of 50 and 70, the disease progresses slowly and culminates in death 10 to 20 years later.

The defects in motor function are due to the progressive loss of dopaminergic neurons in the substantia nigra pars compacta, a population that projects to and innervates neurons in the caudate and putamen (see text). The cause of the progressive deterioration of these dopaminergic neurons is not known. In contrast to other neurodegenerative diseases such as Alzheimer's disease or amyotrophic lateral sclerosis, in Parkinson's disease the spatial distribution of the degenerating neurons is largely restricted to the substantia nigra pars compacta. This spatial restriction, combined with the defined and relatively homogeneous phenotype of the degenerating neurons (i. e., dopaminergic neurons), has provided an opportunity for novel therapeutic approaches to this disorder.

One strategy is so-called gene therapy. Gene therapy refers to the correction of a disease phenotype through the introduction of new genetic information into the affected organism. Although still in its infancy, this approach promises to revolutionize treatment of human disease. One

therapy for Parkinson's disease would be to enhance release of dopamine in the caudate and putamen. In principle, this could be accomplished by implanting cells genetically modified to express tyrosine hydroxylase, the enzyme that converts tyrosine to L-DOPA, which in turn is converted by a nearly ubiquitous decarboxylase into the neurotransmitter dopamine. The feasibility of this approach has been demonstrated by transplanting tissue derived from the midbrain of human fetuses into the caudate and putamen, which produces longlasting symptomatic improvement in a majority of grafted Parkinson's patients. (The fetal midbrain is enriched in developing neurons that express tyrosine hydroxylase and synthesize and release dopamine.) To date, however, ethical, practical, and political considerations have limited use of fetal transplanted tissue. The effects of transplanting nonneuronal cells genetically modified in vitro to express tyrosine hydroxylase are also being studied in patients with Parkinson's disease, an approach that avoids some of these problems.

An alternative strategy to treating Parkinsonian patients involves "neural grafts" using stem cells. Stem cells are self-renewing multipotent progenitors with broad developmental potential (see Chapters 22 and 25). Instead of isolating mature dopaminergic neurons from the fetal midbrain for transplantation, this approach isolates neuronal progenitors at earlier stages of development when these cells are actively proliferating. Critical to this approach is to prospectively identify and isolate stem cells that are multipotent and self-renewing, and to identify the growth factors needed to promote differ-

entiation into the desired phenotype (e.g., dopaminergic neurons). The prospective identification and isolation of multipotent mammalian stem cells has already been accomplished, and several factors likely to be important in differentiation of midbrain precursors into dopamine neurons have been identified. Establishing the efficacy of this approach for Parkinson's patients would increase the possibility of its application to other neurodegenerative diseases. Although therapeutic strategies like these remain experimental, some of them will very likely succeed.

References

BJÖRKLUND, A. AND U. STENEVI (1979) Reconstruction of the nigrostriatal dopamine pathway by intracerebral nigral transplants. Brain Res. 177: 555–560.

MORRISON, S. J., P. M. WHITE, C. ZOCK AND D. J. ANDERSON (1999) Prospective identification, isolation by flow cytometry, and in vivo self-renewal of multipotent mammalian neural crest stem cells. Cell 96(5): 737–49.

OLANOW, C. W., T. B. FREEMAN AND J. H. KORDOWER (1997) Neural transplantion as a therapy for Parkinson's disease. Adv. Neurol. 74: 249–259.

ROEMER, K. AND T. FRIEDMANN (1992) Concepts and strategies for human gene therapy. Eur. J. Biochem. 208: 211–225.

WIDNER, H. AND 9 OTHERS (1992) Bilateral fetal mesencephalic grafting in two patients with Parkinsonism induced by 1-methyl-4-phenyl-1,2,3,6-tetrahydropyridine (MPTP). N. Engl. J. Med. 327: 1556–1563.

YE, W., K. SHIMAMURA, J. L. RUBENSTEIN, M. A. HYNES AND A. ROSENTHAL (1998) FGF and Shh signals control dopaminergic and serotonergic cell fate in the anterior neural plate. Cell 93(5): 755–66.

ZABNER, J. AND 5 OTHERS (1993) Adenovirusmediated gene transfer transiently corrects the chloride transport defect in nasal epithelia of patients with cystic fibrosis. Cell 75: 207–216.

the compacta input to the striatum are *excitation* of the medium spiny neurons that project directly to the internal globus pallidus and *inhibition* of the spiny neurons that project to the external globus pallidus cells in the indirect pathway. Normally, both of these dopaminergic effects serve to decrease the inhibitory outflow of the basal ganglia and thus to increase the excitability of the upper motor neurons (Figure 18.10A). In contrast, when the compacta cells are destroyed, as occurs in Parkinson's disease, the inhibitory outflow of the basal ganglia is abnormally high, and thalamic activation of upper motor neurons in the motor cortex is therefore less likely to occur.

In fact, many of the symptoms seen in Parkinson's disease (and in other *hypokinetic* movement disorders) reflect a failure of the disinhibition normally mediated by the basal ganglia. Thus, Parkinsonian patients tend to have diminished facial expressions and lack "associated movements" such as arm swinging during walking. Indeed, any movement is difficult to initiate and, once initiated, is often difficult to terminate. Disruption of the same circuits also increases the discharge rate of the inhibitory cells in substantia nigra pars reticulata. The resulting increase in tonic inhibition reduces the excitability of the upper motor neurons in the superior colliculus and causes saccades to be reduced in both frequency and amplitude.

Figure 18.10 Summary explanation of hypokinetic disorders such as Parkinson's disease and hyperkinetic disorders like Huntington's disease. In both cases, the balance of inhibitory signals in the direct and indirect pathways is altered, leading to a diminished ability of the basal ganglia to control the thalamic output to the cortex. (A) In Parkinson's disease, the inputs provided by the substantia nigra are diminished (thinner arrow), making it more difficult to generate the transient inhibition from the caudate and putamen. The result of this change in the direct pathway is to sustain the tonic inhibition from the globus pallidus (internal segment) to the thalamus, making thalamic excitation of the motor cortex less likely (thinner arrow from thalamus to cortex). (B) In hyperkinetic diseases such as Huntington's, the projection from the caudate and putamen to the globus pallidus (external segment) is diminished (thinner arrow). This effect increases the tonic inhibition from the globus pallidus to the subthalamic nucleus (larger arrow), making the excitatory subthalamic nucleus less effective in opposing the action of the direct pathway (thinner arrow). Thus, thalamic excitation of the cortex is increased (larger arrow), leading to greater and often inappropriate motor activity. (After DeLong, 1991.)

(A) Parkinson's disease (hypokinetic)

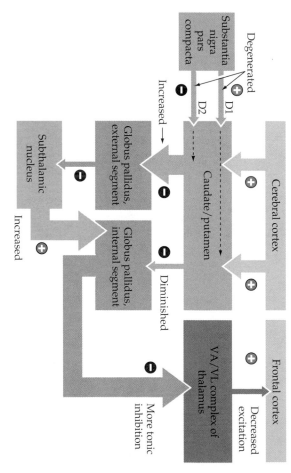

(B) Huntington's disease (hyperkinetic)

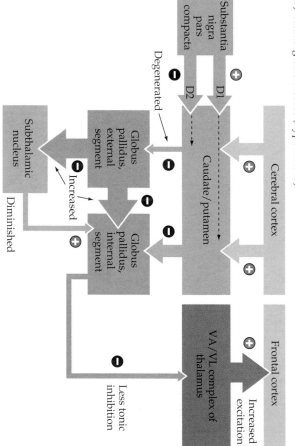

Support for this explanation of hypokinetic movement disorders like Parkinson's disease comes from studies of monkeys in which degeneration of the dopaminergic cells of substantia nigra has been induced by the neurotoxin 1-methyl-4-phenyl-1,2,3,6-tetrahydropyridine (MPTP). Monkeys (or humans) exposed to MPTP develop symptoms that are very similar to those of patients with Parkinson's disease. Furthermore, a second lesion placed in the subthalamic nucleus results in significant improvement in the ability of these animals to initiate movements, as would be expected based on the circuitry of the indirect pathway (see Figure 18.8B).

Similarly, knowledge about the indirect pathway within the basal ganglia helps explain the motor abnormalities seen in Huntington's disease (see Box A). In patients with Huntington's disease, medium spiny neurons that project to the external segment of the globus pallidus degenerate (see Figure 18.9B). In the absence of their normal inhibitory input from the spiny neurons, the external globus pallidus cells become abnormally active; this activity reduces in turn the excitatory output of the subthalamic nucleus to the internal globus pallidus (Figure 18.10B). In consequence, the inhibitory outflow of the basal ganglia is reduced. Without the restraining influence of the basal ganglia, upper motor neurons can be activated by inappropriate signals, resulting in the undesired ballistic and choreic (dancelike) movements that characterize Huntington's disease. Importantly, the basal ganglia may exert a similar influence on other *non-motor* systems with equally significant clinical implications (Box C).

As predicted by this account, GABA agonists and antagonists applied to substantia nigra pars reticulata of monkeys produce symptoms similar to those seen in human basal ganglia disease. For example, intranigral injection of bicuculline, which blocks the GABAergic inputs from the striatal medium spiny neurons to the reticulata cells, increases the amount of tonic inhibition on the upper motor neurons in the deep collicular layers. These animals exhibit fewer, slower saccades, reminiscent of patients with Parkinson's disease. In contrast, injections of the GABA agonist muscimol into substantia nigra pars reticulata decrease the tonic GABAergic inhibition of the upper motor neurons in the superior colliculus, with the result that the injected monkeys generate spontaneous, irrepressible saccades that resemble the involuntary movements characteristic of basal ganglia diseases such as hemiballismus and Huntington's disease (Figure 18.11).

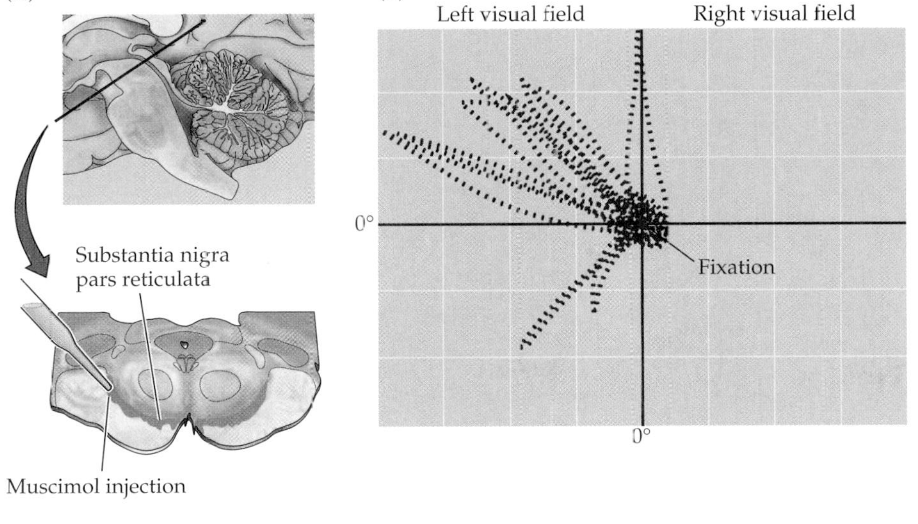

(A)

Substantia nigra pars reticulata

Muscimol injection

(B)

Left visual field   Right visual field

0°

Fixation

0°

Figure 18.11 After the tonically active cells of substantia nigra pars reticulata are inactivated by an intranigral injection of muscimol (A), the upper motor neurons in the deep layers of the superior colliculus are disinhibited and the monkey generates spontaneous irrepressible saccades (B). The cells in both substantia nigra pars reticulata and the deep layers of the superior colliculus are arranged in spatially organized motor maps of saccade vectors (see Chapter 20), and so the direction of the involuntary saccades—in this case toward the upper left quadrant of the visual field—depends on the precise location of the injection site in the substantia nigra.

BOX C
Basal Ganglia Loops and Non-Motor Brain Functions

Traditionally, the basal ganglia have been regarded as motor structures that regulate the initiation of movements. However, they also have functions analogous to their role in motor behavior in other brain systems. Thus, the basal ganglia are also the central structures in anatomical loops similar to the motor loop described in the text, but terminating in non-motor regions of the frontal lobe. These "non-motor" loops include an "executive" loop involving the dorsolateral prefrontal cortex and part of the caudate (see Chapter 26) and a "limbic" loop (see Chapter 29) involving the cingulate cortex and the nucleus accumbens. The similarity of these additional loops to the traditional motor loop suggests that the non-motor regulatory functions of the basal ganglia may be generally the same as what the basal ganglia do in regulating the initiation of movement. For instance, the executive loop may regulate the initiation and termination of cognitive processes such as planning, working memory, and attention. By the same token, the limbic loop

may regulate emotional behavior. Indeed, the deterioration of cognitive and emotional function in both Huntington's disease (see Box A) and Parkinson's disease (see Box B) could be the result of disruption of these non-motor loops.

In fact, a variety of other disorders are now thought to be caused, at least in part, by damage to non-motor components of the basal ganglia. For example, patients with Tourette's syndrome produce inappropriate utterances and obscenities as well as unwanted vocal-motor "tics" and repetitive grunts. These manifestations may be a result of excessive activity in basal ganglia loops that regulate the cognitive circuitry of the prefrontal speech areas. Another example is schizophrenia, which some investigators have argued is associated with excessive activity within the limbic loop, resulting in hallucinations, delusions of persecution, disordered thoughts, and loss of emotional expression. In support of the argument for a basal ganglia contribution to schizophrenia, antipsychotic drugs

such as perphenazine and haloperidol are known to act on the dopaminergic D2 inhibitory receptors on the medium spiny neurons of the striatum's indirect pathway. Still other psychiatric disorders, including obsessive-compulsive disorder, depression, and chronic anxiety, may also involve dysfunctions of the limbic loop. A challenge for future research is therefore to understand more fully the relationships between the clinical problems and other largely unexplored functions of the basal ganglia.

References

ALEXANDER, G. E., M. R. DELONG AND P. L. STRICK (1986) Parallel organization of functionally segregated circuits linking basal ganglia and cortex. Ann. Rev. Neurosci. 9: 357–381.

BHATIA, K. P. AND C. D. MARSDEN (1994) The behavioral and motor consequences of focal lesions of the basal ganglia in man. Brain 117: 859–876.

DREVETS, W. C. AND 6 OTHERS (1997) Subgenual prefrontal cortex abnormalities in mood disorders. Nature 386: 824–827.

GRAYBIEL, A. M. (1997) The basal ganglia and cognitive pattern generators. Schiz. Bull. 23: 459–469.

JENIKE, M. A., L. BAER AND W. E. MINICHIELLO (1990) Obsessive Compulsive Disorders: Theory and Management. Chicago: Year Book Medical Publishers, Inc.

MIDDLETON, F. A. AND P. L. STRICK (2000) Basal ganglia output and cognition: Evidence from anatomical, behavioral, and clinical studies. Brain Cogn. 42: 183–200.

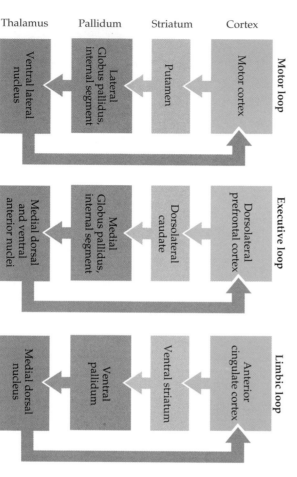

Thalamus	Pallidum	Striatum	Cortex

Motor loop

Ventral lateral nucleus ← Lateral Globus pallidus, internal segment ← Putamen ← Motor cortex

Executive loop

Medial dorsal and ventral anterior nuclei ← Medial Globus pallidus, internal segment ← Dorsolateral caudate ← Dorsolateral prefrontal cortex

Limbic loop

Medial dorsal nucleus ← Ventral pallidum ← Ventral striatum ← Anterior cingulate cortex

Comparison of motor and non-motor basal ganglia loops (only the direct pathways are indicated).

Summary

The contribution of the basal ganglia to motor control is apparent from the deficits that result from damage to the component nuclei. Such lesions compromise the initiation and performance of voluntary movements, as exemplified by the paucity of movement in Parkinson's disease and in the inappropriate "release" of movements in Huntington's disease. The organization of the basic circuitry of the basal ganglia indicates how this constellation of nuclei modulates movement. With respect to motor function, the system forms a loop that originates in almost every area of the cerebral cortex and eventually terminates, after enormous convergence within the basal ganglia, on the upper motor neurons in the motor and premotor areas of the frontal lobe and the superior colliculus. The efferent neurons of the basal ganglia influence the upper motor neurons in the cortex by gating the flow of information through relays in the ventral nuclei of the thalamus. The upper motor neurons in the superior colliculus that initiate saccadic eye movements are controlled by monosynaptic projections from substantia nigra pars reticulata. In each case, the basal ganglia loop regulates movement by a process of disinhibition that results from the serial interaction within the basal ganglia circuitry of two GABAergic neurons. Internal circuits within the basal ganglia system modulate the amplification of the signals that are transmitted through the loop.

ADDITIONAL READING

Reviews

ALEXANDER, G. E. AND M. D. CRUTCHER (1990) Functional architecture of basal ganglia circuits: Neural substrates of parallel processing. Trends Neurosci. 13: 266–271.

DELONG, M. R. (1990) Primate models of movement disorders of basal ganglia origin. Trends Neurosci. 13: 281–285.

GERFEN, C. R. AND C. J. WILSON (1996) The basal ganglia. In *Handbook of Chemical Neuroanatomy,* Vol. 12: *Integrated Systems of the CNS,* Part III. L. W. Swanson, A. Björklund and T. Hokfelt (eds.). New York: Elsevier Science Publishers, pp. 371–468.

GOLDMAN-RAKIC, P. S. AND L. D. SELEMON (1990) New frontiers in basal ganglia research. Trends Neurosci. 13: 241–244.

GRAYBIEL, A. M. AND C. W. RAGSDALE (1983) Biochemical anatomy of the striatum. In *Chemical Neuroanatomy,* P. C. Emson (ed.). New York: Raven Press, pp. 427–504.

HIKOSAKA, O. AND R. H. WURTZ (1989) The basal ganglia. In *The Neurobiology of Eye Movements,* R. H. Wurtz and M. E. Goldberg (eds.). New York: Elsevier Science Publishers, pp. 257–281.

MINK, J. W. AND W. T. THACH (1993) Basal ganglia intrinsic circuits and their role in behavior. Curr. Opin. Neurobiol. 3: 950–957.

WILSON, C. J. (1990) Basal ganglia. In *Synaptic Organization of the Brain.* G. M Shepherd (ed.). Oxford: Oxford University Press, Chapter 9.

Important Original Papers

ANDEN, N.-E., A. DAHLSTROM, K. FUXE, K. LARSSON, K. OLSON AND U. UNGERSTEDT (1966) Ascending monoamine neurons to the telencephalon and diencephalon. Acta Physiol. Scand. 67: 313–326.

BRODAL, P. (1978) The corticopontine projection in the rhesus monkey: Origin and principles of organization. Brain 101: 251–283.

CRUTCHER, M. D. AND M. R. DELONG (1984) Single cell studies of the primate putamen. Exp. Brain Res. 53: 233–243.

DELONG, M. R. AND P. L. STRICK (1974) Relation of basal ganglia, cerebellum, and motor cortex units to ramp and ballistic movements. Brain Res. 71: 327–335.

DIFIGLIA, M., P. PASIK AND T. PASIK (1976) A Golgi study of neuronal types in the neostriatum of monkeys. Brain Res. 114: 245–256.

KEMP, J. M. AND T. P. S. POWELL (1970) The cortico-striate projection in the monkey. Brain 93: 525–546.

KIM, R., K. NAKANO, A. JAYARAMAN AND M. B. CARPENTER (1976) Projections of the globus pallidus and adjacent structures: An autoradiographic study in the monkey. J. Comp. Neurol. 169: 217–228.

KOCSIS, J. D., M. SUGIMORI AND S. T. KITAI (1977) Convergence of excitatory synaptic inputs to caudate spiny neurons. Brain Res. 124: 403–413.

SMITH, Y., M. D. BEVAN, E. SHINK AND J. P. BOLAM (1998) Microcircuitry of the direct and indirect pathways of the basal ganglia. Neurosci. 86: 353–387.

Books

BRADLEY, W. G., R. B. DAROFF, G. M. FENICHEL AND C. D. MARSDEN (eds.) (1991) *Neurology in Clinical Practice.* Boston: Butterworth-Heinemann, Chapters 29 and 77.

KLAWANS, H. L. (1989) *Toscanini's Fumble and Other Tales of Clinical Neurology.* New York: Bantam, Chapters 7 and 10.

Chapter 19

Modulation of Movement by the Cerebellum

Overview

In contrast to the upper motor neurons described in Chapter 17, the efferent cells of the cerebellum do not project directly either to the local circuits of the brainstem and spinal cord that organize movement, or to the lower motor neurons that innervate muscles. Instead—like the basal ganglia—the cerebellum influences movements by modifying the activity patterns of the upper motor neurons. In fact, the cerebellum sends prominent projections to virtually all upper motor neurons. Structurally, the cerebellum has two main components: a laminated cerebellar cortex, and a subcortical cluster of cells referred to collectively as the deep cerebellar nuclei. Pathways that reach the cerebellum from other brain regions (predominantly the cerebral cortex) project to both these components; thus, the afferent axons send branches to both the deep nuclei and the cerebellar cortex. The output cells of the cerebellar cortex project to the deep cerebellar nuclei, which give rise to the main efferent pathways that leave the cerebellum to regulate upper motor neurons in the cerebral cortex, brainstem, and spinal cord. Thus, much like the basal ganglia, the cerebellum is part of a vast loop that receives projections from and sends projections back to the cerebral cortex, brainstem, and spinal cord. The primary function of the cerebellum is evidently to detect the difference, or "motor error," between an intended movement and the actual movement, and, through its projections to the upper motor neurons, to reduce the error. These corrections can be made both during the course of the movement and as a form of motor learning when the correction is stored. When this feedback loop is damaged, as occurs in many cerebellar diseases, the patient makes persistent movement errors whose specific character depends on the location of the damage.

Organization of the Cerebellum

The **cerebellum** can be subdivided into three main parts based on differences in the sources of input (Figure 19.1 and Table 19.1). By far, the largest subdivision in humans is the **cerebrocerebellum**. It occupies most of the lateral cerebellar hemisphere and receives input from many areas of the cerebral cortex. This region of the cerebellum is especially well developed in primates. The cerebrocerebellum is concerned with the regulation of highly skilled movements, especially the planning and execution of complex spatial and temporal sequences of movement (including speech). The phylogenetically oldest part of the cerebellum is the **vestibulocerebellum**. This portion of the cerebellum comprises the caudal lobes of the cerebellum and includes the **flocculus** and the **nodulus**. As its name suggests, the vestibulocerebel-

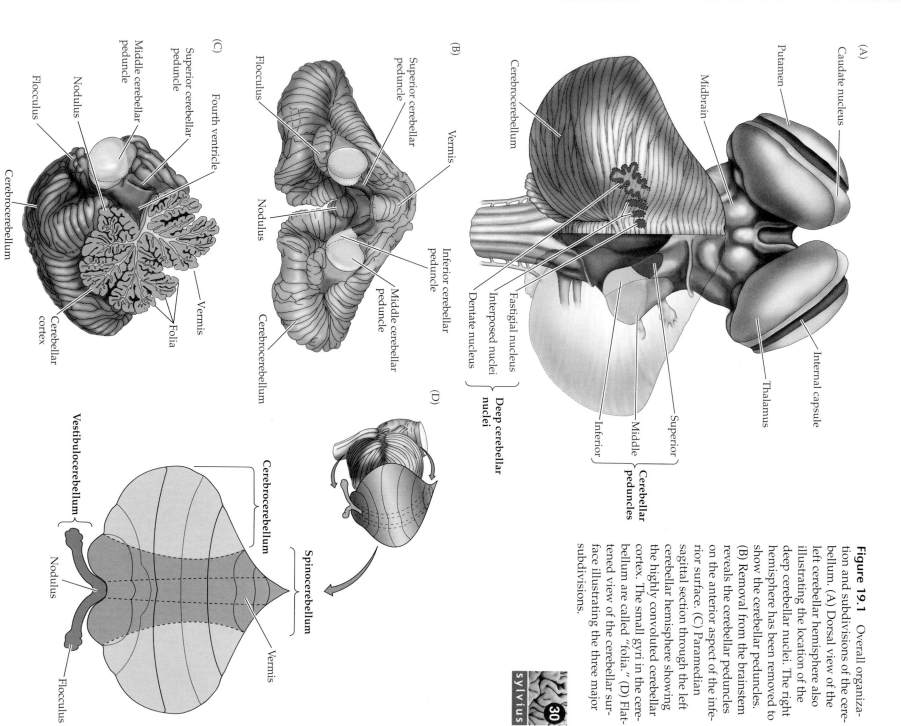

(A)

Caudate nucleus

Putamen

Midbrain

Internal capsule

Thalamus

Cerebrocerebellum

Fastigial nucleus
Interposed nuclei
Dentate nucleus
} **Deep cerebellar nuclei**

Inferior
Middle
Superior
} **Cerebellar peduncles**

(B)

Cerebrocerebellum

Superior cerebellar peduncle

Vermis

Nodulus

Flocculus

Inferior cerebellar peduncle

Middle cerebellar peduncle

Cerebrocerebellum

(C)

Flocculus

Nodulus

Middle cerebellar peduncle

Superior cerebellar peduncle

Fourth ventricle

Cerebrocerebellum

Vermis

Folia

Cerebellar cortex

(D)

Vestibulocerebellum

Cerebrocerebellum

Spinocerebellum

Nodulus

Flocculus

Vermis

sylvius **30**

Figure 19.1 Overall organization and subdivisions of the cerebellum. (A) Dorsal view of the left cerebellar hemisphere also illustrating the location of the deep cerebellar nuclei. The right hemisphere has been removed to show the cerebellar peduncles. (B) Removal from the brainstem reveals the cerebellar peduncles on the anterior aspect of the inferior surface. (C) Paramedian sagittal section through the left cerebellar hemisphere showing the highly convoluted cerebellar cortex. The small gyri in the cerebellum are called "folia." (D) Flattened view of the cerebellar surface illustrating the three major subdivisions.

lum receives input from the vestibular nuclei in the brainstem and is primarily concerned with the regulation of movements underlying posture and equilibrium. The last of the major subdivisions is the **spinocerebellum**. The spinocerebellum occupies the median and paramedian zone of the cerebellar hemispheres and is the only part that receives input directly from the spinal cord. The lateral part of the spinocerebellum is primarily concerned with movements of distal muscles, such as the relatively gross movements of the limbs in walking. The central part, called the **vermis**, is primarily concerned with movements of proximal muscles, and also regulates eye movements in response to vestibular inputs.

The connections between the cerebellum and other parts of the nervous system occur by way of three large pathways called **cerebellar peduncles** (Figures 19.1 to 19.3). The **superior cerebellar peduncle** (or brachium conjunctivum) is almost entirely an efferent pathway. The neurons that give rise to this pathway are in the deep cerebellar nuclei, and their axons project to upper motor neurons in the red nucleus, the deep layers of the superior colliculus, and, after a relay in the dorsal thalamus, the primary motor and premotor areas of the cortex (see Chapter 17). The **middle cerebellar peduncle** (or brachium pontis) is an afferent pathway to the cerebellum; most of the cell bodies that give rise to this pathway are in the base of the pons, where they form the **pontine nuclei** (Figure 19.2). The pontine nuclei receive input from a wide variety of sources, including almost all areas of the cerebral cortex and the superior colliculus. The axons of the pontine nuclei, called

TABLE 19.1
Major Components of the Cerebellum

Cerebellar cortex:
Cerebrocerebellum
Spinocerebellum
Vestibulocerebellum

Deep cerebellar nuclei:
Dentate nucleus
Interposed nuclei
Fastigial rucleus

Cerebellar peduncles
Superior peduncle
Middle peduncle
Inferior peduncle

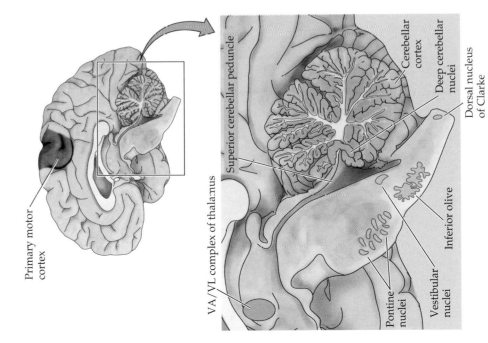

Primary motor cortex

VA/VL complex of thalamus

Superior cerebellar peduncle

Cerebellar cortex

Deep cerebellar nuclei

Dorsal nucleus of Clarke

Pontine nuclei

Vestibular nuclei

Inferior olive

Figure 19.2 Components of the brainstem and diencephalon related to the cerebellum. This sagittal section shows the major structures of the cerebellar system, including the cerebellar cortex, the deep cerebellar nuclei, and the ventroanterior and ventrolateral (VA/VL) complex (which is the target of some of the deep cerebellar nuclei).

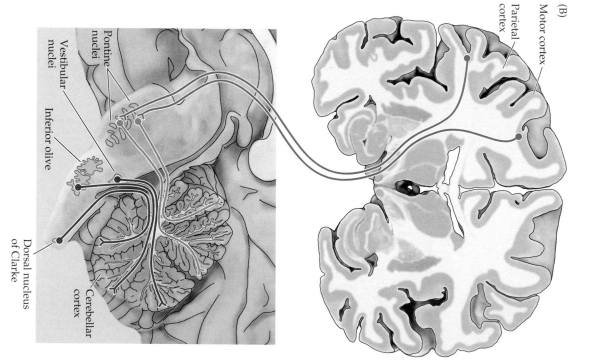

Figure 19.3 Functional organization of the inputs to the cerebellum. (A) Diagram of the major inputs. (B) Idealized coronal and sagittal sections through the human brainstem and cerebrum, showing inputs to the cerebellum from the cortex, vestibular system, spinal cord, and brainstem. The cortical projections to the cerebellum are made via relay neurons in the pons. These axons then cross the midline within the pons and run to the cerebellum via the middle cerebellar peduncle. Axons from the inferior olive, spinal cord, and vestibular nuclei enter via the inferior cerebellar peduncle.

TABLE 19.2
Major Inputs to the Cerebellum (via Inferior and Middle Cerebellar Peduncles)

From cerebral cortex:
Parietal cortex (secondary visual, primary and secondary somatic sensory)
Cingulate cortex (limbic)
Frontal cortex (primary and secondary motor)

Other sources:
Red nucleus
Superior colliculus
Spinal cord (Clarke's column)
Vestibular labyrinth and nuclei
Reticular formation
Inferior olivary nucleus
Locus ceruleus

transverse pontine fibers, cross the midline and enter the cerebellum via the middle cerebellar peduncle (Figure 19.3). Each of the two middle cerebellar peduncles contains over 20 million axons and are thus among the largest pathways in the brain. In comparison, the optic and pyramidal tracts contain only about a million axons. Most of these pontine axons relay information from the cortex to the cerebellum. Finally, the **inferior cerebellar peduncle** (or restiform body) is the smallest but most complex of the cerebellar peduncles, containing multiple afferent and efferent pathways. Efferent pathways in this peduncle project to the vestibular nuclei and the reticular formation; the afferent pathways include axons from the vestibular nuclei, the spinal cord, and several regions of the brainstem tegmentum.

Projections to the Cerebellum

The cerebral cortex is by far the largest source of inputs to the cerebellum, the major destination being the cerebrocerebellum (see Figure 19.3 and Table 19.2). These pathways arise from a somewhat more circumscribed area of the

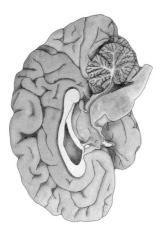

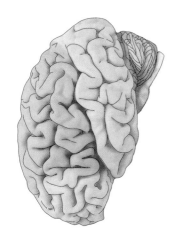

Figure 19.4 Regions of the cerebral cortex that project to the cerebellum (shown in green). The cortical projections to the cerebellum are mainly from the sensory association cortex of the parietal lobe and motor association areas of the frontal lobe.

cortex than do those to the basal ganglia (see Chapter 18). The majority originate in the primary motor and premotor cortices of the frontal lobe, the primary and secondary somatic sensory cortices of the anterior parietal lobe, and the secondary visual regions of the posterior parietal lobe (Figure 19.4). The visual input to the cerebellum originates mostly in association areas concerned with processing moving stimuli (i.e., the cortical targets of the magnocellular stream; see Chapter 12). Indeed, visually guided coordination of ongoing movement is one of the major tasks carried out by the cerebrocerebellum. Most of these cortical pathways relay in the pontine nuclei before entering the cerebellum (see Figure 19.3).

There are also direct sensory inputs to the cerebellum (see Figure 19.3 and Table 19.2). Vestibular axons from the eighth cranial nerve and axons from the vestibular nuclei in the medulla project to the vestibulcerebellum. In addition, relay neurons in the **dorsal nucleus of Clarke** in the spinal cord (a group of relay neurons innervated by proprioceptive axons from the periphery; see Chapter 9) send their axons to the spinocerebellum. The vestibular and spinal inputs provide the cerebellum with information from the labyrinth in the ear, from muscle spindles, and from other mechanoreceptors that monitor the position and motion of the body. The somatic sensory input remains topographically mapped in the spinocerebellum such that there are orderly representations of the body surface within the cerebellum (Figure 19.5). These maps are "fractured," however: Fine-grain electrophysiological analysis indicates that each small area of the body surface is represented multiple times by spatially separated clusters of cells rather than by a continuous

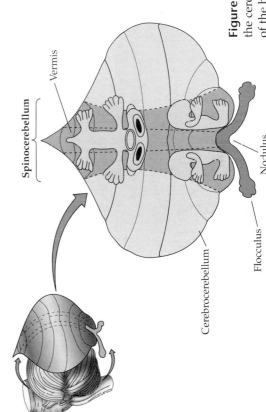

Figure 19.5 Somatotopic maps of the body surface in the cerebellum. The spinocerebellum contains two maps of the body.

Spinocerebellum

Vermis

Cerebrocerebellum

Flocculus

Nodulus

topographic map of the body surface. The vestibular and spinal inputs remain ipsilateral from their point of entry in the brainstem, traveling in the inferior cerebellar peduncle (see Figure 19.3B). This arrangement indicates that, in contrast to most areas of the brain, the right cerebellum is concerned with the right half of the body and the left cerebellum with the left half.

Finally, the entire cerebellum receives modulatory inputs from the **inferior olive** and the locus ceruleus in the brainstem. These nuclei evidently participate in the learning and memory functions served by cerebellar circuitry (see p. 417).

Projections from the Cerebellum

Except for a direct projection from the vestibulocerebellum to the vestibular nuclei, the cerebellar cortex projects to the deep cerebellar nuclei, which project in turn to upper motor neurons in the cortex (via a relay in the thalamus) and spinal cord (via relays in the brainstem) (Figure 19.6 and Table 19.3). There are four major deep nuclei: the **dentate nucleus** (by far the

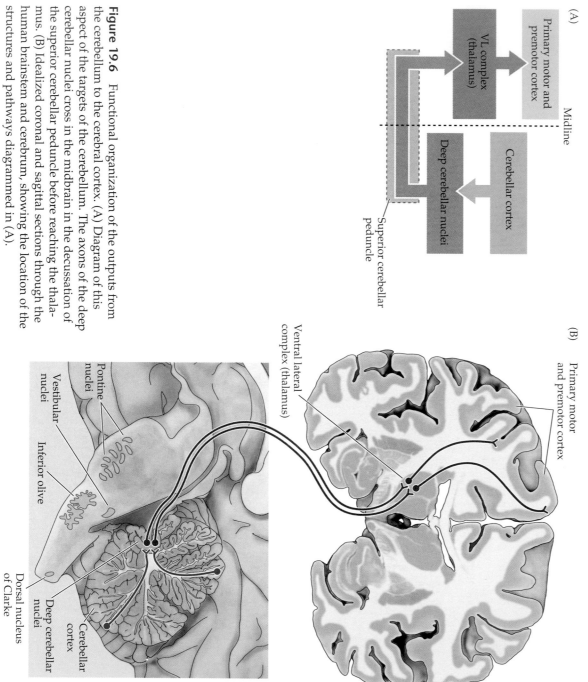

(A)

Primary motor and premotor cortex

VL complex (thalamus)

Midline

Deep cerebellar nuclei

Cerebellar cortex

Superior cerebellar peduncle

(B)

Primary motor and premotor cortex

Ventral lateral complex (thalamus)

Pontine nuclei

Vestibular nuclei

Inferior olive

Dorsal nucleus of Clarke

Deep cerebellar nuclei

Cerebellar cortex

Figure 19.6   Functional organization of the outputs from the cerebellum to the cerebral cortex. (A) Diagram of this aspect of the cerebellum. The axons of the deep cerebellar nuclei cross in the midbrain in the decussation of the superior cerebellar peduncle before reaching the thalamus. (B) Idealized coronal and sagittal sections through the human brainstem and cerebrum, showing the location of the structures and pathways diagrammed in (A).

largest), two **interposed nuclei**, and the **fastigial nucleus**. Each receives input from a different region of the cerebellar cortex. Although the borders are not distinct, in general, the cerebrocerebellum projects primarily to the dentate nucleus, the spinocerebellum to the interposed nuclei, and the vestibulocerebellum to the fastigial nucleus. Axons from the dentate nucleus are destined for the cortex via a projection to the ventral nuclear complex in the thalamus. This pathway must cross the midline if the motor cortex in each hemisphere, which is concerned with contralateral musculature, is to receive information from the cerebellum about the appropriate side of the body. Consequently, the dentate axons exit the cerebellum via the superior cerebellar peduncle, cross at the **decussation of the superior cerebellar peduncle** in the caudal midbrain, and then ascend to the thalamus.

The thalamic nuclei that receive projections from the deep cerebellar nuclei are segregated in two distinct subdivisions of the ventral lateral nuclear complex: the oral, or anterior, part of the posterolateral segment, and a region simply called area X. Both of these thalamic relays project directly to primary motor and premotor association cortices. Thus, the cerebellum has access to the upper motor neurons that organize the sequence of muscular contractions underlying complex voluntary movements. Pathways leaving the deep cerebellar nuclei also project to upper motor neurons in the red nucleus, the superior colliculus, the vestibular nuclei, and the reticular formation (see Table 19.3 and Chapter 17).

Anatomical studies using viruses to trace chains of connections between nerve cells have shown that large parts of the cerebrocerebellum send information back to non-motor areas of the cortex to form "closed loops." That is, a region of the cerebellum projects back to the same cortical area that in turn projects to it. These closed loops run in parallel to "open loops" that receive input from multiple cortical areas and funnel output back to upper motor neurons in specific regions of the motor and premotor cortices (Figure 19.7).

Circuits within the Cerebellum

The ultimate destination of the afferent pathways to the cerebellar cortex is a distinctive cell type called the **Purkinje cell** (Figure 19.8). However, the input from the cerebral cortex to the Purkinje cells is quite indirect. Neurons in the pontine nuclei receive a projection from the cerebral cortex and then relay the information to the contralateral cerebellar cortex. The axons from the pontine nuclei and other sources are called **mossy fibers** because of the appearance of their synaptic terminals. Mossy fibers synapse on **granule cells** in the granule cell layer of the cerebellar cortex (see Figures 19.8 and 19.9). The cerebellar granule cells are widely held to be the most abundant class of neurons in the human brain. They give rise to specialized axons called **parallel fibers** that ascend to the **molecular layer** of the cerebellar cortex. The parallel fibers bifurcate in the molecular layer to form T-shaped branches that relay information via excitatory synapses onto the dendritic spines of the Purkinje cells.

The Purkinje cells present the most striking histological feature of the cerebellum. Elaborate dendrites extend into the molecular layer from a single subjacent layer of these giant nerve cell bodies (called the Purkinje layer). Once in the molecular layer, the Purkinje cell dendrites branch extensively in a plane at right angles to the trajectory of the parallel fibers (Figure 19.8A). In this way, each Purkinje cell is in a position to receive input from a large number of parallel fibers, and each parallel fiber can contact a very large number of Purkinje cells (on the order of tens of thousands). The Purkinje cells receive a direct modulatory input on their dendritic shafts from the **climbing fibers**, all of which arise in the inferior olive (Figure 19.8B). Each Purkinje cell receives

TABLE 19.3
Output Targets of the Cerebellum

Red nucleus
Vestibular nuclei
Superior colliculus
Reticular formation
Motor cortex (via relay in ventral lateral nuclei of thalamus)

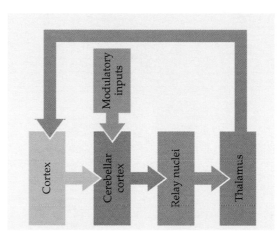

Figure 19.7 Summary diagram of motor modulation by the cerebrocerebellum. The central processing component, the cerebrocerebellar cortex, receives massive input from the cerebral cortex and generates signals that adjust the responses of upper motor neurons to regulate the course of a movement. Note that modulatory inputs also influence the processing of information within the cerebellar cortex. The output signals from the cerebellar cortex are relayed to the thalamus and then back to the motor cortex, where they modulate the motor commands.

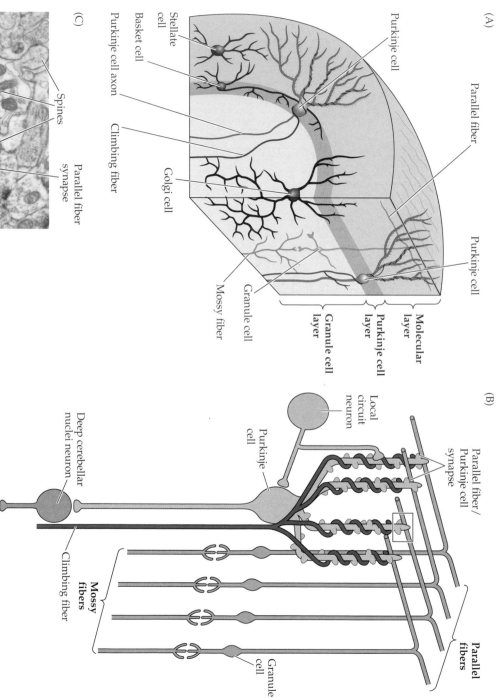

(A)

Stellate cell
Basket cell
Purkinje cell axon
Purkinje cell
Parallel fiber
Golgi cell
Climbing fiber
Granule cell
Mossy fiber
Parallel fiber
Purkinje cell
Molecular layer
Purkinje cell layer
Granule cell layer

(C)

Spines
Parallel fiber synapse
Purkinje cell dendrite

(B)

Parallel fiber/ Purkinje cell synapse
Local circuit neuron
Purkinje cell
Deep cerebellar nuclei neuron
Climbing fiber
Mossy fibers
Granule cell
Parallel fibers

Figure 19.8 Neurons and circuits of the cerebellum. (A) Neuronal types in the cerebellar cortex. Note that the various neuron classes are found in distinct layers. (B) Diagram showing convergent inputs onto the Purkinje cell from parallel fibers and local circuit neurons [boxed region shown at higher magnification in (C)]. The output of the Purkinje cells is to the deep cerebellar nuclei. (C) Electron micrograph showing Purkinje cell dendritic shaft with three spines contacted by synapses from a trio of parallel fibers. (C courtesy of A.-S. La Mantia and P. Rakic.)

numerous synaptic contacts from a single climbing fiber. In most models of cerebellum function, the climbing fibers regulate movement by modulating the effectiveness of the mossy–parallel fiber connection with the Purkinje cells.

The Purkinje cells project in turn to the deep cerebellar nuclei. They are the only output cells of the cerebellar cortex. Since the Purkinje cells are GABAergic, the output of the cerebellar cortex is wholly inhibitory. However, the deep cerebellar nuclei receive excitatory input from the collaterals of the mossy and climbing fibers. The Purkinje cell inhibition of the deep nuclei serves to modulate the level of this excitation (Figure 19.9).

Inputs from local circuit neurons modulate the inhibitory activity of Purkinje cells and occur on both dendritic shafts and the cell body. The most powerful of these local inputs are inhibitory complexes of synapses made around the Purkinje cell bodies by **basket cells** (see Figure 19.8A,B). Another type of local circuit neuron, the **stellate cell**, receives input from the parallel fibers and provides an inhibitory input to the Purkinje cell dendrites. Finally, the molecular layer contains the apical dendrites of a cell type called **Golgi cells**; these neurons have their cell bodies in the granular cell layer. The Golgi cells receive input from the parallel fibers and provide an inhibitory feedback to the cells of origin of the parallel fibers (the granule cells).

This basic circuit is repeated over and over throughout every subdivision of the cerebellum in all mammals and is the fundamental functional module of the cerebellum. Modulation of signal flow through these modules provides the basis for both real-time regulation of movement and the long-term changes in regulation that underlie motor learning. The flow of signals through this admittedly complex intrinsic circuitry is best described in reference to the Purkinje cells (see Figure 19.9). The Purkinje cells receive two types of excitatory input from outside of the cerebellum, one directly from the climbing fibers and the other indirectly via the parallel fibers of the granule cells. The Golgi, stellate, and basket cells control the flow of information through the cerebellar cortex. For example, the Golgi cells form an inhibitory feedback that may limit the duration of the granule cell input to the Purkinje cells, whereas the basket cells provide lateral inhibition that may focus the spatial distribution of Purkinje cell activity. The Purkinje cells modulate the activity of the deep cerebellar nuclei, which are driven by the direct excitatory input they receive from the collaterals of the mossy and climbing fibers.

The modulation of cerebellar output also occurs at the level of the Purkinje cells (see Figure 19.9). This latter modulation may be responsible for

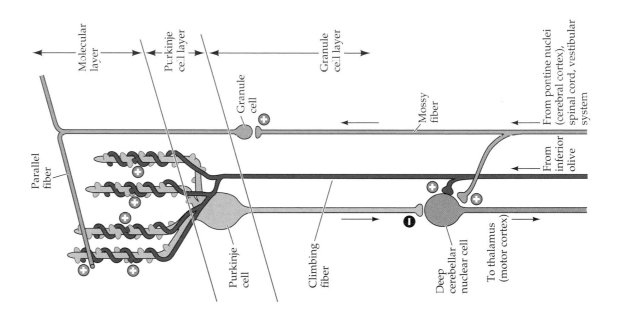

Figure 19.9 Excitatory and inhibitory connections in the cerebellar cortex and deep cerebellar nuclei. The excitatory input from mossy fibers and climbing fibers to Purkinje cells and deep nuclear cells is basically the same. Additional convergent input onto the Purkinje cell from local circuit neurons (basket and stellate cells) and other Purkinje cells establishes a basis for the comparison of ongoing movement and sensory feedback derived from it. The Purkinje cell output to the deep cerebellar nuclear cell thus generates an error correction signal that can modify movements already begun. The climbing fibers modify the efficacy of the parallel fiber–Purkinje cell connection, producing long-term changes in cerebellar output. (After Stein, 1986.)

Parallel fiber

Molecular layer

Purkinje cell layer

Granule cell layer

Granule cell

Mossy fiber

Purkinje cell

Climbing fiber

Deep cerebellar nuclear cell

To thalamus (motor cortex)

From inferior olive

From pontine nuclei (cerebral cortex), spinal cord, vestibular system

BOX A
Prion Diseases

Creutzfeldt-Jakob disease (CJD) is a rare but devastating neurological disorder characterized by cerebellar ataxia, myoclonic jerks, seizures, and the fulminant progression of dementia. The onset is usually in middle age, and death typically follows within a year. The distinctive histopathology of the disease, termed "spongiform degeneration," consists of neuronal loss and extensive glial proliferation, mainly in the cortex of the cerebellum and cerebrum; the peculiar spongiform pattern is due to vacuoles in the cytoplasm of neurons and glia. CJD is the only human disease known to be transmitted by inoculation (either orally or into the bloodstream) *or* inherited through the germline! In contrast to other transmissible diseases mediated by microorganisms such as viruses or bacteria, the agent in this case is a protein called a prion.

Observations dating back some 30 years suggested that CJD was infective. The major clue was a once obscure disease of sheep, called scrapie, which is also characterized by cerebellar ataxia, wasting, and intense itching. The ability to transmit scrapie from one sheep to another strongly suggested an infectious agent. Another clue came from the work of Carlton Gajdusek, a neurologist studying a peculiar human disease called kuru that occurred specifically in a group of New Guinea natives known to practice ritual cannibalism. Like CJD, kuru is a neurodegenerative disease characterized by devastating cerebellar ataxia and subsequent dementia, usually leading to death within a year. The striking similarities in the distinctive histopathology of scrapie and kuru, namely spongiform degeneration, suggested a common pathogenesis and led to the successful transmission of kuru to apes and chimpanzees in the 1960s, confirming that CJD was indeed infectious. The pro-

longed period between inoculation and disease onset (months to years) led Gajdusek to suggest that the transmissible agent was what he called a "slow virus." These extraordinary findings spurred an intensive search for the infectious agent. The transmission of scrapie from sheep to hamsters by Stanley Prusiner at the University of California at San Francisco permitted biochemical characterization of partially purified fractions of scrapie agent from hamster brain. Oddly, he found that the infectivity was extraordinarily resistant to ultraviolet irradiation or nucleases, treatments that degrade nucleic acids. It therefore seemed unlikely that a virus could be the causal agent. Conversely, procedures that modified or degraded proteins markedly diminished infectivity. In 1982, Prusiner coined the term "prion" to refer to the agent causing these transmissible spongiform encephalopathies. He chose the term to emphasize that the agent was a proteinaceous infectious particle (making the abbreviation a little more euphonious in the process.) Subsequently, a half dozen more diseases of animals—including mad cow disease—and four more human diseases have been shown to be caused by prions.

Whether prions contain undetected nucleic acids or are really proteins only remained controversial for some years. Prusiner strongly advocated a "protein only" hypothesis, a revolutionary concept with respect to transmissible diseases. He proposed that the prion is a protein consisting of a modified (scrapie) form (PrP^{Sc}) of the normal host protein (PrP^{C} for "prion protein control"), the propagation of which occurs by a conformational change of endogenous PrP^{C} to PrP^{Sc}, autocatalyzed by PrP^{Sc}. That is, the modified form of the protein (PrP^{Sc}) transforms the normal form (PrP^{C}) into the modified form, much as crystals

form in supersaturated solutions. Differences in the secondary structure of PrP^{C} and PrP^{Sc} evident by optical spectroscopy supported this idea. An alternative hypothesis, however, was that the agent is simply an unconventional nucleic acid–containing virus, and that the accumulation of PrP^{Sc} is an incidental consequence of infection and cell death.

A compelling body of evidence in support of the "protein only" hypothesis has emerged only in the past decade. First, PrP^{C} and scrapie infectivity copurify by a number of procedures, including affinity chromatography using an anti-PrP monoclonal antibody; no nucleic acid has been detected in highly purified preparations, despite intensive efforts. Second, spongiform encephalopathies can be inherited in humans, and the cause is now known to be a mutation (or mutations) in the gene coding for PrP. Third, transgenic mice carrying a mutant PrP gene equivalent to one of the mutations of inherited human prion disease develop a spongiform encephalopathy. Thus, a defective protein is sufficient to account for the disease. Finally, transgenic mice carrying a null mutation for PrP do not develop spongiform encephalopathy when inoculated with scrapie agent, whereas wild-type mice do. These results argue convincingly that PrP^{C} must indeed interact with endogenous PrP^{C} to convert PrP^{C} to PrP^{Sc}, propagating the disease in the process. The protein is highly conserved across mammalian species, suggesting that it serves some essential function, although mice carrying a null mutation of PrP exhibit no detectable abnormalities.

These advances notwithstanding, many questions remain. What is the mechanism by which the conformational transformation of PrP^{C} to PrP^{Sc} occurs? How do mutations at different sites of the same protein culminate in the distinct phenotypes evident in diverse prion diseases of humans? Are conformational changes of proteins a common mechanism of other neurodegenerative diseases? And do these findings suggest a therapy for the dreadful manifestations of spongiform encephalopathies?

Despite these unanswered questions, this work remains one of the most exciting chapters in modern neurological

research, and rightly won Nobel prizes in Physiology or Medicine for both Gajdusek (in 1976) and Prusiner (in 1997).

References

BUELER, H. AND 6 OTHERS (1993) Mice devoid of PrP are resistant to scrapie. Cell 73: 1339–1347.

GAJDUSEK, D. C. (1977) Unconventional viruses and the origin and disappearance of kuru. Science 197: 943–960.

GIBBS, C. J., D. C. GAJDUSEK, D. M. ASHER AND M. P. ALPERS (1968) Creutzfeldt-Jakob disease (spongiform encephalopathy): Transmission to the chimpanzee. Science 161: 388–389.

PRUSINER, S. B. (1982) Novel proteinaceous infectious particles cause scrapie. Science 216: 136–144.

PRUSINER, S. V., M. R. SCOTT, S. J. DEARMOND AND G. E. COHEN (1998) Prion protein biology. Cell 93: 337–348.

RHODES, R. (1997) *Deadly Feasts: Tracking the Secrets of a Terrifying New Plague.* New York: Simon and Schuster.

the motor learning aspect of cerebellar function. According to a model proposed by Masao Ito and his colleagues at Tokyo University, the climbing fibers relay the message of a motor error to the Purkinje cells. This message produces long-term reductions in the Purkinje cell responses to mossy-parallel fiber inputs. This inhibitory effect on the Purkinje cell responses *disinhibits* the deep cerebellar nuclei (for an account of the probable cellular mechanism for this long-term reduction in the efficacy of the parallel fiber synapse on Purkinje cells, see Chapter 25). As a result, the output of the cerebellum to the various sources of upper motor neurons is enhanced, in much the way that this process occurs in the basal ganglia (see Chapter 18).

Cerebellar Circuitry and the Coordination of Ongoing Movement

As expected for a structure that monitors and regulates motor behavior, neuronal activity in the cerebellum changes continually during the course of a movement. For instance, the execution of a relatively simple task like flipping the wrist back and forth elicits a dynamic pattern of activity in both the Purkinje cells and the deep cerebellar nuclear cells that closely follows the ongoing movement (Figure 19.10). Both types of cells are tonically active at rest and change their frequency of firing as movements occur. The neurons respond selectively to various aspects of movement, including extension or contraction of specific muscles, the position of the joints, and the direction of the next movement that will occur. All this information is therefore encoded by changes in the firing frequency of Purkinje cells and deep cerebellar nuclear cells.

As these neuronal response properties predict, cerebellar lesions and disease tend to disrupt the modulation and coordination of ongoing movements. Thus, the hallmark of patients with cerebellar damage is difficulty producing smooth, well-coordinated movements. Instead, movements tend to be jerky and imprecise, a pattern referred to as **cerebellar ataxia** (Box A). Many of these difficulties in performing movements can be explained as disruption of the cerebellum's role in correcting errors in ongoing movements. Normally, the cerebellar error correction mechanism ensures that movements are modified to cope with changing circumstances. As described earlier, the Purkinje cells and the deep cerebellar nuclear cells recognize potential errors by com-

Figure 19.10 Activity of Purkinje cells (A) and deep cerebellar nuclear cells (B) at rest (upper traces) and during movement of the wrist (lower traces). The lines below the action potential records show changes in muscle tension, recorded by electromyography. The durations of the wrist movements are indicated by the colored blocks. Both classes of cells are tonically active at rest. Rapid alternating movements result in the transient inhibition of the tonic activity of both cell types. (After Thach, 1968.)

(A) PURKINJE CELL

At rest

During alternating movement

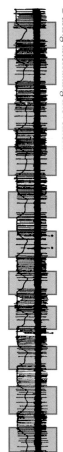

(B) DEEP NUCLEAR CELL

At rest

During alternating movement

paring patterns of convergent activity that are concurrently available to both cell types; the deep nuclear cells then send corrective signals to the upper motor neurons in order to maintain or improve the accuracy of the movement.

As in the case of the basal ganglia, studies of the oculomotor system, saccades in particular, have contributed greatly to understanding the contribution that the cerebellum makes to motor error reduction. For example, cutting part of the tendon to the lateral rectus muscles in one eye of a monkey weakens horizontal eye movements by that eye (Figure 19.11). When a patch is then placed over the normal eye to force the animal to use its weak eye, the saccades performed by the weak eye are *hypometric*; as expected, they fall short of visual targets. Then, over the next few days, the amplitude of the saccades gradually increases until they again become accurate. If the patch is then switched to cover the weakened eye, the saccades performed by the normal eye are *hypermetric*. In other words, over a period of a few days the nervous system corrects the error in the saccades made by the weak eye by increasing the gain in the saccade motor system. Lesions in the vermis of the spinocerebellum (see Figure 19.1) eliminate this ability to reduce the motor error.

Similar evidence of the cerebellar contribution to movement has come from studies of the vestibulo-ocular reflex (VOR) in monkeys and humans. The VOR works to keep the eyes trained on a visual target during head movements (see Chapter 14). The relative simplicity of this reflex has made it possible to analyze some of the mechanisms that enable motor learning as a process of error reduction. When a visual image on the retina shifts its position as a result of head movement, the eyes must move at the same velocity in the opposite direction to maintain a stable percept. In these studies, the adaptability of the VOR to changes in the nature of incoming sensory information is challenged by fitting subjects (either monkeys or humans) with

Figure 19.11 Contribution of the cerebellum to the experience-dependent modification of saccadic eye movements. Weakening of the lateral rectus muscle of the left eye causes the eye to undershoot the target (1). When the experimental subject (in this case a monkey) is forced to use this eye by patching the right eye, multiple saccades must be generated to acquire the target (2). After 5 days of experience with the weak eye, the gain of the saccadic system has been increased and a single saccade is now used to fixate the target. (3) This adjustment of the gain of the saccadic eye movement system depends on an intact cerebellum. (After Optican and Robinson, 1980.)

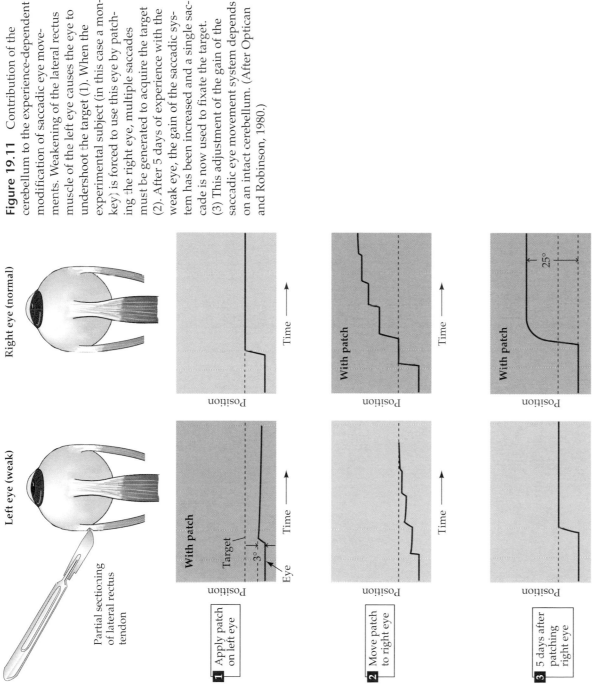

Left eye (weak)

Right eye (normal)

Partial sectioning of lateral rectus tendon

1 Apply patch on left eye

With patch

Target

3°

Eye

Position

Time

2 Move patch to right eye

With patch

Position

Time

With patch

Position

Time

3 5 days after patching right eye

Position

Time

With patch

25°

Position

Time

magnifying or minifying spectacles (Figure 19.12). Because the glasses alter the size of the visual image on the retina, the compensatory eye movements, which would normally have maintained a stable image of an object on the retina, are either too large or too small. Over time, subjects (whether monkeys or humans) learn to adjust the distance the eyes must move in response to head movements to accord with the artificially altered size of the visual field. Moreover, this change is retained for significant periods after the spectacles are removed and can be detected electrophysiologically in recordings from cerebellar Purkinje cells and neurons in the deep cerebellar nuclei. Information that reflects this change in the sensory context of the VOR must therefore be learned and remembered to eliminate the artificially introduced error. Once again, if the cerebellum is damaged or removed, the ability of the VOR to adapt to the new conditions is lost. These observations support the conclusion that the cerebellum is critically important in error reduction during motor learning.

Normal vestibulo-ocular reflex (VOR)

Head and eyes move in a coordinated manner to keep image on retina

VOR out of register

Minifying glasses

Eyes move too far in relation to image movement on the retina when the head moves

After several hours

VOR gain reset

Minifying glasses

Eyes move smaller distances in relation to head movement to compensate

Figure 19.12 Learned changes in the vestibulo-ocular reflex in monkeys. Normally, this reflex operates to move the eyes as the head moves, so that the retinal image remains stable. When the animal observes the world through minifying spectacles, the eyes initially move too far with respect to the "slippage" of the visual image on the retina. After some practice, however, the VOR is reset and the eyes move an appropriate distance in relation to head movement, thus compensating for the altered size of the visual image.

Cerebellar circuitry also provides real-time error correction during ongoing movements. This function is accomplished by changes in the tonically inhibitory activity of Purkinje cells that in turn influence the tonically excitatory deep cerebellar nuclear cells. The resulting effects on the ongoing activity of the deep cerebellar nuclear cells adjust the cerebellar output signals to the upper motor neurons in the cortex and brainstem.

Consequences of Cerebellar Lesions

As predicted from the preceding discussion, patients with cerebellar damage, regardless of the causes or location, exhibit persistent errors in movement. These movement errors are always on the same side of the body as the damage to the cerebellum, reflecting the cerebellum's unusual status as a brain structure in which sensory and motor information is represented ipsilaterally rather than contralaterally. Furthermore, somatic, visual, and other inputs are represented topographically within the cerebellum; as a result, the movement deficits may be quite specific. For example, one of the most common cerebellar syndromes is caused by degeneration in the anterior portion of the cerebellar cortex in patients with a long history of alcohol abuse (Figure 19.13). Such damage specifically affects movement in the lower limbs, which are represented in the anterior spinocerebellum (see Figure 19.5). The consequences include a wide and staggering gait, with little impairment of arm or hand movements. Thus, the topographical organization of the cerebellum allows cerebellar damage to disrupt the coordination of movements performed by some muscle groups but not others.

The implication of these pathologies is that the cerebellum is normally capable of integrating the moment-to-moment actions of muscles and joints throughout the body to ensure the smooth execution of a full range of motor behaviors. Thus, cerebellar lesions lead first and foremost to a lack of coordination of ongoing movements (Box B). For example, damage to the vestibu-

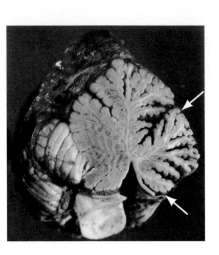

Figure 19.13 The pathological changes in a variety of neurological diseases provide insights about the function of the cerebellum. In this example, chronic alcohol abuse has caused degeneration of the anterior cerebellum (arrows), while leaving other cerebellar regions intact. The patient had difficulty walking but little impairment of arm movements or speech. The orientation of this paramedian sagittal section is the same as Figure 19.1C. (From Victor et al., 1959.)

BOX B
Genetic Analysis of Cerebellar Function

Since the early 1950s, investigators interested in motor behavior have identified and studied strains of mutant mice in which movement is compromised. These mutant mice are easy to spot: The "screen" following induced or spontaneous mutagenesis is simply to look for animals that have difficulty moving. Genetic analysis suggested that some of these abnormal behaviors could be explained by single autosomal recessive or semidominant mutations, in which homozygotes are most severely affected. The strains were given names like *reeler, weaver, lurcher, staggerer,* and *leaner* that reflected the nature of the motor dysfunction they exhibited (see table). The relatively large number of mutations that compromise movement suggested it might be possible to understand some aspects of motor circuits and function at the genetic level.

A common feature of the mutants is ataxia resembling that associated with cerebellar dysfunction in humans. Indeed, all the mutations are associated with some form of cerebellar pathology. The pathologies associated with the *reeler* and *weaver* mutations are particularly striking. In the *reeler* cerebellum, Purkinje cells, granule cells, and interneurons are all displaced from their usual laminar positions, and there are fewer granule cells than normal. In *weaver*, most of the granule cells are lost prior to their migration from the external granule layer (a proliferative region where cerebellar granule cells are generated during development); Purkinje cells and interneurons to carry on the work of the cerebellum. Thus, these mutations causing deficits in motor behavior impair the development and final disposition of the neurons that comprise the major processing circuits of the cerebellum (see Figure 19.8).

Efforts to characterize the cellular mechanisms underlying these motor deficits were unsuccessful, and the molecular identity of the affected genes remained obscure until recently. In the past few years, however, both the *reeler* and *weaver* genes have been identified and cloned.

The *reeler* gene was cloned through a combination of good luck and careful observation. In the course of making transgenic mice by inserting DNA fragments in the mouse genome, investigators in Tom Curran's laboratory created a new strain of mice that behaved much like *reeler* mice and had similar cerebellar pathology. This "synthetic" *reeler* mutation was identified by finding the position of the novel DNA fragment—which turned out to be on the same chromosome as the original *reeler* mutation. Further analysis showed that the same gene had indeed been mutated, and the *reeler* gene was subsequently identified. Remarkably, the protein encoded by this gene is homologous to

Motor Mutations in Mice

Mutation	Inheritance	Chromosome affected	Behavioral and morphological characteristics
reeler (*rl*)	Autosomal recessive	5	Reeling ataxia of gait, dystonic postures, and tremors. Systematic malposition of neuron classes in the forebrain and cerebellum. Small cerebellum, reduced number of granule cells.
weaver (*wv*)	Autosomal recessive	?	Ataxia, hypotonia, and tremor. Cerebellar cortex reduced in volume. Most cells of external granular layer degenerate prior to migration.
leaner (*tg^{la}*)	Autosomal recessive	8	Ataxia and hypotonia. Degeneration of granule cells, particularly in the anterior and nodular lobes of the cerebellum. Degeneration of a few Purkinje cells.
lurcher (*lr*)	Autosomal semidominant	6	Homozygote dies. Heterozygote is ataxic with hesitant, lurching gait and has seizures. Cerebellum half normal size; Purkinje cells degenerate; granule cells reduced in number.
nervous (*nr*)	Autosomal recessive	8	Hyperactivity and ataxia. Ninety percent of Purkinje cells die between 3 and 6 weeks of age.
Purkinje cell degeneration (*pcd*)	Autosomal recessive	13	Moderate ataxia. All Purkinje cells degenerate between the fifteenth embryonic day and third month of age.
staggerer (*sg*)	Autosomal recessive	9	Ataxia with tremors. Dendritic arbors of Purkinje cells are simple (few spines). No synapses of Purkinje cells with parallel fibers. Granule cells eventually degenerate.

(Adapted from Caviness and Rakic, 1978.)

known extracellular matrix proteins such as tenascin, laminin, and fibronectin (see Chapter 22). This finding makes good sense, since the pathophysiology of the *reeler* mutation entails altered cell migration, resulting in misplaced neurons in the cerebral cortex as well as the cerebellar cortex and hippocampus.

Molecular genetic techniques have also led to cloning the *weaver* gene. Using linkage analysis and the ability to clone and sequence large pieces of mammalian chromosomes, Andy Peterson and his colleagues "walked" (i.e., sequentially cloned) several kilobases of DNA in the chromosomal region to find where the *weaver* gene mapped. By comparing normal and mutant sequences within this

region, they determined *weaver* to be a mutation in a K^+ channel that resembles the Ca^{2+}-activated K^+ channels found in cardiac muscle. How this particular molecule influences the development of granule cells or causes their death in the mutants is not yet clear.

The story of the proteins encoded by the *reeler* and *weaver* genes indicates both the promise and the challenge of a genetic approach to understanding cerebellar function. Identifying motor mutants and their pathology is reasonably straightforward, but understanding their molecular genetic basis depends on hard work and good luck.

References

CAVINESS, V. S. JR. AND P. RAKIC (1978) Mechanisms of cortical development: A view from mutations in mice. Ann. Rev. Neurosci. 1: 297–326.

D'ARCANGELO, G., G. G. MIAO, S. C. CHEN, H. D. SOARES, J. I. MORGAN AND T. CURRAN (1995) A protein related to extracellular matrix proteins deleted in the mouse mutation *reeler*. Nature 374: 719–723.

PATIL, N., D. R. COX, D. BHAT, M. FAHAM, R. M. MYERS AND A. PETERSON (1995) A potassium channel mutation in *weaver* mice implicates membrane excitability in granule cell differentiation. Nature Genetics 11: 126–129.

RAKIC, P. AND V. S. CAVINESS JR. (1995) Cortical development: A view from neurological mutants two decades later. Neuron 14: 1101–1104.

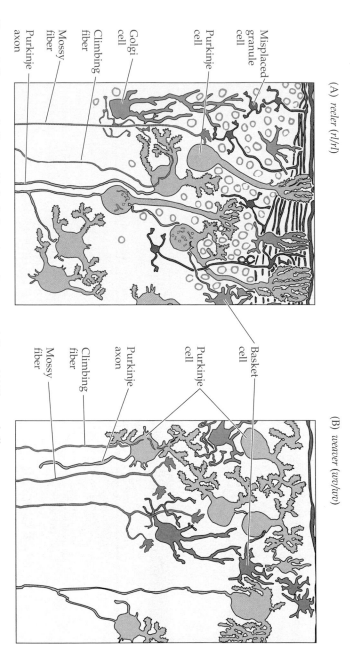

(A) *reeler* (rl/rl)

Misplaced granule cell

Purkinje cell

Golgi cell

Climbing fiber

Mossy fiber

Purkinje axon

(B) *weaver* (wv/wv)

Basket cell

Purkinje cell

Purkinje axon

Climbing fiber

Mossy fiber

The cerebellar cortex is disrupted in both the *reeler* and *weaver* mutations. (A) The cerebellar cortex in homozygous *reeler* mice. The *reeler* mutation causes the major cell types of the cerebellar cortex to be displaced from their normal laminar positions. Despite the disorganization of the cerebellar cortex in *reeler* mutants, the major inputs—mossy fibers and climbing fibers—find appropriate targets. (B) The cerebellar cortex in homozygous *weaver* mice. The granule cells are missing, and the major cerebellar inputs synapse inappropriately on the remaining neurons. (After Rakic, 1977.)

locerebellum impairs the ability to stand upright and maintain the direction of gaze. The eyes have difficulty maintaining fixation; they drift from the target and then jump back with a corrective saccade, a phenomenon called **nystagmus**. Disruption of the pathways to the vestibular nuclei may also result in a loss of muscle tone. In contrast, patients with damage to the spinocerebellum have difficulty controlling walking movements; they have a wide-based gait with small shuffling movements, which represents the inappropriate operation of groups of muscles that normally rely on sensory feedback to produce smooth, concerted actions. The patients also have difficulty performing rapid alternating movements such as the heel-to-shin and/or finger-to-nose tests, a sign referred to as **dysdiadochokinesia**. Over- and underreaching may also occur (called **dysmetria**). During the movement, tremors—called **action** or **intention tremors**—accompany over- and undershooting of the movement due to disruption of the mechanism for detecting and correcting movement errors. Finally, lesions of the cerebrocerebellum produce impairments in highly skilled sequences of learned movements, such as playing a musical instrument. The common denominator of all of these signs, regardless of the site of the lesion, is the inability to perform smooth, directed movements.

Summary

The cerebellum receives input from regions of the cerebral cortex that plan and initiate complex and highly skilled movements; it also receives innervation from sensory systems that monitor the course of the movements. This arrangement enables a comparison of the intended movement with the actual movement and a reduction in the difference, or "motor error." The corrections of motor error produced by the cerebellum occur both in real time and over longer periods, as motor learning. For example, the vestibulo-ocular reflex allows an observer to fixate an object of interest while the head moves; however, lenses that change image size produce a long-term change in the gain of this reflex that depends on an intact cerebellum. Knowledge of cerebellar circuitry suggests that motor learning is mediated by climbing fibers that ascend from the inferior olive to contact the dendrites of the Purkinje cells in the cerebellar cortex. Information provided by the climbing fibers modulates the effectiveness of the second major input to the Purkinje cells, which arrives via the parallel fibers from the granule cells. The granule cells receive information about the intended movement from the vast number of mossy fibers that enter the cerebellum from multiple sources, including the cortico-ponto-cerebellar pathway. As might be expected, the output of the cerebellum from the deep cerebellar nuclei projects to all the major sources of upper motor neurons described in Chapter 17. The effects of cerebellar disease provide strong support for the idea that the cerebellum regulates the performance of movements. Thus, patients with cerebellar disorders show severe ataxias in which the particular movements affected are determined by the site of the lesion.

ADDITIONAL READING

Reviews

ALLEN, G. AND N. TSUKAHARA (1974) Cerebrocerebellar communication systems. Physiol. Rev. 54: 957–1006.

GLICKSTEIN, M. AND C. YEO (1990) The cerebellum and motor learning. J. Cog. Neurosci. 2: 69–80.

LISBERGER, S. G. (1988) The neural basis for learning of simple motor skills. Science 242: 728–735.

STEIN, J. F. (1986) Role of the cerebellum in the visual guidance of movement. Nature 323: 217–221.

THACH, W. T., H. P. GOODKIN AND J. G. KEATING (1992) The cerebellum and adaptive coordination of movement. Ann. Rev. Neurosci. 15: 403–442.

Important Original Papers

ASANUMA, C., W. T. THACH AND E. G. JONES (1983) Distribution of cerebellar terminals and

their relation to other afferent terminations in the ventral lateral thalamic region of the monkey. Brain Res. Rev. 5: 237–265.

BRODAL, P. (1978) The corticopontine projection in the rhesus monkey: Origin and principles of organization. Brain 101: 251–283.

DELONG, M. R. AND P. L. STRICK (1974) Relation of basal ganglia, cerebellum, and motor cortex units to ramp and ballistic movements. Brain Res. 71: 327–335.

ECCLES, J. C. (1967) Circuits in the cerebellar control of movement. Proc. Natl. Acad. Sci. USA 58: 336–343.

McCORMICK, D. A., G. A. CLARK, D. G. LAVOND AND R. F. THOMPSON (1982) Initial localization of the memory trace for a basic form of learning. Proc. Natl. Acad. Sci. USA 79: 2731–2735.

THACH, W. T. (1968) Discharge of Purkinje and cerebellar nuclear neurons during rapidly alternating arm movements in the monkey. J. Neurophysiol. 31: 785–797.

THACH, W. T. (1978) Correlation of neural discharge with pattern and force of muscular activity, joint position, and direction of intended next movement in motor cortex and cerebellum. J. Neurophysiol. 41: 654–676.

VICTOR, M., R. D. ADAMS AND E. L. MANCALL (1959) A restricted form of cerebellar cortical degeneration occurring in alcoholic patients. Arch. Neurol. 1: 579–688.

Books

BRADLEY, W. G., R. B. DAROFF, G. M. FENICHEL AND C. D. MARSDEN (eds.) (1991) *Neurology in Clinical Practice.* Boston: Butterworth-Heinemann, Chapters 29 and 77.

ITO, M. (1984) *The Cerebellum and Neural Control.* New York: Raven Press.

KLAWANS, H. L. (1989) *Toscanini's Fumble and Other Tales of Clinical Neurology.* New York: Bantam, Chapters 7 and 10.

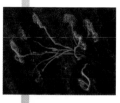

Chapter 20

Eye Movements and Sensory Motor Integration

Overview

Eye movements are, in many ways, easier to study than movements of other parts of the body. This fact arises from the relative simplicity of muscle actions on the eyeball. There are only six extraocular muscles, each of which has a specific role in adjusting eye position. Moreover, there are only four stereotyped kinds of eye movements, each with its own control circuitry. Eye movements have therefore been a useful model for understanding the mechanisms of motor control. Indeed, much of what is known about the regulation of movements by the cerebellum, basal ganglia, and vestibular system has come from the study of eye movements (see Chapters 14, 18, and 19). Here the major features of eye movement control are used to illustrate the principles of sensory motor integration that also apply to more complex motor behaviors.

What Eye Movements Accomplish

Eye movements are important in humans because high visual acuity is restricted to the fovea, the small circular region (about 1.5 mm in diameter) in the central retina that is densely packed with cone photoreceptors (see Chapter 11). Eye movements can direct the fovea to new objects of interest (a process called "foveation") or compensate for disturbances that cause the fovea to be displaced from a target already being attended to.

As demonstrated several decades ago by the Russian physiologist Alfred Yarbus, eye movements reveal a good deal about the strategies used to inspect a scene. Yarbus used contact lenses with small mirrors on them (see Box A) to document (by the position of a reflected beam) the pattern of eye movements made while subjects examined a variety of objects and scenes. Figure 20.1 shows the direction of a subject's gaze while viewing a picture of Queen Nefertiti. The thin, straight lines represent the quick, ballistic eye movements (saccades) used to align the foveas with particular parts of the scene; the denser spots along these lines represent points of fixation where the observer paused for a variable period to take in visual information (little or no visual perception occurs during a saccade, which occupies only a few tens of milliseconds). The results obtained by Yarbus, and subsequently many others, showed that vision is an active process in which eye movements typically shift the view several times each second to selected parts of the scene to examine especially interesting features. The spatial distribution of the fixation points indicates that much more time is spent scrutinizing Nefertiti's eye, nose, mouth, and ear than examining the middle of her cheek or neck. Thus, eye movements allow us to scan the visual field, pausing to focus attention on the portions of the scene that convey the most significant information. As is apparent in Figure 20.1, tracking

Figure 20.1 The eye movements of a subject viewing a picture of Queen Nefertiti. The bust on the left is what the subject saw; the diagram on the right shows the subject's eye movements over a 2-minute viewing period. (From Yarbus, 1967.)

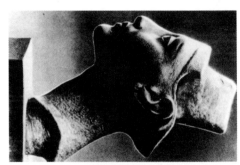

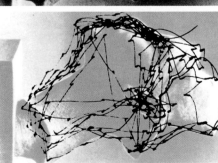

eye movements can be used to determine what aspects of a scene are particularly arresting. Advertisers now use modern versions of Yarbus' method to determine which pictures and scene arrangements will best sell their product.

The importance of eye movements for visual perception has also been demonstrated by experiments in which a visual image is stabilized on the retina, either by paralyzing the extraocular eye muscles or by moving a scene in exact register with eye movements so that the different features of the image always fall on exactly the same parts of the retina (Box A). Stabilized visual images rapidly disappear, for reasons that remain poorly understood. Nonetheless, these observations on motionless images make it plain that eye movements are also essential for normal visual perception.

The Actions and Innervation of Extraocular Muscles

Three antagonistic pairs of muscles control eye movements: the **lateral** and **medial rectus muscles**, the **superior** and **inferior rectus muscles**, and the **superior** and **inferior oblique muscles**. These muscles are responsible for movements of the eye along three different axes: *horizontal*, either toward the nose (adduction) or away from the nose (abduction); *vertical*, either elevation or depression; and *torsional*, movements that bring the top of the eye toward the nose (intorsion) or away from the nose (extorsion). Horizontal movements are controlled entirely by the medial and lateral rectus muscles; the medial rectus muscle is responsible for adduction, the lateral rectus muscle for abduction. Vertical movements require the coordinated action of the superior and inferior rectus muscles, as well as the oblique muscles. The relative contribution of the rectus and oblique groups depends on the horizontal position of the eye (Figure 20.2). In the primary position (eyes straight ahead), both of these groups contribute to vertical movements. Elevation is due to the action of the superior rectus and inferior oblique muscles, while depression is due to the action of the inferior rectus and superior oblique muscles. When the eye is abducted, the rectus muscles are the prime vertical movers. Elevation is due to the action of the superior rectus, and depression is due to the action of the inferior rectus. When the eye is adducted, the oblique muscles are the prime vertical movers. Elevation is due to the action of the inferior oblique muscle, while depression is due to the action of the superior oblique muscle. The oblique muscles are also primarily responsible for torsional movements.

The extraocular muscles are innervated by lower motor neurons that form three cranial nerves: the abducens, the trochlear, and the oculomotor (Figure 20.3). The **abducens nerve** (cranial nerve VI) exits the brainstem from the

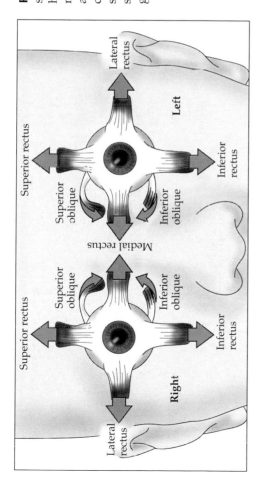

Figure 20.2 The contributions of the six extraocular muscles to vertical and horizontal eye movements. Horizontal movements are mediated by the medial and lateral rectus muscles, while vertical movements are mediated by the superior and inferior rectus and the superior and inferior oblique muscle groups.

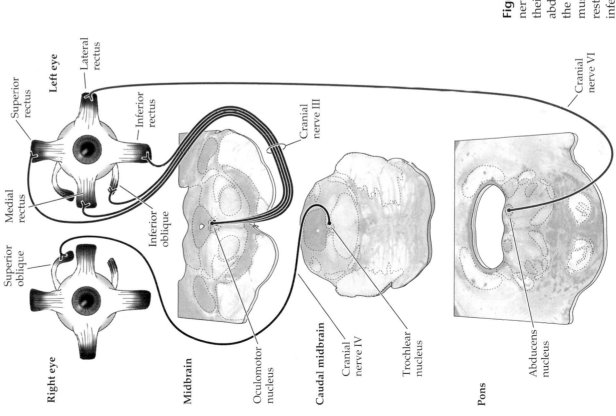

Figure 20.3 Organization of the several cranial nerve nuclei that govern eye movements, showing their innervation of the extraocular muscles. The abducens nucleus innervates the lateral rectus muscle; the trochlear nucleus innervates the superior oblique muscle; and the oculomotor nucleus innervates all the rest of the extraocular muscles (the medial rectus, inferior rectus, superior rectus, and inferior oblique).

BOX A
The Perception of Stabilized Retinal Images

Visual perception depends critically on frequent changes of scene. Normally, our view of the world is changed by saccades, and tiny saccades that continue to move the eyes abruptly over a fraction of a degree of visual arc occur even when the observer stares intently at an object of interest. Moreover, continual drift of the eyes during fixation progressively shifts the image onto a nearby but different set of photoreceptors. As a consequence of these several sorts of eye movements (figure A), our point of view changes more or less continually.

The importance of a continually changing scene for normal vision is dramatically revealed when the retinal

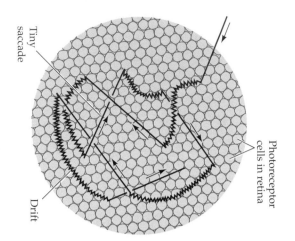

Tiny
saccade

Photoreceptor
cells in retina

Drift

(A) Diagram of the types of eye movements that continually change the retinal stimulus during fixation. The straight lines indicate microsaccades and the curved lines drift; the structures in the background are photoreceptors drawn approximately to scale. The normal scanning movements of the eyes (saccades) are much too large to be shown here, but obviously contribute to the changes of view that we continually experience, as do slow tracking eye movements (although the fovea tracks a particular object, the scene nonetheless changes). (After Pritchard, 1961.)

image is stabilized. If a small mirror is attached to the eye by means of a contact lens and an image reflected off the mirror onto a screen, then the subject necessarily sees the same thing, whatever the position of the eye: Every time the eye moves, the projected image moves exactly the same amount (figure B). Under these circumstances, the stabilized image actually disappears from perception within a few seconds!

A simple way to demonstrate the rapid disappearance of a stabilized retinal image is to visualize one's own retinal blood vessels. The blood vessels, which lie in front of the photoreceptor layer, cast a shadow on the underlying receptors. Although normally invisible, the vascular shadows can be seen by moving a source of light across the eye, a phenomenon first noted by J. E. Purkinje more than 150 years ago. This perception can be elicited with an ordinary penlight pressed gently against the lateral side of the closed eyelid. When the light is wiggled vigorously, a rich network of black blood vessel shadows appears against an orange background. (The vessels appear black because they are shadows.) By starting and stopping the movement, it is readily

apparent that the image of the blood vessel shadows disappears within a fraction of a second after the light source is stilled.

The conventional interpretation of the rapid disappearance of stabilized images is retinal adaptation. In fact, the phenomenon is at least partly of central origin. Stabilizing the retinal image in one eye, for example, diminishes perception through the other eye, an effect known as interocular transfer. Although the explanation of these remarkable effects is not entirely clear, they emphasize the point that the visual system is designed to deal with novelty.

References

BARLOW, H. B. (1963) Slippage of contact lenses and other artifacts in relation to fading and regeneration of supposedly stable retinal images. Q. J. Exp. Psychol. 15: 36–51.

COPPOLA, D. AND D. PURVES (1996) The extraordinarily rapid disappearance of entoptic images. Proc. Natl. Acad. Sci. USA 96: 8001–8003.

HECKENMUELLER, E. G. (1965) Stabilization of the retinal image: A review of method, effects and theory. Psychol. Bull. 63: 157–169.

KRAUSKOPF, J. AND L. A. RIGGS (1959) Interocular transfer in the disappearance of stabilized images. Amer. J. Psychol. 72: 248–252.

(B) Diagram illustrating one means of producing stabilized retinal images. By attaching a small mirror to the eye, the scene projected onto the screen will always fall on the same set of retinal points, no matter how the eye is moved.

Mirrors

Adjustable return path

Mirrors

Contact lens

Screen

Mirror on
contact lens

Light from
projector

pons-medullary junction and innervates the lateral rectus muscle. The **trochlear nerve** (IV) exits from the caudal portion of the midbrain and supplies the superior oblique muscle. In distinction to all other cranial nerves, the trochlear nerve exits from the dorsal surface of the brainstem and crosses the midline to innervate the superior oblique muscle on the contralateral side. The **oculomotor nerve** (III), which exits from the rostral midbrain near the cerebral peduncle, supplies all the rest of the extraocular muscles. Although the oculomotor nerve governs several different muscles, each receives its innervation from a separate group of lower motor neurons within the third nerve nucleus.

In addition to supplying the extraocular muscles, a distinct cell group within the oculomotor nucleus innervates the levator muscles of the eyelid; the axons from these neurons also travel in the third nerve. Finally, the third nerve carries axons from the nearby Edinger-Westphal nucleus that are responsible for pupillary constriction (see Chapter 12). Thus, damage to the third nerve results in three characteristic deficits: impairment of eye movements, drooping of the eyelid (ptosis), and pupillary dilation.

Types of Eye Movements and Their Functions

There are four basic types of eye movements: saccades, smooth pursuit movements, vergence movements, and vestibulo-ocular movements. The functions of each type of eye movement are introduced here; in subsequent sections, the neural circuitry responsible for three of these types of movements is presented in more detail (see Chapters 14 and 19 for further discussion of neural circuitry underlying vestibulo-ocular movements).

Saccades are rapid, ballistic movements of the eyes that abruptly change the point of fixation. They range in amplitude from the small movements made while reading, for example, to the much larger movements made while gazing around a room. Saccades can be elicited voluntarily, but occur reflexively whenever the eyes are open, even when fixated on a target (see Box A). The rapid eye movements that occur during an important phase of sleep (see Chapter 28) are also saccades. The time course of a saccadic eye movement is shown in Figure 20.4. After the onset of a target for a saccade (in this example, the stimulus was the movement of an already fixated target), it takes about 200 ms for eye movement to begin. During this delay, the position of the target with respect to the fovea is computed (that is, how far the eye has to move), and the difference between the initial and intended position, or "motor error" (see Chapter 19), is converted into a motor command that activates the extraocular muscles to move the eyes the correct distance in the appropriate direction. Saccadic eye movements are said to be ballistic because the saccade-generating system cannot respond to subsequent changes in the position of the target during the course of the eye movement. If the target moves again during this time (which is on the order of 15–100 ms), the saccade will miss the target, and a second saccade must be made to correct the error.

Smooth pursuit movements are much slower tracking movements of the eyes designed to keep a moving stimulus on the fovea. Such movements are under voluntary control in the sense that the observer can choose whether or not to track a moving stimulus (Figure 20.5). (Saccades can also be voluntary, but are also made unconsciously.) Surprisingly, however, only highly trained observers can make a smooth pursuit movement in the *absence* of a moving target. Most people who try to move their eyes in a smooth fashion without a moving target simply make a saccade.

The smooth pursuit system can be tested by placing a subject inside a rotating cylinder with vertical stripes. (In practice, the subject is more often seated

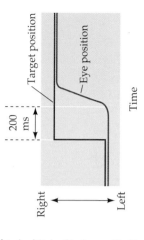

Figure 20.4 The metrics of a saccadic eye movement. The red line indicates the position of a fixation target and the blue line the position of the fovea. When the target moves suddenly to the right, there is a delay of about 200 ms before the eye begins to move to the new target position. (After Fuchs, 1967.)

Figure 20.5 The metrics of smooth pursuit eye movements. These traces show eye movements (blue lines) tracking a stimulus moving at three different velocities (red lines). After a quick saccade to capture the target, the eye movement attains a velocity that matches the velocity of the target. (After Fuchs, 1967.)

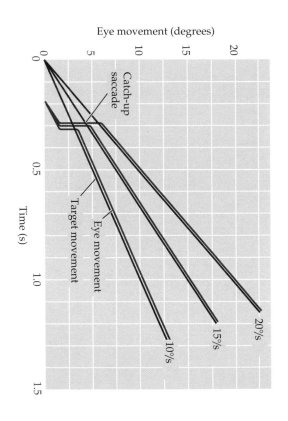

in front of a screen on which a series of horizontally moving vertical bars is presented to conduct this "optokinetic test.") The eyes automatically follow a stripe until they reach the end of their excursion. There is then a quick saccade in the direction opposite to the movement, followed once again by smooth pursuit of a stripe. This alternating slow and fast movement of the eyes in response to such stimuli is called **optokinetic nystagmus.** Optokinetic nystagmus is a normal reflexive response of the eyes in response to large-scale movements of the visual scene and should not be confused with the pathological nystagmus that can result from certain kinds of brain injury (for example, damage to the vestibular system or the cerebellum; see Chapters 14 and 19).

Vergence movements align the fovea of each eye with targets located at different distances from the observer. Unlike other types of eye movements in which the two eyes move in the same direction (**conjugate eye movements**), vergence movements are **disconjugate** (or **disjunctive**); they involve either a convergence or divergence of the lines of sight of each eye to see an object that is nearer or farther away. Convergence is one of the three reflexive visual responses elicited by interest in a near object. The other components of the so-called **near reflex triad** are accommodation of the lens, which brings the object into focus, and pupillary constriction, which increases the depth of field and sharpens the image on the retina (see Chapter 11).

Vestibulo-ocular movements stabilize the eyes relative to the external world, thus compensating for head movements. These reflex responses prevent visual images from "slipping" on the surface of the retina as head position varies. The action of vestibulo-ocular movements can be appreciated by fixating an object and moving the head from side to side; the eyes automatically compensate for the head movement by moving the same distance but in the opposite direction, thus keeping the image of the object at more or less the same place on the retina. The vestibular system detects brief, transient changes in head position and produces rapid corrective eye movements (see Chapter 14). Sensory information from the semicircular canals directs the eyes to move in a direction opposite to the head movement. While the vestibular system operates effectively to counteract rapid movements of the head, it is relatively insensitive to slow movements or to persistent rotation of the head. For example, if the vestibulo-ocular reflex is tested with continuous rotation and without visual cues about the movement of the image (i.e., with eyes closed or in the dark), the compensatory eye movements cease after only about 30 seconds of rotation. However, if the same test is performed with visual cues, eye move-

Eye movement (degrees)

20
15
10
5
0

Catch-up saccade

Eye movement

Target movement

Time (s)
0   0.5   1.0   1.5

20°/s
15°/s
10°/s

ments persist. The compensatory eye movements in this case are due to the activation of the smooth pursuit system, which relies not on vestibular information but on visual cues indicating motion of the visual field.

Neural Control of Saccadic Eye Movements

The problem of moving the eyes to fixate a new target in space (or indeed any other movement) entails two separate issues: controlling the *amplitude* of movement (how far), and controlling the *direction* of the movement (which way). The amplitude of a saccadic eye movement is encoded by the duration of neuronal activity in the lower motor neurons of the oculomotor nuclei. As shown in Figure 20.6, for instance, neurons in the abducens nucleus fire a burst of action potentials prior to abducting the eye (by causing the lateral rectus muscle to contract) and are silent when the eye is adducted. The amplitude of the movement is correlated with the duration of the burst of action potentials in the abducens neuron. With each saccade, the abducens neurons reach a new baseline level of discharge that is correlated with the position of the eye in the orbit. The steady baseline level of firing holds the eye in its new position.

The direction of the movement is determined by which eye muscles are activated. Although in principle any given direction of movement could be specified by independently adjusting the activity of individual eye muscles, the complexity of the task would be overwhelming. Instead, the direction of eye movement is controlled by the local circuit neurons in two **gaze centers** in the reticular formation, each of which is responsible for generating movements along a particular axis. The **paramedian pontine reticular formation (PPRF)** or **horizontal gaze center** is a collection of local circuit neurons near the midline in the pons responsible for generating horizontal eye movements (Figure 20.7). The **rostral interstitial nucleus** or **vertical gaze center** is located in the rostral part of the midbrain reticular formation and is responsible for vertical movements. Activation of each gaze center separately results in movements of the eyes along a single axis, either horizontal or vertical. Activation of the gaze centers in concert results in oblique movements whose trajectories are specified by the relative contribution of each center.

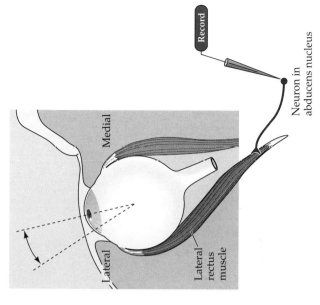

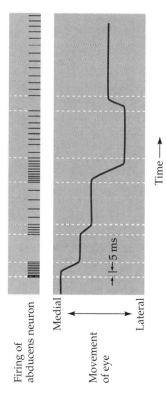

Firing of abducens neuron

Medial ↕ Lateral

Movement of eye

↦ 5 ms

Time →

Figure 20.6 Motor neuron activity in relation to saccadic eye movements. The experimental setup is shown on the right. In this example, an abducens lower motor neuron fires a burst of activity (upper trace) that precedes and extends throughout the movement (solid line). An increase in the tonic level of firing is associated with more lateral displacement of the eye. Note also the decline in firing rate during a saccade in the opposite direction. (After Fuchs and Luschei, 1970.)

Record

Neuron in abducens nucleus

Medial

Lateral

Lateral rectus muscle

An example of how the PPRF works with the abducens and oculomotor nuclei to generate a horizontal saccade to the right is shown in Figure 20.7. Neurons in the PPRF innervate cells in the abducens nucleus on the same side of the brain. There are, however, two types of neurons in the abducens nucleus. One type is a lower motor neuron that innervates the lateral rectus muscle on the same side. The other type, called internuclear neurons, sends their axons across the midline and ascends in a fiber tract called the **medial longitudinal fasciculus**, terminating in the portion of the oculomotor nucleus that contains lower motor neurons innervating the medial rectus muscle. As a result of this arrangement, activation of PPRF neurons on the right side of the brainstem causes horizontal movements of both eyes to the right; the converse is of course true for the PPRF neurons in the left half of the brainstem.

Neurons in the PPRF also send axons to the medullary reticular formation, where they contact inhibitory local circuit neurons. These local circuit neurons, in turn, project to the contralateral abducens nucleus, where they terminate on lower motor neurons and internuclear neurons. In consequence, activation of neurons in the PPRF on the right results in a reduction in the activity of the lower motor neurons whose muscles would oppose movements of the eyes to the right. This inhibition of antagonists resembles the strategy used by local circuit neurons in the spinal cord to control limb muscle antagonists (see Chapter 16).

Although saccades can occur in complete darkness, they are often elicited when something attracts attention and the observer directs the foveas toward the stimulus. How then is sensory information about the location of a target in space transformed into an appropriate pattern of activity in the horizontal and vertical gaze centers? Two structures that project to the gaze centers are demonstrably important for the initiation and accurate targeting

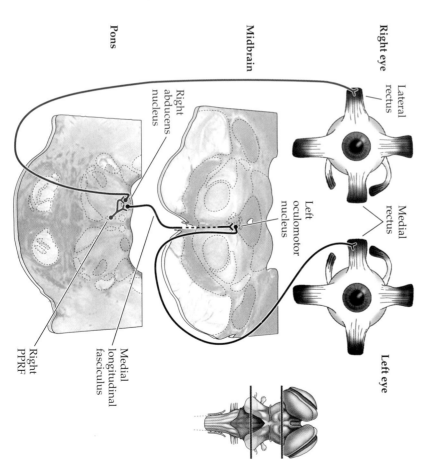

Figure 20.7   Simplified diagram of synaptic circuitry responsible for horizontal movements of the eyes to the right. Activation of local circuit neurons in the right horizontal gaze center (the PPRF; orange) leads to increased activity of lower motor neurons (red and green) and internuclear neurons (blue) in the right abducens nucleus. The lower motor neurons innervate the lateral rectus muscle of the right eye. The internuclear neurons innervate lower motor neurons in the contralateral oculomotor nucleus, which in turn innervate the medial rectus muscle of the left eye.

Right eye

Lateral rectus

Medial rectus

Left eye

Midbrain

Left oculomotor nucleus

Pons

Right abducens nucleus

Medial longitudinal fasciculus

Right PPRF

of saccadic eye movements: the **superior colliculus** of the midbrain, and a region of the frontal lobe that lies just rostral to premotor cortex, known as the **frontal eye field** (**Brodmann's area 8**). Upper motor neurons in both of these structures, each of which contains a topographical motor map, discharge immediately prior to saccades. Thus, activation of a particular site in the superior colliculus or in the frontal eye field produces saccadic eye movements in a specified direction and for a specified distance that is independent of the initial position of the eyes in the orbit. The direction and distance are always the same for a given stimulation site, changing systematically when different sites are activated.

Both the superior colliculus and the frontal eye field also contain cells that respond to visual stimuli; however, the relation between the sensory and motor responses of individual cells is better understood for the superior colliculus. An orderly map of visual space is established by the termination of retinal axons within the superior colliculus (see Chapter 12), and this sensory map is in register with the motor map that generates eye movements. Thus, neurons in a particular region of the superior colliculus are activated by the presentation of visual stimuli in a limited region of visual space. This activation leads to the generation of a saccade that moves the eye by an amount just sufficient to align the foveas with the region of visual space that provided the stimulation (Figure 20.8).

Neurons in the superior colliculus also respond to auditory and somatic stimuli. Indeed, the location in space for these other modalities also is mapped in register with the motor map in the colliculus. Topographically organized maps of auditory space and of the body surface in the superior colliculus can therefore orient the eyes (and the head) in response to a variety of different sensory stimuli. This registration of the sensory and motor maps in the colliculus illustrates an important principle of topographical maps in the motor system, namely to provide an efficient mechanism for sensory motor transformations (Box B).

The functional relationship between the frontal eye field and the superior colliculus in controlling eye movements is similar to that between the motor cortex and the red nucleus in the control of limb movements (see Chapter 17). The frontal eye field projects to the superior colliculus, and the superior colliculus projects to the PPRF on the contralateral side (Figure 20.9). (It also projects to the vertical gaze center, but for simplicity the discussion here is

Figure 20.8 Evidence for sensory motor transformation obtained from electrical recording and stimulation in the superior colliculus. (A) Surface views of the superior colliculus illustrating the location of eight separate electrode recording and stimulation sites. (B) Map of visual space showing the receptive field location of the sites in (A) (white circles), and the amplitude and direction of the eye movements elicited by stimulating these sites electrically (arrows). In each case, electrical stimulation results in eye movements that align the fovea with a region of visual space that corresponds to the visual receptive field of the site. (After Schiller and Stryker, 1972.)

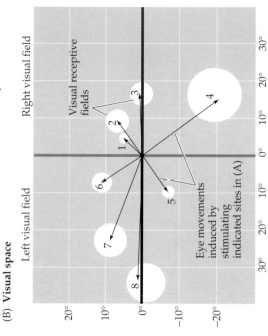

(B) **Visual space**

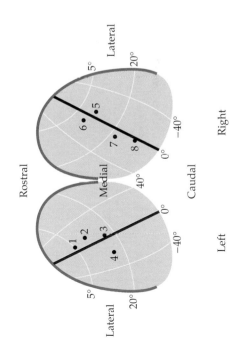

(A) **Superior colliculus**

Figure 20.9 The relationship of the frontal eye field in the right cerebral hemisphere (Brodmann's area 8) to the superior colliculus and the horizontal gaze center (PPRF). There are two routes by which the frontal eye field can influence eye movements in humans: indirectly by projections to the superior colliculus, which in turn project to the contralateral PPRF, and directly by projections to the contralateral PPRF.

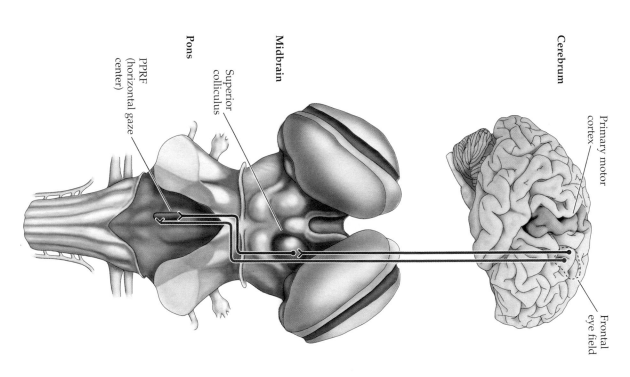

Primary motor cortex

Frontal eye field

Cerebrum

Midbrain

Superior colliculus

Pons

PPRF (horizontal gaze center)

limited to the PPRF.) The frontal eye field can thus control eye movements by activating selected populations of superior colliculus neurons. This cortical area also projects directly to the contralateral PPRF; as a result, the frontal eye field can also control eye movements independently of the superior colliculus. The parallel inputs to the PPRF from the frontal eye field and superior colliculus are reflected in the deficits that result from damage to these structures. Injury to the frontal eye field results in an inability to make saccades to the contralateral side and a deviation of the eyes to the side of the lesion. These effects are transient, however; in monkeys with experimentally induced lesions of this cortical region, recovery is virtually complete in two to four weeks. Lesions of the superior colliculus change the accuracy, frequency, and velocity of saccades; yet saccades still occur, and the deficits also improve with time. These results suggest that the frontal eye fields and the superior colliculus provide complementary pathways for the control of saccades. Moreover, one of these structures appears to be able to compensate (at least partially) for the loss of the other. In support of this interpretation, combined lesions of the frontal eye field and the superior colliculus produce a dramatic and permanent loss in the ability to make saccadic eye movements.

These observations do not, however, imply that the frontal eye fields and the superior colliculus have the same functions. Superior colliculus lesions produce a permanent deficit in the ability to perform very short latency reflex-like eye movements called "express saccades." The express saccades are evidently mediated by direct pathways to the superior colliculus from the retina or visual cortex that can access the upper motor neurons in the colliculus without extensive, and more time-consuming, processing in the frontal cortex (see Box B). In contrast, frontal eye field lesions produce permanent deficits in the ability to make saccades that are not guided by an external target. For example, patients (or monkeys) with a lesion in the frontal eye fields cannot voluntarily direct their eyes *away* from a stimulus in the visual field, a type of eye movement called an "antisaccade." Such lesions also eliminate the ability to make a saccade to the remembered location of a target that is no longer visible.

Finally, the frontal eye fields are essential for systematically scanning the visual field to locate an object of interest within an array of distracting objects (see Figure 20.1). Figure 20.10 shows the responses of a frontal eye field neuron during a visual task in which a monkey was required to foveate

Figure 20.10   (A) Locus of the frontal eye fields on a lateral view of the rhesus monkey brain. (B) Activation of a frontal eye field neuron during visual search for a target. The vertical tickmarks represent action potentials, and each row of tick marks is a different trial. The graphs below show the average frequency of action potentials as a function of time. The change in color form green to purple in each row indicates the time of onset of a saccade toward the target. In the left trace (1), the target (red square) is in the part of the visual field "seen" by the neuron, and the response to the target is similar to the response that would be generated by the neuron even if no distractors (green squares) were present (not shown). In the right trace (3), the target is far from the response field of the neuron. The neuron responds to the distractor in its response field. However, it responds at a lower rate than it would to exactly the same stimulus if the square were not a distractor but a target for a saccade (left trace). In the middle trace (2), the response of the neuron to the distractor has been sharply reduced by the presence of the target in a neighboring region of visual field. (After Schall, 1995.)

(A)

Record

Frontal eye fields

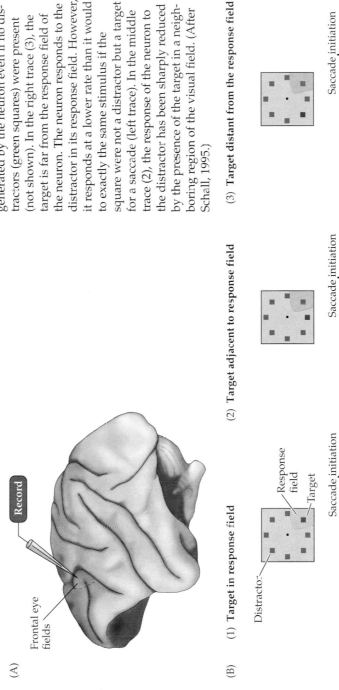

(B)

(1) **Target in response field**

Distractor

Response field

Target

Saccade initiation

Trials

Activity (Hz)

100

50

Time from target (ms)

(2) **Target adjacent to response field**

Saccade initiation

Time from target (ms)

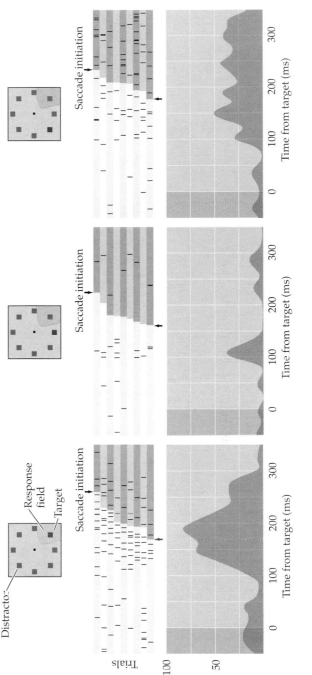

(3) **Target distant from the response field**

Saccade initiation

Time from target (ms)

BOX B
Sensory Motor Integration in the Superior Colliculus

The superior colliculus is a laminated structure in which the differences between the layers provide clues about how sensory and motor maps interact to produce appropriate movements. As discussed in the text, the superficial or "visual" layer of the colliculus receives input from retinal axons that form a topographic map. Thus, each site in the superficial layer is activated maximally by the presence of a stimulus at a particular point of visual space. In contrast, neurons in the deeper or "motor" layers generate bursts of action potentials that command saccades, effectively generating a motor map; thus, activation of different sites generates saccades having different vectors. The visual and motor maps are *in register*, so that visual cells responding to a stimulus in a specific region of visual space are located directly above the motor cells that command eye movements toward that same region (see Figure 20.8).

The registration of the visual and motor maps suggests a simple strategy for how the eyes might be guided toward an object of interest in the visual field. When an object appears at a particular location in the visual field, it will activate neurons in the corresponding part of the visual map. As a result, bursts of action potentials are generated by the subjacent motor cells to command a saccade that rotates the two eyes just the right amount to direct the foveas toward that same location in the visual field. This behavior is called "visual grasp" because successful sensory motor inte-

gration results in the accurate foveation of a visual target.

This seemingly simple model, formulated in the early 1970s when the collicular maps were first found, assumes point to point connections between the visual and motor maps. In practice, however, these connections have been difficult to demonstrate. Neither the anatomical nor the physiological methods available at the time were sufficiently precise to establish these postulated synaptic connections. At about the same time, motor neurons were found to command saccades to nonvisual stimuli; moreover, spontaneous saccades occur in the dark. Thus, it was clear that visual layer activity is not always necessary for saccades. To confuse matters further, animals could be trained *not* to make a saccade when an object appeared in the visual field, showing that the activation of visual neurons is sometimes insufficient to command saccades. The fact that activity of neurons in the visual map is *neither necessary nor sufficient* for eliciting saccades led investigators away from the simple model of direct connections between corresponding regions of the two maps, toward models that linked the layers indirectly through pathways that detoured through the cortex.

Eventually, however, new and better methods resolved this uncertainty. Techniques for filling single cells with axonal tracers showed an overlap between descending visual layer axons and ascending motor layer dendrites, in accord with direct anatomical connec-

tions between corresponding regions of the maps. At the same time, in vitro whole-cell patch clamp recording (see Box A in Chapter 4) permitted more discriminating functional studies that distinguished excitatory and inhibitory inputs to the motor cells. These experiments showed that the visual and motor layers do indeed have the functional connections required to initiate the command for a visually guided saccadic eye movement. A single brief electrical stimulus delivered to the superficial layer generates a prolonged burst of action potentials that resembles the command bursts that normally occur just before a saccade (see figure).

These direct connections presumably provide the substrate for the very short latency reflex-like "express saccades" that are unaffected by destruction of the frontal eye fields. Other visual and nonvisual inputs to the deep layers probably explain why activation of the retina is neither necessary nor sufficient for the production of saccades.

References

LEE, P. H., M. C. HELMS, G. J. AUGUSTINE AND W. C. HALL (1997) Role of intrinsic synaptic circuitry in collicular sensorimotor integration. Proc. Natl. Acad. Sci. USA 94: 13299–13304.

SCHILLER, P. H. AND M. STRYKER (1972) Single-unit recording and stimulation in superior colliculus of the alert rhesus monkey. J. Neurophysiol. 35: 915–924.

SPARKS, D. L. AND J. S. NELSON (1987) Sensory and motor maps in the mammalian superior colliculus. TINS 10(8): 312–317.

a target located within an array of distracting objects. This frontal eye field neuron discharges at different levels to the same stimulus, depending on whether the stimulus is the target of the saccade or a "distractor," and on the location of the distractor relative to the actual target. For example, the differences between the middle and the left and right traces in Figure 20.10 demonstrate that the response to the distractor is much reduced if it is

(A) The superior colliculus receives visual input from the retina and sends a command signal to the gaze centers to initiate a saccade (see text). In the experiment illustrated here, a stimulating electrode activates cells in the visual layer and a patch clamp pipette records the response evoked in a neuron in the subjacent motor layer. The cells in the visual and motor layers were subsequently labeled with a tracer called biocytin. This experiment demonstrates that the terminals of the visual neuron are located in the same region as the dendrites of the motor neuron. (B) The onset of a target in the visual field (top trace) is followed after a short interval by a saccade to foveate the target (second trace). In the superior colliculus, the visual cell responds shortly after the onset of the target, while the motor cell responds later, just before the onset of the saccade. (C) Bursts of excitatory postsynaptic currents (EPSCs) recorded from a motor layer neuron in response to a brief (0.5 ms) current stimulus applied via a steel wire electrode in the visual layer (top; see arrow). These synaptic currents generate bursts of action potentials in the same cell (bottom).

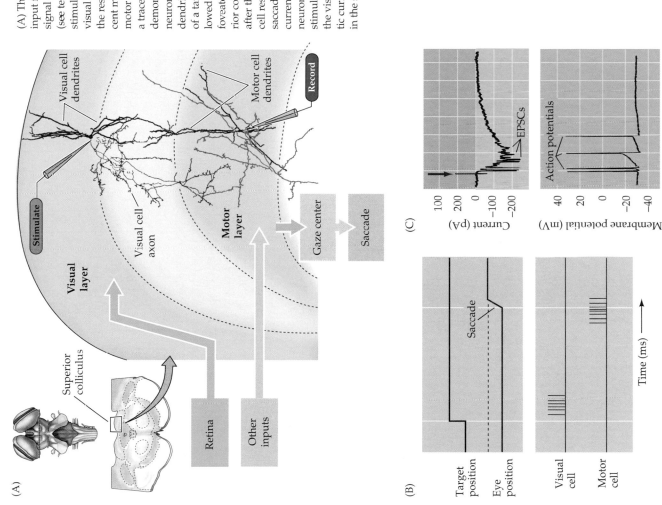

located close to the target in the visual field. Results such as these suggest that lateral interactions within the frontal eye fields enhance the neuronal responses to stimuli that will be selected as saccade targets, and that such interactions suppress the responses to uninteresting and potentially distracting stimuli. These sorts of interactions presumably reduce the occurrence of unwanted saccades to distracting stimuli in the visual field.

Neural Control of Smooth Pursuit Movements

Smooth pursuit movements are also mediated by neurons in the PPRF, but are under the influence of motor control centers other than the superior colliculus and frontal eye field. (The superior colliculus and frontal eye field are exclusively involved in the generation of saccades.) The exact route by which visual information reaches the PPRF to generate smooth pursuit movements is not known (a pathway through the cerebellum has been suggested). It is clear, however, that neurons in the striate and extrastriate visual areas provide sensory information that is essential for the initiation and accurate guidance of smooth pursuit movements. In monkeys, neurons in the middle temporal area (which is largely concerned with the perception of moving stimuli and a target of the magnocellular stream; see Chapter 12) respond selectively to targets moving in a specific direction. Moreover, damage to this area disrupts smooth pursuit movements. In humans, damage of comparable areas in the parietal and occipital lobes also results in abnormalities of smooth pursuit movements. Unlike the effects of lesions to the frontal eye field and the superior colliculus, the deficits are in eye movements made toward the side of the lesion. For example, a lesion of the left parieto-occipital region is likely to result in an inability to track an object moving from right to left.

Neural Control of Vergence Movements

When a person wishes to look from one object to another object that are located at different distances from the eyes, a saccade is made that shifts the direction of gaze toward the new object, and the eyes either diverge or converge until the object falls on the fovea of each eye. The structures and pathways responsible for mediating the vergence movements are not well understood, but appear to include several extrastriate areas in the occipital lobe (see Chapter 12). Information about the location of retinal activity is then relayed through the two lateral geniculate nuclei to the cortex, where the information from the two eyes is integrated. The appropriate command to diverge or converge the eyes, which is based largely on information from the two eyes about the amount of binocular disparity (see Chapter 12), is then sent via upper motor neurons from the occipital cortex to "vergence centers" in the brainstem. One such center is a population of local circuit neurons located in the midbrain near the oculomotor nucleus. These neurons generate a burst of action potentials. The onset of the burst is the command to generate a vergence movement, and the frequency of the burst determines its velocity. There is a division of labor within the vergence center, so that some neurons command convergence movements while others command divergence movements. These neurons also coordinate vergence movements of the eyes with accommodation of the lens and pupillary constriction to produce the near reflex discussed in Chapter 11.

Summary

Despite their specialized function, the systems that control eye movements have much in common with the motor systems that govern movements of other parts of the body. Just as the spinal cord provides the basic circuitry for coordinating the actions of muscles around a joint, the reticular formation of the pons and midbrain provides the basic circuitry that mediates movements of the eyes. Descending projections from higher-order centers in the superior colliculus and the frontal eye field innervate the brainstem gaze centers, providing a basis for integrating eye movements with a variety of sensory information that indicates the location of objects in space. The superior colliculus and the frontal eye field are organized in a parallel as well as a hierarchical fashion, enabling one of these structures to compensate for the loss of the other. Eye movements, like other movements, are also under the control of the basal ganglia and cerebellum (see Chapters 18 and 19); this control ensures the proper initiation and successful execution of these relatively simple motor behaviors, thus allowing observers to interact efficiently with the universe of things that can be seen.

ADDITIONAL READING

Reviews

FUCHS, A. F., C. R. S. KANEKO AND C. A. SCUDDER (1985) Brainstem control of eye movements. Ann. Rev. Neurosci. 8: 307–337.

HIKOSAKA, O AND R. H. WURTZ (1989) The basal ganglia. In *The Neurobiology of Saccadic Eye Movements: Reviews of Oculomotor Research*, Volume 3. R. H. Wurtz and M. E. Goldberg (eds.). Amsterdam: Elsevier, pp. 257–281.

ROBINSON, D. A. (1981) Control of eye movements. In *Handbook of Physiology*, Section 1: *The Nervous System*, Volume II: *Motor Control*, Part 2. V. B. Brooks (ed.). Bethesda, MD: American Physiological Society, pp. 1275–1320.

SCHALL, J. D. (1995) Neural basis of target selection. Reviews in the Neurosciences 6: 63–85.

SPARKS, D. L. AND L. E. MAYS (1990) Signal transformations required for the generation of saccadic eye movements. Ann. Rev. Neurosci. 13: 309–336.

ZEE, D. S. AND L. M. OPTICAN (1985) Studies of adaption in human oculomotor disorders. In *Adaptive Mechanisms in Gaze Control: Facts and Theories*. A Berthoz and G. Melvill Jones (eds.). Amsterdam: Elsevier, pp. 165–176.

Important Original Papers

FUCHS, A. F. AND E. S. LUSCHEI (1970) Firing patterns of abducens neurons of alert monkeys in relationship to horizontal eye movements. J. Neurophysiol. 33: 382–392.

OPTICAN, L. M. AND D. A. ROBINSON (1980) Cerebellar-dependent adaptive control of primate saccadic system. J. Neurophysiol. 44: 1058–1076.

SCHILLER, P. H. AND M. STRYKER (1972) Single unit recording and stimulation in superior colliculus of the alert rhesus monkey. J. Neurophysiol. 35: 915–924.

SCHILLER, P. H., S. D. TRUE AND J. L. CONWAY (1980) Deficits in eye movements following frontal eye-field and superior colliculus ablations. J. Neurophysiol. 44: 1175–1189.

Books

LEIGH, R. J AND D. S. ZEE (1983) *The Neurology of Eye Movements*. Contemporary Neurology Series. Philadelphia: Davis.

YARBUS, A. L. (1967) *Eye Movements and Vision*. Basil Haigh (trans). New York: Plenum Press.

SCHOR, C. M. AND K. J. CIUFFREDA (EDS.) (1983) *Vergence Eye Movements: Basic and Clinical Aspects*. Boston: Butterworth.

Chapter 21

The Visceral Motor System

Overview

The visceral (or autonomic) motor system controls involuntary functions mediated by the activity of smooth muscle fibers, cardiac muscle fibers, and glands. The system comprises two major divisions, the sympathetic and parasympathetic subsystems (the specialized innervation of the gut provides a further semi-independent component and is usually referred to as the enteric nervous system). Although these divisions are always active at some level, the sympathetic system mobilizes the body's resources for dealing with challenges of one sort or another. Conversely, parasympathetic system activity predominates during states of relative quiescence, so that energy sources previously expended can be restored. This continuous neural regulation of the expenditure and replenishment of the body's resources contributes importantly to the overall physiological balance of bodily functions called homeostasis. Whereas the major controlling centers for somatic motor activity are the primary and secondary motor cortices in the frontal lobes and a variety of related brainstem nuclei, the major locus of central control in the visceral motor system is the hypothalamus and the complex (and ill-defined) circuitry that it controls in the brainstem tegmentum and spinal cord. The status of both divisions of the visceral motor system is modulated by descending pathways from these centers to preganglionic neurons in the brainstem and spinal cord, which in turn determine the activity of the primary visceral motor neurons in autonomic ganglia. The autonomic regulation of several organ systems of particular importance in clinical practice (including cardiovascular function, control of the bladder, and the governance of the reproductive organs) is considered in more detail as specific examples of visceral motor control.

Early Studies of the Visceral Motor System

Although humans must always have been aware of involuntary motor reactions to stimuli in the environment (e.g., narrowing of the pupil in response to bright light, constriction of superficial blood vessels in response to cold or fear, increased heart rate in response to exertion), it was not until the late nineteenth century that the neural control of these and other visceral functions came to be understood in modern terms. The researchers who first rationalized the workings of the **visceral motor system** were Walter Gaskell and John Langley, two British physiologists at Cambridge University. Gaskell, whose work preceded that of Langley, established the overall anatomy of the system and carried out early physiological experiments that demonstrated some of its salient functional characteristics (e.g., that the heartbeat of an

443

experimental animal is accelerated by stimulating the outflow of the upper thoracic spinal cord segments). Based on these and other observations, Gaskell concluded in 1866 that "every tissue is innervated by two sets of nerve fibers of opposite characters," and he further surmised that these actions showed "the characteristic signs of opposite chemical processes."

Langley went on to establish the function of **autonomic ganglia** (which harbor the primary visceral motor neurons), defined the terms "preganglionic" and "postganglionic" (see next section), and coined the phrase **autonomic nervous system** (which is basically a synonym for "visceral motor system"; the terms are used interchangeably). Langley's work on the pharmacology of the autonomic system initiated the classical studies indicating the roles of acetylcholine and the catecholamines in autonomic function, and in neurotransmitter function more generally (see Chapter 6). In short, Langley's ingenious physiological and anatomical experiments established in detail the general proposition put forward by Gaskell on circumstantial grounds.

The third major figure in the pioneering studies of the visceral motor system was Walter Cannon at Harvard Medical School, who during the early to mid-1900s devoted his career to understanding autonomic functions in relation to homeostatic mechanisms generally, and to the emotions and higher brain functions in particular (see Chapter 29). He also established the effects of denervation in the visceral motor system, laying some of the basis for much further work on what is now referred to as "neuronal plasticity" (see Chapter 25).

The Sympathetic Division of the Visceral Motor System

Activity of the neurons that make up the sympathetic division of the visceral motor system ultimately prepares individuals for "flight or fight," as Cannon famously put it. Cannon meant that, in extreme circumstances, heightened levels of sympathetic neural activity allow the body to make maximum use of its resources (particularly its metabolic resources), thereby increasing the chances of survival or success in threatening or otherwise challenging situations. Thus, during high levels of sympathetic activity, the pupils dilate and the eyelids retract (allowing more light to reach the retina and the eyes to move more efficiently); the blood vessels of the skin and gut constrict (rerouting blood to muscles, thus allowing them to extract a maximum of available energy); the hairs stand on end (making our hairier ancestors look more fearsome); the bronchi dilate (increasing oxygenation); the heart rate accelerates and the force of cardiac contraction is enhanced (maximally perfusing skeletal muscles and the brain); and digestive and other vegetative functions become quiescent (thus diminishing activities that are temporarily inappropriate) (Figure 21.1). At the same time, sympathetic activity stimulates the adrenal medulla to release epinephrine and norepinephrine into the bloodstream and mediates the release of glucagon and insulin from the pancreas, further enhancing energy mobilizing (or catabolic) functions.

The neurons that drive these effects at the level of the spinal cord are arranged in a column of **preganglionic neurons** that extends from the uppermost thoracic to the upper lumbar segments (T1 to L2 or L3; Table 21.1) and is referred to as the intermediolateral column of the spinal cord gray matter or lateral horn (Figure 21.2). The neurons that innervate the organs in the head and thorax are in the lowest cervical segment and the upper and middle thoracic segments, whereas those that innervate the abdominal and pelvic organs are in the lower thoracic and upper lumbar segments (see Table

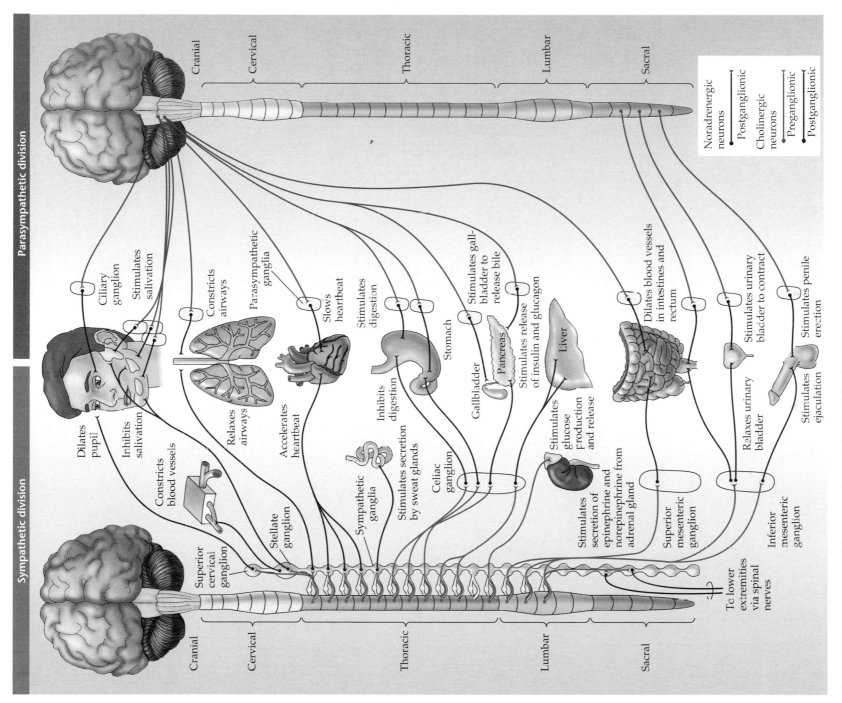

Figure 21.1 Overview of the sympathetic (left side of the figure) and parasympathetic (right side of the figure) divisions of the visceral motor system.

Figure 21.2 Organization of the preganglionic spinal outflow to sympathetic spinal ganglia. (A) General organization of the sympathetic division of the visceral motor system in the spinal cord and the preganglionic outflow to the sympathetic ganglia that contain the primary visceral motor neurons. (B) Cross section of thoracic spinal cord at the level indicated, showing location of the sympathetic preganglionic neurons in the intermediolateral cell column of the lateral horn.

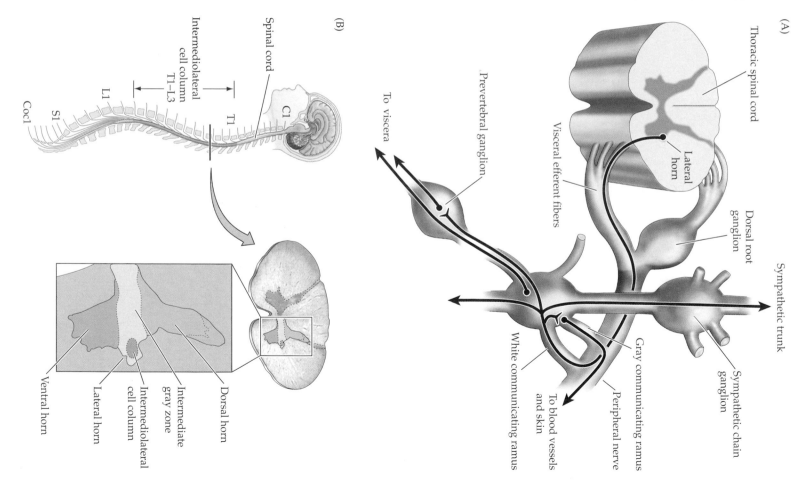

(A)

Thoracic spinal cord

Visceral efferent fibers

Lateral horn

Dorsal root ganglion

Sympathetic chain ganglion

Sympathetic trunk

Gray communicating ramus

Peripheral nerve

White communicating ramus

To blood vessels and skin

Prevertebral ganglion

To viscera

(B)

Intermediolateral cell column T1–L3

Spinal cord

Coc1

S1

L1

T1

C1

Dorsal horn

Intermediate gray zone

Intermediolateral cell column

Lateral horn

Ventral horn

21.1). The axons that arise from these spinal preganglionic neurons typically extend only a short distance, terminating in a series of paravertebral or sympathetic chain ganglia, which, as the name implies, are arranged in a chain that extends along most of the length of the vertebral column (see Figure 21.1). These preganglionic pathways to the ganglia are known as the white communicating rami because of the relatively light color imparted to the rami by the myelinated axons they contain (see Figure 21.2A). Roughly speaking, these preganglionic spinal neurons are comparable to somatic motor interneurons (see Chapter 16).

The neurons in **sympathetic ganglia** are the primary motor neurons of the sympathetic division in that they directly innervate smooth muscles, cardiac muscle, and glands. The **postganglionic axons** arising from these **paravertebral sympathetic chain** neurons travel to various targets in the body wall, joining the segmental spinal nerves of the corresponding spinal segments by way of the gray communicating rami. These rami are another set of short linking nerves, so named because the unmyelinated postganglionic axons give them a somewhat darker appearance than the myelinated preganglionic linking nerves (see Figure 21.2A).

In addition to innervating the sympathetic chain ganglia, the preganglionic axons that govern the viscera extend a longer distance from the spinal cord to sympathetic ganglia that lie in the chest, abdomen, and pelvis in the splanchnic nerves. These **prevertebral ganglia** include sympathetic ganglia in the cardiac plexus, the celiac ganglion, the superior and inferior mesenteric ganglia, and sympathetic ganglia in the pelvic plexus (note that *ganglion* is the singular form, and *ganglia* plural) (see Figure 21.1 and Table 21.1). The postganglionic axons arising from the prevertebral ganglia provide sympathetic innervation to the heart, lungs, gut, kidneys, pancreas, liver, bladder, and reproductive organs (many of these organs also receive some postganglionic innervation from neurons in the sympathetic chain ganglia). Finally, a subset of thoracic preganglionic fibers in the splanchnic nerves innervate the adrenal medulla, which is generally regarded as a sympathetic ganglion modified for a specific endocrine function—namely, the release of catecholamines into the circulation to enhance a widespread sympathetic response to stress. In summary, sympathetic axons contribute to virtually all peripheral nerves, carrying innervation to an enormous range of targets.

Cannon's memorable truism that the sympathetic activity prepares the animal for "fight or flight" notwithstanding, the sympathetic division of the visceral motor system is tonically active to maintain sympathetic target function at appropriate levels whatever the circumstances. Nor should the sympathetic system be thought of as responding in an all-or-none fashion; many specific sympathetic reflexes operate more or less independently, as might be expected from the obvious need to specifically control various organ functions (e.g., the heart during exercise, the bladder during urination, and the reproductive organs during sexual intercourse, as described in later sections).

The Parasympathetic Division of the Visceral Motor System

In contrast to the sympathetic division, the preganglionic outflow from the central nervous system to the ganglia of the parasympathetic division of the visceral motor system stems from neurons whose distribution is limited to the brainstem and the sacral part of the spinal cord (see Figure 21.1). The **cranial preganglionic innervation** arising from the brainstem, which is analogous to the preganglionic sympathetic outflow from the spinal cord, includes the Edinger-Westfall nucleus in the midbrain (which innervates the ciliary ganglion via the oculomotor nerve and mediates the diameter of the pupil in response to light; see Chapter 12), the superior and inferior salivatory nuclei in the pons and medulla (which innervate the salivary glands and tear glands, mediating salivary secretion and the production of tears),

TABLE 21.1
Summary of the Major Functions of the Visceral Motor System

Sympathetic Division

Target organ	Location of preganglionic neurons	Location of ganglionic neurons	Actions
Eye			Pupillary dilation
Lacrimal gland			Tearing
Submandibular and sublingual glands	Upper thoracic spinal cord (C8–T7)	Superior cervical ganglion	Vasoconstriction
Parotid gland			Vasoconstriction
Head, neck (blood vessels, sweat glands, piloerector muscles)			Sweat secretion, vasoconstriction, piloerection
Upper extremity	T3–T6	Stellate and upper thoracic ganglia	Sweat secretion, vasoconstriction, piloerection
Heart	Middle thoracic spinal cord (T1–T5)	Superior cervical and upper thoracic ganglia	Increased heart rate and stroke volume, dilation of coronary arteries
Bronchi, lungs		Upper thoracic ganglia	Vasodilation, bronchial dilation
Stomach		Celiac ganglion	Inhibition of peristaltic movement and gastric secretion, vasoconstriction
Pancreas		Celiac ganglion	Vasoconstriction, insulin secretion
Ascending small intestine, transverse large intestine	Lower thoracic spinal cord (T6–T10)	Celiac, superior, and inferior mesenteric ganglia	Inhibition of peristaltic movement and secretion
Descending large intestine, sigmoid, rectum		Inferior mesenteric hypogastric, and pelvic plexus	Inhibition of peristaltic movement and secretion
Adrenal gland	T9–L2	Cells of gland are modified neurons	Catecholamine secretion
Ureter, bladder	T11–L2	Hypogastric and pelvic plexus	Relaxation of bladder wall muscle and contraction of internal sphincter
Lower extremity	T10–L2	Lower lumbar and upper sacral ganglia	Sweat secretion, vasoconstriction, piloerection

TABLE 21.1
Summary of the Major Functions of the Visceral Motor System (*continued*)

Parasympathetic Division

Target organ	Location of preganglionic neurons	Location of ganglionic neurons	Actions
Eye	Edinger-Westphal nucleus	Ciliary ganglion	Pupillary constriction, accommodation
Lacrimal gland	Superior salivatory nucleus	Pterygopalatine ganglion	Secretion of tears
Submandibular and sublingual glands	Superior salivatory nucleus	Submandibular ganglion	Secretion of saliva, vasodilation
Parotid gland	Inferior salivatory nucleus	Otic ganglion	Secretion of saliva, vasodilation
Head, neck (blood vessels, sweat glands, piloerector muscles)	None	None	None
Upper extremity	None	None	None
Heart	Dorsal motor nucleus of the vagus nerve	Cardiac plexus	Reduced heart rate
Bronchi, lungs	Dorsal motor nucleus of the vagus nerve	Pulmonary plexus	Bronchial constriction and secretion
Stomach	Dorsal motor nucleus of the vagus nerve	Myenteric and submucosal plexus	Peristaltic movement and secretion
Pancreas	Dorsal motor nucleus of the vagus nerve	Pancreatic plexus	Secretion of digestive enzymes
Ascending small intestine, transverse large intestine	Dorsal motor nucleus of the vagus nerve	Ganglia in the myenteric and submucosal plexus	Peristaltic movement and secretion
Descending large intestine, sigmoid, rectum	S3–S4	Ganglia in the myenteric and submucosal plexus	Peristaltic movement and secretion
Adrenal gland	None	None	None
Ureter, bladder	S2–S4	Pelvic plexus	Contraction of bladder wall and inhibition of internal sphincter
Lower extremity	None	None	None

and the dorsal motor nucleus of the vagus nerve in the medulla. The more dorsal part of the nucleus primarily governs glandular secretion via the parasympathetic ganglia located in the viscera of the thorax and abdomen, whereas the more ventral part of the nucleus controls the motor responses of the heart, lungs, and gut elicited by the vagus nerve (e.g., slowing of the heart rate and constriction of the bronchioles). Some preganglionic parasympathetic neurons are also found in the nucleus ambiguus; these innervate parasympathetic ganglia in the submandibular salivary glands and the mediastinum. The location of the parasympathetic brainstem nuclei is shown in Figure 21.3A and B.

The **sacral preganglionic innervation** arises from neurons in the lateral gray matter of the sacral segments of the spinal cord, which are located in much the same position as the sympathetic preganglionics in the intermediolateral column of the thoracic cord (Figure 21.3C,D). The axons from these neurons travel in the splanchnic nerves to innervate parasympathetic ganglia in the lower third of the colon, the rectum, bladder, and reproductive organs.

The **parasympathetic ganglia** innervated by preganglionic outflow from both cranial and sacral levels are in or near the end organs they serve. In this way they are different from the ganglionic targets of the sympathetic system (recall that both the paravertebral chain and prevertebral ganglia are located relatively far from their target organs; see Figure 21.1). Another important anatomical difference between sympathetic and parasympathetic ganglia at the cellular level is that sympathetic ganglion cells tend to have extensive dendritic arbors and are, as might be expected from this arrangement, innervated by a large number of preganglionic fibers. Parasympathetic ganglion cells have few if any dendrites and consequently are each innervated by only one or a few preganglionic axons (see Box B in Chapter 23).

The overall function of the parasympathetic system, as Gaskell, Langley, and later Cannon demonstrated, is generally opposite to that of the sympathetic system, serving to increase metabolic and other resources during periods when the animal's circumstances allow it to "rest and digest." Thus, in contrast to the sympathetic functions enumerated earlier, the activity of the parasympathetic system constricts the pupils, slows the heart rate, and increases the peristaltic activity of the gut. At the same time, diminished activity in the sympathetic system allows the blood vessels of the skin and gut to dilate, the piloerector muscles to relax, and the outflow of catecholamines from the adrenal medulla to decrease.

Although most organs do (as Gaskell surmised) receive innervation from *both* the sympathetic and parasympathetic divisions of the visceral motor system, some receive only sympathetic innervation. These exceptional targets include the sweat glands, the adrenal medulla, the piloerector muscles of the skin, and most arterial blood vessels (see Table 21.1).

The Enteric Nervous System

An enormous number of neurons are specifically associated with the gastrointestinal tract to control its many functions; indeed, more neurons are said to reside in the human gut than in the entire spinal cord. As already noted, the activity of the gut is modulated by both the sympathetic and the parasympathetic divisions of the visceral motor system. However, the gut also has an extensive system of nerve cells in its wall (as do its accessory organs such as the pancreas and gallbladder) that do not fit neatly into the sympathetic or parasympathetic divisions of the visceral motor system (Figure 21.4A). To a surprising degree, these neurons and the complex enteric

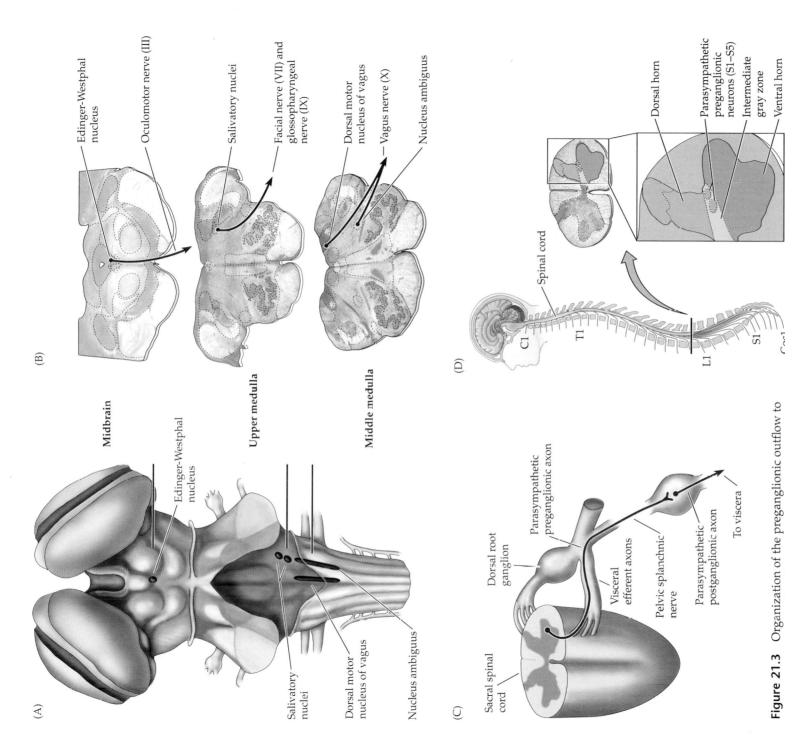

Figure 21.3 Organization of the preganglionic outflow to parasympathetic ganglia. (A) Dorsal view of brainstem showing the location of the nuclei of the cranial part of the parasympathetic division of the visceral motor system. (B) Cross section of the brainstem at the relevant levels [indicated by blue lines in (A)] showing location of these parasympathetic nuclei. (C) Main features of the parasympathetic preganglionics in the sacral segments of the spinal cord. (D) Cross section of the sacral spinal cord showing location of sacral preganglionic neurons.

Figure 21.4 Organization of the enteric component of the visceral motor system. (A) Sympathetic and parasympathetic innervation of the enteric nervous system, and the intrinsic neurons of the gut. (B) Detailed organization of nerve cell plexuses in the gut wall. The neurons of the submucus plexus (Meissner's plexus) are concerned with the secretory aspects of gut function, and the myenteric plexus (Auerbach's plexus) with the motor aspects of gut function (e.g., peristalsis).

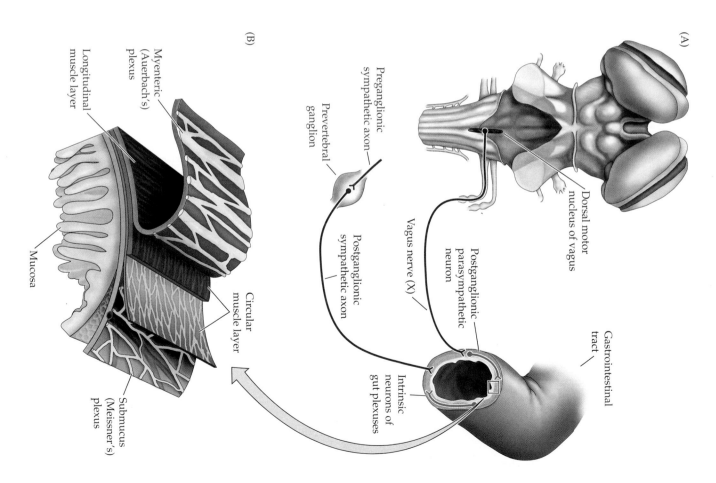

(A)

Gastrointestinal tract

Dorsal motor nucleus of vagus

Vagus nerve (X)

Postganglionic parasympathetic neuron

Preganglionic sympathetic axon

Preverterbral ganglion

Postganglionic sympathetic axon

Intrinsic neurons of gut plexuses

(B)

Longitudinal muscle layer

Myenteric (Auerbach's) plexus

Mucosa

Circular muscle layer

Submucus (Meissner's) plexus

plexuses in which they are found (*plexus* means "network") operate more or less independently according to their own reflex rules; as a result, many gut functions continue perfectly well without sympathetic or parasympathetic supervision (peristalsis, for example, occurs in isolated gut segments in vitro). Thus, most investigators prefer to classify the enteric nervous system as a separate component of the visceral motor system.

The neurons in the gut wall include local and centrally projecting sensory neurons that monitor mechanical and chemical conditions in the gut, local circuit neurons that integrate this information, and motor neurons that influence the activity of the smooth muscles in the wall of the gut and glandular secretions (e.g., of digestive enzymes, mucus, stomach acid, and bile). This complex arrangement of nerve cells intrinsic to the gut is organized into: (1) the myenteric (or Auerbach's) plexus, which is specifically concerned with regulating the musculature of the gut; and (2) the submucus (or Meissner's) plexus, which is located, as the name implies, just beneath the mucus membranes of the gut and is concerned with chemical monitoring and glandular secretion (Figure 21.4B).

As already mentioned, the preganglionic parasympathetic neurons that influence the gut are primarily in the dorsal motor nucleus of vagus nerve in the brainstem and the intermediate gray zone in the sacral spinal cord segments. The preganglionic sympathetic innervation that modulates the action of the gut plexuses derives from the thoraco-lumbar cord, primarily by way of the celiac, superior, and inferior mesenteric ganglia.

Sensory Components of the Visceral Motor System

The visceral motor system clearly requires sensory feedback to control and modulate its many functions. As in the case of somatic sensory modalities (see Chapters 9 and 10), the cell bodies of the visceral afferent fibers lie in the dorsal root ganglia or the sensory ganglia associated with cranial nerves (in this case, the vagus, glossopharyngeal, and facial nerves) (Figure 21.5A). The sensory neurons in the dorsal root ganglia send an axon peripherally to end in sensory receptor specializations, and an axon centrally to terminate in a part of the dorsal horn of the spinal cord near the lateral horn, where the preganglionic neurons of both sympathetic and parasympathetic divisions are located. In addition to making local reflex connections, branches of these visceral sensory neurons also travel rostrally to innervate nerve cells in the brainstem; in this case, however, the target is the nucleus of the solitary tract in the upper medulla (Figure 21.5B). The afferents from viscera in the head and neck that enter the brainstem via the cranial nerves also terminate in the

Figure 21.5 Organization of sensory input to the visceral motor system. (A) Afferent input from the cranial nerves relevant to visceral sensation (as well as afferent input ascending from the spinal cord not shown here) converge on the nucleus of the solitary tract. (B) Cross section of the brainstem showing the location of the nucleus of the solitary tract, which is so-named because of its association with the tract of the myelinated axons that supply it.

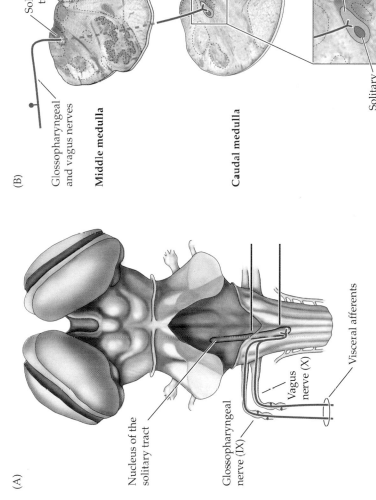
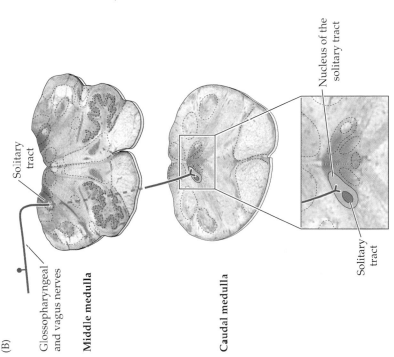

nucleus of the solitary tract (see Figure 21.5B). This nucleus, as described in the next section, integrates a wide range of visceral sensory information and transmits to the hypothalamus and to the relevant motor nuclei in the brainstem tegmentum.

Sensory fibers related to the viscera convey only limited information to consciousness—primarily pain. Nonetheless, the visceral afferent information of which we are *not* aware is essential for the functioning of autonomic reflexes. Specific examples described in more detail later in the chapter include afferent information relevant to cardiovascular control, to the control of the bladder, and to the governance of sexual functions (although sexual reflexes are, exceptionally, not mediated by the nucleus of the solitary tract).

Central Control of the Visceral Motor Functions

The visceral motor system is regulated in part by circuitry in the cerebral cortex: Involuntary visceral reactions such as blushing in response to consciously embarrassing stimuli, vasoconstriction and pallor in response to fear, and autonomic responses to sexual situations make this plain. Indeed, autonomic function is intimately related to emotional experience and expression, as described in Chapter 29. In addition, the hippocampus, thalamus, basal ganglia, cerebellum, and reticular formation all influence the visceral motor system. The major center in the control of the visceral motor system, however, is the hypothalamus (Box A). The hypothalamic nuclei relevant to visceral motor function project to the nuclei in the brainstem that organize many visceral reflexes (e.g., respiration, vomiting, urination), to the cranial nerve nuclei that contain parasympathetic preganglionic neurons, and to the sympathetic and parasympathetic preganglionic neurons in the spinal cord. The general organization of this central autonomic control is summarized in Figure 21.6, and some important clinical manifestations of damage to this descending system are illustrated in Box B.

Although the hypothalamus is the key structure in the overall organization of visceral function, and in homeostasis generally, the visceral motor system continues to function independently if disease or injury impedes the influence of this controlling center. The major subcortical centers for the ongoing regulation of the autonomic function in the absence of hypothalamic control are a series of poorly understood nuclei in the brainstem tegmentum that organize specific visceral functions such as cardiac reflexes, reflexes that control the bladder, and reflexes related to sexual function, as well as other critical autonomic reflexes such as respiration and vomiting.

The afferent information from the viscera that drives these brainstem centers is, as noted already, received by neurons in the nucleus of the solitary tract, which relays these signals to the hypothalamus and to the various autonomic centers in the brainstem tegmentum.

Neurotransmission in the Visceral Motor System

The neurotransmitter functions of the visceral motor system are of enormous importance in clinical practice, and drugs that act on the autonomic system are among the most important in the clinical armamentarium. Moreover, autonomic transmitters have played a major role in the history of efforts to understand synaptic function. Consequently, neurotransmission in the visceral motor system deserves special comment (see also Chapter 6).

Acetylcholine is the primary neurotransmitter of both sympathetic and parasympathetic preganglionic neurons. Nicotinic receptors on autonomic ganglion cells are ligand-gated ion channels that mediate a so-called fast

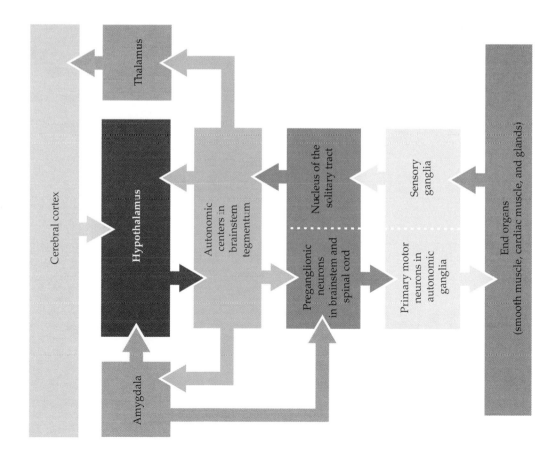

Figure 21.6 Summary of the central control of the visceral motor system. The major organizing center for visceral motor functions is the hypothalamus (see Box A).

EPSP (much like nicotinic receptors at the neuromuscular junction). In contrast, muscarinic acetylcholine receptors on ganglion cells are members of the 7-transmembrane G protein-linked receptor family, and they mediate slower synaptic responses (see Chapters 7 and 8). The primary action of muscarinic receptors in autonomic ganglion cells is to close K^+ channels, making the neurons more excitable and generating a prolonged EPSP. As a result of these two acetylcholine receptor types, ganglionic synapses mediate both rapid excitation and a slower modulation of autonomic ganglion cell activity.

The postganglionic effects of autonomic ganglion cells on their smooth muscle, cardiac muscle, or glandular targets are mediated by two primary neurotransmitters: norepinephrine (NE) and acetylcholine (ACh). For the most part, sympathetic ganglion cells release norepinephrine onto their targets (a notable exception is the cholinergic sympathetic innervation of sweat glands), whereas parasympathetic ganglion cells typically release acetylcholine. As expected from the foregoing account, these two neurotransmitters usually have opposing effects on their target tissue—contraction versus relaxation of smooth muscle, for example.

As described in Chapters 6 to 8, the specific effects of either ACh or NE are determined by the type of receptor expressed in the target tissue, and the downstream signaling pathways to which these receptors are linked.

BOX A
The Hypothalamus

The hypothalamus is located at the base of the forebrain, bounded by the optic chiasm rostrally and the midbrain tegmentum caudally. It forms the floor and ventral walls of the third ventricle and is continuous through the infundibular stalk with the posterior pituitary gland, as illustrated in figure A. Because of its central position in the brain and its proximity to the pituitary, it is not surprising that the hypothalamus integrates information from the forebrain,

brainstem, spinal cord, and various endocrine systems, being particularly important in the central control of visceral motor functions.

The hypothalamus comprises a large number of distinct nuclei, each with its own complex pattern of connections and functions. The nuclei, which are intricately interconnected, can be grouped in three longitudinal regions referred to as *periventricular, medial,* and *lateral.* They can also be grouped along

the anterior–posterior dimension, which are referred to as the *anterior* (or preoptic), *tuberal,* and *posterior* regions (figure B). The anterior periventricular group contains the suprachiasmatic nucleus, which receives direct retinal input and drives circadian rhythms (see Chapter 28). More scattered neurons in the periventricular region (located along the wall of the third ventricle) manufacture peptides known as releasing or inhibiting factors that control the secretion of a variety of hormones by the anterior pituitary. The axons of these neurons project to the median eminence, a region at the junction of the hypothalamus and pituitary stalk, where the peptides are secreted into the portal circulation that supplies the anterior pituitary.

Nuclei in the anterior-medial region include the paraventricular and supraoptic nuclei, which contain the neurosecretory neurons whose axons extend into the posterior pituitary. With appropriate stimulation, these neurons secrete oxytocin or vasopressin (antidiuretic hormone) directly into the bloodstream. Other neurons in the paraventricular nucleus project to the preganglionic neurons of the sympathetic and parasympathetic divisions in the brainstem and spinal cord. It is these cells that are thought to exert hypothalamic control over the visceral motor system and to modulate the activity of the poorly defined nuclei in the brainstem tegmentum that organize specific autonomic reflexes such as respiration and vomiting. The paraventricular nucleus, like other hypothalamic nuclei, receives inputs from the other hypothalamic zones, which are in turn related to the cortex, hippocampus, amygdala, and other central structures that, as noted in the text, are all capable of influencing visceral motor function.

(A) Diagram of the human hypothalamus, illustrating its major nuclei.

Anterior commissure

Paraventricular nucleus

Lateral and medial preoptic nuclei

Anterior nucleus

Suprachiasmatic nucleus

Supraoptic nucleus

Optic chiasm

Infundibular stalk

Anterior pituitary

Fornix

Anterior region

Tuberal region

Lateral-posterior region

Thalamus

Hypothalamic sulcus

Dorsomedial nucleus

Posterior area

Mammillary body

Tuber cinereum

Arcuate nucleus

Ventromedial nucleus

Posterior pituitary

(1)

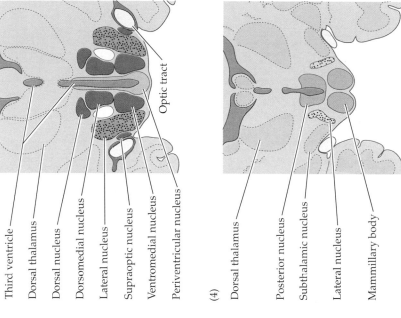

Lateral ventricle
Third ventricle
Anterior commissure
Lateral preoptic nucleus
Medial preoptic nucleus
Suprachiasmatic nucleus

Optic chiasm

(2)

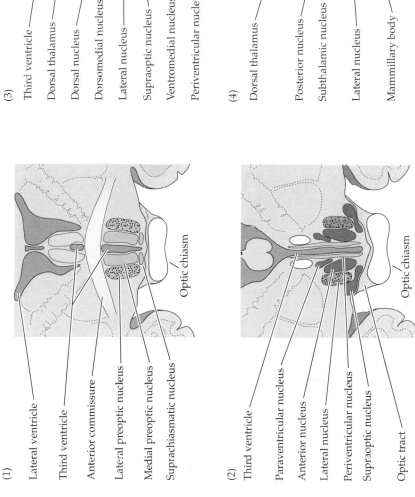

Third ventricle
Paraventricular nucleus
Anterior nucleus
Lateral nucleus
Periventricular nucleus
Supraoptic nucleus
Optic tract

Optic chiasm

(3)

Third ventricle
Dorsal thalamus
Dorsal nucleus
Dorsomedial nucleus
Lateral nucleus
Supraoptic nucleus
Ventromedial nucleus
Periventricular nucleus

Optic tract

(4)

Dorsal thalamus
Posterior nucleus
Subthalamic nucleus
Lateral nucleus
Mammillary body

(B) Coronal sections through the human hypothalamus (see figure A for location of sections 1–4). Color coding of the nuclei illustrates the two dimensions by which hypothalamic nuclei are subdivided (see text). Blue, red, and green illustrate nuclei in the anterior, tuberal, and posterior regions, respectively. The relative shading of these hues illustrates the three mediolateral zones: Lighter shading represents nuclei in the periventricular zone, whereas darker shades represent medial zone nuclei. Nuclei in the lateral zone are stippled. (1) Section through the anterior region illustrating the preoptic and suprachiasmatic nuclei. (2) Rostral tuberal region. (3) Caudal tuberal region. (4) Section through the posterior region illustrating the mammillary bodies.

The medial-tuberal region nuclei (*tuberal* refers to the tuber cinereum, the anatomical name given to the middle portion of the inferior surface of the hypothalamus) include the dorsomedial and ventromedial nuclei, which are involved in feeding, reproductive and parenting behavior, thermoregulation, and water balance. These nuclei receive inputs from structures of the limbic system, as well as from visceral sensory nuclei in the brainstem (e.g., the nucleus of the solitary tract).

Finally, the lateral region of the hypothalamus is really a rostral continuation of the midbrain reticular forma-

tion. Thus, the neurons of the lateral region are not grouped into nuclei, but are scattered among the fibers of the medial forebrain bundle, which runs through the lateral hypothalamus. These cells control behavioral arousal and shifts of attention, especially as related to reproductive activities.

In summary, the hypothalamus regulates an enormous range of physiological and behavioral activities, including control of body temperature, sexual activity, reproductive endocrinology, and attack-and-defense (aggressive) behavior. It is not surprising, then, that this intricate structure is the key control-

ling center for visceral motor activity and for homeostatic functions generally.

References

SAPER, C. B. (1990) Hypothalamus. In *The Human Nervous System*. G. Paxinos (ed.). San Diego: Academic Press, pp. 389–414.

SWANSON, L. W. (1987) The hypothalamus. In *Handbook of Chemical Neuroanatomy, Vol. 5: Integrated Systems of the CNS, Part 1: Hypothalamus, Hippocampus, Amygdala, Retina.* A. Björklund and T. Hökfelt (eds.). Amsterdam: Elsevier, pp. 1–124.

SWANSON, L. W. AND P. E. SAWCHENKO (1983) Hypothalamic integration: Organization of the paraventricular and supraoptic nuclei. Ann. Rev. Neurosci. 6: 269–324.

BOX B
Horner's Syndrome

The characteristic clinical presentation of damage to the pathway that controls the sympathetic division of the visceral motor system to the head and neck is called Horner's syndrome, after the Swiss ophthalmologist who first described this clinical picture in the mid-nineteenth century. The main features, as illustrated in figure A, are decreased diameter of the pupil on the side of the lesion (miosis), a droopy eyelid (ptosis), and a sunken appearance of the affected eye (enophthalmos). Less obvious signs are decreased sweating, increased skin temperature, and flushing on the same side of the face and neck.

All these signs are explained by a loss of sympathetic tone due to damage somewhere along the pathway illustrated in figure B (the parasympathetic pathways are located more medially in the brainstem and are more diffuse; thus they are less frequently and/or obviously affected). The affected route includes the descending axons from the paraventricular nucleus of the hypothalamus that travels through the lateral brainstem to reach the relevant nuclei in the brainstem tegmentum, and ultimately the preganglionic sympathetic neurons in the upper thoracic cord. These preganglionic targets include the neurons in the intermediolateral column in spinal segments T1–T3 that control the dilator muscle of the iris and the tone in smooth muscles of the eyelid

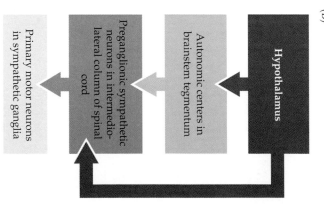

(A)

Ipsilateral pupillary constriction (miosis)

Apparent sinking of eyeball (enophthalmos)

Drooping of eyelid (ptosis)

(B)

Hypothalamus

Autonomic centers in brainstem tegmentum

Preganglionic sympathetic neurons in intermediolateral column of spinal cord

Primary motor neurons in sympathetic ganglia

stab wounds or other traumatic injuries to the head and neck, and tumors of the apex of the lung, thyroid, or cervical lymph nodes.

Horner's syndrome. (A) Major features of the clinical presentation. (B) Diagram of the descending sympathetic pathways in the brainstem that can be interrupted to cause Horner's syndrome. Damage to the preganglionic neurons in the upper thoracic cord, to the superior cervical ganglion, or to the cervical sympathetic trunk can also cause Horner's syndrome (see Figure 21.1).

and globe, the paralysis of which leads to miosis, ptosis, and enophthalmos. The flushing and decreased sweating are likewise the result of diminished sympathetic tone, in this case governed by intermediolateral column neurons in somewhat lower thoracic segments (~T3–T8). Damage to the descending sympathetic pathway in the brainstem will, of course, affect sweating and vascular tone in the rest of the body on the side of the lesion. However, if the damage is to the upper thoracic outflow (as is more typical), the upper thoracic chain, or the superior cervical ganglion, then the manifestations of Horner's syndrome will be limited to the head and neck. Typical causes in these sites are

Peripheral sympathetic targets generally have two subclasses of noradrenergic receptors in their cell membranes, referred to as α and β receptors. Like muscarinic ACh receptors, both α and β receptors and their subtypes belong to the 7-transmembrane G-protein-coupled class of cell surface receptors. The different distribution of these receptors in sympathetic targets allows for a variety of postsynaptic effects mediated by norepinephrine released from postganglionic sympathetic nerve endings (Table 21.2).

TABLE 21.2
Summary of Adrenergic Receptor Types and Some of Their Effects in Sympathetic Targets

Receptor	Tissue	Response
α_1	Smooth muscle of blood vessels, iris, ureter, hairs, uterus, bladder	Contraction of smooth muscle
	Smooth muscle of gut	Relaxation of smooth muscle
	Heart muscle	Positive inotropic effect ($\beta_1 >> \alpha_1$)
	Salivary gland	Secretion
	Adipose tissue	Glycogenolysis, gluconeogenesis
	Sweat glands	Secretion
	Kidney	Na$^+$ reabsorbed
α_2	Adipose tissue	Inhibition of lipolysis
	Pancreas	Inhibition of insulin release
	Smooth muscle of blood vessels	Contraction
β_1	Heart muscle	Positive inotropic effect; positive chronotropic effect
	Adipose tissue	Lipolysis
	Kidney	Renin release
β_2	Liver	Glycogenolysis, gluconeogenesis
	Skeletal muscle	Glycogenolysis, lactate release
	Smooth muscle of bronchi, uterus, gut, blood vessels	Relaxation
	Pancreas	Insulin secretion
	Salivary glands	Thickened secretions

The effects of acetylcholine released by parasympathetic ganglion cells onto smooth muscles, cardiac muscle, and glandular cells also vary according to the subtypes of muscarinic cholinergic receptors found in the peripheral target (Table 21.3). The two major subtypes are known as M1 and M2 recep-

TABLE 21.3
Summary of Cholinergic Receptor Types and Some of Their Effects in Parasympathetic Targets

Receptor	Tissue	Response
Nicotinic	Most parasympathetic targets (and all autonomic ganglion cells)	Relatively fast post-synaptic response
Muscarinic (M1)	Smooth muscles and glands of the gut	Smooth muscle contraction and glandular secretion (relatively slow response)
Muscarinic (M2)	Smooth and cardiac muscle of cardiovascular system	Smooth muscle contraction; some inotropic effect on cardiac muscle
Muscarinic (M3)	Smooth muscles and glands of all targets	Smooth muscle contraction, glandular secretion

tors, M1 receptors being found primarily in the gut and M2 receptors in the cardiovascular system (another subclass of muscarinic receptors, M3, occurs in both smooth muscle and glandular tissues). Muscarinic receptors are coupled to a variety of intracellular signal transduction mechanisms that modify K^+ and Ca^{2+} channel conductances. They can also activate nitric oxide synthase, which promotes the local release of NO in some parasympathetic target tissues (see, for example, the section on autonomic control of sexual function).

In contrast to the relatively restricted responses generated by norepinephrine and acetylcholine released by sympathetic and parasympathetic ganglion cells, respectively, neurons of the enteric nervous system achieve an enormous diversity of target effects by virtue of many different neurotransmitters, most of which are neuropeptides associated with specific cell groups in either the myenteric or submucous plexuses mentioned earlier. The details of these agents and their actions are beyond the scope of this introductory account.

Visceral Motor Reflex Functions

Many examples of specific autonomic functions could be used to illustrate in more detail how the visceral motor system operates. The three outlined here—control of cardiovascular function, control of the bladder, and control of sexual function—have been chosen primarily because of their importance in human physiology and clinical practice.

Autonomic Regulation of Cardiovascular Function

The cardiovascular system is subject to precise reflex regulation so that an appropriate supply of oxygenated blood can be reliably provided to different body tissues under a wide range of circumstances. The sensory monitoring for this critical homeostatic process entails primarily mechanical (barosensory) information about pressure in the arterial system and, secondarily, chemical (chemosensory) information about the level of oxygen and carbon dioxide in the blood. The parasympathetic and sympathetic activity relevant to cardiovascular control is determined by the information supplied by these sensors.

The mechanoreceptors (called baroreceptors) are located in the heart and major blood vessels; the chemoreceptors are located primarily in the carotid bodies, which are small, highly specialized organs located at the bifurcation of the common carotid arteries (some chemosensory tissue is also found in the aorta). The nerve endings in baroreceptors are activated by deformation as the elastic elements of the vessel walls expand and contract. The chemoreceptors in the carotid bodies and aorta respond directly to the partial pressure of oxygen and carbon dioxide in the blood. Both afferent systems convey their status via the vagus nerve to the nucleus of the solitary tract (Figure 21.7), which relays this information to the hypothalamus and the relevant brainstem tegmental nuclei (see earlier).

The afferent information from changes in arterial pressure and blood gas levels reflexively modulates the activity of the relevant visceral motor pathways and, ultimately, of target smooth and cardiac muscles and other more specialized structures. For example, a rise in blood pressure activates baroreceptors that, via the pathway illustrated in Figure 21.7, inhibit the tonic activity of sympathetic preganglionic neurons in the spinal cord. In parallel, the pressure increase stimulates the activity of the parasympathetic preganglionic neurons in the dorsal motor nucleus of the vagus and the nucleus ambiguus that influence heart rate. The carotid chemoreceptors also have some influence, but this is a less important drive than that stemming from the barore-

Figure 21.7 Autonomic control of cardiovascular function.

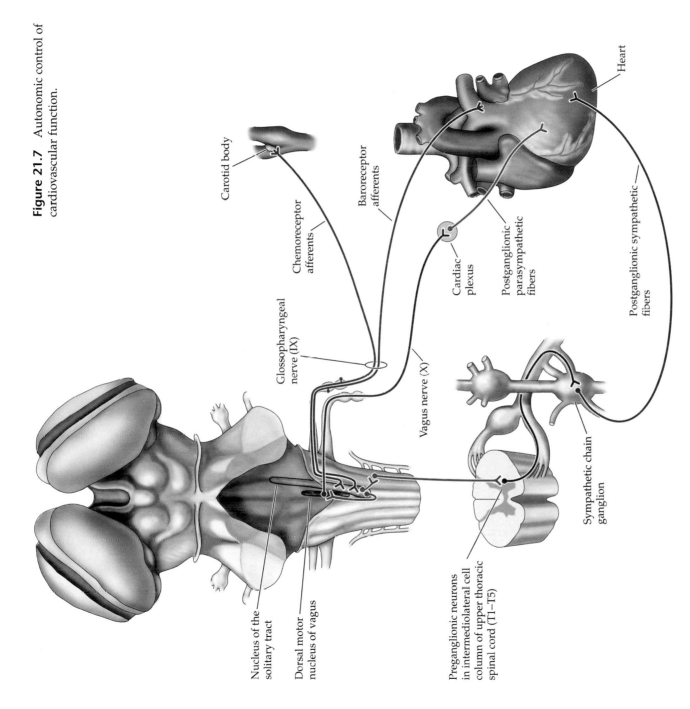

Heart

Carotid body

Baroreceptor afferents

Chemoreceptor afferents

Cardiac plexus

Postganglionic parasympathetic fibers

Glossopharyngeal nerve (IX)

Postganglionic sympathetic fibers

Vagus nerve (X)

Sympathetic chain ganglion

Nucleus of the solitary tract

Dorsal motor nucleus of vagus

Preganglionic neurons in intermediolateral cell column of upper thoracic spinal cord (T1–T5)

ceptors. As a result of this shift in the balance of sympathetic and parasympathetic activity, the stimulatory noradrenergic effects of postganglionic sympathetic innervation on the cardiac pacemaker and cardiac musculature is reduced (an effect abetted by the decreased output of catecholamines from the adrenal medulla and the decreased vasoconstrictive effects of sympathetic innervation on the peripheral blood vessels). At the same time, activation of the cholinergic parasympathetic innervation of the heart decreases the discharge rate of the cardiac pacemaker in the sinoatrial node and slows the ventricular conduction system. These parasympathetic influences are mediated by an extensive series of parasympathetic ganglia in and near the heart, which release acetylcholine onto cardiac pacemaker cells and cardiac muscle fibers. As a result of this combination of sympathetic and parasympathetic effects, heart rate and the effectiveness of the atrial and ventricular myocardial contraction are reduced and the peripheral arterioles dilate, thus lowering the blood pressure.

In contrast to this sequence of events, a drop in blood pressure, as might occur from blood loss, has the opposite effect, inhibiting parasympathetic activity while increasing sympathetic activity. As a result, norepinephrine is released from sympathetic postganglionic terminals, increasing the rate of cardiac pacemaker activity and enhancing cardiac contractility, at the same time increasing release of catecholamines from the adrenal medulla (which further augments these and many other sympathetic effects that enhance the response to this threatening situation). Norepinephrine released from the terminals of sympathetic ganglion cells also acts on the smooth muscles of the arterioles to increase the tone of the peripheral vessels, particularly those in the skin, subcutaneous tissues, and muscles, thus shunting blood away from these tissues to those organs where oxygen and metabolites are urgently needed to maintain function (e.g., brain, heart, and kidneys in the case of blood loss). If these reflex sympathetic responses fail to raise the blood pressure sufficiently (in which case the patient is said to be in shock), the vital functions of these organs begin to fail, often catastrophically.

A more mundane circumstance that requires a reflex autonomic response to a fall in blood pressure is standing up. Rising quickly from a prone position produces a shift of some 300–800 milliliters of blood from the thorax and abdomen to the legs, resulting in a sharp (approximately 40%) decrease in the output of the heart. The adjustment to this normally occurring drop in blood pressure (called orthostatic hypotension) must be rapid and effective, as evidenced by the dizziness sometimes experienced in this situation. Indeed, normal individuals can briefly lose consciousness as a result of blood pooling in the lower extremities, which is the usual cause of fainting among healthy individuals who must stand still for abnormally long periods (the "Beefeaters" who guard Buckingham Palace, for example).

The sympathetic innervation of the heart arises from the preganglionic neurons in the intermediolateral column of the spinal cord, extending from roughly the first through fifth thoracic segments (see Table 21.1). The primary visceral motor neurons are in the adjacent thoracic paravertebral and prevertebral ganglia of the cardiac plexus. The parasympathetic preganglionics, as already mentioned, are in the dorsal motor nucleus of the vagus nerve and the nucleus ambiguus, projecting to parasympathetic ganglia in and around the heart and great vessels.

Autonomic Regulation of the Bladder

The autonomic regulation of the bladder provides a good example of the interplay between the voluntary motor system (obviously, we have voluntary control over urination), and the sympathetic and parasympathetic divisions of the visceral motor system, which operate involuntarily.

The arrangement of afferent and efferent innervation of the bladder is shown in Figure 21.8. The parasympathetic control of the bladder musculature, the contraction of which causes bladder emptying, originates with neurons in the sacral spinal cord segments (S2–S4) that innervate visceral motor neurons in parasympathetic ganglia in or near the bladder wall. Mechanoreceptors in the bladder wall supply visceral afferent information to the spinal cord and to higher autonomic centers in the brainstem (primarily the nucleus of the solitary tract), which in turn project to the various central coordinating centers for bladder function in the brainstem tegmentum and elsewhere.

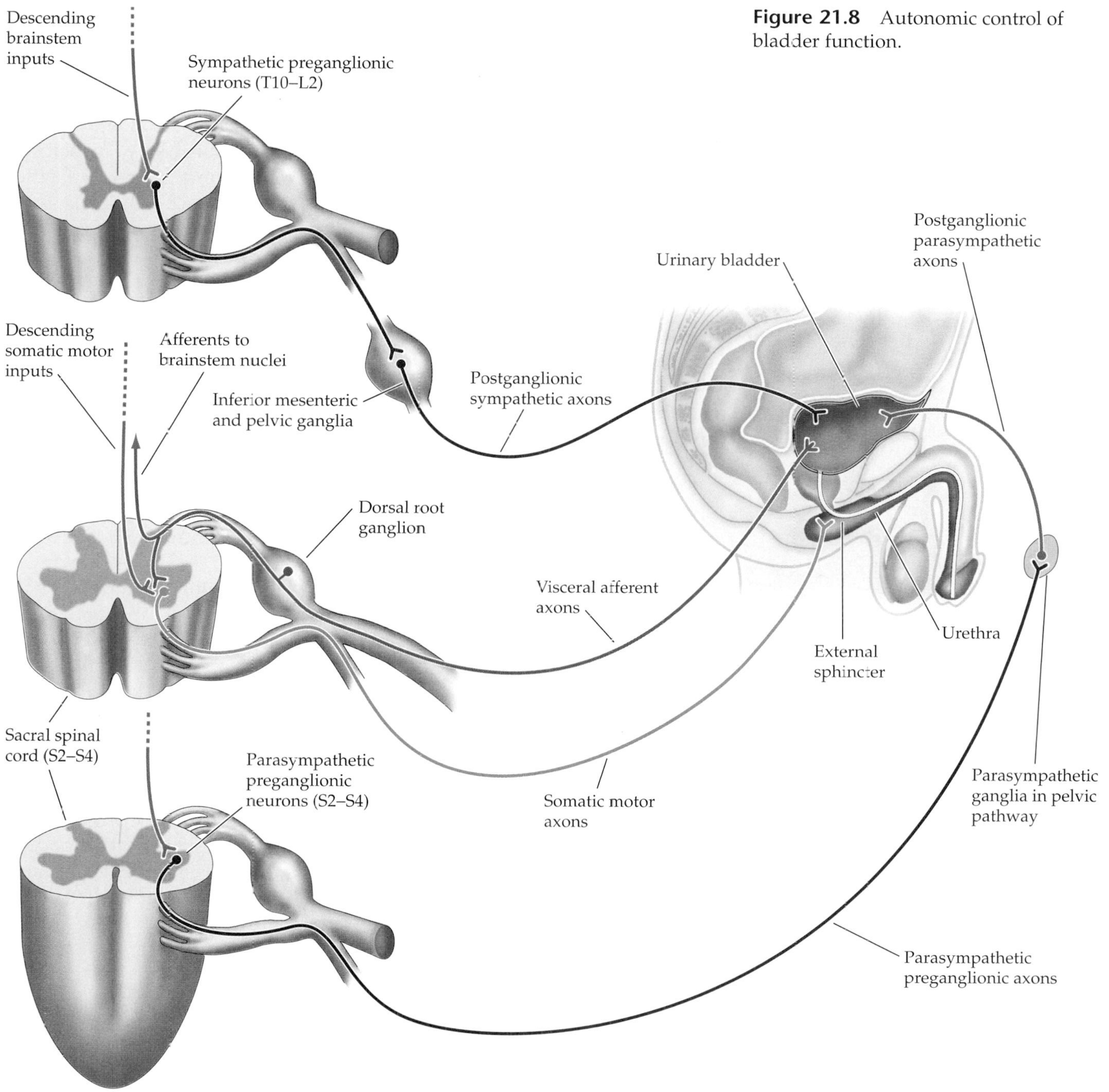

Figure 21.8 Autonomic control of bladder function.

Postganglionic
parasympathetic
axons

Parasympathetic
ganglia in pelvic
pathway

Parasympathetic
preganglionic axons

Urethra

External
sphincter

Urinary bladder

Postganglionic
sympathetic axons

Visceral afferent
axons

Somatic motor
axons

Parasympathetic
preganglionic axons

Sympathetic preganglionic
neurons (T10–L2)

Descending
brainstem
inputs

Inferior mesenteric
and pelvic ganglia

Afferents to
brainstem nuclei

Descending
somatic motor
inputs

Dorsal root
ganglion

Parasympathetic
preganglionic neurons
(S2–S4)

Sacral spinal
cord (S2–S4)

The sympathetic innervation of the bladder originates in the lower thoracic and upper lumbar spinal cord segments (T10–L2), the preganglionic axons running to sympathetic neurons in the inferior mesenteric ganglion and the ganglia of the pelvic plexus. The postganglionic fibers from these ganglia travel in the hypogastric and pelvic nerves to the bladder, where sympathetic activity causes the internal urethral sphincter to close (postganglionic sympathetic fibers also innervate the blood vessels of the bladder, and in males the smooth muscle fibers of the prostate gland). Stimulation of this pathway in response to a modest increase in bladder pressure from the accumulation of urine thus closes the internal sphincter and inhibits the contraction of the bladder wall musculature, allowing the bladder to fill. At the same time, moderate distension of the bladder inhibits parasympathetic activity (which would otherwise contract the bladder and allow the internal sphincter to open). When the bladder is full, afferent activity conveying this information centrally increases parasympathetic tone and decreases sympathetic activity, causing the internal sphincter muscle to relax and the bladder to contract. In this circumstance, the urine is held in check by the voluntary (somatic) motor innervation of the external urethral sphincter muscle (see Figure 21.8).

The voluntary control of the external sphincter is mediated by α-motor neurons of the ventral horn in the sacral spinal cord segments (S2–S4), which cause the striated muscle fibers of the sphincter to contract. During bladder filling (and subsequently, until circumstances permit urination) these neurons are active, keeping the external sphincter closed and preventing bladder emptying. During urination (or "voiding," as clinicians often call this process), this tonic activity is temporarily inhibited, leading to relaxation in the external sphincter muscle. Thus, urination results from the coordinated activity of sacral parasympathetic neurons and temporary inactivity of the α-motor neurons of the voluntary motor system.

The central governance of these events stems from the rostral pons, the relevant pontine circuitry being referred to as the micturition center (*micturition* is also "medicalese" for urination). This phrase implies more knowledge about the central control of bladder function than is actually available. As many as five other central regions have been implicated in the coordination of urinary functions, including the locus coeruleus, the hypothalamus, the septal nuclei, and several cortical regions. The cortical regions primarily concerned with the voluntary control of bladder function include the paracentral lobule, the cingulate gyrus, and the frontal lobes. This functional distribution accords the motor representation of perineal musculature in the medial part of the primary motor cortex (see Chapter 17), and the planning functions of the frontal lobes (see Chapter 26), which are equally pertinent to bodily functions (remembering to stop by the bathroom before going on a long trip, for instance).

Importantly, paraplegic patients, or patients who have otherwise lost descending control of the sacral spinal cord, continue to exhibit autonomic regulation of bladder function, since urination is eventually stimulated reflexively at the level of the sacral cord by sufficient bladder distension. Unfortunately, this reflex is not efficient in the absence of descending motor control, resulting in a variety of problems in paraplegics and others with diminished or absent central control of bladder function. The major difficulty is incomplete bladder emptying, which often leads to chronic urinary tract infections from the culture medium provided by retained urine, and thus the need for an indwelling catheter to ensure adequate drainage.

Autonomic Regulation of Sexual Function

Much like control of the bladder, sexual responses are mediated by the coordinated activity of sympathetic, parasympathetic, and somatic innervation. Although these reflexes differ in detail in males and females, basic similarities allow the two sexes to be considered together, not only in humans but in mammals generally (see Chapter 30). The relevant autonomic effects include: (1) the mediation of vascular dilation, which causes penile or clitoral erection; (2) stimulation of prostatic or vaginal secretions; (3) smooth muscle contraction of the vas deferens during ejaculation or rhythmic vaginal contractions during orgasm in females; and (4) contractions of the somatic pelvic muscles that accompany orgasm in both sexes.

Figure 21.9 Autonomic control of sexual function in the human male.

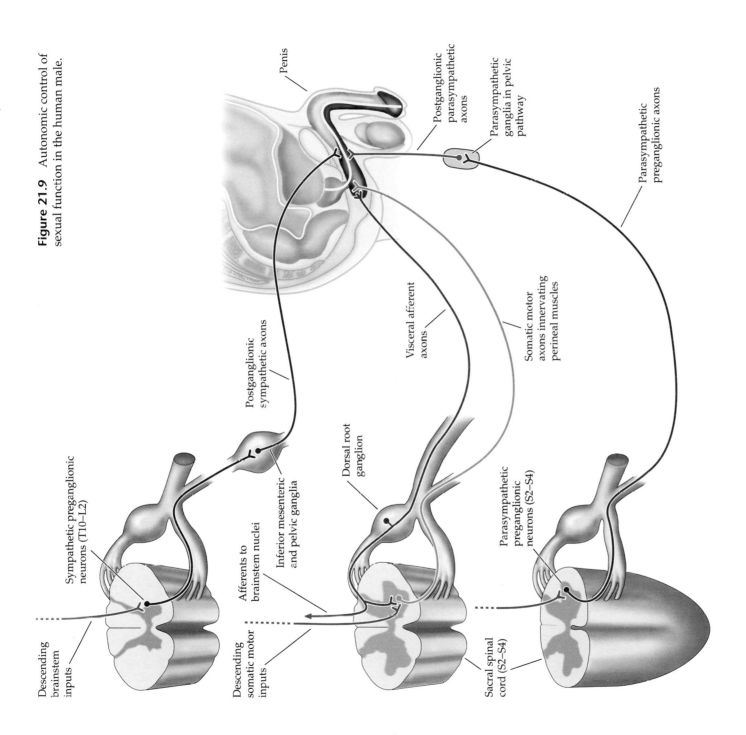

Penis

Postganglionic parasympathetic axons

Parasympathetic ganglia in pelvic pathway

Parasympathetic preganglionic axons

Postganglionic sympathetic axons

Visceral afferent axons

Somatic motor axons innervating perineal muscles

Sympathetic preganglionic neurons (T10–L2)

Dorsal root ganglion

Parasympathetic preganglionic neurons (S2–S4)

Afferents to brainstem nuclei

Inferior mesenteric and pelvic ganglia

Descending brainstem inputs

Descending somatic motor inputs

Sacral spinal cord (S2–S4)

Like the urinary tract, the reproductive organs receive preganglionic parasympathetic innervation from the sacral spinal cord, preganglionic sympathetic innervation from the outflow of the lower thoracic and upper lumbar spinal cord segments, and somatic motor innervation from α-motor neurons in the ventral horn of the lower spinal cord segments (Figure 21.9). The sacral parasympathetic pathway controlling the sexual organs in both males and females originates in the sacral segments S2–S4 and reaches the target organs via the pelvic nerves. Activity of the postganglionic neurons in the relevant parasympathetic ganglia causes dilation of penile or clitoral arteries, and a corresponding relaxation of the smooth muscles of the venous (cavernous) sinusoids, which leads to expansion of the sinusoidal spaces. As a result, the amount of blood in the tissue is increased, leading to a sharp rise in the pressure and an expansion of the cavernous spaces (i.e., erection). The mediator of the smooth muscle relaxation leading to erection is not acetylcholine (as in most postganglionic parasympathetic actions), but nitric oxide (see Chapter 8). The drug sildenafil (Viagra®), for instance, acts by stimulating the activity of guanylate cyclase, which increases the conversion of GTP to cyclic GMP, mimicking the action of NO on the c-GMP pathway, thus enhancing the relaxation of the venous sinusoids and promoting erection in males with erectile dysfunction. Parasympathetic activity also provides excitatory input to the vas deferens, seminal vesicles, and prostate in males, or vaginal glands in females.

In contrast, sympathetic activity causes vasoconstriction and loss of erection. The lumbar sympathetic pathway to the sexual organs originates in the thoraco-lumbar segments (T11–L2) and reaches the target organs via the corresponding sympathetic chain ganglia and the inferior mesenteric and pelvic ganglia, as in the case of the autonomic bladder control.

The afferent effects of genital stimulation are conveyed centrally from somatic sensory endings via the dorsal roots of S2–S4, eventually reaching the somatic sensory cortex (reflex sexual excitation may also occur by local stimulation, as is evident in paraplegics). The reflex effects of such stimulation are increased parasympathetic activity, which, as noted, causes relaxation of the smooth muscles in the wall of the sinusoids and subsequent erection.

Finally, the somatic component of reflex sexual function arises from α-motor neurons in the lumbar and sacral spinal cord segments. These neurons provide excitatory innervation to the bulbocavernosus and ischiocavernosus muscles, which are active during ejaculation in males and mediate the contractions of the perineal (pelvic floor) muscles that accompany orgasm in both male and females.

Sexual functions are governed centrally by the anterior-medial and medial-tuberal zones of the hypothalamus, which contain a variety of nuclei pertinent to visceral motor control and reproductive behavior (see Box A). Although they remain poorly understood, these nuclei act as integrative centers for sexual responses and are also thought to be involved in more complex aspects of sexuality, such as sexual preference and gender identity (see Chapter 30). The relevant hypothalamic nuclei receive inputs from several areas of the brain, including—as one might imagine—the cortical and subcortical structures concerned with emotion and memory (see Chapters 29 and 31).

Summary

Sympathetic and parasympathetic ganglia, which contain the primary visceral-motor neurons that innervate smooth muscles, cardiac muscle, and glands, are controlled by preganglionic neurons in the spinal cord and brainstem. The sympathetic preganglionic neurons that govern ganglion cells in the sympathetic division of the visceral motor system arise from neurons in the thoracic and upper lumbar segments of the spinal cord; parasympathetic preganglionic neurons, in contrast, are located in the brainstem and sacral spinal cord. Sympathetic ganglion cells are distributed in the sympathetic chain (paravertebral) and prevertebral ganglia, whereas the parasympathetic motor neurons are more widely distributed in ganglia that lie within or near the organs they control. Most autonomic targets receive inputs from both the sympathetic and parasympathetic systems, which act in a generally antagonistic fashion. The diversity of autonomic functions is achieved primarily by different types of receptors for the two primary classes of postganglionic autonomic neurotransmitters, norepinephrine in the case of the sympathetic division and acetylcholine in the parasympathetic division. The visceral motor system is regulated by sensory feedback provided by dorsal root and cranial nerve sensory ganglion cells that make local reflex connections in the spinal cord or brainstem and project to the nucleus of the solitary tract in the brainstem, and by descending pathways from the hypothalamus and brainstem tegmentum, the major controlling centers of the visceral motor system (and of homeostasis more generally). The importance of the visceral motor control of organs such as the heart, bladder, and reproductive organs—and the many pharmacological means of modulating autonomic function—have made visceral motor control a central theme in clinical medicine.

ADDITIONAL READING

Reviews

ANDERSSON, K-E. AND G. WAGNER (1995) Physiology of penile erections. Physiol. Rev. 75: 191–236.

BROWN, D. A., F. C. ABOGADIE, T. G. ALLEN, N. J. BUCKLEY, M. P. CAULFIELD, P. DELMAS, J. E. HALEY, J. A. LAMAS AND A. A. SELYANKO (1997) Muscarinic mechanisms in nerve cells. Life Sciences 60(13–14): 1137–44.

COSTA, M. AND S. J. H. BROOKES (1994) The enteric nervous system. Am. J. Gastroenterol. 89: S129–S137.

DAMPNEY, R. A. L. (1994) Functional organization of central pathways regulating the cardiovascular system. Physiol. Rev. 74: 323–364.

GERSHON, M. D. (1981) The enteric nervous system. Ann. Rev. Neurosci. 4: 227–272.

MUNDY, A. R. (1999) Structure and function of the lower urinary tract. In *Scientific Basis of Urology*, A. R. Mundy, J. M. Fitzpatrick, D. E. Neal, and N. J. R. George (eds). Philadelphia: Saunders, pp. 217–242.

PATTON, H. D. (1989) The autonomic nervous system. In *Textbook of Physiology: Excitable Cells and Neurophysiology*, Vol. 1, Section VII: Emotive Responses and Internal Milieu, H. D. Patton, A. F. Fuchs, B. Hille, A. M. Scher, and R. Steiner (eds). Philadelphia: Saunders, pp. 737–758.

PRYOR, J. P. (1999) Male sexual function. In *Scientific Basis of Urology*, A. R. Mundy, J. M. Fitz-

patrick, D. E. Neal and N. J. R. George (eds). Oxford: Isis Medical Media, pp. 243–255.

Important Original Papers

JANSEN, A. S. P., X. V. NGUYEN, V. KARPITSKIY, T. C. METTENLEITER AND A. D. LOEWY (1995) Central command neurons of the sympathetic nervous system: Basis of the fight or flight response. Science 270: 644–646.

LANGLEY, J. N. (1894) The arrangement of the sympathetic nervous system chiefly on observations upon pio-erector nerves J. Physiol. (Lond.) 15: 176–244.

LANGLEY, J. N. (1905) On the reaction of nerve cells and nerve endings to certain poisons chiefly as regards the reaction of striated muscle to nicotine and to curare. J. Physiol. (Lond.) 33: 374–473.

LICHTMAN, J. W., D. PURVES AND J. W. YIP (1980) Innervation of sympathetic neurones in the guinea-pig thoracic chain. J. Physiol. 298: 285–299.

RUBIN, E. AND D. PURVES (1980) Segmental organization of sympathetic preganglionic neurons in the mammalian spinal cord. J. Comp. Neurol. 192: 163–174.

Books

APPENZELLER, O. (1997) *The Autonomic Nervous System: An Introduction to Basic and Clinical Concepts*, 5th Ed. Amsterdam: Elsevier Biomedical Press.

BLESSING, W. W. (1997) *The Lower Brainstem and Bodily Homeostasis*. New York: Oxford University Press.

BRADING, A. (1999) *The Autonomic Nervous System and Its Effectors*. Oxford: Blackwell Science.

BURNSTOCK, G. AND C. H. V. HOYLE (1995) *The Autonomic Nervous System*, Vol. 1: *Autonomic Neuroeffector Mechanism*. London: Harwood Academic.

CANNON, W. B. (1932) *The Wisdom of the Body*. New York: Norton.

FURNESS, J. B. AND M. COSTA (1987) *The Enteric Nervous System*. Edinburgh: Churchill Livingstone.

GABELLA, G. (1976) *Structure of the Autonomic Nervous System*. London: Chapman and Hall.

LANGLEY, J. N. (1921) *The Autonomic Nervous System*. Cambridge, England: Heffer & Sons.

LOEWY, A. D. AND K. M. SPYER (eds.) (1990) *Central Regulation of Autonomic Functions*. New York: Oxford.

PICK, J. (1970) *The Autonomic Nervous System: Morphological, Comparative, Clinical and Surgical Aspects*. Philadelphia: J.B. Lippincott Company.

RANDALL, W. C. (ed.) (1984) *Nervous Control of Cardiovascular Function*. New York: Oxford University Press.

IV

The Changing Brain

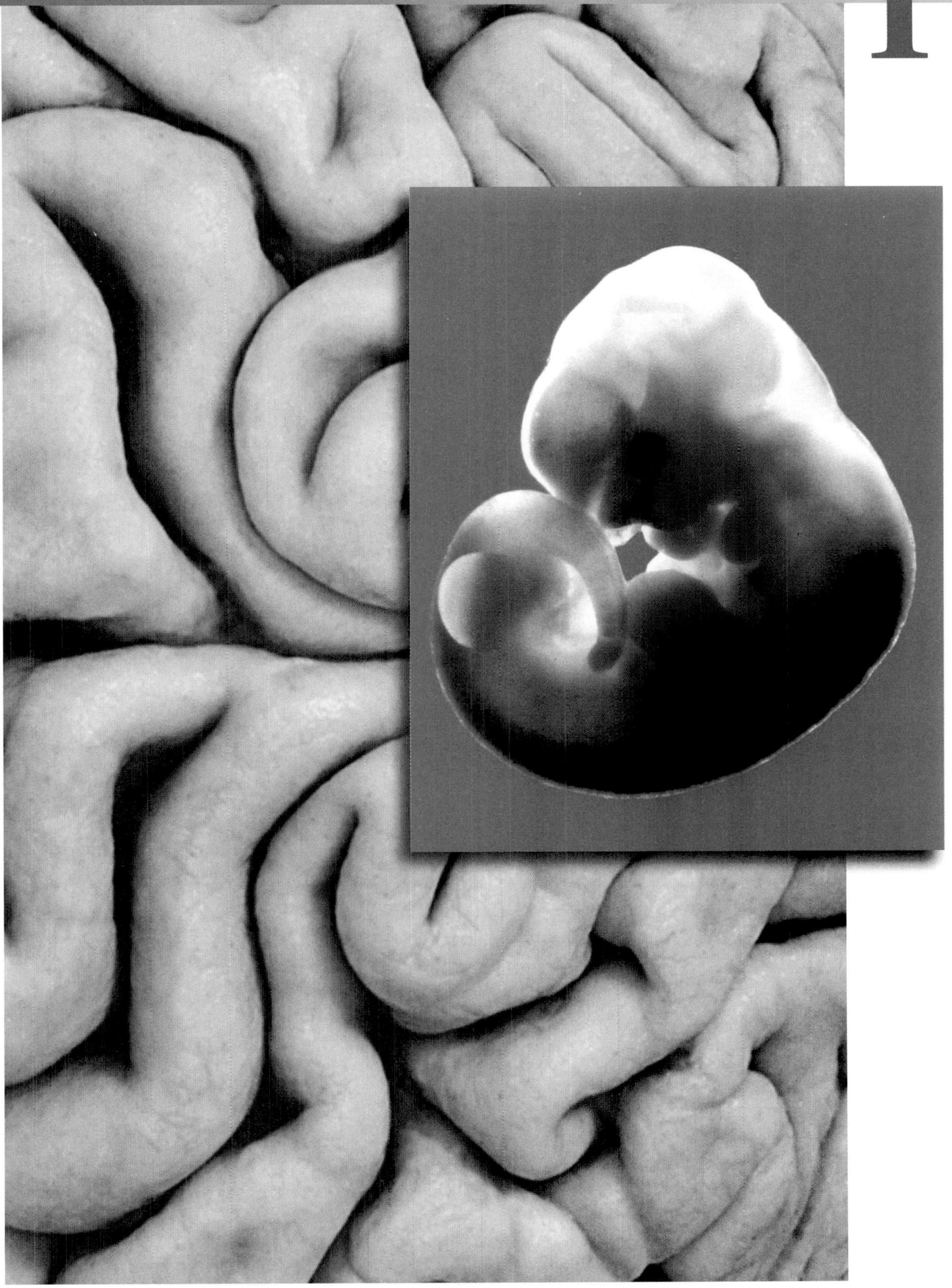

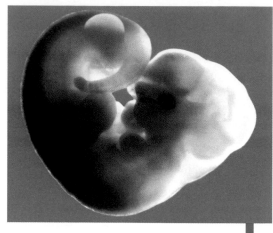

A mammalian embryo in which cells in the developing nervous system responding to the signaling molecule retinoic acid have been labeled by means of a reporter gene. (Courtesy of Anthony-Samuel LaMantia and Elwood Linney.)

Unit IV
The Changing Brain

22   *Early Brain Development*

23   *Construction of Neural Circuits*

24   *Modification of Brain Circuits as a Result of Experience*

25   *Plasticity of Mature Synapses and Circuits*

Although we think of ourselves as the same person throughout life, the structural and functional state of the brain changes dramatically over the human lifespan. The initial development of the nervous system entails the generation and differentiation of neurons, the formation of axonal pathways, and the elaboration of vast numbers of synapses. The circuits that emerge from these processes then mediate an increasingly complex array of behaviors. Subsequent experience during postnatal life continues to shape neural circuits, the related behavioral repertoires, and ultimately cognitive abilities. These changes are most pronounced during developmental windows in early life called critical periods. Even in maturity, however, synaptic connections can be modified as new skills and memories are acquired and older ones are forgotten, and even some new neurons can be generated in a few specialized regions. Some of the mechanisms used during early development are evidently retained and adapted to mediate these ongoing changes in the brain.

Finally, like any other organ system, the brain is subject to disease and traumatic insults that call repair mechanisms into play. Diseases like amyotrophic lateral sclerosis, Parkinson's disease, or Alzheimer's disease all reflect pathologies of processes that normally contribute to neuronal development and to the subsequent maintenance and modification of neural circuitry.

Overview

The elaborate architecture of the adult brain is the final product of genetic instructions, cellular interactions, and eventually interactions between the developing child and the external world. The early development of the nervous system is dominated by events that occur prior to the formation of synapses and are therefore activity-independent. These early events include the establishment of the primordial nervous system in the embryo, the initial generation of neurons from undifferentiated precursor cells, the formation of the major brain regions, and the migration of neurons from the sites of generation to their final positions. These processes set the stage for the subsequent formation of axon pathways and synaptic connections. When any of these processes goes awry—because of genetic mutation, disease, or exposure to drugs or chemicals—the consequences can be disastrous. Indeed, most congenital brain defects result from interference with the normal mechanisms of activity-independent neuronal development. With the advent of powerful new techniques, the cellular and molecular machinery underlying these extraordinarily complex events is beginning to be understood.

The Initial Formation of the Nervous System: Gastrulation and Neurulation

Well before the patch of cells that will eventually become the brain and spinal cord appears, embryonic polarity and the primitive cell layers required for the subsequent formation of the nervous system are established. Critical to this early framework in all vertebrate embryos is the process of **gastrulation**. This invagination of the developing embryo (which starts out as a single sheet of cells) produces the three **germ layers**: the outer layer, or **ectoderm**; the middle layer, or **mesoderm**; and the inner layer, or **endoderm** (Figure 22.1). Gastrulation defines the midline and the anterior-posterior axes of all vertebrate embryos as well.

One key consequence of gastrulation is the formation of the **notochord**, a distinct cylinder of mesodermal cells that extends along the midline of the embryo from anterior to posterior. The notochord forms from an aggregation of mesoderm that invaginates and extends inward from a surface indentation called the **primitive pit**, which subsequently elongates to form the **primitive streak**. As a result of these cell movements during gastrulation, the notochord comes to define the embryonic midline. The ectoderm that lies immediately above the notochord is called the neuroectoderm, and gives rise to the entire nervous system. In addition to specifying the basic topography of the embryo and determining the position of the nervous system, the

Figure 22.1 Neurulation in the mammalian embryo. On the left are dorsal views of the embryo at several different stages of early development; each boxed view on the right is a midline cross section through the embryo at the same stage. (A) During late gastrulation and early neurulation, the notochord forms by invagination of the mesoderm in the region of the primitive streak. (B) As neurulation proceeds, the neural plate begins to fold on itself, forming the neural groove and ultimately the neural tube. The neural plate immediately above the notochord differentiates into the floorplate, whereas the neural crest emerges at the lateral margins of the neural plate (farthest from the notochord). (C) Once the edges of the neural plate meet in the midline, the neural tube is complete. The mesoderm adjacent to the tube then thickens and subdivides into structures called somites—the precursors of the axial musculature and skeleton. (D) As development continues, the neural tube adjacent to the somites becomes the rudimentary spinal cord, and the neural crest gives rise to sensory and autonomic ganglia (the major elements of the peripheral nervous system). Finally, the anterior ends of the neural plate (anterior neural folds) grow together at the midline and continue to expand, eventually giving rise to the brain.

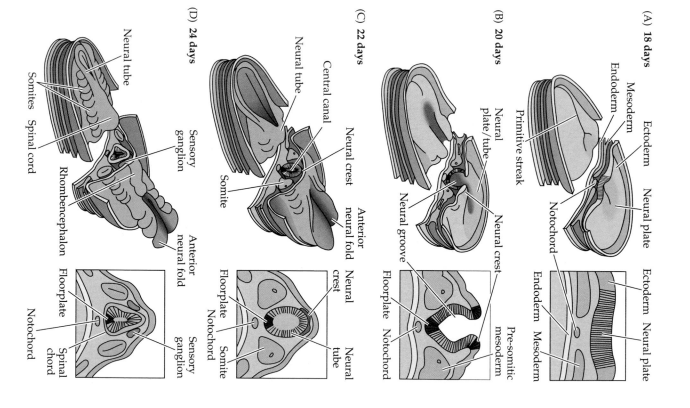

(A) **18 days**

Ectoderm   Neural plate

Mesoderm

Endoderm

Notochord

Primitive streak

Ectoderm   Neural plate

Endoderm   Mesoderm

Pre-somitic mesoderm

(B) **20 days**

Neural plate/tube

Neural crest

Neural groove

Floorplate

Neural crest

Neural tube

Floorplate   Notochord

Somite

(C) **22 days**

Central canal

Neural tube

Anterior neural fold

Neural crest

Somite

Anterior neural fold

Neural crest

Neural tube

Floorplate

Notochord

Somite

Sensory ganglion

(D) **24 days**

Neural tube

Somites   Spinal cord

Rhombencephalon

Sensory ganglion

Anterior neural fold

Floorplate

Notochord

Spinal chord

Sensory ganglion

notochord is required for subsequent neural differentiation (see Figure 22.1). Thus the notochord (along with the primitive pit) sends **inductive signals** to the overlying ectoderm that cause a subset of neuroectodermal cells to differentiate into neural precursor cells. During this process, called **neurulation**, the midline ectoderm that contains these cells thickens into a distinct columnar epithelium called the **neural plate**. The lateral margins of the neural plate then fold inward, eventually transforming the neural plate into a tube. This structure, the **neural tube**, subsequently gives rise to the brain and spinal cord.

The progenitor cells of the neural tube are known as **neural precursor cells**. These precursors are dividing stem cells that produce more precursors and, eventually, nondividing **neuroblasts** that differentiate into neurons. As a result of their proximity to the notochord, the cells at the ventral midline of the neural tube differentiate into a special strip of epithelial-like cells called

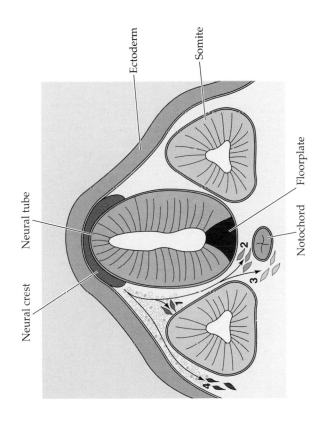

Neural crest

Neural tube

Ectoderm

Somite

Neural crest

Notochord

Floorplate

Figure 22.2 The neural crest. Diagram of a cross section through a developing mammalian embryo at a stage similar to that in Figure 22.1C. The neural crest cells follow four distinct migratory paths that lead to differentiation of distinct cell types and structures. Cells that follow the pathways labeled (1) and (2) give rise to sensory and autonomic ganglia, respectively. The precursors of adrenal neurosecretory cells migrate along pathway (3) and eventually aggregate around the dorsal portion of the kidney. Cells destined to become non-neural tissues (for example, melanocytes) migrate along pathway (4). Each pathway permits the migrating cells to interact with different kinds of cellular environments, from which they receive inductive signals (see Figure 22.11). (After Sanes, 1988.)

the **floorplate**. The position of the floorplate at the ventral midline determines the dorso-ventral polarity of the neural tube and further influences the differentiation of neural precursor cells. Inductive signals from the floorplate lead to the differentiation of cells in the ventral portion of the neural tube that eventually give rise to spinal and hindbrain motor neurons (which are thus closest to the ventral midline). Precursor cells farther away from the ventral midline give rise to sensory neurons within the spinal cord and hindbrain. At the most dorsal limit of the neural tube, a third population of cells emerges in the region where the edges of the folded neural plate join together. Because of their location, this set of precursors is called the **neural crest** (Figure 22.2). The neural crest cells migrate away from the neural tube along specific pathways that expose them to additional inductive signals that influence their differentiation. As a result, neural crest cells give rise to a variety of progeny, including the neurons and glia of the sensory and visceral motor (autonomic) ganglia, the neurosecretory cells of the adrenal gland, and the neurons of the enteric nervous system. They also contribute to variety of non-neural structures such as pigment cells, cartilage, and bone.

The Molecular Basis of Neural Induction

The essential consequence of gastrulation and neurulation for the development of the nervous system is the emergence of a population of neural precursors from a subset of ectodermal cells. Through a variety of experimental manipulations, primarily involving transplantation of different portions of developing embryos, embryologists recognized early on that this process depends on signals arising from cells in the primitive pit and notochord. Because a wide variety of chemical agents and physical manipulations are able to mimic some of the effects of these endogenous signals, their nature remained a mystery for several decades. It is now clear that the generation of cell identity—of which neural induction is but one mechanism—results from the spatial and temporal control of different sets of genes by endogenous signaling molecules. These inducing signals—including those from the primitive pit and notochord—are, not surprisingly, molecules that modulate gene expression.

The increasingly sophisticated effort to understand exactly how these inductive signals work has therefore focused on molecules that can modify

BOX A
Retinoic Acid: Teratogen and Inductive Signal

In the early 1930s, investigators noticed that vitamin A deficiency during pregnancy in animals led to a variety of fetal malformations. The most severe abnormalities affected the developing brain, which was often grossly malformed. At about the same time, experimental studies yielded the surprising finding that *excess* vitamin A caused similar defects. These observations suggested that a family of compounds—the retinoids—are teratogenic. (*Teratogenesis* is the term for

birth defects induced by exogenous agents.) The retinoids include the alcohol form of vitamin A (retinol), the aldehyde form of vitamin A (retinal), and the acid form (retinoic acid). Subsequent experiments in animals confirmed that other retinoids produce birth defects similar to those generated by too much—or too little—vitamin A. The disastrous consequences of exposure to exogenous retinoids during human pregnancy were underscored in the early 1980s when the drug Accutane® (the

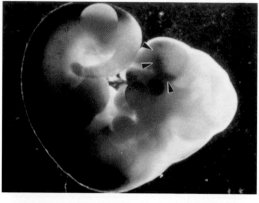

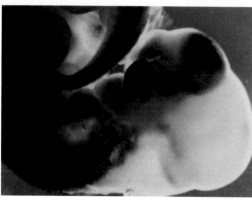

trade name for isoretinoin, or 13-*cis*-retinoic acid) was introduced as a treatment for severe acne. Women who took this drug during pregnancy had an increased number of spontaneous abortions and children born with a range of birth defects. Despite the importance of these several findings, the reasons for the adverse effects of retinoids on fetal development remained obscure until recently.

An important insight into teratogenic potential of retinoids came when embryologists working on limb development in chicks found that retinoic acid mimics the inductive ability of tissues in the limb bud. Still the mystery remained as to just what retinoic acid (or its absence) was doing to influence or compromise development. An important answer came in the mid-1980s, when the receptors for retinoic acid were discovered. These receptors are members of the steroid/thyroid hormone receptor superfamily; when they bind retinoic acid or similar ligands, the receptors act as transcription factors to activate specific genes. Furthermore, careful biochemical analysis showed that retinoic acid was synthesized by embryonic tissues. Subsequent studies have shown that retinoic acid activates gene expression at several sites in

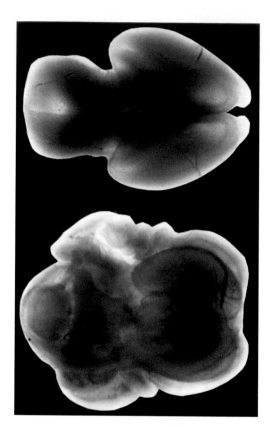

Top panel: At left, retinoic acid activates gene expression in a subset of cells in the normal developing forebrain of a mid-gestation mouse embryo (blue areas indicate β-galactosidase reaction product, an indicator of gene expression in this experiment); at right, after maternal ingestion of a small quantity of retinoic acid (0.00025 mg/g of maternal weight), gene expression is ectopically activated throughout the forebrain. Bottom panel: At left, the brain of a normal mouse at term; at right, the grossly abnormal brain of a mouse whose mother ingested this same amount of retinoic acid at mid-gestation. (Top panels from Anchan et al., 1997; bottom panels from Linney and LaMantia, 1994.)

the embryo including the developing brain (see figure). Among the most important targets for retinoic acid regulation are genes for other inductive signals including sonic hedgehog and Hox genes (see Box B). Thus, an excess or deficiency of retinoic acid can disrupt normal development by eliciting inappropriate patterns of retinoid-induced gene expression.

The role of retinoic acid as both a teratogen and an endogenous signaling molecule implies that the retinoids cause birth defects by mimicking the normal signals that influence gene expression. The story provides a good example of how teratogenic, clinical, cellular, and molecular observations can be combined to explain seemingly bizarre developmental pathology.

References

JOHNSON R. L. AND C. J. TABIN (1997) Molecular models for vertebrate limb development. Cell 90: 979–990

EVANS, R. M. (1988) The steroid and thyroid hormone receptor superfamily. Science 240: 889–895.

LAMANTIA, A.-S., M. C. COLBERT AND E. LINNEY (1993) Retinoic acid induction and regional differentiation prefigure olfactory pathway formation in the mammalian forebrain. Neuron 10: 1035–1048.

LAMMER, E. J. AND 11 OTHERS (1985) Retinoic acid embryopathy. N. Engl. J. Med. 313: 837–841.

SCHARDEIN, J. L. (1993) *Chemically Induced Birth Defects*, 2nd Ed. New York: Marcel Dekker.

THALLER, C. AND G. EICHELE (1987) Identification and spatial distribution of retinoids in the developing chick limb bud. Nature 327: 625–628.

TICKLE, C., B. ALBERTS, L. WOLPERT AND J. LEE (1982) Local application of retinoic acid to the limb bud mimics the action of the polarizing region. Nature 296: 564–565.

WARKANY, J. AND E. SCHRAFFENBERGER (1946) Congenital malformations induced in rats by maternal vitamin A deficiency. Arch. Ophthalmol. 35: 150–169.

patterns of gene expression. An instructive example is **retinoic acid**, a derivative of vitamin A and a member of the steroid/thyroid superfamily of hormones (Box A). Retinoic acid activates a unique class of **transcription factors**—the retinoid receptors—that modulate the expression of a number of target genes. Peptide hormones provide another class of inductive signals, including those that belong to the **fibroblast growth factor (FGF)** and **transforming growth factor (TGF)** families. Another peptide hormone essential for neural induction is **sonic hedgehog (shh)**. These molecules, like retinoic acid, are produced by a variety of embryonic tissues including the notochord, the floorplate, and the neural ectoderm itself; they bind to cell surface receptors, many of which are protein kinases. Some of these molecular signals have been implicated in determining the fates of specific classes of cells in the developing nervous system (Figure 22.3). For example, shh is essential for the differentiation of motor neurons in the ventral spinal cord, whereas a TGF family molecule called *dorsalin* is important for the establishment of dorsal cells in the spinal cord—including the neural crest. Signaling via these peptide hormones activates a cascade of subsequent gene expression in ectodermal cells. In general, if the signaling mediated by any of these molecules is disrupted, the early development of the nervous system is compromised.

A particularly intriguing aspect of molecular signals that influence neural induction is the mechanism by which one class of inductive signals—the bone morphogenetic proteins, or BMPs—cause neural differentiation. As the name suggests, these peptide hormones, which are members of the TGF-β family, elicit osteogenesis from mesodermal cells. If ectodermal cells are exposed to BMP, they assume an epidermal fate. How then does the ectoderm manage to become neuralized, especially since BMPs are produced by the notochord, floorplate, and somites? All of these structures are in position to signal to the neuroectoderm, and therefore to convert it to epidermis. This epidermal fate is evidently avoided in the neural plate by the local activity of other inductive signaling molecules called noggin and chordin. Both of these molecules bind directly to the BMPs and thus prevent their binding to BMP receptors. In this way, the neuroectoderm is "rescued" from becoming epi-

Figure 22.3 Location of some inductive signals in the developing neural tube. Inductive signals are provided by either the notochord, the floorplate, the roofplate and dorsal ectoderm, or the somites. These signals act locally on either the ventral or dorsal neuroepithelium of the developing spinal cord and hindbrain to elicit distinct patterns of gene expression and, ultimately, differentiation of specific classes of neurons. The peptide hormone sonic hedgehog (shh) is the most important ventral signal and is produced by both the notochord and floorplate. In addition, noggin, chordin, and retinoic acid are produced either by the notochord or floorplate. In contrast, a variety of signals including dorsalin and other members of the TGF family as well as noggin and retinoic acid are provided by the roofplate and dorsal ectoderm. These signals influence the differentiation of several dorsal cell types including the neural crest.

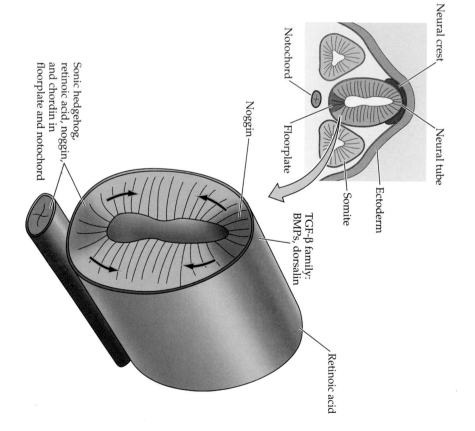

Neural crest

Notochord

Floorplate

Neural tube

Ectoderm

Somite

Noggin

TGF-β family;
BMPs, dorsalin

Sonic hedgehog,
retinoic acid, noggin,
and chordin in
floorplate and notochord

Retinoic acid

dermis. Such negative regulation has led to the conclusion that becoming a neuron is actually the "default" fate for embryonic ectodermal cells.

This and other knowledge about the molecules involved in neural induction has provided a much more informed way of thinking about the etiology and prevention of a number of congenital disorders. Anomalies like **spina bifida** (failure of the posterior neural tube to close completely), **anencephaly** (failure of the anterior neural tube to close at all), and other brain malformations (often accompanied by mental retardation) probably result from defects in inductive signaling or the genes that participate in this process. As already described, excessive intake of vitamin A can impede neural tube closure and differentiation or disrupt later aspects of neuronal differentiation. Embryonic exposure to a variety of other drugs—alcohol and thalidomide are good examples—can also elicit pathological differentiation of the embryonic nervous system by providing inductive signals at inappropriate times or places. Furthermore, dietary insufficiency of substances like **folic acid** can disrupt neural tube formation by compromising cellular mechanisms essential for normal cell division and motility. Because the consequences of disordered neural induction are so severe, pregnant women are well advised to avoid virtually all drugs and dietary supplements—except those deemed safe and prescribed specifically by physicians—especially during the first trimester of pregnancy.

Formation of the Major Brain Subdivisions

Soon after neural tube formation, the forerunners of the major brain regions become apparent as a result of morphogenetic movements that bend, fold, and constrict the neural tube. Initially, the anterior end of the tube forms a crook, giving it the shape of a cane handle (Figure 22.4A). The end of the

(A)

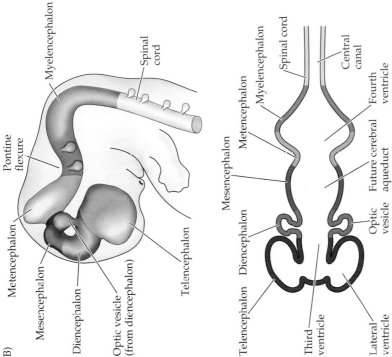

Optic
vesicle

Rhombencephalon

Cervical
flexure

Cranial and
spinal ganglia

Spinal
cord

Cephalic
flexure

Mesencephalon

Prosencephalon

Prosencephalon

Rhombencephalon

Mesencephalon

Spinal cord

(B)

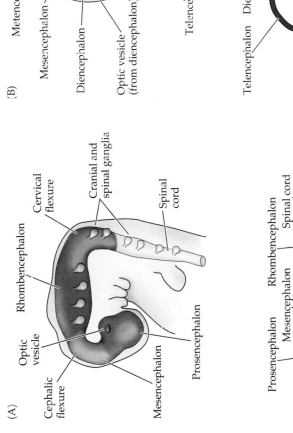

Metencephalon

Mesencephalon

Diencephalon

Optic vesicle
(from diencephalon)

Pontine
flexure

Myelencephalon

Spinal
cord

Telencephalon

Mesencephalon

Diencephalon

Telencephalon

Metencephalon

Myelencephalon

Spinal
cord

Third
ventricle

Lateral
ventricle

Optic
vesicle

Future cerebral
aqueduct

Fourth
ventricle

Central
canal

33

sylvius

(C)

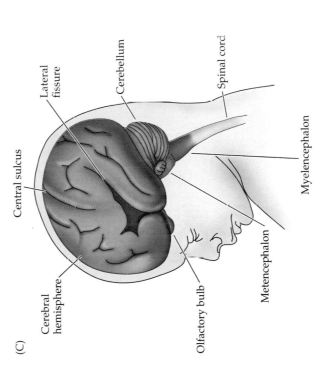

Central sulcus

Lateral
fissure

Cerebellum

Spinal cord

Cerebral
hemisphere

Olfactory bulb

Metencephalon

Myelencephalon

Figure 22.4 Regional specification of the developing brain. (A) Early in gestation the neural tube becomes subdivided into the prosencephalon (at the anterior end of the embryo), mesencephalon, and rhombencephalon. The spinal cord differentiates from the more posterior region of the neural tube. The initial bending of the neural tube at its anterior end leads to a cane shape. Below is a longitudinal section of the neural tube at this stage, showing the position of the major brain regions. (B) Further development distinguishes the telencephalon and diencephalon from the prosencephalon; two other subdivisions—the metencephalon and myelencephalon—derive from the rhombencephalon. These subregions give rise to the rudiments of the major functional subdivisions of the brain, while the spaces they enclose eventually form the ventricles of the mature brain. Below is a longitudinal section of the embryo at the developmental stage shown in (B). (C) The fetal brain and spinal cord are clearly differentiated by the end of the second trimester. Several major subdivisions, including the cerebral cortex and cerebellum, are clearly seen from the lateral surfaces.

Figure 22.5 Sequential gene expression divides the *Drosophila melanogaster* (fruit fly) embryo into regions and segments. (A) Temporal pattern of expression of four genes that influence the establishment of the body plan in *Drosophila*. A series of sections through the anterior-posterior midline of the embryo are shown from early to later stages of development (top to bottom in each row). Initially, expression of the gene *bicoid* (*bcd*) helps define the anterior pole of the embryo. Next, the gene *krüppel* (*kr*) is expressed in the middle and then at the posterior end of the embryo, defining the anterior-posterior axis. Then, the gene *hairy* (*h*) is expressed, which helps to delineate the domains that will eventually form the mature segmented body of the fly. Finally, the gene *wingless* (*wg*) is expressed, further refining the organization of individual segments. (B) The relationship of embryonic segments in the *Drosophila* larva defined by sequential gene expression, shown in (A), to the body plan of the mature fly. (A from Ingham, 1988; B after Gilbert, 1994, and Lawrence, 1992.)

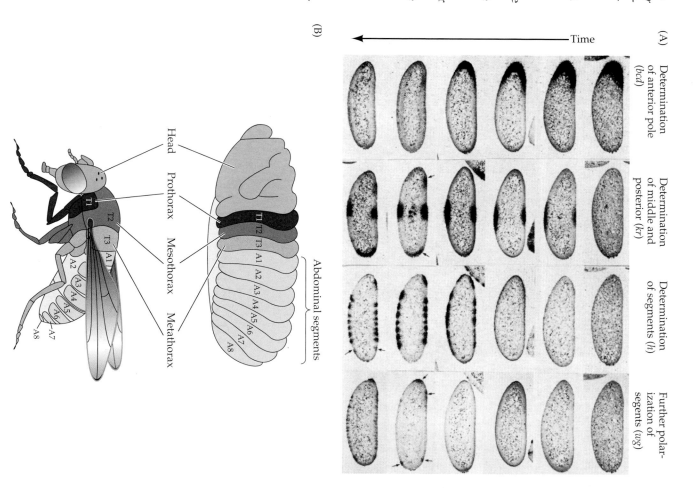

(A)

Determination of anterior pole (*bcd*)

Determination of middle and posterior (*kr*)

Determination of segments (*h*)

Further polarization of segments (*wg*)

(B)

Time

Head

Prothorax

Mesothorax

Metathorax

Abdominal segments

cane nearest the sharper bend, or cephalic flexure, balloons out to form the **forebrain**, or **prosencephalon**. The **midbrain**, or **mesencephalon**, forms as a bulge called the cephalic flexure. The **hindbrain**, or **rhombencephalon**, forms in the long, relatively straight stretch between the cephalic flexure and the more caudal cervical flexure. Caudal to the cervical flexure, the neural tube forms the precursor of the spinal cord. This bending and folding constricts or enlarges the lumen enclosed by the developing neural tube. These lumenal spaces eventually become the ventricles of the mature brain (Figure 22.4B).

Once the primitive brain regions are established in this way, they undergo at least two more rounds of partitioning, each of which produces additional regions in the adult (Figure 22.4C). Thus, the lateral aspects of the rostral prosencephalon forms the **telencephalon**. The two bilaterally symmetric telencephalic vesicles contain the rudiments of the cerebral cortex, hippocampus, basal ganglia, basal forebrain nuclei, and olfactory bulb. The more caudal portion of the prosencephalon forms the **diencephalon**, which contains the rudiments of the thalamus and hypothalamus, as well as a pair of lateral outpocketings (the **optic cups**) from which the neural portion of the retina will form. The rostral part of the rhombencephalon becomes the **metencephalon** and gives rise to the adult cerebellum and pons. Finally, the caudal part of the rhombencephalon becomes the **myelencephalon** and gives rise to the adult medulla.

How can a simple tube of neuronal precursor cells produce such a variety of brain structures? At least part of the answer comes from the observation made early in the twentieth century that much of the neural tube is organized into repeating units called **neuromeres**. This discovery led to the idea that the process of segmentation—used by all animal embryos at the earliest stages of development to establish regional identity in the body—might also establish regional identity in the developing brain. Enthusiasm for this hypothesis was stimulated by observations of the development of the body plan of the fruit fly *Drosophila*. In the fly, early expression of a class of genes called **homeobox genes** (Box B) guides the differentiation of the embryo into distinct segments that give rise to the head, thorax, and abdomen (Figure 22.5). These genes code for DNA-binding proteins that can modulate the expression of other genes. Similar homeobox genes in mammals (referred to as *Hox* genes) have been identified. In some cases their patterns of expression coincide with, or even precede, the formation of morphological features such as the various bends, folds, and constrictions that signify the progressive regionalization of the developing neural tube, particularly in the hindbrain and spinal cord (Box C). More recently, similar genes have been associated with regional differentiation in the brainstem. The patterned expression of homeobox genes, as well as other developmentally regulated transcription factors and signaling molecules, does not by itself determine the fate of a group of embryonic neural precursors. As in the case of neural induction, regionally distinct transcription factor expression contributes to a series of genetic and cellular processes that eventually produce a fully differentiated brain.

Genetic Abnormalities and Altered Human Brain Development

The recent explosion of information about molecules that influence brain development provides a basis for reevaluating the causes of a number of congenital brain malformations as well as various forms of mental retardation. Thus, some forms of **hydrocephalus** (a constriction of the lumen of the neural tube that eventually leads to enlarged ventricles and subsequent cortical atrophy as a result of compression) can be traced to mutations of genes on the X chromosome. Similarly, **fragile X syndrome**, the most common form of congenital mental retardation, is associated with triplet repeats in a subset of genes on the X chromosome.

Beyond these X-linked abnormalities, there are at least two genetic disorders that compromise the nervous system generated by single gene mutations in homeobox-like transcription factors. **Aniridia**, which is characterized by loss of the iris in the eye and mild mental retardation, and **Wardenburg syndrome**, which is characterized by craniofacial abnormalities, spina bifida and hearing loss, are caused by mutations in the *Pax* 6 and

BOX B
Homeotic Genes and Human Brain Development

The notion that particular genes can influence the establishment of distinct regions in an embryo arose from efforts to catalog single-gene mutations that affect development of the fruit fly *Drosophila*. In the 1960s and 1970s, E. B. Lewis at the California Institute of Technology reported a number of mutations that resulted in either the duplication of a distinct body segment or the appearance of an inappropriate structure at an ectopic location in the fly. These genes were called homeotic genes because they were able to convert segments of one sort to those of another (*homeo* is Greek for "similar"). Subsequently, studies by C. Nusslein-Volhard and E. Wieschaus demonstrated the existence of numerous such "master control" genes, each forming part of a cascade of gene expression leading to the distinctive segmentation of the developing embryo. (In 1995, Lewis, Nusslein-Volhard, and Wieschaus shared a Nobel Prize for these discoveries.)

Homeotic genes code for DNA-binding proteins—that is, transcription factors—that bind to a particular sequence of genomic DNA called the "homeobox." Similar genes have been found in most species, including hu-

mans. Using an approach known as cloning by homology, at least four "clusters" of homeobox genes have been identified in virtually all vertebrates that have been examined. The genes of each cluster are closely, but not consecutively, spaced on a single chromosome. Other motifs identified in *Drosophila* have led to the discovery of additional families of DNA-binding proteins, which have again been found in a variety of species.

Importantly, a number of developmental anomalies in mice and humans have been associated with mutations in the homeotic or other developmental control genes initially identified in the fly. Relatively rare diseases like aniridia, Wardenburg's syndrome, and Greig cephalopolysyndactyly syndrome (all disorders that disrupt the nervous system and peripheral structures like the iris or the digits) have been associated with human genes that are homologues of *Drosophila* developmental control genes. In addition, several other developmental disorders including autism and various forms of mental retardation can be associated with mutations or polymorphisms of homeobox genes (see text). Thus, the initial insights into the molecular control

of development gleaned from genetic studies of *Drosophila* have opened new avenues for exploring the molecular basis of developmental disorders in humans.

References

ENGELKAMP, D. AND V. VAN HEYNINGEN (1996) Transcription factors in disease. Curr. Opin. Genet. Dev. 6: 334–42.

GEHRING, W. J. (1993) Exploring the homeobox. Gene 135: 215–221.

GRUSS, P. AND C. WALTHER (1992) *Pax* in development. Cell 69: 719–722.

LEWIS, E. B. (1978) A gene complex controlling segmentation in *Drosophila*. Nature 276: 565–570.

NUSSLEIN-VOLHARD, C. AND E. WIESCHAUS (1980) Mutations affecting segment number and polarity in *Drosophila*. Nature 287: 795–801.

READ, A. P. AND V. E. NEWTON (1997) Waardenburg syndrome. J. Med. Genet. 34: 656–665.

SHIN, S. H., P. KOGERMAN, E. LINDSTROM, R. TOFTGARD AND L. G. BIESECKER (1999) Gli3 mutations in human disorders mimic *Drosophila* cubitus interruptus protein functions and localization. Proc. Natl. Acad. Sci. USA 96: 2880–2884.

Pax 3 genes, respectively, both of which produce transcription factors (see Box B). Finally, developmental disorders like **autism** and other severe learning impairments have in some instances been linked to microdeletions or duplications of specific chromosomal regions. Although the causal links between these genes and the resulting anomalies of brain development are not yet known, such correlations provide a starting point for understanding the molecular pathogenesis of many congenital nervous system disorders.

The Initial Differentiation of Neurons and Glia

Once the neural tube has developed into a rudimentary brain and spinal cord, the generation and differentiation of the permanent cellular elements of the brain—neurons and glia—begin in earnest. The mature human brain contains about 100 billion neurons and many more glial cells, all generated over the course of only a few months from a small population of precursor cells. Except for a few specialized cases (see Chapter 25), the entire neuronal complement of the adult brain is produced during a time window that closes before birth; thereafter, precursor cells disappear, and few if any new neurons can be added to replace those lost by age or injury in most brain regions. The precursor cells are located in the **ventricular zone**, the innermost cell layer surrounding the lumen of the neural tube. The ventricular zone is a region of extraordinary mitotic activity: In humans, it has been estimated that about 250,000 new neurons are generated each minute during the peak of cell proliferation during gestation.

The dividing precursor cells in the ventricular zone undergo a stereotyped pattern of cell movements as they progress through the mitotic cycle (Figure 22.6), leading to the formation of either new stem cells or postmitotic neuroblasts that differentiate into neurons. As cells become postmitotic, they leave the ventricular zone and migrate to their final positions in the developing brain. Knowing when the neurons destined to populate a given brain region are "born"—that is, when they become postmitotic (determined by performing birthdating studies; Box D)—has provided considerable insight into how different regions of the brain are constructed. Thus, different nuclei of the brainstem and thalamus are distinguished by the times when their component neurons are generated. In the cerebral cor-

Figure 22.6 Dividing precursor cells in the vertebrate neuroepithelium (neural plate and neural tube stages) are attached both to the pial (outside) surface of the neural tube and to its ventricular (lumenal) surface. The nucleus of the cell translocates between these two limits within a narrow cylinder of cytoplasm. When cells are closest to the outer surface of the neural tube, they enter a phase of DNA synthesis (the S stage); after the nucleus moves back to the ventricular surface (the G_2 stage), the precursor cells lose their connection to the outer surface and enter mitosis (the M stage). When mitosis is complete, the two daughter cells extend processes back to the outer surface of the neural tube, and the new precursor cells enter a resting (G_1) phase of the cell cycle. At some point a precursor cell generates either another stem cell that will go on dividing and a daughter cell that will not divide further (that is, a neuroblast) or two postmitotic daughter cells.

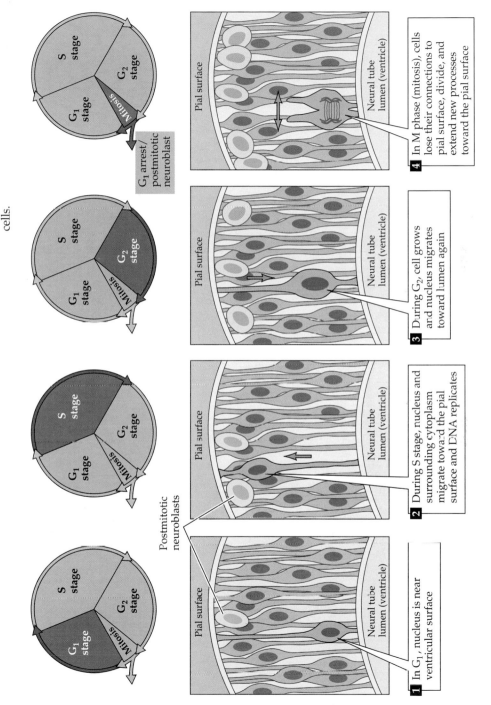

1 In G_1, nucleus is near ventricular surface

2 During S stage, nucleus and surrounding cytoplasm migrate toward the pial surface and DNA replicates

3 During G_2 stage, cell grows and nucleus migrates toward lumen again

4 In M phase (mitosis), cells lose their connections to pial surface, divide, and extend new processes toward the pial surface

Postmitotic neuroblasts

G_1 arrest / postmitotic neuroblast

BOX C
Rhombomeres

An interesting parallel between early embryonic segmentation and early brain development was noticed around the turn of the nineteenth century. Several embryologists reported repeating units in the early neural plate and neural tube, which they called *neuromeres*. In the late 1980s, A. Lumsden, R. Keynes, and their colleagues, as well as R. Krumlauf, R. Wilkinson, and colleagues, noticed further that combinations of homeobox (*Hox*) genes (see Box B) are expressed in banded patterns in

the developing chick nervous system, especially in the hindbrain (the common name for the rhombencephalon and its derivatives). These *Hox* expression domains defined rhombomeres, which in the chick (as well as in most mammals), are a series of seven transient bulges in the developing rhombencephalon corresponding to the neuromeres described earlier. Rhombomeres are sites of differential cell proliferation (cells at rhombomere boundaries divide faster than cells in

the rest of the rhombomere), differential cell mobility (cells from any one rhombomere cannot easily cross into adjacent rhombomeres), and differential cell adhesion (cells prefer to stick to those of their own rhombomere).

Later in development, the pattern of axon outgrowth from the cranial motor nerves also correlates with the earlier rhombomeric pattern. Cranial motor nerves (see Chapter 1) originate either from a single rhombomere or from specific pairs of neighboring rhombomeres

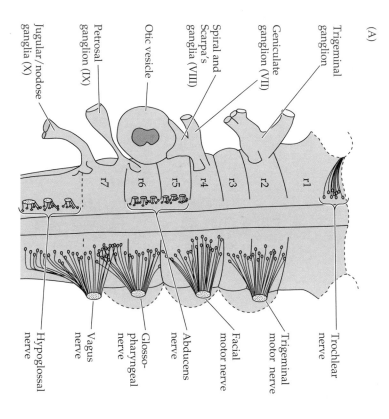

(A)

Trigeminal ganglion

Geniculate ganglion (VII)

Spiral and Scarpa's ganglia (VIII)

Otic vesicle

Petrosal ganglion (IX)

Jugular/nodose ganglia (X)

r7 r6 r5 r4 r3 r2 r1

Trochlear nerve

Trigeminal motor nerve

Facial motor nerve

Abducens nerve

Glosso-pharyngeal nerve

Vagus nerve

Hypoglossal nerve

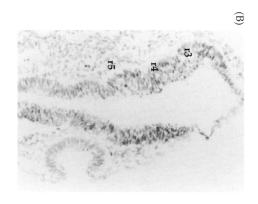

(B)

r3 r4 r5

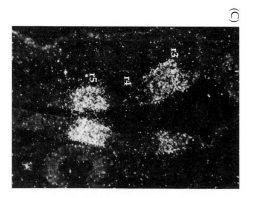

(C)

r3 r4 r5

Rhombomeres in the developing chicken hindbrain and their relationship to the differentiation of the cranial nerves. (A) Diagram of the chick hindbrain, indicating the position of the cranial ganglia and nerves and their rhombomeric origin (rhombomeres denoted as r1 to r7). (B) Section through early chicken hindbrain, showing bulges that will eventually become rhombomeres (in this example, r3 to r5). (C) Differential patterns of transcription factor expression (in this case, *krx 20*, a *Hox*-like gene) define rhombomeres at early stages of development, well before the cranial nerves that will eventually emerge from them are apparent. (A courtesy of Andrew Lumsden; B,C from Wilkinson and Krumlauf, 1990.)

(transplantation experiments indicate that rhombomeres are in fact specified in pairs). Thus, *Hox* gene expression probably represents an early step in the formation of cranial nerves in the developing brain. Mutation or ectopic activation of *Hox* genes in mice alters the position of specific cranial nerves, or prevents their formation. Mutation of the *HoxA-1* gene by homologous recombination—the so-called "knockout" strategy for targeting mutations to specific genes—prevents normal formation of rhombomeres. In these animals, development of the external, middle, and inner ear is also compromised, and cranial nerve ganglia are fused and located incorrectly. Conversely, when the *HoxA-1* gene is expressed in a rhombomere where it is usually not seen, the ectopic expression causes changes in rhombomere identity and subsequent differentiation. It is likely that problems in rhombomere formation are the underlying cause of congenital nervous system defects involving cranial nerves, ganglia, and peripheral structures derived from the cranial neural crest (the part of the neural crest that arises from the hindbrain).

The exact relationship between early patterns of rhombomere-specific gene transcription and subsequent cranial nerve development remains a puzzle. Nevertheless, the correspondence between these repeating units in the embryonic brain and similar iterated units in the development of the insect body (see Figure 22.5) suggests that differential expression of transcription factors in specific regions is essential for the normal development of many species. In a wide variety of animals, spatially and temporally distinct patterns of transcription factor expression coincide with spatially and temporally distinct patterns of differentiation, including the differentiation of the nervous system. The idea that the bulges and folds in the neural tube are segments defined by patterns of gene expression provides an attractive framework for understanding the molecular basis of pattern formation in the developing vertebrate brain.

References

GUTHRIE, S. (1996) Patterning the hindbrain. Curr. Opin. Neurobiol. 6: 41–48.

CARPENTER, E. M., J. M. GODDARD, O. CHISAKA, N. R. MANLEY AND M. CAPECCHI (1993) Loss of *HoxA-1* (*Hox-1.6*) function results in the reorganization of the murine hindbrain. Develop. 118: 1063–1075.

LUMSDEN, A. AND R. KEYNES (1989) Segmental patterns of neuronal development in the chick hindbrain. Nature 337: 424–428.

WILKINSON, D. G. AND R. KRUMLAUF (1990) Molecular approaches to the segmentation of the hindbrain. Trends Neurosci. 13: 335–339.

VON KUPFFER, K. (1906) Die morphogenie des central nerven systems. In *Handbuch der vergleichende und experimentelle Entwicklungslehreder Wirbeltiere*, Vol. 2, 3: 1–272. Fischer Verlag, Jena.

ZHANG, M. AND 9 OTHERS (1993) Ectopic *HoxA-1* induces rhombomere transformation in mouse hindbrain. Develop. 120: 2431–2442.

tex, neurons of the six layers of the cortex are generated in an inside-out manner. The firstborn cells are eventually located in the deepest layers, while later generations of neurons migrate through the older cells and come to lie superficial to them (Figure 22.7). Thus, each cortical layer, like subcortical nuclei, consists of a cohort of cells generated during a specific developmental period. The implication of this fact is that common periods of neurogenesis are important for the development of the cell types and connections that characterize each of the cortical layers.

The Generation of Neuronal Diversity

The neuronal precursor cells in the ventricular zone of the embryonic brain look and act more or less the same. Yet these precursors ultimately give rise to postmitotic cells that are enormously diverse in form and function. The spinal cord, cerebellum, cerebral cortex, and subcortical nuclei (including the basal ganglia and thalamus) each contain several dozen neuronal cell types distinguished by morphology, neurotransmitter content, cell surface molecules, and the types of synapses they make and receive. On an even more basic level, the stem cells of the ventricular zone produce both neurons and glia, cells with markedly different properties and functions. How and when are these different cell types determined?

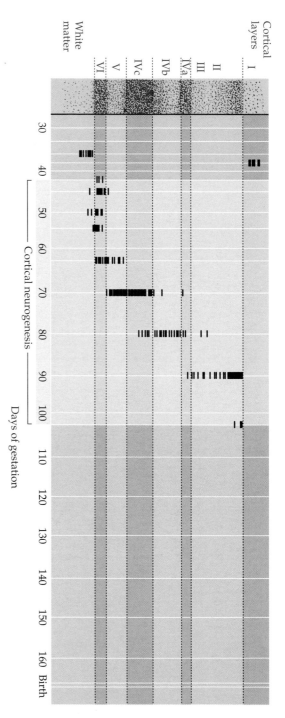

Figure 22.7   Generation of cortical neurons during the gestation of a rhesus monkey (a span of about 165 days). The final cell divisions of the neuronal precursors, determined by maximal incorporation of radioactive thymidine administered to the pregnant mother (see Box D), occur primarily during the first half of pregnancy and are complete on or about embryonic day 105. Each short horizontal line represents the position of a neuron heavily labeled by maternal injection of radiolabeled thymidine at the time indicated by the corresponding vertical line. The numerals on the left designate the cortical layers. The earliest generated cells are found in a transient layer called the subplate (a few of these cells survive in the white matter) and in layer I (the Cajal-Retzius cells). (After Rakic, 1974.)

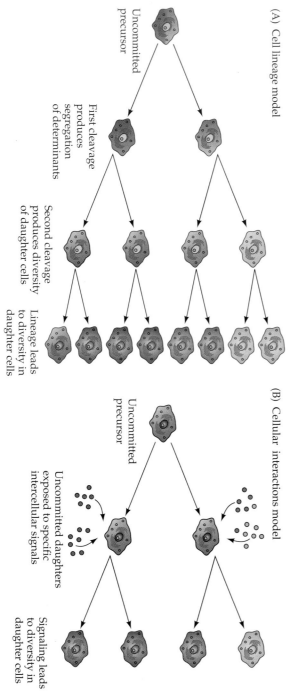

(A) Cell lineage model

Uncommitted precursor

First cleavage produces segregation of determinants

Second cleavage produces diversity of daughter cells

Lineage leads to diversity in daughter cells

(B) Cellular interactions model

Uncommitted precursor

Uncommitted daughters exposed to specific intercellular signals

Signaling leads to diversity in daughter cells

Figure 22.8   Two hypotheses about the generation of cell diversity during embryonic development. (A) Cells acquire diverse fates while still at the precursor stage, relying primarily on information intrinsic to each cell. Subsequent divisions result in the proliferation of these cells, which differentiate according to their lineage. (B) Cells are descended from a pleuripotential precursor, diversity being generated among daughter cells by specific signals from other cells. Experimental evidence in vertebrates favors this second model.

One possibility is that the precursors of different populations of neurons (or neurons and glia) are established very early in development, perhaps at the formation of the neural plate (Figure 22.8A). Separate types of precursor cells would then exist in the ventricular zone, each giving rise to a particular type of cell in the adult. According to this school of thought, a cell's fate is a function of its lineage: Neurons of different types would have distinct "ancestors," as would glial cells. At the other extreme, the ancestral precursor cells might provide essentially no information about eventual phenotype; in this scenario, all subsequent differentiation depends on interactions with other cells in particular brain microenvironments (Figure 22.8B).

There is little evidence that precursor cells are committed irrevocably early on to produce particular types of daughter cells. In fact, in several brain regions precursor cells generate postmitotic daughter cells throughout the course of development that assume a number of different phenotypes (Figure 22.9). In the retina, for example, experiments using lineage-marking techniques have shown that a precursor cell can generate any combination of cell types found in the retina, including photoreceptors, bipolar cells, amacrine cells, ganglion cells, and even glial cells. The developing spinal cord, superior colliculus, cerebellum, and cerebral cortex also appear to lack clearly specified precursor cells. Apparently, cell lineage plays only a minor role in specifying cell fate for at least these components of the developing brain.

The bulk of the evidence favors the view that neuronal differentiation is based primarily on cell-cell interactions (Figure 22.8B). Historically, most experimental approaches to this issue have relied on transplantation strategies, such as moving bits of a particular brain region to a different location in the brain of a host animal to determine whether the transplanted cells acquire the host phenotype or retain their original fate during subsequent development. When very young precursor cells are transplanted, they tend to acquire the host phenotype. Transplants at increasingly older ages, however,

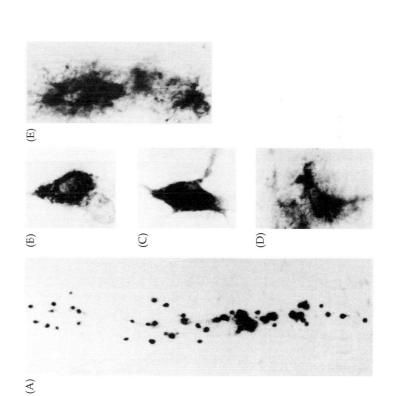

Figure 22.9 Cells derived from the last divisions of a single precursor can assume different fates, implying that lineage has little influence on cellular phenotype. In the chick optic tectum, injection of a replication-incompetent retrovirus at early stages in development inserts a reporter gene into the genome of a single progenitor cell. The progeny of the cell can then be detected using a simple histochemical stain for expression of the reporter gene (in this case for the enzyme β-galactosidase). As shown in (A), the offspring of a single progenitor cell form a narrow column spanning almost the entire thickness of the optic tectum. Both neurons recognized by the shape of their cell body (B and C) and glia recognized by the halo of fine, hairlike processes (D and E) can be derived from the same progenitor. (From Galileo et al., 1990.)

Figure 22.12   Cell signaling during the migration of neural crest cells. The establishment of each precursor type relies on signals provided by one of several specific peptide hormones. The availability of each signal depends on the migratory pathway.

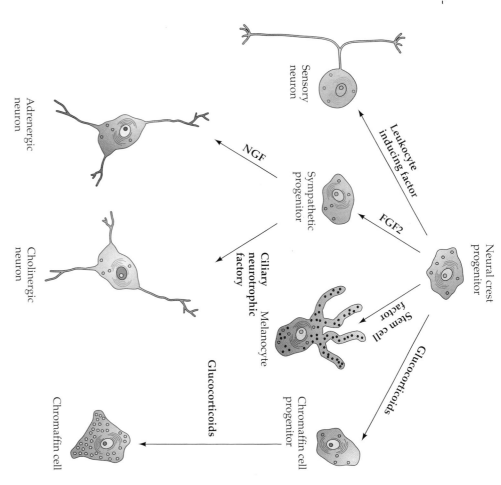

made available from somites, visceral epithelial structures (like the developing dorsal aorta), mesodermally derived mesenchymal cells, and the neural crest cells themselves. Of particular significance is the fact that specific peptide hormone growth factors cause neural crest cells to differentiate into distinct phenotypes (Figure 22.12). These effects depend on the location of the neuronal precursor cell along a migratory pathway, different signals being available at different points. Such position-dependent cues are probably not restricted to the peripheral nervous system; in the cerebellum, for example, different patterns of genes are expressed in migrating granule neurons at different locations, implying the existence of different signals (as yet unknown) along the migratory path.

Thus, neuronal migration involves much more than the mechanics of moving cells from one place to another. As is the case for inductive events during the initial formation of the nervous system, stereotyped movements bring different classes of cells into contact with one another, thereby providing a means of constraining cell-cell signaling to specific times and places.

Summary

The initial development of the nervous system depends on an intricate interplay of cellular movements and inductive signals. In addition to an early reordering of cellular positions as a result of morphogenesis, substantial migration of neuronal precursors is necessary for the subsequent differentiation of distinct classes of neurons, and for the eventual formation of specialized patterns of synaptic connections. The fate of individual precursor cells is not determined simply by their mitotic history; rather, the information required for differentiation arises largely from interactions between the developing cells and their local environment. All of these events are dependent on the same categories of molecular and cellular phenomena: cell-cell signaling, transcriptional regulation, and, ultimately, cell-specific changes in gene expression. The molecules that participate in signaling during early brain development are the same as the signals used by mature cells: hormones, transcription factors, other second messengers (see Chapter 8), and cell adhesion molecules. As might be expected, the identification and characterization of these molecules in the developing brain has begun to explain a variety of congenital neurological defects. These signaling events early in neural development are especially vulnerable to the effects of genetic mutations, and to the actions of the many drugs and toxins that can compromise the elaboration of a normal nervous system.

ADDITIONAL READING

Reviews

ANDERSON, D. J. (1993) Molecular control of cell fate in the neural crest: The sympathoadrenal lineage. Ann. Rev. Neurosci. 16: 129–158.

CAVINESS, V. S. JR. AND P. RAKIC (1978) Mechanisms of cortical development: A view from mutations in mice. Ann. Rev. Neurosci. 1: 297–326.

FRANCIS, N. J. AND S. C. LANDIS (1999) Cellular and molecular determinants of sympathetic neuron development. Ann. Rev. Neurosci. 22: 541–566.

HATTEN, M. E. (1993) The role of migration in central nervous system neuronal development. Curr. Opin. Neurobiol. 3: 38–44.

INGHAM, P. (1988) The molecular genetics of embryonic pattern formation in *Drosophila*. Nature 335: 25–34.

JESSELL, T. M. AND D. A. MELTON (1992) Diffusible factors in vertebrate embryonic induction. Cell 68: 257–270.

KESSLER, D. S. AND D. A. MELTON (1994) Vertebrate embryonic induction: Mesodermal and neural patterning. Science 266: 596–604.

KEYNES, R. AND R. KRUMLAUF (1994) *Hox* genes and regionalization of the nervous system. Ann. Rev. Neurosci. 17: 109–132.

LEWIS, E. M. (1992) The 1991 Albert Lasker Medical Awards. Clusters of master control genes regulate the development of higher organisms. JAMA 267: 1524–1531.

LINNEY, E. AND A. S. LAMANTIA (1994) Retinoid signaling in mouse embryos. Adv. Dev. Biol. 3: 73–114.

RICE, D. S. AND T. CURRAN (1999) Mutant mice with scrambled brains: Understanding the signaling pathways that control cell positioning in the CNS. Genes Dev. 13: 2758–2773.

RUBENSTEIN, J. L. R. AND P. RAKIC (1999) Genetic control of cortical development. Cerebral Cortex 9: 521–523.

SANES, J. R. (1989) Extracellular matrix molecules that influence neural development. Ann. Rev. Neurosci. 12: 491–516.

SELLECK, M. A., T. Y. SCHERSON AND M. BRONNER-FRASER (1993) Origins of neural crest cell diversity. Dev. Biol. 159: 1–11.

ZIPURSKY, S. L. AND G. M. RUBIN (1994) Determination of neuronal cell fate: Lessons from the R7 neuron of *Drosophila*. Ann. Rev. Neurosci. 17: 373–397.

Important Original Papers

ANCHAN, R. M., D. P. DRAKE, C. F. HAINES, E. A. GERWE AND A-S. LAMANTIA (1997) Disruption of local retinoid-mediated gene expression accompanies abnormal development in the mammalian olfactory pathway. J. Comp. Neurol. 379: 171–184.

ANGEVINE, J. B. AND R. L. SIDMAN (1961) Autoradiographic study of cell migration during histogenesis of cerebral cortex in the mouse. Nature 192: 766–768.

BULFONE, A., L. PUELLES, M. H. PORTEUS, M. A. FROHMAN, G. R. MARTIN AND J. L. RUBENSTEIN (1993) Spatially restricted expression of Dlx-1, Dlx-2 (Tes-1), Gbx-2, and Wnt-3 in the embryonic day 12.5 mouse forebrain defines potential transverse and longitudinal segmental boundaries. J. Neurosci. 13: 3155–3172.

EKSIOGLU, Y. Z. AND 12 OTHERS (1996) Periventricular heterotopia: An X-linked dominant epilepsy locus causing aberrant cerebral cortical development. Neuron 16: 77–87.

ERICSON, J., S. MORTON, A. KAWAKAMI, H. ROELINK AND T. M. JESSELL (1996) Two critical periods of *sonic hedgehog* signaling required for the specification of motor neuron identity. Cell 87: 661–673.

GALILEO, D. S., G. E. GRAY, G. C. OWENS, J. MAJORS AND J. R. SANES (1990) Neurons and glia arise from a common progenitor in chicken optic tectum: Demonstration with two retroviruses and cell type-specific antibodies. Proc. Natl. Acad. Sci. USA 87: 458–462.

GRAY, G. E. AND J. R. SANES (1991) Migratory paths and phenotypic choices of clonally related cells in the avian optic tectum. Neuron 6: 211–225.

HAFEN, E., K. BASLER, J. E. EDSTROEM AND G. M. RUBIN (1987) *Sevenless*, a cell-specific homeotic gene of *Drosophila*, encodes a putative transmembrane receptor with a tyrosine kinase domain. Science 236: 55–63.

HEMMATI-BRIVANLOU, A. AND D. A. MELTON (1994) Inhibition of activin receptor signaling promotes neuralization in *Xenopus*. Cell 77: 273–281.

KRAMER, H., R. L. CAGAN AND S. L. ZIPURSKY (1991) Interaction of bride of sevenless membrane-bound ligand and the sevenless tyrosine-kinase receptor. Nature 352: 207–212.

LANDIS, S. C. AND D. L. KEEFE (1983) Evidence for transmitter plasticity *in vivo*: Developmental changes in properties of cholinergic sympathetic neruons. Dev. Biol. 98: 349–372.

LIEM, K. F. JR., G. TREMML AND T. M. JESSELL (1997) A role for the roof plate and its resident TGFβ-related proteins in neuronal patterning in the dorsal spinal cord. Cell 91: 127–138.

McMAHON, A. P. AND A. BRADLEY (1990) The *wnt-1* (*int-1*) protooncogene is required for the development of a large region of the mouse brain. Cell 62: 1073–1085.

NODEN, D. M. (1975) Analysis of migratory behavior of avian cephalic neural crest cells. Dev. Biol. 42: 106–130.

PATTERSON, P. H. AND L. L. Y. CHUN (1977) The induction of acetylcholine synthesis in primary cultures of dissociated rat sympathetic neurons. Dev. Biol. 56: 263–280.

Rakic, P. (1971) Neuron-glia relationship during granule cell migration in developing cerebral cortex. A Golgi and electronmicroscopic study in *Macacus rhesus*. J. Comp. Neurol. 141: 283–312.

Rakic, P. (1974) Neurons in rhesus monkey visual cortex: Systematic relation between time of origin and eventual disposition. Science 183: 425–427.

Sauer, F. C. (1935) Mitosis in the neural tube. J. Comp. Neurol. 62: 377–405.

Spemann, H. and H. Mangold (1924) Induction of embryonic primordia by implantation of organizers from a different species. Trans-

lated into English by V. Hamburger and reprinted in *Foundations of Experimental Embryology*, B. H. Willier and J. M. Oppenheimer (eds.) (1974). New York: Hafner Press.

Stemple, D. L. and D. J. Anderson (1992) Isolation of a stem cell for neurons and glia from the mammalian neural crest. Cell 71: 973–985.

Walsh, C. and C. L. Cepko (1992) Widespread dispersion of neuronal clones across functional regions of the cerebral cortex. Science 255: 434–440.

Yamada, T., M. Placzek, H. Tanaka, J. Dodd and T. M. Jessell (1991) Control of cell pattern in the developing nervous system. Polarizing activity of the floor plate and notochord. Cell 64: 635–647.

Zimmerman, L. B., J. M. De Jesus-Escobar and R. M. Harland (1996) The Spemann organizer signal noggin binds and inactivates bone morphogenetic protein 4. Cell 86: 599–606.

Books

Lawrence, P. A. (1992) *The Making of a Fly: The Genetics of Animal Design*. Oxford: Blackwell Scientific.

Moore, K. L. (1988) *The Developing Human: Clinically Oriented Embryology*, 4th Ed. Philadelphia: W. B. Saunders Company.

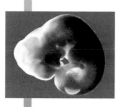

Chapter 23

Overview

Once neurons are generated and have migrated to their final positions, two additional features pertinent to the adult organization of the nervous system must be established. First, different regions must be linked together via axon pathways. Second, orderly synaptic connections must be made among appropriate pre- and postsynaptic partners. The cellular mechanisms that generate axon outgrowth and synapse formation are thus the major determinants of neural circuitry. The directed growth of axons and the recognition of synaptic targets is mediated by a specialization at the tip of growing axons called the growth cone. Growth cones detect and respond to signaling molecules that identify intended pathways and ultimately facilitate correct synaptic partnerships. These signals include surface-bound and diffusible molecules that either attract or repel growing axons. In addition, secreted growth factors promote and maintain stable synapses in appropriate numbers between axons and their targets. As in other instances of intercellular communication, a variety of receptors and second messenger molecules transduce the signals provided to the growth cone and initiate the intracellular events that underlie directed growth and synapse formation. The end results of this dynamic signaling process are a wealth of well-defined peripheral and central axon pathways and detailed synaptic circuitry that allow animals to behave in ever more sophisticated ways as they mature.

The Axonal Growth Cone

Among the many extraordinary features of nervous system development, one of the most fascinating is the ability of growing axons to navigate through a complex cellular embryonic terrain to find appropriate synaptic partners that may be millimeters or even centimeters away. In 1910, Ross G. Harrison, who first observed axons extending in a living tadpole in vitro, noted:

> The growing fibers are clearly endowed with considerable energy and have the power to make their way through the solid or semi-solid protoplasm of the cells of the neural tube. But we are at present in the dark with regard to the conditions which guide them to specific points.

Harrison's observations indicate the central features of axonal growth. First, the energy and power of growing axons reflect the cellular properties of the **growth cone**, a specialized structure at the tip of the extending axon. Growth cones are highly motile structures that explore the extracellular environment, determine the direction of growth, and then guide the extension of the axon in that direction. The primary morphological characteristic of a growth cone is a

Figure 23.1 Photomicrograph of a growth cone at the tip of a sensory ganglion cell axon that is extending in tissue culture. Lamellapodia (flat, sheetlike protrusions) and filopodia (long, fingerlike processes) can be seen arising from the growth cone. These highly motile extensions evidently sample the local environment in order to regulate the speed and direction of axonal growth. (Courtesy of P. Forscher.)

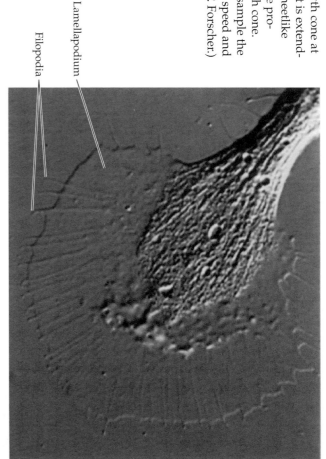

Lamellapodium

Filopodia

sheetlike expansion of the growing axon at its tip called a **lamellapodium.** When examined in vitro, numerous fine processes called **filopodia** rapidly form and disappear from the terminal expansion, like fingers reaching out to touch or sense the environment (Figure 23.1). The cellular mechanisms that underlie these complex searching movements have become a focus of cell biological studies of axon growth and guidance. Such movements are thought to reflect rapid, controlled rearrangement of cytoskeletal elements—particularly molecules related to the actin cytoskeleton—which modulate the changes in growth cone shape and ultimately its course through the developing tissues.

Santiago Ramón y Cajal, Harrison's contemporary, noted that when growth cones move along an established pathway pioneered by other axons, they tend to be simple in shape. In contrast, when a growing axon first extends in a new direction or reaches a region where a choice must be made about the direction to take, the structure (and presumably motility) of its growth cone undergoes dramatic changes. The growth cone flattens and extends numerous filopodia, much as it does in a culture dish, suggesting an active search for appropriate cues to direct subsequent growth. These changes of growth cone shape at "decision points" have been observed in both the peripheral and central nervous system. In the periphery, growth cones of motor neurons undergo shape changes as they enter the primordia of muscles in immature limbs, presumably seeking appropriate targets in the developing musculature. In the central nervous system, growth cones in developing olfactory and optic nerves also change shape when they reach critical points in their trajectories. Of particular functional significance is the decision made by subsets of retinal axons at the optic chiasm. The growth cones of retinal axons slow down and acquire a complex shape as they "choose" whether or not to cross the midline (Figure 23.2).

Non-Diffusible Signals for Axon Guidance

The complex behavior of growth cones during axonal extension suggests the presence of specific cues that cause the growth cone to move in a particular direction. In addition, the growth cone itself must have a specialized array of receptors and transduction mechanisms to respond to these cues. The cues themselves—the "condition(s) which guide…growth cones" referred to by Harrison—remained elusive for more than half a century after his initial observations of axon growth. The identity of some of the relevant molecules has only been established over the last 30 years or so. These signals comprise a large group of molecules that mediate cell-cell contact and motility, including an extensive group of **cell adhesion molecules, or CAMs,** found throughout the embryo. CAMs, which are located on the surface of cells or in organized matrices in the extracellular spaces, interact with cell surface receptors on growth cones; the receptors interact in turn with associated intracellular signaling molecules (e.g., kinases, phosphatases, proteases, and their targets), which are enriched in the cytoplasm of the growth cone (Figure 23.3).

The most remarkable characteristic of cell adhesion molecules is their number and diversity. The array of CAMs at the cell surface provides a daunting menu of signals to individual growth cones, which instruct them to extend along a particular pathway or to interact with a specific target. Most CAMs share several molecular features. First, there are repeated amino acid motifs in the extracellular portion of CAMs. Some of these repeats are homologous to either the immunoglobulin family, the extracellular matrix molecule fibronectin, or epidermal growth factor. There are two families of CAMs, based on sequence homology. The first family is the Ca^{2+}-dependent CAMs, or **cadherins,** which are adhesive only in the presence of Ca^{2+}. The second family is the Ca^{2+}-independent CAMs, which include the **neural cell adhesion molecule, or NCAM.** In general, the transmembrane or intracellular portions of CAMs do not have signaling or enzymatic activity. The information provided by these CAMs is transduced by protein-protein interactions between the CAMs and receptors and cytoplasmic signaling molecules that are not well understood.

The importance of these adhesive interactions is underscored by the pathogenesis of several inherited human conditions that lead to mental retardation and hydrocephalus (an abnormal enlargement of the cerebral ventricles that often causes atrophy of the cerebral cortex). Individuals with these syndromes—X-linked hydrocephalus, MASA (an acronym for *m*ental retardation, *a*phasia, *s*huffling gait, and *a*dducted thumbs), and X-linked spastic paraplegia—are all consequences of mutations in the gene encoding a particular family of CAMs. These mutations can also lead to the absence of the corpus callosum, which connects the two cerebral hemispheres, and of the corticospinal tract, which carries cortical information to the spinal cord. Congenital anomalies such as these (which are fortunately rare) are now understood to arise from errors in the signaling mechanisms normally responsible for axon navigation.

Other molecules that growth cone receptors respond to are located in the **extracellular matrix,** a complex of substances produced by cells but not directly attached to the embryonic cell membranes. These constituents are generally referred to as **extracellular matrix adhesion molecules** and are particularly good substrates for growth cones. The most prominent members of this group are the **laminins, collagens,** and **fibronectin.** An important class of growth cone receptors, known collectively as **integrins,** bind specifically to these molecules (see Figure 23.3).

The binding of laminin, collagen, or fibronectin to integrins triggers a cascade of events within the growth cone that stimulates axon growth and elongation. These changes include fluctuations in levels of intracellular messengers such as calcium and inositol trisphosphate, and the activation of intracellular kinases (see Chapter 8). The role of extracellular matrix molecules in axon guidance is particularly clear in the embryonic periphery.

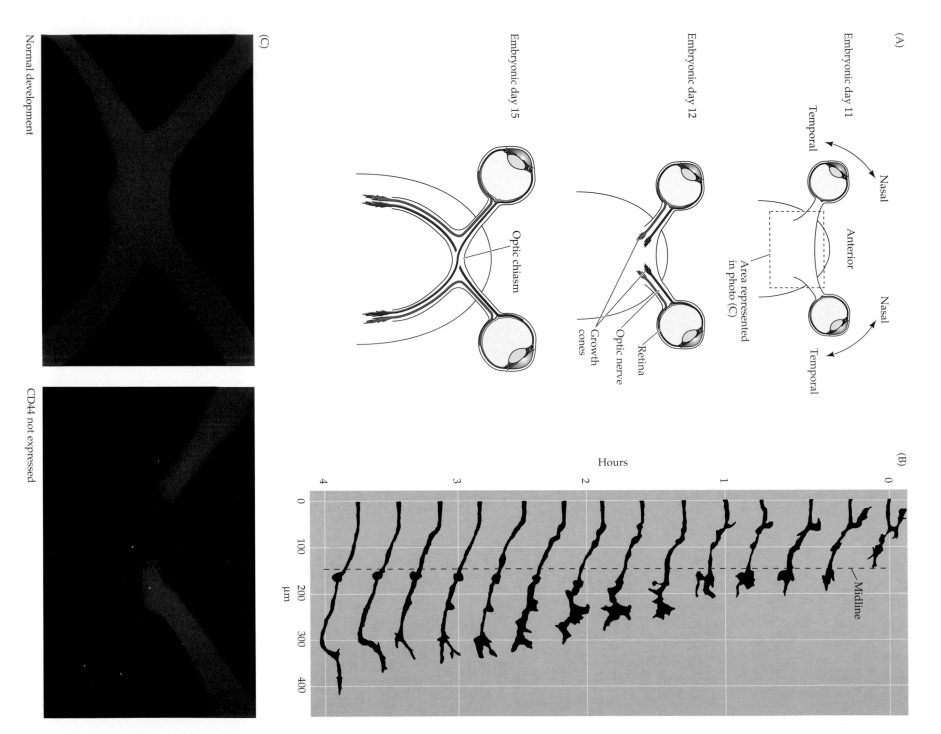

(A)

Embryonic day 11

Temporal

Nasal

Anterior

Nasal

Temporal

Area represented
in photo (C)

Embryonic day 12

Growth
cones

Optic nerve

Retina

Embryonic day 15

Optic chiasm

(B)

Hours

4   3   2   1   0

0

100

200

300

400

μm

Midline

(C)

Normal development

CD44 not expressed

▼ **Figure 23.2**   Growth cone behavior at a decision point (in this case, the optic chiasm). (A) In the embryonic mouse visual system, the growing axons of retinal ganglion cells reach the optic chiasm at about embryonic day 12–13; some temporal axons from each retina remain on the same side of the brain, while most axons cross to the opposite side (see Chapter 12). (B) As growth cones approach the chiasm, they change their speed and shape, as seen in these silhouettes of a living, dye-labeled growth cone of a retinal ganglion cell at various times (shown in hours). When growing toward the chiasm, the growth cone has a tapered, streamlined appearance. At the midline, however, the growth cone slows down and becomes spread out and more complex; after crossing (at about 3 hours in this time-lapse recording), it regains a streamlined shape and advances more rapidly. (C) Cells at the optic chiasm express a number of cues that may elicit these changes. One candidate is a glycoprotein known as CD44. Eliminating either the protein or the cells expressing it interferes with the decussation of the ganglion cell axons and the formation of a normal chiasm. (A and B from Gocemont, 1994; C from Sretavan et al., 1995.)

Axons extending through peripheral tissues grow through loosely arrayed mesenchymal cells that fill the interstices of the embryo; they also grow along the interface of mesenchyme and epidermis (which is identified by an organized sheet of extracellular matrix components called the *basal lamina*). In tissue culture as well as in the embryo, different extracellular matrix molecules have different capacities to stimulate axon growth. Thus, the relative availability of different matrix molecules can influence the speed or direction

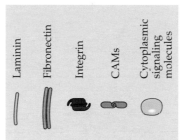
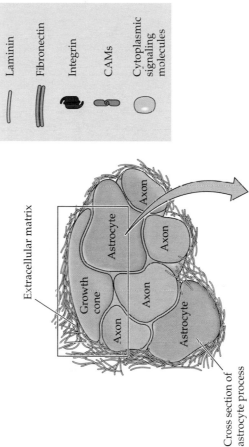
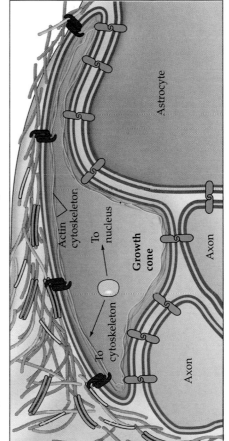

Figure 23.3   Interactions with non-diffusible signaling molecules that enable growth cone navigation through the developing nervous system. In addition to interactions mediated by cell adhesion molecules, growth cones bind to extracellular matrix molecules such as laminin and fibronectin by means of integrins. (After Reichardt et al., 1991.)

of a growing axon. The role of matrix molecules in the central nervous system is less clear. Some of the same molecules are present in the extracellular space but are not organized into orderly substrates like the basal lamina in the periphery, and have therefore been harder to study.

Diffusible Signals for Axon Guidance: Chemoattraction and Repulsion

Another major class of information-bearing molecule that influences axon outgrowth is secreted by target cells, diffusing from the cells of origin to the growing axons. With remarkable foresight, Cajal proposed early in the twentieth century that target-derived signals selectively influenced the movement of axonal growth cones, thereby attracting them to appropriate destinations. In addition to the chemoattraction predicted by Cajal, it was long supposed that there might also be chemorepellent signals that discouraged axon growth toward a particular region. Experiments both in vivo and in vitro have confirmed the basic idea of chemoattraction and repulsion. Until relatively recently, however, the identity of these signals remained uncertain. One problem was the vanishingly small amounts of such factors expressed in the developing embryo. Another was that of distinguishing **tropic** molecules—which *guide* growing axons toward a source—from **trophic** molecules—which *support* the survival and growth of neurons and their processes once an appropriate target has been contacted. These problems were solved by laborious biochemical purification and analysis of attractive or repulsive activities from vertebrate (chick) embryos, and genetic analysis of axon growth in both fruit flies and the tiny roundworm, *C. elegans*; this work eventually led to the cloning of several genes that code for chemotropic factors. Remarkably, the identity and function of chemoattractants and chemorepellents across phyla is highly conserved.

The best-characterized family of chemoattractant molecules is the **netrins** (from the Sanskrit "to guide"), which were isolated and their genes cloned nearly a century after Cajal proposed the existence of such molecules. In chick embryos, the netrins were identified as proteins with chemoattractant activity. In *C. elegans*, netrins were first recognized as the product of a gene that influenced axon growth and guidance (the first such gene was called *Unc-6* for "uncoordinated," which describes the behavioral phenotype of the mutant worms; the cause is misrouted axons as a result of the absence of netrin). The netrins themselves have high homology to extracellular matrix molecules like laminin and in some cases may actually interact with the extracellular matrix to influence directed axon growth. Netrin signals are transduced by specific receptors. Like most cell surface adhesion molecules, netrin receptors have specific repeats in their extracellular domain, a transmembrane domain and an intracellular domain with no enzymatic activity.

Netrins act at a variety of sites in the developing vertebrate nervous system where axons must choose whether to cross the midline or remain ipsilateral (Figure 23.4). Perhaps the best-characterized example of their chemoattractive activity is in the developing spinothalamic tract, the pathway that relays information about pain and temperature from the periphery to the thalamus (see Chapter 10), and thence to the cortex. Netrins are found in the floor plate at the midline of the spinal cord, where the growing spino-thalamic (or commissural) axons cross. In cell culture experiments, these molecules direct the growth of commissural axons. Equally telling, inactivation of netrin genes by homologous recombination in mice disrupts the midline crossing of spinothalamic axons. The netrins and their receptors have also been implicated in midline crossing

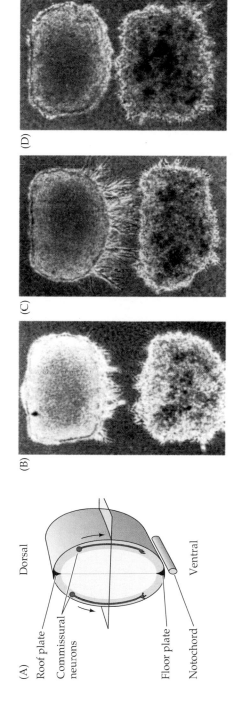

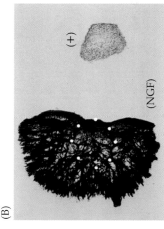

Figure 23.4 Chemotropic molecules (netrins) in the developing spinal cord. (A) The embryonic spinal cord and the cuts used to prepare explants from chick embryos. The commissural neurons send axons to the ventral region of the spinal cord, including a specific region called the floor plate. (B) Axons from commissural neurons in an explant culture (upper piece of tissue) grow directly toward a floor plate explant (lower piece of tissue) about 100 μm away. The axon bundles from the spinal cord explant grow only in the direction of the floor-plate, indicating a diffusible chemotropic factor originating from this site. (C) If the floor plate is replaced by an aggregate of cells from a cell line transfected with the netrin gene, the identical behavior is observed. (D) Control cells that do not contain the netrin gene fail to elicit outgrowth. (A after Serafini et al., 1994; B–D from Kennedy et al., 1994.)

Figure 23.5 Semaphorins act as chemorepulsive cues. (A) In the presence of nerve growth factor (NGF), explant cultures of chick dorsal root ganglia extend halos of neurites that originate from different neuronal subpopulations. (B) Co-culture of a ganglion with non-neuronal cells (+) transfected with the gene for semaphorin III (collapsin) results in asymmetrical growth of the ganglion cell neurites as a result of chemorepulsion. Control cells not transfected with the gene [(−) in panel A] have no effect on the pattern of outgrowth. (From Messersmith et al., 1995.)

at a variety of other sites in the central nervous system, including the optic chiasm, the corpus callosum, and the decussation of the trochlear nerve. Thus, these and other chemoattractant signals help organize the major crossed (or decussated) pathways in the brain.

Most research on axon guidance has focused on molecules that encourage axon outgrowth or attract growing neurons. Constructing the nervous system, however, also entails telling axons where *not* to grow. Two broad classes of chemorepellent molecules have been described. The first is associated with cells that make central nervous system myelin. These molecules may be particularly important after injury to the adult brain, where they inhibit axon growth around a damaged region (see Chapter 25). Molecules belonging to the second class of chemorepellents are active during neural development. These molecules, called **semaphorins** (*semaphor* is the Greek word for signal), are eventually bound to cell surfaces or to the extracellular matrix, where they can prevent the extension of nearby axons (Figure 23.5). Although the semaphorins can be secreted, they can also be anchored to the cell surface. Their receptors, like those for cell surface adhesion molecules, are transmembrane proteins whose cytoplasmic domains have no known catalytic activity.

The most compelling examples of semaphorin chemorepellent activity have come from studies of invertebrates, where mutation or manipulation can cause axons to grow abnormally. Studies with cultured vertebrate neurons have also shown that the semaphorins can cause collapse of growth cones and cessation of axon extension. The activity of these molecules in vivo, however, has been harder to demonstrate. For example, inappropriate growth or targeting of axons is not apparent when single semaphorin genes in mice are deleted. Thus, the nature of semaphorin activity and its functions in the vertebrate nervous system are not understood.

Several other candidate chemoattractants, repellants, and their receptors have been described recently. These include the secreted factor "slit" and its receptor "robo" (named for the phenotypes of *Drosophilia* mutants in which these genes were first identified); both molecules are important for preventing an axon from crossing back over the midline once it has crossed initially in response to a tropic signal like netrin. Although none of these molecules alone can explain the initial choices and resulting trajectories of developing axons, it is clear that they make an important contribution to the orderly construction of axon pathways in the periphery and in the central nervous system.

The Formation of Topographic Maps

In the somatic sensory, visual, and motor systems, neuronal connections are arranged such that neighboring points in the periphery are represented by adjacent locations in the central nervous system (see Chapters 9, 12, and 17). In other systems (e.g., the auditory and olfactory systems), there are also orderly representations of various stimulus attributes like frequency or receptor identity. How do growing axons distribute themselves with such fidelity within target regions in the brain?

In the early 1960s, Roger Sperry, who later did pioneering work on the functional specialization of the cerebral hemispheres (see Chapter 27), articulated the **chemoaffinity hypothesis**, based primarily on work in the visual system of frogs and goldfish. In these animals, the terminals of retinal ganglion cells form a precise topographical map in the optic tectum (which is homologous to the mammalian superior colliculus). When Sperry crushed the optic nerve and allowed it to regenerate (fish and amphibians, unlike mammals, can regenerate axonal tracts in their central nervous system; see Chapter 25), he found that retinal axons reestablished the same pattern of connections in the tectum. Even if the eye was rotated 180°, the regenerating axons grew back to their original tectal destinations (Figure 23.6). Accordingly, Sperry proposed that each tectal cell carries an "identification tag"; he further supposed that the growing terminals of retinal ganglion cells have complementary tags, such that they seek out a specific location in the tectum. In modern parlance, these "chemical" tags are **recognition molecules**, and the "affinity" that they engender is a selective binding of receptor molecules on the growth cone to corresponding molecules on the tectal cells that signal their relative positions.

Further experiments in the amphibian and avian visual systems made the strictest form of the chemoaffinity hypothesis—labeling of each tectal location by a different recognition molecule—untenable. Rather than precise "lock and key" affinity, the behavior of growing axons suggested that there

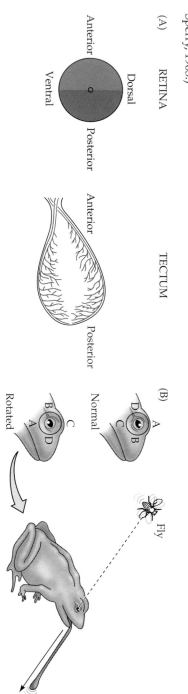

Figure 23.6 The axons of the retinal ganglion cells project to appropriate positions in the optic tectum during both development and regeneration in frogs and other "lower" vertebrates. (A) Posterior retinal axons project to the anterior tectum and anterior retinal axons to the posterior tectum. When the optic nerve of a frog is surgically interrupted, the axons regenerate with the appropriate specificity. (B) Even if the eye is rotated after severing the optic nerve, the axons regenerate to their original position in the tectum. That the topographical visual map in the tectum remains unchanged is evident from the frog's behavior: When a fly is presented above, the frog consistently strikes downward, and vice versa. This outcome indicates a specific matching of retinal neurons to particular regions of the tectum, a phenomenon taken to explain topographical mapping in the mammalian brain as well. (After Sperry, 1963.)

(A) RETINA

Anterior

Dorsal   Ventral

Posterior

TECTUM

Anterior

Posterior

(B) Normal

Rotated

Fly

(B)

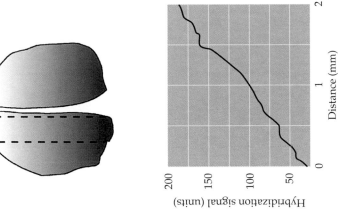

(A)

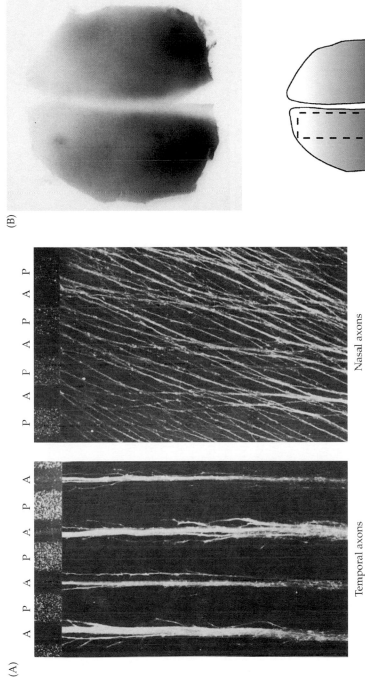

Temporal axons

Nasal axons

Figure 23.7 A possible molecular basis of retinotectal chemoaffinity. (A) An in vitro assay for cell surface molecules that may contribute to topographical specificity in the optic tectum. A set of alternating stripes (90 μm wide) of membranes from anterior (A) and posterior (P) optic tectum of chicks was laid down on a glass coverslip. The posterior membranes have had fluorescent particles added to make the boundaries of the stripes apparent (top of panels). Subsequently, explants of retina were taken from either nasal or temporal retina and placed on the stripes. Temporal axons prefer to grow on anterior membranes and are repulsed by posterior membranes. In contrast, nasal retinal axons grow equally well on both stripes. (B) The messenger RNA for a mammalian homologue of the repellent protein from chick tectum, termed ephrin A–2, is normally distributed in a linear, anterior-to-posterior gradient in the developing mouse tectum, as indicated both by in situ hybridization (top) and measurements of the hybridization signal along the region indicated by the dashed rectangle (middle and bottom). (A from Walter et al., 1987; B from Cheng et al., 1995.)

are gradients of cell surface molecules to which growing axons respond in the tectum to establish the basic axes of the retinotopic map. Normally, axons from the temporal region of the chick retina innervate the anterior pole of the tectum and avoid the posterior pole. In an in vitro assay using cell membranes derived from these two tectal regions, temporal retinal axons, when presented with a choice, grow exclusively on anterior membranes, avoiding membranes derived from the "wrong" region of the tectum (Figure 23.7A). The positive interactions probably are due to increased adhesion of the growth cones to the substrate, whereas the failure to grow into inappropriate regions may result from repulsive interactions that tend to collapse the growth cones.

A likely candidate for the negative guidance signal for temporal axons in the posterior tectum has been purified and its gene cloned. The protein—called RAGS (repulsive axon guidance signal)—is a membrane-anchored ligand for a specific type of tyrosine kinase-coupled receptor; this system of cellular interactions is analogous to the *boss-sevenless* system described for the developing fly eye in Chapter 22. Both the chick protein and a homologous mammalian protein are distributed in a well-defined gradient in the tectum (Figure 23.7B) and could thus contribute to the topographical organization of the visual system.

The RAGS molecule belongs to a family of so-called ephrin or Eph ligands and receptors that are associated with topographic maps in the visual system. Eph ligands are cell adhesion-like molecules that can be either transmembrane proteins or membrane-associated proteins. Eph receptors belong to the single transmembrane domain tyrosine receptor kinase family, and thus can directly transduce a signal from an Eph ligand. The Eph ligands are distributed in a graded fashion in the optic tectum and the lateral geniculate nucleus, while the Eph receptors are found in a complementary gradient in the retina. Disruption of the genes for the Eph ligands or their receptors results in subtle disruptions in the topographic organization of the retino-collicular or retino-thalamic projection. These observations accord with the idea that chemoaffinity operates by a system of gradients in the retina and tectum that give axons and their targets markers of position, rather than a unique lock and key sort of recognition. The Eph receptors and their ligands, RAGS in particular, provide examples of graded information in the retina and tectum that help organize topographic axonal growth in the visual system and other regions of the developing brain.

Selective Synapse Formation

After reaching the correct target or target region, axons must make a further local determination about which particular cells to innervate among a variety of potential synaptic partners. Because of the complexity of brain circuitry, this issue has been studied most thoroughly in the peripheral nervous system, particularly in the innervation of muscle fibers (Box A) and autonomic ganglion cells. Synaptic specificity was first explored by British physiologist John Langley at the end of the nineteenth century. Preganglionic sympathetic neurons located at different levels of the spinal cord innervate cells in sympathetic chain ganglia in a stereotyped and selective manner (Figure 23.8; see also Chapter 21). In the superior cervical ganglion, for example, cells from the highest thoracic level (T1) innervate ganglion cells that project in turn to targets in the eye, whereas neurons from a somewhat lower level (T4) innervate ganglion cells that cause constriction of the blood vessels of the ear. Since the axons of all these neurons run together in the cervical sympathetic trunk to arrive at the ganglion, the mechanisms underlying the differential innervation of the ganglion cells must occur at the level of synapse formation rather than axon guidance to the general vicinity of target cells (see above). Anticipating Sperry by more than 50 years in a different context, Langley concluded that selective synapse formation is based on differential affinities of the pre- and postsynaptic elements.

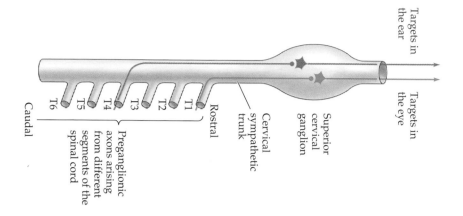

Figure 23.8 Evidence that synaptic connections between mammalian neurons form according to specific affinities between different classes of pre- and postsynaptic cells. In the superior cervical ganglion, preganglionic neurons located in particular spinal cord segments (T1, for example) innervate ganglion cells that project to particular peripheral targets (the eye, for example). The establishment of these preferential synaptic relationships indicates that selective neuronal affinities are a major determinant of neural connectivity.

Modern studies based on intracellular recordings from individual neurons in the superior cervical ganglion have shown, however, that the selective affinities between pre- and postsynaptic neurons are not especially restrictive. Thus, synaptic connections to ganglion cells made by preganglionic neurons of a particular spinal level are preferred, but synaptic contacts from neurons at other levels are not excluded (much like the rules that govern axon guidance). Furthermore, if the innervation to the superior cervical ganglion from a particular spinal level is surgically interrupted, recordings made some weeks later indicate that new connections are established by residual axons arising from what would normally be inappropriate spinal segments. The novel connections also establish a pattern of segmental preferences, as if the system had attempted to achieve the best match it could under the altered circumstances. Despite this relative selectivity during synapse formation, a quite different line of work has shown that *where* a synapse forms on the target cell (at least if the cell is a muscle fiber) is tightly controlled by a set of molecules that are now understood in some detail (see Box A).

From a broader vantage, of course, there are some absolute restrictions to synaptic associations. Thus, neurons do not innervate nearby glial or connective tissue cells, and many instances have been described in which various nerve and target cell types show little or no inclination to establish connections with one another. When synaptogenesis does proceed, however, neurons and their targets in both the central and peripheral nervous systems appear to associate according to a continuously variable system of preferences. Such biases guide the pattern of innervation that arises in development (or reinnervation) without limiting it in any absolute way. The target cells residing in muscles, autonomic ganglia, or elsewhere are certainly not equivalent, but neither are they unique with respect to the innervation they can receive. This relative promiscuity can cause problems following neural injury, since regenerated patterns of innervation are not always appropriate (see Chapter 25).

Trophic Interactions and the Ultimate Size of Neuronal Populations

The formation of synaptic contacts between growing axons and their synaptic partners marks the beginning of a new stage of development. Once synaptic contacts are established, neurons become dependent in some degree on the presence of their targets for continued survival and differentiation; in the absence of synaptic targets, the axons and dendrites of developing neurons atrophy and the innervating nerve cells may eventually die. This long-term dependency between neurons and their targets is referred to as **trophic interaction**. The word *trophic* is taken from the Greek *trophé*, meaning, roughly, "nourishment." Despite this nomenclature, the sustenance provided to neurons by trophic interactions is not the sort derived from metabolites such as glucose or ATP. Rather, the dependence is based on specific signaling molecules called **neurotrophic factors**. Neurotrophic factors, like some other intercellular signaling molecules (mitogens and cytokines, for example), originate from target tissues and regulate neuronal differentiation, growth, and ultimately survival.

Why should neurons depend so strongly on their targets, and what specific cellular and molecular interactions mediate this dependence? The answer to the first part of this question lies in the changing scale of the developing nervous system and the body it serves, and the related need to precisely match the number of neurons in particular populations with the size of their targets. The basic mechanisms by which neurons are initially

BOX A
Molecular Signals That Promote Synapse Formation

Synapses require a precise organization of presynaptic and postsynaptic elements in order to function properly (see Chapters 5–7). At the neuromuscular junction, for example, synaptic vesicles and the related release machinery are located at sites in the nerve terminal called active zones; and, in the postsynaptic muscle cell, acetylcholine receptors and other synapse-specific molecules are localized in high density exactly subjacent to the presynaptic active zones. During the past 25 years, a number of investigators have identified some of the molecular cues that guide the formation of these carefully apposed elements. Their efforts have met with the greatest success at the neuromuscular junction, where a molecule called agrin is now known to be responsible for initiating some of the events that lead to the formation of a fully functional synapse.

Agrin was originally identified as a result of its influence in the reinnervation of frog neuromuscular junctions following damage to the motor nerve. In mature skeletal muscle, each fiber typically receives a single synaptic contact at a highly specialized region called the end plate (see Chapter 5). U. J. McMahan, Josh Sanes, and their colleagues at Harvard and later Stanford and Washington Universities found that regenerating axons reinnervate the original end plate site precisely. In seeking to determine the molecular signals underlying this phenomenon, they took advantage of the fact that each muscle fiber is surrounded by a sheath of extracellular matrix called the basal lamina. When muscle fibers degenerate, they leave the basal lamina behind (as do degenerating axons); moreover, a specific infolding of the basal lamina at the former end plate site allows its continued identification. Remarkably, presynaptic nerve terminals differentiate at these original sites even when the associated muscle fibers are

absent. Equally remarkable is that regenerating muscle fibers form postsynaptic specializations—such as densely packed acetylcholine receptors—at precisely these same basal lamina locations in the absence of nerve fibers! These findings show that the signal(s) guiding synapse formation remain in the extracellular

Using a bioassay based on the aggregation of acetylcholine receptors to analyze the constituents of the basal lamina, McMahan and colleagues isolated and

environment after removal of either nerve or muscle, presumably in the basal lamina "ghost" that surrounds each muscle fiber.

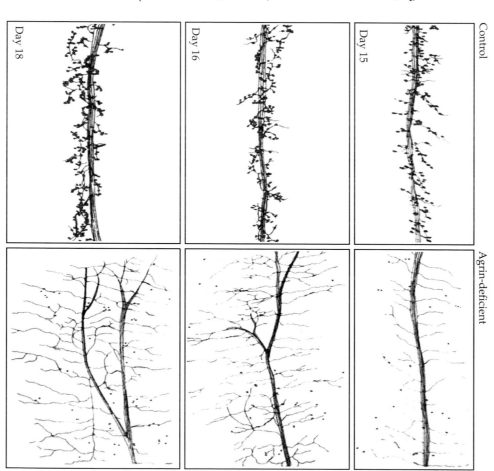

Control

Day 15

Day 16

Day 18

Agrin-deficient

Development of neuromuscular junctions in agrin-deficient mice. Diaphragm muscles from control (left) and agrin-deficient (right) mice at embryonic day 15, 16, and 18 were double-stained for acetylcholine receptors and axons, then drawn with a camera lucida. The developing muscle fibers run vertically. In both control and mutant muscles, an intramuscular nerve (black) and aggregates of AChRs (red) are present by embryonic day 15. In controls, axonal branches and AChR clusters are confined to a band at the central end plate at all stages. Mutant AChR aggregates are smaller, less dense, and less numerous; axons form fewer branches and their synaptic relationships are disorganized. (From Gautam et al., 1996.)

purified agrin. Agrin is a proteoglycan found in both mammalian motor neurons and muscle fibers; it is also abundant in brain tissue. The neuronal form of agrin is synthesized by motor neurons, transported down their axons, and released from growing nerve fibers. Agrin binds to a postsynaptic receptor whose activation leads to a clustering of acetylcholine receptors and, evidently, to subsequent events in synaptogenesis. Support for the role of agrin as an organizer of synaptic differentiation is the finding by Sanes and his collaborators that genetically engineered mice that lack the gene for agrin develop in utero with few neuromuscular junctions (see figure). Importantly, mice lacking only neural agrin were as severely impaired as mice lacking both nerve and muscle agrin. Animals missing the agrin receptor also fail to develop neuromuscular junctions and die at birth. Agrin is therefore one of the first examples of a presynaptically derived molecule that promotes postsynaptic differentiation in target cells.

Because synapse formation requires an ongoing dialogue between pre- and postsynaptic partners, it is likely that postsynaptically derived organizers of presynaptic differentiation also exist. Based on the studies of basal lamina mentioned above, Sanes and his collaborators identified one such group of molecules, the $\beta2$-laminins (originally called s-laminin). Mice lacking s-laminin show deficits in differentiation of motor nerve terminals and, unexpectedly, of terminal-associated glial (Schwann) cells. However, presynaptic defects in $\beta2$-laminin mutants are considerably less severe than postsynaptic defects in agrin mutants, suggesting that additional important retrograde signals remain to be identified.

References

BURGESS, R. W., Q. T. NGUYEN, Y.-J. SON, J. W. LICHTMAN AND J. R. SANES (1999) Alternatively spliced isoforms of nerve- and muscle-derived agrin: Their roles at the neuromuscular junction. Neuron 23: 33–44.

DECHIARA, T. M. AND 14 OTHERS (1996) The receptor tyrosine kinase MuSK is required for neuromuscular junction formation in vivo. Cell 85: 501–512.

GAUTAM, M. AND 6 OTHERS (1996) Defective neuromuscular synaptogenesis in agrin-deficient mutant mice. Cell 85: 525–535.

MCMAHAN, U. J. (1990) The agrin hypothesis. Cold Spring Harbor Symp. Quant. Biol. 50: 407–418.

NOAKES, P. G., M. GAUTAM, J. MUDD, J. R. SANES AND J. P. MERLIE (1995) Aberrant differentiation of neuromuscular junctions in mice lacking s-laminin/laminin $\beta2$. Nature 374: 258–262.

PATTON, B. L., A. Y. CHIU AND J. R. SANES (1998) Synaptic laminin prevents glial entry into the synaptic cleft. Nature 393: 698–701.

SANES, J. R., L. M. MARSHALL AND U. J. MCMAHAN (1978) Reinnervation of muscle fiber basal lamina after removal of myofibers. J. Cell Biol. 78: 176–198.

generated have already been considered in Chapter 22. One important feature of this process, however, has not yet been touched on. A general—and surprising—strategy in the development of vertebrates is the production of an initial surplus of nerve cells (on the order of two- or threefold); the final population is subsequently established by the degeneration of those neurons that fail to interact successfully with their intended targets, a process now known to be mediated by neurotrophic factors.

Evidence that targets play a major role in determining the size of the neuronal populations that innervate them has come from an ongoing series of studies dating from the start of the twentieth century. The seminal observation was that the removal of a limb bud from a chick embryo results, at later embryonic stages, in a striking reduction in the number of nerve cells in the corresponding portions of the spinal cord (Figure 23.9A,B). The interpretation of these experiments is that neurons, in the spinal cord in this case, compete with one another for a resource present in the target (the developing limb) that is available in limited supply. In support of this idea, many neurons that would normally have died can be rescued by augmenting the amount of target available, thereby providing extra trophic support (Figure 23.9C,D). Thus, the size of nerve cell populations in the adult is not fully determined in advance, but is governed in part by idiosyncratic neuron–target interactions in each developing individual.

Importantly, the death of neurons deprived of trophic support from their targets occurs by a different mechanism than the cell death resulting from injury or disease. Trophically deprived neurons degenerate and die through a

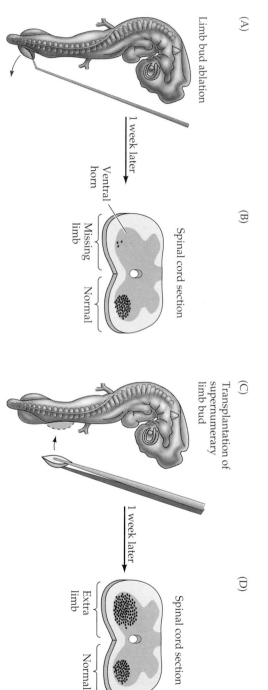

(A) Limb bud ablation

1 week later

(B) Spinal cord section

Ventral horn

Missing limb

Normal

(C) Transplantation of supernumerary limb bud

1 week later

(D) Spinal cord section

Extra limb

Normal

Figure 23.9   Effect of removing or augmenting neural targets on the survival of related neurons. (A) Limb bud amputation in a chick embryo at the appropriate stage of development (about 2.5 days of incubation) depletes the pool of motor neurons that would have innervated the missing extremity. (B) A cross section of the lumbar spinal cord in an embryo that underwent this surgery about a week earlier. The motor neurons (dots) in the ventral horn that would have innervated the hindlimb degenerate almost completely after embryonic amputation; a normal complement of motor neurons is present on the other side. (C) Adding an extra limb bud before the normal period of cell death rescues neurons that normally would have died. (D) Such augmentation leads to an abnormally large number of limb motor neurons (dots) on the side related to the extra limb. (After Hamburger, 1958, 1977, and Hollyday and Hamburger, 1976.)

special process called **apoptosis**, which depends on the active transcription of a host of specific genes that when "turned on" cause neurons or other cells to degenerate. Indeed, the processes underlying apoptosis appear to involve many of the same mechanisms that govern cell differentiation and control of the cell cycle (see Chapter 22). In any event, cell death by apoptosis can properly be regarded as an actively determined state of cell differentiation and may be involved in a variety of neurological disorders in which nerve cells degenerate (Parkinson's disease, for example).

Further Competitive Interactions in the Formation of Neuronal Connections

Once neuronal populations are established by this winnowing, trophic interactions continue to modulate the formation of synaptic connections, a process that begins in embryonic life but extends far beyond birth. Among the problems that must be solved during the establishment of innervation is ensuring that each target cell is innervated by the right number of axons, and that each axon innervates the right number of target cells. Getting these numbers right is another major achievement of trophic interactions between developing nerve and target cells.

Studying synaptic refinement in the complex circuitry of the cerebral cortex or other regions of the central nervous system is a formidable challenge. As a result, many basic ideas about the ongoing modification of developing brain circuitry have again come from simpler, more accessible systems, most notably the vertebrate neuromuscular junction and the innervation of autonomic ganglion cells (Figure 23.10). Adult skeletal muscle fibers and neurons in some classes of autonomic ganglia (parasympathetic neurons) are each innervated by a single axon. Initially, however, each of these target cells receives innervation from several neurons, a condition termed **polyneuronal innervation**. In such cases, inputs are gradually lost during early postnatal development until only one remains. This process of loss is generally referred to as synapse elimination, although the elimination actually refers to a reduction in the number of different axonal *inputs* to the target cells, not to a reduction in the overall number of *synapses* made on the postsynaptic cells. In fact, the overall number of *synapses* in the peripheral nervous system increases steadily during the course of development, as is the case through-

(A) Ganglion cells

(B) Muscle cells

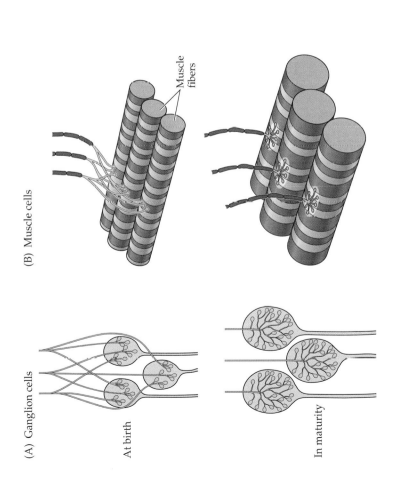

Muscle fibers

At birth

In maturity

Figure 23.10 Major features of synaptic rearrangement during the first few weeks of postnatal life in the mammalian peripheral nervous system. In ganglia comprising neurons without dendrites (A) and in muscles (B), each axon innervates more target cells at birth than in maturity. In both muscles and ganglia, however, the size and complexity of the terminal arbor on each target cell increases. Thus, each axon elaborates more and more terminal branches and synaptic endings on the target cells it will innervate in maturity. The common denominator of this process is not a net loss of synapses, but the focusing by each axon of a progressively increasing amount of synaptic machinery on fewer target cells. (After Purves and Lichtman, 1980.)

out the brain. A variety of experiments have shown that the elimination of some initial inputs to muscle and ganglion cells is a process in which synapses originating from different neurons compete with one another for "ownership" of an individual target cell (Box B).

Importantly, such competition is thought to be modulated by patterns of electrical activity in the pre- and postsynaptic partners. For example, if acetylcholine receptors at the neuromuscular junction are blocked by curare (a potent antagonist of the acetylcholine receptor; see Chapter 7), polyneuronal innervation persists. Blocking presynaptic action potentials in the motor neuron axons (by silencing the nerve with tetrodotoxin, a sodium channel blocker) also prevents the reduction of polyneuronal innervation. Blocking neural activity, therefore, reduces (or delays) competitive interactions and the associated synaptic rearrangements.

The object of this competition is not known. Some of the phenomena of activity-dependent competition in muscles and ganglia (as well as in more complex central nervous system structures) could be explained by postulating that (1) synapses require a certain minimal level of trophic support to persist, (2) the relevant factors are secreted in limited amounts by the postsynaptic (target) cells in response to synaptic activation, and (3) synapses can only avail themselves of trophic support if their activity and that of the target cell coincide. There is, however, little direct evidence for this scenario. Equally plausible is the idea that active synapses provide a destabilizing signal that weakens asynchronously firing inputs. Thus, how activity achieves its effects on synaptic connectivity is a key question that remains unresolved.

To date, the most useful insights into the nature of synaptic rearrangement during development have come from direct observation of this process (Figure 23.11). Using different-colored fluorescent dyes that stain either the presynaptic terminal or the postsynaptic receptors, Jeff Lichtman and his colleagues at Washington University have observed synaptic rearrangement over time in living animals. With this technique they followed the same neuromuscular junction over days, weeks, or longer. These observations have yielded some unexpected results. Competition between synapses arising from different motor neurons does not involve the active displacement of the "losing" input by the eventual "winner." Instead, it appears that the inputs of the two competitors gradually segregate. The "losing" axon then atrophies and retracts from the synaptic site. This is accomplished by a loss of the corresponding postsynaptic specializations associated with the "loser." Neurotransmitter receptors beneath the terminal branches that eventually will be eliminated are also lost. This receptor loss occurs before the nerve terminal has withdrawn and reduces the synaptic strength of the input, which causes a further loss of postsynaptic receptors, leading to further reduction in the strength of the input. The downward spiral of synaptic efficacy presumably results in withdrawal of the presynaptic terminal. The remaining terminals then continue to enlarge and strengthen in place as the end plate region expands during postnatal muscle growth.

A generally similar reorganization is evident in a variety of other peripheral and central nervous system regions. In the peripheral nervous system, the number of presynaptic axons innervating each neuron can also fall, as demonstrated by studies of certain autonomic ganglia. A similar process has been described in the central nervous system. In the cerebellum, each adult Purkinje cell is innervated by a single climbing fiber (see Chapter 19); early in development, however, each Purkinje cell receives multiple climbing fiber inputs. Finally, in the visual cortex, there is initially binocular innervation of cells that is eliminated to establish segregated inputs (see Chapter 24). Clearly, the pattern of synaptic connections that emerges in the adult is not simply a consequence of the biochemical identities of synaptic partners or other determinate developmental rules. Rather, the wiring plan in maturity is the result of a much more flexible process in which neuronal connections are formed, removed, and remodeled according to local circumstances. These interactions guarantee that every target cell is innervated—and continues to be innervated—by the right number of inputs and synapses, and that every innervating axon contacts the right number of target cells with an appropriate number of synaptic endings. The regulation of **convergence** (the number of inputs to a target cell) and **divergence** (the number of connections made by a neuron) in the developing nervous system is another key consequence of trophic interactions among neurons and their targets (see Box B).

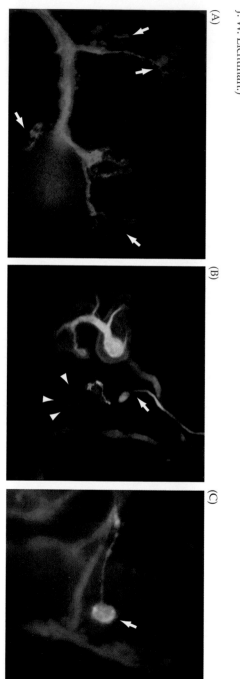

(A)　(B)　(C)

Figure 23.11 Synapse elimination at neuromuscular junctions. (A) Several neuromuscular junctions (arrows) from a mouse fetus (embryonic day 17). The red and green terminals are synapses from two different axons that converge at each of several junctions. (B) A single neuromuscular junction at higher magnification during a late stage of competition in which one of the synaptic inputs is close to elimination (white arrow). The "losing" input has completely segregated from the other axon, and the synaptic area on the muscle fiber that it occupies (labeled red with an acetylcholine antibody) is disappearing as the nerve is being eliminated (arrowheads). (C) This image illustrates the outcome of synaptic competition just after the losing axon (green) has withdrawn, leaving a red axon and its terminal. Note that the "loser" (green axon) has a retraction bulb at the end (arrow), and the "winning" axon (red) is significantly thicker. (Courtesy of J. W. Lichtman.)

BOX B
Why Do Neurons Have Dendrites?

Perhaps the most striking feature of neurons is their diverse morphology. Some classes of neurons have no dendrites at all; others have a modest dendritic arborization; still others have an arborization that rivals the complex branching of a fully mature tree (see Figures 1.1 and 1.2). Why should this be? Although there are many reasons for this diversity, neuronal geometry influences the number of different inputs that a target neuron receives by modulating competitive interactions among the innervating axons.

Evidence that the number of inputs a neuron receives depends on its geometry has come from studies of the peripheral autonomic system, where it is possible to stimulate the full complement of axons innervating an autonomic ganglion and its constituent neurons. This approach is not usually feasible in the central nervous system because of the anatomical complexity of most central circuits. Since individual postsynaptic neurons can also be labeled via an intracellular recording electrode, electrophysiological measurements of the number of different axons innervating a neuron can routinely be correlated with target cell shape. In both parasympathetic and sympathetic ganglia, the degree of preganglionic convergence onto a neuron is proportional to its dendritic complexity. Thus, neurons that lack dendrites altogether are generally innervated by a single input, whereas neurons with increasingly complex dendritic arborizations are innervated by a proportionally greater number of different axons (see figure). This correlation of neuronal geometry and input number holds within a single ganglion, among different ganglia in a single species, and among homologous ganglia across a range of species. Since ganglion cells that have few or no dendrites are initially innervated by several different

inputs (see text), confining inputs to the limited arena of the developing cell soma evidently enhances competition between them, whereas the addition of dendrites to a neuron allows multiple inputs to persist in peaceful coexistence.

A neuron innervated by a single axon will clearly be more limited in the scope of its responses than a neuron innervated by 100,000 inputs (1 to 100,000 is the approximate range of convergence in the mammalian brain). By regulating the number of inputs that neurons receive, dendritic form greatly influences function.

References

HUME, R. I. AND D. PURVES (1981) Geometry of neonatal neurons and the regulation of synapse elimination. Nature 293: 469–471.

PURVES, D. AND R. I. HUME (1981) The relation of postsynaptic geometry to the number of presynaptic axons that innervate autonomic ganglion cells. J. Neurosci. 1: 441–452.

PURVES, D. AND J. W. LICHTMAN (1985) Geometrical differences among homologous neurons in mammals. Science 228: 298–302.

PURVES, D., E. RUBIN, W. D. SNIDER AND J. W. LICHTMAN (1986) Relation of animal size to convergence, divergence and neuronal number in peripheral sympathetic pathways. J. Neurosci. 6: 158–163.

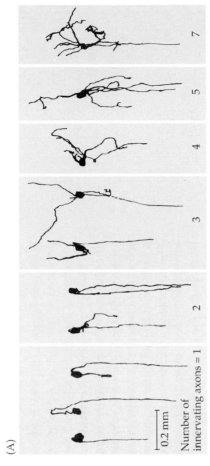

(A)

0.2 mm

Number of innervating axons = 1   2   3   4   5   7

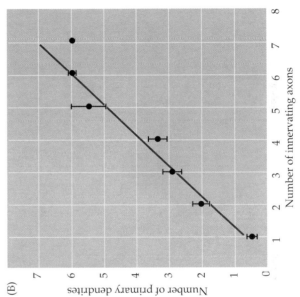

(B)

The number of axons innervating ciliary ganglion cells in adult rabbits. (A) Neurons studied electrophysiologically and then labeled by intracellular injection of a marker enzyme have been arranged in order of increasing dendritic complexity. The number of axons innervating each neuron is indicated. (B) This graph summarizes observations on a large number of cells. There is a strong correlation between dendritic geometry and input number. (After Purves and Hume, 1981.)

Molecular Basis of Trophic Interactions

The two major functions of neurotrophic signaling—the survival of a subset of neurons from a considerably larger population, and the subsequent formation and maintenance of appropriate numbers of connections—can be rationalized in part by the supply and availability of trophic factors. These rules entail several general assumptions about neurons and their targets (which may be other neurons, muscles, or other peripheral structures). First, neurons depend on the availability of some minimum amount of trophic factor for survival, and subsequently for the persistence of appropriate numbers of target connections. Second, target tissues synthesize and make available to developing neurons appropriate trophic factors. Third, targets produce trophic factors in limited amounts; in consequence, the survival of developing neurons (and later, the persistence of neuronal connections) depends on neuronal competition for the available factor. One much-studied trophic molecule, the protein called nerve growth factor (NGF), has provided support for these assumptions. Although the story of nerve growth factor certainly does not explain all aspects of trophic interactions, it has been a useful paradigm for understanding in more detail the manner in which neural targets influence the survival and connections of the nerve cells that innervate them.

NGF was discovered in the early 1950s by Rita Levi-Montalcini and Viktor Hamburger at Washington University. On the basis of experiments involving the survival of motor neurons after removal of developing limb buds (see Figure 23.9), they made an informed guess that the target tissues provided some sort of signal to the relevant neurons, and that limited amounts of this agent explained the apparently competitive nature of nerve cell death. Accordingly, Levi-Montalcini and Hamburger undertook a series of experiments to explore the source and nature of the postulated signal, focusing on dorsal root and sympathetic ganglion neurons rather than the spinal cord neurons. A former student of Hamburger's had earlier removed a limb from a chick embryo and replaced it with a piece of mouse tumor. The surprising outcome of this experiment was that the tumor apparently furnished an even more potent stimulus than the limb, causing an enlargement of the sensory and sympathetic ganglia that normally innervate the appendage. In further experiments, Levi-Montalcini and Hamburger provided evidence that the tumor (a mouse sarcoma) secreted a soluble factor that stimulated the survival and growth of both sensory and sympathetic ganglion cells. Levi-Montalcini then devised a bioassay for the presumed

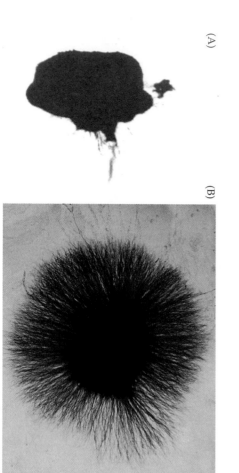

(A)

(B)

Figure 23.12 Effect of NGF on the outgrowth of neurites. (A) A chick sensory ganglion taken from an 8-day-old embryo and grown in organ culture for 24 hours in the absence of NGF. Few, if any, neuronal branches grow out into the plasma clot in which the explant is embedded. (B) A similar ganglion in identical culture conditions 24 hours after the addition of NGF to the medium. NGF stimulates a halo of neurite outgrowth from the ganglion cells. (From Purves and Lichtman, 1985; courtesy of R. Levi-Montalcini.)

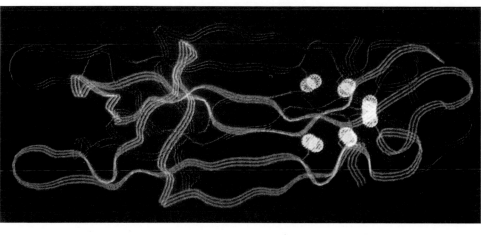

Figure 23.13 The crystal structure of the active subunit of the nerve growth factor complex. The active form of NGF consists of a dimer of identical β subunits. (From McDonald et al., 1991.)

agent and, in collaboration with Stanley Cohen, isolated and characterized the molecule—which had by then been named nerve growth factor for its ability to induce the massive outgrowth of neurites from explanted ganglia (Figure 23.12). (The term "neurite" is used to describe neuronal branches when it is not known whether they are axons or dendrites.) NGF was identified as a protein and was substantially purified from a rich biological source, the salivary glands of the male mouse. Subsequently, its amino acid sequence was determined and the cDNAs encoding NGF cloned in several species; in 1991, the structure of the active moiety of NGF was finally resolved (Figure 23.13).

Support for the idea that NGF is important for neuronal survival in more physiological circumstances emerged from a number of further observations. Depriving developing mice of NGF by the chronic administration of an NGF antiserum or other strategies resulted in adult mice lacking most NGF-dependent neurons (Figure 23.14). Conversely, injection of exogenous NGF into newborn rodents caused enlargement of sympathetic ganglia, an effect opposite that of NGF deprivation. Neurons in ganglia in treated animals were both more numerous and larger; there was also more neuropil between cell bodies, suggesting an overgrowth of axons, dendrites, and other cellular elements. The dramatic influence of NGF on cell survival, together with what was known about the significance of neuronal death in development, suggested that NGF is indeed a target-derived signal that serves to match the number of nerve cells to the number of target cells.

The ability of NGF to support neuronal survival (and of NGF deprivation to enhance cell death) is not in itself unassailable proof of a physiological role for this factor in development. In particular, these observations provided no direct evidence for NGF synthesis by (and uptake from) neuronal targets. This gap was filled by another series of ingenious experiments in several laboratories that showed NGF to be present in sympathetic targets, and to be quantitatively correlated with the density of sympathetic innervation. Furthermore, messenger RNA for NGF was demonstrated in targets innervated by sympathetic and sensory ganglia, but not in the ganglia themselves or in targets innervated by other types of nerve cells. As might be expected from such specificity, the NGF-sensitive neurons were also shown to have receptor molecules for the trophic factor (see next section). Importantly, the NGF message appears only after ingrowing axons have reached their targets; this fact makes it unlikely that secreted NGF acts as a chemotropic (guidance) molecule (like the netrins discussed earlier). Finally, the great majority of sympathetic neurons are lacking in mice in which the gene encoding NGF has been deleted.

In sum, several decades of work in a number of laboratories have shown that NGF mediates cell survival among two specific neuronal populations in birds and mammals (sympathetic and a subpopulation of sensory ganglion cells). These observations include the death of the relevant neurons in the absence of NGF; the survival of a surplus of neurons in the presence of augmented levels of the factor; the presence and production of NGF in neuronal targets; and the existence of receptors for NGF in innervating nerve terminals. Indeed, these observations define the criteria that must be satisfied in order to conclude that a given molecule is indeed a trophic factor.

Although NGF remains the most thoroughly studied neurotrophic factor, it was apparent from the outset that only certain classes of nerve cells respond to NGF. A flurry of work in the last decade has shown that NGF is only one member of a family of related trophic molecules, the **neurotrophins.** At present, there are three well-characterized members of the neurotrophin family in addition to NGF: **brain-derived neurotrophic factor**

Neurotrophin Receptors

Although several neurotrophins are homologous in amino acid sequence and structure, they are very different in their specificity (Figure 23.15). For example, whereas NGF supports the survival of (and neurite outgrowth from) sympathetic neurons, another family member—BDNF—cannot. Conversely, BDNF, but not NGF, can support the survival of certain sensory ganglion neurons, which have a different embryonic origin. NT-3 supports both of these populations, indicating that the specificity of neurotrophins is both distinct and overlapping. Given the diverse systems whose growth and connectivity must be coordinated during neural development, this specificity makes good sense.

The selective actions of the neurotrophins arise from a family of receptor proteins, the **Trk receptors**. **TrkA** is primarily a receptor for NGF, **TrkB** a receptor for BDNF and NT-4/5, and **TrkC** a receptor for NT-3 (Figure 23.16). The expression of a particular Trk receptor subtype therefore confers on that neuron the capacity to respond to the corresponding neurotrophin. Since neurotrophins and Trk receptors are expressed only in certain cell types in the nervous system, the selective binding between ligand and receptor accounts for the specificity of the relevant neurotrophic interactions.

(BDNF), **neurotrophin-3 (NT-3)**, and **neurotrophin 4/5 (NT-4/5)** (Box C). Other trophic factors that are not members of this family have also been described; like the neurotransmitters, it is likely that the number of identified trophic factors will continue to grow.

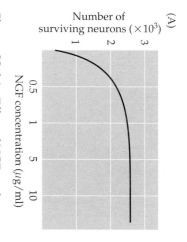

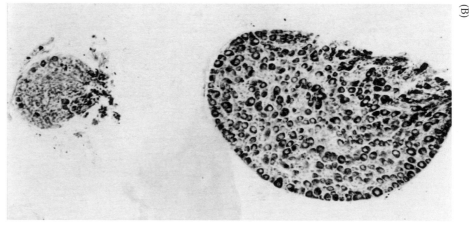

Figure 23.14 Effect of NGF on the survival of sympathetic ganglion cells. (A) The survival of newborn rat sympathetic ganglion cells grown in culture for 30 days evaluated quantitatively as a function of NGF concentration. Dose-response curves such as this one confirm the strict dependence of these neurons on the availability of NGF. (B) Cross section of a superior cervical ganglion from a normal 9-day-old mouse (top) compared to a similar section from a littermate injected daily since birth with NGF antiserum (bottom). The ganglion of the treated mouse shows marked atrophy, with obvious loss of nerve cells. (A after Chun and Patterson, 1977; B from Levi-Montalcini, 1972.)

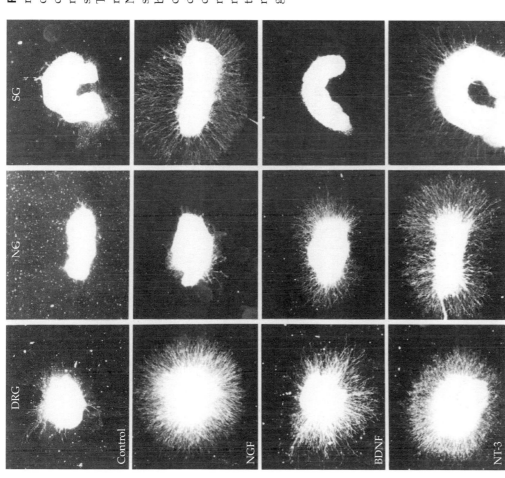

Figure 23.15 The effects of the neurotrophins NGF, BDNF, and NT-3 on the outgrowth of neurites from explanted, dorsal root ganglia (left column), nodose ganglia (middle column), and sympathetic ganglia (right column). The specificities of these several neurotrophins are evident in the ability of NGF to induce neurite outgrowth from sympathetic and dorsal root ganglia, but not from nodose ganglia (which are cranial nerve sensor ganglia that have a different embryological origin from dorsal root ganglia); of BDNF to induce neurite outgrowth from dorsal root and nodose ganglia, but not from sympathetic ganglia; and of NT-3 to induce neurite outgrowth from all three types of ganglia. (From Maisonpierre et al., 1990.)

Trk receptors belong to the superfamily of receptor tyrosine kinases, and their structure and activation closely resemble those of non-neuronal growth factor receptors, such as epidermal growth factor (EGF) receptor. Interestingly, activation of Trk receptors transfected into non-neuronal cells induces

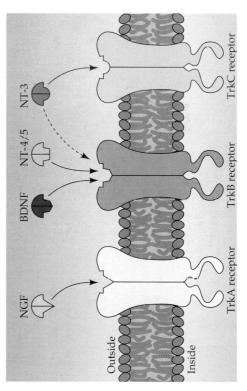

Figure 23.16 The Trk family of receptor tyrosine kinases for the neurotrophins. TrkA is primarily a receptor for NGF, TrkB a receptor for BDNF and NT-4/5, and TrkC a receptor for NT-3. Because of the high degree of structural homology among both the neurotrophins and the Trk receptors, there is some degree of cross-activation between factors and receptors. For example, NT-3 can bind to and activate TrkB under some conditions, as indicated by the dashed arrow. These distinct receptors allow various neurons to respond selectively to the different neurotrophins.

BOX C
The Discovery of BDNF and the Neurotrophin Family

During the 30 years or so that work with NGF showed it to fulfill all the criteria for a target-derived neurotrophic factor (see text), it became clear that NGF affected only a few specific populations of peripheral neurons. It was therefore presumed that other neurotrophic factors must exist that followed similar rules, but supported the survival and growth of other classes of neurons. In particular, whereas NGF was shown to be secreted by the *peripheral* targets of primary sensory and sympathetic neurons, other factors were presumably produced by target neurons in the brain and spinal cord that supported the *central* projections of sensory neurons.

The serendipity of the mouse salivary gland and its extraordinary levels of NGF was not repeated for these additional factors, however, and the hunt for the neurotrophic factors presumed to act in the central nervous system proved to be a long and arduous one. Indeed, it was not until the 1980s that the pioneering work of Yves Barde, Hans Thoenen, and their colleagues succeeded in identifying and purifying a factor from the brain that they named brain-derived neurotrophic factor (BDNF). As with NGF, this factor was purified on the basis of its ability to promote the survival and neurite outgrowth of sensory neurons. However, BDNF is expressed at such vanishingly small levels that over a million-fold purification was necessary before the protein could be identified!

Thereafter, microsequencing and recombinant DNA technology allowed rapid progress even from the scant amounts of purified BDNF protein that were available. By 1989, Barde's group had succeeded in cloning the cDNA for BDNF. Surprisingly—despite its entirely different origin and distinct neuronal specificity—BDNF turned out to be a close relative of NGF. Based on the homologies between the primary structures of NGF and BDNF, the following year six independent laboratories (including Barde's) reported the cloning of a third member of the neurotrophin family, neurotrophin-3 (NT-3). At present, four members of the neurotrophin family have been reported in a variety of vertebrate species (see text).

Experiments on BDNF and other members of the neurotrophin family over last decade have supported the conclusion that the survival and growth of different neuronal populations in both the PNS and CNS is dependent on different neurotrophins, relationships that are mediated by expression of membrane receptors that are specific for each neurotrophin (see figure). However, the dramatic relationship between the survival of neuronal populations and neurotrophins has not been found in the CNS, where BDNF, NT-3, and NT-4/5, as well as their receptors, are primarily expressed. The most striking demonstration of this difference has been in "knockout" mice in which individual genes encoding neurotrophins or Trk receptors have been deleted: While these genetic deletions have led to predictable deficits in the PNS (see text), they have generally had minimal impact on CNS structure and function.

Thus, the part played by neurotrophins in the CNS remains much less certain. One possibility is that these neurotrophins are more involved in regulating neuronal differentiation and phenotype in the CNS than in supporting neuronal survival per se. In this regard, the expression of neurotrophins is tightly regulated by electrical and synaptic activity, suggesting that they may also influence experience-dependent processes during the formation of circuits in the CNS.

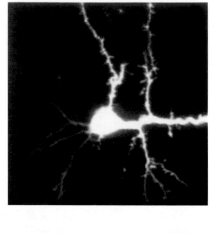

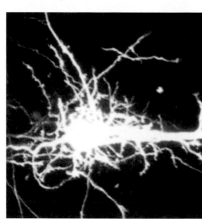

Neurotrophins influence dendritic arbors in the developing cerebral cortex. The cell on the left was transfected with the gene for green fluorescent protein (GFP) alone, the one on the right with GFP plus the gene encoding BDNF. Within a day, BDNF-transfected neurons grow elaborate dendritic branches, reminiscent of the NGF-induced halo in peripheral ganglia (see Figure 23.12B). (From Horch et al., 1999.)

References

HOFER, M. M. AND Y.-A. BARDE (1988) Brain-derived neurotrophic factor prevents neuronal death *in vivo.* Nature 331: 261–262.

HOHN, A., J. LEIBROCK, K. BAILEY AND Y. A. BARDE (1990) Identification and characterization of a novel member of the nerve growth factor/brain-derived neurotrophic factor family. Nature 344: 339–341.

LEIBROCK, J. AND 7 OTHERS (1989) Molecular cloning and expression of brain-derived neurotrophic factor. Nature 341: 149–152.

SNIDER, W. D. (1994) Functions of the neurotrophins during nervous system development: What the knockouts are teaching us. Cell 77: 627–638.

LEWIN, G. R. AND Y.-A. BARDE (1996) Physiology of the neurotrophins. Ann. Rev. Neurosci. 19: 289–317.

cell proliferation, which is the normal response of cells like fibroblasts to EGF. These similarities imply that the intracellular signaling cascades activated downstream of Trk receptors are similar to those of other growth factor receptors. Activation of Trk receptors by neurotrophins leads ultimately to changes in patterns of gene expression in target neurons that underlie the manifestations of trophic interactions. Finally, a second class of receptors that binds all the neurotrophins (called p75) has been identified, but its function is not known.

The Effect of Neurotrophins on the Differentiation of Neuronal Form

The second major role of nerve growth factor and other trophic molecules (in addition to their influence on neuron survival) is to modulate the growth of neuronal branches. In the case of NGF, a compelling indication of this further action is that explanted sensory or sympathetic ganglia exposed to a culture medium containing NGF show a marked outgrowth of neurites within 24 hours (see Figure 23.12). More specific evidence has been obtained using a culture system that distinguishes the local effects of NGF on neurites from effects mediated through the neuronal cell body. In this system, dissociated ganglion cells are placed in the central well of a chamber with three compartments whose NGF concentration can be varied independently (Figure 23.17). If the three compartments contain adequate concentrations of NGF, then neurites from the ganglion cells in the central well extend into both peripheral compartments. However, if NGF is removed from one of the peripheral wells, then the neurites that have grown into that compartment gradually retract. Conversely, if NGF is removed from the central compartment but retained in the peripheral wells, the neurites remain in place. These results indicate that neurites extend or retract as a function of the *local* concentration of NGF. If the effects of NGF on neurites depended only on sufficient trophic factor being supplied to the nerve cell as a whole, then the neurites in the different compartments should grow regardless of which compartments contained NGF. In short, neurite outgrowth can be controlled locally by trophic stimuli and does not simply depend on the overall effects of trophic agents on the parent cell. As a result, some branches of a neuron may extend while others retract, which is what actually happens during the establishment of synaptic connections in normal development.

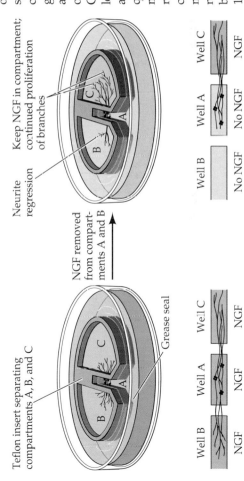

Figure 23.17 Evidence that NGF can influence neurite growth by local action. Three compartments of a culture dish (A, B, C) are separated from one another by a plastic divider sealed to the bottom of the dish with grease. Isolated rat sympathetic ganglion cells plated in compartment A can grow through the grease seal and into compartments B and C. (A magnified view looking down on the compartments is shown below.) Growth into a lateral chamber occurs as long as the compartment contains an adequate concentration of NGF. Subsequent removal of NGF from a compartment causes a local regression of neurites without affecting the survival of cells or neurites in the other compartments. These observations show that neuritic growth can be locally controlled by neurotrophins. (After Campenot, 1981.)

Taken together, the evidence about NGF and the other neurotrophins suggests a general scheme for the regulation of neuronal connections in the nervous system (sometimes referred to as the **neurotrophic hypothesis**). Neuronal targets, whether non-neural cells or other neurons, produce trophic molecules in limited amounts. In embryonic and early postnatal life—and to a more limited extent in maturity—the survival of innervating neurons depends on exposure to a critical amount of these agents. In consequence, neurons sensitive to a particular trophic molecule initially compete with one another, and those that fail in this competition die. Following the establishment of definitive neuronal populations in this way, ongoing trophic dependency is apparent in the growth and retraction of neuronal processes, again as a function of target-derived support. In early postnatal development, this dependence is evident in the continuing growth and rearrangement of the initial connections whereby each neuron comes to innervate an appropriate number of target cells and each target cell comes to be contacted by an appropriate number of axons. Later in life, neural connections made by a fixed number of nerve cells continue to adjust by sprouting and retraction as targets change in size, form, and function during a prolonged period of maturation. In addition to mediating the compensatory adjustments required by growth, competition for trophic molecules allows neuronal branches and their connections to change in response to a variety of other circumstances, including injury (see Chapter 25) and altered patterns of neural activity associated with experience, as described in the following chapter.

Summary

Neurons in the developing brain must integrate a variety of molecular signals in order to determine where to send their axons, whether to live or die, what cells to form synapses on, how many synapses to make, and whether to retain them. Fixed and/or diffusible chemotropic, chemorepulsive, and trophic molecules all regulate the trajectory of growing axons and the synaptic connections they make with target cells. These developmental interactions transpire over weeks, months, and to some extent over the entire lifetime of the animal. In the early stages of development, the most salient effects of trophic agents are on cell survival and differentiation. Once the adult population of neurons is established, trophic signals continue to govern the establishment of neural connections, particularly the extent of axonal and dendritic arborizations. Dysfunction of this range of molecular and cellular interactions has the potential for producing devastating pathology. Defects in the early guidance of axons are responsible for a variety of congenital neurological syndromes, and conditions thought to reflect trophic dysfunction may underlie degenerative diseases such as amyotrophic lateral sclerosis and Parkinson's disease. Understanding the molecular basis of axon guidance, synapse formation, and trophic signaling began a century ago and has now burgeoned into a broad effort that continues to identify additional factors and to illuminate their varied roles in both the developing and adult brain. A further goal that now seems within reach is the application of this knowledge to understanding a spectrum of previously intractable neurological diseases.

ADDITIONAL READING

Reviews

CULOTTI, J. G. AND D. C. MERZ (1998) DCC and netrins. Curr. Op. Cell Biol. 10: 609–613.

LEVI-MONTALCINI, R. (1987) The nerve growth factor 35 years later. Science 237: 1154–1162.

LEWIN, G. R. AND Y. A. BARDE (1996) Physiology of the neurotrophins. Ann. Rev. Neurosci. 19: 289–317.

LICHTMAN, J. W. AND H. COLEMAN (2000) Synapse elimination and indelible memory. Neuron 25: 269–278.

PURVES, D. AND J. W. LICHTMAN (1978) Formation and maintenance of synaptic connections in autonomic ganglia. Physiol. Rev. 58: 821–862.

PURVES, D. AND J. W. LICHTMAN (1980) Elimination of synapses in the developing nervous system. Science 210: 153–157.

PURVES, D., W. D. SNIDER AND J. T. VOYVODIC (1988) Trophic regulation of nerve cell morphology and innervation in the autonomic nervous system. Nature 336: 123–128.

RAPER, J. A. (2000) Semaphorins and their receptors in vertebrates and invertebrates. Curr. Opin. Neurobiol. 10: 88–94.

REICHARDT, L. F. AND K. J. TOMASELLI (1991) Extracellular matrix molecules and their receptors: Functions in neural development. Ann. Rev. Neurosci. 14: 531–570.

RUTISHAUSER, U. (1993) Adhesion molecules of the nervous system. Curr. Opin. Neurobiol. 3: 709–715.

SANES, J. R. AND J. W. LICHTMAN (1999) Development of the vertebrate neuromuscular junction. Ann. Rev. Neurosci. 22: 389–442.

SCHWAB, M. E., J. P. KAPFHAMMER AND C. E. BANDTLOW (1993) Inhibitors of neurite growth. Ann. Rev. Neurosci. 16: 565–595.

SEGAL, R. A. AND M. E. GREENBERG (1996) Intracellular signaling pathways activated by neurotrophic factors. Ann. Rev. Neurosci. 19: 463–489.

SILOS-SANTIAGO, I., L. J. GREENLUND, E. M. JOHNSON JR. AND W. D. SNIDER (1995) Molecular genetics of neuronal survival. Curr. Opin. Neurobiol. 5: 42–49.

TEAR, G. (1999) Neuronal guidance: A genetic perspective. Trends Genet. 15: 113–118.

Important Original Papers

BAIER, H. AND F. BONHOEFFER (1992) Axon guidance by gradients of a target-derived component. Science 255: 472–475.

BALICE-GORDON, R. J. AND J. W. LICHTMAN (1994) Long-term synapse loss induced by focal blockade of postsynaptic receptors. Nature 372: 519–524.

BALICE-GORDON, R. J., C. K. CHUA, C. C. NELSON AND J. W. LICHTMAN (1993) Gradual loss of synaptic cartels precedes axon withdrawal at developing neuromuscular junctions. Neuron 11: 801–815.

BROWN, M. C., J. K. S. JANSEN AND D. VAN ESSEN (1976) Polyneuronal innervation of skeletal muscle in new-born rats and its elimination during maturation. J. Physiol. (Lond.) 261: 387–422.

CAMPENOT, R. B. (1977) Local control of neurite development by nerve growth factor. Proc. Natl. Acad. Sci. USA 74: 4516–4519.

DRESCHER, U., C. KREMOSER, C. HANDWERKER, J. LOSCHINGER, M. NODA AND F. BONHOEFFER (1995) In vitro guidance of retinal ganglion cell axons by RAGS, a 25 kDa tectal protein related to ligands for Eph receptor tyrosine kinases. Cell 82: 359–370.

FREDETTE, B. J. AND B. RANSCHT (1994) T-cadherin expression delineates specific regions of the developing motor axon-hindlimb projection pathway. J. Neurosci. 14: 7331–7346.

FARINAS, I., K. R. JONES, C. BACKUS, X. Y. WANG AND L. F. REICHARDT (1994) Severe sensory and sympathetic deficits in mice lacking neurotrophin-3. Nature 369: 658–661.

KAPLAN, D. R., D. MARTIN-ZANCA AND L. F. PARADA (1991) Tyrosine phosphorylation and tyrosine kinase activity of the *trk* proto-oncogene product induced by NGF. Nature 350: 158–160.

KENNEDY, T. E., T. SERAFINI, J. R. DE LA TORRE AND M. TESSIER-LAVIGNE (1994) Netrins are diffusible chemotropic factors for commissural axons in the embryonic spinal cord. Cell 78: 425–435.

KOLODKIN, A. L., D. J. MATTHES AND C. S. GOODMAN (1993) The semaphorin genes encode a family of transmembrane and secreted growth cone guidance molecules. Cell 75: 1389–1399.

LANGLEY, J. N. (1895) Note on regeneration of pre-ganglionic fibres of the sympathetic. J. Physiol. (Lond.) 18: 280–284.

LEVI-MONTALCINI, R. AND S. COHEN (1956) In vitro and in vivo effects of a nerve growth-stimulating agent isolated from snake venom. Proc. Natl. Acad. Sci. USA 42: 695–699.

LICHTMAN, J. W. (1977) The reorganization of synaptic connexions in the rat submandibular ganglion during post-natal development. J. Physiol. (Lond.) 273: 155–177.

LICHTMAN, J. W., L. MAGRASSI AND D. PURVES (1987) Visualization of neuromuscular junctions over periods of several months in living mice. J. Neurosci. 7: 1215–1222.

LUO, Y., D. RAIBLE AND J. A. RAPER (1993) Collapsin: A protein in brain that induces the collapse and paralysis of neuronal growth cones. Cell 75: 217–227.

MESSERSMITH, E. K., E. D. LEONARDO, C. J. SHATZ, M. TESSIER-LAVIGNE, C. S. GOODMAN AND A. L. KOLODKIN (1995) Semaphorin III can function as a selective chemorepellent to pattern sensory projections in the spinal cord. Neuron 14: 949–959.

OPPENHEIM, R. W., D. PREVETTE AND S. HOMMA (1990) Naturally occurring and induced neuronal death in the chick embryo in vivo requires protein and RNA synthesis: Evidence for the role of cell death genes. Dev. Biol. 138: 104–113.

SPERRY, R. W. (1963) Chemoaffinity in the orderly growth of nerve fiber patterns and connections. Proc. Natl. Acad. Sci. 50: 703–710.

WALTER, J., S. HENKE-FAHLE AND F. BONHOEFFER (1987) Avoidance of posterior tectal membranes by temporal retinal axons. Development 101: 909–913.

Books

LETOURNEAU, P. C., S. B. KATER AND E. R. MACAGNO (EDS.) (1991) *The Nerve Growth Cone.* New York: Raven Press.

LOUGHLIN, S. E. AND J. H. FALLON (EDS.) (1993) *Neurotrophic Factors.* San Diego, CA: Academic Press.

PURVES, D. (1988) *Body and Brain: A Trophic Theory of Neural Connections.* Cambridge, MA: Harvard University Press.

RAMÓN Y CAJAL, S. (1928) *Degeneration and Regeneration of the Nervous System.* R. M. May (ed.). New York: Hafner Publishing.

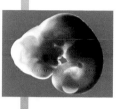

Chapter 24

Modification of Brain Circuits as a Result of Experience

Overview

The rich diversity of human personalities, abilities, and behavior is undoubtedly generated by the uniqueness of individual human brains. These fascinating neurobiological differences among humans derive from both heritable and environmental influences. The first steps in the construction of the brain's circuitry—the establishment of distinct brain regions, the generation of neurons, the formation of major axon tracts, the guidance of growing axons to appropriate targets, and the initiation of synaptogenesis—rely largely on the intrinsic cellular and molecular processes described in the previous chapters. Once the basic patterns of brain connections are established, however, patterns of neuronal activity (i.e., experience) can modify the synaptic circuitry of the developing brain. Neuronal activity generated by interactions with the outside world in postnatal life thus provides a mechanism by which the environment can influence brain structure and function. This activity-mediated influence on the developing brain is most consequential during temporal windows called critical periods. As animals mature, the brain becomes increasingly refractory to the lessons of experience, and the cellular mechanisms that mediate altered neural connectivity become less effective.

Critical Periods

The cellular and molecular mechanisms outlined in Chapters 22 and 23 construct a nervous system of impressive anatomical complexity. These instructions are also sufficient to create some remarkably sophisticated innate or "instinctual" behaviors (see Box A in Chapter 31). For most animals, the behavioral repertoire, including foraging, fighting, and mating strategies, relies on patterns of connectivity established by intrinsic developmental mechanisms. However, the nervous systems of complex ("higher") animals, including humans, clearly adapt to and are influenced by the particular circumstances of an individual's environment. These environmental factors are especially influential in early life, during temporal windows called **critical periods.** In some cases, such as the acquisition of language, instructive influences from the environment are obviously required for the normal development of the behavior (i.e., exposure to the individual's native language). Moreover, some behaviors, such as imprinting in birds (Box A), are expressed only if animals have certain specific experiences during a sharply restricted time in early postnatal (or posthatching) development. On the other hand, critical periods for sensory and motor skills, or complex behaviors such as human language, are longer and much less well delimited.

BOX A
Built-in Behaviors

The idea that animals already possess a set of behaviors appropriate for a world not yet experienced has always been difficult to accept. However, the preeminence of instinctual responses is obvious to any biologist who looks at what animals actually do. Perhaps the most thoroughly studied examples occur in young birds. Hatchlings emerge from the egg with an elaborate set of innate behaviors. First, of course, is the complex behavior that allows the chick to escape from the egg. Having emerged, a variety of additional abilities indicate how much early behavior is "preprogrammed" (see Box A in Chapter 31).

In a series of seminal observations, Konrad Lorenz, working with geese, showed that goslings follow the first large, moving object that they see and hear during their first day of life. Although this object is normally the mother goose, Lorenz found that goslings can imprint on a wide range of animate and inanimate objects presented during this period, including Lorenz himself (see figure). The window for imprinting in goslings is less than a day; if animals are not exposed to an appropriate stimulus during this time, they will never form the appropriate parental relationship. Once imprinting occurs, however, it is irreversible, and geese will continue to follow inappropriate objects (male conspecifics, people, or even inanimate objects). In many mammals, auditory and visual systems are poorly developed at birth, and maternal imprinting relies on olfactory and/or gustatory cues. For example, during the first week of life (but not later), infant rats develop a lifelong preference to odors associated with their mother's nipples. As in birds, this variety of filial imprinting also plays a role in their social development and later sexual preferences.

Imprinting is a two-way street, with parents (especially mothers) rapidly

forming exclusive bonds with their offspring. This phenomenon is especially important in animals like sheep that live in large groups or herds and produce offspring at about the same time of year. Ewes have a critical period 2–4 hours after giving birth during which they imprint on the scent of their own lamb. Following this time, they rebuff approaches by other lambs.

The relevance of this work to primates was underscored in the 1950s by Harry Harlow and his colleagues at the University of Wisconsin. Harlow isolated monkeys within a few hours of birth and raised them in the absence of either a natural mother or a human substitute. In the best-known of these experiments, the baby monkeys had one of two maternal surrogates: a "mother" constructed of a wooden frame covered with wire mesh that supported a nursing bottle, or a similarly shaped object covered with terrycloth. When presented with this choice, the baby monkeys preferred the

Lorenz being followed by imprinted geese. (From Nisbett, 1976; photo courtesy of H. Kacher.)

terrycloth mother and spent much of their time clinging to it, even if the feeding bottle was with the wire mother. Harlow took this to mean that newborn monkeys have a built-in need for maternal care and have at least some innate idea of what a mother should be like. More recently, a number of other endogenous behaviors have been carefully studied in infant monkeys, including a naïve monkey's fear reaction to the presentation of certain objects (e.g., a snake) and the "looming" response (fear elicited by the rapid approach of any formidable object). Most of these built-in behaviors have analogs in human infants.

Taken together, these observations make plain that many complicated behaviors, emotional responses, and other predilections are well established in the nervous system prior to any significant experience, and that the need for certain kinds of early experience for normal development is predetermined. These built-in behaviors and their neural substrates have presumably evolved to give newborns a better chance of surviving in a predictably dangerous world.

References

HARLOW, H. F. (1959) Love in infant monkeys. Sci. Am. 2(9): 68–74.

HARLOW, H. F. AND R. R. ZIMMERMAN (1959) Affectional responses in the infant monkey. Science 130: 421–432.

LORENZ, K. (1970) Studies in Animal and Human Behaviour. Translated by R. Martin. Cambridge, MA: Harvard University Press.

MACFARLANE, A.J. (1975) Olfaction in the development of social preferences in the human neonate. Ciba Found. Symp. 33: 103–17.

SCHAAL, B.E., H. MONTAGNER, E. HERTLING, D. BOLZONI, A. MOYSE AND R. QUICHON (1980) Les stimulations olfactives dans les relations entre l'enfant et la mere. Reproduction, Nutrition, Developement 20(3b): 843–858.

TINBERGEN, N. (1953) Curious Naturalists. Garden City, NY: Doubleday.

Despite the fact that critical periods vary widely in both the behaviors affected and their duration, they all share some basic properties. A critical period is defined as the time during which a given behavior is especially susceptible to, and indeed requires, specific environmental influences to develop normally. Once this period ends, the behavior is largely unaffected by subsequent experience (or even by the complete absence of the relevant experience). Conversely, failure to be exposed to appropriate stimuli during the critical period is difficult or impossible to remedy subsequently.

While psychologists and ethologists (that is, biologists who study the natural behavior of animals) have long recognized that early postnatal or posthatching life is a period of special sensitivity to environmental influences, their studies of critical periods focused on behavior. Work in the latter part of the twentieth century has increasingly examined the underlying changes in the relevant brain circuits and their mechanisms.

The Development of Language: A Critical Period in Humans

Many animals communicate by means of sound, and some (humans and songbirds are examples) learn these vocalizations. There are, in fact, provocative similarities in the development of human language and birdsong (Box B). Most animal vocalizations, like alarm calls in mammals and birds, are innate, and require no experience to be correctly produced. For example, quails raised in isolation or deafened at birth so that they never hear conspecifics nonetheless produce the full repertoire of species-specific vocalizations. In contrast, humans obviously require extensive postnatal experience to produce and decode speech sounds that are the basis of language.

Importantly, this linguistic experience, to be effective, must occur in early life. The requirement for hearing and practicing during a critical period is apparent in studies of language acquisition in congenitally deaf children. Whereas most babies begin producing speechlike sounds at about 7 months (babbling), congenitally deaf infants show obvious deficits in their early vocalizations, and such individuals fail to develop language if not provided with an alternative form of symbolic expression (such as sign language; see Chapter 27). If, however, these deaf children are exposed to sign language at an early age (from approximately six months onward), they begin to "babble" with their hands just as a hearing infant babbles audibly. This suggests that, regardless of the modality, early experience shapes language behavior (Figure 24.1). Children who have acquired speech but subsequently lose their hearing before puberty also suffer a substantial decline in spoken language, presumably because they are unable to hear themselves talk and thus lose the opportunity to refine their speech by auditory feedback.

Examples of pathological situations in which normal children were never exposed to a significant amount of language make much the same point. In one well-documented case, a girl was raised by deranged parents until the age of 13 under conditions of almost total language deprivation. Despite intense subsequent training, she never learned more than a rudimentary level of communication. This and other examples of so-called "feral children" starkly define the importance of early experience. In contrast to the devastating effects of deprivation on children, adults retain their ability to speak and comprehend language even if decades pass without exposure or speaking. In short, the normal acquisition of human speech is subject to a critical period: The process is sensitive to experience or deprivation during a restricted period of life (before puberty) and is refractory to similar experience or deprivations in adulthood.

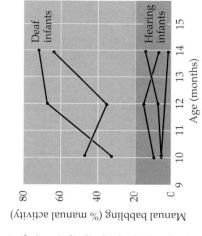

Figure 24.1 Manual "babbling" in deaf infants raised by deaf, signing parents compared to manual babble in hearing infants. Babbling was judged by scoring hand positions and shapes that showed some resemblance to the components of American Sign Language. In deaf infants, meaningful hand shapes increase as a percentage of manual activity between ages 10 and 14 months. Hearing children raised by hearing, speaking parents do not produce similar hand shapes. (After Petito and Marentette, 1991.)

BOX B
Birdsong

Anyone witnessing language development in a child cannot help but be amazed at how quickly learning takes place. This facility contrasts with the adult acquisition of a new language, which can be a painfully slow process that never produces complete fluency. In fact, many learned behaviors are acquired during a period in early life when experience exerts an especially potent influence on subsequent behavior. Particularly well characterized is the sensitive period for learning courtship songs by oscine songbirds such as canaries and finches. In these species, the quality of early sensory exposure is the major determinant of subsequent perceptual and behavioral capabilities. Furthermore, developmental periods for learning these and other behaviors are restricted during postnatal life, suggesting that the nervous system changes in some manner to become refractory to further experience. Understanding how critical periods are regulated has many implications, not least the possibility of reactivating this enhanced learning capacity in adults. Nonetheless, such periods are often highly specialized for the acquisition of species-typical behaviors and are not merely times of general enhanced learning.

Avian song learning illustrates the interactions between intrinsic and environmental factors in this developmental process. Many birds sing to attract mates, but oscine songbirds are special in that their courtship songs are dependent on auditory and vocal experience. The sensitive period for song learning comprises an initial stage of sensory acquisition, when the juvenile bird listens to and memorizes the song of a nearby adult male tutor (usually of its own species), and a subsequent stage of vocal learning, when the young bird matches its own song to the now-memorized tutor model via auditory feedback. This sensory motor learning stage ends with the onset of sexual maturity, when songs become acoustically stable, or crystallized. In all species studied to date, young songbirds are especially impressionable during the first two months after hatching and then become refractory to further exposure to tutor song as they age. The impact of this early experience is profound, and the memory it generates can remain intact for months, and perhaps years, before the onset of the vocal practice phase. Even constant exposure to other songs after sensory acquisition during the sensitive period ends does not affect this memory: The songs heard during sensory acquisition, but not later, are those that the bird vocally mimics. Early auditory experience is crucial to the bird's Darwinian success. In the absence of a tutor, or if raised only in the presence of another species, birds produce highly abnormal "isolate" songs, or songs of the foster species, neither of which succeeds in attracting females of their own kind.

Two other features of song learning indicate an intrinsic predisposition for this specialized form of vocal learning. First, juveniles often need to hear the tutor song only 10 or 20 times to then vocally mimic it many months later. Second, when presented with a variety of songs played from tape recordings that include their own and other species' songs, juvenile birds preferentially copy the song of their own species, even with no external reinforcement. These observa-

On a more subtle level, the phonetic structure of the language an individual hears during early life shapes both the perception and production of speech. Many of the thousands of human languages and dialects use appreciably different repertoires of speech elements called phonemes to produce spoken words (examples are the phonemes "ba" and "pa" in English). Very young human infants can perceive and discriminate between differences in all human speech sounds, and are not innately biased towards the phonemes characteristic of any particular language. However, this universal appreciation does not persist. For example, adult Japanese speakers cannot reliably distinguish between the /r/ and /l/ sounds in English, presumably because this phonemic distinction is not present in Japanese. Nonetheless, 4-month-old Japanese infants can make this discrimination as reliably as 4-month-olds raised in English-speaking households (as indicated by increased suckling frequency or head turning in the presence of a novel stimulus). By 6 months of age, however, infants show preferences for phonemes in their

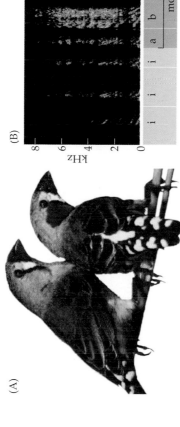

(A)

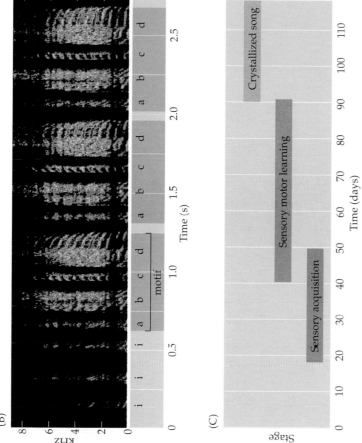

(A) A pair of zebra finches (the male is on the right), a species that has been the subject of many song acquisition studies. (B) Spectrogram of the typical adult song. The male's song comprises characteristic repeating elements including introductory notes (i), and single or multi-note syllables (a–d). Syllables are grouped into motifs; both syllable structure and order are learned in this species. (C) Chronology of song acquisition in the zebra finch. (Courtesy of Rich Mooney.)

tions show that juveniles are not really "naïve," but are innately biased to learn the songs of their own species over those of others. In short, intrinsic factors make the nervous system of oscine birds especially sensitive to songs that are species-typical. It is likely that similar biases influence human language learning.

References

DOUPE, A. AND P. KUHL (1999) Birdsong and human speech: Common themes and mechanisms. Ann. Rev. Neurosci. 22: 567–631.

MOONEY, R. (1999) Sensitive periods and circuits for learned birdsong. Curr. Op. Neurobiol. 9: 121–127.

native language over those in foreign languages, and by the end of their first year no longer respond to phonetic elements peculiar to non-native languages. The ability to perceive these phonemic contrasts evidently persists for several more years, as evidenced by the fact that children can learn to speak a second language without accent and with fluent grammar until about age 7 or 8. After this age, however, performance gradually declines no matter what the extent of practice or exposure (Figure 24.2).

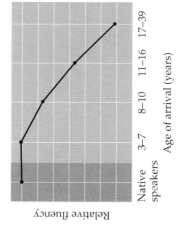

Figure 24.2 A critical period for learning language is shown by the decline in language ability (fluency) of non-native speakers of English as a function of their age upon arrival in the United States. The ability to score well on tests of English grammar and vocabulary declines from approximately age 7 onward. (After Johnson and Newport, 1989.)

A number of changes in the developing brain could explain these observations. One possibility is that experience acts selectively to preserve the circuits in the brain that perceive phonemes and phonetic distinctions. The absence of exposure to non-native phonemes would then result in a gradual atrophy of the connections representing those sounds, accompanied by a declining ability to distinguish between them. In this formulation, circuits that are used are retained, whereas those that are unused get weaker (and eventually disappear). Alternatively, experience could promote the growth of rudimentary circuitry pertinent to the experienced sounds.

The reality, however, is considerably more complex than either of these scenarios suggest. Experiments by Patricia Kuhl and her colleagues have demonstrated that as a second language is acquired, the brain gradually groups sounds according to their similarity with phonemes in the native language. For example, when asked to categorize a continuous spectrum of artificial phonemes between /r/ and /l/, native English speakers, but not Japanese speakers, tend to perceive sounds as all sounding like either /r/ or /l/, a phenomenon that Kuhl has likened to a "perceptual magnet." Related but varying sounds (defined by their audiographic spectrum) are evidently grouped together and eventually perceived as representing the same phoneme. Without ongoing experience during the critical period, this process fails to occur. Interestingly, the "baby-talk" or "parentese" used by adults speaking to young children actually emphasizes these phonetic distinctions compared to normal speech among adults. Thus, learning language during the critical period for its development entails an amplification and reshaping of innate biases by appropriate postnatal experience.

Critical Periods in Visual System Development

Although critical periods for language and other distinctively human behaviors are in some ways the most compelling examples, it is difficult if not impossible to study the underlying changes in brain circuits. A much deeper understanding of the changes in circuitry that accompany critical periods has come from studies of the developing visual system. In an extraordinarily influential series of experiments, David Hubel and Torsten Wiesel found that depriving an experimental animal of normal visual experience during a restricted period of early postnatal life irreversibly alters neuronal connections (and functions) in the visual cortex. These observations provided the first evidence that the brain translates the effects of early experience (that is, patterns of neural activity) into permanently altered wiring.

To understand these experiments and their implications, it is important to review the organization and development of the mammalian visual system. Information from the two eyes is first integrated in the primary visual (striate) cortex, where most afferents from the lateral geniculate nucleus of the thalamus terminate (see Chapter 12). In some mammals—carnivores, anthropoid primates, and humans—the afferent terminals form an alternating series of eye-specific domains in cortical layer IV called **ocular dominance columns** (Figure 24.3). As already noted in Chapter 12, ocular dominance columns can be visualized by injecting tracers, such as radioactive proline, into one eye; the tracer is then transported along the visual pathway to specifically label the geniculocortical terminals (i.e., synaptic terminals in the visual cortex) corresponding to that eye (Box C). In the adult macaque monkey, the domains representing the two eyes are stripes of about equal width (0.5 mm) that occupy roughly equal areas of layer IV of the primary visual cortex.

Electrical recordings confirm that the cells within layer IV of macaques respond strongly or exclusively to stimulation of either the left or the right eye, while neurons in layers above and below layer IV integrate inputs from one or the other eye or are being recorded, detailed assessment of ocular dominance can be made at the level of individual cells (see Figure 12.13). In these studies, Hubel and Wiesel assigned neurons to one of seven ocular dominance categories. Group 1 cells were defined as being driven only by stimulation of the contralateral eye; group 7 cells were driven entirely by the ipsilateral eye. Neurons driven equally well by either eye were assigned to group 4. Using this approach, they found that the ocular dominance distribution across the cortical layers in primary visual cortex is roughly Gaussian in a normal adult (cats were used in these experiments). Most cells were activated to some degree by both eyes, and about a quarter were more activated by either the contralateral or ipsilateral eye (Figure 24.4A).

Hubel and Wiesel then asked whether this normal distribution of ocular dominance could be altered by visual experience. When they simply closed one eye of a kitten early in life and let the animal mature to adulthood (which takes about 6 months), a remarkable change was observed. Electrophysiological recordings now showed that very few cells could be driven from the deprived eye; that is, the ocular dominance distribution had shifted such that all cells were driven by the eye that had remained open (Figure 24.4B).

Effects of Visual Deprivation on Ocular Dominance

As described in Chapter 12, if an electrode is passed at a shallow angle through the cortex while the responses of individual neurons to stimulation of the left and right eyes and respond to visual stimuli presented to either eye. Ocular dominance is thus apparent in two related phenomena: the degree to which individual cortical neurons are driven by stimulation of one eye or the other, and domains (stripes) in cortical layer IV in which the majority of neurons are driven exclusively by one eye or the other. The clarity of these eyes can be manipulated led to a series of experiments that greatly clarified the neurobiological processes underlying critical periods.

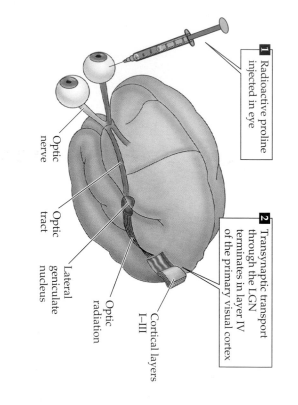

1 Radioactive proline injected in eye

Optic nerve

Optic tract

Lateral geniculate nucleus

2 Transsynaptic transport through the LGN terminates in layer IV of the primary visual cortex

Optic radiation

Cortical layers I–III

3 Terminations are visible as bright bands on the autoradiogram

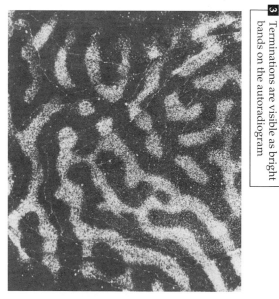

Figure 24.3 Ocular dominance columns (which in most anthropoid primates are really stripes or bands) in layer IV of the primary visual cortex of an adult macaque monkey. Diagram indicates the labeling procedure (see also Box C); following transsynaptic transport, the pattern of geniculocortical terminations related to that eye is visible as a series of bright stripes in this autoradiogram of a section through layer IV in the plane of the cortex (that is, as if looking down on the cortical surface). The dark areas are the zones occupied by geniculocortical terminals related to the other eye. The pattern of the human ocular dominance column is shown in Figure 12.10. (From LeVay, Wiesel and Hubel, 1980.)

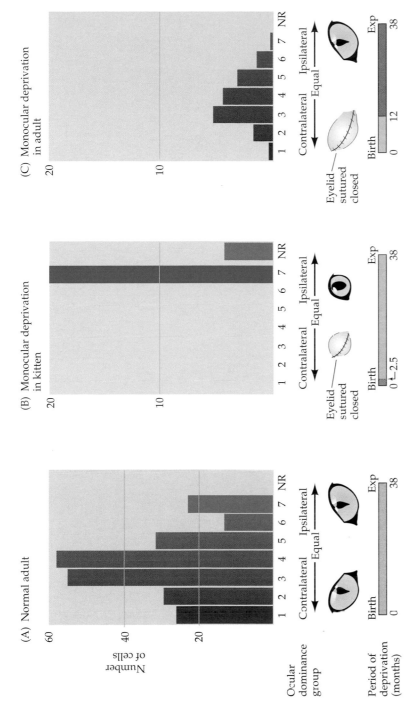

Figure 24.4 Effect of early closure of one eye on the distribution of cortical neurons driven by stimulation of both eyes. (A) Ocular dominance distribution of single unit recordings from a large number of neurons in the primary visual cortex of normal adult cats. Cells in group 1 were activated exclusively by the contralateral eye; cells in group 7 by the ipsilateral eye. Diagrams below these graphs indicate procedure, and bars indicate duration of deprivation (purple). "Exp" = time when experimental observations were made. (B) Following closure of one eye from 1 week after birth until 2.5 months of age (indicated by the bar underneath the graph), no cells could be activated by the deprived (contralateral) eye. Some cells could not be activated by either eye (NR). Note that the closed eye is opened at the time of the experimental observations, and that the recordings are not restricted to any particular cortical layer. (C) A much longer period of monocular deprivation in an adult cat has little effect on ocular dominance (although overall cortical activity is diminished). In this case, the contralateral eye was closed from 12 to 38 months of age. (A after Hubel and Weisel, 1962; B after Wiesel and Hubel, 1963; C after Hubel and Wiesel, 1970.)

Recordings from the retina and lateral geniculate layers related to the deprived eye indicated that these more peripheral stations in the visual pathway worked quite normally. Thus, the absence of cortical cells that responded to stimulation of the closed eye was not a result of retinal degeneration or a loss of retinal connections to the thalamus. Rather, the deprived eye had been functionally disconnected from the visual cortex. Consequently, such animals are behaviorally blind in the deprived eye. This "cortical blindness," or amblyopia, is permanent (see next section). Even if the formerly deprived eye is subsequently left open indefinitely, little or no recovery occurs.

Remarkably, the same manipulation—closing one eye—had no effect on the responses of cortical cells in the visual cortex of an adult cat. If one eye of a mature cat was closed for a year or more, both the ocular dominance distribution and the animal's visual behavior were indistinguishable from normal when tested through the reopened eye (Figure 24.4C). Thus, sometime

BOX C
Transneuronal Labeling with Radioactive Amino Acids

Unlike many brain structures, ocular dominance columns are not easily visible by means of conventional histology. Thus, the striking cortical patterns evident in cats and monkeys were not seen until the early 1970s, when the technique of anterograde tracing using radioactive amino acids was introduced. In this approach, an amino acid commonly found in proteins (usually proline) is radioactively tagged and injected into the area of interest. Neurons in the vicinity take up the label from the extracellular space and incorporate it into newly made proteins. Some of these proteins are involved in the maintenance and function of the neuron's synaptic terminals; thus, they are shipped via anterograde transport from the cell body to nerve terminals, where they accumulate. After a suitable interval, the tissue is fixed, and sections are made, placed on glass slides, and coated with a sensitive photographic emulsion. The radioactive decay of the labeled amino acids in the proteins causes silver grains to form in the emulsion. After several months of exposure, a heavy concentration of silver grains accumulates over the regions that contain synapses originating from the

injected site. For example, injections into the eye will heavily label the terminal fields of retinal ganglion cells in the lateral geniculate nucleus.

Transneuronal transport takes this process a step further. After tagged proteins reach the axon terminals, a fraction is actually released into the extracellular space, where the proteins are degraded into amino acids or small peptides that retain their radioactivity. An even smaller fraction of this pool of labeled amino acids is taken up by the postsynaptic neurons, incorporated again into proteins, and transported to synaptic terminals of the second set of neurons. Because the label passes from the presynaptic terminals of one set of cells to the postsynaptic target cells, the process is called transneuronal transport. By such transneuronal labeling, the chain of connections originating from a particular structure can be visualized. In the case of the visual system, proline injections into one eye label appropriate layers of the lateral geniculate nucleus (as well as

other retinal ganglion cell targets such as the superior colliculus), and subsequently the terminals in the visual cortex of the geniculate neurons receiving inputs from that eye. Thus, when sections of the visual cortex are viewed with dark-field illumination to make the silver grains glow a brilliant white against the unlabeled background, ocular dominance columns in layer IV are easily seen (see Figure 24.3).

References

COWAN, W. M., D. I. GOTTLIEB, A. HENDRICKSON, J. L. PRICE AND T. A. WOOLSEY (1972) The autoradiographic demonstration of axonal connections in the central nervous system. Brain Res. 37: 21–51

GRAFSTEIN, B. (1971) Transneuronal transfer of radioactivity in the central nervous system. Science 172: 177–179.

GRAFSTEIN, B. (1975) Principles of anterograde axonal transport in relation to studies of neuronal connectivity. In *The Use of Axonal Transport for Studies in Neuronal Connectivity*, W. M. Cowan and M. Cuénod (eds.). Amsterdam: Elsevier, pp. 47–68.

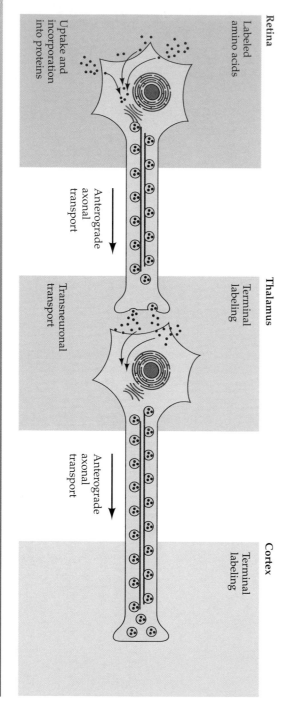

Retina

Labeled amino acids

Uptake and incorporation into proteins

Anterograde axonal transport

Transneuronal transport

Thalamus

Terminal labeling

Anterograde axonal transport

Cortex

Terminal labeling

Transneuronal transport. A neuron in the retina is shown taking up a radioactive amino acid, incorporating it into proteins, and moving the proteins down the axons and across the extracellular space between neurons. This process is repeated in the thalamus, and eventually label accumulates in the thalamocortical terminals in layer IV of the primary visual cortex.

between the time a kitten's eyes open (about a week after birth) and a year of age, visual experience determines how the visual cortex is wired with respect to eye dominance. In fact, further experiments showed that eye closure is effective only if the deprivation occurs during the first 3 months of life. In keeping with the ethological observations described earlier in the chapter, Hubel and Weisel called this period of susceptibility to visual deprivation the critical period for the development of ocular dominance. During the height of the critical period (about 4 weeks of age in the cat), as little as 3 to 4 days of eye closure profoundly alters the ocular dominance profile of the striate cortex (Figure 24.5). Similar experiments in the monkey have shown that the same phenomenon occurs in primates, although the critical period is longer (up to about 6 months of age).

A key advance arising from Hubel and Weisel's work was to show that visual deprivation causes changes in cortical connectivity (Figure 24.6). The implications of altered circuitry was amply confirmed by complementary anatomical studies. In monkeys, a central aspect of circuitry—the alternating stripelike patterns of geniculocortical axons representing the two eyes that form ocular dominance columns—is already present at birth (Figure 24.6A). As in the case of language development, in which infants exhibit early preferences for speech sounds, the visual cortex is not a blank slate on which the effects of experience are later inscribed. Nevertheless, animals deprived of vision in one eye from birth develop abnormal patterns of ocular dominance stripes in the visual cortex (Figure 24.6B). The open-eye stripes are substantially wider than normal, whereas the stripes representing the deprived eye are correspondingly diminished. The absence of cortical neurons that respond to the deprived eye in electrophysiological studies is not simply a result of the relatively inactive inputs withering away. If this were the case, one would expect to see areas of layer IV devoid of any thalamic innervation. Instead, inputs from the active (open) eye take over some of the territory that formerly belonged to the inactive

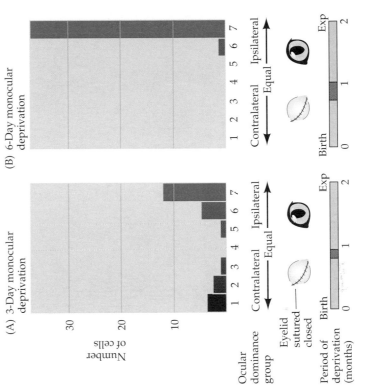

(A) 3-Day monocular deprivation

(B) 6-Day monocular deprivation

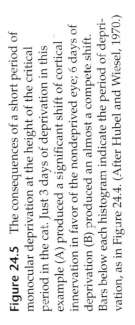

Figure 24.5 The consequences of a short period of monocular deprivation at the height of the critical period in the cat. Just 3 days of deprivation in this example (A) produced a significant shift of cortical innervation in favor of the nondeprived eye; 6 days of deprivation (B) produced an almost a compete shift. Bars below each histogram indicate the period of deprivation, as in Figure 24.4. (After Hubel and Weisel, 1970.)

(closed) eye. Hubel and Wiesel interpreted these results as demonstrating a competitive interaction between the two eyes during the critical period (see Chapter 23). At birth, the cortical representation of both eyes starts out equal, and in a normal animal, this balance is retained if both eyes experience roughly comparable levels of visual stimulation. When, however, an imbalance in visual experience is induced by monocular deprivation, the active eye gains a competitive advantage and replaces many of the synaptic inputs from the closed eye, such that few if any neurons can be driven by the deprived eye (see Figure 24.4B).

The idea that a competitive imbalance underlies the altered distribution of inputs after deprivation has been confirmed by closing *both* eyes shortly after birth, thereby equally depriving all visual cortical neurons of normal experience. The arrangement of ocular dominance recorded some months later is, by either electrophysiological or anatomical criteria, much more normal than if just one eye is closed. Although several peculiarities in the response properties of cortical cells are apparent, roughly normal proportions of neurons representing the two eyes are present. Because there is no imbalance in the visual activity of the two eyes (both sets of related cortical inputs being deprived), both eyes retain their territory in the cortex. If disuse atrophy of the closed-eye inputs were the main effect of deprivation, then binocular deprivation would cause the visual cortex to be largely unresponsive.

Experiments using techniques that label individual axons from the lateral geniculate nucleus terminating in layer IV have shown in greater detail what happens to the arborizations of individual neurons after visual deprivation (Figure 24.7). As noted, monocular deprivation causes a loss of cortical territory related to the deprived eye, with a concomitant expansion of the open eye's territory. At the level of single axons, these changes are reflected in an increased extent and complexity of the arborizations related to the open eye, and a decrease in the size and complexity of the arborizations related to the deprived eye. Individual neuronal arborizations can be substantially altered after as little as one week of deprivation, and perhaps even less. This latter finding highlights the ability of developing thalamic and cortical neurons to rapidly remodel their connections—actually making and breaking synapses—in response to environmental changes.

(A)

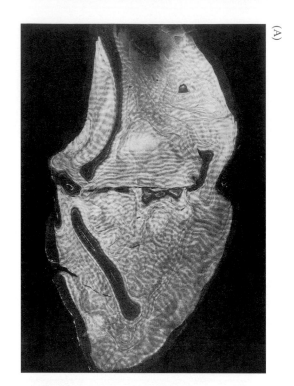

(B)

Figure 24.6 Effect of monocular deprivation on ocular dominance columns in the macaque monkey. (A) In normal monkeys, ocular dominance columns seen as alternating stripes of roughly equal width are already present at birth. (B) The picture is quite different after monocular deprivation. This dark-field autoradiograph shows a reconstruction of several sections through layer IV of the primary visual cortex of a monkey whose right eye was sutured shut from 2 weeks of age to 18 months, when the animal was sacrificed. Two weeks before death, the normal (left) eye was injected with radiolabeled amino acids (see Box C). The columns related to the nondeprived eye (white stripes) are much wider than normal, whereas as those related to the deprived eye are shrunken. (A from Horton and Hocking, 1999; B from Hubel et al., 1977.)

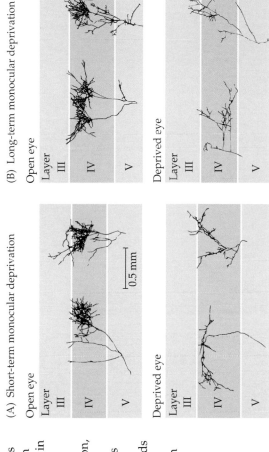

Figure 24.7 Terminal arborizations of lateral geniculate nucleus axons in the visual cortex can change rapidly in response to mcnocular deprivation during the critical period. (A) After only a week of monocular deprivation, axons from the deprived eye have greatly reduced numbers of branches compared with those from the open eye. (B) Deprivation for longer periods does not result in appreciably larger changes. Numbers on the left of each figure indicate cortical layers. (After Antonini and Stryker, 1993.)

Critical Periods, Cortical Plasticity, and Amblyopia in Humans

These developmental phenomena in the visual system of experimental animals accord with clinical problems in children who have experienced similar deprivation. The loss of acuity, diminished stereopsis, and problems with fusion that arise from early deficiencies of visual experience is called **amblyopia** (from the Greek meaning "dim sight").

In humans, amblyopia is most often the result of **strabismus**—a misalignment of the two eyes due to improper control by the eye muscles and referred to colloquially as "lazy eye." Depending on the muscles affected, the misalignment can produce a convergent strabismus called esotropia ("cross-eyed") or divergent strabismus called exotropia ("wall-eyed"). These alignment errors are surprisingly common, affecting about 5% of children. Since such misalignment produces double vision, the response of the visual system in some of these individuals is to suppress input from one eye by mechanisms that are not understood. Functionally, however, the suppressed eye eventually has very low acuity and may render the child effectively blind in that eye.

Another common cause of visual deprivation in humans is cataracts. Cataracts, which can be caused by several congenital conditions, render the lens opaque. Diseases such as onchocerciasis or "river blindness" (a parasitic infection caused by the nematode *Onchocerca volvulus*) and trachoma (caused by *Chlamydia trachomatis*, a small, bacteria-like organism) affect millions of people in undeveloped tropical regions, often inducing corneal opacity in one or both eyes. A cataract in one eye is functionally equivalent to monocular deprivation in experimental animals; left untreated in children, this defect also results in an irreversible effect on the visual acuity of the deprived eye. If the corneal opacity is removed before about 4 months of age, however, the consequences of monocular deprivation are largely avoided (although other binocular abilities such as stereopsis may be compromised). As expected from Hubel and Weisel's work, bilateral cataracts, which are similar to binocular deprivation in experimental animals, produce less dramatic deficits even if treatment is delayed. Apparently, unequal competition during the critical period for normal vision (e.g., that caused by monocular deprivation) is more deleterious than the overall visual deprivation that occurs with binocular deprivation.

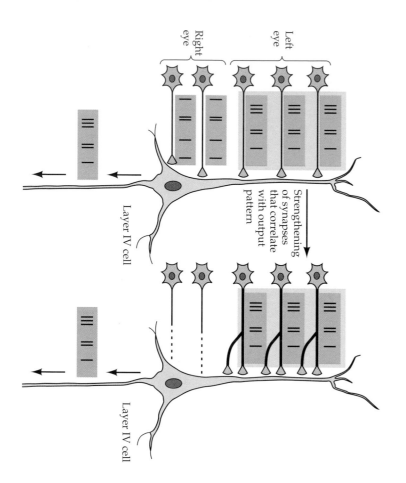

In keeping with the findings in experimental animals, the visual abilities of individuals monocularly deprived of vision as adults (by cataracts, for example) are unaffected, even after decades, when vision is restored. Nor is there any evidence of anatomical change in this circumstance. For instance, a patient whose eye was surgically removed in adulthood showed normal ocular dominance columns when his brain was examined postmortem many years later (see Figure 12.10). In short, the visual system of humans exhibits much the same critical period for visual cortical development and behavior as experimental animals.

Mechanisms by which Neuronal Activity Affects the Development of Neural Circuits

How are differences in patterns of neural activity translated into changes in neural circuitry? In 1949, the psychologist D. O. Hebb hypothesized that coordinated activity of a presynaptic terminal and a postsynaptic neuron strengthens the synaptic connections between them. Hebb's postulate, as it has come to be known, was originally formulated to explain the cellular basis of learning and memory (see Chapter 25), but has been widely applied to situations that involve long-term modifications in synaptic strength, including those that occur during cortical development. In this context, Hebb's postulate implies that synaptic terminals strengthened by correlated activity will be retained or sprout new branches, whereas those that are persistently weakened by uncorrelated activity will eventually lose their hold on the postsynaptic cell (Figure 24.8; see also Chapter 23). In the visual system, the action potentials of the thalamocortical inputs related to one eye are presumably better correlated with each other than with the activity related to the other eye—at least in layer IV. If sets of correlated inputs tend to dominate the activity of groups of locally connected postsynaptic cells, this relationship would exclude uncorrelated inputs. Thus, patches of cortex occu-

Figure 24.8   Representation of Hebb's postulate as it might operate during development of the visual system. The cell represents a postsynaptic neuron in layer IV of the primary visual cortex. Early in development, inputs from the two eyes converge on single postsynaptic cells. The two sets of presynaptic inputs, however, have different patterns of electrical activity (represented by the short vertical bars). In the example here, the three left eye inputs are better able to fire the postsynaptic cell; as a result, their activity is highly correlated with the postsynaptic cell's activity. According to Hebb's postulate, these synapses are therefore strengthened. The inputs from the right eye carry a different pattern of activity that is less well correlated with the majority of the activity elicited in the postsynaptic cell. These synapses gradually weaken and are eventually eliminated (right-hand side of figure), while the correlated inputs form additional synapses.

pied exclusively by inputs representing one eye or the other could arise. In this scenario, ocular dominance column rearrangements in layer IV are generated by cooperation between inputs carrying *similar* patterns of activity, and competition between inputs carrying *dissimilar* patterns.

Monocular deprivation, which dramatically changes ocular dominance columns, clearly alters both the levels and patterns of neural activity between the two eyes. To test the role of correlated activity in driving the competitive postnatal rearrangement of cortical connections, it is necessary to create a situation in which activity levels in each eye remain the same but the correlations between the two eyes are altered. This circumstance can be created in experimental animals by cutting one of the extraocular muscles in one eye. As already mentioned, this condition, in which the two eyes can no longer be aligned, is called strabismus. The major consequence of strabismus is that corresponding points on the two retinas are no longer stimulated by objects in the same location in visual space at the same time. As a result, differences in the visually evoked patterns of activity between the two eyes are far greater than normal. Unlike monocular deprivation, however, the overall amount of activity in each eye remains roughly the same; only the correlation of activity arising from corresponding retinal points is changed.

The effects of bilateral strabismus provide an illustration of the basic accuracy of Hebb's postulate. The anatomical pattern of ocular dominance columns in layer IV of cats is sharper than normal, implying that the asynchronous patterns of activity have accentuated the normal separation of inputs from the two eyes. In addition, the ocular asynchrony prevents the binocular convergence that normally occurs in cells above and below layer IV: Ocular dominance histograms from strabismic animals show that most cells in *all* layers are driven exclusively by one eye *or* the other (Figure 24.9). Evidently, strabismus not only accentuates the competition between the two sets of thalamic inputs in layer IV, but also prevents binocular interactions in the other layers, which are mediated by local connections originating from cells in layer IV. These observations in experimental animals have important implications for children with strabismus. Unless the ocular deviation is corrected during the critical period (by patching the good eye or surgically altering the mechanics of the extraocular muscles), a strabismic child may ultimately have poor binocular fusion, diminished depth perception, and degraded acuity; in other words, he or she will become amblyopic (see previous section).

Even before visual experience exerts these effects, innate mechanisms have ensured that the basic outlines of a functional system are present. These intrinsic mechanisms establish the general circuitry required for vision, but allow modifications to accommodate the individual requirements that occur with changes in head size or eye alignment. Normal visual experience evidently validates the initial wiring, preserving, augmenting, or adjusting the normal arrangement. In the case of abnormal experience, such as monocular deprivation, the mechanisms that allow these adjustments result in more dramatic anatomical (and ultimately behavioral) changes, such as those that occur in amblyopia. The eventual decline of this capacity to remodel cortical (and subcortical) connections is presumably the cellular basis of critical periods in a variety of neural systems, including the development of language. By the same token, these differences in plasticity as a function of age presumably provide a neurobiological basis for the general observation that human behavior is much more susceptible to normal or pathological modification early in development than later on. Although many cellular and molecular mechanisms have been proposed to explain these effects (see Chapter

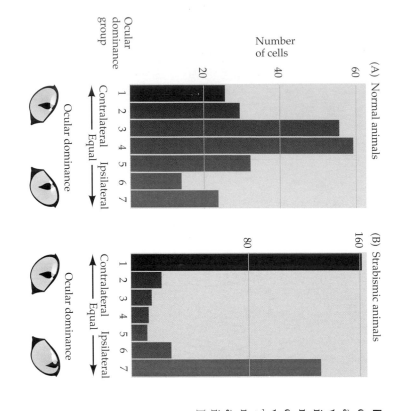

Figure 24.9 Ocular dominance histograms obtained by electrophysiological recordings in normal adult cats (A) and adults cats in which strabismus was induced during the critical period (B). The data in (A) is the same as that shown in Figure 24.3A. The number of binocular cells is sharply decreased as a consequence of strabismus; most of the cells are driven exclusively by stimulation of one eye or the other. This enhanced segregation of the inputs presumably results from the greater discrepancy in the patterns of activity between the two eyes as a result of surgically interfering with normal conjugate vision. (After Hubel and Wiesel, 1965.)

25), the mechanisms responsible for creating and eventually terminating critical periods remain largely unknown.

Evidence for Critical Periods in Other Sensory Systems

Although the neural basis of critical periods has been most thoroughly studied in the mammalian visual system, similar phenomena exist in a number of sensory systems, including the auditory, somatic sensory and olfactory systems. In the auditory system, experiments on the role of auditory experience and neural activity in owls (who use auditory information to localize prey) indicate that neural circuits for auditory localization are similarly shaped by experience. Thus, deafening an owl or altering neural activity during early postnatal development compromises the bird's ability to localize sounds and can alter the neural circuits that mediate this capacity. In the somatic sensory system as well, somatic sensory maps can be changed by experience during a critical period of postnatal development. In mice or rats, for instance, the anatomical patterns of "whisker barrels" in the somatic sensory cortex (see Chapter 9) can be altered by abnormal sensory experience during a narrow window in early postnatal life. And, as already mentioned, behavioral studies in the olfactory system indicate that exposure to maternal odors for a limited period can change the ability to respond to such odorants, a change that can persist throughout life. Clearly, the phenomenon of critical periods is general.

Summary

An individual animal's history of interaction with its environment—its "experience"—helps to shape neural circuits and thus determines subsequent behavior. In some cases, experience functions primarily as a switch to activate innate behaviors. More often, however, experience during a specific time in early life (referred to as a "critical period") determines the adult

behavioral repertoire. Critical periods influence behaviors as diverse as maternal bonding and the acquisition of language. Although it is possible to define the behavioral consequences of critical periods for these complex functions, their biological basis has been more difficult to understand. The most accessible and thoroughly studied example of a critical period is the one pertinent to the establishment of normal vision. These studies show that experience is translated into distinct patterns of neuronal activity that influence the function and connectivity of the relevant neurons. In the visual system, and other systems as well, competition between inputs with different patterns of activity is an important determinant of adult connectivity. Correlated patterns of activity in afferent axons tend to stabilize connections (and conversely). When normal patterns of activity are disturbed (experimentally in animals or by pathology in humans) during a critical period in early life, the connectivity in the visual cortex is altered, as is visual function. If not reversed before the end of the critical period, these structural alterations of brain circuitry are difficult or impossible to change. In normal development, the influence of activity on neural connectivity presumably enables the maturing brain to store the vast amounts of information that reflect the specific experience of the individual.

ADDITIONAL READING

Reviews

KATZ, L. C. AND C. J. SHATZ (1996) Synaptic activity and the construction of cortical circuits. Science 274: 1133–1138.

KNUDSEN, E. I. (1995) Mechanisms of experience-dependent plasticity in the auditory localization pathway of the barn owl. J. Comp. Physiol. 184(A): 305–321.

SHERMAN, S. M. AND P. D. SPEAR (1982) Organization of visual pathways in normal and visually deprived cats. Physiol. Rev. 62: 738–855.

WIESEL, T. N. (1982) Postnatal development of the visual cortex and the influence of environment. Nature 299: 583–591.

Important Original Papers

ANTONINI, A. AND M. P. STRYKER (1993) Rapid remodeling of axonal arbors in the visual cortex. Science 260: 1819–1821.

CABELLI, R. J., A. HOHN AND C. J. SHATZ (1995) Inhibition of ocular dominance column formation by infusion of NT-4/5 or BDNF. Science 267: 1662–1666.

HORTON, J. C. AND D. R. HOCKING (1999) An adult-like pattern of ocular dominance columns in striate cortex of newborn monkeys prior to visual experience. J. Neurosci. 16: 1791–1807.

HUBEL, D. H. AND T. N. WIESEL (1965) Binocular interaction in striate cortex of kittens reared with artificial squint. J. Neurophysiol. 28: 1041–1059.

HUBEL, D. H. AND T. N. WIESEL (1970) The period of susceptibility to the physiological effects of unilateral eye closure in kittens. J. Physiol. 206: 419–436.

HUBEL, D. H., T. N. WIESEL AND S. LEVAY (1977) Plasticity of ocular dominance columns in monkey striate cortex. Phil. Trans. R. Soc. Lond. B. 278: 377–409.

KUHL, P. K., K. A. WILLIAMS, F. LACERDA, K. N. STEVENS AND B. LINDBLOM (1992) Linguistic experience alters phonetic perception in infants by 6 months of age. Science 255: 606–608.

LEVAY, S., T. N. WIESEL AND D. H. HUBEL (1980) The development of ocular dominance columns in normal and visually deprived monkeys. J. Comp. Neurol. 191: 1–51.

RAKIC, P. (1977) Prenatal development of the visual system in the rhesus monkey. Phil. Trans. R. Soc. Lond. B. 278: 245–260.

STRYKER, M. P. AND W. HARRIS (1986) Binocular impulse blockade prevents the formation of ocular dominance columns in cat visual cortex. J. Neurosci. 6: 2117–2133.

WIESEL, T. N. AND D. H. HUBEL (1965) Comparison of the effects of unilateral and bilateral eye closure on cortical unit responses in kittens. J. Neurophysiol. 28: 1029–1040.

Books

CURTISS, S. (1977) Genie: A Psycholinguistic Study of a Modern-Day "Wild Child." New York: Academic Press.

HUBEL, D. H. (1988) Eye, Brain, and Vision. Scientific American Library Series. New York: W. H. Freeman.

PURVES, D. (1994) Neural Activity and the Growth of the Brain. Cambridge, UK: Cambridge University Press.

Plasticity of Mature Synapses and Circuits

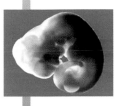

Overview

The capacity of the nervous system to change—generally referred to as neural plasticity—is obvious during the development of neural circuits. However, the adult brain must also possess at least some plasticity to learn new skills, establish new memories, and respond to injury throughout life. Although the mechanisms responsible for ongoing changes in the adult brain are not completely understood, altered neural connectivity in maturity appears to rely primarily on carefully regulated changes in the strength of extant synapses. Experiments carried out in a variety of animals, ranging from sea slugs to primates, have shown that synaptic strength can be altered over periods that range from milliseconds to months. The molecular mechanisms underlying these changes are post-translational modifications of existing proteins and, in the case of longer-lasting effects, changes in gene expression. To some extent, changes in synaptic circuitry can also occur by localized formation of new axon terminals and dendritic processes. More extensive changes occur when the adult nervous system is damaged by trauma or disease, although regeneration of connections in the brain and spinal cord is sharply limited. Modest optimism regarding this unfortunate clinical situation is warranted by the recent finding that new neurons can be generated throughout life in several brain regions, providing the raw material for entirely novel circuitry.

Mechanisms of Synaptic Plasticity in Relatively Simple Invertebrates

An obvious obstacle to exploring change in the brains of humans and other mammals is the enormous numbers of neurons and the complexity of synaptic connections in the mammalian central nervous system. As a consequence, it is difficult to unambiguously attribute a behavioral modification to changes in the properties of specific neurons or synapses. One way to circumvent this dilemma is to examine plasticity in far simpler nervous systems. The assumption in this strategy is that plasticity is so fundamental that its essential cellular and molecular underpinnings are likely to be conserved in the nervous systems of very different organisms.

One of the most successful examples of this approach has been that of Eric Kandel and colleagues at Columbia University using the marine mollusk *Aplysia californica* (Figure 25.1A). This sea slug has only a few tens of thousands of neurons, many of which are quite large (up to 1 mm in diameter) and in stereotyped locations within the ganglia that make up the animal's nervous system (Figure 25.1B). These attributes make it practical to monitor the electrical and chemical signaling of specific, identifiable nerve

Figure 25.1 Short-term sensitization of the *Aplysia* gill withdrawal reflex. (A) Diagram of the animal. (B) The abdominal ganglion of *Aplysia*. The cell bodies of many of the neurons involved in gill withdrawal can be recognized by their size, shape, and position within this ganglion. (C) Changes in the gill withdrawal behavior due to habituation and sensitization. The first time that the siphon is touched, the gill contracts vigorously. Repeated touches elicit smaller gill contractions due to habituation. Subsequently pairing a siphon touch with an electrical shock to the tail restores a large and rapid gill contraction, due to short-term sensitization. (After Kandel and Schwartz, 1982.)

(A)

(B)

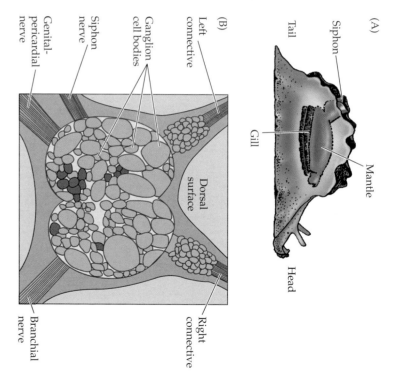

(C)

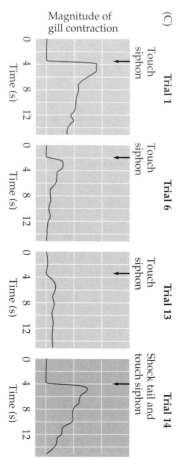

Magnitude of gill contraction

Trial 1 — Touch siphon

Trial 6 — Touch siphon

Trial 13 — Touch siphon

Trial 14 — Shock tail and touch siphon

cells, and to define the synaptic circuits involved in mediating the limited behavioral repertoire of *Aplysia*.

An elementary form of behavioral plasticity in *Aplysia* (and many other species, including humans) is **sensitization**. Sensitization is a process that allows an animal to generalize an aversive response elicited by a noxious stimulus to a variety of other, non-noxious stimuli. In *Aplysia*, a light touch to the animal's siphon results in gill withdrawal, a response that gradually habituates (becomes less strong) with repeated stimulation. After several repetitions, the animal no longer bothers to withdraw the gill after being touched. However, if touching the siphon is paired with a strong electrical stimulus to the animal's tail, then the siphon stimulus again elicits a strong withdrawal of the gill (Figure 25.1C). Thus, the noxious stimulus to the tail sensitizes the gill withdrawal reflex to light touch. Even after a single stimulus to the tail, the gill withdrawal reflex remains enhanced for at least an hour (called short-term sensitization). With repeated pairing of tail and siphon stimuli, this behavior can be altered for days or weeks. Such a long-lasting change in the gill withdrawal reflex is an example of long-term sensitization.

Although hundreds of neurons are ultimately involved in producing this simple behavior, the activities of only a few different types of neurons can account for gill withdrawal and its plasticity during sensitization. These critical neurons include mechanosensory neurons that innervate the siphon, motor neurons that innervate muscles in the gill, and interneurons that receive inputs from a variety of sensory neurons (Figure 25.2A). Touching

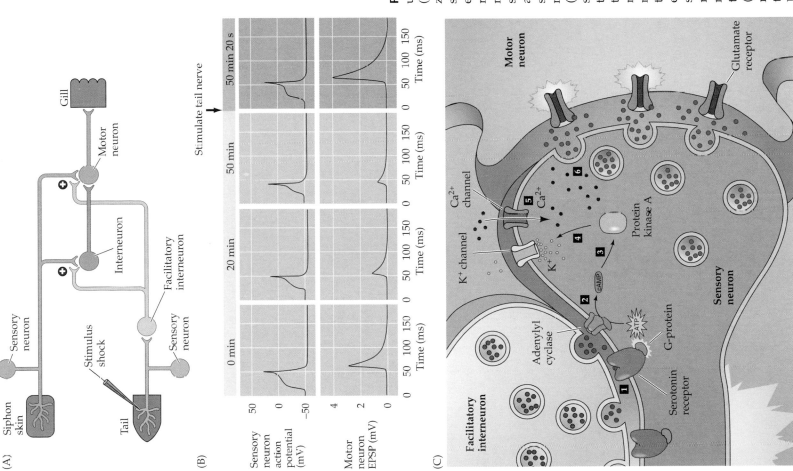

Figure 25.2 Synaptic mechanisms underlying short-term sensitization. (A) Neural circuitry involved in sensitization. Normally, touching the siphon skin activates sensory neurons that excite interneurons and gill motor neurons, yielding a contraction of the gill muscle. A shock to the animal's tail stimulates facilitatory interneurons that alter synaptic transmission between the siphon sensory neurons and gill motor neurons, resulting in sensitization. (B) Changes in synaptic efficacy at the sensory motor synapse during short-term sensitization. Prior to sensitization, activating the siphon sensory neurons causes an EPSP to occur in the gill motor neurons. Activation of the serotonergic facilitatory interneurons enhances release of transmitter from the sensory neurons onto the motor neurons, increasing the EPSP in the motor neurons and causing the motor neurons to more strongly excite the gill muscle. (C) Mechanism of presynaptic enhancement underlying short-term sensitization. See text for explanation. (After Kandel and Schwartz, 1982.)

the siphon activates mechanosensory neurons that innervate the siphon. These neurons in turn form excitatory synapses on both the interneurons and the motor neurons; thus, touching the siphon increases the probability that both these postsynaptic targets will produce action potentials. The interneurons form excitatory synapses on motor neurons, further increasing the likelihood of the motor neurons firing action potentials. When the motor neurons are activated by the summed synaptic excitation of the sensory neurons and interneurons, they excite the muscle cells of the gill, producing gill withdrawal.

Synaptic activity in this circuit is modified during sensitization. The tail shock that evokes sensitization activates additional sensory neurons that innervate the tail. These sensory neurons in turn excite interneurons that release the neurotransmitter serotonin onto the presynaptic terminals of the sensory neurons of the siphon (see Figure 25.2A). Serotonin produces a prolonged enhancement of transmitter release from the siphon sensory neuron terminals, leading to increased synaptic excitation of the motor neurons (Figure 25.2B). This simple form of behavioral plasticity is therefore due to recruitment of additional synaptic elements that change synaptic transmission in the gill withdrawal circuit.

The probable mechanism of enhanced transmission during short-term sensitization is shown in Figure 25.2C. Serotonin released by the facilitatory interneurons binds to G-protein-linked receptors on the presynaptic terminals of the siphon sensory neurons (step 1), which stimulates production of the second messenger, cAMP (step 2). cAMP activates protein kinase A (PKA; step 3), which then phosphorylates several proteins, probably including K$^+$ channels (step 4). The net effect of the action of PKA is to reduce the probability that the K$^+$ channels open during a presynaptic action potential. This effect prolongs the presynaptic action potential, thereby opening more presynaptic Ca^{2+} channels (step 5). Finally, the enhanced influx of Ca^{2+} into the presynaptic terminals increases the amount of transmitter released onto motor neurons during a sensory neuron action potential (step 6). In summary, short-term sensitization of gill withdrawal is mediated by a signal transduction cascade that involves neurotransmitters, second messengers, and ion channels. This cascade ultimately enhances synaptic transmission between the sensory and motor neurons within the gill withdrawal circuit.

The same mechanisms that mediate short-term sensitization underlie long-term sensitization. During long-term sensitization, however, this circuitry is affected for up to several weeks. The duration of this form of plasticity is evidently due to changes in gene expression and thus protein synthesis. With repeated training (that is, additional tail shocks), the serotonin-activated, cAMP-dependent protein kinase involved in short-term sensitization now phosphorylates—and thereby activates—the transcriptional activator CREB (see Chapter 8). CREB binding to the cAMP responsive elements (CREs) increases the rate of transcription of downstream genes. Although the changes in genes and gene products that follow CRE activation have been difficult to sort out, two consequences of gene activation have been identified. Some cAMP-dependent protein kinases no longer require serotonin to be activated but remain persistently active. Gene activation also increases the number of synapses between the sensory and the motor neurons. Such structural increases are not seen following short-term sensitization and may represent an anatomical basis for the long-lasting change in overall strength of the relevant connections.

Work on *Aplysia* and other invertebrates such as the fruit fly (Box A) has led to several generalizations about the neural mechanisms underlying plasticity in the adult nervous system that presumably extend to mammals and other vertebrates. First, behavioral plasticity can clearly arise from changes in the efficacy of synaptic transmission. Second, these changes in synaptic function can be either short-term effects that rely on post-translational modification of existing synapses and synaptic proteins, or long-term changes that require changes in gene expression, new protein synthesis, and perhaps even growth of new synapses (or the elimination of existing ones). The following sections explore the evidence for these generalizations in neuronal circuits and synapses of the mammalian nervous system.

Mechanisms of Short-Term Synaptic Plasticity in the Mammalian Nervous System

Evidence for synaptic plasticity in the mammalian nervous system is widespread and also occurs on timescales ranging from milliseconds to days, weeks, or longer. Although these changes occur throughout the brain, short-term forms of plasticity that last for minutes or less have been studied in greatest detail at peripheral neuromuscular synapses.

Repeated activation of the neuromuscular junction triggers several sorts of change that vary in both time course and direction (Figure 25.3). **Synaptic facilitation**, which is a transient increase in synaptic strength, occurs when two or more action potentials invade the presynaptic terminal in close succession. Facilitation results in more neurotransmitter being released by each succeeding action potential, causing the postsynaptic end plate potential

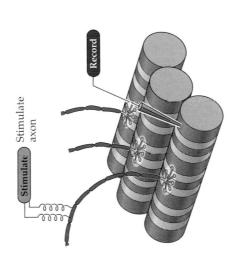

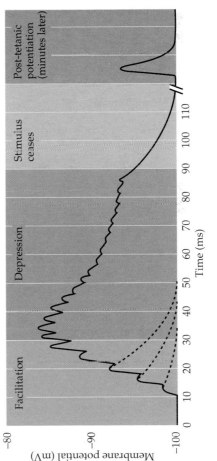

Figure 25.3 Short-term plasticity at the neuromuscular synapse. Electrical recording of EPPs elicited in a muscle fiber by a train of electrical stimuli applied to the presynaptic motor nerve. Facilitation of the EPP occurs at the beginning of the stimulus train and is followed by depression of the EPP. After the train of stimuli ends, EPPs are larger than before the train. This phenomenon is called post-tetanic potentiation. (After Katz, 1966.)

BOX A
Genetics of Learning and Memory in the Fruit Fly

As part of a renaissance in the genetic analysis of simple organisms in the mid-1970s, several investigators recognized that the genetic basis of learning and memory might be effectively studied in the fruit fly, *Drosophila melanogaster*. In the intervening quarter-century, this approach has yielded some fundamental insights. Although learning and memory has certainly been one of the more difficult problems tackled by *Drosophila* geneticists, their efforts have been surprisingly successful. A number of genetic mutations have been discovered that alter learning and memory, and the identification of these genes has provided a valuable framework for studying the cellular mechanisms of these processes.

The initial problem in this work was to develop behavioral tests that could identify abnormal learning and/or memory defects in large populations of flies. This challenge was met by Seymour Benzer and his colleagues Chip Quinn and Bill Harris at California Institute of Technology, who developed the olfactory and visual learning tests that have become

the basis for most subsequent analyses of learning and memory in the fruit fly (see figure). Behavioral paradigms pairing odors or light with an aversive stimulus allowed Benzer and colleagues to assess associative learning in flies. The design of ingenious testing apparatus controlled for non-learning-related sensory cues that had previously complicated such behavioral testing. Moreover, the apparatus allowed large numbers of flies to be screened relatively easily, expediting the analysis of mutagenized populations.

These studies led to the identification of an ever-increasing number of single gene mutations that disrupt learning and/or memory in flies. The behavioral and molecular studies of the mutants (given whimsical but descriptive names like *dunce*, *rutabaga*, and *amnesiac*) suggested that a central pathway for learning and memory in the fly is signal transduction mediated by the cyclic nucleotide cAMP. Thus, the gene products of the *dunce*, *rutabaga*, and *amnesiac* loci are, respectively, a phosphodiesterase (which degrades cAMP), an adenylyl cyclase

(which converts ATP to cAMP), and a peptide transmitter that stimulates adenylyl cyclase. This conclusion about the importance of cAMP has been confirmed by the finding that genetic manipulation of the CREB transcription factor also interferes with learning and memory in normal flies.

These observations in *Drosophila* accord with conclusions reached in studies of *Aplysia* (see text), and have emphasized the importance of cAMP-mediated mechanisms of learning and memory in a wide range of additional species.

References

QUINN, W. G., W. A. HARRIS AND S. BENZER (1974) Conditioned behavior in *Drosophila melanogaster*. Proc. Natl. Acad. Sci. USA 71: 708–712.

TULLY, T. (1996) Discovery of genes involved with learning and memory: An experimental synthesis of Hirshian and Benzerian perspectives. Proc. Natl. Acad. Sci. USA 93: 13460–13467.

WEINER, J. (1999) *Time, Love, Memory: A Great Biologist and His Quest for the Origins of Behavior*. New York: Knopf.

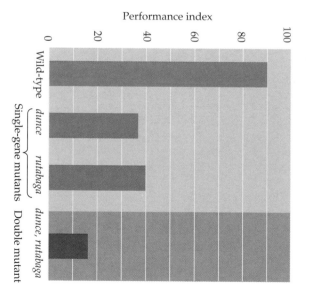
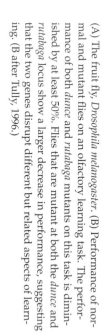

Performance index

(A) The fruit fly, *Drosophila melanogaster*. (B) Performance of normal and mutant flies on an olfactory learning task. The performance of both *dunce* and *rutabaga* mutants on this task is diminished by at least 50%. Flies that are mutant at both the *dunce* and *rutabaga* locus show a larger decrease in performance, suggesting that the two genes disrupt different but related aspects of learning. (B after Tully, 1996.)

(EPP) to increase progressively. Much evidence suggests that synaptic facilitation is the result of prolonged elevation of presynaptic calcium levels following synaptic activity. Although the entry of Ca²⁺ into the presynaptic terminal occurs within a millisecond or two after an action potential invades (see Chapter 5), the mechanisms that return calcium to resting levels are much slower. Thus, when action potentials arrive close together in time, calcium builds up within the terminal and allows more neurotransmitter to be released by a subsequent presynaptic action potential. A high-frequency burst of presynaptic action potentials (colloquially referred to as a "tetanus") can yield even more prolonged elevation of presynaptic calcium levels, causing another form of synaptic enhancement called **post-tetanic potentiation (PTP)**. PTP is delayed in its onset and typically persists for some minutes after the train of stimuli ends. The difference in duration distinguishes PTP from synaptic facilitation. PTP is thought to arise from calcium-dependent processes that make more synaptic vesicles available for transmitter release.

Synaptic transmission also can be diminished following repeated synaptic activity. Such **synaptic depression** occurs when many presynaptic action potentials occur in rapid succession and depends on the amount of neurotransmitter that has been released (see Figure 25.3). Depression arises because of the progressive depletion of the pool of synaptic vesicles available for fusion in this circumstance. During synaptic depression, the strength of the synapse declines until this pool can be replenished via the mechanisms involved in recycling of synaptic vesicles (see Chapter 5).

During repeated synaptic activity, these various forms of plasticity can interact in complex ways. For example, at the neuromuscular synapse, repeated activity first facilitates synaptic transmission; then depletion of synaptic vesicles allows depression to dominate and weaken the synapse (see Figure 25.3). After the stimulus train ends, the invasion of the terminal by another action potential causes enhanced transmitter release (i.e., pos-tetanic potentiation).

These forms of short-term plasticity are observed at virtually all chemical synapses and continually modify synaptic strength. Thus, the efficacy of chemical synaptic transmission changes dynamically as a consequence of the recent history of synaptic activity.

Mechanism of Long-Term Synaptic Plasticity in the Mammalian Nervous System

Facilitation, depression, and post-tetanic potentiation can briefly modify synaptic transmission. These mechanisms cannot, however, provide the basis for memories or other manifestations of behavioral plasticity that persist for weeks, months, or years. As might be expected, many synapses in the mammalian central nervous system exhibit long-lasting forms of synaptic plasticity that are more plausible substrates for enduring changes in brain function. Because of their duration, these forms of synaptic plasticity in mammals are widely believed to be cellular correlates of learning and memory. Thus, a great deal of effort has gone into understanding how they are generated.

Some patterns of synaptic activity in the CNS produce a long-lasting increase in synaptic strength known as **long-term potentiation (LTP)**, whereas other patterns of activity produce a long-lasting decrease in synaptic strength, known as **long-term depression (LTD)**. LTP and LTD are broad terms that describe only the direction of change in synaptic efficacy. In general, these different forms of synaptic plasticity are produced by different histories of activity, and are mediated by different complements of intracellular signal transduction pathways in the nerve cells involved.

Long-Term Synaptic Potentiation

LTP has been most thoroughly studied in the mammalian hippocampus, an area of the brain that is especially important in the formation and/or retrieval of some forms of memory (see Chapter 31). In humans, functional imaging shows that the human hippocampus is activated during certain kinds of memory tasks, and that damage to the hippocampus results in an inability to form certain types of new memories. In rodents, hippocampal neurons fire action potentials only when an animal is in certain locations. Such "place cells" appear to encode spatial memories, an interpretation supported by the fact that hippocampal damage prevents rats from developing proficiency in spatial learning tasks (Figure 25.4). Although many other brain areas are involved in the complex process of memory formation, storage, and retrieval, these observations have led many investigators to study this particular form of synaptic plasticity in the hippocampus.

Work on LTP began in the early 1970s, when Timothy Bliss and his colleagues at Mill Hill in England discovered that a few seconds of high-frequency electrical stimulation can enhance synaptic transmission in the rabbit hippocampus for days or even weeks. More recently, however, progress in understanding the mechanism of LTP has relied heavily on in vitro studies of slices of living hippocampus. The arrangement of neurons allows the hippocampus to be sectioned such that most of the relevant circuitry is left intact. In such preparations, the cell bodies of the pyramidal neurons lie in a single densely packed layer that is readily apparent (Figure 25.5). This layer is divided into several distinct regions, the major ones being CA1 and CA3. "CA" refers to Cornu Ammon, a Latin name for Ammon's horn—the ram's horn that resembles the shape of the hippocampus. The dendrites of pyramidal cells in the CA1 region form a thick band (the stratum radiatum), where they receive synapses from Schaffer collaterals, the axons of pyramidal cells

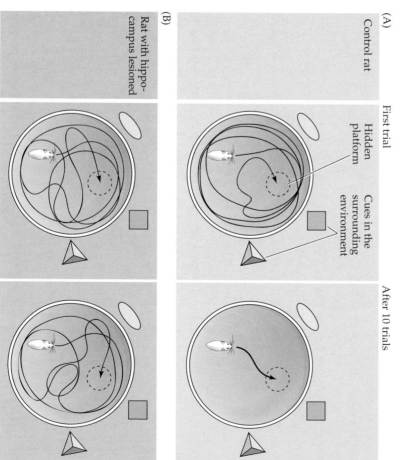

Figure 25.4 Spatial learning in rodents. (A) Rats are placed in a circular arena (about the size and shape of a child's wading pool) filled with cloudy water. The arena itself is featureless, but the surrounding environment contains the positional cues such as windows, doors, light fixtures, and so on. A small platform is located just below the surface. As rats search for this resting place, the pattern of their swimming (indicated by the traces in the figure) is monitored by a video camera. After a few trials, normal rats swim directly to the platform on each trial. (B) The swimming patterns of rats with impaired spatial memories—induced by hippocampal lesions—indicate a seeming inability to remember where the platform is located. (After Schenk and Morris, 1985.)

(A)   Control rat   First trial   After 10 trials

Hidden platform

Cues in the surrounding environment

(B)   Rat with hippocampus lesioned

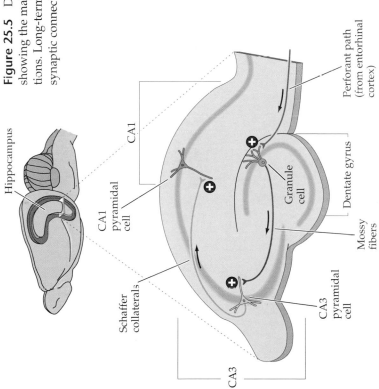

Figure 25.5 Diagram of a section through the rodent hippocampus showing the major regions, excitatory pathways, and synaptic connections. Long-term potentiation has been observed at each of the three synaptic connections shown here.

in the CA3 region. Much of the work on LTP has focused on the synaptic connections between the Schaffer collaterals and CA1 pyramidal cells. Electrical stimulation of Schaffer collaterals generates excitatory postsynaptic potentials (EPSPs) in the postsynaptic CA1 cells (Figure 25.6). If the Schaffer collaterals are stimulated only two or three times per minute, the size of the evoked EPSP in the CA1 neurons remains constant. However, a brief, high-frequency train of stimuli to the same axons causes LTP, which is evident as a long-lasting increase in EPSP amplitude. LTP also occurs at many other synapses, both within the hippocampus and in a variety of other brain regions, including the cortex, amygdala, and cerebellum.

LTP of the Schaffer collateral synapse exhibits several properties that make it an attractive neural mechanism for information storage. First, LTP is *state-dependent*: The degree of depolarization of the postsynaptic cell determines whether or not LTP occurs (Figure 25.7). If a single stimulus to the Schaffer collaterals—which would not normally elicit LTP—is paired with strong depolarization of the postsynaptic CA1 cell, the size of the EPSP is increased. The increase occurs only if the paired activities of the presynaptic and postsynaptic cells are tightly linked in time, such that the strong postsynaptic depolarization occurs within about 100 ms of presynaptic transmitter release. Recall that a requirement for coincident activation of presynaptic and postsynaptic elements is the hallmark of the Hebbian postulate (see Chapter 24), an early attempt to establish a theoretical framework of the synaptic changes underlying learning and memory.

LTP also exhibits the property of *input specificity*: When LTP is induced by the stimulation of one synapse, it does not occur in other, inactive synapses that contact the same neuron (see Figure 25.6). Thus, LTP is input-specific in the sense that it is restricted to activated synapses rather than to all of the synapses on a given cell (Figure 25.8A). This feature of LTP is consistent with its involvement in memory formation. If activation of one set of synapses led to all other synapses—even inactive ones—being potentiated, it

Figure 25.6 Long-term potentiation of Schaffer collateral-CA1 synapses. (A) Arrangement for recording synaptic transmission; two stimulating electrodes (1 and 2) each activate separate populations of Schaffer collaterals, thus providing test and control synaptic pathways. (B) Left: Synaptic responses recorded in a CA1 neuron in response to single stimuli of synaptic pathway 1, minutes before and one hour after a high-frequency train of stimuli. The high-frequency stimulus train increases the size of the EPSP evoked by a single stimulus. Right: Responses produced by stimulating synaptic pathway 2, which did not receive high-frequency stimulation, is unchanged. (C) The time course of changes in the amplitude of EPSPs evoked by stimulation of pathways 1 and 2. High-frequency stimulation of pathway 1 causes a prolonged enhancement of the EPSPs in this pathway 1 (purple). This potentiation of synaptic transmission in pathway 1 persists for several hours, while the amplitude of EPSPs produced by pathway 2 (orange) remains constant. (After Malinow et al., 1989.)

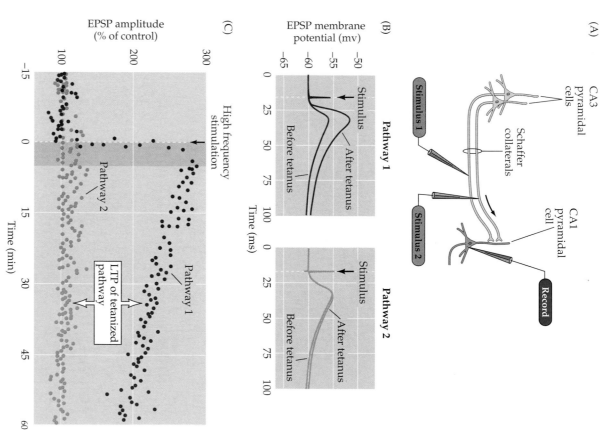

(A)

(B) EPSP membrane potential (mv)

Pathway 1   Pathway 2

Stimulus   Before tetanus   After tetanus

(C) EPSP amplitude (% of control)

High frequency stimulation

Pathway 2

Pathway 1

LTP of tetanized pathway

Time (min)

would be difficult to selectively enhance particular sets of inputs, as is presumably required for learning and memory (see also Box B).

Another important property of LTP is *associativity* (Figure 25.8B). As noted, weak stimulation of a pathway will not by itself trigger LTP. However, if one pathway is weakly activated at the same time that a neighboring pathway onto the same cell is strongly activated, both synaptic pathways undergo LTP. This selective enhancement of conjointly activated sets of synaptic inputs is often considered a cellular analog of associative or classical conditioning. More generally, associativity is expected in any network of neurons that links one set of information with another.

Although there is clearly a gap between understanding LTP of hippocampal synapses and understanding behavioral plasticity, this form of synaptic plasticity provides a plausible neural mechanism for long-lasting changes in a part of the brain that is known to be involved in the formation of certain kinds of memories.

Figure 25.7 Pairing presynaptic and postsynaptic activity causes LTP. Single stimuli applied to a Schaffer collateral synaptic input evokes EPSPs in the postsynaptic CA1 neuron. These stimuli alone do not elicit any change in synaptic strength. However, when the CA1 neuron's membrane potential is briefly depolarized (by applying current pulses through the recording electrode) in conjunction with the Schaffer collateral stimuli, there is a persistent increase in the EPSPs. (After Gustafsson et al., 1987.)

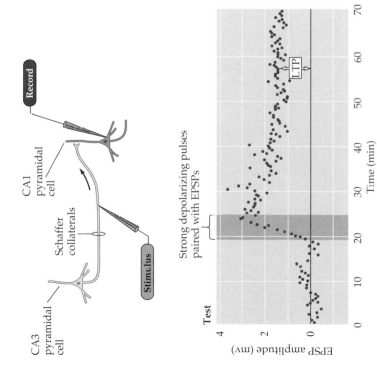

Molecular Mechanisms Underlying LTP

Despite the fact that LTP was discovered more than 30 years ago, its molecular mechanism is still not well understood. One advance in this effort occurred in the mid-1980s, when the unique properties of the NMDA type of glutamate receptor were first appreciated (see Chapter 7). At about the same time, it was discovered that antagonists of NMDA receptors prevent LTP, but have no effect on the synaptic response evoked by low-frequency stimulation of the Schaffer collaterals. The biophysical properties of the NMDA receptor channels pro-

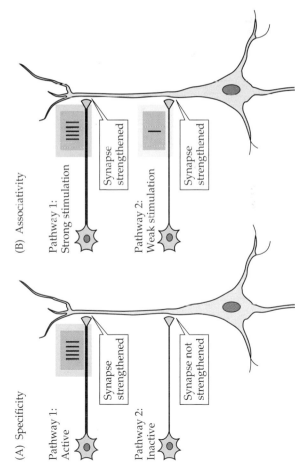

Figure 25.8 Properties of LTP at a CA1 pyramidal neuron receiving synaptic inputs from two independent sets of Schaffer collateral axons. (A) Strong activity initiates LTP at active synapses (pathway 1) without initiating LTP at nearby inactive synapses (pathway 2). (B) Weak stimulation of pathway 2 alone does not trigger LTP. However, when the same weak stimulus to pathway 2 is activated together with strong stimulation of pathway 1, both sets of synapses are strengthened.

Figure 25.9 The NMDA receptor channel can open only during depolarization of the postsynaptic neuron from its normal resting level. Depolarization expels Mg²⁺ from the NMDA channel, allowing current to flow into the post-synaptic cell. This leads to Ca²⁺ entry, which in turn triggers LTP. (After Nicoll, Malenka and Kauer, 1988.)

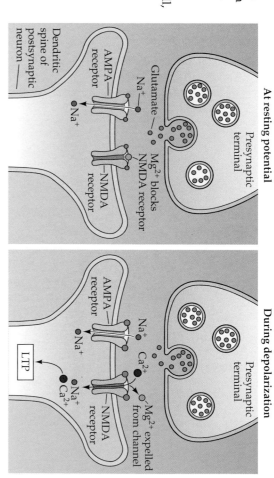

At resting potential

Presynaptic terminal

Glutamate
Na⁺

AMPA receptor

Mg²⁺ blocks NMDA receptor

NMDA receptor

Na⁺

Dendritic spine of postsynaptic neuron

During depolarization

Presynaptic terminal

AMPA receptor

Na⁺

Na⁺
Ca²⁺

Mg²⁺ expelled from channel

NMDA receptor

Na⁺
Ca²⁺

LTP

vided considerable insight into the induction of LTP (Figure 25.9). As described in Chapter 7, the NMDA receptor channel is permeable to Ca²⁺, but is blocked by physiological concentrations of Mg²⁺. Thus, during low-frequency synaptic transmission, glutamate released by the Schaffer collaterals binds to both NMDA-type and AMPA/kainate-type glutamate receptors. While both types of receptors bind glutamate, if the postsynaptic neuron is at its normal resting membrane potential, the NMDA channels will be blocked by Mg²⁺ ions and no current will flow. Because blockade of the NMDA channel by Mg²⁺ is voltage-dependent, the function of the synapse changes markedly when the postsynaptic cell is depolarized. Thus, Mg²⁺ is expelled from the NMDA channel during high-frequency stimulation (as in Figure 25.6), or when the cell is depolarized directly (as in Figure 25.7), allowing Ca²⁺ to enter the postsynaptic neuron. The resulting increase in Ca²⁺ concentration within the dendritic spines of the postsynaptic cell turns out to be the critical trigger for LTP: The NMDA receptor thus behaves like a molecular "and" gate: The channel opens (to induce LTP) only when glutamate is bound to NMDA receptors *and* the postsynaptic cell is depolarized to relieve the Mg²⁺ block of the NMDA channel. Thus, the receptor can "detect" the coincidence of two events.

These properties of the NMDA receptor can account, in principle, for both the specificity and the associativity of LTP (see Figure 25.8). When only one group of synaptic inputs is strongly stimulated, LTP is confined to the active synapses because glutamate opens NMDA channels only at these sites. With respect to associativity, a weakly stimulated input releases glutamate, but cannot sufficiently depolarize the postsynaptic cell to relieve the Mg²⁺ block. If neighboring inputs are strongly stimulated, however, they provide the "associative" depolarization necessary to relieve the block. LTP induced by the pairing of synaptic input with depolarization (see Figure 25.7) may work similarly: The synaptic input releases glutamate, while the coincident depolarization relieves the Mg²⁺ block of the NMDA receptor.

Several sorts of observations have confirmed that a primary signal for LTP induction is a rise in the concentration of Ca²⁺ in the postsynaptic CA1 neuron. Imaging studies, for instance, have shown that activation of NMDA receptors causes increases in postsynaptic Ca²⁺ levels. Furthermore, injection of Ca²⁺ chelators blocks LTP induction, whereas elevation of Ca²⁺ levels in postsynaptic neurons potentiates synaptic transmission. A likely scenario is that Ca²⁺ ions entering through NMDA receptors stimulate one or more Ca²⁺-activated protein kinases in the postsynaptic neuron. At least two protein kinases have been implicated in LTP induction: Ca²⁺/calmodulin-dependent protein kinase (CaMKII) and protein kinase C (PKC; see Chapter 8). CaMKII seems to play an especially important role: This enzyme is the most abundant postsynaptic protein at Schaffer collateral synapses, and inhibition or genetic deletion of CaMKII prevents LTP. The downstream targets of these kinases are not yet known.

Despite much effort, the additional mechanism(s) responsible for prolonged strengthening of synaptic transmission during LTP are not fully resolved. The most popular school of thought is that LTP arises from changes in the sensitivity of the postsynaptic cell to glutamate, either by rapid addition of new receptors to the potentiated synapses or by increased current flow through receptors already present. Several lines of evidence support a role for such postsynaptic mechanisms in LTP (Box B). However, other evidence suggests that LTP causes a sustained increase in transmitter release by presynaptic terminals, perhaps by modification of the proteins involved in exocytosis (see Chapter 5). Because LTP clearly is triggered by the actions of Ca²⁺ within the postsynaptic neuron, this hypothesis requires that a retrograde signal (perhaps NO) spread from the postsynaptic region to the presynaptic terminals (Figure 25.10). Which of these schemes is correct remains to be determined; indeed, elements of both could operate.

Like synaptic plasticity in *Aplysia*, LTP has an early phase that involves post-translation mechanisms, such as activation of protein kinases, and a later phase that entails changes in gene expression and synthesis of new proteins. Thus, blocking protein synthesis prevents LTP measured several hours after a stimulus but does not affect LTP measured at earlier times. Presum-

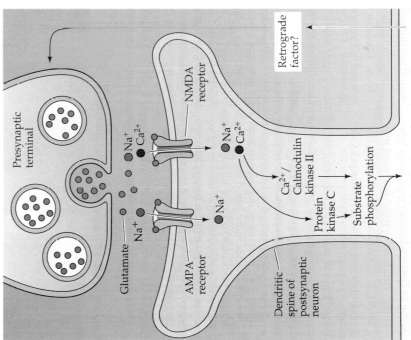

Figure 25.10 Mechanisms underlying LTP. During glutamate release, the NMDA channel opens only if the postsynaptic cell is sufficiently depolarized. The Ca²⁺ ions that enter the cell through the channel activate postsynaptic protein kinases. These kinases may act postsynaptically to increase the sensitivity to glutamate. Alternatively, a retrograde signal may be released that acts on the presynaptic terminal to enhance neurotransmitter release.

Despite these intriguing observations, the mechanism, purpose, and significance of the reorganization of sensory and motor maps that occurs in adult cortex are not known. Clearly, limited changes in cortical circuitry can occur in the adult brain, even though the basic features of cortical organization—such as ocular dominance columns and the broader topographical organization of inputs from the thalamus—remain fixed (see Chapter 24). If a greater degree of cortical plasticity were possible, recovery from brain injury would be far more vigorous and effective than centuries of clinical observation have shown it to be.

Given their rapid and reversible character, most of these changes in cortical function probably reflect alterations in the strength of synapses already present.

Recovery from Neural Injury

These various observations on adult plasticity indicate that normal experience can alter the strength of existing synapses and might even elicit some local remodeling of synapses and circuits. More extensive growth and remodeling are stimulated by nervous system injury.

Traumatic injury, interruption of blood supply, and degenerative diseases all can damage axons in peripheral nerves, or neuronal cell bodies and synapses in the more complex circuitry of the brain or spinal cord. When peripheral nerves are injured, the damaged axons regenerate vigorously and can regrow over distances of many centimeters or more. Under favorable circumstances, these regenerated axons can also reestablish synaptic connections with their targets in the periphery. In contrast, CNS axons typically fail to regenerate (Figure 25.16). As a result, axonal damage in the retina, spinal cord, or the rest of the brain leads to permanent blindness, paralysis, and other disabilities. What explains this difference in the regeneration of peripheral nerves compared to axonal regeneration in the brain or spinal cord?

Successful regeneration in peripheral nerves depends on two critical conditions. First, the injured neuron must respond to axon interruption by initiating a program of gene expression that can support axon elongation. Many of the genes involved in the outgrowth of axons over long distances during embryonic development (see Chapter 23) are not normally expressed in adult neurons. Interruption of axons reactivates expression of

Figure 25.16 Different responses to injury in the peripheral (A) and central (B) nervous systems. Damage to a peripheral nerve leads to series of cellular responses, collectively called Wallerian degeneration (after Augustus Waller, the nineteenth century English physician who first described these phenomena). *Distal* to the site of injury, axons disconnected from their cell bodies degenerate, and invading macrophages remove the cellular debris. Schwann cells that formerly ensheathed the axons proliferate, align to form longitudinal arrays, and increase their production of neurotrophic factors that can promote axon regeneration. Schwann cell surfaces and the extracellular matrix also provide a favorable substratum for the extension of regenerating axons. In the CNS, the removal of myelin debris is relatively slow, and the myelin membranes produce inhibitory molecules that can block axon growth (see Chapter 23). Astrocytes at the site of injury also interfere with regeneration. *Proximal* to the injury, neuron cell bodies react to peripheral nerve injury by inducing expression of growth-related genes, including those for major components of axonal growth cones. Following CNS injury, however, neurons typically fail to activate these growth-associated genes.

some of these genes in the peripheral nervous system, but not in the adult CNS. Axons damaged in the long tracts of the brain or spinal cord, particularly at sites far from their cell bodies, rarely re-express these genes. Second, once a damaged neuron initiates a genetic program that can support axon

(A) Peripheral nervous system

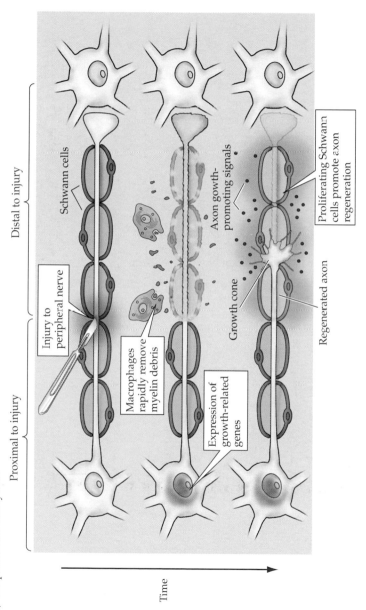

Proximal to injury — Distal to injury

Schwann cells

Injury to peripheral nerve

Macrophages rapidly remove myelin debris

Expression of growth-related genes

Axon growth-promoting signals

Growth cone

Regenerated axon

Proliferating Schwann cells promote axon regeneration

Time

(B) Central nervous system

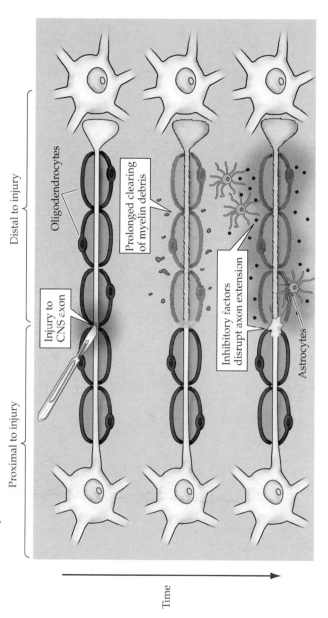

Proximal to injury — Distal to injury

Oligodendrocytes

Injury to CNS axon

Prolonged clearing of myelin debris

Inhibitory factors disrupt axon extension

Astrocytes

Time

regrowth, the emerging growth cones must encounter an environment that can support and guide the regrowing axons. In peripheral nerves, damage or degeneration triggers changes that produce a favorable environment for axon elongation. Schwann cells and other non-neuronal cells respond to axonal injury by elaborating cell adhesion molecules, extracellular matrix components, and an array of neurotrophins and other signals that promote axon growth (see Chapter 23). Equally important, damaged peripheral nerves are invaded by macrophages that rapidly remove fragments of degenerating axons and myelin that might otherwise inhibit the growth of regenerating axons.

In contrast, damage to axonal tracts in the adult CNS triggers a very different set of changes. As axons and their myelin sheaths break down, the remnants are not cleared efficiently and can persist for many weeks, posing a substantial impediment to regeneration. This inhibition appears to reflect the activity of a protein called Nogo that blocks axon extension by interacting with advancing growth cones (see Box D). Nogo is produced by oligodendrocytes, the glia that normally form myelin sheaths around CNS axons. To make matters worse, astrocytes reacting to CNS injury express additional inhibitors of axon extension. As a consequence, even if a central neuron initiates a genetic program for regeneration, growth cones emerging from the site of a lesion in the adult CNS encounter an array of circumstances that impede recovery.

The role of the axonal environment in regeneration of CNS axons was explored by Albert Aguayo and his co-workers at McGill University in the 1980s. They grafted segments of peripheral nerve into sites in the CNS, such as optic nerve, spinal cord, or other locations, and then determined whether neurons were able to regenerate axons through the peripheral grafts. Their studies showed that at least some CNS axons are able to take advantage of the more supportive growth environment of the peripheral nerve, regenerating over distances of many centimeters and in some cases restoring appropriate synaptic connections (see Box D).

This demonstration that CNS axons can sometimes regenerate successfully into a peripheral nerve graft sparked intensive efforts by many labs to produce a similarly supportive environment for axon growth within the long tracts of the brain or spinal cord. For example, Martin Schwab and his collaborators showed that implanting cells engineered to secrete antibodies against inhibitory proteins, including Nogo, alleviated some of the inhibitory properties of CNS myelin in experimental animals. Another approach was to introduce cells that provide a more supportive environment for regenerating axons in the damaged CNS. Schwann cells, neural stem cells (see next section), and specialized glial cells from the olfactory nerve all can be grown in tissue culture and introduced into the brains or spinal cords of experimental animals, where they modestly improve axon regrowth and, in some cases, functional recovery.

In summary, regeneration in adults is held in check by ongoing suppression of genes required for effective axon elongation. Injury to the peripheral nervous system readily induces expression of this genetic program, while interruption of mammalian CNS axons does not. Once CNS neurons have activated these genes, in principle regrowth could be enhanced by removal or neutralization of inhibitory molecules, and by the introduction of cells that provide a more supportive growth environment. These strategies, however, have not been proven clinically useful. Why this patently maladaptive state of affairs has persisted in evolution is much debated (Box D).

Generation of Neurons in the Adult Brain

It has long been known that mature, differentiated neurons do not divide (see Chapter 22). It does not follow, however, that *all* the neurons that make up the adult brain are produced during embryonic development, even though this interpretation has generally been assumed. The merits of this assumption were questioned in the 1980s, when Fernando Nottebohm and colleagues at Rockefeller University demonstrated the production of new neurons in the brains of adult songbirds. They showed that labeled DNA precursors injected into adult birds could be found subsequently in fully differentiated neurons, indicating that the neurons had undergone their final round of cell division after the labeled precursor was injected. Moreover, the new neurons were able to extend dendrites and project long axons to establish appropriate connections with other brain nuclei. Production of new neurons was apparent in many parts of the birds' brains, but was especially prominent in areas involved in song production (see Box B in Chapter 24). These observations showed that the adult brain can generate at least some new nerve cells and incorporate them into neural circuits (see also Chapter 15).

The production of new neurons in the adult brain has now been examined in mice, rats, monkeys and, finally, humans. In all cases, however, the new nerve cells in the mammalian CNS have been restricted to just two regions of the brain: (1) The granule cell layer of the olfactory bulb; and (2) the dentate gyrus of the hippocampus. Furthermore, the new nerve cells are primarily local circuit neurons or interneurons. New neurons with long distance projections have not been seen. Each of these populations in the olfactory bulb and hippocampus is apparently generated from nearby sites near the surface of the lateral ventricle. As in bird brains, the newborn nerve cells extend axons and dendrites and become integrated into functional synaptic circuits. Evidently, a limited production of new neurons occurs continually in a few specific loci.

If neurons cannot divide (see Chapter 22), how does the adult brain generate these nerve cells? The answer emerged with the discovery that the subventricular zone that produces neurons during development retains some **neural stem cells** in the adult. The term "stem cells" refers to a population of cells that are self-renewing—each cell can divide symmetrically to give rise to more cells like itself, but also can divide asymmetrically, giving rise to a new stem cell plus one or more differentiated cells. Over the past decade, several research groups have isolated stem cells from the adult brain that can reproduce in large numbers in cell culture. Such cells can then be induced to differentiate into neurons and glial cells, when exposed to appropriate signals. Many of these same signals mediate neuronal differentiation in normal development. Adult stem cells can be isolated not only from the anterior subventricular zone (near the olfactory bulb) and dentate gyrus, but from many other parts of the forebrain, cerebellum, midbrain, and spinal cord, although they do not apparently produce any new neurons in these sites. Inhibitory signals in these regions may prevent stem cells from generating neurons.

Why the generation of neurons is so limited in the adult brain is not known. This peculiar limitation is presumably related to the reasons discussed in Box D. Nevertheless, the fact that new neurons can be generated in a few regions of the adult brain suggests that this phenomenon can occur throughout the adult CNS. The ability of newly generated neurons to integrate into at least some synaptic circuits adds to the mechanisms available for plasticity in the adult brain. Thus, many investigators have begun to explore the potential applications of stem cell technology for the repair of circuits damaged by traumatic injury or degenerative disease.

BOX D
Why Aren't We More Like Fish and Frogs?

The central nervous system of adult mammals, including humans, recovers only poorly from injury. As indicated in the text, once severed, major axon tracts (such as those in the spinal cord) never regenerate. The devastating consequences of these injuries—e.g., loss of movement and the inability to control basic bodily functions—has led many neuroscientists to seek ways of restoring the connections of severed axons. There is no a priori reason for this biological failure, since "lower" vertebrates—e.g., lampreys, fish, and frogs—*can* regenerate a severed spinal cord or optic nerve. Even in mammals, the inability to regenerate axonal tracts is a special failing of the central nervous system; peripheral nerves can and do regenerate in adult animals, including humans. Why, then, not the central nervous system?

At least a part of the answer to this puzzle apparently lies in the molecular

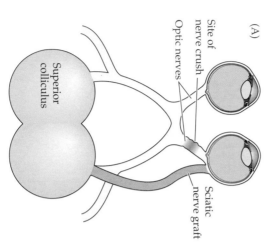

(A)

Superior
colliculus

Site of
nerve crush

Optic nerves

Sciatic
nerve graft

cues that promote and inhibit axon outgrowth. In mammalian peripheral nerves, axons are surrounded by a basement membrane (a proteinaceous extracellular layer composed of collagens, glycoproteins, and proteoglycans) secreted in part by Schwann cells, the glial cells associated with peripheral axons. After a peripheral nerve is crushed, the axons within it degenerate; the basement membrane around each axon, however, persists for months. One of the major components of the basement membrane is laminin, which (along with other growth-promoting molecules in the basement membrane) forms a hospitable environment for regenerating growth cones. The surrounding Schwann cells also react by releasing neurotrophic factors, which further promote axon elongation (see text). This peripheral environment is so favorable to regrowth that even neurons from the central nervous

system can be induced to extend into transplanted segments of peripheral nerve. Albert Aguayo and his colleagues at the Montreal General Hospital found that grafts derived from peripheral nerves can act as "bridges" for central neurons (in this case, retinal ganglion cells), allowing them to grow for over a centimeter (figure A); they even form a few functional synapses in their target tissues (figure B).

These several observations suggest that the failure of central neurons to regenerate is not due to an intrinsic inability to sprout new axons, but rather to something in the local environment that prevents growth cones from extending. This impediment could be the absence of growth-promoting factors—such as the neurotrophins—or the presence of molecules that actively prevent axon outgrowth. Studies by Martin Schwab and his colleagues point to the

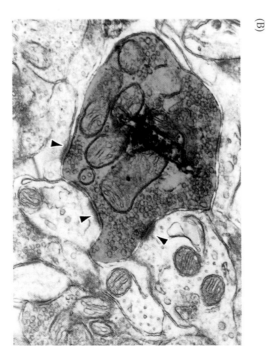

(B)

Implantation of a section of peripheral nerve into the central nervous system facilitates the extension of central axons. (A) Mammalian retinal ganglion neurons, which do not normally regenerate following a crush injury, will grow for many millimeters into a graft derived from the sciatic nerve. (B) If the distal end of the graft is inserted into a normal target of retinal ganglion cells, such as the superior colliculus, a few regenerating axons invade the target and form functional synapses, as shown in this electron micrograph (arrowheads). The dark material is an intracellularly transported label that identifies particular synaptic terminals as originating from a regenerated retinal axon. (A after So and Aguayo, 1985; B from Bray et al., 1991.)

latter possibility. Schwab found that central nervous system myelin contains an inhibitory component that causes growth cone collapse in vitro and prevents axon growth in vivo. This component, recognized by a monoclonal antibody called IN-1, is found in the myelinated portions of the central nervous system but is absent from peripheral nerves. IN-1 also recognizes molecules in the optic nerve and spinal cord of mammals, but is missing in the same sites in fish, which do regenerate these central tracts. Nogo-A, the primary antigen recognized by the IN-1 antibody, is secreted by oligodendrocytes, but not by Schwann cells in the peripheral nervous system. Most dramatically, the IN-1 antibody increases the extent of spinal cord regeneration when provided at the site of injury in rats with spinal cord damage. All this implies that the human central nervous system differs from that of many "lower" verte-

brates in that humans and other mammals present an unfavorable molecular environment for regrowth after injury. Why this state of affairs occurs is not known. One speculation is that the extraordinary amount of information stored in mammalian brains puts a premium on a stable pattern of adult connectivity.

At present there is only one modestly helpful treatment for CNS injuries such as spinal cord transection. High doses of a steroid, methylprednisolone, immediately after the injury prevents some of the secondary damage to neurons resulting from the initial trauma. Although it may never be possible to fully restore function after such injuries, enhancing axon regeneration, blocking inhibitory molecules and providing additional trophic support to surviving neurons could in principle allow sufficient recovery of motor control to give afflicted individuals a better quality of life than

they now enjoy. The best "treatment," however, is to prevent such injuries from occurring, since there is now very little that can be done after the fact.

References

BRAY, G. M., M. P. VILLEGAS-PEREZ, M. VIDAL-SANZ AND A. J. AGUAYO (1987) The use of peripheral nerve grafts to enhance neuronal survival, promote growth and permit terminal reconnections in the central nervous system of adult rats. J. Exp. Biol. 132: 5–19.

SCHNELL, L. AND M. E. SCHWAB (1990) Axonal regeneration in the rat spinal cord produced by an antibody against myelin-associated neurite growth inhibitors. Nature 343: 269–272.

VIDAL-SANZ, M., G. M. BRAY, M. P. VILLEGAS-PEREZ, S. THANOS AND A. J. AGUAYO (1987) Axonal regeneration and synapse formation in the superior colliculus by retinal ganglion cells in the adult rat. J. Neurosci. 7: 2894–2909.

Summary

The adult nervous system exhibits plasticity in a variety of circumstances. Studies of behavioral plasticity in several invertebrates and of the neuromuscular junction suggest that modification of synaptic strength is responsible for much of the ongoing change in synaptic function in adults. Synapses exhibit many forms of plasticity that occur over a broad temporal range. At the shortest times (seconds to minutes), facilitation, post-tetanic potentiation, and depression provide rapid but transient modifications based on alterations in Ca^{2+} signaling and synaptic vesicle pools at recently active synapses. Longer-lasting forms of synaptic plasticity such as LTP and LTD are also based on Ca^{2+} and other intracellular second messengers. In these more enduring forms of plasticity, protein phosphorylation and changes in gene expression greatly outlast the period of synaptic activity and can yield persistent changes in synaptic strength (hours to days or longer). Different brain regions evidently use one or more of these strategies to learn new behaviors and acquire new memories. Neuronal damage can also induce plastic changes. Peripheral neurons can regenerate axons following damage, though the capacity of CNS axons to regenerate is limited. In addition, stem cells are present in certain regions of the adult brain, allowing the production of some new neurons in selected sites. These various forms of adult plasticity can modify the function of the mature brain and provide some hope for improving the limited ability of the CNS to recover successfully from trauma and neurological disease.

ADDITIONAL READING

Reviews

BARRES, B. A. (1999) A new role for glia: Generation of neurons! Cell 97: 667–670.

BLISS, T. V. P. AND G. L. COLLINGRIDGE (1993) A synaptic model of memory: Long-term potentiation in the hippocampus. Nature 361: 31–39.

GAGE, F. H. (2000) Mammalian neural stem cells. Science 287: 1433–1438.

GOLDBERG, J. L. AND B. A. BARRES (2000) Nogo in nerve regeneration. Nature 403: 369–370.

ITO, M. (1989) Long-term synaptic depression. Ann. Rev. Neurosci. 12: 85–102.

KANDEL, E. R. AND J. H. SCHWARTZ (1982) Molecular biology of learning: Modulation of transmitter release. Science 218: 433–443.

KAUER, J. A., R. C. MALENKA AND R. A. NICOLL (1988) A persistent postsynaptic modification mediates long-term potentiation in the hippocampus. Neuron 1: 911–917.

KEMPERMANN, G. AND F. H. GAGE. (1999) New nerve cells for the adult brain. Sci. Am. 280 (May): 48–53.

MALENKA, R. C. AND R. A. NICOLL (1999) Long-term potentiation—a decade of progress? Science 285: 1870–1874.

MERZENICH, M. M., G. H. RECANZONE, W. M. JENKINS AND K. A. GRAJSKI (1990) Adaptive mechanisms in cortical networks underlying cortical contributions to learning and nondeclarative memory. Cold Spring Harbor Symp. Quant. Biol. 55: 873–887.

QIU, J., D. CAI AND M. T. FILBIN (2000) Glial inhibition of nerve regeneration in the mature mammalian CNS. Glia 29: 166–174.

SANES, J. R. AND J. W. LICHTMAN (1999) Can molecules explain long-term potentiation? Nature Neurosci. 2: 597–604.

Important Original Papers

ALVAREZ, P. S. ZOLA-MORGAN AND L. R. SQUIRE (1995) Damage limited to the hippocampal region produces long-lasting memory impairment in monkeys. J. Neurosci. 15: 3796–3807.

BLISS, T. V. P. AND A. R. GARDNER-MEDWIN (1973) Long-lasting potentiation of synaptic transmission in the dentate area of the unanaesthetized rabbit following stimulation of the perforant path. J. Physiol. 232: 357–374.

BLISS, T. V. P. AND T. LØMO (1973) Long-lasting potentiation of synaptic transmission in the dentate area of the anaesthetized rabbit following stimulation of the perforant path. J. Physiol. 232: 331–356.

BREGMAN, B. S., E. KUNKEL-BAGDEN, L. SCHNELL, H. N. DAI, D. GAO AND M. E. SCHWAB (1995) Recovery from spinal cord injury mediated by antibodies to neurite growth inhibitors. Nature 378: 498–501.

COLLINGRIDGE, G. L., S. J. KEHL AND H. MCLENNAN (1983) Excitatory amino acids in synaptic transmission in the Schaffer collateral-commissural pathway of the rat hippocampus. J. Physiol. 334: 33–46.

ENGERT, F. AND T. BONHOEFFER (1999) Dendritic spine changes associated with hippocampal long-term synaptic plasticity. Nature 399: 66–70.

ERIKSSON, P. S. AND 6 OTHERS (1998) Neurogenesis in the adult human hippocampus. Nature Medicine 4: 1313–1317.

FAGGIN, B. M., K. T. NGUYEN AND M. A. L. NICOLELIS (1997) Immediate and simultaneous sensory reorganization at cortical and subcortical levels of the somatosensory system. PNAS 94: 9428–9433.

FINCH, E. A. AND G. J. AUGUSTINE (1998) Local calcium signaling by IP3 in Purkinje cell dendrites. Nature 396: 753–756.

GILBERT, C. D. AND T. N. WIESEL (1992) Receptive field dynamics in adult primary visual cortex. Nature 356: 150–152.

GOLDMAN, S. A. AND F. NOTTEBOHM (1983) Neuronal production, migration, and differentiation in a vocal control nucleus of the adult female canary brain. Proc. Natl. Acad. Sci. USA 80: 2390–2394.

JENKINS, W. M., M. M. MERZENICH, M. T. OCHS, E. ALLARD AND T. GUIC-ROBLES (1990) Functional reorganization of primary somatosensory cortex in adult owl monkeys after behaviorally controlled tactile stimulation. J. Neurophysiol. 63: 82–104.

KATZ, B. AND R. MILEDI (1968) The role of calcium in neuromuscular facilitation. J. Physiol. (Lond.) 195: 481–492.

KELSO, S. R. AND T. H. BROWN (1986) Differential conditioning of associative synaptic enhancement in hippocampal brain slices. Science 232: 85–87.

KEMPERMANN, G., H. G. KUHN AND F. H. GAGE (1997) More hippocampal neurons in adult mice living in an enriched environment. Nature 386: 493–495.

LASHLEY, K. S. (1950) In search of the engram. Symp. Soc. Exp. Biol. 4: 454–482.

LINDEN, D. J., M. H. DICKINSON, M. SMEYNE AND J. A. CONNOR (1991) A long-term depression of AMPA currents in cultured cerebellar Purkinje neurons. Neuron 7: 81–89.

MALENKA, R. C., J. A KAUER, R. S. ZUCKER AND R. A. NICOLL (1988) Postsynaptic calcium is sufficient for potentiation of hippocampal synaptic transmission. Science 242: 81–84.

MALINOW, R., H. SCHULMAN, AND R. W. TSIEN (1989) Inhibition of postsynaptic PKC or CaMKII blocks induction but not expression of LTP. Science 245: 862–866.

MCDONALD, J. W. AND 7 OTHERS (1999) Transplanted embryonic stem cells survive, differentiate and promote recovery in injured rat spinal cord. Nature Medicine 5: 1410–1412.

MERZENICH, M. M., R. J. NELSON, M. P. STRYKER, M. S. CYNADER, A. SCHOPPMANN AND J. M. ZOOK (1984) Somatosensory cortical map changes following digit amputation in adult monkeys. J. Comp. Neurol. 224: 591–605.

MULKEY, R. M., C. E. HERRON AND R. C. MALENKA (1993) An essential role for protein phosphatases in hippocampal long-term depression. Science 261: 1051–1055.

NEUMANN, S. AND C. J. WOOLF (1999) Regeneration of dorsal column fibers into and beyond the lesion site following adult spinal cord injury. Neuron 23: 83–91.

NICOLELIS, M. A. L., R. C. S. LIN, D. J. WOODWARD AND J. K. CHAPIN (1993) Induction of immediate spatiotemporal changes in thalamic networks by peripheral block of ascending cutaneous information. Nature 361: 533–536.

O'KEEFE, J. (1990) A computational theory of the hippocampal cognitive map. Prog. Brain Res. 83: 301–312.

RAMON-CUETO, A., M. I. CORDERO, F. F. SANTOS-BENITO AND J. AVILA (2000) Functional recovery of paraplegic rats and motor axon regeneration in their spinal cords by olfactory ensheathing glia. Neuron 25: 425–35.

SAKURAI, M. (1987) Synaptic modification of parallel fibre-Purkinje cell transmission in in vitro guinea-pig cerebellar slices. J. Physiol. (Lond) 394: 463–480.

SILVA, A. J., R. PAYLOR, J. M. WEHNER AND S. TONEGAWA (1992) Impaired spatial learning in alpha-calcium-calmodulin kinase II mutant mice. Science 257: 206–211.

SQUIRE, L. R., J. G. OJEMANN, F. M MIEZEN, S. E. PETERSEN, T. O. VIDEEN AND M. E. RAICHLE (1995) Activation of the hippocampus in normal humans: A functional anatomical study of memory. Proc. Natl. Acad. Sci. USA 89: 1837–1841.

Books

BAUDRY, M. AND J. D. DAVIS (1991) Long-Term Potentiation: A Debate of Current Issues. Cambridge, MA: MIT Press.

LANDFIELD, P. W. AND S. A. DEADWYLER (eds.) (1988) Long-Term Potentiation: From Biophysics to Behavior. New York: A. R. Liss.

V

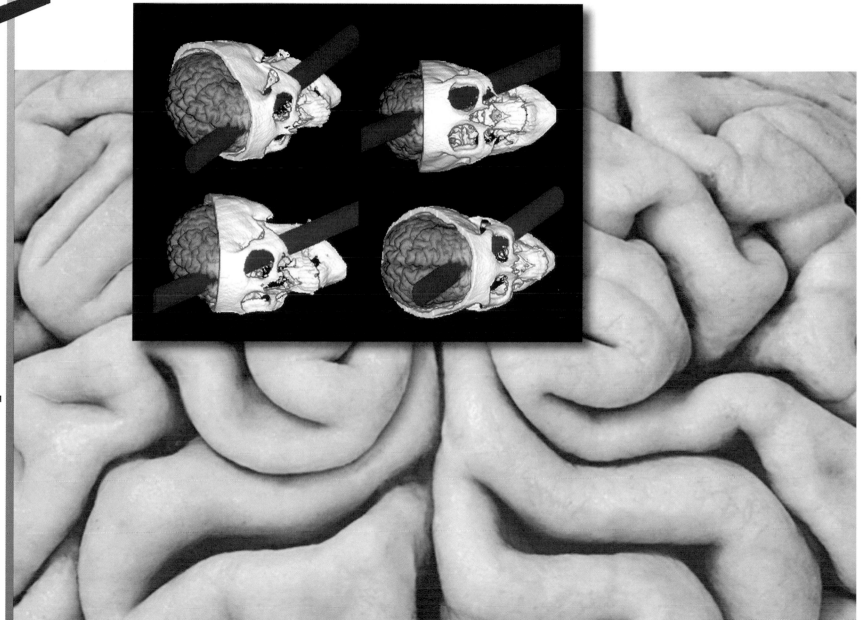

Unit V

Complex Brain Functions

26 *The Association Cortices*

27 *Language and Lateralization*

28 *Sleep and Wakefulness*

29 *Emotions*

30 *Sex, Sexuality, and the Brain*

31 *Human Memory*

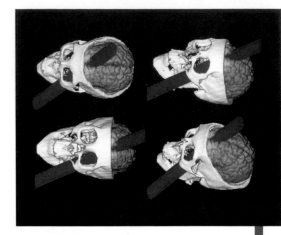

The function of the frontal cortex was first suggested by a dramatic nineteenth century accident in which a tamping rod was driven through the frontal part of the brain of a railroad worker named Phineas P. Gage. Remarkably, Gage survived, and his behavioral deficits stimulated much early thinking about complex brain functions. The illustration here is a reconstruction of the trajectory of the rod based on Gage's skull, which is housed in the Warren Museum at Harvard Medical School. (Courtesy of H. Damasio.)

The awareness of physical and social circumstances, the ability to have thoughts and feelings (emotions), to be sexually attracted to others, to express these things to our fellow humans by language, and to store such information in memory certainly rank among the most intriguing functions of the human brain. Given their importance in our daily lives—and for human culture generally—it is not surprising that much of the human brain is devoted to these and other complex mental functions. The intrinsic interest of these aspects of human behavior is unfortunately equaled by the difficulty—both technical and conceptual—involved in unraveling their neurobiological underpinnings. Nonetheless, a good deal of progress has been made in deciphering the structural and functional organization of the relevant brain regions. Especially important has been the steady accumulation of case studies during the last century or more that, by the signs and symptoms resulting from damage to specific brain regions, have indicated much about the neural basis of complex brain functions. The advent of noninvasive brain imaging techniques has recently provided another way of understanding some of these abilities in normal human subjects as well as in neurological patients. Finally, complementary electrophysiological experiments in nonhuman primates have begun to elucidate the cellular correlates of some of these functions. Taken together, these observations have established a rapidly growing body of knowledge about these more complex aspects of the human brain. This domain of investigation has come to be called "cognitive neuroscience," a field that promises to loom ever larger in the new century.

Chapter 26

The Association Cortices

Overview

The association cortices include most of the cerebral surface of the human brain and are largely responsible for the complex processing that goes on between the arrival of input in the primary sensory cortices and the generation of behavior. The diverse functions of the association cortices are loosely referred to as "cognition," which literally means the process by which we come to know the world ("cognition" is perhaps not the best word to indicate this wide range of neural functions, but it has already become part of the working vocabulary of neurologists and neuroscientists). More specifically, cognition refers to the ability to attend to external stimuli or internal motivation, to identify the significance of such stimuli, and to plan meaningful responses to them. Given the complexity of these tasks, it is not surprising that the association cortices receive and integrate information from a variety of sources, and that they influence a broad range of cortical and subcortical targets. Inputs to the association cortices include projections from the primary and secondary sensory and motor cortices, the thalamus, and the brainstem. Outputs from the association cortices reach the hippocampus, the basal ganglia and cerebellum, the thalamus, and other association cortices. Insight into how the association areas work has come primarily from observations of human patients with damage to one or another of these regions. Noninvasive brain imaging of normal subjects, functional mapping at neurosurgery, and electrophysiological analysis of comparable brain regions in nonhuman primates have generally confirmed these clinical impressions. Together, these studies indicate that, among other functions, the parietal association cortex is especially important for attending to complex stimuli in the external and internal environment, that the temporal association cortex is especially important for identifying the nature of such stimuli, and that the frontal association cortex is especially important for planning appropriate behavioral responses to the stimuli.

The Association Cortices

The preceding chapters have considered in some detail the parts of the brain responsible for encoding sensory information and commanding movements. But these regions account for only a fraction (perhaps a fifth) of the cerebral cortex (see Figure 26.1). The consensus has long been that much of the remaining cortex is concerned with attending to complex stimuli, identifying the relevant features of such stimuli, recognizing the related objects, and planning appropriate responses (as well as storing aspects of this information). Collectively, these abilities are referred to as **cognition**, and it is evi-

Figure 26.1 Lateral and medial views of the human brain, showing the extent of the association cortices in blue. The primary sensory and motor regions of the cortical mantle. The primary cortices occupy a relatively small fraction of the total area of the cortical mantle. The remainder of the neocortex—defined by exclusion as the association cortices—is the seat of human cognitive ability. The term "association" refers to the fact that these regions of the cortex integrate (associate) information derived from other brain regions.

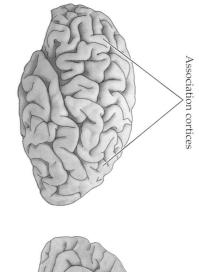

Association cortices

Primary sensory and motor areas

An Overview of Cortical Structure

Before delving into a more detailed account of the functions of these cortical regions, it is important to have a general understanding of cortical structure and the organization of its canonical circuitry. Most of the cortex that covers the cerebral hemispheres is **neocortex**, defined as cortex that has six cellular layers, or laminae. Each layer comprises more or less distinctive populations of cells based on their different densities, sizes, shapes, inputs, and outputs. The laminar organization and basic connectivity of the human cerebral cortex are summarized in Figure 26.2A and Table 26.1. Despite an overall uniformity, regional differences based on these laminar features have long been apparent (Box A), allowing investigators to identify about 50 subdivisions of the cerebral cortex (Figure 26.2B). These histologically defined subdivisions are referred to as **cytoarchitectonic areas**, and over the years, a zealous band of neuroanatomists has painstakingly mapped these areas in humans and some of the more widely used laboratory animals.

Early in the twentieth century, cytoarchitectonically distinct regions were identified with little or no knowledge of their functional significance. Even-

dently the association cortices in the parietal, temporal, and frontal lobes that make cognition possible. (The association cortex of the occipital lobe is equally important in cognition; its functions, however, are largely concerned with vision, and much of what is known about these areas has been discussed in Chapter 12.) The primary sensory and motor cortices occupy a relatively limited portion of the cortical mantle, the majority of the cortex being devoted to more integrative functions. These other areas of the cerebral cortex are referred to collectively as the **association cortices** (Figure 26.1).

TABLE 26.1
The Major Connections of the Neocortex

Direct cortical inputs	Targets of cortical output
Thalamus	Thalamus
	Caudate and putamen (striatum)
Brainstem nuclei	Pontine nuclei (to cerebellum)
Other cortical regions	Spinal cord
	Brainstem nuclei

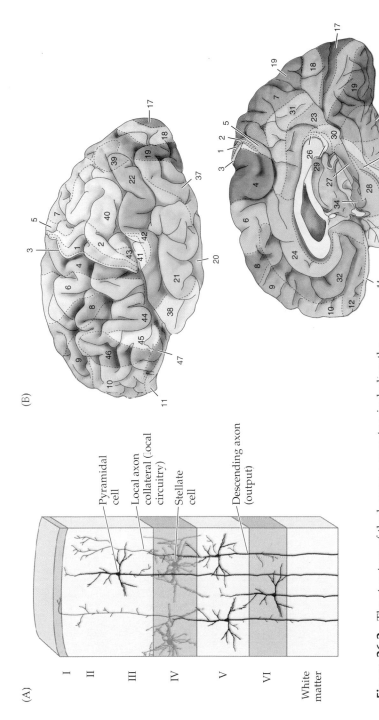

(A)

I
II
III
IV
V
VI
White matter

Pyramidal cell

Local axon collateral (local circuitry)

Stellate cell

Descending axon (output)

(B)

Figure 26.2 The structure of the human neocortex, including the association cortices. (A) A summary of the cellular composition of the six layers of the neocortex. (B) Based on variations in the thickness, cell density, and other histological features of the six neocortical laminae, the human brain can be divided into about 50 cytoarchitectonic areas, in this case those recognized by the neuroanatomist Korbinian Brodmann in his seminal monograph in 1909. (See Box A for additional detail.)

tually, however, studies of patients in whom one or more of these cortical areas had been damaged, supplemented by electrophysiological mapping in both laboratory animals and neurosurgical patients, added to this knowledge. This work showed that many of the regions neuroanatomists had identified on histological grounds are also functionally distinct. Thus, cytoarchitectonic areas can also be distinguished by the physiological response properties of their constituent cells, and often by their patterns of local and long-distance connections.

Despite significant variations among different cytoarchitectonic areas, the circuitry of all cortical regions has some common features (Figure 26.3). First, each cortical layer has a primary source of inputs and a primary output target. Second, each area has connections in the vertical axis (called *columnar* or *radial* connections) and connections in the horizontal axis (called *lateral* connections). Third, cells with similar functions tend to be arrayed in radially aligned groups that span all of the cortical layers and receive inputs that are often segregated into radial or columnar bands. Finally, interneurons within specific cortical layers give rise to extensive local axons that extend horizontally in the cortex, often linking functionally similar groups of cells. The particular circuitry of any cortical region is a variation on this canonical pattern of inputs, outputs, and vertical and horizontal patterns of connectivity.

Figure 26.3   Canonical neocortical circuitry. Green arrows indicate outputs to the major targets of each of the neocortical layers in humans; white arrow indicates thalamic input (primarily to layer IV); dark purple arrows indicate input from other cortical areas; and light purple arrows indicate input from the brainstem modulatory systems to each layer.

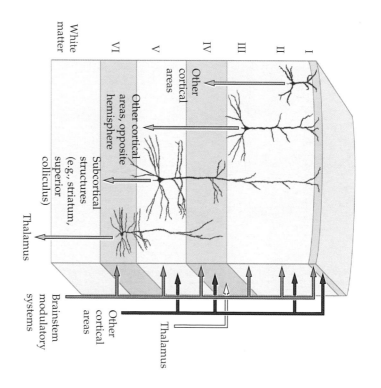

Specific Features of the Association Cortices

These generalizations notwithstanding, the connectivity of the association cortices is appreciably different from primary and secondary sensory and motor cortices, particularly with respect to inputs and outputs. For instance, three thalamic nuclei that are not involved in relaying primary motor or sensory information provide much of the subcortical input to the association cortices: the **pulvinar** projects to the parietal association cortex, the **lateral posterior nuclei** to the temporal association cortex, and the **medial dorsal nuclei** to the frontal association cortex. Unlike the thalamic nuclei that receive peripheral sensory information and project to primary sensory cortices, the input to these three nuclei comes from other regions of the cortex. In consequence, the signals coming into the association cortices via the thalamus reflect sensory and motor information that has *already* been processed in the primary sensory and motor areas of the cerebral cortex, and is being fed back to the association regions. The primary sensory cortices, in contrast, receive thalamic information that is more directly related to peripheral sense organs (see, for example, Chapter 9). Similarly, much of the thalamic input to primary motor cortex is derived from the thalamic nuclei related to the basal ganglia and cerebellum rather than to other cortical regions (see Unit III).

A second major difference in the sources of innervation to the association cortices is their enrichment in *direct* projections from other cortical areas, called **corticocortical connections** (see Figure 26.3). Indeed, these connections form the majority of the input to the association cortices. Ipsilateral corticocortical connections arise from primary and secondary sensory and motor cortices, and from other association cortices within the same hemisphere. Corticocortical connections also arise from both corresponding and noncorresponding cortical regions in the opposite hemisphere via the corpus callosum and anterior commissure, which together are referred to as **inter-hemispheric connections**. In the association cortices of humans and other pri-

BOX A
A More Detailed Look at Cortical Lamination

Much knowledge about the cerebral cortex is based on descriptions of differences in cell number and density throughout the cortical mantle. Nerve cell bodies, because of their high metabolic rate, are rich in basophilic substances (RNA, for instance), and therefore tend to stain darkly with reagents such as cresyl violet acetate. These so-called Nissl stains (named after F. Nissl, who first described this technique when he was a medical student in nineteenth-century Germany) provide a dramatic picture of brain structure at the histological level. The most striking feature revealed in this way is the distinctive lamination of the cortex in humans and other mammals (see figure). In humans, there are three to six cortical layers, which are usually designated by Roman numerals, with letters for laminar subdivisions (layers IVa, b, and c in the visual cortex, for example).

Each of the cortical laminae in the so-called *neocortex* (which covers the bulk of the cerebral hemispheres and is defined by six layers) has characteristic functional and anatomical features (see Figures 26.2 and 26.3). For example, cortical layer IV is typically rich in stellate neurons with locally ramifying axons; in the primary sensory cortices, these neurons receive input from the thalamus, the major sensory relay from the periphery. Layer V, and to a lesser degree layer VI, contain pyramidal neurons whose axons typically leave the cortex. The generally smaller pyramidal neurons in layers II and III (which are not as distinct as their Roman numeral assignments suggest) have primarily corticocortical connections, and layer I contains mainly neuropil. Korbinian Brodmann, who early in the twentieth century devoted his career to an analysis of brain regions distinguished in this way, described about 50 distinct cortical regions, or cytoarchitectonic areas (see Figure 26.2B). These structural features of the cerebral cortex

continue to figure importantly in discussions of the brain, particularly in structural/functional correlation of intensely studied regions such as the primary sensory and motor cortices.

Not all of the cortical mantle is six-layered neocortex. The hippocampus, for example, which lies deep in the temporal lobe and has been implicated in acquisition of memories (see Chapter 31), has only three or four laminae. The hippocampal cortex is regarded as evolutionarily more primitive, and is therefore called *archicortex* to distinguish it from the six-layered neocortex. Another type of cortex, called *paleocortex*, generally has

three layers and is found on the ventral surface of the cerebral hemispheres and along the parahippocampal gyrus in the medial temporal lobe.

The functional significance of different numbers of laminae in neocortex, archicortex, and paleocortex is not known, although it seems likely that the greater number of layers in neocortex reflects more complex information processing than in archi- or paleocortex. The general similarity of neocortical structure across the entire cerebrum clearly suggests that there is a common denominator of cortical operation, but no one has yet deciphered what it is.

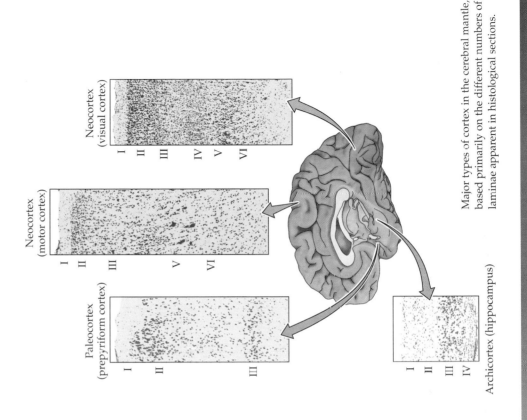

Major types of cortex in the cerebral mantle, based primarily on the different numbers of laminae apparent in histological sections.

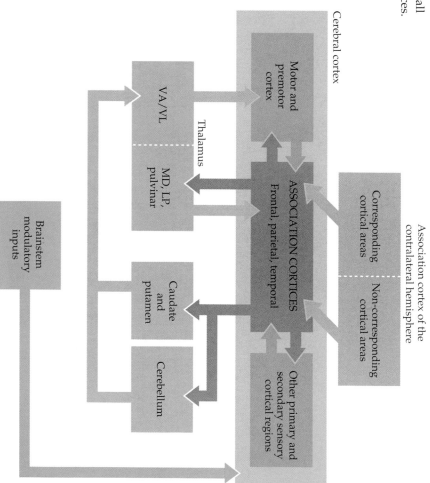

Cerebral cortex

Association cortex of the contralateral hemisphere

Motor and premotor cortex

VA/VL

Thalamus

MD, LP, pulvinar

ASSOCIATION CORTICES
Frontal, parietal, temporal

Caudate and putamen

Cerebellum

Brainstem modulatory inputs

Corresponding cortical areas

Non-corresponding cortical areas

Other primary and secondary sensory cortical regions

Figure 26.4 Summary of the overall connectivity of the association cortices.

mates, corticocortical connections often form segregated bands in which interhemispheric projection bands are interdigitated with bands of ipsilateral corticocortical projections.

The third source of innervation to the association areas is diffuse inputs from the dopaminergic, noradrenergic, and serotonergic nuclei in the brainstem reticular formation, as well as cholinergic nuclei in the brainstem and basal forebrain. These sources project to different cortical layers and, among other functions, determine mental status along a continuum that ranges from deep sleep to high alert (see Chapter 28).

The general wiring plan for the association cortices is summarized in Figure 26.4. Despite the high degree of interconnectivity, extensive inputs and outputs of the association cortices should not be taken to imply that everything is simply connected to everything else in these regions. On the contrary, each association cortex is defined by a distinct, if overlapping, subset of thalamic, corticocortical, and subcortical connections. It is nonetheless difficult to conclude much about the role of these different cortical areas based solely on connectivity (this information is, in any event, quite limited for the human association cortices; most of the evidence comes from anatomical tracing studies in nonhuman primates, supplemented by the limited pathway tracing that can be done in human brain tissue postmortem). As a result, inferences about the function of human association areas continue to depend critically on observations of patients with cortical lesions. Damage to the association cortices in the parietal, temporal, and frontal lobes, respectively, results in specific cognitive deficits that indicate much about the operations and purposes of each of these regions.

Lesions of the Parietal Association Cortex: Deficits of Attention

In 1941, the British neurologist W. R. Brain reported three patients with unilateral parietal lobe lesions in whom the primary problem was varying degrees of perceptual difficulty. Brain described their peculiar deficiency in the following way:

> Though not suffering from a loss of topographical memory or an inability to describe familiar routes, they nevertheless got lost in going from one room to another in their own homes, always making the same error of choosing a right turning instead of a left, or a door on the right instead of one on the left. In each case there was a massive lesion in the right parieto-occipital region, and it is suggested that this ... resulted in an inattention to or neglect of the left half of external space.

> The patient who is thus cut off from the sensations which are necessary for the construction of a body scheme may react to the situation in several different ways. He may remember that the limbs on his left side are still there, or he may periodically forget them until reminded of their presence. He may have an illusion of their absence, i.e. they may 'feel absent' although he knows that they are there; he may believe that they are absent but allow himself to be convinced by evidence to the contrary; or, finally, his belief in their absence may be unamenable to reason and evidence to the contrary and so constitute a delusion.

> W. R. Brain, 1941 (*Brain* 64: pp. 257 and 264)

This description is generally considered the first account of the link between parietal lobe lesions and deficits in attention or perceptual awareness. Based on a large number of patients studied since Brain's pioneering work, these deficits are now referred to as **contralateral neglect syndrome**.

The hallmark of contralateral neglect is an inability to perceive and attend to objects, or even one's own body, in a part of space, despite the fact that visual acuity, somatic sensation, and motor ability remain intact. Affected individuals fail to report, respond to, or even orient to stimuli presented to the side of the body (or visual space) opposite the lesion (Figure 26.5). They may also have difficulty performing complex motor tasks on the neglected side, including dressing themselves, reaching for objects, writing, drawing, and, to a lesser extent, orienting to sounds. The signs of neglect can be as subtle as a temporary lack of contralateral attention that rapidly improves as the patient recovers, or they can be as profound as permanent denial of the existence of the side of the body and extrapersonal space opposite the lesion. Since Brain's original description of contralateral neglect and its relationship to lesions of the parietal lobe, it has been generally accepted that the parietal cortex, particularly the inferior parietal lobe, is the primary cortical region (but not the only region) governing attention (Figure 26.6).

Importantly, contralateral neglect syndrome is specifically associated with damage to the *right* parietal cortex. The unequal distribution of this particular cognitive function between the hemispheres is thought to arise because the right parietal cortex mediates attention to both left and right halves of the body and extrapersonal space, whereas the left hemisphere mediates attention only to the right. Thus, left parietal lesions tend to be compensated by the intact right hemisphere. In contrast, when the right parietal cortex is damaged, there is little or no compensatory capacity in the left hemisphere to mediate attention to the left side of the body or extrapersonal space.

This interpretation has been confirmed by noninvasive imaging of parietal lobe activity during specific attention tasks carried out by normal sub-

Figure 26.5 Characteristic performance on visuospatial tasks by individuals suffering from contralateral neglect syndrome. In (A), the patient was asked to draw a house by copying the figure on the left; on the right is the subject's imitation. In (B), the patient was asked to bisect the line. (C) Sequential self-portraits made by the artist Anton Raederscheidt over the period of his recovery from a right parietal stroke, showing progressive improvement in his attention to the left side of his mirror image. (A,B adapted from Posner and Raichle, 1994; C from Jung, 1974.)

(A) "Draw a house"

Model   Patient's copy

(B) "Bisect the line"

(C)

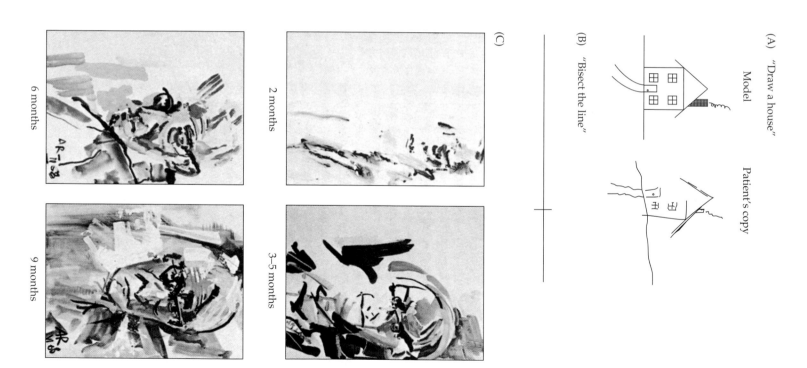

2 months

3–5 months

6 months

9 months

jects. Such studies show that blood flow is increased in *both* the right and left parietal cortices when subjects are asked to perform tasks in the *right* visual field requiring selective attention to distinct aspects of a visual stimulus such as its shape, velocity, or color. However, when a similar challenge is pre-

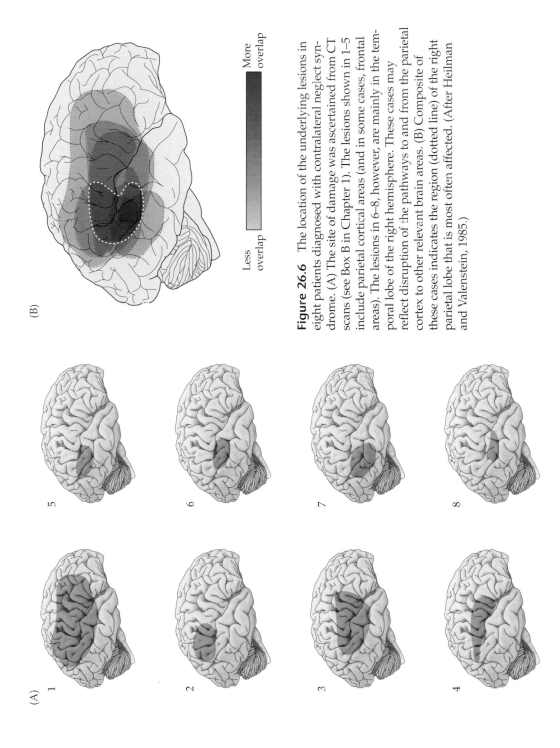

Figure 26.6 The location of the underlying lesions in eight patients diagnosed with contralateral neglect syndrome. (A) The site of damage was ascertained from CT scans (see Box B in Chapter 1). The lesions shown in 1–5 include parietal cortical areas (and in some cases, frontal areas). The lesions in 6–8, however, are mainly in the temporal lobe of the right hemisphere. These cases may reflect disruption of the pathways to and from the parietal cortex to other relevant brain areas. (B) Composite of these cases indicates the region (dotted line) of the right parietal lobe that is most often affected. (After Heilman and Valenstein, 1985.)

sented in the left visual field, only the *right* parietal cortex is activated. Moreover, when normal subjects are asked to maintain a state of general alertness, the right parietal cortex is especially active (Figure 26.7). There is also evidence of increased activity in the right frontal cortex during such tasks (see also Figure 26.6). This latter observation suggests that, at least to some extent, regions outside the parietal lobe also contribute to attentive behavior, and perhaps to some aspects of the pathology of neglect syndromes. Overall, however, metabolic mapping is consistent with the clinical fact that contralateral neglect typically arises from a right parietal lesion, and it endorses the broader idea that there is hemispheric specialization for attention, just as there is for a number of other cognitive functions (see Chapter 27).

Interestingly, patients with contralateral neglect are not simply deficient in their attentiveness to the left visual field, but to the left sides of objects generally. For example, in tasks in the *right* visual field, patients more readily identify novel stimuli on the right side of any object, wherever it happens to be in the *right* visual field. This observation suggests that attentiveness to objects is predicated on their "centroids," and that the relevant frame of reference in neglect is at least in part object-based.

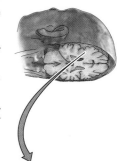

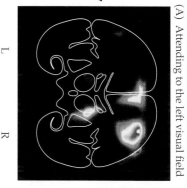

(A) Attending to the left visual field

L R

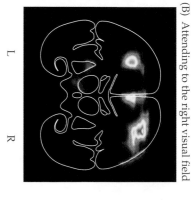

(B) Attending to the right visual field

L R

Figure 26.7 In confirmation of the impressions derived from neurological patients with parietal lobe damage, the right parietal cortex of normal subjects is highly active during tasks requiring attention. (A) A subject has been asked to attend to objects in the left visual field; only the right parietal cortex is active. (B) When attention is shifted from the left visual field to the right, the right parietal cortex remains active, but activity is apparent in the left parietal cortex as well. This arrangement implies that damage to the left parietal lobe does not generate right-sided hemineglect because the right parietal lobe also serves this function. (After Posner and Raichle, 1994.)

Lesions of the Temporal Association Cortex: Deficits of Recognition

Clinical evidence from patients with lesions of the association cortex in the temporal lobe indicate that one of the major functions of this part of the brain is recognition and identification of stimuli that are attended to, particularly complex stimuli. Thus, damage to either temporal lobe can result in difficulty recognizing, identifying, and naming different categories of objects. These disorders, collectively called **agnosias** (from the Greek for "not knowing"), are quite different from the neglect syndromes. As noted, patients with right parietal lobe damage often deny awareness of sensory information in the left visual field (and are less attentive to the left sides of objects generally), despite the fact that the sensory systems are intact (for instance, an individual with contralateral neglect syndrome typically withdraws his left arm in response to a pinprick, even though he may not admit the arm's existence). Patients with agnosia, on the other hand, acknowledge the presence of a stimulus, but are unable to report what it is. These latter disorders have both a lexical aspect (a mismatching of verbal or cognitive symbols with sensory stimuli; see Chapter 27) and a mnemonic aspect (a failure to recall stimuli when confronted with them again; see Chapter 31).

One of the most thoroughly studied agnosias following damage to the temporal association cortex in humans is the inability to recognize and identify faces. This disorder, called **prosopagnosia** (*prosopo-*, from the Greek for "face" or "person"), was recognized by neurologists in the late nineteenth century and remains an area of intense investigation. After damage to the temporal lobes, typically the right temporal lobe, patients are often unable to identify familiar individuals by their facial characteristics, and in some cases cannot recognize a face at all. Nonetheless, such individuals are perfectly aware that some sort of visual stimulus is present and can describe particular aspects or elements of it without difficulty.

An example is the case of L.H., a patient described by the neuropsychologist N. L. Etcoff and colleagues in 1991. (The use of initials is standard neurological practice in published reports is standard practice.) This 40-year-old minister and social worker had sustained a severe head injury as the result of an automobile accident when he was 18. After recovery, L.H. could not recognize familiar faces, report that they were familiar, or answer questions about faces from memory. He was nonetheless able to lead a fairly normal and productive life. He could still identify other common objects, could discriminate subtle shape differences, and could recognize the sex, age, and even the "likability" of faces. Moreover, he could identify particular people by nonfacial cues such as voice, body shape, and gait. The only other category of visual stimuli he had trouble recognizing was animals and their expressions, though these impairments were not as severe as for human faces. Noninvasive brain imaging showed that L.H.'s prosopagnosia was the result of damage to the right temporal lobe.

More recently, imaging studies in normal subjects have confirmed that the inferior temporal lobe mediates face recognition and that nearby regions are responsible for categorically different recognition functions (Figure 26.8). In general, lesions of the right temporal lobe lead to agnosia for faces and objects, whereas lesions of the corresponding regions of the left temporal lobe tend to result in difficulties with language-related material (recall that the primary auditory cortex is on the superior aspect of the temporal lobe; as described in the following chapter, the cortex adjacent to the auditory cortex in the left temporal lobe is specifically concerned with language). The lesions that typically cause recognition deficits are in the inferior temporal lobe in or near the fusiform gyrus (those that cause language-related problems in the left temporal lobe tend to be on the lateral surface of the lobe). Consistent with these conclusions, electrical stimulation in subjects whose temporal lobes are being mapped for neurosurgery (typically removal of an epileptic focus) may have a transient prosopagnosia as a consequence of this abnormal activation of the relevant regions of the right temporal lobe.

Prosopagnosia and related agnosias involving objects are specific instances of a broad range of functional deficits that have as their hallmark the inability to recognize a complex sensory stimulus as familiar, and to identify and name that stimulus as a meaningful entity in the environment. Depending on the laterality, location, and size of the lesion in temporal cortex, agnosias can be as specific as for human faces, or as general as an inability to name most familiar objects.

Figure 26.8 Functional brain imaging of temporal lobes during face recognition. (A) Face stimulus presented to a normal subject at time indicated by arrow. Graph shows activity change in the relevant area of the right temporal lobe. (B) Location of *f*MRI activity in right inferior frontal lobe. (Courtesy of Greg McCarthy.)

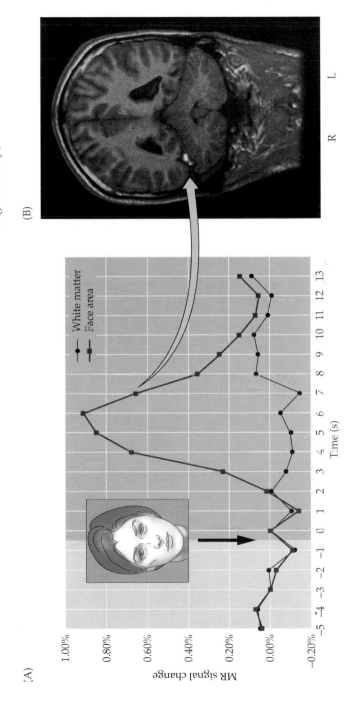

Lesions of the Frontal Association Cortex: Deficits of Planning

The functional deficits that result from damage to the human frontal lobe are diverse and devastating, particularly if both hemispheres are involved. This broad range of clinical effects stems from the fact that the frontal cortex has a wider repertoire of functions than any other neocortical region (consistent with the fact that the frontal lobe comprises a greater number of cytoarchitectonic areas). The particularly devastating nature of the behavioral deficits after frontal lobe damage reflects the role of this part of the brain in maintaining what is normally thought of as an individual's "personality." The frontal cortex integrates complex perceptual information from sensory and motor cortices, as well as from the parietal and temporal association cortices. The result is an appreciation of self in relation to the world that allows behaviors to be planned and executed normally. When this ability is compromised, the afflicted individual often has difficulty carrying out complex behaviors that are appropriate to the circumstances. These deficiencies in the normal ability to match ongoing behavior to present or future demands are, not surprisingly, interpreted as a change in the patient's "character."

The case that first called attention to the consequences of frontal lobe damage was Phineas Gage, a worker on the Rutland and Burlington Railroad in mid-nineteenth-century Vermont. A conventional way of blasting a rock in that era was to tamp powder into a hole with a heavy metal rod. Gage, the popular and respected foreman of the crew, was undertaking this procedure one day in 1848 when his tamping rod sparked the powder, setting off an explosion that drove the rod, which was about a meter long and 4 or 5 centimeters in diameter, through his left orbit (eye socket), destroying much of the frontal part of his brain in the process (see the illustration on the page introducing this Unit). Gage, who never lost consciousness, was promptly taken to a local doctor, who treated his wound. An infection set in, presumably destroying additional frontal lobe tissue, and Gage was an invalid for several months. Eventually, he recovered and was—to outward appearances—well again. Those who knew Gage, however, were profoundly aware that he was not the "same" individual that he had been before. A temperate, hardworking, and altogether decent person had, by virtue of this accident, been turned into an inconsiderate, intemperate lout who could no longer cope with normal social intercourse or the kind of practical planning that had allowed Gage the social and economic success he enjoyed before.

The physician who had looked after Gage until his death in 1863 summarized his impressions of Gage's personality as follows:

[Gage was] fitful, irreverent, indulging at times in the grossest profanity (which was not previously his custom), manifesting but little deference for his fellows, impatient of restraint or advice when it conflicts with his desires, at times pertinaciously obstinate, yet capricious and vacillating, devising many plans of future operations, which are no sooner arranged than they are abandoned in turn for others appearing more feasible. A child in his intellectual capacity and manifestations, he has the animal passions of a strong man. Previous to his injury, although untrained in the schools, he possessed a well-balanced mind, and was looked upon by those who knew him as a shrewd, smart businessman, very energetic and persistent in executing all his plans of operation. In this regard his mind was radically changed, so decidedly that his friends and acquaintances said he was 'no longer Gage'.

J. M. Harlow, 1868 (*Publications of the Massachusetts Medical Society* 2: pp. 339–340)

Another classical case of frontal lobe deficits was a patient followed for many years by the neurologist R. M. Brickner during the 1920s and '30s. Joe A., as Brickner referred to his patient, was a stockbroker who underwent bilateral frontal lobe resection because of a large tumor at age 39. After the operation, Joe A. had no obvious sensory or motor deficits; he could speak and understand verbal communication and was aware of people, objects, and temporal order in his environment. He acknowledged his illness and retained a high degree of intellectual power, as judged from an ongoing ability to play an expert game of checkers. Nonetheless, Joe A.'s personality had undergone a dramatic change. A restrained, modest man, he became boastful of professional, physical, and sexual prowess, showed little restraint in conversation, and was unable to match the appropriateness of what he said to his audience. Like Gage, his ability to plan for the future was largely lost, as was much of his earlier initiative and creativity. Even though he retained the ability to learn complex procedures, he was unable to return to work and had to rely on his family for support and care.

Sadly, these effects of damage to the frontal lobes have also been documented by the many thousands of frontal lobotomies (or "leukotomies") performed in the 1930s and '40s as a means of treating mental illness. The rise and fall of this "psychosurgery" provides a compelling example of the frailty of human judgment in medical practice, and of the conflicting approaches of neurologists, neurosurgeons, and psychiatrists in that era to the treatment of mental disease (Box B).

"Attention Neurons" in the Monkey Parietal Cortex

These clinical and pathological observations clearly indicate distinct cognitive functions for the parietal, temporal, and frontal lobes. They do not, however, provide much insight into how the nervous system represents this information in nerve cells and their interconnections. The apparent functions of the association cortices implied by these clinical observations stimulated a number of informative electrophysiological studies in nonhuman primates, particularly macaque (usually rhesus) monkeys.

As in humans, a wide range of cognitive abilities in monkeys are mediated by the association cortices of the parietal, temporal, and frontal lobes (Figure 26.9A). Moreover, these functions can be tested using behavioral paradigms that assess attention, identification, and planning capabilities, the broad functions respectively assigned to the parietal, temporal, and frontal association cortices in humans. Needless to say, it is far more practical to study neuronal behavior in relation to cognitive functions in experimental animals. Using implanted electrodes, recordings can be made from single neurons in the brains of awake, behaving monkeys to assess the activity of individual cells in the association cortices as various cognitive tasks are performed (Figure 26.9B).

An example is neurons apparently related to the attentive functions of the parietal cortex. These particular studies of cellular electrophysiology and behavior take advantage of the fact that eye movements provide an excellent indicator of attentive behavior in primates. Thus, fixation of the eyes on a target the monkey has been trained to attend to can be used to identify attention-sensitive neurons in this part of the cortex (Figure 26.10). As might be expected from the clinical evidence in humans, some neurons in specific regions of the parietal cortex of the rhesus monkey increase their rate of firing when the animal fixates on a target of interest, and they maintain their activity for the duration of the eye fixation. Responses of this sort are also

(A)

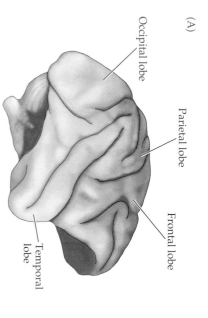

Occipital lobe

Parietal lobe

Frontal lobe

Temporal lobe

(B)

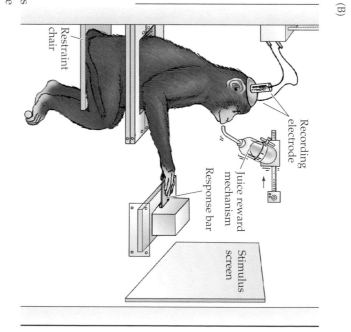

Restraint chair

Recording electrode

Juice reward mechanism

Response bar

Stimulus screen

Figure 26.9 Recording from single neurons in the brain of an awake, behaving rhesus monkey. (A) Lateral view of the rhesus monkey brain showing the parietal (red), temporal (green), and frontal (blue) cortices. The occipital cortex is shaded purple. (B) The animal is seated in a chair and gently restrained. Several months before data collection begins, a recording well is placed through the skull using a sterile surgical technique. For electrophysiological recording experiments, a tungsten microelectrode is inserted through the dura and arachnoid, and into the cortex. The screen and the response bar in front of the monkey are for behavioral testing. In this way, individual neurons can be monitored while the monkey performs specific cognitive tasks.

observed when the monkey simply fixates on a test stimulus that has been associated with a food reward. When attention to the stimulus flags, eye movements resume, and the firing of the neurons falls to baseline levels. Thus, the monkey parietal cortex contains neurons that respond specifically when the animal attends to a behaviorally meaningful stimulus.

(A)

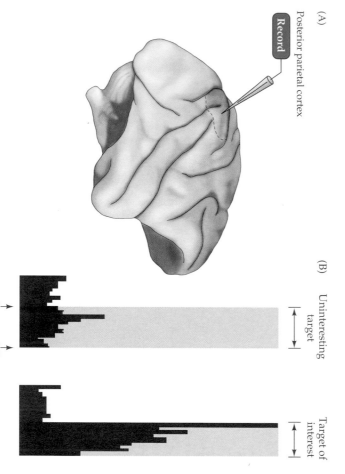

Record

Posterior parietal cortex

(B)

Uninteresting target

On — Off

Target of interest

Figure 26.10 Selective activation of neurons in the parietal cortex of a rhesus monkey during the fixation of a significant visual target (in this case, a spot of light associated with a food reward). (A) Region of recording. (B) Although the baseline level of activity of the neuron being studied here changes little in response to an uninteresting target (left), it increases its firing rate tenfold during eye fixation on a target of interest (right). The histograms indicate action potential frequency per unit time. (After Lynch et al., 1977.)

BOX B
Psychosurgery

The consequences of frontal lobe destruction have been all too well documented by a disturbing yet fascinating episode in twentieth-century medical practice. During the period from 1935 through the 1940s, neurosurgical destruction of the frontal lobe (frontal lobotomy or leukotomy) was a popular treatment for certain mental disorders. More than 20,000 of these procedures were performed, mostly in the United States.

Enthusiasm for this approach to mental disease grew from the work of Egas Moniz, a respected Portuguese neurologist, who, among other accomplishments, did pioneering work on cerebral angiography before becoming the leading advocate of psychosurgery. Moniz recognized that the frontal lobes were important in personality structure and behavior, and concluded that interfering with frontal lobe function might alter the course of mental diseases such as schizophrenia and other chronic psychiatric disorders. He also recognized that destroying the frontal lobe would be relatively easy to do and, with the help of Almeida Lima, a

neurosurgical colleague, introduced a simple surgical procedure for indiscriminately destroying most of the connections between the frontal lobe and the rest of the brain (see figure).

In the United States, the neurologist Walter Freeman at George Washington University School of Medicine, in collaboration with neurosurgeon James Watts, became an equally strong advocate of this approach. Freeman devoted his life to treating a wide variety of mentally disturbed patients in this way. He popularized a form of the procedure that could be carried out under local anesthesia and traveled widely across the United States to demonstrate this technique and encourage its use.

Although it is easy in retrospect to be critical of this zealotry in the absence of either evidence or sound theory, it is important to remember that effective psychotropic drugs were not then available, and patients suffering from many of the disorders for which leucotomies were done were confined under custodial conditions that were at best dismal, and at

worst brutal. Rendering a patient relatively tractable, albeit permanently altered in personality, no doubt seemed the most humane of the difficult choices that faced psychiatrists and others dealing with such patients in that period.

With the advent of increasingly effective psychotropic drugs in the late 1940s and the early 1950s, frontal lobotomy as a psychotherapeutic strategy rapidly disappeared, but not before Moniz was awarded the Nobel Prize for Physiology or Medicine in 1949. The history of this instructive episode in modern medicine has been compellingly told by Eliot Valenstein, and his book on the rise and fall of psychosurgery should be read by anyone contemplating a career in neurology, neurosurgery, or psychiatry.

References

BRICKNER, R. M. (1932) An interpretation of function based on the study of a case of bilateral frontal lobectomy. Proceedings of the Association for Research in Nervous and Mental Disorders 13: 259–351.

BRICKNER, R. M. (1952) Brain of patient A after bilateral frontal lobectomy: Status of frontal lobe problem. Arch. Neurol. Psychiatry 68: 293–313.

FREEMAN, W. AND J. WATTS (1942) *Psychosurgery: Intelligence, Emotion and Social Behavior Following Prefrontal Lobotomy for Mental Disorders.* Springfield, IL: Charles C. Thomas.

MONIZ, E. (1937) Prefrontal leukotomy in the treatment of mental disorders. Am. J. Psychiatry 93: 1379–1385

VALENSTEIN, E. S. (1986) *Great and Desperate Cures: The Rise and Decline of Psychosurgery and Other Radical Treatments for Mental Illness.* New York: Basic Books.

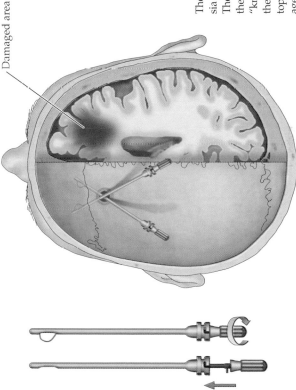

Damaged area

The surgical technique for frontal leukotomy under local anesthesia described and advocated by Egas Moniz and Almeida Lima. The "leukotome" was inserted into the brain at approximately the angles shown. When the leukotome was in place, a wire "knife" was extended and the handle rotated. The right side of the figure depicts a horizontal slice of the brain (parallel to the top of the skull) with Moniz's estimate of the extent of the damage done by the procedure. (After Moniz, 1937.)

"Recognition Neurons" in the Monkey Temporal Cortex

In keeping with human deficits of recognition following temporal lobe lesions, neurons with responses that correlate with the recognition of specific stimuli are present in the temporal cortex of rhesus monkeys (Figure 26.11). The behavior of these neurons in the vicinity of the inferior temporal gyrus is generally consistent with one of the major functions ascribed to the human temporal cortex—namely, the recognition and identification of complex stimuli. For example, some neurons in the inferior temporal gyrus of the rhesus monkey cortex respond specifically to the presentation of a monkey face.

(A)

Inferior temporal cortex

Record

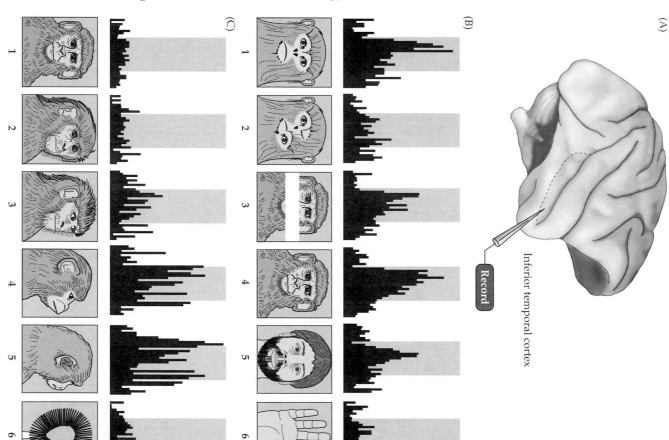

(B)

1

2

3

4

5

6

(C)

1

2

3

4

5

6

Figure 26.11 Selective activation of face cells in the inferior temporal cortex of a rhesus monkey. (A) Region of recording. (B) The neuron being recorded from here responds selectively to faces seen from the front. Scrambled parts of faces (stimulus 2) or faces with parts omitted (stimulus 3) do not elicit a maximal response. The cell responds best to different monkey faces, as long as they are complete and viewed from the front (stimulus 4); the cell also responds to a bearded human face (stimulus 5), although not quite as robustly. An irrelevant stimulus (in this case a hand; stimulus 6) does not elicit a response. (C) In this case, the neuron being recorded from responds to profiles of faces. A face viewed from the front (stimulus 1), 30° (stimulus 2), or 60° (stimulus 3) is not as effective as a true profile (stimulus 4). The cell will respond to profiles of different monkeys (stimulus 5), but is unresponsive to an irrelevant stimulus (in this case a brush; stimulus 6). (After Desimone et al., 1984.)

These cells are generally selective; thus, some respond only to the frontal view of a face and others only to profiles (Figure 26.11B,C). Furthermore, the cells are not easily deceived. When parts of faces or generally similar objects are presented, such cells typically fail to respond.

In principle, it is unlikely that such "face cells" are tuned to specific faces or objects, and no cells have so far been found that are selective for a particular face. However, it is not hard to imagine that populations of neurons differentially responsive to various features of faces or other objects could act in concert to enable the recognition of such complex sensory stimuli.

"Planning Neurons" in the Monkey Frontal Cortex

In further confirmation of the human clinical evidence about the function of the frontal association cortices, neurons that appear to be specifically involved in planning have been identified in the frontal cortices of rhesus monkeys.

The behavioral test used to study cells in the monkey frontal cortex is called the **delayed response task** (Figure 26.12A). Variants of this task are used to assess frontal lobe function in a variety of situations, including the clinical evaluation of frontal lobe function in humans (Box C). In the delayed response task, the monkey watches an experimenter place a food morsel in one of two wells; both wells are then covered. Subsequently, a screen is lowered for an interval of a few seconds to several minutes (the delay). When the screen is raised, the monkey gets only one chance to uncover the well containing food and receive the reward. Thus, the animal must decide that he wants the food, remember where it is placed, recall that the cover must be removed to obtain the food, and keep all this information available during the delay so that it can be used to get the reward. The monkey's ability to carry out this task is diminished or abolished if the area anterior to the motor region of the frontal cortex—called the prefrontal cortex—is destroyed bilaterally (in accord with clinical findings in human patients).

Some neurons in the prefrontal cortex, particularly those in and around the principal sulcus (Figure 26.12B), generate a response that is correlated with the delayed response task; that is, they are maximally active during the period of the delay, as if their firing represented the information maintained from the presentation part of the trial (i.e., the cognitive information needed to guide behavior when the screen is raised; Figure 26.12C,D). Such neurons return to a low level of activity during the actual motor phase of the behavior, suggesting that they represent working memory and planning rather than the actual movement itself. Delay-specific neurons in the prefrontal cortex are also active in monkeys that have been trained to perform a variant of the delayed response task in which the response is to internally generated memories. Evidently, these neurons are equally capable of using stored information to guide behavior. Thus, if a monkey is trained to associate eye movements to a particular target with a delayed reward, the delay-associated neurons in the prefrontal cortex will fire during the delay, even if the monkey moves his eyes to the appropriate region of the visual field in the absence of the target.

The existence of delay-specific neurons in the frontal cortex of rhesus monkeys, as well as attention-specific cells in the parietal cortex and recognition-specific cells in the temporal cortex, generally supports the functions of these cortical areas inferred from clinical evidence in humans. Nonetheless, functional localization, whether inferred by examining human patients or by recording single neurons in monkeys, is an imprecise business. The observations summarized here are only a rudimentary guide to thinking about how

(A)

CUE

1 Food is placed in randomly selected well visible to monkey

Food morsel

Empty dish

DELAY

2 Screen is lowered and food covered for a standard time

RESPONSE

3 Screen is raised and monkey uncovers well containing food

(B)

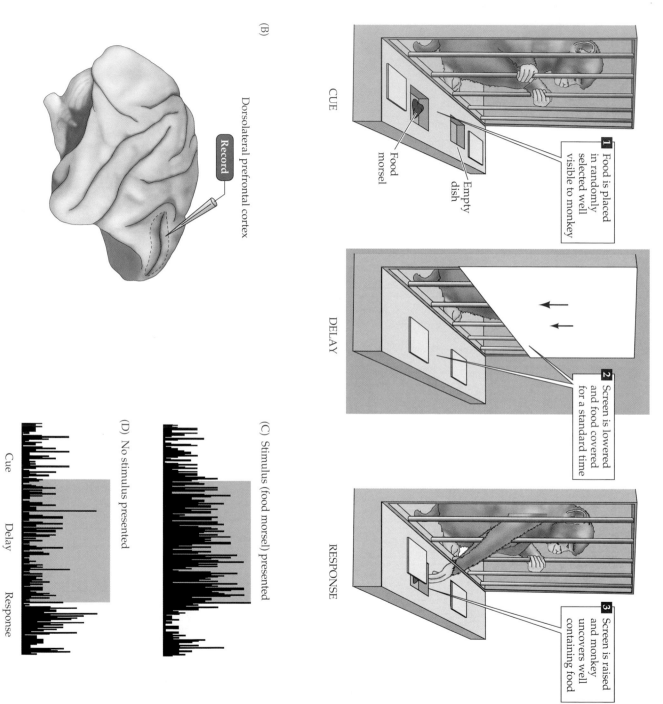

Record

Dorsolateral prefrontal cortex

(C) Stimulus (food morsel) presented

(D) No stimulus presented

Cue   Delay   Response

Figure 26.12 Activation of neurons near the principal sulcus of the frontal lobe during delayed response task. (A) Illustration of task. The experimenter randomly varies the well in which the food is placed. The monkey watches the morsel being covered, and then the screen is lowered for a standard time. When the screen is raised, the monkey is allowed to uncover only one well to retrieve the food. Normal monkeys learn this task quickly, usually performing at a level of 90% correct after less than 500 training trials. (B) Region of recording. (C) Activity of a delay-specific neuron in the prefrontal cortex of a rhesus monkey recorded during the delayed response task shown in (A). The histograms show the number of action potentials during the cue, delay, and response periods. The neuron begins firing when the screen is lowered and remains active throughout the delay period. (D) When no food is presented, but the screen is still lowered and raised, the same neuron is relatively silent. (After Goldman-Rakic, 1987.)

BOX C
Neuropsychological Testing

Long before PET scanning and functional MRI were used to evaluate normal and abnormal cognitive function, several "low-tech" methods proved to be reliable means of assessing these abilities in human subjects. From the late 1940s onward, psychologists and neurologists developed a battery of behavioral tests—generally called neuropsychological tests—to evaluate the integrity of cognitive function and to help localize lesions.

One of the most frequently used measures is the Wisconsin Card Sorting Task. In this test, the examiner places four cards with symbols that differ in number, shape, or color before the subject, who is given a set of response cards with similar symbols on them. The subject is then asked to place an appropriate response card in front of the stimulus card based on a sorting rule established, but not stated, by the examiner (i.e., sort by color, number, or shape; see figure). The examiner then indicates whether the response is "right" or "wrong." After 10 consecutive correct responses, the examiner changes the sorting rule simply by saying "wrong." The subject must then ascertain the new sorting rule and perform 10 cor-

rect trials. The sorting rule is then changed again, until six cycles have been completed.

In 1963, the neuropsychologist Brenda Milner at the Montreal Neurological Institute showed that patients with frontal lobe lesions show consistently poor performance in the Wisconsin Card Sorting Task. By comparing patients with known brain lesions as a result of surgery for epilepsy or tumor, Milner was able to demonstrate that this impairment is fairly specific for frontal lobe damage. Particularly striking is the inability of frontal lobe patients to use previous information to guide subsequent behavior. A widely accepted explanation for the sensitivity of the Wisconsin Card Sorting Task to frontal lobe deficits is the "planning" aspect of this test. To respond correctly, the subject must retain information about the previous trial, which is then used to guide behavior on future trials. Processing this sort of information is characteristic of frontal lobe function.

A variety of other neuropsychological tests have been devised to evaluate the functional integrity of other cognitive functions. These include tasks in which a

patient is asked to identify familiar faces in a series of pictures, and others in which "distractors" interfere with the patient's ability to attend to salient stimulus features. An example of the latter is the Stroop Interference Test, in which patients are asked to read the names of colors presented in color-conflicting print (for example, the word "green" printed in red ink). This sort of challenge evaluates both attention and identification abilities.

The simplicity, economy, and accumulated experience with such tests continue to make them a valuable means of evaluating cognitive functions.

References

BERG, E. A. (1948) A simple objective technique for measuring flexibility in thinking. J. Gen. Psychol. 39: 15–22.

LEZAK, M. D. (1995) *Neuropsychological Assessment*, 3rd Ed. New York: Oxford University Press.

MILNER, B. (1963) Effects of different brain lesions on card sorting. Arch. Neurol. 9: 90–100.

MILNER, B. AND M. PETRIDES (1984) Behavioural effects of frontal-lobe lesions in man. Trends Neurosci. 4: 403–407.

Sort by number

Sort by shape

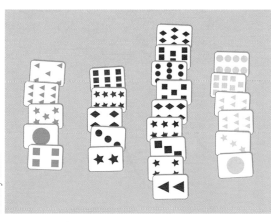

Sort by color

BOX D
Brain Size and Intelligence

The fact that so much of the brain is occupied by the association cortices raises a fundamental question: does more of it provide individuals with greater cognitive ability? Humans and other animals obviously vary in their talents and predispositions for a wide range of cognitive behaviors. Does a particular talent imply a greater amount of neural space in the service of that function?

Historically, the most popular approach to the issue of brain size and behavior in humans has been to relate the overall size of the brain to a broad index of performance, conventionally measured in humans by "intelligence" tests. This way of studying the relationship between brain and behavior has caused considerable trouble. In general terms, the idea that the size of brains from different species reflects intelligence represents a simple and apparently valid idea (see figure). The ratio of brain weight to body weight for fish is 1:5000; for reptiles it is about 1:1500; for birds, 1:220; for most mammals, 1:180, and for humans, 1:50. If intelligence is defined as the full spectrum of cognitive performance, surely no one would dispute that a human is more intelligent than a mouse, or that this difference is explained in part by the 3000-fold difference in the size of the brains of these species. Does it follow, however, that relatively small differences in the size of the brain among related species, strains, genders, or individuals—which often persist even after correcting for differences in body size—are also a valid measure of cognitive abilities? Certainly no issue in neuroscience has provoked a more heated debate than the notion that alleged differences in brain size among races—or the demonstrable differences in brain size between men and women—reflect differences in performance. The passion attending this controversy has been generated not only by the scientific

issues involved, but also by the spectre of racism or misogyny.

Nineteenth-century enthusiasm for brain size as a simple measure of human performance was championed by some remarkably astute scientists (including Darwin's cousin Francis Galton and the French neurologist Paul Broca), as well as others whose motives and methods are now suspect (see Gould, 1978, 1981 for a fascinating and authoritative commentary). Broca, one of the great neurologists of his day and a gifted observer, not only thought that brain size reflected intelligence, but was of the opinion (as was just about every other nineteenth-century male scientist) that white European males had larger and better-developed brains than anyone else. Based on what was known about the human brain in the late nineteenth century, it was perhaps reasonable for Broca to consider it an organ, like the liver or the lung, having a largely homogeneous function. Ironically, it was Broca himself who laid the groundwork for the modern view that the brain is a heterogeneous collection of highly interconnected but functionally discrete systems (see Chapter 27). Nonetheless, the simplistic nineteenth-century approach to brain size and intelligence has persisted in some quarters well beyond its time.

There are at least two reasons why measures such as brain weight or cranial capacity are not easily interpretable indices of intelligence, even though small observed differences may be statistically valid. First is the obvious difficulty of defining and accurately measuring intelligence among animals, particularly among humans with different educational and cultural backgrounds. Second is the functional diversity and connectional complexity of the brain. Imagine assessing the relationship between body size and athletic ability, which might be considered the somatic analogue of intelligence. Body weight, or any other global

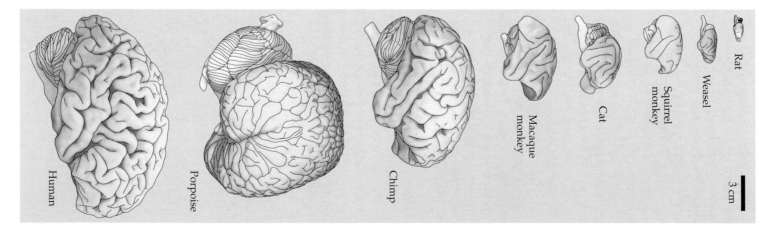

Rat
Weasel
Squirrel monkey
Cat
Macaque monkey
Chimp
Human
Porpoise
3 cm

measure of somatic phenotype, would be a woefully inadequate index of athletic ability. Although the evidence would presumably indicate that bigger is better in the context of sumo wrestling or basketball, more subtle somatic features would no doubt be correlated with extraordinary ability in Ping Pong, gymnastics, or figure skating. The diversity of somatic function vis-à-vis athletic ability confounds the interpretation of any simple measure such as body size.

The implications of this analogy for the brain are straightforward. Any program that seeks to relate brain weight, cranial capacity, or some other measure of overall brain size to individual performance ignores the reality of the brain's

functional diversity. Thus, quite apart from the political or ethical probity of attempts to measure "intelligence" by brain size, by the yardstick of modern neuroscience (or simple common sense), this approach will inevitably generate more heat than light. A more rational approach to the issue, which has become feasible in the last few years, is to relate the size of measurable regions of known function (the primary visual cortex, for example) to the corresponding functions (visual performance), as well as to cellular features such as synaptic density and dendritic arborization. These correlations have greater promise for functional validity, and less pretense of judgment and discrimination.

References

BROCA, P. (1861) Sur le volume et al forme du cerveau suivant les individus et suivant les races. Bull. Soc. Anthrop. 2: 139–207, 301–321.

GALTON, F. (1883) *Inquiries into Human Faculty and Its Development.* London: Macmillan.

GOULD, S. J. (1978) Morton's ranking of races by cranial capacity. Science 200: 503–509.

GOULD, S. J. (1981) *The Mismeasure of Man.* New York: W. W. Norton and Company.

GROSS, B. R. (1990) The case of Phillipe Rushton. Acad. Quest. 3: 35–46.

SPITZKA, E. A. (1907) A study of the brains of six eminent scientists and scholars belonging to the American Anthropometric Society, together with a description of the skull of Professor E. D. Cope. Trans. Amer. Phil. Soc. 21: 175–308.

WALLER, A. D. (1891) *Human Physiology.* London: Longmans, Green.

Summary

The majority of the human cerebral cortex is devoted to tasks that transcend encoding primary sensations or commanding motor actions. Collectively, the association cortices mediate these cognitive functions of the brain—broadly defined as the ability to attend to, identify, and act meaningfully in response to complex external or internal stimuli. Descriptions of patients with cortical lesions, functional brain imaging of normal subjects, and behavioral and electrophysiological studies of nonhuman primates have established the general purpose of the major association areas. Thus, parietal association cortex is involved in attention and awareness of the body and the stimuli that act on it; temporal association cortex is involved in the recognition and identification of highly processed sensory information; and frontal association cortex is involved in guiding complex behavior by planning responses to ongoing stimulation (or remembered information), matching such behaviors to the demands of a particular situation. More than any other brain regions, the association areas support the mental processes that make us human.

complex cognitive information is represented and processed in the brain, and how the relevant brain areas and their constituent neurons contribute to such important but still ill-defined qualities as personality, intelligence (Box D), or other cognitive functions that define what it means to be a human being.

ADDITIONAL READING

Reviews

BEHRMANN, M. (1999) Spatial frames of reference and hemispatial neglect. In *The Cognitive Neurosciences*, 2nd Ed. M. Gazzaniga (ed.). Cambridge, MA: MIT Press, pp. 651–666.

DAMASIO, A. R. (1985) The frontal lobes. In *Clinical Neuropsychology*, 2nd Ed. K. H. Heilman and E. Valenstein (eds.). New York: Oxford University Press, pp. 409–460.

DAMASIO, A. R., H. DAMASIO AND G. W. VAN HOESEN (1982) Prosopagnosia: Anatomic basis and behavioral mechanisms. Neurology 32: 331–341.

DESIMONE, R. (1991) Face-selective cells in the temporal cortex of monkeys. J. Cog. Neurosci. 3: 1–8.

FILLEY, C. M. (1995) *Neurobehavioral Anatomy*, Ch. 8, Right hemisphere syndromes. Boulder: University of Colorado Press, pp. 113–130.

GOLDMAN-RAKIC, P. S. (1987) Circuitry of the prefrontal cortex and the regulation of behavior by representational memory. In *Handbook of Physiology*, Section 1, *The Nervous System*. Vol. 5, *Higher Functions of the Brain*, Part I. F. Plum (ed.), Bethesda: American Physiological Society, pp. 373–417.

HALLIGAN, P. W. AND J. C. MARSHALL (1994) Toward a principled explanation of unilateral neglect. Cog. Neuropsych. 11(2): 167–206.

LÀDAVAS, E., A. PETRONIO AND C. UMILTA (1990) The deployment of visual attention in the intact field of hemineglect patients. Cortex 26: 307–317.

MACRAE, D. AND E. TROLLE (1956) The defect of function is visual agnosia. Brain 77: 94–110.

POSNER, M. I. AND S. E. PETERSEN (1990) The attention system of the human brain. Annu. Rev. Neurosci. 13: 25–42.

VALLAR, G. (1998). Spatial hemineglect in humans. Tr. Cog. Sci.: 2(3), 87–96.

Important Original Papers

BRAIN, W. R. (1941) Visual disorientation with special reference to lesions of the right cerebral hemisphere. Brain 64: 224–272.

DESIMONE, R., T. D. ALBRIGHT, C. G. GROSS AND C. BRUCE (1984) Stimulus-selective properties of inferior temporal neurons in the macaque. J. Neurosci. 4: 2051–2062.

ETCOFF, N. L., R. FREEMAN AND K. R. CAVE (1991) Can we lose memories of faces? Content specificity and awareness in a prosopagnosic. J. Cog. Neurosci. 3: 25–41.

FUNAHASHI, S., M. V. CHAFEE AND P. S. GOLDMAN-RAKIC (1993) Prefrontal neuronal activity in rhesus monkeys performing a delayed anti-saccade task. Nature 365: 753–756.

FUSTER, J. M. (1973) Unit activity in prefrontal cortex during delayed-response performance: Neuronal correlates of transient memory. J. Neurophysiol. 36: 61–78.

GESCHWIND, N. (1965) Disconnexion syndromes in animals and man. Parts I and II. Brain 88: 237–294.

HARLOW, J. M. (1868) Recovery from the passage of an iron bar through the head. Publications of the Massachusetts Medical Society 2: 327–347.

MOUNTCASTLE, V. B., J. C. LYNCH, A. GEORGOPOULOS, H. SAKATA AND C. ACUNA (1975) Posterior parietal association cortex of the monkey: Command function from operations within extrapersonal space. J. Neurophys. 38: 871–908.

Books

BRICKNER, R.M. (1936) *The Intellectual Functions of the Frontal Lobes*. New York: Macmillan.

DAMASIO, A. R. (1994) *Descartes' Error: Emotion, Reason and the Human Brain*. New York: Grosset/Putnam.

DEFELIPE, J. AND E. G. JONES (1988) *Cajal on the Cerebral Cortex: An Annotated Translation of the Complete Writings*. New York: Oxford University Press.

GAREY, L. J. (1994) *Brodmann's "Localisation in the Cerebral Cortex."* London: Smith-Gordon. (Translation of K. Brodmann's 1909 book, *Leipzig: Verlag von Johann Ambrosius Barth*.)

HEILMAN, H. AND E. VALENSTEIN (1985) *Clinical Neuropsychology*, 2nd Ed. New York: Oxford University Press, Chapters 8, 10, and 12.

KLAWANS, H. L. (1988) *Toscanini's Fumble, and Other Tales of Clinical Neurology*. Chicago: Contemporary Books.

KLAWANS, H. L. (1991) *Newton's Madness*. New York: Harper Perennial Library.

POSNER, M. I. AND M. E. RAICHLE (1994) *Images of Mind*. New York: Scientific American Library.

SACKS, O. (1987) *The Man Who Mistook His Wife for a Hat*. New York: Harper Perennial Library.

SACKS, O. (1995) *An Anthropologist on Mars*. New York: Alfred A. Knopf.

Chapter 27

Language and Lateralization

Overview

One of the most remarkable features of complex cortical functions in humans is the ability to associate arbitrary symbols with specific meanings to express thoughts and emotions to ourselves and others by means of language. Indeed, the achievements of human culture rest largely upon this skill, and a person who for one reason or another fails to develop a facility for language as a child is severely incapacitated. Studies of patients with damage to specific cortical regions indicate that linguistic abilities of the human brain depend on the integrity of several specialized areas of the association cortices in the temporal and frontal lobes. In the vast majority of people, these primary language functions are located in the left hemisphere: the linkages between speech sounds and their meanings are mainly represented in the left temporal cortex, and the circuitry for the motor commands that organize the production of meaningful speech is mainly found in the left frontal cortex. Despite this left-sided predominance, the emotional (affective) content of language is governed largely by the right hemisphere. Studies of congenitally deaf individuals have shown further that the cortical areas devoted to sign language are the same as those that organize spoken and heard communication. The regions of the brain devoted to language are, therefore, specialized for symbolic representation and communication, rather than heard and spoken language as such. Understanding functional localization and hemispheric lateralization of language is especially important in clinical practice. The loss of language is such a devastating blow that neurologists and neurosurgeons make every effort to identify and preserve those cortical areas involved in its comprehension and production.

Language Is Both Localized and Lateralized

It has been known for more than a century that two more or less distinct regions in the frontal and temporal association cortices of the left cerebral hemisphere are especially important for normal human language. That language abilities are localized is expected; ample evidence of the localization of other cognitive functions was reviewed in Chapter 26. The unequal representation of language functions in the two cerebral hemispheres is, however, especially clear in this domain. Although functional lateralization has already been introduced in the unequal functions of the parietal lobes in attention and of the temporal lobes in recognizing different categories of objects, it is in studies of language that this controversial concept was first proven. Such functional asymmetry is referred to as **hemispheric lateralization** and, because language is so important, has given rise to the misleading idea that one hemisphere in humans is actually "dominant" over the other—namely, the

587

BOX A
Do Apes Have Language?

Over the centuries, theologians, natural philosophers, and a good many modern neuroscientists have argued that language is uniquely human, this extraordinary behavior setting us qualitatively apart from our fellow animals. The gradual accumulation of evidence during the last 75 years demonstrating highly sophisticated systems of communication in species as diverse as bees, birds, and whales has made this point of view increasingly untenable, at least in a broad sense (see Box B in Chapter 24). Until recently, however, human language *has* appeared unique in its semantic aspect, i.e., in the human ability to associate specific meanings with arbitrary symbols, ad infinitum. In the dance of the honeybee described so beautifully by Karl von Frisch, for example, each symbolic movement made by a foraging bee that returns to the hive encodes only a single meaning, whose expression and appreciation has been hardwired into the nervous system of the actor and the respondents.

A series of controversial studies in great apes, however, have indicated that the rudiments of the human symbolic communication are indeed evident in the behavior of our closest relatives. Although early efforts were sometimes patently misguided (some initial attempts to teach chimpanzees to speak were without merit simply because they lack the necessary vocal apparatus), modern work on this issue has shown that if chimpanzees are given the means to communicate symbolically, they demonstrate some surprising talents. While techniques have varied, most psychologists who study chimps have used some form of manipulable symbols that can be arranged to express ideas in an interpretable manner. For example, chimps can be trained to manipulate tiles or other symbols such as the gestures of sign language to represent words and syntactical constructs, allowing them to communicate simple demands, questions, and even spontaneous expressions. The most remarkable results have come from increasingly sophisticated work with chimps using symbolic keyboards (figure A). With appropriate training, chimps can choose from as many as 400 different symbols to construct expressions, allowing the researchers to have something resembling a rudimentary conversation with their charges. The more accomplished of these animals are alleged to have "vocabularies" of several thousand words or phrases, equivalent to the speech abilities of a child of about 3 or 4 years of age.

Given the challenge this work presents to some long-held beliefs about the superiority and evolutionary independence of human language, it is not surprising that these claims continue to stir up a great deal of debate and are not universally accepted. Nonetheless, the issues raised certainly deserve careful consideration by anyone interested in human language abilities, and how our remarkable symbolic skills may have evolved from the capabilities of our ancestors. The pressure for the evolution of some form of symbolic communication in great apes seems clear enough. Ethologists studying chimpanzees in the wild have described extensive social communication based on gestures, the manipulation of objects, and facial expressions. This intricate social intercourse is likely to be the antecedent of human language; one need only think of the importance of gestures and facial expressions as ancillary aspects of our own speech to appreciate this point. (The sign language studies described later in the text are also pertinent here.) Whether the regions of the temporal, parietal, and frontal cortices that support human language also serve these symbolic functions in the brains of great apes (figure B) is an

(A)

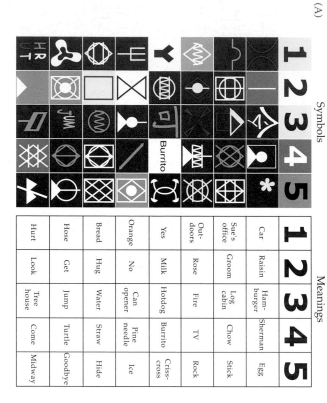

Symbols

1	2	3	4	5

Meanings

1	2	3	4	5
Car	Raisin	Hamburger	Sherman	Egg
Sue's office	Groom	Log cabin	Chow	Stick
Yes	Milk	Hotdog	Burrito	Criss-cross
Out-doors	Rose	Fire	TV	Rock
Orange	No	Can opener	Pine needle	Ice
Bread	Hug	Water	Straw	Hide
Hose	Get	Jump	Turtle	Goodbye
Hurt	Look	Tree house	Come	Midway

Section of keyboard showing lexical symbols used to study symbolic communication in great apes (From Savage-Rumbaugh et al., 1998.)

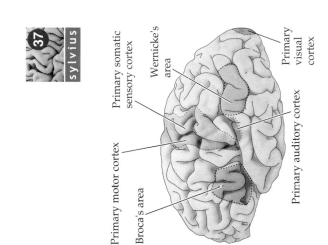

Figure 27.1 Diagram of the major brain areas involved in the comprehension and production of language. The primary sensory, auditory, visual, and motor cortices are indicated to show the relation of Broca's and Wernicke's language areas to less specialized areas that are nonetheless involved in the comprehension and production of speech.

Primary motor cortex
Broca's area
Primary somatic sensory cortex
Wernicke's area
Primary visual cortex
Primary auditory cortex

sylvius 37

hemisphere in which the major capacity for language resides. The true significance of lateralization, however, lies in the efficient subdivision of complex functions between the hemispheres, rather than in any superiority of one hemisphere over the other. Indeed, a safe presumption is that every region of the brain is doing *something important!*

The representation of language in the association cortices is clearly distinct from the circuitry concerned with the motor control of the mouth, tongue, larynx, and pharynx, the structures that produce speech sounds; it is also distinct from the circuits underlying the auditory perception of spoken words and the visual perception of written words in the primary auditory and visual cortices, respectively (Figure 27.1). The neural substrate for language transcends these essential motor and sensory functions in that its main concern is with a system of symbols—spoken and heard, written and read (or, in the case of sign language, gestured and seen). The essence of language, then, is symbolic representation. The obedience to grammatical rules and the use of appropriate emotional tone are recognizable regardless of the particular mode of representation and expression (a point that is especially important in comparing human language and the communicative abilities of great apes, which suggests how language may have evolved in the brains of our prehominid ancestors; Box A).

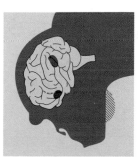

(B)

The brains of great apes are remarkably similar to those of humans, including regions that, in humans, support language. The areas comparable to Broca's area and Wernicke's area are indicated.

important question that remains to be tackled.

Although much uncertainty remains, only someone given to extraordinary anthropocentrism would, in light of this

evidence, continue to argue that symbolic communication is a uniquely human attribute.

References

CERUTTI, D. AND D. RUMBAUGH (1993) Stimulus relations in comparative primate perspective. Psychological Record 43: 811–821.

GOODALL, J. (1990) *Through a Window: My Thirty Years with the Chimpanzees of Gombe.* Boston: Houghton Mifflin Company.

GRIFFIN, D. R. (1992) *Animal Minds.* Chicago: The University of Chicago Press.

HELTNE, P. G. AND L. A. MARQUARDT (EDS.) (1989) *Understanding Chimpanzees.* Cambridge, MA: Harvard University Press.

MILES, H. L. W. AND S. E. HARPER (1994) "Ape language" studies and the study of human language origins. In *Hominid Culture in Primate Perspective,* D. Quiatt and J. Itani (eds.). Niwot, CO: University Press of Colorado, pp. 253–278.

SAVAGE-RUMBAUGH, S., J. MURPHY, R. A. SEVCIK, K. E. BRAKKE, S. L. WILLIAMS AND D. M. RUM-

BAUGH (1993) *Language Comprehension in Ape and Child.* Monographs of the Society for Research in Child Development, Serial No. 233, Vol. 58, Nos. 3, 4.

SAVAGE-RUMBAUGH, S., S. G. SHANKER, AND T. J. TAYLOR (1998) *Apes, Language, and the Human Mind.* New York: Oxford University Press.

TERRACE, H. S. (1983) Apes who "talk": Language or projection of language by their teachers? In *Language in Primates: Perspectives and Implications,* J. de Luce and H. T. Wilder (eds.). New York: Springer-Verlag, pp. 19–42.

WHITEN, A., J. GOODALL, W. C. MCGREW, T. NISHIDA, V. REYNOLDS, Y. SUGIYAMA, C. E. G. TUTIN, R. W. WRANGHAM AND C. BOESCH (1999) Cultures in chimpanzees. Nature 399: 682–685.

VON FRISCH, K. (1993) *The Dance Language and Orientation of Bees* (Transl. by Leigh E. Chadwick). Cambridge, MA: Harvard University Press.

WALLMAN, J. (1992) *Aping Language.* New York: Cambridge University Press.

Aphasias

The distinction between language and the related sensory and motor capacities on which it depends was first apparent in patients with damage to specific brain regions. Clinical evidence of this sort showed that the ability to move the muscles of the mouth, tongue, larynx, and pharynx can be compromised without abolishing the ability to use spoken language to communicate (even though a motor deficit may make communication difficult). Similarly, damage to the auditory pathways can impede the ability to hear without interfering with language functions per se. Damage to other quite specific regions of the brain, however, can compromise essential language functions while leaving the sensory and motor components of verbal communication intact. These latter syndromes, collectively referred to as **aphasias**, diminish or abolish the ability to comprehend or to produce language, while sparing the ability to perceive the relevant stimuli and to produce words. Missing in these patients is the capacity to recognize or employ the *symbolic* value of words, thus depriving such individuals of the linguistic understanding, grammatical organization, and intonation that distinguishes language from nonsense (Box B).

The localization of language function to a specific region (and to a degree hemisphere) of the cerebrum is usually attributed to the French neurologist Paul Broca and the German neurologist Carl Wernicke, who made their seminal observations in the late 1800s. Both Broca and Wernicke examined the brains of individuals who had become aphasic and later died. Based on correlations of the clinical picture and the location of the brain damage, Broca suggested that language abilities were localized in the ventroposterior region of the frontal lobe (Figures 27.1 and 27.2). More importantly, he observed that the loss of the ability to produce meaningful language—as opposed to the ability to move the mouth and produce words—was usually associated with damage to the left hemisphere. "*On parle avec l'hemisphere gauche,*" Broca concluded. The preponderance of aphasic syndromes associated with damage to the left hemisphere has supported his claim that one speaks with the left hemisphere, a conclusion amply confirmed by a variety of modern studies using functional imaging (with some important caveats, discussed later in the chapter).

Although Broca was basically correct, he failed to grasp the limitations of thinking about language as a unitary function localized in a single cortical region. This issue was better appreciated by Wernicke, who distinguished between patients who had lost the ability to comprehend language and those who could no longer produce language. Wernicke recognized that

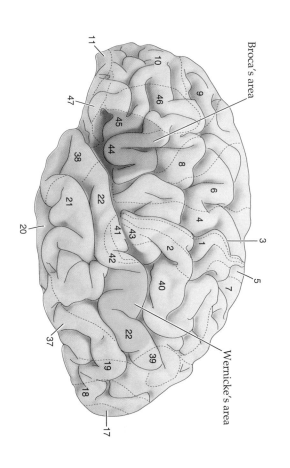

Figure 27.2 The relationship of the major language areas to the classical cytoarchitectonic map of the cerebral cortex. As discussed in Chapter 26, about 50 histologically distinct regions (cytoarchitectonic areas) have been described in the human cerebral cortex. Whereas primary sensory and motor functions are sometimes coextensive to these areas, more general cognitive functions like attention, identification, and planning typically encompass several cytoarchitectonic areas in one or more cortical lobes. The language functions described by Broca and Wernicke are associated with at least three of the cytoarchitectonic areas defined by Brodmann (area 22, at the junction of the parietal and temporal lobes [Wernicke's area]; and areas 44 and 45, in the ventral and posterior region of the frontal lobe [Broca's area]).

Broca's area

Wernicke's area

BOX B
Words, Syntax, and Meaning

When Samuel Johnson (figure A) compiled his *Dictionary of English Language* in 1755 under the sponsorship of Oxford University, he defined only 43,500 entries; the current *Oxford English Dictionary*, a lineal descendant of Johnson's

Samuel Johnson

seminal work and most recently revised in the 1980s, contains over 500,000 definitions! This difference is not the result of an increase in the number of English words since the eighteenth century, but rather an indication of the difficulty collecting the enormous number of words we use in daily communication. The average college-educated English speaker typically has a working vocabulary of more than 100,000 words. The enormous ambiguity of language indicates further that there is far more to a lexicon—be it in a dictionary or in the left temporal cortex—than simply the definitions of words. Even when the meaning of a word is known, it must be used according to the rules of grammar to produce effective communication, and a given word must always be understood in a proper context—as figure B illustrates.

From both the point of view of neuroscience and linguistics, two related questions about words and grammar are especially germane in relation to the

main themes of this chapter. First, what is the nature of the neural machinery that allows us to learn language? And second, why do we have such a profound drive to learn language? The major twentieth-century figure who has grappled with these questions is linguist Noam Chomsky, who, while not interested in brain structure, has argued that the complexity of language is such that it cannot simply be learned. Chomsky therefore proposed that language must be predicated on a "universal grammar" laid down in the evolution of our species. Whereas this argument is undoubtedly correct (the basic language machinery, like all aspects of brain circuitry that support adult behavior, is indeed constructed during the normal development of each individual as a result of inheritance; see Chapters 23 and 24), Chomsky's eschewing of neurobiology avoids the central question of how, in evolutionary or developmental terms, this machinery comes to be and how it actually encodes words and grammar. Whatever the mechanisms

eventually prove to be, much of the language we use is obviously learned by making neuronal linkages between arbitrary symbols and the objects, concepts, and interrelationships they signify in the real world. As such, human language provides a rich source for understanding how the relevant parts of the human cortex and their constituent neurons work to produce the enormous facility for making associations, which appears to be a fundamental aspect of all cortical functions.

References

CHOMSKY, N. (1975) *Reflections on Language.* New York: Pantheon/Random House.

CHOMSKY, N. (1980) *Rules and Representations.* New York: Columbia University Press.

CHOMSKY, N. (1981) Knowledge of language: Its elements and origins. Philos. Trans. Roy. Soc. Lond. B 295: 223-234.

MILLER, G. A. (1991) *The Science of Words.* New York: Scientific American Library.

PINKER, S. (1994) *The Language Instinct.* New York: W. Morrow and Co.

WINCHESTER, S. (1998) *The Professor and the Madman.* New York: Harper Perennial.

(B)

The importance of context. When a person says "I'm going to our house on the lake," the meaning of the expression obviously depends on usage and context, rather than on the literal structure of the sentence uttered. This example indicates the enormous complexity of the task we all accomplish when becoming fluent in language. (From Miller, 1991.)

some aphasic patients do not understand language but retain the ability to produce utterances with reasonable grammatical and emotional content. He concluded that lesions of the posterior and superior temporal lobe on the left side tend to result in a deficit of this sort. In contrast, other patients continue to comprehend language but lack the ability to organize or control the linguistic content of their response. Thus, they produce nonsense syllables, transposed words, and generally utter grammatically incomprehensible phrases. These deficits are associated with damage to the posterior and inferior region of the left frontal lobe, the area that Broca had emphasized as an important substrate for language (see Figures 27.1 and 27.2).

As a consequence of these early observations, two rules about the localization of language have been taught ever since. The first is that lesions of the left frontal lobe in a region referred to as **Broca's area** affect the ability to *produce* language efficiently. This deficiency is called **motor** or **expressive aphasia** and is also known as **Broca's aphasia**. (Such aphasias must be specifically distinguished from **dysarthria**, which is the inability to move the muscles of the face and tongue that mediate speaking.) The deficient motor-planning aspects of expressive aphasias accord with the complex motor functions of the posterior frontal lobe and its adjacency to the primary motor cortex already discussed (see Chapters 16 and 26). The second rule is that damage to the left temporal lobe causes difficulty *understanding* spoken language, a deficiency referred to as **sensory** or **receptive aphasia**. (Deficits of reading and writing—alexias and agraphias—are separate disorders that arise from damage to related but different brain areas.) Receptive aphasia, also known as **Wernicke's aphasia**, generally reflects damage to the auditory association cortices in the posterior temporal lobe, a region referred to as **Wernicke's area**. A final broad category of language deficiency syndromes is **conduction aphasia**. These disorders arise from lesions to the pathways connecting the relevant temporal and frontal regions, such as the arcuate fasciculus in the subcortical white matter that links Broca's and Wernicke's areas. Interruption of this pathway may lead to an inability to produce appropriate responses to heard communication, even though the communication is understood.

In a classic Broca's aphasia the patient can't express himself appropriately because the organizational aspects of language (its grammar) have been disrupted, as shown in the following example reported by Howard Gardner (who is the interlocutor). The patient was a 39-year-old Coast Guard radio operator named Ford who had suffered a stroke that affected his left posterior frontal lobe.

'I am a sig…no…man…uh, well…again.' These words were emitted slowly, and with great effort. The sounds were not clearly articulated; each syllable as uttered harshly, explosively, in a throaty voice. With practice, it was possible to understand him, but at first I encountered considerable difficulty in this. 'Let me help you,' I interjected. 'You were a signal…' 'A sig-nal man…right,' Ford completed my phrase triumphantly. 'Were you in the Coast Guard?' 'No, er, yes, yes,…ship…Massachu…chusetts…Coastguard…years.' He raised his hands twice, indicating the number nineteen. 'Oh, you were in the Coast Guard for nineteen years.' 'Oh…boy…right…right,' he replied. 'Why are you in the hospital, Mr. Ford?' Ford looked at me strangely, as if to say, Isn't it patently obvious? He pointed to his paralyzed arm and said, 'Arm no good,' then to his mouth and said, 'Speech…can't say…talk, you see.'

Howard Gardner, 1974. (*The Shattered Mind: The Person after Brain Damage*; pp. 60–61.)

In contrast, the major difficulty in Wernicke's aphasia is putting together objects or ideas and the words that signify them. Thus, in a Wernicke's aphasia,

TABLE 27.1
Characteristics of Broca's and Wernicke's Aphasias

Broca's aphasia[a]	*Wernicke's aphasia[b]*
Halting speech	Fluent speech
Repetitive (perseveration)	Little repetition
Disordered syntax	Syntax adequate
Disordered grammar	Grammar adequate
Disordered structure of individual words	Contrived or inappropriate words

[a]Also called motor, expressive, or production aphasia
[b]Also called sensory or receptive aphasia

speech is fluent and well structured, but makes little or no sense because words and meanings are not correctly linked, as is apparent in the following example (again from Gardner). The patient in this case was a 72-year-old retired butcher who had suffered a stroke affecting his left posterior temporal lobe.

> Boy, I'm sweating, I'm awful nervous, you know, once in a while I get caught up, I can't get caught up, I can't mention the tarripoi, a month ago, quite a little, I've done a lot well, I impose a lot, while, on the other hand, you know what I mean. I have to run around, look it over, trebbin and all that sort of stuff. Oh sure, go ahead, any old think you want. If I could I would. Oh, I'm taking the word the wrong way to say, all of the barbers here whenever they stop you it's going around and around, if you know what I mean, that is tying and tying for repucer, repuceration, well, we were trying the best that we could while another time it was with the beds over there the same thing...

> Ibid., p. 68.

The major differences between these two classical aphasias are summarized in Table 27.1.

Despite the validity of Broca's and Wernicke's original observations, the classification of language disorders is considerably more complex. An effort to refine the nineteenth-century categorization of aphasias was undertaken by the American neurologist Norman Geschwind during the 1950s and early 1960s. Based on data from a large number of patients and on the better understanding of cortical connectivity gleaned by that time from animal studies, Geschwind concluded correctly that several other regions of the parietal, temporal, and frontal cortices are critically involved in human linguistic capacities. Basically, he showed that damage to these additional areas results in identifiable, if more subtle, language deficits. His clarification of the definitions of language disorders has been largely confirmed by functional brain imaging in normal subjects, and remains the basis for much contemporary clinical work on the aphasias.

A Dramatic Confirmation of Language Lateralization

Until the 1960s, observations about language localization and lateralization were based primarily on patients with brain lesions of varying severity, location, and etiology. The inevitable uncertainties of clinical findings allowed more than a few skeptics to argue that language function (or other complex cognitive functions) might not be lateralized (or even localized) in the brain. Definitive evidence supporting the inferences from neurological observations came from studies of patients whose corpus callosum and anterior commissure had been severed as a treatment for medically intractable epileptic seizures. (Recall that a certain fraction of severe epileptics are refractory to medical treatment, and that interrupting the connection between the two hemispheres remains an effective way of treating epilepsy in highly selected patients; see Box C in Chapter 25). In such patients, investigators could assess the function of the two cerebral hemispheres *independently*, since the major axon tracts that connect them had been interrupted. The first studies of these so-called **split-brain patients** were carried out by Roger Sperry and his colleagues at the California Institute of Technology in the 1960s and 1970s, and established the hemispheric lateralization of language beyond any doubt; this work also demonstrated many other functional differences between the left and right hemispheres and continues to stand as an extraordinary contribution to the understanding of brain organization.

To evaluate the functional capacity of each hemisphere in split-brain patients, it is essential to provide information to one side of the brain only. Sperry, Michael Gazzaniga (who was a key collaborator in this work), and others devised several simple ways to do this, the most straightforward of which was to ask the subject to use each hand independently to identify objects without any visual assistance (Figure 27.3A). Recall from Chapter 9 that somatic sensory information from the right hand is processed by the left hemisphere, and vice versa. By asking the subject to describe an item being manipulated by one hand or the other, the language capacity of the relevant hemisphere could be examined. Using the left hemisphere, split-brain patients were able to name objects held in the right hand without difficulty. In contrast, and quite amazingly, an object held in the left hand could not be named! Using the right hemisphere, subjects could produce only an indirect description of the object that relied on rudimentary words and phrases rather than the precise lexical symbol for the object (for instance, "a round thing" instead of "a ball"), and some could not provide any verbal account of what they held in their left hand. Observations using special techniques to present visual information to the hemispheres independently (a method called *tachistoscopic presentation*; see Figure 27.3B) showed further that the left hemisphere can respond to written commands, whereas the right hemisphere can respond only to nonverbal stimuli (e.g., pictorial instructions, or, in some cases, rudimentary written commands). These distinctions reflect broader hemispheric differences summarized by the statement that the left hemisphere in most humans is specialized for processing verbal and symbolic

(A)

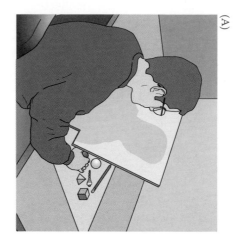

(B)

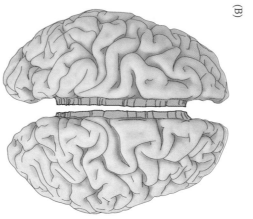

Some left hemisphere functions	Some right hemisphere functions
Analysis of right visual field	Analysis of left visual field
Stereognosis (right hand)	Stereognosis (left hand)
Lexical and syntactic language	Emotional coloring of language
Writing	Spatial abilities
Speech	Rudimentary speech

Figure 27.3 Confirmation of the linguistic specialization of the left hemisphere in the vast majority of humans obtained by studying individuals in whom the connections between the right and left hemispheres have been surgically divided. (A) Single-handed, vision-independent stereognosis can be used to evaluate the language capabilities of each hemisphere in split-brain patients. Objects held in the right hand, which provides somatic sensory information to the left hemisphere, are easily named; objects held in the left hand, however, cannot be identified by name in these patients. (B) Schematic representation of some of the different functional abilities of the left and right hemispheres, deduced from a variety of behavioral tests in split-brain patients. Simple visual instructions can also be given independently to the right or left hemisphere in these individuals. Since the left visual field is perceived by the right hemisphere (and vice versa; see Chapter 12), a briefly presented (tachistoscopic) instruction in the left visual field is appreciated only by the right brain (assuming that the individual maintains his or her gaze straight ahead). A wide range of functions can be evaluated using this tachistoscopic method, even in normal subjects.

material important in communication, whereas the right hemisphere is specialized for visuospatial and emotional processing (Figure 27.3B).

The ingenious work of Sperry and his colleagues on split-brain patients put an end to the century-long controversy about language lateralization; in most individuals, the left hemisphere is unequivocally the seat of the major language functions (although see Box C). It would be wrong to imagine, however, that the right hemisphere has no language capacity. As noted, in some individuals the right hemisphere can produce rudimentary words and phrases, and it is normally the source of emotional coloring of language. Moreover, the right hemisphere in at least some split-brain patients understands language to a modest degree, since it can respond to simple visual commands presented tachistoscopically. Consequently, Broca's conclusion that we speak with our left brain is not strictly correct; it would be more accurate to say that one speaks very much better with the left hemisphere than with the right, and that the contributions of the two hemispheres to language are markedly different.

Anatomical Differences between the Right and Left Hemispheres

The clear differences in language function between the left and right hemispheres have naturally inspired neurologists and neuropsychologists to find a structural correlate of this behavioral asymmetry. One hemispheric difference that has received much attention over the years was identified in the late 1960s by Norman Geschwind and his colleagues at Harvard Medical School, who found an asymmetry in the superior aspect of the temporal lobe known as the **planum temporale** (Figure 27.4). This area was significantly larger on the left

Figure 27.4 Asymmetry of the right and left human temporal lobes. (A) The superior portion of the brain has been removed as indicated to reveal the dorsal surface of the temporal lobes in the right-hand diagram (which presents a dorsal view of the horizontal plane). A region of the surface of the temporal lobe called the planum temporale is significantly larger in the left hemisphere of most (but far from all) individuals. (B) Measurements of the planum temporale in 100 adult and infant brains. The mean size of the planum temporale is expressed in arbitrary planimetric units to get around the difficulty of measuring the curvature of the gyri within the planum. The asymmetry is evident at birth and persists in adults at roughly the same magnitude (on average, the left planum is 57% larger in infants and 50% larger in adults). (C) An MR image in the frontal plane, showing this asymmetry (arrows) in a normal adult subject.

(A)

Frontal and parietal lobes removed

Right planum temporale

Left planum temporale

(B)

Planum temporale measurements of 100 adult and 100 infant brains		
	Left hemisphere	Right hemisphere
Infant	20.7	11.7
Adult	37.0	18.4

(C)

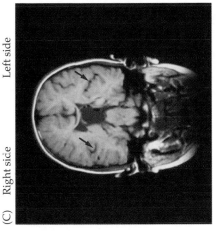

Right side   Left side

BOX C
Handedness

Approximately 9 out of 10 people are right-handed, a proportion that appears to have been stable over thousands of years and across all cultures in which handedness has been examined. Handedness is usually assessed by having individuals answer a series of questions about preferred manual behaviors, such as "Which hand do you use to write?"; "Which hand do you use to throw a ball?"; or "Which hand do you use to brush your teeth?" Each answer is given a value, depending on the preference indicated, providing a quantitative measure of the inclination toward right- or left-handedness. Anthropologists have determined the incidence of handedness in ancient cultures by examining artifacts; the shape of a flint ax, for example, can indicate whether it was made by a right- or left-handed individual. Handedness in antiquity has also been assessed by examining the incidence of figures in artistic representations who are using one hand or the other. Based on this evidence, our species appears always to have been a right-handed one. Moreover, handedness is probably not peculiar to humans; many studies have demonstrated paw preference in animals ranging from

mice to monkeys that is, at least in some ways, similar to human handedness.

Whether an individual is right- or left-handed has a number of interesting consequences. As will be obvious to left-handers, the world of human artifacts is in many respects a right-handed one. Implements such as scissors, knives, coffee pots, and power tools are constructed for the right-handed majority. Books and magazines are also designed for right-handers (compare turning this page with your left and right hands), as are golf clubs and guitars. Perhaps as a consequence of this bias, the accident rate for left-handers in all categories (work, home, sports) is higher than for right-handers. The rate of traffic fatalities among left-handers is also greater than that among right-handers. However, there are also some advantages to being left-handed. For example, an inordinate number of international fencing champions have been left-handed. The reason for this fact is ultimately obvious: since the majority of any individual's opponents will be right-handed, the average fencer, right- or left-handed, is less prepared to parry thrusts from left-handers.

One of the most hotly debated questions about the consequences of handed-

ness in recent years has been whether being left-handed entails a diminished life expectancy. No one disputes the fact that there is currently a surprisingly small number of left-handers among the elderly (see figure). These data have come from studies of the general population and have been supported by information gleaned from *The Baseball Encyclopedia*, in which longevity and other characteristics of a large number of healthy left- and right-handers have been recorded because of interest in the U.S. national pastime. Two explanations of this peculiar finding have been put forward. Stanley Coren and his collaborators at the University of British Columbia have argued that these statistics reflect a higher mortality rate among left-handers, partly as a result of increased accidents, but also because of other data that show left-handedness to be associated with a variety of pathologies. In this regard, Coren and others have suggested that left-handedness may arise because of developmental problems in the pre- and/or perinatal period. If true, then a further rationale for decreased longevity would have been identified. An alternative explanation, however, is that the diminished number of left-handers among the elderly is primarily a reflection of sociological factors—namely, a greater acceptance of left-handed children today compared to earlier in the twentieth century. In this view, there are fewer older left-handers now because

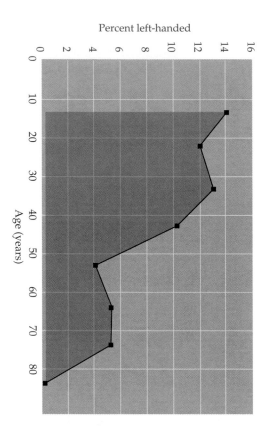

The percentage of left-handers in the normal population as a function of age (based on more than 5000 individuals). Taken at face value, these data indicate that right-handers live longer than left-handers. Another possibility, however, is that the paucity of elderly left-handers at present may simply reflect changes over the decades in the social pressures on children to become right-handed. (From Coren, 1992.)

parents, teachers, and others encouraged right-handedness in earlier generations. Although this controversy continues, the weight of the evidence favors the sociological explanation.

The relationship between handedness and other lateralized functions—language in particular—has long been a source of confusion. It is unlikely that there is any direct relationship between language and handedness, despite much speculation to the contrary. The most straightforward evidence on this point comes from the results of the Wada test (the injection of sodium amytal into one carotid artery to determine the hemisphere in which language function is located; see text on this page). The large number of such tests carried out for clinical purposes indicate that about 97% of humans, including the majority of left-handers, have their major language functions in the left hemisphere (although it should be noted that right hemispheric dominance for language is much more common among left-handers). Since most left-handers have language function on the side of the brain opposite the control of their preferred hand, it is hard to argue for any strict relationship between these two lateralized functions.

In all likelihood, handedness, like language, is first and foremost an example of the advantage of having any specialized function on one side of the brain or the other to make maximum use of the available neural circuitry in a brain of limited size.

References

BAKAN, P. (1975) Are left-handers brain damaged? New Scientist 67: 200–202.

COREN, S. (1992) *The Left-Hander Syndrome: The Causes and Consequence of Left-Handedness.* New York: The Free Press.

DAVIDSON, R. J. AND K. HUGDAHL (EDS.) (1995) *Brain Asymmetry.* Cambridge, MA: MIT Press.

SALIVE, M. E., J. M. GURALNIK AND R. J. GLYNN (1993) Left-handedness and mortality. Am. J. Pub. Health 83: 265–267.

side in about two-thirds of human subjects studied postmortem. A similar difference has been described in higher apes, but not in other primates.

Because the planum temporale is near (although not congruent with) the regions of the temporal lobe that contain cortical areas essential to language, it has been widely assumed that this leftward asymmetry reflects the greater involvement of the left hemisphere in language. Nonetheless, these anatomical differences in the two hemispheres of the brain, which are recognizable at birth, have by no means been proven to be an anatomical correlate of the lateralization of language function. The fact that a detectable planum asymmetry is present in only 67% of human brains, whereas the preeminence of language in the left hemisphere is evident in 97% of the population, argues that this association should be regarded with caution. In fact, recent studies using noninvasive imaging have indicated that there is less variability in the planum temporale than originally described. The anatomical underpinning of left/right differences in hemispheric language abilities, if any, remains uncertain, as it does for the lateralized hemispheric functions described in Chapter 26.

Mapping Language Function

The pioneering work of Broca and Wernicke, and later Geschwind and Sperry, clearly established differences in hemispheric function. Several techniques have since been developed that allow hemispheric attributes to be assessed in neurological patients with an intact corpus callosum, and in normal subjects.

One method that has long been used for the assessment of language lateralization was devised in the 1960s by Juhn Wada at the Montreal Neurological Institute. In the so-called Wada test, a short-acting anesthetic (e.g., sodium amytal) is injected into the left carotid artery; this procedure transiently "anesthetizes" the left hemisphere and thus tests the functional capabilities of the affected half of the brain. If the left hemisphere is indeed dominant for language, then the patient becomes transiently aphasic while carrying out an ongoing verbal task like counting (the anesthetic is rapidly diluted by the circulation, but not before its local effects can be observed).

Less invasive (but less definitive) ways to test the cognitive abilities of the two hemispheres include positron emission tomography, functional magnetic resonance imaging (see Box C in Chapter 1), and the sort of tachistoscopic presentation used so effectively by Sperry and his colleagues (even when the hemispheres are normally connected, subjects show delayed verbal responses and other differences when the right hemisphere receives the instruction). These techniques have all confirmed the hemispheric lateralization of language functions. More importantly, they have provided valuable diagnostic tools to determine in preparation for neurosurgery which hemisphere is "eloquent"; although most individuals have the major language functions in the left hemisphere, a few—about 3% of the population—do not (the latter are much more often left-handed; see Box C).

Once the appropriate hemisphere is known, the neurosurgeon can map language functions more definitively by electrical stimulation of the cortex during the surgery. By the 1930s, the neurosurgeon Wilder Penfield and his colleagues at the Montreal Neurological Institute had already carried out a more detailed localization of cortical capacities in a large number of patients (see Chapter 9). Penfield used electrical mapping techniques adapted from neurophysiological work in animals to delineate the language areas of the cortex prior to the removal of brain tissue in the treatment of tumors or epilepsy. Such intraoperative mapping guaranteed that the cure would not be worse than the disease and has been widely used ever since.

As a result, considerable new information about language localization emerged. Penfield's observations, together with more recent studies performed by George Ojemann and his group at the University of Washington, have generally confirmed the conclusions inferred from postmortem correlations and other approaches: a large region of the perisylvian cortex of the left hemisphere is clearly involved in language production and comprehension (Figure 27.5). A surprise, however, was the variability in language localization from patient to patient in such studies. Thus, Ojemann found that the brain regions involved in language are only approximately those indicated by most textbook treatments, and that their exact locations differ unpredictably among individuals. Indeed, bilingual patients do not necessarily use the same bit of cortex for storing the names of the same objects in two different languages! Moreover, although single neurons in the temporal cortex in and around Wernicke's area respond preferentially to spoken words,

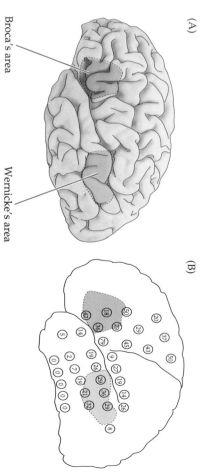

(A)

Broca's area

Wernicke's area

(B)

Figure 27.5 Cortical mapping of the language areas in the left cerebral cortex during neurosurgery. (A) Location of the classical language areas. (B) Evidence for the variability of language representation among individuals. This diagram summarizes data from 117 patients whose language areas were electrically mapped at the time of surgery. The number in each circle indicates the percentage of the patients who showed interference with language in response to stimulation at that site. Note also that many of the sites that elicited interference fall outside the classic language areas. (B after Ojemann et al., 1989.)

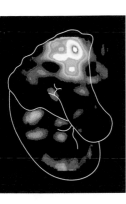

Passively viewing words

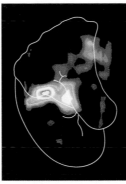

Listening to words

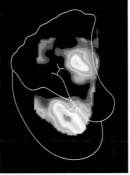

Speaking words

Generating words

Figure 27.6 Language-related regions of the left hemisphere mapped by positron emission tomography (PET) in a normal human subject. Language tasks such as listening to words and generating words elicit activity in Broca's and Wernicke's areas, as expected. However, there is also activity in primary and association sensory and motor areas for both active and passive language tasks. These observations indicate that language processing involves cortical regions other than the classic language areas. (From Posner and Raichle, 1994.)

they do not show preferences for a particular word. Rather, a wide range of words can elicit a response in any given neuron.

Despite these advances, neurosurgical studies are complicated by their intrinsic difficulty and to some extent by the fact that the brains of the patients in whom they are carried out are not normal. The advent of positron emission tomography in the 1980s, and more recently functional magnetic resonance imaging, has allowed the investigation of the language regions in normal subjects by noninvasive imaging. Recall that these techniques reveal the areas of the brain that are active during a particular task because the related electrical activity increases local metabolic activity and therefore local blood flow. The results of this approach, particularly in the hands of Marc Raichle, Steve Petersen, and their colleagues at Washington University in St. Louis, have challenged excessively rigid views of localization and lateralization of linguistic function. Although high levels of activity occur in the expected regions, large areas of both hemispheres are activated in word recognition or production tasks (Figure 27.6).

Finally, Hanna Damasio and her colleagues at the University of Iowa have recently shown that distinct regions of the temporal cortex are activated by tasks in which subjects named particular people, animals, or tools (Figure 27.7). This arrangement helps explain the clinical finding that when a relatively limited region of the temporal lobe is damaged (usually by a stroke on the left side), language deficits are sometimes restricted to a particular category of objects. These studies are also consistent with Ojemann's electrophysiological studies, indicating that language is organized according to categories of meaning rather than individual words. Taken together, such studies are rapidly augmenting the information available about how language is organized.

More on the Role of the Right Hemisphere in Language

Since exactly the same cytoarchitectonic areas exist in the cortex of both hemispheres, a puzzling issue remains. What do the comparable areas in the right hemisphere actually do? In fact, language deficits often *do* occur following damage to the right hemisphere. Most obvious is an absence of the normal emotional and tonal components—called **prosodic** elements—of language, which impart additional meaning (often quite critical) to verbal communication (see Chapter 29). These deficiencies, referred to as **aprosodias,** are associated with right-hemisphere lesions of the cortical areas that correspond to Broca's and Wernicke's areas in the left hemisphere. The aprosodias emphasize that although the left hemisphere (or, better put, distinct cortical regions within that hemisphere) figures prominently in the comprehension and production of language for most humans, other regions, including areas in the right hemisphere, are needed to generate the full richness of everyday speech

In summary, whereas the classically defined regions of the left hemisphere operate more or less as advertised, a variety of more recent studies

Figure 27.7 Different regions in temporal lobe are activated by different word categories using PET imaging. Dotted lines show location of the relevant temporal regions in these horizontal views. Note the different patterns of activity in the temporal lobe in response to each stimulus category. (After Damasio et al., 1996.)

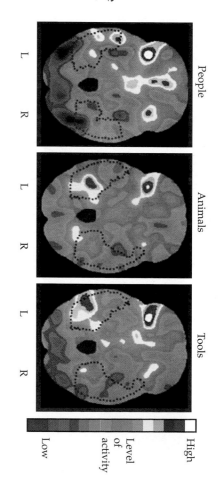

People   Animals   Tools

L R   L R   L R

Level of activity: High — Low

have shown that other left- and right-hemisphere areas clearly make a significant contribution to generation and comprehension of language.

Sign Language

The account so far has already implied that language localization and lateralization does not simply reflect brain specializations for hearing and speaking; the language regions of the brain appear to be more broadly organized for processing symbols. Strong support for this conclusion has come from studies of sign language in individuals deaf from birth.

American Sign Language has all the components (e.g., grammar and emotional tone) of spoken and heard language. Based on this knowledge, Ursula Bellugi and her colleagues at the Salk Institute examined the localization of sign language in patients who had suffered localized lesions of either the left or right hemisphere. All these individuals were prelingually deaf, had been signing throughout their lives, had deaf spouses, were members of the deaf community, and were right-handed. The patients with left-hemisphere lesions, which in each case involved the language areas of the frontal and/or temporal lobes, had measurable deficits in sign production and comprehension when compared to normal signers of similar age (Figure 27.8). In contrast, the patients with lesions in approximately the same areas in the right hemisphere did not have sign "aphasias." Instead, as predicted from other studies of subjects with normal hearing, visuospatial and other abilities (e.g., emotional processing and the emotional tone of signing) were impaired. Although the number of subjects studied was necessarily small (deaf signers with lesions of the language areas are understandably difficult to find), the capacity for signed and seen communication is evidently represented predominantly in the left hemisphere, in the same areas as spoken language. This evidence confirms that the language regions of the brain are specialized for the representation of symbolic communication, rather than for heard and spoken language per se.

The capacity for seen and signed communication, like its heard and spoken counterpart, emerges in early infancy. Careful observation of babbling in hearing (and, eventually, speaking) infants shows the production of a predictable pattern of sounds related to the ultimate acquisition of spoken language. Thus, babbling represents an early behavior that prefigures true language, indicating that an innate capacity for language imitation is a key part of the learning process. The congenitally deaf offspring of deaf, signing parents "babble" with their hands in gestures that are apparently the forerun-

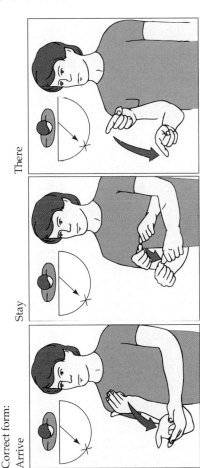

Figure 27.8 Signing deficits in congenitally deaf individuals *who* had learned sign language from birth and later suffered lesions of the language areas in the left hemisphere. Left hemisphere damage produced signing problems in these patients analogous to the aphasias seen after comparable lesions in hearing, speaking patients. In this example, the patient (upper panels) is expressing the sentence "We arrived in Jerusalem and stayed there." Compared to a normal control (lower panels), he cannot properly control the spatial orientation of the signs. The direction of the correct signs and the aberrant direction of the "aphasic" signs are indicated in the upper left-hand corner of each panel. (After Bellugi et al., 1989.)

ners of signs (see Figure 24.1). Like babbling, the amount of manual "babbling" increases with age until the child begins to form accurate, meaningful signs. These observations indicate that the strategy for acquiring the rudiments of symbolic communication from parental or other cues—regardless of the means of expression—is similar in deaf and hearing individuals. These developmental facts are also pertinent to the possible antecedents of language in the nonverbal communication of great apes (see Box A).

Summary

Neurological, neuropsychological, and electrophysiological methods have all been used to localize linguistic function in the human brain. This effort began in the nineteenth century by correlating clinical signs and symptoms with the location of brain lesions determined postmortem. In the twentieth century, additional clinical observations together with studies of split-brain patients, mapping at neurosurgery, sodium amytal anesthesia of a single hemisphere, and noninvasive imaging techniques such as PET and fMRI have greatly extended knowledge about the localization of language. Together, these various approaches show that the perisylvian cortices of the left hemisphere are especially important for normal language in the vast majority of humans. The right hemisphere also contributes to language, most obviously by giving it emotional tone. The similarity of the deficits after comparable brain lesions in congenitally deaf patients and their speaking counterparts strongly supports the idea that the cortical representation of language is independent of the means of its expression or perception (spoken and heard, versus gestured and seen). The specialized language areas that have been identified to date are evidently the major components of a widely distributed set of brain regions that allow humans to communicate effectively by means of symbols and concepts.

ADDITIONAL READING

Reviews

BELLUGI, U., H. POIZNER AND E. S. KLIMA (1989) Language, modality, and the brain. Trends Neurosci. 12: 380–388.

DAMASIO, A. R. (1992) Aphasia. New Eng. J. Med. 326: 531–539.

DAMASIO, A. R. AND H. DAMASIO (1992) Brain and language. Sci. Amer. 267 (Sept.): 89–95.

DAMASIO, A. R. AND N. GESCHWIND (1984) The neural basis of language. Annu. Rev. Neurosci. 7: 127–147.

ETCOFF, N. L. (1986) The neurophysiology of emotional expression. In *Advances in Clinical Neuropsychology*, Volume 3, G. Goldstein and R. E. Tarter (eds.). New York: Quantum, pp. 127–179.

LENNEBERG, E. H. (1967) Language in the context of growth and maturation. In *Biological Foundations of Language*. New York: John Wiley and Sons, pp. 125–395.

OJEMANN, G. A. (1983) The intrahemispheric organization of human language, derived with electrical stimulation techniques. Trends Neurosci. 4: 184–189.

OJEMANN, G. A. (1991) Cortical organization of language. J. Neurosci. 11: 2281–2287.

SPERRY, R. W. (1974) Lateral specialization in the surgically separated hemispheres. In *The Neurosciences: Third Study Program*, F. O. Schmitt and F. G. Worden (eds.). Cambridge, MA: The MIT Press, pp. 5–19.

SPERRY, R. W. (1982) Some effects of disconnecting the cerebral hemispheres. Science 217: 1223–1226.

Important Original Papers

CREUTZFELDT, O., G. OJEMANN AND E. LETTICH (1989) Neuronal activity in the human temporal lobe. I. Response to Speech. Exp. Brain Res. 77: 451–475.

CARAMAZZA, A. AND A. E. HILLIS (1991) Lexical organization of nouns and verbs in the brain. Nature 349: 788–790.

DAMASIO, H., T. J. GRABOWSKI, D. TRANEL, R. D. HICHWA AND A. DAMASIO (1996) A neural basis for lexical retrieval. Nature 380: 499–505.

EIMAS, P. D., E. R. SIQUELAND, P. JUSCZYK AND J. VIGORITO (1971) Speech perception in infants. Science 171: 303–306.

GAZZANIGA, M. S. (1998) The split brain revisited. Sci. Amer. 279 (July): 50–55.

GAZZANIGA, M.S., R. B. LURY AND G. R. MANGUN (1998) Ch. 8, Language and the Brain. In *Cognitive Neuroscience: The Biology of the Mind*. New York: W.W. Norton and Co., pp. 289–321.

GAZZANIGA, M. S. AND R. W. SPERRY (1967) Language after section of the cerebral commissures. Brain 90: 131–147.

GESCHWIND, N. AND W. LEVITSKY (1968) Human brain: Left-right asymmetries in temporal speech region. Science 161: 186–187.

OJEMANN, G. A. AND H. A. WHITAKER (1978) The bilingual brain. Arch. Neurol. 35: 409–412.

PETERSEN, S. E., P. T. FOX, M. I. POSNER, M. MINTUN AND M. E. RAICHLE (1988) Positron emission tomographic studies of the cortical anatomy of single-word processing. Nature 331: 585–589.

PETITTO, L. A. AND P. F. MARENTETTE (1991) Babbling in the manual mode: Evidence for the ontogeny of language. Science 251: 1493–1496.

WADA, J. A., R. CLARKE AND A. HAMM (1975) Cerebral hemispheric asymmetry in humans: Cortical speech zones in 100 adult and 100 infant brains. Arch. Neurol. 32: 239–246.

WESTBURY, C. F., R. J. ZATORRE AND A. C. EVANS (1999) Quantifying variability in the planum temporale: A probability map. Cerebral Cortex 9: 392–405.

Books

GARDNER, H. (1974) *The Shattered Mind: The Person After Brain Damage*. New York: Vintage.

LENNEBERG, E. (1967) *The Biological Foundations of Language*. New York: Wiley.

PINKER, S. (1994) *The Language Instinct: How the Mind Creates Language*. New York: William Morrow and Company.

POSNER, M. I. AND M. E. RAICHLE (1994) *Images of Mind*. New York: Scientific American Library.

Chapter 28

Sleep and Wakefulness

Overview

Sleep—which is defined behaviorally by the normal suspension of consciousness and electrophysiologically by specific brain wave criteria—consumes fully a third of our lives. Sleep occurs in all mammals, and probably all vertebrates. We crave sleep when deprived of it, and, to judge from animal studies, continued sleep deprivation can ultimately be fatal. Surprisingly, however, this peculiar state is not the result of a simple diminution of brain activity; rather, sleep is a series of precisely controlled brain states, and in some of these the brain is as active as it is when people are awake. The sequence of sleep states is governed by a group of brainstem nuclei that project widely throughout the brain and spinal cord. The reason for high levels of brain activity during some phases of sleep, the significance of dreaming, and the basis of the restorative effect of sleep are all topics that remain poorly understood. The clinical importance of sleep is obvious from the prevalence of sleep disorders (insomnias). Each year about 40 million Americans suffer from chronic sleep disorders and an additional 20 million experience occasional sleeping problems.

Why Do Humans and Many Other Animals Sleep?

To feel rested and refreshed upon awaking, most adults require 7–8 hours of sleep, although this number varies among individuals (Figure 28.1A). As a result, a substantial fraction of our lives is spent in this mysterious state. For infants, the requirement is much higher (about 16 hours a day), and teenagers need on average about 9 hours of sleep. As people age, they tend to sleep more lightly and for shorter times, although often needing about the same amount of sleep as in early adulthood (Figure 28.1B). Getting too little sleep creates a "sleep debt" that must be repaid in the following days. In the meantime, judgment, reaction time, and other functions are impaired. Drivers who fall asleep at the wheel are estimated to cause some 56,000 traffic accidents annually and 1,500 highway deaths.

Sleep (or at least a physiological period of quiescence) is a highly conserved behavior that occurs in animals ranging from fruit flies to humans (Box A). This prevalence not withstanding, why we sleep is not well understood. Since animals are particularly vulnerable while sleeping, there must be advantages that outweigh this considerable disadvantage. Shakespeare characterized sleep as "nature's soft nurse," noting the restorative nature of sleep. From a perspective of energy conservation, one function of sleep is to replenish brain glycogen levels, which fall during the waking hours. In keeping with this idea, humans and many other animals sleep at night. Since it is

(A)

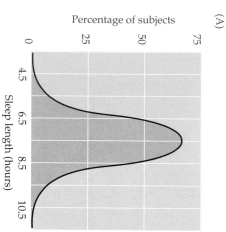

Percentage of subjects

Sleep length (hours)

(B)

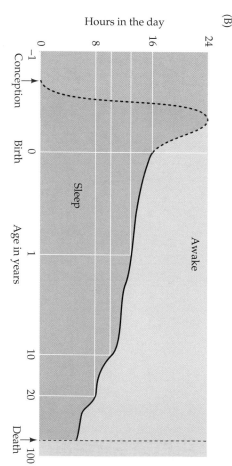

Hours in the day

Conception Birth Age in years Death

Sleep

Awake

Figure 28.1 The duration of sleep. (A) The duration of sleep each night in adults is normally distributed with a mean of 7.5 hours with a standard deviation of about 1.25 hours. Thus, about two-thirds of the population sleeps between 6.25 and 8.75 hours each night. (B) The duration of daily sleep as a function of age. (After Hobson, 1989.)

generally colder at night, more energy would have to be expended to keep warm were we nocturnally active. Furthermore, body temperature has a 24-hour cycle, reaching a minimum at night and thus reducing heat loss. As might be expected, human metabolism measured by oxygen consumption decreases during sleep.

Whatever the reasons for sleeping, in mammals sleep is evidently necessary for survival. For instance, rats completely deprived of sleep die in a few weeks (Figure 28.2). Sleep-deprived rats lose weight despite increasing food intake, and progressively fail to regulate body temperature. They also develop infections, suggesting an impairment of the immune system.

(A) Experimental setup

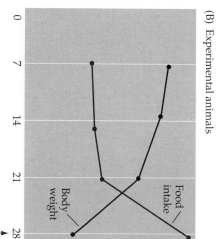

Onset of non-REM sleep in experimental rat triggers floor movement

EEG Experimental rat

EEG Control rat

Feeder

Experimental rat

Control rat

Feeder

Feeder

Motor

Gears to rotate cage floor

(B) Experimental animals

Days of sleep deprivation

Body weight

Food intake

Death

Figure 28.2 The consequences of total sleep deprivation in rats. (A) In this apparatus, an experimental rat is kept awake because the onset of sleep (detected electroencephalographically) triggers movement of the cage floor. The control rat can thus sleep intermittently, whereas the experimental animal cannot. (B) After two to three weeks of sleep deprivation, the experimental animals begin to lose weight, fail to control their body temperature, and eventually die. (After Bergmann et al., 1989.)

BOX A
Styles of Sleep in Different Species

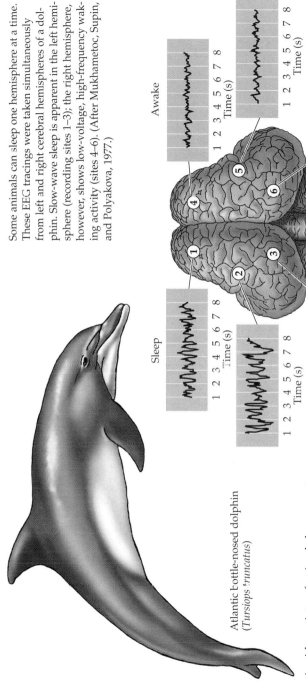

Atlantic bottle-nosed dolphin
(*Tursiops truncatus*)

A wide variety of animals have a rest-activity cycle that often (but not always) occurs in a daily (circadian) rhythm. Even among mammals, however, the organization of sleep depends very much on the species in question. As a general rule, predatory animals can indulge, as humans do, in long, uninterrupted periods of sleep that can be nocturnal or diurnal, depending on the time of day when the animal acquires food, mates, cares for its young, and deals with life's other necessities. The survival of animals that are preyed upon, however, depends much more critically on continued vigilance. Such species—as diverse as rabbits and giraffes—sleep during short intervals that usually last no more than a few minutes. Shrews, the smallest mammals, hardly sleep at all.

An especially remarkable solution to the problem of maintaining vigilance during sleep is shown by dolphins and seals, in whom sleep alternates between the two cerebral hemispheres (see figure). Thus, one hemisphere can exhibit the electroencephalographic signs of wakefulness, while the other shows the characteristics of sleep (see Box C and Figure 28.5). In short, although periods of rest are evidently essential to the proper functioning of the brain, and more generally to normal homeostasis, the manner in which rest is obtained depends on the particular needs of each species.

Some animals can sleep one hemisphere at a time. These EEG tracings were taken simultaneously from left and right cerebral hemispheres of a dolphin. Slow-wave sleep is apparent in the left hemisphere (recording sites 1–3); the right hemisphere, however, shows low-voltage, high-frequency waking activity (sites 4–6). (After Mukhametoc, Supin, and Polyakova, 1977.)

Sleep

Awake

1 2 3 4 5 6 7 8
Time (s)

References

ALLISON, T. AND D. V. CICCHETTI (1976) Sleep in mammals: Ecological and constitutional correlates. Science 194: 732–734.

ALLISON, T. H. AND H. VAN TWYVER (1970) The evolution of sleep. Natural History 79: 56–65.

ALLISON, T., H. VAN TWYVER AND W. R. GOFF (1972) Electrophysiological studies of the echidna, *Tachyglossus aculeatus*. Arch. Ital. Biol. 110: 145–184.

In humans, lack of sleep leads to impaired memory and reduced cognitive abilities, and, if the deprivation persists, mood swings and even hallucinations. The longest documented period of voluntary sleeplessness is 264 hours (approximately 11 days), a record achieved without any pharmacological stimulation. The young man involved recovered after a few days, during which he slept only somewhat more than normal, and seemed none the worse for wear.

The Circadian Cycle of Sleep and Wakefulness

Human sleep occurs with circadian (*circa* = about, and *dia* = day) periodicity, and biologists interested in circadian rhythms have explored a number of questions about this daily cycle. What happens, for example, when individuals are prevented from sensing the cues they normally have about night and day? This question has been answered by placing volunteers in an environment (caves or bunkers have sometimes been used) without external cues about time (Figure 28.3). During a five-day period of acclimation that included social interactions, meals at normal times, and temporal cues (radio, TV), the subjects arose and went to sleep at the usual times and maintained a 24-hour sleep-wake rhythm. After removing these cues, however, the subjects awakened later each day, and the cycle of sleep and wakefulness gradually lengthened to about 28 hours instead of the normal 24. When the volunteers were returned to a normal environment, the 24-hour cycle was rapidly restored. Thus, humans (and many other animals; see Box B) have an internal "clock" that continues to operate in the absence of any external information about the time of day; under these conditions, the clock is said to be "free running."

Presumably, circadian clocks evolved to maintain appropriate periods of sleep and wakefulness in spite of the variable amount of daylight and darkness in different seasons and at different places on the planet. To synchronize physiological processes with the day-night cycle (called photoentrainment), the biological clock must detect decreases in light levels as night approaches. The receptors that sense these light changes are, not surprisingly, in the outer

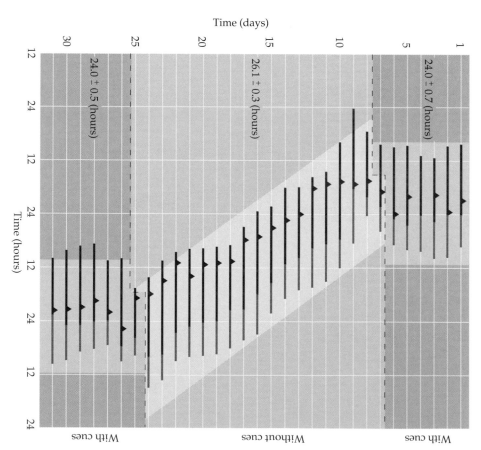

Figure 28.3 Rhythm of waking (blue lines) and sleeping (red lines) of a volunteer in an isolation chamber with and without cues about the day-night cycle. Numbers represent the mean ± standard deviation of a complete waking/sleeping cycle during each period (blue triangles represent times when the rectal temperature was maximum). (After Aschoff, 1965, as reproduced in Schmidt et. al., 1983.)

nuclear layer of the retina; although removing the eye abolishes photoentrainment. The detectors are not, however, the rods or cones. Rather, these poorly understood cells lie within the ganglion and amacrine cell layers of the primate and murine retinas, and project to the **suprachiasmatic nucleus (SCN)** of the hypothalamus, the site of the circadian control of homeostatic functions generally (Figure 28.4A). These peculiar retinal photoreceptors contain a novel photopigment called melanopsin. Perhaps the most convincing evidence of the SCN's role as a sort of master biological clock is that its removal in experimental animals abolishes their circadian rhythm of sleep and waking. The SCN also governs other functions that are synchronized with the sleep-wake cycle, including body temperature (see Figure 28.3), hor-

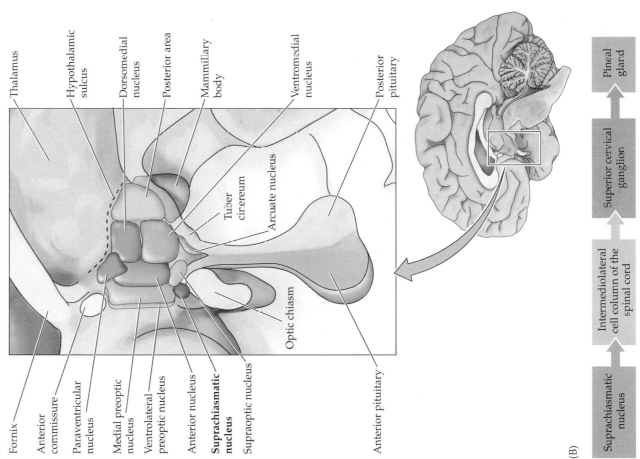

(A)

- Fornix
- Anterior commissure
- Paraventricular nucleus
- Medial preoptic nucleus
- Ventrolateral preoptic nucleus
- Anterior nucleus
- **Suprachiasmatic nucleus**
- Supraoptic nucleus
- Anterior pituitary
- Thalamus
- Hypothalamic sulcus
- Dorsomedial nucleus
- Posterior area
- Mammillary body
- Ventromedial nucleus
- Posterior pituitary
- Tuber cirereum
- Arcuate nucleus
- Optic chiasm

(B)

- Suprachiasmatic nucleus
- Intermediolateral cell column of the spinal cord
- Superior cervical ganglion
- Pineal gland

Figure 28.4 Anatomical underpinnings of circadian rhythms. (A) The hypothalamus, showing the location of the suprachiasmatic nucleus (SCN), which in mammals is the primary "biological clock." The name "suprachiasmatic" derives from the location of the nucleus just above the optic chiasm. (B) Diagram of the pathway from the suprachiasmatic nucleus to the pineal gland.

BOX B
Molecular Mechanisms of Biological Clocks

Virtually all plants and animals adjust their physiology and behavior to the 24-hour day-night cycle under the governance of circadian clocks. Molecular biological studies have now indicated much about the genes and proteins that make up the machinery of these clocks, a story that began nearly 30 years ago.

In the early 1970s, Ron Konopka and Seymour Benzer, working at the California Institute of Technology, discovered three mutant strains of fruit flies whose circadian rhythms were abnormal. Further analysis showed the mutants to be alleles of a single locus, which Konopka and Benzer called the *period* or *per* gene. In the absence of normal environmental cues (that is, in constant light or dark), wild-type flies have periods of activity geared to a 24-hour cycle; *per*^{s} mutants have 19 hour rhythms, *per*^{1} mutants have 29-hour rhythms, and *per*^{0} mutants have no apparent rhythm. About 10 years later, Michael Young at Rockefeller University and Jeffrey Hall and Michael Rosbash at Brandeis University independently cloned the first of the three *per* genes. Cloning a gene does not necessarily reveal its function, however, and so it was in this case. Nonetheless, the gene product PER, a nuclear protein, is found in many *Drosophila* cells pertinent to the production of the fly's circadian rhythms. Moreover, normal flies show a circadian variation in the amount of *per* mRNA and PER protein, whereas *per*^{0} flies, which lack a circadian rhythm, do not show this circadian rhythmicity of gene expression.

Many of the genes and proteins responsible for circadian rhythms in fruit flies have now been discovered in mammals. In mice, the circadian clock arises from the temporally regulated activity of proteins (in capital letters) and genes (in italics) including CRY (*cryptochrome*),

CLOCK (C) (*Circadian locomotor output cycles kaput*), BMAL1 (B) (*brain and muscle, ARNT-like*), PER1 (*Period1*), PER2 (*Period2*), PER3 (*Period3*), and vasopressin preprepropressophysin (VP) (*clock controlled*

genes; ccg). These genes and their proteins give rise to transcription/translation autoregulatory feedback loops with both excitatory and inhibitory components (see figure). The key points to

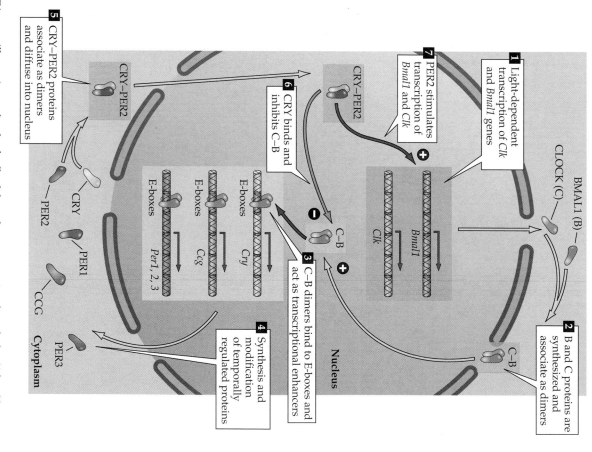

1 Light-dependent transcription of *Clk* and *Bmal1* genes

2 B and C proteins are synthesized and associate as dimers

3 C–B dimers bind to E-boxes and act as transcriptional enhancers

4 Synthesis and modification of temporally regulated proteins

5 CRY–PER2 proteins associate as dimers and diffuse into nucleus

6 CRY binds and inhibits C–B

7 PER2 stimulates transcription of *Bmal1* and *Clk*

CLOCK (C)

BMAL1 (B)

CRY–PER2

C–B

Cry

Ccg

Per1, 2, 3

E-boxes

Clk

Bmal1

Nucleus

Cytoplasm

CRY

PER2

PER1

CCG

PER3

Diagram illustrating molecular feedback loop that governs circadian clocks. (After Okamura et al., 1999.)

understanding this system are: (1) that the concentrations of BMAL1 (B) and the three PER proteins cycle in counterpoint; (2) that PER2 is a positive regulator of the *Bmal1* loop; and (3) that CRY is a negative regulator of the *period* and *cryptochrome* loops. The two positive components of this loop are influenced, albeit indirectly, by light or temperature.

At the start of the day, the transcription of *Clk* and *Bmal1* commences, and the proteins CLOCK (C) and BMAL1 (B) are synthesized in tandem. When the concentrations of C and B increase sufficiently, they associate as dimers and bind to regulatory DNA sequences (E-boxes) that act as a circadian transcriptional enhancers of the genes *Cry, Per 1, 2, and 3* and *CCG*. As a result, the proteins PER1, 2, and 3, CRY, and proteins such as VP are produced. These proteins then diffuse from the nucleus into the cytoplasm, where they are modified.

Although the functions of PER1 and PER3 remain to be elucidated, when the cytoplasmic concentrations of PER2 and CRY increase, they associate as CRY–PER2, and diffuse back into the nucleus. Here, PER2 stimulates the synthesis of C, and B, and CRY binds to C–B dimers, inhibiting their ability to stimulate the synthesis of the other genes. The complete time course of these feedback loops is 24 hours.

References

DUNLAP, J. C. (1993) Genetic analysis of circadian clocks. Ann. Rev. Physiol. 55: 683–728.

KING, D. P. AND J. S. TAKAHASHI (2000) Molecular mechanism of circadian rhythms in mammals. Ann. Rev. Neurosci. 23: 713–742.

HARDIN, P. E., J. C. HALL AND M. ROSBASH (1990) Feedback of the *Drosophila* period gene product on circadian cycling of its messenger RNA levels. Nature 348: 536–540.

OKAMURA, H. AND 8 OTHERS (1999) Science 286: 2531–2534.

SHEARMAN, L. P. AND 10 OTHERS (2000) Interacting molecular loops in the mammalian circadian clock. Science 288: 1013–1019.

TAKAHASHI, J. S. (1992) Circadian clock genes are ticking. Science 258: 238–240.

VITATERNA, M. H. AND 9 OTHERS (1994) Mutagenesis and mapping of a mouse gene, *clock*, essential for circadian behavior. Science 264: 719–725.

mone secretion, urine production, and changes in blood pressure. The cellular mechanisms of circadian control are summarized in Box B.

Activation of the superchiasmatic nucleus evokes responses in neurons whose axons descend to the preganglionic sympathetic neurons in the lateral horn of the spinal cord (Figure 28.4B). These cells, in turn, modulate neurons in the superior cervical ganglia whose postganglionic axons project to the **pineal gland** (pineal means shaped like a pinecone) in the midline near the dorsal thalamus. The pineal gland synthesizes the sleep promoting neurohormone melatonin (*N*-acetyl-5-methoxytryptamine) from tryptophan, and secretes it into the bloodstream to help modulate the brainstem circuits that ultimately govern the sleep-wake cycle (see p. 615 ff.). Predictably, melatonin synthesis increases as light decreases and reaches it maximal level between 2:00 and 4:00 A.M. In the elderly, the pineal gland calcifies and less melatonin is produced, perhaps explaining why older people sleep fewer hours and are more often afflicted with insomnia.

Stages of Sleep

The normal cycle of sleep and wakefulness implies that, at specific times, various neural systems are being activated while others are being turned off. A key to the neurobiology of sleep is therefore to understand the various stages of sleep. For centuries—indeed up until the 1950s—most people who thought about sleep considered it a unitary phenomenon whose physiology was essentially passive and whose purpose was simply restorative. In 1953, however, Nathaniel Kleitman and Eugene Aserinksy showed, by means of electroencephalographic (EEG) recordings from normal subjects, that sleep actually comprises different stages that occur in a characteristic sequence (Figures 28.5 and 28.6).

Humans descend into sleep in stages that succeed each other over the first hour or so after retiring (Figure 28.5). These characteristic stages are defined primarily by electroencephalographic criteria (Box C). Initially, during "drowsiness," the frequency spectrum of the electroencephalogram (EEG) is shifted toward lower values and the amplitude of the cortical waves slightly increases. This drowsy period, called **stage I sleep**, eventually gives way to light or **stage II sleep**, which is characterized by a further decrease in the frequency of the EEG waves and an increase in their amplitude, together with intermittent high-frequency spike clusters called **sleep spindles**. Sleep spindles are periodic bursts of activity at about 10–12 Hz that generally last 1 or 2 seconds and arise as a result of interactions between thalamic and cortical neurons. In **stage III sleep**, which represents moderate to deep sleep, the number of spindles decreases, whereas the amplitude of low-frequency waves increases still more. In the deepest level of sleep, **stage IV sleep**, the predominant EEG activity consists of low frequency (1–4 Hz), high-amplitude fluctuations called **delta waves**, the characteristic slow waves for which this phase of sleep is named. The entire sequence from drowsiness to deep stage IV sleep usually takes about an hour.

These four sleep stages are called **non-rapid eye movement (non-REM) sleep**, and its most prominent feature is the **slow-wave (stage IV) sleep**. It is most difficult to awaken people from slow-wave sleep; hence it is considered to be the deepest stage of sleep. Following a period of slow-wave sleep, however, EEG recordings show that the stages of sleep reverse to reach a quite different state called **rapid eye movement, or REM, sleep**. In REM sleep, the EEG recordings are remarkably similar to that of the awake state (see Figure 28.5). After about 10 minutes in REM sleep, the brain typically cycles back through the non-REM sleep stages. Slow-wave sleep usually occurs again in the second period of this continual cycling, but not during the rest of the night (see Figure 28.6). On average, four additional periods of REM sleep occur, each having longer durations.

In summary, the typical 8 hours of sleep experienced each night actually comprise several cycles that alternate between non-REM and REM sleep, the brain being quite active during much of this supposedly dormant, restful time. For reasons that are not clear, the amount of REM sleep each day decreases from about 8 hours at birth to 2 hours at 20 years to only about 45 minutes at 70 years of age.

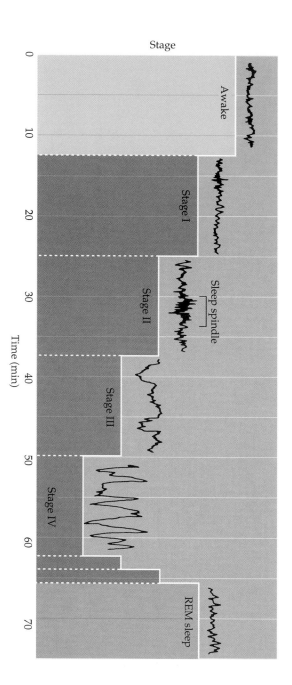

Figure 28.5 EEG recordings during the first hour of sleep. The waking state with the eyes open is characterized by high-frequency (15–60 Hz), low-amplitude activity (~30 μV). This pattern is called beta activity. Descent into stage I non-REM sleep is characterized by decreasing EEG frequency (4–8 Hz) and increasing amplitude (50–100 μV), called theta waves. Descent into stage II non-REM sleep is characterized by 10–15 Hz oscillations (50–150 μV) called spindles, which occur periodically and last for a few seconds. Stage III non-REM sleep is characterized by slower waves at 2–4 Hz (100–150 μV). Stage IV sleep is defined by slow waves (also called delta waves) at 0.5–2 Hz (100–200 μV). After reaching this level of deep sleep, the sequence reverses itself and a period of rapid eye movement sleep, or REM sleep, ensues. REM sleep is characterized by low-voltage, high-frequency activity similar to the EEG activity of individuals who are awake. (Adapted from Hobson, 1989.)

Figure 28.6 Physiological changes in a male volunteer during the various sleep states in a typical 8-hour period of sleep (A). The duration of REM sleep increases from 10 minutes to up to 50 minutes in the first cycle to up to 50 minutes in the final cycle; note that slow-wave (stage IV) sleep is attained only in the first two cycles. (B) The upper panels show the electro-oculogram (EOG) and the lower panels show changes in various muscular and autonomic functions. Movement of neck muscles was measured using an electromyogram (EMG). Other than the few slow eye movements approaching stage I sleep, all other eye movements evident in the EOG occur in REM sleep. The greatest EMG activity occurs during the onset of sleep and just prior to awakening. The heart rate (beats per minute) and respiration (breaths per minute) slow in non-REM sleep, but increase almost to the waking levels in REM sleep. Finally, penile erection occurs only during REM sleep. (After Schmidt et al., 1983.)

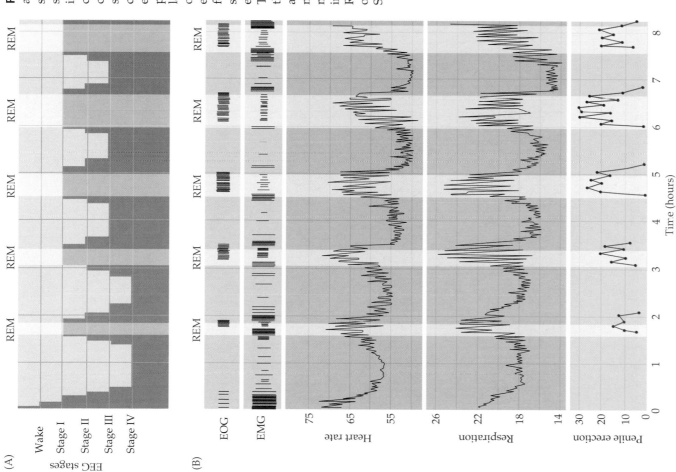

Physiological Changes in Sleep States

A variety of physiological changes take place during the different stages of sleep (see Figure 28.6). Periods of non-REM sleep are characterized by decreases in muscle tone, heart rate, breathing, blood pressure, and metabolic rate. All these parameters reach their lowest values during slow-wave sleep. In non-REM sleep, body movements are reduced compared to wakefulness, although it is common to change sleeping position (tossing and turning). Periods of REM sleep, in contrast, are characterized by increases in blood pressure, heart rate, and metabolism to levels almost as high as those found in the awake state. In addition, REM sleep, as the name implies, is characterized by rapid, rolling eye movements, paralysis of large muscles,

BOX C
Electroencephalography

Although electrical activity recorded from the exposed cerebral cortex of a monkey was reported in 1875, it was not until 1929 that Hans Berger, a psychiatrist at the University of Jena, first made scalp recordings of this activity in humans. Since then, the electroencephalogram, or EEG, has received mixed press, touted by some as a unique opportunity to understand human thinking and denigrated by others as too complex and poorly resolved to allow anything more than a superficial glimpse of what the brain is actually doing. The truth probably lies somewhere in between. Certainly no one disputes that electroencephalography has provided a valuable tool to both researchers and clinicians, particularly in the fields of sleep physiology and epilepsy.

The major advantage of electroencephalography, which involves the application of a set of electrodes to standard positions on the scalp, is its great simplicity. Its most serious limitation is its poor spatial resolution, allowing localization of an active site only to within several centimeters. Four basic EEG phenomena have been defined in humans (albeit somewhat arbitrarily). The alpha rhythm is typically recorded in awake

and the twitching of fingers and toes. Penile erection also occurs during REM sleep, a fact that is clinically important in determining whether a complaint of impotence has a physiological or psychological basis. Interestingly, REM sleep is found only in mammals (and juvenile birds)

Despite the similar EEG recordings obtained in REM sleep and wakefulness, the two conditions are clearly not equivalent brain states. Unlike wakefulness, REM sleep is characterized by dreaming, visual hallucinations, increased emotion, lack of self-reflection, and a lack of volitional control. Since most muscles are inactive during REM sleep, the motor responses to dreams

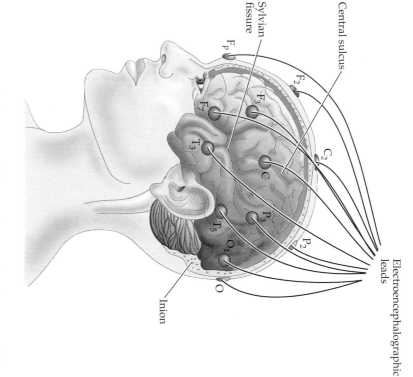

The electroencephalogram represents the voltage recorded between two electrodes applied to the scalp. Typically, pairs of electrodes are placed in 19 standard positions distributed over the head. Letters indicate position (F = frontal, P = parietal, T = temporal, O = occipital, C = central). The recording obtained from each pair of electrodes is somewhat different because each samples the activity of a population of neurons in a different brain region.

subjects with their eyes closed. By definition, the frequency of the alpha rhythm is 8–13 Hz, with an amplitude that is typically 10–50 mV. Lower-amplitude beta activity is defined by frequencies of 14–60 Hz and is indicative of mental activity and attention. The theta and delta waves, which are characterized by frequencies of 4–7 Hz and less than 4 Hz, respectively, imply drowsiness, sleep, or one of a variety of pathological conditions (these slow waves in normal individuals are the signature of stage IV non-REM sleep). Far and away the most obvious component of these various oscillations is the alpha rhythm. Its prominence in the occipital region—and its modulation by eye opening and closing—implies that it is somehow linked to visual processing, as was first pointed out in 1935 by the British physiologist E. D. Adrian. In fact, evidence from very large numbers of subjects suggests that at least several different regions of the brain have their own characteristic rhythms; for example, within the alpha band (8–13 Hz), one rhythm, the classic alpha rhythm, is associated with visual cortex, one (the mu rhythm) with the sensory motor cortex around the central sulcus, and yet another (the kappa rhythm) with the auditory cortex.

In the 1940s, Edward W. Dempsey and Robert Morrison showed that these EEG rhythms depend in part on activity in the thalamus, since thalamic lesions can reduce or abolish the oscillatory cortical discharge (although some oscillatory activity remains even after the thalamus has been inactivated). At about the same time, H. W. Magoun and G. Moruzzi showed that the reticular activating system in the brainstem is also important in modulating EEG activity. For example, activation of the reticular formation changes the cortical alpha rhythm to beta activity, in association with greater behavioral alertness. In the 1960s, Per Andersen and his colleagues in Sweden further advanced these studies by showing that virtually all areas of the cortex participate in these oscillatory rhythms, which reflect a feedback loop between neurons in the thalamus and cortex (see text).

The cortical origin of EEG activity has been clarified by animal studies, which have shown that the source of the current that causes the fluctuating scalp potential is primarily the pyramidal neurons and their synaptic connections in the deeper layers of the cortex. (This conclusion was reached by noting the location of electrical field reversal upon passing an electrode vertically through the cortex from surface to white matter.) In general, oscillations come about either because membrane voltage of thalamo-cortical cells fluctuates spontaneously, or as a result of the reciprocal interaction of excitatory and inhibitory neurons in circuit loops (see Figure 28.10). The oscillations of the EEG are thought to arise from the latter mechanism.

Despite these intriguing observations, the functional significance of these cortical rhythms is not known. The purpose of the brain's remarkable oscillatory activity is a puzzle that has now defied electroencephalographers and neurobiologists for more than 60 years.

References

ADRIAN, E. D. AND K. YAMAGIWA (1935) The origin of the Berger rhythm. Brain 58: 323–351.

ANDERSEN, P. AND S. A. ANDERSSON (1968) *Physiological Basis of the Alpha Rhythm.* New York: Appleton-Century-Crofts.

CATON, R. (1875) The electrical currents of the brain. Brit. Med. J. 2: 278.

DA SILVA, F. H. AND W. S. VAN LEEUWEN (1977) The cortical source of the alpha rhythm. Neurosci. Letters 6: 237–241.

DEMPSEY, E. W. AND R. S. MORRISON (1943) The electrical activity of a thalamocortical relay system. Amer. J. Physiol. 138: 283–296.

NIEDERMEYER, E. AND F. L. DA SILVA (1993) *Electroencephaiography: Basic Principles, Clinical Applications, and Related Fields.* Baltimore: Williams & Wilkins.

NUÑEZ, P. L. (1981) *Electric Fields of the Brain: The Neurophysics of EEG.* New York: Oxford University Press.

are relatively minor (sleepwalking actually occurs during non-REM sleep and is not accompanied or motivated by dreams). This relative paralysis arises from increased activity in GABAergic neurons in the pontine reticular formation that contact lower motor neuron circuitry in the spinal cord. Similarly, activity of descending inhibitory projections from the pons to the dorsal column nuclei causes a diminished response to somatic sensory stimuli during REM sleep. Taken together, these observations have led to the aphorism that non-REM sleep is characterized by an inactive brain in an active body, whereas REM sleep is characterized by an active brain in an inactive body.

The Possible Functions of REM Sleep and Dreaming

Despite this wealth of descriptive information about the stages of sleep, the functional purposes of the various sleep states are not known. Whereas most sleep researchers accept the idea that the purpose of non-REM sleep is at least in part restorative, the function of REM sleep remains a matter of considerable controversy.

A possible clue about the purposes of REM sleep is the prevalence of dreams during these epochs of the sleep cycle. The occurrence of dreams can be tested by waking volunteers during either non-REM or REM sleep and asking them if they were dreaming. Subjects awakened from REM sleep recall elaborate, vivid, hallucinogenic and emotional dreams, whereas subjects awakened during non-REM sleep report fewer dreams, which, when they occur, are more conceptual, less vivid and less emotion-laden.

Dreams have been studied in a variety of ways, perhaps most notably within the psychoanalytic framework of revealing unconscious thought processes considered to be at the root of neuroses. Sigmund Freud's *The Interpretation of Dreams*, published in 1900, speaks eloquently to the complex relationship between conscious and unconscious mentation. It is by no means agreed upon, however, that dreams have the deep significance that Freud and others have given them, and the psychoanalytic interpretation of dreams has recently fallen into disfavor. Nevertheless, most people probably give some credence to the significance of dream content, at least privately. In more recent studies of dreams, about 65% are associated with sadness, apprehension, or anger; 20% with happiness or excitement; and, somewhat surprisingly, only 1% with sexual feelings or acts.

Adding to the uncertainty about the purposes of REM sleep and dreaming is the fact that deprivation of REM sleep in humans for as much as two weeks has little or no obvious effect on behavior. Such studies have been done by waking volunteers whenever their EEG recordings showed the characteristic signs of REM sleep. Although the subjects in these experiments compensate for the lack of REM sleep by having more of it after the period of deprivation has ended, they suffer no obvious adverse effects. Similarly, patients taking certain antidepressants (MAO inhibitors) have little or no REM sleep, yet show no obvious ill effects, even after months or years of treatment. The apparent innocuousness of REM sleep deprivation contrasts markedly with the effects of total sleep deprivation (see earlier). The implication of these several findings is that we can get along without REM sleep, but need non-REM sleep in order to survive.

Several general hypotheses about dreams and REM sleep have been advanced. Francis Crick (of DNA fame) and Grahame Mitchison suggested that dreams act as an "unlearning" mechanism, whereby certain modes of neural activity are erased by random activation of cortical connections. The hypothesis is based on the idea that the human brain represents information by the activity of sets of neuronal networks that are widely distributed and overlapping. In computers, neural network architectures are subject to unwanted patterns of activity that can indeed degrade rather than enhance the information content of the system. By analogy, these "parasitic" modes of activity might be unwanted thoughts or erroneous information, which, if not expunged, could become the basis for obsession, paranoia, or other pathologies of thought that prevent the "system" from working as efficiently as it should. In a different vein, Michel Jouvet proposed that dreaming reinforces behaviors not commonly encountered during the awake state (aggression, fearful situations) by rehearsing them while dreaming. Yet another hypothesis is that REM sleep and dreams are involved in the transfer of memories between the hippocampus and neocortex. Finally, it has been suggested that dreaming is simply an incidental consequence of REM sleep. None of these ideas are generally accepted.

In short, the questions of why we have REM sleep and why we dream remain unanswered, as are questions about consciousness per se (Box D).

Neural Circuits Governing Sleep

From the descriptions of the physiological changes that occur during sleep, it is clear that periodic excitatory and inhibitory changes occur in many neural circuits. What follows is a brief overview of the still incompletely understood circuits and their interactions that govern sleeping and wakefulness.

One of the first clues about the circuits involved in the sleep-wake cycle was provided in 1949 by Horace Magoun and Giuseppe Moruzzi. They found that electrically stimulating a group of cholinergic neurons that lies near the junction of the pons and midbrain causes a state of wakefulness and arousal (the name "reticular activating system" was therefore given to this region of the brainstem) (Figure 28.7A). This observation implied that wakefulness requires a special mechanism, not just the presence of adequate sensory experience. About the same time, the Swiss physiologist Walter Hess found that stimulating the thalamus with low-frequency pulses in an awake animal produced a slow-wave sleep as measured by cortical EEG activity (Figure 28.7B). These important experiments showed that sleep entails a patterned interaction between the thalamus and cortex.

The saccade-like rapid eye movements that define REM sleep arise because, in the absence of external visual stimuli, endogenously generated signals from the **pontine reticular formation** are transmitted to the motor region of the superior colliculus. As described in Chapter 20, collicular neurons project to the **paramedialpontine reticular formation** (PPRF), which coordinates timing and direction of eye movements. REM sleep is also characterized by EEG waves that originate in the pontine reticular formation and propagate through the lateral geniculate nucleus of the thalamus to the occipital cortex. These **pontine-geniculo-occipital (PGO) waves** therefore provide a useful marker for the beginning of REM sleep; they also indicate yet another neural network by which brainstem nuclei can activate the cortex. As already noted, the function of these eye movements is not known.

Human MRI and PET studies have also been used to compare the activity in the awake state and in REM sleep. Activity in the amygdala, parahip-

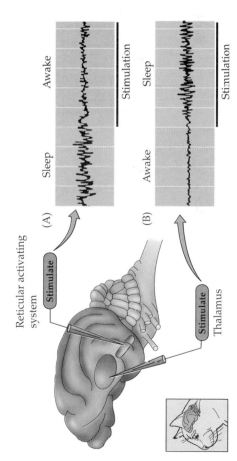

Figure 28.7 Activation of specific neural circuits triggers sleep and wakefulness. (A) Electrical stimulation of the cholinergic neurons near the junction of pons and midbrain (the reticular activating system) causes a sleeping cat to awaken. (B) Electrical stimulation of the thalamus causes an awake cat to fall asleep. Graphs show EEG recordings before and during stimulation.

BOX D
Consciousness

As the text explains, the mechanisms of sleep and wakefulness determine mental status at any moment on a continuum that normally ranges from stage IV sleep to high alert. There is, however, another way that "wakefulness" has been considered, namely from the perspective of *consciousness* as such. Although the brainstem circuits and projections supporting consciousness are beginning to be understood, these neurological aspects of consciousness are—not surprisingly—insufficient to satisfy philosophers, theologians, and neuroscientists interested in the more general issues that the phenomenon of consciousness raises.

The common concern of these diverse groups is what it takes to generate consciousness in the first place, whether in humans, other animals, or machines. Despite a longstanding debate about consciousness in other animals, it would be foolish to assert that we alone possess this obviously useful biological property. From a purely logical vantage, it is impossible, strictly speaking, to state whether *any* being is conscious other than ourselves; as philosophers have long pointed out, we must inevitably take the consciousness of others on faith (or perhaps on the basis of common sense). Nonetheless, it is reasonable to assume that animals with brains structured much like ours (other primates and indeed mammals generally) have much the same ability as we do to be self-aware. Reflection on the past and planning for the future made possible by self-awareness are surely evolutionary advantages. At what phylogenetic level this assumption of consciousness becomes untenable is, of course, unknown. But another reasonable supposition would be that consciousness is present in animals in proportion to the complexity of their brains.

The question of whether machines can ever be conscious is a much more contentious issue, but also subject to common sense informed by some knowledge of neurobiology. If one rejects dualism (the Cartesian proposition that consciousness or "mind" is an entity really subject to neurobiological investigation, these fascinating questions about consciousness are not really subject to neurobiological investigation. Although a number of contemporary scientists have advocated the idea that neurobiology will reveal the "basis" of consciousness in the near future (Nobel Laureates seem especially prone to such pontification and include such outspoken individuals as John Eccles, Francis Crick, and Gerald Edelman), that is not likely to happen. As information grows about the nature of other animals, about computers, and indeed about the brain, the question "What is consciousness?" may simply fade from center stage in much the same way that the question "What is life?" (which stirred up similar debate early in the twentieth century) has become less and less frequently asked as biologists and others recognized it as an ill-posed problem that admits no definite answer.

A great deal of literature on the subject notwithstanding, these fascinating questions about consciousness are not really subject to neurobiological investigation. Although a number of contemporary scientists have advocated the idea that neurobiology will reveal the "basis" of consciousness in the near future (Nobel Laureates seem especially prone to such pontification and include such outspoken individuals as John Eccles, Francis Crick, and Gerald Edelman), that is not likely to happen. As information grows about the nature of other animals, about computers, and indeed about the brain, the question "What is consciousness?" may simply fade from center stage in much the same way that the question "What is life?" (which stirred up similar debate early in the twentieth century) has become less and less frequently asked as biologists and others recognized it as an ill-posed problem that admits no definite answer.

beyond the ken of physics, chemistry, and biology, and not therefore subject to the rules of these disciplines), it follows that a structure could be built by sufficiently wise agents that either mimicked our own consciousness by being effectively isomorphic with brains, or achieved consciousness using physically different elements (e.g., computer elements) in sufficiently biological ways to allow consciousness. There are, of course, many who feel otherwise. An interesting argument in this respect was put forward by the philosopher John Searle in part to rebut those who imagine that present-day computers, because their operations in some ways resemble mental processes, can be considered to have the rudiments of consciousness. His famous "Chinese Room" analogy describes a cubicle in which workers are handed English letters that they then translate into Chinese characters. The workers themselves have no knowledge of English or Chinese, but simply a set of rules that enables the characters to be efficiently translated. The output of the room is sensible statements in Chinese. Yet the workers have no knowledge of the meaning of the information they are dealing with or of the room's larger purpose. Searle uses this image to emphasize that meaningful output from a computer, however sophisticated, cannot provide evidence for consciousness within it. Despite clever arguments like this that deflate simplistic assertions that extant machines exhibit a rudimentary form of consciousness, Searle does not

dispute the argument that nothing *in principle* stands in the way of constructing conscious entities.

References

CHURCHLAND, P. M. AND P. S. CHURCHLAND (1990) Could a machine think? Sci. Am. 262 (Jan.): 32–37.

CRICK, F. (1995) *The Astonishing Hypothesis: The Scientific Search for the Soul.* New York: Touchstone.

CRICK, F. AND C. KOCH (1998) Consciousness and neuroscience. Cerebral Cortex 8: 97–107.

PENROSE, R. (1996) *Shadows of the Mind: A Search for the Missing Science of Consciousness.* Oxford: Oxford University Press.

SEARLE, J. R. (1992) *The Rediscovery of the Mind.* Cambridge, MA: MIT Press.

SEARLE, J. R. (2000) Consciousness. Ann. Rev. Neurosci. 23: 557–578.

TONONI, G. AND G. EDELMAN (1998) Consciousness and complexity. Science 282: 1846–1851.

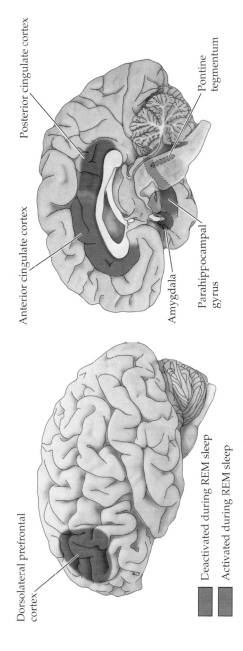

Figure 28.8 Diagram showing cortical regions whose activity is increased or decreased during REM sleep. (After Hobson et al., 1998.)

pocampus, pontine tegmentum, and anterior cingulate cortex are all increased in REM sleep, whereas activity in the dorsolateral prefrontal and posterior cingulate cortices is decreased (Figure 28.8). The increase in limbic system activity, coupled with a marked decrease in the influence of the frontal cortex during REM sleep, presumably explains some characteristics of dreams (e.g., their emotionality and the often inappropriate social content; see Chapter 26 for the normal role of the frontal cortex in determining behavior that is appropriate to circumstances in the waking state).

Most investigators now agree that a key component of the reticular activating system is a group of **cholinergic nuclei** near the **pons-midbrain junction**, which project to thalamocortical neurons. The relevant neurons in these nuclei are characterized by high discharge rates during waking and in REM sleep, and by quiescence during non-REM sleep. When stimulated, they cause "desynchronization" of the electroencephalogram (that is, a shift of EEG activity from high-amplitude, synchronized waves to lower-amplitude, higher-frequency, desynchronized ones) (see Figure 28.7A). These features imply that activity of cholinergic neurons in the reticular activating system is a primary cause of wakefulness and REM sleep, and that their relative inactivity is important for producing non-REM sleep.

Activity of these neurons is not, however, the only cellular basis of wakefulness; also involved are **noradrenergic neurons** of the **locus coeruleus** and **serotonergic neurons** of the **raphe nuclei**. The cholinergic and monoaminergic networks responsible for the awake state are periodically inhibited by neurons in the ventrolateral preoptic nucleus (VLPO) of the hypothalmus (see Figure 28.4). Thus, activation of VLPO neurons contributes to the onset of sleep, and lesions of VLPO neurons produce insomnia. These complex interactions and effects are summarized in Table 28.1. Both monoaminergic and cholinergic systems are active during the waking state and suppress REM sleep. Thus, decreased activity of the monoaminergic and cholinergic systems leads to the onset of non-REM sleep. In REM sleep, the monoaminergic and serotonin neurotransmitter levels markedly decrease, while the cholinergic levels increase to approximately the levels found in the awake state.

With so many systems and transmitters involved in the different phases of sleep, it is clear that a variety of drugs can influence the sleep cycle.

TABLE 28.1
Summary of the Cellular Mechanisms that Govern Sleep and Wakefulness

Brainstem nuclei responsible	Neurotransmitter involved	Activity state of the relevant brainstem neurons
WAKEFULNESS		
Cholinergic nuclei of pons-midbrain junction	Acetylcholine	Active
Locus coeruleus	Norepinephrine	Active
Raphe nuclei	Serotonin	Active
NON-REM SLEEP		
Cholinergic nuclei of pons-midbrain junction	Acetylcholine	Decreased
Locus coeruleus	Norepinephrine	Decreased
Raphe nuclei	Serotonin	Decreased
REM SLEEP ON		
Cholinergic nuclei of pons-midbrain junction	Acetylcholine	Active (PGO waves)
Raphe nuclei	Serotonin	Inactive
Locus coeruleus	Norepinephrine	Active
REM SLEEP OFF		
Locus coeruleus	Norepinephrine	Active

More generally, these effects on mental status are achieved by modulating the rhythmicity of interactions between the thalamus and the cortex. Thus, the activity of several ascending systems from the brainstem decreases both the rhythmic bursting of the thalamocortical neurons and the related synchronized activity of cortical neurons (hence the diminution and ultimate disappearance of high-voltage, low-frequency slow waves during waking and REM sleep).

To appreciate how different sleep states reflect modulation of thalamocortical cell activity, some understanding of the electrophysiological responses of these neurons is useful. As might be expected from the summary in Table 28.1, thalamocortical neurons receive ascending projections from the locus coeruleus (noradregeneric), raphe nuclei (serotonin), and reticular activating system (acetylcholine), and, as their name implies, project to cortical pyramidal cells. The primary characteristic of thalamocortical neurons is that they can be in one of two stable electrophysiological states (Figure 28.9): an intrinsic **oscillatory state**, and a **tonically active state** that is generated when neurons are depolarized, as occurs when the reticular activating system generates wakefulness (see Figure 28.7). In the tonic firing state, thalamocortical neurons transmit information to the cortex that matches the spike trains encoding peripheral stimuli. In contrast, when thalamocortical neurons are in the oscillatory/bursting mode, the neurons in the thalamus become synchronized with those in the cortex, essentially "disconnecting" the cortex from the outside world. During slow-wave sleep, when EEG recordings show the lowest frequency and the highest amplitude, this disconnection is maximal.

The oscillatory state of thalamocortical neurons can be transformed into the tonically active state by activating the cholinergic or monoaminergic projections from the brainstem nuclei (Figure 28.10). Moreover, the oscillatory

Thalamocortical Interactions

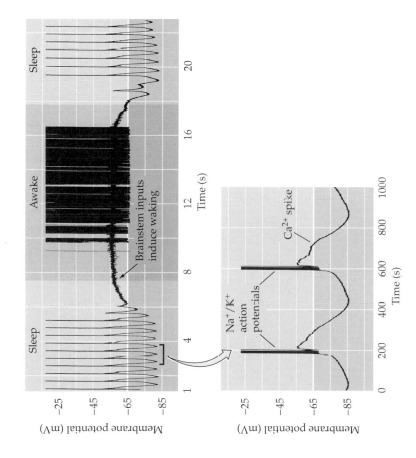

Figure 28.9 Recordings from a thalamocortical neuron, showing the *oscillatory* mode corresponding to a sleep state, and the *tonically active mode* corresponding to an awake state. An expanded view of oscillatory phase is shown below. Bursts of action potentials are evoked only when the thalamocortical neuron is hyperpolarized sufficiently to activate low-threshold calcium channels. These bursts account for the spindle activity in EEG recordings in stages II and III sleep. Depolarizing the cell either by injecting current or by stimulating the reticular activating system transforms this oscillatory activity into a tonically active mode. (After McCormick and Pape, 1990.)

state is stabilized by hyperpolarizing the relevant thalamic cells. Such hyperpolarization can occur as a consequence of stimulation by GABAergic neurons in the thalamic reticular nucleus. These neurons receive ascending information from the brainstem and descending projections from cortical neurons, and they contact the thalamocortical neurons. When neurons in the reticular nucleus undergo a burst of activity, they cause thalamocortical neurons to generate short bursts of action potentials, which in turn generate spindle activity in cortical EEG recordings (indicating a lighter sleep state; see Figure 28.5).

These admittedly complex interactions between the cortex and the relevant subcortical systems are summarized in Figure 28.11. In brief, the control of sleep and wakefulness depends on brainstem and hypothalamic modulation of the thalamus and cortex. It is this thalamocortical loop that generates the EEG signature of mental function along the continuum of deep sleep to high alert. The major components of the brainstem modulatory system are the cholinergic nuclei of the pons-midbrain junction, the noradrenergic cells of the locus coeruleus in the pons, the serotonergic raphe nuclei, and GABAergic neurons in the VLPO. All of these nuclei can exert direct as well as indirect effects on the overall cortical acitivity that determines sleep and wakefulness.

Sleep Disorders

An estimated 20% of the U.S. population experience during their lifetime some kind of sleep disorder. These problems range from simply annoying to life threatening. The most common problems are insomnia, sleep apnea, "restless legs" syndrome, and narcolepsy.

Insomnia, or the inability to sleep, has many causes. Short-term insomnia can arise from stress, jet lag, or simply drinking too much coffee. These prob-

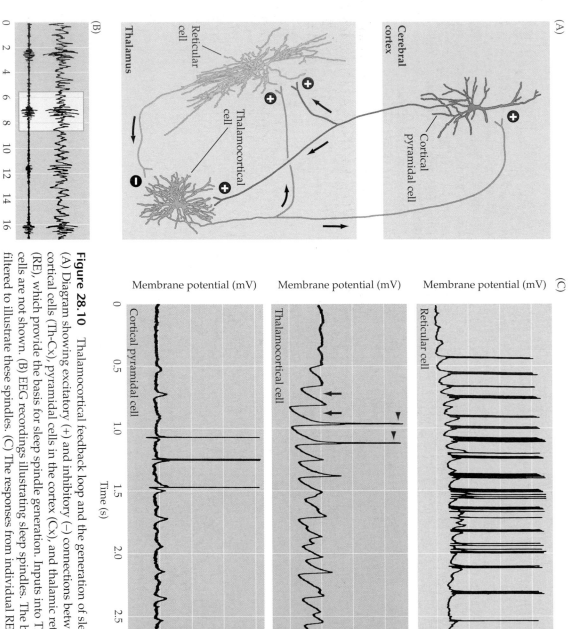

Figure 28.10 Thalamocortical feedback loop and the generation of sleep spindles. (A) Diagram showing excitatory (+) and inhibitory (–) connections between thalamocortical cells (Th-Cx), pyramidal cells in the cortex (Cx), and thalamic reticular cells (RE), which provide the basis for sleep spindle generation. Inputs into Th-Cx and RE cells are not shown. (B) EEG recordings illustrating sleep spindles. The bottom trace is filtered to illustrate these spindles. (C) The responses from individual RE, Th-Cx, and Cx cells during the generation of the middle spindle [box in (B)]. Note that when the GABAergic RE cells are in the bursting mode they hyperpolarize the Th-Cx cells (arrows), causing them to burst (arrowheads) (see Figure 28.9). This bursting behavior is apparent as spikes in Cx cells, and as spindles in EEG recordings. (After Steriade et al., 1997.)

lems can usually be prevented by improving sleep habits, avoiding stimulants like caffeine at night, and in some cases taking sleep-promoting medications. More serious insomnia is associated with psychiatric disorders such as depression that presumably affect the balance between the cholinergic, adrenergic, and serotinergic systems that control the onset and duration of the sleep cycles. Long-term insomnia is a particular problem in the elderly because they sleep less, are subject to more depression, and frequently take medications that affect the relevant neurotransmitter systems.

Sleep apnea refers to interrupted breathing during sleep that affects about 18 million Americans, most often obese males in middle age. A person suffering from sleep apnea may awaken hundreds of times during the night,

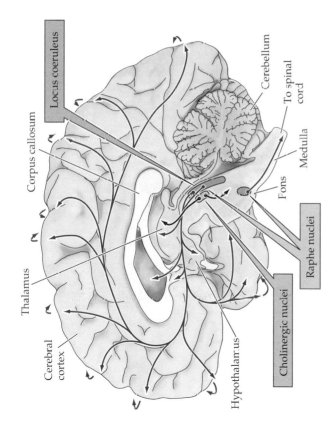

Figure 28.11 Summary of subcortical/cortical interactions that generate wakefulness and sleep. A variety of brainstem nuclei using several different neurotransmitters determine mental status on a continuum that ranges from deep sleep to a high level of alertness. These nuclei, which include the cholinergic nuclei of the pons-midbrain junction, the locus coeruleus, and the raphe nuclei, all have widespread ascending and descending connections (arrows) to other regions, which explain their numerous effects. Curved arrows along the perimeter of the cortex indicate the innervation of lateral cortical regions not shown in this plane of section.

with the result that they have little or no slow-wave sleep and spend less time in REM sleep (Figure 28.12). Consequently, these individuals are chronically tired in the daytime and often suffer from depression that exacerbates the problem. In some high-risk individuals, sleep apnea may even lead to sudden death from respiratory arrest during sleep. The underlying problem is that the airway collapses during breathing, thus blocking air flow. In normal sleep, breathing slows and muscle tone decreases throughout the body, including the tone of the pharynx. If the brainstem circuitry regulating commands to the chest wall or to pharyngeal muscles is affected, or if the airway is compressed because of the weight on it during sleep, the pharynx tends to collapse as the muscles relax during breathing. As a result, oxygen levels decrease and the ensuing reflex to inspire more air awakens people suffering from this disorder.

A third sleep disorder is **"restless legs syndrome,"** a familial disorder causing unpleasant crawling, prickling, or tingling sensations in the legs and

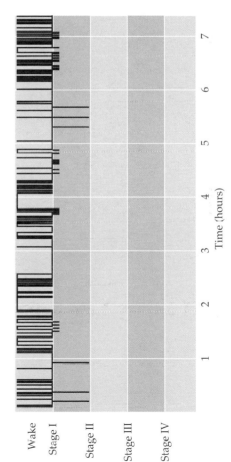

Figure 28.12 Sleep apnea. The sleep pattern of a patient with obstructive sleep apnea. In this condition, patients awake frequently and never descend into stages III or IV sleep. The brief descents below stage I in the record represent short periods of REM sleep. (After Carskadon and Dement, 1989, based on data from G. Nino-Murcia.)

feet, and an urge to move them about for relief. This problem is surprisingly common, affecting about 12 million (mostly elderly) Americans. The result is constant leg movement during the day and fragmented sleep patterns at night. The neurobiology of this particular problem is not understood, but the problem often can be relieved by drugs that inhibit the release of dopamine.

The sleep disorder that is best understood in neurobiological terms is **narcolepsy**, a chronic disorder that affects about 250,000 people in the United States. Individuals with narcolepsy have frequent "REM sleep attacks" during the day, in which they enter REM sleep from wakefulness without going through non-REM sleep. Such individuals may become cataplectic during these episodes, and can fall down (cataplexy refers to a temporary loss of muscle control). Insight into the causes of narcolepsy have come from studies of dogs suffering from a genetic disorder similar to the human disease. In these animals, narcolepsy is caused by a mutation of the hypocretin receptor 2 gene (*Hcrtr2*). Hypocretins (or orexins) are neuropeptides homologous to secretin, which is found exclusively in cells in the tuberal region of the hypothalamus (see Figure 28.4). These hypothalamic cells project to the reticular formation, the locus coeruleus and the dorsal raphe nucleus, all of which control various aspects of sleep (see Table 28.1). The evidence in dogs suggests that the *Hcrtr2* mutation cause a hyperexcitability of the neurons that generate REM sleep, and/or impairment of the circuits that inhibit REM sleep.

Summary

All animals have a restorative cycle of rest following activity, but only mammals divide the period of rest into distinct phases of non-REM and REM sleep. Why mammals (and many other animals) need a restorative phase of suspended consciousness accompanied by decreased metabolism and lowered body temperature is not known. Even more mysterious is why the human brain is periodically active during sleep at levels not appreciably different from the waking state (that is, the neural activity during REM sleep). Despite the electroencephalographic similarities, the psychological states of wakefulness and REM sleep are obviously different. The highly organized sequence of human sleep states is actively generated by nuclei in the brainstem, most importantly the cholinergic nuclei of the pons-midbrain junction, the noradrenergic cells of the locus coeruleus, and the serotonergic neurons of the raphe nuclei. The activity of the relevant cell groups controls the degree of mental alertness on a continuum from deep sleep to waking attentiveness. These brainstem systems are in turn influenced by a circadian clocks located in the suprachiasmatic nucleus and VLPO of the hypothalamus. The clock adjusts periods of sleep and wakefulness to appropriate durations during the 24-hour cycle of light and darkness that is fundamental to life on Earth.

ADDITIONAL READING

Reviews

HOBSON, J. A. (1990) Sleep and dreaming. J. Neurosci. 10: 371–382.

HOBSON, J. A., R. STICKGOLD AND E. F. PACE-SCHOTT (1998) The neuropsychology of REM sleep dreaming. NeuroReport 9: R1–R14.

LU J., M. A. GRECO, P. SHIROMANI AND C. B. SAPER (2000) Effect of lesions of the ventrolateral preoptic nucleus on NREM and REM sleep. J. Neurosci. 20: 3830–3842.

McCARLEY, R. W. (1995) Sleep, dreams and states of consciousness. In Neuroscience in Medicine, P. M. Conn (ed.). Philadelphia: J. B. Lippincott, pp. 535–554.

McCORMICK, D. A. (1989) Cholinergic and noradrenergic modulation of thalamocortical processing. Trends Neurosci. 12: 215–220.

McCORMICK, D. A. (1992) Neurotransmitter actions in the thalamus and cerebral cortex. J. Clin. Neurophysiol. 9: 212–223.

POSNER, M. I. AND S. DEHAENE (1994) Attentional networks. Trends Neurosci. 17: 75–79.

PROVENCIO, I. AND 5 OTHERS (2000) A novel human opsin in the inner retina. J. Neurosci. 20: 600–605.

SAPER, C. B. AND F. PLUM (1985) Disorders of consciousness. In Handbook of Clinical Neurology, Volume 1 (45): Clinical Neuropsychology, J. A. M. Frederiks (ed.). Amsterdam: Elsevier Science Publishers, pp. 107–128.

STERIADE, M. (1992) Basic mechanisms of sleep generation. Neurol. 42: 9–18.

STERIADE, M. (1999) Coherent oscillations and short-term plasticity in corticothalamic networks. TINS 22: 337–345.

STERIADE, M., D. A. McCORMICK AND T. J. SEJNOWSKI (1993) Thalamocortical oscillations in the sleeping and aroused brain. Science 262: 679–685.

Important Original Papers

ASERINSKY, E. AND N. KLEITMAN (1953) Regularly occurring periods of eye motility, and concomitant phenomena, during sleep. Science 118: 273–274.

ASCHOFF, J. (1965) Circadian rhythms in man. Science 148: 1427–1432.

CRICK, F. AND G. MITCHISON (1983) The function of dream sleep. Science 304: 111–114.

DEMENT, W. C. AND N. KLEITMAN (1957) Cyclic variation in EEG during sleep and their relation to eye movements, body motility and dreaming. Electroenceph. Clin. Neurophysiol. 9: 673–690.

MORUZZI, G. AND H. W. MAGOUN (1949). Brain stem reticular formation and activation of the EEG. Electroenceph. Clin. Neurophysiol. 1: 455–473.

ROFFWARG, H. P., J. N. MUZIO AND W. C. DEMENT (1966) Ontogenetic development of the human sleep-dream cycle. Science 152: 604–619.

Book

HOBSON, J. A. (1989) *Sleep*. New York: Scientific American Library.

LAVIE, P. (1996). *The Enchanted World of Sleep*. (Transl. by A. Barris). New Haven: Yale University Press.

Chapter 29

Emotions

Overview

The subjective feelings and associated physiological states known as emotions are essential features of normal human experience. Moreover, some of the most devastating psychiatric problems involve emotional (affective) disorders. Although everyday emotions are as varied as happiness, surprise, anger, fear, and sadness, they share some common characteristics: All emotions are expressed through both visceral motor changes and stereotyped somatic motor responses, especially movements of the facial muscles. These responses accompany subjective experiences that are not easily described, but which are much the same in all human cultures. Emotional expression is closely tied to the visceral motor system, and therefore entails the activity of all of the central brain structures that govern the preganglionic neurons in the brainstem and spinal cord. Historically, the neural centers that coordinate emotional responses have been grouped under the rubric of the limbic system. More recently, however, several brain regions in addition to the classical limbic system have been shown to play a pivotal role in emotional processing, including the amygdala and several cortical areas in the orbital and medial aspects of the frontal lobe. This broader constellation of cortical and subcortical regions encompasses not only the central components of the visceral motor system but also regions in the forebrain and diencephalon that motivate lower motor neuronal pools concerned with the somatic expression of emotional behavior. Effectively, the concerted action of these diverse brain regions constitutes an emotional motor system.

Physiological Changes Associated with Emotion

The most obvious signs of emotional arousal involve changes in the activity of the visceral motor (autonomic) system (see Chapter 21). Thus, increases or decreases in heart rate, cutaneous blood flow (blushing or turning pale), piloerection, sweating, and gastrointestinal motility can all accompany various emotions. These responses are brought about by changes in activity in the sympathetic, parasympathetic, and enteric components of the visceral motor system, which govern smooth muscle, cardiac muscle, and glands throughout the body. As discussed in Chapter 21, Walter Cannon argued that intense activity of the sympathetic division of the visceral motor system prepares the animal to fully utilize metabolic and other resources in challenging or threatening situations. Conversely, activity of the parasympathetic division (and the enteric division) promotes a building up of metabolic reserves. Cannon further suggested that the natural opposition of the expenditure and storage of resources is reflected in a parallel opposition of the emotions associated with these different physiological states. As Cannon pointed out, "The desire for food and drink, the relish of taking them, all the pleasures of the table are naught in the presence of anger or great anxiety."

Activation of the visceral motor system, particularly the sympathetic division, was long considered an all-or-nothing process. Once effective stimuli engaged the system, it was argued, a widespread discharge of all of its components ensued. More recent studies have shown that the responses of the autonomic nervous system are actually quite specific, with different patterns of activation characterizing different situations and their associated emotional states. Indeed, emotion-specific expressions produced voluntarily can elicit distinct patterns of autonomic activity. For example, if subjects are given muscle-by-muscle instructions that result in facial expressions recognizable as anger, disgust, fear, happiness, sadness, or surprise without being told which emotion they are simulating, each pattern of facial muscle activity is accompanied by specific and reproducible differences in visceral motor activity (as measured by indices such as heart rate, skin conductance, and skin temperature). Moreover, autonomic responses are strongest when the facial expressions are judged to most closely resemble actual emotional expression and are often accompanied by the subjective experience of that emotion! One interpretation of these findings is that when voluntary facial expressions are produced, signals in the brain engage not only the motor cortex but also some of the circuits that produce emotional states. Perhaps this relationship helps explain how good actors can be so convincing. Nevertheless, we are quite adept at recognizing the difference between a contrived facial expression and the spontaneous smile that accompanies a pleasant emotional state (Box A).

This evidence, along with many other observations, indicates that one source of emotion is sensory drive from muscles and internal organs. This input forms the sensory limb of reflex circuitry that allows rapid physiological changes in response to altered conditions. However, physiological responses can also be elicited by complex and idiosyncratic stimuli mediated by the forebrain. For example, an anticipated tryst with a lover, a suspenseful episode in a novel or film, stirring patriotic or religious music, or dishonest accusations can all lead to autonomic activation and strongly felt emotions. The neural activity evoked by such complex stimuli is relayed from the forebrain to autonomic and somatic motor nuclei via the hypothalamus and brainstem reticular formation, the major structures that coordinate the expression of emotional behavior (see next section).

In summary, emotion and motor behavior are inextricably linked. As William James put it more than a century ago:

> What kind of an emotion of fear would be left if the feeling neither of quickened heart-beats nor of shallow breathing, neither of trembling lips nor of weakened limbs, neither of goose-flesh nor of visceral stirrings, were present, it is quite impossible for me to think ... I say that for us emotion dissociated from all bodily feeling is inconceivable.
>
> William James, 1893 (*Psychology*: p. 379)

The Integration of Emotional Behavior

In 1928, Phillip Bard reported the results of a series of experiments that pointed to the **hypothalamus** as a critical center for coordination of both the autonomic and somatic components of emotional behavior (see Box A in Chapter 21). Bard removed both cerebral hemispheres (including the cortex, underlying white matter, and basal ganglia) in a series of cats. When the anesthesia had worn off, the animals behaved as if they were enraged. The angry behavior occurred spontaneously and included the usual autonomic correlates of this emotion: increased blood pressure and heart rate, retraction of the nictitating membranes (the thin connective tissue sheets associated with feline eyelids), dilation of the pupils, and erection of the hairs on the back and tail. The cats also exhibited somatic motor components of anger, such as arching the back, extending the claws, lashing the tail, and snarling. This behavior was called **sham rage** because it had no obvious target. Bard showed that a complete response occurred as long as the caudal hypothalamus was intact (Figure 29.1). Sham rage could not be elicited, however, when the brain was transected at the junction of the hypothalamus and midbrain (although some uncoordinated components of the response were still apparent). Bard suggested that whereas the subjective experience of emotion might depend on an intact cerebral cortex, the expression of coordinated emotional behaviors does not necessarily depend on cortical processes. He also emphasized that emotional behaviors are often directed toward self-preservation (a point made by Charles Darwin in his classic book on the evolution of emotion), and that the functional importance of emotions in all mammals is consistent with the involvement of phylogenetically older parts of the nervous system.

Complementary results were reported by Walter Hess, who showed that electrical stimulation of discrete sites in the hypothalamus of awake, freely moving cats could also lead to a rage response, and even to subsequent attack behavior. Moreover, stimulation of other sites in the hypothalamus caused a defensive posture that resembled fear. In 1949, a share of the Nobel Prize in Physiology or Medicine was awarded to Hess "for his discovery of the functional organization of the interbrain [hypothalamus] as a coordinator of the activities of the internal organs." Experiments like those of Bard and Hess led to the important conclusion that the basic circuits for organized behaviors accompanied by emotion are in the diencephalon and the brainstem structures connected to it. Furthermore, their work emphasized that the control of the involuntary motor system is not entirely separable from the control of the voluntary pathways.

The routes by which the hypothalamus and other forebrain structures influence the visceral and somatic motor systems are complex. The major targets of the hypothalamus lie in the **reticular formation**, the tangled web of nerve cells and fibers in the core of the brainstem. This structure contains over 100 identifiable cell groups, including some of the nuclei that control the brain states associated with sleep and wakefulness described in the previous chapter. Other important circuits in the reticular formation control cardiovascular function, respiration, urination, vomiting, and swallowing, as described in Chapter 21. The reticular neurons receive hypothalamic input and feed into both somatic and autonomic effector systems in the brainstem and spinal cord. Their activity can therefore produce widespread visceral motor and somatic motor responses, often overriding reflex function and sometimes involving almost every organ in the body (as implied by Cannon's dictum about the sympathetic preparation of the animal for fight or flight).

In addition to the hypothalamus, other sources of descending projections from the forebrain to the brainstem reticular formation contribute to the expression of emotional behavior. Collectively, these additional centers in the forebrain are considered part of the **limbic system**, which is described in the following section. These descending influences on the expression of somatic and visceral motor behavior arise outside of the classic motor cortical areas in the frontal lobe.

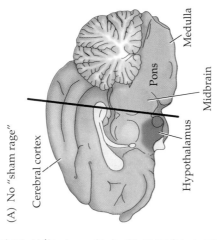

(A) No "sham rage"

Cerebral cortex

Medulla

Pons

Midbrain

Hypothalamus

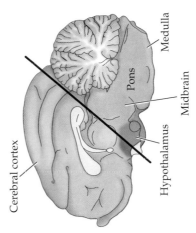

(B) "Sham rage" remains

Cerebral cortex

Medulla

Pons

Midbrain

Hypothalamus

Figure 29.1 Midsagittal view of a cat's brain, illustrating the regions sufficient for the expression of emotional behavior. (A) Transection through the midbrain, disconnecting the hypothalamus and brainstem, abolishes "sham rage." (B) The integrated emotional responses associated with "sham rage" survive removal of the cerebral hemispheres as long as the caudal hypothalamus remains intact. (After LeDoux, 1987.)

BOX A
Facial Expressions: Pyramidal and Extrapyramidal Contributions

In 1862, the French neurologist and physiologist G.-B. Duchenne de Boulogne published a remarkable treatise on facial expressions. His work was the first to systematically examine the contributions of small groups of cranial muscles to the expressions that communicate the rich experience of human emotion. Duchenne reasoned that "one would be able, like nature herself, to paint the expressive lines of the emotions of the soul on the face of man." In so doing, he sought to understand how the coordinated contractions of groups of muscles express distinct, pan-cultural emotional states. To achieve this goal, he pioneered the use of transcutaneous electrical stimulation (called "faradization" after the British chemist and physicist Michael Faraday) to activate single muscles and small groups of muscles in the face, dorsal surface of the

head, and neck. Duchenne also documented the faces of his subjects with another technological innovation: photography (see figure A). His seminal contribution was the identification of muscles and muscle groups, such as the obicularis oculi, that cannot be activated by force of the will, but only "put into play by the sweet emotions of the soul." Duchenne concluded that the emotion-driven contraction of these muscle groups surrounding the eyes, together with the zygomaticus major, communicates the genuine experience of happiness, joy and laughter. The smile characteristic of these emotional states has therefore been termed the "Duchenne smile" by subsequent investigators.

In normal individuals, such as the Parisian shoemaker illustrated here (figure A), the difference between a forced smile (produced by voluntary contrac-

tion or electrical stimulation of facial muscles) and a spontaneous, "emotional" smile testifies to the convergence of descending motor signals from different forebrain centers onto premotor and motor neurons in the brainstem that control the facial musculature. In contrast to the Duchenne smile, the contrived smile of volition (sometimes called a "pyramidal smile") is driven by the motor cortex, which communicates with the brainstem and spinal cord via the pyramidal tracts. The Duchenne smile is motivated by accessory motor areas in the prefrontal cortex and ventral parts of the basal ganglia that access brainstem nuclei via multisynaptic, "extrapyramidal" pathways through the brainstem reticular formation.

Studies of patients with specific neurological injury to these separate descending systems of control have further differentiated the forebrain centers responsible for control of the muscles of facial expression (figure B). Patients with unilateral facial paralysis due to damage of descending pathways from the motor cortex (upper motor neuron

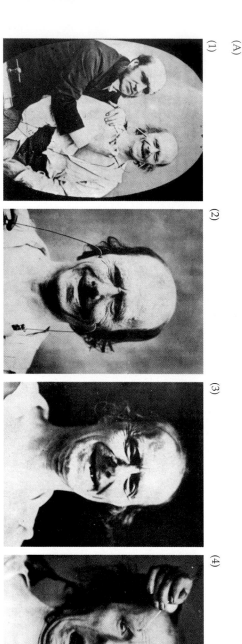

(A)

(1)　　　　(2)　　　　(3)　　　　(4)

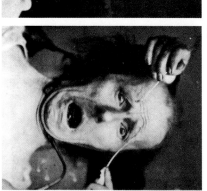

(A) Duchenne and one of his subjects undergoing "faradization" of the muscles of facial expression (1). Bilateral electrical stimulation of the zygomaticus major mimicked a genuine expression of happiness (2), although closer examination shows insufficient contraction of the obicularis oculi (surrounding the eyes) compared to spontaneous laughter (3). Stimulation of the brow and neck produced an expression of "terror mixed with pain, torture … that of the damned" (4); however, the subject reported no discomfort or emotional experience consistent with the evoked contractions.

syndrome; see Chapter 17) are unable to move their lower facial muscles on one side, either voluntarily or in response to commands, a condition called voluntary facial paresis (figure B, left panels). Nonetheless, many such individuals produce symmetrical *involuntary* facial movements when they laugh, frown, or cry in response to amusing or distressing stimuli. In such patients, pathways from regions of the forebrain other than the classical motor cortex in the frontal lobe remain available to activate facial movements in response to stimuli with emotional significance. A much less common form of neurological injury, called emotional facial paresis, demonstrates the opposite set of impairments, i.e., loss of the ability to express emotions by using the muscles of the face without loss of volitional control (figure B, right panels). Such individuals are able to produce symmetrical pyramidal smiles, but fail to display spontaneous emotional expressions involving the facial musculature contralateral to the lesion. These two systems are diagrammed in figure C.

References

DUCHENNE DE BOULOGNE, G.-B. (1862) *Mecanisme de la Physionomie Humaine.* Paris: Editions de la Maison des Sciences de l'Homme. Edited and translated by R. A. Cuthbertson (1990). Cambridge: Cambridge University Press.

TROSCH, R. M., G. SZE, L. M. BRASS AND S. G. WAXMAN (1990) Emotional facial paresis with striatocapsular infarction. J. Neurol. Sci. 98:195–201.

WAXMAN, S. G. (1996) Clinical observations on the emotional motor system. In *Progress in Brain Research*, Vol. 107. G. Holstege, R. Bandler and C. B. Saper (eds). Amsterdam: Elsevier, pp. 595–604.

(B)

"Pyramidal smile"

Duchenne smile

Voluntary facial paresis

Emotional facial paresis

(C)

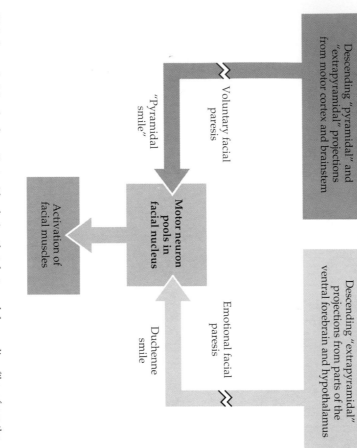

VOLITIONAL MOVEMENT

Descending "pyramidal" and "extrapyramidal" projections from motor cortex and brainstem

NEURAL SYSTEMS FOR EMOTION

Descending "extrapyramidal" projections from parts of the ventral forebrain and hypothalamus

Voluntary facial paresis

"Pyramidal smile"

Emotional facial paresis

Duchenne smile

Motor neuron pools in facial nucleus

Activation of facial muscles

(B) Left panels: Mouth of a patient with a lesion that destroyed descending fibers from the right motor cortex displaying voluntary facial paresis. When asked to show her teeth, the patient was unable to contract the muscles on the left side of her mouth (upper left), yet her spontaneous smile in response to a humorous remark is nearly symmetrical (lower left). Right panels: Face of a child with a lesion in the left forebrain that interrupted descending pathways from nonclassical motor cortical areas, producing emotional facial paresis. When asked to smile volitionally, the contractions of the facial muscles are nearly symmetrical (upper right). In spontaneous response to a humorous comment, however, the right side of the patient's face fails to express emotion (lower right). (C) The complementary deficits demonstrated in figure B are explained by selective lesions of one of two anatomically and functionally distinct sets of descending projections that motivate the muscles of facial expression.

(A)

NEURAL SYSTEMS FOR EMOTION

Somatic motor system	**Visceral motor system**
Volitional	*Nonvolitional*
	Nonvolitional

(B)

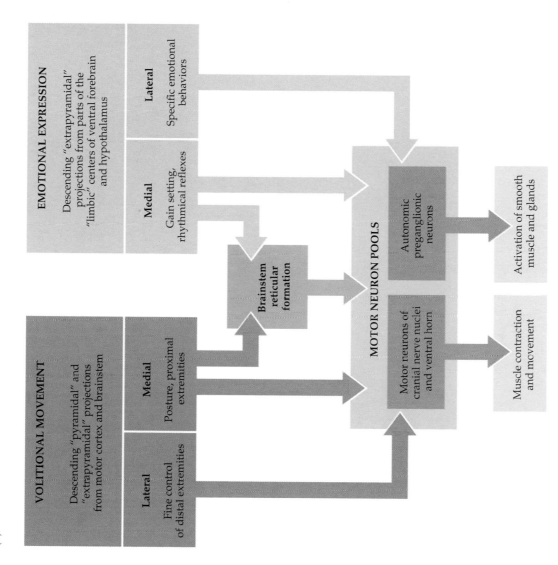

VOLITIONAL MOVEMENT

Descending "pyramidal" and "extrapyramidal" projections from motor cortex and brainstem

Lateral	**Medial**
Fine control of distal extremities	Posture, proximal extremities

EMOTIONAL EXPRESSION

Descending "extrapyramidal" projections from parts of the "limbic" centers of ventral forebrain and hypothalamus

Medial	**Lateral**
Gain setting, rhythmical reflexes	Specific emotional behaviors

Brainstem reticular formation

MOTOR NEURON POOLS

Motor neurons of cranial nerve nuclei and ventral horn	Autonomic preganglionic neurons
Muscle contraction and movement	Activation of smooth muscle and glands

Thus, the descending control of emotional expression entails two parallel systems that are anatomically and functionally distinct (Figure 29.2). The voluntary motor component described in detail in Chapters 17 through 19 comprises the classical motor areas of the frontal lobe and related circuitry in the basal ganglia and cerebellum. The descending pyramidal and extrapyramidal projections from the motor cortex and brainstem ultimately con-

Figure 29.2 Components of the nervous system that organize emotional experience and expression. (A) The neural systems that process emotion include the visceral motor system and forebrain centers that govern the nonvolitional expression of somatic motor behavior. (B) Diagram of the descending systems that control the relevant visceral and somatic motor effectors. Functionally and anatomically distinct centers in the forebrain govern the expression of emotional behavior. Motor cortical areas in the frontal lobe give rise to descending projections that, together with secondary projections arising in the brainstem, are organized into medial and lateral components, as described in Chapter 17. "Limbic" centers in the ventral forebrain and hypothalamus also give rise to medial and lateral descending projections. For both systems of descending projections, the lateral components elicit specific behaviors (e.g., volitional digit movements and involuntary facial expressions), while the medial components provide support for the display of such behaviors. The descending projections of both systems terminate in several integrative centers in the brainstem reticular formation, as well as the motor neuronal pools of the brainstem and spinal cord. In addition, the limbic forebrain centers innervate components of the visceral motor system that govern preganglionic autonomic neurons in the brainstem and spinal cord.

vey the impulses responsible for voluntary somatic movements. In addition to the descending systems that govern volitional movements, several cortical and sub-cortical structures in the ventral forebrain, including related circuitry in the ventral part of the basal ganglia and hypothalamus, give rise to separate descending projections that run parallel to the pathways of the volitional motor system. These descending projections of the ventral forebrain terminate on visceral motor centers in the brainstem reticular formation, preganglionic autonomic neurons, and certain somatic premotor and motor neuronal pools that also receive projections from the volitional motor component. The two types of facial paresis illustrated in Box A underscore this dual nature of descending motor control.

In short, the somatic and visceral activities associated with unified emotional behavior are mediated by the activity of both the somatic and visceral motor neurons, which integrate parallel, descending inputs from a constellation of forebrain sources. The remaining sections of the chapter are devoted to the organization and function of the forebrain centers that specifically govern the experience and expression of emotional behavior.

The Limbic System

Attempts to understand the effector systems that control emotional behavior have a long history. In 1937, James Papez first proposed that specific brain circuits are devoted to emotional experience and expression (much as the occipital cortex is devoted to vision, for instance). In seeking to understand what parts of the brain serve this function, he began to explore the medial aspects of the cerebral hemisphere. In the 1850s, Paul Broca had used the term "limbic lobe" to refer to the part of the cerebral cortex that forms a rim (*limbus* is Latin for rim) around the corpus callosum on the medial face of the hemispheres (Figure 29.3). Two prominent components of this region are the **cingulate gyrus**, which lies above the corpus callosum, and the **hippocampus**, which lies in the medial temporal lobe.

For many years, these structures, along with the olfactory bulbs, were thought to be concerned primarily with the sense of smell. Papez, however, speculated that the function of the limbic lobe might be more related to the emotions. He knew from the work of Bard and Hess that the hypothalamus

Figure 29.3 The so-called limbic lobe includes the cortex on the medial aspect of the cerebral hemisphere that forms a rim around the corpus callosum, including the cingulate gyrus (lying above the corpus callosum) and the parahippocampal gyrus.

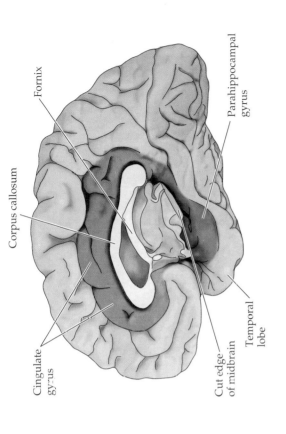

Corpus callosum

Fornix

Parahippocampal gyrus

Cingulate gyrus

Cut edge of midbrain

Temporal lobe

influences the expression of emotion; he also knew, as everyone does, that emotions reach consciousness, and that higher cognitive functions affect emotional behavior. Ultimately, Papez showed that the cingulate cortex and hypothalamus are interconnected via projections from the **mammillary bodies** (part of the posterior hypothalamus) to the **anterior nucleus of the dorsal thalamus**, which projects in turn to the **cingulate gyrus.** The cingulate gyrus (and a lot of other cortex as well) projects to the hippocampus. Finally, he showed that the hippocampus projects via the **fornix** (a large fiber bundle) back to the hypothalamus. Papez suggested that these pathways provided the connections necessary for cortical control of emotional expression, and they became known as the "Papez circuit."

Over time, the circuitry initially described by Papez has been revised to include parts of the **orbital and medial prefrontal cortex, ventral parts of the basal ganglia, the mediodorsal nucleus of the thalamus** (a different thalamic nucleus than the one emphasized by Papez), and a large nuclear mass in the temporal lobe anterior to the hippocampus, called the **amygdala.** This set of structures, together with the hippocampus and cingulate cortex, is generally referred to as the **limbic system** (Figure 29.4). Thus, whereas some of the structures that Papez originally described (the hippocampus, for example) now appear to have little to do with emotional behavior, the amygdala, which was hardly mentioned by Papez, clearly plays a major role in the experience and expression of emotional behavior (Box B).

About the same time that Papez proposed that some of these structures were important for the integration of emotional behavior, Heinrich Klüver and Paul Bucy were carrying out a series of experiments on rhesus monkeys in which they removed a large part of both medial temporal lobes, thus destroying much of the limbic system. They reported a set of abnormal behaviors in these animals that is now known as the Klüver-Bucy syndrome (Box C). Among the most prominent changes was visual agnosia: The animals appeared to be unable to recognize objects, although they were not blind, a deficit similar to that seen in human patients following lesions of the temporal cortex (see Chapter 26). In addition, the monkeys displayed bizarre oral behaviors. For instance, these animals would put objects into their mouths that normal monkeys would not. They exhibited hyperactivity and hypersexuality, approaching and making physical contact with virtually anything in their environment; most importantly, they showed marked changes in emotional behavior. Because they had been caught in the wild,

the monkeys had typically reacted with hostility and fear to humans before their surgery. Postoperatively, however, they were virtually tame. Motor and vocal reactions generally associated with anger or fear were no longer elicited by the approach of humans, and the animals showed little or no excitement when the experimenters handled them. Nor did they show fear when presented with a snake—a strongly aversive stimulus for a normal rhesus monkey. Klüver and Bucy concluded that this remarkable change in behavior was at least partly due to the interruption of the pathways described by Papez. Interestingly, a similar syndrome has been described in humans who have suffered bilateral damage of the temporal lobes.

When it was later demonstrated that the emotional disturbances of the Klüver-Bucy syndrome could be elicited by removal of the amygdala alone, attention turned more specifically to the role of this structure in the control of emotional behavior.

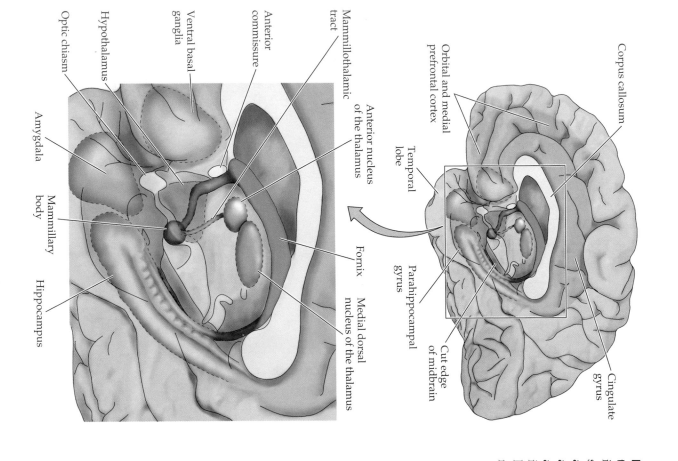

Figure 29.4 Modern conception of the limbic system. Two especially important components of the limbic system not initially emphasized in anatomical accounts of limbic circuitry are the orbital-medial prefrontal cortex and the amygdala. Although some investigators consider the concept of a limbic system to be outmoded, no newer idea has emerged to replace it.

Emotions **633**

BOX B
The Anatomy of the Amygdala

The amygdala is a complex mass of gray matter buried in the anterior-medial portion of the temporal lobe, just rostral to the hippocampus (figures A, B). It comprises multiple, distinct subnuclei and is richly connected to nearby cortical areas on the medial aspect of the hemispheric surface. The amygdala (or amygdaloid complex, as it is often called) contains three functional subdivisions, each of which has a unique set of connections with other parts of the brain (figure C). The medial group of subnuclei has extensive connections with the olfactory bulb and the olfactory cortex. The basolateral group, which is especially large in humans, has major connections with the cerebral cortex, especially the orbital and medial prefrontal cortex. The central and anterior group of nuclei is characterized by connections with the brainstem and hypothalamus and with visceral sensory structures, such as the nucleus of the solitary tract.

(A) Sagittal view of the brain, illustrating the location of the amygdala in the temporal lobe. The line indicates the level of the section in (B). (B) Coronal section through the forebrain at the level of the amygdala.

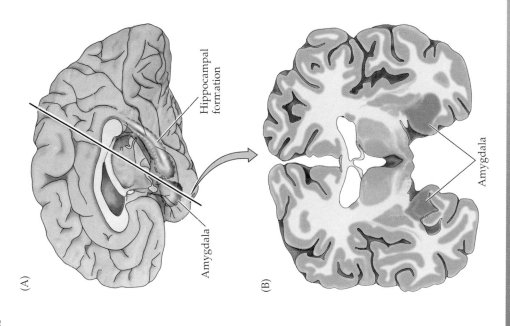

Hippocampal formation

Amygdala

(A)

Amygdala

(B)

The Importance of the Amygdala

Experiments first performed in the late 1950s by John Downer at University College London vividly demonstrated the importance of the amygdala in aggressive behavior. Downer removed one amygdala in rhesus monkeys, at the same time transecting the optic chiasm and the commissures that link the two hemispheres (see Chapter 27). In so doing, he produced an animal with a single amygdala that had access only to visual inputs from the eye on the same side of the head. Downer found that the animals' behavior depended on which eye was used to view the world. When the eye on the side of the intact amygdala was covered, the monkeys behaved in some respects like the monkeys described by Klüver and Bucy; for example, they were relatively placid in the presence of humans. If, however, they were allowed to see only with the eye on the side of the intact amygdala, they reverted to their normal fearful and often aggressive behavior. Thus, in the absence of the amygdala, a monkey does not interpret the significance of the visual stimulus presented by an approaching human in the same way as a normal animal. Importantly,

only visual stimuli presented to the eye on the side of the ablation produced this abnormal state; thus if the animal was touched on either side, a full aggressive reaction occurred, implying that somatic sensory information about both sides of the body had access to the remaining amygdala. Taken together, these results show that the amygdala mediates processes that invest sensory experience with emotional significance.

To better understand the role of the amygdala in evaluating stimuli, and to define more precisely the specific circuits and mechanisms involved, several other animal models of emotional behavior have been developed. One of the most useful is based on conditioned fear responses in rats. Conditioned fear develops when an initially neutral stimulus is repeatedly paired with an inherently aversive one. Over time, the animal begins to respond to the neutral stimulus with behaviors similar to those elicited by the threatening stimulus (i.e., it learns to attach a new meaning to it). Studies of the parts of the brain involved in the development of conditioned fear in rats

---

The amygdala thus links cortical regions that process sensory information with hypothalamic and brainstem effector systems. Cortical inputs provide information about highly processed visual, somatic sensory, visceral sensory, and auditory stimuli. These pathways from sensory cortical areas distinguish the amygdala from the hypothalamus, which receives relatively unprocessed visceral sensory inputs. The amygdala also receives sensory input directly from some thalamic nuclei, the olfactory bulb, and the nucleus of the solitary tract in the brainstem.

Physiological studies have confirmed this convergence of sensory information. Thus, many neurons in the amygdala respond to visual, auditory, somatic sensory, visceral sensory, gustatory, and olfactory stimuli. Moreover, highly complex stimuli (faces, for instance) are often required to evoke a response.

In addition to sensory inputs, the prefrontal cortical connections of the amygdala give it access to more cognitive neocortical circuits, which integrate the emotional significance of sensory stimuli and guide complex behavior.

Projections from the amygdala to the hypothalamus and brainstem (and possibly as far as the spinal cord) allow it to play an important role in the expression of emotional behavior by influencing activity in both the somatic and visceral motor efferent systems.

Reference

PRICE, J. L., F. T. RUSSCHEN AND D. G. AMARAL (1987) The limbic region II: The amygdaloid complex. In *Handbook of Chemical Neuroanatomy*, Vol. 5, *Integrated Systems of the CNS*, Part I, *Hypothalamus, Hippocampus, Amygdala, Retina*. A. Björklund and T. Hökfelt (eds.). Amsterdam: Elsevier, pp. 279–388.

(C)

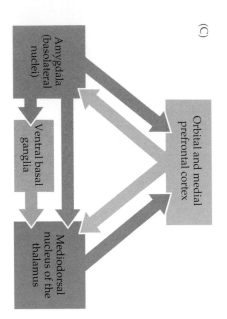

Amygdala (basolateral nuclei)

Ventral basal ganglia

Mediodorsal nucleus of the thalamus

Orbital and medial prefrontal cortex

(C) Connections between the amygdala (specifically, the basolateral group of nuclei) and the orbital and medial prefrontal cortex. The amygdala participates in a "triangular" circuit linking the amygdala, the thalamic mediodorsal nucleus (directly and indirectly via the ventral parts of the basal ganglia), and the orbital and medial prefrontal cortex. These complex interconnections allow direct interactions between the amygdala and the prefrontal cortex, as well as indirect modulation via the circuitry of the ventral basal ganglia.

sylvius

42

BOX C
The Reasoning Behind an Important Discovery*

Paul Bucy explains why he and Heinrich Klüver removed the temporal lobes in monkeys:

When we started out, we were not trying to find out what removal of the temporal lobe would do, or what changes in behavior of the monkeys it would produce. What we found out was completely unexpected! Heinrich had been experimenting with mescaline. He had even taken it himself and had experienced hallucinations. He had written a book about mescaline and its effects. Later Heinrich gave mescaline to his monkeys. He gave everything to his monkeys, even his lunch! He noticed that the monkeys acted as though they experienced paraesthesias in their lips. They licked, bit and chewed their lips. So he came to me and said, "Maybe we can find out where mescaline has its actions in the brain." So I said, "OK."

We began by doing a sensory denervation of the face, but that didn't make any difference to the mescaline-induced behavior. So we tried motor denervation. That didn't make any difference, either. Then we had to sit back and think hard about where to look. I said to Heinrich, "This business of licking and chewing the lips is not unlike what you see in cases of temporal lobe epilepsy. Patients chew and smack their lips inordinately. So, let's take out the uncus." Well, we could just as well take out the whole temporal lobe, including the uncus. So we did.

We were especially fortunate with our first animal. This was an older female.... She had become vicious—absolutely nasty. She was the most vicious animal you ever saw; it was dangerous to go near her. If she didn't hurt you, she

would at least tear your clothing. She was the first animal on which we operated. I removed one temporal lobe.... The next morning my phone was ringing like mad. It was Heinrich, who asked, "Paul, what did you do to my monkey? She is tame!" Subsequently, in operating on non-vicious animals, the taming effect was never so obvious.

That stimulated our getting the other temporal lobe out as soon as we could evaluate her. When we removed the other temporal lobe, the whole syndrome blossomed.

*Excerpt from an interview of Bucy by K. E. Livingston in 1981. K. E. Livingston (1986) Epilogue: Reflections on James Wenceslas Papez, According to Four of his Colleagues. In *The Limbic System: Functional Organization and Clinical Disorders.* B. K. Doane and K. E. Livingston (eds.). New York: Raven Press.

---

have begun to shed some light on this process. Joseph LeDoux and his colleagues at New York University trained rats to associate a tone with a foot shock delivered shortly after onset of the sound. To assess the animals' responses, they measured blood pressure and the length of time the animals crouched without moving (a behavior called "freezing"). Before training, the rats did not react to the tone, nor did their blood pressure change when the tone was presented. After training, however, the onset of the tone caused a marked increase in blood pressure and prolonged periods of behavioral freezing. Using this paradigm, LeDoux worked out the neural circuitry that established the association between the tone and fear (Figure 29.5). First, he demonstrated that the medial geniculate nucleus is necessary for the development of the conditioned fear response. This result is not surprising, since all auditory information that reaches the forebrain travels through the medial geniculate nucleus of the dorsal thalamus (see Chapter 13). He went on to show, however, that the responses were still elicited if the connections between the medial geniculate and auditory cortex were severed, leaving only a projection between the medial geniculate and the amygdala. Furthermore, if the part of the medial geniculate that projects to the amygdala was also destroyed, the fear responses were abolished. Subsequent work in LeDoux's laboratory established that projections from the

amygdala to the midbrain reticular formation are critical in the expression of freezing behavior, and that projections from the amygdala to the hypothalamus control the rise in blood pressure.

Since the amygdala is a site where neural activity produced by both tones and shocks can be processed, it is reasonable to suppose that the amygdala is also the site where learning about fearful stimuli occurs. These results, among others, have led to the broader hypothesis that the amygdala participates in establishing associations between neutral sensory stimuli, such as a mild auditory tone or the sight of inanimate object in the environment, and other stimuli that have some primary reinforcement value (Figure 29.6). The neu-

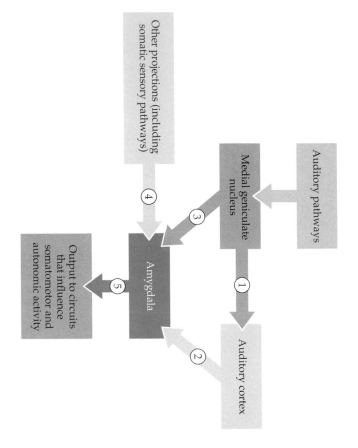

Figure 29.5 Pathways in the rat brain that mediate the association of auditory and aversive stimuli. Information processed by the auditory centers in the brainstem is relayed to the auditory cortex via the medial geniculate nucleus (1). The amygdala receives auditory information from the cortex (2) and from one subdivision of the medial geniculate (3). The amygdala also receives sensory information about other modalities, including pain (4). Thus, the amygdala is in a position to associate diverse sensory inputs, leading to new behavioral and autonomic responses to the stimulus (5).

Other projections (including somatic sensory pathways)

Medial geniculate nucleus

Auditory pathways

Auditory cortex

Amygdala

Output to circuits that influence somatomotor and autonomic activity

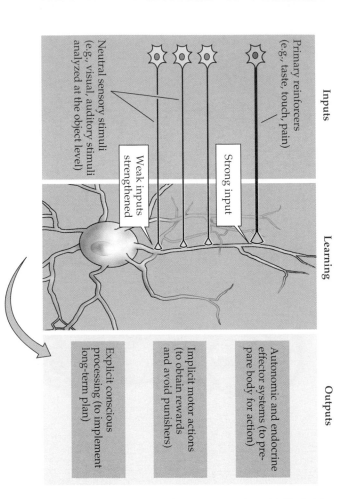

Inputs

Primary reinforcers (e.g., taste, touch, pain)

Neutral sensory stimuli (e.g., visual, auditory stimuli analyzed at the object level)

Weak inputs strengthened

Strong input

Learning

Outputs

Autonomic and endocrine effector systems (to prepare body for action)

Implicit motor actions (to obtain rewards and avoid punishers)

Explicit conscious processing (to implement long-term plan)

Figure 29.6 Model of associative learning in the amygdala relevant to emotional function. Neutral sensory inputs are relayed to principal neurons in the amygdala by projections from "higher order" sensory processing areas that represent objects (e.g., faces), rather than elementary components of sensory stimuli. If these sensory inputs depolarize amygdalar neurons at the same time as inputs that represent other sensations with primary reinforcing value, then associative learning occurs by strengthening synaptic linkages between the previously neutral inputs and the neurons of the amygdala (see Chapter 25 for possible mechanisms). The output of the amygdala then informs a variety of integrative centers responsible for the somatic and visceral motor expression of emotion, and for modifying behavior relevant to seeking rewards and avoiding punishment. (After Rolls, 1999.)

BOX D
Fear and the Human Amygdala: A Case Study

Studies of fear conditioning in rodents show that the amygdala plays a critical role in the association of an innocuous auditory tone with an aversive mechanical sensation. Does this finding imply that the human amygdala is similarly involved in the experience of fear and the expression of fearful behavior? Recent reports of one extraordinary patient support the idea that the amygdala is indeed a key brain center for the experience of fear.

The patient (S.M.) suffers from a rare, autosomal recessive condition called Urbach-Wiethe disease. The disorder caused the bilateral calcification and atrophy of the anterior-medial temporal lobes. As a result, the amygdala in each of S.M.'s hemispheres is extensively damaged, with little or no detectable injury to the hippocampal formation or nearby temporal neocortex. She has no motor or sensory impairment, and no notable deficits in intelligence, memory, or language function. However, when asked to rate the intensity of emotion in a series of photographs of facial expressions, she cannot recognize the emotion of fear (figure A). Indeed, S.M.'s ratings of emotional content in fearful facial expressions were five standard deviations below the ratings of control patients who had suffered damage outside of the anterior-medial temporal lobe.

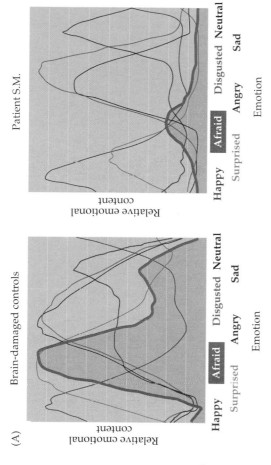

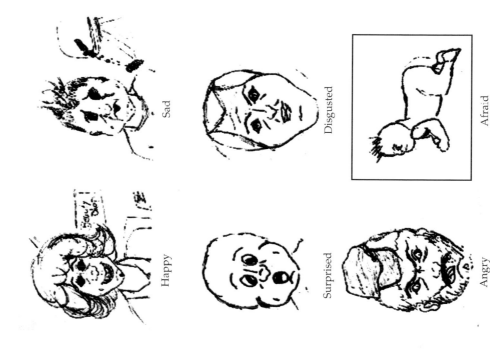

(A) Patients with brain damage outside of the anterior-medial temporal lobe and patient S.M. rated the emotional content of a series of facial expressions. Each colored line represents the intensity of the emotions judged in facial expressions. S.M. recognized happiness, surprise, anger, disgust, sadness and neutral qualities in facial expressions compared to controls. However, she failed to recognize the appropriate amount of fear (orange lines). (B) Sketches made by S.M. when asked to draw facial expressions of emotion.

tral sensory stimuli can be communicated centrally via the special sensory afferent systems, or internal stimuli derived from activation of visceral sensory receptors. The stimuli with primary reinforcement value include sensory stimuli that are inherently rewarding, such as the sight, smell, and taste of food, or stimuli with negative values such as an aversive taste, loud sounds, or painful mechanical stimulation. The associative learning process itself is probably a Hebbian-like mechanism (see Chapters 24 and 25) that strengthens the connections relaying the information about the neutral stimulus, provided that they activate the postsynaptic neurons in the amygdala at the same time as inputs pertaining to the primary reinforcer. The discovery that long-term potentiation (LTP) can be evoked in the amygdala provides further support for this hypothesis. Indeed, the acquisition of conditioned fear in rats is blocked by infusion into the amygdala of NMDA antagonists, which prevents the induction of LTP. Finally, the behavior of patients with selective damage to the anterior-medial temporal lobe indicates that the amygdala plays a similar role in the human experience of fear (Box D).

The Relationship between Neocortex and Amygdala

As these observations on the limbic system (and the amygdala in particular) make plain, understanding the neural basis of emotions requires understanding the role of the cerebral cortex. In animals like the rat, most behavioral responses are highly stereotyped. In more complex brains, however, individual experience is increasingly influential in determining responses to special and even idiosyncratic stimuli. Thus in humans, a stimulus that evokes fear or sadness in one person may have little or no effect on the emotions of another. Although the pathways underlying such responses are not well understood, the amygdala and its interconnections with an array of neocortical areas in the prefrontal cortex and several subcortical structures appear to be especially important in the higher order processing of emotion. In addition to its connections with the hypothalamus and brainstem centers that regulate autonomic function, the amygdala has significant connections with several cortical areas in the orbital and medial aspects of the frontal

The investigators next asked S.M. (and the brain-damaged control subjects) to draw facial expressions of the same set of emotions from memory. Although the subjects obviously differed in artistic abilities and the detail of their renderings, S.M. (who has a background in art) produced skillful pictures of each emotion, except for fear (figure B). At first, she could not produce a sketch of a fearful expression and, when prodded to do so, explained that "she did not know what an afraid face would look like." After several failed attempts, she produced the sketch of a cowering figure with hair standing on end, evidently because she knew these clichés about the expression of fear. In short, S.M. has a severely limited concept of fear and, consequently, fails to recognize the emotion of fear in facial expressions. Studies of other individuals with bilateral destruction of the amygdala are consistent with this account. As might be expected, S.M.'s disability also limits her ability to experience fear in situations where this emotion is appropriate.

Despite the adage "have no fear," to truly live without fear is to be deprived of a crucial neural mechanism that facilitates and promotes appropriate social behavior, making advantageous decisions in critical circumstances, and, ultimately, survival.

References

ADOLPHS, R., D. TRANEL, H. DAMASIO AND A. R. DAMASIO (1995) Fear and the human amygdala. J. Neurosci. 15: 5879–5891.

BECHARA, A., H. DAMASIO, A. R. DAMASIO AND G. P. LEE (1999) Differential contributions of the human amygdala and ventromedial prefrontal cortex to decision-making. J. Neurosci. 19: 5473–5481.

BOX E
Affective Disorders

Whereas some degree of disordered emotion is present in virtually all psychiatric problems, in affective (mood) disorders the essence of the disease is an abnormal regulation of the feelings of sadness and happiness. The most severe of these afflictions are major depression and manic depression. (Manic depression is also called "bipolar disorder," since such patients experience alternating episodes of depression and euphoria.) Depression, the most common of the major psychiatric disorders, has a lifetime incidence of 10–25% in women and 5–12% in men. For clinical purposes, depression (as distinct from bereavement or neurotic unhappiness) is defined by a set of standard criteria. In addition to an abnormal sense of sadness, despair, and bleak feelings about the future (depression itself), these criteria include disordered eating and weight control, disordered sleeping (insomnia or hypersomnia), poor concentration, inappropriate guilt, and diminished sexual interest.

The personally overwhelming quality of major depression has been compellingly described by patient/authors such as William Styron, and by afflicted psychologists such as Kay Jamison. But the depressed patient's profound sense of despair has been nowhere better expressed than by Abraham Lincoln, who during a period of depression wrote:

I am now the most miserable man living. If what I feel were equally distributed to the whole human family, there would not be one cheerful face on earth. Whether I shall ever be better, I cannot tell; I awfully forebode I shall not. To remain as I am is impossible. I must die or be better, it appears to me.

Indeed, about half the suicides in this country occur in individuals with clinical depression.

Not many decades ago, depression and mania were considered disorders that arose from circumstances or a neu-

rotic inability to cope. It is now universally accepted that these conditions are neurobiological disorders. Among the strongest lines of evidence for this consensus are studies of the inheritance of these diseases. For example, the concordance of affective disorders is high in monozygotic compared to dizygotic twins. It has also become possible to study the brain activity of patients suffering from affective disorders by non-invasive brain imaging (see figure). In at least one condition, unipolar depression, abnormal patterns of blood flow are apparent in the "triangular" circuit interconnecting the amygdala, the mediodorsal nucleus of the thalamus, and the orbital and medial prefrontal cortex (see Box B). Of particular interest is the significant correlation of abnormal blood flow in the amygdala and the clinical severity of depression, as well as the observation that the abnormal blood flow pattern in the prefrontal cortex returns to normal when the depression has abated.

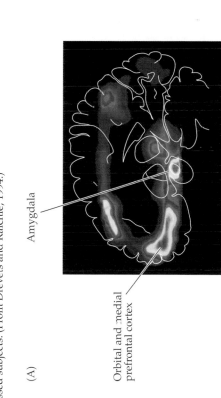

(A)

Amygdala

Orbital and medial prefrontal cortex

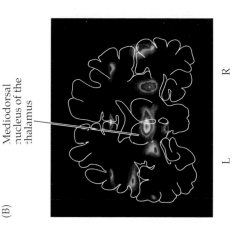

(B)

Mediodorsal nucleus of the thalamus

L   R

Areas of increased blood flow in the left amygdala, orbital and medial prefrontal cortex (A), and in a location in the left medial thalamus consistent with the mediodorsal nucleus (B) from a sample of patients diagnosed with unipolar clinical depression. The "hot" colors indicate statistically significant increases in blood flow, compared to a sample of nondepressed subjects. (From Drevets and Raichle, 1994.)

Despite evidence for a genetic predisposition and an increasing understanding of the brain areas involved, the cause of these conditions remains unknown. The efficacy of a large number of drugs that influence catecholaminergic and serotonergic neurotransmission strongly implies that the basis of the disease(s) is ultimately neurochemical. The majority of patients (about 70%) can be effectively treated with one of a variety of drugs (including tricyclic antidepressants, monoamine oxidase inhibitors, and selective serotonin reuptake inhibitors), which are among the most widely prescribed agents worldwide. Most successful are drugs that selectively block the uptake of serotonin without affecting the uptake of other neurotransmitters. Three such inhibitors—fluoxetine (Prozac®), sertraline (Zoloft®), and paroxetine (Paxil®)—are especially effective in treating depression and have few of the side effects of the older, less specific drugs. Perhaps the best indicator of the success of these drugs has been their wide acceptance: Although Prozac® was approved for clin-

ical use only in the late 1980s, it is now the second-largest selling drug in the United States.

Most depressed patients who use drugs such as Prozac® report that they lead fuller lives and are more energetic and organized. Based on such information, Prozac® is sometimes used not only to combat depression but also to "treat" individuals who have no definable psychiatric disorder. This abuse raises important social questions, and Prozac® has been compared by some to "Soma," the mythical drug routinely administered to the inhabitants of Aldous Huxley's *Brave New World*.

References

BREGGIN, P. R. (1994) *Talking Back to Prozac: What Doctors Won't Tell You about Today's Most Controversial Drug*. New York: St. Martin's Press.

DREVETS, W. C. AND M. E. RAICHLE (1994) PET imaging studies of human emotional disorders. In *The Cognitive Neurosciences*, M. S. Gazzaniga (ed.). Cambridge, MA: MIT Press, pp. 1153–1164.

FREEMAN, P. S., D. R. WILSON AND F. S. SIERLES (1993) *Psychopathology*. In *Behavior Science for Medical Students*, F. S. Sierles (ed.). Baltimore: Williams and Wilkins, pp. 239–277.

GREENBERG, P. E., L. E. STIGLIN, S. N. FINKELSTEIN AND E. R. BERNDT (1993) The economic burden of depression in 1990. J. Clin. Psychiatry 54: 405–424.

JAMISON, K. R. (1995) *An Unquiet Mind*. New York: Alfred A. Knopf.

JEFFERSON, J. W. AND J. H. GRIEST (1994) Mood disorders. In *Textbook of Psychiatry*, J. A. Talbott, R. E. Hales and S. C. Yudofsky (eds.). Washington: American Psychiatric Press, pp. 465–494.

ROBINS, E. (1981) *The Final Months: A Study of the Lives of 134 Persons Who Committed Suicide*. New York: Oxford University Press.

STYRON, W. (1990) *Darkness Visible: A Memoir of Madness*. New York: Random House.

WONG, D. T. AND F. P. BYMASTER (1995) Development of antidepressant drugs: Fluoxetin (Prozac®) and other selective serotonin uptake inhibitors. Adv. Exp. Med. Biol. 363: 77–95.

WONG, D. T., F. P. BYMASTER AND E. A. ENGLEMAN (1995) Prozac® (fluoxetine, Lilly 110140), the first selective serotonin uptake inhibitor and an antidepressant drug: Twenty years since its first publication. Life Sci. 57(5): 411–441.

WURTZEL, E. (1994). *Prozac Nation: Young and Depressed in America*. Boston: Houghton-Mifflin.

lobe (see Box B). These cortical fields associate information from every sensory modality (including information about visceral activities) and can thus integrate a variety of inputs pertinent to moment-to-moment experience. In addition, the amygdala projects to the thalamus (specifically, the mediodorsal nucleus), which projects in turn to these same cortical areas. Moreover, the amygdala innervates neurons in the ventral portions of the basal ganglia that receive the major cortico-striatal projections from the regions of the prefrontal cortex thought to process emotions. Considering all these seemingly arcane anatomical connections, the amygdala emerges as a nodal point in a network that links together the cortical (and subcortical) brain regions involved in emotional processing.

Clinical evidence concerning the significance of this circuitry linked through the amygdala has come from functional imaging studies of patients suffering from unipolar depression (Box E), in which this set of interrelated forebrain structures displayed abnormal patterns of cerebral blood flow, especially in the left hemisphere. More generally, the amygdala and its connections to the prefrontal cortex and basal ganglia are likely to influence the selection and initiation of behaviors aimed at obtaining rewards and avoiding punishments (recall that the process of motor program selection and initiation is an important function of basal ganglia circuitry; see Chapter 18). The parts of the prefrontal cortex interconnected with the amygdala are also

involved in organizing and planning future behaviors; thus, the amygdala may provide emotional input to overt (and covert) deliberations of this sort (see "The Interplay of Emotion and Reason" below).

Cortical Lateralization of Emotional Functions

Since functional asymmetries of complex cortical processes are commonplace (see Chapters 26 and 27), it should come as no surprise that the two hemispheres make different contributions to the governance of emotion.

Emotionality is lateralized in the cerebral hemispheres in at least two ways. First, as discussed in Chapter 27, the right hemisphere is especially important for the expression and comprehension of the affective aspects of speech. Thus, patients with damage to the supra-Sylvian portions of the posterior frontal and anterior parietal lobes on the right side may lose the ability to express emotion by modulation of their speech patterns (recall that this loss of emotional expression is referred to as **aprosody** or **aprosodia**, and that similar lesions in the left hemisphere give rise to Broca's aphasia). Patients with aprosodia tend to speak in a monotone, no matter what the circumstances or meaning of what is said. For example, one such patient, a teacher, had trouble maintaining discipline in the classroom. Because her pupils (and even her own children) couldn't tell when she was angry or upset, she had to resort to adding phrases such as "I am angry and I mean it" to indicate the emotional significance of her remarks. The wife of another patient felt her husband no longer loved her because he could not imbue his speech with cheerfulness or affection. Although such patients cannot express emotion in speech, they nonetheless experience normal emotional feelings.

A second way in which the hemispheric processing of emotionality is asymmetrical concerns mood. Both clinical and experimental studies indicate that the left hemisphere is more importantly involved with what can be thought of as positive emotions, whereas the right hemisphere is more involved with negative ones. For example, the incidence and severity of depression (see Box E) is significantly higher in patients with lesions of the left anterior hemisphere compared to any other location. In contrast, patients with lesions of the right anterior hemisphere are often described as unduly cheerful. These observations suggest that lesions in the left hemisphere result in the loss of positive feelings, facilitating depression, whereas lesions of the right hemisphere result in the loss of negative feelings, leading to inappropriate optimism.

Hemispheric asymmetry related to emotion is also apparent in normal individuals. For instance, auditory experiments that introduce sound into one ear or the other indicate a right-hemisphere superiority in detecting the emotional nuances of speech. Moreover, when facial expressions are specifically presented to either the right or the left visual hemifield, the depicted emotions are more readily and accurately identified from the information in the left hemifield (that is, the hemifield perceived by the right hemisphere; see Chapters 12 and 27). Finally, kinematic studies of facial expressions show that most individuals more quickly and fully express emotions with the left facial musculature than with the right (recall that the left lower face is controlled by the right hemisphere, and vice versa) (Figure 29.7). Taken together, this evidence is consistent with the idea that the right hemisphere is more intimately concerned with both the perception and expression of emotions than is the left hemisphere. However, it is important to remember that, as in the case of other lateralized behaviors (language, for instance), both hemispheres participate in processing emotion.

Figure 29.7 Asymmetrical smiles on some famous faces. Studies of normal subjects show that facial expressions are often more quickly and fully expressed by the left facial musculature than the right, as is apparent in many of these examples. Since the left lower face is governed by the right hemisphere, some psychologists have suggested that the majority of humans are "left-faced," in the same general sense that most of us are right-handed. (After Moscovitch and Olds, 1982; images from Microsoft® Encarta Encyclopedia 98.)

The Interplay of Emotion and Reason

The experience of emotion—even on a subconscious level—has a powerful influence on the neural faculties responsible for making rational decisions. Evidence for this statement has come principally from studies of patients with damage to parts of the orbital and medial prefrontal cortex, as well as patients with injury or disease involving the amygdala (see Box D). Such patients have in common an impairment in emotional processing, especially emotions engendered by complex personal and social situations, and an inability to make advantageous decisions (see also Chapter 26).

Antonio Damasio and his colleagues at the University of Iowa have suggested that such decision making entails the rapid evaluation of a set of possible outcomes with respect to the future consequences associated with each course of action. It seems plausible that the generation of conscious or subconscious mental images that represent the consequences of each contingency triggers emotional states that involve either actual alterations of somatic and visceral motor function, or the activation of neural representations of such activity. Whereas William James proposed that we are "afraid because we tremble," Damasio and his colleagues suppose a vicarious representation of motor action and sensory feedback in the neural circuits of the frontal and parietal lobes. It is these vicarious states, according to Damasio, that give mental representations of contingencies the emotional valence that helps an individual to identify favorable or unfavorable outcomes.

Experimental studies of fear conditioning have suggested just such a linking role for the amygdala in associating sensory stimuli with aversive consequences. Indeed, the patient described in Box D showed an inability to recognize and experience fear, together with impairment in rational decision making. Similar evidence of the emotional influences on decision making have also come from studies of patients with lesions in the orbital and medial prefrontal cortex. These clinical observations suggest that the amygdala and prefrontal cortex, as well as their striatal and thalamic connections, are not only involved in processing emotions, but also participate in the complex neural processing responsible for what we consider rational thinking.

Summary

The word "emotion" covers a wide range of states that have in common the association of visceral motor responses, somatic behavior (e.g., facial expressions), and powerful subjective feelings. The visceral motor responses are mediated by the visceral motor nervous system, which is itself regulated by inputs from many other parts of the brain. The organization of the somatic motor behavior associated with emotion is governed by circuits in the limbic system, which includes the hypothalamus, the amygdala, and several regions of the cerebral cortex. Although a good deal is known about the neuroanatomy and transmitter chemistry of the different parts of the limbic system, there is still a dearth of information about how this complex circuitry mediates specific emotional states. Similarly, neuropsychologists and neurologists are only now coming to appreciate the important role of emotional processing in other complex brain functions, such as decision making. A variety of other evidence indicates that the two hemispheres are differently specialized for the governance of emotion, the right hemisphere being the more important in this regard. The prevalence and social significance of human emotions and their disorders ensure that the neurobiology of emotion will be an increasingly important theme in modern neuroscience.

ADDITIONAL READING

Reviews

APPLETON, J. P. (1993) The contribution of the amygdala to normal and abnormal emotional states. Trends Neurosci. 16: 328–333.

CAMPBELL, R. (1986) Asymmetries of facial action: Some facts and fancies of normal face movement. In *The Neuropsychology of Face Perception and Facial Expression*, R. Bruyer (ed.). Hillsdale, NJ: Erlbaum, pp. 247–267.

DAVIS, M. (1992) The role of the amygdala in fear and anxiety. Ann. Rev. Neurosci. 15: 353–375.

LEDOUX, J. E. (1987) Emotion. In *Handbook of Physiology*, Section 1, *The Nervous System*, Vol. 5, F. Blum, S. R. Geiger, and V. B. Mountcastle (eds.). Bethesda, MD: American Physiological Society, pp. 419–459.

SMITH, O. A. AND J. L. DEVITO (1984) Central neural integration for the control of autonomic responses associated with emotion. Annu. Rev. Neurosci. 7: 43–65.

Important Original Papers

BARD, P. (1928) A diencephalic mechanism for the expression of rage with special reference to the sympathetic nervous system. Am. J. Physiol. 84: 490–515.

DOWNER, J. L. DE C. (1961) Changes in visual agnostic functions and emotional behaviour following unilateral temporal pole damage in the "split-brain" monkey. Nature 191: 50–51.

EKMAN, P., R. W. LEVENSON AND W. V. FRIESEN (1983) Autonomic nervous system activity distinguishes among emotions. Science 221: 1208–1210.

KLÜVER, H. AND P. C. BUCY (1939) Preliminary analysis of functions of the temporal lobes in monkeys. Arch. Neurol. Psychiat. 42: 979–1000.

MACCLEAN, P. D. (1964) Psychosomatic disease and the "visceral brain": Recent developments bearing on the Papez theory of emotion. In *Basic Readings in Neuropsychology*, R. L. Isaacson (ed.). New York: Harper & Row, Inc., pp. 181–211.

PAPEZ, J. W. (1937) A proposed mechanism of emotion. Arch. Neurol. Psychiat. 38: 725–743.

ROSS, E. D. AND M-M. MESULAM (1979) Dominant language functions of the right hemisphere? Prosody and emotional gesturing. Arch. Neurol. 36: 144–148.

Books

APPLETON, J. P. (ed.) (1992) *The Amygdala: Neurobiological Aspects of Emotion, Memory and Mental Dysfunction*. New York: Wiley-Liss.

CORBALLIS, M. C. (1991) *The Lopsided Ape: Evolution of the Generative Mind*. New York: Oxford University Press.

DAMASIO, A. R. (1994) *Descartes' Error: Emotion, Reason, and the Human Brain*. New York: Avon Books.

DARWIN, C. (1890) *The Expression of Emotion in Man and Animals*, 2nd Ed. In *The Works of Charles Darwin*, Vol. 23, 1989. London: William Pickering.

HELLIGE, J. P. (1993) *Hemispheric Asymmetry: What's Right and What's Left*. Cambridge, MA: Harvard University Press.

HOLSTEGE, G., R. BANDLER AND C. B. SAPER (eds.) (1996) *Progress in Brain Research*, Vol. 107. Amsterdam: Elsevier.

JAMES, W. (1890) *The Principles of Psychology*, Vols. 1 and 2. New York: Dover Publications (1950).

ROLLS, E. T. (1999) *The Brain and Emotion*. Oxford: Oxford University Press.

Chapter 30

Sex, Sexuality, and the Brain

Overview

"Vive la difference." "Isn't that just like a (wo)man?" "It's on the Y chromosome." These expressions denote pleasure (or displeasure) with phenotypic sexual differences—how females and males look and behave. These sex-related differences in the phenotypic expression of genotype are called sexual dimorphisms. While some of the behavioral distinctions involved may be rooted in cultural or social norms, the majority of gender differences arise because the nervous systems of females and males are in some respects different. In the rat, the animal in which most experimental work has been done, several structures in female and male brains differ in the number, size, and connectivity of their constituent neurons. In humans and other primates, structural differences are less obvious and, as a result, more controversial. In both rodents and humans, sexually dimorphic brain structures tend to cluster around the third ventricle in the anterior hypothalamus and are an integral part of the system that governs visceral motor behavior. Some sexual dimorphisms, however, are apparent in cerebral cortical structures, implying differences in more complex regulatory behaviors. The development of these differences depends on the early influence of hormones on maturing brain circuits, especially estrogens, an influence that apparently continues to some extent throughout life. The functional consequences of sexual dimorphisms in rodents are beginning to be understood. Although the significance of such differences in humans is much less clear, the influence of hormones on neural circuitry provides a plausible basis for the wide variety of human sexual behavior.

Sexually Dimorphic Behavior

Many animal behaviors differ between the sexes and are therefore referred to as **sexually dimorphic** (*dimorphic* means having two forms). Most of these sexually dimorphic behaviors are part of the reproductive repertoire. A good example is apparent in songbirds. In many species, the male produces complex song, whereas the female does not. The production of song in male birds arises from the activity of specific brain nuclei whose growth and connectivity depend on the presence of testosterone during a critical period of development (see Box B in Chapter 24). In rodents, many sexually dimorphic behaviors are also associated with reproduction. Examples are priming of the genitalia for sexual intercourse, and a stereotypical position assumed while having sex (lordosis for females, mounting for males). Just as courting and behaviors associated with the sex act can be dimorphic, other reproductive behaviors such as building nests, caring for the young, foraging for

BOX A
The Development of Male and Female Phenotypes

The presence of two X chromosomes, or conversely an X and a Y chromosome, in the cells of an embryo sets in motion events that establish phenotypic sex, including the sexually dimorphic development of the brain. The relevant neural effects are determined by the production of estrogens or androgens, which depends in turn on the presence of either female or male gonads.

The early stages of human embryonic development follow a plan that produces common precursors for the gonads. At about the sixth week of gestation, the primordial gonads have formed from somatic mesenchyme tissue, near the developing kidneys. Cells in the gonads differentiate into supporting and hormone-producing cells (the cells that will become ova and sperm migrate from the yolk sac). Attached to the primordial gonads are two sets of tubes that are the progenitors of the internal genitalia, the Müllerian and Wolffian ducts. Developing simultaneously is an undifferentiated structure called the urogenital groove, the progenitor of the external genitalia.

The primary genetic influence on the development of the typical male phenotype is the sex-determining region on the Y chromosome, the *Sry* gene. When this region of the chromosome is activated during development, it turns on the production of a protein called testicular determining factor (TDF). TDF then instructs the testes to begin developing. Once activated, the primordial gonads in the male begin to produce testosterone (elaborated by the Leydig cells). Müllerian-inhibiting hormone is also secreted by cells of the testes, which prevents the Müllerian ducts from developing and allows the Wolffian ducts to develop into the epididymis, vas deferens, and seminal vesicles. Under these same influences, the tissue around the urogenital groove becomes the penis and scrotum.

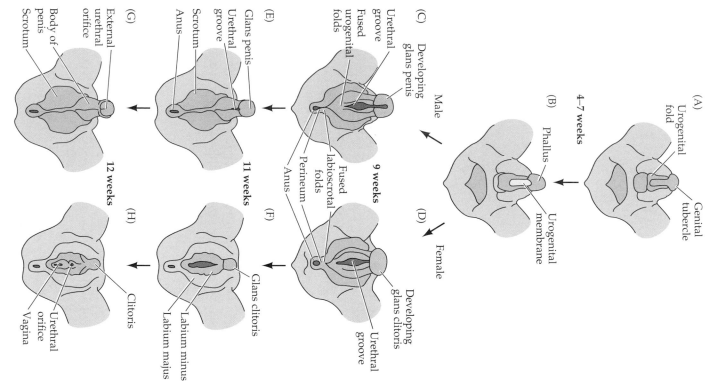

(A)
Urogenital fold
Genital tubercle

4–7 weeks

(B)
Phallus
Urogenital membrane

Male →

Female ↘

(C)
Urethral groove
Developing glans penis

(D)
Developing glans clitoris

9 weeks

(E)
Glans penis
Urethral groove
Scrotum
Anus
Fused labioscrotal folds
Perineum
Anus
Fused urogenital folds

(F)
Glans clitoris
Labium minus
Labium majus
Urethral groove

11 weeks

(G)
External urethral orifice
Body of penis
Scrotum

(H)
Clitoris
Urethral orifice
Vagina

12 weeks

Development of human female and male external genitalia. A and B show the indifferent stage during weeks 4–7 of gestation. D, F, and H show differentiation in the female genitalia at weeks 9, 11, and 12, respectively; C, E, and G show differentiation into male genitalia at the same intervals. (After Moore, 1977.)

However, androgens alone are not sufficient for male differentiation. Ken Korach's group at the National Institute for Environmental Health Sciences has demonstrated that estrogens are also needed for the hormonal differentiation of the testes. Thus, XY mice lacking estrogen receptors develop testes but there is disruption of spermatogenesis and degeneration of the seminiferous tubules, leading eventually to sterility. The presence of TDF and the consequent production of androgens early in life lead to the differentiation of the male body and brain, but estrogens are essential for the full development of the male phenotype.

In XX embryos, the absence of TDF, testosterone, and Müllerian-inhibiting hormone allows the indifferent gonad to differentiate into an ovary, the Wolffian ducts degenerate, and the Müllerian ducts develop into the oviducts, uterus, and cervix. The tissue around the urogenital groove becomes the clitoris, labia, and vagina. Thus, development of the female phenotype depends on the absence of androgens during early development. However, development of the female phenotype also depends on estrogens; lack of both the α and β estrogen receptors results in ovaries that resemble testes in that they have structures resembling seminiferous tubules, including Sertoli-like cells and expression of Müllerian inhibiting substance. Nonetheless, it is primarily the early lack of androgens that leads to the differentiation of the female body and brain.

References

COUSE, J. F. AND 6 OTHERS (1999) Postnatal sex reversal of the ovaries in mice lacking estrogen receptors alpha and beta. Science 286: 2328–2331.

EDDY, E. M. AND 7 OTHERS (1996) Targeted disruption of the estrogen receptor gene in male mice causes alteration of spermatogenesis and infertility. Endocrinology 137: 4796–4805.

JOHNSON, M. H. AND B. J. EVERITT (1988) Essential Reproduction, 3rd Ed. Oxford: Blackwell Scientific, pp. 1–34.

KOOPMAN, P., J. GUBBAY, N. VIVIAN, P. GOODFELLOW AND R. LOVELL-BADGE (1991) Male development of chromosomally female mice transgenic for Sry. Nature 351: 117–121.

SINCLAIR, A. H. AND 9 OTHERS (1990) A gene from the human sex-determining region encodes a protein with homology to a conserved DNA-binding motif. Nature 346: 240–242.

food, nursing, and so on can take two different forms in females and males. In humans, the different behaviors of males and females can be far more subtle, including one's sense of sexual identity, the choice of a sexual partner, and behaviors that are not related directly to sexual or reproductive function, such as spatial thinking and use of language.

In both human and animal examples, behavioral differences are based on the details of the underlying neural circuitry. Accordingly, neurobiologists have long looked for differences between the brains of females and males that might explain sexually dimorphic behaviors and, as described in subsequent sections, have found many examples. These differences in the nervous system, like the behavioral differences they give rise to, are also referred to as sexually dimorphic. Bear in mind, however, that while brain differences in animals like rodents often have two distinct forms, in human females and males these neural differences probably vary along a continuum.

What Is Sex?

Roughly speaking, sex can be considered in terms of three categories: genotypic sex, phenotypic sex, and gender. **Genotypic sex** refers specifically to an individual's two sex chromosomes. Most people have either two X chromosomes (genotypic female) or an X and a Y chromosome (genotypic male). **Phenotypic sex** refers to an individual's sex as determined by their internal and external genitalia, expression of secondary sex characteristics, and behavior. If everything proceeds according to plan during development (Box A), the XX genotype leads to a person with ovaries, oviducts, uterus, cervix, clitoris, labia, and vagina—i.e., a phenotypic female. By the same token, the XY genotype leads to a person with testicles, epididymis, vas deferens, seminal vesicles, penis, and scrotum—a phenotypic male. **Gender** refers more broadly to an individual's subjective perception of their sex and their sexual orientation, and is therefore harder to define than genotypic or phenotypic sex. Generally speaking, gender identity entails self-appraisal according to the traits most often associated with one sex or the other (called gender traits), and these can be influenced to some degree by cultural norms. Sexual orientation also entails self-appraisal in the context of culture. For purposes of understanding the neurobiology of sex, it is helpful to think of genotypic sex as largely immutable, phenotypic sex as modifiable (by developmental processes, hormone treatment, and/or surgery), and gender as a more complex construct that is determined culturally as well as biologically.

Clearly, then, genotypic sex, phenotypic sex, and gender are not always aligned. Variations in alignment can be minor, or they can challenge the usual definitions of female and male and lead to psychosocial conflicts and sexual dysfunction (see Box B). Genetic variations include individuals who are XO (Turner's syndrome), XXY (Klinefelter's syndrome), or XYY. Each of these genotypes has its own particular phenotype. Other genetic variations arise from mutations in genes coding for hormone receptors or for the hormones themselves. For instance, a metabolic disorder that leads to overactive adrenals during maturation, called congenital adrenal hyperplasia (CAH), causes abnormally high levels of circulating androgens and hence, along with severe salt imbalance, an ambiguous sexual phenotype. In addition to having a large clitoris and fused labia at birth, females with CAH typically exhibit "tomboyish" behavior as children and tend to form homosexual relationships as adults. High levels of circulating androgens from the adrenals may cause sexually dimorphic brain circuitry to have a male rather than female organization, leading to more aggressive play and the eventual choice of a female sexual partner.

An example of a mutation in a gene responsible for hormone receptors is androgen insensitivity syndrome (AIS), also called testicular feminization. The receptor deficiency leads to the development of the internal genitalia of a male and the external genitalia of a female in an individual who is genotypically XY. Thus, people with androgen insensitivity syndrome look like females and self-identify as female, even though they have a Y chromosome. Since they are generally not aware of their condition until puberty, when they fail to menstruate, they see themselves and are experienced by others as female. Thus, their gender identity matches external sexual phenotype, but not genotype. Although this syndrome is relatively uncommon (about 1 in 4000 births), there are some well-known examples of individuals thought or known to have had AIS (e.g., Joan of Arc and Wallis Simpson, the woman for whom King Edward of England gave up his throne).

Another variation in the alignment of genotype, phenotype, and gender is genotypic males who are phenotypic females early in life, but whose sexual phenotype changes at puberty. As infants and children, these individuals are phenotypic females because they lack an enzyme, 5-α-reductase, that promotes the early development of male genitalia (see Box A). Such children have somewhat ambiguous but generally female-appearing genitalia (they have labia with an enlarged clitoris, and undescended testes). As a result, they are generally raised as females. At puberty, however, when the testicular secretion of androgen becomes high, the clitoris develops into a penis and the testes descend, changing these individuals into phenotypic males. In the Dominican Republic and Haiti, where this congenital syndrome has been thoroughly studied in a particular pedigree, the condition is referred to colloquially as "testes-at-twelve." Such individuals generally exhibit male gender behavior at puberty, and most eventually live as males.

The term used to describe all these variations is "intersexuality." Taken together, these individuals make up approximately 1–2% of all live births. In addition to the more clearly defined categories of Klinefelter's syndrome, Turner's syndrome, CAH, AIS, and 5-α-reductase deficiency, subtle permutations and combinations of genes, hormones, and environment give rise to a large number of biological and behavioral possibilities. In all these permutations and combinations, the relevant brain circuitry established early in development generally determines sexual behavior and identity (Box B).

Hormonal Influences on Sexual Dimorphism

The development of sexual dimorphisms in the central nervous system is ultimately an outcome of genotypic sex. Genotype normally determines the phe-

BOX B
The Case of John/Joan

In the early 1960s, identical XY twins were born to a Canadian couple. When the twins were 7 months old, the parents had them circumcised. Unfortunately, the surgeon performing the operation burned one of the twins' penises with an electrocautery knife so severely that the penis was, in essence, destroyed. The consensus conveyed to the parents by the local physicians was that the disfigured twin would be unable to have a typical heterosexual life, would be shunned by his peers, and would suffer in a variety of other ways. Given this dire prognosis, the parents decided to consult John Money, an eminent sex researcher at Johns Hopkins University, to decide what should be done. After meeting with the family, Money advocated that they surgically reassign the child's sex and raise the little boy as a girl. The parents consented, and at age 17 months the child's testes were removed and his scrotum reshaped to resemble a vulva. The little boy, called "John" in Money's medical records, became known as "Joan," and the parents did everything they could to raise the infant as a typical female.

Although Money's published reports were optimistic, subsequent interviews with the family, including "John" himself

(who is now nearly 40), indicated that the truth was far more complex, and indeed deeply problematic. In a detailed follow-up of the case, Milton Diamond of the University of Hawaii and Keith Sigmundson describe the struggle that "Joan" suffered. From an early age, the child would not wear dresses, urinated standing up, always felt that something was wrong, and eventually refused to comply with the hormone treatments that were initiated at puberty. At 14 "Joan" demanded to know the truth, and was reluctantly told of the circumstances by "her" equally frustrated parents. Ironically, "Joan" was greatly relieved to understand why "she" had always been subject to such deeply conflicting feelings, which had sometimes made "her" life so miserable that suicide was contemplated. Upon understanding what had happened, "Joan" immediately reverted to male dress and behavior and eventually underwent surgery to be reconfigured as a phenotypic male. "John" eventually married, adopted his wife's children, and has lived a conventional life as a father and husband.

This case underscores the fact that, in the words of Diamond and Sigmundson, "...the evidence [is] overwhelming that normal humans are not psychosocially

neutral at birth but are, in keeping with their mammalian heritage, predisposed and biased to interact with environmental, familial, and social forces in either a male or female mode."

Cases like this raise serious moral and ethical questions about the assignment of sex when, for one reason or another, that option is open. Since there is often no way to know at birth how the brain has been shaped by early exposure to hormones, there is in many instances insufficient information to know with what sex the child, or adult, will ultimately identify. In the case of John/Joan, however, a grievous mistake was made by ignoring the overwhelming influence of a brain whose sex has been already determined.

References

COLAPINTO, J. (2000) *As Nature Made Him: The Boy Who Was Raised as a Girl.* New York: Harper Collins.

DIAMOND, M. AND H. K. SIGMUNDSON (1997) Sex reassignment at birth: Long-term review and clinical implications. Arch. Ped. Adolesc. Med. 151: 298–304.

DREGER, A. D. (1998) "Ambiguous sex" or ambiguous medicine? The Hastings Center Report 28: 24–35.

notype of the gonads; and the gonads, in turn, are responsible for producing most of the circulating sex hormones (see Box A). Because the gonads are dimorphic, the production of sex steroids during development is itself dimorphic: The testes synthesize androgens and the ovaries estrogens. The resulting differences in circulating hormones lead to a variety of differential effects on the development of XX and XY individuals (as well as other genotypes), including physical appearance, response to pharmacological treatments, susceptibility to certain diseases, and the development of the brain.

In general, the establishment of phenotypic dimorphisms in rodents (and presumably humans) is generated by different levels of hormones at specific times in XX and XY individuals. Males have an early surge of testosterone, and females a later surge of estrogens. Paradoxically, the testosterone surge in males has the same effect as an estrogen surge, because neurons contain an enzyme (aromatase) that converts testosterone to **estradiol**, a form of estrogen (Figure 30.1). Thus, the surge of testosterone in developing males is, ultimately, converted into a surge of estradiol. Although testosterone is popularly considered the "male" hormone and estrogen the "female" hormone, the active agent in the brains of both males and females is estradiol. Once the conversion of testosterone to estradiol has occurred, estradiol can influence gene tran-

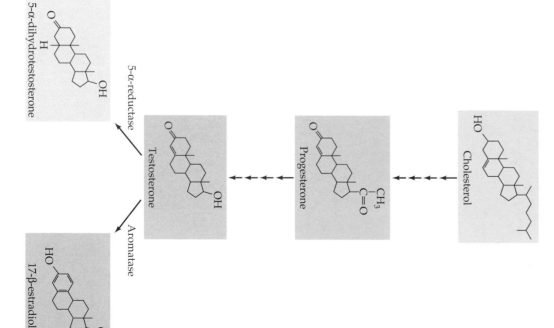

Figure 30.1 All sex steroids are synthesized from cholesterol. Cholesterol is first converted to progesterone, the common precursor, by four enzymatic reactions (represented by the four arrows). Progesterone can then be converted into testosterone via another series of enzymatic reactions; testosterone in turn is converted to 5-α-dihydrotestosterone via 5-α-reductase, or to 17-β-estradiol via an aromatase. 17-β-estradiol mediates most known hormonal effects in the brains of both female and male rodents.

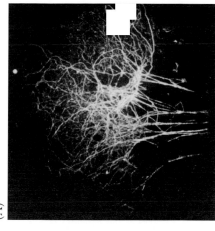

(A)

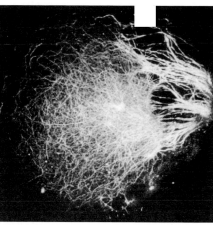

(B)

Figure 30.2 Estrogen causes exuberant outgrowth of neurites in hypothalamic explants from newborn mice. (A) Control explant showing only a few silver-impregnated processes growing from the explant. (B) An estradiol-treated explant has many more neurites growing from its center. (From Torand-Allerand, 1978.)

scription by binding with intracellular receptors (the α- and β-estrogen receptors) that in turn bind to hormone-responsive regions of DNA (Box C).

It is important during development, therefore, to sequester all exogenous sources of circulating hormones that might interfere with sexual differentiation. One such source is the maternal blood supply, which is rich in estrogens produced by the mother's gonads and placenta. To counter potential interference from this source, mammals have a circulating protein called **α-fetoprotein** that binds circulating estrogens. As a result, the female brain is kept from early exposure to estrogens in large amounts since they are complexed by α-fetoprotein; the male brain, however, is exposed to estrogens because the early testosterone surge is not affected by α-fetoprotein and can thus be aromatized to estradiol once inside neurons.

In sum, the way that estrogens—the ultimate agents of both female and male sexual differentiation—stimulate sexually dimorphic patterns of development is by binding to estrogen receptors. These receptors, which are transcription factors activated by estrogen binding, can then influence gene transcription and subsequently the development of an array of targets, including sexually dimorphic neural circuits.

The Effect of Sex Hormones on Neural Circuitry

During development, and to some degree throughout life, estradiol stimulates brain dimorphisms by increasing size, nuclear volume, dendritic length, dendritic branching, dendritic spine density, and synaptic connectivity of the sensitive neurons. One of the first demonstrations of such effects was provided by Dominique Torand-Allerand at Columbia University, who noted the striking consequences of adding estrogens to fetal hypothalamic explants (Figure 30.2). Estradiol can also stimulate neurons to increase the number of synaptic contacts the neurons receive. For example, during periods of high circulating estrogen in the estrous cycle of female rodents—or after administration of estrogens—there is an increase in the density of spines and synapses on the apical dendrites of pyramidal neurons (Figure 30.3). These sorts of changes in the sensitive neurons presumably underlie changes in neuronal circuitry and ultimately differences in sexual behavior.

Hormonally generated differences in brain circuits leading to differences in reproductive behaviors in females and males have been documented in experimental animals by administering testosterone (or estrogens) to females, or by depriving males of testosterone by castrating them at birth. Geoffrey Raisman and Pauline Field, then working at Oxford University, found a greater number of synapses on spines in the preoptic region of the hypothalamus in normal female rats compared to the equivalent region in males. This difference was linked with ovulatory behavior and was found to be directly under the influence of hormones during development. Thus, castrating males within 12 days of birth increased the density of these synapses to female levels, producing a surge of luteinizing hormone (which, with ovaries, would lead to ovulation); administration of testosterone to developing females, in contrast, led to a reduction of preoptic spine synapses to male levels and an abolishment of the surge of luteinizing hormone.

Subsequently, Roger Gorski and his colleagues at the University of California at Los Angeles discovered a nucleus in the male rodent hypothalamus that is essentially missing in the female; logically enough, they called this structure the sexually dimorphic nucleus (SDN). The nucleus is responsible for the different sexual postures of rodents during coitus, and develops under the influence of hormones. Thus, Gorski found that the SDN in male

| Hormone levels | Dendritic spine density | Dendritic morphology |

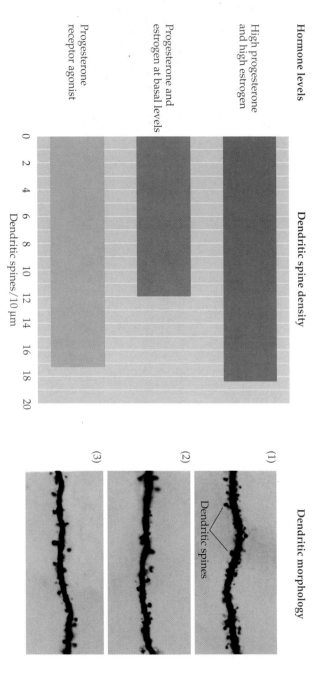

Figure 30.3 Changes in the dendrites of rat hippocampal neurons following various hormonal regimes. Left: Dendritic spine density under each of the indicated conditions (recall that dendritic spines, which are small extensions from the dendritic shaft, are sites of synapses). Right: Tracings of representative apical dendrites from hippocampal pyramidal neurons. (1) After administration of progesterone and estrogen in high dosage. (2) After administration of progesterone and estrogen at basal levels. (3) After administration of a progesterone receptor antagonist. (After Woolley and McEwen, 1992.)

rats could be reduced in size to that of females by castration within the first two weeks after birth. Similarly, the size of the female SDN could be increased to that of the male by early administration of androgens. Female rodents given testosterone exhibited mounting behavior; male rodents deprived of testosterone exhibited behaviors receptive to mounting.

Thus, the development of sexually dimorphic structures, at least in the rodent brain, is under the control of circulating sex hormones.

Central Nervous System Dimorphisms Related to Reproductive Behaviors

As these examples show, the actions of sex steroids on neurons provide powerful mechanisms for the production of behavioral differences between females and males. In experimental animals at least, the consequences in the central nervous system are evident in sexually dimorphic circuits and behaviors ranging from the control of motor responses to aspects of cognition. This section briefly reviews examples that are specifically related to reproductive behavior; the following section considers sexual dimorphisms related to cognitive behavior.

A good example of sexual dimorphism related to motor control of a reproductive behavior is the difference in size of a nucleus in the lumbar segment of the rat spinal cord called the **spinal nucleus of the bulbocavernosus**. The motor neurons of this nucleus innervate two striated muscles of the perineum, the bulbocavernosus and ischiocavernosus (Figure 30.4A). In males, the bulbocavernosus and the ischiocavernosus attach to the penis and play a role in both urination and copulation. In females, the bulbocavernosus and the ischiocavernosus are much smaller and attach to the base of the clitoris; they are used to construct the opening of the vagina. Marc Breedlove and his colleagues first showed that the spinal nucleus containing the motor neurons that innervate the bulbocavernosus is much smaller in female rats compared to males (Figure 30.4B,C). They next demonstrated that the development of this dimorphism in the spinal cord depends on the maintenance

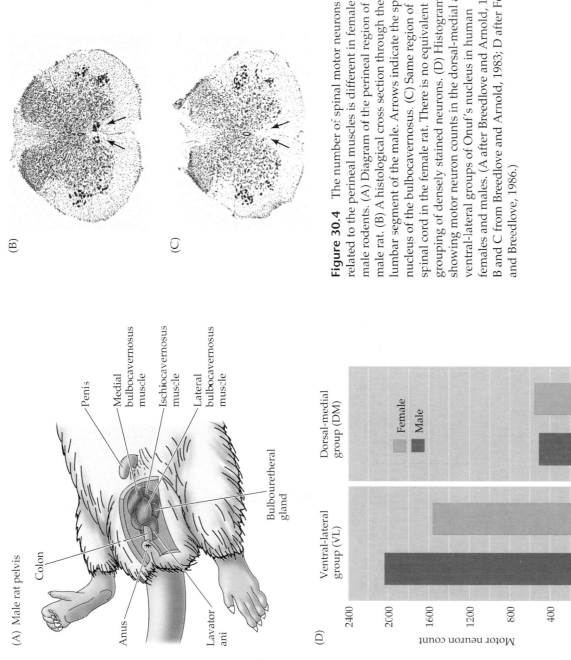

Figure 30.4 The number of spinal motor neurons related to the perineal muscles is different in female and male rodents. (A) Diagram of the perineal region of a male rat. (B) A histological cross section through the fifth lumbar segment of the male. Arrows indicate the spinal nucleus of the bulbocavernosus. (C) Same region of the spinal cord in the female rat. There is no equivalent grouping of densely stained neurons. (D) Histograms showing motor neuron counts in the dorsal-medial and ventral-lateral groups of Onuf's nucleus in human females and males. (A after Breedlove and Arnold, 1984; B and C from Breedlove and Arnold, 1983; D after Forger and Breedlove, 1986.)

of target muscles by circulating androgens. Since developing males have high levels of circulating sex steroids whereas females do not, these muscles largely degenerate in developing female rats, leaving the motor neurons to atrophy in the absence of trophic support (see Chapter 23).

In humans, the spinal cord structure that corresponds to the spinal nucleus of the bulbocavernosus in rats is called **Onuf's nucleus**, which consists of two cell groups in the sacral cord (a dorsal-medial and a ventral-lateral group). Although the dorsal-medial group is not sexually dimorphic, human females have fewer neurons in the ventral-lateral group than males (Figure 30.4D). In contrast to rodents, the female perineal muscles in humans remain relatively large throughout life, but are nonetheless smaller than in the male. The difference in nuclear size in humans, like rats, presumably reflects the difference in the number of muscle fibers the motor neurons must innervate.

BOX C
The Actions of Sex Hormones

Sex hormones, which include progestagens, androgens, and estrogens, are all steroids derived from a common precursor, cholesterol (see Figure 30.1). Despite the tendency to speak of these hormones as female or male, it is not really correct to think of estrogens as female and androgens as male; females and males synthesize both estrogens and androgens, and estrogens are the effective agonist in both sexes (see text). What is important is the receptors available to bind the two steroids when they are in the circulation. The brain has receptors for all these sex steroids, but the distribution of each receptor type is slightly different in females and males. For example, a higher level of estrogen receptors occurs in the arcuate, dorsomedial nucleus, and ventrolateral nucleus of the hypothalamus of females than in males.

Because sex steroids are lipids, they do not need special membrane receptors to enter cells; they simply diffuse through the lipid bilayer of the membrane. However, neurons and other cells have the capacity to select, concentrate, and retain specific steroids by means of receptors and binding proteins in the cytoplasm and nucleus. Different areas of the adult brain have different steroid receptor patterns, with overlapping distributions of receptor types. Thus, particular brain regions can be targets for the actions of different classes of steroids. For instance, estradiol receptors are sparsely distributed in the neocortex of the rat, but are prevalent in preoptic and hypothalamic areas and the anterior pituitary (figure A). Conversely, whereas receptors for 5-α-dihydrotestosterone (5-DHT) are found only in certain nuclei in the septum and hypothalamus, both estradiol and 5-DHT receptors are abundant in the frontal, prefrontal, and cingulate areas of the cortex. Some neurons express receptors for more than one steroid. Thus, all neurons with proges-

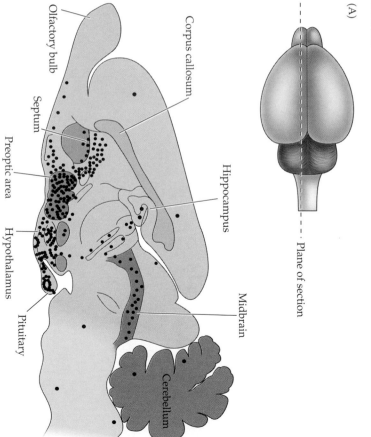

(A)

Olfactory bulb

Corpus callosum

Septum

Preoptic area

Hippocampus

Hypothalamus

Pituitary

Midbrain

Cerebellum

Plane of section

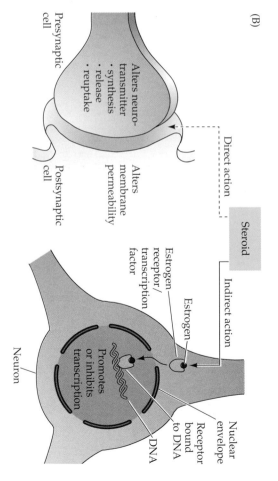

(B)

Presynaptic cell

Postsynaptic cell

Alters neurotransmitter
· synthesis
· release
· reuptake

Alters membrane permeability

Direct action

Steroid

Indirect action

Estrogen receptor/ transcription factor

Estrogen

Promotes or inhibits transcription

Nuclear envelope

Receptor bound to DNA

DNA

Neuron

(A) Distribution of estradiol-sensitive neurons in a sagittal section of the rat brain. Animals were given radioactively labeled estradiol; dots represent regions where the label accumulated. In the rat, most estradiol-sensitive neurons are located in the preoptic area, hypothalamus, and amygdala. (B) Steroids have direct and indirect effects on neurons. Dashed line shows direct effects of hormones on the pre- or postsynaptic membrane, which alter neurotransmitter release and affect neurotransmitter receptors. Solid line shows indirect effects of hormones, which act at the level of the nucleus to alter protein synthesis. (A after McEwen, 1976; B after McEwen et al., 1978.)

terone receptors also express estrogen receptors. As a result, hormones can have a synergistic effect; for example, certain reproductive behaviors brought on by progesterone can be enhanced if estrogen is given first.

Steroids can have a direct effect on neural activity by altering the permeability of the membrane to neurotransmitters and their precursors, or by altering the functioning of the neurotransmitter receptors (figure B). This type of effect has a latency to onset of seconds to minutes. As a consequence of these actions, sex steroids can modulate the efficacy of neural signaling.

Sex steroids can also have an indirect effect on neural activity by forming noncovalent bonds with steroid receptors or by indirectly affecting other signaling pathways. Binding to a steroid receptor causes a conformational change that allows the receptor to bind to specific DNA recognition elements called hormone responsive elements. Steroid receptor coactivators, members of a family of coactivators that modulate the activity of steroid receptors, can enhance the effects of steroids by opening up chromatin structure, or by stabilizing the preinitiation complex at the promoter. Consequently, hormones can alter gene expression, leading to changes in the synthesis of specific proteins (see figure B). Such hormonal actions have a latency to onset of minutes to hours.

Most sexually dimorphic differences in the brains of females and males are thought to arise by the indirect actions of hormones on gene expression.

References

BROWN, T. J., J. YU, M. GAGNON, M. SHARMA AND N. J. MACLUSKY (1996) Sex differences in estrogen receptor and progestin receptor induction in the guinea pig hypothalamus and preoptic area. Brain Res. 725: 37–48.

MCEWEN, B. S., P. G. DAVIS, B. S. PARSONS AND D. W. PFAFF (1979) The brain as a target for steroid hormone action. Ann. Rev. Neurosci. 2: 65–112.

ROWAN, B. G., N. L. WEIGEL AND B. W. O'MALLEY (2000) Phosphorylation of steroid receptor coactivator-1: Identification of the phosphorylation sites and phosphorylation through the mitogen activated protein kinase pathway. J. Biol. Chem. 275: 4475–4483.

TSAI, M.-J. AND B. W. O'MALLEY (1994) Molecular mechanisms of action of steroid/thyroid receptor superfamily members. Ann. Rev. Biochem. 63: 451–486.

A variety of reproductive behaviors in both non-human primates and humans are governed by the hypothalamus, including mating, priming, and parenting behaviors (Figure 30.5). In rhesus monkeys, electrophysiological recordings from hypothalamic neurons during sexual activity show that neurons of the medial preoptic area of the anterior hypothalamus fire during different components of the sexual act. Such recordings have been carried out on male monkeys sitting in a flexible restraining chair that allows the male to gain access to a receptive female by pressing a bar, which brings the female close enough to allow mounting by the male. In this way, the responses of hypothalamic neurons can be correlated with "desire" (number of bar presses) and mating behavior (contact, mounting, intromission, thrusting). Neurons in the medial preoptic area of the male hypothalamus fire rapidly before sexual behavior but decrease their activity upon contact with the female and mating (Figure 30.6). In contrast, neurons in the dorsal anterior hypothalamus begin firing at the onset of mating and continue to fire vigorously during intercourse. Although these studies do not speak to sexual dimorphism, they provide direct evidence that the anterior hypothalamus helps to regulate aspects of sexual behavior.

These and other studies of rodents and non-human primates have stimulated a variety of observations in the human hypthalamus. The most thoroughly documented examples of sexually dimorphic hypothalamic nuclei in humans have been described by Laura Allen and Roger Gorski at the University of California at Los Angeles and supported by Dick Swaab and his colleagues at the Netherlands Institute for Brain Research. There are four cell groupings within the anterior hypothalamus of humans, collectively called the **interstitial nuclei of the anterior hypothalamus** (INAH; the nuclei are numbered 1–4 from dorsolateral to ventromedial) (Figure 30.7). The original studies by Allen and Gorski reported that nuclei 1–3 of the INAH can be

Figure 30.5 Organization of the components of the hypothalamus involved in regulating sexual functions. (A) The human hypothalamus, illustrating the location of the anterior hypothalamic area and other nuclei in which sexual dimorphisms have been observed in either humans or experimental animals. (B) Diagram of the major relationships of the anterior hypothalamus with other brain regions. Blue arrows denote neural connections; yellow arrows denote hormonal links; purple arrow denotes a combination of hormonal and neural connections. Although this information comes largely from studies of rodents, it is reasonable to assume that these interactions are characteristic of mammals.

(A)

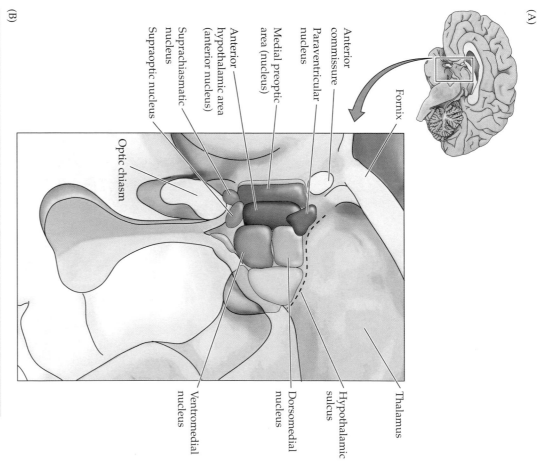

Anterior commissure

Paraventricular nucleus

Medial preoptic area (nucleus)

Anterior hypothalamic area (anterior nucleus)

Suprachiasmatic nucleus

Supraoptic nucleus

Fornix

Optic chiasm

Hypothalamic sulcus

Thalamus

Dorsomedial nucleus

Ventromedial nucleus

(B)

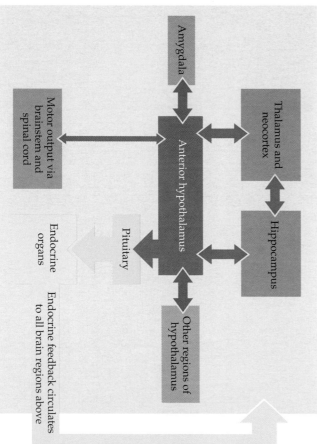

Amygdala

Thalamus and neocortex

Motor output via brainstem and spinal cord

Anterior hypothalamus

Hippocampus

Pituitary

Other regions of hypothalamus

Endocrine organs

Endocrine feedback circulates to all brain regions above

Figure 30.6 Many neurons in the primate hypothalamus are actively associated with sexual behavior. This example shows a histogram of neuronal activity recorded in the medial preoptic area in a male monkey exposed to a receptive female (see text). The firing rate of the neuron changes during different phases of sexual activity. (After Oomura et al., 1983.)

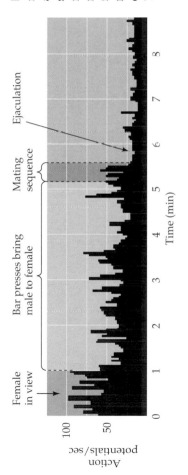

more than twice as large in males as they are in females. These studies were extended by William Byne who reported, along with Swaab and colleagues, that only INAH-3 is consistently sexually dimorphic in adults. INAH-1 and INAH-2 change in size over time, which may account for differences between the findings of various investigators. INAH-1 is the same size in females and males up until 2–4 years of age; it then becomes larger in males

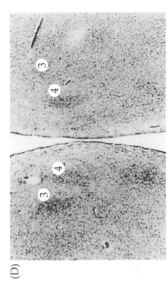

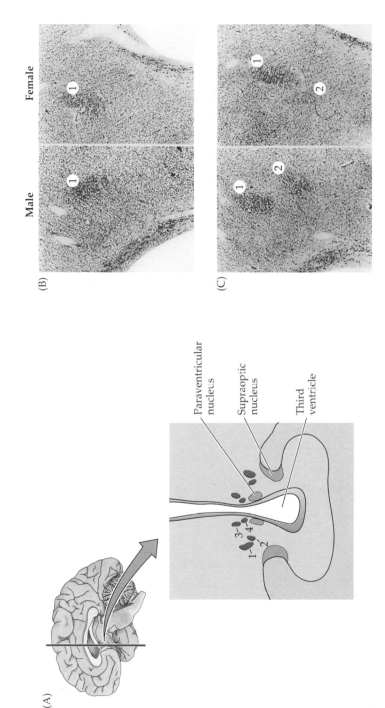

Figure 30.7 Sexual dimorphisms in the interstitial nuclei of the human anterior hypothalamus (INAH). (A) Diagrammatic coronal section through the anterior hypothalamus. The four interstitial nuclei of the anterior hypothalamus (red) are indicated by the numbers 1–4. (B–D) Micrographs showing the interstitial nuclei from a male (left column) and a female (right column). The male examples were taken from the left side of the brain, female examples from the right side at the same level. (B) INAH-1. (C) INAH-1 and -2. Note that INAH-2 is less compact in the female. (D) INAH-3 and -4. INAH-4 is well represented in both the male and female, whereas INAH-3 is clearly less distinct in the female. (B–D from Allen et al., 1989.)

until approximately 50 years of age, when it decreases in size in both sexes. Although generally larger in males, INAH-2 is larger in females of child-bearing age than in prepubescent and postmenopausal females. Such changes in nuclear size with age suggest that in humans, as in rodents, these hypothalamic dimorphisms may be related to levels of circulating sex steroids.

One aspect of human reproduction in which these nuclei have been implicated is the choice of a sexual partner. In addition to heterosexual behavior, some people express sexual interest in both females and males (**bisexuality**), and some only in members of their own phenotypic sex (**homosexuality**). Still other people are interested in the opposite sex but have a gender identity that is at odds with their phenotypic sex (**transgenderism**). Based on experimental work in animals and evidence that relatively simple sexual behaviors are influenced by brain dimorphisms, explaining these more complex behaviors in the same general way has been an attractive possibility. To investigate this issue, Simon LeVay, then working at the Salk Institute, compared the INAH nuclei of heterosexual males and homosexual males. LeVay first confirmed Allen and Gorski's findings that the INAH nuclei are sexually dimorphic. He went on to discover that INAH-3 is more than twice as large in male heterosexuals as in male homosexuals (Figure 30.8A). LeVay suggested that this difference is related to sexual orientation.

Other researchers have also concluded that dimorphisms of the hypothalamic nuclei are related to sexual orientation and gender identity. Swaab and Michel Hofman examined the suprachiasmatic nucleus of the hypothalamus, which lies just above the optic chiasm in both rodents and humans and generates circadian rhythms (see Figure 30.5A and Chapter 28). This nucleus is also involved in reproductive behavior. In examining the suprachiasmatic nuclei of females, heterosexual males, and homosexual males, Swaab and Hofman found the volume of the suprachiasmatic nucleus to be almost twice as large in male homosexuals compared to male heterosexuals (Figure 30.8B). There was no difference, however, between the size of the suprachiasmatic nucleus in females and heterosexual males. Like LeVay, they suggested that the difference in nuclear size between homosexual and heterosexual men might be related to sexual orientation. This same group has also reported a dimorphism that may be related to gender identity. In comparing male-to-female transgendered individuals to heterosexual males, they found that another hypothalamic structure, the bed nucleus of the stria terminalis, is smaller in transgendered males, being closer in size to that of females.

Taken together, this evidence suggests a plausible explanation of the continuum of human sexuality: Small differences in the relevant brain structures generate significant differences in sexual identity and behavior. In analogy to the rodent, these brain dimorphisms are probably established by the early influence of hormones acting on the brain nuclei that mediate various aspects of sexuality. For instance, low levels of circulating androgens in a male early in life could lead to a relatively "feminine" brain in genotypic males, whereas high levels of circulating androgens in females could lead to a relatively "masculine" brain in genotypic females.

As attractive as this hypothesis may be (and it should be emphasized that it remains unproven), the development of sexuality in humans is probably a good deal more complicated than this scenario implies. Although LeVay's findings support the idea that homosexuality is related to "feminization" of the male brain (recall that INAH-3 in gay males is smaller than in straight

Figure 30.8 Brain dimorphisms in heterosexual and homosexual human males. (A) Micrographs showing difference in INAH-3 between heterosexual and homosexual males. Arrowheads outline the nucleus. (B) The suprachiasmatic nucleus may also differ between homosexual and heterosexual males. Note that the suprachiasmatic nucleus of homosexual males appears to be larger (left histogram) and to contain more neurons (right histogram) than that of heterosexual males with or without AIDS. (A from LeVay, 1991; B after Swaab and Hofman, 1990.)

(A) INAH-3

Heterosexual male

Homosexual male

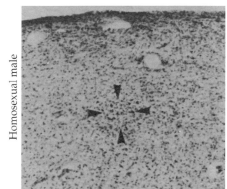

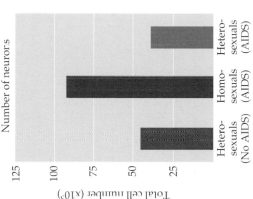

(B) Suprachiasmatic nucleus

Volume

Number of neurons

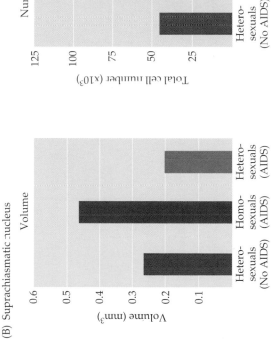

males), Swaab and Hofman's data on the size of the suprachiasmatic nucleus undermine the interpretation that the male homosexual's brain is simply "feminized" by a lack of androgens early in development. Whereas they found a difference in the volume of the suprachiasmatic nucleus between homosexual and heterosexual males, in contrast to LeVay they found no difference in the volume of the nucleus between females and heterosexual males. In addition, the development of the INAH-1 dimorphism occurs between 2 and 4 years of age—long after the first testosterone surge in human males. These discrepancies suggest that the development of sexually dimorphic nuclei in humans does not depend solely on early hormone levels. Even adult neural circuits have some plasticity (see next section and Chapter 25), leaving open the possibility that behavior, experience, and changes in circulating hormone levels generate dimorphisms at later stages. In accord with this suggestion, Breedlove and colleagues have reported that the posterodorsal nucleus of the medial amygdala has a greater volume in male rats than in female, but that castration of adult males and androgen treatment of adult females reverses this effect. Thus, the question of whether we are simply born "that way" with respect to sexuality remains difficult to answer with any great confidence. Like most developmental events, a combination of intrinsic and extrinsic factors are probably involved.

Despite these uncertainties, work over the last decade has placed human sexuality on a much firmer biological footing. This is a welcome advance over the not-too-distant past when unusual sexual behavior was commonly explained in social, Freudian, or, worse yet, moralistic terms.

Brain Dimorphisms Related to Cognitive Function

Many differences in sexual behavior obviously involve the cognitive functions mediated by the cerebral cortex and functions mediated by subcortical structures other than (or in addition to) the hypothalamus (e.g., the basal ganglia and amygdala—see Chapters 18 and 29). Evidence for sexually dimorphic cognitive behavior comes mainly from clinical observations. For example, neurologists have reported that females suffer aphasia less often than males after damage to the left hemisphere. This observation led to the suggestion that language functions are to some degree differently represented in females and males. To explore this issue, Doreen Kimura at the University of Western Ontario studied language ability in right-handed patients with unilateral lesions of the left cerebral cortex. She found that females were more likely to suffer aphasia if the damage was to the anterior left hemisphere, while males were more likely to suffer aphasia if the damage was located posteriorly. Kimura suggested that language areas of the female brain are more anteriorly represented than the same areas in males.

Another brain structure relevant to the cognitive abilities of females and males is the corpus callosum, the large fiber bundle that connects the right and left cerebral hemispheres (see Chapters 1 and 27). In the most complete study to date, Laura Allen and her colleagues found that the corpus callosum, while not different in size between females and males, does differ in shape; a particular region of the callosum called the splenium is more bulbous in females than in males (Figure 30.9A,B). The other fiber bundle that connects the cerebral hemispheres, the anterior commissure, is also sexually dimorphic, its midsagittal surface area being about 12% larger than in males (Figure 30.9C). Since both the anterior commissure and the corpus callosum mediate the interhemispheric transfer of information, these findings raise the possibility that there may be some differences between females and males in the way information is integrated across the hemispheres.

The performance of tasks that depend more on one hemisphere than the other has also been examined in females and males. One simple test for visuospatial differences entails how well girls and boys are able to identify shapes after feeling them with the right or left hand while blindfolded. Both sexes perform equally well with either hand up to about 6 years of age. Thereafter, boys start scoring better when they use their left hand, whereas girls continue to score equally well with either hand up to 13 years of age, when they also begin to do better with their left hand. This study suggests that boys develop the right hemispheric lateralization of visuospatial skills earlier than girls.

The idea that females and males develop lateralized functions at different rates is supported by studies of the development of the prefrontal cortex of non-human primates. Removing the prefrontal cortex before 15 to 18 months of age does not affect motor-planning functions in female rhesus monkeys, although the same lesion in male monkeys at this age diminishes these skills. In a similar vein, human females and males use different strategies to navigate in an unfamiliar environment. Males trying to find their way out of

(A)

(B)

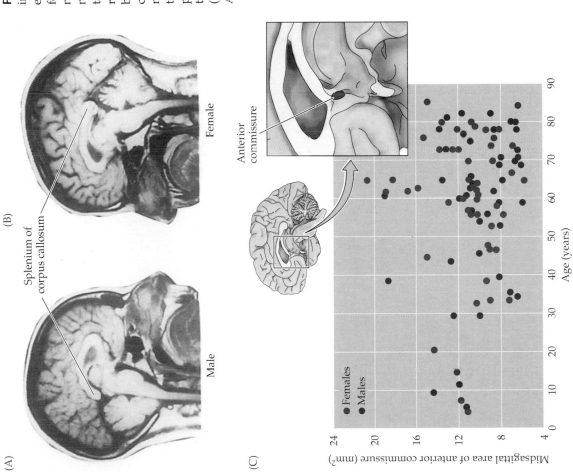

Male

Female

Splenium of
corpus callosum

Figure 30.9 Structural dimorphisms in the human brain that might differently influence cognitive behaviors in females and males. The magnetic resonance images show the body and splenium of the human corpus callosum taken from the midsagittal plane of a representative male (A) and female (B) brain. (C) Size differences of the anterior commissures of human females and males measured at autopsy. Although the differences in the means of the two populations are statistically significant, the distributions obviously overlap. (A,B from Allen et al., 1991; C after Allen and Gorski, 1991.)

(C)

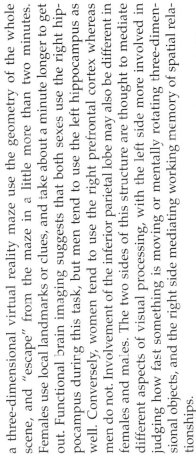

Anterior
commissure

a three-dimensional virtual reality maze use the geometry of the whole scene, and "escape" from the maze in a little more than two minutes. Females use local landmarks or clues, and take about a minute longer to get out. Functional brain imaging suggests that both sexes use the right hippocampus during this task, but men tend to use the left hippocampus as well. Conversely, women tend to use the right prefrontal cortex whereas men do not. Involvement of the inferior parietal lobe may also be different in females and males. The two sides of this structure are thought to mediate different aspects of visual processing, with the left side more involved in judging how fast something is moving or mentally rotating three-dimensional objects, and the right side mediating working memory of spatial relationships.

Although none of these studies is in itself compelling, there probably are some differences in cognitive function in males and females, and in the way these abilities are represented in the brain.

(A) Female rat

Location of nipples on ventrum

Primary somatic sensory cortex

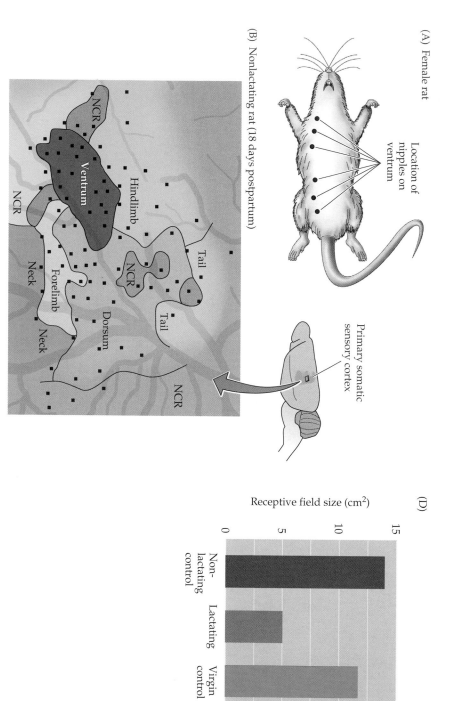

(B) Nonlactating rat (18 days postpartum)

NCR

Ventrum

Hindlimb

Tail

NCR

Tail

NCR

Forelimb

Neck

Dorsum

Neck

NCR

(C) Lactating rat (19 days postpartum)

HL

Ventrum

NCR

Hindlimb

Forelimb

NCR

Tail

NCR

Dorsum

Neck

(D)

Receptive field size (cm²)

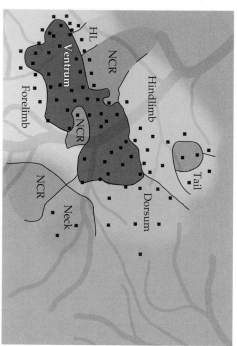

Non-lactating control

Lactating

Virgin control

Figure 30.10 Changes in the cortical representation of the chest wall in the rat primary somatic sensory cortex during lactation. (A) Ventrum of the female rat; dots mark the position of nipples. (B) Diagram of somatic sensory cortex in a nonlactating control rat, showing the amount of cortex normally activated by stimulation of the ventrum. Squares mark electrode penetrations; colors signify the estimated representation. (C) Similar diagram from a 19-day postpartum, lactating rat. Note the expansion of the representation of the ventrum. NCR = no cutaneous response. (D) Histogram of receptive field sizes of single neurons in nonlactating control, lactating, and virgin control rats. The receptive field sizes of neurons in lactating mothers are decreased. (B–C after Xerri et al., 1994.)

Hormone-Sensitive Brain Circuits in Adult Animals

One of the many problems associated with studies of differences in female and male brain structures is that the relevant regions are usually studied long after the related behaviors are established. As mentioned earlier, there is growing evidence that some brain circuits continue to change over the course of an individual's life, depending on experience and on the hormonal milieu. A good example is the changes in the brain circuits of adult rats in conjunction with parenting behavior (Figure 30.10). Michael Merzenich and his colleagues have shown that the cortical representation of the ventrum (chest wall) is altered in the somatic sensory cortex of lactating females. As determined by electrophysiological mapping, the representation of the ventrum is approximately twice as large in nursing females as in non-lactating controls. Moreover, the receptive fields of the neurons representing the skin of the ventrum in lactating females are decreased in size by about a third. Both the increase in cortical representation and the decrease in receptive field size indicate that changes in behavior are reflected in changes of cortical circuitry.

Another example of adult plasticity under hormonal control is the altered connections between cells of the female rat hypothalamus after giving birth. In females prior to pregnancy, the relevant hypothalamic neurons are isolated from each other by thin astrocytic processes. Under the influence of the hormonal environment prevailing during birth and lactation, the glial processes retract and the oxytocin- and vasopressin-secreting neurons become electrically coupled by gap junctions (Figure 30.11). Whereas these neurons fire independently before the female gives birth, during lactation they fire synchronously, releasing pulses of oxytocin into the maternal circulation. These surges of oxytocin cause the contraction of smooth muscles in the mammary glands, and hence milk ejection. Interestingly, the changes are thought to be mediated by olfactory cues, since the lactating circuits can be induced in virgin females simply by placing them in the vicinity of pups.

These examples suggest that the sexual circuits of the brain are malleable not only during development, but to some degree throughout life as the hormonal milieu changes.

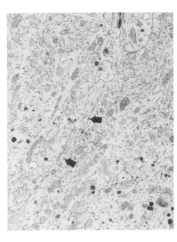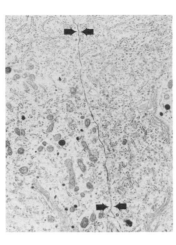

Figure 30.11 Changes in neurons of the rat supraoptic nucleus during lactation. Top: Before birth, the relevant neurons and their dendrites are isolated from each other by astrocytic processes (arrows). Bottom: During nursing of the young, the astrocytic processes withdraw, and neurons and their dendrites show close apposition (arrow pairs) that allows electrical synapses to form between adjacent neurons (see Chapter 5). (From Modney and Hatton, 1990.)

Summary

Differences in female and male behaviors ranging from copulation to cognition are linked to differences in brain structure. Although the neural basis for these sexual dimorphisms is much clearer in experimental animals, the evidence for sex-related differences in the human brain is now substantial. The region of the brain in which the most clear-cut structural dimorphisms occur is the anterior hypothalamus, which governs reproductive behavior. In rats and monkeys, the nuclei in this region play a role not only in the mechanics of sex, but also in desire, parenting, and sexual preferences. In the rodent, sexual dimorphisms of both behavior and the underlying circuitry develop primarily as a result of hormonal action on neurons during early development. On the strength of this knowledge about sexual development in experimental animals, neurobiological explanations for a range of human sexual behaviors have been proposed. Such models remain controversial because the sexual dimorphisms of the human brain and their functional significance are neither fully established nor well understood. Nonetheless, from what is now known it seems likely that a deeper understanding of how sex hormones influence the developing (and to some extent the adult) brain will eventually explain the fascinating continuum of human sexuality.

ADDITIONAL READING

Reviews

BLACKLESS, M., A. CHARUVASTRA, A. DERRYCK, A. FAUSTO-STERLING, K. LAUZANNE AND E. LEE (2000) How sexually dimorphic are we? Review and synthesis. Am. J. Human Biol. 12: 151–166.

MACLUSKY, N. J. AND F. NAFTOLIN (1981) Sexual differentiation of the central nervous system. Science 211: 1294–1302.

McEWEN, B.S. (1999) Permanence of brain sex differences and structural plasticity of the adult brain. PNAS 96: 7128–7130.

SMITH, C. L AND B. W. O'MALLEY (1999) Evolving concepts of selective estrogen receptor action: From basic science to clinical applications. Trends Endocrinol. Metab. 10: 299–300.

SWAAB, D. F. (1992) Gender and sexual orientation in relation to hypothalamic structures. Horm. Res. 38(Suppl. 2): 51–61.

SWAAB, D. F. AND M. A. HOFMAN (1984) Sexual differentiation of the human brain: A historical perspective. In *Progress in Brain Research*, Vol. 61. G. J. De Vries (ed.). Amsterdam: Elsevier, pp. 361–374.

Important Original Papers

ALLEN, L. S., M. HINES, J. E. SHRYNE AND R. A. GORSKI (1989) Two sexually dimorphic cell groups in the human brain. J. Neurosci. 9: 497–506.

ALLEN, L. S., M. F. RICHEY, Y. M. CHAI AND R. A. GORSKI (1991) Sex differences in the corpus callosum of the living human being. J. Neurosci. 11: 933–942.

BREEDLOVE, S. M. AND A. P. ARNOLD (1981) Sexually dimorphic motor nucleus in the rat lumbar spinal cord: Response to adult hormone manipulation, absence in androgen-insensitive rats. Brain Res. 225: 297–307.

BYNE, W., M. S. LASCO, E. KEMETHER, A. SHINWARI, L. JONES AND S. TOBET (2000) The interstitial nuclei of the human anterior hypothalamus: Assessment for sexual variation in volume and neuronal size, density and number. Brain Res. 856: 254–258.

COOKE, B. M., G. TABIBNIA AND S. M. BREEDLOVE (1999) A brain sexual dimorphism controlled by adult circulating androgens. PNAS 96: 7538–7540.

FREDERIKSE, M.E., A. LU, E. AYLWARD, P. BARTA AND G. PEARLSON (1999) Sex differences in the inferior parietal lobule. Cerebral Cortex 9: 896–901.

GORSKI, R. A., J. H. GORDON, J. E. SHRYNE AND A. M. SOUTHAM (1978) Evidence for a morphological sex difference within the medial preoptic area of the rat brain. Brain Res. 143: 333–346.

GRON, G., A. P. WUNDERLICH, M. SPITZER, R. TOMCZAK AND M. W. RIEPE (2000) Brain activation during human navigation: Gender different neural networks as substrate of performance. Nature Neurosci. 3: 404–408.

LEVAY, S. (1991) A difference in hypothalamic structure between heterosexual and homosexual men. Science 253: 1034–1037.

MEYER-BAHLBURG, H. F. L., A. A. EHRHARDT, L. R. ROSEN AND R. S. GRUEN (1995) Prenatal estrogens and the development of homosexual orientation. Dev. Psych. 31: 12–21.

MODNEY, B. K. AND G. I. HATTON (1990) Motherhood modifies magnocellular neuronal interrelationships in functionally meaningful ways. In *Mammalian Parenting*, N. A. Krasnegor and R. S. Bridges (eds.). New York: Oxford University Press, pp. 306–323.

RAISMAN, G. AND P. M. FIELD (1973) Sexual dimorphism in the neuropil of the preoptic area of the rat and its dependence on neonatal androgen. Brain Res. 54: 1–29.

SWAAB, D.F. AND E. FLIERS (1985) A sexually dimorphic nucleus in the human brain. Science 228: 1112–5.

WOOLLEY, C. S. AND B. S. McEWEN (1992) Estradiol mediates fluctuation in hippocampal synapse density during the estrous cycle in the adult rat. J. Neurosci. 12: 2549–2554.

XERRI, C., J. M. STERN AND M. M. MERZENICH (1994) Alterations of the cortical representation of the rat ventrum induced by nursing behavior. J. Neurosci. 14: 1710–1721.

ZHOU, J.-N., M. A. HOFMAN, L. J. G. GOOREN AND D. F. SWAAB (1995) A sex difference in the human brain and its relation to transsexuality. Nature 378: 68–70.

Books

FAUSTO-STERLING, A. (2000) *Sexing the Body.* New York: Basic Books.

GOY, R. W. AND B. S. McEWEN (1980) *Sexual Differentiation of the Brain.* Cambridge, MA: MIT Press.

LEVAY, S. (1993) *The Sexual Brain.* Cambridge, MA: MIT Press.

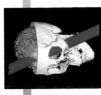

Chapter 31

Human Memory

Overview

One of the most intriguing of the brain's complex functions is the ability to store information provided by experience and to retrieve much of it at will. Without this ability, many of the cognitive functions discussed in the preceding chapters cannot operate. Learning is the name given to the process by which new information is acquired by the nervous system, and memory to the mechanism of storage and/or retrieval of that information. Equally fascinating (and important) is the normal ability to forget information. Pathological forgetfulness, or amnesia, has been especially instructive about the neurological underpinnings of memory; amnesia is defined as the inability to learn new information or to retrieve information that has already been acquired. The importance of memory in daily life has made understanding these several phenomena one of the major challenges of modern neuroscience, a challenge that has only begun to be met. The mechanisms of plasticity that provide plausible cellular and molecular bases for some aspects of information storage have been considered in Chapters 23 through 25. The present chapter summarizes the broader organization of human memory, surveys the major clinical manifestations of memory disorders, and considers the implications of these disorders for ultimately understanding human memory in more detailed terms.

Qualitative Categories of Human Memory

Humans have at least two qualitatively different systems of information storage, which are generally referred to as **declarative memory** and **procedural memory** (Figure 31.1; see also Box A). Declarative memory is the storage (and retrieval) of material that is available to consciousness and can therefore be expressed by language (hence, "declarative"). Examples of declarative memory are the ability to remember a telephone number, a song, or the images of some past event. Procedural memory, on the other hand, is not available to consciousness, at least not in any detail. Such memories involve skills and associations that are, by and large, acquired and retrieved at an unconscious level. Remembering how to dial the telephone, how to sing the song, how to efficiently inspect a scene, or to make the myriad associations that occur continuously are all examples of memories that fall in this category. It is difficult or impossible to say how we do these things, and we are not conscious of any particular memory during their occurrence. In fact, thinking about such activities may actually inhibit the ability to perform them smoothly and efficiently (thinking about how to stroke a tennis ball or trying to remember the name associated with a familiar face often makes matters worse).

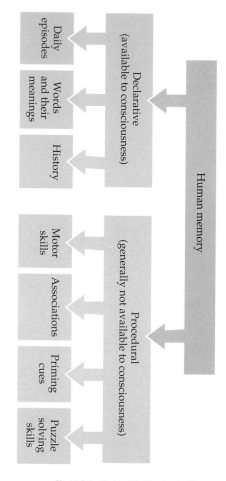

Figure 31.1 The major qualitative categories of human memory. Declarative memory includes those memories that can be brought to consciousness and expressed as remembered events, images, sounds, and so on. Procedural memory includes motor skills, cognitive skills, simple classical conditioning, priming effects, and other information that is acquired and retrieved unconsciously.

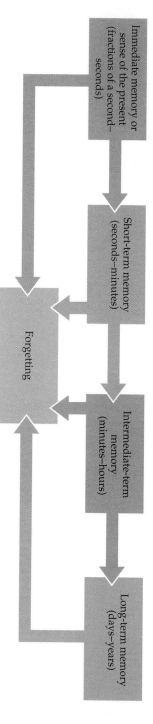

Figure 31.2 The major temporal categories of human memory.

Temporal Categories of Memory

In addition to the types of memory defined by the nature of what is remembered, memory can also be categorized according to the *time* over which it is effective. Although the details are still debated by both psychologists and neurobiologists, three temporal classes of memory are generally accepted (Figure 31.2). The first of these is **immediate memory**. By definition, immediate memory is the routine ability to hold ongoing experiences in mind for a few seconds. The capacity of this register is very large, involves all modalities (visual, verbal, tactile, and so on), and provides the ongoing sense of a "present."

Short-term memory, the second temporal category, is the ability to hold information in mind for seconds to minutes once the present moment has passed. A conventional way of testing the integrity of (declarative) short-term memory at the bedside is to present a string of randomly ordered digits, which the patient is then asked to repeat; surprisingly, the normal "digit span" is only 7–9 numbers. A special sort of (procedural) short-term memory is called **working memory**, which refers to the ability to hold information in mind long enough to carry out sequential actions. An example is searching for a lost object; working memory allows the hunt to proceed efficiently, avoiding places already inspected. A particular advantage of working memory is that it can be readily examined in experimental animals (see Chapter 26).

The third temporal category is **long-term memory** and entails the retention of information in a more permanent form of storage for days, weeks, or even a lifetime. There is general agreement that the so-called **engram** (i.e., the physical embodiment of the long-term memory in neuronal machinery) depends on long-term changes in the efficacy of transmission of the relevant synaptic connections, and/or the actual growth and reordering of such con-

BOX A
Phylogenetic Memory

A category of information storage not usually considered in standard accounts is memories that arise from the experience of the species over the eons, established by natural selection acting on the cellular and molecular mechanisms of neural development. Such stored information does not depend on postnatal experience, but on what a given species has typically encountered in its environment. These "memories" are no less consequential than those acquired by individual experience and are likely to have much underlying biology in common with the memories established during an individual's lifetime.

Information about the experience of the species, as expressed by endogenous or "instinctive" behavior, can be quite sophisticated, as is apparent in examples collected by ethologists in a wide range of animals, including primates. The most thoroughly studied instances of such behaviors are those occurring in young birds. Hatchlings arrive in the world with an elaborate set of innate behaviors. First is the complex behavior that allows the young bird to emerge from the egg. Having hatched, a variety of additional behaviors indicate how much of its early life is dependent on inherited information. Hatchlings of precocial species "know" how to preen, peck, gape their beaks, and carry out a variety of other complex acts immediately. In some species, hatchlings automatically crouch down in the nest when a hawk passes overhead but are oblivious to the overflight of an innocuous bird. Konrad Lorenz and Niko Tinbergen used handheld silhouettes to explore this phenomenon in naïve herring gulls (see figure). "It soon became obvious," wrote Tinbergen, "that . . . the reaction was mainly one to shape. When the model had a short neck so that the head protruded only a little in front of

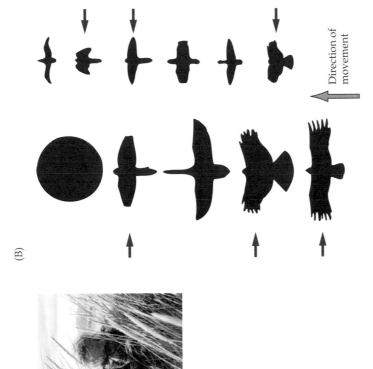

(A)

(B)

↓ ↓ ↓

← Direction of movement

↑ ↑ ↑

(A) Niko Tinbergen at work. (B) Silhouettes used to study alarm reactions in hatchlings. The shapes that were similar to the shadow of the bird's natural predators (red arrows) when moving in the appropriate direction elicited escape responses (crouching, crying, seeking cover); silhouettes of songbirds and other innocuous species (or geometrical forms) elicited no obvious response. (After Tinbergen, 1969.)

the line of the wings, it released alarm, independent of the exact shape of the dummy." Evidently, the memory of what the shadow of a predator looks like is built into the nervous system of this species. Examples in primates include the innate fear that newborn monkeys have of snakes and looming objects.

Despite the relatively scant attention paid to this aspect of memory, it is probably the most important component of the stored information in the brain that determines whether or not an individual survives long enough to reproduce. It may be that the mechanisms by which such associations are made (between the shadow of a predator and the appropriate motor and emotional response in the hatchling, for instance) are similar, or even identical, to those that promote appropriate associations by virtue of postnatal experience.

References

TINBERGEN, N. (1969) *Curious Naturalists.* Garden City, NY: Doubleday.

TINBERGEN, N. (1953) *The Herring Gull's World.* New York: Harper & Row.

LORENZ, K. (1970) *Studies in Animal and Human Behaviour.* (Translated by R. Martin.) Cambridge, MA: Harvard University Press.

DUKAS, R. (1998) *Cognitive Ecology.* Chicago: University of Chicago Press.

TABLE 31.1
The Fallibility of Human Memory[a]

(A) Initial list of words	(B) Subsequent test list
candy	taste
sour	point
sugar	sweet
bitter	
good	
taste	
tooth	
nice	
honey	
soda	
chocolate	
heart	
cake	
eat	
pie	

[a] After hearing the words in list A read aloud, subjects were asked to identify which of the items in list B had also been on list A. See text for the results.

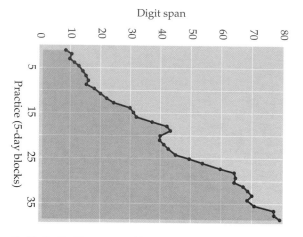

Figure 31.3 Increasing the digit span by practice (and the development of associational strategies). During many months involving one hour of practice a day for 3–5 days a week, this subject increased his digit span from 7 to 79 numbers. Random digits were read to him at the rate of one per second. If a sequence was recalled correctly, one digit was added to the next sequence. (After Ericsson et al., 1980.)

nections. As discussed in Chapter 25, there is good reason to think that both these varieties of synaptic change occur.

Evidence for a continual transfer of information from short-term to long-term memory storage is apparent in the phenomenon of **priming**. Priming is typically demonstrated by presenting subjects with a set of items to which they are exposed under false pretenses. For example, a list of words can be given with the instruction that the subjects are to identify some feature that is actually extraneous to the experiment (e.g., whether the words are verbs, adjectives, or nouns). Later—typically the next day—the same individuals are given a different test in which they are asked to fill in the missing letters of words with whatever letters come to mind. The test list actually includes fragments of words that were presented the day before, mixed among fragments of words that were not. Subjects fill in the letters to make the words that were presented the day before at a much higher rate than expected by chance, even though they have no specific memory of the words seen earlier; moreover, they are faster at filling in letters to make words that were previously seen than new words. Priming shows that information previously presented is influential, even though we are entirely unaware of its effect on subsequent behavior. The significance of priming is well known—at least intuitively—to advertisers, teachers, spouses, and others who have reason to influence the way we think and act.

Despite the prevalence of such transfer, the information stored in this process is not particularly reliable. Consider, for instance, the list of words in Table 31.1A. If the list is read to a group of students who are immediately thereafter asked to identify which of several *new* items (see Table 31.1B) were on the original list, and which were not, the result is surprising. Typically, about half the students report that the word "sweet" was included in the list in Table 31.1A, and moreover that they are quite certain about it! The mechanism of such erroneous "recall" is presumably the strong associations that have previously been made between the words on the list and the word "sweet," which bias the witnesses to think that "sweet" was a member of the original set. Clearly, memories, even those we feel quite certain about, are often false.

The Importance of Association in Information Storage

As mentioned, the normal human capacity for remembering relatively meaningless information is surprisingly limited (a string of about 7–9 numbers or other arbitrary items). This capacity can, however, be increased dramatically. For example, a college student who for some months spent an hour each day practicing the task of remembering randomly presented numbers was able to increase his reiterative ability up to a string of about 80 digits (Figure 31.3). He did this by making subsets of the string of numbers he was given signify dates or times at track meets (he was a competitive runner)—in essence, giving meaningless items a meaningful context. This same strategy of association is used by most professional "mnemonists," who amaze audiences by apparently prodigious feats of memory. Similarly, a good chess player can remember the position of many more pieces on a briefly examined board than a poor player, presumably because the posi-

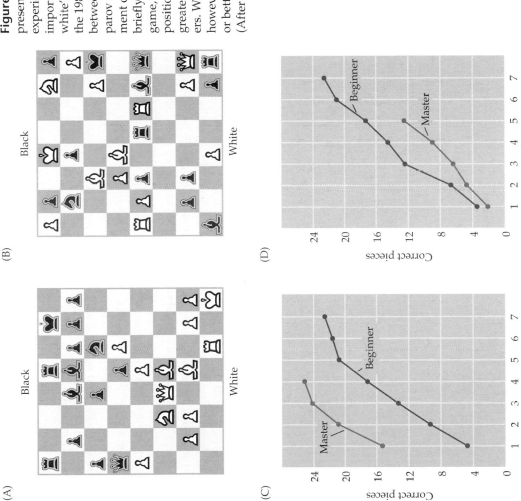

Figure 31.4 The retention of briefly presented information depends on past experience, context, and its perceived importance. (A) Board position after white's twenty-first move in game 10 of the 1985 World Chess Championship between A. Karpov (white) and G. Kasparov (black). (B) A random arrangement of the same 28 pieces. (C,D) After briefly viewing the board from the real game, master players reconstruct the positions of the pieces with much greater efficiency than beginning players. With a randomly arranged board, however, beginners perform as well as or better than accomplished players. (After Chase and Simon, 1973.)

tions have much more significance for individuals who know the intricacies of the game than for neophytes (Figure 31.4). Thus, the capacity of short-term memory very much depends on what the information in question means to the individual, and how readily it can be associated with information that has already been stored.

The ability of humans to remember *significant* information in the normal course of events is, in fact, enormous. Consider Arturo Toscanini, the late conductor of the NBC Philharmonic Orchestra, who allegedly kept in his head the complete scores of more than 250 orchestral works, as well as the music and librettos for about 100 operas. Once, just before a concert in St. Louis, the first bassoonist approached Toscanini in some consternation because he had just discovered that one of the keys on his bassoon was broken. After a minute or two of deep concentration, the story goes, Toscanini turned to the alarmed bassoonist and informed him that there was no need for concern, since that note did not appear in any of the bassoon parts for the evening's program. A parallel example of a prodigious quantitative memory is the mathematician Alexander Aitken. After an undistinguished career in elementary school, Aitken was, at the age of 13, greatly taken with the manipulation of numbers. For the next four years he undertook, as a personal

BOX B
Savant Syndrome

A fascinating developmental anomaly of human memory is rare individuals who, until recently, were referred to as "*idiots savants*" (the current literature tends to discuss this phenomenon in terms of the less pejorative phrase "savant syndrome"). Savants are people who, for a variety of poorly understood reasons (typically brain damage in the perinatal period), are severely restricted in most mental activities but extraordinarily competent and mnemonically capacious in one particular domain. The grossly disproportionate skill compared to the rest of their restricted mental life can be striking. Indeed, these individuals—whose special talent may be in calculation, history, art, language, or music—are usually diagnosed as severely retarded.

Many striking examples could be cited, but a summary of one such case suffices to make the point. The individual whose history is summarized here was given the fictitious name "Christopher" in a detailed study carried out by psychologists Neil Smith and Ianthi-Maria Tsimpli. Christopher was discovered to be severely brain damaged at just a few weeks of age (perhaps as the result of rubella during his mother's pregnancy, or anoxia during birth; the record is uncertain in this respect). Christopher had been institutionalized since childhood because he was unable

to care for himself, could not find his way around, had poor hand-eye coordination, and a variety of other deficiencies. Tests on standard IQ scales were low, consistent with his general inability to cope with daily life (scores on the Wechsler Scale were, on different occasions, 42, 67, and 52). Despite this severe mental incapacitation, Christopher took an intense interest in books from the age of about three, particularly those providing factual information and lists (e.g., telephone directories and dictionaries). At about six or seven he began to read technical papers that his sister sometimes brought home from work, and he soon showed a surprising proficiency in foreign languages. His special talent in the acquisition and use of language, an area where savants are often especially limited, grew rapidly. As an early teenager, Christopher could translate from—and communicate in—a variety of languages including Danish, Dutch, Finnish, French, German, modern Greek, Hindi, Italian, Norwegian, Polish, Portuguese, Russian, Spanish, Swedish, Turkish, and Welsh (in which his skills were described as ranging from rudimentary to fluent). This extraordinary level of linguistic accomplishment is all the more remarkable since he had no formal training in language even at the elementary school level, and cannot play tic-tac-toe or checkers because he

cannot grasp the rules needed to make moves in these games.

The neurobiological basis for such extraordinary individuals is not understood. It is fair to say, however, that savants are unlikely to have an ability in their areas of expertise that exceeds the competency of normally intelligent individuals who focus passionately on a particular subject (see examples in the text). Presumably the savant's intense interest in a particular cognitive domain is due to one or more brain regions that continue to work reasonably well. Whether because of social feedback or self-satisfaction, savants clearly spend a great deal of their mental time and energy practicing the skill they can exercise more or less normally. The result is that the relevant associations they make become especially rich, as Christopher demonstrates.

References

MILLER, L. K. (1989) *Musical Savants: Exceptional Skill in the Mentally Retarded.* Hillsdale, New Jersey: Lawrence Erlbaum Associates.

SMITH, N. AND I.-M. TSIMPLI (1995) *The Mind of a Savant: Language Learning and Modularity.* Oxford, England: Basil Blackwell Ltd.

HOWE, M. J. A. (1989) *Fragments of Genius: The Strange Feats of Idiots Savants.* Routledge, New York: Chapman and Hall.

challenge, to master mental calculation. He began by memorizing the value of π to 1000 places, and could soon do calculations in his head with such facility that he became a local celebrity. When asked for the squares of three-digit numbers, he was able to give these almost instantly. The square roots for each were produced to five significant digits in 2 to 3 seconds; the squares of four-digit numbers allegedly took him only about 5 seconds. Aitken went on to become a professor of mathematics at Edinburgh and was eventually elected a Fellow of the Royal Society for his contributions to numerical mathematics, statistics, and matrix algebra. At the age of 30 or so, he began to lose his enthusiasm for "mental yoga," as he called his penchant. In part, his waning enthusiasm stemmed from the realization that the advent of calculators was making his prowess obsolete (it was then 1930). He also discovered that the last 180 digits of π that he had memorized as a boy were wrong; he had taken the values from the published work of another mental calculator, who erred in an era when there was no way to check the correct value. In fact, Aitken's feat has long since been superseded. In 1981, an Indian mnemonist memorized the value of π to 31,811 places, only to have a Japanese mnemonist increase this record to 40,000 places a few years later!

Toscanini's and Aitken's mental processes in these feats were not rote learning, but a result of the fascination that aficionados bring to their special interests (Box B). Although few can boast the mnemonic prowess of such talented individuals, the human ability to remember the things that deeply interest us—whether baseball statistics, soap opera plots, or the details of brain structure—is amazing.

Forgetting

Some years ago, a poll showed that 84% of psychologists agreed with the statement that "everything we learn is permanently stored in the mind, although sometimes particular details are not accessible." The 16% who thought otherwise should get the higher marks. Common sense indicates that, were it not for forgetting, our brains would be impossibly burdened with the welter of useless information that is briefly encoded in our immediate memory "buffer." In fact, the human brain is very good at forgetting. In addition to the unreliable performance on tests such as the example in Table 31.1, Figure 31.5 shows that the memory of the appearance of a penny (an icon seen thousands of times since childhood) is uncertain at best, and that people gradually forget what they have seen over the years (TV shows, in this case). Clearly we forget things that have no importance, and unused memories deteriorate over time.

The ability to forget unimportant information may be as critical for normal mentation as retaining information that is significant. One reason for this presumption is rare individuals who have difficulty with the normal

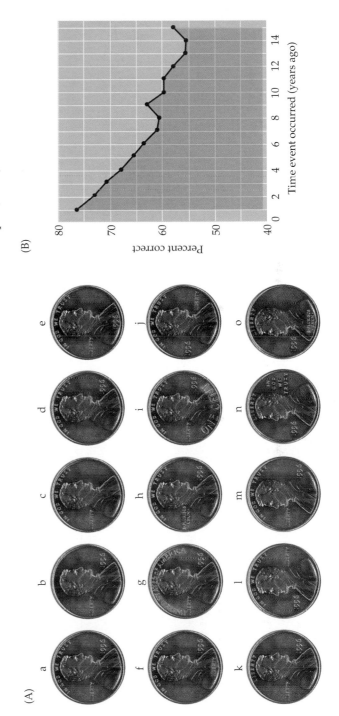

Figure 31.5 Forgetting. (A) Different versions of the "heads" side of a penny. Despite innumerable exposures to this familiar design, few people are able to pick out (a) as the authentic version. Clearly, repeated information is not necessarily retained. (B) The deterioration of long-term memories was evaluated in this example by a multiple-choice test in which the subjects were asked to recognize the names of television programs that had been broadcast for only one season during the past 15 years. Forgetting of stored information that is no longer used evidently occurs gradually and progressively over the years (chance performance = 25%). (A after Rubin and Kontis, 1983; B after Squire, 1989.)

erasure of information. Perhaps the best-known case is a subject studied over several decades by the Russian psychologist A. R. Luria, who referred to the subject simply as "S." Luria's description of an early encounter gives some idea why S, then a newspaper reporter, was so interesting:

I gave S a series of words, then numbers, then letters, reading them to him slowly or presenting them in written form. He read or listened attentively and then repeated the material exactly as it had been presented. I increased the number of elements in each series, giving him as many as thirty, fifty, or even seventy words or numbers, but this too, presented no problem for him. He did not need to commit any of the material to memory; if I gave him a series of words or numbers, which I read slowly and distinctly, he would listen attentively, sometimes ask me to stop and enunciate a word more clearly, or, if in doubt whether he had heard a word correctly, would ask me to repeat it. Usually during an experiment he would close his eyes or stare into space, fixing his gaze on one point; when the experiment was over, he would ask that we pause while he went over the material in his mind to see if he had retained it. Thereupon, without another moment's pause, he would reproduce the series that had been read to him.

A. R. Luria (1987), *The Mind of a Mnemonist*, pp. 9–10

S's phenomenal memory, however, did not always serve him well. He had difficulty ridding his mind of the trivial information that he tended to focus on, sometimes to the point of incapacitation. As Luria put it:

Thus, trying to understand a passage, to grasp the information it contains (which other people accomplish by singling out what is most important) became a tortuous procedure for S, a struggle against images that kept rising to the surface in his mind. Images, then, proved an obstacle as well as an aid to learning in that they prevented S from concentrating on what was essential. Moreover, since these images tended to jam together, producing still more images, he was carried so far adrift that he was forced to go back and rethink the entire passage. Consequently, a simple passage—a phrase, for that matter—would turn out to be a Sisyphean task.

A. R. Luria (1987), *The Mind of a Mnemonist*, p. 113

TABLE 31.2
Causes of Amnesia

Causes	Examples	Site of damage
Vascular occlusion of both posterior cerebral arteries	Patient R.B. (Box C)	Bilateral medial temporal lobe, the hippocampus in particular
Midline tumors	—	Medial thalamus bilaterally (hippocampus and other related structures if tumor is large enough)
Trauma	Patient N.A. (Box C)	Bilateral medial temporal lobe
Surgery	Patient H.M. (Box C)	Bilateral medial temporal lobe
Infections	Herpes simplex encephalitis	Bilateral medial temporal lobe
Vitamin B$_1$ deficiency	Korsakoff's syndrome	Medial thalamus and mammillary bodies
Electroconvulsive therapy (ECT) for depression	—	Uncertain

Although forgetting is a normal and apparently essential mental process, it can also be pathological, a condition called **amnesia**. Some of the causes of amnesia are listed in Table 31.2. An inability to establish new memories is called **anterograde amnesia**, whereas difficulty retrieving previously established memories is called **retrograde amnesia**. Anterograde and retrograde amnesia are often present together, but can be dissociated under various circumstances. Amnesias following bilateral lesions of the temporal lobe and diencephalon have given particular insight into where and how at least some categories of memory are formed and stored (see next section and Box C).

Brain Systems Underlying Declarative and Procedural Memories

Three extraordinary clinical cases of amnesia have been especially revealing about the brain systems responsible for the short-term storage of declarative information and are now familiar to neurologists and neurobiologists as patients H.M., N.A., and R.B. (see Box C). Taken together, these cases provide dramatic evidence of the importance of midline diencephalic and medial temporal lobe structures—the **hippocampus**, in particular—in establishing new declarative memories (Figure 31.6). These patients also demonstrate

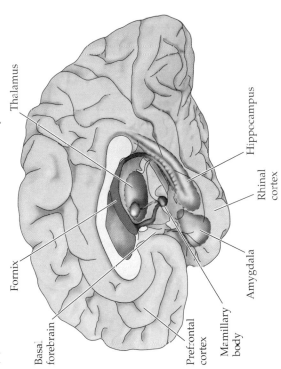

(A) Brain areas associated with declarative memory disorders

Thalamus

Fornix

Basal forebrain

Prefrontal cortex

Mamillary body

Amygdala

Rhinal cortex

Hippocampus

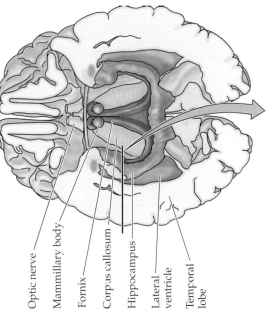

(B) Ventral view of hippocampus and related structures with part of temporal lobes removed

Optic nerve

Mammillary body

Fornix

Corpus callosum

Hippocampus

Lateral ventricle

Temporal lobe

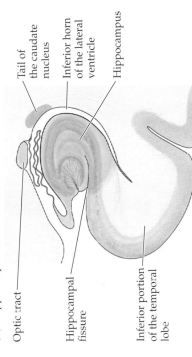

(C) Hippocampus in coronal section

Optic tract

Hippocampal fissure

Inferior portion of the temporal lobe

Tail of the caudate nucleus

Inferior horn of the lateral ventricle

Hippocampus

Figure 31.6 Brain areas that, when damaged, tend to give rise to declarative memory disorders. By inference, declarative memory is based on the physiological activity of these structures. (A) Studies of amnesic patients have shown that the formation of declarative memories depends on the integrity of a subset of limbic circuits (see Chapter 29), particularly those of the hippocampus and its subcortical connections to the mammillary bodies and dorsal thalamus. (B) Diagram showing the location of the hippocampus in a cutaway view in the horizontal plane. (C) The hippocampus as it would appear in a histological section in the coronal plane, at approximately the level indicated by the green line in (B).

BOX C
Clinical Cases That Reveal the Anatomical Substrate for Declarative Memories

The Case of H.M.

At the age of 27, H.M., who had suffered minor seizures since age 10 and major seizures since age 16, underwent surgery to correct his increasingly debilitating epilepsy. A high school graduate, H.M. had been working as a technician in a small electrical business until shortly before the time of his operation. His attacks involved generalized convulsions with tongue biting, incontinence, and loss of consciousness (all typical of grand mal seizures). Despite a variety of medications, the seizures remained uncontrolled and increased in severity. A few weeks before his surgery, H.M. became unable to work and had to quit his job.

On September 1, 1953, a bilateral medial temporal lobe resection was carried out in which the amygdala, uncus, hippocampal gyrus, and anterior two-thirds of the hippocampus were removed. At the time, it was unclear that bilateral surgery of this kind would cause a profound memory defect. Severe amnesia was evident, however, upon H.M.'s recovery from the operation, and his life was changed radically.

The first formal psychological exam of H.M. was conducted nearly 2 years after the operation, at which time a profound memory defect was still obvious. Just before the examination, for instance, H.M. had been talking to the psychologist; yet he had no recollection of this experience a few minutes later, denying that anyone had spoken to him. He gave the date as March 1953 and seemed oblivious to the fact that he had undergone an operation, or that he had become incapacitated as a result. Nonetheless, his score on the Wechsler-Bellevue Intelligence Scale was 112, a value not significantly different from his preoperative IQ. Various psychological tests failed to

reveal any deficiencies in perception, abstract thinking, or reasoning; he seemed highly motivated and, in the context of casual conversation, normal. Importantly, he also performed well on tests of the ability to learn new skills, such as mirror writing or puzzle solving (that is, his ability to form procedural memories was intact). Moreover, his early memories were easily recalled, showing that the structures removed during H.M.'s operation are not a permanent repository for such information. On the Wechsler Memory Scale (a specific test of declarative memory), however, he performed very poorly, and he could not recall a preceding test-set once he had turned his attention to another part of the exam. These deficits, along with his obvious inability to recall events in his daily life, all indicate a profound loss of short-term declarative memory function.

During the subsequent decades, H.M. has been studied extensively, primarily by Brenda Milner and her colleagues at the Montreal Neurological Institute. His memory deficiency has continued unabated, and, according to Milner, he has little idea who she is in spite of their acquaintance for nearly 50 years. Sadly, he has gradually come to appreciate his predicament. "Every day is alone," H.M. reports, "whatever enjoyment I've had and whatever sorrow I've had."

The Case of N.A.

N.A. was born in 1938 and grew up with his mother and stepfather, attending public schools in California. After a year of junior college, he joined the Air Force. In October of 1959 he was assigned to the Azores as a radar technician and remained there until December 1960, when a bizarre accident made him a celebrated neurological case. N.A. was

assembling a model airplane in his barracks room while his roommate, unbeknownst to him, was making thrusts and parries with a miniature fencing foil behind N.A.'s chair. N.A. turned suddenly and was stabbed through the right nostril. The foil penetrated the cribriform plate (the structure through which the olfactory nerve enters the brain) and took an upward course into the left forebrain. N.A. lost consciousness within a few minutes (presumably because of bleeding in the region of brain injury) and was taken to a hospital. There he exhibited a right-sided weakness and paralysis of the right eye muscles innervated by the third cranial nerve. Exploratory surgery was undertaken and the dural tear repaired. Gradually he recovered and was sent home to California. After some months, his only general neurological deficits were some weakness of upward gaze and mild double vision. He retained, however, a severe anterograde amnesia for declarative memories. MRI studies first carried out in 1986 showed extensive damage to the thalamus and the medial temporal lobe, mostly on the right side; the mammillary bodies also appeared to be missing bilaterally. The exact extent of his lesion, however, is not known, as N.A. remains alive and well.

N.A.'s memory from the time of his injury over 40 years ago to the present has remained impaired, and like H.M. he fails badly on formal tests of new learning ability. His IQ is 124, and he shows no defects in language skills, perception, or other measures of intelligence. He can also learn new procedural skills quite normally. His amnesia is not as dense as that of H.M. and is more verbal than spatial. He can, for example, draw accurate diagrams of material presented to him earlier. Nonetheless, he loses track of his possessions, forgets what he has done, and tends to forget who has come to visit him. He has only vague impressions of political, social, and sporting events that have occurred since his injury. Watching television is difficult because he tends to forget the storyline during commercials. On the other hand, his memory for events prior to 1960 is extremely good; indeed, his lifestyle tends to reflect the 1950s.

The Case of R.B.

At the age of 52, R.B. suffered an ischemic episode during cardiac bypass surgery. Following recovery from anesthesia, a profound amnesic disorder was apparent. As in the cases of H.M. and N.A., his IQ was normal (111), and he showed no evidence of cognitive defects

other than memory impairment. R.B. was tested extensively for the next five years, and, while his amnesia was not as severe as that of H.M. or N.A., he consistently failed the standard tests of the ability to establish new declarative memories. When R.B. died in 1983 of congestive heart failure, a detailed examination of his brain was carried out. The only significant finding was bilateral lesions of the hippocampus—specifically, cell loss in the CA1 region that extended the full rostral–caudal length of the hippocampus on both sides. The amygdala, thalamus, and mammillary bodies, as well as the structures of the basal forebrain, were normal. R.B.'s case is particularly important because it suggests that hippocampal lesions alone can result in profound anterograde amnesia for declarative memory.

References

CORKIN, S. (1984) Lasting consequences of bilateral medial temporal lobectomy: Clinical course and experimental findings in H.M. Semin. Neurol. 4: 249–259.

CORKIN, S., D. G. AMARAL, R. G. GONZÁLEZ, K. A. JOHNSON AND B. T. HYMAN (1997) H. M.'s medial temporal lobe lesion: Findings from MRI. J. Neurosci. 17: 3964-3979.

HILTS, P. J. (1995) *Memory's Ghost: The Strange Tale of Mr. M. and the Nature of Memory.* New York: Simon and Schuster.

MILNER, B., S. CORKIN AND H.-L. TEUBER (1968) Further analysis of the hippocampal amnesic syndrome: A 14-year follow-up study of H.M. Neuropsychologia 6: 215–234.

SCOVILLE, W. B. AND B. MILNER (1957) Loss of recent memory after bilateral hippocampal lesions. J. Neurol. Neurosurg. Psychiat. 20: 11–21.

SQUIRE, L. R., D. G. AMARAL, S. M. ZOLA-MORGAN, M. KRITCHEVSKY AND G. PRESS (1989) Description of brain injury in the amnesic patient N.A. based on magnetic resonance imaging. Exp. Neurol. 105: 23–35.

TEUBER, H. L., B. MILNER AND H. G. VAUGHN (1968) Persistent anterograde amnesia after stab wound of the basal brain. Neuropsychologia 6: 267–282.

ZOLA-MORGAN, S., L. R. SQUIRE AND D. AMARAL (1986) Human amnesia and the medial temporal region: Enduring memory impairment following a bilateral lesion limited to the CA1 field of the hippocampus. J. Neurosci. 6: 2950–2967.

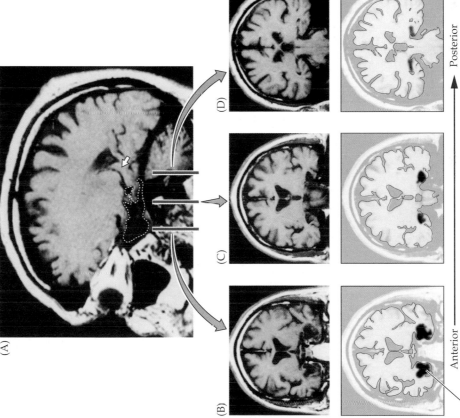

MRI images of the brain of patient H.M. (A) Sagittal view of the right hemisphere; the area of the anterior temporal lobectomy is indicated by the white dotted line. The intact posterior hippocampus is the banana-shaped object indicated by the white arrow. (B–D) Coronal sections at approximately the levels indicated by the red lines in (A). Image (B) is the most rostral and is at the level of the amygdala. The amygdala and the associated cortex are entirely missing. Image (C) is at the level of the rostral hippocampus; again, this structure and the associated cortex have been removed. Image (D) is at the caudal level of the hippocampus; the posterior hippocampus appears intact, although somewhat shrunken. Outlines below give a clearer indication of the parts of H.M.'s brain that have been ablated (black shading). (From Corkin et al., 1997.)

Posterior

Anterior

Damaged area

that there is a different anatomical substrate for anterograde and retrograde amnesia, since in each of these individuals, memory for events *prior* to the precipitating injury was retained. The devastating deficiency is (or was in the case of R.B.), the inability to establish new memories. Retrograde amnesia—the loss of memory for events preceding an injury or illness—is more typical of the generalized lesions associated with head trauma and neurodegenerative disorders, such as Alzheimer's disease (Box D). Although a degree of retrograde amnesia can occur with the more focal lesions that cause anterograde amnesia, the long-term storage of memories is presumably distributed throughout the brain (see next section). Thus, the hippocampus and related diencephalic structures indicated in Figure 31.6 form and consolidate declarative memories that are ultimately stored elsewhere. Finally, it is important to emphasize that H.M., N.A., and R.B. had no problems establishing (or recalling) procedural memories, indicating that procedural memories are laid down by means of a different anatomical substrate.

Other causes of amnesia have also provided some insight into the parts of the brain relevant to various aspects of memory (see Table 31.2), **Korsakoff's syndrome**, for example, occurs in chronic alcoholics as a result of thiamine (vitamin B_1) deficiency. In such cases, loss of brain tissue occurs bilaterally in the mammillary bodies and the medial thalamus, for reasons that are not well understood.

The Long-Term Storage of Information

Revealing though they have been, clinical studies of amnesic patients have provided relatively little insight into the long-term storage of information in the brain (other than to indicate quite clearly that such information is *not* stored in the midline diencephalic structures that are affected in anterograde amnesia). Nonetheless, a good deal of circumstantial evidence implies that the cerebral cortex is the major long-term repository for many aspects of memory.

One line of evidence comes from observations of patients undergoing electroconvulsive therapy (ECT). Individuals with severe depression are often treated by the passage of enough electrical current through the brain to cause the equivalent of a full-blown seizure (this procedure being done under anesthesia in well-controlled circumstances). This remarkably useful treatment was discovered because depression in epileptics was perceived to be alleviated after a spontaneous seizure. However, ECT often causes both anterograde and retrograde amnesia. The patients typically do not remember the treatment itself or the events of the preceding days, and their recall of events of the previous 1–3 years can also be affected. Animal studies (rats tested for maze learning, for example) have confirmed the amnesic consequences of ECT. The memory loss usually clears over a period of weeks to months. However, to mitigate this side effect (which may be the result of excitotoxicity; see Box B in Chapter 6), ECT is often delivered to only one hemisphere at a time. The nature of amnesia following ECT supports the conclusion that long-term memories are widely stored in the cerebral cortex, since this is the part of the brain predominantly affected by this therapy.

Since different cortical regions have different cognitive functions (see Chapters 26 and 27), it is not surprising that these sites store information that reflects the cognitive function of the relevant part of the brain. For example, the lexicon that links speech sounds and their symbolic significance is located in the association cortex of the superior temporal lobe, since damage to this area typically results in an inability to link words and meanings (Wernicke's aphasia; see Chapter 27). Presumably, the widespread con-

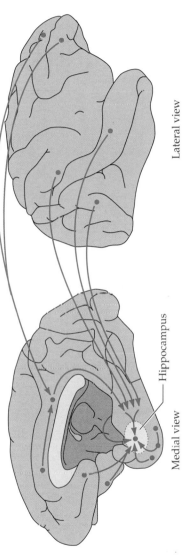

(A) Afferent connections of the hippocampal region

Lateral view

Medial view

Hippocampus — Hippocampal region

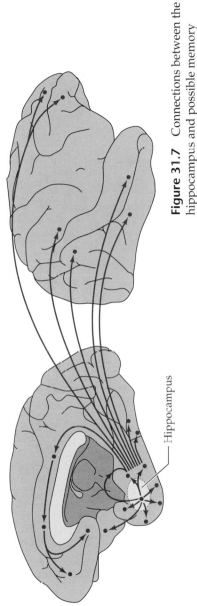

(B) Efferent connections of the hippocampal region

Hippocampus

Figure 31.7 Connections between the hippocampus and possible memory storage sites. The rhesus monkey brain is shown because these connections are much better documented in subhuman primates than in humans. Projections to this region are shown in (A); the efferent projections from the hippocampus are shown in (B). Projections from numerous cortical areas converge on the hippocampus and the related structures known to be involved in human memory; most of these sites also send projections to the same cortical areas. Medial and lateral views are shown, the latter rotated 180° for clarity. (After Van Hoesen, 1982.)

nections of the hippocampus to the language areas serve to consolidate declarative information in these and other language-related cortical sites (Figure 31.7). By the same token, the inability of patients with temporal lobe lesions to recognize objects and/or faces suggests that such memories are stored in that location. Similarly, frontal lobe syndromes imply that memories about appropriate behaviors in a given social context and future plans reside in the frontal cortex (see Chapter 26).

With respect to procedural learning, the motor skills gradually acquired through practice are evidently stored in the basal ganglia, cerebellum, and premotor cortex (see Chapters 17–19). Thus lesions of these sites cause a loss of the ability to make complex coordinated movements that can be considered a sort of "motor amnesia." This scheme for long-term information storage is diagrammed in Figure 31.8, although the generality of the diagram only emphasizes the rudimentary state of present thinking about exactly how and where long-term memories are stored. A reasonable guess is that each complex memory is instantiated in the activity of an extensive

Acquisition and storage of declarative information

Long-term storage (a variety of cortical sites: Wernicke's areas for the meanings of words, temporal cortex for the memories of objects and faces, etc.)

Short-term memory storage (hippocampus and related structures)

Acquisition and storage of procedural information

Long-term memory storage (cerebellum, basal ganglia, premotor cortex, and other sites related to motor behavior)

Short-term memory storage (sites unknown but presumably widespread)

Figure 31.8 Summary diagram of the acquisition and storage of declarative versus procedural information.

BOX D
Alzheimer's Disease

Dementia is a syndrome characterized by failure of recent memory and other intellectual functions that is usually insidious in onset but steadily progresses. Alzheimer's disease (AD) is the most common dementia, accounting for 60–80% of cases in the elderly. It affects 5–10% of the population over the age of 65, and as much as 45% of the population over 85. The earliest sign is typically an impairment of recent memory function and attention, followed by failure of language skills, visual-spatial orientation, abstract thinking, and judgment. Inevitably, alterations of personality accompany these defects.

The tentative diagnosis of Alzheimer's disease is based on these characteristic clinical features, and can only be confirmed by the distinctive cellular pathology evident on post-mortem examination of the brain. The histopathology consists of three principal features: (1) collections of intraneuronal cytoskeletal filaments called *neurofibrillary tangles*; (2) extracellular deposits of an abnormal protein in a matrix called amyloid in so-called *senile plaques*; and (3) a diffuse loss of neurons. These changes are most apparent in neocortex, limbic structures (hippocampus, amygdala, and their associated cortices), and selected brainstem nuclei (especially the basal forebrain nuclei).

Although the vast majority of AD cases arise sporadically, the disorder is inherited in an autosomal dominant pattern in a tiny fraction (<1%) of patients. Identification of the mutant gene in a few families with an early-onset autosomal dominant form of the disease provides considerable insight into the kinds of processes that go awry in Alzheimer's. Investigators had long suspected that the mutant gene responsible for familial AD might reside on chromo-

some 21, primarily because Down's syndrome has similar clinical and neuropathologic features, but with a much earlier onset (at about age 30 in most cases). Down's syndrome is typically caused by an extra copy of chromosome 21 (trisomy 21).

The prominence of amyloid deposits in AD further suggested that a mutation of a gene encoding amyloid precursor protein is somehow involved. The gene for amyloid precursor protein (APP) was cloned by D. Goldgaber and colleagues, and found to reside on chromosome 21. This discovery eventually led to the identification of mutations of the APP gene in almost 20 families with the early-onset autosomal dominant form of AD. Importantly, only a few of the early-onset families, and none of the late-onset families, exhibited these particular

mutations. The mutant genes underlying two additional autosomal dominant forms of AD have been subsequently identified (*presenilin 1* and *presenilin 2*). Thus, mutation of any one of these three genes appears to be sufficient to cause a heritable form of AD.

The most common form of Alzheimer's, however, occurs late in life, and although the relatives of affected individuals are at a greater risk, the disease is clearly not inherited in any simple sense. The central role of APP in the families with the early-onset form of the disease nonetheless suggested that APP might somehow be linked to the chain of events culminating in the "spontaneous" forms of Alzheimer's disease. In particular, biochemists Warren Strittmatter and Guy Salvesen theorized that pathologic deposition of proteins com-

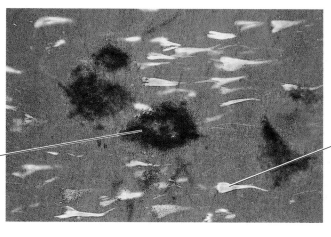

Neurofibrillary tangle

Amyloid plaque

Histological section of the cerebral cortex from a patient with Alzheimer's disease, showing characteristic amyloid plaques and neurofibrillary tangles. Elements of these features of the disease are indicated in the relevant diagrams. (From Roses, 1995; courtesy of Gary W. Van Hoesen.)

plexed with a derivative of APP might be responsible. To test this idea, they immobilized a recombinant form of the APP derivative on nitrocellulose paper and searched for proteins in the cerebrospinal fluid of patients with Alzheimer's disease that bound with high affinity. One of the proteins they detected was apolipoprotein E, a molecule that normally chaperones cholesterol through the bloodstream.

This discovery was especially provocative in light of a discovery made by Margaret Pericak-Vance, Allen Roses, and their colleagues at Duke University Medical Center, who found that affected members of some families with the late-onset form of the inherited disease exhibited an association with genetic markers on chromosome 19. This finding was of particular interest because a gene encoding an isoform of apolipoprotein E (the $\varepsilon 4$ allele) is located in the same region of chromosome 19 implicated by the family studies. As a result, they began to explore the association of the different alleles of apolipoprotein E with affected members in families with a late-onset but inherited form of Alzheimer's disease.

There are three major alleles of apolipoprotein E, $\varepsilon 2$, $\varepsilon 3$, and $\varepsilon 4$. The frequency of allele $\varepsilon 3$ in the general population is 0.78, and the frequency of allele $\varepsilon 4$ is 0.14. The frequency of the $\varepsilon 4$ allele in late-onset familial AD patients, however, is 0.52, almost four times higher than the general population. Thus, the inheritance of the $\varepsilon 4$ allele is a risk factor for late-onset AD. In fact, people homozygous for $\varepsilon 4$ are about 8 times more likely to develop AD compared to individuals homozygous for $\varepsilon 3$. Among individuals in late-onset Alzheimer's families with no copies of $\varepsilon 4$, only 20% develop AD by age 75 compared to 90% of individuals with two copies of $\varepsilon 4$. An increased association of the $\varepsilon 4$ allele has also been shown in the sporadic form of AD, an especially important discovery because this category constitutes by far the most common form of the disease.

It is not known whether the $\varepsilon 4$ allele of ApoE itself is responsible for the increased risk, or whether it is linked to another gene on chromosome 19 that is the real culprit. The fact that ApoE binds avidly to amyloid plaques in AD brains favors the idea that the $\varepsilon 4$ allele of ApoE itself is the problem. However, in contrast to the mutations of APP or presenilin 1 and presenilin 2 that cause familial forms of AD, inheriting the $\varepsilon 4$ form of ApoE is *not* sufficient to cause AD; rather, inheriting this gene simply increases the risk of developing AD. Moreover, some of the individuals with early-onset forms of familial AD do *not* have the $\varepsilon 4$ allele. Thus, a variety of related molecular problems appear to underlie AD.

A possible common denominator of these problems at the cellular level is the "amyloid cascade" hypothesis. A prominent constituent of the amyloid plaques is an abnormal cleavage product of APP called amyloid-β peptide (or β-A4). The cascade hypothesis proposes that accumulation of β-A4 is critical to the pathogenesis of AD. Opponents, however, argue that extracellular deposition of β-A4 may not be a key event in the pathogenesis of AD because the density of the β-A4 plaques correlates only poorly with severity of the dementia (the degree of dementia being much better correlated with the density of neurofibrillary tangles). Moreover, a transgenic mouse model of AD based on a *presenilin 1* mutation exhibits neurodegeneration without amyloid plaque formation.

Clearly, AD has a complex pathology and probably reflects a variety of related molecular and cellular abnormalities. It is unlikely that this important problem will be understood without a great deal more research, much hyperbole in the lay press notwithstanding.

References

CHUI, D. H., H. TANAHASHI, K. OZAWA, S. IKEDA, F. CHECLER, O. UEDA, H. SUZUKI, W. ARAKI, H. INOUE, K. SHIROTANI, K. TAKAHASHI, F. GALLYAS AND T. TABIRA (1999) Transgenic mice with Alzheimer *presenilin 1* mutations show accelerated neurodegeneration without amyloid plaque formation. Nature Med. 5: 560–564.

CITRON, M., T. OLTERSDORF, C. HAAS, L. MCCONLOGUE, A. Y. HUNG, P. SEUBERT, C. VIGO-PELFREY, I. LIEBERBURG AND D. J. SELKOE (1992) Mutation of the β-amyloid precursor protein in familial Alzheimer's disease increases β-protein production. Nature 360: 672–674.

CORDER, E. H., A. M. SAUNDERS, W. J. STRITTMATTER, D. E. SCHMECHEL, P. C. GASKELL, G. W. SMALL, A. D. ROSES, J. L. HAINES AND M. A. PERICAK-VANCE (1993) Gene dose of apolipoprotein E type 4 allele and the risk of Alzheimer's disease in late onset families. Science 261: 921–923.

GOLDGABER, D., M. I. LERMAN, O. W. MCBRIDE, U. SAFFIOTTI AND D. C. GAJDUSEK (1987) Characterization and chromosomal localization of a cDNA encoding brain amyloid of Alzheimer's disease. Science 235: 877–880.

LI, T. ET AL. (2000) Photoactivated gamma-secretase inhibitors directed to the active site covalently label presenilin 1. Nature 405: 689–694.

MURRELL, J., M. FARLOW, B. GHETTI AND M. D. BENSON (1991) A mutation in the amyloid precursor protein associated with hereditary Alzheimer's disease. Science 254: 97–99.

ROGAEV, E. I., R. SHERRINGTON, E. A. ROGAEVA, G. LEVESQUE, M. IKEDA, Y. LIANG, H. CHI, C. LIN, K. HOLMAN, T. TSUDA, L. MAR, S. SORBI, B. NACMIAS, S. PIACENTINI, L. AMADUCCI, I. CHUMAKOV, D. COHEN, L. LANNFELT, P. E. FRASER, J. M. ROMMENS AND P. H. ST. GEORGE-HYSLOP (1995) Familial Alzheimer's disease in kindreds with missense mutations in a gene on chromosome 1 related to the Alzheimer's disease type 3 gene. Nature 376: 75–778.

SHERRINGTON, R., E. I. ROGAEV, Y. LIANG, ET AL. (1995) Cloning of a gene bearing missense mutations in early-onset familial Alzheimer's disease. Nature 375: 754–760.

Figure 31.9 Brain size as a function of age. The human brain reaches its maximum size (measured by weight in this case) in early adult life and decreases progressively thereafter. This decrease presumably represents the gradual loss of neural circuitry in the aging brain, which presumably underlies the progressively diminished memory function in older individuals. (After Dekaban and Sadowsky, 1978.)

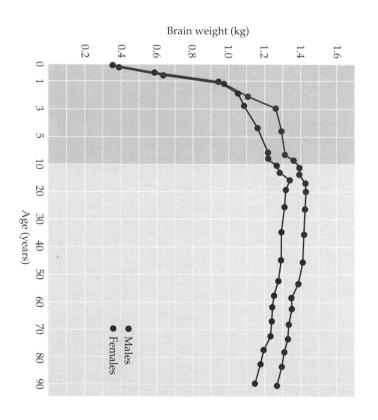

network of neurons whose triggering depends on synaptic weightings that have been molded and modified by experience.

Memory and Aging

Although outward appearance obviously changes with age, we tend to imagine that the brain is resistant to the ravages of time. Unfortunately, the evidence suggests that this optimistic view is unjustified. From early adulthood onward, the average weight of the normal human brain, as determined at autopsy, steadily decreases (Figure 31.9). In elderly individuals, this effect can be observed with noninvasive imaging as a slight but nonetheless significant shrinkage of the brain. Counts of synapses in the cerebral cortex generally decrease in old age (although the number of neurons probably does not change very much), suggesting that it is mainly the connections between neurons (i.e., neuropil) that are lost as we grow old (consistent with the idea that the networks of connections that represent memories—i.e., the engrams—gradually deteriorate).

These several observations accord with the difficulty many older people have in making associations (e.g., remembering names or the details of recent experiences) and with declining scores on tests of memory as a function of age. The normal loss of some memory function with age means that there is a large gray zone between individuals undergoing normal aging and patients suffering from age-related dementias such as Alzheimer's disease (see Box D).

Summary

Human memory entails a number of biological strategies and anatomical substrates that are unlikely to be explained in terms of any particular cellular or molecular mechanism. Primary among these are a system for memories that can be expressed by means of language and can be made available to the conscious mind (declarative memory), and a separate system that concerns skills and associations that are essentially prelinguistic, operating at a largely unconscious level (procedural memory). Based on evidence from amnesic patients and knowledge about normal patterns of neural connections in the human brain, the hippocampus and associated midline diencephalic and medial temporal lobe structures are critically important in laying down new declarative memories, although not in storing them (a process that occurs primarily in the association cortices). In contrast, procedural memories for motor and other unconscious skills depends on the integrity of the premotor cortex, basal ganglia, and cerebellum, and is not affected by lesions that impair the declarative memory system. The common denominator of these categories of stored information is generally thought to be alterations in the strength and number of the synaptic connections in the cerebral cortices that mediate associations between stimuli and the behavioral responses to them.

ADDITIONAL READING

Reviews

BUCKNER, R. L. (2000) Neuroimaging of memory. In *The New Cognitive Neurosciences*, M. Gazzaniga (ed.). Cambridge, MA: MIT Press, pp. 817–840.

CABEZA, R. (In press) Function neuroimaging of cognitive aging. In *Handbook of Functional Neuroimaging of Cognition*, R Cabeza and A. Kingstone (eds.). Cambridge, MA: MIT Press.

ERICKSON, C. A., B. JAGADEESH AND R. DESIMONE (2000) Learning and memory in the inferior temporal cortex of the macaque. In *The New Cognitive Neurosciences*, M. Gazzaniga (ed.). Cambridge, MA: MIT Press, pp. 743–752.

MISHKIN, M. AND T. APPENZELLER (1987) The anatomy of memory. Sci. Amer. 256(6): 80–89.

PETRI, H. AND M. MISHKIN (1994) Behaviorism, cognitivism, and the neuropsychology of memory. Amer. Sci. 82: 30–37.

SCHACTER, D. L. AND R. L. BUCKNER (1998) Priming and the brain. Neuron 20: 185–195

SQUIRE, L. R. AND B. J. KNOWLTON (2000) The medical temporal lobe, the hippocampus, and the memory systems of the brain. In *The New Cognitive Neurosciences*, M. Gazzaniga (ed.). Cambridge, MA: MIT Press, pp. 765–779.

SQUIRE, L. R. (1992) Memory and hippocampus: A synthesis from findings with rats, monkeys, and humans. Psych. Rev. 99: 195–231.

THOMPSON, R. F. (1986) The neurobiology of learning and memory. Science 223: 941–947.

ZOLA-MORGAN, S. M. AND L. R. SQUIRE (1993) Neuroanatomy of memory. Ann. Rev. Neurosci. 16: 547–563.

Important Original Papers

GOBET, F. AND H. A. SIMON (1998) Expert chess memory: Revisiting the chunking hypothesis. Memory 6: 225–255.

SCOVILLE, W. B. AND B. MILNER (1957) Loss of recent memory after bilateral hippocampal lesions. J. Neurol. Neurosurg. Psychiat. 20: 11–21.

SQUIRE, L. R. (1989) On the course of forgetting in very long-term memory. J. Exp. Psychol. 15: 241–245.

ZACKS, R. T., L. HASHER AND K. Z. H. LI (1999) Human memory. In *The Handbook of Aging and Cognition*. F. I. M. Craik and T. A. Salthouse (eds.). Mahweh, New Jersey: Lawrence Erlbaum Associates, pp. 293–357.

ZOLA-MORGAN, S. M. AND L. R. SQUIRE (1990) The primate hippocampal formation: Evidence for a time-limited role in memory storage. Science 250: 288–290.

Books

BADDELEY, A. (1982) *Your Memory: A User's Guide*. New York: Macmillan.

CRAIK, F. I. M. AND T. A. SALTHOUSE (1999) *The Handbook of Aging and Cognition*. Mahweh, New Jersey: Lawrence Erlbaum Associates.

DUKAS, R. (1998) *Cognitive Ecology: The Evolutionary Ecology of Information: Processing and Decision Making*. Chicago: University of Chicago Press.

GAZZANIGA, M.S. (2000) *The New Cognitive Neurosciences*, 2nd Ed. Cambridge MA: MIT Press.

GAZZANIGA, M. S., R. B. IVRY AND G. R. MANGUN (1998) *Cognitive Neuroscience: The Biology of the Mind*. New York: W. W. Norton & Company.

LURIA, A. R. (1987) *The Mind of a Mnemonist*. Translated by Lynn Solotaroff. Cambridge, MA: Harvard University Press.

NEISSER, U. (1982) *Memory Observed: Remembering in Natural Contexts*. San Francisco: W. H. Freeman.

PENFIELD, W. AND L. ROBERTS (1959) *Speech and Brain Mechanisms*. Princeton, NJ: Princeton University Press.

SAPER, C. B. AND F. PLUM (1985) *Handbook of Clinical Neurology, Vol. 1(45): Clinical Neuropsychology*, P. J. Vinken, G. S. Bruyn and H. L. Klawans (eds.). New York: Elsevier, pp. 107–128.

SMITH, S. B. (1983) *The Great Mental Calculators: The Psychology, Methods, and Lives of Calculating Prodigies, Past and Present*. New York: Columbia University Press.

SQUIRE, L. R. (1987) *Memory and Brain*. New York: Oxford University Press, pp. 202–223.

ZECHMEISTER, E. B. AND S. E. NYBERG (1982) *Human Memory: An Introduction to Research and Theory*. Monterey, CA: Brooks/Cole Publishing.

Glossary

acetylcholine Neurotransmitter at motor neuron synapses, in autonomic ganglia and a variety of central synapses; binds to two types of receptors—ligand-gated ion channels (nicotinic receptors) and G-protein-coupled receptors (muscarinic receptors).

achromatopsia, cerebral Loss of color vision as a result of damage to extrastriate visual cortex.

action potential The electrical signal conducted along axons (or muscle fibers) by which information is conveyed from one place to another in the nervous system.

activation The time-dependent opening of ion channels in response to a stimulus, typically membrane depolarization.

adaptation The phenomenon of sensory receptor adjustment to different levels of stimulation; critical for allowing sensory systems to operate over a wide dynamic range.

adenylyl cyclase Membrane-bound enzyme that can be activated by G-proteins to catalyze the synthesis of cyclic AMP from ATP.

adhesion molecules see cell adhesion molecules.

adrenaline see epinephrine.

adrenal medulla The central part of the adrenal gland that, under visceral motor stimulation, secretes epinephrine and norepinephrine into the bloodstream.

adrenergic Refers to synaptic transmission mediated by the release of norepinephrine or epinephrine.

adult The mature form of an animal, usually defined by the ability to reproduce.

afferent An axon that conducts action potentials from the periphery toward the central nervous system.

agnosia The inability to name objects.

alpha (α) motor neurons Neurons in the ventral horn of the spinal cord that innervate skeletal muscle.

amacrine cells Retinal neurons that mediate lateral interactions between bipolar cell terminals and the dendrites of ganglion cells.

amblyopia Diminished visual acuity as a result of the failure to establish appropriate visual cortical connections in early life.

amnesia The pathological inability to remember or establish memories; retrograde amnesia is the inability to recall existing memories, whereas anterograde amnesia is the inability to lay down new memories.

amphetamine A synthetically produced central nervous system stimulant with cocaine-like effects; drug abuse may lead to dependence.

ampullae The juglike swellings at the base of the semicircular canals that contain the hair cells and cupulae (see also cupulae).

amygdala A nuclear complex in the temporal lobe that forms part of the limbic system; its major functions concern autonomic, emotional, and sexual behavior.

androgen insensitivity syndrome A condition in which, due to a defect in the gene that codes for the androgen receptor, testosterone cannot act on its target tissues.

anencephaly A congenital defect of neural tube closure, in which much of the brain fails to develop.

anosmia Loss of the sense of smell.

anterior Toward the front; sometimes used as a synonym for rostral, and sometimes as a synonym for ventral.

anterior commissure A small midline fiber tract that lies at the anterior end of the corpus callosum; like the callosum, it serves to connect the two hemispheres.

anterior hypothalamus Region of the hypothalamus containing nuclei that mediate sexual behaviors; not to be confused with region in rodent called the medial preoptic area, which lies anterior to hypothalamus and also contains nuclei that mediate sexual behavior (most notably the sexually dimorphic nucleus).

anterograde A movement or influence acting from the neuronal cell body toward the axonal target.

anterolateral pathway (anterolateral system) Ascending sensory pathway in the spinal cord and brainstem that carries information about pain and temperature to the thalamus.

antiserum Serum harvested from an animal immunized to an agent of interest.

aphasia The inability to comprehend and/or produce language as a result of damage to the language areas of the cerebral cortex (or their white matter interconnections).

apoptosis Cell death resulting from a programmed pattern of gene expression; also known as "programmed cell death."

aprosodia The inability to infuse language with its normal emotional content.

arachnoid mater One of the three coverings of the brain that make up the meninges; lies between the dura mater and the pia mater.

areflexia Loss of reflexes.

association cortex Defined by exclusion as those neocortical regions that are not involved in primary sensory or motor processing.

associativity In the hippocampus, the enhancement of a weakly activated group of synapses when a nearby group is strongly activated.

astrocytes One of the three major classes of glial cells found in the central nervous system; important in regulating the ionic milieu of nerve cells and, in some cases, transmitter reuptake.

astrotactin A cell surface molecule that causes neurons to adhere to radial glial fibers during neuronal migration.

athetosis Slow, writhing movements seen primarily in patients with disorders of the basal ganglia.

ATPase pumps Membrane pumps that use the hydrolysis of ATP to translocate ions against their electrochemical gradients.

atrophy The physical wasting away of a tissue, typically muscle, in response to disuse or other causes.

attention The selection of a particular sensory stimulus or mental process for further analysis.

auditory meatus Opening of the external ear canal.

auditory space map Topographic representation of sound source location, as occurs in the inferior colliculus.

autonomic nervous system The components of the nervous system (peripheral and central) concerned with the regulation of smooth muscle, cardiac muscle, and glands (see also visceral motor system).

axon The neuronal process that carries the action potential from the nerve cell body to a target.

axoplasmic transport The process by which materials are carried from nerve cell bodies to their terminals (anterograde transport), or from nerve cell terminals to the neuronal cell body (retrograde transport).

baroreceptors Sensory receptors in the visceral motor system that respond to changes in blood pressure.

basal ganglia A group of nuclei lying deep in the subcortical white matter of the frontal lobes that organize motor behavior. The caudate and putamen and the globus pallidus are the major components of the basal ganglia; the subthalamic nucleus and substantia nigra are often included.

basal lamina (basement membrane) A thin layer of extracellular matrix material (primarily collagen, laminin, and fibronectin) that surrounds muscle cells and Schwann cells. Also underlies all epithelial sheets.

basilar membrane The membrane that forms the floor of the cochlear duct, on which the cochlear hair cells are located.

basket cells Inhibitory interneurons in the cerebellar cortex whose cells bodies are located within the Purkinje cell layer and whose axons make basketlike terminal arbors around Purkinje cell bodies.

binocular Referring to both eyes.

biogenic amines The bioactive amine neurotransmitters; includes the catecholamines (epinephrine, norepinephrine, dopamine), serotonin, and histamine.

bipolar cells Retinal neurons that provide a direct link between photoreceptor terminals and ganglion cell dendrites.

bisexuality Sexual attraction to members of both the opposite and the same phenotypic sex.

blastomere A cell produced when the egg undergoes cleavage.

blastula An early embryo during the stage when the cells are typically arranged to form a hollow sphere.

blind spot The region of visual space that falls on the optic disk; due to the lack of photoreceptors in the optic disk, objects that lie completely within the blind spot are not perceived.

blood-brain barrier A diffusion barrier between the brain vasculature and the substance of the brain formed by tight junctions between capillary endothelial cells.

bouton (synaptic bouton) A swelling specialized for the release of neurotransmitter that occurs along or at the end of an axon.

bradykinesia Pathologically slow movement.

brain-derived neurotrophic factor (BDNF) One member of a family of neutrophic factors, the best-known constituent of which is nerve growth factor.

brainstem The portion of the brain that lies between the diencephalon and the spinal cord; comprises the midbrain, pons, and medulla.

Broca's aphasia Difficulty producing speech as a result of damage to Broca's area in the left frontal lobe.

Broca's area An area in the left frontal lobe specialized for the production of language.

CA1 A region of the hippocampus that shows a robust form of long-term potentiation.

CA3 A region of the hippocampus containing the neurons that form the Schaffer collaterals.

cadherins A family of calcium-dependent cell adhesion molecules found on the surfaces of growth cones and the cells over which they grow.

calcarine sulcus The major sulcus on the medial aspect of the occipital lobe; the primary visual cortex lies largely within this sulcus.

cAMP response element binding protein (CREB) A protein activated by cyclic AMP that binds to specific regions of DNA, thereby increasing the transcription rates of nearby genes.

cAMP response elements (CREs) Specific DNA sequences that bind transcription factors activated by cAMP (see also cAMP response element binding protein).

carotid bodies Specialized tissue masses found at the bifurcation of the carotid arteries in humans and other mammals that respond to the chemical composition of the blood (primarily the partial pressure of oxygen and carbon dioxide).

catecholamine A term referring to molecules containing a catechol ring and an amino group; examples are the neurotransmitters epinephrine, norepinephrine, and dopamine.

cauda equina The collection of segmental ventral and dorsal roots that extend from the caudal end of the spinal cord to their exit from the spinal canal.

caudal Posterior, or "tailward."

caudate nucleus One of the three major components of the basal ganglia (the other two are the globus pallidus and putamen).

cell adhesion molecules A family of molecules on cell surfaces that cause them to stick to one another (see also fibronectin and laminin).

central nervous system The brain and spinal cord of vertebrates (by analogy, the central nerve cord and ganglia of invertebrates).

central pattern generator Oscillatory spinal cord or brainstem circuits responsible for programmed, rhythmic movements such as locomotion.

central sulcus A major sulcus on the lateral aspect of the hemispheres that forms the boundary between the frontal and parietal lobes. The anterior bank of the sulcus contains the primary motor cortex; the posterior bank contains the primary sensory cortex.

cerebellar ataxia A pathological inability to make coordinated movements associated with lesions to the cerebellum.

cerebellar cortex The superficial gray matter of the cerebellum.

cerebellar peduncles The three bilateral pairs of axon tracts (inferior, middle, and superior cerebellar peduncles) that carry information to and from the cerebellum.

cerebellum Prominent hindbrain structure concerned with motor coordination, posture, and balance. Composed of a three-layered cortex and deep nuclei; attached to the brainstem by the cerebellar peduncles.

cerebral aqueduct The portion of the ventricular system that connects the third and fourth ventricles.

cerebral cortex The superficial gray matter of the cerebral hemispheres.

cerebral peduncles The major fiber bundles that connect the brainstem to the cerebral hemispheres.

cerebrocerebellum The part of the cerebellar cortex that receives input from the cerebral cortex via axons from the pontine relay nuclei.

cerebrospinal fluid A normally clear and cell-free fluid that fills the ventricular system of the central nervous system; produced by the choroid plexus in the third ventricle.

cerebrum The largest and most rostral part of the brain in humans and other mammals, consisting of the two cerebral hemispheres.

c-fos Cellular Feline Osteosarcoma gene product; a transcription factor that binds as a heterodimer, thus activating gene transcription.

chemical synapses Synapses that transmit information via the secretion of chemical signals (neurotransmitters).

chemoaffinity (chemoaffinity hypothesis) The idea that nerve cells bear chemical labels that determine their connectivity.

chemotaxis The movement of a cell up (or down) the gradient of a chemical signal.

chemotropism The growth of a part of a cell (axon, dendrite, filopodium) up (or down) a chemical gradient.

chimera An experimentally generated embryo (or organ) comprising cells derived from two or more species (or other genetically distinct sources).

cholinergic Referring to synaptic transmission mediated by acetylcholine.

chorea Jerky, involuntary movements of the face or extremities associated with damage to the basal ganglia.

choreoathetosis The combination of jerky, ballistic, and writhing movements that characterizes the late stages of Huntington's disease.

choroid plexus Specialized epithelium in the ventricular system that produces cerebrospinal fluid.

chromosome Nuclear organelle that bears the genes.

ciliary body Circular band of muscle surrounding the lens; contraction allows the lens to round up during accommodation.

cingulate cortex Cortex of the cingulate gyrus that surrounds the corpus callosum; important in emotional and visceral motor behavior.

cingulate gyrus Prominent gyrus on the medial aspect of the hemisphere, lying just superior to the corpus callosum; forms a part of the limbic system.

cingulate sulcus Prominent sulcus on the medial aspect of the hemisphere.

circadian rhythms Variations in physiological functions that occur on a daily basis.

circle of Willis Arterial anastomosis on the ventral aspect of the midbrain; connects the posterior and anterior cerebral circulation.

cisterns Large, cerebrospinal-fluid-filled spaces that lie within the subarachnoid space.

class A taxonomic category subordinate to phylum; comprises animal orders.

climbing fibers Axons that originate in the inferior olive, ascend through the inferior cerebellar peduncle, and make terminal arborizations that invest the dendritic tree of Purkinje cells.

clone The progeny of a single cell.

cochlea The coiled structure within the inner ear where vibrations caused by sound are transduced into neural impulses.

cognition A general term referring to higher order mental processes; the ability of the central nervous system to attend, identify, and act on complex stimuli.

collapsin A molecule that causes collapse of growth cones; a member of the semaphorin family of signaling molecules.

colliculi The two paired hillocks that characterize the dorsal surface of the midbrain; the superior colliculi concern vision, the inferior colliculi audition.

competition The struggle among nerve cells, or nerve cell processes, for limited resources essential to survival or growth.

concha A component of the external ear.

conduction aphasia Difficulty producing speech as a result of damage to the connection between Wernicke's and Broca's language areas.

conduction velocity The speed at which an action potential is propagated along an axon.

conductive hearing loss Diminished sense of hearing due reduced ability of sounds to be mechanically transmitted to the inner ear. Common causes include occlusion of the ear canal, perforation of the tympanic membrane, and arthritic degeneration of the middle ear ossicles. Contrast with sensorineural hearing loss.

cone opsins The three distinct photopigments found in cones; the basis for color vision.

cones Photoreceptors specialized for high visual acuity and the perception of color.

congenital adrenal hyperplasia Genetic deficiency that leads to overproduction of androgens and a resultant masculinization of external genitalia in genotypic females.

conjugate The paired movements of the two eyes in the same direction, as occurs in the vestibulo-ocular reflex (see also vergence movements and vestibulo-ocular reflex).

conspecific Fellow member of a species.

contralateral On the other side.

contralateral neglect syndrome Neurological condition in which the patient does not acknowledge or attend to the left visual hemifield or the left half of the body. The syndrome typically results from lesions of the right parietal cortex.

contrast The difference, usually expressed in terms of a percentage in luminance, between two territories in the visual field (can also apply to color when specified as spectral contrast).

convergence Innervation of a target cell by axons from more than one neuron.

cornea The transparent surface of the eyeball in front of the lens; the major refractive element in the optical pathway.

coronal Referring to a plane through the brain that runs parallel to the coronal suture (the mediolateral plane). Synonymous with frontal plane.

corpus callosum The large midline fiber bundle that connects the cortices of the two cerebral hemispheres.

corpus striatum General term applied to the caudate and putamen; name derives from the striated appearance of these basal ganglia nuclei in sections of fresh material.

cortex The superficial mantle of gray matter covering the cerebral hemispheres and cerebellum, where most of the neurons in the brain are located.

cortico-cortical connections Connections made between cortical areas in the same hemisphere or between the two hemispheres via the cerebral commissures (the corpus callosum and the anterior commissure).

corticospinal tract Pathway carrying motor information from the primary and secondary motor cortices to the brain stem and spinal cord.

co-transmitters Two or more types of neurotransmitters within a single synapse; may be packaged into separate populations of synaptic vesicles or co-localized within the same synaptic vesicles.

cranial nerve ganglia The sensory ganglia associated with the cranial nerves; these correspond to the dorsal root ganglia of the spinal segmental nerves.

cranial nerve nuclei Nuclei in the brainstem that contain the neurons related to cranial nerves III–XII.

cranial nerves The 12 pairs of nerves arising from the brainstem that carry sensory information toward (and sometimes motor information away from) the central nervous system.

CREB see cAMP response element binding protein.

crista The hair cell-containing sensory epithelium of the semicircular canals.

critical period A restricted developmental period during which the nervous system is particularly sensitive to the effects of experience.

cuneate nuclei Sensory relay nuclei that lie in the lower medulla; they contain the second-order sensory neurons that relay mechanosensory information from peripheral receptors in the upper body to the thalamus.

cupulae Gelatinous structures in the semicircular canals in which the hair cell bundles are embedded.

cytoarchitectonic areas Distinct regions of the neocortical mantle identified by differences in cell size, packing density, and laminar arrangement.

decerebrate rigidity Excessive tone in extensor muscles as a result of damage to descending motor pathways at the level of the brainstem.

declarative memory Memories available to consciousness that can be expressed by language.

decussation A crossing of fiber tracts in the midline.

deep cerebellar nuclei The nuclei at the base of the cerebellum that relay information from the cerebellar cortex to the thalamus.

delayed response genes Genes that are synthesized de novo after a cell is stimulated; usually refers to transcriptional activator proteins that are synthesized after preexisting transcription factors are first activated by an inducing stimulus.

delayed response task A behavioral paradigm used to test cognition and memory.

delta waves Slow (<4 Hz) electroencephalographic waves that characterize stage IV (slow-wave) sleep.

dendrite A neuronal process arising from the cell body that receives synaptic input.

denervation Removal of the innervation to a target.

dentate gyrus A region of the hippocampus; so named because it is shaped like a tooth.

depolarization The displacement of a cell's membrane potential toward a less negative value.

dermatome The area of skin supplied by the sensory axons of a single spinal nerve.

determination Commitment of a developing cell or cell group to a particular fate.

dichromatic Referring to the majority of mammals (and most color-blind humans), which have only two instead of three cone pigments to mediate color vision.

diencephalon Portion of the brain that lies just rostral to the midbrain; comprises the thalamus and hypothalamus.

differentiation The progressive specialization of developing cells.

dihydrotestosterone A more potent form of testosterone that masculinizes the external genitalia.

disinhibition Arrangement of inhibitory and excitatory cells in a circuit that generates excitation by the transient inhibition of a tonically active inhibitory neuron.

disjunctive eye movements Movements of the two eyes in opposite directions (see also vergence movements).

distal Farther away from a point of reference (the opposite of proximal).

divergence The branching of an axon to innervate multiple target cells.

dopamine A catecholamine neurotransmitter.

dorsal Referring to the back.

dorsal column nuclei Second-order sensory neurons in the lower medulla that relay mechanosensory information from the spinal cord to the thalamus; comprises the cuneate and gracile nuclei.

dorsal columns Major ascending tracts in the spinal cord that carry mechanosensory information from the first-order sensory neurons in dorsal root ganglia to the dorsal column nuclei; also called the posterior funiculi.

dorsal horn The dorsal portion of the spinal cord gray matter; contains neurons that process sensory information.

dorsal root ganglia The segmental sensory ganglia of the spinal cord; contain the first-order neurons of the dorsal column/medial lemniscus and spinothalamic pathways.

dorsal roots The bundle of axons that runs from the dorsal root ganglia to the dorsal horn of the spinal cord, carrying sensory information from the periphery.

dura mater The thick external covering of the brain and spinal cord; one of the three components of the meninges, the other two being the pia mater and arachnoid mater.

dynorphins A class of endogenous opioid peptides.

dysarthria Difficulty producing speech as a result of damage to the primary motor centers that govern the muscles of articulation; distinguished from aphasia, which results from cortical damage.

dysmetria Inaccurate movements due to faulty judgment of distance. Characteristic of cerebellar pathology.

dystonia Lack of muscle tone.

early inward current The initial electrical current, measured in voltage clamp experiments, that results from the voltage-dependent entry of a cation such as Na⁺ or Ca²⁺; produces the rising phase of the action potential.

ectoderm The most superficial of the three embryonic germ layers; gives rise to the nervous system and epidermis.

Edinger-Westphal nucleus Midbrain nucleus containing the autonomic neurons that constitute the efferent limb of the pupillary light reflex.

efferent An axon that conducts information away from the central nervous system.

electrical synapses Synapses that transmit information via the direct flow of electrical current at gap junctions.

electrochemical equilibrium The condition in which no net ionic flux occurs across a membrane because ion concentration gradients and opposing transmembrane potentials are in exact balance.

electrogenic Capable of generating an electrical current; usually applied to membrane transporters that create electrical currents while translocating ions.

embryo The developing organism before birth or hatching.

end plate current (EPC) Postsynaptic current produced by neurotransmitter release and binding at the motor end plate.

end plate potential (EPP) Depolarization of the membrane potential of skeletal muscle fiber, caused by the action of the transmitter acetylcholine at the neuromuscular synapse.

endocrine Referring to the release of signaling molecules whose effects are made widespread by distribution in the general circulation.

endocytosis A budding off of vesicles from the plasma membrane, which allows uptake of materials in the extracellular medium.

endoderm The innermost of the three embryonic germ layers.

endogenous opioids Peptides in the central nervous system that have the same pharmacological effects as morphine and other derivatives of opium.

endolymph The potassium-rich fluid filling both the cochlear duct and the membranous labyrinth; bathes the apical end of the hair cells.

endorphins One of a group of neuropeptides that are agonists at opioid receptors, virtually all of which contain the sequence Tyr-Gly-Gly-Phe.

end plate The complex postsynaptic specialization at the site of nerve contact on skeletal muscle fibers.

engram The term used to indicate the physical basis of a stored memory.

enkephalins A general term for endogenous opioid peptides.

ependyma The epithelial lining of the canal of the spinal cord and the ventricles.

ependymal cells Epithelial cells that line the ventricular system.

epidermis The outermost layer of the skin; derived from the embryonic ectoderm.

epigenetic Referring to influences on development that arise from factors other than genetic instructions.

epinephrine (adrenaline) Catecholamine hormone and neurotransmitter that binds to α- and β-adrenergic G-protein-coupled receptors.

epineurium The connective tissue surrounding axon fascicles of a peripheral nerve.

epithelium Any continuous layer of cells that covers a surface or lines a cavity.

equilibrium potential The membrane potential at which a given ion is in electrochemical equilibrium.

estradiol One of the biologically important C_{18} class of steroid hormones capable of inducing estrous in females.

eukaryote An organism that contains cells with nuclei.

excitatory postsynaptic potential (EPSP) Neurotransmitter-induced postsynaptic potential change that depolarizes the cell, and hence increases the likelihood of initiating a postsynaptic action potential.

explant A piece of tissue maintained in culture medium.

external segment A subdivision of the globus pallidus.

extracellular matrix A matrix composed of collagen, laminin, and fibronectin that surrounds most cells (see also basal lamina).

extrafusal muscle fibers Fibers of skeletal muscles; a term that distinguishes ordinary muscle fibers from the specialized intrafusal fibers associated with muscle spindles.

face cells Neurons in the temporal cortex of rhesus monkeys that respond specifically to faces.

facilitation The increased transmitter release produced by an action potential that follows closely upon a preceding action potential.

family A taxonomic category subordinate to order; comprises genera.

fasciculation The aggregation of neuronal processes to form a nerve bundle; also refers to the spontaneous discharge of motor units after muscle denervation.

α-fetoprotein A protein that actively sequestors circulating estrogens.

fetus The developing mammalian embryo at relatively late stages when the parts of the body are recognizable.

fibrillation Spontaneous contractile activity of denervated muscle fibers.

fibroblast growth factor (FGF) A peptide growth factor, originally defined by its mitogenic effects on fibroblasts; also acts as an inducer during early brain development.

fibronectin A large cell adhesion molecule that binds integrins.

filopodium Slender protoplasmic projection, arising from the growth cone of an axon or a dendrite, that explores the local environment.

fissure A deep cleft in the brain; distinguished from sulci, which are shallower cortical infoldings.

flexion reflex Polysynaptic reflex mediating withdrawal from a painful stimulus.

floorplate Region in the ventral portion of the developing spinal cord; important in the guidance and crossing of growing axons.

folia The name given to the gyral formations of the cerebellum.

forebrain The anterior portion of the brain that includes the cerebral hemispheres (includes the telencephalon and diencephalon).

fornix An axon tract, best seen from the medial surface of the divided brain, that interconnects the hypothalamus and hippocampus.

fourth ventricle The ventricular space that lies between the pons and the cerebellum.

fovea Area of the retina specialized for high acuity in the center of the macula; contains a high density of cones and few rods.

foveola Capillary and rod-free zone in the center of the fovea.

frontal lobe One of the four lobes of the brain; includes all the cortex that lies anterior to the central sulcus and superior to the lateral fissure.

G-protein-coupled receptors (metabotropic receptors) A large family of neurotransmitter or hormone receptors, characterized by seven transmembrane domains; the binding of these receptors by agonists leads to the activation of intracellular G-proteins.

G-proteins Term for two large groups of proteins—the heterotrimeric G-proteins and the small-molecule G-proteins—that can be activated by exchanging bound GDP for GTP.

gamma (γ) motor neurons Class of spinal motor neurons specifically concerned with the regulation of muscle spindle length; these neurons innervate the intrafusal muscle fibers of the spindle.

ganglion (plural, ganglia) Collections of hundreds to thousands of neurons found outside the brain and spinal cord along the course of peripheral nerves.

ganglion cell   A neuron located in a ganglion.

gap junction   A specialized intercellular contact formed by channels that directly connect the cytoplasm of two cells.

gastrula   The early embryo during the period when the three embryonic germ layers are formed; follows the blastula stage.

gastrulation   The cell movements (invagination and spreading) that transform the embryonic blastula into the gastrula.

gender identification   Self-perception of one's alignment with the traits associated with being a phenotypic female or male in a given culture.

gene   A hereditary unit located on the chromosomes; genetic information is carried by linear sequences of nucleotides in DNA that code for corresponding sequences of amino acids.

genome   The complete set of an animal's genes.

genotype   The genetic makeup of an individual.

genotypic sex   Sexual characterization according to the complement of sex chromosomes; XX is a genotypic female, and XY is a genotypic male.

genus   A taxonomic division that comprises a number of closely related species within a family.

germ cell   The egg or sperm (or the precursors of these cells).

germ layers   The three primary layers of the developing embryo from which all adult tissues arise: ectoderm, mesoderm, and endoderm.

glia (neuroglial cells)   The support cells associated with neurons (astrocytes, oligodendrocytes, and microglia in the central nervous system; Schwann cells in peripheral nerves; and satellite cells in ganglia).

globus pallidus   One of the three major nuclei that make up the basal ganglia in the cerebral hemispheres; relays information from the caudate and putamen to the thalamus.

glomeruli   Characteristic collections of neuropil in the olfactory bulbs; formed by dendrites of mitral cells and terminals of olfactory receptor cells, as well as processes from local interneurons.

glutamate-glutamine cycle   A metabolic cycle of glutamate release and resynthesis involving both neuronal and glial cells.

G$_{olf}$ A G-protein found uniquely in olfactory receptor neurons.

Golgi tendon organs   Receptors located in muscle tendons that provide mechanosensory information to the central nervous system about muscle tension.

gracile nuclei   Sensory nuclei in the lower medulla; these second-order sensory neurons relay mechanosensory information from the lower body to the thalamus.

gradient   A systematic variation of the concentration of a molecule (or some other agent) that influences cell behavior.

granule cell layer   The layer of the cerebellar cortex where granule cell bodies are found. Also used to refer to cell-rich layers in neocortex and hippocampus.

gray matter   General term that describes regions of the central nervous system rich in neuronal cell bodies and neuropil; includes the cerebral and cerebellar cortices, the nuclei of the brain, and the central portion of the spinal cord.

growth cone   The specialized end of a growing axon (or dendrite) that generates the motive force for elongation.

gyri   The ridges of the infolded cerebral cortex (the valleys between these ridges are called sulci).

hair cells   The sensory cells within the inner ear that transduce mechanical displacement into neural impulses.

helicotrema   The opening at the apex of the cochlea that joins the scala vestibuli and scala tympani.

Hensen's node   see primitive pit.

heterotrimeric G-proteins   A large group of proteins consisting of three subunits (α, β, and γ) that can be activated by exchanging bound GDP with GTP resulting in the liberation of two signaling molecules—αGTP and the $\beta\gamma$-dimer.

higher-order neurons   Neurons that are relatively remote from peripheral targets.

hindbrain   see rhombencephalon.

hippocampus   A cortical structure in the medial portion of the temporal lobe; in humans, concerned with short-term declarative memory, among many other functions.

histamine   A biogenic amine neurotransmitter derived from the amino acid histidine.

homeobox genes   A set of master control genes whose expression establishes the early body plan of developing organisms (see also homeotic mutant).

homeotic mutant   A mutation that transforms one part of the body into another (e.g., insect antennae into legs).

homologous   Technically referring to structures in different species that share the same evolutionary history; more generally, referring to structures or organs that have the same general anatomy and perform the same function.

homosexuality   Sexual attraction to an individual of the same phenotypic sex.

horizontal cells   Retinal neurons that mediate lateral interactions between photoreceptor terminals and the dendrites of bipolar cells.

horseradish peroxidase   A plant enzyme widely used to stain nerve cells (after injection into a neuron, it generates a visible precipitate by one of several histochemical reactions).

Huntington's disease   An autosomal dominant genetic disorder in which a single gene mutation results in personality changes, progressive loss of the control of voluntary movement, and eventually death. Primary target is the basal ganglia.

hydrocephalus   Enlarged cranium as a result of increased cerebrospinal fluid pressure (typically due to a mechanical outflow blockage).

hyperalgesia   Increased perception of pain.

hyperkinesia   Excessive movement.

hyperpolarization   The displacement of a cell's membrane potential toward a more negative value.

hypokinesia   A paucity of movement.

hypothalamus   A collection of small but critical nuclei in the diencephalon that lies just inferior to the thalamus; governs reproductive, homeostatic, and circadian functions.

imprinting   A rapid and permanent form of learning that occurs in response to early experience.

inactivation   The time-dependent closing of ion channels in response to a stimulus, such as membrane depolarization.

inducers   Chemical signals originating from one set of cells that influence the differentiation of other cells.

induction   The ability of a cell or tissue to influence the fate of nearby cells or tissues during development by chemical signals.

inferior colliculi (singular, colliculus) Paired hillocks on the dorsal surface of the midbrain; concerned with auditory processing.

inferior olive (inferior olivary nucleus) Prominent nucleus in the medulla; a major source of input to the cerebellum.

infundibulum The connection between the hypothalamus and the pituitary gland; also known as the pituitary stalk.

inhibitory postsynaptic potential (IPSP) Neurotransmitter-induced postsynaptic potential change that tends to decrease the likelihood of a postsynaptic action potential.

innervate Establish synaptic contact with a target.

innervation Referring to all the synaptic contacts of a target.

input The innervation of a target cell by a particular axon; more loosely, the innervation of a target.

input elimination The developmental process by which the number of axons innervating some classes of target cells is diminished.

instructive A developmental influence that dictates the fate of a cell rather than simply permitting differentiation to occur.

insula The portion of the cerebral cortex that is buried within the depths of the lateral fissure.

integral membrane proteins Proteins that possess hydrophobic domains that are inserted into membranes.

integration The summation of excitatory and inhibitory synaptic conductance changes by postsynaptic cells.

integrins A family of receptor molecules found on growth cones that bind to cell adhesion molecules such as laminin and fibronection.

intention tremor Tremor that occurs while performing a voluntary motor act. Characteristic of cerebellar pathology.

internal arcuate tract Mechanosensory pathway in the brainstem that runs from the dorsal column nuclei to form the medial lemniscus.

internal capsule Large white matter tract that lies between the diencephalon and the basal ganglia; contains, among others, sensory axons that run from the thalamus to the cortex and motor axons that run from the cortex to the brainstem and spinal cord.

interneuron Technically, a neuron in the pathway between primary sensory and primary effector neurons; more generally, a neuron that branches locally to innervate other neurons.

interstitial nuclei of the anterior hypothalamus (INAH) Four cell groups located slightly lateral to the third ventricle in the anterior hypothalamus of primates; thought to play a role in sexual behavior.

intrafusal muscle fibers Specialized muscle fibers found in muscle spindles.

invertebrate An animal without a backbone (includes about 97% of extant animals).

in vitro Referring to any biological process studied outside of the organism (literally, "in glass").

in vivo Referring to any biological process studied in an intact living organism (literally "in life").

ion channels Integral membrane proteins possessing pores that allow certain ions to diffuse across cell membranes, thereby conferring selective ionic permeability.

ion exchangers Membrane transporters that translocate one or more ions against their concentration gradient by using the electrochemical gradient of other ions as an energy source.

ionotropic (ionotropic receptors) Receptors in which the ligand binding site is an integral part of the receptor molecule.

ion pumps see transporters.

ipsilateral On the same side of the body.

iris Circular pigmented membrane behind the cornea; perforated by the pupil.

ischemia Insufficient blood supply.

kinocilium A true ciliary structure which, along with the stereocilia, comprises the hair bundle of vestibular and fetal cochlear hair cells in mammals (it is not present in the adult mammalian cochlear hair cell).

Korsakoff's syndrome An amnesic syndrome seen in chronic alcoholics.

labyrinth Referring to the internal ear; comprises the cochlea, vestibular apparatus, and the bony canals in which these structures are housed.

lamellipodia The leading edge of a motile cell or growth cone, which is rich in actin filaments.

laminae (singular, lamina) Cell layers that characterize the neocortex, hippocampus, and cerebellar cortex. The gray matter of the spinal cord is also arranged in laminae.

laminin A large cell adhesion molecule that binds integrins.

late outward current The delayed electrical current, measured in voltage clamp experiments, that results from the voltage-dependent efflux of a cation such as K⁺. Produces the repolarizing phase of the action potential.

lateral columns The lateral regions of spinal cord white matter that convey motor information from the brain to the spinal cord.

lateral (Sylvian) fissure The cleft on the lateral surface of the brain that separates the temporal and frontal lobes.

lateral geniculate nucleus (LGN) A nucleus in the thalamus that receives the axonal projections of retinal ganglion cells in the primary visual pathway.

lateral olfactory tract The projection from the olfactory bulbs to higher olfactory centers.

lateral posterior nucleus A thalamic nucleus that receives its major input from sensory and association cortices and projects in turn to association cortices, particularly in the parietal and temporal lobes.

lateral superior olive (LSO) The auditory brainstem structure that processes interaural intensity differences and, in humans, mediates sound localization for stimuli greater than 3 kHz.

learning The acquisition of novel behavior through experience.

lens Transparent structure in the eye whose thickening or flattening in response to visceral motor control allows light rays to be focused on the retina.

lexical The quality of associating a symbol (e.g., a word) with a particular object, emotion, or idea.

lexicon Dictionary. Sometimes used to indicate region of brain that stores the meanings of words.

ligand-gated ion channels Term for a large group of neurotransmitter receptors that combine receptor and ion channel functions into a single molecule.

limb bud The limb rudiment of vertebrate embryos.

limbic lobe Cortex that lies superior to the corpus callosum on the medial aspect of the cerebral hemispheres; forms the cortical component of the limbic system.

limbic system Term that refers to those cortical and subcortical structures concerned with the emotions; the most prominent components are the cingulate gyrus, the hippocampus, and the amygdala.

lobes The four major divisions of the cerebral cortex (frontal, parietal, occipital, and temporal).

local circuit neurons General term referring to neurons whose activity mediates interactions between sensory systems and motor systems; interneuron is often used as a synonym.

locus coeruleus A small brainstem nucleus with widespread adrenergic cortical and descending connections; important in the governance of sleep and waking.

long-term Lasting days, weeks, months, or longer.

long-term depression A persistent weakening of synapses based on recent patterns of activity.

long-term memory Memories that last days, weeks, months, years, or a lifetime.

long-term potentiation (LTP) A persistent strengthening of synapses based on recent patterns of activity.

lower motor neuron Spinal motor neuron; directly innervates muscle (also referred to as α or primary motor neuron).

lower motor neuron syndrome Signs and symptoms arising from damage to α motor neurons; these include paralysis or paresis; muscle atrophy, areflexia, and fibrillations.

macroscopic Visible with the naked eye.

macroscopic currents Ionic currents flowing through large numbers of ion channels distributed over a substantial area of membrane.

macula The central region of the retina that contains the fovea (the term derives from the yellowish appearance of this region in ophthalmoscopic examination); also, the sensory epithelia of the otolith organs.

magnocellular A component of the primary visual pathway specialized for the perception of motion; so named because of the relatively large cells involved.

mammal An animal the embryos of which develop in a uterus and the young of which begin to suckle at birth (technically, a member of the class Mammalia).

mammillary bodies Small prominences on the ventral surface of the diencephalon; functionally, part of the caudal hypothalamus.

map The ordered projection of axons from one region of the nervous system to another, by which the organization of the body (or some function) is reflected in the organization of the nervous system.

mechanoreceptors Receptors specialized to sense mechanical forces.

medial Located nearer to the midsagittal plane of an animal (the opposite of lateral).

medial dorsal nucleus A thalamic nucleus that receives its major input from sensory and association cortices and projects in turn to association cortices, particularly in the frontal lobe.

medial geniculate complex The major thalamic relay for auditory information.

medial lemniscus Axon tract in the brainstem that carries mechanosensory information from the dorsal column nuclei to the thalamus.

medial longitudinal fasciculus Axon tract that carries excitatory projections from the abducens nucleus to the contralateral oculomotor nucleus; important in coordinating conjugate eye movements.

medial superior olive (MSO) The auditory brainstem structure that processes interaural time differences and serves to compute the horizontal location of a sound source.

medium spiny neuron The principal projection neuron of the caudate and putamen.

medulla The caudal portion of the brainstem, extending from the pons to the spinal cord.

medullary pyramids Longitudinal bulges on the ventral aspect of the medulla that signify the corticospinal tracts at this level of the neuraxis.

Meissner's corpuscles Encapsulated cutaneous mechanosensory receptors specialized for the detection of fine touch and pressure.

membrane conductance The reciprocal of membrane resistance. Changes in membrane conductance result from, and are used to describe, the opening or closing of ion channels.

meninges The external covering of the brain; includes the pia, arachnoid, and dura mater.

Merkel's disks Encapsulated cutaneous mechanosensory receptors specialized for the detection of fine touch and pressure.

mesencephalon see midbrain.

mesoderm The middle of the three germ layers; gives rise to muscle, connective tissue, skeleton, and other structures.

mesopic Light levels at which both the rod and cone systems are active.

metabotropic (metabotropic receptors) Refers to receptors that are indirectly activated by the action of neurotransmitters or other extracellular signals, typically through the aegis of G-protein activation.

Meyer's loop That part of the optic radiation that runs in the caudal portion of the temporal lobe.

microglial cells One of the three main types of central nervous system glia; concerned primarily with repairing damage following neural injury.

microscopic currents Ionic currents flowing through single ion channels.

midbrain (mesencephalon) The most rostral portion of the brainstem; identified by the superior and inferior colliculi on its dorsal surface, and the cerebral peduncles on its ventral aspect.

middle cerebellar peduncle Large white matter tract that carries axons from the pontine relay nuclei to the cerebellar cortex.

miniature end plate potential (MEPP) Small, spontaneous depolarization of the membrane potential of skeletal muscle cells, caused by the release of a single quantum of acetylcholine.

mitral cells The major output neurons of the olfactory bulb.

mnemonic Having to do with memory.

modality A category of function. For example, vision, hearing, and touch are different sensory modalities.

molecular layer The layer of the cerebellar cortex containing the apical dendrites of Purkinje cells, parallel fibers from granule cells, a few local circuit neurons, and the synapses between these elements.

monoclonal antibody An antibody molecule raised from a clone of transformed lymphocytes.

morphine A plant alkaloid that gives opium its analgesic properties.

morphogen A molecule that influences morphogenesis.

morphogenesis The generation of animal form.

morphology The study of the form and structure of organisms; or, more commonly, the form and structure of an animal or animal part.

motor Pertaining to movement.

motor cortex The region of the cerebral cortex lying anterior to the central sulcus concerned with motor behavior; includes the primary motor cortex in the precentral gyrus and associated cortical areas in the frontal lobe.

motor neuron By usage, a nerve cell that innervates skeletal muscle. Also called primary or α motor neuron.

motor neuron pool The collection of motor neurons that innervates a single muscle.

motor system A broad term used to describe all the central and peripheral structures that support motor behavior.

motor unit A motor neuron and the skeletal muscle fibers it innervates; more loosely, the collection of skeletal muscle fibers innervated by a single motor neuron.

mucosa Term referring the mucus membranes lining the nose, mouth, gut, and other epithelial surfaces.

muscarinic receptors A group of G-protein-coupled acetylcholine receptors activated by the plant alkaloid muscarine.

muscle spindle Highly specialized sensory organ found in most skeletal muscles; provides mechanosensory information about muscle length.

muscle tone The normal, ongoing tension in a muscle; measured by resistance of a muscle to passive stretching.

myelin The multilaminated wrapping around many axons formed by oligodendrocytes or Schwann cells.

myelination Process by which glial cells wrap axons to form multiple layers of glial cell membrane that increase axonal conduction velocity.

myotatic reflex (stretch reflex) A fundamental spinal reflex that is generated by the motor response to afferent sensory information arising from muscle spindles.

myotome The part of each somite that contributes to the development of skeletal muscles.

Na⁺/K⁺ transporter (or Na⁺ pump) A type of ATPase transporter in the plasma membrane of most cells that is responsible for accumulating intracellular K^+ and extruding intracellular Na^+.

nasal (nasal division) Referring to the region of the visual field of each eye in the direction of the nose.

near reflex Reflexive response induced by changing binocular fixation to a closer target; includes convergence, accommodation, and pupillary constriction.

neocortex The six-layered cortex that forms the surface of most of the cerebral hemispheres.

Nernst equation A mathematical relationship that predicts the equilibrium potential across a membrane that is permeable to only one ion.

nerve A collection of peripheral axons that are bundled together and travel a common route.

nerve growth factor (NGF) A neurotrophic protein required for survival and differentiation of sympathetic ganglion cells and certain sensory neurons. Preeminent member of the neurotrophin family of growth factors.

netrins A family of diffusible molecules that act as attractive or repulsive cues to guide growing axons.

neural cell adhesion molecule (N-CAM) Molecule that helps bind axons together and is widely distributed in the developing nervous system. Structurally related to immunoglobin.

neural crest A group of progenitor cells that forms along the dorsum of the neural tube and gives rise to peripheral neurons and glia (among other derivatives).

neural plate The thickened region of the dorsal ectoderm of a neurula that gives rise to the neural tube.

neural tube The primordium of the brain and spinal cord; derived from the neural ectoderm.

neurite A neuronal branch (usually used when the process in question could be either an axon or a dendrite, such as the branches of isolated nerve cells in tissue culture).

neuroblast A dividing cell, the progeny of which develop into neurons.

neurogenesis The development of the nervous system.

neuroglial cells see glia.

neuroleptics A group of antipsychotic agents that cause indifference to stimuli by blocking brain dopamine receptors.

neuromere A segment of the rhombencephalon (synonym for rhombomere).

neuromuscular junction The synapse made by a motor axon on a skeletal muscle fiber.

neuron Cell specialized for the conduction and transmission of electrical signals in the nervous system.

neuronal geometry The spatial arrangement of neuronal branches.

neuron-glia cell adhesion molecule (Ng-CAM) A cell adhesion molecule, structurally related to immunoglobin molecules, that promotes adhesive interactions between neurons and glia.

neuropeptides A general term describing a large number of peptides that function as neurotransmitters or neurohormones.

neuropil The dense tangle of axonal and dendritic branches, and the synapses between them, that lies between neuronal cell bodies in the gray matter of the brain and spinal cord.

neurotransmitter Substance released by synaptic terminals for the purpose of transmitting information from one nerve cell to another.

neurotrophic factors A general term for molecules that promote the growth and survival of neurons.

neurotrophic hypothesis The idea that developing neurons compete for a limited supply of trophic factors secreted by their targets.

neurotrophins A family of trophic factor molecules that promote the growth and survival of several different classes of neurons.

neurula The early vertebrate embryo during the stage when the neural tube forms from the neural plate; follows the gastrula stage.

neurulation The process by which the neural plate folds to form the neural tube.

nociceptors Cutaneous and subcutaneous receptors (usually free nerve endings) specialized for the detection of harmful (noxious) stimuli.

nodes of Ranvier Periodic gaps in the myelination of axons where action potentials are generated.

non-rapid eye movement (non-REM) sleep Collectively, those phases of sleep characterized by the absence of rapid eye movements.

norepinephrine (noradrenaline) Catecholamine hormone and neurotransmitter that binds to α- and β-adrenergic receptors, both of which are G-protein-coupled receptors.

notochord A transient, cylindrical structure of mesodermal cells underlying the neural plate (and later the neural tube) in vertebrate embryos. Source of important inductive signals for spinal cord.

nucleus (plural, nuclei) Collection of nerve cells in the brain that are anatomically discrete, and which typically serve a particular function.

nucleus proprius Region of the dorsal horn of the spinal cord that receives information from nociceptors.

nystagmus Literally, a nodding movement. Refers to repetitive movements of the eyes normally elicited by large-scale movements of the visual field (optokinetic nystagmus). Nystagmus in the absence of appropriate stimuli usually indicates brainstem or cerebellar pathology.

occipital lobe The posterior lobe of the cerebral hemisphere; primarily devoted to vision.

ocular dominance columns The segregated termination patterns of thalamic inputs representing the two eyes in primary visual cortex of some mammalian species.

odorants Molecules capable of eliciting responses from receptors in the olfactory mucosa.

olfactory bulb Olfactory relay station that receives axons from cranial nerve I and transmits this information via the olfactory tract to higher centers.

olfactory epithelium Pseudostratified epithelium that contains olfactory receptor cells, supporting cells, and mucus-secreting glands.

olfactory receptor neurons Bipolar neurons in olfactory epithelium that contain receptors for odorants.

olfactory tracts see lateral olfactory tract.

oligodendrocytes One of three classes of central neuroglial cells; their major function is to elaborate myelin.

ontogeny The developmental history of an individual animal; also used as a synonym for development.

Onuf's nucleus Sexually dimorphic nucleus in the human spinal cord that innervates striated perineal muscles mediating contraction of the bladder in males, and vaginal constriction in females.

opioid Any natural or synthetic drug that has pharmacological actions similar to those of morphine.

opsins Proteins in photoreceptors that absorb light (in humans, rhodopsin and the three specialized cone opsins).

optic chiasm The junction of the two optic nerves on the ventral aspect of the diencephalon, where axons from the nasal parts of each retina cross the midline.

optic cup see optic vesicle.

optic disk The region of the retina where the axons of retinal ganglion cells exit to form the optic nerve.

optic nerve The nerve (cranial nerve II) containing the axons of retinal ganglion cells; extends from the eye to the optic chiasm.

optic radiation Portion of the internal capsule that comprises the axons of lateral geniculate neurons that carry visual information to the striate cortex.

optic tectum The first central station in the visual pathway of many vertebrates (analogous to the superior colliculus in mammals).

optic tract The axons of retinal ganglion cells after they have passed through the region of the optic chiasm en route to the lateral geniculate nucleus of the thalamus.

optic vesicle The evagination of the forebrain vesicle that generates the retina and induces lens formation in the overlying ectoderm.

optokinetic eye movements Movements of the eyes that compensate for head movements; the stimulus for optokinetic movements is large-scale motion of the visual field.

optokinetic nystagmus Repeated reflexive responses of the eyes to ongoing large-scale movements of the visual scene.

orbital (and medial prefrontal) cortex Division of the prefrontal cortex that lies above the orbits in the most rostral and ventral extension of the sagittal fissure; important in emotional processing and rational decision-making.

order A taxonomic category subordinate to class; comprises animal families.

orientation selectivity A property of many neurons in visual cortex in which they respond to edges presented over a narrow range of orientations.

oscillopsia An inability to fixate visual targets while the head is moving as a result of vestibular damage.

ossicles The bones of the middle ear

otoconia The calcium carbonate crystals that rest on the otolithic membrane overlying the hair cells of the sacculus and utricle.

otolithic membrane The gelatinous membrane on which the otoconia lie and in which the tips of the hair bundles are embedded.

otoliths Dense calcific structures (literally "ear stones"); important in generating the vestibular signals pertinent to balance.

outer segment Portion of photoreceptors made up of membranous disks that contain the photopigment responsible for initiating phototransduction.

oval window Site where the middle ear ossicles transfer vibrational energy to the cochlea.

overshoot The peak, positive-going phase of an action potential, caused by high membrane permeability to a cation such as Na$^+$ or Ca^{2+}.

oxytocin A 9-amino-acid neuropeptide that is both a putative neurotransmitter and a neurohormone.

Pacinian corpuscle Encapsulated mechanosensory receptor specialized for the detection of high-frequency vibrations.

Papez's circuit System of interconnected brain structures (mainly cingulate gyrus, hippocampus, and hypothalamus) in the medial aspect of the telencephalon and diencephalon described by James Papez. Participates in emotional processing, short-term declarative memory, and autonomic functions.

paracrine Term referring to the secretion of hormone-like agents whose effects are mediated locally rather than by the general circulation.

parallel fibers The bifurcated axons of cerebellar granule cells that synapse on dendritic spines of Purkinje cells.

paralysis Complete loss of voluntary motor control.

paramedian pontine reticular formation (PPRF) Neurons in the reticular formation of the pons that coordinate the actions of motor neurons in the abducens and oculomotor nuclei to generate horizontal movements of the eyes; also known as the "horizontal gaze center."

parasympathetic nervous system A division of the visceral motor system in which the effectors are cholinergic ganglion cells located near target organs.

paresis Partial loss of voluntary motor control; weakness.

parietal lobe The lobe of the brain that lies between the frontal lobe anteriorly, and the occipital lobe posteriorly.

Parkinson's disease A neurodegenerative disease of the substantia nigra that results in a characteristic tremor at rest and a general paucity of movement.

parvocellular Referring to the component of the primary visual pathway specialized for the detection of detail and color; so named because of the relatively small cells involved.

passive current flow Current flow across neuronal membranes that does not entail the action potential mechanism.

patch clamp An extraordinarily sensitive voltage clamp method that permits the measurement of ionic currents flowing through individual ion channels.

periaqueductal gray matter Region of brainstem gray matter that contains, among others, nuclei associated with the modulation of pain perception.

perilymph The potassium-poor fluid that bathes the basal end of the cochlear hair cells.

perineurium The connective tissue that surrounds a nerve fascicle in a peripheral nerve.

peripheral nervous system All nerves and neurons that lie outside the brain and spinal cord.

permissive An influence during development that permits differentiation to occur but does not specifically instruct cell fate.

phasic Transient firing of action potentials in response to a prolonged stimulus; the opposite of tonic.

phenotype The visible (or otherwise discernible) characteristics of an animal that arise during development.

phenotypic sex The visible body characteristics associated with sexual behaviors.

phospholipase A2A G-protein-activated enzyme that hydrolizes membrane phospholipids at the inner leaflet of the plasma membrane to release fatty acids such as arachadonic acid.

phospholipase CA G-protein-activated enzyme that hydrolizes membrane phospholipids at the inner leaflet of the plasma membrane to release a diacylglycerol and an inositol phosphate such as inositol trisphosphate (IP3).

photopic vision Vision at high light levels that is mediated entirely by cones.

phylogeny The evolutionary history of a species or other taxomonic category.

phylum A major division of the plant or animal kingdom that includes classes having a common ancestry.

pia mater The innermost of the three layers of the meninges, which is closely applied to the surface of the brain.

pigment epithelium Pigmented coat underlying the retina important in the normal turnover of photopigment in rods and cones.

pineal gland Midline neural structure lying on the dorsal surface of the midbrain; important in the control of circadian rhythms (and, incidentally, considered by Descartes to be the seat of the soul).

pinna A component of the external ear.

pituitary gland Endocrine structure comprising an anterior lobe made up of many different types of hormone-secret-

ing cells, and a posterior lobe that secretes neuropeptides produced by neurons in the hypothalamus.

placebo An inert substance that when administered may, because of the circumstances, have physiological effects.

planum temporale Region on the superior surface of the temporal lobe posterior to Heschl's gyrus; notable because it is larger in the left hemisphere in about two-thirds of humans.

plasticity Term that refers to structural or functional changes in the nervous system.

polarity Referring to a continually graded organization along one of the major axes of an animal.

polymodal Responding to more than one sensory modality.

polyneuronal innervation A state in which neurons or muscle fibers receive synaptic inputs from multiple, rather than single, axons.

pons One of the three components of the brainstem, lying between the midbrain rostrally and the medulla caudally.

pontine-geniculate-occipital (PGO) waves Characteristic encephalographic waves that signal the onset of rapid eye movement sleep.

pontine relay nuclei Collections of neurons in the pons that receive input from the cerebral cortex and send their axons across the midline to the cerebellar cortex via the middle cerebellar peduncle.

pore A structural feature of membrane ion channels that allows ions to diffuse through the channel.

pore loop An extracellular domain of amino acids, found in certain ion channels, that lines the channel pore and allows only certain ions to pass.

postcentral gyrus The gyrus that lies just posterior to the central sulcus; contains the primary somatic sensory cortex.

posterior Toward the back; sometimes used as a synonym for caudal or dorsal.

postganglionic Referring to axons that link visceral motor neurons in autonomic ganglia to their targets.

postsynaptic current (PSC) The current produced in a postsynaptic neuron by the binding of neurotransmitter released from a presynaptic neuron.

postsynaptic Referring to the component of a synapse specialized for transmitter reception; downstream at a synapse.

postsynaptic potential (PSP) The potential change produced in a postsynaptic neuron by the binding of neurotransmitter released from a presynaptic neuron.

post-tetanic potentiation (PTP) An enhancement of synaptic transmission resulting from high-frequency trains of action potentials.

precentral gyrus The gyrus that lies just anterior to the central sulcus; contains the primary motor cortex.

prefrontal cortex Cortical regions in the frontal lobe that are anterior to the primary and association motor cortices; thought to be involved in planning complex cognitive behaviors and in the expression of personality and appropriate social behavior.

preganglionic Referring to neurons and axons that link visceral motor neurons in spinal cord and brainstem to autonomic ganglia.

premotor cortex Motor association areas in the frontal lobe anterior to primary motor cortex; thought to be involved in planning or programming of voluntary movements.

pre-proproteins The first protein translation products synthesized in a cell. These polypeptides are usually much larger than the final, mature peptide, and often contain signal sequences that target the peptide to the lumen of the endoplasmic reticulum.

presynaptic Referring to the component of a synapse specialized for transmitter release; upstream at a synapse.

pretectum A group of nuclei located at the junction of the thalamus and the midbrain; these nuclei are important in the pupillary light reflex, relaying information from the retina to the Edinger-Westphal nucleus.

prevertebral (prevertebral ganglia) Sympathetic ganglia that lie anterior to the spinal column (distinct from the sympathetic chain ganglia).

primary auditory cortex The major cortical target of the neurons in the medial geniculate nucleus.

primary motor cortex A major source of descending projections to motor neurons in the the spinal cord and cranial nerve nuclei; located in the precentral gyrus (Brodmann's area 4) and essential for the voluntary control of movement.

primary neuron A neuron that directly links muscles, glands, and sense organs to the central nervous system.

primary sensory cortex Any one of several cortical areas receiving the thalamic input for a particular sensory modality.

primary visual cortex see striate cortex.

primary visual pathway (retinogeniticulocortical pathway) Pathway from the retina via the lateral geniculate nucleus of the thalamus to the primary visual cortex; carries the information that allows conscious visual perception.

primate An order of mammals that includes lemurs, tarsiers, marmosets, monkeys, apes, and humans (technically, a member of this order).

priming A phenomenon in which the memory of an initial exposure is expressed unconsciously by improved performance at a later time.

primitive pit The thickened anterior end of the primitive streak; an important source of inductive signals during early development.

primitive streak Axial thickening in the ectoderm of the gastrulas of reptiles, birds, and mammals; the mesoderm forms by the ingression of cells at this site.

procedural memory Unconscious memories such as motor skills and associations.

production aphasia Aphasia that derives from cortical damage to those centers concerned with the motor aspects of speech.

promoter DNA sequence (usually within 35 nucleotides upstream of the start site of transcription) to which the RNA polymerase and its associated factors bind to initiate transcription.

proproteins Partially processed forms of proteins containing peptide sequences that play a role in the correct folding of the final protein.

proprioceptors Sensory receptors (usually limited to mechanosensory receptors) that sense the internal forces acting on the body; muscle spindles and Golgi tendon organs are the preeminent examples.

prosencephalon The part of the brain that includes the diencephalon and telencephalon (derived from the embryonic forebrain vesicle).

prosody (adjective, prosodic) The emotional tone or quality of speech.

prosopagnosia The inability to recognize faces; usually associated with lesions to the right inferior temporal cortex.

proteoglycan Molecule consisting of a core protein to which one or more long, linear carbohydrate chains (glycosaminoglycans) are attached.

proximal Closer to a point of reference (the opposite of distal).

psychotropic Referring to drugs that alter behavior, mood, and perception.

pulvinar A thalamic nucleus that receives its major input from sensory and association cortices and projects in turn to association cortices, particularly in the parietal lobe.

pupil The perforation in the iris that allows light to enter the eye.

pupillary light reflex The decrease in the diameter of the pupil that follows stimulation of the retina.

Purkinje cell The large principal projection neuron of the cerebellar cortex that has as its defining characteristic an elaborate apical dendrite.

putamen One of the three major nuclei that make up the basal ganglia.

pyramidal tract White matter tract that lies on the ventral surface of the medulla and contains axons descending from motor cortex to the spinal cord.

pyriform cortex Component of cerebral cortex in the temporal lobe pertinent to olfaction; so named because of its pearlike shape.

radial glia Glial cells that contact both the luminal and pial surfaces of the neural tube, providing a substrate for neuronal migration.

ramus Branch; typically applied to the white and gray communicating rami that carry visceral motor axons to the segmental nerves.

raphe nuclei A collection of serotonergic nuclei in the brainstem tegmentum; important in the governance of sleep and waking.

rapid eye movement (REM) sleep Phase of sleep characterized by low-voltage, high-frequency electroencephalographic activity accompanied by rapid eye movements.

receptive field Region of a receptor surface (e.g., the body surface or the retina) that causes a sensory nerve cell (or axon) to respond.

receptor A molecule specialized to bind any one of a large number of chemical signals, preeminently neurotransmitters.

receptor neuron A neuron specialized for the transduction of energy in the environment into electrical signals.

receptor potential The membrane potential change elicited in receptor neurons during sensory transduction.

5-α-reductase Enzyme that converts testosterone to dihydrotestosterone.

reflex A stereotyped (involuntary) motor response elicited by a defined stimulus.

refractory period The brief period after the generation of an action potential during which a second action potential is difficult or impossible to elicit.

remodeling Change in the anatomical arrangement of neural connections.

reserpine An antihypertensive drug that is no longer used due to side effects such as behavioral depression.

resting potential The inside-negative electrical potential that is normally recorded across all cell membranes.

reticular activating system Region in the brainstem tegmentum that, when stimulated, causes arousal; involved in modulating sleep and wakefulness.

reticular formation A network of neurons and axons that occupies the core of the brainstem, giving it a reticulated appearance in myelin-stained material; major functions include control of respiration and heart rate, posture, and state of consciousness.

retina Laminated neural component of the eye that contains the photoreceptors (rods and cones) and the initial processing machinery for the primary (and other) visual pathways.

retinoic acid A derivative of vitamin A that acts as an inducer during early brain development.

retinotectal system The pathway between ganglion cells in the retina and the optic tectum of vertebrates.

retrograde A movement or influence acting from the axon terminal toward the cell body.

reversal potential The membrane potential of a post-synaptic neuron (or other target cell) at which the action of a given neurotransmitter causes no net current flow.

rhodopsin The photopigment found in rods.

rhombencephalon The part of the brain that includes the pons, cerebellum, and medulla (derived from the embryonic hindbrain vesicle).

rhombomere Segment of the developing rhombencephalon.

rising phase The initial, depolarizing, phase of an action potential, caused by the regenerative, voltage-dependent influx of a cation such as Na$^+$ or Ca^{2+}.

rods Photoreceptors specialized for operating at low light levels.

rostral Anterior, or "headward."

rostral interstitial nucleus Neurons in the midbrain reticular formation that coordinate the actions of neurons in the oculomotor nuclei to generate vertical movements of the eye; also known as the "vertical gaze center."

saccades Ballistic, conjugate eye movements that change the point of foveal fixation.

sacculus The otolith organ that detects linear accelerations and head tilts in the vertical plane.

sagittal Referring to the anterior-posterior plane of an animal.

saltatory conduction Mechanism of action potential propagation in myelinated axons; so named because action potentials "jump" from one node of Ranvier to the next due to generation of action potentials only at these sites.

Scarpa's ganglion The ganglion containing the bipolar cells that innervate the semicircular canals and otolith organs.

Schaffer collaterals The axons of cells in the CA3 region of hippocampus that form synapses in the CA1 region.

Schwann cells Neuroglial cells in the peripheral nervous system that elaborate myelin (named after the nineteenth-century anatomist and physiologist Theodor Schwann).

sclera The external connective tissue coat of the eyeball.

scotoma A defect in the visual field as a result of pathological changes in some component of the primary visual pathway.

scotopic Referring to vision in dim light, where the rods are the operative receptors.

second-order neurons Projection neurons in a sensory pathway that lie between the primary receptor neurons and the third-order neurons.

segment One of a series of more or less similar anterior-posterior units that make up segmental animals.

segmentation The anterior-posterior division of animals into roughly similar repeating units.

semaphorins A family of diffusible, growth-inhibiting molecules (see also collapsin).

semicircular canals The vestibular end organs within the inner ear that sense rotational accelerations of the head.

sensitization Increased sensitivity to stimuli in an area surrounding an injury. Also, a generalized aversive response to an otherwise benign stimulus when it is paired with a noxious stimulus.

sensorineural hearing loss Diminished sense of hearing due to damage of the inner ear or its related central auditory structures. Contrast with conductive hearing loss.

sensory Pertaining to sensation.

sensory aphasia Difficulty in communicating with language that derives from cortical damage to those areas concerned with the comprehension of speech.

sensory ganglia see dorsal root ganglia.

sensory system Term sometimes used to describe all the components of the central and peripheral nervous system concerned with sensation.

sensory transduction Process by which energy in the environment is converted into electrical signals by sensory receptors.

serotonin A biogenic amine neurotransmitter derived from the amino acid tryptophan.

sexually dimorphic Having two different forms depending on genotypic or phenotypic sex.

short-term memory Memories that last from seconds to minutes.

silver stain A classical method for visualizing neurons and their processes by impregnation with silver salts (the best-known technique is the Golgi stain, developed by the Italian anatomist Camillo Golgi in the late nineteenth century).

size principle The orderly recruitment of motor neurons by size to generate increasing amounts of muscle tension.

sleep spindles Bursts of electroencephalographic activity, at a frequency about 10–14 Hz and lasting a few seconds; spindles characterize the initial descent into non-REM sleep.

small molecule neurotransmitters Referring to the non-peptide neurotransmitters such as acetylcholine, the amino acids glutamate, aspartate, GABA, and glycine, as well as the biogenic amines.

smooth pursuit eye movements Slow, tracking movements of the eyes designed to keep a moving object aligned with the fovea.

soma (plural, somata) The cell body.

somatic cells Referring to the cells of an animal other than its germ cells.

somatic sensory cortex That region of the cerebral cortex concerned with processing sensory information from the body surface, subcutaneous tissues, muscles, and joints; located primarily in the posterior bank of the central sulcus and on the postcentral gyrus.

somatic sensory system Components of the nervous system involved in processing sensory information about the mechanical forces active on both the body surface and on deeper structures such as muscles and joints.

somatotopic maps Cortical or subcortical arrangements of sensory pathways that reflect the organization of the body.

somites Segmentally arranged masses of mesoderm that lie alongside the neural tube and give rise to skeletal muscle, vertebrae, and dermis.

species A taxonomic category subordinate to genus; members of a species are defined by extensive similarities, including the ability to interbreed.

specificity Term applied to neural connections that entail specific choices between neurons and their targets.

spina bifida A congenital defect in which the neural tube fails to close at its posterior end.

spinal cord The portion of the central nervous system that extends from the lower end of the brainstem (the medulla) to the cauda equina.

spinal ganglia see dorsal root ganglia.

spinal nucleus of the bulbocavernosus Sexually dimorphic collection of neurons in the lumbar region of the rodent spinal cord that innervate striated perineal muscles.

spinal shock The initial flaccid paralysis that accompanies damage to descending motor pathways.

spinal trigeminal tract Brainstem tract carrying fibers from the trigeminal nerve to the spinal nucleus of the trigeminal complex (which serves as the relay for painful stimulation of the face).

spinocerebellum Region of the cerebellar cortex that receives input from the spinal cord, particularly Clarke's column in the thoracic spinal cord.

spinothalamic pathway see anterolateral pathway.

spinothalamic tract Ascending white matter tract carrying information about pain and temperature from the spinal cord to the VP nuclear complex in the thalamus; also referred to as the anterolateral tract.

split-brain patients Individuals who have had the cerebral commissures divided in the midline to control epileptic seizures.

sporadic Cases of a disease that apparently occur at random in a population; contrasts with familial or inherited.

stem cells Undifferentiated cells from which other cells, including neurons, can be derived.

stereocilia The actin-rich processes that, along with the kinocilium, form the hair bundle extending from the apical surface of the hair cell; site of mechanotransduction.

stereopsis The perception of depth that results from the fact that the two eyes view the world from slightly different angles.

strabismus Developmental misalignment of the two eyes; may lead to binocular vision being compromised.

stria vascularis Specialized epithelium lining the cochlear duct that maintains the high potassium concentration of the endolymph.

striate cortex Primary visual cortex in the occipital lobe (also called Brodmann's area 17). So named because the prominence of layer IV in myelin-stained sections gives this region a striped appearance.

striatum (neostriatum) see corpus striatum.

striola A line found in both the sacculus and utricle that divides the hair cells into two populations with opposing hair bundle polarities.

subarachnoid space The cerebrospinal fluid—filled space over the surface of the brain that lies between the arachnoid and the pia.

substance P An 11-amino acid neuropeptide; the first neuropeptide to be characterized.

substantia nigra Nucleus at the base of the midbrain that receives input from a number of cortical and subcortical structures. The dopaminergic cells of the substantia nigra send their output to the caudate/putamen, while the GABAergic cells send their output to the thalamus.

subthalamic nucleus A nucleus in the ventral diencephalon that receives input from the caudate/putamen and participates in the modulation of motor behavior.

sulci (singular, sulcus) The infoldings of the cerebral hemisphere that form the valleys between the gyral ridges.

summation The addition in space and time of sequential synaptic potentials to generate a larger than normal postsynaptic response.

superior colliculus Laminated structure that forms part of the roof of the midbrain; plays an important role in orienting movements of the head and eyes.

suprachiasmatic nucleus Hypothalamic nucleus lying just above the optic chiasm that receives direct input from the retina; involved in light entrainment of circadian rhythms.

Sylvian fissure see lateral fissure.

sympathetic nervous system A division of the visceral motor system in vertebrates comprising, for the most part, adrenergic ganglion cells located relatively far from the related end organs.

synapse Specialized apposition between a neuron and its target cell for transmission of information by release and reception of a chemical transmitter agent.

synaptic cleft The space that separates pre- and postsynaptic neurons at chemical synapses.

synaptic depression A short-term decrease in synaptic strength resulting from the depletion of synaptic vesicles at active synapses.

synaptic vesicle recycling A sequence of budding, and fusion reactions that occurs within presynaptic terminals to maintain the supply of synaptic vesicles.

synaptic vesicles Spherical, membrane-bound organelles in presynaptic terminals that store neurotransmitters.

syncytium A group of cells in protoplasmic continuity.

target (neural target) The object of innervation, which can be either non-neuronal targets, such as muscles, glands, and sense organs, or other neurons.

taste buds Onion-shaped structures in the mouth and pharynx that contain taste cells.

tectorial membrane The fibrous sheet overlying the apical surface of the cochlear hair cells; produces a shearing motion of the stereocilia when the basilar membrane is displaced.

tectum A general term referring to the dorsal region of the brainstem (tectum means "roof").

tegmentum A general term that refers to the central gray matter of the brainstem.

telencephalon The part of the brain derived from the anterior part of the embryonic forebrain vesicle; includes the cerebral hemispheres.

temporal (temporal division) Referring to the region of the visual field of each eye in the direction of the temple.

temporal lobe The hemispheric lobe that lies inferior to the lateral fissure.

terminal A presynaptic (axonal) ending.

tetraethylammonium A quaternary ammonium compound that selectively blocks voltage-sensitive K^+ channels; eliminates the delayed K^+ current measured in voltage clamp experiments.

tetrodotoxin An alkaloid neurotoxin, produced by certain puffer fish, tropical frogs, and salamanders, that selectively blocks voltage-sensitive Na^+ channels; eliminates the initial Na^+ current measured in voltage clamp experiments.

thalamus A collection of nuclei that forms the major component of the diencephalon. Although its functions are many, a primary role of the thalamus is to relay sensory information from lower centers to the cerebral cortex.

thermoreceptors Receptors specialized to transduce changes in temperature.

threshold The level of membrane potential at which an action potential is generated.

tight junction A specialized junction between epithelial cells that seals them together, preventing most molecules from passing across the cell sheet.

tip links The filamentous structures that link the tips of adjacent stereocilia; thought to mediate the gating of the hair cell's transduction channels.

tonic Sustained activity in response to an ongoing stimulus; the opposite of phasic.

tonotopy the topographic mapping of frequency across the surface of a structure, which originates in the cochlea and is preserved in ascending auditory structures, including the auditory cortex.

transduction see sensory transduction.

transforming growth factor (TGF) A class of peptide growth factors that acts as an inducer during early development.

transgenderism Gender identification with the opposite phenotypic sex.

transmitter see neurotransmitter.

transporters (active transporters) Cell membrane molecules that consume energy to move ions up their concentration gradients, thus restoring and/or maintaining normal concentration gradients across cell membranes.

trichomatic Referring to the presence of three different cone types in the human retina, which generate the initial steps in color vision by differentially absorbing long, medium, and short wavelength light.

tricyclic antidepressants A class of antidepressant drugs named for their three-ringed molecular structure; thought to act by blocking the reuptake of biogenic amines.

trigeminal ganglion The sensory ganglion associated with the trigeminal nerve (cranial nerve V).

Trk receptors The receptors for the neurotrophin family of growth factors.

trophic The ability of one tissue or cell to support another; usually applied to long-term interactions between pre- and postsynaptic cells.

trophic factor (agent) A molecule that mediates trophic interactions.

trophic interactions Referring to the long-term interdependence of nerve cells and their targets.

trophic molecules see trophic factor.

tropic An influence of one cell or tissue on the direction of movement (or outgrowth) of another.

tropic molecules Molecules that influence the direction of growth or movement.

tropism Orientation of growth in response to an external stimulus.

tuning curve Referring to a common physiological test in which the receptive field properties of neurons are gauged against a varying stimulus such that maximum sensitivity or maximum responsiveness can be defined by the peak of the tuning curve.

tympanic membrane The eardrum.

undershoot The final, hyperpolarizing phase of an action potential, typically caused by the voltage-dependent efflux of a cation such as K^+.

upper motor neuron A neuron that gives rise to a descending projection that controls the activity of lower motor neurons in the brainstem and spinal cord.

upper motor neuron syndrome Signs and symptoms that result from damage to descending motor systems; these include paralysis, spasticity, and a positive Babinski sign.

utricle The otolith organ that senses linear accelerations and head tilts in the horizontal plane.

vasopressin A 9-amino-acid neuropeptide that acts as a neurotransmitter, as well as a neurohormone.

ventral Referring to the belly; the opposite of dorsal.

ventral horn The ventral portion of the spinal cord gray matter; contains the primary motor neurons.

ventral posterior complex Group of thalamic nuclei that receives the somatic sensory projections from the dorsal column nuclei and the trigeminal nuclear complex.

ventral posterior lateral nucleus Component of the ventral posterior complex of thalamic nuclei that receives brainstem projections carrying somatic sensory information from the body (excluding the face).

ventral posterior medial nucleus Component of the ventral posterior complex of thalamic nuclei that receives brainstem projections related to somatic sensory information from the face.

ventral roots The collection of nerve fibers containing motor axons that exit ventrally from the spinal cord and contribute the motor component of each segmental spinal nerve.

ventricles The fluid-filled spaces in the vertebrate brain that represent the lumen of the embryonic neural tube.

ventricular zone The sheet of cells closest to the ventricles in the developing neural tube.

vergence movements Disjunctive movements of the eyes (convergence or divergence) that align the fovea of each eye with targets located at different distances from the observer.

vertebrate An animal with a backbone (technically, a member of the subphylum Vertebrata).

vesicle Literally, a small sac. Used to refer to the organelles that store and release transmitter at nerve endings. Also used to refer to any of the three dilations of the anterior end of the neural tube that give rise to the three major subdivisions of the brain.

vestibulocerebellum The part of the cerebellar cortex that receives direct input from the vestibular nuclei or vestibular nerve.

vestibulo-ocular reflex Involuntary movement of the eyes in response to displacement of the head. This reflex allows retinal images to remain stable while the head is moved.

visceral (noun, viscera) Referring to the internal organs of the body cavity.

visceral motor system The component of the motor system (also known as the autonomic nervous system) that motivates and governs visceral motor behavior.

visceral nervous system Synonymous with autonomic nervous system.

visual field The area in the external world normally seen by one or both eyes (referred to, respectively, as the monocular and visual binocular fields).

vital dye A reagent that stains cells when they are alive.

voltage clamp A method that uses electronic feedback to control the membrane potential of a cell, simultaneously measuring transmembrane currents that result from the opening and closing of ion channels.

voltage-gated Term used to describe ion channels whose opening and closing is sensitive to membrane potential.

Wallerian degeneration The process by which the distal portion of a damaged axon segment degenerates; named after Augustus Waller, a nineteenth-century physician and neuroanatomist.

Wernicke's aphasia Difficulty comprehending speech as a result of damage to Wernicke's language area.

Wernicke's area Region of cortex in the superior and posterior region of the left temporal lobe that helps mediate language comprehension. Named after the nineteenth-century neurologist, Carl Wernicke.

white matter A general term that refers to large axon tracts in the brain and spinal cord; the phrase derives from the fact that axonal tracts have a whitish cast when viewed in the freshly cut material.

working memory Memories held briefly in mind that enable a particular task to be accomplished (e.g., efficiently searching a room for a lost object).

Illustration Credits

Chapter 1   The Organization of the Nervous System

Figure 1.4   JONES, E. G. AND M. W. COWAN (1983) The nervous tissue. In *The Structural Basis of Neurobiology*, E. G. Jones (ed.) New York: Elsevier, Chapter 8.

Chapter 2   Electrical Signals of Nerve Cells

Figures 2.6 & 2.7   HODGKIN, A. L. AND B. KATZ (1949) The effect of sodium ions on the electrical activity of the giant axon of the squid. J. Physiol. (Lond.) 108: 37–77.

Chapter 3   Voltage-Dependent Membrane Permeability

Figures 3.1, 3.2, 3.3 & 3.4   HODGKIN, A. L. AND A. F. HUXLEY (1952a) Currents carried by sodium and potassium ions through the membrane of the giant axon of Loligo. J. Physiol. 116: 449–472.   **Figure 3.5**   ARMSTRONG, C. M. AND L. BINSTOCK (1965) Anomalous rectification in the squid giant axon injected with tetraethylammonium chloride. J. Gen. Physiol. 48: 859–872. MOORE, J. W., M. P. B-AUSTEN, N. C. ANDERSON AND T. NARAHASHI (1967) Basis of tetrodotoxin's selectivity in blockage of squid axons. J. Gen. Physiol. 50: 1401–1411.   **Figures 3.6 & 3.7**   HODGKIN, A. L. AND A. F. HUXLEY (1952b) The components of membrane conductance in the giant axon of Loligo. J. Physiol. 116: 473–496.   **Figure 3.8**   HODGKIN, A. L. AND A. F. HUXLEY (1952d) A quantitative description of membrane current and its application to conduction and excitation in nerve. J. Physiol. 116: 507–544.   **Figure 3.10**   HODGKIN, A. L. AND W. A. RUSHTON (1938) The electrical constarts of a crustacean nerve fibre. Proc. R. Soc. Lond. 133: 444–479.

Chapter 4   Channels and Transporters

Figure 4.1B,C   BEZANILLA, F. AND A. M. CORREA (1995) Single-channel properties and gating of Na+ and K– channels in the squid giant axon. In *Cephalopod Neurobiology*, N. J. Abbott, R. Williamson and L. Maddock (eds.). New York: Oxford University Press, pp. 131–151.   **Figure 4.1D**   VANDERBERG, C. A. AND F. BEZANILLA (1991) A sodium channel model based on single channel, macroscopic ionic and gating currents in the squid giant axor. Biophys. J. 60: 1511–1533.   **Figure 4.1E**   CORREA, A. M. AND F. BEZANILLA (1994) Gating of the squid sodium channel at positive potentials. II. Single channels reveal two open states. Biophys. J. 66: 1864–1878.   **Figure 4.2B–D**   AUGUSTINE, C. K. AND F. BEZANILLA (1990) Phosphorylation modulates potassium concuctance and gat:ng current of perfused giant axons of squid. J. Gen. Physiol. 95: 245–271.   **Figure 4.2E**   PEROZO, E., D. S.

JONG AND F. BEZANILLA (1991) Single-channel studies of the phosphorylation of K+ channels in the squid giant axon. II. Nonstationary conditions. J. Gen. Physiol. 98: 19–34.   **Figure 4.10**   HODGKIN, A. L. AND R. D. KEYNES (1955) Active transport of cations in giant axons from Sepia and Loigo. J. Physiol. 128: 28–60. LINGREL, J. B., J. VAN HUYSSE, W. O'BRIEN, E. JEWELL-MOTZ, R. ASKEW AND P. SCHULTHEIS (1994) Structure-function studies of the Na, K-ATPase. Kidney Internat. 45: S32–S39.   **Figure 4.11**   RANG, H. P. AND J. M. RICHIE (1968) On the electrogenic sodium pump in mammalian non-myelinated nerve fibres and its activation by various external cations. J. Physiol. 196: 183–221.   **Figure 4.12**   LINGREL, J. B., J. VAN HUYSSE, W. O'BRIEN, E. JEWELL-MOTZ, R. ASKEW AND P. SCHULTHEIS (1994) Structure-function studies of the Na, K-ATPase. Kidney Internat. 45: S32–S39.

Chapter 5   Synaptic Transmission

Figure 5.2B   FURSHPAN, E. J. AND D. D. POTTER (1959) Transmission at the giant motor synapses of the crayfish. J. Physiol. (Lond.) 145: 289–325.   **Figure 5.4**   FATT, P. AND B. KATZ (1952) Spontaneous subthreshold activity at motcr nerve endings. J. Physiol. (Lord.) 117: 109–128.   **Figure 5.5**   BOYD, I. A. AND A. R. MARTIN (1955) Spontaneous subthreshold activity at mammalian neuromuscular junctions. J. Physiol. 132: 61–73.   **Figure 5.6**   HEUSER, J. E., T. S. REESE, M. J. DENNIS, Y. JAN, L. JAN AND L. EVANS (1979) Synaptic vesicle exocytosis captured by quick freezing and correlated with quantal transmitter release. J. Cell Biol. 81: 275–300.   **Figure 5.7**   HEUSER, J. E. AND T. S. REESE (1973) Evidence for recycling of synaptic vesicle membrane during transmitter release at the frog neuromuscular junction. J. Cell Biol. 57: 315–344.   **Figure 5.8**   AUGUSTINE, G. J. AND R. ECKERT (1984) Divalent cations differentially support transmitter release at the squid giant synapse. J. Physiol. 346: 257–271.   **Figure 5.9A**   SMITH, S. J., J. BUCHANAN, L. R. OSSES, M. P. CHARLTON AND G. J. AUGUSTINE (1993) The spatial distribution of calcium signals in squid presynaptic terminals. J. Physiol. (Lond.) 472: 573–593.   **Figure 5.9B**   MILEDI, R. (1973) Transmitter release induced by injection of calcium ions into nerve terminals. Proc. R. Soc. Lond. B 183: 421–425.   **Figure 5.9C**   ADLER, E. M. ADLER, G. J. AUGUSTINE, M. P. CHARLTON AND S. N. DUFFY (1991) Alien intracellular calcium chelators attenuate neurotransmitter release at the squid giart synapse. J. Neurosci. 11: 1496–1507.   **Figure 5.10**   JESSELL, T. M. AND E. R. KANDEL (1993) Synaptic transm:ssion:

A bicirectional and self-modifiable form of cell-cell communication. Cell 72/Neuron 10(Supp.): 1–30.

Chapter 6   Neurotransmitters

Figure 6.7   PETERS, A., PALAY, S. L. AND H. WEBSTER (1991) *The Fine Structure of the Nervous System: Neurons and Their Supporting Cells*, Third Ed. Oxford University Press, New York.

Chapter 7   Neurotransmitter Receptors and Their Effects

Figure 7.3   TAKEUCHI, A. AND N. TAKEUCHI (1960) On the permeability of end-plate membrane during the action of transmitter. J. Physiol. 154: 52–67.

Chapter 9   The Somatic Sensory System

Figure 9.3   DARIAN-SMITH, I. (1984) The sense of touch: Performance and peripheral neural processes. In *Handbook of Physiology: The Nervous System*, Vol. III., J. M. Brookhart and V. B. Mountcastle (eds.). Bethesda, MD: American Physiological Society, pp. 739–788.   **Figure 9.4**   WEINSTEIN, S. (1969) Neuropsychological studies of the phantom. In *Contributions to Clinical Neuropsychology*, A. L. Benton (ed.). Chicago: Aldine Publishing Company, pp. 73–106.   **Figure 9.5**   MATTHEWS, P. B. C. (1964) Muscle spindles and their motor control. Physiol. Rev. 44: 219–288.   **Figure 9.7**   BRODAL, P. (1992) The Central Nervous System: Structure and Function. New York: Oxford University Press, p. 151. JONES, E. G. AND D. P. FRIEDMAN (1982) Projection pattern of functional components of thalamic ventrobasal complex on monkey somatosensory cortex. J. Neurophys. 48: 521–544.   **Figure 9.8**   PENFIELD, W. AND T. RASMUSSEN (1950) *The Cerebral Cortex of Man: A Clinical Study of Localization of Function*. New York: Macmillan, p. 14. CORSI, P. (1991) *The Enchanted Loom: Chapters in the History of Neuroscience*, P. Corsi (ed.). New York: Oxford University Press.   **Figure 9.9**   KAAS, J. H. (1989) The functional organization of somatosensory cortex in primates. Ann. Anat. 175: 509–518.

Chapter 10   Pain

Figure 10.1   FIELDS, H. L. (1987) *Pain*. New York: McGraw-Hill.   **Figure 10.2**   FIELDS, H. L. (ed.) (1990) *Pain Syndromes in Neurology*. London: Butterworths.   **Box B**   Solonen, K. A. (1962) The phantom phenomenon in amputated Finnish war veterans. Acta Orthop. Scand. Suppl. 54: 1–37.

Chapter 11   Vision: The Eye

Figure 11.3A–C   HILFER, S. R. AND J. W. YANG (1980) Accumulation of CPC-precipitable material at apical cell surfaces during formation of the optic cup. Anat. Rec. 197: 423–433. **Box A, Figure D** WESTHEIMER, G. (1974) In *Medical Physiology*, 13th Ed. V. B. Mountcastle (ed.) St. Louis: Mosby. **Figure 11.5** SCHNAPF, J. L. AND D. A. BAYLOR (1987) How photoreceptors respond to light. Sci. Am. 256: 40–47.

Chapter 12 Central Visual Pathways
Figure 12.10B J. HORTON AND E. T. HEDLEY-WHYTE (1984) Mapping of cytochrome oxidase patches and ocular dominance columns in human visual cortex. Philo Trans. 304: 255–172. **Figure 12.14A** WATANABE, M., AND R. W. RODIECK, (1989) Parasol and midget ganglion cells of the primate retina. J. Comp. Neurol. 289: 434–454. **Figure 12.15A** MAUNSELL, J. H. R. AND W.T. NEWSOME (1987) Visual processing in monkey extrastriate cortex. Ann. Rev. Neurosci. 10: 363–401. **Figure 12.15B** FELLEMAN, D. J. AND D. C. VAN ESSEN (1991) Distributed hierarchical processing in primate cerebral cortex. Cereb. Cortex 1: 1–47. **Box B, Figure A** WANDELL, B. A. (1995) *Foundations of Vision*. Sunderland, MA: Sinauer Associates. **Box B, Figure C** *Super Stereogram* (1994) San Francisco: Cadence Books, p. 40.

Chapter 13 The Auditory System
Figure 13.5 DALLOS, P. (1992) The active cochlea. J. Neurosci. 12: 4575–4585. VON BÉKÉSY, G. (1960) Experiments in Hearing. New York: McGraw-Hill. **Figure 13.7A** LINDEMAN, H. (1973) Anatomy of the otolith organs. Adv. Otorhinolaryngol. 20: 405–433. **Figure 13.7B** HUDSPETH, A. J. (1983) The hair cells of the inner ear. Sci. Amer. 248: 54–64. **Figure 13.7D** PICKLES, J. O., S. D. COMIS AND M. P. OSBORNE (1984) Cross-links between stereocilia in the guinea pig organ of Corti, and their possible relation to sensory transduction. Hear. Res. 15: 103–112. **Figure 13.8A,B** LEWIS, R. S. AND A. J. HUDSPETH (1983) Voltage- and ion-dependent conductances in solitary vertebrate hair cells. Nature 304: 538–541. **Figure 13.8C** PALMER, A. R. AND I. J. RUSSELL (1986) Phase-locking in the cochlear nerve of the guinea-pig and its relation to the receptor potential of inner hair cells. Hear. Res. 24: 1–15. **Figure 13.9** KIANG, N. Y. AND E. C. MOXON (1972) Physiological considerations in artificial stimulation of the inner ear. Ann. Otol. Rhinol. Laryngol. 81: 714–730. **Figure 13.10** KIANG, N.Y.S. (1984) Peripheral neural processing of auditory information. In *Handbook of Physiology: A Critical, Comprehensive Presentation of Physiological Knowledge and Concepts, Section 1: The Nervous System, Vol. III: Sensory Processes, Part 2*, J. M. Brookhart, V. B. Mountcastle, I. Darian-Smith and S. R. Geiger (eds.). Bethesda, MD: American Physiological Society, pp. 639–674. **Figure 13.12** JEFFRESS, L. A. (1948) A place theory of sound localization. J. Comp. Physiol. Psychol. 41: 35–39.

Chapter 14 The Vestibular System
Figure 14.3 LINDEMAN, H. H. (1973) Anatomy of the otolith organs. Adv. Otorhinolaryngol. 20: 405–433. **Figure 14.6** GOLDBERG, J. M.

AND C. FERNÁNDEZ (1976) Physiology of peripheral neurons innervating otolith organs of the squirrel monkey, Parts 1, 2, 3. J. Neurophys. 39: 970–1008. **Figure 14.9** GOLDBERG, J. M. AND C. FERNÁNDEZ (1971) Physiology of peripheral neurons innervating semicircular canals of the squirrel monkey, Parts 1, 2, 3. J. Neurophys. 34: 635–684.

Chapter 15 The Chemical Senses
Figure 15.2 PELOSI, P. (1994) Odorant-binding proteins. Crit. Rev. Biochem. Mol. Biol. 29: 199–228. **Figure 15.3** CAIN, W. S. AND J. F. GENT (1986) Use of odor identification in clinical testing of olfaction. In *Clinical Measurement of Taste and Smell*, H. L. Meiselman and R. S. Rivlin (eds.), New York: Macmillan, pp. 170–186. **Figure 15.4** MURPHY, C. (1986) Taste and smell in the elderly. In *Clinical Measurement of Taste and Smell*, H. L. Meiselman and R. S. Rivlin (eds.), New York: Macmillan, pp. 343–371. **Figure 15.5A** ANHOLT, R. R. H. (1987) Primary events in olfactory reception. Trends Biochem. Sci. 12: 58–62. **Figure 15.5B** FIRESTEIN, S., F. ZUFALL AND G. M. SHEPHERD (1991) Single odor-sensitive channels in olfactory receptor neurons are also gated by cyclic nucleotides. J. Neurosci. 11: 3565–3572. **Figure 15.7B** GETCHELL, M. L. (1986) In *Neurobiology of Taste and Smell*, T. E. Finger and W. L. Silver (eds). New York: John Wiley and Sons, p. 112. **Figure 15.8A** LAMANTIA, A.-S., S. L. POMEROY AND D. PURVES (1992) Vital imaging of glomeruli in the mouse olfactory bulb. J. Neurosci. 12: 976–988. **Figure 15.8B,C** POMEROY, S. L., A.-S. LAMANTIA AND D. PURVES (1990) Postnatal construction of neural activity in the mouse olfactory bulb. J. Neurosci. 10: 1952–1966. **Figure 15.9** RUBIN, B. D. AND L. C. KATZ (1999) Optical imaging of odorant representations in the mammalian olfactory bulb. Neuron 23: 499–511. **Figure 15.11C** ROSS M. H., L. J. ROMMELL AND G. I. KAYE (1995) *Histology, A Text and Atlas*. Baltimore: Williams and Wilkins. **Figure 15.16** COMETTO-MUNIZ, J. E. AND W. S. CAIN (1990) Thresholds for odor and nasal pungency. Physiol. Behav. 48: 719–725.

Chapter 16 Lower Motor Circuits and Motor Control
Figure 16.2 BURKE, R. E., P. L. STRICK, K. KANDA, C. C. KIM AND B. WALMSLEY (1977) Anatomy of medial gastrocnemius and soleus motor nuclei in cat spinal cord. J. Neurophys. 40: 667–680. **Figure 16.5** BURKE, R. E., D. N. LEVINE, M. SALCMAN AND P. TSAIRIS (1974) Motorunits in cat soleus muscle: Physiological, histochemical and morphological characteristics. J. Physiol. (Lond.) 238: 503–514. **Figure 16.6** WALMSLEY, B., J. A. HODGSON AND R. E. BURKE (1978) Forces produced by medial gastrocnemius and soleus muscles during locomotion in freely moving cats. J. Neurophys. 41: 1203–1216. **Figure 16.8** MONSTER, A. W. AND H. CHAN (1977) Isometric force production by motor units of extensor digitorum communis muscle in man. J. Neurophys. 40: 1432–1443. **Figure 16.10** HUNT, C. C. AND S. W. KUFFLER (1951) Stretch receptor discharges during mus-

cle contraction. J. Physiol. (Lond.) 113: 298–315. **Figure 16.11B** PATTON, H. D. (1965) Reflex regulation of movement and posture. In *Physiology and Biophysics*, 19th Ed., T. C. Ruch and H. D. Patton (eds.), Philadelphia: Saunders, pp. 181–206. **Figure 16.14** PEARSON, K. (1976) The control of walking. Sci. Am. 235: 72–86.

Chapter 19 Modulation of Movement by the Cerebellum
Figure 19.9 STEIN, J. F. (1986) Role of the cerebellum in the visual guidance of movement. Nature 323: 217–221. **Figure 19.10** THACH, W. T. (1968) Discharge of Purkinje and cerebellar nuclear neurons during rapidly alternating arm movements in the monkey. J. Neurophys. 31: 785–797. **Figure 19.11** OPTICAN, L. M. AND D. A. ROBINSON (1980) Cerebellar-dependent adaptive control of primate saccadic system. J. Neurophys. 44: 1058–1076. **Box B** RAKIC, P. (1977) Genesis of the dorsal lateral geniculate nucleus in the rhesus monkey: Site and time of origin, kinetics of proliferation, routes of migration and pattern of distribution of neurons. J. Comp. Neuro. 176: 23–52.

Chapter 20 Eye Movements and Sensory Motor Integration
Figure 20.1 YARBUS, A. L. (1967) *Eye Movements and Vision*, Basil Haigh, trans. New York: Plenum Press. **Figure 20.4** FUCHS, A. F. (1967) Saccadic and smooth pursuit eye movements in the monkey. J. Physiol. (Lond.) 191: 609–631. **Figure 20.6** FUCHS, A. F. AND E. S. LUSCHEI (1970) Firing patterns of abducens neurons of alert monkeys in relationship to horizontal eye movements. J. Neurophys. 33: 382–392. **Figure 20.8** SCHILLER, P. H. AND M. STRYKER (1972) Single unit recording and stimulation in superior colliculus of the alert rhesus monkey. J. Neurophys. 35: 915–924.

Chapter 22 Early Brain Development
Figure 22.2 SANES, J. R. (1989) Extracellular matrix molecules that influence neural development. Ann. Rev. Neurosci. 12: 491–516. **Figure 22.5A** INGHAM, P. (1988) The molecular genetics of embryonic pattern formation in *Drosophila*. Nature 335: 25–34. **Figure 22.5B** GILBERT, S. F. (1994) *Developmental Biology*, 4th Ed. Sunderland, MA: Sinauer Associates. **Figure 22.9** GALILEO, S. S., GRAY, G. E., OWENS, G. C., MAJORS, J., AND SANES, J. R. (1990) Neurons and glia arise from a common progenitor in chicken optic tectum: Demonstration with two retroviruses and cell type-specific antibodies. Proc. Natl. Acad. Sci. USA 87: 458–462.

Chapter 23 Construction of Neural Circuits
Figure 23.2A,B GODEMONT, P., L. C. WANG AND C. A. MASON (1994) Retinal axon divergence in the optic chiasm: Dynamics of growth cone behavior at the midline. J. Neurosci 14: 7024–7039. **Figure 23.2C** SRETAVAN, D. W., L. FENG, E. PURE AND L. F. REICHARDT (1994) Embryonic neurons of the developing optic chiasm express L1 and CD44, cell surface molecules with opposing effects on retinal axon

growth. Neuron 12: 957–975. **Figure 23.3** REICHARDT, L. F. AND K. J. TOMASELLI (1991) Extracellular matrix molecules and their receptors: Functions ir: neural development. Ann. Rev. Neurosci. 14: 531–570. **Figure 23.4A** SERAFINI, T., T. E. KENNEDY, M. J. GALKO, C. MIRZAYAN, T. M. JESSELL, M. TESSIER-LAVIGNE (1994) The netrins define a family of axon outgrowth-promoting proteins homologous to C. elegans UNC-6. Cell 78: 409–424. **Figure 23.4B–D** KENNEDY, T. E., T. SERAFINI, J. R. DE LA TORRE AND M. TESSIER-LAVIGNE (1994) Netrins are diffusible chemotropic factors for commissural axons in the embryonic spinal cord. Cell 78: 425–435. **Figure 23.5** MESSERSMITH, E. K., E. D. LEONARDO, C. J. SHATZ, M. TESSIER-LAVIGNE, C. S. GOODMAN AND A. L. KOLODKIN (1995) Semaphorin III can function as a selective chemcrepellent to pattern sensory projections in the spinal cord. Neuron. 14: 949–959. **Figure 23.6** SPERRY, R. W. (1963) Chemoaffinity in the orderly growth of nerve fiber patterns and connections. Proc. Natl. Acad. Sci. USA 50: 703–710. **Figure 23.7A** WALTER, J., S. HENKE-FAHLE AND F. BONHOEFFER (1987) Avoidance of posterior tectal membranes by temporal retinal axons. Development. 101: 909–913. **Figure 23.7B** CHENG, H. J., M. NAKAMOTO, A. D. BERGEMANN AND J. G. FLANAGAN (1995) Complementary gradients in expression and binding of ELF-1 and Mek4 in development of the topographic retinotectal projection map. Cell 82: 371–381. **Figure 23.9** HOLLYDAY, M. AND V. HAMBURGER (1976) Reduction of the naturally occurring motor neuron loss by enlargement of the periphery. J. Comp. Neurol. 170: 311–320 . HOLLYDAY, M. AND V. HAMBURGER (1958) Regression versus peripheral controls of differentiation in motor hypoplasia. Am. J. Anat. 102: 365–409. HAMBURGER, V. (1977) The developmental history of the motor neuron. The F. O. Schmitt Lecture in Neuroscience. 1970. Neurosci. Res. Prog. Bull. 15, Suppl. III: 1–37. **Figure 23.10** PURVES, D. AND J. W. LICHTMAN (1980) Elimination of synapses in the developing nervous system. Science 210: 153–157. **Figure 23.12** PURVES, D. AND J. W. LICHTMAN (1985) *Principles of Neural Development.* Sunderland, MA: Sinauer Associates, Inc. **Figure 23.13** McDONALD, N. Q., R. LAPATTO, J. MURRAY-RUST, J. GUNNING, A. WLODAWER AND T. L. BLUNDELL (1991) New protein fold revealed by a 2.3-Å resolution crystal structure of nerve growth factor. Nature 354: 411–414. **Figure 23.14A** CHUN, L. L. AND P. H. PATTERSON (1977) Role of nerve growth factor in the development of rat sympathetic neurons in vitro. III; Effect on acetylcholine production. J. Cell Biol. 75: 712–718. **Figure 23.15** MAISONPIERRE, P. C., L. BELLUSCIO, S. SQUINTO, N.Y. IP, M.E. FURTH, R.M. LINDSAY AND G.D. YANCOPOULOS (1990) Neurotrophin-3: A neurotrophic factor related to NGF and BDNF. Science 247: 1446–1451. **Figure 23.17** CAMPENOT, R. B. (1981) Regeneration of neurites on long-term cultures of sympatic neurons deprived of nerve growth factor. Science 214: 579–581.

nucleus in the rat lumbar spinal cord: Response to adult hormone manipulation, absence in androgen-insensitive rats. Brain Res. 225: 297–307. **Figure 30.4B,C** BREEDLOVE, S. M. AND A. P. ARNOLD (1983) Hormonal control of a developing neuromuscular system. II. Sensitive periods for the androgen-induced masculinization of the rat spinal nucleus of the bulbocavernosus. J. Neurosci. 3: 424–432. **Figure 30.4D** FORGER, N. G. AND S. M. BREEDLOVE (1986) Sexual dimorphism in human and canine spinal cord: Role of early androgen. Proc. Natl. Acad. Sci. USA 83: 7527–7531. **Figure 30.6** OOMURA, Y., H. YOSHIMATSU AND S. AOU (1983) Medial preoptic and hypothalamic neuronal activity during sexual behavior of the male monkey. Brain Res. 266: 340–343. **Figure 30.7B–D** ALLEN, L. S., M. HINES, J. E. SHYRNE AND R. A. GORSKI (1989) Two sexually dimorphic cell groups in the human brain. J. Neurosci. 9: 497–506. **Figure 30.8A** LEVAY, S. (1991) A difference in hypothalamic structure between heterosexual and homosexual men. Science 253: 1034–1037. **Figure 30.8B** SWAAB, D. F. AND M. A. HOFFMAN (1990) An enlarged suprachiasmatic nucleus in homosexual men. Brain Res. 537: 141–148. **Figure 30.9A,B** ALLEN, L. S., M. F. RICHEY, Y. M. CHAI AND R. A. GORSKI (1991) Sex differences in the corpus callosum of the living human being. J. Neurosci. 11: 933–942. **Figure 30.9C** ALLEN, L. S. AND R. A. GORSKI (1991) Sexual dimorphism of the anterior commissure and massa intermedia of the human brain. J. Comp. Neurol. 312: 97–104. **Figure 30.10B–C** XERRI, C., J. M. STERN AND M. M. MERZENICH (1994) Alterations of the cortical representation of the rat ventrum induced by nursing behavior. J. Neurosci. 14: 1710–1721. **Box A** MOORE, K. L. (1977) *The Developing Human*, 2nd Ed. Philadelphia: W. B. Saunders, p. 219. **Box C** McEWEN, B. S. (1976) Interactions between hormones and nerve tissue. Sci. Am. 235: 48–58; McEWEN, B. S., P. G. DAVIS, B. PARSONS AND D. W. PFAFF (1979) The brain as a target for steroid hormone action. Ann. Rev. Neurosci. 2: 65–112.

Chapter 31   Human Memory
Figure 31.3 ERICSSON, K. A., W. G. CHASE, AND S. FALOON (1980) Acquisition of a memory skill. Science. 208: 1181–1182. **Figure 31.4** CHASE W. G. AND H. A. SIMON (1973) *The Mind's Eye in Chess in Visual Information Processing*, W. G. Chase, ed. New York: Academic Press, pp. 215–281. **Figure 31.5A** RUBIN, D. C. AND T. C. KONTIS (1983) A schema for common cents. Mem. Cog. 11: 335–341. **Figure 31.5B** SQUIRE, L. R. (1989) On the course of forgetting in very long-term memory. J. Exp. Psychol. 15: 241–245. **Figure 31.7** VAN HOESEN, G. W. (1982) The parahippocampal gyrus. Trends Neurosci. 5: 345–350. **Figure 31.9** DEKABAN, A. S. AND D. SADOWSKY (1978) Changes in brain weights during the span of human life: Relation of brain weights to body heights and body weights. Ann. Neurol. 4: 345–356.

Chapter 24   Modification of Brain Circuits as a Result of Experience
Figure 24.1 PETITTO, L. A. AND P. F. MARENTETTE (1991) Babbling in the manual mode: Evidence for the ontogeny of language. Science 251: 1493–1496. **Figure 24.2** JOHNSON, J. S. AND E. I. NEWPORT (1989) Critical period effects in second language learning: the influences of maturational state on the acquisition of English as a second language. Cogn. Psychol. 21. **Figure 24.4A** HUBEL, D. H. AND T. N. WIESEL (1962) Receptive fields, binocular interaction and functional architecture in the cat's visual cortex. J. Physiol. 160: 106–154. **Figure 24.4B** HUBEL, D. H. AND T. N. WIESEL (1963) Receptive fields of cells in striate cortex of very young, visually inexperienced kittens. J. Neurophys. 26: 994–1003.

Chapter 25   Plasticity of Mature Synapses and Circuits
Figure 25.3 KATZ, B. (1966) *Nerve, Muscle and Synapse.* New York: McGraw Hill. **Figure 25.4** SCHENK, F. AND R. G. MORRIS (1985) Dissociation between components of spatial memory in rats after recovery from the effects of retrohippocampal lesions. Exp. Brain Res. 58: 11–28. **Figure 25.9** NICOLL, R. A., J. A. KAUER AND R. C. MALENKA (1988) The current excitement in long-term potentiation. Neuron. 1: 97–103. **Box C** DYRO, F. M. (1989) *The EEG Handbook.* Boston: Little, Brown and Company.

Chapter 26   The Association Cortices
Figure 26.10 LYNCH, J. C., V. B. MOUNTCASTLE, W. H. TALBOT AND T. C. YIN (1977) Parietal lobe mechanisms for directed visual attention. J. Neurophys. 40: 362–369.

Chapter 27   Language and Lateralization
Figure 27.5B OJEMANN, G. A., I. FRIED AND E. LETTICH (1989) Electrocorticographic (EcoG) correlates of language. Electroencephalo. Clin. Neurophys. 73: 453–463. **Figure 27.6** POSNER, M. I. AND M. E. RAICHLE (1994) *Images of Mind.* New York: Scientific American Library.

Chapter 28   Sleep and Wakefulness
Figure 28.2 BERGMANN, B. M., C. A. KUSHIDA, C. A. EVERSON, M. A. GILLILAND, W. OBERYMEYER AND A. RECHTSCHAFFEN (1989) Sleep deprivation in the rat: II. Methodology. Sleep 12: 5–12.

Chapter 29   Emotions
Figure 29.7 MOSCOVITCH, M. AND J. OLDS (1982) Asymmetries in spontaneous facial expression and their possible relation to hemispheric specialization. Neuropsych. 20: 71–81.

Chapter 30   Sex, Sexuality, and the Brain
Figure 30.2 TORAND-ALLERAND, C. D. (1978) Gonadal hormones and brain development. Cellular aspects of sexual differentiation. Amer. Zool. 18: 553–565. **Figure 30.3** WOOLLEY, C. S. AND B. S. McEWEN (1992) Estradiol mediates fluctuation in hippocampal synapse density during the estrous cycle in the adult rat. J. Neurosci. 12: 2549–2554. **Figure 30.4A** BREEDLOVE, S. M. AND A. P. ARNOLD (1984) Sexually dimorphic motor

Index

ABCR gene, 236
Abducens nerve, 15, 428, 429, 431
  characteristics of, 18–19
  saccadic eye movements and, 433, 434
Abducens nuclei, 16, 311, 429
Aberrations, spherical and chromatic, 225
Accessory nerve, 15, 18–19
Accessory olfactory nuclei, 329
Accommodation, 225
Accurane®, 474
Acetylcholine (ACh), 101, 105
  chemical structure, 121
  congenital myasthenic syndromes and, 108
  discovery of, 117
  functional features of, 127
  organophosphates and, 128–129
  packaging in synaptic vesicles, 126
  sites of action, 117–118, 127
  synthesis of, 127–128
  visceral motor system and, 454–455, 459, 467
Acetylcholine receptors
  muscarinic, 153, 157, 169, 172, 455, 459–460
  myasthenic syndromes and, 108, 158
  in neuromuscular junctions, dynamics of, 141–147
  nicotinic, 151–153, 157, 454–455, 459
  See also Cholinergic receptors
Acetylcholinesterase, 108, 128, 158
Acetyl coenzyme A, 127, 128
ACh. See Acetylcholine
Achromatopsia, 272
Acids, taste responses, 335–336, 337
Aconitine, 90
Acquired hearing loss, 277
Acromelic acid, 157
Across-neuron hypothesis, 340, 341
Action potentials, 5
  axon myelination and, 72–74
  conduction velocity, 70
  defined, 43
  diversity in, 55
  electrophysiological recording of, 9–10
  end plate potentials and, 142
  Hodgkin-Huxley model, 64–65
  ion channels and, 80–81
  ionic basis of, 50–51, 52, 54, 56
  long-distance signaling, 65, 67, 69–72
  measuring, 57–59
  in motor neurons, 354
  observing with microelectrodes, 43–44
  overview of, 76
  passive current flow in, 69, 70–72, 74
  phases of, 54, 56, 65
  refractory period, 72
  regenerative quality of, 65
  saltatory propagation, 73, 74
  stimulating current and, 44–45
  summation of postsynaptic potentials and, 149–150
  threshold potential and, 66
  voltage-dependent ionic currents in, 59–61
  voltage-dependent membrane conductance in, 61–64
Action tremors, 425
Active tactile exploration, 199
Active transporters, 45
  ion gradients and, 89, 91
  overview of, 77, 96–97
  types of, 91
Acupuncture
  induced analgesia, 139
  placebo effect and, 219
Adaptation
  in olfactory receptor neurons, 324
  in taste cells, 338, 340
Addiction, 138
Adenosine, 135
  purinergic receptors and, 161
Adenosine triphosphate (ATP), 120
  chemical structure, 121
  as neurotransmitter, 127, 135
  purinergic receptors and, 161
  Na$^+$/K$^+$ pump and, 93, 94, 95
Adenylyl cyclase, 171, 174
Adrenal gland, 448
Adrenaline. See Epinephrine
Adrenal medulla, 447
Adrenergic receptors, 162, 169
  dopamine and, 133
  sympathetic system and, 458, 459
Adrenocorticotropin, 135
Adrian, E. D., 613
Adrian, Edgar, 328
Affective disorders, 640–641
Afferent neurons, 8
α-Agatoxins, 157
Age-related macular degeneration, 236
Aggressive behavior, amygdala and, 634–635
Aging, memory and, 580
Agnosias, 574–575, 632
Agrin, 504–505
Aguayo, Albert, 558, 560
Aitken, Alexander, 669–671
Alcohol / Alcohol abuse
  addiction, 138
  cerebellar cortex and, 422
  congenital neurological disorders and, 476
  GABA receptors and, 160
  Korsakoff's syndrome and, 676
Allen, Laura, 655, 657
Alpha rhythm, 612–613
Alzheimer's disease, 678–679
Amacrine cells, 2, 228, 229, 237
Amanita, 156
Amblyopia, 526, 530, 532
Aδ mechanosensitive nociceptors, 210
Aδ mechanothermal nociceptors, 210
American Sign Language, 600
Ametropia, 226
Amiloride, 335, 341
Amine neurotransmitters. See Biogenic amines
Aminergic neurons, 137
Amino acid neurotransmitters, 121, 126
Amino acids
  radioactive, in transneuronal labeling, 524, 527
  taste perception of, 338
Aminoglycoside antibiotics, 277
2-Amino-5-phosphono-valerate (APV), 156–157
4-Aminopyridine (4-AP), 105
Ammon's horn, 542
Amnesia
  causes of, 672
  clinical cases, 674–675
  defined, 665
  electroconvulsive therapy and, 676
  Korsakoff's syndrome, 676
  types of, 673
amnesiac gene, 540
Amoore, John, 319
AMPA, 154
AMPA receptors, 153, 182, 183, 546, 548
  in long-term depression and potentiation, 549, 551, 552
  properties of, 154, 155, 156, 157
Amphetamine, 137
Amplitude, of sound waves, 275
Ampullae, 298, 299, 306
Amputation
  functional topographic remapping and, 553
  phantom sensations, 216–217
Amsler grid, 236
Amygdala, 30
  anatomy and connectivity of, 634–635
  association of sensory experience and emotional behavior, 635–639
  decision making and, 644
  declarative memory and, 674
  kindling seizures in, 554
  Klüver-Bucy syndrome and, 633
  limbic system and, 632
  long-term potentiation in, 639
  neocortex and, 639, 641–642
  olfactory processing and, 317, 318, 329, 330
  sexual dimorphism in, 659
  sleep and, 615, 617
  somatic sensory input, 207
  taste processing and, 331, 332
Amyloid cascade hypothesis, 679
Amyloid-β peptide, 679
Amyloid plaques, 678, 679
Amyloid precursor protein, 678, 679

Amyotrophic lateral sclerosis, 367
Analgesia
  actions of analgesics, 211
  capsaicin and, 342
  opioid peptides and, 139
Anamirta cocculus, 157
Andersen, Per, 613
Androgen insensitivity syndrome, 648
Androgens
  congenital adrenal hyperplasia, 648
  in sexual development, 646–647, 650
  sexual identity and, 658
5α-Androst-16-en-3-one, 319
Anencephaly, 476
Anesthetics, somatic sensory receptive fields and, 553
Angiotensin II, 122
Aniridia, 479–480
Aδ nociceptors, 209–210, 211
Anomalous trichromacy, 241, 242
Anopsias, 258
Anosmia, 320
Antennal lobes, 322, 328
Anterior cerebral arteries, 35, 36, 37, 39
Anterior chamber (eye), 223
Anterior cingulate cortex, 617
Anterior circulation, 35, 37
Anterior commissure, 24, 30, 31, 456, 568
  sexual dimorphism in, 654
Anterior communicating artery, 35, 36
Anterior cranial fossa, 33, 34
Anterior hypothalamus, sexual dimorphism in, 655, 657–658, 663
Anterior inferior cerebellar artery (AICA), 37, 39
Anterior nucleus of the dorsal thalamus, 632
Anterior pituitary, sex hormone receptors in, 654
Anterior (position), 12, 13
Anterior spinal artery, 35
Anterior ventral cochlear nucleus, 289
Anterograde amnesia, 673, 674–675
Anterograde tracing, 527
Anterolateral column (spinal cord), 20
Anterolateral system, 201, 212. See also Spinothalamic pathway
Anteroventral cochlear nucleus, 288, 290, 291, 292

Anti-anxiety drugs, 137, 160
Antidepressants, 137, 614
Antihistamines, 133
Antihypertensive agents, 137
Antipsychotic drugs, 134, 137
Antisaccades, 437
Antiseizure medications, 555
Anxiety disorders, 137
Anxiety reducing drugs, 137, 160
Apamin, 90
β-A4 peptide, 679
Apes, language function in, 588–589
Aphasias, 590, 592–593
Aplysia, synaptic plasticity in, 535–539
Apolipoprotein E, 679
Apoptosis, 506
Aprosodia, 599, 642
Aprosody. See Aprosodia
APV, 156–157
Apyrase, 135
Aqueous humor, 223–224
Arachnoid granulations, 33, 34

Arachnoid mater, 33, 34
Arachnoid trabeculae, 33, 34
Arachnoid villi, 33, 34
Archicortex, 29, 569
Arcuate fasciculus, 592
Arcuate nucleus, 456, 654
Areca catechu, 157
Arecoline, 157
Aromatase, 650
Arousal responses, evoked by pain, 216
Arrestin, 234
Arteries. See Blood supply
Artificial consciousness, 616
Aserinsky, Eugene, 609
Aspartate, 119, 120, 121
Aspirin, 211
Associational systems, 10
Association cortex lesions
  frontal, 576–577
  parietal, 574–575
  temporal, 571–573
Association cortices, 207, 393
  cognition and, 565–566
  feature of connectivity in, 568, 570
  language representation in, 587
  overview of, 565, 566, 585
Associative learning, amygdala and, 635–639
Astrocytes, 7, 12
  blood-brain barrier and, 38
  inhibition of axonal regeneration, 558, 561
Ataxia, 16
ATP. See Adenosine triphosphate
ATPase pumps, 91
Atropine, 153, 157
Attention deficits, contralateral neglect syndrome, 571–573
Attention-specific neurons, 577–578
Audible spectrum, 276, 278
Auditory cortex. See Primary auditory cortex
Auditory localization, 533
Auditory meatus, 278
Auditory nerve, 277, 289
  in auditory pathways, 288
  characteristics of, 18–19
  labyrinth and, 298
  phase-locking, 287–288
  tuning of, 287
  See also Cranial nerve VIII
Auditory space map, 292
Auditory stimuli
  audible spectrum, 276, 278
  conditioned fear response and, 636–637, 638, 639
  physical properties of sound, 275–276
Auditory system
  audible spectrum, 276, 278
  auditory nerve tuning and timing, 287–288
  auditory thalamus, 293–294
  basilar membrane mechanics, 280–282
  critical period in, 533
  ear anatomy, 278–282
  inferior colliculus, 292–293
  major pathways, 289
  mechanoelectrical sound transduction, 282–286
  monoaural pathway to the lateral lemniscus, 292
  overview of, 275, 276–277, 295–296
  parallel circuitry in, 288

  primary auditory cortex, 294–295
  sound localization and, 288, 290–292
  volley theory of information transfer, 287
Auerbach's plexus, 452, 453
Autism, 480
Autoimmune diseases
  multiple sclerosis and, 75
  myasthenia gravis, 158
Autonomic ganglia, 12, 444, 454–455, 466. See also Parasympathetic ganglia; Sympathetic ganglia
Autonomic nervous system, 11, 12, 444. See also Visceral motor system
Autonomic neurons, 4
Autosomal dominant retinitis pigmentosa, 231
Autosomal recessive retinitis pigmentosa, 231
Autostereograms, 264–265
Axonal growth
  chemoaffinity hypothesis, 500–502
  diffusible signals and, 498–500
  formation of topographic maps, 500–502
  growth cones, 493–494
  non-diffusible signals and, 494–495, 497–498
  selective synapse formation, 502–503
Axon myelination
  action potentials and, 72–74
  multiple sclerosis and, 75
Axons, 2, 5
  action potentials (see Action potentials)
  cell death in multiple sclerosis, 75
  fast and slow transport, 125, 126
  nodes of Ranvier, 72
  passive current flow, 65, 67, 68–69, 70–72, 74
  refractory period, 72
  regeneration and, 556–558, 560–561
  in squid, 53
  tracts, 11–12
Axon tracts, 30

Babbling, language acquisition and, 600–601
Babinski sign, 386, 387
Baclofen, 157
Balance, 314, 371
Banded krait, 156
Barbiturates, 131
Bard, Phillip, 626–627
Barde, Yves, 514
Barnard, Eric, 85
Baroreceptor afferents, 461
Basal forebrain, 32
Basal forebrain nuclei, 30, 31
Basal ganglia
  amygdala and, 641
  anatomy of, 29–30, 31
  Alzheimer's disease and, 678
  component nuclei and associated structures, 391
  discharges prior to movement, 395–396
  disinhibition circuitry, 396–397, 398, 399, 400–402
  emotional behavior and, 631
  eye movement studies, 397–399
  limbic system and, 632
  modulation of movement and, 391, 394–396, 397–399, 402, 407
  motor control and, 349
  neurons of, 394

non-motor circuits, 406
procedural memory and, 677, 681
projections from, 392, 396–397
projections to, 391–396
subcortical loop formed by, 391
system circuits, 400–405
ventricular space, 32
Basal ganglia syndromes
hemiballismus, 402, 405
Huntington's disease, 400
hypokinetic movement disorders, 405
nigrostriatal dopaminergic neurons and, 402–404
Parkinson's disease, 402–405
simulating with GABA agonists and antagonists, 405
Basal lamina, 497, 504
Basement membrane, 560
Basilar artery, 35, 36, 37
Basilar membrane, 295
auditory nerve tuning and, 287
mechanics of, 280–282
outer hair cells and, 287
sound distortion and, 285
Basket cells, 416, 417
Batrachotoxin, 90
BDNF. See Brain-derived neurotrophic factor
Bed nucleus of the stria terminalis, 658
Beecher, Henry, 217, 219
Behavior
built-in, 520
critical periods and, 519, 520, 521
instinctive, 667
See also Sexual behavior
Behavioral deficits, frontal lobe damage and, 576–577
Behavioral plasticity, 536–539
Békésy, Georg von, 281
Belladonna, 157
Bell pepper, 319
Bellugi, Ursula, 600
Belt areas, 294
Benadryl®, 133
Benperidol, 137
Benzaldehyde, 319
Benzer, Seymour, 540, 608
Benzodiazepines, 131, 137, 160
Berger, Hans, 612
Beta-blockers, 162
Beta waves, 613
Betel nut, 157
Betz cells, 376
γ Bias, 357
bicoid (bcd) gene, 478
Bicuculline, 157, 405
Bifocal lenses, 227
Binocular vision, 273
effect of visual disorders on, 263
field of view, 255–257
stereopsis, 262–263
visual cortex neurons and, 261–263
Biogenic amine neurotransmitters, 137
Biogenic amines, 122
overview of, 131, 133–135
packaging in synaptic vesicles, 126
rate of secretion, 126
Bipolar retinal cells, 228, 229, 237, 243
Bipolar spiral ganglion cells, 288
Birds
auditory localization, 533

imprinting, 520
phylogenetic memory, 667
sexually dimorphic behavior in, 645
song learning, 522–523
Birth defects, related to neural development, 476
Bisexuality, 658
Bitemporal hemianopsia, 260
Bitter compounds, taste responses, 335, 337, 338
Black widow spider venom, 114
Bladder
autonomic regulation of, 462–464
referred pain, 215
visceral motor system and, 448, 449
voluntary control, 464
Blind spot, 251, 254
Bliss, Timothy, 542
Blood-brain barrier, 38, 39
Blood pressure, autonomic regulation of, 462
Blood supply
blood-brain barrier, 38, 39
to brain, 35–37, 39
to spinal cord, 35
strokes and, 33, 39
subarachnoid space and, 34
Bmal1 gene, 608, 609
BMAL1 protein, 508, 609
Bombesin, 122
Bone morphogenetic proteins, 475–476
boss gene, 487
boss-sevenless gene, 502
Botulinum toxin, 108–109, 114
Bowman's gland, 32?
Brachial plexus anesthesia, 216
Brachium conjuctivum. See Superior cerebellar peduncle
Brachium pontis. See Middle cerebellar peduncle
Bradykinin, 211
Brair, 11
anatomical planes, 12, 13
axonal regeneration and, 556, 557–558, 560–561
blood supply to, 35–37, 39
brain weight : body weight ratios, 584
generation of new neurons in, 559
injury and excitotoxicity, 130
number of neurons in, 3
sensory-motor talents and space allocation, 383
sex hormone receptors in, 654–655
sexual dimorphisms in, 651–652, 655, 657–659, 660–661, 663
strokes and, 33
Brain, W. R., 571
Brain anatomy
blood-brain barrier, 38, 39
external, 20–25
imaging technology, 26–27, 28–29
internal, 25, 29–34
major arteries, 35–37, 39
meninges, 33–34
overview of, 39
ventricular system, 31–33
Brain anatomy, external
dorsal and ventral surfaces, 22–23
lateral surface, 21–22
midline saggital surface, 23–25
overview of, 20–21

Brain anatomy, internal
axon tracts, 30–31
cerebral hemispheres, 25, 29–30
diencephalon, 30
meninges, 33–34
ventricular system, 31–33
Brain circuitry modification
activity-dependent mechanisms, 531–533
avian song learning, 522–523
imprinting, 520
language acquisition, 521–524
ocular dominance, 524, 525–526, 528–529, 530–531
Brain-derived neurotrophic factor (BDNF), 511–512, 514
Brain development
formation of major subdivisions, 476–479
genetic abnormalities, 479–480
homeotic genes and, 479, 480
neurogenesis, 480–481, 483
neuronal differentiation, 483–485, 487–488, 489–490
neuronal migration, 488–490
overview of, 491
rhombomeres, 482–483
See also Neural development
Brain/gut peptides, 135
Brain imaging
computer tomography, 26, 27
functional imaging techniques, 28–29
magnetic resonance imaging, 26–27
optical imaging of visual cortex, 266–267
positron emission tomography, 28–29, 599
X-rays, 26
Brain modules, 206
Brain size, intelligence and, 584–585
Brainstem, 11
amygdala and, 634
anatomy of, 13, 15–17, 18, 24
assessing integrity of, 308–309
auditory system and, 288
in balance and posture, 370–375
blood supply to, 37, 39
dorsal column–medial lemniscus pathway and, 202
emotional behavior and, 626, 627, 630–631
local motor neurons and, 350
motor control centers in, 369, 370–375
output to cerebellum, 412, 414
output to spinal cord, 370, 371, 373–375
output to striatal medium spiny neurons, 395
parasympathetic system and, 447, 450, 451
primary motor cortex projections to, 376, 388
reticular activating system of, 615, 617
sensory ganglia, 12
serotonin and, 134
significance to clinical neuroanatomy, 15, 16
sleep and, 615, 617, 618, 619
smiling and, 628
trigeminal somatic sensory pathway and, 203
visceral sensory neurons and, 453–454
See also Medulla; Midbrain; Pons
Brainstem nuclei
Alzheimer's disease and, 678
sound localization and, 288, 290–292

Brainstem tegmentum, visceral motor system and, 454
Brave New World (Huxley), 641
Breedlove, Marc, 652, 659
Brewster, David, 264
Brickner, R. M., 577
Brightman, Milton, 38
Brightness, perception of, 244–245
Broca, Paul, 584, 631
Broca's aphasia, 592
Broca's area, 592
Brodmann, Korbinian, 569
Brodmann area 1, 204, 205
Brodmann area 2, 204, 205
Brodmann area 3a, 204–205
Brodmann area 3b, 204
Brodmann area 4, 348, 375
Brodmann's area 6, 348
Brodmann's area 8. *See* Frontal eye field
Brodmann's area 17. *See* Primary visual cortex
Brodmann's area V1. *See* Primary visual cortex
Bromodeoxyuridine (BrDu), 486
Bronchi, visceral motor system and, 448, 449
Bucy, Paul, 632–633, 636
Bulbocavernosus muscles, 466
Bulbocavernosus, spinal nucleus of, 652–653
α-Bungarotoxin, 152, 156
Bungarus multicinctus, 156
Butyl mercaptan, 321
Byne, William, 657

Cadherins, 495
Caenorhabditis elegans, 498
Caffeine, 161
Cajal, Santiago Ramón y. *See* Ramón y Cajal, Santiago
Calbindin, 173
Calcarine sulcus, 24, 258
Calcineurin, 178
Calcium channels
  diversity in, 83
  functions of, 83
  types of, 172, 173, 174
Ca²⁺/calmodulin-dependent protein kinase
  GTP-binding protein regulation of, 172
  in hair cells, 285, 286
  intracellular, 86
  in phototransduction, 230, 232, 234
  in presynaptic terminals, 107, 108, 109, 110
  structure of, 88
  in taste transduction, 336, 337
  GABA receptors and, 161
  genetic disorders of, 92–93
  type II (CaMKII), 112, 176, 547, 550
  type II calmodulin kinase IV, 176
Calcium ions (Ca²⁺)
  in chemical synapses, 101–102
  co-transmitter release and, 123
  cytosolic concentration, 172
  intracellular ion channels and, 86
  in intracellular signal transduction, 172, 173, 174
  long-term potentiation and, 546–547
  long-term synaptic depression and, 183, 550–552
  in mechanoelectrical sound transduction, 285, 286
  neurotransmitter release and, 107, 109, 111
  NMDA receptors and, 154–155
  in odorant signal transduction, 323, 324
  in phototransduction, 230, 232, 233–234
  in synaptic facilitation, 541
  synaptic vesicle fusion and, 113
  synaptotagmin and, 111
  vestibular hair cells and, 304, 305

Calcium pumps, 91, 172
Calcium response element (CaRE), 179
Calmodulin, 172
CaMKII. *See* Ca²⁺/calmodulin-dependent protein kinase type II
cAMP. *See* Cyclic AMP
cAMP-dependent protein kinase A (PKA), 174, 176, 538
cAMP Response element binding protein (CREB), 179–180, 538
cAMP Response element (CRE), 179, 538
CaMs. *See* Cell adhesion molecules
Canal reuniens, 298
Cancer, myasthenic syndromes and, 108
Cannon, Walter, 444, 625–626
CAPS, 112
Capsaicin, 341, 342, 343
Carbamazepine, 555
Carbidopa, 133
Cardiac plexus, 461
Cardiovascular system, autonomic control of, 460–462
CaRE. *See* Calcium response element
Carotid body, 461
Cartesian dualism, 616
Carvone, 319
Cataracts, 530
Caudal (position), 12, 13
Caudate nucleus, 29, 31, 391, 392, 410
  anticipatory discharges, 395
  in basal ganglia circuitry, 394, 396, 399, 401
  cortical input to, 392–393, 395
  in disinhibition, 396, 399
  in Huntington's disease, 400
  in hypokinetic and hyperkinetic diseases, 400, 403, 404

Cell adhesion molecules (CAMs), 495
Cell-associated signaling molecules, 167–168
Cell-attached patch clamping, 78
Cell-impermeant signaling molecules, 167, 168
Cell-permeant signaling molecules, 167, 168
Cellular receptors, in intracellular signal transduction, 168–170, 185
Center-surround mechanisms
  detection of differences in luminance, 242–243, 245–246
  in light adaptation, 247–249

Central lateral nucleus, 216
Central nervous system
  anatomy of (*see* Neuroanatomy)
  axonal regeneration and, 556, 557–558, 560–561
  components of, 11
  neurotrophins and, 514
  sexual dimorphism and, 652–653, 655–660
  subdivisions of, 13, 14, 18
Central pattern generator
  evidence of autonomy in, 364–365
  in leeches and lampreys, 362
  in locomotion, 366
Central sulcus, 21, 23, 24
Cephalic flexure, 478
Cereal hairs, 195
Cerci, 195
Cerebellar ataxia, 418, 419, 423
Cerebellar cortex, 25, 410
  effects of alcohol abuse on, 422
  excitatory and inhibitory connections, 416, 417
  inputs to, 412
  layers of, 415, 416
  neurons of, 416
  outputs from, 414
Cerebellar lesions, consequences of, 419, 422, 425
Cerebellar peduncles, 410, 411–412
Cerebellar vermis, 314
Cerebellum, 11, 13, 14, 23, 24, 25
  circuits of, 415–417, 419
  coordination and correction of movement, 420–421, 425
  genetic analysis of function, 423–424
  lesions and, 419, 422, 425
  long-term synaptic depression in, 551–552
  major components of, 410, 411
  major pathways of, 411–412
  motor control and, 348–349
  motor learning and, 409, 417, 419, 420–421, 425
  neuronal migration and differentiation in, 490
  neurons of, 416
  organization of, 409–412
  primary function of, 409
  prion diseases, 418–419
  procedural memory and, 677, 681
  projections from, 414–415
  projections to, 412–414
  remodeling of synaptic connections through time, 508
  somatotopic mapping and, 413–414
  ventricular space, 32
Cerebral achromatopsia, 272
Cerebral aqueduct, 24, 32
Cerebral arteries, 35, 36, 37, 39
Cerebral cortex, 11–12
  aging and, 680
  amygdala and, 634
  anatomy of, 25, 29
  association cortices and, 568, 570, 585
  cellular organization, 4
  corticospinal tract and, 377
  cytoarchitectonic areas, 566–567, 569
  EEG rhythms and, 613
  epilepsy and, 554–555
  extrastriate visual areas, 270–272, 273
  input from cerebellum, 414

laminar organization, 481, 483, 486, 566, 568
long-term memory and, 676–677, 680
modularity in, 205
neuronal migration in, 488
output to basal ganglia, 392–395
output to cerebellum, 412–413
pain perception and, 216–217
patterns of connectivity in, 567
plasticity in, 553–556
sleep and, 615, 617, 618–619
structure of, 20
taste processing and, 340, 341
V4 area, 270
ventricular space, 32
vestibular system inputs, 315
visceral motor system and, 454
visual processing, 251
in voluntary bladder control. 464
in wakefulness and sleep, 615, 617, 618–619
Cerebral hemispheres, 13
anatomical differences between, 595, 597
external anatomy, 21–22, 23–24
internal anatomy, 25, 29–30
lateralization of emotion processing, 642, 644
lateralization of language function, 587, 589, 593–595, 597–598
sexual dimorphism and, 660
during sleep in dolphins and seals, 605
Cerebral peduncles, 15, 22, 23, 377
Cerebrocerebellum, 409, 410
closed loops to the cortex, 415
consequences of lesions to, 425
projections from, 415
visual coordination of movement, 413
Cerebrospinal fluid, 32–33
Cerebrum, 11, 14
descending pain modulatory systems, 218
vestibular thalamocortical system and. 314
Cervical enlargement, 14, 19
Cervical nerves, 14, 18
Cervical spinal cord, 17, 18
Cesium chloride, 335
C fibers, 135–136
in pain perception, 211
c-fos, 181
cGMP. See Cyclic GMP
cGMP-dependent protein kinase, 174
Channel-linked receptors, 168, 169. See also Ligand-gated ion channels
Channelopathies, 92–93
Characteristic frequency, 287
Charybdotoxin, 90
Chemical signaling
signal amplification, 166, 167
strategies of, 165–166
types of, 165
See also Intracellular signal transduction
Chemical synapses, 5–6, 100
neuromuscular, 103–105
overview of, 99
short-term plasticity, 539, 541
structure of, 101
transmission at, 101–105
Chemoaffinity hypothesis, 500–502
Chemoattraction, in axonal growth, 498–499, 500

Chemorepulsion, in axonal growth, 499, 500
Chest wall, cortical representation in rats, 662, 663
Children, language acquisition, 521–524
Chimpanzees, language and, 588–589
Chlamydia trachomatis, 530
Chlordiazepoxide, 137, 160
Chloride channels
diversity in, 83
structure of, 88
Chloride (CL⁻) ions, GABA receptors and, 159–160
Chlorpromazine, 137
Cholecystokinin octapeptide, 122
Cholesterol, sex hormone synthesis from, 650
Choline, 127–128
Choline acetyltransferase, 127
Cholinergic nerves/nuclei
acetylcholine metabolism in, 128
inputs to association cortices, 570
in wakefulness and sleep, 617, 619
Cholinergic receptors
acetylcholinesterase and, 128
muscarinic, 153, 157, 169, 172, 455, 459–460
myasthenia gravis and, 158
neurotoxins and, 152, 156, 157
nicotinic, 151–153, 157, 454–455, 459
parasympathetic system and, 454–455, 459–460
structure of, 152–153
Chomsky, Noam, 591
Chondodendron tomentosum, 157
Chorda tympani, 339, 340
Chordin, 475
Choreiform movement, 400
Choroid, retinal, 223, 224, 238
Choroid plexus, 32
Chromatic aberration, 225
Chronic pain, 215, 216
Ciliary body, 223
Ciliary ganglion, 253
Ciliary muscle, 224, 225
Ciliary processes, 223
Cingulate cortex, 632
Cingulate gyrus, 24, 464, 631, 632, 633
Cingulate sulcus, 24
Circadian clock, 622
brain pathways in, 607, 609
under "free running" conditions, 606
molecular mechanisms of, 608–609
photoentrainment, 606–607
Circle of Willis, 35, 36, 37
Circuits. See Brain circuitry modification; Neural circuits
Circumvallate papillae, 333, 334
Cisterns, 34
Clasp-knife phenomenon, 387
Clathrin, 111, 112, 113
CLL 112
Climbing fibers, 415–416, 419, 551
Clitoris, 465–466, 647, 652
Clk gene, 608, 609
CLOCK protein, 608, 609
Clonus, 386, 387
Clostridial toxins, 108–109, 114
Clostridium, 108–109
CL1 protein, 114
Coated vesicles, 106

Cocaine
addiction and, 138
dopamine and, 133
Coccygeal nerve, 14, 18
Cochlea, 276–277, 278, 295, 298
function of, 279, 280–282
labyrinth and, 297
mechanoelectrical sound transduction, 282–286
otoacoustical emissions, 287
structure of, 279–280
topographical map of, 294
types of hair cells in, 286–287
Cochlear hair cells, 295
acquired hearing loss and, 277
auditory nerve phase locking and, 287
mechanoelectrical sound transduction, 282–286
sound distortion and, 285
structure of, 282–283
types of, 286–287
Cochlear implants, 287
Cochlear nerve, 278
Cochlear nucleus
in auditory pathways, 277, 288, 295
monoaural pathway to the lateral lemniscus, 292
in sound localization, 288, 290, 291, 292
Cognition
association cortices and, 565–566
defined, 565
neuropsychological testing and, 583
sexual dimorphism in, 660–661
Cohen, Stanley, 511
Coincidence detectors, 291
Cole, Kenneth, 58
Collagens, 495
Colliculospinal tract, 371
Color blindness, 241, 242, 272
Color constancy, 240
Color contrast, 240
Color vision
cerebral achromatopsia, 272
cone system and, 239, 241–242
importance of context in, 240
P retinal ganglion cells and, 268–269
Combination tones, 285
Communication, temporal structure of sound and, 294–295
Complex cells. 261
Complexin, 111, 112
Computers, consciousness and, 616
Computer tomography, 26, 27
Concha. 278
Conditioned fear response, 635–637, 639
Conduction aphasia, 592
Conduction velocity, action potentials, 70
Conductive hearing loss, 279, 295
Cone cells/system, 228, 229
anatomical distribution, 237–238
color vision, 239, 241–242
connectivity to ganglion cells, 237
M and P retinal ganglion cells, 268–269
photopigment in, 232
responsiveness to luminance, 235, 237
retinitis pigmentosa, 231
sensitivity of phototransduction in, 235
spatial resolution of, 234–235
See also Photoreceptors
Cone snail toxins, 157

Congenital adrenal hyperplasia, 648
Congenital disorders, related to neural development, 476
Congenital stationary night blindness, 93
Conjugate eye movements, 432
Conotoxins, 157
Consciousness, 616
Constrictor muscle, 253
Contact lenses, 227
Contralateral neglect syndrome, 571–573
Convergence, in neural circuit construction, 660
Copper/zinc superoxide dismutase, 367
Cornea, 223, 224–225, 225
Cornu Ammon, 542
Coronal sections, 12, 13
Corpus callosum, 22, 24, 25, 30, 31, 568, 633, 660
Corpus striatum
anticipatory discharges, 395–396
in basal ganglia disinhibitory circuitry, 396–397, 398, 401
cortical inputs, 392–395
dopamine and, 133
functional subdivision by inputs, 393–395
medium spiny neurons in, 391–392, 395
rostrocaudal bands, 394–395
subdivisions of, 391
Cortical blindness, 526
Cortical motor maps
creation of, 379–381
information content of, 380
space allocation and sensory motor talents, 383
Cortical pyramidal neurons, 2, 394. *See also* Pyramidal neurons
Corticobulbar tract, 376, 378
Corticocortical connections, 568, 570
Corticoreticulospinal tract, 373, 374
Corticospinal tracts, 20, 373–375, 376, 377
Corticostriatal pathway, 392–393
Co-transmitters, 122–123, 135
Cranial cavity, 33
Cranial motor nerves, 15, 18
Cranial nerve I, 18–19. *See also* Olfactory nerve
Cranial nerve II, 18–19. *See also* Optic nerve
Cranial nerve III, 15, 18–19. *See also* Oculomotor nerve
Cranial nerve IV, 15, 16, 18–19, 429, 431. *See also* Trochlear nerve
Cranial nerve V, 15, 18–19, 341. *See also* Trigeminal ganglion; Trigeminal nerve
Cranial nerve VI, 15, 18–19, 428, 429, 431. *See also* Abducens nerve
Cranial nerve VII, 15, 18–19, 331, 334. *See also* Facial nerve
Cranial nerve VIII, 15, 18–19, 288, 310
labyrinth and, 298
output to the cerebellum, 413
See also Auditory nerve; Vestibular nerve
Cranial nerve IX, 15, 18–19
taste system and, 331, 334, 340
See also Glossopharyngeal nerve
Cranial nerve X, 15, 18–19
taste system and, 331, 334, 340
trigeminal chemosensory system and, 341

See also Vagus nerve
Cranial nerve XI, 15, 18–19. *See also* Accessory nerve
Cranial nerve XII, 15, 18–19. *See also* Hypoglossal nerve
Cranial nerve ganglia, 12
Cranial nerve nuclei, 13, 15–16, 18
Cranial nerves
development of, 482–483
innervation of extraocular muscles, 428, 429, 431
motor deficits and, 378
taste system and, 331–332
visceral afferent fibers and, 453
See also individual nerves
Cranial sensory ganglia, 13
Cranial sensory nerves, 13, 15
CRE. *See* cAMP Response element
CREB. *See* cAMP Response element binding protein
Cremasteric reflex, 386, 388
Cresyl violet acetate, 569
Creutzfeldt-Jakob disease, 418
Cribriform plate, 318, 321, 326
Crick, Francis, 614, 616
Crickets, 195
Crista, 306
Critical periods
activity-dependent mechanisms in, 531–532
auditory localization, 533
in avian song learning, 522–523
imprinting, 520
language acquisition and development, 524–526, 528–529, 530–531, 534
overview of, 519, 521, 533–534
response to odorants, 520, 533
somatic sensory mapping, 533
in visual development, 521–524
Crossed extension reflex, 361
Cross-eyedness, 530
CRY protein, 608, 609
cryptochrome gene, 608, 609
Cuneate nucleus, 200, 202
Cuneate tract, 200, 202
Cupula, 306–307
Curare, 104, 157
Curran, Tom, 423
Curtis, David, 130–131
Cutaneous sensory receptors, 189–190, 192–193
Cyclic AMP (cAMP), 173, 174
Cyclic AMP-mediated second messenger pathways
in learning and memory, 540
in odorant signal transduction, 323, 324
overview of, 174
in short-term sensitization, 538
Cyclic GMP (cGMP), 173, 174
in phototransduction, 232–233, 234
Cyclic nucleotides, 173, 174
Cyclooxygenase, 211
Cysteine string protein, 112

2,4-D. *See* 2,4-Dichlorophenoxyacetic acid
DAG. *See* Diacylglycerol
Damasio, Antonio, 643–644
Damasio, Hanna, 599

dance gene, 540
Darwin, Charles, 627
Deadly night shade, 157
Deaf children, language acquisition in, 600–601
Decerebrate rigidity, 314, 386, 388
Decibels, 275
Decision making, emotion and, 643–644
Declarative memory, 680–681
brain systems underlying, 673, 676
clinical cases of amnesia in, 674–675
defined, 665
Decussation, 202
pyramidal decussation, 374, 377
Deep cerebellar nuclei, 409, 410
components of, 414–415
coordination and correction of movement, 419–420, 422
excitatory and inhibitory connections, 417
motor learning and, 421
projections from, 414, 416
Deep tendon reflex. *See* Muscle stretch reflex
Deep cerebellar nuclei, 409, 410, 414, 417, 419
Purkinje cells and, 414, 417, 419
Delayed response genes, 181
Delayed response task, 581, 582
Delay-specific neurons, 581, 582
Delta waves, 610, 613
Dementia, Alzheimer's disease, 678–679
Dempsey, Edward, 613
Demyelination, 75
Dendrites, 2, 3
impact on neuronal innervation, 509
Dendrotoxin, 90
Dentate gyrus, 559
Dentate nucleus, 410, 414–415
Deoxyhemoglobin/oxyhemoglobin ratios, 266
Deoxyribonucleic acid (DNA)
promoters, 178
transcriptional activator proteins, 178–181
Depolarization, 44, 66
Depression, 640, 641
electroconvulsive therapy, 676
Dermatomes, 201
Desipramine, 137
Deuteranomalous trichromacy, 241
Deuteranopia, 241
Diacylglycerol (DAG), 173, 174–175, 176
in long-term synaptic depression, 183
in taste transduction, 338
Diamond, Milton, 649
Diazepam, 137, 160
2,4-Dichlorophenoxyacetic acid (2,4-D), 118
Dichromacy, 241, 242
Dieldrin, 157
Diencephalon, 13, 14, 479
anatomy of, 24, 25
descending pain modulatory systems, 218
memory and, 673, 676
retinal ganglion projections to, 252
ventricular space, 32
Digenea simplex, 157
5-α-Dihydrotestosterone, 650, 654
Dihydroxyphenylalanine (DOPA), 131, 133
Dilantin®, 555
Dimethylsulfide, 319
Dinoflagellates, 90

Diphenhydramine, 133
Diphenyl trichloroethane (DTT), 128
Diplopia, 262
Disconjugate eye movements, 432
Disinhibition, 397
 basal ganglia circuitry in, 396–397, 398, 399, 400–402
Disjunctive eye movements, 432
Dissociated sensory loss, 212
Divergence, in neural circuit construction, 508
DNA. *See* Deoxyribonucleic acid
DOC2, 112
Dolphins, sleep in, 605
DOPA. *See* Dihydroxyphenylalanine
DOPA carboxylase, 133
Dopamine, 120, 131
 chemical structure, 121
 distribution in brain, 134
 drug addiction and, 138
 G-protein-coupled receptors and, 171
 overview of, 133
 Parkinson's disease and, 403
Dopamine β-hydroxylase, 133
Dopamine receptors, 137, 162
Dopaminergic neurons, 133
 Parkinson's disease and, 402–404
Dopaminergic nuclei, 570
Dorsal cochlear nucleus, 289
Dorsal column-medial lemniscus system, 199–202
Dorsal column nuclei, 202, 374
Dorsal columns (spinal cord), 20, 202
Dorsal horn (spinal cord), 19, 20
 nociceptive axons in, 212
Dorsalin, 475
Dorsal lateral geniculate nucleus, 252
Dorsal motor nucleus of vagus, 451, 452, 461
Dorsal nucleus of Clarke, 412, 413
Dorsal (position), 12, 13
Dorsal root ganglia, 12
 mechanosensory pathway, 199, 201–202
 nociceptive axons, 212
 visceral afferent fibers and, 453
Dorsal roots (spinal cord), 18, 20, 199
Dorsal thalamus, 24, 25, 397
 limbic system and, 632
 ventricular space, 32
 See also Ventral anterior nucleus; Ventral lateral nucleus
Dorsolateral tract of Lissauer, 212
Dorsomedial nucleus, 456, 457, 654
Double vision, 262
Downer, John, 634
Down's syndrome, 678
Dreams, 614–615, 617
Drosophila
 eye development in, 487
 genetic basis of learning and memory, 540
 homeobox genes, 479, 480
 patterned gene expression in, 478, 479, 480
Drug addiction, 138
Drugs
 tolerance, 138
 withdrawal syndromes, 138
DTT. *See* Diphenyl trichloroethane
Dualism, 616
Duchenne de Boulogne, G.-B., 628
Duchenne smile, 628

Dura mater, 33, 34
Dutchman's breeches, 157
Dynamin, 111, 112, 113
Dynorphins, 136, 139, 220–221
Dysarthria, 592
Dysdiadochokinesia, 425
Dysmetria, 425

Ear
 external, 278–279
 inner, 279–286
 middle, 279
 See also Auditory system
Eardrum. *See* Tympanic membrane
Eccles, John, 616
Ectoderm, 471
Ecto-5'nucleotidase, 135
Edelman, Gerald, 616
Edinger-Westphal nucleus, 16, 253, 255, 431, 447, 451
EE cells, 294
Effector enzymes, 171–172, 185
Efferent neurons, 8–9
Ehrlich, Paul, 38
EI cells, 294
Ejaculation, 466
Electrical signals
 ion movements in, 45–51
 overview of, 43–45
Electrical synapses, 99–101
Electrochemical equilibrium
 basic dynamics of, 46–47
 with more than one permeant ion, 49–51
 Nernst equation, 47–49
Electroconvulsive therapy, 676
Electroencephalography
 measurements during sleep, 610
 principle waves, 612–613
Electrophysiological studies
 of monkey association cortices, 577
 myotatic spinal reflex, 9–10
Embolic strokes, 33
Emmetropia, 226, 227
Emotional facial paresis, 629
Emotion/Emotional behavior
 affective disorders, 640–641
 amygdala and, 635–639, 641–642
 associating sensory experience with, 635–639
 facial expressions, 626, 628–629
 hemispheric lateralization, 642
 limbic system and, 631–633
 neural systems involved in, 625, 626–627, 630–631, 639, 641–642
 overview of, 625, 644
 physiological changes associated with, 625–626
 rational thought and, 643–644
 sham rage, 627
 visceral motor system and, 454, 625–626
Encapsulated sensory receptors, 189, 192–193
Endocrine signaling, 165, 166
Endocytosis, 106
Endoderm, 471
Endogenous behavior, 667
Endogenous opioids, 220–221
Endolymph
 cochlear, 286
 vestibular, 297–298
Endolymphatic duct, 298

Endorphins, 135, 136, 220–221
 amino acid sequences, 122, 221
Endosomes, 106
End plate current (EPC), 142, 143–147
End plate potentials (EPPs), 103–105
 dynamics of, 142, 143, 146, 147
 myasthenia gravis and, 158
End plates, 103
Engrams, 666, 668
Enkephalins, 136, 220–221
Enophthalmos, 458
Ensemble hypothesis, 340, 341
Enteric nervous system, 12, 450, 452–453, 460
Enteric plexuses, 450, 452, 453
Entorhinal complex, 318
Entorhinal cortex, 329
Enzyme-linked receptors, 168–169
Eph ligands, 502
Eph receptors, 502
Epididymis, 646
Epiglottis, taste buds, 334
Epilepsy, 93, 554–555
Epinephrine, 120, 131
 adrenergic receptors and, 162
 chemical structure, 121
 overview of, 133
Episodic ataxias, 93
EPPs. *See* End plate potentials
Equilibrium potential, 47–49
Erabutoxin, 156
Erection, penile or clitoral, 465–466
ERKs. *See* Extracellular signal-regulated kinases
Esophagus
 referred pain, 215
 taste buds, 334
Esotropia, 530
Estradiol, 650–651
 receptors, 554
Estrogen, 654, 655
 α-fetoprotein and, 651
 in sexual development, 647, 650, 651
Estrogen receptors, 651, 654, 655
Etcoff, N. L., 574
Ethacrynic acid, 277, 286
Ethanol, 319, 343
Ethyl acetate, 319
Ethyl mercaptan, 321
Eustachian tube, 278
Evarts, Ed, 381
Excitatory postsynaptic potentials (EPSPs)
 dynamics of, 147–148, 149
 long-term depression, 550
 long-term potentiation, 543
 summation of, 149–150
Excitotoxicity, 130
Exotropia, 530
Expressive aphasia, 592
Express saccades, 437
External auditory meatus, 278
External segment, globus pallidus, 401, 402, 405
Extracellular matrix, axonal growth and, 495, 497–498
Extracellular matrix adhesion molecules, 495, 497–498
Extracellular signal-regulated kinases (ERKs), 177–78
Extrafusal muscle fibers, 350

Extraocular muscles
actions of, 428, 429
innervation of, 428, 429, 431
Extrapyramidal pathways, 628
Eye
anatomy of, 223–224
formation of retinal images, 224–225, 227
Horner's syndrome and, 458
light-mediated growth, 226–227
muscular control, 353
near reflex triad, 432
pupillary light reflex, 252–253, 255
retina structure and function, 227–230
visceral motor system and, 448, 449
See also entries at Visual
Eye development, 228
in *Drosophila*, 487
Eyelids, 431
Eye movements
basal ganglia and, 397–399
brainstem motor control centers and, 369
cerebellar contributions, 420–421
characteristics and significance of, 427–428
conjugate and disconjugate, 432
extraocular muscles, 428–429, 431
functions of, 431–433
neural control of, 433–441
reflex responses to head rotation, 308–309, 316
in REM sleep, 615
saccades, 398, 399, 420, 431, 433–439, 615
smooth pursuit movements, 431–432, 440
vergence movements, 432, 440
vestibulo-ocular reflex, 310–312, 420–421, 422, 432–433

Face
diagnosing motor deficits, 378
pain and temperature pathways, 212, 213, 214
trigeminal mechanosensory pathway, 202–203
Face recognition, deficits in, 574–575
Facial expressions, 626, 628–629
asymmetry in, 642, 643
Facial nerve, 15, 16
characteristics of, 18–19
parasympathetic system and, 451
taste neural coding and, 339, 340
taste system and, 331, 334
Facial nucleus, 378
Facial paresis, 629
Familial ALS, 367
Familial epilepsy, 554–555
Familial hemiplegic migraine, 92
Familial infantile myasthenia, 108
Faradization, 628
Far cells, 263
Farsightedness, 226
Fast axonal transport, 126
Fast fatigable motor units, 352, 353
Fast fatigue-resistant motor units, 352, 353
Fastigial nucleus, 410, 415
Fatt, Paul, 103
Fear, 635–639
Feedback cycles
in action potentials, 65
in postural control, 373

Feedforward mechanisms, in postural control, 372–373
Female sexual development, 646, 647
Fentanyl, 136
α-Fetoprotein, 651
Fibroblast growth factors, 475, 487–488
Fibronectin, 495
Field, Pauline, 651
Filopodia, 494
First-order neurons, mechanosensory pathway, 199, 201–202
First pain, 210–211
Fish, Mauthner cells, 312–313
Fisher, C. Miller, 33
Fissures, 20
Flexion reflex, 361
Flocculus, 409, 410
Floorplate, 487
Floor plate, 473
Fluoxetine, 137, 641
Folia, 25
Folia, diagnosing motor deficits, 378
Foliate papillae, 333, 334
Folic acid, 476
Forebrain, 13
Forgetting. *See* Functional magnetic resonance imaging
Alzheimer's disease and, 678
anatomy of, 24
basal nuclei, 30, 31, 678
development of, 478, 479
emotional behavior and, 626, 627, 631
olfactory processing and, 317
ventricular space, 32
Forgetting, 671–673. *See also* Amnesia
Fornix, 24, 30, 31, 632
Fourth ventricle, 24, 32, 33
Fovea, 224, 238, 255–256, 258
Foveola, 238
Fragile X syndrome, 479
Freeman, Walter, 579
Free nerve endings, 189, 190
Free sensory receptors, 189
Frequency, of sound waves, 275
Freud, Sigmund, 614
Frisch, Karl von, 588
Fritsch, G. Theodor, 376
Frontal association cortex
language function and, 587
planning and behavioral deficits, 576–577
thalamic connectivity to, 568
Frontal cortex
delay-specific neurons in monkeys, 581, 582

memory and, 677
projections to basal ganglia, 393
taste processing and, 331, 332
Frontal eye field, 435–439, 440–441
Frontal lobe, 23–24
anatomy of, 21–22
language function and, 590, 592
neuropsychological testing, 587, 589
planning and behavioral deficits, 576–577
somatic sensory input, 207
in voluntary bladder control, 464
Frontal lobotomies, 577, 579
Frontal operculum, 331, 332
Frontal (position), 12
Functional brain imaging, 28–29
Functional magnetic resonance imaging
(fMRI), 29, 599
Fungiform papillae, 333, 334

Fused tetanus, 354
Fusiform gyrus, 575

GABA. *See* Gamma-aminobutyric acid
GABA aminotransferase, 131
GABAergic neurons, 405
in inhibitory basal ganglia circuits, 396–397
GABA receptors
in wakefulness and sleep, 613, 619
inhibitory postsynaptic potentials and, 148–149
neurotoxins and, 157
properties of, 159–161
Gage, Phineas, 576
Gain, stretch reflex and, 357, 358, 386
Gajdusek, Carlton, 418, 419
Gamma-aminobutyric acid (GABA), 120
chemical structure, 121
functional features of, 127
synthesis, release, and reuptake, 130–131, 131
Ganglia, 11, 12
Ganglion cells, retinal. *See* Retinal ganglion cells
Gap junctions
in electrical synapses, 99
structure and function of, 100, 101
GAPs. *See* GTPase-activating proteins
Gardner, Howard, 592, 593
Gaskell, Walter, 443–444
Gastrocnemius, 353, 372
Gastrointestinal tract, enteric nervous system, 450, 452–453
Gastrulation, 471
Gate theory of pain, 220
Gaze centers, 433–434, 435–436, 440
Gazzaniga, Michael, 594
Gender, 648
Gender identity, 659
Gene expression
long-term potentiation and, 547
long-term sensitization and, 538
modulated in neural development, 473, 475–476
retinoic acid and, 474–475
transcriptional activator proteins, 178–181
Generalized anxiety disorder, 137
Generalized epilepsy with febrile seizures, 93
Gene therapy, 403
Genetic disorders
affecting ion channels, 92–93
in brain development, 479–480
Geniculocortical axons, 524, 528
Genitalia, 646–647
Genotypic sex, 647, 648
Gentamycin, 277
Geosmin, 319
Germ layers, 471
Geschwind, Norman, 593, 595
Giant neurons, squid, 53
Gill withdrawal reflex, 535–539
Glaucoma, 224
Glial cells
axons tracts and, 12
differentiation of, 480–481, 483
generation of diverse cell types, 483–485, 487–488
glutamate-glutamine cycle and, 129

in neuronal migration, 488–489
types of, 7–8
Globus pallidus, 29, 31, 391
basal ganglia circuitry and, 394, 396, 398, 401
basal ganglia output to, 392
external segment of, 401, 402, 405
Huntington's disease and, 405
in hypokinetic and hyperkinetic diseases, 404
inhibitory projections from medium spiny neurons, 396–397
internal division of, 397–398, 401, 402
projections from, 397
Glomeruli, 326, 327
in hawk moths, 322
responsiveness to odorants, 329, 330
Glossopharyngeal nerve, 15, 452
characteristics of, 18–19
control of cardiovascular function and, 461
nucleus of the solitary tract and, 453
parasympathetic system and, 451
taste system and, 331, 334, 340
trigeminal chemosensory system and, 341
Glucocorticoid hormones, nuclear receptors, 180–181
Glutamate, 119, 120, 131
chemical structure, 121
excitotoxicity and, 130
functional features of, 127
G-protein-coupled receptors and, 171
long-term potentiation and, 546, 547
long-term synaptic depression and, 551, 552
silent synapses and, 548
synthesis and removal, 129
taste perception of, 338
Glutamate-glutamine cycle, 129
Glutamate receptors
excitatory postsynaptic potentials and, 148
long-term potentiation and, 545–547
metabotropic, 157, 159
neurotoxins and, 157
properties of, 15–157
retinal ganglion cells and, 243, 245
silent synapses, 548–549
types of, 153–154
Glutamatergic synapses, silent, 548–549
Glutamic acid decarboxylase, 131
Glutaminase, 129
Glutamine, 129
Glutamine synthetase, 129
Glycine, 119, 120
chemical structure, 121
functional features of, 127
as neurotransmitter, 130, 131
NMDA receptors and, 154–155
synthesis of, 132
Glycine receptors, neurotoxins and, 157
Goldbager, D., 678
Goldman equation, 50
Goldmann, Edwin, 38
Golgi, Camillo, 1–2
Golgi cells, 416, 417
Golgi tendon organs, 190, 358–361
in proprioception, 198, 199
Gonads, 646, 647, 650
Gorski, Roger, 651, 655, 657

G-protein-coupled receptors, 169
catecholaminergic, 162
overview of, 151
peptide receptors, 162
purinergic, 161
in taste transduction, 336, 337, 338
See also Metabotropic receptors
G-proteins. *See* GTP-binding proteins
Gracile nucleus, 200, 202
Gracile tract, 200, 202
Grammar, 591
Granisetron, 161
Granule cells
cerebellar, 415, 416, 425
olfactory, 326, 329, 559
Graybiel, Ann, 393–394
Gray communicating rami, 446, 447
Gray matter, 12
spinal cord, 19
See also Periaqueductal gray matter
Great apes, language and, 588–589
Greig cephalopolysyndactyly, 480
Growth cones
morphological characteristics, 493–494
non-diffusible signals and, 494–495, 497–498
shape changes, 494
GTPase-activating proteins (GAPs), 171
GTP-binding proteins
activation of ion channels, 172
classes of, 170–171
effector enzyme targets, 171–172
in G-protein-coupled receptors, 169
metabotropic receptors and, 151
termination of signaling by, 171
Guanylyl cyclase, 171, 174, 175
Gurdon, John, 85
Gustatory cortex, 331, 332
Gustatory nucleus, 332
Gustatory selectivity, 338
Gustatory system. *See* Taste system
Gustducin, 338
Gut, enteric nervous system, 450, 452–453
Gyri, 20

Hair cells. *See* Cochlear hair cells; Vestibular hair cells
hairy (*h*) gene, 478
Hall, Jeffrey, 608
Haloperidol, 137, 406
Hamburger, Viktor, 510
Hand, mechanoreceptors in, 192–193
Handedness, 596–597
Hanig, Deiter, 333
Haptics, 199
Harlow, Harry, 520
Harris, Bill, 540
Harrison, Ross G., 493
Hawk moths, 322
Hctr2 gene, 622
Head, visceral motor system and, 448
Head movements/position
brainstem motor control centers and, 369
detection by otolith organs and vestibular nerve, 300–303, 306
detection of angular acceleration, 306–307, 310
proprioception and, 198
translational and rotational planes of, 300
vestibulospinal pathway, 313–314

Hearing. *See* Auditory system
Hearing loss
acquired, 277
conductive, 279, 295
monoaural, 288
sensorineural, 279
Heart
autonomic control of, 460–462
referred pain, 215
sympathetic and parasympathetic innervation, 462
visceral motor system and, 448, 449
Heat-activated ion channels, 86
Hebb, D. O., 531
Hebb's postulate, 531–532, 543
Helicotrema, 280
Helmholtz, Herman von, 254
Hemiballismus, 402, 405
Hemispheric lateralization
of emotional processing, 642
of language function, 587, 589, 593–595, 597–598
sexual dimorphism and, 660
Hemorrhagic strokes, 33
Henbane, 157
Henneman, Elwood, 353
Heroin, 138
Hertz, 275
Hess, Walter, 615, 627
Heteronomous hemianopsia, 260
Heterotrimeric G-proteins, 170
Heuser, John, 105, 106
Hikosaka, Okihide, 397
Hindbrain, 478. *See also* Rhombencephalon
Hippocampal gyrus, declarative memory and, 674
Hippocampus, 22, 30, 31, 661
cortex of, 29
laminar structure, 569
limbic system and, 631
long-term synaptic depression in, 550–551
long-term synaptic potentiation in, 542–543
memory and, 673, 674, 676, 677, 681
neural stem cells and, 559
olfactory processing and, 318, 330
somatic sensory input, 207
Substance P and, 135
ventricular space, 32
His, Wilhelm, 488
Histamine, 120, 131
chemical structure, 121
distribution in brain, 134
functional features of, 127
overview of, 133–134
release during tissue damage, 211
synthesis of, 133, 135
Histamine methyltransferase, 133
Histidine, 135
Histidine decarboxylase, 133
Hitzig, Eduard, 376
Hodgkin, Alan, 51, 52, 54, 59, 60–65, 77, 79, 80
Hofman, Michel, 659
Hofman, Michel, 480
Homeobox genes, 479, 480, 482, 483
Homeobox, 480
Homonymous quadrantanopsia, 259
Homosexuality, 648, 658–659
Homunculus, 205
Horizontal cells, 228–229
in light adaptation, 247–248

Horizontal gaze center. *See* Paramedian pontine reticular formation

Horizontal (position), 12, 13

Hormone responsive elements, 655

Hormones
neurotransmitters and, 118, 120
See also individual hormones; Sex hormones

Hormone-secreting neurons, 101

Horner's syndrome, 458

Horseradish peroxidase (HRP), 106

Horseradish peroxidase (HRP), 479, 482, 483

Hox genes, 475, 479, 482, 483

Huxley, Andrew, 59, 60–65, 77, 79, 80

Hydrocephalus, 479

Hydrogen cyanide, 321

5-Hydroxytryptamine. *See* Serotonin

5-Hydroxytryptophan, 135

Hyperalgesia, 211

Hyperglycinemia, 131

Hypermetric saccades, 420

Hyperopia, 387

Hyperpolarization, 44

Hypertonia, 387

Hypocretins, 622

Hypoglossal nerve, 15, 18–19

Hypoglossal nucleus, 378

Hypokinetic movement disorders, 404–405.
See also Parkinson's disease

Hypometric saccades, 420

Hypothalamic sulcus, 456

Hypothalamus, 22, 24, 25
amygdala and, 634, 637
circadian clock and, 607, 609
control of visceral motor system, 454, 456–457
electrical synapses in, 101
emotional behavior and, 626–627, 631
histamine and, 133
limbic system and, 632
location, 456
median eminence of, 456
modification in adults by sex hormones, 663
narcolepsy and, 622
nuclei of, 456–457
olfactory processing and, 317, 318, 330
rate of neurotransmitter release, 126
retinal ganglion projections to, 255
sex hormone receptors in, 654
sexual behavior and, 655, 656, 658, 659
sexual dimorphism in, 651–652, 655, 657–658, 663
suprachiasmatic nucleus, 255, 456, 457, 608, 609, 658, 659
taste processing and, 331, 332
ventricular space, 32
visual system and, 253
wakefulness and, 617, 619

Hypotonia, 386, 387

Hudspeth, A. J., 285

Huntingtin gene, 400

Huntington, George, 400

Huntington's disease, 349, 400, 405

Huxley, Aldous, 641

Hubel, David, 206, 260–261, 524, 525–526, 528–529

5-HT receptors. *See* Serotonin receptors

5-HT. *See* Serotonin

HRP. *See* Horseradish peroxidase

4-Hydroxyoctanoic acid lactone, 319

Ia afferents, 355
muscle spindles and, 198
muscle tone and, 387

Ib afferents, Golgi tendons and, 199

Ib inhibitory interneurons, 360

Ib sensory axons, 359, 360

Ibotenic acid, 157

Idiots savants, 670

Imaging technology. *See* Brain imaging

Immediate early gene, 181

Immediate memory, 666

Immunoglobulins, olfactory receptor neurons and, 323

Impedance, 279

Imprinting, 520

INAH. *See* Interstitial nuclei of the anterior hypothalamus

Incus, 278

Inderol®, 162

Inductive signals, in neural development, 472, 473, 475–476

Infants
imprinting and built-in behaviors, 520
language acquisition, 521–524

Inferior cerebellar peduncle, 16, 410, 412

Inferior colliculus, 16, 24, 25
in auditory processing, 277, 289, 292–293, 295–296
ventricular space, 32

Inferior division, visual field, 255, 256–257

Inferior homonymous quadrantanopsia, 259

Inferior oblique muscle, 428

Inferior olive, 15, 23
action potential in, 55
climbing fibers and, 415
inputs to cerebellum, 412
outputs to the cerebellum, 414

Inferior parietal lobe, 661

Inferior rectus muscle, 428

Inferior salivatory nuclei, 447

Inferior temporal gyrus, 580–581

Inferior temporal lobe, 575

Inferotemporal cortex, 272, 273

Infundibular stalk, 15, 22, 23, 24, 25, 456

Inhibitory postsynaptic potentials (IPSPs)
dynamics of, 148–149
summation of, 149–150

Injuries
axonal regeneration and, 556–558, 560–561
excitotoxicity and, 130
pain and, 211

IN-1 monoclonal antibody, 561

Inner ear
cochlear function, 279, 280–282
cochlear structure, 279–280
mechanoelectrical sound transduction, 282–286

Inner hair cells, 286, 295
acquired hearing loss and, 277

Innervation
axonal growth, 493–502
competitive interactions and refinements, 506–508
by motor neurons, 349–350, 351–352
neuronal dendritic geometry and, 509
neurotrophins and, 510–516
polyneural, 506, 507
reciprocal, 355

selective synapse formation, 502–503, 504–505
topographic map formation, 500–502
trophic interactions, 503, 505–506, 510–512

Inositol triphosphate (IP$_3$), 173, 174–175
long-term synaptic depression and, 183, 551

Inositol triphosphate (IP$_3$) receptor, 174
in taste transduction, 337, 338

Insects
somatic sensory system in, 195
temporal olfactory coding, 328

Inside-out patch clamping, 78

Insomnia, 617, 619–620

Instinctive behavior, 667

Insula, 21, 22
taste system and, 331, 332

Integrins, 168, 495

Intelligence, brain size and, 584–585

Intention tremors, 425

Interaural intensity differences, 290, 291–292

Interaural time differences, 290–291

Interhemispheric connections, 568

Intermediolateral column (spinal cord), 444, 445, 446–447

Internal arcuate tract, 200, 202

Internal capsule, 30, 31, 252, 410
corticospinal tract and, 377
injury to, 375

Internal carotid arteries, 35, 36, 37

Internal division, globus pallidus, 397–398, 401, 402

Interneurons, 4
myotatic spinal reflex, 9, 10
in neural circuits, 9
in sensitization, 537–538
tetanus toxin and, 109

Internuclear neurons, 434

Interocular transfer, 430

Interposed nuclei, 410, 415

Intersexuality, 648–649

Interstitial nuclei of the anterior hypothalamus (INAH), 655, 657–658, 659

Interventricular foramen, 32

Interventricular formation, 18

Intervertebral receptors, 169–170

Intracellular signal transduction
control of cell behavior, 166
defined, 165–166
examples of, 181–184
GTP-binding proteins and, 170–172
in long-term synaptic depression, 182–184, 551–552
NGF/TrkA pathway, 181–182
overview of, 165, 185
protein kinases and protein phosphatases in, 175–178
regulation of gene expression, 178–181
second messengers, 172–175
in short-term sensitization, 538
signal amplification, 166, 167
signaling molecules, 167–168

Intracortical microstimulation, 379

Intrafusal muscle fibers, 198, 350

Intralaminar complex, arousal responses and, 216

Invertebrates, synaptic plasticity in, 535–539

Ion channels

action potentials and, 80–81
activation by GTP-binding proteins, 172
characteristics of, 77, 79–80
cyclic nucleotide-gated, 174
diversity of, 81–82
expression in *Xenopus* oocytes, 85
genetic diseases of, 92–93
heat-activated, 86
overview of, 77, 96–97
patch clamp methods, 78–79
in phototransduction, 230, 232, 234
physical and molecular structure, 86–89
pores, 86–87, 88–89
selective permeability and, 45
selectivity filters, 87
stretch-activated, 86
subunit proteins, 82
toxins affecting, 90
in vestibular hair cells, 305
See also Ligand-gated ion channels;
    Voltage-gated ion channels
Ion concentration gradients
active transporters and, 89, 91
Ionic conductances, 61–64, 65, 143, 147,
    162–163
Ionic currents
in electrical synapses, 100
membrane conductance and, 61–64, 65
membrane depolarization experiment, 59
voltage-dependent, 59–61
β-Ionone, 319
Ionotropic receptors, 141, 163
cholinergic, 151–153
GABAergic, 159–160
glutamatergic, 153–157
overview of, 151
purinergic, 161
serotonergic, 161
structure of, 154
subunits, 154
See also Ligand-gated ion channels
Ion transporters, 45. *See also* Active
    transporters
IP₃. *See* Inositol triphosphate
Ipratropium, 153
Iris, 223, 224, 225
Irritants, trigeminal chemosensory system
    and, 341
Ischemia, excitotoxicity and, 130
Ischiocavernosus muscles, 466, 652
2-Isobutyl-3-methoxypyrazine, 319
Isoretinoin, 474
Ito, Masao, 419

Jackson, John Hughlings, 379
James, William, 626, 643
Jamison, Kay, 640
Johnson, Samuel, 551
Joint receptors, 190, 198, 199
Joro spider, 157

Joro toxin, 157
Jouvet, Michel, 624
Julesa, Bela, 264
Juverile myoclonic epilepsy, 554–555

Kaas, Jon, 553
Kainate, 157
Kainate receptors, 153, 154, 155, 157
Kainic acid, 154
Kanamycin, 277
Kandel, Eric, 535
Kappa rhythm, 613
Karnovsky, Morris, 38
Katz, Bernard, 51, 32, 54, 103, 105, 109
Keynes, Richard, 93, 482
Kimura, Doreen, 660
Kindling, 554
Kinesin, 126
Kinocilium, 282–283, 298, 299, 306
Kleitman, Nathaniel, 609
Klinefelter's syndrome, 648
Klüver, Heinrich, 632–633, 636
Klüver-Bucy syndrome, 632–633, 636
Knee-jerk reflex, 202. *See also* Myotatic reflex;
    Stretch reflex
Konopka, Ron, 608
Korach, Ken, 647
Korsakoff's syndrome, 676
Krumlauf, R., 482
krüppel (krp) gene, 478
Kuffler, Stephen, 242, 245
Kuhl, Patricia, 524
Kuru, 418
Kuypers, Hans, 373
Kytril®, 161

Labeled line hypothesis, 340–341
Labia, 647
Labyrinth, 297–298, 312. *See also* Otolith
    organs; Semicircular canals
Lacrimal gland, 448, 449
Lambert-Eaton myasthenic syndrome
    (LEMS), 108
Lamellapodium, 494
Laminated sheets, 20
Laminins, 495, 505
    axonal regeneration and, 560
Lampreys, 362, 363
Land, Edwin, 240
Land Mondrians, the, 240
Langley, John N., 141, 443–444, 502
Language
grammar and brain structure, 591
prosodic elements, 599
signed, 600–601
as symbolic representation, 589
working vocabulary of English speakers,
    591
Language acquisition/development, 521–524
infant babbling and, 600–601
Language function
in apes, 588–589
aphasias, 590, 592–593
handedness and, 597
hemispheric lateralization, 587, 589,
    593–595, 597–598
localization of, 587, 592, 598–599
mapping of, 597–599
overview of, 587, 601–602
right hemisphere in, 587, 599–600

Large dense-core vesicles, 126
Large intestine, visceral motor system and,
    448, 449
Laser therapy, for macular degeneration, 236
Lateral column (spinal cord), 20
Lateral corticospinal tract, 374, 376, 377
Lateral fissure, 21, 22
Lateral geniculate nucleus
magnocellular and parvocellular layers,
    269
transneuronal labeling, 527
visual deprivation and, 526, 529
visual pathways and, 253, 258, 260, 261,
    269, 273
Lateral horns (spinal cord), 19, 20
Lateral lemniscus, 292, 295
auditory pathway and, 289
Lateral olfactory tract, 329
Lateral (position), 12
Lateral posterior nuclei, 568
Lateral premotor cortex, 348, 384
Lateral rectus muscle, 428, 431
Lateral rubrospinal tract, 374
Lateral superior olive (LSO), 289, 291, 292
Lateral ventricles, 31–32
Lateral vestibulospinal tract, 314
α-Latrotoxin, 114
Laurent, Gilles, 328
Lazy eye, 530
Learning
defined, 665
genetic studies in *Drosophila*, 540
long-term synaptic plasticity and, 542
Learning disorders, genetic basis, 480
LeDoux, Joseph, 636
Leeches, 362, 363
Left cerebral hemisphere
emotion processing and, 642
language function and, 587, 590, 597–598
Left homonymous hemianopsia, 260
Left superior quadrantanopsia, 260
Lens
development, 228
in image formation, 224, 225
refractive errors, 226, 227
Lenticulostriate arteries, 36, 37
Lesions
cerebellar, 419, 422, 425
in extrastriate regions, 271–272
frontal association cortex, 576–577
parietal association cortex, 571–573
spinal cord dorsal columns, 202
temporal association cortex, 574–575
See also Neural injuries; Strokes
Leucine enkephalin, 122, 221
Leukotomies, 577, 579
Leukotrienes, 211
Leutinizing hormone-releasing hormone, 122
Levator muscles, 431
Levi-Montalcini, Rita, 510–511
Levodopa, 133
Lewis, E. B., 480
Librium®, 137, 160
Lichtman, Jeff, 507–508
Ligand-gated ion channels, 83, 86, 163, 168,
    169
cholinergic, 151–153
glutamatergic, 153–157
at neuromuscular junctions, dynamics of,
    141–147

overview of, 151
purinergic, 161
serotonergic, 161
structure of, 154
subunits, 154
Light adaptation, 233–234, 247–249
Lima, Almeida, 579
Limbic lobe, 24–25
Limbic system, 25
Alzheimer's disease and, 678
components of, 627, 631–632
emotional behavior and, 627, 631–633, 644
See also Amygdala
Lincoln, Abraham, 640
Lindstrom, Jon, 158
Lithium chloride, 335
Llinás, Rodolfo, 109
Lobes, 21
Lobsters, stomatogastric ganglion, 364–365
Local circuit neurons, 9
central pattern generators, 366
in cerebellar circuitry, 416
in motor control, 347–348
muscle stretch reflex, 355–357
patterning of connections, 370
saccadic eye movements and, 434
spatial organization, 369
Local motor neuron syndrome, 366–367
Locomotion
in leeches and lampreys, 362–363
local neuronal circuitry and, 361, 366
See also Movement/Motor control
Locus coeruleus, 133, 414, 617, 618, 619
Loewi, Otto, 117
Long circumferential arteries, 37, 39
Longitudinal fissure, 23
Long-term memory, 666, 668
Long-term sensitization, 538
Long-term synaptic depression (LTD),
182–184, 541, 549
Long-term synaptic potentiation (LTP), 541,
639
AMPA receptors and, 549
associativity of, 544
hippocampus and, 542–543
input specificity of, 543–544
memory and, 542, 543, 544
molecular mechanisms of, 545–547
phases of, 547, 550
state-dependency of, 543
Lorenz, Konrad, 520, 667
Loudness, 275
Lou Gehrig's disease, 367
Lower motor neurons
amyotrophic lateral sclerosis and, 367
brainstem and, 350
Golgi tendon organ and, 360
innervation of skeletal muscles, 349, 350,
351–352
locomotion and, 365, 366
modulation by upper motor neurons, 347,
357–358
motor neuron pools, 349
motor units, 351–353
muscle stretch reflex and, 355–357
overview of, 347
regulation of muscle force, 353–354
spatial organization in spinal cord, 349,
369

syndrome from damage to, 366–367
types of, 349–350
Low-threshold mechanoreceptors, 192–193,
198
LSO. See Lateral superior olive
LTD. See Long-term synaptic depression
LTP. See Long-term synaptic potentiation
Lucas, D. R., 130
Lumbar enlargement, 14, 19
Lumbar nerves, 14, 18
Lumbar puncture, 18
Luminance
detecting differences in, 242–243, 245–246
perception of, 244–245
Luminance contrast, 245–246
Lumsden, A., 482
Lungs, visceral motor system and, 448, 449
Luria, A. R., 672
Luteinizing hormone, 651

Macaque monkeys. See Monkeys
Macroglomerular complex, 322
Macrophages, 8, 558
Macroscopic currents, 79, 80
Macular degeneration, 235, 236
Macula (retina), 236, 252
Macular sparing, 259–260
Macula (vestibular), 300, 302
Mad cow disease, 418
Magnesium (Mg^{2+}) ions, NMDA receptors
and, 155, 546, 548
Magnetic resonance imaging (MRI), 26–27
Magnocellular layer, lateral geniculate
nucleus, 269
Magnocellular stream, 269, 272, 273
Magoun, Horace, 615
Major depression, 640
Male sexual development, 646–647
Malleus, 278
Mammillary bodies, 15, 22, 23, 31, 456, 457,
676
limbic system and, 632, 633
Mammillothalamic tract, 633
Mandibular nerve, 202
Manduca sexta, 322
Manic depression, 640
Mannitol, 38
MAO inhibitors, 137, 614
MAPK. See Mitogen-activated protein kinase
Mapping. See Somatotopic maps; Topo-
graphic maps
Marine molluscs. See Aplysia
Mariotte, Edmé, 254
MASA, 495
Mast cells, histamines and, 133–134
Mauthner cells, 312–313
Maxillary nerve, 202
M cells, 268–269
McMahan, U. J., 504
Mechanoreceptors, 189
dorsal root ganglia and, 199
mechanosensory discrimination and,
194, 197
in proprioception, 197–199
receptive fields, 194–197
sensory stimuli and, 190, 192
types of, 192–193
Mechanosensory discrimination, 194, 197
Mechanosensory neurons, in sensitization,
537–538

Mechanosensory pathways
dorsal column-medial lemniscus system,
199–202
overview of, 189
trigeminal system, 200, 202–203
Medial dorsal nuclei, 568
Medial gastrocnemius, 350
Medial geniculate complex, 289
Medial geniculate nucleus, 636
Medial lemniscus
mechanosensory systems and, 200, 202
spinothalamic tract and, 213
Medial longitudinal fasciculus, 310, 311, 313,
314, 434
Medial nucleus of the trapezoid body
(MNTB), 291, 292
Medial (position), 12
Medial prefrontal cortex, 632, 633
depression and, 640
emotional behavior and, 643
Medial premotor cortex, 348, 384
Medial rectus muscle, 428
Medial superior olive (MSO), 289, 290–291
Medial temporal lobe, memory and, 681
Medial thalamus, Korsakoff's syndrome and,
676
Medial vestibular nucleus, 311, 313, 314
Median eminence of the hypothalamus, 456
Median sagittal sections, 12
Mediodorsal nucleus of the thalamus, 640,
641
Medium spiny neurons
anticipatory discharges, 395
in basal ganglia circuitry, 394
cortical inputs, 391, 395
in disinhibition circuitry, 396–397, 401, 402
dopaminergic influences, 402
function of, 391–392
Huntington's disease and, 405
noncortical inputs, 395
output from, 392, 402
spatial distribution in corpus striatum,
395
Medulla, 13, 14, 15, 23, 24, 25, 200
blood supply to, 37
corticospinal tract and, 377
in descending pain modulatory systems,
218
dorsal column nuclei, 16
internal organization, 17
mechanosensory pathway, 202
output to the cerebellum, 413
somatic sensory system and, 191
spinothalamic tract and, 213
ventricular space, 32
vestibular thalamocortical system and,
314
vestibulo-ocular reflex and, 311
Medullary arteries, 35
Medullary pyramids, 15, 376
Medullary reticular formation, 372
Meissner's corpuscles, 190, 192–193
Meissner's plexus, 452, 453
Melanocyte-stimulating hormone, 135
Melanopsin, 608
Melatonin, 609
Melzack, Ronald, 216, 220
Membrane conductance
in action potentials, 65

activation, 62–63
defined, 61–62
inactivation, 63
membrane potential and, 63–64
Membrane permeability
dynamics of postsynaptic receptors and, 141–147
membrane conductance and, 61–62
with more than one permeant ion, 49–51
resting potential and, 51–52
selective, 45–47
voltage-dependent, 57–65
Membrane potential
depolarization, 44, 66
dynamics of postsynaptic receptors and, 141, 142, 143, 144, 145, 146, 147
equilibrium, 47–49
hyperpolarization, 44
ionic flux and, 48–49
ionic movements and, 45–51
membrane conductance and, 63–64
passive current flow and, 68–69
resting, 43, 44, 51–52
Na⁺/K⁺ pump and, 95
threshold, 44, 66
voltage clamping and, 58
voltage-dependent ionic currents and, 59–61
voltage-dependent membrane permeability and, 57–59
See also Action potentials
Membranous labyrinth, 297
Memory
aging and, 680
Alzheimer's disease, 678–679
association and information storage, 668–671
brain systems underlying, 673, 676
clinical cases of amnesia in declarative memory, 674–675
fallibility of, 668
forgetting and, 671–673
genetic studies in Drosophila, 540
long-term information storage, 676–677, 680
long-term synaptic plasticity and, 542
long-term synaptic potentiation and, 542, 543, 544
overview of, 665, 680–681
phylogenetic, 667
priming and, 668
qualitative categories of, 665
savant syndrome, 670
temporal categories of, 666
Meninges, 33–34
Menstrual cycles, synchronization of, 320–321
Mental illness
affective disorders, 640–641
psychosurgery and, 577, 579
See also Psychiatric disorders
Meperidine, 136
MEPPs. See Miniature end plate potentials
Merkel's disks, 190, 193
Merzenich, Michael, 553, 663
Mesencephalic reticular formation, 372
Mesencephalon, 32, 478. See also Midbrain
Mesoderm, 471
Mesopic vision, 235
Metabotropic receptors, 141, 163, 169
catecholaminergic, 162

GABAergic, 161
glutamatergic, 159, 551–552
long-term synaptic depression and, 551–552
muscarinic cholinergic receptors, 153
overview of, 151
peptide receptors, 162
serotonergic, 161
structure and function of, 159
subunits, 159
Metencephalon, 32, 479
Methadone, 136
Methionine enkephalin, 121, 122, 221
Methylprednisolone, 561
Meyer's loop, 259
Mice. See Rodents
Microelectrodes, 43–44
Microglial cells, 7
Microscopic currents, 79–80
Microtubules, 126
Micturition center, 464
Midbrain, 13, 14, 15, 17, 410
anatomy of, 24, 25
auditory processing in, 292–293
blood supply to, 37
corticospinal tract and, 377
descending pain modulatory systems, 218
development of, 478
mechanosensory system and, 200
in pain modulation, 220
projections to basal ganglia, 393
spinothalamic tract and, 213
trigeminal mechanosensory system and, 200
ventricular space, 32
vestibulo-ocular reflex and, 311
See also Mesencephalon
Midbrain dopamine system, 138
Middle cerebellar peduncle, 15, 16, 410, 411–412
Middle cerebral artery, 35, 36, 37
Middle cranial fossa, 33, 34
Middle ear, 279
Middle temporal area, 270
Midget ganglion cells, 237
Midline diencephalic structures, memory and, 673, 676, 681
Migraines, 92
Miledi, Ricardo, 85, 109
Milner, Brenda, 583, 674
Miniature end plate potentials (MEPPs), 104–105, 158
Miosis, 458
Mitchison, Graeame, 614
Mitogen-activated protein kinase (MAPK), 177–178
Mitral cells, 326, 329
Mnemonists, 668, 669–671
MNTB. See Medial nucleus of the trapezoid body
Modules, in brain organization, 206
Molecular layer, cerebellar cortex, 415
Molecular signaling, 165–166. See also Intracellular signal transduction
Mondrian, Piet, 240
Money, John, 649
Moniz, Egas, 579
Monkeys
attention-specific neurons, 577–578
delay-specific neurons, 581, 582

electrophysiological studies of association cortices, 577
hypothalamic control of sexual behavior, 655
recognition-specific neurons, 580–581
Monoamine oxidase (MAO), 133, 134
Monoaminergic system, sleep and, 617
Monomeric G-proteins, 170–171
Monosodium glutamate, 334, 337
Monosynaptic reflexes, 355–357
Morphine, 136
Morrison, Robert, 613
Moruzzi, Giuseppe, 615
Mossy fibers, 415, 416, 417, 419
Motor aphasia, 592
Motor axons, 4
Motor cortex, 21–22
basal ganglia circuitry and, 396, 397, 398, 401
in disinhibitory circuitry, 398, 401
emotional behavior and, 630–631
in hypokinetic and hyperkinetic diseases, 404
smiling and, 628–629
Motor deficits, diagnosing, 378
Motor learning, cerebellum and, 409, 417, 419, 420–421, 425
Motor maps, 435, 438
Motor neuron pools, 349
Motor neurons
action potential in, 55
agrin and, 505
amyotrophic lateral sclerosis and, 367
botulinum toxin and, 109
competitive synapse interactions and refinements, 506–508
development of, 473
myotatic spinal reflex and, 9, 10
in sensitization, 537–538
in spinal cord, 20
See also Lower motor neurons; α Motor neurons; γ Motor neurons; Upper motor neurons
α Motor neurons, 347
amyotrophic lateral sclerosis and, 367
Golgi tendon organ and, 360
injury to, 367
innervation of muscle fibers, 350, 351–352
maintenance of muscle length, 357
modulation of, 357–358
in motor units, 351–352
muscle tone and, 387
in reflex sexual function, 466
stretch reflex and, 355, 356
in voluntary bladder control, 464
γ Motor neurons, 198
modulation of, 357–358
muscle spindles and, 349–350, 357
muscle tone and, 387
stretch reflex gain, 357, 358
Motor nuclei, in brainstem, 15
Motor subsystems, 347–349
Motor systems, 10, 11
Motor units
components of, 351–352
regulation of muscle force, 353
size principle, 353
types of, 352–353
Mountcastle, Vernon, 206

Movement disorders
  Huntington's disease, 400, 405
  hypokinetic, 404–405
  Parkinson's disease, 402–404
Movement/Motor control
  afferent activity and, 367
  amyotrophic lateral sclerosis and, 357–361
  basal ganglia and, 391, 394–396, 397–399, 402, 407
  cerebellum and, 409, 413, 419–422, 425
  direction of force, 381
  encoding of intention, 385
  Golgi tendon organ and, 358–361
  locomotion, 361, 362–363, 365, 366
  motor neuron-muscle relationships, 349–350
  motor units and, 351–353
  muscle force regulation, 353–354
  neural circuits responsible for, 347–349
  premotor cortex and, 385
  primary motor neurons and, 381, 383–384
  self-initiated, 385
  sensory motor talents and cortical space allocation, 383
  striatal rostrocaudal bands and, 394–395
  upper motor neurons and, 369, 388
  visually-guided, encoding of, 383
  See also Reflexes
Movement selection, 385, 395–396
M retinal ganglion cells, 268–269
MSO. See Medial superior olive
Mucus, nasal, 323
Müllerian ducts, 646, 647
Müllerian-inhibiting hormone, 646, 647
Multiple sclerosis, 75
Mu rhythm, 613
Muscarine, 153, 156
Muscarinic cholinergic receptors, 153, 169
  effects in parasympathetic targets, 504
  ion channel activation, 172
  neurotoxins and, 157
  in visceral motor system, 455, 459–460
Muscimol, 157, 405
Muscle fibers
  action potentials and contraction, 354
  basal lamina sheath, 504
  competitive synapse interactions and refinements, 506–508
  fused tetanus, 354
  intrafusal and extrafusal, 350
Muscle fields, 381
Muscle force
  Golgi tendon organ and, 360
  regulation of, 353–354
Muscles
  cortical motor maps of, 379–381
  Golgi tendon organ, 358–361
  innervation by motor neurons, 349–350, 351–352
  maintenance of tone and length, 357, 387
  motor units, 351–353
  regulation of force, 353–354
  stretch reflex, 355–357
  See also Skeletal muscles
Muscle spindles, 190, 349–350
  clonus and, 387
  modulation of, 357
  in proprioception, 198–199
  stretch reflex, 355
  structure of, 198

Muscle stretch reflex, 355–357
  clasp-knife phenomenon and, 387
Muscle tone, 357
  decerebrate rigidity and, 387
  overview of, 387
Mushroom body, 328
Mushroom toxins, 156, 157
Myasthenic syndromes, 108, 158
Myelencephalon, 32, 479
Myelin, 72
  inhibition of axonal regeneration, 561
  oligodendrocytes and, 7
  See also Axon myelination
Myenteric plexus, 452, 453
Myopia, 226
Myotatic reflex, 9–10, 202. See also Muscle stretch reflex
Myotonia, 92

Naloxone, 219
Narcolepsy, 622
Nasal division, visual field, 255, 256–257
Nasal mucosa, 323
Nathans, Jeremy, 241
Natural toxins, 156–157
NCAMs. See Neural cell adhesion molecules
Near cells, 263
Nearsightedness, 432
Near reflex triad, 432
Neck, visceral motor system and, 448, 449
Neher, Erwin, 79
Neocortex, 25
  Alzheimer's disease and, 678
  amygdala and, 639, 641–642
  cytoarchitectonic areas, 566–567
  in hypokinetic and hyperkinetic diseases, 404
  laminar organization, 566
  major connections of, 566
  patterns of connectivity in, 567
  projections to basal ganglia, 392–393
  sex hormone receptors in, 654
Nernst equation, 47–49
  expansion of, 50
Nerve gas, 128
Nerve growth factor (NGF), 510–511, 514
  neurite growth and, 515
  NGF/TrkA pathway, 181–182
  TrkA receptor, 512
Nerves, 11
Nervous system
  anatomical planes, 12–13
  anatomy of (see Neuroanatomy)
  cellular components, 1–3
  development of (see Neural development)
Netrins, 498–499
Neural cell adhesion molecules (NCAMs), 168, 495
Neural circuits, 8–10. See also Brain circuitry
  modification; Neuronal connections
Neural crest, 473
Neural crest cells, 488, 490
Neural development
  formation of major brain subdivisions, 476–479
  gastrulation, 471
  genetic abnormalities, 479–480
  homeotic genes and, 479, 480
  induction in, 472, 473, 475–476
  neurogenesis, 480–481, 483, 486

neuronal differentiation, 483–485, 487–488, 489–490
  neuronal migration, 488–490
  notochord formation and neurulation, 471–473
  overview of, 471, 490–491
  patterned gene expression and, 479, 480
  related congenital disorders, 476
  retinoic acid and, 475–476
  rhombomeres, 482–483
  teratogens and, 474, 476
Neural differentiation, 475–476
Neural grafting, 403
Neural induction, 472, 473, 475–476
Neural injuries
  axonal regeneration and, 556–558, 560–561
  dissociated sensory loss, 212
  excitotoxicity and, 130
  in extrastriate regions, 271–272
  visual field deficits, 258–260
  See also Lesions; Strokes
Neural injury, in extrastriate regions, 271–272
Neural plasticity
  axonal regeneration and, 556–558, 560–561
  behavioral plasticity in invertebrates, 535–539
  epilepsy and, 554–555
  generation of neurons in adult brain, 559
  long-term synaptic depression, 541, 550–552
  long-term synaptic potentiation, 541, 542–547, 550
  neural stem cells and, 559
  overview of, 535, 561
  short-term synaptic plasticity, 539, 541
  See also Synaptic plasticity
Neural plate, 472–473
Neural precursor cells, 472–473, 481, 483, 485, 487
Neural stem cells, 559
Neural systems, 10–12
Neural tube
  congenital disorders, 476
  development of, 472–473
  formation of major brain subdivisions, 476, 477, 478
  neuromere organization, 479, 482
Neurexins, 112, 114
Neurites, 511
Neuroanatomy
  blood-brain barrier, 38, 39
  blood vessels, 35–37, 39
  brain, external, 20–25
  brain, internal, 25, 29–34
  brain imaging technology, 26–27, 28–29
  cellular components, 1–3
  central nervous system subdivisions, 13, 14, 18
  cranial nerves, 18–19
  meninges, 33–34
  neural circuits, 8–10
  neural systems, 10–12
  neuroglial cells, 6–8
  neurons, 3–6
  overview of, 39
  spinal cord, 14, 18–20
  terminology, 12–13
  ventricular system, 31–33

Neuroblasts, 472, 486
  differentiation of, 487–488
  migration of, 488
Neuroectoderm, 471–472, 475–476
Neuroendocrine cells, 126
Neurofibrillary tangles, 678
Neurogenesis, 480–481, 483, 486
Neuroglial cells, 2, 6–8. See also Glial cells
Neurokinin A, 136
Neurological disorders, abnormal neuronal migration and, 489
Neuromeres, 479, 482
Neuromuscular junctions
  dynamics of postsynaptic receptors, 141–147
  short-term synaptic plasticity, 539, 541
  synapse formation in, 504–505
Neuromuscular synapses, 4
  botulism and, 108
  competitive interactions and refinements, 506–508
  formation of, 504–505
  myasthenic syndromes and, 108, 158
  quantal transmission at, 103–105
  short-term plasticity, 539, 541
Neuronal birthdating, 481, 486
Neuronal connections, 516
  axonal growth, 493–502
  competitive interactions and refinements, 506–508
  convergence and divergence in, 508
  neuronal dendritic geometry and, 509
  neurotrophic hypothesis, 516
  neurotrophins and, 510–516
  selective synapse formation, 502–503, 504–505
  topographic map formation, 500–502
  trophic interactions, 503, 505–506, 510–512
Neuronal injury
  excitotoxicity and, 130
  See also Neural injuries
Neuronal migration, 488–490
Neuronal plasticity, 444
Neuronal populations, size determination, 505–506
Neurons, 2
  afferent and efferent, 8–9
  apoptosis and, 506
  birthdating, 481, 486
  dendritic geometry and innervation, 509
  electrical characteristics, 43
  extracellular and intracellular ion concentrations, 50
  generation of diverse cell types, 483–485, 487–488
  giant squid neurons, 53
  migration, 488–490
  morphological features and functions, 3–6
  in neural circuits, 8–9
  neurogenesis, 480–481, 483, 486
  neurotransmitter synthesis, 123–126
  numbers in brain, 3
  recording passive and active electrical signals in, 44
  reticular theory of communication, 1
  staining of, 569
  summation of postsynaptic potentials and, 149–150
  synaptic terminals, 3–5
  transneuronal labeling, 524, 527

transneuronal transport, 527
trophic interactions, 503, 505–506, 510–512
Neuropeptide GAMMA, 136
Neuropeptide K, 136
Neuropeptide receptors, 162
Neuropeptides, 120, 121, 122
  biological activity and, 135
  categories of, 135–136, 139
  fast axonal transport, 126
  functional features of, 127
  packaging in synaptic vesicles, 126
  synthesis of, 124, 125–126
Neuropeptide-Y, 122
Neuropil, 8
  olfactory, 328–329
Neuropsychological testing
  delayed response task, 581, 582
  overview of, 583
Neurosurgery
  intraoperative mapping, 598
  See also Psychosurgery
Neurotensin, 122
α-Neurotoxin, 156
Neurotoxins
  affecting presynaptic terminals, 108–109
  effect on neurotransmitter release, 109, 114
  postsynaptic receptors and, 152, 156–157
  See also Toxins
Neurotransmitter receptors, 6
  catecholaminergic, 162
  cholinergic, 151–153
  classes of, 141, 150–151
  dynamics of, 141–147
  excitatory and inhibitory postsynaptic potentials, 147–149
  GABAergic, 159–161
  glutamatergic, 153–157, 159
  myasthenic syndromes and, 108, 158
  neurotoxins and, 156–157
  overview of, 141, 162–163
  peptide receptors, 162
  purinergic, 161
  serotonergic, 161
  in visceral motor system, 454–455, 458–460, 467
  See also specific types
Neurotransmitter release
  in chemical synapses, 102–103
  co-transmitters, 122–123
  differences in rates, 126
  myasthenic syndromes and, 108
  proteins involved in, 111–113, 115
  quantal, 103–105
  role of calcium in, 107, 109, 110
  from synaptic vesicles, 105–106
  toxins affecting, 109, 114
Neurotransmitters, 6
  in chemical synapses, 101, 102–103
  co-transmitters, 122–123
  criteria for defining, 119
  discovery of, 117
  enteric nervous system and, 460
  functional features of, 127
  hormones and, 118, 120
  ligand-gated channels and, 86
  localization of action, 118
  major categories of, 120–122
  overview of, 117, 139–140
  packaging in synaptic vesicles, 126

"putative," 118, 119
removal from synaptic cleft, 126–127
synthesis of, 123–126
visceral motor system and, 454, 455, 458, 459, 467
See also Neurotransmitter release
Neurotrophic factors
  axonal regeneration and, 560
  described, 503
Neurotrophic hypothesis, 516
Neurotrophin-3 (NT-3), 512
Neurotrophin 4/5 (NT 4/5), 512
Neurotrophin receptors, 512–513, 515
Neurotrophins
  discovery of, 510–511, 514
  members of, 511–512
  nerve growth factor, 510–511
  neurite growth and, 515–516
  receptors, 512–513, 515
Neurottoxins
  affecting ion channels, 90
  affecting presynaptic terminals, 108–109
  clostridial, 108–109, 114
  effect on neurotransmitter release, 109, 114
Neurulation, 472–473. See also Neural induction
Newborns, imprinting and built-in behaviors, 520
Newhouse, J. P., 130
NGF. See Nerve growth factor
NGF/TrkA pathway, 181–182
Nicotiana tabacum, 156
Nicotine, 138, 151, 156
Nicotinic cholinergic receptors, 151–153
  effects in parasympathetic targets, 459
  neurotoxins and, 152, 157
  in visceral motor system, 454–455
Night vision, 231
Nissl, F., 569
Nissl stains, 569
Nitric oxide, 173, 174, 175, 466
Nitric oxide synthase, 175, 460
NMDA, 154
NMDA agonists, fear response and, 639
NMDA receptors
  in long-term potentiation, 545–547
  in long-term synaptic depression, 550
  properties of, 154–157
  in silent synapses, 548–549
Nó, Rafael Lorente de, 206
Nociception. See Pain perception
Nociceptive axons
  in flexion reflexes, 361
  nociceptor types, 209–210
  in pain perception, 210–211
  spinal trigeminal tract and, 212, 213, 214
  spinothalamic tract and, 212, 213
  trigeminal chemoreception and, 341, 342, 343
  types of, 209–210
Nociceptors, 189
  dorsal root ganglia and, 199
  sensitization and, 211
Nodes of Ranvier, 72
Nodulus, 409, 410
Noggin, 475
Nogo protein, 558, 561
Noise, acquired hearing loss and, 277
2-trans-6-cis-Nonadienal, 319

Non-rapid eye movement (non-REM) sleep
 brain activity in, 617
 dreaming in, 614
 physiological characteristics, 611
 purpose of, 614
 stages of, 610
Non-receptor tyrosine kinases, 176
Non-REM sleep. *See* Non-rapid eye movement sleep
Noradrenaline. *See* Norepinephrine
Noradrenergic nuclei
 inputs to the association cortices, 570
 in wakefulness and sleep, 617, 619
Noradrenergic receptors, 458
Norepinephrine, 120, 131
 adrenergic receptors and, 162
 antihypertensive agents and, 137
 cardiovascular function and, 462
 chemical structure, 121
 distribution in brain, 134
 G-protein-coupled receptors and, 171
 overview of, 133
 visceral motor system and, 455, 458, 467
Notochord, 471–472
Nottebohm, Fernando, 559
nSec-1, 111
NSF protein, 111, 112
NT-3. *See* Neurotrophin-3
NT 4/5. *See* Neurotrophin 4/5
Nuclear bag fibers, 198
Nuclear chain fibers, 198
Nuclear receptors, 180–181
Nuclei, 11
Nuclei of lateral lemniscus, 289, 292
Nucleus accumbens, 138
Nucleus ambiguus, 16, 451, 462
Nucleus of the solitary tract, 453, 454
 amygdala and, 634
 control of cardiovascular function and, 460, 461
 taste processing and, 340, 341
Nucleus proprius, 212
Nystagmus
 optokinetic, 432
 physiological, 308–309

Obicularis oculi, 628
Oblique muscles, 428, 429, 431
Occipital lobe, 21, 22
 alpha rhythm and, 613
 cognition and, 566
 eye movements and, 440
 visual field retinotopic representation, 258
 in visual processing, 270
Ocular dominance
 activity-dependent modification, 532
 effects of visual deprivation on, 525–526, 528–529
 ocular dominance columns and, 524–525
 optical imaging of, 267
Ocular dominance columns, 261, 267
 activity-dependent modification, 532
 effects of visual deprivation on, 525–526, 528–529
 in ocular dominance, 524–525
Oculomotor nerve, 15, 434, 440
 characteristics of, 18–19
 eye movements and, 428, 429, 431
 parasympathetic system and, 451

Oculomotor nuclei, 16, 311, 429
Odorant receptors, 324–325
Odorants, 317
 categories of, 319–320
 glomeruli responsiveness to, 329, 330
 human sensitivity to, 319, 320
 pheromones, 321, 322
Odorant signal transduction, 323–324
Off-center retinal ganglion cells, detection of differences in luminance, 242–243, 245–246
Ohm's Law, 62
Ojemann, George, 598
Olfaction. *See* Olfactory perception
Olfactory bulb, 22, 23, 317, 318
 amygdala and, 634
 cellular organization, 4
 central projections of, 329–330
 functional organization of, 326, 327–329
 neural stem cells and, 559
 olfactory nerve and, 327–328
 ventricular space, 32
Olfactory cilia, 321, 323
Olfactory cortex, 634
Olfactory epithelium, 317, 318, 326, 327
 structure and function of, 321, 323
 surface area of, 319
Olfactory nerve, 18–19, 22, 327–328
Olfactory perception
 in hawk moths, 322
 in humans, 319–320
Olfactory receptor genes, 325
Olfactory receptor molecules, 324–325
Olfactory receptor neurons, 4, 326, 343, 344
 adaptation, 324
 in odorant signal transduction, 323–324
 olfactory coding and, 327
 projections to glomeruli, 329
 regeneration of, 323
 responses to chemical stimuli, 325, 327
 structure and function of, 321, 323
Olfactory signal transduction, 323–324
Olfactory system
 central projections, 329–330
 critical period in, 520, 533
 odorant receptors, 324–325
 odorant signal transduction, 323–324
 olfactory bulb, 326, 327–329
 olfactory receptor neurons, 321, 323
 organization of, 317–319
 overview of, 317, 343, 344
 space coding, 327
 temporal coding, 327, 328
Olfactory tract, 22, 23, 317, 318
Olfactory transduction, 343
Olfactory tubercle, 318, 329
Oligodendrocytes, 7, 12
 in myelination, 72
 Nogo protein secretion, 558, 561
Olney, John, 130
Ommatidia, 487
On-center retinal ganglion cells, detection of differences in luminance, 242–243, 245–246

Onchocerca volvulus, 530
Onchocerciasis, 530
Ondansetron, 161

Onuf's nucleus, 653
Oocytes, ion channel expression, 85
Ophthalmic nerve, 202
Opiate receptors, 162
Opioid peptides, 135, 136, 139
Opioid receptors, 162
Opioids, endogenous, 220–221
Opium, 136
Opsins, 232, 234
Optic chiasm, 15, 22, 23, 31, 251–252, 253, 456, 457
 axonal growth and, 494, 496
 visual field deficits and, 258, 259, 260
Optic cup, 227, 228, 479
Optic disk, 224, 251, 252, 254
Optic nerve, 15, 224, 228, 251, 253
 characteristics of, 18–19
 visual field deficits and, 258, 260
Optic papilla, 251
Optic radiation, 252, 253
 visual field deficits and, 259, 260
Optic system. *See* Visual system
Optic tectum
 Eph ligands in, 502
 retinotopic map formation, 500–502
Optic tract, 15, 252, 253
 visual field deficits and, 258, 259, 260
Optic vesicle, 227, 228
Optokinetic nystagmus, 432
Orbital frontal cortex
 olfactory processing and, 318, 330
 taste processing and, 332
Orbital prefrontal cortex, 632, 633
 depression and, 640
 emotional behavior and, 643
Orb weaver spider, 157
Orexins, 622
Organ of Corti, 277, 280
Organophosphates, 128–129
Orgasm, 466
Orientation-selective neurons, 260–261
 columnar organization, 263, 265, 267
Orthostatic hypotension, 462
Oscillatory state, of thalamocortical neurons, 618–619

Oscillopsia, 312
Oscine songbirds, 522–523
Ossicles, 279
Otoacoustical emissions, 287
Otoconia, 300, 301
Otolithic membrane, 300, 301
Otolith organs, 297, 316
 detection of head movements, 300–302
 inputs to the vestibular nuclei, 314
 See also Sacculus; Utricle
Ototoxic drugs, acquired hearing loss and, 277
Ouabain, 95
Outer hair cells, 286–287, 295
 acquired hearing loss and, 277
Outside-out patch clamping, 78
Oval window, 278, 279–280
Overshoot phase, action potential, 54
Owls, auditory localization, 533
2-Oxoglutarate, 129
Oxyhemoglobin. *See* Deoxyhemoglobin/oxyhemoglobin ratios
Oxytocin, 122, 457, 663

Pacinian corpuscles, 190, 193
Pain modulation
actions of analgesics, 211
capsaicin and, 342
physiological basis, 220–221
placebo effect, 219–220
Pain perception
arousal responses, 216
categories of, 210–211
central pathways in, 212–214
central regulation of, 217, 219
cerebral cortex and, 216–217
chronic, 215, 216
fibers involved in, 210–211
gate theory, 220
hyperalgesia, 211
modulatory descending systems, 216, 218
nociceptors, 209–210
overview of, 209
phantom, 216
placebo effect, 219–220
referred, 215
sensitization, 211
spinal trigeminal tract in, 212, 213, 214
spinothalamic tract in, 212, 213
thalamus and, 214, 216
Pain relief. *See* Pain modulation
Paleocortex, 29, 569
Pancreas, 448, 449
Panic attacks, 137
Papez, James, 631
Papez circuit, 632
Papillae, 333, 334
Papilledema, 251
Paracentral lobule, 464
Paracrine signaling, 165, 166
Parahippocampal gyrus, 22, 23
Parahippocampus, sleep and, 615, 617
Parallel fibers, 415, 416, 417, 419
in long-term synaptic depression, 182–184, 551, 552
Paralysis
genetic disorders of ion channels, 92
from ion channel poisoning, 90
Paramedian circumferential artery, 39
Paramedian pontine reticular formation (PPRF), 433, 434, 435, 436, 440, 615
Paramedian sagittal sections, 12
Paraplegics
autonomic bladder regulation, 464
reflex sexual excitation, 466
Parasympathetic ganglia, 449, 450, 451, 466
muscarinic receptors of targets, 459–460
neurotransmitters released on targets, 455
Parasympathetic preganglionic neurons
acetylcholine and, 454
location of, 449, 453, 462, 466
organization of, 451
parasympathetic ganglia and, 450
in regulation of sexual responses, 465
Parasympathetic system
bladder regulation, 462, 464
control of cardiovascular function and, 460–462
emotional behavior and, 625
function of, 449, 450
ganglia in, 12
neurotransmission in, 454–455, 459–460
organization of, 445, 447, 450, 451

in regulation of sexual responses, 464, 465–466
Paraventricular nucleus, 456, 457
Paravertebral sympathetic chain, 447, 466
Parietal association cortex, 565
contralateral neglect syndrome, 571–573
thalamic connectivity to, 568
Parietal cortex
attention-specific cells in monkeys, 577–578
projections to basal ganglia, 393
Parietal lobe, 21, 22
contralateral neglect syndrome, 571–573
eye movements and, 440
visual processing, 251
in visual processing, 270, 272, 273
Parieto-occipital sulcus, 24
Parkinson, James, 403
Parkinson's disease
basal ganglia and, 349, 403–404
disruption of disinhibition pathways and, 403–404
dopamine and, 133
loss of nigrostriatal dopaminergic neurons in, 402–404, 405
novel therapeutic approaches, 403
Parotid gland, 448, 449
Paroxetine, 641
Pars caudalis, 214
Pars compacta. *See* Substantia nigra pars compacta
Pars interpolaris, 214
Pars reticulata. *See* Substantia nigra pars reticulata
Parvocellular layer of the lateral geniculate nucleus, 269
Parvocellular stream, 269, 272, 273
Passive current flow, 65, 67, 68–69
in action potentials, 69, 70–72, 74
Patch clamp methods, 78–79
Patrick, Jim, 158
Pax genes, 479–480
Paxil®, 641
P cells, 268–269
Pelvic splanchnic nerve, 451
Perfield, Wilder, 379, 380, 598
Penis, 646, 652
erection, 465–466
Pentadecalactone, 319
Pentobarbital, 160
Peptidases, 135
Peptide neurotransmitters. *See* Neuropeptides
Peptide receptors, 62
Peptide toxins, 90
Periaqueductal gray matter, 220, 221
Pericak-Vance, Margaret, 679
Periglomerular cells, 326
Perilymph
cochlear, 286
vestibular, 298
Perineal muscles, 653
period genes, 608, 609
Peripheral nerves
axonal regeneration and, 556–557, 558, 560
in spinal cord, 18
Peripheral nervous system
competitive synapse interactions and refinements, 506–508

neuronal migration in, 488
neurotrophins and, 510–511, 514
organization of, 12
overview of, 11
Peripheral neurons, 4
Peripheral taste system, 334–335
Perisylvian cortex, 598
Perphenazine, 406
PER proteins, 608, 609
Petersen, Steve, 599
Peterson, Andy, 424
Petit mal epilepsy, 555
Phantom limbs, 197, 216–217
functional topographic remapping and, 553
Phantom pain, 216
Phasic sensory receptors, 192
Phenelzine, 133, 137
Phenobarbital, 160, 555
Phenotypic sex, 646–647, 647–648
Phenylethanolamine-N-methyltransferase, 133
Phenylthiocarbamide, 335
Phenytoin, 555
Pheromones, 321, 322
β-Philanthotoxin, 157
Phosphatidylinositol bisphosphate (PIP₂), 173, 174, 175
Phosphatidylserine, 176
Phosphodiesterases, 174
in phototransduction, 232, 233
Phospholipase C, 171
in taste transduction, 337, 338
Phosphorylation
regulation by protein kinases and phosphatases, 175–178
of tyrosine hydroxylase, 184
Photoentrainment, 606–607
Photopic vision, 235, 237
Photopigments, 232
in color vision, 239, 241–242
Photoreceptors, 229, 249
anatomical distribution of, 237–238
color vision and, 239, 241–242
light adaptation, 233–234
macular degeneration and, 236
M and P retinal ganglion cells, 268–269
phototransduction, 230, 232–234, 235
in retinal structure and function, 229–230
retinitis pigmentosa and, 231
signal amplification, 233
types of, 228
See also Cone cells/system; Rod cells/system
Phototransduction, 230, 232–234, 249
Phylogenetic memory, 667
Physiological nystagmus, 308–309
Pia mater, 33, 34
Picrotoxin, 157
Pigment epithelium, 227, 228, 229, 230, 236, 238
Pineal gland, 24, 609
Pinna, 278–279
PIP₂. *See* Phosphatidylinositol bisphosphate
Pitch (rotation), 300
Pitch (sound), 275
Pituitary gland, 24, 25, 456
Pituitary stalk. *See* Infundibular stalk
PKA. *See* cAMP-dependent protein kinase A

PKC. See Protein kinase C
Placebo effect, 219–220
Place cells, 542
Planning
  deficits in, 576–577
  delay-specific neurons in monkeys, 581, 582
Plant toxins, 90
  curare, 104, 157
Planum temporale, 595–597
Plasma membrane
  passive current flow and, 68, 69
  See also entries at Membrane
Point of fixation, 255
Poisson statistics, 105
Polymodal nociceptors, 210
  capsaicin and, 342
  trigeminal chemoreception and, 341, 342, 343
Polyneural innervation, 506, 507
Pons, 13, 14, 15, 16, 17, 23, 24, 25
  blood supply to, 37
  corticospinal tract and, 377
  descending projections from vestibular nuclei, 314
  serotonin and, 134
  taste processing and, 332
  trigeminal mechanosensory system and, 200
  ventricular space, 32
  vestibulo-ocular reflex and, 311
Pons-midbrain junction, 617, 619
Pontine nuclei, 411–412, 415
  inputs to cerebellum, 412
Pontine reticular formation, 372, 615
Pontine tegmentum, sleep and, 617
Pores, of ion channels, 86–87, 88–89
Positron emission tomography (PET), 28–29, 599

Postcentral gyrus, 21, 22, 23
Posterior (position), 12, 13
Posterior cerebral arteries, 35, 36, 37
Posterior chamber (eye), 223
Posterior circulation, 37
Posterior communicating artery, 35, 36
Posterior cranial fossa, 33, 34
Posterior funiculi, 202
Posterior inferior cerebellar artery (PICA), 16, 37, 39
Posterior speech cortex, 295
Posterior spinal artery, 35
Posterior ventral cochlear nucleus, 289
Posterodorsal nucleus of the medial amygdala, 659
Posteroventral cochlear nucleus, 288
Postganglionic axons, 447
Postganglionic neurons
  conductance changes and, 143, 145, 162–163
  dynamics of neurotransmitter receptors, 141–147
  in electrical synapses, 100
  excitotoxicity and, 130
  summation of postsynaptic potentials and, 149–150
  synapse formation and, 149–150
Postsynaptic current (PSC), 147
Postsynaptic potentials (PSPs)
  effect of postsynaptic current on, 147

  excitatory and inhibitory, 147–148, 149
  summation of, 149–150
Postsynaptic receptors
  classes of, 150–151
  dynamics of, 141–147
  neurotoxins and, 156–157
  See also Neurotransmitter receptors
Postsynaptic specialization, 3–4
Post-tetanic potentiation (PTP), 541
Postural control, 388
  brainstem motor control centers and, 369, 372–373
  feedforward and feedback mechanisms, 372–373
  vestibulospinal pathway in, 313–314
Potassium channels
  action potentials and, 80–81
  diversity in, 83
  functional states, 83
  GABA receptors and, 161
  genetic disorders of, 93
  muscarinic receptors and, 172
  properties and functions of, 83, 84
  structure of, 86–87
  in taste transduction, 336, 337, 338
  toxins affecting, 90
Potassium ions (K+)
  action potentials and, 50–51, 54, 56, 57, 59, 61, 62–65, 76
  dynamics of postsynaptic acetylcholine receptors and, 144–147
  electrochemical equilibrium, 46–50
  in mechanoelectrical sound transduction, 285, 286
  membrane conductance and, 62–64, 65
  release during tissue damage, 211
  resting membrane potential and, 51–52
  tetraethylammonium ions and, 61
  voltage-dependent current, 61
PPRF. See Paramedian pontine reticular formation

Precentral gyrus, 21–22, 23
Prefrontal cortex
  amygdala and, 641–642
  decision making and, 643, 644
  delay-specific neurons, 581, 582
  depression and, 640
  emotional behavior and, 643
  sexual dimorphism related to cognitive function, 660, 661
Preganglionic autonomic neurons, 631. See also Parasympathetic preganglionic neurons; Sympathetic preganglionic neurons
Preganglionic visceral motor neurons, 12, 19–20
Premotor cortex, 374, 376, 384–385, 388
  in basal ganglia disinhibitory circuitry, 401
Preoptic area, sex hormone receptors and, 677, 681
Preoptic nuclei, 456, 457
Preoptic area, sex hormone receptors in, 654
Pre-proenkephalin A, 136
Pre-proopiomelanocortin, 136
Pre-propeptides, 125
  proteolytic processing of, 136
Presbycusis, 277
Presbyopia, 227

presenilin genes, 678, 679
Presynaptic neurons, synapse formation and, 502–503
Presynaptic proteins
  overview of, 111–113
  toxins affecting, 114
Presynaptic terminals, 3–4, 5
  calcium channels in, 107, 109
  in chemical synapses, 101–102
  co-transmitters, 122–123
  diseases and disorders affecting, 108–109
  in electrical synapses, 100
  proteins associated with, 111–113, 115
Pretectum, 252
  visual system and, 253
P retinal ganglion cells, 268–269
Prevertebral ganglia, 447, 466
Primary auditory cortex, 289, 294–295, 296, 394

Primary motor cortex, 348
  cerebellar input to, 414
  descending pathways, 376, 377, 388
  direct and indirect influence on movement, 376, 388
  functional organization, 376, 379–384
  injury to, 375
  lateral corticospinal tract and, 374
  mapping of body musculature, 379–381
  overview of, 375
  projections from, 370, 373–375, 376, 378
  reorganization in amputees, 216
  sensory motor talents and space allocation, 213
Primary sensory endings, 198
Primary sensory neurons, 15
Primary somatic sensory cortex, 191, 204, 206, 207, 374
  mechanosensory pathways and, 200
  spike-triggered averaging of neurons, 381
  thalamic input to, 568
  spinothalamic tract and, 213
  trigeminal pain and temperature system, 213

Primary visual cortex
  activity-dependent modification, 531–532
  binocularity and stereopsis, 261–263
  columnar organization, 263, 265, 267
  critical period and, 525–526, 528–529, 530, 531, 534
  functional organization, 260–263
  mapping of orientation preference, 263, 265, 267
  ocular dominance and, 525–526, 528–529
  ocular dominance columns, 267, 524–525
  optical imaging of functional domains, 267
  orientation-selective neurons, 260–261
  output to basal ganglia, 394
  retinal ganglion input to, 252
  transneuronal labeling, 524, 527
  visual field retinotopic representation, 258
  See also Striate cortex
Primary visual pathway
  magnocellular and parvocellular streams, 269
  overview of, 251, 273
  retinal ganglion projections, 251–252
  striate cortex functional organization, 260–263
  See also Visual pathways

Priming, 668
Primitive pit, 471, 472
Primitive streak, 471
Principal nucleus of the trigeminal complex, 203, 213
Prion diseases, 418–419
Procedural memory, 665, 681
Progesterone, 650
Progesterone receptors, 654–655
Promoters, 178
Propanolol, 162
Propeptides, 125, 135
Proprioception, 197–199
Proprioceptors, 197–199, 201
Prosencephalon, 32, 478, 479. *See also* Forebrain
Prosopagnosia, 574–575
Prostaglandins, 211
Prostate, referred pain, 215
Protanomalous trichromacy, 241
Protanopia, 241
Protein kinase C (PKC), 176, 547
long-term synaptic depression and, 551–552
Protein kinases, 168–169
in intracellular signal transduction, 175–178
Protein phosphatases, in intracellular signal transduction, 175–176, 178
Protein tyrosine kinases, 176–177. *See also* Tyrosine kinases
Prozac®, 137, 641
Prusiner, Stanley, 418, 419
Psychiatric disorders
affective, 640–641
biogenic amine neurotransmitters and, 137
non-motor basal ganglia loops and, 406
See also Mental illness
Psychosurgery, 577, 579
Psychotherapeutic drugs, categories of, 137
Psychotropic drugs, biogenic amine neurotransmitters and, 137
Ptc gene, 335
Ptosis, 431, 458
Pulvinar, 568
Pupil, 224, 225–226
Horner's syndrome and, 458
Pupillary light reflex, 252–253, 255
Purinergic receptors, 161
Purines, as neurotransmitters, 135
Purkinje, J. E., 430
Purkinje cells, 2, 4
action potential in, 55
in cerebellar circuitry, 415–416, 417, 419
coordination and correction of movement, 419–420, 422
long-term synaptic depression and, 182–184, 551–552
loss of innervating fibers through time, 508
motor learning and, 417, 419, 421
Purkinje layer, 415
Putamen, 29, 31, 391, 410
anticipatory discharges, 395
in basal ganglia circuitry, 394, 399, 401
cortical inputs, 392–393, 395
in Huntington's disease, 400
in hypokinetic and hyperkinetic diseases, 404

inhibitory projections from, 396–397
Parkinson's disease and, 403
Pyramidal decussation, 374, 377
Pyramidal neurons, 2, 4, 394, 569
EEG rhythms and, 613
hippocampal, 542–543
long-term potentiation and, 543, 546–547
long-term synaptic depression and, 550
pyriform cortex, 330
Pyramidal smile, 628
Pyramidal tracts, 628
Pyridoxal phosphate, 131
Pyriform cortex, 22
olfactory processing and, 317, 318, 329–330

Quadrantanopia, 260
Quantal transmission, 103–105
Quanylate cyclase, 234
Quinine, taste and, 333, 339
Quinn, Chip, 540
Quisqualate, 157
Quisqualis indica, 157

Rab 3, 112
Rabphilin, 112
Radial glia, 488–489
Raederscheidt, Anton, 572
RAGS protein, 501–502
Raichle, Marc, 599
Raisman, Geoffrey, 651
Rami, gray and white communicating, 446, 447
Ramón y Cajal, Santiago, 1–2, 488, 494, 498
Random dot stereograms, 264
Raphe nuclei, 218
in wakefulness and sleep, 617, 618, 619
Rapid eye movement (REM) sleep
brain activity in, 615, 617
deprivation studies, 614
dreaming in, 614
EEG characteristics, 610
eye movements in, 615
narcolepsy and, 622
physiological characteristics, 611–613
pontine-geniculo-occipital waves, 615
possible functions of, 614–615
Rapidly adapting sensory receptors, 192
Ras, 171
Rasmussen's encephalitis, 554
Rats. *See* Rodents
Raviola, Elio, 226
Reason, emotion and, 643–644
Receptive aphasia, 592
Receptive fields
of retinal ganglion cells, 242, 243, 245–246
of somatic sensory receptors, 194, 195, 196–197
Receptor kinases, in neuronal differentiation, 487
Receptor molecules, 141. *See also* Cellular receptors; Neurotransmitter receptors
Receptor tyrosine kinases, 176
Reciprocal innervation, 355
Recognition deficits, 574–575
Recognition molecules, 500
Recognition-specific neurons, 580–581
Rectum, visceral motor system and, 448, 449
Rectus muscles
actions of, 428, 429

innervation of, 428, 429, 431
vestibulo-ocular reflex and, 310–311
Red nucleus, 369, 370, 371, 376, 388, 415
Red tide, 90
5-α-Reductase deficiency, 648
reeler mouse mutation, 423–424, 486
Reese, Tom, 38, 105, 106
Referred pain, 215
Reflexes
flexion reflex, 361
Golgi tendon organ, 358–361
muscle stretch reflex, 355–357
Reflex sexual function, 464–466
Refractive errors, in vision, 226–227
Refractory period, of action potentials, 72
Regeneration, of axons, 556–558, 560–561
Reissner's membrane, 280
Relay neurons, outputs to the cerebellum, 413
REM sleep. *See* Rapid eye movement sleep
Reserpine, 137
Respiratory epithelium, 323
Resting membrane potential, 43, 44, 51–52
Restless legs syndrome, 621–622
Reticular activating system, 615, 617, 618
Reticular formation, 369, 374
amygdala and, 637
cerebellar projections to, 415
descending pain modulatory systems, 218
description of, 371
EEG rhythms and, 613
emotional behavior and, 626, 627, 631
gaze centers and, 433
in postural control, 372–373, 388
primary motor cortex projections to, 376, 388
projections to the spinal cord, 370
smiling and, 628
Reticular nucleus, thalamus, 619
Reticular theory, 1
Reticulospinal tract, 371
Retina, 223, 224
blind spot, 254
development of, 227, 228, 485
distribution of photoreceptors in, 237–238
Eph receptors in, 502
formation of images on, 224–225, 227
functional organization, 227–230
neuronal types in, 4, 228
optic disk, 251, 254
retinitis pigmentosa, 231
visual field, 255–257
visual field deficits, 258
Retinal, 232, 234, 474
Retinal adaptation, 430
Retinal axons
growth cones, 494, 496
topographic map formation in optic tectum, 500–502
Retinal bipolar cells, 2
Retinal circuits, 250
for detecting differences in luminance, 242–243, 245–246
in light adaptation, 247–249
Retinal ganglion cells, 2, 228, 229, 249–250
central projections of, 251–253, 255, 273
connections to photoreceptors, 237
detecting differences in luminance, 242–243, 245–246
light adaptation, 247, 248
luminance contrast and, 245–246

M and P populations, 268–269
receptive fields, 242, 243, 245–246
visual field representation and, 257–258
Retinal images
formation of, 224–225, 227
stabilized, 430
Retinal neurons, 4
functional organization, 228–230
types of, 228
See also Photoreceptors; Retinal ganglion cells
Retinal pigment epithelium, 227, 229, 230, 236, 238
Retinex theory, 240
Retinitis pigmentosa, 231
Retinogeniculostriate pathway, 252, 273
Retinohypothalamic pathway, 255
Retinoic acid, 474–475
Retinoid receptors, 474, 475
Retinoids, 475
Retinol, 474
Retinotopic maps, 255–258, 500–502
Retrograde amnesia, 673, 676
Reversal potential, 143, 144, 145, 147, 148
Rexed's laminae, 201–202, 212, 220
Rhesus monkeys. See Monkeys
Rhinal fissure, 23
Rhodopsin, 169
in phototransduction, 232, 233
recycling of, 234
retinitis pigmentosa and, 231
Rhombencephalon, 32, 478, 479, 482
Rhombomeres, 482–483
Right cerebral hemisphere
emotion processing and, 642, 644
language function and, 587, 599–600
RIM, 112
Rising phase, action potential, 54
River blindness, 530
RNA polymerase, 178
Rod bipolar cells, 237
Rod cells/system, 228, 229
ABCR gene, 236
anatomical distribution, 237–238
connectivity to ganglion cells, 237
photopigment in, 232
responsiveness to luminance, 235, 237
sensitivity of phototransduction in, 235
spatial resolution of, 235
See also Photoreceptors
Rodents
genetic analysis of cerebellar function, 423–424
hormone-sensitive brain circuits in adults, 662, 663
sexually dimorphic behavior in, 645, 652–653
somatotopic maps in, 205
Roll (rotation), 300
Roses, Allen, 679
Rostral interstitial nucleus, 433
Rostral medulla
endogenous opioids and, 221
epinephrine and, 133
opiate-sensitivity, 220
pain modulation and, 220
Rostral pons, 464
Rostral (position), 12, 13
Rostrocaudal bands, 394–395, 397
Round window, 278, 279–280

Rubrospinal tract, 370, 371, 374
Ruffini's corpuscles, 190, 193
Ruggero, M., 285
rutabaga gene, 540
Ryanodine receptor, 174

Saccades, 398, 399, 430
antisaccades, 437
cerebellar contributions, 437
express, 437
hypometric and hypermetric, 420
neural control of, 433–439
overview of, 431
in REM sleep, 615
Saccharin, taste responses, 335
Saccular macula, 299, 301
Sacculus, 297
detection of head movements, 300–302
electrical resonance and, 305
hair cells in, 298, 299, 305
Sacral nerves, 14, 18
Sacral parasympathetic neurons
in bladder regulation, 462, 464
in sexual responses, 465–466
Sacral preganglionic innervation, 450
Sagittal (position), 12, 13
Sakmann, Bert, 79
Salicylic acid, 211
Salivary nuclei, 447, 451
Salt, taste responses, 335, 337
Saltatory propagation, of action potentials, 73, 74
Salvesen, Guy, 678
Sanes, Joshua, 504, 505
Sarin, 128
Savant syndrome, 670
Saxitoxin, 90
Scala media, 280
Scala tympani, 280, 286
Scala vestibuli, 280
Scarpa's ganglion, 298, 307, 310, 311
Schaffer collaterals
in hippocampus, 542–543
long-term potentiation and, 543, 546, 547
long-term synaptic depression and, 550
Schiller, Peter, 243
Schizophrenia, 406
Schwab, Martin, 558, 560–561
Schwann cells, 7, 11, 193
axonal regeneration and, 556, 560
Sclera, 223, 224
Scopolamine, 153, 157
Scorpion toxins, 90
Scotomas, 254, 258
Scotopic vision, 235, 237
Scrapie, 418
Scrotum, 646
SDN. See Sexually dimorphic nucleus
Seals, sleep in, 605
Searle, John, 616
Sea slugs, gill withdrawal reflex, 535–539
Secondary auditory cortex, 294, 295
Secondary somatic sensory cortex (SII), 207
Second language acquisition, 523–524
Second messengers
ligand-gated channels and, 86
overview of, 172–175
regulation of gene expression, 178–181
regulation of protein phosphorylation, 175–178

See also Cyclic AMP-mediated second messenger pathways
Second-order neurons, 202, 203
Second pain, 210–211
Sedatives, 133
Segmental ganglia, 362
Segmental nerves, 18
Segmental reflexes, Rexed's laminae and, 202
Seizures, 93. See also Epilepsy
Selective permeability, 45–47
resting potential and, 51–52. See also Membrane permeability
Selective serotonin reuptake inhibitors (SSRIs), 137, 641
Selectivity filters, in ion channels, 87
Selegiline, 133
Self-awareness, 616
Semaphorins, 499
Semicircular canals, 278, 297, 298, 316
functional organization of, 306–307
Seminal vesicles, 646
Senile plaques, 678, 679
Sensitization
in invertebrates, 536–539
Sensorineural hearing loss, 279
Sensory aphasia, 592
Sensory fibers, of visceral motor system, 453–454
Sensory ganglia, 12, 199
dermatomes and, 201
Sensory maps, 435, 438. See also Somatotopic maps; Topographic maps
Sensory motor integration
eye movements and, 427
in superior colliculus, 435, 438–439
Sensory motor talents, cortical space allocation and, 383
Sensory neurons
development of, 473
myotatic spinal reflex, 9, 10
Sensory stimuli, 190, 192
Sensory systems, 10. See also individual systems
Sensory transduction, 189–190, 192
Septal forebrain nuclei, 30
Septum, sex hormone receptors in, 654
Septum pellucidum, 32
Serine hydroxymethyltransferase, 131
Serotonergic neurons, 134
in wakefulness and sleep, 617, 619
Serotonergic nuclei, 570
Serotonin, 120, 131
chemical structure, 121
distribution in brain, 134
functional features of, 127
overview of, 134–135
release during tissue damage, 211
in sensitization, 538
synthesis of, 135
uptake blockers, 137
Serotonin receptors, 161
Ser/Thr kinases, 176
Ser/Thr phosphatases, 176, 178
Sertraline, 641
sevenless gene, 487
Sex
categories of meaning, 647–648
reassigning, 649

Sex development
  human, 646–647
  intersexual variations, 648–649
Sex hormones
  effects on neural circuitry and activity, 651–652, 655, 663
  plasticity in adult neural circuits, 662, 663
  receptors in brain, 654–655
  sexual dimorphism and, 649–651
  sexual identity and, 658
Sexual behavior
  dimorphic, 645, 647, 652–653, 655–660
  hypothalamus and, 655, 656, 658, 659
Sexual dimorphism
  behavioral, 645, 647, 652–653, 655–660
  central nervous system and, 652–653, 655–660
  in cognitive function, 660–661
  overview of, 645, 663
  sex hormones and, 649–651
Sexually dimorphic nucleus (SDN), 651–652
Sexual orientation, 658–659
Sexual responses, autonomic regulation of, 464–466
Sham rage, 627
Sherrington, Charles, 347, 379
Shock, dopamine and, 133
Short circumferential artery, 39
Short-term memory, 666, 668
Short-term sensitization, 536, 537, 538
Sigmoid, visceral motor system and, 448, 449
Sigmundson, Keith, 549
Signal amplification
  in intracellular signal transduction, 166, 167
  in phototransduction, 233
Signaling molecules, 167–168
  in neuronal differentiation, 487–488
Signal transduction pathways. See Intracellular signal transduction
Sign language, 600–601
Sildenafil, 466
Silent synapses, 548–549
Simple cells, 261
Single-photon emission computerized tomography (SPECT), 29
Skeletal muscles
  Golgi tendon organ, 358–361
  innervation of, 349–350, 351–352
  local motor neurons and, 347, 349–350
  maintenance of tone and length, 357
  motor units, 351–352
  regulation of force, 353–354
  stretch reflex, 355–357
  See also Muscles
Skin
  mechanoreceptors in, 192–193
  sensory receptors in, 189–190, 192
  tactile discrimination and, 199
Sleep
  circadian clock and, 606–607, 609
  in different species, 605
  disorders in, 619–622
  dreaming, 614–615
  human requirements, 603
  neural circuits governing, 615, 617, 618–619
  overview of, 603
  physiological changes in, 611–613
  possible functions of, 614–615

possible reasons for, 603–604
  sleep deprivation, 604–605
  stages of, 609–610
Sleep apnea, 620–621
Sleep disorders, 619–622
Sleep spindles, 610, 619, 620
Sleepwalking, 613
Slow axonal transport, 125
Slowly adapting sensory receptors, 192
Slow motor units, 352, 353
Slow-wave sleep, 610, 618
Small cell carcinoma, Lambert-Eaton myasthenic syndrome and, 108
Small clear-core vesicles, 126
Small G-proteins, 170–171
Small intestine, visceral motor system and, 448, 449
Small-molecule neurotransmitters, 120–123
  packaging in synaptic vesicles, 126
  rate of secretion, 126
  removal from synaptic cleft, 127
  synthesis of, 124, 125
Smiling, 628–629, 642, 643
Smith, Neil, 670
Smooth pursuit eye movements, 431–432, 440
Snake venom, 156
Snapin, 111, 112
SNAP proteins, 111, 112, 113, 114
SNARE proteins, 109, 111, 113, 114
SOD1 gene, 367
Sodium channels
  characteristics of, 79–80
  diversity of, 82–83
  functional states, 82
  genetic disorders of, 93
  GTP-binding protein regulation of, 172
  inactivation, 81
  in phototransduction, 232
  structure of, 88
  in taste transduction, 336, 337, 338
  toxins affecting, 90
Sodium chloride, taste responses, 335, 339, 340
Na⁺/Ca²⁺ exchanger, 91, 172
Na⁺/H⁺ exchanger, 91
Sodium ions (Na⁺)
  action potentials and, 50–51, 53, 54, 56, 57, 59, 60–61, 62–65, 70, 76
  dynamics of postsynaptic acetylcholine receptors and, 144–147
  electrochemical equilibrium, 49–50
  membrane conductance and, 62–64, 65
  in phototransduction, 232
  tetrodotoxin and, 61
  voltage-dependent current, 60–61
Na⁺/K⁺ pump, 91
  electrogenic properties, 95
  functional properties of, 93–95
  molecular structure of, 95–96
Soleus, 350, 353
Somatic motor division, 11
Somatic motor system, 12
  See also Motor neurons
Somatic sensory cortex, 4, 22
  functions of regions, 204–205
  modification by sex-hormones, 662, 663
  modularity in, 206
  plasticity in, 555–556
  regions of, 204

somatotopic mapping in, 204, 205
  trigeminal chemosensory system and, 341
Somatic sensory receptive fields
  dynamic aspects of, 196–197
  in mechanosensory discrimination, 194, 197
  response to anesthetics, 555
  topographic mapping in insects, 195
Somatic sensory receptors
  functional groups, 189
  major classes of, 190
  mechanosensory discrimination and, 194, 197
  phasic and tonic, 192
  receptive fields, 194–197
  sensory stimuli and, 190, 192
  sensory transduction in, 189–190
  See also Mechanoreceptors
Somatic sensory system, 191
  active tactile exploration, 199
  critical period in, 533
  dermatomes, 201
  descending projections, 207
  dorsal column-medial lemniscus pathway, 199–202
  dynamic information processing in, 196–197
  higher-order cortical fields, 207
  in insects, 195
  mechanosensory discrimination, 194, 197
  mechanosensory processing in, 189
  overview of, 189, 207
  plasticity in, 553–556
  receptive fields, 194–197
  sensory receptors, 189–190, 192–193, 195
  thalamic components, 202, 203
  trigeminal mechanosensory system, 202–203
Somatostatin-14, 122
Somatotopic maps
  cerebellum and, 413–414
  distortions in, 205
  plasticity in, 553, 554
  in somatic sensory cortex, 204, 205
  See also Topographic maps
Somites, 201
Songbirds
  generation of new neurons in, 559
  song learning in, 522–523
Sonic hedgehog, 475
  sonic hedgehog gene, 475
Sorge, W. A., 285
Sound
  audible spectrum, 276, 278
  distortion, 285
  physical properties of, 275–276
  temporal structure and communication, 294–295
  See also Auditory stimuli; Auditory system
Sound localization
  external ear and, 278–279
  neural basis of, 288, 290–292
Sour compounds, taste responses, 335–336, 337
Space coding, olfactory system and, 327
Spasticity, 386, 387, 388
Spatial memory, long-term synaptic potentiation and, 542
Sperry, Roger, 500, 594

Spherical aberration, 225
Spike-triggered averaging, 381
Spina bifida, 476
Spinal anesthesia, 18
Spinal cord, 11
　anatomical planes, 12
　anatomy of, 14, 18–20
　blood supply to, 35
　corticospinal tract and, 377
　descending projections to, 314, 371, 373–375
　development of, 478
　intermediolateral column, 444, 445, 446–447
　local circuit neurons and (*see* Local circuit neurons)
　locomotion and, 361, 366
　lower motor neurons and (*see* Lower motor neurons)
　mechanosensory pathway, 199–202
　outputs to the cerebellum, 413
　parasympathetic preganglionic neurons and, 447, 450
　reflexes and, 9, 10, 355–357, 361
　upper motor neuron pathways and, 370, 373–375
　ventricular space, 32
　vestibular nuclei inputs, 32
Spinal cord injury, 202
　dissociated sensory loss, 212
Spinal cord reflex, 355–357, 386
Spinal lemniscus, spinothalamic tract and, 213
Spinal nerves, 18
　dermatomes and, 201
Spinal nucleus of the bulbocavernosus, 652–653
Spinal nucleus of the trigeminal complex, 203, 213
Spinal shock, 386
Spinal trigeminal nucleus, 16
Spinal trigeminal tract, 212, 213, 214
Spinocerebellum, 410, 411, 413
　consequences of lesions to, 425
　eye movement and, 420
　projections from, 415
Spinothalamic pathway, 201, 212, 213, 498
Spiral ganglion, 280
Splenium, 660
Split-brain patients, language function studies in, 593–595
Spongiform degeneration, 418
Squid neurons
　action potential, 55
　axon ion channels, 81
　description of, 53
　extracellular and intracellular ion concentrations, 50
　resting membrane potential, 51–52
SRY gene, 646
Stage I sleep, 610
Stage II sleep, 610
Stage III sleep, 610
Stage IV sleep, 610
Stapes, 278
Stargardt disease, 236
Status epilepticus, 130
Stellate neurons, 416, 417, 569
Stem cells, 483, 486

in neural development, 472
　neural grafting and, 403
Stereocilia, 282, 283, 298, 299
Stereograms, 264–265
Stereopsis, 262–263
Stereoscopy, 264
Steroid receptor coactivators, 655
Steroid / thyroid hormone receptors, 474
Stimulants, 137
Stomach, visceral motor system and, 448, 449
Stomatogastric ganglion, 364–365
Strabismus, 263, 530, 532
Stretch-activated ion channels, 86
Stretch reflex, 355–357
　clasp-knife phenomenon and, 387
Striate cortex
　columnar organization, 260, 265, 267
　functional organization, 260–263, 273
　optical imaging of functional domains, 267
　remodeling of synaptic connections through time, 508
　See also Primary visual cortex
Striatum, 29, 31. *See also* Corpus striatum
Stria vascularis, 286
Striola, 299, 301–302
Striosomes, 395
Strittmatter, Warren, 678
Strokes, 33, 39
　excitotoxicity and, 130
　internal capsule and, 30
Stroop Interference Test, 583
Strychnine, 157
　taste perception, 333
Strychnos nux-vomica, 157
Styron, William, 640
Subarachnoid hemorrhage, 35
Subarachnoid space, 33, 34
Subcutaneous sensory receptors, 189–190, 192, 193
Subdural hemorrhage, 35
Sublingual glands, 448, 449
Submandibular glands, 448, 449
Submucus plexus, 452, 453
Substance dependence, 138
Substance P, 122, 135–136, 211
Substantia gelatinosa, 212
Substantia nigra pars compacta, 25, 391
　dopamine and, 133
　in hypokinetic and hyperkinetic diseases, 404
　inputs to striatal medium spiny neurons, 395
　outputs to basal ganglia, 393
　Parkinson's disease and, 402–404, 405
Substantia nigra pars reticulata
　in basal ganglia disinhibitory circuitry, 396–397, 399, 401–402
　basal ganglia output to, 392
　hypokinetic movement disorders and, 404, 405
　inhibitory projections from medium spiny neurons, 396–397
　projections from, 397, 398
　saccadic eye movements and, 398
Subthalamic nuclei, 391, 392, 396, 401–402, 404
Succinic semialdehyde dehydrogenase, 131
Sugars, taste responses, 335, 337, 338, 339, 340

Suicide, 640
Sulci, 20
Summation, of synaptic potentials, 149–150
Superficial reflexes, 386, 388
Superior cerebellar peduncle, 410, 411, 415
Superior cervical ganglion, 458, 502–503
Superior colliculus, 16, 24, 25, 369, 370, 371, 397, 398, 415
　in basal ganglia circuitry, 396, 399
　REM eye movements and, 615
　retinal ganglion projections to, 255
　saccadic eye movements and, 435–437, 440–441
　sensory motor integration in, 435, 438–439
　ventricular space, 32
　visual system and, 253
Superior division, of visual field, 255, 256–257
Superior homonymous quadrantanopsia, 259
Superior oblique muscle, 428, 431
Superior olivary complex, 277, 295
Superior rectus muscle, 428
Superior salivatory nuclei, 447
Superior temporal association cortex, 676
Superior temporal gyrus, 294
Supertasters, 335
Supporting cells, 2
Supporting cells, 335
Supraoptic nucleus, 456, 457
Suprachiasmatic nucleus, 255, 456, 457
　circadian clock and, 607, 609
　sexual orientation and, 658, 659
SV2, 112
Swaab, Dick, 655, 658, 659
Sweet compounds, taste responses, 335, 337, 338
Sylvian fissure, 21, 22
Sympathetic chain, 447, 466
Sympathetic ganglia, 448, 466
　dopamine and, 133
　neurotransmitters released on targets, 455
　noradrenergic receptors of targets, 458
　norepinephrine and, 133
Sympathetic preganglionic neurons
　acetylcholine and, 454
　innervation targets, 447
　location of, 448, 453, 466
　organization of, 444, 445, 446–447
　in regulation of sexual responses, 465
Sympathetic system
　actions of, 444
　bladder regulation, 463–464
　control of cardiovascular function, 460–462
　emotional behavior and, 625
　functions of, 448
　ganglia in, 12
　Horner's syndrome, 458
　innervation targets, 447
　neurotransmission in, 454–455, 458–459
　organization of, 444, 445, 446–447
　regulation of sexual responses, 464, 465, 466
Synapse elimination, 507
Synapse formation
　competitive interactions and refinements, 506–508
　molecular signals in, 504–505
　selective, 502–503
　on target cells, 503

Synapses, 5–6, 115
  chemical, 101–105
  classes of, 189
  discovery of, 1–2
  electrical, 99–101
  excitatory and inhibitory synaptic potentials, 147–148, 149
  neuromuscular, 103–105
  neurotoxins and, 90, 108–109, 114, 152, 156–157
  silent, 548–549
  synaptic terminals, 3–5
  See also Synaptic transmission
Synapsin, 112, 113
Synaptic cleft
  in chemical synapses, 101, 102
  removal of neurotransmitters from, 126–127
Synaptic depression, 541
  long-term, 541, 550–552
Synaptic facilitation, 539, 541
Synaptic plasticity
  epilepsy and, 554–555
  in invertebrates, 535–539
  long-term depression, 541, 550–552
  long-term potentiation, 541, 542–547, 550
  overview of, 535, 552, 561
  short-term, 539, 541
  See also Neural plasticity
Synaptic potentials
  excitatory and inhibitory, 147–148, 149
  summation of, 149–150
  See also Postsynaptic potentials
Synaptic terminals, 3–5
Synaptic transmission, 5–6, 115
  at chemical synapses, 101–105
  co-transmitters, 122–123
  diseases and disorders affecting, 108–109, 114
  dynamics of postsynaptic receptors, 141–147
  at electrical synapses, 99–101
  excitatory and inhibitory postsynaptic potentials, 147–149
  excitotoxicity and, 130
  local recycling of synaptic vesicles, 106–107
  at neuromuscular synapses, 103–105
  quantal, 103–105
  transmitter secretion, 105–106, 107, 109, 111–113, 115
Synaptic vesicle cycle, 106–107
  proteins associated with, 111, 113, 115
Synaptic vesicles, 6, 101
  ATP in, 135
  co-transmitters, 122–123
  local recycling, 106–107
  myasthenic syndromes and, 108
  packaging neurotransmitters in, 126
  proteins associated with, 111–113, 115
  release of neurotransmitters from, 105–106, 107
  types of, 126
Synaptobrevin, 111, 112, 113, 114
Synaptophysin, 112
Synaptotagmin, 111, 112, 113, 114
Synein, 112
Syntaphilin, 111, 112
Syntaxin, 111, 112, 113, 114
Synthetic sweeteners, taste responses, 335

Tachistoscopic presentation, 594
Tactile perception
  active exploration, 199
  context of stimuli, 197
  mechanosensory discrimination, 194, 197
Tail-flip escape reflex, 312–313
Tartini, Giuseppe, 285
Tastants
  idiosyncratic responses to, 335–336
  taste perception, 332–334
Taste buds, 330, 331
  distribution on tongue, 333
  innervation of, 335
  structure of, 333, 334
  taste transduction and, 334–335
Taste cells, 330, 331, 333, 334–335
  adaptation, 338, 340
  gustatory selectivity, 338
  taste transduction, 336–338, 340
Taste perception, 332–334, 343–344
  idiosyncratic responses, 335–336
Taste pores, 333, 335
Taste receptors, 336–338
Taste system
  central projections, 331–332
  gustatory selectivity, 338
  neural coding in, 339, 340–341
  organization of, 330–332
  overview of, 317, 343–344
  peripheral, 334–335
  taste cell adaptation, 338, 340
  taste receptors and taste transduction, 336–338, 340
Taste transduction, 336–338, 340, 343–344
Tectorial membrane, 280, 282
Tectum, 25
Tegmentum, 25
Telencephalon, 32, 479
Temperature perception
  spinal trigeminal tract, 212, 213, 214
  spinothalamic tract, 212, 213
Temporal association cortex, 565
  language function and, 587
  memory and, 676
  recognition deficits, 574–575
  thalamic connectivity to, 568
Temporal coding, olfactory system and, 327, 328
Temporal cortex
  language function and, 599
  projections to basal ganglia, 393
  recognition-specific neurons in monkeys, 580–581
Temporal division, of visual field, 255, 256, 257
Temporal lobe, 21, 22
  language function and, 592
  left-right asymmetry in, 595–597
  memory and, 673, 677, 681
  olfactory processing and, 329–330
  recognition deficits and, 574–575
  visual field deficits and, 259
  in visual processing, 251, 270, 272, 273
Teratogenesis, 475
Teratogens
  alcohol, 476
  retinoid, 474–475
  thalidomide, 476
Terminal ganglion, 195
"Testes-at-twelve" syndrome, 648
Testicular determining factor, 646, 647

Testicular feminization, 648
Testosterone, 646, 650, 651
Tetanus, fused, 354
Tetanus toxin, 108, 109, 114
Tetraethylammonium ion, 61
Tetrodotoxin, 61, 80, 90, 109
Thalamocortical neurons
  electrophysiological states, 618
  in sleep, 618–619
Thalamocortical pathways, 315
Thalamus, 31, 410
  amygdala and, 641
  auditory processing and, 289, 293–294
  basal ganglia circuitry and, 396, 397, 398, 401
  connectivity to cortical regions, 568
  depression and, 640
  descending pain modulatory systems and, 218
  EEG rhythms and, 613
  in hypokinetic and hyperkinetic diseases, 404
  inputs from the cerebellum, 415
  Korsakoff's syndrome and, 676
  limbic system and, 632
  mechanosensory components, 202, 203
  mediodorsal nucleus of, 640, 641
  motor pathways and, 397
  nociceptive components, 214, 216
  olfactory processing and, 318, 330
  retinal ganglion projections to, 252
  sleep and, 615, 618–619
  somatic sensory system and, 191
  taste processing and, 331, 332, 340, 341
  trigeminal chemosensory system and, 341
  trigeminal pain and temperature system, 213
  vestibular system and, 314, 315
  See also Dorsal thalamus; Ventral posterior lateral nucleus; Ventral posterior medial nucleus
Thalidomide, 476
Theophylline, 161
Thermoceptors, 189
  dorsal root ganglia and, 199
Theta waves, 613
Thiamine deficiency, Korsakoff's syndrome and, 676
Third-order neurons, 202
Third ventricle, 31, 32
Thoenen, Hans, 514
Thoracic nerves, 14, 18
Threshold potential, 44, 66
Thrombotic strokes, 33
Thyroid hormone, nuclear receptors, 181
Thyrotropin releasing hormone, 122
Tight junctions
  blood-brain barrier and, 38
  cochlear hair cells and, 286
  vestibular hair cells and, 298
Tinbergen, Niko, 667
Tip links, 283, 285
Tissue damage, substance released following, 211
Tobacco, 156
Tolerance, 138
Tomosyn, 111, 112
Tongue
  taste buds, 334–335
  taste perception, 333–334

Tonically active state, of thalamocortical neurons, 618
Tonic inhibition
basal ganglia motor circuits and, 397, 398, 400–401
movement disorders and, 404, 405
Tonic sensory receptors, 192
Tonotopy, 277, 281–282
Topographic maps
of auditory space, 292
cochlear, 294
cortical motor maps, 379–381, 383
distortions in, 205
plasticity in, 553, 554
retinotopic, 255–258, 500–502
sensory motor integration in superior colliculus, 435, 438
See also Somatotopic maps
of sensory neuron receptive fields, 195
Torand-Allerand, Dominique, 651
Toscanini, Arturo, 669
Tourette's syndrome, 406
Toxins. See Neurotoxins
α-Toxins, 90
β-Toxins, 90
Trachoma, 530
Tracts, 11–12
Tranquilizing drugs, 160
Transcription, regulation of, 178–181
Transcriptional activator proteins, 178–181
Transcription factors, 178–181, 185, 475
Transducin, 232, 233
Transforming growth factors (TGFs), 475, 487–488
Transgenderism, 658
7-Transmembrane receptors, 169
Transneuronal labeling, 524, 527
Transneuronal transport, 527
Transverse pontine fibers, 412
Tranylcypromine, 133
Trazodone, 137
Tremors, 425
2,3,6-Trichloroanisole, 319
Trichromacy, 241
Tricyclic antidepressants, 137
Trigeminal brainstem complex, 137
Trigeminal chemosensory system, 317, 341, 343, 344
Trigeminal ganglion, 200, 202–203, 213
Trigeminal lemniscus, 200, 203, 213, 214
Trigeminal mechanosensory system, 200
Trigeminal nerve, 15, 16
characteristics of, 18–19
subdivisions of, 202–203
trigeminal chemosensory system and, 341
Trigeminal nucleus, 341
Trigeminal pain and temperature system, 212, 213, 214
Trigeminal somatic sensory system, 202–203
Trigeminothalamic tract, 200, 203, 213, 214
Triplet repeat diseases, Huntington's disease, 400
Trk receptors, 169, 512–513, 515
Trochlear nerve, 15, 16, 18–19, 431. See also Cranial nerve IV
Trochlear nucleus, 429
Trophic interactions, 503, 505–506
neurotrophins, 510–516
refinement of synaptic connections, 506–508

Trophic molecules, 498, 516
Tropic molecules, 498
Tropical spastic paraparesis, 75
Tryptophan, 134, 135
Tryptophan-5-hydroxylase, 134
Tsimpli, Ianthe-Maria, 670
Tuber cinereum, 456
δ-Tubocurarine, 157
Tuffed cells, 326
Tuning curves, 287
Tunnel vision, 231
Turner's syndrome, 648
Two-point discrimination, 194, 197
Tyler, Chris W., 265
Tympanic canal, 280
Tympanic membrane, 278, 279
Tyrosine, 133
Tyrosine hydroxylase, 131, 184, 403
Tyrosine kinase receptors, 176
NGF/TrkA pathway, 181–182
overview of, 168–169
Tyrosine kinases, 176–177

Unc-6 gene, 498
Uncus, 22, 23
Undershoot phase, action potential, 54, 65
Unipolar depression, 640, 641
Upper motor neuron pathways
brainstem motor control centers, 370–375
damage syndromes, 386, 388
direct and indirect projections to the spinal cord, 373–375
overview of, 369, 388
patterning of connections in spinal cord, 370
Upper motor neurons
basal ganglia and, 397
cerebellum and, 348, 415
cranial nerve motor nuclei and, 378
modulation of local motor neurons, 347, 357–358
modulation of stretch reflex gain, 357
movement and, 348
premotor cortex and, 384–385
saccadic eye movements and, 398, 399
Upper motor neuron syndrome, 386, 388, 628–629
Upper thoracic chain, 458
Urbach-Wiethe disease, 638–639
Ureter
referred pain, 215
visceral motor system and, 215
Urethral sphincter muscles, 463, 464
Urinary system, referred pain, 215
Urination, 464
Urogenital groove, 646, 647
Utricle, 297
detection of head movements, 300–302
electrical resonance and, 305
hair cells in, 33, 298, 299
Utricular macula, 299, 301

Vagina, 647, 652
Vagus nerve, 15, 16, 452, 462
cardiovascular function and, 461
characteristics of, 18–19
enteric system and, 452
nucleus of the solitary tract and, 453
parasympathetic system and, 451

Valenstein, Eliot, 579
Valium®, 137, 160
Valproic acid, 555
Vascular endothelium, 38
Vas deferens, 646
Vasoactive intestinal peptide, 122
Vasopressin, 118, 120, 122, 457
VA/VL complex of thalamus. See Ventral anterior nucleus; Ventral lateral nucleus
V4 cortex, 270
Ventral anterior nucleus, 392, 397
basal ganglia circuitry and, 396, 398, 401
in hypokinetic and hyperkinetic diseases, 404
Ventral column (spinal cord), 20
Ventral corticospinal tract, 374, 376
Ventral horns (spinal cord), 20
Ventral lateral nucleus, 392, 397, 414
mechanosensory system and, 202, 203
pain perception and, 214, 218
trigeminal chemosensory system and, 341
trigeminal pain and temperature system, 213
Ventral parts of basal ganglia, 632
Ventral (position), 12, 13
Ventral posterior lateral (VPL) nucleus, 191, 202, 203
mechanosensory system and, 200
pain perception and, 214, 218
spinothalamic tract and, 213
trigeminal mechanosensory system and, 200
Ventral posterior medial (VPM) nucleus, 191, 203
pain perception and, 214, 218
taste processing and, 331, 332
trigeminal chemosensory system and, 341
trigeminal pain and temperature system, 213
Ventral roots (spinal cord), 18, 20
Ventral-tegmental area, 138
Ventricular system, 31–33
Ventricular zone, 481, 483, 485
neural stem cells and, 559
Ventrolateral column (spinal cord), 20
Ventrolateral nucleus of the hypothalamus, 654
Ventrolateral preoptic nucleus (VLPO), 617, 619
Ventromedial nucleus, 456, 457
Veratridine, 90
Vergence center, 440
Vergence eye movements, 432, 440
Vermis, 410, 411
Vertebral arteries, 35, 36, 37
Vertical gaze center, 433
Vestibular hair cells
adaptation, 304
electrical tuning, 304–305
mechanoelectrical transduction in, 298
orientations of, 299, 306
otolithic, 300–302
semicircular canals, 306–307
Vestibular labyrinth, 297–298, 312
Vestibular nerve, 278
characteristics of, 18–19
detection of angular acceleration, 307, 310
labyrinth and, 298

preganglionic parasympathetic neurons and, 450, 453
taste system and, 331, 334, 340
trigeminal chemosensory system and, 341

response to otolith organs, 302–303, 306
Scarpa's ganglion and, 310
spontaneous firing in, 298–299, 307
Vestibular nerve ganglion, 310
Vestibular nuclei, 16, 369, 370–371, 372, 388
disruption of pathways to, 425
inputs to, 310, 415
output to cerebellum, 412, 413
postural control and, 313–314
thalamocortical system and, 314
vestibulo-ocular reflex and, 310–312
vestibulospinal pathway and, 313–314
Vestibular system
caloric testing of, 308–309
descending pathways, 314
detection of angular acceleration, 306–307, 310
detection of displacement and linear acceleration, 300–303, 306
labyrinth structure, 297–298
otolith organs, 300–302
overview of, 297, 316
semicircular canals, 306–307
thalamocortical pathway, 315
vestibular hair cells, 298–299, 304–305
vestibular nerve, 298–299, 302–303, 306, 307, 310
vestibulo-ocular reflex, 310–312
vestibulospinal pathway, 313–314
Vestibule, 278
Vestibulocerebellum, 409, 410, 411
consequences of lesions to, 422, 425
inputs to, 413
Vestibulocochlear nerve, 15
Vestibulo-ocular reflex, 310–312, 420–421, 422, 432–433
Vestibulospinal pathway, 312, 313–314, 371
Viagra®, 466
Visceral motor division, 11
Visceral motor system
autonomic regulation by, 443, 460–466
cardiovascular function and, 460–462
central control of, 454, 467
early studies of, 443–444
emotional behavior and, 625–626, 631, 644
enteric nervous system, 450, 452–453
functions of, 448–449
Horner's syndrome, 458
hypothalamus and, 454, 456–457
major controlling centers in, 443
neurotransmission in, 454–455, 458–460
overview of, 443, 466–467
parasympathetic division, 445, 447, 449, 450
regulation of, 467
sensory components, 453–454
sexual responses and, 464–466
sympathetic division, 443, 444, 445, 446–447, 448
Visceral pain, 215
Visual cortex. See Primary visual cortex; Striate cortex
Visual disorders/deficits, 530–531
in color vision, 239, 241, 242
congenital stationary night blindness, 93
from damaged extrastriate regions, 271–272
effects on binocular vision, 263
glaucoma, 224
light-mediated eye growth and, 226–227

macular degeneration, 235, 236
myopia, 226
refractive errors, 226, 227
retinitis pigmentosa, 231
visual field deficits, 258–260
Visual field
binocular, 255–257
blind spot, 254
deficits, 258–260
description of, 255
extrastriate representations, 270–271
retinal ganglion cell projections, 257–258
retinotopic representation of, 255–258
Visual grasp, 438
Visual pathways
central projections of retinal ganglion cells, 251–253, 255
extrastriate visual areas, 270–272, 273
interactions between systems, 272
magnocellular and parvocellular streams, 269
M and P retinal ganglion cell populations, 268–269
overview of, 251, 273
pupillary light reflex, 252–253, 255
striate cortex functional organization, 260–263
visual field representation, 255–258
See also Primary visual cortex; Striate cortex
Visual perception
eye movement characteristics, 427–428
magnocellular and parvocellular streams in, 269
stabilized retinal images and, 430
Visual system
activity-dependent modification, 531–532
alpha rhythms and, 613
anatomical distribution of photoreceptors, 237–238
color vision, 239, 240, 241–242
critical period, 525–526, 528–529, 530, 531, 534
detecting differences in luminance, 242–243, 245–246
eye anatomy, 223–224
formation of retinal images, 224–225, 227
light adaptation, 233–234, 247–249
luminance contrast, 245–246
ocular dominance, 525–526, 528–529, 530–531
ocular dominance columns, 524–525
overview of, 223, 249–250
perception of luminance, 244–245
phototransduction, 230, 232–234
retina structure and function, 227–230
transneuronal labeling, 524, 527
visual deficiencies, 530–531
Vitamin A, 475
Vitamin B6, 131
Vitamin B6, deficiency, 676
Vitreal "floaters," 224
Vitreous humor, 224
VLPO. See Ventrolateral preoptic nucleus
Vocabulary, 591
Volhard-Nusslein, C., 480
Voltage clamping, 58, 59
Voltage-gated ion channels
action potentials and, 80–81
characteristics of, 77, 79–80

defined, 81
diversity of, 81, 82–83
functions of, 83
genetic disorders of, 92–93
GTP-binding protein regulation of, 172
in hair cells, 285, 286
overview of, 77, 96
structure of, 88–89
in taste transduction, 336, 337–338
voltage sensors, 81, 89
Voltage sensors, 81, 89
Voluntary facial paresis, 629
Vomeronasal organs, 321
VP protein, 608, 609
VR-1 receptors, 342

Wada, Juhn, 397
Wada test, 597
Wakefulness, cellular basis of, 615, 617
Wall-eyedness, 530
Wardenburg syndrome, 479–480
Wasp venom, 157
Watkins, Jeffrey, 130–131
Watts, James, 579
Waveform, of sound, 275
weaver mouse mutation, 423, 424
Wernicke, Carl, 590, 592
Wernicke's aphasia, 592–593
Wernicke's area, 295
White communicating rami, 446, 447
White matter, 12
spinal cord, 19, 20
Whole-cell patch clamping, 78
Wieschaus, E., 480
Wiesel, Torsten, 206, 226, 260–261, 524, 525–526, 528–529
Wilkinson, R., 482
Willis, Thomas, 158
wingless (wg) gene, 478
Wisconsin Card Sorting Task, 589
Withdrawal syndromes, 138
Wolffian ducts, 646, 647
Woolsey, Clinton, 380
Words, 585
Working memory, 666
Wurtz, Robert, 397

Xanthines, 161
X chromosome, X-linked abnormalities, 479
Xenopus, ion channel expression in oocytes, 85
X-linked congenital stationary night blindness, 93
X-linked hydrocephalus, 495
X-linked retinitis pigmentosa, 231
X-linked spastic paraplegia, 495
X-ray technology, 26

Yarbus, Alfred, 427
Yaw, 300
Y chromosome, SRY gene, 646
Young, John Z., 53
Young, Michael, 608
Young, Thomas, 240, 241

Zofran®, 161
Zoloft®, 641
Zonule fibers, 224, 225
Zygomaticus major, 628

About the Book

Editor: Andrew D. Sinauer

Project Editor: Joyce Zymeck

Production Manager: Christopher Small

Book Layout and Production: Joan Gemme

Art Editing and Illustration Program: S. Mark Williams, Pyramis Studios, Inc.

Book Design: Joan Gemme

Cover Design: S. Mark Williams

Subject Indexer: Grant Hackett

Cover Manufacturer: Henry Sawyer Company, Inc.

Book Manufacturer: The Courier Companies, Inc.

(continued from inside front cover)

ICON	STRUCTURE(S) OF INTEREST	MODULE	SELECTION
13	General organization of the somatic sensory system	Sectional anatomy Spinal cord and brainstem	Structure Mode: Somatosensory system Cross Sectional Atlas tab
14	Organization of the mechanosensory pathways	Pathways Spinal cord and brainstem	Pathways tab: Mechanosensory pathway Cross Sectional Atlas tab
15	Location of the human somatosensory cortex	Surface anatomy Sectional anatomy Animations	Structure Mode: Major cortical areas Structure Mode: Somatosensory system Sagittal, coronal, and axial MRI sections
16	Cytoarchitecture of the somatosensory cortex	Surface anatomy	Structure Mode: Brodmann's areas
17	Major pathways for pain and temperature sensation	Pathways Spinal cord and brainstem	Pathways tab: Anterolateral system Cross Sectional Atlas tab
18	Nociceptive components of thalamus and cortex	Sectional anatomy	Structure Mode: Somatosensory system
19	Major components of the visual system	Sectional anatomy Animations	Structure Mode: Visual system Sagittal, coronal, and axial MRI sections
20	Location of the primary visual (striate) cortex	Sectional anatomy Surface anatomy	Structure Mode: Visual system Structure Mode: Major cortical areas
21	Major components of the auditory system	Sectional anatomy Surface anatomy	Structure Mode: Auditory system Cross Sectional Atlas tab
22	Structures involved in the vestibulo-ocular reflex	Spinal cord and brainstem	Cross Sectional Atlas tab
23	Location of the vestibular nuclei	Spinal cord and brainstem	Cross Sectional Atlas tab
24	Central components of the olfactory system	Sectional anatomy Sectional anatomy	Structure Mode: Chemical senses Structure Mode: Limbic system
25	Central components of the human taste system	Sectional anatomy Spinal cord and brainstem	Structure Mode: Chemical senses Cross Sectional Atlas tab
26	Brainstem structures that project to the spinal cord	Spinal cord and brainstem	Cross Sectional Atlas tab
27	Location of the motor and premotor cortices	Surface anatomy Sectional anatomy	Structure Mode: Major cortical areas Structure Mode: Motor systems
28	Organization of the corticospinal system	Pathways Spinal cord and brainstem	Pathways tab: Corticospinal tract Cross Sectional Atlas tab
29	Components of the human basal ganglia	Sectional anatomy Sectional anatomy Animations	Structure Mode: Basal ganglia Structure Mode: Motor systems Sagittal, coronal, and axial MRI sections